STEEL DESIGNERS'
MANUAL

STEEL DESIGNERS' MANUAL

SEVENTH EDITION

The Steel Construction Institute

Edited by

Buick Davison
Department of Civil & Structural Engineering,
The University of Sheffield

Graham W. Owens
Consultant, The Steel Construction Institute

WILEY-BLACKWELL

A John Wiley & Sons, Ltd., Publication

This edition first published 2012 © 2012 by Steel Construction Institute

Blackwell Publishing was acquired by John Wiley & Sons in February 2007. Blackwell's publishing program has been merged with Wiley's global Scientific, Technical and Medical business to form Wiley-Blackwell.

Registered office: John Wiley & Sons, Ltd, The Atrium, Southern Gate, Chichester, West Sussex, PO19 8SQ, UK

Editorial offices: 9600 Garsington Road, Oxford, OX4 2DQ, UK
The Atrium, Southern Gate, Chichester, West Sussex, PO19 8SQ, UK
2121 State Avenue, Ames, Iowa 50014-8300, USA

For details of our global editorial offices, for customer services and for information about how to apply for permission to reuse the copyright material in this book please see our website at www.wiley.com/wiley-blackwell.

Library of Congress Cataloging-in-Publication Data

Steel designers' manual / the Steel Construction Institute ; edited by Buick Davison, Graham W. Owens.. – 7th ed.
    p. cm.
  Includes bibliographical references and index.
  ISBN-13: 978-1-4051-8940-8 (hardback)
  ISBN-10: 1-4051-8940-1 ()
  ISBN-13: 978-1-1192-4986-3 (paperback)
  I. Davison, Buick. II. Owens, Graham W. (Graham Wynford) III. Steel Construction Institute (Great Britain)
  TA685.G67 2011
  624.1'821–dc23

2011028324

A catalogue record for this book is available from the British Library.

This book is published in the following electronic formats: ePDF 9781444344844; ePub 9781444344851; Mobi 9781444344868

Set in 10/12 pt Times Ten by Toppan Best-set Premedia Limited
Printed and bound by CPI Group (UK) Ltd, Croydon, CR0 4YY

4   2020

Although care has been taken to ensure, to the best of our knowledge, that all data and information contained herein are accurate to the extent that they relate to either matters of fact or accepted practice or matters of opinion at the time of publication, the Steel Construction Institute assumes no responsibility for any errors in or misinterpretations of such data and or information or any loss or damage arising from or related to their use.

Extracts from the British Standards are reproduced with the permission of BSI. Complete copies of the standards quoted can be obtained by post from BSI Sales, Linford Wood, Milton Keynes, MK14 6LE.

Contents

CONNECTION DESIGN

Introduction to the seventh edition

At the instigation of the Iron and Steel Federation, the late Bernard Godfrey began work in 1952 on the first edition of the *Steel Designers' Manual*. As principal author, he worked on the manuscript almost continuously for a period of two years. On many Friday evenings he would meet with his co-authors, Charles Gray, Lewis Kent and W.E. Mitchell, to review progress and resolve outstanding technical problems. A remarkable book emerged. Within approximately 900 pages it was possible for the steel designer to find everything necessary to carry out the detailed design of most conventional steelwork. Although not intended as an analytical treatise, the book contained the best summary of methods of analysis then available. The standard solutions, influence lines and formulae for frames could be used by the ingenious designer to disentangle the analysis of the most complex structure. Information on element design was intermingled with guidance on the design of both overall structures and connections. It was a book to dip into rather than read from cover to cover. However well one thought one knew its contents, it was amazing how often a further reading would give some useful insight into current problems. Readers forgave its idiosyncrasies, especially in the order of presentation. How could anyone justify slipping a detailed treatment of angle struts between a very general discussion of space frames and an overall presentation on engineering workshop design?

The book was very popular. It ran to four editions with numerous reprints in both hard and soft covers. Special versions were also produced for overseas markets. Each edition was updated by the introduction of new material from a variety of sources. However, the book gradually lost the coherence of its original authorship and it became clear in the 1980s that a more radical revision was required.

After 36 very successful years, it was decided to rewrite and reorder the book, while retaining its special character. This decision coincided with the formation of the Steel Construction Institute and it was given the task of co-ordinating this activity.

A complete restructuring of the book was undertaken for the fifth edition, with more material on overall design and a new section on construction. The analytical material was condensed because it is now widely available elsewhere, but all the design data were retained in order to maintain the practical usefulness of the book as a day-to-day design manual. Allowable stress design concepts were replaced by limit state design encompassing BS 5950 for buildings and BS 5400 for bridges. Design examples are to the more appropriate of these two codes for each particular application.

The fifth edition was published in 1992 and proved to be a very worthy successor to its antecedents. It also ran to several printings in both hard and soft covers; an international edition was also printed and proved to be very popular in overseas markets. The sixth edition of 2003 maintained the broad structure introduced in 1992, reflecting its target readership of designers of structural steelwork of all kinds, and included updates to accommodate changes in the principal design codes, BS5400 and BS5950.

This seventh edition, while maintaining the same overall structure, has required a more radical review of the content. The most significant changes are:

- The adoption of the Eurocodes, presenting relevant parts of their background in the chapters on element and connection design and using them for all worked examples.
- Recognition of the growing importance of light steel and secondary steelwork with separate chapters on types of light steel structure, the detailed design of light gauge elements and secondary steelwork.
- Recognition of both the greater importance of sustainability to the built environment and the associated need for holistic, integrated approaches to design and construction.
- A revised approach to analysis, recognising the growing importance of computer methods.

Because all these changes introduced more material into an already very large text book, it was decided to concentrate on building structures, removing all references to bridges.

Introduction: Chapters 1–3

An introduction to both design to the Eurocodes and the need for integrated design.

- Introduction – design to the Eurocodes (Chapter 1)
- Integrated design for successful steel construction (Chapter 2)
- Loading to the Eurocodes (Chapter 3).

Design synthesis: Chapters 4–9

A description of the processes by which design solutions are formed for a wide range of steel structures.

- Single storey buildings (Chapter 4)
- Multi-storey buildings (Chapter 5)
- Industrial steelwork (Chapter 6)
- Special steel structures (Chapter 7)
- Light steel structures (Chapter 8)
- Secondary steelwork (Chapter 9)

Applied metallurgy: Chapters 10–11

Background material sufficient to inform designers of the important issues inherent in the production and use of steel and methods of accounting for them in practical design and construction.

- Applied metallurgy of steel (Chapter 10)
- Failure processes (Chapter 11)

Analysis: Chapters 12 and 13

A resumé of analytical methods for determining the forces and moments in structures subject to static or dynamic loads, both manual and computer-based.

Comprehensive tables for a wide variety of beams and frames are given in the Appendix.

- Analysis (Chapter 12)
- Structural vibrations (Chapter 13)

Element Design: Chapters 14–24

A comprehensive treatment of the design of steel elements, singly, in combination or acting compositely with concrete.

- Local buckling and cross-section classification (Chapter 14)
- Tension members (Chapter 15)
- Columns and struts (Chapter 16)
- Beams (Chapter 17)
- Plate girders (Chapter 18)
- Members with compression and moments (Chapter 19)
- Trusses (Chapter 20)
- Composite slabs (Chapter 21)
- Composite beams (Chapter 22)
- Composite columns (Chapter 23)
- Light gauge elements (Chapter 24)

Connection design: Chapters 25–28

The general basis of design of connections is surveyed and amplified by consideration of specific connection methods.

- Bolting assemblies (Chapter 25)
- Welds and design for welding (Chapter 26)

- Joint design and simple connection (Chapter 27)
- Moment connections (Chapter 28)

Foundations: Chapters 29 and 30

Relevant aspects of sub-structure design for steel construction.

- Foundations and holding down systems (Chapter 29)
- Steel piles (Chapter 30)

Construction: Chapters 31–36

Important aspects of steel construction about which a designer must be informed in order to produce structures which can be economically fabricated and erected, and which will have a long and safe life.

- Design for movement (Chapter 31)
- Tolerances (Chapter 32)
- Fabrication (Chapter 33)
- Erection (Chapter 34)
- Fire protection and fire engineering (Chapter 35)
- Corrosion and corrosion prevention (Chapter 36)

A comprehensive collection of data of direct use to the practising designer is compiled into a series of Appendices.

Throughout the book, Eurocode notation has been adopted.

By kind permission of the British Standards Institution, references are made to British Standards, including the Eurocodes, throughout the manual. The tables of fabrication and erection tolerances in Chapter 32 are taken from the *fifth edition of the National Structural Steelwork Specification.* Both these sources are used with the kind permission of the British Constructional Steelwork Association, the publishers.

Finally, we would like to thank all the contributors and acknowledge their hard work in updating the content and their co-operation in compiling this latest edition. All steelwork designers are indebted to their efforts in enabling this manual to be maintained as the most important single source of information on steel design.

Buick Davison and Graham W. Owens

Contributors

David G. Brown

David Brown graduated from the University of Bradford in 1982 and worked for several years for British Rail, Eastern Region, before joining a steelwork contractor as a designer, and then technical director. He joined the Steel Construction Institute in 1994 and has been involved with connections, frame design, Eurocodes and technical training.

Michael Burdekin

Michael Burdekin graduated from Cambridge University in 1961. After fifteen years of industrial research and construction experience, during which he was awarded a PhD from Cambridge University, he was appointed Professor of Civil and Structural Engineering at UMIST in 1977. He retired from this post in December 2002 and is now an Emeritus Professor of the University of Manchester. His specific expertise is the field of welded steel structures, particularly in materials behaviour and the application of fracture mechanics to fracture and fatigue failure.

Katherine Cashell

Dr Katherine Cashell is a Senior Engineer at the Steel Construction Institute and a Chartered Member of the Institution of Civil Engineers. Previous to this, she worked as a research assistant at Imperial College London and a Design Engineer at High Point Rendel Consultants.

Kwok-Fai Chung

Professor K F Chung obtained a bachelor degree from the University of Sheffield in 1984 and a doctoral degree from Imperial College in 1988. He joined the Steel Construction Institute in 1989 and worked as a research engineer for six years on steel, steel-concrete composite and cold-formed steel structures as well as structural fire engineering. After practising as a structural engineer in a leading consultant firm in Hong Kong for approximately a year, he joined the Hong Kong Polytechnic University in 1996 as an Assistant Professor and was promoted to a full Professor in 2005. He has published about 150 technical papers in journals and conferences together with five SCI design guides. Moreover, he has taught about 30 professional courses to practising engineers in Hong Kong, Singapore, Malaysia and Macau. He was Chairman of the Editorial Board and Chief Editor of the Proceedings of the IStructE Centenary Conference 2008. Currently, he is the Founding President of the Hong Kong Constructional Metal Structures Association and Advisor to the Macau Society of Metal Structures.

Graham Couchman

Graham Couchman graduated from Cambridge University in 1984 and completed a PhD in composite construction from the Swiss Federal Institute of Technology in Lausanne in 1994. He has experience of construction, design and research, specialising in composite construction and light gauge construction. He first joined SCI in 1995 then, after a brief spell at BRE, became Chief Executive of SCI in 2007. He is currently chairman of the European committee responsible for Eurocode 4.

Buick Davison

Dr Buick Davison is a Senior Lecturer in the Department of Civil and Structural Engineering at the University of Sheffield. In addition to his wide experience in teaching and research of steel structures, he is a Chartered Engineer and has worked in consultancies on the design of buildings and stadia.

David Deacon

David Deacon qualified as a Coating Technologist in 1964 and after working for the British Iron and Steel Research Association and the Burma Castrol Group, he started in consultancy of coatings for iron and steel structures in 1970.

His first major consultancy was the protective coatings for London's Thames Barrier, which is now some 28 years old and a recent major survey has extended the life of the coatings to first major maintenance from 25 to 40 years.

His consultancy and advisory activities has taken him to over 50 countries worldwide on a range of projects; he is currently working on the refurbishment of the Cutty Sark iron frame, the Forth Rail and Road Bridges and numerous other structures.

He has given many papers on his specialist coatings subjects and is a co-author of the book 'Steelwork Corrosion Control'. He is a Past President of the Institute of Corrosion and last year was awarded a unique Lifetime Achievement Award by the Institute of Corrosion.

David Dibb-Fuller

David Dibb-Fuller started his career with the Cleveland Bridge and Engineering Company in London. His early bridge related work gave a strong emphasis to heavy fabrication; in later years he moved on to building structures. As technical director for Conder Southern in Winchester, his strategy was to develop close links between design for strength and design for production. He moved on to become a partner with Gifford and Partners in Southampton until his retirement; he remains a consultant to the partnership.

Richard Dobson

Richard Dobson graduated from the University of Cambridge in 1980. For the first eight years of his professional career, he worked for consulting engineers in areas of bridge design and off-shore steel jacket design for the North Sea and other parts of the world. Twenty-four years ago Richard joined CSC (UK) Ltd, designing and developing software solutions for structural engineers. For the last 12 years, Richard has been the technical director at CSC overseeing the global development of CSC's range of software products – Fastrak, Orion and Tedds.

Leroy Gardner

Dr Leroy Gardner is a Reader in Structural Engineering at Imperial College London and a Chartered Civil and Structural Engineer. He leads an active research group in the area of structural steelwork, teaches at both undergraduate and postgraduate levels and carries out specialist advisory work for industry. He has co-authored two textbooks and over 100 technical papers, and is a member of the BSI committee responsible for Eurocode 3.

Jeff Garner

Jeff Garner has over 30 years industrial experience in fabrication and welding. He is a professional Welding Engineer with a Masters degree in Welding Engineering. Working in a range of key industry sectors including petrochemical, nuclear, steel making, railways and construction he has acted as a consulting welding engineer, delivered welding technology training courses and provided representation on a number of British and European welding code/standards committees. Jeff joined the British Constructional Steelwork Association in 2008 as Welding and Fabrication Manager, responsible for providing welding and fabrication technology support throughout the UK steel construction industry. In 2011 he moved to edf.

Martin Heywood

Martin Heywood has worked at the SCI since 1998. He currently holds the title 'Associate Director Construction Technology' and has responsibility for a portfolio of projects involving light gauge steel, modern methods of construction, floor vibrations and building envelope systems. Previously, Martin worked for several years in the SCI's Codes and Standards division where he authored the SCI's Guide to the amendments to BS 5950-1:2000 and BS 5950 worked examples. Prior to joining the SCI, Martin worked for 3 years in civil engineering contracting and obtained a PhD in structural dynamics from Birmingham University.

Roger Hudson

Roger Hudson studied metallurgy at Sheffield Polytechnic whilst employed by BISRA. He also has a Masters degree from the University of Sheffield. In 1968, he joined the United Steel Companies at Swinden Laboratories in Rotherham to work on the corrosion of stainless steels. The laboratories later became part of British Steel where he was responsible for the Corrosion Laboratory and several research projects. He became principal technologist for Corus . He is a member of several technical and international standards committees, has written technical publications, and has lectured widely on the corrosion and protection of steel in structures. He has had a longstanding professional relationship with the Institute of Corrosion.

Mark Lawson

Mark Lawson is part-time Professor of Construction Systems at the University of Surrey and a Specialist Consultant to the Steel Construction Institute. In 1987, he joined the newly formed SCI as Research Manager for steel in buildings, with particular reference to composite construction, fire engineering and cold-formed steel. A graduate of Imperial College, and the University of Salford, where he worked in the field of cold-formed steel, Mark Lawson spent his early career at Ove Arup and Partners and the Construction Industry Research and Information Association. He is a member of the Institutions of Civil and Structural Engineers and the American Society of Civil Engineers.

Ian Liddell

Ian Liddell was a Founding Partner of Buro Happold in 1976. He has been responsible for a wide range of projects with special innovatory structural engineering including Sydney Opera House, the Millennium Dome, Mannheim Gridshell Roof, and the concept and scheme for Phoenix Stadium Retractable Roof. He is one of the world's leading experts in the field of lightweight tension and fabric structures. He is a Royal Academy of Engineering Visiting Professor at the University of Cambridge and was awarded the Institution of Structural Engineers Gold Medal in 1999.

Allan Mann

Allan Mann graduated from Leeds University and gained a PhD there. Since then he has over 40 years of experience in steel structures of all kinds over the commercial, industrial and nuclear sectors. He also has extensive experience in roller coasters and large observation wheels. Allan has authored a number of papers, won a number of prizes and been closely associated with the Institution of Structural Engineers throughout his career.

Fergus McCormick

Fergus McCormick is a specialist in cable, long-span, dynamic and moving structures and wind engineering and is a sector specialist in Sports Stadia. His past projects include the BA London Eye, the City of Manchester Stadium and the Infinity Footbridge. For Buro Happold he has been Structural Leader for Astana Stadium Retractable Roof; Kirkby Stadium, Everton Football Club; Aviva Stadium and currently leads the structural team for the London 2012 Olympic Stadium.

David Moore

Dr David Moore is the Director of Engineering at the British Constructional Steelwork Association. Dr Moore has over 20 years experience of research and specialist advisory work in the area of structural engineering and has published over 70 technical papers on a wide range of subjects. He has also made a significant contribution to a number of specialised design guides and best practice guides for the UK and European steel industry. Many of these publications are used daily by practising structural engineers and steelwork contractors.

David Nethercot

Since graduating from the University of Wales, Cardiff, David Nethercot has completed forty years of teaching, research and specialist advisory work in the area of structural steelwork. The author of over 400 technical papers, he has lectured frequently on post-experience courses; he is a past Chairman of the BSI Committee responsible for BS 5950, and is a frequent contributor to technical initiatives associated with the structural steelwork industry. Since 1999 he has been head of the Department of Civil and Environmental Engineering at Imperial College. He is a past president of IStructE, received the 2008 Charles Massonnet prize from ECCS and was awarded a Gold Medal by the IStructE in 2009.

Graham W. Owens

Dr Graham Owens has 45 years' experience in designing, constructing, teaching and researching in structural steelwork. After six years' practical experience and 16 years at Imperial College, he joined the Steel Construction Institute at its formation in 1986. He was Director from 1992 until his retirement in 2008. He was President of the Institution of Structural Engineers in 2009. He continues some consultancy interests in wave energy, New Nuclear Build and Education.

Roger Pope

Dr Roger Pope is a consulting engineer who specialises in steel construction. His career in steel and steel construction began with the Steel Company of Wales in 1964. He is currently chairman of CB/203 the technical committee responsible for British Standards dealing with the design and execution of steel structures. He is also chairman of the Codes, Standards and Regulations Committee established by the Steel Construction Industry Sector under the auspices of the UK Government.

Alan Rathbone

Alan Rathbone is Chief Engineer at CSC. His previous experience includes design, research and advice on reinforced concrete and masonry, together with the design of volumetric building systems, using all main structural materials. He has worked with his co-contributor, Richard Dobson, for almost 25 years in the development of software solutions for structural engineers. He is a member of the BSI Committee CB/203 which is responsible for many of the steelwork design codes. A long-time member of the BCSA/SCI Connections Group, he has made a significant contribution to their efforts in producing the 'Green Book' series and is currently the Chairman. He is a Fellow of the Institution of Civil Engineers.

John Roberts

John Roberts graduated from the University of Sheffield in 1969 and was awarded a PhD there in 1972 for research on the impact loading of steel structures. His professional career includes a short period of site work with Alfred McAlpine plc, following which he has worked as a consulting engineer, since 1981 with Allott & Lomax/Babtie Group/Jacobs. He is an Executive Director of Operations at Jacobs Engineering UK Limited and has designed many major steelwork structures. He was President of the Institution of Structural Engineers in 1999-2000 and has served as a council member of both the Steel Construction Institute and the BCSA.

Alan Rogan

Alan Rogan, sponsored by British Steel (now TATA), obtained a PhD at the University of the West of England, Bristol, focusing on the steel sector. He then went on to be the Corus sponsored Reader at Oxford Brookes University in the school of Architecture. Alan is currently Managing Director of Metek Building Systems, a leading company involved in the design and construction of Light steel framing. The company, under Alan's leadership, are now involved in the building of up to 7 storey high buildings. Alan has over 40 years of steel construction experience from bridge building to structures of all sizes and shapes. Alan continues to have a close relationship with TATA through a manufacturing agreement between the companies at Llanwern Steelworks, South Wales.

Michael Sansom

Michael Sansom has 19 years experience of environmental and sustainability work in consultancy, research and research management roles in the construction sector.

After completion of his PhD in nuclear waste disposal at Cardiff University in 1995, he worked for the UK Construction Industry Research and Information Association (CIRIA) managing construction research projects. He then worked for a large US consultancy working on contaminated land investigation and remediation.

Michael joined the SCI in 1999 where he now leads the Sustainability Division. He is involved in a range of sustainable construction activities including life cycle assessment, carbon foot-printing, BREEAM assessments and operational carbon emissions assessment and reduction.

Ian Simms

Ian Simms is currently the manager responsible for the fire engineering and composite construction departments at the Steel Construction Institute. Ian joined the SCI in April 1998 to work as a specialist in Fire Engineering, after completing his PhD at the University of Ulster.

Andrew Smith

Andy Smith worked for the Steel Construction Institute for four years after graduating from the University of Cambridge. He was involved in many projects relating to floor vibration and steel-concrete composite construction, and regularly presented courses on both these subjects. He is the lead author of SCI P354 on the design of floors for vibration and participated in an ECSC funded European project on the subject. Andy now lives and works as a consulting structural engineer in Canada.

Colin Taylor

Colin Taylor graduated from Cambridge in 1959. He started his professional career in steel fabrication, initially in the West Midlands and subsequently in South India. After eleven years he moved into consultancy where, besides practical design, he became involved with drafting work for British Standards and later for Eurocodes. Moving to the Steel Construction Institute on its formation, he also became involved with BS EN 1090-2 Execution of Steel Structures.

Colin contributed to all 9 parts of BS 5950, including preparing the initial drafts for Parts 1 and 2, compiled the 1989 revision of BS 449-2 and then contributed to all the 17 parts of the ENV stage of Eurocode 3, besides other Eurocodes. His last job before retiring from SCI was compiling and correcting BS 5950-1:2000.

Since then Colin has worked as a consultant for SCI and for DCLG and continued on numerous BSI committees and as convenor for EN 1993-6 Crane supporting structures. His main interest now is serving in local government as an elected member of two councils.

Richard Thackray

Richard Thackray graduated from Imperial College London with a degree and PhD in the field of Materials Science, and joined the University of Sheffield as Corus Lecturer in Steelmaking in 2003. His particular expertise is in the areas of clean steel production, continuous casting, and the processing of next generation high strength steels. He is currently the Chairman of the Iron and Steel Society, a division of the Institute of Materials, Minerals and Mining.

Mark Tiddy

Mark Tiddy is the Technical Manager of Cooper & Turner, the UK's leading manufacturer of fasteners used in the construction industry. He spent the first part of his career as a graduate metallurgist in the steel industry and for the past twenty five years has worked in the fastener industry. He is the UK expert on CEN/TC 185/WG6 the European committee responsible for structural bolting.

Andrew Way

After graduating from the University of Nottingham, Andrew Way has spent his career in the field of steel construction design and research. In 1996 he joined the Steel Construction Institute where he now holds the position of Manager of Light Gauge Construction. He is a Chartered Engineer and is also responsible for the management of the SCI Assessed third party verification scheme.

Richard White

Richard White is an Associate Director of Ove Arup and Partners. He has worked in their Building Engineering office in London since graduating from Newcastle University in 1980. During that time he has both contributed to and led the structural design of a wide range of building projects. His experience includes the design and construction of offices, airports, railway stations and art galleries. These projects embrace a broad range of primary structural materials and incorporate a variety of secondary steel components including architecturally expressed metalwork.

Richard was a member of the structures working party for the British Council for Offices Design Guide 2000 and 2005 and is a member of the Institutions of Civil and Structural Engineers.

Erica Wilcox

Erica Wilcox worked as a civil and geotechnical engineering designer at Arup for many years, after graduating from the University of Bristol. Her experience covers a wide range of geotechnical design, including the design of numerous embedded retaining walls. Her research MSc at the University of Bristol covered the monitoring and back analysis of a steel sheet pile basement.

John Yates

John Yates was appointed to a personal chair in mechanical engineering at the University of Sheffield in 2000, after five years as a reader in the department. He graduated from Pembroke College, Cambridge in 1981 in metallurgy and materials science and then undertook research degrees at Cranfield and the University of Sheffield, before several years of post-doctoral engineering and materials research. His particular interests are in developing structural integrity assessment tools based on the physical mechanisms of fatigue and fracture. He is the honorary editor of *Engineering Integrity* and an editor of the international journal *Fatigue and Fracture of Engineering Materials and Structures*. John moved to the University of Manchester in August 2010 as Professor of Computational Mechanics and Director of the Centre for Modelling and Simulation.

Ralph B.G. Yeo

Ralph Yeo graduated in metallurgy at Cardiff and Birmingham and lectured at the University of the Witwatersrand. In the USA he worked on the development of weldable high-strength and alloy steels with International Nickel and US Steel and on industrial gases and the development of welding consumables and processes at Union Carbide's Linde Division. Commercial and general management activities in the UK, mainly with the Lincoln Electric Company, were followed by twelve years as a consultant and expert witness with special interest in improved designs for welding.

Commercial and general management activities in the UK, mainly with the Lincoln Electric Company, were followed by twelve years as a consultant and expert witness with special interest in improved designs for welding before retirement and occasional lectures on improvements in design for welding.

Andrew Oldham

The illustrations in this Seventh edition of the manual are the careful work of senior CAD technician Andrew Oldham. Andrew has over fifteen years experience in architectural, civil and structural drawing working with Arup, Hadfield Cawkwell and Davidson, and SKM in Sheffield. The editors gratefully acknowledge his excellent contribution to the much improved quality of the illustrations in this edition.

Chapter 1
Introduction – designing to the Eurocodes

GRAHAM COUCHMAN

1.1 Introduction

For more than twenty years, the design of steel framed buildings in the UK, including those where composite (steel and concrete) construction is used, has generally been in accordance with the British Standard BS 5950. This first appeared in 1985 to replace BS 449 and introduced designers to the concept of limit state design. However, BS 5950 was withdrawn in March 2010 and replaced by the various parts of the Structural Eurocodes.

Bridge design in the UK has generally been in accordance with BS 5400, which was also introduced in the early 1980s and was also replaced in 2010.

The Structural Eurocodes are a set of structural design standards, developed by the European Committee for Standardisation (CEN) over the last 30 years, to cover the design of all types of structures in steel, concrete, timber, masonry and aluminium. In the UK, they are published by BSI under the designations BS EN 1990 to BS EN 1999. Each of the ten Eurocodes is published in several parts, and each part is accompanied by a National Annex that adds certain UK-specific provisions to go alongside the CEN document when it is implemented in the UK.

In England, implementation of these Standards for building design is achieved through Approved Document A to the Building Regulations. In Scotland and Northern Ireland, corresponding changes will be made to their regulations. It is expected that adoption of the Eurocodes by building designers will increase steadily from 2010 onwards.

As a public body, the Highways Agency is committed to specifying the Eurocodes for the design of all highway bridges as soon as it is practicable to do so. British Standard information reflected in the numerous BDs and BAs will be effectively replaced, and a comprehensive range of complementary guidance documents will be produced.

Steel Designers' Manual, Seventh Edition. Edited by Buick Davison and Graham W. Owens.
© 2012 Steel Construction Institute. Published 2012 by Blackwell Publishing Ltd.

1.2 Creation of the Eurocodes

The Structural Eurocodes were initiated by the Commission of the European Communities as a set of common structural design standards to provide a means for eliminating barriers to trade. Their scope was subsequently widened to include the EFTA countries, and their production was placed within the control of CEN. CEN had been founded in 1961 by the national standards bodies in the European Economic Community and EFTA countries. Its Technical Committees and Sub-Committees managed the actual process of bringing appropriate state-of-the-art technical content together to form the Eurocodes.

The size of the task, and indeed the difficulty of reaching agreement between the member states, is evident not only from the length of time it has taken to produce the final ENs, but also from the need for interim ENV documents (pre-norms) and a mechanism to allow national variations. ENVs appeared in the early 1990s and were intended to be useable documents that would permit feedback from 'real use'. In the UK this did not really happen as most designers, largely driven by commercial pressures, did not change 'until they had to'.

During the past fifteen years or so, the ENVs have been developed into EN documents. For each Eurocode part, a Project Team of experts was formed and duly considered national comments from the various member states. The UK was well represented on all the major Project Teams, which is a very positive reflection of our national expertise and ensured that the UK voice was heard.

1.3 Structure of the Eurocodes

There are ten separate Structural Eurocodes, as noted in Table 1.1.

Each Eurocode comprises a number of Parts, which are published as separate documents, and each Part consists of:

- the main body of text
- normative annexes
- informative annexes.

Table 1.1 The structural Eurocodes

EN	Eurocode
EN 1990	Eurocode: Basis of structural design
EN 1991	Eurocode 1: Actions on structures
EN 1992	Eurocode 2: Design of concrete structures
EN 1993	Eurocode 3: Design of steel structures
EN 1994	Eurocode 4: Design of composite steel and concrete structures
EN 1995	Eurocode 5: Design of timber structures
EN 1996	Eurocode 6: Design of masonry structures
EN 1997	Eurocode 7: Geotechnical design
EN 1998	Eurocode 8: Design of structures for earthquake resistance (depending on the location)
EN 1999	Eurocode 9: Design of Aluminium Structures

CEN publish the full text of each Eurocode Part in three languages (English, French and German) with the above EN designations. National standards bodies may translate the text into other languages but may not make any technical changes. The information given in each part is thus the same for each country in Europe.

To allow national use, the EN document is provided with a front cover and foreword by each national standards body, and published nationally using an appropriate prefix (for example EN 1990 is published by BSI as BS EN 1990). The text may be followed by a National Annex (see below), or a National Annex may be published separately.

The structure of the various Eurocodes and their many parts has been driven by logic. Thus the design basis in EN 1990 applies irrespective of the construction material or the type of structure. For each construction material, requirements that are independent of structural form are given in General Parts (one for each aspect of design) and form-specific requirements are given in other Parts. Taking Eurocode 3 as an example, indeed one where this philosophy was taken to an extreme, the resultant parts are given in Table 1.2.

Within each part the content is organised considering physical phenomena, rather than structural elements. So whereas BS 5950-1-1 includes Sections such as 'compression members with moments', EN 1993-1-1 contains Sections such as 'buckling resistance of members' and 'resistance of cross-sections'. The design of a structural element will therefore entail referring to numerous Sections of the document.

Another key aspect of the formatting logic is that there can be no duplication of rules, i.e. they can only be given in one document. A consequence of dividing the Eurocodes into these separate parts, and not allowing the same rule to appear in more than one part, is that when designing a steel structure many separate Eurocode documents will be required. Logical yes, but not very user friendly.

The Eurocode Parts contain two distinct types of rules, known as Principles and Application Rules. The former must be followed if a design is to be described as compliant with the code. The latter are given as ways of satisfying the Principles, but it may be possible to use alternative rules and still achieve this compliance (although how this would work in practice is yet to be seen).

Within the text of each Eurocode Part, provision is made for national choice in the setting of some factors and the adoption of some design methods (e.g. when guidance is given in an Informative Annex, a national view may be taken on the applicability of that 'information'). The choices are generally referred to as Nationally Determined Parameters (NDPs), and details are given in the National Annex to the Part. In many cases the annex will simply tell the designer to use the recommended value/option (i.e. there will be no national deviation). In addition, the National Annex may give references to publications that contain so-called non-contradictory complementary information (NCCI) that will assist the designer (see below).

This facility for national variations was adopted to provide regulatory bodies with a means of maintaining existing national levels of safety. The guidance given in a

Table 1.2 The parts of Eurocode 3 (titles given are informative, not necessarily as published)

EN	Eurocode
EN 1993-1-1	Eurocode 3: Design of Steel Structures – Part 1-1: General rules and rules for buildings
EN 1993-1-2	Eurocode 3: Design of Steel Structures – Part 1-2: General rules – structural fire design
EN 1993-1-3	Eurocode 3: Design of Steel Structures – Part 1-3: General rules – cold formed thin gauge members and sheeting
EN 1993-1-4	Eurocode 3: Design of Steel Structures – Part 1-4: General rules – structures in stainless steel
EN 1993-1-5	Eurocode 3: Design of Steel Structures – Part 1-5: General rules – strength and stability of planar plated structures without transverse loading
EN 1993-1-6	Eurocode 3: Design of Steel Structures – Part 1-6: General rules – strength and stability of shell structures
EN 1993-1-7	Eurocode 3: Design of Steel Structures – Part 1-7: General rules – design values for plated structures subjected to out of plane loading
EN 1993-1-8	Eurocode 3: Design of Steel Structures – Part 1-8: General rules – design of joints
EN 1993-1-9	Eurocode 3: Design of Steel Structures – Part 1-9: General rules – fatigue strength
EN 1993-1-10	Eurocode 3: Design of Steel Structures – Part 1-10: General rules – material toughness and through thickness assessment
EN 1993-1-11	Eurocode 3: Design of Steel Structures – Part 1-11: General rules – design of structures with tension components
EN 1993-1-12	Eurocode 3: Design of Steel Structures – Part 1-12: General rules – supplementary rules for high strength steels
EN 1993-2	Eurocode 3: Design of Steel Structures – Part 2: Bridges
EN 1993-3-1	Eurocode 3: Design of Steel Structures – Part 3-1: Towers, masts and chimneys – towers and masts
EN 1993-3-2	Eurocode 3: Design of Steel Structures – Part 3-2: Towers, masts and chimneys – chimneys
EN 1993-4-1	Eurocode 3: Design of Steel Structures – Part 4-1: Silos, tanks and pipelines – silos
EN 1993-4-2	Eurocode 3: Design of Steel Structures – Part 4-2: Silos, tanks and pipelines – tanks
EN 1993-4-3	Eurocode 3: Design of Steel Structures – Part 4-3: Silos, tanks and pipelines – pipelines
EN 1993-5	Eurocode 3: Design of Steel Structures – Part 5: Piling
EN 1993-6	Eurocode 3: Design of Steel Structures – Part 6: Crane supporting structures

National Annex applies to structures that are to be constructed within the country of its publication (so, for example, a UK designer wishing to design a structure in Germany will need to comply with the National Annexes published by DIN, not those published by BSI). The National Annex (NA) is therefore an essential document when using a Eurocode Part.

1.4 Non-contradictory complementary information – NCCI

The Eurocodes are design standards, not design handbooks. This is reflected in their format, as discussed above. They omit some design guidance where it is considered to be readily available in textbooks or other established sources. It is also accepted that they cannot possibly cover everything that will be needed when carrying out a design. For building design, the SCI has been collating some of the additional information that designers will need, some of which was contained in BS 5950. For highway bridges, the Highways Agency has led a thorough examination of the Eurocodes to determine what additional requirements will be needed to ensure bridges will continue to be safe, economic, maintainable, adaptable and durable.

The Eurocode format allows so-called non-contradictory complementary information (NCCI) to be used to assist the designer when designing a structure to the Eurocodes. According to CEN rules, a National Annex cannot contain NCCI but it may give references to publications containing NCCI. As the name suggests, any guidance that is referenced in the National Annex must not contradict the principles of the Eurocode.

The steel community has established a website that will serve as an up-to-date repository for NCCI, primarily aimed at steel and composite building design (www.steel-ncci.co.uk). This is referenced in the appropriate National Annexes. Additionally, BSI is publishing NCCI guidance in the form of Published Documents (PDs). These documents are only informative and do not have the status of a Standard. They include a number of background documents to the UK National Annexes. It is the intention of the Highways Agency that most additional guidance will be published as BSI PDs.

1.5 Implementation in the UK

The stated aim of the Eurocodes is to be mandatory for European public works, and become the 'default standard' for private sector projects. However, for buildings, this aim must be understood within the context of the UK regulatory system, which does not oblige designers to adopt any particular solution contained within an Approved Document. At the time of writing, Approved Document A to the Building Regulations (England and Wales) states that 'These Eurocodes ... when used in conjunction with their National Annexes and when approved by the Secretary of State, are intended to be referenced in this Approved Document as practical guidance on meeting the Part A Requirements'. Approved Document A will be updated in due course to make specific reference to the Eurocodes. Regulations in Scotland and Northern Ireland will also be updated to refer to the Eurocodes.

This means that, in principle, British Standards can be widely used to design buildings for the foreseeable future, assuming clients and / or insurers are happy with the use of 'out-of-date' guidance. Over time, the information contained within these BSs will suffer from a lack of maintenance (they will not be maintained beyond 2010).

For bridges the situation is somewhat different, because as a public body the Highways Agency will specify Eurocodes for the design of all highway structures as soon as it is practicable to do so. The Agency has indeed been preparing for the introduction of the Eurocodes for some time.

1.6 Benefits of designing to the Eurocodes

At a purely technical level, the benefits of the Eurocodes can only be due to either more 'accurate' rules, or rules that cover a broader scope than previous standards and therefore facilitate the use of a broader range of solutions. Benefits will there-fore be greater in countries where existing national rules were either out-of-date or limited in their scope. Not surprisingly, this means that the technical benefits to the UK are limited (a few examples are however given below).

However, to consider only the technical benefits is to miss the point of the Eurocodes. They were always intended to be more about removing trade barriers than advancing the state-of-the-art beyond the best practice present in the best national standards. Indeed, the Eurocodes are now a permanent feature so a fun-damental benefit they possess is that they represent the future. Designers rely heavily on design guides, training courses and software. In future these will all be based on the Eurocodes, indeed one could envisage significant improvements in software given the vastly improved ratio between development cost and sales value that pan-national rules will bring.

A further conclusion from the studies carried out by the Highways Agency, and one which may be of general application, was that the less prescriptive approaches adopted in the Eurocodes will allow greater scope for innovation and encourage designers to use advanced analysis techniques.

Some specific benefits of the Eurocodes are considered below.

1.6.1 Clarity and style

In the UK we tend to be rather pragmatic in our approach to design, and this is often considered by our continental colleagues, mistakenly, to show a lack of rigour. One of the manifestations of our approach is our liking for lookup tables, perhaps based on empirical information. The Eurocodes tend to be much more transparent in the way they present the physics behind aspects of structural behaviour. Whilst less user friendly, this can only help to reduce the instances of information being used out-of-context, by enabling the intelligent user to appreciate better what the code writers were considering.

One of the conclusions from the extensive studies initiated by the Highways Agency was that the clauses were expressed in a more 'mathematical' style than found in British Standards. It was also noted that the design principles are generally clear (although it was also noted that it is not always obvious how they should be satisfied). Having climbed a 'significant learning curve' the trial designers found use of the codes to be different but not necessarily more difficult.

1.6.2 Scope

The scope of rules given in BS 5950 for the design of buildings is, not surprisingly, comparable with that of EN 1993. The most obvious exceptions to this cover-all statement come from the world of composite steel-concrete construction. EN 1994 extends the scope that was included in BS 5950-3-1 to include continuous beams and, perhaps more usefully, composite columns. In terms of the latter, both filled tubes and encased open sections are covered. There is also considerably more guidance given on connection design and behaviour than can be found in BS 5950, although this has, of course, been well covered in the past for UK designers by the SCI-BCSA 'Green Books'.

Clearly the scope covered by the various parts of EN 1993 enables a consistent design approach to be adopted for buildings, bridges, silos, masts, etc.

1.6.3 Technical improvements

As noted above, it would be unreasonable to expect the rules in the Eurocodes to represent a significant technical improvement over the content of current British Standards. According to the Highways Agency, trial calculations of highway bridges revealed that the Eurocodes 'would make little difference to common forms of bridges and highway structures in terms of member sizes' and, compared on a like-for-like basis, the Eurocodes generally resulted in sectional resistances that were within 10% of the results from the British Standards. Whilst the author is not aware of such thorough comparisons of building designs, a similar conclusion would be expected.

Perhaps the biggest benefit will come from the option to use lower load factors. The load combination equations given in EN 1990 mean that in the absence of wind loading, gravity loads on beams can be reduced. The factors for dead and imposed load drop from 1.4 and 1.6 to 1.25 and 1.5 respectively. Loads may therefore be reduced by between 5 and 10%.

In terms of member performance there are some significant gains to be had in a number of areas. Often it becomes clear that the price of having 'simple' rules in BS 5950 was their conservatism.

1.7 Industry support for the introduction of the Eurocodes

Whilst recognising that it would be inappropriate to 'abandon' the British Standards prematurely, the steel sector (in terms of the combined forces of Tata, SCI and

BCSA) has nevertheless been busy preparing design guides since late 2007. This ensures that as the Eurocodes become available to use, designers have the help of high quality guidance. Ten guides were published in 2009, including explanatory texts, worked examples and design data (section and member properties). These and further publications will cover key aspects of both steel and composite design for buildings and bridges.

In addition to traditional design guides, comprehensive information is available through sector websites. Access Steel (www.access-steel.com) offers guidance on project initiation, scheme development and detailed design. Whilst initially populated with only harmonised information, new UK-specific information (reflecting the National Annexes) is being frequently added. The site includes many interlinked modules on the detailed design of elements, with step-by-step guidance, full supporting information and worked examples, to give a thorough understanding of how the Eurocodes should be used.

NCCI may be found on www.steel-ncci.co.uk. This website is referenced in the National Annexes to various parts of Eurocode 3 and 4 and serves as a listing of non-contradictory complementary information for the design of steel and composite structures. The NCCI references are associated with relevant clauses of the codes and provide links to other resources. Where the NCCI is a public electronic resource, hyperlinks are provided.

SCI and others have offered a range of courses covering design to various Eurocodes for some time. Uptake has been steadily increasing as design offices realise that the codes will not go away, and indeed begin to be asked to price for Eurocode-compliant designs.

All the major software houses have been developing Eurocode-compliant tools for some time, waiting for the right time to put them on the market. Similar to the purchase of design guides and attendance at courses, there is something of a 'chicken and egg' situation here – designers will only start using the Eurocodes once help is available, but there is no real market for that help until designers start to use the codes. Changes to the Building Regulations will force this situation in due course.

1.8 Conclusions

The production of the Eurocodes has occupied a very substantial part of the careers of some leading engineers across Europe. SCI has been associated with Eurocode development since its inception in 1986, and indeed the author first used an ENV version of Eurocode 4 to design one of the buildings at Sizewell B in the mid- to late eighties.

This lengthy gestation period has only led to minor technical improvements over existing British Standards (indeed in some cases it has been a regressive technical step), but that is to be expected given the advanced state of most British Standards. For some other countries, the technical advance is much more significant, for example in countries that had no national code for composite construction and no resources to develop one.

The great benefit of the Eurocodes is, however, the opportunities they provide from the removal of trade barriers, and the fact that in the future, disparate national experts and software houses will all be pulling together for common advancement.

The steel construction sector in the UK is well prepared for the time when designers will need to use the Eurocodes in earnest, with a multitude of design guides, software, courses, etc. already available and more to come in the coming years.

Chapter 2
Integrated design for successful steel construction

GRAHAM COUCHMAN and MICHAEL SANSOM

2.1 Client requirements for whole building performance, value and impact

This section considers a broad range of building performance requirements. Aspects of sustainability and economy are considered in greater detail in Sections 2.2 and 2.3.

2.1.1 Client priorities

As engineers or scientists with a particular interest we may sometimes forget that the average building owner and/or occupier has no interest in the beauty of a nice steelwork connection! They are only interested in one thing, that being whether or not the building as a whole is fit for the purpose for which it was intended. Fitness may of course include aesthetic requirements. In this chapter we will consider what exactly clients want from their buildings, and how the choice of a well designed steel frame can help meet these numerous and varied needs.

In order that a building may be described as truly fit for purpose it must:

- provide the performance that is needed 'now'
- facilitate provision of the performance that may be needed in the future, i.e. future-proof the building
- do so in the most (financially) economical way
- do so in the most environmentally friendly way.

These bullets resonate with the three dimensions of sustainability; social, economic and environmental. Each of them is considered in some detail below.

Steel Designers' Manual, Seventh Edition. Edited by Buick Davison and Graham W. Owens.
© 2012 Steel Construction Institute. Published 2012 by Blackwell Publishing Ltd.

2.1.2 Current building performance

For a given type of building the performance requirements will be governed by two things, namely regulatory requirements and specific owner / occupier requirements. Buildings to be constructed in England and Wales must conform with the requirements of the Building Regulations.[1] The stated aim of these is to ensure the health and safety of people in and around buildings. They are also concerned with energy conservation and making buildings more accessible and convenient for people with disabilities. Approved Documents (ADs) provide practical guidance on how to meet the requirements of the Building Regulations, and include:

- Part A – structure
- Part B – fire safety
- Part L – conservation of fuel and power
- Part E – resistance to passage of sound.

The ADs include lists of reference documents that have been approved as fit for the purpose of supporting this practical guidance. AD A, published in 2004, explicitly notes that the Eurocodes are on their way and will in due course be listed. An updated version of AD A, listing the Eurocodes, was prepared in late 2009 but at the time of writing it looks unlikely to 'appear' until 2013. The regulatory systems in Scotland and Northern Ireland are similar.

The nature of the UK regulatory system is such that the guidance given in the ADs and their references does not have to be used, although a designer's insurance provider may think otherwise. For certain types of building other regulations may need to be satisfied, for instance schools must comply with the rules in various Building Bulletins, and hospitals must satisfy Health Technical Memoranda.

In addition to satisfying regulatory requirements a building owner may impose their own requirements for building performance. These may be driven by the function for which the building is to be used, or perhaps issues associated with the image of the client's brand. It is important that consideration is given to the type of occupancy (now and in the future) so that appropriate floor loads are considered in the design. Values are given in EN 1991-1-1[2] ('Eurocode 1') for characteristic values of imposed loads (noting that loads are called actions in the Eurocodes) as well as densities and self-weights. Some key values are given below:

- Category A, domestic and residential floors: 1.5 to $2.0 \, \text{kN/m}^2$
- Category B, offices: 2.0 to $3.0 \, \text{kN/m}^2$

The higher values in both ranges are 'recommended', meaning that they may vary from country to country as stated in the relevant National Annex. The highest recommended value is $5.0 \, \text{kN/m}^2$, which applies to shopping areas and public areas subject to large crowds, amongst other things. Other loads (actions) are stated in other parts of Eurocode 1, for instance EN 1991-1-4 covers wind loading.

Clearly there may be a potential conflict in some cases. Designing floors for the relatively light loading associated with residential occupancy will result in an economical solution, but would not readily permit future building adaption to enable other uses (that might be associated with higher floor loads). The need for stiffness under these loads will also vary depending on use, floors in certain parts of hospitals being a well known and obvious example where avoiding 'bounciness' is particularly important.

Having designed the structure so that it will stand up and not deform excessively under the expected level of loading, the next most important requirement is that it will provide the necessary resistance to fire. Regulations require that the building will stand up for a certain period (between 30 minutes and 2 hours) depending on the building height and type of occupancy, and the provision of active fire protection. For steel framed buildings this is achieved by fire protecting the structure (which keeps steel temperatures sufficiently low) and/or so-called fire engineering (columns and beams are sized so that they can resist the load even in a state where their strength is reduced due to a temperature increase). In addition to regulatory requirements the owner may decide to use an active fire protection system (sprinklers) to protect the building and its contents.

In addition to structural performance under a range of actions including fire, it is important that the building is designed to be 'comfortable'. This means considering thermal and acoustic performance. The provision of adequate daylighting is also necessary but not something that the steel designer has much control over (and so not discussed here). Structural performance is covered in detail in other chapters of this Manual. Aspects of 'comfort' are considered below.

2.1.2.1 Thermal performance

To achieve thermal comfort it will be necessary for the designer to consider the heating of the building when external temperatures are low, and the cooling of the building when temperatures are high. The relative importance of these clearly depends on the external climate, which is likely to change over time. Regulatory requirements, which clearly reflect the current UK climate, may be found in Part L of the Building Regulations (L1 covers residential buildings and L2 covers other types). The control of internal temperatures must be achieved in a way that minimises materials and energy consumption and emissions in order to achieve the most sustainable building solution.

Ways of keeping the internal temperatures higher than those externally are well understood and covered by established regulations and published guidance. Strategies include insulation of the external walls, roof and ground floor (U-values), and achievement of airtightness. Consideration should also be given to promoting solar gain in a controlled manner at appropriate times of the year.

Less well understood in the UK, because traditionally it has not been a major concern, are ways of controlling overheating in summer. Strategies include:

Figure 2.1   Prestige building façade with brise soleil

- control solar gain
- provide thermal mass
- control ventilation
- insulate and make airtight the envelope
- control internal gains.

To control solar gain the designer must think about the orientation of windows, and indeed the internal configuration of rooms, e.g. preventing excessive solar gain in rooms occupied during the day. It is acknowledged that other constraints will often compromise choices. The choice of glass and/or provision of brise soleil (see Figure 2.1), will also affect gain.

Thermal mass is a 'hot' topic in debates about the relative benefits of 'lightweight steel structures' and 'heavyweight concrete structures'. The right amount of thermal mass may undoubtedly be beneficial as it serves as a heat sink to absorb energy during the day. However, provision of thermal mass will only ever be beneficial as part of an overall strategy, as the energy absorbed must then be purged during the cooler night time, using ventilation which is appropriate (for example secure) and controlled (not by infiltration through a leaky envelope).

The amount of thermal mass will ideally reflect not only the external climate (how many consecutive hot days are typical) but also the building occupancy pattern. For example too much mass may not provide the responsiveness needed in a home that is unoccupied during the day and on cold days needs to heat up rapidly just before the occupants arrive home from work. For a dwelling that is unoccupied during the day, absorbing heat which is subsequently radiated back into the living space during the night may also be less than desirable for temperatures in bedrooms. Clearly some careful planning is needed.

A further problem with the provision of thermal mass is exposing it. Studies have shown that blockwork walls with a plasterboard finish only provide half the effective mass of the same wall with 'wet' plaster. There may also be a conflict when trying to achieve 'integrated design'; floors should have an exposed soffit to maximise their effectiveness as a heat sink, but this may be in direct conflict with the use of a false ceiling to improve acoustic or aesthetic performance.

Completely divorced from the building construction, but very important when controlling internal temperatures, is the performance of appliances. Studies have shown that the number of degree hours above 27°C (which is a recognised indicator of when a building becomes uncomfortable) could be reduced by 25% by using low energy consumption appliances. In time one would imagine that this will happen by default as other sorts of appliances will no longer be available.

2.1.2.2 Acoustic performance

The level of acoustic performance required depends on the building and room types. Normally the main concern is ensuring that noise from public areas does not impact on the comfort of those occupying residential spaces, but rules also exist for ensuring a minimum level of noise in public areas (e.g. to ensure conversations in a restaurant cannot be overheard). Regulations governing acoustic performance include Building Regulations Part E, Building Bulletin 93 (Acoustic design of schools)[3] and Health Technical Memorandum 08-01.[4]

Where one room is separated from another room, sound can travel either directly through the separating element (direct transmission), or around the separating element through adjacent building elements (flanking transmission). Sound insulation for both routes is controlled by the following three characteristics:

- mass
- isolation
- sealing.

Direct transmission depends upon the properties of the separating wall or floor and can be estimated from laboratory measurements. Flanking transmission is more difficult to predict because it is influenced by the details of the junctions between the building elements and the quality of construction on site. In certain circumstances flanking transmission can account for the passage of more sound than direct transmission. It is therefore important that the junctions between separating elements are detailed and built correctly to minimise flanking sound transmission. Figure 2.2 shows typical construction details.

Transmission of airborne sound across a solid wall or a single skin partition will obey what is known as the mass law. The insulation of a solid element will increase

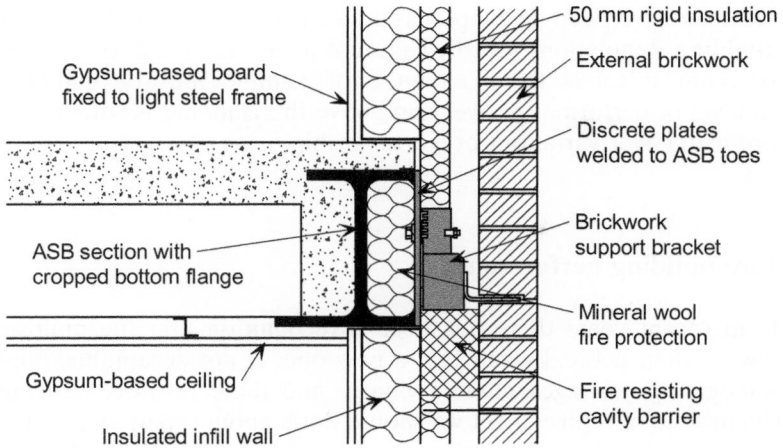

Figure 2.2 Flanking sound transmission

by approximately 5 dB per doubling of mass. However, lightweight framed construction achieves far better standards of airborne sound insulation than the mass law would suggest because of the presence of a cavity which provides a degree of isolation between the various layers of the construction. It has been demonstrated that the sound insulations of individual elements within a double skin partition tend to combine together in a simple cumulative linear relationship. The overall performance of a double skin partition can therefore generally be determined by simply adding together the sound insulation ratings of its constituent elements. In this way, two comparatively lightweight partitions can be combined to give much better insulation than the mass law alone would suggest. This is the basis of many lightweight partition systems. The width of the cavity between separate layers is important, and should be at least 40 mm. It will of course be recognised that the need to isolate elements to avoid acoustic transmission may be in conflict with the need to tie elements together to achieve adequate structural performance.

It is important to provide adequate sealing around floors and partitions because even a small gap can lead to a marked deterioration in acoustic performance. Joints between walls and between walls and ceilings should be sealed with tape or caulked with sealant. Where walls abut profiled metal decks, or similar elements, mineral wool packing and acoustic sealants may be required to fill any gaps. Where there are movement joints at the edges of walls, special details are likely to be necessary.

Ideally, wall linings should not be penetrated by services. This is particularly important for separating walls between dwellings. Where service penetrations do occur in sensitive locations, particular attention should be given to the way in which these are detailed.

SCI has published guidance on the detailing of both hot-rolled and cold-formed steel structures, including indicative acoustic performance values.

Robust Details have been developed as a means of complying with the requirements of Building Regulations Part E. Their use avoids the need for site testing, as in order to become a Robust Detail a number of examples must be built and tested, and show a level of performance over and above the Building Regulation requirements in order to cover variations of workmanship on site.

2.1.3 Future building performance

Thankfully in recent years there has been a recognition that the philosophy of 'throw it away when it breaks and buy a new one' is not acceptable. Population growth, demographic changes, climate change and finite resources are just four reasons why instead of replacement we should think about repair and/or modification. This applies to buildings as much as it applies to anything else, indeed it is very important in terms of the built environment given the materials and energy that are used to build and then operate.

It has already been noted that the right choice for the designer is not always obvious. Designing a new residential building for the prescribed level of imposed loads for that application will mean that the building would not readily lend itself to future conversion for office use, where certain rooms would be subject to much higher loading. Designing the building for the highest possible level of loading (to cover all possible future uses) would clearly be uneconomic, wasting materials and the client's money. Some of the other choices are less obvious.

Design for deconstruction is frequently mentioned as a methodology for reducing the materials used in construction. It clearly makes sense to be able to dismantle a building when it has reached the end of its useful life, and reuse or recycle the components. Steel framed buildings are well placed for this given their similarity to Meccano. To ease deconstruction it must be relatively easy to separate the building components. However, if we think about one of the most structurally efficient forms of construction currently used in buildings, with composite steel and concrete beams and floor slabs, the efficiency comes from the fact that the materials act together. By tying together steel beams with concrete slabs, via shear studs welded to the beam and embedded in the concrete, the tensile strength of the steel and compressive strength of the concrete are used to best effect. The profile shape and formed embossments of steel decking enable it to transfer shear with the concrete to which it is attached, so that it acts as external reinforcement. So by intimately tying together the steel and concrete it is possible to use less of the materials. Composite solutions are also an extremely effective way of forming long spans, which allow greater internal flexibility because there are fewer columns or load bearing walls. So is it better to adopt a solution that can be easily dismantled, e.g. non-composite, or is it better to use fewer materials in the first place and provide a building which could have a longer useful life because it is more adaptable?

2.1.4 Building economy

The cost of the structural frame is a relatively minor part of the overall building cost (less than 10%). More costly items are the building services and cladding, either of which might typically cost three times as much as the frame. Other parts of this Manual explain how to design and detail cost-effective frames, but it makes no sense to save a proportion of the frame cost if it compromises the services or the cladding.

The cost of cladding can be reduced by reducing the overall height of a building. This can be achieved by making the effective depth of structural floor zones less. This could either mean the depth of the structural zone is literally less, or effectively less because it allows services to occupy the same zone. Steel solutions such as so-called slim floors address the first solution (by making the steel beams occupy the same space as the concrete slab), whereas solutions such as cellular beams enable services to pass through the beam webs so avoiding them taking up more space below the structural zone. Some slim floor solutions allow services to run within the depth of the slab, at least in one direction. So the choice of floor solution will affect the cost of the cladding, and the ease of installation and therefore cost of the services, as well as the cost of the structure. Interface detailing, and its impact on building economy, is considered in more detail in Section 2.3. Reducing the envelope area of a building also reduces fabric heat loss.

Another aspect of economy for the designer to consider is the cost of time spent on site. For some applications this is vitally important. A number of years ago there was a well publicised example of a McDonalds' drive-through that was erected in 24 hours using a modular light steel framed solution that was constructed off-site (and before the clock started). Off-site volumetric solutions have also served the needs of the education sector very well, allowing student accommodation to be completed on site within the summer vacation. Whilst both these examples are rather niche, steel solutions in general allow time to be saved on site. The SCI cost comparison study,[5] which was last updated in 2004, shows that the ease and consequent speed of erection of steel solutions can translate into real financial benefits.

2.1.5 Environmental impact

Clients are increasingly demanding buildings with lower environmental impacts. Uncertainty remains however about how to quantify these impacts robustly, prompting the important question 'lower than what?'

Although legislation is an important driver, many clients are voluntarily taking steps to procure 'green' buildings that go beyond regulatory compliance. As well as yielding whole life savings, e.g. through operational energy efficiency, such decisions are increasingly seen as enhancing company reputation and brand value.

The environmental impacts of buildings and construction are numerous and diverse and the construction industry, in general, has a poor environmental profile. The industry is the largest consumer of non-renewable resources and the largest generator of waste, while the operation of buildings is responsible for around a half of the UK's total carbon dioxide emissions. Not surprisingly therefore, it is in these areas where measures, both voluntary and mandatory, have been introduced to effect change.

It is important to distinguish between embodied and operational environmental impacts. Embodied impacts refer to the impacts associated with the winning, processing, transporting, and erecting of construction products and materials. Operational impacts relate to the impacts arising from heating, lighting, cooling and maintaining the building.

Regulatory requirements have, to date, focussed on the operational environmental impacts of buildings, mainly via Part L of the Building Regulations. Historically this has made good sense since the relative importance of building operational impacts has been much greater than the embodied impacts. Based on the 2006 Building Regulations for example, the ratio of embodied to operational impacts of a commercial building over a 60-year design life is estimated to be of the order of 1:6.

This ratio is changing, however, and as building operational impacts are reduced through regulation, by a combination of building fabric and services improvements and the introduction of zero- and low-carbon technologies, the relative importance of the embodied impacts becomes greater. It is therefore highly likely that in the future, the Building Regulations will address embodied as well as operational impacts of buildings. Unfortunately, quantification of embodied impacts is far more complex and controversial than operational impacts and this may hinder their inclusion within the Building Regulations for some time.

The prominence of climate change and carbon within the sustainability agenda has meant that environmental impact assessment has mainly focussed on embodied energy and/or carbon. For the case of operational impacts such as heating, cooling and lighting, carbon is a sensible proxy for environmental impact and is relatively easy to quantify. Embodied impacts are often far more diverse and although carbon remains an important metric, there are many other impacts that should be considered. These include water extraction, mineral resource extraction, toxicity, waste, eutrophication and ozone creation. This raises the important question about how these different impacts can be objectively compared.

The most widely used and highly regarded tool for quantifying the environmental impacts of construction is life cycle assessment (LCA). Despite being conceptually quite straightforward, LCA can be very complex with many important, often material-specific, assumptions than can significantly influence the outcome.

In the UK, the leading environmental assessment methodology for construction materials and products is BRE's Environmental Profiles. Although frequently criticised by the industry, particularly for its lack of transparency, the methodology is founded on LCA principles and is used to derive the Green Guide to Specification[6] ratings that are used in BREEAM[7] and the Code for Sustainable Homes[8] to asses the embodied environmental impact of construction products.

The embodied carbon impact of the structure typically represents 15% of the total embodied impact of a building. Although designers should be aware of the environmental impacts of the structural materials they specify, their choice should not be at the cost of the wider sustainability aspects of the building that the structure supports. This, and a broad range of sustainability issues, is addressed in more detail in Section 2.2.

2.2 Design for sustainability

Many of the issues considered above may be wrapped up under the broad heading of 'sustainability'. The purpose of this section is to consider sustainability in a more detailed and focused way, with specific references to regulations, guidance, tools and themes.

Before the engineer can begin to consider how to design a sustainable building he or she has to understand what a sustainable building actually is. This is arguably a greater challenge than any structural design problem they will encounter in their career! Engineers, in general, work with algorithms and absolute values that yield black or white answers, albeit with factors of safety to provide a degree of conservatism that covers, for example, variations in material properties and necessary simplifications in design models. Sustainable development is not like this; it is a concept, not an absolute that can be defined by algorithms. Work is underway to develop more robust metrics for sustainability but its complexity will preclude the establishment of an agreed set of metrics within the short to medium term.

The concept of sustainable development is simple but the detail is complex. Despite numerous attempts, the early (1987) Brundtland[9] definition is still hard to beat 'development that meets the needs of the present without compromising the ability of future generations to meet their own needs'. Central to this definition is the consumption of non-renewal resources and the environmental impacts (to air, ground and water) arising from human activities.

This definition raises an important point of relevance to sustainable construction; namely that it is not physically possible to construct a building, no matter how small, without depleting some resources and without generating some impact on the environment. The challenge to the engineer is to fulfil their traditional role of providing robust, safe, fit for purpose and economic buildings, but with the minimum impact in terms of non-renewable resource use and environmental impact.

The first and arguably most significant decision therefore is whether or not a proposed building is really needed. It is noted that such decisions are generally not within the remit of the structural engineer. Now let us assume that the need for a building has been established, the challenge for the design team is to ensure that the building is fit for purpose, is affordable (however defined) and that these objectives are met in the most sustainable way possible. There is no single solution to this challenge and it is more likely that the client will set a specific target, or set of targets, that are reasonably measurable.

Much of the sustainable construction agenda has, to date, been dominated by inter-material claims. This has been unfortunate since it has distracted designers from delivering more sustainable buildings. Although the environmental impact of construction materials is clearly an important part of sustainability, the product of primary consideration is the building itself. Be it a school, a hospital, an office building or a home, how successfully the building fulfils its intended function is paramount. Buildings that achieve this are likely to be cherished by users and hence last a long time; a key feature of a sustainable building. Both concrete and steel framed buildings can equally achieve high BREEAM ratings (see Section 2.2.4). Therefore it is how different materials, usually in conjunction with one another, can enable the construction of sustainable buildings that is key.

Sustainable development is generally recognised as having three interdependent dimensions; that is economic, social and environmental. Although the challenge is to consider and balance these dimensions holistically, the focus or priority is to reduce the environmental impact while simultaneously addressing the economic and social dimensions. In recognition of this, by far the majority of effort to date has focussed on understanding, quantifying and setting targets for the environmental impacts of construction and buildings.

2.2.1 The social dimension

The social dimension of sustainable construction is diverse. Planners are particularly concerned with social impacts of the built environment and achieve this by addressing spatial development issues associated with new construction and infrastructure. All national and local planning polices including Local Development Frameworks and Regional Spatial Strategies have sustainability at the core. Architects generally address social issues at a more local scale including master planning and at the individual site or building level.

The quality of buildings and the built environment is a key social sustainability consideration. Good design (and construction) yields buildings that can enhance the quality of life of their users. Social attributes of a sustainable building include issues such as security, indoor air quality, thermal comfort, safety and access.

The social considerations of sustainable construction are however much more than just the physical building itself. Other issues that should be considered include:

- the welfare of construction workers particularly their health and safety, skills and training
- local impacts of construction work including congestion, noise, and disruption
- the social impact of the construction materials supply chains
- corporate social responsibility (CSR) of contractors and material suppliers.

The social dimension of sustainable construction is probably the least well understood, particularly when it comes to its assessment and metrics. Work is underway within the European standardisation body CEN to develop a framework for the assessment of social performance of buildings but it is likely to be some time before robust assessment methodologies are available.

2.2.2 The economic dimension

Construction is estimated to contribute around 10% of the UK's gross domestic product, and as such is a cornerstone of the national economy. Design, R&D, codes and standards and product manufacture are areas in which the UK is a world leader. The contribution of these activities to the sustainability of the UK economy should not be underestimated.

At the project level, economic considerations are generally dictated by the prevailing market conditions. The construction industry has notoriously been fixated with initial capital cost rather than whole life cost and whole life value. Although there are examples of decision-making based on whole life cost/value, the industry still has some way to go to make this commonplace. This bias towards initial cost is frequently counter to holistic sustainable construction decision-making.

Building energy efficiency is a good example of where decisions based on whole life costing can yield both environmental and direct economic benefits particularly for building owner occupiers. For many other environmental impacts however there are economic costs for their mitigation.

Material, construction and energy costs are relatively easy to manage, albeit they are subject to significant fluctuations. There are other elements of sustainability for which an economic value could be attributed. An obvious example is carbon but others could include building quality, flexibility, adaptability and reusability of buildings and components. Although such attributes are recognised qualitatively, their economic quantification is more difficult to define. A further complication of such assessments is that traditional whole life costing using treasury discount rates yields only marginal net present value benefit over typical building design lives.

As for social impacts, work is underway at CEN to develop a framework for the assessment of the economic performance of buildings.

2.2.3 The environmental dimension

Reducing the environmental impact of construction and buildings is the greatest sustainability challenge facing the industry and this is reflected in the vast number of initiatives undertaken over recent years.

It is beyond the scope of this chapter to address environmental sustainability in detail. Instead, the fundamental aspects of sustainable building design will be summarised and some key areas where the structural steel designer can play a part will be highlighted. Assessment methods for measuring the environmental sustainability of buildings are also addressed.

2.2.4 Sustainable building assessment

There are many schemes for assessing the sustainability of buildings although, in the main, they focus solely on environmental aspects.

Within the UK, BREEAM is the leading building environmental assessment scheme and is currently the *de facto* national standard. Originally launched in 1990, the methodology has been adapted and updated to reflect the improved understanding of the subject. Although the core BREEAM methodology is consistent, different versions are available for different generic building types, including offices, industrial, schools, hospitals, etc. BREEAM is a voluntary tool, however many publicly-funded buildings are now required to have a minimum BREEAM rating. The residential version of BREEAM (EcoHomes)[10] has been developed into the Code for Sustainable Homes. Since 2008, it is mandatory for new homes to have a Code rating although it is not required that buildings be assessed against the Code (if not assessed they are given a zero rating).

Under BREEAM, buildings are awarded credits according to their performance within nine environmental categories. The number of credits available in each category does not necessarily reflect the relative importance of the issues being assessed. Credits achieved are then added together using a set of environmental weightings to produce a single overall score for the building. Table 2.1 summarises the issues under each of the nine categories.

2.2.5 Fundamentals of sustainable building design and construction

The complexity of sustainable construction and buildings requires the designer to consider a diverse range of design issues. These can include regulatory compliance, for example Part L of the Building Regulations, or specific client requirements, for example achievement of a certain BREEAM rating.

Below are listed the fundamentals of sustainable building design that are currently most commonly considered in UK construction projects. This is by no means a comprehensive list and no priority or importance should be inferred from the listings.

2.2.5.1 Location

- Select a suitable site and consider its proximity to the public transport network and local amenities.
- Where possible, redevelop a brownfield site.
- Consider opportunities to reuse any existing buildings on the site.
- Maximising recycling of existing materials on site, for example by using the ICE's Demolition Protocol.[11]

Table 2.1 Categories and associated issues considered by BREEAM (information based on BREEAM Offices 2008)

Category	Performance issues assessed
Management	Building commissioning and user guide; Considerate Constructors scheme and measuring of construction site impacts; building security
Health and well being	Natural and artificial lighting; view out; glare control; indoor air quality and VOC emissions; lighting and thermal zoning and controls; acoustic performance; potential for natural ventilation
Energy	Reduction in operational CO_2 emissions; sub-metering; external lighting; zero- and low-carbon technologies; energy efficient lifts and transport systems
Transport	Accessibility to the public transport network; proximity to local amenities; provision of cyclist facilities; travel plan for building users; pedestrian and cyclist safety
Water	Reduction in water consumption; water metering; leak detection; proximity detection shut-off in toilet facilities;
Materials	Environmental impact of construction materials; reuse of existing facades; responsible sourcing of construction materials; robust design
Waste	Construction site waste management; use of recycled/secondary aggregates; provision of storage for recyclable waste streams; specification of floor finishes by the building occupant
Land use and ecology	Development of brownfield sites; remediation and development of contaminated sites; preservation and enhancement of the ecological value of the site; minimising the impact on the surrounding biodiversity
Pollution	Low GWP refrigerant services; prevention of refrigerant leaks; low boiler NOx emissions; minimising flood risk; minimising surface water run-off from buildings; minimising night time light pollution; minimising the impact of noise from the building on its neighbours

- Site the building and use landscaping to develop a favourable micro-climate.
- Site the building to enhance the ecological value and biodiversity of the site.
- Strike a balance between tempering the wind and using it to promote natural ventilation and/or as a source of renewable energy.

2.2.5.2 Operational energy efficiency

- Orientate and space buildings to take advantage of solar gain, natural lighting and any appropriate zero- and low-carbon technologies.
- Incorporate the right amount of thermal mass within the building.
- Use glazing, rooflights, etc. to allow solar gain.
- Provide shading to prevent overheating and glare.
- Insulate the building well to reduce fabric heat losses.
- Incorporate a suitable degree of air-tightness.

- Consider the integration of appropriate zero- and low-carbon technologies within or near to the building.
- The HVAC systems should be selected to match the building response and the pattern of occupancy.
- Use natural ventilation wherever possible.
- Design buildings and services that have the capacity to meet predicted future climate change scenarios.
- Provide good controls, sub-metering and operating instructions for the building users.
- Undertake seasonal commissioning of the building.
- Provide appropriate and controllable ventilation to ensure good indoor air quality.

2.2.5.3 Materials

- Specify responsibly-sourced materials.
- Consider the whole life environmental impacts of construction materials using tools such as the Green Guide to Specification.[6]
- Assess buildings holistically, for example by taking account of the structural efficiency of materials, recyclability, etc.
- Specify reused, recycled or recyclable materials and products.
- Minimise the use of primary aggregates.
- Consider whether Modern Methods or off-site construction offer advantages in terms of speed, waste, etc.
- Specify inert and low emissions finishes.
- Where possible use prefabricated and standardised components to minimise waste and enable subsequent reuse.
- Use tools such as the Design Quality Indicators[12] (DQIs) to ensure good design.

2.2.5.4 On site

- Adopt Site Waste Management plans.
- Use the Considerate Constructors scheme.[13]
- Monitor and minimise on-site construction impacts including waste, transport, CO_2 emissions, productivity, etc.
- Minimise local impacts from construction activities such as traffic congestion, noise and dust.

2.2.5.5 Buildings in use

- Design flexible buildings that can be adapted as the user requirements change over time, e.g. long spans and non-loading bearing internal walls that can be reconfigured.

- Design robust buildings.
- Design buildings to minimise ongoing maintenance requirements.
- Consider site and building security, for example using the Secured by Design[14] Principles.

2.2.5.6 Buildings at their end-of-life

- Design buildings that can be easily deconstructed.
- Design buildings using components and materials that can be easily reclaimed, segregated and either reused again or recycled – this should include the sub-structure and particularly any piled foundations.

2.2.6 What the structural engineer can do

Delivering sustainable buildings requires an integrated effort by the design team with early involvement and interaction of all relevant parties. It is important that the sustainability requirements of the project are clearly set out and agreed with the client early in the design process and that all parties are fully aware of the impacts of their design choices and decisions.

It may appear that the structural engineer's role is relatively minor when many of the above criteria are considered. However there are many aspects of the structural design that have a bearing on the sustainability of the building. Issues that need to be considered and balanced by the structural engineer include the following.

2.2.6.1 Structural efficiency and environmental impact

The engineer has an important role to play in designing an efficient structure with a low environmental impact. It is important that the structural efficiency of different materials is considered and that whole building assessments are undertaken. For example a lighter superstructure may require smaller foundations.

In terms of environmental impact, it is important that recycling and the recyclability of structural materials are properly taken into account in any assessments undertaken.

2.2.6.2 Design for climate change

Climate change predictions require the designer to future-proof buildings over their design life. Issues for the structural engineer to consider include:

- changing weather patterns including increased wind loading and storm events
- ground movements resulting from extended wet and dry periods, impacting on the building substructure
- prolonged hot periods causing overheating
- designing new buildings such that low- and zero-carbon technologies can be easily retrofitted, if necessary, in the future.

2.2.6.3 Low- and zero-carbon technologies

New and existing buildings are increasingly likely to require the integration of low- and zero-carbon and other green technologies. The structural engineer has a role to consider how the structural form of the building can be optimised so that the performance of such technologies, e.g. wind and solar, can be maximised. There are also loading and vibration issues that the structural engineer should consider, for example green roofs and roof-mounted wind turbines.

2.2.6.4 Adapting and extending existing buildings

Structural engineers have a key role in considering how existing buildings can be structurally extended and/or adapted to increase their longevity. Façade retention, internal reconfiguration of floors and walls, and horizontal and vertical extensions all require specialist structural design expertise.

2.2.6.5 Adaptable buildings

In addition to refurbishing existing buildings, structural engineers have a role to play in designing new buildings that are adaptable to changing needs of the building users, i.e. future-proofing the building. For example by providing a degree of redundancy in the structure or providing long clear spans that allow internal walls to be reconfigured. Both of these examples may require heavier potentially over-designed structures and therefore the engineer has to weigh up the relative benefits.

Building services are likely to have a much shorter life than the structure and therefore providing a structure that is flexible to alternative servicing strategies is a further important consideration for the structural engineer. Examples include the provision of holes through floor slabs and the use of cellular steel beams which provide flexible servicing options through the web openings. This integration of services within the structural zone can reduce floor-to-floor heights with knock-on benefits in terms of façade costs and reduced heat loss through the envelope.

2.2.6.6 *Impact of the structure on the building servicing strategy*

In partnership with the architect and the services (M&E) engineer, the structural engineer has a role to play in how the structural form of the building is coordinated with the appropriate servicing strategy. This is far too complex a subject to deal with here. However, key issues include:

- Natural ventilation – the structural form can contribute to the effectiveness of cross or stack natural ventilation where this is appropriate.
- Thermal mass – key issues relating to the provision of optimum levels of thermal mass in naturally ventilated buildings relate to its position and degree of exposure. Where floor soffits are exposed their appearance is often a key consideration.
- Thermal bridging – attention should be paid to fabric heat losses caused by thermal bridging. Particular issues include thermal bridging at structural interfaces, e.g. floor to wall junctions, eaves, penetrations of the building envelope such as balcony supports, fixings and structural elements.
- Solar shading – provision of appropriate external solar shading such as brise soleil to limit unwanted solar gains.

2.2.6.7 *Design for deconstruction*

Designers are increasingly required to consider the fate of buildings when they come to the end of their useful life. Where possible buildings should be designed so that they can easily be deconstructed to facilitate reuse of the building or its components or, as a last resort, so that the building materials can be easily segregated for recycling.

In the case of steel structures, sections should be standardised where possible, clearly labelled, in terms of their material properties, and straightforward bolted connections used in preference to welds.

2.3 Design for overall economy

The steel frame does not exist in isolation, it sits on a foundation and supports cladding, services etc., so it should not be designed in isolation. Greatest overall economy will be achieved if the interfaces are given due consideration, and this can only happen if the designer recognises and understands two fundamental ideas:

- Geometric deviations occur on site. Whilst the positions of components will hopefully be within tolerance, they will certainly not be in the nominal positions 'anticipated' on the drawings. Tolerances vary depending on the component and/or material, to reflect what can realistically be achieved during manufacture and erection. Whilst it might be possible to achieve tighter tolerances than 'normal', there will almost certainly be a cost associated with this.

● The most efficient use of space (for example the depth of the floors) is often achieved by combining functions. This is illustrated with some of the examples given below.

More detail on many of the issues covered in this section may be found in the SCI publication *Design for Construction*.[15]

2.3.1 Building services

Although due consideration should be given to the provision of daylight to internal spaces, which may impact on the floor plans, in general it will be desirable to use long span floor solutions so the need for load-bearing internal walls and or columns is minimised. In this way the adaptability of the internal space will be maximised, helping to 'future proof' the building. A potential downside with long span solutions is however their depth; typical composite beams will have a span-to-depth ratio of around 18 to 22. Increased depth may mean a greater area (and therefore cost) of cladding, or fewer floors if the total height of the building is limited.

A range of steel and composite solutions is available. Some of these ease the problem of deeper 'down stand' beams by enabling the services to be passed within the structural depth of the floor. One of the most common modern products is cellular beams, see Figure 2.3, which are formed by welding together (at around mid depth) top and bottom 'Tees' which have been cut from rolled open sections. The cutting and rejoining process includes forming holes in the web, which are normally circular but can be elongated. The top and bottom 'Tees' can be formed from different rolled sections, so that the bottom flange can be considerably bigger than the top flange. This asymmetry is useful when, as is often the case, the beams are composite and so exploit the presence of a concrete upper flange. A similar end product can be achieved using welded sections with holes simply cut into the web plate. Other solutions adopted during the past twenty years or more include things like tapered beams, where the beam depth varies as a function of the structural need at a particular point in the span, so that services can pass under the shallower sections. Truss solutions also allow services to be 'woven' between the internal members. A downside of the more complex solutions such as trusses is that the cost of fire protection may increase.

The depth of the beams may also be reduced by making them continuous, or perhaps semi-continuous, by adopting moment-resisting end connections. Deflections, which often govern long span beams, can be greatly reduced with relatively little end continuity. Downsides to this approach are that column sizes will probably increase due to moment transfer from the beams, frame design may be more complicated, and connection details may be more complicated.

Figure 2.3 Service integration with cellular beams (Courtesy of Westok)

Rather than adopting long span downstand beams it may be more appropriate in some buildings to use a so-called 'slim floor' solution. The steel beams are integrated within the depth of the concrete slab, rather than sitting underneath it. The slab may be formed from precast concrete planks, or may adopt profiled steel decking and in-situ concrete to form a composite slab. 'Deep decking' has been developed for this application, with a profile that is over 200 mm deep and sits on the bottom flange of the steel beams (which must have a wide bottom flange to facilitate this). Unpropped spans of around 6 m can be achieved. The shape of the decking means that very useful voids are formed within the slab, which are big enough to allow services to pass within the slab depth and through holes in the beam webs, see Figure 2.4. This solution therefore saves on floor depth firstly by integrating steel beams and concrete within the same zone and secondly by locating (some of) the services in this zone.

One of the great things about design is that various issues must be weighed up in order to identify the best solution for a given situation. The same is true when considering how best to combine the structural floor and services. Whilst long span solutions with services passing through the beams are attractive, for a building that is highly serviced and where the services are likely to be replaced frequently it may be better to simply hang them underneath the structural floor for ease of fitting and removal.

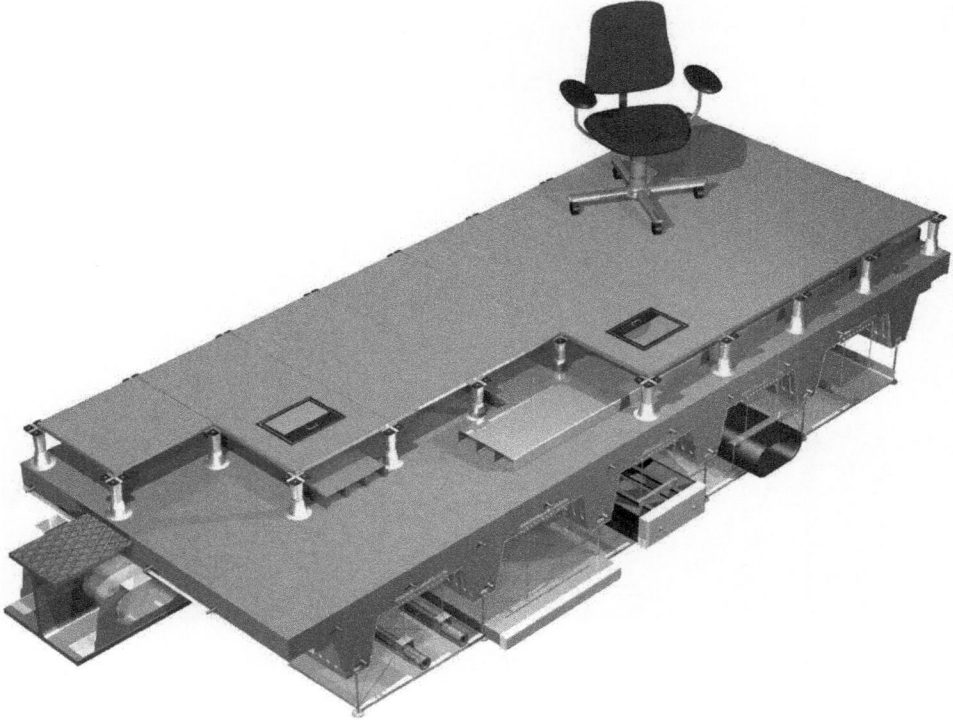

Figure 2.4   Slim floor construction with services

In addition to considering their integration within common space, building services and the structure may interact as part of the building's heating and ventilation strategy. As discussed above, there has been much debate in recent times, some of it rather misguided, about the need to provide thermal mass in order to limit internal temperatures during hot periods. When choosing the optimum building solution the designer must consider potentially conflicting requirements for aesthetics (exposing the thermal mass of the floor slabs may mean exposing the soffits), structural performance and internal temperature control. More exotic solutions such as water-cooled slabs may also be considered.

2.3.2 Building envelope

A wide range of cladding solutions may be used with steel framed buildings, depending on required aesthetics, cost, and durability. One of the biggest issues for designers to consider is that the tolerances within which the steel frame can be fabricated and erected will normally be considerably (perhaps ten times) greater than those associ-

Figure 2.5   Nice reflections in flat rainscreen surface

ated with the cladding. These are not artificial limitations; it is not possible to control the position of steel frame members to within say 5 mm because within the span of an element there is no provision to adjust things on site (the ends of a beam can be adjusted to bring them within tolerance but the mid-span position relative to the ends is a function of the fabrication deviations and load present), and indeed the steel frame members will move as weight is added (e.g. during concreting of the floors) or even depending on the position of the sun (causing differential thermal expansion). The National Structural Steelwork Specification recognises these limitations. On the other hand larger deviations in the position of cladding panels may affect the integrity of seals between panels, the aesthetics of a plane of cladding etc. To avoid increased cost, delays and general frustration on site the designer should adopt details such as adjustable brackets that will allow deviations to be accommodated and maintain the various building components within tolerance. Figure 2.5 shows the attractive architectural finish that is achievable with a carefully adjusted envelope.

The designer should also give due consideration to how a building will deform under load. Some types of cladding are brittle – glazing being the most obvious example. Problems will arise if brittle cladding is fixed to flexible parts of the supporting structure via rigid connections. For example, a 20 m span beam might be expected to deflect by 50 mm or so at mid span under imposed load, so it would be advisable to connect brittle cladding to stiffer parts of the structure and/or make provision in the bracketry for this amount of frame movement. When heavyweight cladding such as brickwork is used, consideration should also be given to how this will affect the frame. Significant load applied eccentrically to the central part of a beam will cause torsional moments.

Another key consideration when designing and detailing the structure-to-cladding interface is the minimisation of cold bridging. This occurs when metal (or other conductive material) penetrates an insulation layer. There is potential conflict between the need to positively fix the external skin and the desire to avoid any thermal bridging from the skin back to the insides of the building. Common examples of this are at the foundations (see below) and when balconies are present. A number of specific materials and products have been developed for balcony connections that combine the necessary qualities of structural resistance and resistance to the transmission of heat.

2.3.3 Foundations

Perhaps the biggest issue for designers to consider when detailing foundation interfaces is the tolerances that can be achieved for many types of foundation which are normally considerably greater than those that are needed at the base of the steel frame. As shown in Figure 2.6, provision must be made to accommodate deviations in both line and level. This is normally done using shims to allow for differing distances between the positions required for the undersides of the column baseplates and the upper surface of the concrete slab or footings. Deviations in the line of the foundations are best allowed for by using holding-down bolts in sleeves (which are filled with grout only once the columns have been finally positioned). Alternatives such as cast-in column bases or post-drilled bolts may also be considered.

When detailing column base connections the designer should also give consideration both to buildability, including minimising risks during erection, and cost. In terms of buildability four-bolt connections should always be used so that the column can be landed on shims (to ensure the right level) and then held by the four bolts as its position (line and verticality) is adjusted. Whilst two bolts might suffice for the final condition they would not be advisable during the temporary state. Base connections should be kept simple to keep costs down – whilst moment-resisting bases may appear to save money on the steelwork by allowing member sizes to be reduced simply at the expense of thicker baseplates, this misses the fact that the concrete base may need considerably more reinforcement, and in addition to the material costs this will inevitably make erection more difficult. It should also be recognised during the design process that a particular construction sequence may be dictated by site access, so the designer may not be free to choose which columns can be erected first to form the basic (permanently) braced unit to which other steelwork is subsequently attached.

A different kind of issue also affects the foundations interface, namely that of thermal bridging. Clearly column bases must be placed on something that is stiff

Figure 2.6   Holding down bolts and baseplate

and strong, and has been brought to the right level (perhaps using shims to make up for any deviations). However, it may also be necessary to isolate the column bases from the cold external environment, so that the columns do not represent significant thermal bridges penetrating the insulation layer at the base of the building. Structural design cannot be considered in isolation, the 'thermal design' of the building must also be considered, and if this is done in an integrated manner it will result in the most effective final solution.

2.4 Conclusions

Integrated design is essential if the three aspects of sustainability, namely social, economic and environmental, are to be addressed in an effective way.

Clients want the building to perform 'now', clearly in terms of standing up against loads but also in terms of thermal and acoustic performance etc. They would also like the building to meet future performance requirements, which may include a desire for it to be adaptable to suit different needs. They will want these things to be achieved with the greatest economy and lowest environmental impact.

In order to achieve the most cost-effective solution it is important that the structural engineer puts the frame in the context of the whole building. Cladding and services are considerably more costly items than the frame itself, and the frame should be designed recognising this, for example by facilitating services routing. Getting the interfaces right is key to achieving an economic solution.

It is also important to note that the relative importance of 'embodied' and 'operational' considerations is changing. In the past the operational impact of a building swamped other considerations, but as building efficiencies improve (driven primarily by regulations) embodied considerations are becoming much more important.

Some compromises will clearly need to be made. Designing the floors to carry relatively high levels of loading will offer the greatest flexibility in terms of future use. However, such an option will also maximise initial material use, and cost. Designing a structure that can be dismantled will facilitate reuse or recycling, but combining materials in composite solutions will often reduce material use because of the structural efficiencies implicit in such forms of construction.

Despite the fact that the steel frame is a relatively small part of the overall building cost, and structural engineering is only one of a broad range of design skills that are needed, this chapter identifies a number of areas where decisions made by the structural engineer are key to developing a good, integrated solution.

References to Chapter 2

1. Building Regulations – see www.planningportal.gov.uk
2. British Standards Institution (2002) *EN 1991-1-1:2002 Eurocode 1: Actions on structures – Part 1-1: General actions – Densities, self-weight, imposed loads for buildings*. London, BSI.
3. Department for Education and Skills (2003) *Building Bulletin 93: Acoustic Design of Schools*. London, The Stationery Office. http://www.ribabookshops.com/search/Department+for+Education+and+Skills+/
4. Department of Health. *Health Technical Memorandum 08-01*. London, The Stationery Office.
5. Steel Construction Institute (2004) *Comparative Structure Cost of Modern Commercial Buildings* (2nd edn), SCI Publication P137. Ascot, SCI.
6. Building Research Establishment *Green Guide to Specification*, www.bre.co.uk/greenguide
7. Building Research Establishment *BRE Environmental Assessment Method*, www.breeam.org
8. Code for Sustainable Homes, www.planningportal.gov.uk
9. United Nations World Commission on Environment and Development (1987) *Our Common Future*. Oxford, Oxford University Press.
10. Building Research Establishment *Ecohomes 2006 – The environmental rating for homes – The Guidance – 2006/Issue 1.2*. Watford, BRE.
11. Institution of Civil Engineers (2008) *Demolition Protocol*. London, ICE.
12. Design Quality Indicators, www.dqi.org.uk
13. Considerate Constructors scheme, www.ccscheme.org.uk/
14. Secured by Design, www.securedbydesign.com/
15. Steel Construction Institute (1997) *Design for Construction*, SCI Publication P178. Ascot, SCI.

Chapter 3
Loading to the Eurocodes

DAVID G. BROWN

This chapter covers the determination of actions, and the combination of those actions at the ultimate limit state and the serviceability limit state. Actions are found in the BS EN 1991 series of the Eurocodes, and the combinations of actions are found in BS EN 1990. In all cases, it is particularly important to refer to the National Annex for that part. In this chapter, the influence of the UK National Annex is reflected – if a structure were to be constructed elsewhere, the National Annex for that country would be required.

3.1 Imposed loads

Imposed loads are taken from BS EN 1991-1-1 and its National Annex. In addition to imposed loads, BS EN 1991-1-1 also covers densities of construction materials and densities of stored materials. For designers, the key interest will be imposed floor loads, minimum imposed roof loads and horizontal loads on parapets and barriers.

3.1.1 Imposed floor loads

BS EN 1991-1-1 identifies four categories of use:

A areas for domestic and residential activities
B office areas
C areas where people may congregate
D shopping areas.

All categories have additional sub-categories – the UK NA provides a large number of sub-categories with examples. For each category of loaded area, imposed

Steel Designers' Manual, Seventh Edition. Edited by Buick Davison and Graham W. Owens.
© 2012 Steel Construction Institute. Published 2012 by Blackwell Publishing Ltd.

Table 3.1 Imposed floor loads (common categories)

Category	Example	q_k (kN/m^2)
A1	All areas within self-contained single family dwellings or modular student accommodation Communal areas (including kitchens) in blocks of flats with limited use that are no more than 3 storeys, and only 4 dwellings per floor are accessible from a single staircase	1.5
A2	Bedrooms and dormitories except those in A1 and A3	1.5
A3	Bedrooms in hotels and motels; hospital wards; toilet areas	2.0
B1	General office use other than in B2	2.5
B2	Office areas at or below ground floor level	3.0
C31	Corridors, hallways, aisles which are not subjected to crowds or wheeled vehicles and communal areas in blocks of flats not covered by A1	3.0
C51	Areas susceptible to large crowds	5.0
C52	Stages in public assembly areas	7.5
D	Areas in general retail shops and department stores	4.0

loads are given both as a point load, Q_k and a uniformly distributed load, q_k. For most design purposes, the point load has little effect other than for local checking. The uniformly distributed load will be used for general design.

Imposed loads for the commonest categories are given in Table 3.1.

The UK NA to BS EN 1991-1-1 should be consulted for the full list of categories and imposed loads. Note that despite the imposed loads given in Table 3.1 (and in the equivalent previous British Standard), common practice has been conservatively to take an imposed floor load as that for crowds, even in a commercial office or similar construction. There is no need for such conservatism in design.

Imposed loads due to storage are given in Tables NA.4 and NA.5 of the UK NA and these should be consulted if storage of goods is envisaged. Although the UK NA offers guidance on specific types of storage, such as libraries, paper storage, file rooms etc, a note to Table NA.5 very sensibly encourages designers to liaise with clients to determine more specific load values.

BS EN 1991-1-1 is very specific about the imposed load due to moveable partitions, which should be added to the imposed loads on floors. The Standard identifies moveable partitions in three categories, with an equivalent uniformly distributed load as given in Table 3.2.

3.1.2 Imposed load reductions

The imposed loads may be reduced, depending on the extent of the area, and the imposed load in columns may be reduced depending on the number of storeys. The reduction factors are given in the UK NA.

Table 3.2 Categories of moveable partition and equivalent UDL

Self-weight of moveable partition	q_k (kN/m²)
≤1.0 kN/m of partition length	0.5
≤2.0 kN/m of partition length	0.8
≤3.0 kN/m of partition length	1.2

3.1.2.1 Reduction factor for area

The reduction factor for area, α_A is given as:

$$\alpha_A = 1.0 - A/1000 \quad \text{but} \geq 0.75$$

The greatest reduction is when the area is at least $250\,\text{m}^2$.

3.1.2.2 Reduction factor for several storeys

The reduction factor for loads from several storeys, α_n depends on the number of storeys, n as given in Table 3.3.

The application of the reduction factor for several storeys has a considerable benefit when designing columns, and is recommended.

The UK NA clarifies that load reductions based on area may be applied if $\alpha_A < \alpha_n$, but cannot be used at the same time as α_n. Thus for a two or three storey structure with a large area, the reduction due to area, α_A, may be 0.75, and may be used in preference to the reduction due to several storeys, α_n, which is 0.9 or 0.8. However, both these reductions may not be used at the same time.

A very important proviso is contained in Clause 3.3.2(2) of BS EN 1991-1-1 that restricts the use of the α_n reduction factor. If the imposed floor load is an accompanying action (i.e. not the leading variable action) when calculating the ultimate limit state (ULS) loads (see Section 3.5.2) it would have a ψ factor applied. Clause 3.3.2(2) requires that *either* the ψ factor is applied *or* the α_n reduction factor, but not both at the same time. It is clear that both factors are addressing the reduced probability of the characteristic load being present on all loaded areas, and cannot be used simultaneously.

Table 3.3 Reduction factor for imposed loads from several storeys

Number of storeys, n	Reduction factor, α_n
$1 \leq n \leq 5$	$1.1 - n/10$
$5 < n \leq 10$	0.6
$n > 10$	0.5

Table 3.4 Horizontal loads on partition walls and parapets

Category of loaded area*	Sub-category	Example	q_k (kN/m)
B and C1	(iii)	Areas not susceptible to overcrowding in office and institutional buildings	0.74
	(iv)	Restaurants and cafes	1.5
C2, C3, C4, D	(viii)	All retail areas	1.5
C5	(x)	Theatres, cinemas, discotheques, bars, auditoria, shopping malls, assembly areas, studios	3.0
F and G	(xv)	Pedestrian areas in car parks including stairs, landings, ramps, edges or internal floors, footways, edges of roofs	1.5

*Categories of loaded areas are given in the UK NA, Table National Annex.2

3.1.3 Loads on parapets

Horizontal loads on parapets and partition walls acting as barriers are taken from Table NA.8 of the UK NA. The load application should be assumed to act no higher than 1.2 m above the adjacent surface. Table NA.8 specifies loads for different categories of loaded areas. The more significant categories are given in Table 3.4. Note that Table 3.4 (and Table NA.8 in the UK NA) do not include loads on partition walls and parapets in grandstands and stadia – the appropriate certifying authority must be consulted.

3.2 Imposed loads on roofs

In previous British Standards, the minimum imposed roof load was rather obscured and appeared as a minimum load as part of the determination of snow loads. Under the Eurocode loading system, imposed roof loads are identified quite separately and distinct from snow loads. Imposed loads on roofs are given in the UK NA to BS EN 1991-1-1, and depend on the roof slope. A point load, Q_k is given, which would be used for local checking of roof materials and fixings, and a uniformly distributed load, q_k, to be applied vertically, as given in Table 3.5.

Table 3.5 Imposed loads on roofs

Roof slope, α	q_k (kN/m^2)
$\alpha < 30°$	0.6
$30° \leq \alpha < 60°$	$0.6[(60 - \alpha)/30]$
$\alpha \geq 60°$	0

The imposed loads in Table 3.5 are for roofs not accessible except for normal maintenance and repair.

Clause 3.3.2(1) of BS EN 1991-1-1 states that

'On roofs (particularly for category H roofs), imposed loads, need not be applied in combination with either snow loads and/or wind actions.'

The intent is that:

(a)   imposed roof loads and wind should not be combined
(b)   imposed roof loads and snow should not be combined
(c)   that snow loads and wind loads should be combined.

3.3 Snow loads

Snow loads should be determined from BS EN 1991-1-3 and its National Annex.

According to BS EN 1991-1-3, and the Clause NA.2.2 of the NA, three design situations should be considered:

1.   undrifted snow
2.   drifted snow (partial removal of snow from one slope)
3.   exceptional snow drifts, which should be treated as accidental actions.

The possibility of 'exceptional snow on the ground', without exceptional snow drifts is specifically excluded as a design situation by NA.2.6.

Both undrifted snow and drifted snow are considered as persistent design situations, and should be combined with other actions using expressions 6.10 or 6.10a and 6.10b of BS EN 1990, as described in Section 3.5.2. The exceptional snow drift case is an accidental action, and should be considered in combination with other actions using expression 6.11b.

3.3.1 Undrifted snow

Snow loads on roofs are given by:

$$s = \mu_i C_e \, C_t \, s_k$$

where:

μ_i is a snow load shape coefficient
C_e is the exposure coefficient
C_t is the thermal coefficient
s_k is the characteristic value of snow load on the ground.

C_e and C_t are recommended as 1.0 (NA.2.15 and NA.2.16)

s_k is to be determined by reference to the characteristic ground snow map in the UK NA, given for a height of 100 m above sea level. The characteristic ground snow map is divided into numbered zones, and the characteristic value s_k determined as:

$$s_k = [0.15 + (0.1Z + 0.05)] + \left(\frac{A - 100}{525}\right), \text{ in which}$$

A is the site altitude in metres and
Z is the zone number from the characteristic ground snow map.

Much of England and Wales is Zone 3. At a site altitude of 150 m, the characteristic value of snow load on the ground, s_k is therefore given by:

$$s_k = [0.15 + (0.1 \times 3 + 0.05)] + \left(\frac{150 - 100}{525}\right) = 0.60 \text{ kN/m}^2$$

Some areas of England and much of Scotland are classed as Zone 4. Some areas of mainland Scotland are classed as Zone 5. BS EN and the NA only cover sites up to 1500 m above sea level. Specialist advice from the Meteorological Office should be sought for sites at higher altitudes.

The snow load shape coefficient for roofs is given by Figure 3.1. For roof slopes less than 30° pitch, $\mu_1 = 0.8$. If snow is prevented from sliding off the roof by a parapet or similar, the snow load shape coefficient should not be taken as less than 0.8.

BS EN 1991-1-3 gives snow load shape coefficients for cylindrical roofs in Clause 5.3.5, which depends on the angle of the tangent to the roof. Where the tangent is steeper than 60°, the snow load shape coefficient is taken as zero, and for all zones where the tangent is less than 60°, a uniform snow load shape coefficient is calculated based on the rise:span ratio, but with a minimum value of 0.8 according to Figure 5.6 of the Standard. A maximum coefficient of 2.0 is specified in the National Annex.

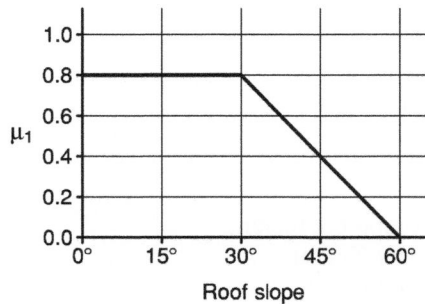

Figure 3.1   Snow load shape coefficient – uniform snow (adapted from Figure 5.1 of BS EN 1991-1-3)

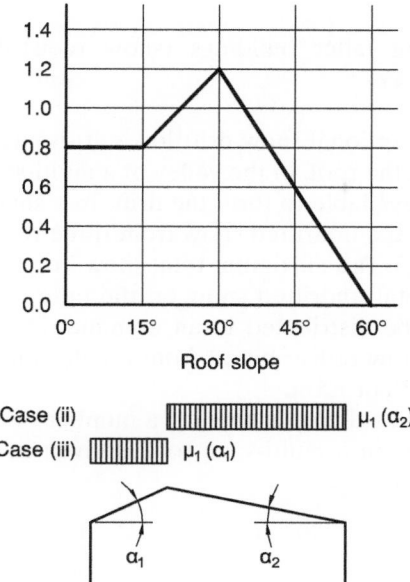

Figure 3.2   Snow load shape coefficient – drifted snow

3.3.2 Drifted snow

The drifted snow condition applies to duo-pitched roofs, and assumes that the snow is completely removed from one slope at a time. Note that the snow load shape coefficients are subtly different from the 'uniform' snow load shape coefficients, and are given in Figure 3.2. For roof slopes less than 15°, $\mu_1 = 0.8$.

The UK National Annex gives snow load shape coefficients for snow drifts on cylindrical roofs. These are given in Table NA.2, and assume that the snow load shape coefficient is zero for one side of the roof, and dependent on the equivalent roof slope on the other. The drifted snow arrangement need only be considered when the angle of a line between the eaves and the crown of the cylindrical roof is greater than 15°, meaning this need not be considered for many low rise roofs.

3.3.3 Exceptional snow drifts

The UK National Annex specifies in Clause NA.2.3 and NA.2.18 that Annex B of BS EN 1991-1-3 should be used to determine exceptional snow drift loads. These snow drift loads occur:

- in valleys of multi-span roofs
- behind parapets

- behind obstructions
- on buildings abutting taller buildings (snow redistributed from the taller building).

When considering exceptional snow drift loads, it should be assumed that there is no snow elsewhere on the roof. In the valley of a multi-span roof, there are limits to the amount of snow available to form the drift. In a single valley, the maximum drift should be taken as the undrifted snow from three roof slopes (assuming they are approximately equal). The maximum total snow in a number of valleys should be no more than the total undrifted snow on the entire roof, although this total volume of snow could be distributed in an asymmetric way. It is generally only necessary to consider snow redistributed from a taller building to a lower roof if the buildings are closer than 1.5 m.

Annex B1 gives snow load coefficients for a number of different situations. The common case of a valley in a multi-span roof is given in Figure 3.3 (taken from Figure B1 of BS EN 1991-1-3).

3.3.3.1 Wind actions

Wind actions are taken from BS EN 1991-1-4 and its National Annex. The UK National Annex is a substantial document and makes significant changes to the core Eurocode – reference to the UK NA is crucial. Designers should ensure that the

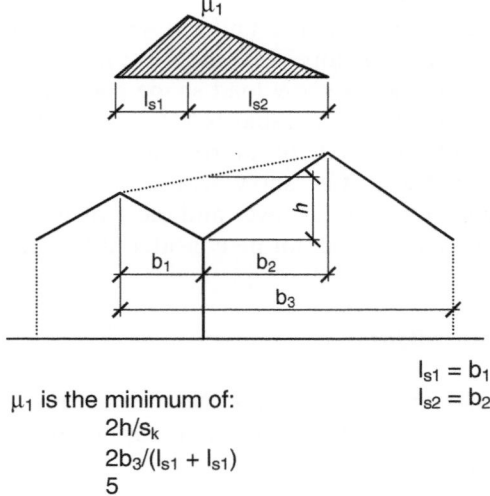

μ_1 is the minimum of:
$2h/s_k$
$2b_3/(l_{s1} + l_{s1})$
5

$l_{s1} = b_1$
$l_{s2} = b_2$

Figure 3.3 Shape coefficients and drift lengths for exceptional snow drifts in the valleys of multi-span roofs

latest standard and NA are used; Amendment No. 1 of January 2011 made significant changes to the pressure coefficients on roofs.

PD 6688-1-4 contains background information and additional guidance. This Published Document contains information on buildings with re-entrant corners, recessed bays, internal wells, irregular faces, inset faces and inset storeys – none of which is supplied in the Eurocode. The PD also provides directional pressure coefficients for walls.

3.3.3.2 General approach

Designers can choose several alternative routes to calculate wind actions. More design effort will generally lead to reduced loads. This section provides a general introduction – more detail is given later. There are four stages in calculating wind actions:

1. calculation of the peak velocity pressure
2. calculation of the structural factor
3. determination of external and internal pressure coefficients
4. calculation of wind forces.

Pressures are calculated as the product of the peak velocity pressure, the structural factor and pressure coefficients. External and internal pressure coefficients are given in the Eurocode. External coefficients given in the Eurocode are for wind within a 45° sector each side of the normal to the face. PD 6688-1-4 gives directional coefficients for walls, at 15° intervals, but use of directional coefficients is not recommended in this Manual – the standard coefficients are appropriate for orthodox structures. Coefficients are given for elements with loaded areas of up to $1\,\mathrm{m}^2$ and loaded areas of over $10\,\mathrm{m}^2$ with logarithmic interpolation for areas between the two. The UK NA simplifies this, and states that the coefficients for $10\,\mathrm{m}^2$, known as $C_{pe,10}$ be used for any loaded area larger than $1\,\mathrm{m}^2$.

The Eurocode also provides force coefficients, which are useful in calculating the overall loads on structures, and include friction effects. Force coefficients are convenient when calculating loads for bracing systems. For portal frames and similar structures loads on elements are required, which are a function of the internal pressures and the external pressures.

3.3.3.3 Calculation of the peak velocity pressure

The peak velocity pressure is based on:

- the fundamental value of the basic wind velocity
- the site altitude

- the height of the structure
- a roughness factor which allows for the ground roughness upwind of the site
- an orography factor, which allows for the effects of hills, if the orography is significant
- the terrain, which may be country or town; in town terrain, the peak velocity pressure is reduced by upwind obstructions
- the distance from the sea, and from the edge of town.

Four alternative approaches to calculating the peak velocity pressure are given below. The different approaches demand different levels of information about the site, and involve different levels of calculation effort. The approaches are summarised in Table 3.6.

Approach 1

The peak velocity pressure can be calculated based on 12 directions around the site. In each direction the preceding factors are calculated, and a peak velocity pressure calculated. The maximum peak velocity pressure can then be used to calculate pressures on elements. This approach demands no knowledge of the orientation of the structure.

Table 3.6 Comparison of approaches to determine the peak velocity pressure.

	Approach 1	Approach 2	Approach 3	Approach 4
Peak velocity pressure	Calculated in 12 directions at 30°	Most onerous factors from 360° around the site	Calculated in 4 quadrants of 90°	Calculated in 4 quadrants of 90° at ±45° to the normal
Building orientation	Not required	Not required	Not required	Required
Outcome	Generally the least onerous result	The most onerous result	Generally a less onerous pressure than Approach 2	Generally a less onerous pressure than Approach 2, but specific to each building face
Calculation effort	Significant	Least effort	Modest effort	Modest effort
Application	Both orthogonal directions use the same peak velocity pressure	Both orthogonal directions use the same peak velocity pressure	Both orthogonal directions use the same peak velocity pressure	Action on each face may use a different peak velocity pressure
Comments	Least onerous peak velocity pressure, but significant calculation effort	Simple, conservative	Recommended approach	Useful for asymmetric structures or where dominant openings are significant

Approach 2

The most conservative approach is to calculate the peak velocity pressure taking the most conservative value of each factor. This simple approach will result in a single peak velocity, which can be used to calculate pressures on elements. This approach demands no knowledge of the orientation of the structure.

Approach 3

The approach recommended in this Manual is to consider quadrants of 90°, and determine the most onerous value of each factor that contributes to the peak velocity pressure. This will result in four values of the peak velocity pressure. The maximum of these four is used to calculate pressures on elements. This approach demands no knowledge of the orientation of the structure. Any orientation of the quadrants can be chosen, as long as the full 360° are captured by the four quadrants.

Approach 4

If the orientation of the structure is known, it may be advantageous to choose the four quadrants (of Approach 3) oriented with respect to the building faces. Each 90° quadrant is formed from 45° each side of the normal to the face. This approach demands knowledge of the building's orientation, but means that wind actions on each building face can be considered with an associated peak velocity pressure. This may be useful if the structure is asymmetric, or there are particular details, such as a dominant opening, where a detailed knowledge of the actions on a specific face is important. For most regular structures, this is not necessary, and Approach 3 will be entirely satisfactory.

It is important to note that in each approach, the influence of factors from 360° around the site must be 'captured' within the determination of the peak velocity pressure. Figure 3.4 illustrates the comparison between approaches 3 and 4.

3.3.3.4 Procedure to calculate the peak velocity pressure

1. Determine the fundamental basic wind velocity (before altitude correction), $v_{b,map}$ from the windspeed map given as Figure NA.1.
2. Calculate the altitude factor, c_{alt}
   The altitude factor depends on the height, z. The definition of z is not entirely clear – according to the NA it is either z_s as defined in Figure 6.1 of BS EN 1991-1-4 or z_e of Figure 7.4. According to Reference 1, z_e should always be used when calculating the velocity pressure. Accordingly, in this Manual, it is recommended that z_e be used, which is taken as the height of the structure. In some

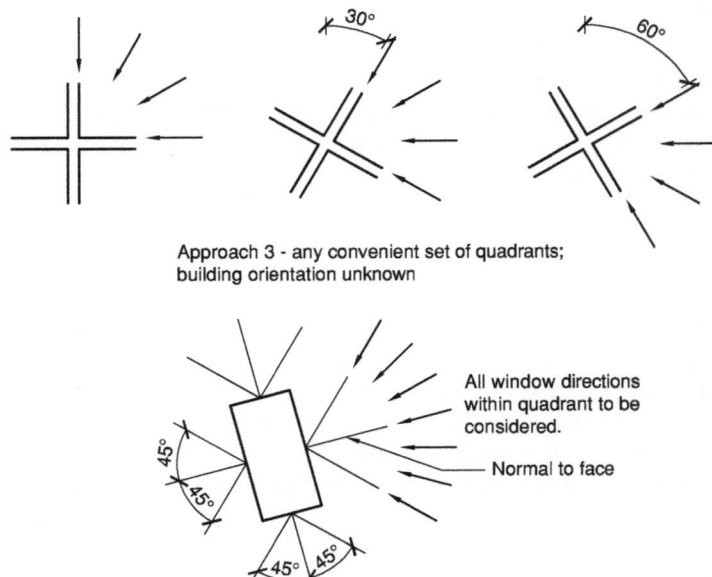

Approach 3 - any convenient set of quadrants;
building orientation unknown

All window directions
within quadrant to be
considered.

Normal to face

Approach 4 - quadrants at ±45° to the normal
to each face; building orientation known

Figure 3.4   Comparison of Approaches 3 and 4

circumstances (for tall buildings where the height exceeds the breadth) the peak velocity pressure may be calculated at a number of heights – see Figure 7.4 of BS EN 1991-1-4. z is the height to the top of the element under consideration.

Two equations are given in the UK NA. For buildings less than 10 m in height, and conservatively for all buildings:

$$c_{alt} = 1 + 0.001\,A \tag{NA.2a}$$

For buildings over 10 m in height:

$$c_{alt} = 1 + 0.001\,A(10/z)^{0.2} \tag{NA.2b}$$

The conservatism in using expression NA.2.a for any building increases with building height and site altitude. For a 40 m building at an altitude of 200 m the c_{alt} factors are 1.2 and 1.15 from NA.2a and NA.2b respectively. Using NA.2a in this case leads to an 8% increase in the peak velocity pressure. For a 40 m building at an altitude of 100 m, the conservatism that results from using NA.2a is 4.5%. The recommendation in this Manual is to use NA.2b when the structure is over 10 m.

3. Calculate the fundamental basic wind velocity $v_{b,0} = v_{b,map} c_{alt}$

4. Determine the directional factor, c_{dir}

   c_{dir} is given in Table NA.1 of the UK NA, at 30° intervals, clockwise, from North. This allows the peak velocity pressure to be calculated in 12 directions, if required (see Approach 1 above). Alternatively, C_{dir} can be simply and conservatively assumed to be 1.0

   Determine the Season factor, c_{season}

   c_{season} is given in Table NA.2 of the UK National Annex, for various specific sub-periods within a year. To use a factor of less than 1.0 would imply absolute confidence that the structure would be exposed to wind for specific months, or parts of a year. In this Manual, it is recommended that c_{season} be taken as 1.0
5. Calculate the basic wind velocity, $v_b = c_{season}\, c_{dir}\, v_{b,0}$
6. Calculate the basic wind pressure, $q_b = 0.613\, v_b^2$
7. Determine the peak velocity pressure. There are two approaches possible at this stage – the general method, which must be used if orography is significant and z is larger than 50 m, or the simplified method which may be used if orography is not significant, and when orography is significant but z is less than 50 m.

   Both approaches demand a knowledge of the upwind terrain, which according to the National Annex may be one of three types:

   Country terrain, which covers areas not directly exposed to the open sea, but only covered with low vegetation and isolated obstacles;

   Town terrain, which covers urban and suburban areas and areas such as permanent forest;

   Sea terrain, which covers costal areas directly exposed to the open sea.

Simplified method

The NA provides two Figures – NA.7 for sites in country terrain and NA.8 for sites in town terrain. Closely-spaced buildings in towns cause the wind to behave as if the ground level was raised to a displacement height, h_{dis}. Annex A.5 of BS EN 1991-1-4 gives the procedure to determine h_{dis}. For sites in the country, $h_{dis} = 0$. The calculation of h_{dis} requires knowledge of the average height of upwind obstructions, and the distance to the upwind obstructions. A.5 notes that in towns, the average height of the upwind obstructions may be taken as 15 m. There is no information on the distance to the obstruction, but previous practice has been to assume 30 m for structures in urban areas, in the absence of any detailed information.

Figure NA.7 gives an exposure factor, $c_e(z)$ at heights z (or at $z - h_{dis}$ in towns), depending on the distance from the sea. Figure NA.8 gives a correction factor, $c_{e,T}$ for sites in town terrain, depending on the upwind distance to the edge of town, and the height $z - h_{dis}$.

The peak velocity pressure $q_p(z)$, is then given by:

$q_p(z) = c_e(z)\, q_b$ for sites in country terrain

$q_p(z) = c_e(z)\, c_{e,T}\, q_b$ for sites in town terrain

The procedure when following the simplified method is given below.

Simplified method (when orography is not significant ($c_o = 1.0$)):

Choose terrain category:

	Country	**Town**
Step 1		calculate h_{dis} from Annex A.5
Step 2	From NA.7, select $c_e(z)$ for distance from sea, and height z $(h_{dis} = 0)$	From NA.7, select $c_e(z)$ for distance from sea, and height $(z - h_{dis})$
Step3		From NA.8, select $c_{e,T}$ for distance from edge of town and height $(z - h_{dis})$
Step 4	$q_p(z) = c_e(z)\, q_b$	$q_p(z) = c_e(z)\, c_{e,T}\, q_b$

Simplified method (when orography is significant but $z < 50$ m):
Follow steps 1 to 4 above, to calculate $q_p(z)$.

Step 5	Calculate the orography factor, $c_o(z)$, from Annex A.3 of BS EN 1991-1-4
Step 6	Calculate $q_p(z) = q_p(z)\, [(c_o(z) + 0.6)/1.6]^2$

General method

The general method is slightly more involved than the simplified method. It must be used when the orography is significant and $z > 50$ m. The general method involves the determination of an orography factor $c_o(z)$, a roughness factor $c_r(z)$ and a turbulence intensity, $I_v(z)$. The roughness factor and the turbulence intensity are taken from Figures NA.3 and NA.5 respectively, and depend on the distance from the sea and the height z (or $z - h_{dis}$) in towns. In towns, a correction factor is applied to the roughness factor and the turbulence intensity taken from Figures NA.4 and NA.6 respectively. Both correction factors depend on the distance from the edge of town and the height $z - h_{dis}$.

The procedure when following the general method is given below:

General method (when orography is significant and $z > 50$ m):
Choose terrain category:

	Country	**Town**
Step 1	Calculate the orography factor, $c_o(z)$, from Annex A.3 of BS EN 1991-1-4	
Step 2	Determine $c_r(z)$ from NA.3, for distance from sea, and height z $(h_{dis} = 0)$	Determine $c_r(z)$ from NA.3, for distance from sea, and height $(z - h_{dis})$
Step 3		Determine c_{rT} from NA.4 for distance from edge of town and height $(z - h_{dis})$

Step 4 Determine $I_v(z)_{,\text{flat}}$ from NA.5 for distance from sea and height z ($h_{\text{dis}} = 0$) Determine $I_v(z)_{,\text{flat}}$ from NA.5 for distance from sea and height ($z - h_{\text{dis}}$)

Step 5 Determine $k_{1,T}$ from NA.6 for distance from edge of town and height ($z - h_{\text{dis}}$)

Step 6 Calculate the mean wind velocity, v_m from $v_m = c_r(z)\, c_o(z)\, v_b$

Step 7 Calculate the peak velocity pressure, $q_p(z)$ from
$$q_p(z) = (1 + (3\, I_v(z)_{,\text{ flat}}/c_o(z)))^2 \times 0.613\, v_m^2$$

3.3.3.5 Wind forces

Wind forces may be calculated using force coefficients, or by calculation of pressures on surfaces.

In both cases, a structural factor $c_s c_d$ is applied, although for the majority of traditional low-rise or framed buildings this may be conservatively set as 1.0. The National Annex allows the structural factor to be separated into a size factor, c_s and a dynamic factor, c_d, which is generally recommended as smaller forces will result from considering each factor separately. Note that the $c_s c_d$ factor is *not* applied to internal pressures – it can only be applied when using overall force coefficients (expressions 5.3 and 5.4) and when using external pressure coefficients (expression 5.5), according to Clause 5.3.

The size factor is obtained from Table NA.3, which depends on the width of the building (or element), the height of the building (or element), and the selection of a zone. Zones are given in Figures NA.7 and NA.8 for country and town locations respectively. The size effect factor accounts for the non-simultaneous action of gusts over external surfaces. It may be applied to individual components and to loads on the overall structure.

The dynamic factor is obtained from Figure NA.9. For this factor, a degree of structural damping, δ_s must be determined by selecting an appropriate category from Table F.2 of BS EN 1991-1-4. Steel buildings have a δ_s value of 0.05, according to this table. The dynamic factor, c_d can then be determined based on the height of the structure and the height:breadth ratio.

Although it is simple and generally conservative to take $c_s c_d = 1.0$, the separation into the individual factors will generally be beneficial in reducing wind forces, and is recommended.

3.3.3.6 Overall force coefficients

Overall force coefficients are given in Table NA.4. They are most useful for calculating overall loads when verifying overturning or sliding, or when designing the vertical bracing system.

3.3.3.7 *External pressure coefficients*

External pressure coefficients are given in the Eurocode, for walls and for roofs.

Funnelling occurs when the wind blows between buildings, depending on their spacing. The notes to Table NA.4 specify the coefficients to be taken and the circumstances when funnelling occurs.

Pressure coefficients are given for loaded areas of $1\,m^2$ and $10\,m^2$. The UK NA states that the values for $10\,m^2$ should be used for all loaded areas greater than $1\,m^2$ – this will be the situation for almost all structural design.

Pressure coefficients fall into a number of zones, with higher suctions next to corners, such as the vertical corner of a wall, or adjacent to the eaves and ridge of a duopitch roof. Figure 3.5 shows the typical zones for a wall, and Figure 3.6 shows the typical zones for a duopitch roof.

The extent of the zones depends on e, which is the smaller of $2h$ or b, where h is the building height and b is the cross-wind breadth. The demarcation between roof zones does not correspond to the zones on the walls, which complicates the assessment of actions on individual frames in typical structures. Some engineering judgement is required to identify the most onerous combinations of actions on the most heavily loaded frame. In many low rise industrial structures, the penultimate frame is likely to be the most critical frame.

If calculating overall loads using pressure coefficients, the National Annex allows a factor of 0.85 (from Clause 7.2.2(3) of BS EN 1991-1-4) to be applied to all the horizontal components of both walls and roof, due to the lack of correlation between the maximum forces calculated for the windward and leeward faces. For pitched roofs, it is not clear how the separation of the horizontal components of force should be accomplished, or if this reduction may be applied when determining the bending moments around a portal frame.

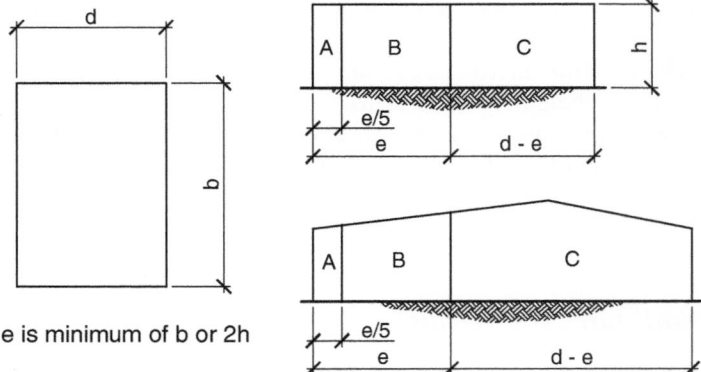

Figure 3.5 Typical wall zones

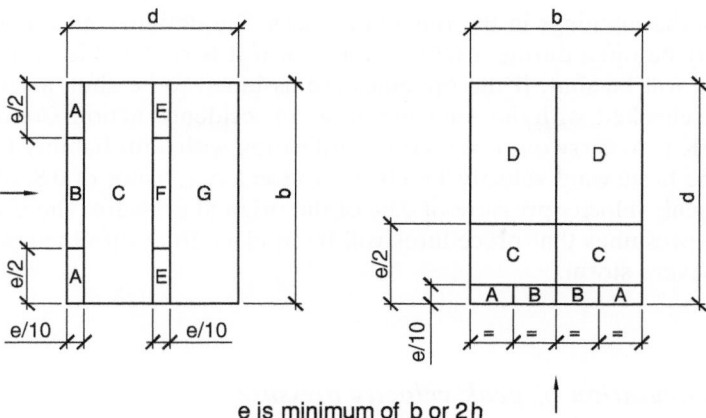

e is minimum of b or 2h

Figure 3.6 Typical roof zones

3.3.3.8 Internal pressure coefficients

Internal pressure coefficients are given in Clause 7.2.9 of BS EN 1991-1-4. For buildings without dominant openings, the value of the internal pressure coefficient is given by Figure 7.13 of the Eurocodes Standard, which offers an opportunity to calculate the internal pressure coefficient based on the opening ratio in the face under consideration. The opening ratio μ is given by:

$$\mu = \frac{\sum \text{area of opening } s \text{ where } c_{pe} \text{ is negative or } -0.0}{\sum \text{area of all openings}}$$

Note 2 to Figure 7.13 of the Eurocodes Standards advises that where it is not possible, or not considered justified to estimate μ for a particular case then c_{pi} should be taken as the more onerous of +0.2 and −0.3. The National Annex gives the permeability of a limited selection of forms of construction. In the absence of more information, which may be forthcoming in due course, a conservative approach is to determine wind loads with both +0.2 and −0.3.

3.3.3.9 Dominant openings

Dominant openings increase the internal pressure or suction dramatically, as high as 75% or 90% of the c_{pe} value at the opening, depending of the size of the opening compared to the openings in the remaining faces. The designer must decide if the openings may be open during a severe storm, or if it is reasonable to assume that the openings will be shut. If the openings are assumed to be shut, a second ULS case must be checked, with the wind actions as an accidental action. Common practice in the UK is to carry out this second verification with a probability factor, c_{prob}, applied to the basic wind velocity. Practice is to use a c_{prob} factor of 0.8, which leads to reduced peak velocity pressure of 0.64 of the original pressure. The use of a c_{prob} factor of 0.8 presumes that procedures will be in place to ensure the openings are closed in a severe storm.

3.3.3.10 Calculation of peak velocity pressure

A worked example is included at the end of the chapter.

3.4 Accidental actions

Three common design situations are treated as accidental design situations:

- drifted snow, determined using Annex B of BS EN 1991-1-3 (see Section 3.3.3)
- opening of a dominant opening assumed to be shut at ULS (see Section 3.3.3.9)
- robustness requirements of BS EN 1991-1-7 and its National Annex.

3.4.1 Robustness

Robustness requirements are to avoid disproportionate collapse. The design calculations for this case are carried out independently of the 'normal' design checks, and substantial permanent deformation is anticipated – generally elements and connection components are designed based on ultimate strengths.

In the UK, the requirement for robustness in a structure is defined by the Building Regulations.

BS EN 1991-1-7 and the UK NA identify and permit three strategies to limit the effect of localised failure:

1. the design of 'key elements'
2. design to limit failure to a modest area – defined in the Building Regulations as the smaller of 15% of the floor area or $70\,m^2$, and limited to the immediate adjacent storeys
3. following prescriptive rules that provide acceptable robustness for the structure.

The strategy to be adopted depends on the classification of the structure, as defined in the Building Regulations and Annex A of BS EN 1991-1-7. Structures are of consequence Class 1, 2a, 2b or 3. Examples of Class 1 buildings are single occupancy structures and agricultural buildings. Examples of Class 2a structures are offices not exceeding 4 storeys. Class 2b structures include offices and residential buildings greater than four storeys but less than 15 storeys. Class 3 structures include larger structures, stadia and those open to significant numbers of the public.

For Class 1 structures, no special provisions are needed, other than good practice.

For Class 2a structures, horizontal ties must be provided. These should be located around the perimeter of the structure and internally, at right angle directions to tie columns together. Annex A1 of BS EN 1991-1-7 states that at least 30% of the ties should be located within the close vicinity of the grid lines of the columns and the walls. Ties may comprise rolled steel sections, steel bar reinforcement in concrete slabs, or steel mesh reinforcement and profiled steel sheeting in composite floors (if directly connected to the steel beams with shear connectors), or a combination of these.

Each continuous tie, including its end connections, should be capable of sustaining a design tensile load as given by:

Internal ties $T_i = 0.8(g_k + \psi q_k)sL$ or 75 kN whichever is the greater

Perimeter ties $T_p = 0.4(g_k + \psi q_k)sL$ or 75 kN whichever is the greater

where:
    s is the spacing of the ties
    L is the length of the tie
    ψ is the combination coefficient relevant to the accidental design action being considered (ψ_1 or ψ_2). In the absence of any other guidance, the larger ψ factor should be selected.

If $g_k = 3.5\,\text{kN/m}^2$, $q_k = 5\,\text{kN/m}^2$, $s = 3\,\text{m}$ and $L = 7\,\text{m}$, and assuming $\psi_1 = 0.5$, then for an internal tie:

$$T_i = 0.8(g_k + \psi q_k)sL = 0.8 \times (3.5 + 0.5 \times 5) \times 3 \times 7 = 101 \text{ kN}$$

Note that if using expression 6.10 to calculate the ULS forces, the end shear reaction is equal to:

$0.5 \times (1.35 \times 3.4 + 1.5 \times 5) \times 3 \times 7 = 128\,\text{kN}$, demonstrating that for orthodox construction, the tie force is generally less than the end shear force.

For Class 2b structures, horizontal and vertical ties are required, and the damage resulting from the notional removal of any column remains within the limits described above. If this cannot be satisfied, the element must be designed as a 'key element'.

Vertical tying requirements affect the design of any splices in the columns. Splices should be capable of resisting an accidental design tensile force equal to the largest design vertical permanent and variable load reaction applied to the column from any one storey.

'Key elements' should be capable of sustaining an accidental design action ($34\,kN/m^2$ is recommended) applied in horizontal and vertical directions (in one direction at a time) to the member and any attached components. The Eurocode points out that the ultimate strength of the attached components, such as cladding, should be considered as such components may have failed under the applied load, rather than transferring load to the member being considered.

For Class 3 structures, in addition to the requirements for Class 2b structures, a systematic risk assessment of the building should be undertaken taking into account both foreseeable and unforeseeable hazards. Generally this will involve making specific provision for hazards.

3.4.2 Actions during execution

Actions during execution (construction) are given in BS EN 1991-1-6 and its National Annex. One common design situation is the effect of wind on a partially erected structure – the designer is referred to a BCSA publication No. 39/05, *Guide to steel erection in windy conditions.*[2]

A second common design situation is the construction loads to be considered when casting concrete, which are covered by Clause 4.11.2 of BS EN 1991-1-6. The recommendation[3] is to apply a construction load of $0.75\,kN/m^2$ over the *whole* area of concrete, in addition to the self-weight. A *further* construction load of 10% of the slab self-weight, but not less than $0.75\,kN/m^2$, is applied over the 'working area' which is an area $3\,m \times 3\,m$, situated to give the most onerous effect.

3.5 Combinations of actions

Combinations of actions are given in BS EN 1990, with important factors given in the UK National Annex. BS EN 1990 covers both ultimate limit state (ULS) and serviceability limit state (SLS), although for the SLS, onward reference is made to the material codes (for example BS EN 1993-1-1 for steelwork) to identify which expression should be used and what SLS limits should be observed.

3.5.1 Ultimate limit states

Four ultimate limit states are identified in BS EN 1990:

- EQU, which should be verified when considering overturning or sliding
- STR, which concerns strength of the structure

- GEO, covering strength of the ground and is used in foundation design
- FAT, covering fatigue failure.

Of these, STR is of most interest to structural designers and is described in detail in this Manual. BS EN 1990 should be consulted for details of the remaining three ultimate limit states.

Design situations are classified as:

Persistent:	normal conditions of use
Transient:	temporary conditions applicable to the structure e.g. loads applied during execution
Accidental:	exceptional conditions applicable to the structure or to its exposure e.g. fire, explosion or the consequence of localised failure
Seismic:	conditions that are applicable to the structure during a seismic event.

The most common design situation is the Persistent situation. The Accidental situation covers situations such as exceptional drifted snow and robustness requirements. Design situations during construction or refurbishment are transient situations.

3.5.2 Combinations of actions – persistent and transient

Combinations of effects of actions (excluding fatigue, seismic and accidental design situations) are given by expression 6.10 of BS EN 1990, where the design value of the combination of actions is given by:

$$\sum_{j\geq1}\gamma_{G,j}G_{k,j} \; "+" \; \gamma_{Q,1}Q_{k,1} \; "+" \; \sum_{i\geq1}\gamma_{Q,i}\psi_{0,i}Q_{k,i} \tag{6.10}$$

Since pre-stressing is largely irrelevant in steelwork design, it has been removed from expression 6.10 above.

In expression 6.10:

- the permanent actions, G are multiplied by a partial factor, γ_G
- a 'leading' or 'main' variable action Q is identified, which is multiplied by a partial factor, γ_Q
- any other variable actions are multiplied by the partial factor, γ_Q, but also multiplied by a combination factor, ψ_0

The ψ_0 factors are specific to the type of action. Thus wind actions have a specific ψ_0 value, and imposed floor loads another. The subscript is an important identifier – ψ_0 values should be used, not ψ_1 or ψ_2.

If there is more that one variable action, each must be identified in turn as the 'leading' or 'main' variable action, with the other variable actions being combined, but each with its specific ψ_0 value.

The symbol '+' merely means 'combined with', presuming that the combination makes the design effect more onerous.

For STR and EQU ULS *only*, the design value of the combination of actions may be found from the more onerous of the pair of expressions (6.10a) and (6.10b) given below:

$$\sum_{j \geq 1} \gamma_{G,j} G_{k,j} \,"+"\, \gamma_{Q,1} \psi_{0,1} Q_{k,1} \,"+"\, \sum_{i \geq 1} \gamma_{Q,i} \psi_{0,i} Q_{k,i} \tag{6.10a}$$

$$\sum_{j \geq 1} \xi_j \gamma_{G,j} G_{k,j} \,"+"\, \gamma_{Q,1} Q_{k,1} \,"+"\, \sum_{i \geq 1} \gamma_{Q,i} \psi_{0,i} Q_{k,i} \tag{6.10b}$$

The use of this pair of expressions is recommended in this Manual – the design value of the combination of actions will be smaller that that determined from expression 6.10. Comparing 6.10a and 6.10b with 6.10:

- expression 6.10a includes a ψ_0 factor on the 'leading' variable action, making it less onerous that of 6.10
- expression 6.10b includes a ξ factor on the permanent actions, making it less onerous that of 6.10.

For these reasons, the pair of expressions 6.10a and 6.10b will produce a less onerous design value than expression 6.10. If using expressions 6.10a and 6.10b, both expressions should be checked to identify the most onerous of the pair. In practice, expression 6.10b will be the critical expression for most orthodox loading.

Values for γ_G, γ_Q, ψ_0 and ξ should be taken from the UK NA. Table 3.7 gives the partial factors for actions.

Table 3.8 gives the ψ_0 factors for buildings. Note that ψ_1 and ψ_2 factors are also included in Table 3.8.

Taking an example with one permanent action, G and two variable actions – imposed floor loads in office areas Q_f and snow loads Q_s, the design value of the combination of actions according to expression 6.10 is:

$$\sum_{j \geq 1} \gamma_{G,j} G_{k,j} \,"+"\, \gamma_{Q,1} Q_{k,1} \,"+"\, \sum_{i \geq 1} \gamma_{Q,i} \psi_{0,i} Q_{k,i}$$

which with the imposed floor load as the leading variable action becomes:

$$1.35 \times G \,"+"\, 1.5 \times Q_f \,"+"\, 1.5 \times 0.5 \times Q_s$$

and with the snow load as the leading variable action becomes:

$$1.35 \times G \,"+"\, 1.5 \times Q_s \,"+"\, 1.5 \times 0.7 \times Q_f$$

Table 3.7   Partial factors for actions

Ultimate Limit State	Permanent Actions $\gamma_{G,j}$		Leading or Main Variable Action $\gamma_{Q,1}$	Accompanying Variable Action $\gamma_{Q,i}$
	Unfavourable	Favourable		
EQU	1.1	0.9	1.5	1.5
STR	1.35	1.0	1.5	1.5

Note: When variable actions are favourable Q_k should be taken as zero

Table 3.8   Values of ψ factors for buildings

Action	ψ_0	ψ_1	ψ_2
Imposed loads in buildings, category (see EN 1991-1-1)			
Category A: domestic, residential areas	0.7	0.5	0.3
Category B: office areas	0.7	0.5	0.3
Category C: congregation areas	0.7	0.7	0.6
Category D: shopping areas	0.7	0.7	0.6
Category E: storage areas	1.0	0.9	0.8
Category H: roofs	0.7	0	0
Snow loads on buildings (see EN 1991-3)			
– for sites located at altitude H > 1 000 m a.s.l.	0.70	0.50	0.20
– for sites located at altitude H ≤ 1 000 m a.s.l.	0.50	0.20	0
Wind loads on buildings (see (EN 1991-1-4)	0.5	0.2	0
Temperature (non-fire) in buildings (see EN 1991-1-5)	0.6	0.5	0

When using expression 6.10a, the design value of the combination of actions is given by:

$$\sum_{j\geq1}\gamma_{G,j}G_{k,j}\ "+"\ \gamma_{Q,1}\psi_{0,1}Q_{k,1}\ "+"\ \sum_{i\geq1}\gamma_{Q,i}\psi_{0,i}Q_{k,i}$$

which with the imposed floor load as the leading variable action becomes:

$$1.35\times G\ "+"\ 1.5\times0.7\times Q_f\ "+"\ 1.5\times0.5\times Q_s$$

and with the snow load as the leading variable action becomes:

$$1.35\times G\ "+"\ 1.5\times0.5\times Q_s\ "+"\ 1.5\times0.7\times Q_f$$

When using expression 6.10b, the design value of the combination of actions is given by:

$$\sum_{j\geq1}\xi_j\gamma_{G,j}G_{k,j}\ "+"\ \gamma_{Q,1}Q_{k,1}\ "+"\ \sum_{i\geq1}\gamma_{Q,i}\psi_{0,i}Q_{k,i}$$

which with the imposed floor load as the leading variable action becomes:

$$0.925 \times 1.35 \times G \text{ "+" } 1.5 \times Q_f \text{ "+" } 1.5 \times 0.5 \times Q_s$$

and with the snow load as the leading variable action becomes:

$$0.925 \times 1.35 \times G \text{ "+" } 1.5 \times Q_s \text{ "+" } 1.5 \times 0.7 \times Q_f$$

When checking uplift under permanent and wind actions $Q_{w,up}$ with expression 6.10 the design value of the combination of actions is given by:

$$\sum_{j \geq 1} \gamma_{G,j} G_{k,j} \text{ "+" } \gamma_{Q,1} Q_{k,1} \text{ "+" } \sum_{i \geq 1} \gamma_{Q,i} \psi_{0,i} Q_{k,i}$$

which becomes:

$$1.0 \times G \text{ "+" } 1.5 \times Q_{w,up}$$

Table 3.9 illustrates the use of the expression 6.10 and the pair of expressions 6.10a and 6.10b. The example assumes the permanent action is $3.5 \, \text{kN/m}^2$; the imposed floor load is $5 \, \text{kN/m}^2$ and the snow load is $0.8 \, \text{kN/m}^2$.

Table 3.9 illustrates that the use of expression 6.10 is generally conservative, and that of the pair of expressions 6.10a and 6.10b, 6.10b is the most onerous.

For a beam which carries only permanent actions and imposed office floor loads (i.e. $\psi_0 = 0.7$), it is simple to demonstrate that if $\xi = 0.925$, $\gamma_G = 1.35$ and $\gamma_G = 1.5$ expression 6.10b will be the most onerous of the pair, unless the unfactored permanent action, G is more than $4.45 \times$ the unfactored variable action, Q.

Table 3.9 The use of the expression 6.10 and the pair of expressions 6.10a and 6.10b

The table assumes the permanent action is $3.5 \, \text{kN/m}^2$; the imposed floor load is $5 \, \text{kN/m}^2$ and the snow load is $0.8 \, \text{kN/m}^2$.

Expression	Leading variable action	Permanent action	Leading variable action	Other variable action	Total
		Factored loads (kN/m^2)			
6.10	imposed floor load	4.73	7.50	0.60	12.83
6.10	snow load	4.73	1.20	5.25	11.18
6.10a	imposed floor load	4.73	5.25	0.60	10.58
6.10a	snow load	4.73	0.60	5.25	10.58
6.10b	imposed floor load	4.37	7.50	0.60	12.47
6.10b	snow load	4.37	1.20	5.25	10.92

3.5.3 Combinations of actions – accidental

The combination of actions for accidental design situations is given by expression 6.11b in BS EN 1990 as:

$$\sum_{j\geq 1} G_{k,j} \; "+" \; A_d \; "+" \; \psi_{1,1}Q_{k,1} \; "+" \; \sum_{i\geq 1} \psi_{2,i}Q_{k,i} \tag{6.11b}$$

Note that the prestressing term has been removed from expression 6.11b as found in the Eurocode. In addition, the UK NA to BS EN 1990 specifies that the combination factor ψ_1 be used for the leading variable action in expression 6.11b.

The common accidental design situations are:

- drifted snow
- robustness (the avoidance of disproportionate collapse)
- checking the accidental opening of the dominant opening.

3.5.4 Serviceability limit state

BS EN 1990 gives three expressions which may be used when checking the serviceability state, and directs the designer to the appropriate material standard. For steel structures, BS EN 1993-1-1 in turn refers to the National Annex. The UK NA states that the characteristic combination of actions should be used at SLS. The characteristic combination of actions is given as expression 6.14b in BS EN 1990:

$$\sum_{j\geq 1} G_{k,j} \; "+" \; Q_{k,1} \; "+" \; \sum_{i>1} \psi_{0,i}Q_{k,i} \tag{6.14b}$$

The UK National Annex states that serviceability deflections should be based on unfactored variable actions, and that permanent actions need not be included. In the UK therefore, expression 6.14b reduces to:

$$Q_{k,1} \; "+" \; \sum_{i>1} \psi_{0,i}Q_{k,i}$$

The UK NA also provides certain deflection limits for members. Vertical deflection limits are given in Table 3.10.

Table 3.10 Suggested limits for vertical deflections

Vertical deflection	
Cantilevers	Length/180
Beams carrying plaster or other brittle finish	Span/360
Other beams (except purlins and sheeting rails)	Span/200
Purlins and sheeting rails	To suit the characteristics of the particular cladding

Table 3.11 Suggested limits for horizontal deflections

Horizontal deflection	
Tops of columns in single-storey buildings except portal frames	Height/300
Columns in portal frame buildings, not supporting crane runways	To suit the characteristics of the particular cladding
In each storey of a building with more than one storey	Height of that storey/300

Horizontal deflections are given in Table 3.11.

In some circumstances, such as when aesthetics are particularly significant, the absolute deflection under permanent and variable actions may be important. The designer should consider these if necessary and make an appropriate calculation.

References to Chapter 3

1. Cook N. (2007) *Designers' Guide to EN 1991-1-4. Eurocode 1: Actions on structures, general actions Part 1-4. Wind actions.* London, Thomas Telford.
2. British Constructional Steelwork Association (2005) *Guide to steel erection in windy conditions*, Publication 39/05. London, BCSA.
3. Brettle M. E and Brown D. G. (2009) *Steel Building Design: Concise Eurocodes.* SCI publication P362. Ascot, Steel Construction Institute.

		Job No.		Sheet 1 of 4		Rev	
The Steel Construction Institute		Job Title	Steel Designers' Manual				
Silwood Park, Ascot, Berks SL5 7QN Telephone: (01344) 623345 Fax: (01344) 622944		Subject	Wind actions to BS EN 1991-1-4 Calculation of peak velocity pressure				
		Client		Made by	DGB	Date	Dec 2009
CALCULATION SHEET		SCI		Checked by		Date	

Calculation of peak velocity pressure

This example demonstrates the calculation of the peak velocity pressure for the hypothetical site shown in Figure 1, on the eastern edge of Norwich. This example demonstrates Approaches 1, 2 and 3, none of which require knowledge of the building orientation.

References to expressions, tables and figures from the UK National Annex are preceded by 'NA' – all other references are to BS EN 1991-1-4

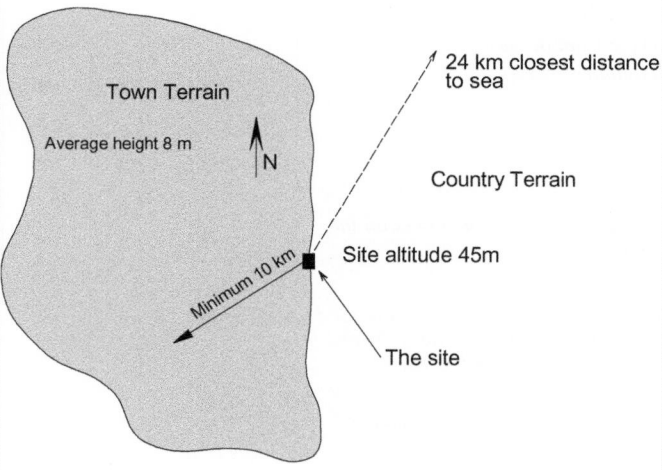

Figure 1   Details of the site

Details of the site are:

Site altitude	*45 m above sea level*
Building height	*27 m*

Terrain category:

From 0° to 180° is country terrain

From 180° to 360° is town terrain – the average height of the upwind obstruction (h_{ave}) is 8 m and the spacing to the upwind obstructions (x) is 30 m. Within this sector, the site is at least 10 km inside the town

The closest distance from the sea is 24 km, at a bearing of 30° from the site.

Example Concept Design	Sheet 2 of 4	Rev

Approach 1

This approach is the simplest, but the most conservative.

$v_{b,map} = 22.5\,m/s$

$c_{alt} = 1 + 0.001\,A\,(10/z)^{0.2}$

$c_{alt} = 1 + 0.001 \times 45 \times (10/27)^{0.2} = 1.037$

$v_{b,0} = v_{b,map} \times c_{alt}$

$v_{b,0} = 22.5 \times 1.037 = 23.3\,m/s$

$c_{dir} = 1.0$ (the maximum from any direction)

$c_{season} = 1.0$

$v_b = c_{season}\,c_{dir}\,v_{b,0}$

$v_b = 1.0 \times 1.0 \times 23.3 = 23.3\,m/s$

$q_b = 0.613 v_b^2$

$q_b = 0.613 \times 23.3^2 \times 10^{-3} = 0.33\,kN/m^2$

Terrain: country, when wind blowing from the east

Distance from the sea: 24 km minimum

height $z = 27\,m$

$c_e(z) = 3.1$

$q_p(z) = c_e(z)\,q_b$

$q_p(z) = 3.1 \times 0.33 = 1.02\,kN/m^2$

This peak velocity pressure may be used to determine forces on the structure in each orthogonal direction.

Approach 2

This approach demands knowledge of the upwind terrain all around the site. Examples of two directions, 330° and 60° are shown below, and then the full details of each direction in tabular format.

330°	60°
$v_{b,0} = 23.3\,m/s$ (as above)	$v_{b,0} = 23.3\,m/s$ (as above)
$c_{dir} = 0.82$	$c_{dir} = 0.73$
$c_{season} = 1.0$	$c_{season} = 1.0$
$v_b = c_{season}\,c_{dir}\,v_{b,0}$	$v_b = c_{season}\,c_{dir}\,v_{b,0}$
$v_b = 0.82 \times 1.0 \times 23.3$	$v_b = 0.73 \times 1.0 \times 23.3$
$= 19.1\,m/s$	$= 17.0\,m/s$
$q_b = 0.613 v_b^2$	$q_b = 0.613 v_b^2$
$q_b = 0.613 \times 19.1^2 \times 10^{-3}$	$q_b = 0.613 \times 17.0^2 \times 10^{-3}$
$= 0.22\,kN/m^2$	$= 0.18\,kN/m^2$
Terrain: town	Terrain: country

$h_{ave} = 8\,m$

$x = 30\,m$ (the spacing to the upwind obstructions)

$2 \times h_{ave} < x < 6 \times h_{ave}$

$2 \times 8 < 30 < 6 \times 8$

$16 < 30 < 48$

Right column references:

Figure NA.1

NA.2b

NA.1

Table NA.1

Table NA.2

4.1

4.10 and NA.2.18

NA.2.11

Figure NA.7

Table NA.1

Table NA.2

4.1

4.10 and NA.2.18

Annex A.5

Example Concept Design		Sheet 3 of 4	Rev

Therefore,

$h_{dis} = min(1.2\,h_{ave} - 0.2x;\ 0.6\,h)$

$h_{dis} = min(1.2 \times 8 - 0.2 \times 30;\ 0.6 \times 27)$

$h_{dis} = min(3.6;\ 16.2) = 3.6\,m$

Distance from the sea = 42 km	*Distance from the sea = 25 km*
$c_e(z) = 2.94$ (at $z - h_{dis} = 23.4\,m$)	$c_e(z) = 3.10$ (at $z = 27\,m$)

Interpolated from Figure NA.7

Distance inside town = 10 km

$c_{e,T} = 0.88$ (at $z - h_{dis} = 23.4\,m$) — *Figure NA.8*

$q_p(z) = c_e(z)\ c_{e,T}\ q_b$ — *NA.2.17*

$q_p(z) = 2.94 \times 0.88 \times 0.22 = 0.57\,kN/m^2$

$q_p(z) = c_e(z)\ q_b$

$q_p(z) = 3.1 \times 0.18 = 0.56\,kN/m^2$

Bearing	0	30	60	90	120	150	180	210	240	270	300	330
$v_{b,0}$ (m/s)	23.33	23.33	23.33	23.33	23.33	23.33	23.33	23.33	23.33	23.33	23.33	23.33
c_{dir}	0.78	0.73	0.73	0.74	0.73	0.80	0.85	0.93	1.00	0.99	0.91	0.82
c_{season}	1.0	1.0	1.0	1.0	1.0	1.0	1.0	1.0	1.0	1.0	1.0	1.0
v_b (m/s)	18.20	17.03	17.03	17.26	17.03	18.66	19.83	21.70	23.33	23.10	21.23	19.13
q_b (kN/m^2)	0.20	0.18	0.18	0.18	0.18	0.21	0.24	0.29	0.33	0.33	0.28	0.22
Distance from sea (km)	32	24	25	27	34	47	73	>100	>100	>100	66	42
$c_e(z)$	3.07	3.1	3.1	3.08	3.06	3.04	3.00	2.88	2.88	2.88	2.91	2.94
$c_{e,T}$								0.88	0.88	0.88	0.88	0.88
$q_p(z)$ (kN/m^2)	0.61	0.56	0.56	0.57	0.55	0.64	0.72	0.73	0.84	0.84	0.72	0.57

Following Approach 2 results in a maximum peak velocity pressure of $0.84\,kN/m^2$, compared to $1.02\,kN/m^2$ from Approach 1. The peak velocity pressure of $0.84\,kN/m^2$ may be used to determine forces on the structure in each orthogonal direction.

Approach 3

In Approach 3, the most onerous values of any factor are taken from any direction within the chosen quadrant. Quadrants may be chosen judiciously to produce the lowest peak velocity pressure. The lowest peak velocity pressure found by Approach 3 will never be smaller than that from Approach 2, but is generally less than the peak velocity pressure from Approach 1.

Two quadrants are demonstrated in detail, and then the results presented in summary form.

Assume that the quadrants are 0° to 90° inclusive, 90° to 180° inclusive, etc, and taking the quadrant from 90° to 180° as an example.

With reference to the Table in Approach 2:

$v_{b,0} = 23.3\,m/s$

maximum c_{dir} from within the quadrant 90° to 180° = 0.85 (at 180°) — *Table NA.1*

$c_{season} = 1.0$ — *Table NA.2*

$v_b = c_{season}\ c_{dir}\ v_{b,0}$ — *4.1*

$v_b = 0.85 \times 1.0 \times 23.3 = 19.8\,m/s$

$q_b = 0.613 v_b^2$ $q_b = 0.613 \times 19.8^2 \times 10^{-3} = 0.24\,kN/m^2$ — *4.10 and NA.2.18*

Terrain: country — *NA.2.11*

Closest distance from the sea within the sector 90° to 180° = 27 km (at 90°)

height $z = 27\,m$

Example Concept Design	Sheet 4 of 4	Rev

$c_e(z) = 3.08$

Interpolated from Figure NA.7
Na.2.17

$q_p(z) = c_e(z)\, q_b$
$q_p(z) = 3.08 \times 0.24 = 0.74\, kN/m^2$

Taking the quadrant from 210° to 300° as a second example

$v_{b,0} = 23.3\, m/s$

maximum c_{dir} from within the quadrant 210° to 300° = 1.0 (at 240°)

Table NA.1

$c_{season} = 1.0$

Table NA.2

$v_b = c_{season}\, c_{dir}\, v_{b,0}$
$v_b = 1.0 \times 1.0 \times 23.3 = 23.3\, m/s$

4.1

$q_b = 0.613 v_b^2$ $q_b = 0.613 \times 23.3^2 \times 10^{-3} = 0.33\, kN/m^2$

4.10 and NA.2.18

Terrain: Town

NA.2.11

$h_{dis} = 3.6\, m$

Annex A.5

Closest distance from the sea within the sector 210° to 300° = 66 km (at 300°)

$c_e(z) = 2.91$ *(at $z - h_{dis} = 23.4\, m$)*

Interpolated from Figure NA.7

$c_{e,T} = 0.88$ *(at $z - h_{dis} = 23.4\, m$)*

Interpolated from Figure NA.8

$q_p(z) = c_e(z)\, c_{e,T}\, q_b$
$q_p(z) = 2.91 \times 0.88 \times 0.33$
$\quad\quad = 0.85\, kN/m^2$

NA.2.17

The summary results are shown below:

Sector	peak velocity pressure (kN/m^2)
30° to 120° inclusive	0.57
120° to 210° inclusive	0.88
210° to 300° inclusive	0.85
300° to 30° inclusive	0.86

When the quadrants are chosen as above, the maximum peak velocity pressure of $1.00\, kN/m^2$ may be used to determine forces on the structure in each orthogonal direction.

Sector	peak velocity pressure (kN/m^2)
0° to 90° inclusive	0.63
90° to 180° inclusive	0.74
180° to 270° inclusive	1.00
270° to 0° inclusive	1.00

When the quadrants are chosen as above, the maximum peak velocity pressure of $0.88\, kN/m^2$ may be used to determine forces on the structure in each orthogonal direction.

In both these examples of the application of Approach 3, the resulting maximum peak velocity pressure is less than that from Approach 1, but more than that from Approach 2. The example demonstrates the beneficial effects of judicious choice of quadrants.

Chapter 4
Single-storey buildings

Compiled by GRAHAM W. OWENS

4.1 The roles for steel in single-storey buildings

It is estimated that around 50% of the hot-rolled constructional steel used in the UK is fabricated into single-storey buildings, being some 40% of the total steel used in them. The remainder is light-gauge steel, cold-formed into purlins, rails, cladding and accessories. Over 90% of single-storey non-domestic buildings have steel frames, demonstrating the dominance of steel construction for this class of building. These relatively light, long-span, durable structures are simply and quickly erected, and developments in steel cladding have enabled architects to design economical buildings of attractive appearance to suit a wide range of applications and budgets.

The traditional image was a dingy industrial shed, with a few exceptions such as aircraft hangars and exhibition halls. Changes in retailing and the replacement of traditional heavy industry with electronics-based products have led to a demand for increased architectural interest and enhancement.

Clients expect their buildings to have the potential for easy change of layout several times during the building's life. This is true for both institutional investors and owner-occupiers. The primary feature is therefore flexibility of planning, which, in general terms, means as few columns as possible consistent with economy. The ability to provide spans up to 60 m, but most commonly around 30 m, gives an extremely popular structural form for the supermarkets, do-it-yourself stores and the like which are now surrounding towns in the UK. The development of steel cladding in a wide variety of colours and shapes has enabled distinctive and attractive forms and house styles to be created.

Improved reliability of steel-intensive roofing systems has contributed to their acceptability in buildings used by the public and, perhaps more importantly, in 'high-tech' buildings requiring controlled environments.

The structural form will vary according to span, aesthetics, integration with services, cost and suitability for the proposed activity. A cement manufacturing building will clearly have different requirements from a warehouse, food processing plant or computer factory.

Steel Designers' Manual, Seventh Edition. Edited by Buick Davison and Graham W. Owens.
© 2012 Steel Construction Institute. Published 2012 by Blackwell Publishing Ltd.

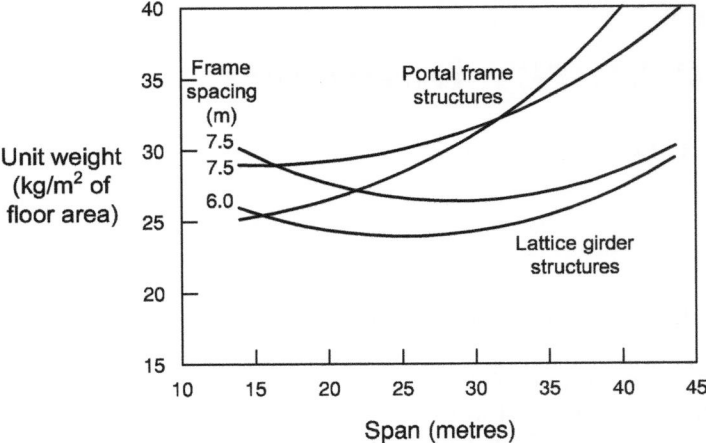

Figure 4.1   Comparison of bare frame weights for portal and lattice structures

The growth of the leisure industry has provided a challenge to designers, and buildings vary from the straightforward requirement of cover for bowls, tennis, etc., to an exciting environment which encourages people to spend days of their holidays indoors at water centres and similar controlled environments suitable for year-round recreation.

In all instances the requirement is to provide a covering to allow a particular activity to take place; the column spacing is selected to give as much freedom of use of the space consistent with economy. The normal span range will be from 12 m to 50 m, but larger spans are feasible for hangars and enclosed sports stadia.

Figure 4.1 shows how steel weight varies with structural form and span.[1]

4.2 Design for long term performance

4.2.1 General design factors

The development of a design solution for a single-storey building, such as a large enclosure or industrial facility, is more dependent on the activity being performed and future requirements for the space than in other buildings such as commercial and residential buildings. Although industrial buildings are primarily functional, they are commonly designed with strong architectural involvement dictated by planning requirements and client 'branding'.

The following overall design requirements should be considered at the concept design stage of industrial buildings and large enclosures, depending on the building form and use:

- space use, for example, specific requirements for handling of materials or components in a production facility

- flexibility of space for current and future use
- speed of construction
- environmental performance, including services requirements and thermal performance
- aesthetics and visual impact
- acoustic isolation, particularly in production facilities
- access and security
- sustainability considerations
- design life and maintenance requirements, including end-of-life issues.

To enable the concept design to be developed, it is necessary to review these considerations based on the type of single-storey building. For example, the requirements for a distribution centre will be different from a manufacturing facility. A review of the importance of various design issues is presented in Table 4.1 for common building types.

Table 4.1 Important design factors for single-storey buildings

Type of single-storey buildings	Space requirements	Flexibility of use	Speed of construction	Access and security	Standardisation of components	Environmental performance	Aesthetics and visual impact	Acoustic isolation	Sustainability considerations	Design life, maintenance and re-use
High bay warehouses	✓✓	✓✓	✓✓	✓✓	✓✓	✓	✓		✓	
Manufacturing facility	✓✓	✓✓	✓	✓	✓		✓	✓✓	✓	✓
Distribution centres	✓✓	✓✓	✓✓	✓✓	✓✓	✓	✓		✓	✓
Retail superstores	✓✓	✓✓	✓	✓✓	✓✓	✓✓	✓✓		✓	✓
Storage / cold storage	✓	✓	✓	✓✓	✓	✓✓	✓		✓	✓✓
Office and light manufacturing	✓	✓	✓	✓	✓	✓✓	✓	✓✓	✓	✓
Processing facility	✓		✓	✓✓	✓	✓		✓✓	✓	✓
Leisure centres	✓	✓✓	✓	✓	✓	✓✓	✓✓	✓	✓✓	✓
Sports halls	✓✓	✓✓	✓	✓	✓	✓✓	✓✓		✓✓	✓
Exhibition halls	✓✓	✓✓	✓	✓	✓	✓✓	✓✓	✓✓	✓✓	✓
Aircraft hangars	✓✓	✓	✓	✓✓	✓	✓	✓	✓	✓	✓

Legend: No tick = Not important ✓ = important ✓✓ = very important

4.2.2 Minimising the energy-in-use

Significant changes have been introduced to the Building Regulations (England & Wales) – Approved Document Part L for Non-Domestic Buildings[2] since it was recognised that a significant proportion of the UK's emission of greenhouse gases arises from energy used in the day-to-day use of building structures.

The significant changes are outlined below:

- increased thermal insulation standards by use of improved U-values
- introduction of an air leakage index
- introduction of as-built inspections
- consideration of whole building design by integrating the building fabric with heating, cooling and air-conditioning requirements
- monitoring of material alterations to an existing structure
- requirement for operating log books and energy consumption meters.

Improved U-values

Approved Document Part L embraces significant changes to the insulation requirements of the building fabric that will impact on the use of fuel and power.

The reduction in the U-values of both roofs and walls is detailed in Table 4.2.

As a consequence of the above, the insulation thickness in composite, glass fibre and rock fibre cladding systems has significantly increased. This has an effect on the dead load that is applied to the supporting structure and results in some nominal increases in section sizes of the primary and secondary steelwork of the building.

Air leakage index

It is now mandatory for buildings to be designed and constructed so that an air leakage index will not be exceeded. This index is set at $5\,m^3/hr/m^2$ of external surface at a pressure difference of 50 pascals.

To achieve the above, care must be taken to specify and effectively construct sealed junctions between all neighbouring elements.

The air leakage index is quantified by in situ air pressurisation tests, using fans on buildings with a floor area that exceeds $1000\,m^2$.

Table 4.2 U-values in Part L of the Building Regulations

	2003	2006	2010
Walls	0.35	0.25	0.20
Roof	0.20	0.16	0.13

Should the building fail the test, remedial measures must be undertaken, and further tests carried out until the building is deemed to comply, to the satisfaction of building control.

'As-built' inspections

In order that the revised insulation standards are adhered to, it is necessary to ensure that the provision of insulation within the fabric itself does not exhibit zones that could compromise the specification. For example, there should be no significant leakage paths between panels; thermal bridges should be minimised; insulation should be continuous, dry and as 'uncompressed' as possible.

One method of ascertaining compliance is the use of infra-red thermography, where the external fabric of the building is 'photographed' using specialist equipment. By this method, areas of the external fabric that are 'hot' – implying poor insulation and heat loss from the building – are shown as colours at the red end of the spectrum. Conversely, areas that are 'cold' are shown at the blue end of the spectrum.

Whole building design

As the title suggests, the building should be designed taking due account of how the constituent parts interact with each other. To this end, insulation, air tightness, windows, doors and rooflights, heating systems and the like should not be viewed as individual elements, but rather as elements that, in some way, have influence on the in-service performance of each other.

Material alterations

Material alteration is intended to cover substantial works to existing buildings that were designed and constructed prior to the changes in Approved Document Part L.

For example, should major works be required to roof or wall cladding, it is necessary to provide insulation to achieve the U-value of a new building. This includes the need to make provision for improving the building's airtightness.

The replacement of doors and windows also entails the use of products that meet the requirements of new buildings, as does the installation of new heating systems.

Building log books and energy meters

The building owner is initially provided with details of all the products that constitute the building, including a forecast of annual energy consumption based on the building design specification. The owner in turn is required to keep detailed maintenance records and the like to ensure that the products within the building are properly maintained, and, when replaced, fully comply with the new requirements.

Energy meters should be provided to ensure that comparisons of energy consumption can be made with those forecast at design stage. These meters in turn provide any new owner/tenant with detailed information on which to base future energy forecasts.

4.3 Anatomy of structure

A typical single-storey building, consisting of cladding, secondary steel frame structure and associated bracing is shown in Figure 4.2.

4.3.1 Cladding

Cladding is required to be weathertight, to provide insulation, to have penetrations for daylight and access, to be aesthetically pleasing, and to last the maximum time with a minimum of maintenance consistent with the budget.

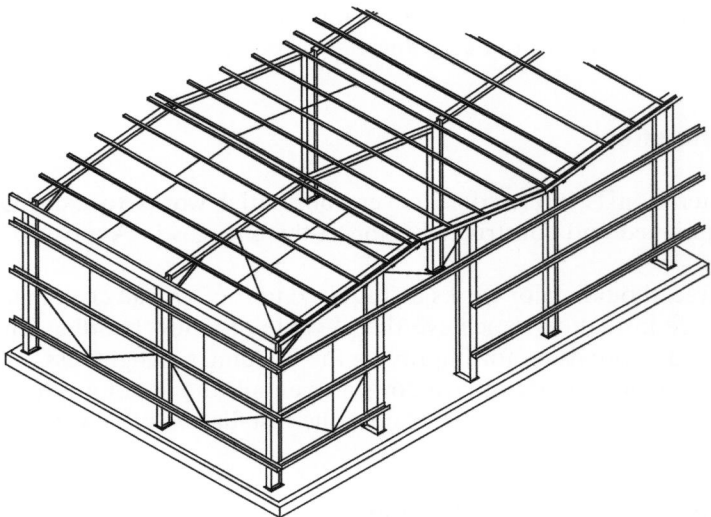

Figure 4.2 Structural form for portal-frame building (some rafter bracing omitted for clarity)

The cladding has also to withstand the applied loads of snow, wind, and foot traffic during fixing and maintenance. It must also provide the necessary lateral stability to the supporting purlin and siderail systems. Occasionally it will form part of the lateral stability of the structure in the form of a stressed-skin diaphragm.

The requirements for the cladding to roofs and walls are somewhat different.

The ability of the roof to remain weathertight is clearly of paramount importance, particularly as the demand for lower roof pitches increases, whereas aesthetic considerations tend to dictate the choice of walling.

Over the past 30 years, metal cladding has been the most popular choice for both roofs and walls, comprising a substrate of either steel or aluminium.

Cladding with a steel substrate tends to be more economical from a purely cost point of view and, coupled with a much lower coefficient of thermal expansion than its aluminium counterpart, has practical advantages. However, the integrity of the steel substrate is very much dependent on its coatings to maintain resistance to corrosion. In some 'sensitive' cases, the use of aluminium has been deemed to better serve the specification.

4.3.1.1 Cladding materials

Typical external and internal coatings for steel substrates manufactured by Tata are detailed below.

Substrate – steel

- Galvatite, hot-dipped zinc-coated steel to BS EN 10326: 2004. Grade Fe E220G with a Z275 zinc coating.
- Galvalloy, hot-dipped alloy-coated steel substrate (95% zinc, 5% aluminium) to BS EN 10326: 2004. Grade S220 GD+ZA with a ZA255 alloy coating.

Coatings – external

- Colorcoat Prisma – a 50 μm organic coating provided with a 25-year guarantee.
- Colorcoat HPS200 Ultra – 200 μm coating applied to the weatherside of the sheet on Galvalloy, above. Provides superior durability, colour stability and corrosion resistance.
- Colorcoat PVDF – 27 μm, stoved fluorocarbon, coating on Galvatite, above. Provides excellent colour stability.
- Colorcoat Silicon Polyester – an economic coating on Galvatite, above. Provides medium term durability for worldwide use.

Coatings – internal

- Colorcoat Lining Enamel – 22 μm coating, 'bright white', with an easily cleaned surface.
- Colorcoat HPS200 Plastisol – 200 μm coating, used in either a corrosive environment or one of high internal humidity.
- Colorcoat Stelvetite Foodsafe – 150 μm coating, comprising a chemically inert polymer for use in cold stores and food processing applications.

The full range of Colors products can be viewed at http://www.colorcoat-online.com/en/products

The reader should note that there is an increasing move towards lifecycle costing of buildings in general, on which the cladding element has a significant influence. A cheaper cladding solution at the outset of a project may result in a smaller initial outlay for the building owner. Over the life of the building, however, running costs could offset (and possibly negate) any savings that may have accrued at procurement stage. A higher cladding specification will reduce not only heating costs but also the carbon footprint of the building.

4.3.1.2 Types of cladding

The construction of the external skin of a building can take several forms, the most prevalent being the following.

Single-skin trapezoidal sheeting

Single-skin sheeting is widely used in agricultural and industrial structures where no insulation is required. It can generally be used on roof slopes as low as 4°, providing the laps and sealants are as recommended by the manufacturers for shallow slopes. The sheeting is fixed directly to the purlins and side rails, as illustrated in Figure 4.3 and provides positive restraint to them. In some cases, insulation is suspended directly beneath the sheeting.

Double-skin system

Double skin or built-up roof systems usually use a steel liner tray that is fastened to the purlins, followed by a spacing system (plastic ferrule and spacer or rail and bracket spacer), insulation and the outer profiled sheeting. Because the connection between the outer and inner sheets may not be sufficiently stiff, the liner tray and fixings must be chosen so that they alone will provide the required level of

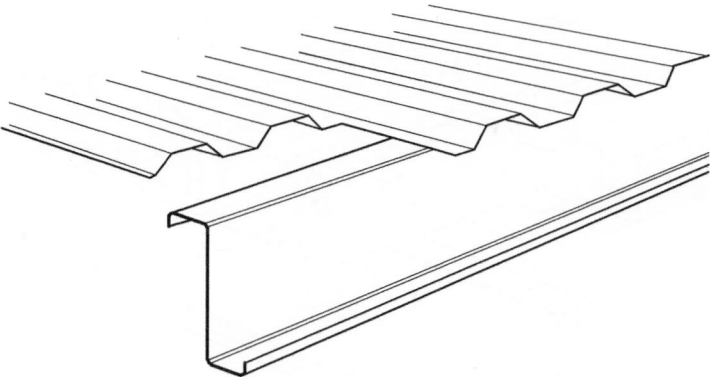

Figure 4.3   Single-skin trapezoidal sheeting

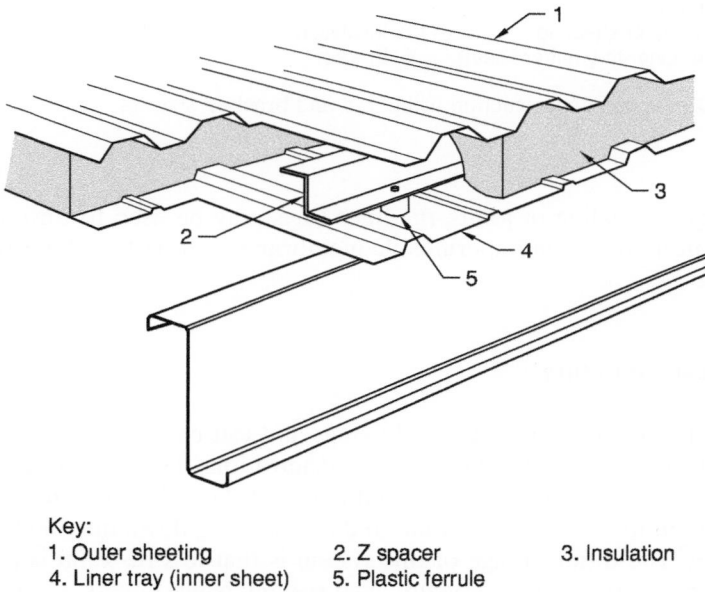

Key:
1. Outer sheeting   2. Z spacer   3. Insulation
4. Liner tray (inner sheet)   5. Plastic ferrule

Figure 4.4   Double-skin construction using plastic ferrule and Z spacers

restraint to the purlins. This form of construction using plastic ferrules is shown in Figure 4.4.

As insulation depths have increased, there has been a move towards 'rail and bracket' solutions as they provide greater lateral restraint to the purlins. This system is illustrated in Figure 4.5.

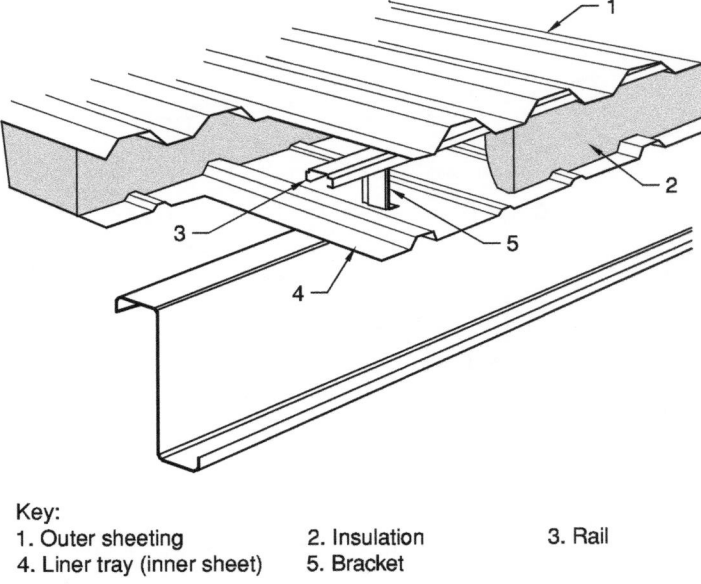

Key:
1. Outer sheeting   2. Insulation   3. Rail
4. Liner tray (inner sheet)   5. Bracket

Figure 4.5   Double-skin construction using 'rail and bracket' spacers

With adequate sealing of joints, the liner trays may be used to form an airtight boundary. Alternatively, an impermeable membrane on top of the liner tray should be provided.

Standing seam sheeting

Standing seam sheeting has concealed fixings and can be fixed in lengths of up to 30 m. The advantages are that there are no penetrations directly through the sheeting that could lead to water leakage, and fixing of the roof sheeting is rapid. The fastenings are in the form of clips that hold the sheeting down but allow it to move longitudinally. The disadvantage of this system is that less restraint is provided to the purlins than with a conventionally fixed system. Nevertheless, a correctly fixed liner tray should provide adequate restraint.

Composite or sandwich panels

Composite or sandwich panels are formed by creating a foam insulation layer between the outer and inner layer of sheeting. Composite panels have good spanning capabilities due to the composite action of the core with the steel sheets. Both standing seam (see Figure 4.6) and direct fixing systems are available. These will

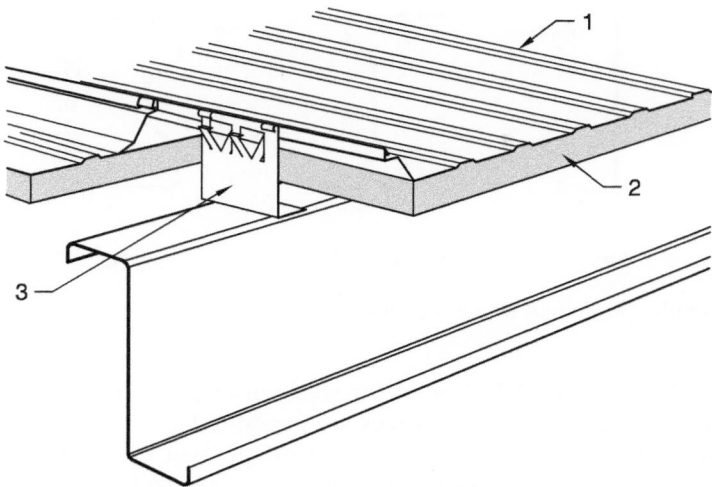

Figure 4.6   Standing seam panels with liner trays. 1 Outer skin; 2 insulation; 3 proprietary cleat

clearly provide widely differing levels of restraint to the purlins. The manufacturers should be consulted for more information.

4.3.2 Secondary elements

In a normal single-storey building the cladding is supported on secondary members which transmit the loads back to main structural steel frames. The spacing of the frames, determined by the overall economy of the building, is normally in the range 5–8 m, with 6 m and 7.5 m being the most common.

A combination of cladding performance, erectability and the restraint requirements for economically-designed main frames dictates that the purlin and rail spacing should be within the range 1.5–2 m.

For this range the most economic solution has proved to be cold-formed light-gauge sections of proprietary shape and size produced to order on computer numerically controlled (CNC) rolling machines. These have proved to be extremely efficient since the components are delivered to site pre-engineered to the exact requirements which minimises fabrication and erection times and eliminates material wastage. Because of the high volumes, manufacturers have been encouraged to develop and test a range of materially-efficient sections. These fall into three main categories: Zed, modified Zed and Sigma sections. Figure 4.7 illustrates the range.

The Zed section was the first shape to be introduced. It is material-efficient but the major disadvantage is that the principal axes are inclined to the web. If subject

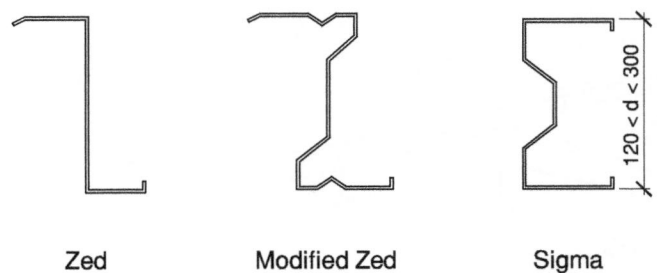

Zed   Modified Zed   Sigma

Figure 4.7   Popular purlin and frame sections

to unrestrained bending in the plane of the web, out-of-plane displacements occur; if these are restrained, out-of-plane forces are generated.

More complicated shapes have to be rolled rather than press braked. This is a feature of the UK, where the market is supplied by relatively few manufacturers and the volumes produced by each allow advanced manufacturing techniques to be employed, giving competitive products and service.

As roof pitches become lower, modified Zed sections have been developed with the inclination of the principal axis considerably reduced, so enhancing overall performance. Stiffening has been introduced, improving material efficiency.

The Sigma shape, in which the shear centre is approximately coincident with the load application line, has advantages. One manufacturer now produces, using rolling, a third-generation product of this configuration, which is economical.

4.3.3 Primary structure

The basic structural form of a single-storey building may be of various generic types, as shown in Figure 4.8. The figure shows a conceptual cross-section through each type of building.

The basic design concepts for each structural type are described below.

Simple roof beam, supported on columns

The span will generally be modest, up to approximately 15 m. The roof beam may be pre-cambered. Bracing will be required in the roof and all elevations, to provide in-plane and longitudinal stability.

Portal frame

A portal frame is a rigid frame with moment-resisting connections to provide stability in-plane. A portal frame may be single-bay or multi-bay. The members are gener-

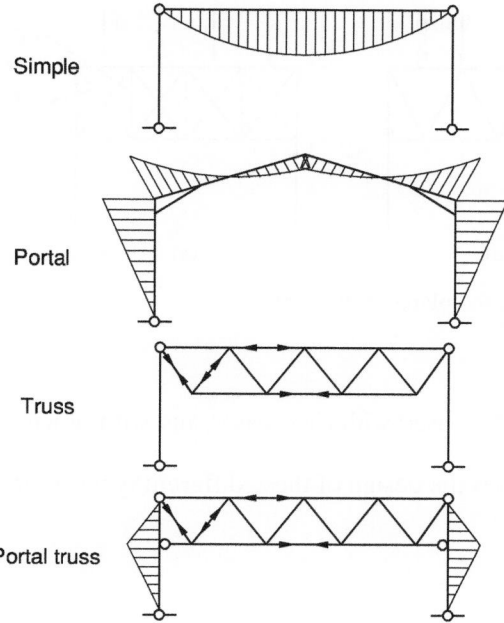

Simple

Portal

Truss

Portal truss

Figure 4.8   Structural concepts

ally plain rolled sections, with the resistance of the rafter enhanced locally with a haunch. In many cases, the frame will have pinned bases.

Stability in the longitudinal direction is provided by a combination of bracing in the roof, across one or both end bays, and vertical bracing in the elevations. If vertical bracing cannot be provided in the elevations (due to industrial doors, for example) stability is often provided by a rigid frame within the elevation.

Trusses

Truss buildings generally have roof bracing and vertical bracing in each elevation to provide stability in both orthogonal directions. The trusses may take a variety of forms, with shallow or steep external roof slopes.

A truss building may also be designed as rigid in-plane, although it is more common to provide bracing to stabilise the frame.

Other forms of construction

Built–up columns (two plain beams, connected to form a compound column) are often used to support heavy loads, such as cranes. These may be used in portalised

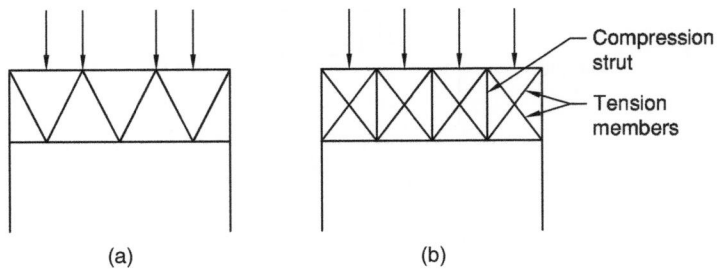

Figure 4.9   Bracing in the plane of the roof

structures, but are often used with rigid bases, and with bracing to provide in-plane stability.

Section 4.4 discusses the design of these different types of primary frame in more detail.

4.3.4 Resistance to sway forces

Most of the common forms provide resistance to sidesway forces in the plane of the frame. It is essential also to provide resistance to out-of-plane forces; these are usually transmitted to the foundations by a combination of horizontal and vertical girders. The horizontal girder in the plane of the roof can be of two forms, as shown in Figure 4.9. Type (a) is formed from members, often tubes, capable of carrying tension or compression. One of the benefits is in the erection stage as the braced bay can be erected first and lined and levelled to provide a square and safe springboard for the erection of the remainder.

Type (b) uses less material but more members are required. The diagonals are tension-only members (wire rope has been used) and the compression is taken in the orthogonal strut which has the shortest possible effective length. It may be possible to use the purlins, strengthened where necessary, for this purpose.

Similar arrangements must be used in the wall to carry the forces down to foundation level. If the horizontal and vertical girders are not in the same bay, care must be taken to provide suitable connecting members.

Bracing is discussed in more detail in Section 4.7.

4.4 Loading

Chapter 3 presents the general application of BS EN 1991. The sections below summarise the issues that are specific to single-storey buildings.

4.4.1 External gravity loads

The dominant gravity loading is snow.

The trend towards both curved and multi-span pitched and curved roof structures, with eaves and gable parapets may add to the number of load combinations that the designer must recognise. Careful consideration needs to be given to the possibility of drifting in the valleys of multi-span structures and adjacent to parapets, in addition to drifting at positions of abrupt changes in height. The drift condition must be allowed for not only in the design of the frame itself, but also in the design of the purlins that support the roof cladding, since the intensity of loading at the position of maximum drift is often far in excess of the minimum basic uniform snow load.

In practice, the designer will usually design the purlins for the uniform load case, thereby arriving at a specific section depth and gauge. In the areas subject to drift, the designer will maintain that section and gauge and reduce the purlin spacing locally. In some instances, however, it may be possible to maintain purlin depth but increase purlin gauge in the area of the drift. An increase in purlin gauge implies a stronger purlin, which in turn implies that the spacing of the purlins may be increased over that of a thinner gauge. However, there is the possibility on site that purlins which appear identical to the eye, but are of different gauge, may not be positioned in the location that the designer envisaged. As such, the practicality of the site operations should also be considered, thereby minimising the risk of construction errors.

Over the years, the calculation of drift loading and associated purlin design has been made relatively straightforward by the major purlin manufacturers, a majority of whom offer state of the art software to facilitate rapid design, invariably free of charge.

4.4.2 Wind loads

Wind loads matter for single-storey buildings. They are inherently complex and likely to influence the final design of most buildings. As discussed in Chapter 3, the designer needs to make a careful choice between a fully rigorous, complex assessment of wind loads and the use of simplifications which ease the design process but make the wind loads more conservative.

4.4.3 Internal gravity loads

Traditionally, service loads for lighting, etc. were reasonably included in the global 'snow' loading. As service requirements have increased, it has become necessary to consider carefully whether additional provision is needed.

Most purlin manufacturers can provide proprietary clips for hanging limited point loads to give flexibility of layout. Where services and sprinklers are required, it is normal to design the purlins for a global service load of $0.1-0.2\,kN/m^2$ with a reduced value for the main frames to take account of likely spread. Particular items of plant must be treated individually. The specifying engineer should make a realistic assessment of the load as the elements are sensitive, and while the loads may seem low, locally they may represent a significant percentage of the total.

4.4.4 Cranes

The most common form of craneage is the overhead type running on beams supported by the columns. The beams are carried on cantilever brackets or, in heavier cases, by providing dual columns.

In addition to the self-weight of the cranes and their loads, the effects of acceleration and deceleration have to be considered. For simple cranes, this is by a quasi-static approach with increased loads.

For heavy, high-speed or multiple cranes the allowances should be specially calculated with reference to the manufacturers.

The constant movement of a crane gives rise to a fatigue condition. This is, however, restricted to the local areas of support, that is, the crane beam itself, the support brackets and their connection to the main column. It is not normal to design the whole frame for fatigue as the stress levels due to the crane travel are relatively low.

4.5 Common types of primary frame

In this section, the common structural forms introduced in Section 4.3 are discussed in more detail, highlighting any special aspects of design, including the use of the Eurocodes.

4.5.1 Beam and column

General

The building cross-section shown in Figure 4.10 is undoubtedly the simplest framing solution which can be used for single-storey buildings. It is used predominantly in spans of up to 15 m. Where flat roof construction is acceptable, the frame comprises standard hot-rolled sections with simple or moment-resisting joints.

Flat roofs are notoriously difficult to weatherproof, since deflections of the horizontal cross-beam induce ponding of rainwater on the roof, which tends to penetrate

Figure 4.10 Simplest single-storey structure

the laps of traditional cladding profiles and, indeed, any weakness of the exterior roofing fabric. To counteract this, either the cross-member is cambered to provide the required fall across the roof, or the cladding itself is laid to a predetermined fall, again facilitating drainage of surface water off the roof.

Due to the need to control excessive deflections, the sections tend to be somewhat heavier than those required for strength alone, particularly if the cross-beam is designed as simply-supported. In its simplest form, the cross-beam is designed as spanning between columns, which, for gravity loadings, are in direct compression apart from a small bending moment at the top of the column due to the eccentricity of the beam connection. The cross-beam acts in bending due to the applied gravity loads, the compression flange being restrained either by purlins, which support the roof sheet, or by a proprietary roof deck, which may span between the main frames and which must be adequately fastened. The columns are treated as vertical cantilevers for in-plane wind loads.

Resistance to lateral loads is achieved by the use of a longitudinal wind girder, usually situated within the depth of the cross-beam. This transmits load from the top of the columns to bracing in the vertical plane, and thence to the foundation. The bracing is generally designed as a pin-jointed frame, in keeping with the simple joints used in the main frame. A typical layout is shown in Figure 4.11 (purlins are omitted for clarity).

Buildings which employ the use of beam-and-column construction often have brickwork cladding in the vertical plane. With careful detailing, the brickwork can be designed to provide the vertical sway bracing, acting in a similar manner to the shear walls of a multi-storey building.

Resistance to lateral loading can also be achieved either by the use of rigid connections at the column/beam joint or by designing the columns as fixed-base cantilevers. The latter point is covered in more detail in the following subsection relating to the truss and stanchion framing system.

Design to the Eurocodes

Detailed design of beam and column structures to the Eurocodes presents no particular difficulty. Stability of the single-storey frame may be considered as a special

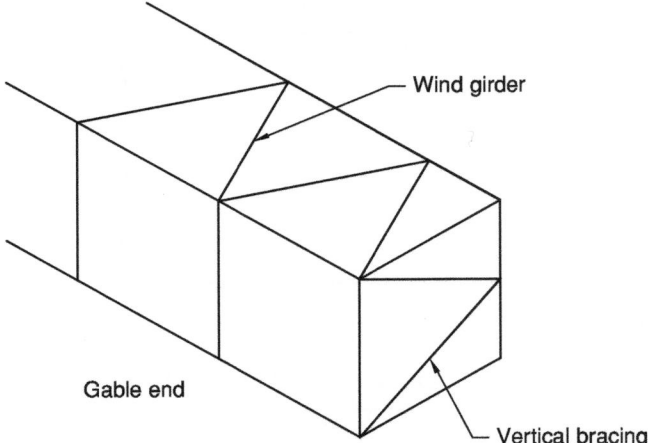

Figure 4.11   Simple wind bracing system

case of multi-storey frame stability discussed in Chapter 5. Beams, columns and connections are addressed in 16, 17, 25 and 27.

4.5.2 Truss and stanchion

General

The truss and stanchion system is essentially an extension of the beam-and-column solution, providing an economic means of increasing the useful span.

Typical truss shapes are shown in Figure 4.12.

Members of lightly-loaded trusses are generally hot-rolled angles as the web elements, and either angles or structural tees as the boom and rafter members, the latter facilitating ease of connection without the use of gusset plates. More heavily loaded trusses comprise universal beam and column sections and hot-rolled channels, with connections invariably employing the use of heavy gusset plates.

Special considerations

In some instances there may be a requirement for alternate columns to be omitted for planning requirements. In this case, load transmission to the foundations is effected by the use of long-span eaves beams carrying the gravity loads of the intermediate truss to the columns: lateral loading from the intermediate truss is transmitted to points of vertical bracing, or indeed vertical cantilevers by means of

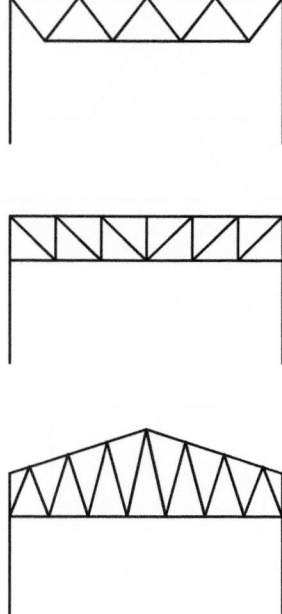

Figure 4.12 Truss configurations

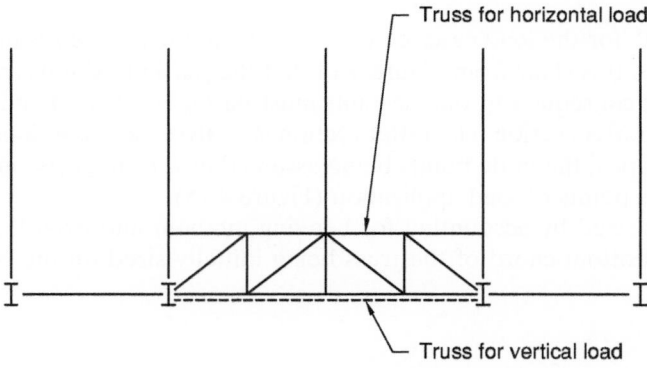

Figure 4.13 Additional framing where edge column is omitted

longitudinal bracing as detailed in Figure 4.13. The adjacent frames must be designed for the additional loads.

Considering the truss and stanchion frame shown in Figure 4.14, the initial assumption is that all joints are pinned, that is, they have no capacity to resist bending moment. The frame is modelled in a structural analysis package or by hand

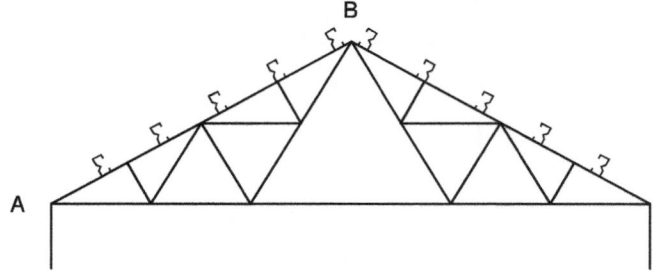

Figure 4.14   Truss with purlins offset from nodes

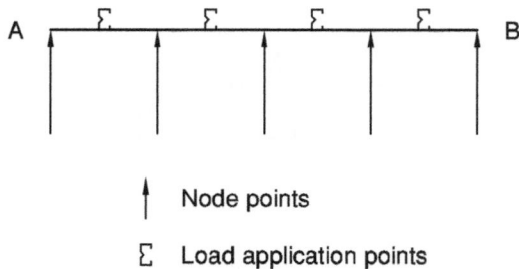

↑   Node points

Ʃ   Load application points

Figure 4.15   Rafter analysed for secondary bending

calculation, and, for the load cases considered, applied loads are assumed to act at the node points. It is clear from Figure 4.14 that the purlin positions and nodes are not coincident; consequently, due account must be taken of the bending moment induced in the rafter section. The rafter section is analysed as a continuous member from eaves to apex, the node points being assumed as the supports, and the purlin positions as the points of load application (Figure 4.15).

The rafter is sized by accounting for bending moment and axial loads, the web members and bottom chord of the truss being initially sized on the basis of axial load alone.

Analysis

Use of structural analysis packages allows the engineer to rapidly analyse any number of load combinations. Typically, dead load, live load and wind load cases are analysed separately, and their factored combinations are then investigated to determine the worst loading case for each individual member. Most software packages provide an envelope of forces on the truss for all load combinations, giving maximum tensile and compressive forces in each individual member, thus facilitating rapid member design.

Out-of-plane bracing

Under gravity loading the bottom chord of the truss will be in tension and the rafter chords in compression. In order to reduce the slenderness of the compression members, lateral restraint must be provided along their length, which in the present case is provided by the purlins which support the roof cladding. In the case of load reversal, the bottom chord is subject to compression and must be restrained. A typical example of restraint to the bottom chord is the use of ties, which run the length of the building at a spacing governed by the slenderness limits of the compression member; they are restrained by a suitable end bracing system. Another solution is to provide a compression strut from the chord member to the roof purlin, in a similar manner to that used to restrain compression flanges of rolled sections used in portal frames. The sizing of all restraints is directly related to the compressive force in the primary member, usually expressed as a percentage of the compressive force in the chord. Care must be taken in this instance to ensure that, should the strut be attached to a thin-gauge purlin, bearing problems in thin-gauge material are accounted for. Examples of restraints are shown in Figure 4.16.

Connections

Connections are initially assumed as pins, thereby implying that the centroidal axes of all members intersecting at a node point are coincident. Practical considerations frequently dictate otherwise, and it is quite common for member axes to be eccentric to the assumed node for reasons of fit-up and the physical constraints that are inherent in the truss structure. Such eccentricities induce secondary bending stresses of the node points, which must be accounted for not only by local bending and axial load checks at the ends of all constituent members, but also in connection design. Typical truss joints are illustrated in Figure 4.17.

It is customary to calculate the net bending moment at each node point due to any eccentricities, and proportion this moment to each member connected to the node in relation to member stiffness.

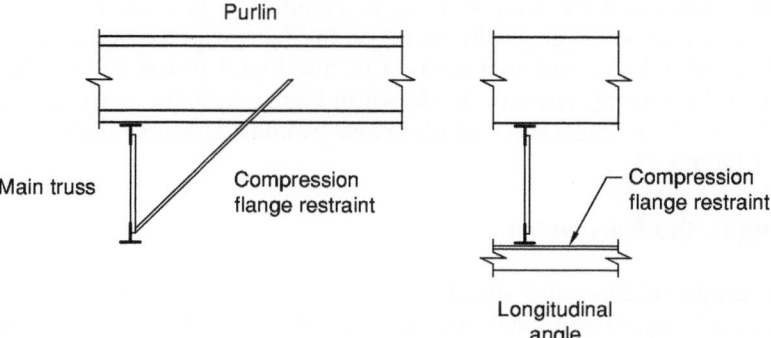

Figure 4.16   Restraints to bottom chord members

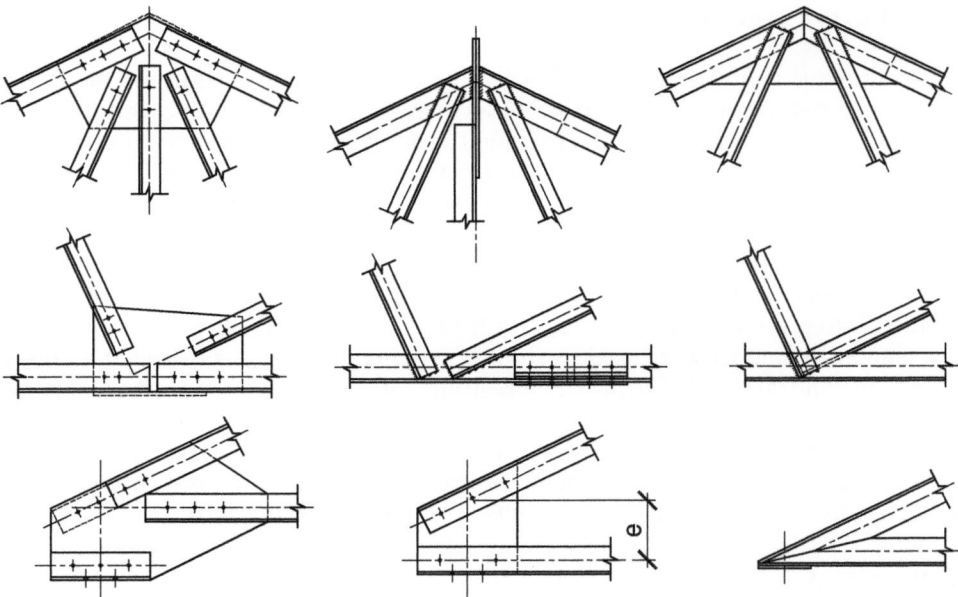

Figure 4.17   Typical joints in trusses

In heavily-loaded members, secondary effects may be of such magnitude as to require member sizes to be increased quite markedly above those required when considering axial load effects alone. In such instances, consideration should be given to the use of gusset plates, which can be used to ensure that member centroids are coincident at node points, as shown in Figure 4.18. Types of truss connections are very much dependent on member size and loadings. For lightly-loaded members, welding is most commonly used with bolted connections in the chords if the truss is to be transported to site in pieces and then erected. In heavily-loaded members, using gusset plates, either bolting, welding or a combination of the two may be used. The type of connection is generally based on the fabricator's own reference.

Longitudinal stability and transmission of horizontal forces to the foundations are provided by bracing systems. As shown in Figure 4.19, these are usually at the ends of the building, unless dictated otherwise by considerations of thermal expansion, see Chapter 31.

Designing to the Eurocodes

Detailed design of truss and stanchion structures to the Eurocodes presents no particular difficulty. Overall stability may be addressed by similar techniques to those adopted for multi-storey buildings in Chapter 5. Trusses, stanchions and connections are addressed in Chapters 20, 16, 25, 27 and 28.

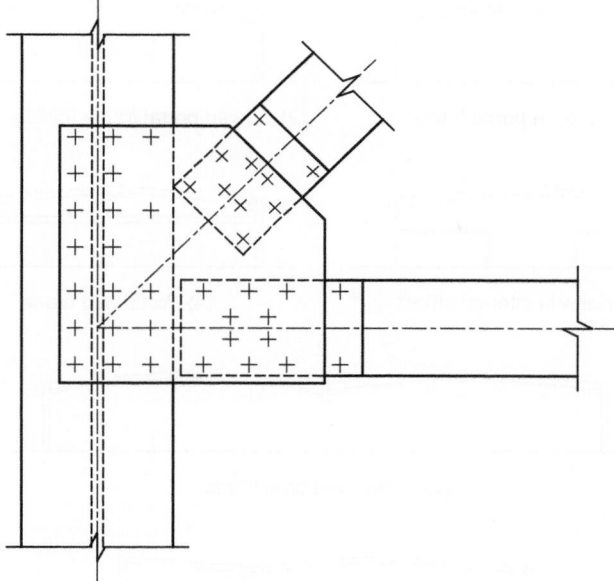

Figure 4.18   Ideal joint with all member centroids coincident

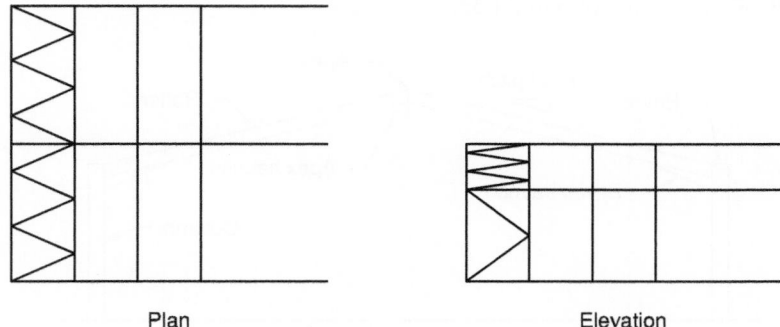

Plan   Elevation

Figure 4.19   Gable-ended bracing systems

4.5.3 Portal frames

General

By far the most common structural form for single-storey buildings is the portal frame. Various configurations of portal frame can be designed using the same structural concept, as shown in Figure 4.20.

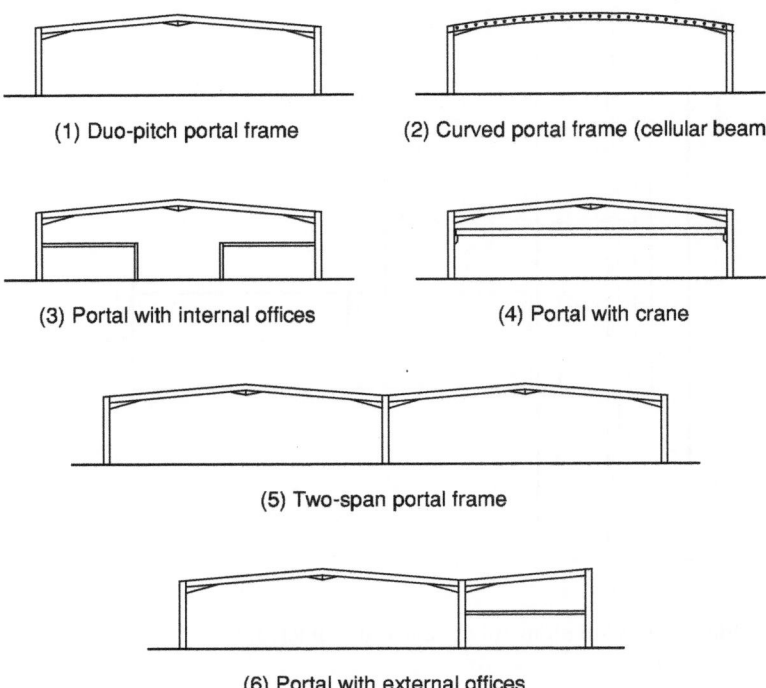

(1) Duo-pitch portal frame

(2) Curved portal frame (cellular beam)

(3) Portal with internal offices

(4) Portal with crane

(5) Two-span portal frame

(6) Portal with external offices

Figure 4.20   Various types of portal frame

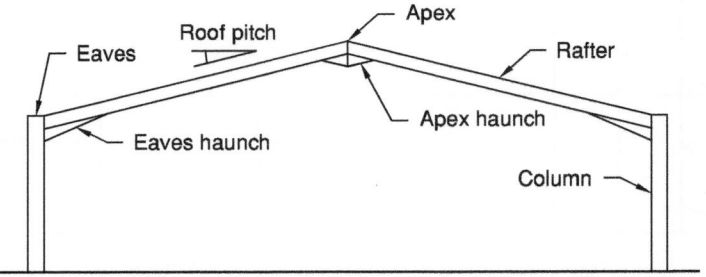

Figure 4.21   Single-span symmetric portal frame

Pitched roof portal frame

A single-span symmetrical portal frame (as illustrated in Figure 4.21) is typically of the following proportions:

- A span between 15 m and 50 m (25 m to 35 m is the most efficient).
- An eaves height (base to rafter centreline) of between 5 m and 10 m (7.5 m is commonly adopted). The eaves height is determined by the specified clear height between the top of the floor and the underside of the haunch.

- A roof pitch between 5° and 10° (6° is commonly adopted).
- A frame spacing between 5 m and 8 m (the greater frame spacings being used in longer span portal frames).
- Members are I-sections rather than H-sections, because they must carry significant bending moments and provide in-plane stiffness.
- Sections are generally S275. Only where deflections are not critical can the use of higher strength steel be justified.
- Haunches are provided in the rafters at the eaves to enhance the bending resistance of the rafter and to facilitate a bolted connection to the column.
- Small haunches are provided at the apex, to facilitate the bolted connection.

The eaves haunch is typically cut from the same size rolled section as the rafter, or one slightly larger, and is welded to the underside of the rafter. The length of the eaves haunch is generally 10% of the span. The length of the haunch means that the hogging bending moment at the 'sharp' end of the haunch is approximately the same as the maximum sagging bending moment towards the apex, as shown in Figure 4.22.

The end frames of the building are generally called gable frames. They may be designed as lighter frames, supported on gable posts. However, gable frames are frequently identical to the internal frames, even though they experience lighter loads. If future extension to the building is envisaged, the use of such gable frames reduces the impact of the structural works.

A typical gable frame is shown in Figure 4.23.

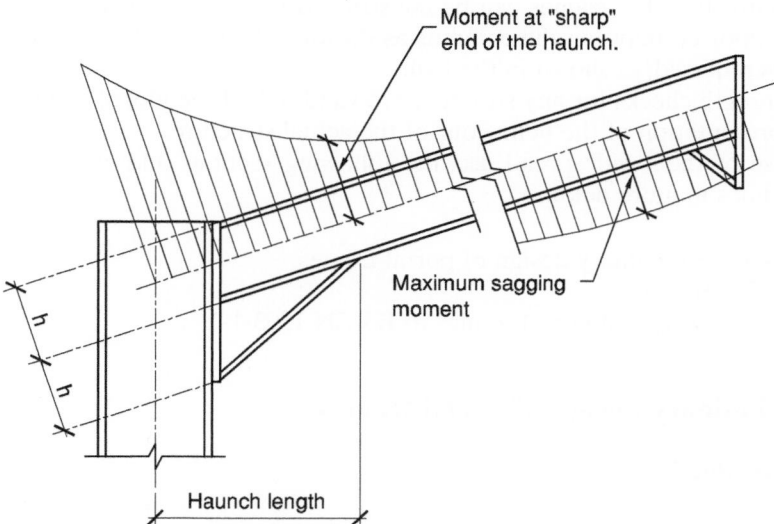

Figure 4.22 Rafter bending moment and haunch length

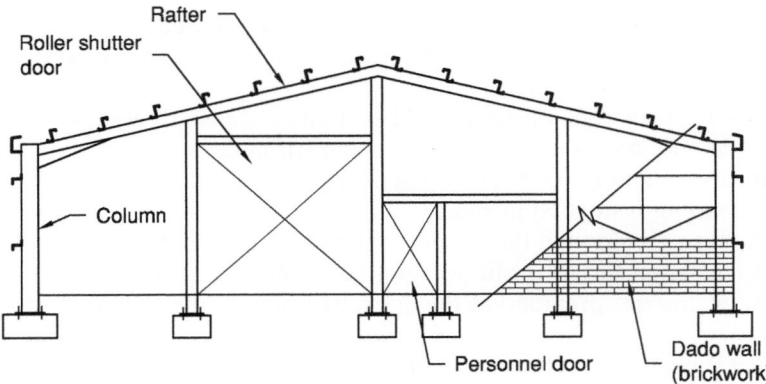

Figure 4.23   Typical details of an end gable of a portal frame building

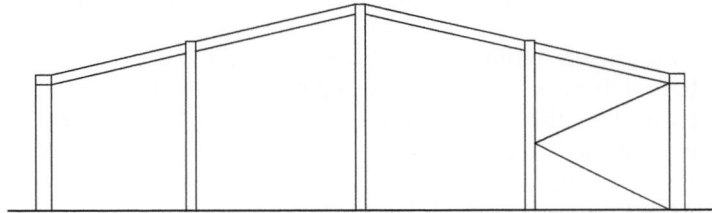

Figure 4.24   Gable frame (not a portal frame)

Alternatively, gable frames can be constructed from columns and short rafters, simply supported between the columns as shown in Figure 4.24. In this case, gable bracing is required, as shown in the figure.

The strength checks for any structure are valid only if the global analysis gives a good representation of the behaviour of the actual structure.

Because of their widespread use, the following sections address the design of portal frames in more detail:

Section 4.6   Preliminary design of portal frames
Section 4.7   Bracing
Section 4.8   Design of portal frames to BS EN 1993-1-1.

4.6 Preliminary design of portal frames

4.6.1 Introduction

There are two alternative methods for determining the size of columns and rafters of single-span portal frames at the preliminary design stage. Further detailed calculations will be required at the final design stage. Both methods are conservative relative to detailed design at the ultimate limit state, but it should be noted that neither method takes account of:

- stability at the ultimate limit state
- deflections at the serviceability limit state.

Further checks will therefore be required, which may necessitate increasing the size of the members in some cases.

4.6.2 Method 1 – tabulated member sizes

4.6.2.1 Basic assumptions

The publication *Portal frames*[3] presents tables that permit a rapid determination of member size to be made for estimating purposes. The span range is 15 to 40 m. A reformatted version of the tables from the publication is presented in Table 4.3 here. The assumptions made in creating this table are as follows:

- the roof pitch is 6°
- the steel grade is S275
- the rafter load is the total factored dead load (including self-weight) and factored imposed load
- the haunch length is 10% of the span of the frame
- a column is treated as restrained when torsional restraints are provided along its length (these columns are therefore lighter than the equivalent unrestrained columns).

A column is treated as unrestrained if no torsional restraint can be provided in its length.

The member sizes given by the tables are suitable for rapid preliminary design, or at the estimating stage. However, where strict deflection limits are specified, it may be necessary to increase the member sizes. Where an asterisk (*) is shown in the table, a suitable section size has not been calculated.

4.6.2.2 Example: using Method 1

Consider a frame of 30 m span, height to eaves of 7 m, 6° roof pitch, pinned bases and total load of 11.3 kN/m.

From Table 4.3, assuming eaves height = 8 m, span = 30 m, rafter loading = 12 kN/m.

Table 4.3 Symmetrical single-span portal frame with 6° roof pitch

	Rafter load (kN/m)	Eaves height (m)	Span of frame (m)					
			15	20	25	30	35	40
Rafter	8	6	254 × 102 × 22 UB	356 × 127 × 33 UB	406 × 140 × 39 UB	406 × 140 × 46 UB	406 × 178 × 60 UB	457 × 191 × 67 UB
	8	8	254 × 102 × 22 UB	356 × 127 × 33 UB	406 × 140 × 39 UB	406 × 178 × 54 UB	457 × 191 × 67 UB	457 × 191 × 74 UB
	8	10	254 × 102 × 22 UB	356 × 127 × 33 UB	406 × 140 × 39 UB	406 × 178 × 54 UB	457 × 191 × 67 UB	457 × 191 × 74 UB
	8	12	*	356 × 127 × 33 UB	406 × 140 × 39 UB	406 × 178 × 54 UB	457 × 191 × 67 UB	457 × 191 × 74 UB
Restrained column	8	6	305 × 165 × 40 UB	356 × 171 × 51 UB	457 × 191 × 67 UB	533 × 210 × 82 UB	533 × 210 × 92 UB	610 × 229 × 113 UB
	8	8	305 × 165 × 40 UB	356 × 171 × 51 UB	457 × 191 × 67 UB	533 × 210 × 82 UB	610 × 229 × 113 UB	610 × 229 × 113 UB
	8	10	305 × 165 × 40 UB	406 × 178 × 54 UB	457 × 191 × 67 UB	533 × 210 × 82 UB	610 × 229 × 101 UB	686 × 254 × 125 UB
	8	12	*	406 × 178 × 54 UB	457 × 191 × 67 UB	533 × 210 × 82 UB	610 × 229 × 101 UB	686 × 254 × 125 UB
Unrestrained column	8	6	356 × 171 × 51 UB	457 × 191 × 67 UB	533 × 210 × 82 UB	533 × 210 × 92 UB	610 × 229 × 113 UB	686 × 254 × 125 UB
	8	8	406 × 178 × 60 UB	533 × 210 × 82 UB	610 × 229 × 101 UB	610 × 229 × 113 UB	686 × 254 × 125 UB	762 × 267 × 147 UB
	8	10	457 × 191 × 67 UB	533 × 210 × 92 UB	610 × 229 × 113 UB	686 × 254 × 125 UB	762 × 267 × 147 UB	762 × 267 × 173 UB
	8	12	*	610 × 229 × 101 UB	686 × 254 × 125 UB	762 × 267 × 147 UB	762 × 267 × 173 UB	838 × 292 × 194 UB

Member								
Rafter	10	6	305 × 102 × 25 UB	356 × 127 × 33 UB	406 × 140 × 46 UB	406 × 178 × 60 UB	457 × 191 × 67 UB	533 × 210 × 82 UB
	10	8	305 × 102 × 25 UB	356 × 127 × 33 UB	406 × 140 × 46 UB	406 × 178 × 60 UB	457 × 191 × 74 UB	533 × 210 × 82 UB
	10	10	305 × 102 × 25 UB	406 × 140 × 39 UB	406 × 140 × 46 UB	406 × 178 × 60 UB	457 × 191 × 74 UB	533 × 210 × 82 UB
	10	12	*	406 × 140 × 39 UB	406 × 140 × 46 UB	457 × 191 × 67 UB	457 × 191 × 74 UB	533 × 210 × 92 UB
Restrained column	10	6	356 × 171 × 45 UB	406 × 178 × 60 UB	457 × 191 × 74 UB	533 × 210 × 92 UB	610 × 229 × 113 UB	686 × 254 × 125 UB
	10	8	356 × 171 × 45 UB	406 × 178 × 60 UB	533 × 210 × 82 UB	533 × 210 × 92 UB	610 × 229 × 113 UB	686 × 254 × 125 UB
	10	10	356 × 171 × 45 UB	406 × 178 × 60 UB	533 × 210 × 82 UB	610 × 229 × 101 UB	610 × 229 × 113 UB	686 × 254 × 140 UB
	10	12	*	406 × 178 × 60 UB	533 × 210 × 82 UB	610 × 229 × 101 UB	686 × 254 × 125 UB	686 × 254 × 140 UB
Unrestrained column	10	6	406 × 178 × 54 UB	457 × 191 × 74 UB	533 × 210 × 92 UB	610 × 229 × 101 UB	686 × 254 × 125 UB	686 × 254 × 125 UB
	10	8	457 × 191 × 67 UB	533 × 210 × 92 UB	610 × 229 × 113 UB	686 × 254 × 125 UB	686 × 254 × 140 UB	762 × 267 × 173 UB
	10	10	457 × 191 × 74 UB	610 × 229 × 101 UB	686 × 254 × 125 UB	762 × 267 × 147 UB	762 × 267 × 173 UB	838 × 292 × 194 UB
	10	12	*	610 × 229 × 113 UB	686 × 254 × 140 UB	762 × 267 × 173 UB	838 × 292 × 194 UB	914 × 305 × 224 UB

Table 4.3 (*Continued*)

	Rafter load (kN/m)	Eaves height (m)	Span of frame (m)					
			15	20	25	30	35	40
Rafter	12	6	305 × 102 × 28 UB	406 × 140 × 39 UB	406 × 178 × 54 UB	457 × 191 × 67 UB	533 × 210 × 82 UB	533 × 210 × 92 UB
	12	8	305 × 102 × 28 UB	406 × 140 × 39 UB	406 × 178 × 54 UB	457 × 191 × 67 UB	533 × 210 × 82 UB	533 × 210 × 92 UB
	12	10	305 × 102 × 28 UB	406 × 140 × 39 UB	406 × 178 × 54 UB	457 × 191 × 67 UB	533 × 210 × 82 UB	610 × 229 × 101 UB
	12	12	*	406 × 140 × 39 UB	406 × 178 × 54 UB	457 × 191 × 67 UB	533 × 210 × 82 UB	610 × 229 × 101 UB
Restrained column	12	6	356 × 141 × 45 UB	457 × 191 × 67 UB	533 × 210 × 82 UB	610 × 229 × 101 UB	686 × 254 × 125 UB	686 × 254 × 140 UB
	12	8	356 × 171 × 45 UB	457 × 191 × 67 UB	533 × 210 × 82 UB	610 × 229 × 101 UB	686 × 254 × 125 UB	762 × 267 × 147 UB
	12	10	356 × 171 × 51 UB	457 × 191 × 67 UB	533 × 210 × 92 UB	610 × 229 × 113 UB	686 × 254 × 125 UB	762 × 267 × 147 UB
	12	12	*	457 × 191 × 67 UB	533 × 210 × 92 UB	610 × 229 × 113 UB	686 × 254 × 125 UB	762 × 267 × 147 UB
Unrestrained column	12	6	406 × 178 × 60 UB	533 × 210 × 82 UB	610 × 229 × 101 UB	610 × 229 × 113 UB	686 × 254 × 125 UB	762 × 267 × 147 UB
	12	8	457 × 191 × 74 UB	610 × 229 × 101 UB	610 × 229 × 113 UB	686 × 254 × 140 UB	762 × 267 × 173 UB	838 × 292 × 176 UB
	12	10	533 × 210 × 82 UB	610 × 229 × 113 UB	686 × 254 × 140 UB	762 × 267 × 173 UB	838 × 292 × 194 UB	914 × 305 × 224 UB
	12	12	*	686 × 254 × 125 UB	762 × 267 × 173 UB	836 × 292 × 176 UB	914 × 305 × 224 UB	914 × 305 × 253 UB

Rafter	14	6	356 × 127 × 33 UB	406 × 140 × 46 UB	406 × 178 × 60 UB	457 × 191 × 74 UB	533 × 210 × 82 UB	610 × 229 × 101 UB
	14	8	356 × 127 × 33 UB	406 × 140 × 46 UB	406 × 178 × 60 UB	457 × 191 × 74 UB	533 × 210 × 92 UB	610 × 229 × 101 UB
	14	10	356 × 127 × 33 UB	406 × 140 × 46 UB	406 × 178 × 60 UB	457 × 191 × 74 UB	533 × 210 × 92 UB	610 × 229 × 101 UB
	14	12	*	406 × 140 × 46 UB	406 × 178 × 60 UB	533 × 210 × 82 UB	533 × 210 × 92 UB	610 × 229 × 113 UB
Restrained column	14	6	356 × 171 × 51 UB	457 × 191 × 74 UB	533 × 210 × 92 UB	610 × 229 × 113 UB	686 × 254 × 140 UB	762 × 267 × 147 UB
	14	8	406 × 178 × 54 UB	457 × 191 × 74 UB	533 × 210 × 92 UB	610 × 229 × 113 UB	686 × 254 × 140 UB	762 × 267 × 173 UB
	14	10	406 × 178 × 54 UB	457 × 191 × 74 UB	610 × 229 × 101 UB	686 × 254 × 125 UB	686 × 254 × 140 UB	762 × 267 × 173 UB
	14	12	*	457 × 191 × 74 UB	610 × 229 × 101 UB	686 × 254 × 125 UB	762 × 267 × 147 UB	762 × 267 × 173 UB
Unrestrained column	14	6	457 × 191 × 67 UB	533 × 210 × 82 UB	610 × 229 × 101 UB	686 × 254 × 125 UB	686 × 254 × 140 UB	762 × 267 × 173 UB
	14	8	533 × 210 × 82 UB	610 × 229 × 101 UB	686 × 254 × 125 UB	762 × 267 × 147 UB	762 × 267 × 173 UB	838 × 292 × 176 UB
	14	10	533 × 210 × 92 UB	686 × 254 × 125 UB	762 × 267 × 147 UB	762 × 267 × 173 UB	838 × 292 × 194 UB	914 × 305 × 224 UB
	14	12	*	686 × 254 × 140 UB	762 × 267 × 173 UB	838 × 292 × 194 UB	914 × 305 × 224 UB	914 × 305 × 289 UB

Table 4.3 (*Continued*)

	Rafter load (kN/m)	Eaves height (m)	Span of frame (m)					
			15	20	25	30	35	40
Rafter	16	6	356 × 127 × 33 UB	406 × 140 × 46 UB	457 × 191 × 67 UB	533 × 210 × 82 UB	533 × 210 × 92 UB	610 × 229 × 113 UB
	16	8	356 × 127 × 33 UB	406 × 140 × 46 UB	457 × 191 × 67 UB	533 × 210 × 82 UB	610 × 229 × 101 UB	610 × 229 × 113 UB
	16	10	356 × 127 × 33 UB	406 × 140 × 46 UB	457 × 191 × 67 UB	533 × 210 × 82 UB	610 × 229 × 101 UB	610 × 229 × 113 UB
	16	12	*	406 × 140 × 46 UB	457 × 191 × 67 UB	533 × 210 × 82 UB	610 × 229 × 101 UB	686 × 254 × 125 UB
Restrained column	16	6	406 × 178 × 54 UB	533 × 210 × 82 UB	610 × 229 × 101 UB	686 × 254 × 125 UB	762 × 267 × 147 UB	762 × 267 × 173 UB
	16	8	406 × 178 × 54 UB	533 × 210 × 82 UB	610 × 229 × 101 UB	686 × 254 × 125 UB	762 × 267 × 147 UB	838 × 292 × 176 UB
	16	10	406 × 178 × 60 UB	533 × 210 × 82 UB	610 × 229 × 101 UB	686 × 254 × 125 UB	762 × 267 × 173 UB	838 × 292 × 176 UB
	16	12	*	533 × 210 × 82 UB	610 × 229 × 101 UB	686 × 254 × 125 UB	762 × 267 × 173 UB	838 × 292 × 194 UB
Unrestrained column	16	6	457 × 191 × 67 UB	533 × 210 × 92 UB	610 × 229 × 113 UB	686 × 254 × 125 UB	762 × 267 × 147 UB	762 × 267 × 173 UB
	16	8	533 × 210 × 82 UB	610 × 229 × 113 UB	686 × 254 × 140 UB	762 × 267 × 173 UB	838 × 292 × 176 UB	914 × 305 × 201 UB
	16	10	610 × 229 × 101 UB	686 × 254 × 125 UB	762 × 267 × 173 UB	838 × 292 × 194 UB	914 × 305 × 224 UB	914 × 305 × 253 UB
	16	12	*	686 × 254 × 140 UB	838 × 292 × 176 UB	914 × 305 × 224 UB	914 × 305 × 253 UB	914 × 305 × 289 UB

Select

Rafter $457 \times 191 \times 67$UB Grade S275.
Restrained column $610 \times 229 \times 101$UB Grade S275.

4.6.3 Method 2 – design charts / graphs

Design charts / graphs available in a number of publications, assist in the rapid estimation of horizontal base force, moments in the rafters and the columns, and the position of the rafter hinge. These charts require slightly more work than the tables mentioned above, but are much more flexible and accurate for a particular design case.

Charts / graphs for portal frames with pinned bases devised by A D Weller are given in *Introduction to steelwork design to BS 5950-1:2000*.[4] Charts for the design of portal frames with bases having various degrees of restraint, devised by Surtees and Yeap, have been published in *The Structural Engineer*.[5]

The graphs for portal frames with pinned bases are reproduced in Figure 4.25 to Figure 4.28. They are based on the following assumptions:

- plastic hinges form in the column at the bottom of the haunch and near the apex in the rafter
- the rafter depth is approximately frame span / 55

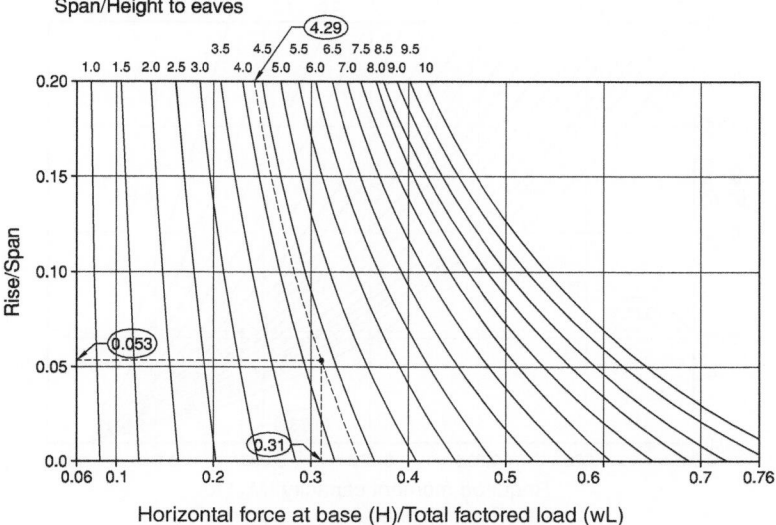

Figure 4.25   Graph 1. Horizontal force ratio at base

- the haunch depth below the rafter is approximately the same as the rafter depth
- the haunch length is approximately 10% of the frame span
- the moment in the rafter at the top of the eaves haunch $\leq 0.87\, M_p$, i.e. the haunch region remains elastic
- wind loading does not control design.

The chosen sections must be checked separately for stability.

The notation for the graphs is as follows:

H is the horizontal base reaction
w is the factored load (dead + imposed) per unit length on the rafter
L is the span of the frame
M_{pr} is the required plastic moment resistance of the rafter
M_{pl} is the required plastic moment resistance of the column
P^N is the distance of the point of maximum moment in the rafter from the column.

Rise is the difference between the apex and eaves height. The graphs cover the range of span/height to eaves between 1 and 10, and a rise/span ratio of 0 to 0.2 (i.e. flat to 22°). Interpolation is permissible but extrapolation is not.

In Figure 4.25, Graph 1 gives the horizontal force at the foot of the frame as a proportion of the total factored load wL.

In Figure 4.26, Graph 2 gives the value of the required moment resistance of the rafters as a proportion of wL^2.

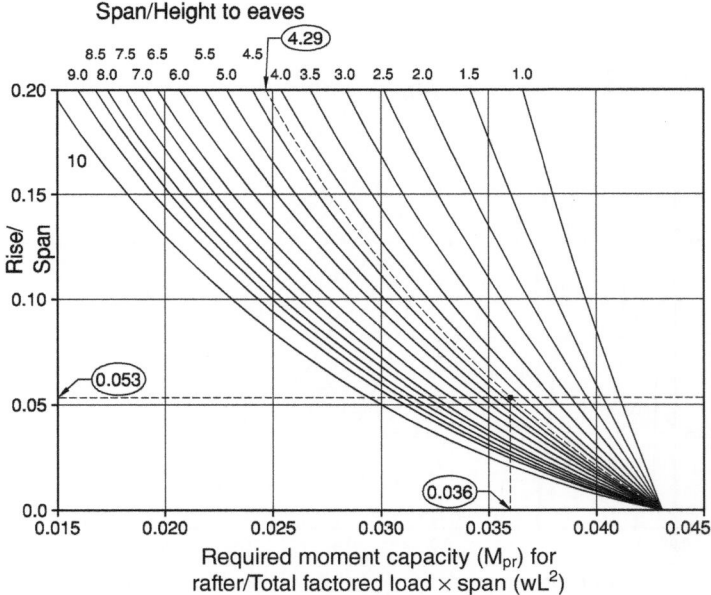

Figure 4.26 Graph 2. Moment capacity ratio required in rafter

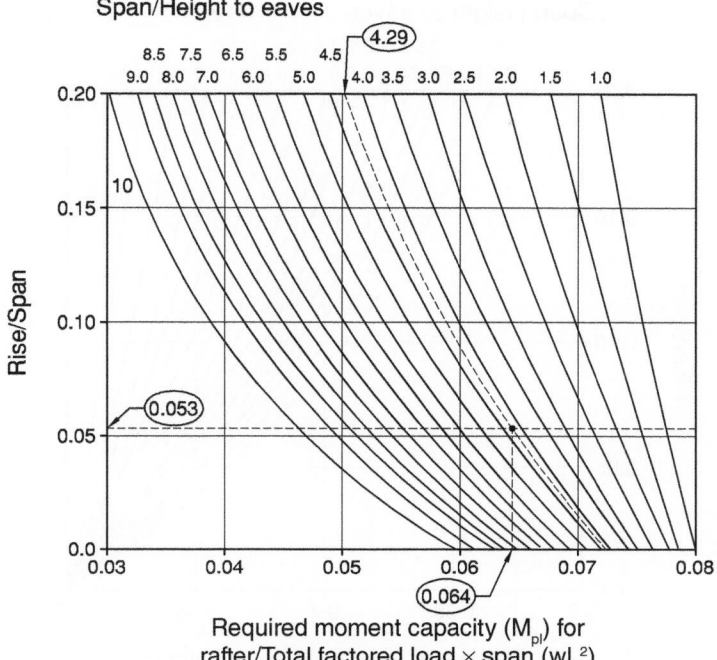

Figure 4.27   Graph 3. Moment capacity ratio required in column

In Figure 4.27, Graph 3 gives the value of the required moment resistance of the columns as a proportion of wL^2.

In Figure 4.28, Graph 4 gives the position of the rafter hinge as a proportion of the span L.

4.6.3.1 *Method of use*

- Determine the ratio span/height to eaves (based on the intersection of the centre-lines of the members).
- Determine the ratio rise/span.
- Calculate wL and wL^2.
- Look up the values from the graphs, as follows:
  - horizontal reaction H = value from Graph 1 $\times$ wL
  - rafter: M_{pr} = value from Graph 2 $\times$ wL^2
  - column: M_{pl} = value from Graph 3 $\times$ wL^2
  - distance to the position of maximum moment in the rafter l' = value from Graph 4 $\times$ L.

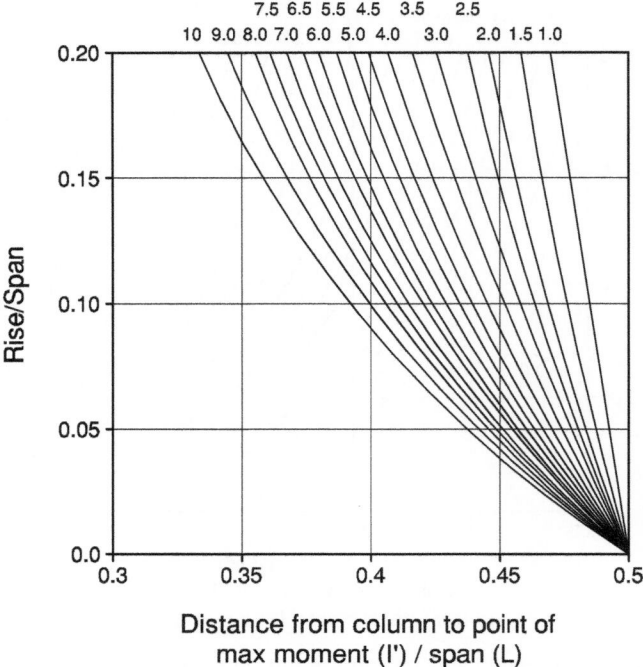

Figure 4.28   Graph 4. Distance to position of maximum moment in rafter

4.6.3.2 *Example using design charts/graphs*

Consider the same example as in Section 4.6.3.1, i.e. span 30 m, height to eaves of 7 m, 6° roof pitch, pinned bases and total load of 11.3 kN/m.

Span/height to eaves	L/h	=	30/7	=	4.23
Rise/span	hr/L	=	1.58/30	=	0.053
Vertical load	wL	=	11.3×30	=	339 kN
	wL^2	=	11.3×30^2	=	10170 kNm

From the graphs:

Required moment capacity of rafter	=	$0.036 \times 10170 = 366$ kNm
Required moment capacity of column	=	$0.064 \times 10170 = 651$ kNm

Based on these required capacities, select:

Rafter	$457 \times 191 \times 67$ UB in S275 steel	M_{cx}	=	405 kNm
Column	$533 \times 210 \times 101$ UB in S275 steel	M_{cx}	=	692 kNm

4.7 Bracing

4.7.1 General

Bracing is required to resist lateral loads, principally wind loads, and the destabilising effects of the imperfections defined in Section 5.3 of BS EN 1993-1-1. This bracing must be correctly positioned and have adequate strength and stiffness to justify the assumptions made in the analysis and member checks.

Section 5.3 of BS EN 1993-1-1 allows the imperfections to be described either as geometrical imperfections or as equivalent horizontal forces.

The equivalent horizontal forces, which cause the forces in the bracing, do not increase the **total** load on the **whole** structure, because they form a self-equilibrating load case.

All the examples in this section relate to portal frame construction because of its popularity. However, the principles may equally be applied to other structural forms.

4.7.2 Vertical bracing

The primary functions of vertical bracing in the side walls of buildings are:

- to transmit the horizontal loads, acting on the end of the building, to the ground
- to provide a rigid framework to which side rails may be attached so that they can in turn provide stability to the columns
- to provide temporary stability during erection.

The bracing system will usually take the form of:

- circular hollow sections in a V pattern
- circular hollow sections in a K pattern
- crossed flats (within a cavity wall), considered to act in tension only
- crossed hot-rolled angles.

The bracing may be located at:

- one or both ends of the building, depending on the length of the structure
- at the centre of the building (but this is rarely done due to the need to begin erection from one braced bay at, or close to, the end of the building)
- in each portion between expansion joints (where these occur).

Where the side wall bracing is not in the same bay as the plan bracing in the roof, an eaves strut is required to transmit the forces from the plan bracing into the wall bracing.

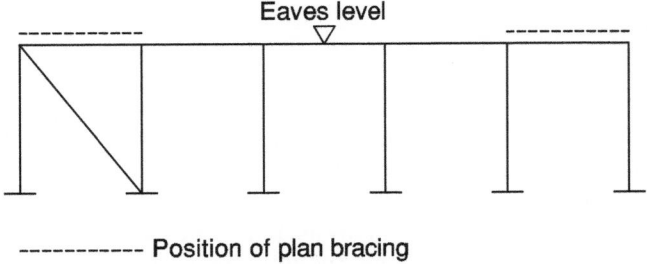

Figure 4.29   Single diagonal bracing for low frames using circular hollow sections

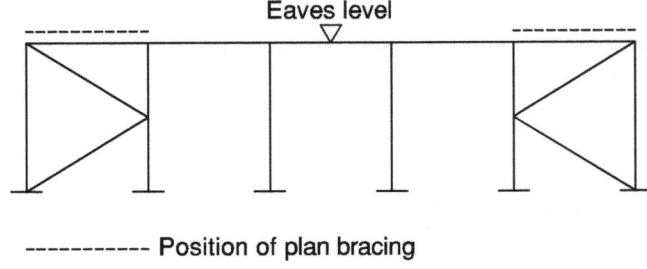

Figure 4.30   K bracing arrangement for taller frames using circular hollow sections

Bracing using circular hollow sections

Circular hollow sections are very efficient in compression, which eliminates the need for cross bracing. Where the height to eaves is approximately equal to the spacing of the frames, a single bracing member at each location is economic (Figure 4.29). Where the eaves height is large in relation to the frame spacing, a K-brace is often used (Figure 4.30).

Bracing using angle sections or flats

Cross braced angles or flats (within a masonry cavity wall) may be used as bracing (as shown in Figure 4.31). In this case, it is assumed that only one of the diagonal members acts in tension under wind load.

Combining bracing with frame restraint

As shown in Figure 4.32, it may be possible to modify the vertical bracing geometry, thereby enabling it also to form the basis of the torsional restraint system to the bottom of the haunch.

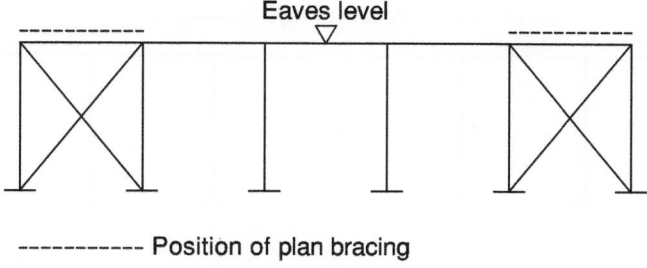

Figure 4.31   Typical cross bracing system using angles or flats as tension members

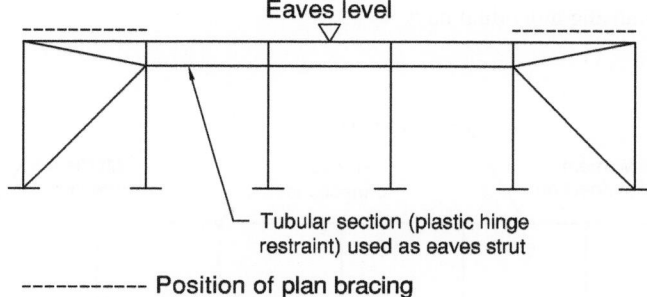

Figure 4.32   Bracing arrangement using the hollow section member as a restraint and as an eaves strut

Portalised braced bays

Where it is difficult or impossible to brace the frame vertically by conventional bracing, a portalised structure can be provided. There are two basic possibilities:

- a portal structure in one or more bays (Figure 4.33)
- a hybrid portal/pinned structure down the full length of the side (Figure 4.34).

The advantage of the first approach is that the conventional portal structure can be determined relatively early. It has the disadvantage that additional members are required and that openings in the side of the building may be restricted.

The second approach provides a lighter and much more open structure. Although in practice its actual stiffness is perhaps less than calculated (due to the flexibility of the internal struts), it is a method that has been used successfully.

In the design of both systems, it is suggested that:

- The bending resistance of the portalised bay (not the main portal frame) is checked using an elastic frame analysis.

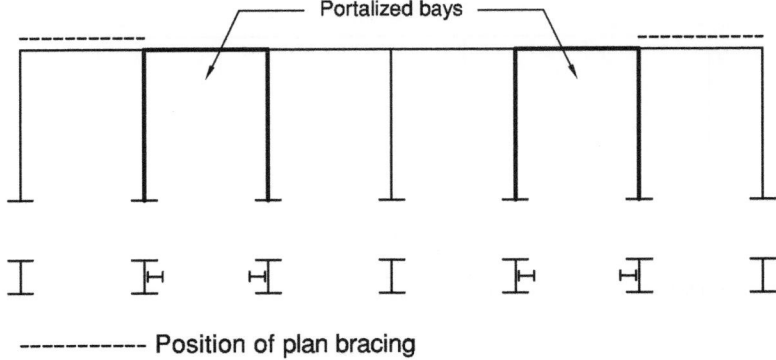

Figure 4.33   Portalising individual bays

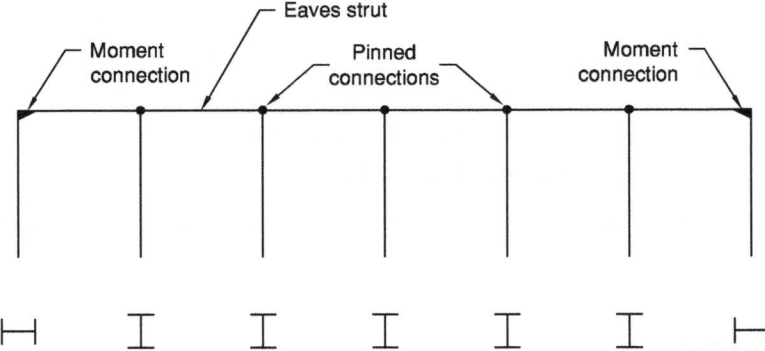

Figure 4.34   Hybrid portal along the full length of the building

- Deflection under the notional horizontal forces is restricted to $h/1000$.
- The stiffness is assured by restricting serviceability deflections to a maximum of $h/360$, where h is the height of the portalised bay.

Bracing to restrain longitudinal crane surge

If a crane is directly supported by the frame, the longitudinal surge force will be eccentric to the column, and will tend to cause the column to twist, unless additional restraint is provided. A horizontal truss at the level of the girder top flange or, for lighter cranes, a horizontal member on the inside face of the column flange tied into the wall bracing, may be adequate to provide the necessary restraint.

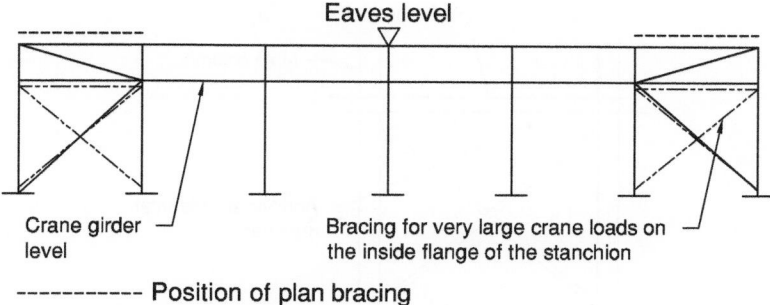

Figure 4.35   Elevation showing position of additional bracing in the plane of the crane girder

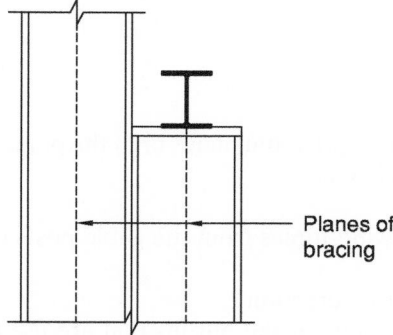

Figure 4.36   Detail showing position of additional bracing in the plane of the crane girder

Table 4.4   Bracing requirements for crane girders

Factored longitudinal force	Bracing requirement
Small (<15 kN)	Use wind bracing
Medium (15–30 kN)	Use horizontal bracing to transfer force to plane of bracing
Large (>30 kN)	Provide additional bracing in the plane of the longitudinal force

For large horizontal forces, additional bracing should be provided in the plane of the crane girder (Figure 4.35 and Figure 4.36). The criteria given in Table 4.4 were developed by Fisher and are reproduced in reference 4.

If the bracing is attached directly to the column, it will tend to attract vertical load and, for heavily loaded crane girders, it may be necessary to provide an additional horizontal member to prevent fatigue failure of the connection (Figure 4.37).

106   *Single-storey buildings*

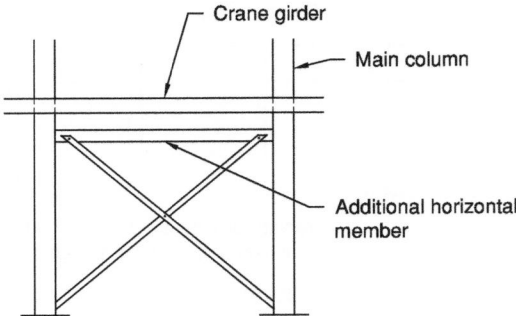

Figure 4.37 Alternative configuration of additional bracing to prevent fatigue failure

4.7.3 Plan bracing

General

Plan bracing is placed in the horizontal plane, or in the plane of the roof. The primary functions of the plan bracing are:

- to transmit horizontal wind forces from the gable posts to the vertical bracing in the walls
- to provide stability during erection
- to provide a stiff anchorage for the purlins that are used to restrain the rafters.

In order to transmit the wind forces efficiently, the plan bracing should connect to the top of the gable posts wherever possible.

The purlins are not usually designed to resist axial forces due to wind loading.

Bracing using circular hollow sections

In modern construction, circular hollow section bracing members are generally used in the roof and are designed to resist both tension and compression. Many arrangements are possible, depending on the spacing of the frames and the positions of the gable posts. A typical arrangement is shown in Figure 4.38.

It is good practice to provide an eaves tie along the length of the building.

4.7.4 Bracing to inner flanges

Bracing is also required to restrain the inner flanges of the primary frames where they are in compression. Under gravity loads, this will occur near the eaves; under uplift conditions, it will occur near mid-span.

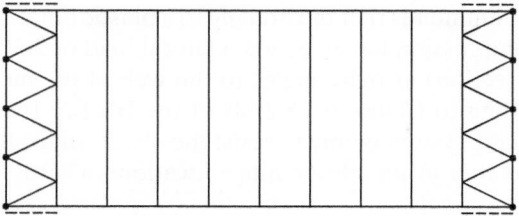

- Position of gable posts
---- Location of vertical bracing

Figure 4.38   Plan view showing both end bays braced (using circular hollow section members)

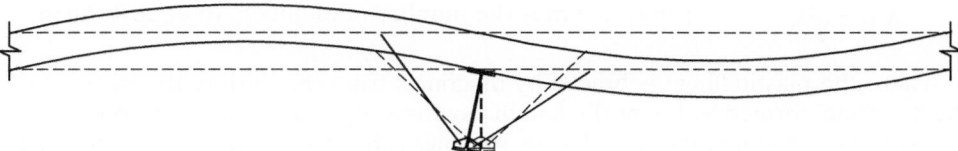

Figure 4.39   Effect of purlin flexibility on bracing

Bracing to the inner flanges is often most conveniently formed by diagonal struts from the purlins or sheeting rails to small stiffeners welded to the inner flange and web. Light cold formed angles are commonly used. If flats are used as diagonals, only one of the pair will be effective, so the strength may be impaired.

The effectiveness of such bracing depends on the stiffness of the system, especially the stiffness of the purlins. The effect of purlin flexibility on the bracing is shown in Figure 4.39. Where the proportions of the members, purlins and spacings differ from proven previous practice, the effectiveness should be checked. This can be done using the formula given in Section 4.7.5, or other methods, such as may be found in bridge codes for U-frame action.

The compression flange of eaves and valley haunches must be braced at the column-haunch connection, unless stability analysis or test data prove it to be unnecessary. This restraint is needed, because the haunch stability checks are based on full torsional restraint at this point.

4.7.5 Bracing at plastic hinges

Section 6.3.5.2 of BS EN 1993-1-1 recommends that bracing should be provided to both tension and compression flanges at or within $0.5h$ of the calculated plastic hinges, where h is the depth of the member.

BS EN 1993-1-1 recommends that the bracing to a plastic hinge should be designed assuming that the compression flange exerts a lateral load of 2.5% (plastic moment resistance/depth of section) at right angles to the web of the member.

In addition, according to Clause 6.3.5.2 5B of the BS EN 1993-1-1, it should be checked that the bracing system is able to resist the effects of local forces Q_m applied at each stabilised member at the plastic hinge locations, where:

$$Q_m = 1.5\alpha_m \frac{N_{f,Ed}}{100}$$

where:

$N_{f,Ed}$ is the axial force in the compressed flange of the stabilised member at the plastic hinge location

$\alpha_m = \sqrt{0.5\left(1+\frac{1}{m}\right)}$ in which m is the number of members to be restrained.

Where the plastic hinge is braced by diagonals from the purlins, the stiffness of the 'U-frame' formed by the purlin and diagonals is especially important. Where the proportions of the members, purlins or spacings differ from proven previous practice, the effectiveness should be checked. In the absence of other methods, the stiffness check may be based on the work of Horne and Ajmani as presented in Reference 7. Thus, the support member (the purlin or sheeting rail) should have $I_{y,s}$ such that:

$$\frac{I_{y,s}}{I_{y,f}} \geq \frac{f_y}{190\times10^3} \frac{L(L_1+L_2)}{L_1 L_2}$$

where:

f_y is the yield strength of the frame member

$I_{y,s}$ is the second moment of area of the supporting member (purlin or sheeting rail) about the axis parallel to the longitudinal axis of the frame member (i.e. the purlin major axis in normal practice)

$I_{y,f}$ is the second moment of area of the frame member about the major axis

L is the span of the purlin or sheeting rail

L_1 and L_2 are the distances either side of the plastic hinge to the eaves (or valley) or points of contraflexure, whichever are the nearest to the hinge.

Hinges that form, rotate then cease, or even unload and rotate in reverse, must be fully braced. However, hinges that occur in the **collapse** mechanism but rotate only above ULS need not be considered as plastic hinges for ULS checks. These hinges are easily identified by elastic-plastic or graphical analysis, but are not shown by virtual-work (rigid-plastic) analysis. However, it is important to note that the mathematics of the analysis can calculate the presence of hinges that form and then disappear at the same load factor. This indicates that no rotation takes place and, therefore, no **hinge** occurs. In these cases, it is not necessary to provide the usual restraint associated with plastic hinges; only the restraints for normal elastic stability are required.

Analysis cannot account for all of the section tolerances, residual stresses and material tolerances. Care should be taken to brace points where these effects could affect the hinge positions, e.g. the shallow end of the haunch instead of the top of the column. Wherever the bending moments come close to the plastic moment capacity, the possibility of a hinge should be considered.

4.8 Design of portal frames to BS EN 1993-1-1

4.8.1 Scope

This section guides the designer through all the steps involved in the detailed design of portal frames to Eurocode 3, taking due account of the role of computer analysis with commercially available software.

It primarily focuses on a single-span portal frame with pinned or fixed column bases. A concluding section outlines the issues involved in multi-bay portal structures.

4.8.2 Second-order effects in portal frames

The strength checks for any structure are valid only if the global analysis gives a good representation of the behaviour of the actual structure.

When any frame is loaded, it deflects and its shape under load is different from the undeformed shape. The deflection causes the axial loads in the members to act along different lines from those assumed in the analysis, as shown diagrammatically in Figure 4.40 and Figure 4.41. If the deflections are small, the consequences are

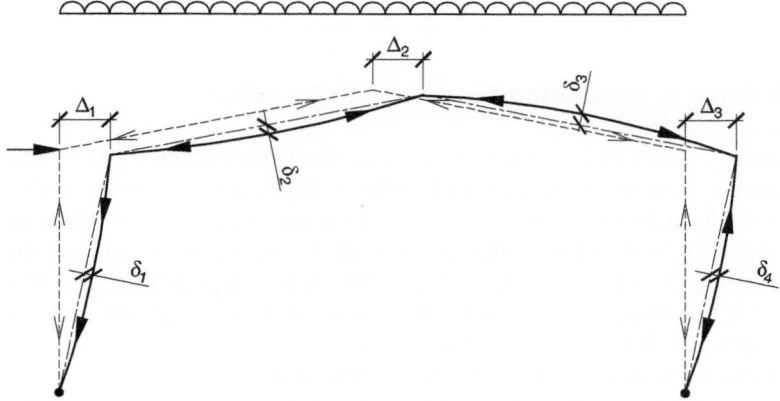

Figure 4.40   Asymmetric or sway mode deflection

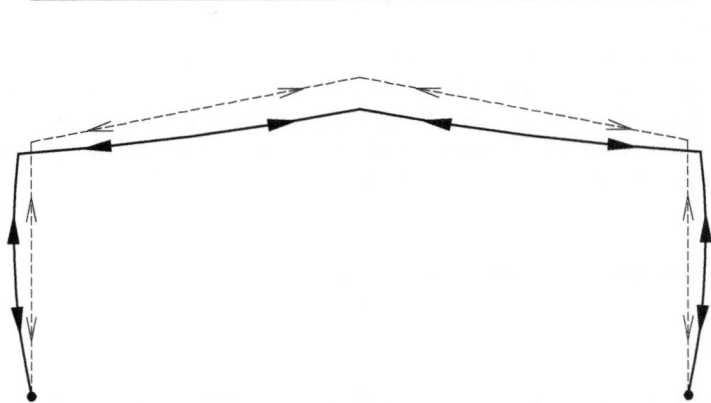

Figure 4.41   Symmetric mode deflection

very small and a first-order analysis (neglecting the effect of the deflected shape) is sufficiently accurate. However, if the deflections are such that the effects of the axial load on the deflected shape are large enough to cause significant additional moments and further deflection, the frame is said to be sensitive to second-order effects. These second-order effects, or *P*-delta effects, can be sufficient to reduce the resistance of the frame.

There are two categories of second-order effects:

- effects of deflections within the length of members, sometimes called *P*-δ (*P*-little delta) effects
- effects of displacements of the intersections of members, sometimes called *P*-Δ (*P*-big delta) effects.

4.8.3 Methods of accounting for second-order effects

Second-order effects increase not only the deflections but also the moments and forces beyond those calculated by first-order analysis. Second-order analysis is the term used to describe analysis methods in which the effects of increasing deflection under increasing load are considered explicitly in the solution, so that the results include the *P*-Δ (*P*-big delta) and *P*-δ (*P*-little delta) effects described above. The results will differ from the results of first-order analysis by an amount dependent on the magnitude of the *P*-Δ and *P*-δ effects.

The effects of the deformed geometry are assessed in BS EN 1993-1-1 by calculating the alpha crit (α_{cr}) factor. The limitations to the use of first-order analysis are defined in BS EN 1993-1-1, Section 5.2.1 (3) as:

For elastic analysis: $\alpha_{cr} \geq 10$
For plastic analysis: $\alpha_{cr} \geq 15$

Details of the calculation of α_{cr} are given below.
When a second-order analysis is required, there are two main methods to proceed:

- Rigorous second-order analysis (i.e. in practice, using an appropriate second-order software).
- Approximate second-order analysis (i.e. hand calculations using first-order analysis with magnification factors). Although the modifications involve approximations, they are sufficiently accurate within the limits given by BS EN 1993-1-1. In this method, also known as 'modified first-order analysis', the α_{cr} factor is used to determine a sway factor. This factor will be used to calculate an equivalent horizontal force, to take into account the second-order effects by running first-order calculations.

Determination of α_{cr}

BS EN 1993-1-1 Clause 5.2.1 (4) B suggests:

$$\alpha_{cr} = \left(\frac{H_{Ed}}{V_{Ed}}\right)\left(\frac{h}{\delta_{H,Ed}}\right)$$

However, because of the second-order effects due to the rafter compression (P-δ), this simple check for α_{cr} is unconservative for portal frames. A more accurate formula, using a new factor $\alpha_{cr,est}$ has been developed by J. Lim and C. King[7] and is detailed below.

For each load case, an estimate of the elastic critical buckling load factor may be obtained as follows.

For frames in which the rafters are straight between the columns:

$$\alpha_{cr,est} = \alpha_{cr,s,est}$$

For frames with pitched rafters:

$$\alpha_{cr,est} = \min(\alpha_{cr,s,est}, \alpha_{cr,r,est}).$$

where:
   $\alpha_{cr,s,est}$ is the estimate of α_{cr} for sway buckling mode
   $\alpha_{cr,r,est}$ is the estimate of α_{cr} for rafter snap-through buckling mode.

The parameters required to calculate the sway parameter $\alpha_{cr,s,est}$ for a portal frame are shown in Figure 4.42. As can be seen, δ_{EHF} is the lateral deflection at the top of each column when subjected to an arbitrary lateral load H_{EHF}. (The magnitude of the total lateral load is arbitrary, as it is simply used to calculate the sway stiffness

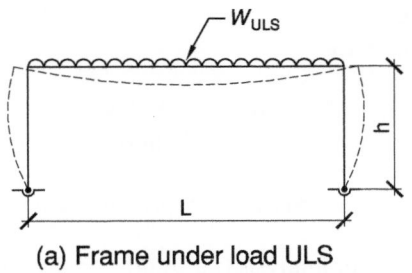

(a) Frame under load ULS

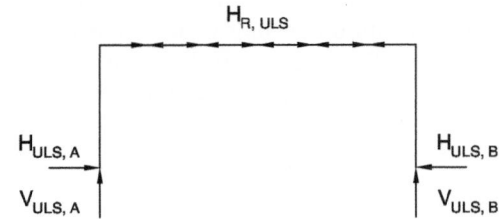

(b) Reactions and axial force in rafter at ULS

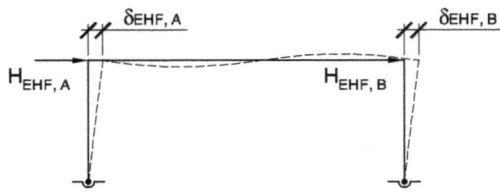

(c) Horizontal deflection at top of columns

Figure 4.42   Parameters required to estimate alpha crit

H_{EHF}/δ_{EHF}.) The horizontal load applied at the top of each column should be proportional to the vertical reaction. Thus, for an individual column:

$$\frac{H_{EHF,i}}{V_{ULS,i}} = \frac{H_{EHF}}{V_{ULS}}$$

where:

H_{EHF} is the sum of all the equivalent horizontal forces at column tops

V_{ULS} is the sum of all factored vertical reactions at ULS calculated from first-order plastic analysis

$H_{EHF,i}$ is equivalent horizontal force at top of column i (there are two columns in a single-span portal, three in a two-span portal, etc.)

$V_{ULS,i}$ is factored vertical reaction at ULS at column i, calculated from first-order plastic analysis.

An estimate of α_{cr} can then be obtained from:

$$\alpha_{cr,s,est} = 0.8\left\{1-\left(\frac{N_{R,ULS}}{N_{R,cr}}\right)_{max}\right\}\left\{\left(\frac{h_i}{V_{ULS,i}}\right)\left(\frac{H_{EHF,i}}{\delta_{EHF,i}}\right)\right\}_{min}$$

where:

$\left(\dfrac{N_{R,ULS}}{N_{R,cr}}\right)_{max}$ is the maximum ratio in any rafter

$N_{R,ULS}$ is the axial force in rafter (see Figure 4.42(b))

$N_{R,cr} = \dfrac{\pi^2 E I_r}{L^2}$ is the Euler load of the full span of the rafter (assumed pinned)

I_r is the in-plane second moment of area of the rafter

$\delta_{EHF,i}$ is the horizontal deflection of column top (see Figure 4.42(c))

$\left\{\left(\dfrac{h_i}{V_{ULS,i}}\right)\left(\dfrac{H_{EHF,i}}{\delta_{EHF,i}}\right)\right\}_{min}$ is the minimum value for columns 1 to n (n = the number of columns).

It is also necessary to evaluate the factor associated with $\alpha_{cr,r,est}$ snap-through buckling.

For frames with rafter slopes not steeper than 1:2 (26°), $\alpha_{cr,r}$ may be taken as:

$$\alpha_{cr,r,est} = \left(\frac{D}{L}\right)\left(\frac{55.7(4+L/h)}{\Omega-1}\right)\left(\frac{I_c+I_r}{I_r}\right)\left(\frac{275}{f_{yr}}\right)(\tan 2\theta_r)$$

This has to be checked because it is possible to design portals of 3 spans or more with stiff outer bays that provide horizontal support to the rafters of the inner spans. Then the rafters of the inner spans can act as arches with the horizontal reaction provided by the outer bays. Where this arching action works, the rafters will support more vertical load than if they were acting only as beams. This check is used to ensure that the rafters are not so flexible that they 'snap through'.

Note that where $\Omega \leq 1$, $\alpha_{cr,r} = \infty$

where:

D is cross-section depth of rafter

L is span of bay

h is mean height of column from base to eaves or valley

I_c is in-plane second moment of area of the column (taken as zero if the column is not rigidly connected to the rafter, or if the rafter is supported on a valley beam)

I_r is in-plane second moment of area of the rafter

f_{yr} is nominal yield strength of the rafters in N/mm^2

θ_r is roof slope if roof is symmetrical, or else $\theta_r = \tan^{-1}(2h_r/L)$

h_r is height of apex of roof above a straight line between the tops of columns

Ω is arching ratio, given by $\Omega = W_r/W_0$

W_0 is value of W_r for plastic failure of rafters as a fixed ended beam of span L
W_r is total factored vertical load on rafters of bay.

If the two columns or two rafters of a bay differ, the mean value of I_c should be used.

4.8.4 Serviceability limit state

The serviceability limit state (SLS) analysis should be performed using the SLS load cases, to ensure that the deflections are acceptable at the 'working loads'.

No specific deflection limits are set in BS EN 1993-1-1. According to BS EN 1993-1-1, Clause 7.2 and BS EN 1990 – Annex A1.4, deflection limits should be specified for each project and agreed with the client. The relevant National Annex to BS EN 1993-1-1 may specify limits for application in individual countries. Where limits are specified they have to be satisfied.

The SLS is normally assessed by a first-order analysis. It should check that there is no plasticity, simply to validate the deflection calculation.

4.8.5 Analysis for the ultimate limit state

General

At the ultimate limit state, the methods of frame analysis fall broadly into two types: elastic analysis and plastic analysis. The term *plastic analysis* is used to cover both rigid-plastic and elastic-plastic analysis. Plastic analysis commonly results in a more economical frame because it allows relatively large redistribution of bending moments throughout the frame, due to plastic hinge rotations. These plastic hinge rotations occur at sections where the bending moment reaches the plastic moment at loads below the full ULS loading. The rotations are normally considered to be localised at 'plastic hinges'.

A typical 'plastic' bending moment diagram for a symmetrical portal under symmetrical vertical loads is shown in Figure 4.43. This shows the position of the plastic hinges for the plastic collapse mechanism. The first hinge to form is normally adjacent to the haunch (shown in the column in this case). Later, depending on the proportions of the portal frame, hinges form just below the apex at the point of maximum sagging moment.

In practice, most load combinations will be asymmetric because they include either equivalent horizontal forces (see Section 4.8.3) or wind loads. A typical loading diagram and bending moment diagram is shown in Figure 4.44.

A typical bending moment diagram resulting from an elastic analysis of a frame with pinned bases is shown in Figure 4.45. In this case, the maximum moment is higher, and the structure has to be designed for this higher moment regime.

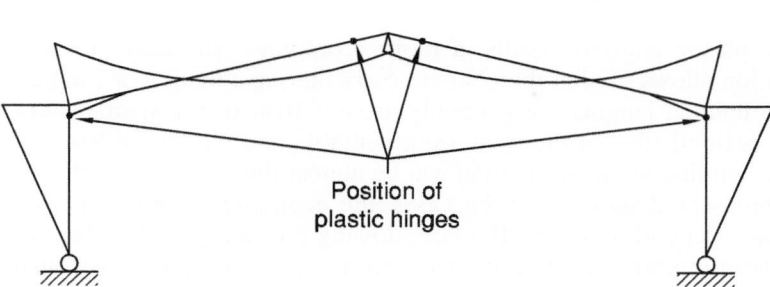

Figure 4.43   Bending moment diagram resulting from the plastic analysis of a symmetrical portal frame under symmetrical loading

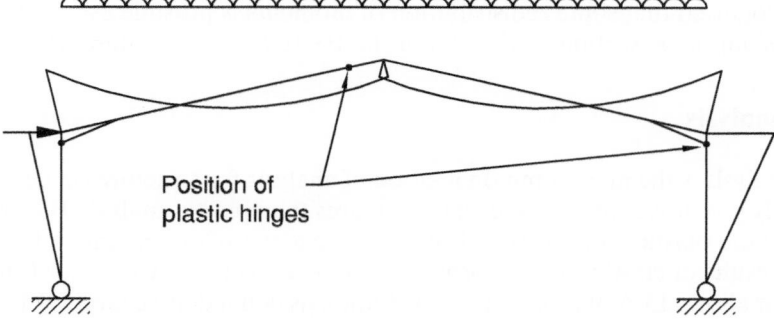

Figure 4.44   Bending moment diagram resulting from plastic analysis of a symmetrical portal frame under asymmetric loading

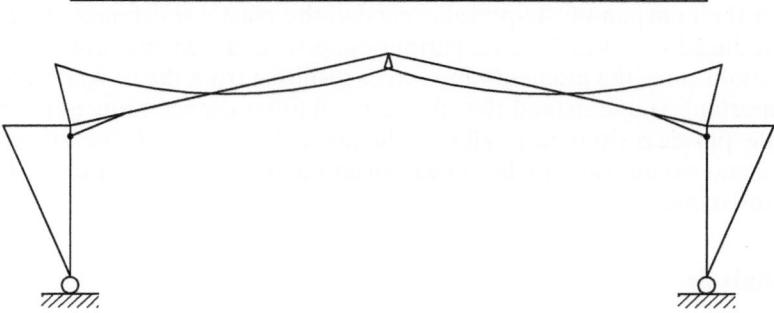

Figure 4.45   Bending moment diagram resulting from the elastic analysis of a symmetrical portal frame under symmetrical loading

Elastic vs. plastic analysis

Generally, plastic analysis results in more economical structures because plastic redistribution allows smaller members to carry the same loads. For frames analysed plastically, haunch lengths are generally around 10% of the span. Where haunch lengths of around 15% of the span are acceptable and the lateral loading is small, the elastic bending moment diagram will be almost the same as the plastic collapse bending moment diagram. In such cases, the economy of the design of slender frames could depend on the method of allowing for second-order effects. The simplest of these methods are approximate, so it is possible that elastic analysis will allow the most economical frames in certain cases.

Where deflections (SLS) govern design, there is no advantage in using plastic analysis for the ULS. The economy of plastic analysis also depends on the bracing system, because plastic redistribution imposes additional requirements on the bracing of members. The overall economy of the frame might, therefore, depend on the ease with which the frame can be braced.

It is recognised that some redistribution of moments is possible even with elastic design assumptions. Section 5.4.1.4 (B) of the BS EN 1993-1-1 allows 15%.

Elastic analysis

Elastic analysis is the most common method of analysis for structures in general but will usually give less economical portal structures than plastic analysis. BS EN 1993-1-1 allows the plastic cross-sectional resistance, e.g. the plastic moment, to be used with the results of elastic analysis, provided the section class is Class 1 or Class 2. In addition, it allows 15% of moment redistribution as defined in Section 5.4.1.4 (B) of BS EN 1993-1-1. To make full use of this in portal design, it is important to recognise the spirit of the Clause, which was written with continuous horizontal beams of uniform depth in mind. Thus, in a haunched portal rafter, up to 15% of the bending moment at the shallow end of the haunch could be redistributed, if the moment exceeded the plastic resistance of the rafter and the moments and forces resulting from redistribution could be carried by the rest of the frame. Alternatively, if the moment at the midspan of the portal exceeded the plastic resistance, this moment could be reduced by up to 15% redistribution, provided that the remainder of the structure could carry the moments and forces resulting from the redistribution.

It is important to understand that the redistribution cannot reduce the moment to below the plastic resistance. To allow reduction below the plastic resistance would be illogical and would result in dangerous assumptions in the calculation of member buckling resistance.

Plastic analysis

In practice, plastic analysis is almost always carried out by suitable commercial software using elastic/perfectly-plastic assumption.

The elastic/perfectly-plastic method applies loads in small increments and puts plastic hinges into the structure as they form with increasing load. It assumes that the members deform as linear elastic elements past first yield at M_y and right up to the full plastic moment M_p. The subsequent behaviour is assumed to be perfectly plastic without strain hardening. It is possible to predict hinges that form, rotate then cease, or even unload or reverse. The final mechanism will be the true collapse mechanism and will be identical to the lowest load factor mechanism that can be found by the traditional rigid-plastic methods. Figure 4.46 shows the development of the plastic mechanism as the loads increase.

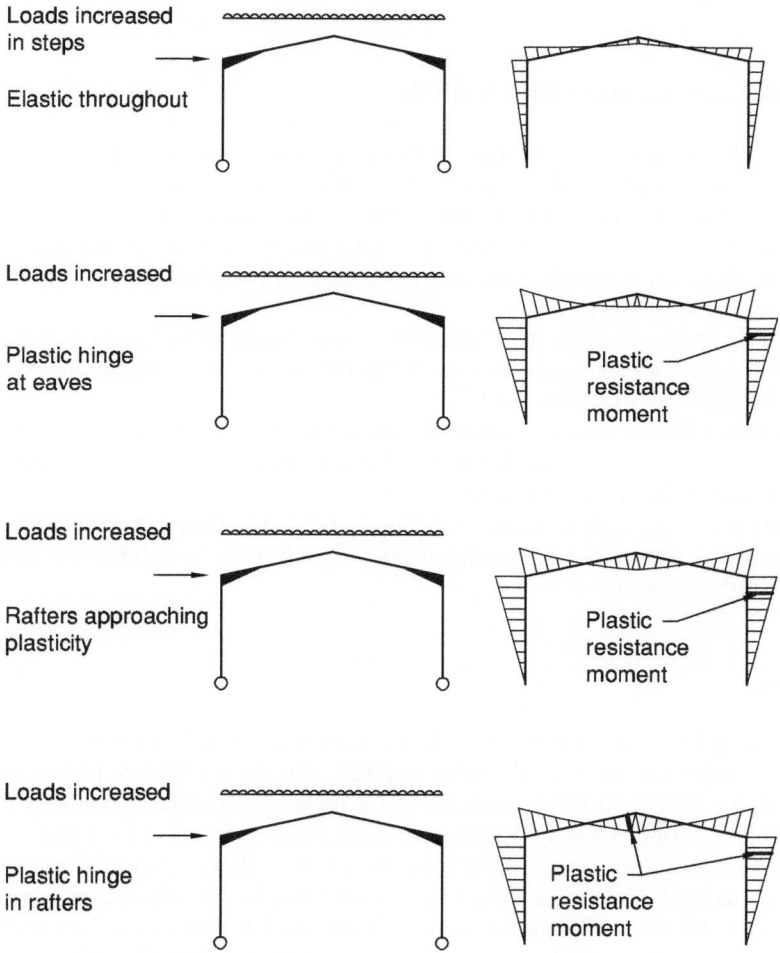

Figure 4.46   Elastic-perfectly plastic method

The method has the following advantages:

- The true collapse mechanism is identified.
- All plastic hinges are identified, including any that might form and cease. Any hinges that cease would not appear in the collapse mechanism but would need restraint.
- Hinges forming at loads greater than ULS can be identified. Where appropriate, the cost of member restraint at these positions could be reduced. This may produce economies in structures where the member resistance is greater than necessary, as occurs when deflections govern the design or when oversize sections are used.
- The true bending moment diagram at collapse, or at any stage up to collapse, can be identified.

First-order and second-order analysis

For either plastic analysis of frames, or elastic analysis of frames, the choice of first-order or second-order analysis depends on the in-plane flexibility of the frame, characterised by the calculation of the α_{cr} factor (see Section 4.8.3). When a second-order analysis is required, a modified first-order method can be useful if full second-order analysis is not available. This method is slightly different for elastic and plastic analysis, and is detailed below.

For elastic frame analysis, the 'amplified sway moment method' is the simplest one for introducing second-order effects for elastic frame analysis; the principle is given in BS EN 1993-1-1, Clause 5.2.2.

A first-order linear elastic analysis is first carried out; then the *horizontal loads* H_{Ed} *(e.g. wind) and equivalent loads* $V_{Ed}\ \phi$ *due to imperfections* are amplified by a 'factor' to ascertain the second-order effects.

For portal frames with shallow roof slopes, provided that the axial compression in the beams or rafters is not significant and $\alpha_{cr} \geq 3,0$ the 'amplification factor' can be calculated according to:

$$\left(\frac{1}{1 - 1/\alpha_{cr}} \right)$$

where α_{cr} may be calculated in accordance with Section 4.8.3 of this publication.

For first-order elastic/plastic frame analysis, the design philosophy is to derive loads that are amplified to account for the effects of deformed geometry (second-order effects). Application of these amplified loads through a first-order analysis gives bending moments, axial forces and shear forces that include the second-order effects approximately. The amplification is calculated by a method that is sometimes known as the Merchant-Rankine method. This provides an equivalent method for plastic analysis to the method for elastic frames in BS EN 1993-1-1, Clause 5.2.2 (4), 'For frames where the first sway buckling mode is predominant, first-order elastic analysis should be carried out with subsequent amplification of relevant action effects (e.g. bending moments) by appropriate factors.' Because, in plastic analysis, the plastic hinges limit the moments resisted by the frame, the amplification is performed *on the actions that are applied to the first-order analysis* (i.e. all actions and not only those related to wind and imperfections) instead of the action effects that are calculated by the analysis.

The method places frames into one of two categories:

- Category A: Regular, symmetric and mono-pitched frames
- Category B: Frames that fall outside of Category A but excluding tied portals.

For each of these two categories of frame, a different amplification factor should be applied to the loads. The method has been verified for frames that satisfy the following criteria:

1. Frames in which $\dfrac{L}{h} \leq 8$ for any span
2. Frames in which $\alpha_{cr} \geq 3$

where:

     L is span of frame (see Figure 4.47)
     h is the height of the lower column at either end of the span being considered (see Figure 4.47)
     α_{cr} is the elastic critical buckling load factor (calculated either exactly using software or estimated from the first sway mode (see Section 4.6.3).

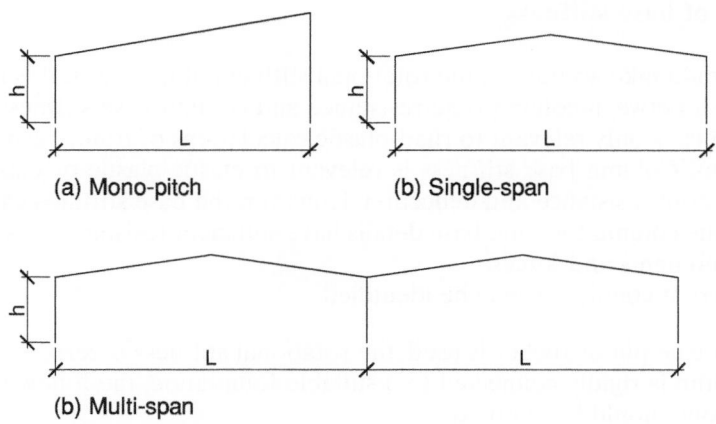

(a) Mono-pitch   (b) Single-span

(b) Multi-span

Note:
h is measured from the intersection of the rafter centreline and the column centreline, ignoring any haunch

Figure 4.47   Examples of Category A

Other frames should be designed using second-order elastic-plastic analysis software.

Amplification factors are determined as follows:

- Category A: Regular, symmetric and asymmetric pitched and mono-pitched frames

  Regular, symmetric and mono-pitched frames are either single-span frames or multi-span frames in which there is only a small variation in height (h) and span (L) between the different spans; variations in height and span of the order of 10% may be considered as being sufficiently small.

  In the traditional industrial application of this approach, for such frames first-order analysis may be used if all forces and moments are amplified for simplicity and safety by $\left(\dfrac{1}{1-1/\alpha_{cr}}\right)$, even though this is conservative for the column axial forces. The factor α_{cr} may be calculated in accordance with Section 4.8.3.

  From the foregoing, it is clear that there are significant advantages to adopting second-order elastic/plastic analysis, which will take direct account of all these effects.

- Category B: Frames that fall outside of Category A and excluding tied portals.

  For frames that fall outside of Category A, first-order analysis may be used if all the applied loads are amplified by $\left(\dfrac{1.1}{1-1/\alpha_{cr}}\right)$

  where α_{cr} may be calculated in accordance with Section 4.8.3.

Treatment of base stiffness

Analysis should take account of the rotational stiffness of the bases. It is important to distinguish between column base resistance and column base stiffness. Column base resistance is only relevant to rigid-plastic calculations of frame resistance, not to deflections. Column base stiffness is relevant to elastic-plastic or elastic frame analysis for both resistance and deflection. However, the base stiffness can only be included if the column foot and base details have sufficient resistance to sustain the calculated moments and forces.

Four different conditions may be identified:

1. Where a true pin or rocker is used, the rotational stiffness is zero.
2. If a column is rigidly connected to a suitable foundation, the following recommendations should be adopted:
   - Elastic global analysis:
     Ultimate Limit State calculations: stiffness of the base as equal to the stiffness of the column.
     Service Limit State calculations: base can be treated as rigid to determine deflections under serviceability loads.

- Plastic global analysis:
  Any base moment capacity between zero and the plastic moment capacity of the column may be assumed, provided that the foundation is designed to resist a moment equal to this assumed moment capacity, together with the forces obtained from the analysis.
- Elastic – plastic global analysis:
  The assumed base stiffness must be consistent with the assumed base moment capacity, but should not exceed the stiffness of the column.
  Where a 'dummy member' is used in the analysis to simulate base fixity, its potential effect on the base reaction must be recognised. The base reaction must be taken as the axial force in the column, which equals the sum of the reactions at the base and the pinned end of the dummy member.

3. Nominally semi-rigid column bases
   A nominal base stiffness of up to 20 % of the column may be assumed in elastic global analysis, provided that the foundation is designed for the moments and forces obtained from this analysis.

4. Nominally pinned bases
   If a column is nominally pin-connected to a foundation that is designed assuming that the base moment is zero, the base should be assumed to be pinned when using elastic global analysis to calculate the other moments and forces in the frame under ultimate limit state loading.

   The stiffness of the base may be assumed to be equal to the following proportion of the column stiffness:
   - 10% when checking frame stability or determining in-plane effective lengths
   - 20% when calculating deflections under serviceability loads.

4.8.6 Element design

It is necessary to consider separately: element design adjacent to plastic hinges and element design away from plastic hinges.

4.8.6.1 Stable lengths adjacent to plastic hinges

Eurocode 3 introduces four types of stable length, L_{stable}, L_m, L_k and L_s. Each is discussed below. All references are to BS EN 1993-1-1:2000.

L_{stable} **(Clause 6.3.5.3(1)B)**

This is the basic stable length for a uniform beam segment under linear moment and without 'significant' axial compression. This simple base case is of limited use in the verification of practical portal frames.

In this context, 'significant' may be related to the determination of α_{cr} in BS EN1993-1-1:2000 Clause 5.2.1 4(B) Note 2B. The axial compression is not significant if it is less than:

$$\frac{A_{fy}}{11\bar{\lambda}^2}$$

where:

$\bar{\lambda}$ is the in-plane, non-dimensional slenderness of the rafter or column, based on its system length with pinned ends.

L_m (Appendix BB3.1.1)

This is a stable length between the torsional restraint at the plastic hinge and the adjacent lateral restraint. It takes account of both member compression and the distribution of moments along the member. Different expressions are available for:

- uniform members (Equation BB.5)
- three flange haunches (Equation BB.9)
- two flange haunches (Equation BB.10).

There are two stable lengths between torsional restraints, L_k and L_s, which recognise the stabilising effects of intermediate restraints to the **tension** flange.

L_k (Appendix BB.3.1.2 (1)B)

This applies to the stable length between a plastic hinge location and the adjacent torsional restraint in the situation where the member is subject to a constant moment, providing the spacing of the restraints to either the tension or compression flange is not greater than L_m. Conservatively, this limit may also be applied to a non-uniform moment.

L_s (Appendix BB3.1.2 (2)B) and (3)B

This applies to the stable length between a plastic hinge location and the adjacent torsional restraint in the situation where the member is subject to axial compression and linear moment gradient, providing the spacing of the restraints to either the tension or compression flange is not greater than L_m.

Different C factors and different equations are used for linear moment gradients and uniform members (Equation BB.7) and non-linear moment gradients with uniform members (Equation BB.8).

Where the segment varies in cross-section along its length, i.e. in a haunch, two different approaches are adopted:

- for both linear and non-linear moments on three-flange haunches – Equation BB.11
- for both linear and non-linear moments on two-flange haunches – Equation BB.12.

The flowchart in Figure 4.48 summarises the practical application of the different stable length formulae for any member segment adjacent to a plastic hinge. (In the absence of a plastic hinge, the member segment is verified by conventional elastic criteria using Equations 6.61 and 6.62).

4.8.6.2 Element design remote from plastic hinges

Away from plastic hinges, rafters and columns must be checked in accordance with Chapter 19.

The objectives are to:

- minimise the number of torsional restraints
- ensure the elements are stable under combined in-plane and out-of-plane effects.

A worked example follows which is relevant to Chapter 4.

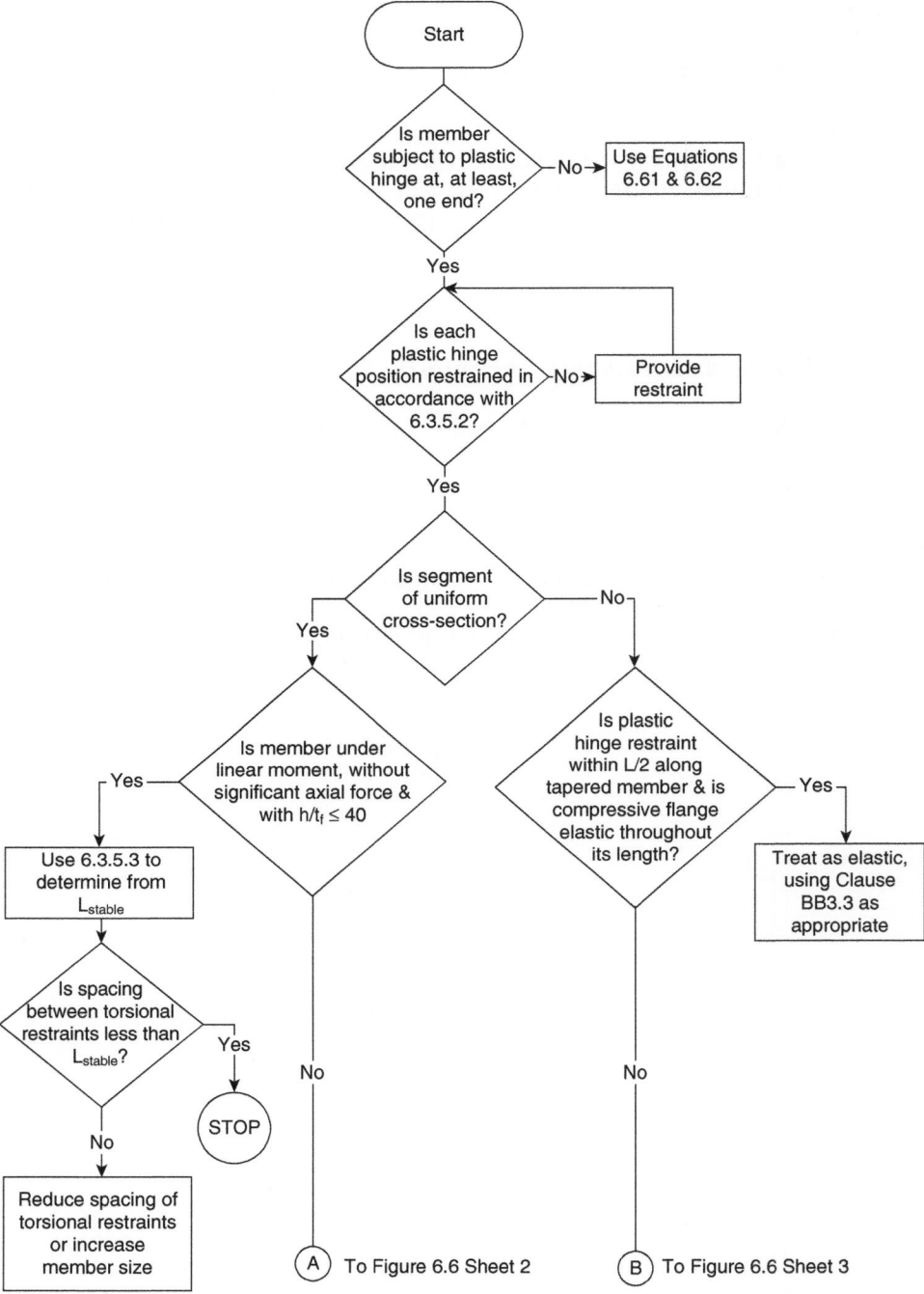

Figure 4.48 Decision tree for selecting appropriate stable length criteria for any segment in a portal frame – Sheet 1

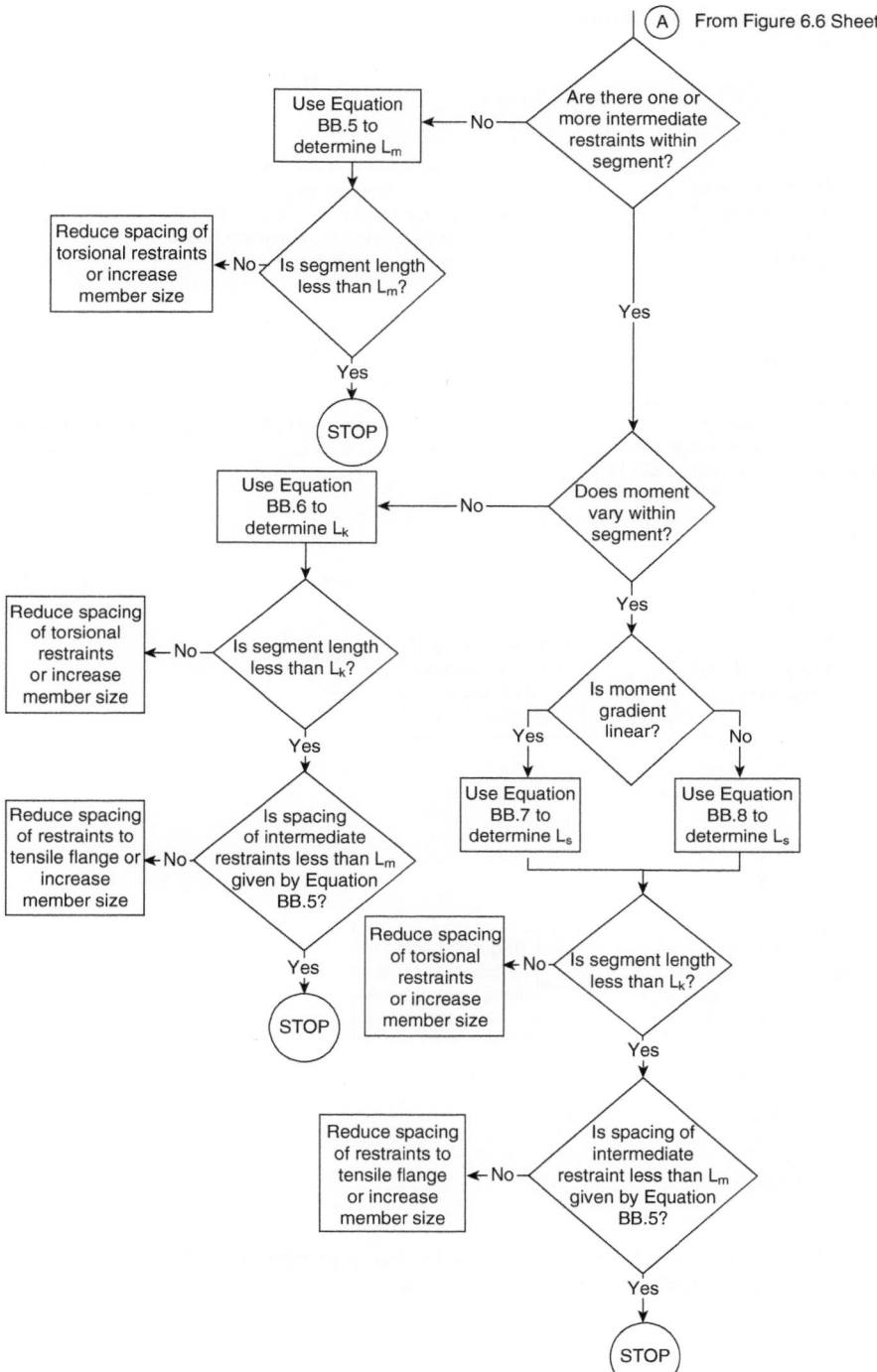

Figure 4.48 *(Continued)* Decision tree for selecting appropriate stable length criteria for any segment in a portal frame – Sheet 2

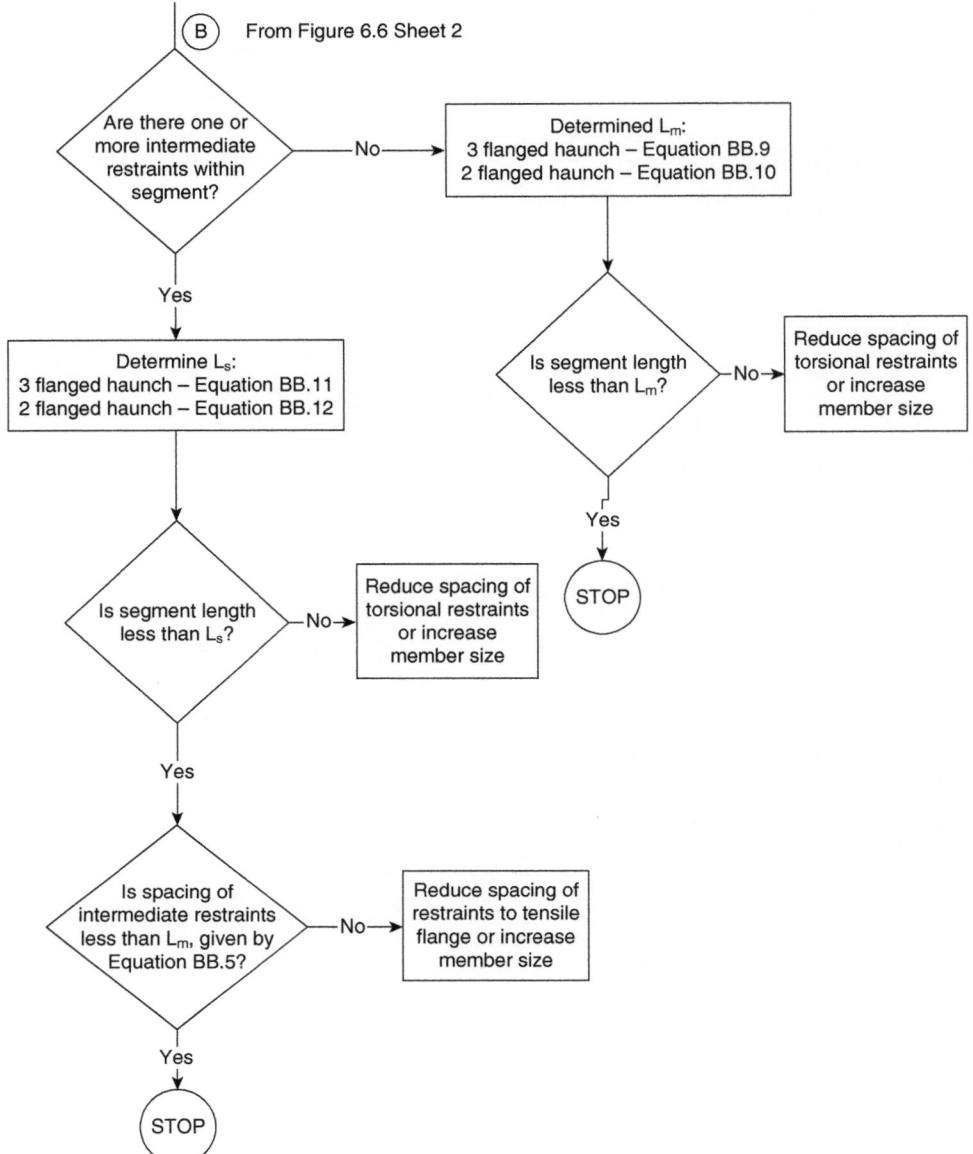

Figure 4.48   (Continued) Decision tree for selecting appropriate stable length criteria for any segment in a portal frame – Sheet 3

References to Chapter 4

1. Horridge J.F. (1985) Design of Industrial Buildings, *Civil Engineering* Steel Supplement, November.
2. The Building Act 1984: the Building Regulations and the Building (Approved Inspectors etc.) Regulations 2000 Building Regulations 2000, schedule 1, part L approved documents L1A, L1B, L2A, L2B multi-foil insulation.
3. Todd A.J. (1996) *Portal frames*. British Steel Structural Advisory Service.
4. Way A.G.J. and Salter P.R. (2003) *Introduction to steelwork design to BS 5950-1:2000 (P325)*, The Steel Construction Institute, Ascot.
5. Surtees J.O. and Yeap S.H. (1996) Load strength charts for pitched roof, haunched steel portal frames with partial base restraint, *The Structural Engineer*, Vol. 74, No. 1, January.
6. Davies J.M. and Raven G.K. (1986) Design of cold formed purlins. In: *Thin Walled Metal Structures in Buildings*, 151–60. IABSE Colloquium, Zurich, Switzerland.
7. Lim J., King C.M., Rathbone A.J. *et al.* (2005) Eurocode 3: The in-plane stability of portal frames, *The Structural Engineer*, Vol. 83, No. 21, November.

		Job No.		Sheet 1 of 6		Rev	
The Steel Construction Institute		Job Title	Worked Example Portal Frame Design				
Silwood Park, Ascot, Berks SL5 7QN Telephone: (01344) 623345 Fax: (01344) 622944		Subject	Chapter 4				
		Client		Made by	LRN	Date	Nov 2006
CALCULATION SHEET				Checked by	GWO	Date	

Portal Frame Design   12567b.wmf

Example: Design of portal frame using plastic analysis

This example introduces the design of a portal frame for a single-storey building, using a plastic method of global analysis. The frame uses hot-rolled I-sections for rafters and columns.

This example presents the overall frame geometry (including restraint position), definition of loads and selection of load combination.

Frame Geometry

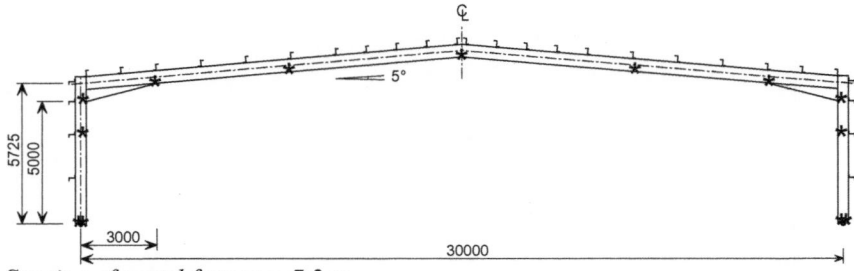

Spacing of portal frames = 7.2 m

* *– torsional restraint*

Portal Frame Geometry

The cladding to the roof and walls is supported by purlins and side-rails as indicated. The positions and spacing of the purlins and side rails were chosen as follows:

- *Torsional restraints are to be provided at both ends of the haunch (i.e. one at the column end and one at the rafter end).*
- *A torsional restraint is to be provided at an intermediate position along the rafter.*
- *The spacing of the purlins and side-rails should generally be about 1800 mm for lateral stability of the rafters and columns. The spacing will be less in regions where the frame members are close to their plastic moment of resistance.*
- *Uniform spacings are adopted within the above constraints.*

Portal Frame Design	Sheet 2 of 6	Rev

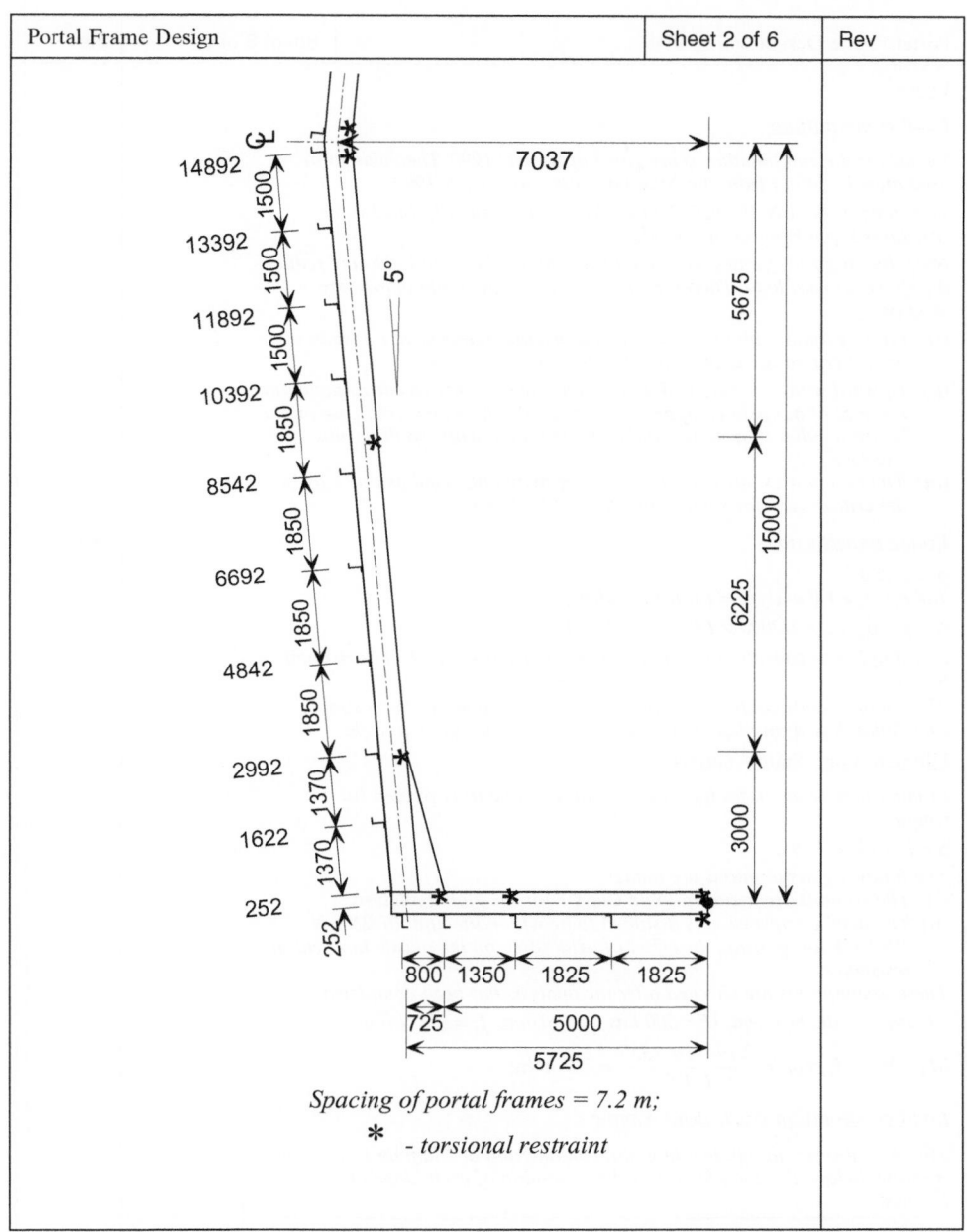

Spacing of portal frames = 7.2 m;

* *- torsional restraint*

Portal Frame Design	Sheet 3 of 6	Rev

Loads

Load combinations

Values for the combination ψ are given in BS EN 1990. The value to be used must be found from the National Annex to BS EN 1990.

The value in BS EN 1990:2002 Table A1.1 is 0.7 generally, but 1.0 for structures supporting storage loads.

Note that in portal frames with small roof slopes, the wind load may reduce the effects of roof load. Therefore the critical design combinations are usually:

(i)   *Gravity loads without wind, causing sagging moments at midspan of the rafter and hogging moment in the haunches.*

(ii) *Upward wind pressure with minimum gravity loads, causing maximum reversal of moment compared with case (i). The worst wind case might be from either transverse wind or longitudinal wind, so both must be checked.*

(iii) *The design must also be checked for gravity plus wind as this may be the critical case for some geometries of building.*

BS EN1993-1-1 Cl 5.3.2(3)

Frame imperfections

$\phi_0 = 1/200$
Taking $\alpha_h = 1.0$ and $\alpha_m = 1.0$ for simplicity,
$\phi = \phi_0 \cdot \alpha_h \cdot \alpha_m = 1/2000 \times 1.0 \times 1.0 = 1/200$
It is simplest to consider the frame imperfections as equivalent horizontal forces.

The column loads could be calculated by a frame analysis, but a simple calculation based on plan areas is suitable for single-storey portals.

Ultimate Limit State Analysis

In this example the bases have been assumed to be truly pinned for simplicity.

Steel grade is S355.

The following assumptions are made:
(i)   *The sections are assumed to be Class 1 for the global analysis.*
(ii) *The axial compression is assumed to be within the limit in BS EN 1993-1-1 for ignoring the effect of axial force on the plastic moment of resistance.*
These assumptions are checked after the analysis has been completed.

Column plastic moment: IPE 500 has $t_f < 40\,mm$, $f_y = 355\,N/mm^2$

$$M_p = W_{pl,y} \cdot f_y / \gamma_{MO} = \frac{2.194 \cdot 10^6 \cdot 355}{1 \cdot 10^6} = 779 \text{ kNm}$$

Load combination No.1: dead + snow

The second-order design bending moment diagram at Ultimate Limit State is shown below. The load factors at the formation of each hinge are as follows:

Load factor Fraction of ULS	Hinge number	Member	Position (m)
0.898	1	Right hand column	5.000
1.070	2	Left hand rafter	12.041

A mechanism is not formed until the second hinge has formed at a load factor of 1.07. Therefore the section sizes are suitable for this load combination.

Portal Frame Design	Sheet 4 of 6	Rev

Source: 12433a. wmf

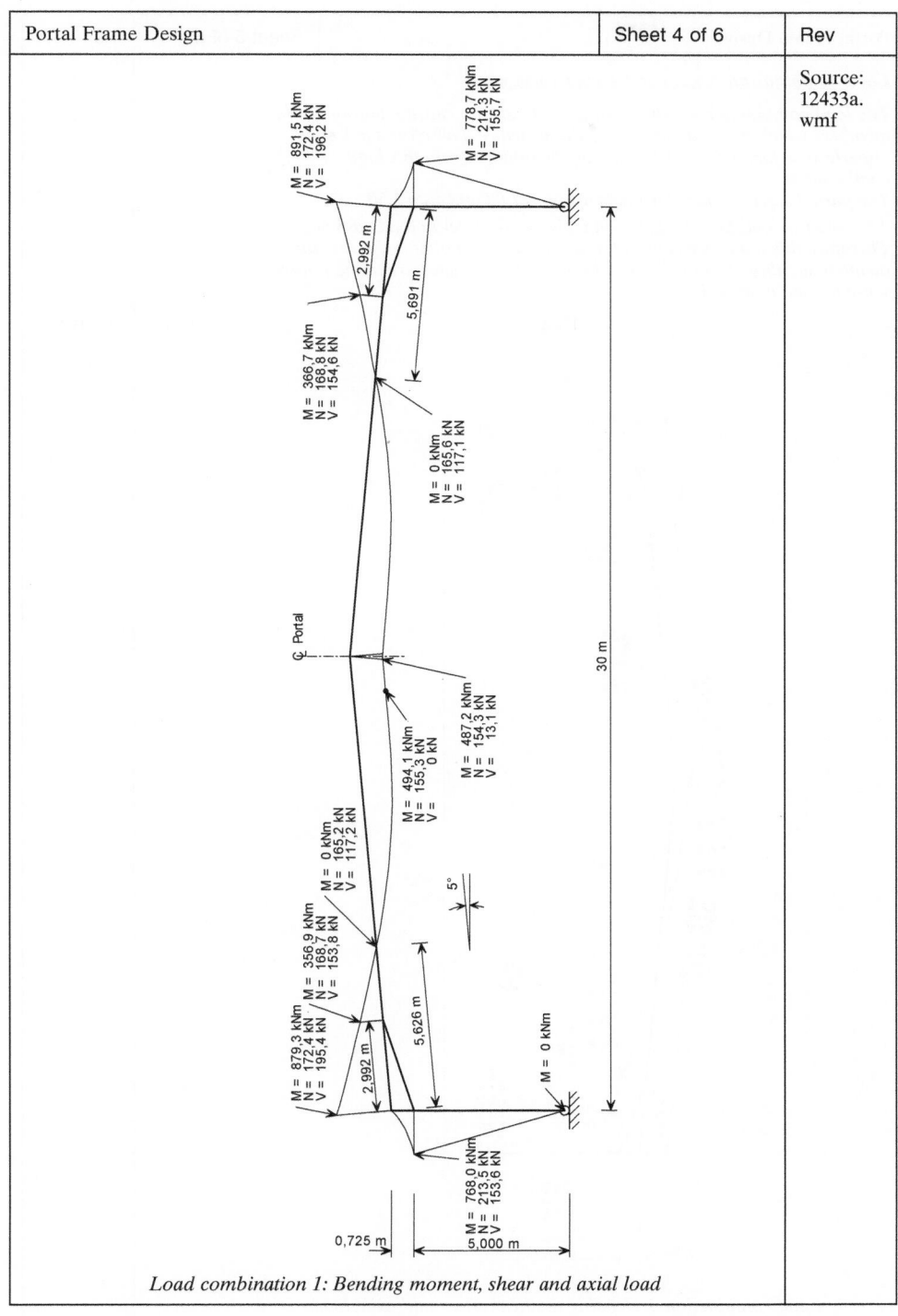

Load combination 1: Bending moment, shear and axial load

Portal Frame Design	Sheet 5 of 6	Rev

Load combination No.2: dead + transverse wind

The load combination results in an uplift load case causing tension in the members which does not destabilise the structure. Therefore the frame imperfection factor EC 3 Cl 5.3.2 may be omitted from this load combination.

The partial safety factors for loads are $\gamma_G = 1.00$ and $\gamma_Q = 1.50$

The collapse load factor = 6.22, which is greater than in load case no.1. Therefore this load case is not critical for cross-sectional resistance, but member stability must be checked because the moments are in the opposite sense to load case no.1.

Source: 12495.wmf

Load combination 2: Bending moment, shear and axial load

Portal Frame Design	Sheet 6 of 6	Rev

Load combination No.3: dead + longitudinal wind

In this case the wind loads applied to the structure result in a net upward force (except LH Column) on the roof as in load case no. 2.

The collapse load factor = 2.69, which is greater than in load case no.1. Therefore this load case is not critical for cross-sectional resistance, but member stability for this case must be checked because the moments are in the opposite sense to load case no.1.

Source: 12496. wmf

6,107 m

M = 433.35 kNm
N = 109.4 kN
V = 160,8 kN

M = 362,9 kNm
N = 107,4 kN
V = 94,0 kN

M = 0 kNm
N = 107,5 kN
V = 51,2 kN

M = 174,1 kNm
N = 110,6 kN
V = 71,4 kN

M = 0 kNm
N = 110,6 kN
V = 41,1 kN

M = 114,1 kNm
N = 112,3 kN
V = 9,8 kN

M = 114,1 kNm
N = 112,3 kN
V = 9,8 kN

30 m

M = 0 kNm
N = 110,6 kN
V = 41,1 kN

M = 175,5 kNm
N = 110,0 kN
V = 71,6 kN

M = 362,9 kNm
N = 10,4 kN
V = 94,0 kN

M = 0 kNm
N = 107,6 kN
V = 51,2 kN

6,107 m

2,992 m

5°

M = 433,35 kNm
N = 109,4 kN
V = 100,8 kN

0,725 m   5,000 m

Load combination 3: Bending moment, shear and axial load

Chapter 5
Multi-storey buildings

Compiled[1] by BUICK DAVISON

5.1 Introduction

Commercial buildings, such as offices, shops and mixed residential-commercial buildings, account for 20% of construction output in the EU, representing over 20 million square metres of floor space per year. The commercial sector demands buildings that are rapid to construct, of high quality, flexible and adaptable in application, and energy efficient in use. Steel and composite construction has achieved over 60% market share in this sector in some countries of Europe where the benefits of long spans, speed of construction, service integration, improved quality and reduced environmental impact have been recognised. These attributes are briefly outlined below.

Speed of construction

All steel construction uses pre-fabricated components that are rapidly installed on site. Short construction periods leads to savings in site preliminaries, earlier return on investment and reduced interest charges. Time-related savings can easily amount to 3–5% of the overall project value, reducing the client's requirements for working capital and improving cash flow. In many inner city projects, it is important to reduce disruption to nearby buildings and roads. Steel construction, especially highly pre-fabricated systems, dramatically reduces the impact of the construction operation on the locality.

Flexibility and adaptability

Long spans allow the space to be arranged to suit open plan offices, different layouts of cellular offices and variations in office layout throughout the height of the building. Where integrated beam construction (see 5.7.1) is used, the flat soffit gives

---

[1]This chapter is based on a series of access-steel documents. The source documents are available at www.access-steel.com.

Steel Designers' Manual, Seventh Edition. Edited by Buick Davison and Graham W. Owens.
© 2012 Steel Construction Institute. Published 2012 by Blackwell Publishing Ltd.

complete flexibility of layout allowing all internal walls to be relocated, leading to fully adaptable buildings.

Service integration

Steel and composite structures can be designed to reduce the overall depth of the floor zone by integrating major services within the depth of the structure, or by achieving the minimum structure depth. This is important in cases in which the building height is restricted for planning reasons, or in renovation projects. A 300 mm reduction in floor-floor depth can lead to equivalent savings of 20 to 30 Euros per square metre of floor area.

Quality and safety

Off-site prefabrication improves quality by factory-controlled production, and is less dependent on site trades and the weather. Working in a controlled manufacturing environment is substantially safer than working on site. The use of pre-fabricated components reduces site activity for frame construction by up to 75%, thereby substantially contributing to overall construction safety.

Environmental benefits

Many of the intrinsic properties of steel usage in construction have significant environmental benefits. For example, the steel structure is 100% recyclable, repeatedly and without any degradation, the speed of construction and reduced disruption of the site gives local environmental benefits and the flexibility and adaptability of steel structures maximise the economic life of the building as they can accommodate radical changes in use.

5.2 Costs and construction programme

5.2.1 Construction costs

A breakdown of construction costs for a typical office building is approximately as below:

- Foundations   5–15%
- Super-structure and floors   10–15%
- Cladding and roofing   15–25%
- Services (mechanical and electrical)   15–25%
- Services (sanitation, and other services etc.)   5–10%
- Finishes, partitioning and fitments   10–20%
- Preliminaries (site management)   10–15%

Preliminaries represent the costs of the site management and control facilities, including cranes, storage and equipment. Site preliminaries can vary with the scale of the project and a figure of 15% of the total cost is often allowed for steel intensive construction reducing to 10% for higher levels of prefabrication. The super-structure or framework cost is rarely more than 10% of the total, but it has an important effect on other costs. For example, as discussed above, a reduction of 100 mm in the ceiling-floor zone can lead to a 2.5% saving in cladding cost or 0.5% saving in overall building cost.

5.2.2 Cost of ownership/occupancy

It is estimated that the total cost of running a building during a 60-year design life may be 3 to 5 times the cost of initial construction. Major components in the longer term costs include direct running costs of heating, lighting, air conditioning etc., refurbishing the interior, minor redecoration (every 3–5 years), major refitting (every 10–20 years), replacing the services (every 15–20 years), possibly re-cladding the building after 25–30 years.

The European Directive[1] on the energy performance of buildings requires that office buildings have an energy performance certificate comparing energy use with benchmarks. Many modern buildings are designed with energy saving measures in mind, including double-skin façades, exposed thermal mass and chimneys for natural ventilation, photovoltaics in roofing etc.

5.2.3 Construction programme

A typical construction programme for a medium-sized office building is shown in Figure 5.1. One of the advantages of steel construction is that the initial period of site preparation and foundation construction is used for procurement of the basic steel sections and their off-site fabrication into a 'kit of parts' for rapid erection on site. The installation of the primary structure and floors takes approximately 20–

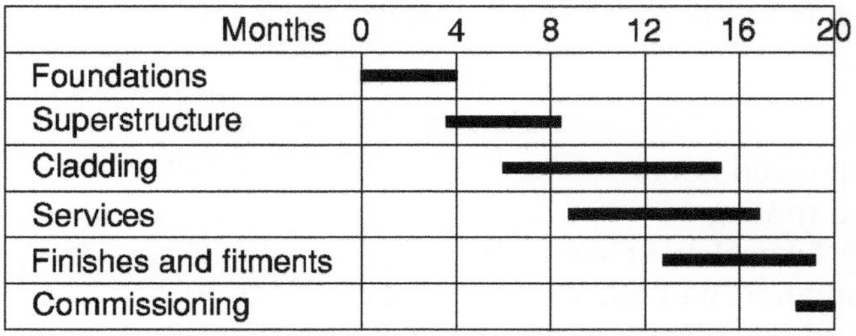

Months	0	4	8	12	16	20
Foundations						
Superstructure						
Cladding						
Services						
Finishes and fitments						
Commissioning						

Figure 5.1   Typical construction programme for a medium-sized office building

Table 5.1 Typical imposed loads for offices (kN/m²)

Application	Imposed Loading	Ceiling, Services, etc.	Self-weight*
Offices – general	2.5 + 1.0⁺	0.7	3.5
Offices – speculative	3.5 + 1.0⁺	1.0	3.5
Corridors and circulation	5.0	0.7	4.0
Library	7.5	0.7	4.0
Storage areas	3.5 + 1.0⁺	0.7	3.5

+Includes partitions
*Self-weight of a typical composite floor

25% of the total construction period but its completion permits an early start on cladding and servicing. It is for these reasons that steel construction leads to considerable advantage in terms of speed of construction, as it is a prefabricated and essentially 'dry' form of construction.

Typical time-related cost savings are 2% to 4% of the total costs i.e. a very significant proportion of the superstructure cost. Furthermore, in renovation projects or major building extensions, speed of construction and reduced disruption to the occupants or adjacent buildings can be even more important. Typical loading values for office buildings are shown in Table 5.1.

For a steel solution to be effective, it is essential that the overall design acknowledges the value that it brings to both the final outcome and the speed and simplicity of construction. The detailed steel design may be carried out by either a design team within the steelwork contractor or by an independent engineer.

Another key issue is to ensure that there are clear responsibilities for all the interfaces between specialist packages, e.g. between structure and cladding. Interface costs are significant and need to be accounted for properly as procurement is initiated.

5.3 Understanding the design brief

5.3.1 Design decisions

The development of any proposal for a multi-storey building requires a complex series of design decisions that are interrelated. Some interaction of the important early decisions is almost inevitable as the client's needs themselves are likely to change as the design develops. Design decisions must begin with the clearest possible understanding of the client requirements and of local conditions or regulations (which may be grouped under the heading of 'Planning').

Planning requirements are likely to define the overall height and plot ratio for the building and the need for set-backs to allow light to neighbouring buildings. The principal choices that need to be made, in close consultation with the client, are:

- The depth of the floor zone and the overall structure/service interaction strategy. Is it possible to provide an additional floor within overall height limitations? If expensive cladding is required, is it economic to reduce the floor zone?
- The need/justification for special investment in structure for prestigious public circulation areas.
- The provision of some 'loose fit' between structure and services, to permit future adaptability.
- The benefit of using longer span structure, at negligible extra cost, in order to enhance flexibility of layout.

Based on the design brief, a concept design is then prepared and is reviewed by the design team and client. It is this early interactive stage where the important decisions are made that influence the cost and value of the final project. Close involvement with the client is essential.

Once the concept design is agreed, the detailed design of the building and its components may be completed with less interaction with the client. Connections and interfaces between the components are often detailed by the fabricator or specialist designer but the lead architect should have an understanding of the form of these details.

5.3.2 Client requirements

Client requirements may be defined firstly by general physical aspects of the building, for example, the number of potential occupants, planning modules, floor-floor zones (3.6 to 4.2 m), floor-ceiling zone (2.7 to 3 m typically) etc. The floor-floor zone is a key parameter, which is influenced by planning requirements on overall building height, cladding cost etc. Minimum floor loadings and fire resistance periods are defined in national regulations, but the client may wish to specify higher requirements.

Other client requirements may be defined under servicing, Information Technology and other communication issues. In most inner city projects, air conditioning or comfort cooling is essential, because noise limits the use of natural ventilation. In suburban or more rural sites, natural ventilation may be preferred. Client requirements for design of the primary building services are likely to be a fresh air supply of 8–12 litres/sec per person, internal temperatures 22C ± 2C, cooling load of 40–70W/m^2 C and thermal insulation (walls) U < 0.3W/m^2 C.

Data communications are normally placed under a raised access floor to facilitate cabling access by the user and future modifications.

Floor loadings are presented in national Regulations or in EN 1991-1-1, and minimum values can be increased by client requirements. Floor loading has three basic components:

- imposed loading, including partitions
- ceiling and services, and a raised floor
- self-weight of the structure.

Imposed loading is dependent on the use of the building and design loads range from 2.5 to 7.5 kN/m² . An additional load of 1 kN/m² is often allowed for lightweight partitions and a total load of 0.7 to 1.0 kN/m² is allowed for ceiling, services and raised floor. For perimeter beams, is it necessary to include the loading from façade walls and internal finishes which can range from 3–5 kN/m for lightweight cladding to 8–10 kN/m for brickwork and 10–15 kN/m for precast concrete. The self-weight of a typical composite floor is 2.5 to 4 kN/m² which is only about 50–70% of that of a 200 mm deep reinforced concrete flat slab. The weight of a hollowcore concrete slab and concrete topping is 3.5 to 4.5 kN/m² .

5.3.3 Building location and ground conditions

The greatest uncertainties and therefore the greatest risks for any specific building project are unforeseen ground conditions and an adequate site investigation is essential to minimise these risks. Although the forms of substructure are generally similar for both steel and concrete superstructures, a steel superstructure will be less than half the mass of a concrete superstructure, making it easier to build on poor ground.

Building on brownfield sites has specific technical problems, including obstruction of existing foundations and services and attachment to adjacent buildings (and the legal rights of these owners or users). Piling can be problematical between existing foundations. The greater economic spanning capability of steel construction can enable a more flexible layout for the superstructure. It may therefore be possible to design around existing foundations, or indeed re-use them. Façade retention schemes often require use of a temporary external or internal steel structure to support the façade. This structure can be relatively complex and must be integrated into the final design.

Building in congested city centres brings with it logistical problems associated with lack of storage for materials and equipment, requirement for 'just-in-time' delivery of major components to site and reduction in noise and vibration levels, which may affect nearby properties. The speed of erection is directly related to the number of cranes and individual pieces that can be lifted. For most steelwork projects, a rate of 20–30 pieces of steelwork per day can be erected per tower crane.

Steel structures can be designed to span over railway lines, existing buildings or rivers taking advantage of the relatively light weight of the structure, and the ability to design full-height bracing in the walls to act as 'deep beams'. Building over tunnels requires consideration of reduction of ground movements to an acceptable minimum (often as low as 5 mm), reduction of loading due to self-weight of the building and limitations on the method and time of working.

Building adjacent to rivers or canals affects the design of basements and below ground works, both in the temporary and permanent conditions. Sheet piling is often used to provide temporary stability, and de-watering or ground freezing may be required to prevent water ingress.

Construction below ground requires careful consideration of ground water pressures, ground conditions and potential movement, temporary stability of the

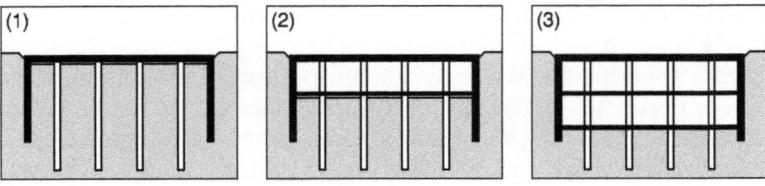

Key:
Stage 1: Piling and cast slab at ground level
Stage 2: Excavation below slab and install basement
Stage 3: Complete basement sub-structure

Figure 5.2   Stages in top-down construction

earthworks and water-tightness of the permanent basement. Sheet piled walls can be designed to resist earth and water pressure in the temporary and permanent conditions, and can be designed to be sufficiently water-tight for applications in below-ground car parking.

'Top-down' construction may be used in major projects in which the ground floor and basement structure is used to resist horizontal forces due to earth pressure as the earth is excavated. This form of construction using sheet piling is illustrated in Figure 5.2.

5.4 Structural arrangements to resist sway

5.4.1 Introduction

Within EN 1993-1-1,[2] there are different options for resisting sway loads, ensuring sway stability, addressing imperfections (both overall and local) and using simple, semi-continuous or continuous construction. Although EN 1993-1-1 provides a comprehensive treatment for a very wide range of frames, the resulting diversity of approaches can be perplexing. Section 5.4.3 explains a range of simplified approaches to both choice of framing and method of analysis. It should be noted that multi-storey frames are three dimensional structures with (usually) orthogonal horizontal grids, that is, with primary and secondary beams in two directions at 90°. Resistance to horizontal forces and the achievement of satisfactory sway stability need to be considered separately in these two principal directions and different solutions may be appropriate in the two directions.

This guidance is aimed at low and medium rise multi-storey buildings for all sectors of application, primarily commercial and residential buildings. Low and medium rise buildings are defined as those where the requirements to resist horizontal loads and achieve adequate sway stability do not have a major impact on structural form. For buildings of normal proportions in non-seismic regions, the upper limit may be taken as 10 storeys.

5.4.2 Principal concepts, definitions and significance for design

5.4.2.1 Braced and unbraced frames

As shown in Figure 5.3, a **braced frame** resists horizontal actions by one of two means:

- attachment of the steel structure to a stiff, concrete core, usually encompassing lifts, vertical services and stairs
- the use of bracing to resist horizontal actions.

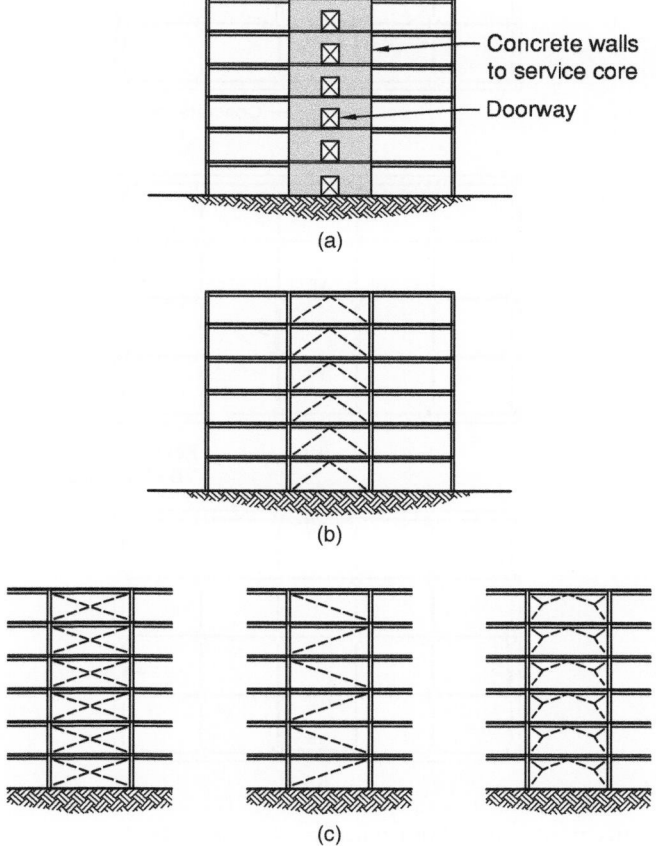

Figure 5.3   Types of braced frame: (a) stiff concrete core, (b) inverted V bracing, (c) alternative types of triangulated bracing

Where cross-bracing cannot be accommodated, it may be replaced by bracing in the form of stability portals.

As shown in Figure 5.4, to be effective the stiff core or braced bays need to be positioned more or less symmetrically about the overall plan of the building. Where the building is divided into sections by expansion joints, each section should be considered as a separate building. Floors act as diaphragms to transfer the overall horizontal actions to the stiff cores or braced bays.

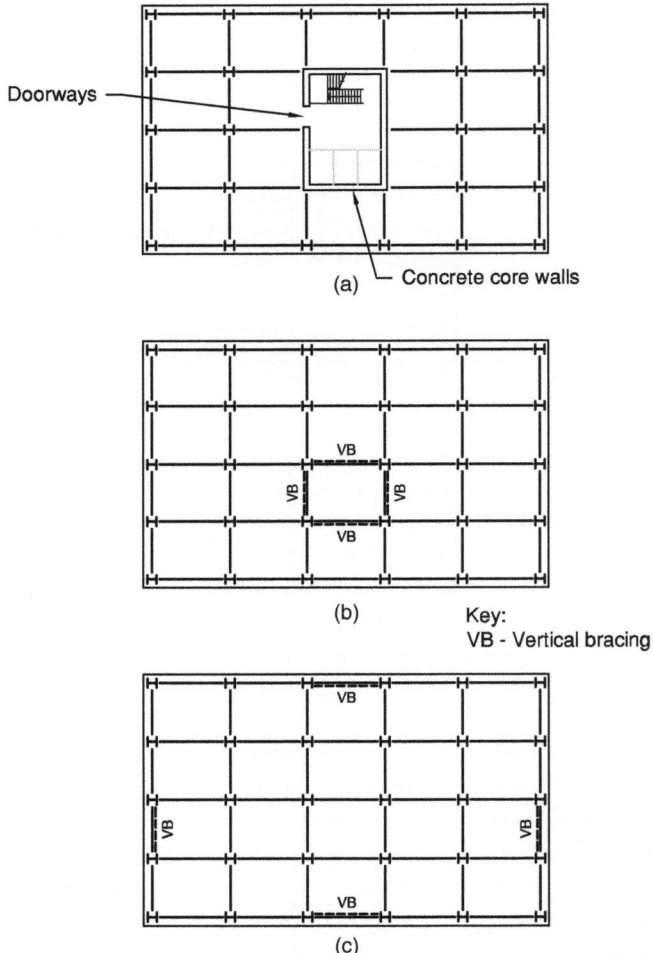

Figure 5.4   Effective positioning of resistance to sway forces: (a) Concrete core surrounding stairs, lifts, service shafts etc., (b) Stiff core of cross-braced panels, (c) Stiff panels not grouped as a core

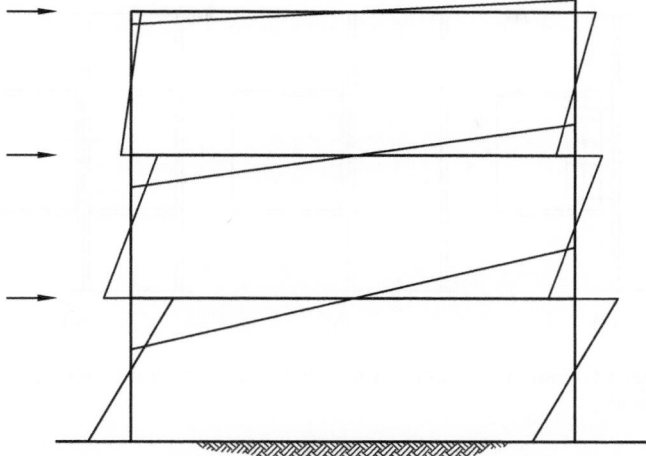

Figure 5.5   Bending moments in columns and beams from sway forces in a multi-storey single-bay un-braced frame

As shown in Figure 5.5, **unbraced frames** resist horizontal action by column and beam bending moments and shear forces. It is noteworthy that the beam moments arising from sway action are at their greatest where they frame into the columns. Clearly the connections in such frames must be capable of transferring substantial moments between beams and columns.

This frame action may be distributed between all the parallel frames in the building; see Section 5.4.3.3 for more details, or concentrated on selected 'stability portals' as explained in Section 5.4.3.4.

5.4.2.2 Sway stiffness of frames

The sway stiffness of a frame affects the choice of analysis method. First-order analysis may be used for frames satisfying the stiffness requirements in EN 1993-1-1 Clause 5.2. A design procedure to allow first-order analysis is given in Reference 3. The effects of deformations should be considered for all other frames. Design procedures are given in Reference 4. All frames sway to some extent; they may be categorised as either **Non-sway frames** or **Sway frames** depending on the magnitude of the effects of deformations. In the latter, the effects of deformed geometry, that is, second-order effects should be included in the analysis. It is likely that a frame that is lightly braced will behave as a sway frame.

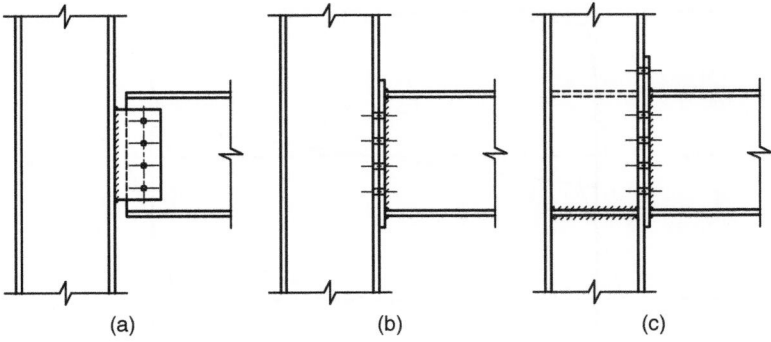

Figure 5.6 Types of beam to column connection: (a) nominally pinned, (b) semi-rigid, (c) rigid

5.4.2.3 *Simple, semi-continuous and continuous construction*

This classification of types of construction is based on the types of beam-to-column connection, illustrated in Figure 5.6 and explained in greater detail in Chapter 27.

In **simple construction** (in which the beam-to-column connections are nominally pinned), the beam-to-column connections only transmit shear from the beam ends to the columns. The proportions of the connections must permit sufficient end rotation of the beams for them to be designed safely as simply supported.

In **semi-continuous construction**, the semi-rigid beam to column connections additionally transmit significant bending moments from the beam ends to the columns. However, the local rotation within the connection associated with these moments is sufficiently large for it to influence the overall distribution of moments within the frame. Global analysis of semi-rigid construction requires appropriate modelling of connection behaviour and is therefore complex.

In **continuous construction** the beam-to-column connections are effectively rigid. No account need be taken of local connection rotations in the global analysis. They must be designed to transmit the full beam end moments and shear forces into the columns.

5.4.2.4 *Imperfections*

The effects of imperfections arising from residual stresses and geometrical imperfections are included in the analysis and design of multi-storey steel frames through the use of equivalent imperfections detailed in clause 5.3 of EN 1993-1-1. For many practical structures, the global imperfections are most effectively addressed by replacing them with systems of equivalent horizontal forces, as discussed in EN

1993-1-1 clause 5.3.2(7). See Reference 5 for detailed guidance on imperfections and Reference 6 for the design of vertical bracing.

Member imperfections are addressed within the buckling resistance of members, EN1993-1-1 clause 5.3.4(1).

5.4.3 Guidance on economical choice of frame for a specific building

The choices available to the designer are presented in ascending order of complexity and generally, in descending order of economy, that is, with the simplest and most economic first. Each principal direction of the column grid should be considered separately. Different choices may be appropriate in the two directions. For **unbraced frames**, sway stability can only be provided by frame action. It is therefore essential that continuous construction is adopted. (The use of semi-continuous construction to provide sway stability is also possible but more complex.) Continuous construction may be adopted for all frames or may be concentrated in discrete 'stability portals' with simple construction adopted for all other frames.

For **braced frames**, the designer may chose between simple or continuous construction on grounds of economy noting that:

- Where strength of the structure governs design, simple construction should **always** be adopted.
- Where stiffness (serviceability) governs design, greater economy will still generally be achieved with simple construction but if beam depth is severely limited, there may be benefits in considering continuous construction. It is recommended that alternative schemes are developed, discussed with the steelwork contractor and costed before the final choice is made.

5.4.3.1 Braced frames in simple construction

These frames use simple beam-to-column connections, as illustrated in Figure 5.6(a).

The design concept and assumptions are explained in Reference 7.

Advantages

- Simple, economic beam-to-column connections.
- Minimum column sizes and masses.
- Beams under sagging moments only, making best simple use of composite construction.
- Simple analysis of a determinate steel structure, thus facilitating optimised choice of beam and column elements.

Disadvantages

- None for small scale construction, if bracing can be accommodated and strength governs floor design.
- Could lead to uneconomic beams if serviceability governs floor design.

If these frames have stiff bracing i.e. with $\alpha_{cr} \geq 10$, they may be designed using first-order analysis. The stiff bracing may be provided either by a concrete core or triangulated steel bracing, as shown in Figure 5.3. As frames increase in height and number of storeys, bracing sizes increase and may become uneconomic or difficult to accommodate within architectural layouts. Simple guidance to ensure that triangulated steel bracing is proportioned to achieve $\alpha_{cr} \geq 10$ is provided in Reference 8. Where a concrete core is provided, it is assumed that the concrete core provides sufficient sway stiffness to ensure $\alpha_{cr} \geq 10$ for most practical low and medium rise buildings (if the designer is in doubt, an eigenvalue analysis may be carried out). First-order analysis may be performed manually; the primary benefit from computer analysis is the organisation of load cases, and the determination of governing actions within the structure.

Where the triangulated bracing to achieve $\alpha_{cr} \geq 10$ cannot architecturally or economically be accommodated, it may be possible to design a lighter bracing system. Practical guidance on the selection of minimum equivalent horizontal forces may be found in Reference 9. The α_{cr} may be determined approximately and manually by EN 1993-1-1 clause 5.2.1 4(B) for certain classes of structure. However, an appropriate computer analysis will provide both an accurate value of α_{cr} and assist in the organisation of load cases and the determination of governing actions within the structure.

It is theoretically possible to design a frame with α_{cr} less than 3.0, using a second-order analysis in accordance with EN 1993-1-1 clause 5.2.1 However, the resulting frame will be so flexible laterally that it is likely to fail practical sway deflection criteria for conventional construction. This approach should therefore only be utilised for special forms of construction that are beyond the scope of this chapter.

5.4.3.2 Braced frames with continuous construction

Where floor stiffness governs design, for example where beam depths are severely restricted, it *may* be more economic to adopt continuous construction in braced frames.

Bracing considerations are the same as for braced frames in simple construction. It will be necessary to carry out overall frame analysis to determine internal forces and moments in columns, beams, bracing and connections. Plastic design may be useful for these frames.

Advantages

- Continuous beam-to-column connections substantially increase stiffness of floor systems to ensure serviceability in presence of long spans and/or restricted beam depth.

Disadvantages

- Columns, especially external columns, increase in mass substantially to resist bending moments.
- Continuous beam-to-column connections are costly.
- Governing beam moments are likely to be at supports, substantially reducing the benefits from composite construction.
- Global analysis is complex, making it difficult to optimise element sizes.

Computer analysis is essential:

- to determine internal forces and moments in beams and columns
- to determine the distribution of sway action between the bracing and the frame
- to determine α_{cr}.

Where full second-order computer analysis is available, it is probably more effective in design effort to use it rather than use the amplification approach of EN 1993-1-1 clause 5.2.1 (3).

5.4.3.3 Unbraced sway frames

Where it is not architecturally possible to accommodate bracing, it is essential to adopt continuous or semi-rigid construction to provide sway stability and to resist sway forces. (Semi-rigid construction is complex and is therefore beyond the scope of this manual).

Designing a highly redundant structure is complex. Selecting appropriate initial sizes is essential for achieving an economical final outcome.

- Use Reference 10 to size the columns. These charts make approximate allowance for the coincidental effects of axial loads and moments.
- Size beams for $\dfrac{wl^2}{12}$, where w is the distributed load per unit length and l is the span of the beam.

- Initially, carry out separate first-order analysis for horizontal and vertical forces and use EN 1993-1-1 clause 5.2.1(4)B to estimate α_{cr} and check that the structure complies with horizontal deflection limits, see Reference 11.

Second-order effects must be included in the analysis. If $\alpha_{cr} \geq 3.0$, EN 1993-1-1 clauses 5.2.2 (5)B and (6)B permit the use of an amplified moment and force approach to horizontal actions. This may be more convenient than a full second-order analysis.

Advantages

- Permits architectural arrangements without any triangulated bracing.

Disadvantages

- Columns, especially external columns, increase in mass substantially to resist bending moments.
- Beam-to-column connections are complex and costly.
- Governing beam moments are likely to be at supports, substantially reducing the benefits of composite construction.
- Global analysis is complex, making it difficult to optimise element sizes.

Complete analysis is essential for similar reasons to Section 5.4.3.2 above.

5.4.3.4 Discrete stability frames

As shown in Figure 5.7(a), it may be architecturally desirable to use discrete 'strong' frames or 'portalised' frames to provide all the sway stability and resistance to sway forces. Providing thought is given to compatibility of sway stiffness, it may also be possible to combine such stability frames with the use of braced frames, as shown in Figure 5.7(b). In both these circumstances, it is possible to apply different design approaches for the different parts of the overall structure:

- Use the techniques of Section 5.4.3.3 above for the stability frames. These frames must provide sway stiffness and sway resistance to the entire building. Therefore their beams and columns need to be very stiff to resist the sway effects of the complete structure.
- Use the simple braced frame concepts of Section 5.4.3.1 for the remainder of the structure.

A full global analysis of such a hybrid arrangement should be carried out to determine α_{cr}.

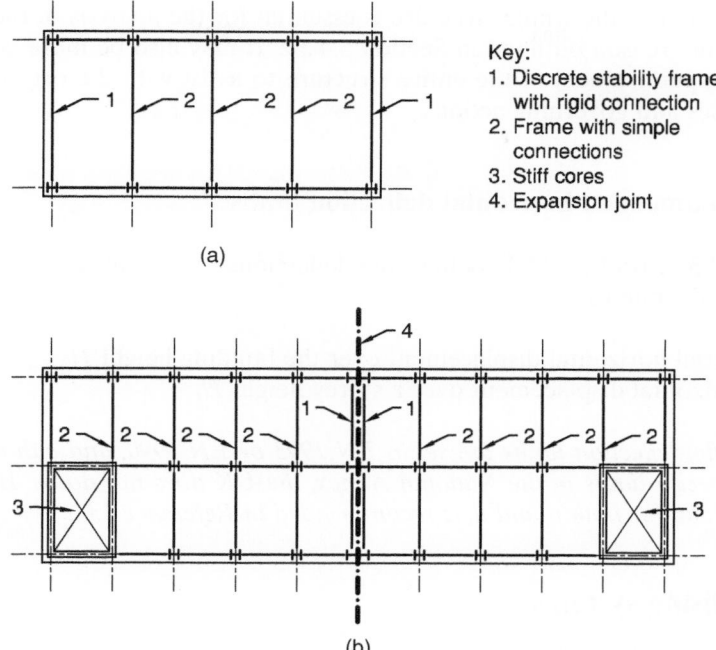

Key:
1. Discrete stability frame with rigid connection
2. Frame with simple connections
3. Stiff cores
4. Expansion joint

(a)

(b)

Figure 5.7 Examples of stability portals: (a) small, long span structure with no internal columns and no stiff cores, (b) small, long span structure with no internal columns and no stiff cores

Advantages

- Permit architectural arrangements without any triangulated bracing.
- Reduces the number of beam-to-column connections and columns that are required to carry large bending moments. This will generally lead to a greater economy than if the frame action is distributed throughout the entire structure which requires all beam-to-column connections to carry significant moments.

Disadvantages

- Columns, especially external columns, in 'portalised' frames increase in mass substantially.
- Beam-to-column connections in portalised frames are complex and costly.
- Careful consideration needs to be given to compatibility of sway stiffness between 'portalised' frames, simple frames and braced frames (if the latter exist).
- Global analysis is complex, making it difficult to optimise element sizes.

Consideration of the whole structure is essential for the analysis of the stability portals, for the reason outlined in Section 5.4.3.2. It may also be more convenient to use a computer model of the entire structure to assist with the organisation of the load cases and governing actions.

5.4.4 Recommended horizontal deflection limits

Annex A1.4.3 of BS EN 1990[12] defines the deflections to be considered at the serviceability limit state as:

u is the overall horizontal displacement over the building height H
u_i is the horizontal displacement over a storey height H_i

No specific deflection limits are set in EN 1993 or EN 1990, and although some countries specify limits in the National Annex, the UK does not do so. However, a value of $H/300$ for both u and u_i is recommended in Reference 13.

5.5 Stabilising systems

5.5.1 Introduction

Horizontal actions take the form of wind loads, stabilising forces and, in some regions, seismic actions. Various generic forms of stabilising system may be used depending on the height and horizontal scale of the building, as follows:

Medium rise buildings (4–8 storeys):	Braced bays around the cores or in the façade or concrete core
Tall buildings (8–20 storeys):	Concrete core or steel plated core
Ultra-tall buildings (20+ storeys):	External bracing or 'Frame-tube' concept

As discussed in Section 5.4.3.4, frame action may be used as an alternative to braced bays. However, it increases the size of columns and substantially increases the cost of connections. For economy, rigidly-jointed frames should be avoided where permitted by the building's architecture. For low- or medium-rise buildings, semi-rigid frames may offer some economy and their use is covered by the Eurocodes.

5.5.2 Types of bracing

Bracing may be in various generic forms, such as X, K and V forms, as in Figure 5.8. Where X-bracing is used, the bracing members may be designed to act only in tension (the members are slender elements that buckle at low compressive forces, thus not being effective in compression). Where K or V bracing is used, the bracing members must be able to carry compression forces.

Flat plates or angles may be used for X-bracing but tubular or H-sections are generally used for K- or V-bracing.

The nature of the forces in the bracing members is illustrated in Figure 5.9 (the magnitude of the forces depends on the panel geometry). As shown, compression forces are ignored in redundant bracing systems.

It is usually possible to accommodate the bracing within the width of the walls, thus minimising its effect on architectural detailing. X-bracing is the least intrusive. Inverted V- or single bracing may be required to accommodate doorways in walls.

An illustration of X-bracing is shown in Figure 5.10. The bracing members are usually structural hollow sections, for V- and K-bracing (angles may also be used

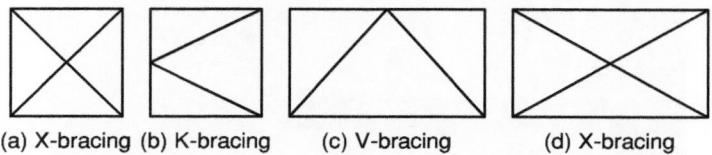

(a) X-bracing (b) K-bracing   (c) V-bracing   (d) X-bracing

Figure 5.8   Different types of bracing

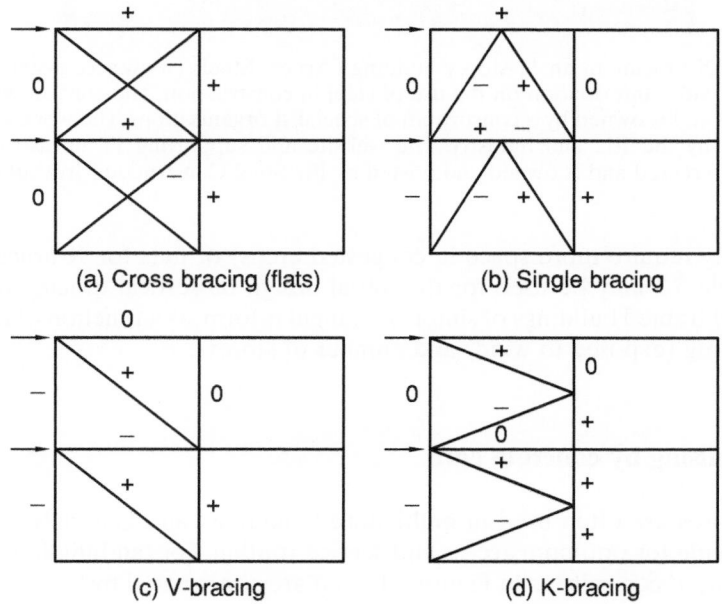

Compressive forces positive, tensile forces negative

Figure 5.9   Forces in X-, K- and V-bracing

Figure 5.10   X-bracing in an 11-storey building (Access Steel) (www.access-steel.com provides information on the use of steel in construction. The content was created
and is owned by a consortium of specialist organisations. The work was funded
by the EU and industry. The website and supporting IT infrastructure was
created and is owned and hosted by the Steel Construction Institute (SCI))

but generally require more space in congested areas) or flats for X-bracing. Table
5.2 and Table 5.3 may be used for the initial design of vertical bracing in V or X
form in steel framed buildings of simple rectangular form, as a function of the length
of the building (exposed to wind) and number of storeys.

5.5.3 Stabilising by concrete core

Concrete cores are often used in multi-storey buildings, and generally located on
plan to provide for optimum access and service routing. For tall buildings, they are
normally placed centrally as in Figure 5.11 and are constructed by 'slip forming', in
advance of the installation of the surrounding steelwork. Their size is determined
by the number of lifts, the service risers and zones for toilets and circulation. The
core area typically represents about 5–7% of the plan area for medium rise buildings
but can increase to 12–20% for high rise buildings.

Table 5.2 Sizes of tubular bracing members in V form (diameter × thickness)

Number of Storeys	Length of Building (m)			
	20	30	40	50
4	100 × 10	120 × 8	120 × 12.5	150 × 8
6	120 × 8	120 × 12.5	150 × 10	150 × 12.5
8	120 × 12.5	150 × 10	150 × 16	2 × 150 × 8
12	2 × 120 × 8	2 × 120 × 12.5	2 × 150 × 10	2 × 150 × 12.5

Bracing is at both ends of the building; 2× means two braced bays at each end
Floor-floor height is 4 m and wind load is 1 kN/m²

Table 5.3 Size of flat bracing members in X form (plate depth × thickness)

Number of Storeys	Length of Building (m)			
	20	30	40	50
2	120 × 10	150 × 12	150 × 15	200 × 20
4	150 × 15	2 × 150 × 12	2 × 200 × 15	2 × 200 × 20
6	2 × 200 × 12	2 × 200 × 20	2 × 220 × 20	2 × 220 × 22
8	2 × 200 × 15	2 × 220 × 20	2 × 250 × 20	2 × 250 × 25

Bracing is at both ends of the building; 2× means two braced bays at each end
Floor-floor height is 4 m and wind load = 1 kN/m²

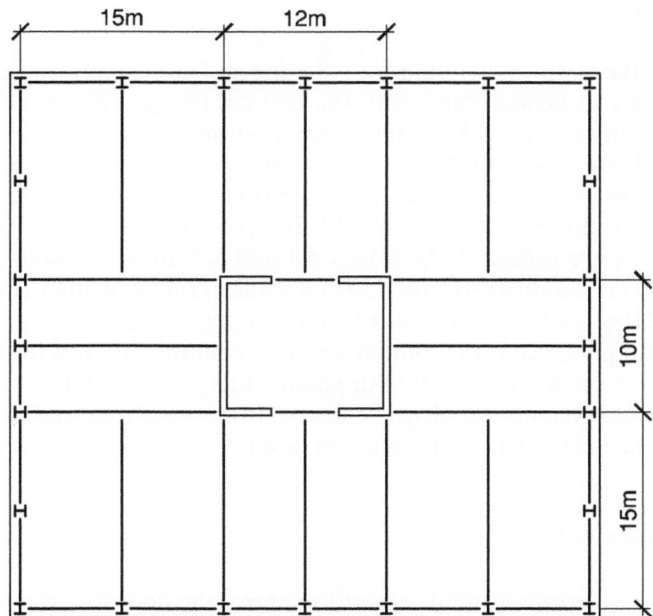

Figure 5.11 Example of a concrete core in a steel framed building

Steel beams are attached to the concrete core by various methods:

- casting a steel plate within the wall attached by shear connectors to which the beam connection is subsequently made on-site
- creating holes or pockets into the wall into which the beams are placed.

The steel plate method allows for positioning of welded attachments, taking into account site tolerances. The thickness of the concrete wall is typically 200 to 300 mm. Sufficient reinforcement is placed to resist over-turning effects due to wind action. Concrete lintels above openings often require heavy reinforcement.

5.6 Columns

The columns and other vertical load-bearing elements of the structure are generally designed to have the minimum impact on the useable space of the building and therefore are of the minimum size possible. The size of the columns clearly depends on the height of the building and the floor area supported. There is also an advantage in using higher grade steel and considering an integrated fire-resistant design (see Sections 5.6.2 and 5.6.3). The various options for columns are illustrated in Figure 5.12.

5.6.1 H-sections

H-sections are the simplest solution for columns and are usually orientated so that the larger (primary) beams frame into the column flange. This makes connection detailing considerably easier. The same column serial size is normally chosen at all floor levels, although the weight of the section can be varied, to simplify column splices. For economy and convenience of erection, columns are placed in lengths equivalent to 2 or 3 times the floor height. For concept design, two tables for HE and UC columns are presented in Tables 5.4 and 5.5. In these tables, an imposed load of 4kN/m^2 is assumed together with a total permanent load (including self-weight) of 4kN/m^2 and the floor-floor height is taken as 4 m.

More detailed guidance on the initial sizing of columns is given in Reference 10.

H-columns are usually provided with passive fire protection, most commonly in the form of board systems for visual reasons. However, if required architecturally, they may also be protected by intumescent paints.

5.6.1.1 Column splices

Column splices are normally made about 1 m above the floor for ease of installation of the bolts. Four splice configurations are illustrated in Figure 5.13. For non-machined ends to the columns, axial loads are transferred through splice plates with multiple bolts. In practice, a modern cold saw can give the equivalent of a machined

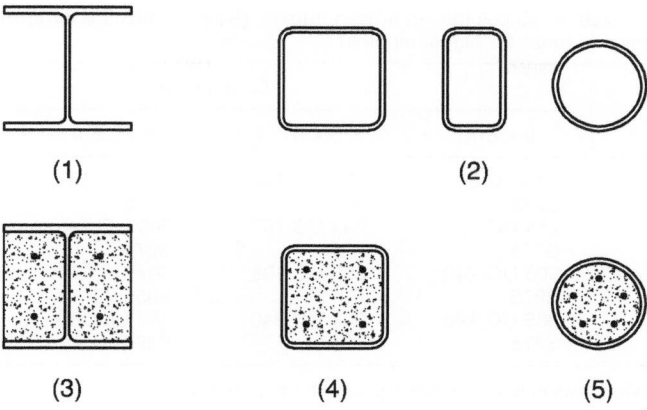

Key:
(1) HE or UC section
(2) Structural hollow sections
(3) Partially encased H section
(4) Composite square hollow section column
(5) Composite circular hollow section column

Figure 5.12   Column options

Table 5.4   Typical sizes of HE columns in braced frames. (Sizes shown are for lowest length of column, with reduction in mass for higher lengths)

Number of Storeys	Column Grid			
	$6 \times 6\,m$	$6 \times 9\,m$	$6 \times 12\,m$	$6 \times 15\,m$
4	HE 220 B	HE 280 B	HE 240 M	HE 260 M
6	HE 280 B	HE 240 M	HE 260 B	HE 300 M
8	HE 300 B	HE 260 M	HE 300 M	HE 320 M
10	HE 240 M	HE 300 M	HE 320 M	HD 400 x 347

All in S355 steel; Imposed load = $3\,kN/m^2$ plus $1\,kN/m^2$ for partitions

end to the column without the need for further machining. Countersunk bolts may be used where the flange is sufficiently thick, and where the bolt heads would affect the finishes to the columns. An end plate detail may be used for lightly loaded columns. Column lengths of 8 to 12 m are most economic, representing 2 or 3 storey heights.

5.6.2 Partially encased H-sections

Partial encasement between the flanges of the columns increases both their compressive resistance and their fire resistance. Typically, partially encased columns can achieve 60 or 90 minutes fire resistance, depending on the amount of bar

Table 5.5 Typical sizes of UC columns in braced frames. (Sizes shown are for lowest length of column, with reduction in mass for higher lengths)

Number of Storeys	Column Grid			
	6 × 6 m	6 × 9 m	6 × 12 m	6 × 15 m
4	203 UC 86 S275	254 UC 132 S275	254 UC 167 S275	305 UC 198 S275
6	254 UC 132 S275	254 UC 167 S275	305 UC 198 S275	305 UC 240 S355
8	305 UC 240 S275	305 UC 198 S275	305 UC 240 S355	356 UC 235 S355
10	305 UC 198 S275	305 UC 240 S355	356 UC 340 S355	356 UC 340 S355

Steel grade as shown; Imposed load = 3 kN/m² plus 1 kN/m² partitions

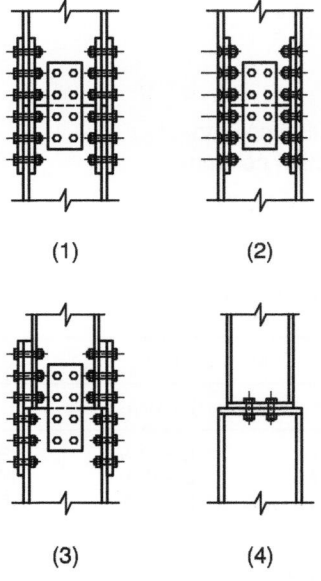

Key:
(1) Splice connection - shear transferred through bolts
(2) End bearing connection - with countersunk bolts
(3) Splice connection - dissimilar column sizes
(4) End plate connection - dissimilar column sizes

(1)   (2)

(3)   (4)

Figure 5.13   Example of column splice details

reinforcement.[14] Design Table 5.6 for partially encased HE sections is given below for concept design.

5.6.3 Concrete-filled structural hollow sections

Concrete-filled circular and square hollow sections are architecturally very important and achieve excellent composite properties due to confinement of the concrete infill. Concrete-filled tubes also achieve excellent fire resistance because compression is

Table 5.6 Typical sizes of partially encased sections in braced frames

Number of Storeys	Column Grid			
	6 × 6 m	6 × 9 m	6 × 12 m	6 × 15 m
4	HE 240A	HE 240B	HE 280B	HE 300B
6	HE 240B	HE 280B	HE 340B	HE 400B
8	HE 280B	HE 340B	HE 450B	–
10	HE 300B	HE 400B	–	–

All in S355 steel; Imposed load = 3 kN/m² plus 1 kN/m² for partitions

Table 5.7 Typical sizes of concrete-filled hollow sections in braced frames

Number of Storeys	Column Grid			
	6 × 6 m	6 × 9 m	6 × 12 m	6 × 15 m
4	219 × 10	219 × 12.5	273 × 12.5	323 × 12.5
6	219 × 12.5	273 × 16	323 × 16	355 × 16
8	273 × 12.5	323 × 16	355 × 20	406 × 16
10	323 × 12.5	355 × 16	406 × 20	457 × 20

Diameter (mm) × thickness (mm)
All in S355 steel; Imposed load = 3 kN/m² plus 1 kN/m² for partitions

transferred to the cooler concrete and its bar reinforcement. For fire resistance purposes, the minimum amount of bar reinforcement should generally satisfy the limits in EN 1994-1-2. No reinforcement is generally required for 60 minutes fire resistance. Larger diameter tubes can be concrete-filled from the base but most smaller tubes are filled from their top. Typical sizes for concept design are shown in Table 5.7.

5.7 Floor systems

5.7.1 Introduction

Figure 5.14 illustrates a range of popular floor systems. In most cases a grillage of beams is required with an overlying concrete slab either in the form of a composite slab or precast units.

5.7.2 Composite slabs

Composite slabs are relatively shallow floors which use steel decking as both the temporary support to the wet concrete during construction, and as a composite element to resist imposed loads on the completed floor system. The slab spans directly between support beams. Two generic forms of composite slabs exist – shallow slabs (100 to 180 mm depth typically) and deep slabs (280 to 350 mm depth

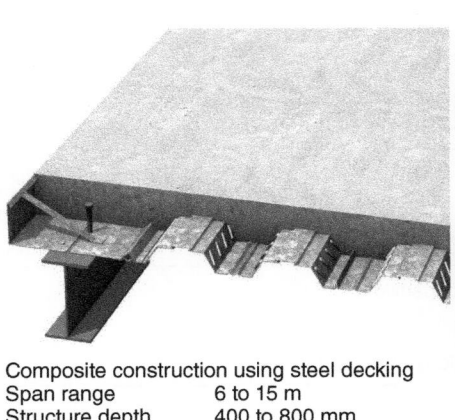

Composite construction using steel decking
Span range   6 to 15 m
Structure depth   400 to 800 mm

(a)

Cellular beams in composite construction
Span range   9 to 18 m
Structure depth   600 to 1000 mm

(b)

Integrated beams with deep decking
Span range   5 to 9 m
Structure depth   300 to 350 mm

(c)

Fabricated or rolled beams with large web openings
Span range   9 to 20 m
Structure depth   600 to 1200 mm

(d)

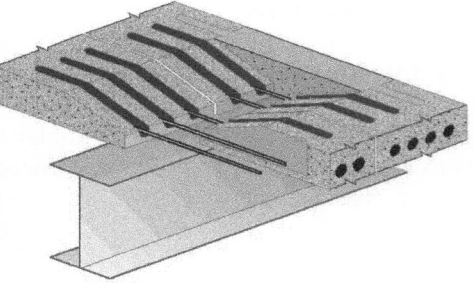

Integrated beams supporting precast
concrete slabs
Span range   5 to 9 m
Structure depth   300 to 400 mm

(e)

Steel beams supporting precast concrete
slabs
Span range   5 to 10 m
Structure depth   500 to 900 mm

(f)

Figure 5.14   Construction systems and their ranges of application

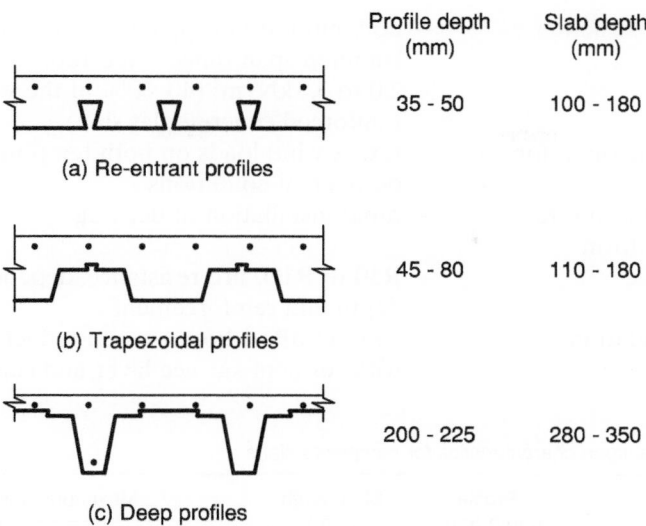

	Profile depth (mm)	Slab depth (mm)
(a) Re-entrant profiles	35 - 50	100 - 180
(b) Trapezoidal profiles	45 - 80	110 - 180
(c) Deep profiles	200 - 225	280 - 350

Figure 5.15   Various forms of composite slab. (It should be noted that not all products are currently available throughout Europe)

Table 5.8   Range of application of composite slabs

Deck shape	Profile depth (mm)	Slab depth (mm)	Span range (m)
1. Re-entrant profiles	35–50	100–150	2–3.5
2. Trapezoidal profiles	45–80	120–180	2.5–4.5
3. Deep profiles	200–225	280–350	4.5–8.5

typically). Deck profile depths range from 35 to 80 mm for shallow decks, and 200 to 225 mm for deep decks. Cross-sections through typical composite slabs are shown in Figure 5.15. Both trapezoidal and re-entrant deck profiles may be used; the re-entrant profile is widely available across Europe. Embossments are rolled into the deck profile to improve composite action with the in-situ concrete.

The range of applications of composite slabs is presented in Table 5.8. Composite slabs may be designed to act as part of a composite beam system by using shear connectors.

5.7.2.1 Application benefits of composite slabs

The benefits of composite slabs in construction may be summarised as:

- speed of construction   — no temporary propping required in the common span range – see Table 5.9
- light weight   — 2.0 to 3.5 kN/m^2 (40–60% of the weight of a reinforced concrete flat slab)
- transfers horizontal forces — resists wind loads on both temporary and permanent conditions
- decking acts as a safe working platform — rapid installation of decking
- fire resistance   — R30 to R120 fire resistance, depending on slab depth and reinforcement
- acoustic insulation   — 56 to 60 dB airborne sound reduction achieved with resilient surface layer and ceiling.

Table 5.9 Typical span characteristics for composite slabs

Slab type	Profile depth (mm)	Slab depth (mm)	Maximum span (m)		
			Un-propped		Propped
			t = 0.9 mm	t = 1.2 mm	
	50	100	3.2	3.5	3.6
		120	2.9	3.2	4.2
		150	2.7	3.0	4.5
	60	120	3.2	3.6	4.0
		150	2.8	3.2	4.2
	200	300	5.5	5.0	7.0
		350			8.0

t = steel thickness (grade S350)
Imposed loading = 3 kN/m^2; plus 1 kN/m^2; for partitions etc.

5.7.2.2 Design aspects

The structural design of a composite slab in the temporary condition depends on the self-weight of the slab and a temporary construction load of 0.75 to 1.5 kN/m^2, representing the loads applied during concreting. Composite action depends on the shear-bond strength or adherence between the concrete and steel profile, which provides the resistance to subsequent imposed load. Normally composite action is sufficiently good that it is the construction condition which controls the design of the slab. Fire resistance can be achieved by using various sizes of mesh reinforcement.

Table 5.9 presents the span characteristics of composite slabs in multi-span applications. Decks of 50 and 60 mm depth require secondary floor beams at spacings to suit the structural grid. They are usually used unpropped (without temporary propping) for both speed and simplicity of construction, in which case the steel deck

Table 5.10 Typical fire resistance requirements for composite slabs

Slab type	Fire resistance	Minimum slab depth (mm)	Minimum reinforcement
	R60	100	A142
	R90	120	A193
	R120	130	A252
	R60	130	A142
	R90	140	A193
	R120	160	A252
	R60	280	16 mm bar
	R90	300	25 mm bar
	R120	320	32 mm bar

All data are for maximum span in unpropped construction
The reinforcement requirements are dependent on national regulations
A142 = 142 mm^2/m; reinforcement in slab
Imposed loading as = 3 kN/m^2; plus 1 kN/m^2; for partitions etc.

thickness is typically 0.9 to 1.2 mm. For propped construction, thinner steel (as low as 0.7 mm) may be used. Deeper decks may avoid the use of secondary beams, especially when propped during construction. Individual manufacturers provide comprehensive design tables for their products.

The requirements for fire resistance in terms of minimum slab depth and reinforcement size are presented in Table 5.10, based on the results of fire tests. The amount of reinforcement is dependent on national regulations, whereas the minimum slab depth is obtained from EN1994-1-2. For spans or slab depths outside these ranges, the fire engineering method in EN1994-1-2 may be used.

5.7.3 Secondary beams

A grillage of floor beams comprises secondary beams which support the floor slab directly, and primary beams to which the secondary beams are attached. Both secondary and primary beams may be designed to act compositely with the slab. The spacing of the secondary beams is dependent on the spanning characteristics of the slab, and a spacing of 2.5 m to 3.6 m is commonly used in composite construction. For square column grids, the secondary beams support less load than the primary beams and, therefore, will be lighter or shallower.

For rectangular column grids, two generic floor configurations exist:

- primary beams supporting shorter span secondary beams – see Figure 5.16(b)
- long span secondary beams spanning directly to columns, or attached to shorter span primary beams – see Figure 5.16(c).

In the first case, the secondary beams are shallower than the primary beams, whereas in the second case, the beam depths can be designed to be approximately equal. The choice of system is frequently determined by structure/services

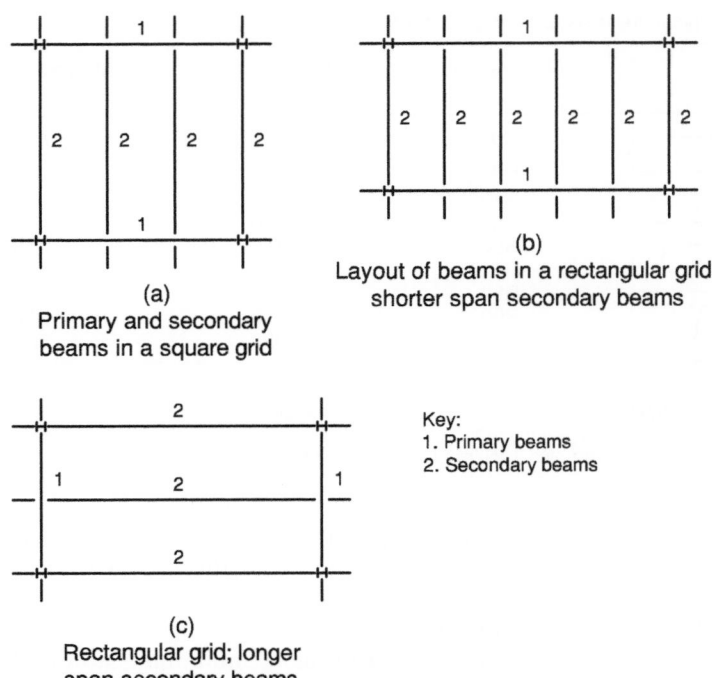

(a)
Primary and secondary
beams in a square grid

(b)
Layout of beams in a rectangular grid,
shorter span secondary beams

(c)
Rectangular grid; longer
span secondary beams

Key:
1. Primary beams
2. Secondary beams

Figure 5.16   Alternative arrangements for secondary and primary composite beams

integration strategy. Where no such constraints exist, the latter solution is usually more economical.

Long span secondary beams can be manufactured with multiple regular openings, which are called 'cellular beams' (see Figure 5.17). These beams may be fabricated by cutting and re-welding I- or H-profiles in a circular wave form. They are similar in form to castellated beams, which have a hexagonal opening form. Cellular beams are often asymmetric in cross-section by re-welding different beam sizes, which leads to optimum performance in composite design. Alternatively, such beams may be formed as plate girders; fabrication costs are higher but it is possible to optimise both the top and bottom flanges and the web, saving mass compared with an equivalent rolled section.

The benefits of long span secondary beams include structural efficiency, with span/depth ratios of 20–25 for composite design and service integration with regular holes of up to 70% of beam depth.

5.7.3.1 Design aspects

The structural design of secondary beams depends on the floor grillage and the opportunity for service integration. Two generic cases are considered below: rolled

Figure 5.17   Long span cellular beams

Table 5.11   Sizes of composite secondary beams using IPE/HE sections (S235 steel)

Rolled steel beam	Maximum span of beam				
	6 m	7.5 m	9 m	10.5 m	12 m
Minimum weight	IPE 270A	IPE 300	IPE 360	IPE 400	IPE 500
Minimum depth	HE 220A	HE 240A	HE 280A	HE 320A	HE340B

Imposed load = 3 kN/m² plus 1 kN/m² for partitions etc
Slab depth = 130 mm; Beam spacing = 3 m

steel beams (IPE/HPE in Table 5.11 or UB/UC in Table 5.12); and cellular beams of asymmetric section, in Table 5.13 and Table 5.14. In all cases the beams act compositely with a composite slab and are designed to EN 1994-1-1.[15] Various beams sizes are presented in the tables.

These tables assume a typical imposed load for offices and the self-weight as determined by the slab depth and span and beam size. S275 steel is commonly used for secondary beams, whose design is limited by deflection. However, cellular beams often use S355 steel, as their design is often controlled by shear in the web-post.

Elongated openings may be created in cellular beams by cutting out the web-post between openings. The optimum position for these longer openings is close to mid-span, as illustrated in Figure 5.18.

Table 5.12   Sizes of composite secondary beams using UB/UC sections (S275 steel)

Rolled steel beam	Maximum span of beam				
	6 m	7.5 m	9 m	10.5 m	12 m
Minimum weight	254 × 146 × 31 kg/m	305 × 127 × 42 kg/m	356 × 171 × 51 kg/m	406 × 178 × 60 kg/m	457 × 191 × 74 kg/m
Minimum depth	203 × 203 × 46 kg/m	203 × 203 × 71 kg/m	254 × 254 × 89 kg/m	305 × 305 × 97 kg/m	305 × 305 × 158 kg/m

Imposed load = 3 kN/m^2 plus 1 kN/m^2 for partitions etc
Slab depth = 130 mm; Beam spacing = 3 m

Table 5.13   Sizes of composite cellular beams as secondary beams (IPE/HE sections in S355 steel)

Cellular beam	Maximum span of beam (m)				
	12	13.5	15	16.5	18
Opening diameter (mm)	300	350	400	450	500
Beam depth (mm)	460	525	570	630	675
Top chord	IPE 360	IPE 400	IPE 400	IPE 450	IPE 500
Bottom chord	HE 260A	HE 300A	HE 340B	HE 360B	He 400M

Imposed load = 3 kN/m^2 plus 1 kN/m^2 for partitions etc
Slab depth = 130 mm; Beam spacing = 3 m

Table 5.14   Sizes of composite cellular beams as secondary beams (UC sections in S355 steel)

Cellular beam	Maximum span of beam (m)				
	12	13.5	15	16.5	18
Opening diameter (mm)	300	350	400	450	450
Beam depth (mm)	415	490	540	605	625
Top chord	305 UC 54	356 UC 67	406 UC 67	457 UC 67	457 UC 82
Bottom chord	254 UC 89	305 UC 54	305 UC 137	356 UC 153	356 UC 287

Imposed load = 3 kN/m^2 plus 1 kN/m^2 for partitions etc
Slab depth = 130 mm; Beam spacing = 3 m

5.7.4 Primary beams

Primary beams support secondary beams, and by their loading, tend to be heavier or deeper than secondary beams of the same span. They are subject to one or more point loads at a spacing given by the span of the floor slab. Primary beams can be of two generic forms:

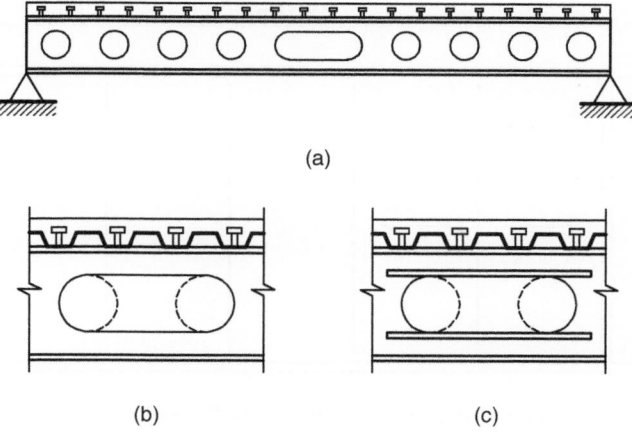

(a)

(b)   (c)

Key:
(a) Cellular beams with elongated openings
(b) Elongated opening (in mid-span)
(c) Stiffened opening

Figure 5.18   Schematic of long span cellular beam with elongated opening

- hot-rolled steel sections (using IPE or UB sections)
- fabricated sections (from welded plates).

For long spanning primary beams (span > 12 m), illustrated in Figure 5.19, it is possible to form large rectangular openings in the webs of the sections close to mid-span where the shear forces are low. Fabricated beams are often used as primary beams because they can be designed efficiently as composite asymmetric sections. Cellular beams can also be used as primary beams, although they are less efficient for this case because of the higher shear forces acting on primary beams. Primary beams should generally connect to the flanges of columns both for stiffness and for efficiency in fabrication of the connections.

The benefits of long span primary beams are as follows:

- Primary beams can be designed   — either hot-rolled or fabricated beams
  for a range of spans   may be used.
- Fabricated sections are efficient   — they are 'tailor-made' for their span
                                       and load.
- Service integration   — large web openings can be formed
                                       close to mid-span.
- Saving on fire protection costs   — heavier sections can achieve up to 30
                                       minutes fire resistance without
                                       protection.

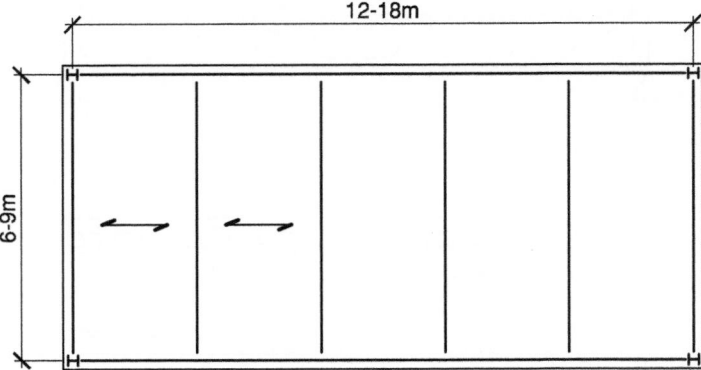

Figure 5.19   Layout of long span primary beams of 12 to 18 m span

5.7.4.1 Design aspects

The structural design of primary beams depends on the size and layout of beams in the floor grillage. Tables 5.15 and 5.16 give typical sizes of primary beams for various column spacings in orthogonal directions.

Table 5.15   Sizes of composite primary beams using IPE sections

Span of secondary beams (m)	Maximum span of primary beam (m)				
	6	7.5	9	10.5	12
6	IPE 360	IPE 400	IPE 450	IPE 550	IPE 600R
7.5	IPE 400	IPE 450	IPE 550	IPE 600R	IPE 750 × 137
9	IPE 450	IPE 500	IPE 600	IPE 750 × 137	IPE 750 × 173

Imposed load = 3 kN/m^2 plus 1 kN/m^2 for partitions etc

Table 5.16   Sizes of composite primary beams using UB sections

Span of secondary beams (m)	Maximum span of primary beam (m)				
	6	7.5	9	10.5	12
6	305 × 127 × 42 kg/m	356 × 171 × 57 kg/m	406 × 178 × 74 kg/m	457 × 191 × 98 kg/m	533 × 210 × 122 kg/m
7.5	356 × 171 × 45 kg/m	406 × 178 × 67 kg/m	457 × 191 × 89 kg/m	533 × 210 × 122 kg/m	610 × 229 × 140 kg/m
9	406 × 178 × 54 kg/m	457 × 191 × 74 kg/m	533 × 210× 101 kg/m	610 × 229 × 140 kg/m	610 × 305 × 179 kg/m

Imposed load = 3 kN/m^2 plus 1 kN/m^2 for partitions etc

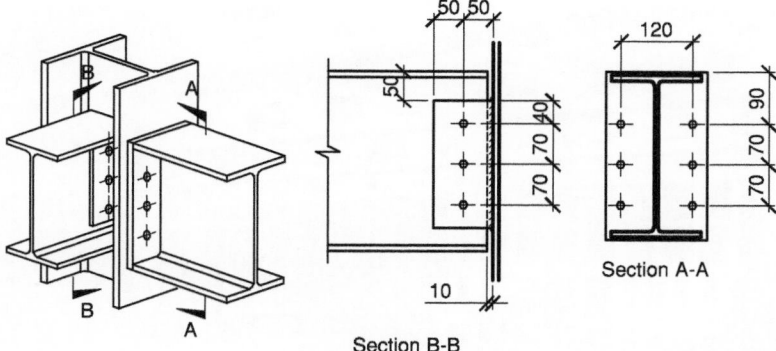

Figure 5.20   End plate connection of primary beam to column and fin plate connection of secondary beam to column

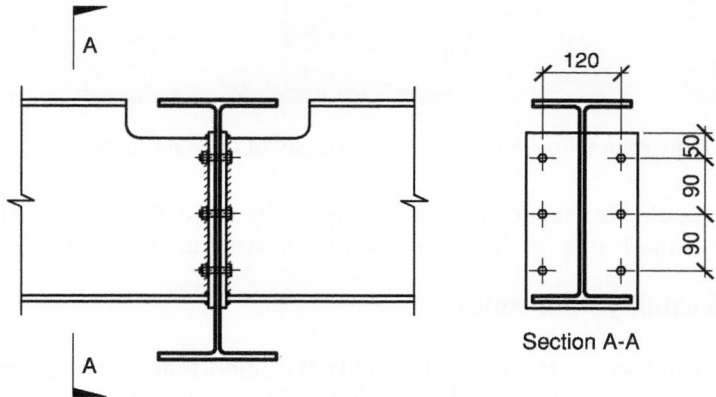

Figure 5.21   Beam-beam connection showing notch at top flange

Primary beams should be connected to column flanges, for example by end plate details as shown in Figure 5.20. Extended end plates may be used to increase the stiffness of the connection and reduce deflections of the beam.

Secondary beams may be connected to primary beams by end plate details but the top flange should be notched where beams are of the same level, as shown in Figure 5.21. Fin plate or double angle cleats may alternatively be used. This latter arrangement leads to improved economy in fabrication as all secondary beams only have to be cut to length and bolted and all the welding is concentrated on the smaller number of primary beams.

For fabricated beams, a variety of section sizes is possible. For efficient structural design, the span/depth ratio of composite primary beams is in the range of 15–18. However, the depth of the section can be increased in order to achieve the maximum

Figure 5.22   Long span fabricated beam with a variety of opening shapes

size of web opening for service integration (typically up to 70% of the depth of the section). An example of a fabricated primary beam structure is shown in Figure 5.22.

5.7.5 Serviceability limit states

In addition to checks on strength and stability, it is important to check floor systems for deflection, vibration and sound transmission. Detailed guidance on vibrations is provided in Chapter 13 and Reference 16. and will not be covered here.

No specific deflection limits are set in EN 1993-1-1 but clause 7.2 states that the serviceability criteria, including deflection limits, should be specified and agreed with the client for each project. Verification should be based on criteria set to ensure that deformations do not adversely affect appearance, comfort of users, functioning of the structure or cause damage to finishes or non-structural members. Although in some countries the National Annex to EN 1993-1-1 specifies the limits, this is not the case in the UK. Reference 17 recommends the limits below for deflections defined in Figure 5.23.

Beams generally	w_{max} not checked
	$w_2 + w_3 < L/200$
Beams carrying brittle finishes	w_{max} not checked
	$w_2 + w_3 < L/360$

(NB w_2 is usually ignored as it is negligible for steel beams and unpropped steel construction).

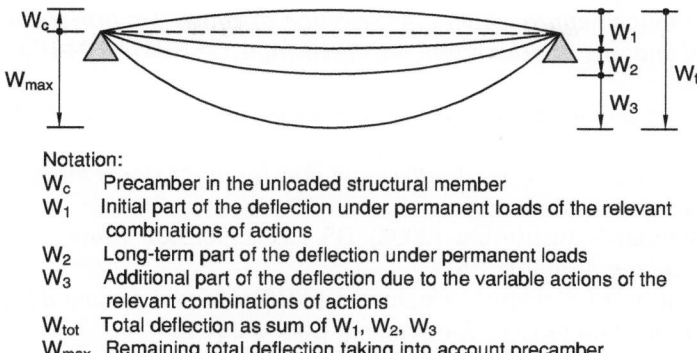

Notation:

W_c Precamber in the unloaded structural member
W_1 Initial part of the deflection under permanent loads of the relevant
          combinations of actions
W_2 Long-term part of the deflection under permanent loads
W_3 Additional part of the deflection due to the variable actions of the
          relevant combinations of actions
W_{tot} Total deflection as sum of W_1, W_2, W_3
W_{max} Remaining total deflection taking into account precamber

Figure 5.23   Definitions of vertical deflections

The acoustic performance of separating floors is dependent on both maximising airborne sound reduction and minimising impact sound transmission. It is often control of impact sound which is more problematical, and a resilient layer on top of the floor is required to reduce direct sound transfer. Acoustic performance in residential construction is an inexact but demanding requirement. It depends not only on detailed design but also significantly on the quality of construction. The UK has adopted a series of Robust Details[18] to ensure a relatively 'fail-safe' method of meeting the national Building Regulations[19] without the need for pre-completion testing to demonstrate compliance with the regulations. The same details are relevant also to European practice.[20]

References to Chapter 5

1. Directive 2002/91/EC of the European Parliament and of the Council 16 December 2002 On the energy performance of buildings. *Official Journal of the European Communities*, 4.1.2003.
2. British Standards Institution (2005) BS EN 1993-1-1:2005 Eurocode 3: *Design of steel structures – Part 1-1: General rules and rules for buildings.* London, BSI.
3. *Flow Chart: Simple method for the design of no-sway braced frames*, SF015a-EN-EU, www.access-steel.com
4. *Flow chart: Frame analysis*, SF002a-EN-EU, www.access-steel.com
5. *NCCI: Simplified approaches to the selection of equivalent horizontal forces for the global analysis of braced and unbraced frames*, SN047a-EN-EU, www.steel-ncci.co.uk
6. *Flowchart: Vertical bracing design*, SF007a-EN-EU, www.access-steel.com
7. *NCCI: 'Simple Construction' – concept and typical frame arrangements*, SN020a-EN-EU, www.steel-ncci.co.uk
8. *NCCI: 'Simple Construction' – concept and typical frame arrangements*, SN020a-EN-EU, www.steel-ncci.co.uk

9. *NCCI: Simplified approaches to the selection of equivalent horizontal forces for the global analysis of braced and unbraced frames*, SN047a-EN-EU, www.steel-ncci.co.uk

10. *NCCI: Sizing guidance – non-composite columns-UC sections*, SN012a-EN-GB, www.steel-ncci.co.uk

11. *NCCI: Vertical and horizontal deflection limits for multi-storey buildings* SN034a-EN-EU, www.steel-ncci.co.uk

12. British Standards Institution (2002) BS EN 1990:2002 *Eurocode – Basis of structural design*. London, BSI.

13. The Institution of Structural Engineers (2010) *Manual for the design of steelwork building structures to Eurocode 3*. London, ICE.

14. British Standards Institution (2005) BS EN 1994-1-2:2005 *Eurocode 4 – Design of composite steel and concrete structure Part 1-2: General rules – Structural fire design*. London, BSI.

15. British Standards Institution (2004) BS EN 1994-1-1:2004 *Eurocode 4 – Design of composite steel and concrete structure Part 1-1: General rules and rules for buildings*. London, BSI.

16. *NCCI: Vibrations*, SN036a_EN-EU, www.access-steel.com

17. *NCCI: Vertical and horizontal deflection limits for multi-storey buildings* SN034a-EN-EU, www.access-steel.com

18. www.robustdetails.com

19. The Building Regulations (2000), Approved Document E – Resistance to the passage of sound, 2003 edn. incorporating 2004 amendments. RIBA Bookshops, London (download from www.planningportal.gov.uk).

20. Scheme development: Acoustic performance in residential construction with light steel framing, SS032a-EN-EU, www.acess-steel.com

Chapter 6
Industrial steelwork

JOHN ROBERTS and ALLAN MANN

6.1 Introduction

6.1.1 Range of structures and scale of construction

Structural steelwork for heavy industrial use is characterised by its function, which is primarily concerned with the support, protection and operation of plant and equipment. In scale, the steelwork ranges from simple supports for single tanks, motors or similar equipment, to some of the largest integrated steel structures, for example, complete electric power-generating facilities.

Whereas conventional single- and multi-storey structures provide environmental protection to space enclosed by walls and roof and, for multi-storey buildings, support of suspended floor areas, these features never dominate in industrial steelwork. For many industrial structures, the framework does also support an envelope giving weather protection, sometimes to high standards of thermal and sound insulation. However, it is not unusual for envelopes simply to provide rain shielding, whilst some plant and equipment is able to function and operate effectively without any weather protection at all. Generally, accommodating plant is key, and wherever cladding and roofing is provided, the wall and roof profiles are designed to fit around and suit the industrial plant and equipment rather than the other way around. Most industrial steelwork structures have some areas of conventional floor construction but this is not a primary requirement and the flooring is incidental, being provided mostly for maintenance access. Likewise, for maintenance, industrial structures frequently require a plethora of ladders and platforms.

Floors are provided to allow access to and around installations, being arranged to suit particular operational features. They are, therefore, unlikely to be constructed at constant vertical spacings or be laid out on plan in any repetitive pattern. Steelwork designers must be particularly careful not to neglect the importance of two factors. First, floors cannot automatically be assumed to provide a horizontal wind girder or diaphragm to distribute lateral loadings to vertically braced or framed bays; openings, missing sections or changes in levels can easily destroy this in-plane

Steel Designers' Manual, Seventh Edition. Edited by Buick Davison and Graham W. Owens.
© 2012 Steel Construction Institute. Published 2012 by Blackwell Publishing Ltd.

function. Secondly, column design can be hampered by the lack of spaced two-directional lateral supports that are commonly available in normal multi-storey structures via the incoming beams. Floors require further consideration regarding the choice of construction (see Section 6.2.3) and for their loading requirements (see Section 6.3).

6.1.2 Design management

Steelwork design and detailing for industrial structures is dominated by accommodation of plant and the design process is primarily affected by:

- loading
- space requirements, including avoidance of clashes with frame connections (piping runs are a particular issue)
- framing arrangements to allow for installation and access to the plant in the first place and for possible removal in later life
- fabrication and erection.

Full information on the plant is crucial for the modern trend of trying to build a complete CAD model of the structure prior to fabrication (Chapter 2). So, gaining and controlling (changeable) information is a major activity within the general management of the design process, which is one of information gathering followed by an iterative evolution of the design as the plant manufacturer's own design gradually evolves (usually in parallel or behind the structural design). (Sections 6.1.2.3 and 6.3.2. provide more detail.)

6.1.2.1 Plant loading

Unlike many other structures, the loading applied to industrial structures is dominated by bespoke plant loading. Such loading is often heavy and applied as point loads. Plant loading can be indeterminate, especially if a large item is spread over several supporting beams where there is interaction between the support stiffness and the plant item stiffness. Plant loads may have inertial components (as in crane braking) and they may have dynamic components (interaction between plant motion and the supporting structure's own natural frequency). Plant loads are typically applied in three directions: gravity down and two orthogonal horizontal directions such as those from pipe thrust restraint points or loads from changes in pipe direction.

As well as the dominant plant loads, there will often be a requirement for a blanket uniformly distributed loading (u.d.l.) to cover surface loading and simulate the hanging of all sorts of ducts and pipework. Configuring the steelwork to support such local loads, both in size and attachment capability, is part of the design process.

Having an appreciation of the overall construction process is vital since plant may often be installed after the framework is complete. This necessitates the definition of access routes (with appropriate strength) and later in the life of the building perhaps strengthened 'lay down' areas to cope with placing equipment removed for maintenance.

In some circumstances exceptional loads may apply, linked to the plant safety case. In nuclear plant, this may be coping with seismic design or extremes of weather. In chemical plant, it may be dealing with blast loads. Since UK design can be for plant in any part of the world, local conditions need to be understood and these may include seismic design (over a range of earthquakes). Where this applies, seismic design may be critical since plant mass can be very high.

Lateral loads are a particular problem. These arise as defined by plant suppliers but there are many internal structures which have no naturally defined horizontal stability loads (such as wind) and so an artificial lateral load linked to plant weight may be required to simulate sway loads. Overall, plant support structures need to be robust and substantially stable laterally.

For both vertical and lateral loads, it is equally important to agree the stiffness demands required of the supporting structure. Section 6.3 gives more detailed advice on loading and stiffness demands.

6.1.2.2 *Structural layout and arrangements*

The needs for plant access and maintenance can have a significant effect on structural layout, often requiring the placement of bracing in zones which are less than ideal. Within the plant, floors and roofs may have many large penetrations, some of which are large enough to require secondary support systems and others small enough to be accommodated by the type of floor adopted. There may be a need for holes through beams and there are issues with potential clashes between pipe runs and the zones required by the structural engineer for connection sizing. The later needs for maintenance (and indeed installation) may necessitate the provision of overhead lifting beams.

6.1.2.3 *Information management*

A key requirement for a successful industrial project is acquisition and control of relevant design information. Professionally, the steelwork designer is advised to be proactive in seeking information and setting out what it is they need to know. Plant design is likely to evolve over the project development life, and the supplied information is likely to be delivered in stages, starting as very general and gradually narrowing down to be more specific. Thus, at the beginning, for example, the only information might be that a crane is required. This may later change into a crane of certain capacity but only in the final stages will the mode of operation and the number of wheels and so on be defined. Likewise, it is naïve to believe that the demand for penetrations will be accurately known early on. Rather the design will have to progress, starting with a concept that has the ability to accept floor holes and gradually moving towards one of greater refinement. This can even be a design that facilitates late site changes.

Consequently, the structural steelwork designer must take an open-minded approach to the contract information and it is wise for the project planning team to agree contingency loading margins to limit whole scale redesign at late stages. To assist in this, the most important advice to the steelwork designer for any industrial purpose is to ensure reasonable familiarity with the entire process or operation involved. Existing facilities can be visited and the plant designers and operators will normally be willing to give a briefing. This then provides an opportunity to describe the form of structure envisaged to the plant designer of the new structure at an early stage, so avoiding later misunderstandings.

In all cases, an initial design strategy needs to be set out which defines the stages of information supply, the accuracy of that information and the contingency measures to be adopted in recognition of consequences of later change.

6.1.2.4 Fabrication and erection

Where plants are large and there is considerable repetition, for example of long span roof beams, it will be part of the design process to optimise the interaction between design, fabrication, transport and erection to achieve the best overall concept and cost. This has implications for the procurement route and contractual arrangements, with early contractor involvement advisable.

There is always interaction between the design concept and the later needs of erection, but in heavy industrial plants in particular, the bay location of permanent bracing may not be best for starting the erection process and squaring up. Attention should therefore be given to all columns and their bases and foundations so as to assure that columns can be stable and free standing, for erection purposes. An integrated program may have to be made between framework erection and plant installation and this can include detailing in temporary steelwork simply to aid plant positioning, or it might include provision of temporary stability bracing.

6.1.3 Types of structure

6.1.3.1 Power station structures

Industrial steelwork for electrical generating plants varies considerably, depending on station size and the fuel being used. These variations are most marked in boiler house structures in coal- or oil-fired stations. Gas-fired stations only require fairly straightforward structures for housing. However, in all stations there are turbine halls largely independent of fuel type. These halls are long sheds, usually including an electric overhead travelling (EOT) crane. Other plant structures commonly

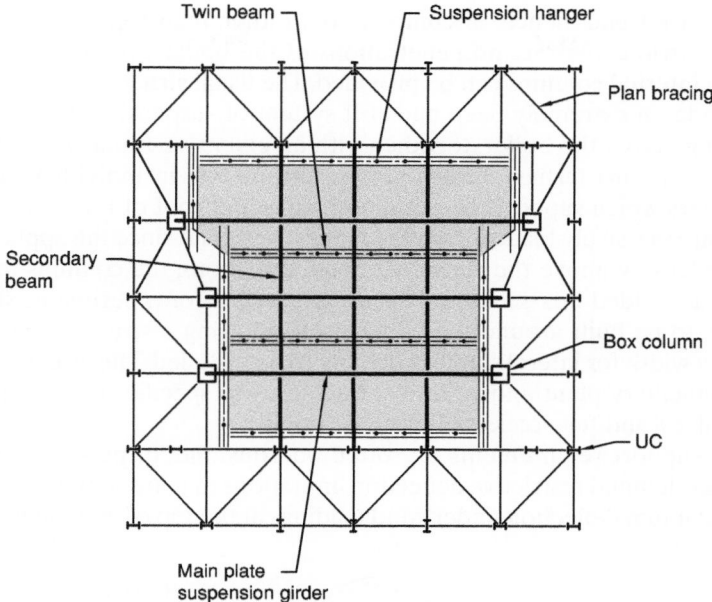

Figure 6.1   Plan of boiler house framing

required are: mechanical annexes, electrical switchgear buildings, pump-houses, conveyors, coal hoppers, bunkers and silos.

Boiler houses

These coal- or oil-fired plants, shown in Figure 6.1, have to solve one overriding design criterion and, as a result, can be considered exercises in pure structural design. Boilers are often huge single pieces of plant with typical dimensions of 20 m × 20 m × 60 m high for a single 500 MW coal boiler. Where poor quality coal is burnt or higher capacities are required, the dimensions can be even larger, up to about 25 m × 25 m × 80 m high for 900 MW size sets. As may be anticipated, the weights of plant structural support steel required are equally massive, typically in the range of 7,000–10,000 t for the plant sizes noted above.

Boilers are always top-suspended from their supporting structures and not built directly from foundation level upwards, nor carried by a combination of top and bottom support. This is because boiler thermal expansion prevents dual support systems, and because insurmountable buckling and stability problems would arise within the boiler thin-walled tube structure of the casing if the boiler was bottom-supported and, hence, in compression, rather than top-suspended and in tension. For obvious reasons, no penetrations of the boiler can be acceptable and therefore no internal columns can be provided. The usual structural system is therefore to provide an extremely deep and stiff system of suspension girders (plate or box) spanning across the boiler, together with an extensive framework of primary, secondary and (twin) tertiary beams above, terminating in individual suspension rods or hangers which support the perimeter walls and roof of the boiler itself.

Columns are massively loaded from the highest level and since the applied loading can be considerably above the capacity of rolled sections, the columns are usually constructed as welded box-sections. It is usual practice for a perimeter strip, some 5–10 m wide, to be built around the boiler itself, allowing a structural grid with an adequate bay width for lateral stability bracing to be installed. The grid also provides support for ancillary plant and equipment adjacent to specific zones and for pipework and valves, and for access walkways or floors.

Pipework support requirements are often onerous and, in particular, pipework designers may demand restrictive deflection limitations that are sometimes set as low as 50 mm maximum deflection under wind loadings at the top of 90 m high structures.

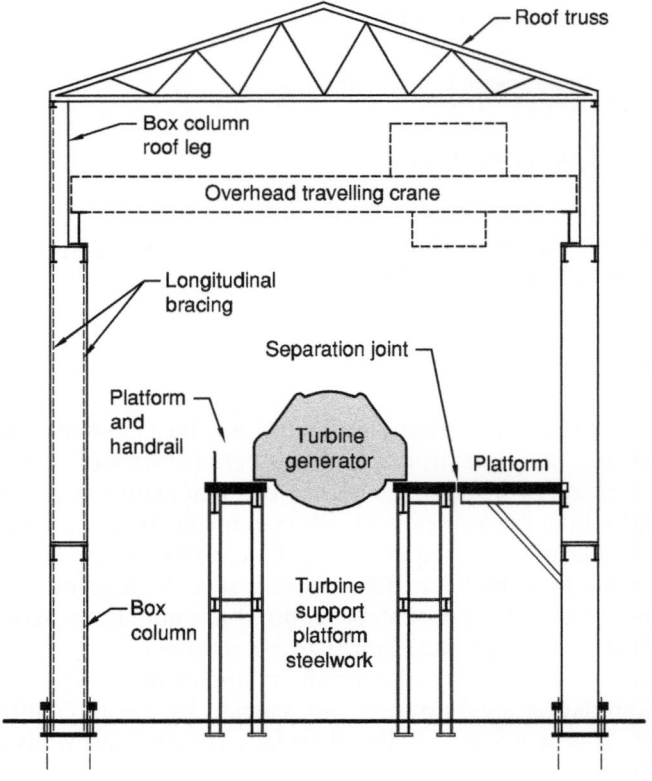

Figure 6.2   Cross-section through turbine hall

Turbine halls

These structures, shown in Figure 6.2, support and house turbo-generating machines operating on steam produced by the boiler, thus converting heat into mechanical energy. Turbo generators are linear, built around a single rotating shaft, typically some 25 m long for 500 MW units. The function of a steel-framed turbine hall is to protect and allow access to the generator, to support steam supply and condensed water return pipework and numerous other items of ancillary plant and equipment. Heavy crane capacity is usually installed, since generators are working machines that require routine servicing as well as major overhaul and repairs. Turbo generator block design (i.e. direct support of the rotating plant) is specialised, but the essential requirement is to inhibit dynamic interaction between the generator and its support system. Linked to this, there will be a need for complete structural separation between the system and the main turbine hall envelope.

Zones on adjoining suspended floors are set aside for strip-down and servicing and specific 'lay down' loading areas are needed in the design of these areas to cater for heavy point and distributed loads. The halls themselves are often long, so requiring consideration of thermal expansion and multiple wall bracing panels. Other typical design issues include crane girder design and the provision and support of large sliding doors.

6.1.3.2 Process plant steelwork

Steelwork for process and manufacturing plants and maintenance facilities varies across a wide spectrum of industrial uses. Here it is considered to be steelwork intimately connected with the support and operation of plant and equipment, rather than merely being a steel-framed envelope constructed over a process plant.

Although processes and plant vary widely, the essential features of this type of industrial steelwork are common to many applications and conveniently examined by reference to some typical examples. There are many similarities with the power station boiler house and the turbine hall steelwork described earlier.

Assembly/maintenance plants

Substantial overhead services to the various assembly lines characterise these plants. Thus, it is normal to incorporate a heavy-duty and closely-spaced grid roof structure capable of supporting hanging loads, which will also support the roof covering. Reasonably large spans are needed to allow flexibility in arranging assembly line layouts without them being constrained by column locations. Automation of the assembly process brings with it stiffness requirements, especially when robots are used for precise operations such as welding and bonding. Open trusses in two directions are likely to satisfy most of these requirements, providing structural depth for deflection control and a zone above bottom boom level that can be used for service runs and a close grid, with adequate scope for hanging load support. Building plans are normally regular with rectangular plan forms and uniformly regular roof profiles. Long span roofs are not uncommon and their design requires extra care in coping with deflection, perhaps defining precamber and in assuring member temporary stability on erection.

Petrochemical plants and comparable 'external' industrial processes

Petrochemical plants tend to be open structures with little or no weather protection. The steelwork required for them is dedicated to providing support to plant and pipework and support for access walkways, gangways, stairs and ladders. Plant layouts are relatively static over long periods of use and the steelwork is relatively economic in relation to the equipment costs. It is, therefore, normal to design the steelwork in a layout exactly suited to the plant and equipment without regard for uniformity of the structural grid. Benefits can still be gained from standardisation of sizes or members and from maintaining an orthogonal grid to avoid connection problems. But, overall, the steelwork is simple to design yet detailing-intensive. Access floors and interconnecting walkways and stairways must be carefully designed for all weather conditions; use of grid flooring is almost universal. Protection systems for the steelwork should acknowledge both the threat from the potentially corrosive local environments due to liquid or gaseous emissions, plus the requirements to prevent closing down the facility for routine maintenance of the selected protection system.

Careful account needs to be taken of wind loading in terms of the loads that occur on an open structure, not least because the plant itself will feature all sorts of flanges and protuberances, so the applicable drag factors are very uncertain. Consequently, a robust assessment is required. Moreover, the lack of well-defined horizontal diaphragms from conventional floors complicates load paths to ground.

6.1.3.3 Conveyors, handling and stacking plants

Many industrial processes need bulk or continuous handling of materials with a typical sequence as follows:

- unloading from bulk delivery or direct from mining or quarrying work
- transportation from bulk loading area to short-term storage (stacking or holding areas)
- reclaim from short-term storage and transport to process plant.

Structural steelwork for the industrial plant utilised in these operations is effectively part of a piece of working machinery (Figure 6.3), so fatigue-resistant design may well be a requirement, especially in the direct support of moving parts. More widely, design and construction standards must recognise the dynamic nature of the

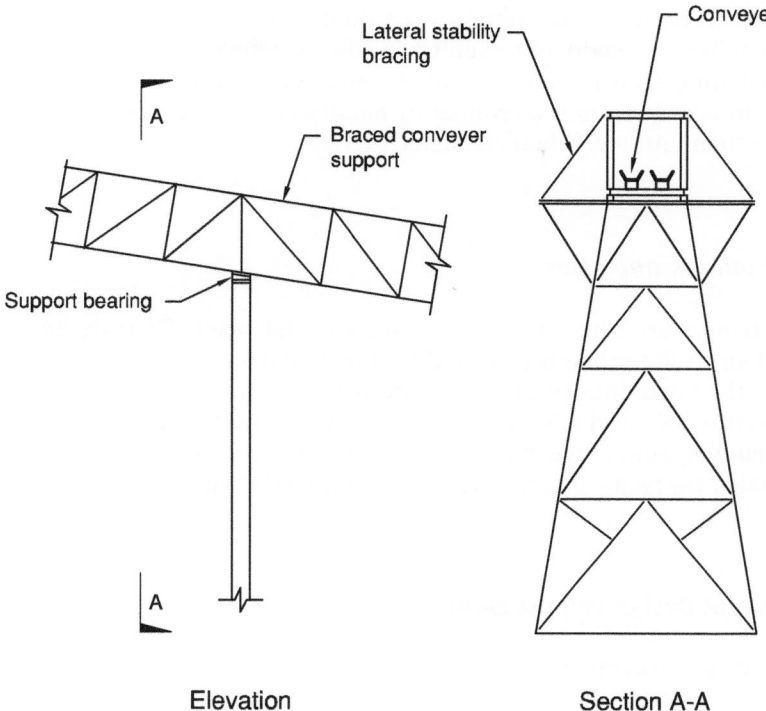

Lateral stability
bracing

Conveyer

A

Braced conveyer
support

Support bearing

A

Elevation

Section A-A

Figure 6.3   Typical details of conveyor support

loadings and must particularly cater for out-of-balance running, overload conditions and plant fault or machinery failure conditions, any of which can cause stresses and deflections significantly higher than those resulting from normal operation. Most designers would adopt slightly lower factors of safety on loading for these conditions, but decisions need to be based on engineering judgement taking account of the probability of occurrence. The plant designer himself may well be unaware that such design decisions can be made for the support steelwork, and frequently provide single maximum loading parameters that could incorporate a combination of all such plausible events rather than a separate tabulation. The steelwork designer who takes the trouble to understand the operation of the plant can therefore ask for the appropriate information and use it to best advantage.

Towers and long conveyor supports often support only light loading so their structural form can be that of light braced frameworks where angles are the dominant member. On long span supports, deflection has to be considered for proper running of conveyors and specifying precamber might be appropriate.

Foundation levels can be set at constant heights or, if variations have to occur, then modular steps above or below a standard height should be adopted. The route should utilise standard plan angles between straight sections and uniform vertical

sloping sections between horizontal runs. Common base plate details and foundation bolt details can be adopted, even where this may be uneconomic for the initial
layout installation. Consideration should be given to allowing the supporting structure to be broken down into conveniently handled sections for transport and erection, rather than into individual elements.

6.1.3.4 Bunkers and silos

These containers are often needed within industrial plant. Their design is rather
specialised since the pressures exerted by the contained materials during storage
and during the conditions of discharge are complex. A design feature can also be
the need to provide special linings to minimise wear and to accommodate agitators
or other handling/discharge mechanical items. The units can often be purchased
from specialist suppliers and are not considered further here.

6.1.4 General design requirements

6.1.4.1 Design procedure

In common with all structural design, the starting point is to establish the functional
demands on the structure generally, in this case the plant loads and the geometric
constraints which govern the steel location for plant supports. The second step is to
determine a structural anatomy, assuring that there are clear load paths to ground
for gravity and horizontal loads; then carry out the detail design. This anatomy also
has to be configured to take account of joints, transport, erection needs and the
needs of later plant installation. Behind all this, there must also be a strategy of
design evolution. Section 6.1.2.3 has pointed out the probable evolution of plant
information so the structural design needs to grow in parallel. It is vital to be aware
of the potential inaccuracy of the plant information being used for design at any
stage, and to avoid carrying out designs at an inappropriately advanced level (it may
periodically be necessary to assess structural cost).

Possibilities of coping with the evolution of design are the sizing of members on
lower grade steel to start with (the cost penalties for use of higher grade are not
that great later on) and the use of contingency loadings (provided that the plant
designers have not also allowed contingency margins already in their own estimates). Since plant and operational costs are high in relation to steel costs, it is not
sensible overall management to absolutely minimise tonnage if this risks incurring
an overall project delay to accommodate inevitable changes.

The structural design engineer also needs to be proactive in seeking information
on tolerances at support locations and on finishes to ensure these are compatible
with plant operation. Corrosion risks and durability can be severe constraints within

certain types of plant so members will frequently be galvanised (Chapter 36 provides guidance).

6.2 Anatomy of structure

In setting out the structure, the steelwork engineer needs to liaise with the plant process designer since not all structural requirements are recognised as important by others. Examples which might need explaining are as follows:

- The physical space requirements of bracing members and the fact that they cannot be dispensed with or moved locally to give clearances.
- The actual size of finished steelwork taking account of splice plates, bolt heads and fittings projecting from the section sizes noted on drawings.
- The fact that a steel structure is not 100% stiff and that all loads cause deflections.
- The fact that a steel structure may interact with any dynamic loading and that dynamic overload multipliers calculated or allowed on the assumption of a fully rigid or infinite mass support are not always appropriate.

6.2.1 Gravity load paths

Vertical loadings on industrial steelwork can be extraordinarily heavy; some individual pieces of plant have a mass of 10,000 t or more. Furthermore, by their very nature, these loadings generally act as discrete point or line loads rather than as uniformly distributed loadings. For these reasons, compounded by the general uncertainty of loading in magnitude and position, gravity load paths must be established at an early stage to provide a simple, logical and well-defined system. The facility should exist to cater for new load locations within the general area of the equipment without the need to alter all existing main structural element locations. Typically, this means provision of a layered system of beams or trusses with known primary span directions and spacings, combined with secondary (and, in complex layouts, tertiary) beams as well, which actually provide direct vertical support to the plant.

Whilst simplicity and structural layout uniformity are always attractive, the non-uniform loadings and layout of plant mean that supporting columns may have to be positioned in other than a completely regular grid so as to provide the most direct and effective load path to foundation level. It is certainly preferable to compromise on a layout that gives short spans and direct, simple routes for gravity loads to columns rather than proceed with designing on a regular grid of columns, only to end up with a large range of member sizes and several types of beams, girders or trusses (Figure 6.4).

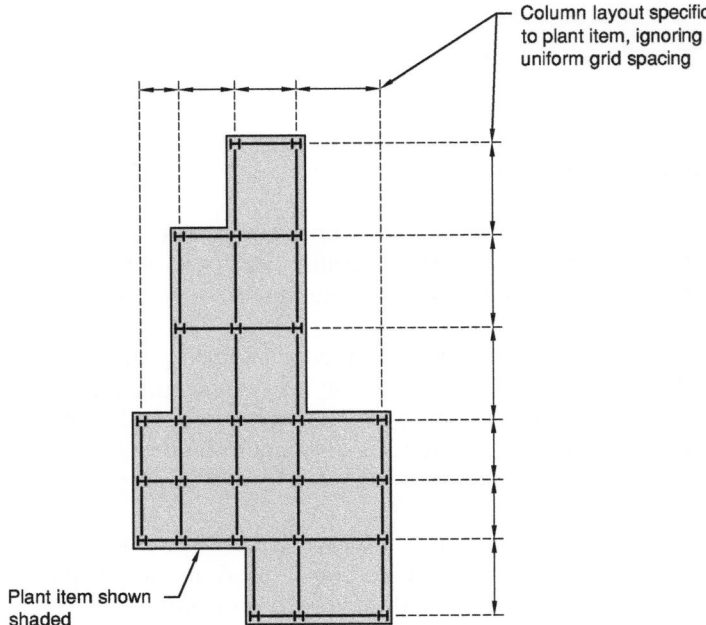

Column layout specific
to plant item, ignoring
uniform grid spacing

Plant item shown
shaded

Figure 6.4 Support steelwork for an unsymmetrical plant item

Similarly, although it is clearly preferable for columns to run consistently down
to the foundations, interference at low levels by further plant or equipment is quite
common, which may make it preferable to transfer vertical loading to an offset
column at height rather than use much larger spans at all higher levels in the structure. A conscious effort should be made to be familiar with all aspects of the industrial process and to have close liaison with the plant designers to make them aware
of the importance of an early and inviolate scheme for column locations. These twin
actions are both necessary to overcome this problem. The layout difficulties will be
compounded when different plant designers are handling the equipment layout at
differing levels in the building.

If a widely variable layout of potential loading attachment points exists, then
consideration can be given to hanging or top-supporting certain types of plant. A
typical example of this would be for an assembly line structure (or maintenance
plant), where the overhead system of conveyors/tooling can sensibly be supported
on a deep truss roof with a regular column grid. If this solution is contrasted with
the multitude of columns and foundations that would be needed to support such
plant from below, and the prospect of having to reposition these supports during
the design stage as more accurate information on the plant becomes available
and is reckoned with, then it can be an appropriate method of establishing a gravity
load path.

6.2.2 Sway load paths

For two fundamental reasons, sway load paths require particular consideration in industrial steelwork. First, the plant or equipment itself can produce significant lateral loading. Such lateral loadings can exist with their points of application differing from the usual cladding and floor intersection locations, and the route such loads take to ground has to be fully defined. For example, where there are crane girders, it is normal to provide dedicated lateral surge girders and thence bracing to ground. Second, many industrial steelwork structures lack a regular and complete floor capable of providing a competent horizontal diaphragm. Hence the lateral load transfer system must be considered carefully as a whole and at a very early stage in the design process (Figure 6.5).

Naturally, there are many different methods of achieving lateral stability. Where floor construction is reasonably complete and regular throughout the building, then the design can be based on using horizontal diaphragms transferring load back to vertical stiff elements at intervals along the building length or width. However, where large openings or penetrations exist in otherwise conventional concrete floors, then it is important to design both the floor itself and the connections between the floor and supporting steelwork for the actual forces acting, rather than simply relying on the provision of an effective diaphragm, as may justifiably occur in the absence of such openings. To assure a reasonable width of floor along each

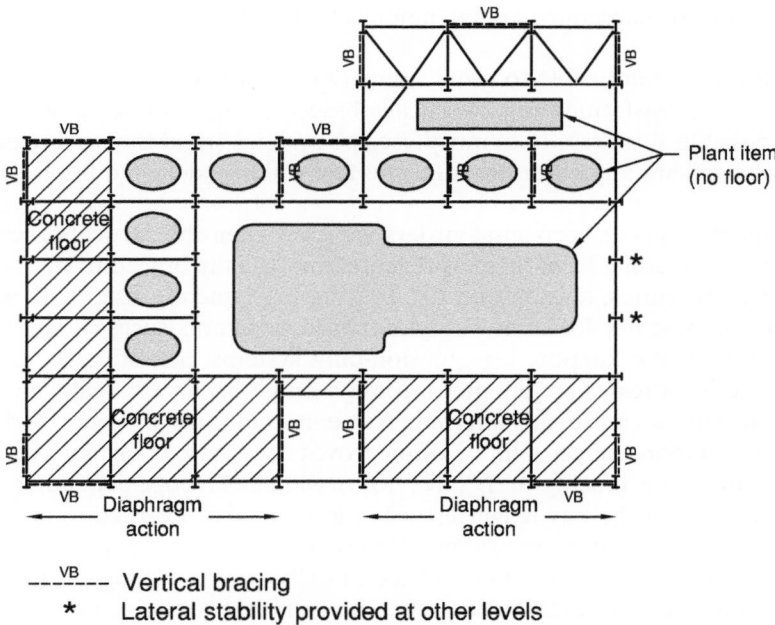

Figure 6.5 Establishment of sway load paths

external building face, it can be worthwhile deliberately influencing the layout to allow, for example, a width of the order of 10% of the horizontal spacing between braced bays or other vertical stiff elements.

When concrete floors, as described above, are wholly or sensibly absent, then other types of horizontal girders or diaphragms can be developed. If solid plate or open mesh steel flooring is used, then it is possible, at least in theory, to design such flooring as a horizontal diaphragm, usually by incorporating steel beams as 'flange' members of an idealised girder, where the floor steel acts as the web. However, in practice, this is usually inadvisable as the flooring plate fixings are rarely found to be adequate for load transfer and indeed it may be a necessary criterion that some or all of the plates can be removable for operational purposes. A further factor is that any line of beams used as a 'flange' must be checked for additional direct compression or tension loadings, with their end connections also having to be designed for axial force transfer.

In the absence of concrete floor construction, it is normal to provide plan bracing. The influence this bracing will have on plant penetrations, pipework routing and many other factors must be considered early in liaison with the plant engineers. Naturally, the design considerations of steel beams serving a dual function within plan bracing must not be neglected. Indeed, it may be preferable to separate totally the lateral load restraint steelwork provided on plan from any other steelwork in the horizontal plane. This will avoid clashes of purpose and clearly signal to the plant designers the steelwork function, and as a result hopefully prevent misuse or abuse at a later stage. Many instances exist where plan bracing members have been removed owing to subsequent plant modifications, to the detriment of building stability.

Where it proves impossible to provide any type of horizontal plan bracing, then each and every cross-frame can be vertically braced or rigid-framed down to foundation. If possible, it is best not to mix these two systems (rigid frames and bracing) since they have markedly differing stiffnesses and will thus deflect differently under loading.

Where diaphragms or horizontal girders are used, then the vertical braced bays receiving lateral loading from them as reactions are usually braced in steelwork. In conventional structures, tension-only 'X' bracing is frequently used and, whereas this may be satisfactory for some straightforward structures such as tank support frames and conveyor support legs, tension-only systems lack stiffness (certainly when high design stresses are adopted). Tension-only systems are really unsuitable when bracing stresses frequently alternate between conditions of tension and 'slack'. Consequently, a more robust solution often proves necessary, designing a combined tension/compression bracing in 'N', 'K', 'M' or similar layouts, depending on the relative geometry of the bay height to width and on what obstruction the bracing members cause to plant penetrations (Reference 1 provides general advice on bracing system design).[1] It is sound advice initially to provide significantly more bracing than may be considered necessary, for example, by bracing in two, three or more bays on one line both for stiffness and robustness. When later developments in the plant and equipment layout mean that perhaps one or more panels must be altered or even removed, it is then still possible to provide lateral stability by rechecking or redesigning the bracing, without a major change in the global bracing location.

When the side vertical members are deep plate girders, it is common to provide dual sets of bracing, one set aligned with each flange. This can also be essential in crane support systems to assure a direct load path to ground from crane girder longitudinal surge.

One particular aspect of vertical bracing design that requires care is in the evaluation of uplift forces in the tension legs of braced frames. Where plant and equipment provide a significant proportion of the total dead load, then it is important that *minimum* dead weights of the plant items are used in stability calculations. For virtually every other design requirement, it is likely that rounding-up or contingency additions to loadings will have been made, especially at the early stages of the design. It is also important to establish whether part of the plant loading has any variability associated with changes in content and, if so, to deduct this when examining global stability. In some cases, for example in hoppers, silos or tanks, this is obvious; but boilers or turbines which normally operate on steam may have their weights expressed in a hydraulic test condition when flooded with water. The structural engineer has to be aware of these plant design features in order to seek out the correct data for design.

6.2.3 Floors

Regular and continuous suspended concrete floors are unusual in industrial steelwork structures but flooring types that are used can consist of any of the following:

- in situ concrete cast on to removable formwork (this concrete may also be composite)
- in situ concrete cast on to metal deck formwork (this concrete may also be composite)
- fully precast flooring (no topping)
- precast concrete units with in situ concrete topping
- raised pattern 'Durbar' solid plate
- flat solid steel plate
- open grid steel flooring.

6.2.3.1 Selection

The selection of floor type depends on the functional requirements and anticipated usage of the floor areas. Functional requirements include strength (for u.d.l. and point loads), stiffness, ease of forming holes, wear resistance, slip resistance, ease of cleaning and perhaps resistance to chemical spillage. The achievement of tolerances may also be important.

Solid steel plates are often used where transit or infrequent access is required and where floors must be removable for future access to plant or equipment. They are normally used only internally (at least in the UK) to avoid problems with wet surfaces. Open grid flooring is used inside for similar functions as steel plate, especially where air flow through the floor is important. It is also used for stair treads and landings. Consideration should be given to making, say, landings and access strips from solid plate at intervals to assist in promoting a feeling of security among users. Open grid flooring is also in common use externally due to its excellent performance in wet weather.

Concrete floors are used where heavy-duty non-removable floor areas are necessary and where floors have to be strong, supporting either u.d.l. or point loads (for the latter reason, pot type floors are of restricted value). In modern construction, MEWPs (Mobile Elevating Work Platforms) are frequently used for overhead installations and the flooring type has to be capable of supporting their wheel loads.

A particular advantage of both precast flooring and metal deck permanent formwork is that, in many industrial structures, the floor zones are irregular in plan and elevation and therefore deploying cheap repetitive formwork is often impracticable. Metal decking can overcome this problem and, when used in conjunction with in situ concrete (including composite construction), is especially adaptable. Even where formwork can sensibly be used, the early installation of major plant items alongside the steel frame erection can complicate the formwork supports and harm the construction programme.

6.2.3.2 Holes and penetrations

Floors must be able to accept holes, openings and plant penetrations on a random layout and must often accept them very late in the design stage or as an alteration after construction. This provides significant problems for certain flooring, particularly precast concrete. In situ concrete can accept most types of openings in a convenient manner prior to construction but it may be prudent deliberately to allow for randomly positioned holes, up to a certain size, by over-sizing reinforcement in both directions to act as trimming around holes, within the specified units without extra reinforcement.

Metal deck formwork is perhaps not as adaptable as regards large openings and penetrations, since it is usually one-way spanning, especially where the formwork is of a type that can also act as reinforcement. If there is sufficient depth of concrete above the top of the metal deck profile, then conventional reinforcing bars can be used to trim openings. Many designers use bar reinforcement as a matter of course, with metal deck formwork to overcome this problem and also to overcome fire protection problems that sometimes occur when unprotected metal decks are used as reinforcement.

Steel-plate flooring should be designed to span two ways where possible, adding to its adaptability in coping with openings, since it can then be altered to span in one direction locally if required (albeit with deflection penalties). Open-grid flooring is less adaptable in this respect as it only spans one way and therefore openings will usually need special trimming via support steelwork. Steel-plate floors are fixed to supporting steelwork below by countersunk set screws; by countersunk bolts where access to nuts on the underside is practicable for removal of plates, or by welding where plates are permanent features and unlikely to require replacement following damage. Proprietary clip fixings are used for open-grid flooring plates to secure them to supporting flanges.

6.2.3.3 Finishes

Except in particularly aggressive environments, floor areas are usually left unfinished in industrial structures. For concrete floors, hard-trowelled finishes, floated finishes and ground surfaces can all be used. Selection depends on the use and wear that will occur (especial attention is required if fork lift trucks are to be used). Steel plate (solid or open-grid) is normally supplied with either paint or hot-dip galvanised finishes, depending on the corrosiveness of the environment; further guidance is given in BS 4592: Part 5[2] and by floor plate manufacturers. Ladders and handrails are normally supplied galvanised.

6.2.3.4 Information flow

As part of project management and information gathering, there is a need to gather information about the need for holes probably starting with crude requirements and gradually refining these as the design stages progress. Early agreement with plant and services engineers is vital to establish the likely maximum random opening required, and any structure-controlled restraint on location. Other policy matters that should be agreed early are the treatment of hole edges, edges of floor areas, transition treatments between different floor constructions and plant plinth or plant foundation requirements.

This last item is extremely important as many plant items have fixings or bearings directly onto steel and the exact interface details and limit of supply of structural steelwork must be agreed. Tolerances of erected structural steelwork are sometimes much larger than anticipated by plant and equipment designers and some method of local adjustment in both position and level must often be provided. Where plant sits onto areas of concrete flooring, then plinths are usually provided to raise equipment above floor level for access and pipe or cable connections. It is convenient to cast plinths later than the main floor, but adequate connection, for example by means of dowelled vertical bars and a scabbled or hacked surface should always be provided. It is usually more practical to drill and fix subsequently all dowel bars or anchor bolts for small-scale steelwork and for holding-down plant items, than to attempt casting them direct into the concrete floor.

6.2.4 Main and secondary beams

6.2.4.1 Layout

The plan arrangement of main and secondary beams in industrial steelwork structures is frequently dictated by the layout of the main items of plant; primary beams clearly need to align with columns. Thus, a sequence of design decisions often occurs in which main beam locations dictate the column locations and not vice versa. If major plant is located at more than one level, then some compromise on column position and hence beam layout, may be needed. Since large plant items usually impose a line or point loading (which can create high shear and local bearing issues), there are clear advantages in placing main or secondary beams directly below plant support positions. Brackets, plinths or bearings may be fitted directly to steelwork and, for major items of plant, this is preferable to the alternative of allowing the plant to sit directly on a concrete or steel floor. Where plant or machinery requires a local floor zone around its perimeter for access or servicing, it is common practice to leave out the flooring below the plant either for later access or because the plant protrudes below the support level.

6.2.4.2 Stiffness

Deflection requirements between support points should be ascertained. They may well control the beam design since stringent limits, for example relative deflections of 1 in 1000 of support spans, may apply. Additionally, when piped services are connected to the plant, then total deflections of the support structure relative to the beam-to-column intersections may also be restricted. Relative deflections can best be controlled by using deep beams (in lower-grade steel if necessary) and total deflections by placing columns as closely as possible to the support positions. Where beams support handrails, there may be torsional stiffness requirements (to take the torque from imposed lateral loads on the rails).

In some industrial plants used for delicate purposes, vibration levels must be absolutely minimised. This requires that the support structure's natural frequency be assessed and design constrained within certain limits.

6.2.4.3 Detailing

It is preferable to avoid the necessity for load-bearing stiffeners at support points unless the plant dimensions are fixed before steelwork design and detailing take

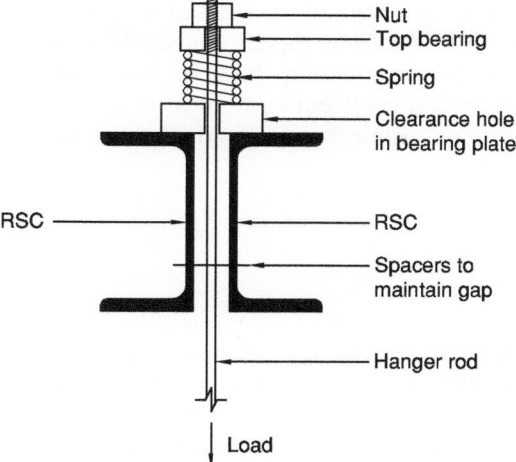

Figure 6.6   Detail at hanger support

place. Where this is not possible, then stiffened zones to prevent secondary bending of top flanges (or bottom if supports are hung) should be provided, even if the design requirements do not require load-bearing stiffeners. This then allows a measure of tolerance for aligning the support positions without causing local over-stressing problems.

Care may be required on the format of connections to ensure they do not clash with pipe runs (alternatively the piping designers should be advised early on if, for example, beam haunches are envisaged). Equally, liaison with piping/duct designers should be made if holes are required through beams. The forming of tailor made holes is, of course, best avoided, though cell-form beams can be used to advantage.

When hanger supports are needed, then pairs of beams or channels are a convenient solution which allows for random hanger positioning in the longitudinal direction (Figure 6.6). However, for minor items, there are many proprietary support systems that can be installed to carry the loads. Not infrequently, hanger supports have springs/bearings to minimise variations in support conditions due to plant temperature changes, or to avoid the plant stiffness interactions described elsewhere.

The method of installation or removal of major plant items frequently requires that beams above or, less commonly, below the plant must be designed to cater for hoisting or jacking up the installed plant sections. Often this requires the installation of dedicated lifting beams or jacking points. When main or secondary beams are specifically designed for infrequent but heavy lifting operations, it is good practice to fit a lifting connection to the beam to give positive location to the lifting position and to allow it to be marked with a safe working load.

The walls of industrial structures are often faced with cladding supported on proprietary cold-rolled sheeting rail systems. A detail requirement, especially for tall buildings with high wind suction, is to ensure that the rail thickness is adequate to take the fastener loads. It is normal to clad and fix from one side only using self-tap and drill screws whose pull out capacity is sensitive to rail gauge, hence rail selection is based not just on bending capacity but also on material thickness. Likewise, cold-rolled purlins are frequently used in roofs and a check should be made on their capacity to sustain hanger loads if these are a requirement of the building.

6.2.5 Columns

Column location in industrial buildings will, as for beams, be dictated by the needs of the plant support. Although regular grid layouts are desirable, it is sometimes impossible to avoid an irregular layout which reflects the plant and equipment location. Obviously, some degree of regularity is of considerable benefit in standardising as many horizontal members as possible; a common method of achieving this is to lay out the columns on a line-grid basis with a uniform spacing between lines. This compromise will allow standard lengths for beam or similar components in one direction, while giving the facility to vary spans and provide direct plant support at least in the other direction. If possible, the line-grids should be set out perpendicular to the longer direction of the structure.

When vertical loadings are high and the capacity of rolled sections is exceeded, several types of built-up columns are available. Where bending capacity is also of importance, or perhaps dominant, large plate I-sections are appropriate, for example, in frameworks where rigid frame action is required in one direction (in such cases, the members are effectively 'vertical beams' rather than 'columns'). To provide high capacity, it is not difficult to fabricate welded plate girders with relatively thick webs and flanges. However, if high vertical loads do truly dominate, then fabricated box columns are often employed (Figure 6.7). Design of box columns is principally constrained by practical fabrication and erection considerations. Internal access during fabrication is usually necessary for the fitting of internal stiffeners and, similarly, internal access may be needed during erection for making splice connections between column lengths or for beam-to-column connections. Preferred minimum dimensions are of the order of 1 m with absolute minimum dimensions of about 900 mm. Whenever possible, column plates should be sized to avoid the necessity of longitudinal and transverse stiffeners to control plate buckling. The simpler fabrication that results from the use of thick plates without stiffeners should lead to overall economies and the increased member weight is not a serious penalty to pay in columns. The same argument does not apply to long-span box girders where increases in self-weight may well be of overriding importance. Under most conditions of internal exposure, no paint protection is necessary to the box interior. If erection access is needed, then simple internal fixed ladders should be detailed. Transverse diaphragms are necessary at intervals (say 3–4 times the minimum column dimension) to assist in maintaining a straight, untwisted profile, and also at splices and at major beam-to-column intersections even if, as is usually the case, rigid connections

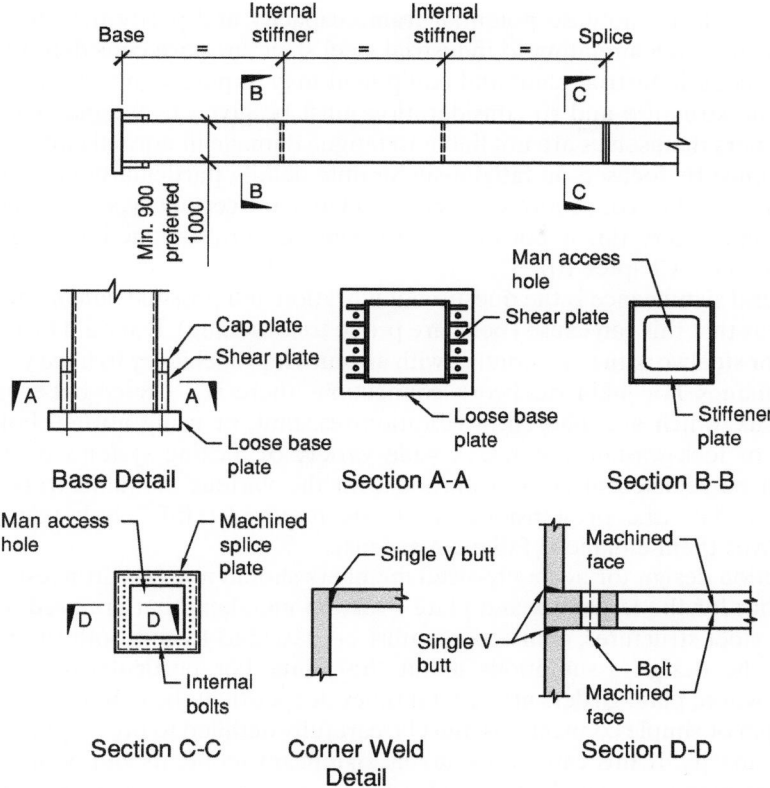

Figure 6.7   Typical details of box columns

are not being used. Diaphragms should be welded to all four box sides and be provided with manhole cut-outs when internal access is needed.

Where really substantial columns are used (either in weight or height), special design attention should be given to their bases. Fixed bases onto pile caps can be used but even if bases are nominally 'pinned', they should be carefully designed to provide stability during erection.

6.2.6 Connections

Connections between structural elements are similar to those used in general structural practice but, due to high loading, they often need to be substantial. Even when the loading is not defined as high, it is false economy to jeopardise available frame strength by skimping on connection capacity. As a generality, there ought to be some correlation between the end connection and the size of the member joined, partly so as to be able to mobilise potential frame capacity and partly for 'robustness'. Specific requirements relating to industrial steel structures are considered below.

On occasions, industrial plant and equipment may impose significant load variations on the structure and so consideration must be given to fatigue. Since basic steel members themselves are not liable to fatigue damage in normal circumstances, attention must be focused on fatigue-susceptible details, particularly those relating to welded and other connections. Specific guidance for certain types of structure is available and, where this is not directly relevant, general fatigue design guidance can be used (see Chapter 10).

Of general significance is the question of vibration and possible damage to bolted connections that this can cause (bolts are prone to loosening). It should be common practice for steelwork in close contact with any moving machinery to have vibration-resistant fixings. For main steelwork connections there is a choice between using HSFG bolts, which are inherently vibration-resistant, or using normal bolts with lock-nuts or lock-washer systems. A wide variety of locking systems is available which can be selected after consultation with the various manufacturers. Where fatigue is a real issue, pre-tensioned bolts are required (HSFG bolts) since their preload gives them enhanced fatigue resistance.

Connection design for normally-sized members should not vary from established practice, but for the large box and plate I-section members that are used in major industrial steel structures, connections must be designed to suit both the member type and the design assumptions about the joints. For particularly deep beam members, where plate girders are several times deeper than the column dimensions, assumed pin or simple connections must be carefully detailed to prevent inadvertent moment capacity. If this care is not taken, significant moments can be introduced into column members even by notional simple connections due to the relative scale of the beam depth. If fatigue conditions do exist, then such secondary moments would have to be accounted for.

In certain cases, it will be necessary to load a column centrally to restrict bending on it. A typical example is where deep suspension girders on power station boilers apply very high vertical loadings to their supporting columns. Here, a rocker cap plate detail is often used to assure centroidal load transfer into the column (Figure 6.8). Conventional connections on smaller-scale members would not usually require such a precise connection, as load eccentricities would be allowed for in the design.

Because members and their connections can be very large, it is even more important to account for ease of site assembly during connection detailing than it is for more standard frames.

6.2.7 Bracing, stiff walls or cores

Section 6.2.2 discusses the particular features of industrial steel structures in relation to achieving a horizontal or sway load path, and describes the various methods by which horizontal loads can be satisfactorily transferred to braced bays or other

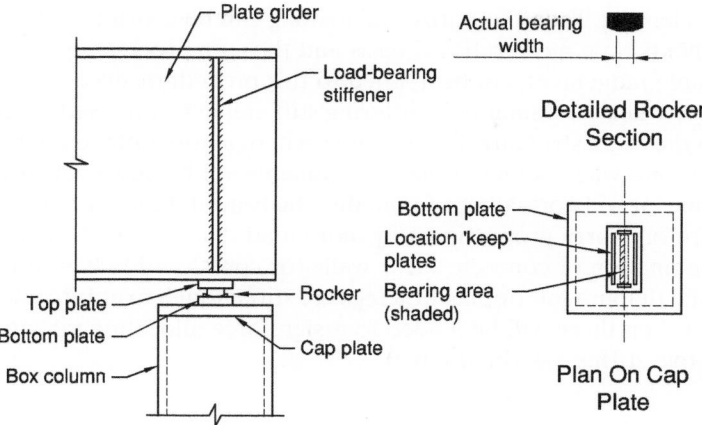

Figure 6.8   Rocker bearing – plate girder to box column

vertically stiff elements. General design requirements and some practical suggestions are also given in Section 6.2.2 for braced steel bay design.

The layout on plan of vertically stiff elements is often difficult, even in conventional and regularly framed structures. General guiding principles for best positioning are that the centre of resistance of the bracing system in any direction should be coincident with the centre of action of the horizontal forces in that direction. In practice this means that the actions and resistances should be evaluated, initially qualitatively, in the two directions perpendicular to the structural frame layout.

Another desirable feature which is also common to many structures is that the braced bays or stiff cores should be located centrally on plan rather than at the extremities. This is to allow for expansion and contraction of the structure away from the core without undue restraint from the stiff bays, and applies equally to a single structure or to an independent part of a structure separated by movement joints from other parts. It is difficult to achieve this ideal in a regular and uniform structure and almost impossible in a typical, highly irregular, industrial steelwork structure. Fortunately, steelwork buildings are reasonably tolerant of temperature movements and rarely suffer distress from what may be considered to be a less than ideal stiff bay layout (Section 6.4).

The procedure for design should be as follows. First a basic means of transferring horizontal loading to foundation level must be decided, identifying clear load paths, and guidance on this is given in Section 6.2.2. Next, in either or both directions, where discrete braced bays or stiff walls or cores are being utilised, an initial geometric apportionment of the total loading should be made by an imaginary division of the structure on plan, into sections that terminate centrally between the vertically stiff structural elements. The loadings thus obtained are used to design each stiff element.

When this process has been completed and, if the means exist by horizontal diaphragm or adequate plan bracing to force equal horizontal deflections onto each element, a second-stage appraisal may be needed to investigate the relative stiffness of each stiff element. Then the horizontal loading can be distributed between vertical stiff elements on a more rational basis and the step process repeated.

Considerable judgement can be applied to this procedure since it is usually only of significance when fundamentally differing stiff elements are used together in one direction on the same structure. For example, where a horizontal diaphragm or plan bracing exists and where some of the stiff elements are braced steelwork and some are rigid frames, it will normally be found that the braced frames are relatively stiffer and will therefore carry proportionately more load than the rigid frames. Similarly, where a combination of concrete shear walls (or cores) and braced frames is used and horizontal diaphragms or plan bracings are stiff enough to enforce their uniform displacement then there will be a need to assign force allocation via an assessment of their relative stiffnesses (Figure 6.9).

6.2.8 Fire protection

Many industrial structures have no fire protection at all. This is partly due to the limited occupancy, partly because the members can be massive and a fire engineering solution will give adequate fire life. Nevertheless, because the plant value can be very significant, it should always be checked what protection is required.

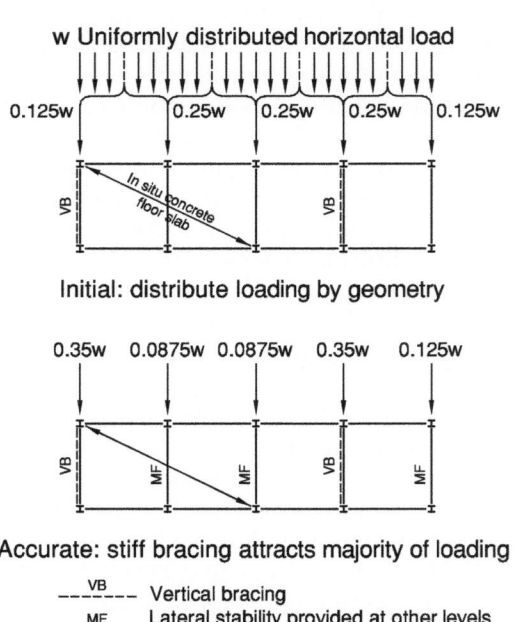

Figure 6.9 Apportioning horizontal loads to vertical stiff elements

6.3 Loading

6.3.1 General

Three headline difficulties exist in defining appropriate loadings on industrial structures. First the actual weight and, particularly, details of the position and method of load application of items of plant or equipment, must be established. Even for routine or replicated plant, this information is often difficult to obtain in a form that suits the structural engineer. To compound this, when the plant or equipment is being custom-built, the problem becomes one of timing; information may just not be available early enough for the structural design program.

The second difficulty that frequently occurs is in the choice of a general, uniformly distributed imposed loading for any remaining floor areas not occupied by items of plant or equipment. Guidance from codes of practice must be used carefully, when applied only to circulation spaces between all the known plant items, because a more appropriate loading may be that associated with plant installation or later maintenance.

The third general difficulty lies with defining design loads where there is an element of dynamic interaction between plant and its supporting structure. For common plant like cranes, codes advise dynamic enhancement factors (multiples of the defined plant dead loadings) but assessing loading is much more complex where there is support interaction with rotary or reciprocating machinery.

6.3.2 Quality of design information

At any stage of the design process, it is vital to be aware of the accuracy of the current information and to avoid carrying out designs at an inappropriately advanced level. Experience shows that provided loadings are often only approximate at the preliminary design stage and it is not at all unusual for them to be altered later, or for new loadings to be added at new locations.

There is also a need to thoroughly understand the nature of information. Frequently, loadings are given as a single all-up value, the components of which may not all act together or have only a very small probability of so doing. Alternatively, the maximum loadings may represent a peak testing condition or a fault overload condition, whereas normal operating loadings may be considerably less in value. Worthwhile and justifiable savings in steelwork can be made if a statistically-based examination of the expected frequency and duration of such unusual conditions is made, leading to the adoption of reduced load factors without reducing the overall plant safety. Comparisons with, for example, wind loading can be used to establish load factors on a reasonably logical basis.

Structural designers should query the exact method and route of plant installation to ensure that the temporary hoisting, jacking, rolling or set-down loads are catered for by the steel framework. Experience suggests that such information is not provided as a matter of course. Where, to ease problems, plant installation occurs coincident with steel erection, then the method of removing plant during the building lifespan may be the most significant temporary loading for the framework.

6.3.3 Process plant and equipment loading

Most plant installations are purpose-designed and the layout and loadings provided for them initially will be estimated or approximate values. A separate problem derives from the interaction between plant and support stiffness. The loading distribution given by plant design engineers may automatically assume fully stiff (zero deflection) supports. For example, when deep-walled tanks, bunkers or silos span across several beams, it is likely they will impose uniform deflection with corresponding load distribution but when the stiffness of the supporting structure is not uniform in relation to supported plant, then significant redistribution of applied loads can take place. As the supporting structure deflects, it causes a redistribution related to the (indeterminate) stiffness of plant items overhead spanning across the structural beams. Where the plant support positions, the plant loads, and the structural steel layout are symmetrical, then engineering judgement can be applied without quantitative evaluation. In extreme cases, a plant-structure interaction analysis may be required to establish the loadings accurately. Either way, it is essential to explore the support structure stiffness requirements as well as its strength requirements.

6.3.4 Uniformly distributed loadings (u.d.l.)

It is necessary to define a u.d.l. live loading on all floors. This has to cover general circulation (which is light) but also rather indefinable amounts to cater for occasional heavy tooling, lay down, storage and maintenance loads. Below the floor there may well be a network of relatively light cabling and ductwork. In some dirty plants it has even been known for dust build up loads to be significant.

By definition, items of fixed plant can be treated as dead load in accordance with BS 6399: Part 1[3] when specific location and loads are known but, at the early stages of design, it is usual to adopt a relatively large *imposed* loading for floors simply to cater for such fixed items of plant or equipment, the location and magnitude of which may not even be known at this design stage. The choice of what imposed loading to use at this stage can be assisted by the following guidance:

- very light industrial processes $7.5 \, kN/m^2$
- medium/average industrial processes $10–15 \, kN/m^2$
- very heavy industrial processes $20–30 \, kN/m^2$

One factor which will influence the choice within these values is the timing of final plant and equipment design data release, in relation to the steelwork design and fabrication detailing process. If a second-stage design is possible, then a lower imposed load can be allowed since local variations can then, if needed, be accommodated to account for specific items of fixed plant which exceed the imposed plan loading allowance.

Under these conditions, it can also be worthwhile, particularly for designers with previous experience of the industrial process, to design columns and foundations for a lower imposed loading than beams. In certain layouts with long span main beams or girders set at wide spacings, this preliminary reduction can also be used for beam members. (Note: although it is quite possible for any of these high loads to occur locally at any point on a floor as imposed load, it is most unlikely that the intensity will occur over all the floor all at the same time and over several floors all at the same time). It should be stressed that these proposals are not intended to contradict or override the particular reduced loading clauses in BS EN 1991-1-1: 2002 Clause 6.2.1(4)[4] but are a practical suggestion for the preliminary design stage.

When detailed plant layouts with location and loading data do become available, fixed plant and equipment can be considered as dead load and subject therefore to the appropriate load factor. Remaining zones of floor space without major items of plant should be allocated an imposed load which reflects only the access and potential use of the floor and may, therefore, very well be reduced from the preliminary imposed loading, often in the range of 5–10 kN/m^2. Specific laydown areas for removal, replacement or maintenance of heavy plant are the only likely exceptions to this range of loading.

6.3.5 Lateral loadings on structures supporting plant

Lateral loads on structures supporting plant derive from loads imposed by the plant itself plus 'notional' sway loads simply related to plant weight. These may be either equivalent horizontal loads derived from elastic sway where vertical loads are placed eccentrically on rigid frames, or sway linked to the fact that columns are never erected truly vertical. Either way, the sway magnitude is a function of the supported plant weight. It is particularly important to presume a horizontal loading for stability in the absence of any definable loading such as plant or wind action.

Defined lateral loadings from plant on the structure derive from three sources. These are considered separately, although there is a common theme throughout that the cause of loading is linked to the operating process undergoing a change in regime. Many of the actions giving rise to horizontal loads also cause vertical loads or at least vertical loading components which must be incorporated into the design. However, whereas vertical loadings are readily understood, horizontal components can sometimes cause confusion. This is because equilibrium considerations dictate that lateral force components are in balance and, consequentially, may be ignored by the plant designer or not be fully defined. This is not satisfactory for the support designer as the balancing components may act a considerable distance apart and even act at different levels. Thus, in all cases, the total load path must be examined in detail since loads may well be transmitted along the supporting structure and so must be accounted for along the load path by the steelwork designer. The three causes of horizontal loading are as follows:

- temperature-induced restraints
- restrained rotational or linear motion (crane surge laterally or longitudinally is an obvious source)
- restraint against hydraulic or gaseous pressures.

1. Where plant undergoes a significant change in temperature, plant designers will typically assume that the structure is fully rigid, i.e. stiff enough to prevent free thermal expansion and contraction. They will then design the plant itself for such forces and the additional stresses caused. This is a safe upper bound procedure since such forces generated in both structure and plant represent maxima, with any support deflection reducing forces in both elements. In spite of the apparent conservatism of this approach, it is frequently adopted for convenience on small items and to avoid complexities in interconnection between plant items and with piped and ducted services on larger items.

   The alternative approach, common on major plant subjected to significant thermal variation such as boilers and ovens, is to assume completely free supports with zero restraint against expansion or contraction. This is a lower bound solution which needs some rational assessment of any forces that could possibly result from bearing or guide misalignment, malfunction or simple inefficiency. Where large plant items are involved, the forces even at such guides or bearings can be significant and, as a minimum, longitudinal friction forces derived from the thermal movement should be allowed for (plant normal loading $\times$ μ) in the structural support design.

2. The lateral forces from constant speed, rotating machinery, which are usually relatively low, are generally balanced by an essentially equal and opposite set of forces coming from the source of motive power. Nevertheless, their point of application must be considered and a load path established which transfers them back to balance each other or, alternatively, down to foundation level in the conventional way. The structural steel designer must have a completely clear and unambiguous understanding of the source and effect of all the moving plant forces, to ensure that all of them are accounted for in the crucial interface between plant and equipment.

   The start-up forces, when inertia of the plant mass is being overcome, and the fault or 'jamming' loads which can apply when rotating or linear machinery is brought to a rapid halt, need consideration. They are obviously of short duration and so can justifiably be treated as special cases with a lower factor of safety. At the same time, the load combinations that can actually co-exist should be established to avoid any loss of design economy.

3. Pressure pipe loadings (hydraulic or gas), significant in many plant installations, occur wherever there are changes in direction of pipework and act on any associated pipework supports or restraints. Thermal change must also be considered. Pipework designers may well combine the effects to provide a schedule of the total forces acting at each support position. Very occasionally there may be force fit loadings imposed by the plant designer, i.e. deliberately introduced to counteract loadings in service. The validity of these will be linked to assumptions on structural support stiffness.

   As well as establishing force levels, it is equally important in all cases to define the line of action of all the plant force components since these will normally be positioned at some eccentricity to the supporting steelwork.

6.3.6 Inertial and dynamic loading

Inertial loading is that which derives from acceleration or braking (as in the motion of crane girders and their crabs) and is defined as mass × acceleration. Inertial loading is a form of dynamic loading but, more generally, dynamic loading is the equivalent load which results from the relationship between the input motion of a piece of plant (which may be unbalanced rotation or something moving to and fro) and the natural frequency of the support structure. If the frequency of the motion matches the frequency of the structure, a condition of resonance can exist where the equivalent forces on the structure become exceedingly large, limited only by damping.

The assessment of dynamic loads is very specialised (see Chapter 13) and is best carried out by appropriately experienced engineers. In very simple cases, some multiple of the normal static loads may be adequate but this must be considered with care. The steelwork designer should be aware of three key facts:

1. The plant designer cannot define the loads on his own in the absence of some knowledge of the support structure stiffness (or frequency).
2. There is always a need to be careful whenever any plant is in motion (to guard against fatigue). As part of project management, the structural engineer needs to confirm, or not, whether there is any reciprocating plant or plant in motion to be supported.
3. Quite commonly, the vibratory plant may be isolated from the structure via support on some form of springs. However, it will still remain necessary for the plant designer to define the force at the base of the spring.

Particular care is required on longer span floors (see Chapter 13) and, if more accurate force evaluations are warranted, a plant-structure interaction dynamic analysis must be undertaken.

6.3.7 Wind loads

Wind loadings on fully enclosed industrial structures do not differ from wind loadings on conventional structures. The only special consideration that must be given applies to the assessment of pressure or force coefficients on irregular or unusual-shaped buildings. A number of sources give guidance on this topic and specific advice can be sought.[5] As many industrial buildings can be tall, care should be taken in the design of their cladding (and cladding fixings) and support rails against high local wind suctions. There may often be a need to consider the effects of dominant openings for industrial sheds with large doors.

On partly or wholly open structures with exposed plant or equipment, great care must be exercised in dealing with wind loading. It must be appreciated that frictional drag on such items with their numerous protuberances, platforms and ladders and so on will be high and not definable with accuracy. It is frequently the case that the total wind loads are higher than on fully clad buildings of the same size and this is due to two causes. First, small structural elements attract a higher force than equivalent exposed areas which form part of a large facade. Secondly, repetitive structural elements of plant items which are nominally shielded from any particular wind direction are not actually shielded in each wind direction and each element is subjected individually to a wind load. The procedure for carrying out this assessment is not covered in detail in BS 6399: Part 2[6] but guidance is available from the references listed in that document. The steelwork designer should be careful to define responsibility for the design of the plant itself (under wind loads) as distinct from the wind forces the plant transmits to the supporting framework (but as part of project information gathering he will require the definition of local plant wind loadings onto the structure). When assessing the effect of wind on frameworks, consideration should be given to both the 'face on' case and to wind blowing 'across the diagonal'.

Loading on particularly large individual pieces of plant or equipment exposed to wind can be calculated by considering them to be small buildings and deriving overall force coefficients that relate to their size and shape. For smaller or more complex shapes, such as ductwork, conveyors and individual smaller plant items, it is more sensible to take a conservative and easy to apply rule-of-thumb and use a net pressure coefficient $C_p = 2.0$ applied to the projected exposed area. The point of application of wind loadings from plant items on to the structure may be different from the vertical loading transfer points if sliding bearings or guides are being used. The designer should always take care to extract the point of load application since this may be eccentric to supports.

6.3.8 Safety case loading

Many industrial structures pose hazards if they fail. Examples are petrochemical plants, refineries, mills, stores for hazardous materials and so on. As such, the overall Safety Case for the plant may define certain unusual loading cases for the structural steel. Examples may be seismic; blast (which may be from internal hazard or terrorist bomb); fire or severe corrosion. Dropped loads or severe impact cases may be other examples. In the nuclear industry, extremes of weather such as wind and rain are considered. It is the responsibility of the persons charged with developing the Case to define the load intensities and functional requirements to the steelwork designer.

Defining the functional requirements or design standards is crucial since, under these rare conditions, it is often the case that the structure merely has to 'survive'. Thus, under high seismic loading, blast or impact, advantage may be taken of steelwork's inherent ductility to absorb the finite amount of energy imposed. Whilst this is sound strategy, it is normally implicit that the member support connections need to be substantial and frequently must carry the full plastic moment of whatever member they join. Nowadays it is possible to carry out sophisticated analyses which will characterise the system performance right through the elastic stages into post-elastic behaviour, but such advanced analyses are best dealt with by engineers trained in the appropriate skills (Chapters 13 and 35).

6.4 Thermal effects

This section deals with thermal effects from environmental factors; plant induced thermal effects have been considered as a special aspect of the lateral loadings in Section 6.3.5. Conventional guidance on the provision of structural expansion or contraction joints is often inappropriate and impractical to implement, and indeed joints frequently fail to perform as intended.

The key to avoiding damage or problems from thermal movements is to consider carefully the detailing of vulnerable finishes (for example, brickwork, blockwork, concrete floors, large glazing areas and similar rigid or brittle materials). Provided that conventional guidance is followed in the movement provisions for these materials, then structural joints in steel frames can usually be avoided unless there is particularly severe restraint between foundations and low-level steelwork. This arises because most expansion is in the roof and the bending moments induced in supporting columns from imposed roof expansion are inversely proportional to (height).[2] This implies that long but tall buildings may be adequate; equally more care is required with lower buildings having the same length.

Many industrial structures with horizontal dimensions of 100–200 m or more, have been constructed without thermal movement joints, usually with lightweight, non-brittle cladding and roofing, and a lack of continuous suspended concrete floors. Part of the success can be put down to the fact that although the external temperature varies, the internal temperature fluctuates over a much smaller range, especially if the plant has a high thermal mass, so the actual steel temperature lags considerably behind the air temperature.

When high restraint does exist close to foundations, or vulnerable plant or finishes are present, a thermal analysis can be carried out on the framework to examine induced stresses and deflections and to evaluate options such as the introduction of joints or altering the structural restraints. Where steelwork is externally exposed in the UK, the conditions vary locally but a minimum range of -5°C to +35°C suffices for an initial sensitivity study. For many of the likely erection conditions, a median temperature of 15°C can be assumed, and a range of ±20°C can thus be examined (though this is likely to be conservative). Where the effects of initial investigations based on these values highlight a potential problem, more specific consideration can be given to the actual characteristic minimum and maximum temperatures and the likely variation of internal temperature. (Advice on UK air temperatures is available from the Meteorological Office).

6.5 Crane girder/lifting beam design

6.5.1 Crane girders

Crane girders are very common in industrial buildings and they need to be reliably designed, otherwise the effect on the plant process might be quite severe. Loads are applied to the girders via the crane wheels, with the normal load path being crane wheel, crane rail, rail fixings, top flange and thence along the beam to the supports. The loads applied act in three directions. Vertical load is obvious but the peak load occurs with the crab positioned to the end of its side travel with full pick up value. Two cases arise – maximum bending and maximum shear. The maximum bending case will occur somewhere around girder midspan but exactly where will depend on the number of wheels and the wheelbase of the crane. Once that information is known, the peak bending can be easily evaluated. Likewise, the peak shear can be worked out using the same information. Lateral bending is linked to the side surge of the combined weight of the crab and lifted weight, whilst longitudinal loads are developed by crane acceleration and braking and, again, these are normally expressed as a percentage of wheel loads. Codes will define the percentages that must be used for the three directional loadings. Another type of force that can exist is referred to as crabbing: this occurs when the whole crane twists in plan relative to its two girders (say if one side drives faster than the other) and this twisting imposes a force couple on the rails and their supporting girders. Again Codes will define force levels.

Peak top flange stresses derive from vertical load combined with lateral load and, to carry the latter, it is quite common to use an asymmetrical girder with either the top flange capped with an inverted channel or with a side surge girder spanning between adjacent columns (standard handbooks give the section properties of gantry girders). It is normal for the girder to be treated as simply supported.

The top flange clearly has to take the stresses but it also has to be wide enough to accommodate the rail (which is normally defined by the crane supplier) plus the crane retention clips. Various proprietary clips exist and these retain the rail, carry the loads and permit adjustment to achieve alignment. It should be obvious that the rail will not be perfectly aligned above the web so the heavy vertical loads will tend to cause girder overturning which should be resisted by the end connections to supporting columns (transmission can be via the top flange or surge girder). A separate problem related to vertical load transfer can exist in welded plate girders. Experience has shown that it is not good design practice to assume that vertical load transfer can be made via a perfect fit between the underside of the top flange and the web. There is a history of fatigue failures here when fillet welds are used since, in practice, such welds carry both longitudinal shear and vertical loading. To overcome this problem, a butt weld should be specified.[7,8]

Apart from the case of damage to fillet welds (and rail clips) it is not usual to have to design parts of the girder for fatigue, though service duty of the crane should always be checked. Nevertheless it is not regarded as good practice to support crane girders on column brackets unless the cranes are very light duty, say less than 5 tonnes. In other cases, a detail such as Figure 6.10 is common.

The vertical stiffness of girders is important and usually restricted to around span/1000. Equally, it is normally necessary to restrict the horizontal displacement between the two opposite girders, especially where the structure support form can spread (say if it is a pitched portal frame). In assessing lateral displacement (and stress), it is normal to carry all the horizontal loads on one girder alone. Where a check is made on spread, it is common to take 'full surge + half wind' or 'full

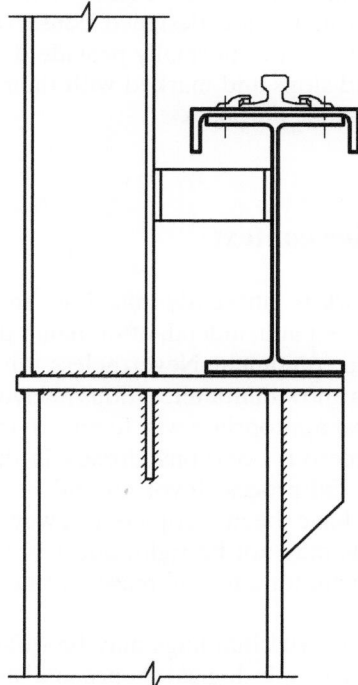

Figure 6.10   Preferred support for medium and heavy duty crane girders

wind + half surge' as the design cases. Maintaining the alignment of the rails is crucial. To facilitate this, the detailing of the girder end supports should allow for lateral shimming (with eccentricities allowed for in the stress calculations) and, additionally, the rail should be adjustable on the flange top. The NSSS specifies tolerances for crane girder fabrication.

Crane girder detailing will usually include an end stop (to loads and level to match the crane buffers). It will also be common to detail some power supply attachments, for example bus bars. It will also be normal to detail in some means of access such as caged ladders. The designer should always query the manner of installing overhead cranes, as this may affect the overall detailing of the structure.

6.5.2 Lifting beams

Lifting beams normally have a trolley running along their lower flanges and the design case is general beam bending plus cross bending of the lower flanges due to wheel load application. Cross-bending can be greatest if the wheels are able to approach a free end. As the cross flange bending is sensitive to wheel numbers and positions, it is vital these data are defined. Applying a load to the lower flange when supports are from the top flange also puts the (thin) web into tension. The connection supports from the top flange are designed conservatively to guard against fatigue failure. The applicable Codes normally provide design rules. Lifting beams should be provided with end stops and marked with their safe working load. They will require testing for insurance purposes.

6.6 Structure in its wider context

Industrial structural steelwork is inherently inflexible, being purpose-designed for a particular function or process and, indeed, often being detailed to suit quite specific items of major plant and equipment. Nevertheless, it is important to try to cater for at least local flexibility to allow minor alterations in layout, upgrading or replacement of plant items. The most appropriate way to ensure this is to repeat the advice that has been given on numerous occasions already in this chapter. The designer must understand the industrial process involved and be aware of both structural and layout solutions that have been adopted elsewhere for similar processes. Previous structural solutions may not be right, but it is preferable to be aware of them and positively reject them for a logical reason, rather than reinvent the wheel at regular intervals.

General robustness in industrial buildings may be difficult to achieve alone by the normal route of adopting simple, logical shapes and structural forms, with well-defined load-paths and frequent effective bracing or other stability provisions. As a supplement to these provisions, it is sensible to ensure reasonable margins on element and connection design. Typically, planning for a 60–80% capacity utilisation at the initial design stages will be appropriate. Thereafter, even when these allowances are reduced during the final design and checking stages, as so frequently occurs, adequate spare capacity may still exist to ensure that no individual element or joint can disproportionately weaken the overall building strength. As has been pointed out, to assure robustness, the destabilising effect of heavy masses on their own should not be overlooked. Where buildings are divided into sections, each section should be tested for 'robustness' on its own.

References to Chapter 6

1. Ji T. (2003) Concepts for designing stiffer structures, *Journal of the Institution of Structural Engineers*, Vol 81, No 21, 4th Nov.
2. British Standards Institution (2006) BS 4592:5 *Industrial type flooring and stair treads. Solid plates in metal and GRP Specification*. London, BSI.
3. British Standards Institution (1996) BS 6399 *Loading for buildings Part I: Code of practice for dead and imposed loads*. London, BSI.
4. British Standards Institution (1991) BS EN 1991-1-1: 2002 Clause 6.2.1(4).
5. Cook N.J. (1985) *The designer's guide to wind loading of building structures. Part 1*. Cambridge, Butterworths.
6. British Standards Institution (1995) BS 6399 *Loading for buildings. Part 2: Code of practice for wind loads*. London, BSI.
7. Senior A.G. and Gurney, T.R. (1963) The Design and Service life of the upper part of welded crane girders, *Journal of the Institution of Structural Engineers*, Vol 41, No 10, 1st Oct.
8. Kuwamura H. and Hanzawa M. (1987) Inspection and repair of fatigue cracks in crane runway girders, *Journal of Structural Engineering, ASCE*, 113, No. 11, Nov., 2181–95.

Further reading for Chapter 6

Masterton, G.T. Power and Industry (2008) Centenary Special Edition, *Journal of the Institution of Structural Engineers*, Vol 86, No.14, 21st July.
Morris, J.M. (2000) An overview of the Design of the AAT Air Cargo handling Terminal, Hong Kong, *Journal of the Institution of Structural Engineers*, Vol 78, No. 16, 15th Aug.
Luke, S.J. McElligott, M. and Everett, M. (1998) General Electric Aircraft Engine Services GE 90 Test Cell, *Journal of the Institution of Structural Engineers*, Vol 76, No. 7, 7th April.
Luke, S.J. and Corp, H.L. (1996) British Airways Heavy Maintenance Hanger, Cardiff Wales Airport: Structural System, *Journal of the Institution of Structural Engineers*, Vol 74, No. 14, 16th July.

Ford, R.F. and Lilley, C. (1996) Alusaf Hillside Aluminium Smelter, Richards Bay, *Journal of the Institution of Structural Engineers*, Vol 74, No. 19, 1st Oct.

Krige, G.J. (1996) The Design of Shaft Steelwork Towers at Reef Intersection in Deep Mines, *Journal of the Institution of Structural Engineers*, Vol 74, No. 19, 1st Oct.

Zai, J., Cao, J and Bell, A.J. (1994) The Fatigue Strength of Box Girders in Overhead Travelling Cranes. *Journal of the Institution of Structural Engineers*, Vol 72, No. 23, 6th Dec.

Bloomer, D.A. (1993) Stainless Steel Ventilation Stack: WEP Sellafield, *Journal of the Institution of Structural Engineers*, Vol 71, No. 16, 17th Aug.

Tvieito, G., Froyland, T. and Wilson, R.A. (1992) The Development of Kvaerner Govan Shipyard, Govan, Glasgow, *Journal of the Institution of Structural Engineers*, Vol 70, No. 2, 21st Jan.

Jordan, G.W. and Mann, A.P. (1990) THORP Receipt and Storage – design and construction. *Journal of the Institution of Structural Engineers*, Vol 68, No. 1, 9th Jan.

Forzey, E.J. and Prescott, N.J. (1989) Crane supporting girders in BS 15 – a general review, *Journal of the Institution of Structural Engineers*, Vol 67, No. 11, 6th June.

Dickson, H. (1964) Power Stations as a Structural Problem, *Journal of the Institution of Structural Engineers*, 45, No. 5, 1st May.

Fisher, J.M. and Buckner, D.R. (1979) *Light and Heavy Industrial Buildings.* American Institute of Steel Construction, Chicago, USA.

Mann, A.P. and Brotton, D.M. (1989) The design and construction of large steel framed buildings. *Proc. Second East Asia Pacific Conference on Structural Engineering and Construction*, Chiang Mai, Thailand, 2nd Jan, 1342–7.

Chapter 7
Special steel structures

IAN LIDDELL and FERGUS McCORMICK

7.1 Introduction

This chapter aims to open to the reader the possibilities and opportunities of designing and delivering 'special' structures in steel. These are ones which are not the baseline 'beam and stick' or common portal frame structures which form much of the volume of structures our industry engages with but the special structures that excite and create drama, interest and intrigue within the built environment. The chapter offers a state-of-the-art commentary to illustrate the versatility of steel for use in these special and non-standard designs.

Special structures are generally those with complex 3D loadpaths or non-regular geometry, or those that are required to have a longer than normal span or have a lower than normal self-weight. Typical modern special structures include:

- 3-dimensional grids based on regular solids, e.g. space frames
- forms generated by equilibrium such as:
  - surface-stressed tension structures and cable nets
  - compression structures whose form is derived from inverted hanging chain models
- deployable or moving structures
- curved forms such as domes and vaults with geometries defined by circles, ellipses, spheres etc.
- curved forms with irregular geometry generated by spline curves using CAD methods
- irregular forms with organic or chaotic or random geometries.

There is always a place for the most efficient, effective form-driven engineering solutions. However, trends in architecture and the possibilities arising from new technology have caused changes in, and an expansion of, types of special structures. The restrictions of having to use defined regular geometries with their attendant

---

Steel Designers' Manual, Seventh Edition. Edited by Buick Davison and Graham W. Owens.
© 2012 Steel Construction Institute. Published 2012 by Blackwell Publishing Ltd.

benefits of symmetry and repetition have gone and the space frame systems that peaked in the 60s have more or less disappeared and have been replaced by 'free-form' surfaces and chaotic frames.

Driven by the development of 3D CAD programs in the 1990s that used NURBS, non-uniform B splines, (which are cubic functions for fitting curves through a number of points) architects and engineers are able to evolve forms with doubly curved surfaces that they can manipulate to suit their tastes. Steel has become the material of choice for this new architecture that is made with curving surfaces or with surprisingly jagged, twisted or falling-over forms.

The forms developed with these tools have driven advances in fabrication processes and here computational methods have enabled changes. Forms based on these strong aesthetic drivers would be less structurally efficient and thus more materially costly than ones based purely on structural engineering and production drivers, but the problem can be managed by linking the form development software to analysis programs and work is proceeding on automatically improving their structural efficiency.

All solutions with complex geometries impose a greater amount of detailed engineering input in both the design and the construction phases than standard building frames, thus requiring specialist engineers and steel contractors. Some of the key issues for analysis, design, fabrication and construction are described with reference to case studies, which illustrate the issues in practice.

7.2 Space frame structures: 3-dimensional grids based on regular solids

3D space grids have a long history of development and, for a while, were extensively used, their benefit being a lightweight, open, structural system that can be constructed from standardized repetitive elements. The best known is Mero prefabricated 3D frames, begun and developed by Max Mengeringhausen in the 1930s. This used square cut tubes with special end fittings that could join into spherical nodes at predetermined angles. Initially this was used for constructing multi-layer space frames based on regular solids. CNC machining of the nodes allowed the range of geometries to be extended to include cylindrical and spherical domes and the like. Structures were delivered to site as straight members and nodes and were highly efficient in the use of material.

An application of this system with greater complexity than a simple flat roof was the football stadium roof at Split, 1979. A curved sheet of a 2-layer space grid was created to form an arch with about 210 m span along each side of the pitch covering the main stands. During the 1960s and 70s these were the high-tech forms of the time. Since that time, systematic space grids have fallen out of use because modern methods of design and manufacture allow greater freedom in the development of geometrical forms. However, an outstanding recent example of space frame technology is the Eden Domes (see Jones *et al.*, 2001[1]).

Case study: Eden Project Domes, Cornwall, UK

The domes represent a modern structure developed from regular geometries (Figures 7.1 and 7.2). The form is based on spheres constructed with a grid based on the subdivided pentagons of a regular dodecahedron, a solid with twelve faces. The resulting pattern of hexagons and pentagons is reminiscent of the skeleton of certain zooplankton called radiolaria and was thought to provide an appropriate link to the natural world. A single-layer grid would carry external loads with large bending moments in the members. To make it more structurally efficient, and to enable prefabrication of the components, a second grid layer to the space frame was added to form a 'hex-tri-hex' grid. This was constructed with members and nodes developed from the Mero system with cup nodes for the outer layer and spherical

Figure 7.1   Aerial view of the Eden Domes (courtesy of Grimshaw)

Figure 7.2   View from inside Dome looking at steel structure (courtesy of Grimshaw)

nodes for the inner. The ends of the members are finished square to the axis while the nodes are machined and drilled to the correct angles.

7.3 Lightweight tension steel cable structures

7.3.1 Introduction

Lightweight tension structures comprise a family of designs whose common approach is the use of steel cables often within form-found structures. The solutions are attractive for special structures such as roofs and bridges for a number of reasons. Cables have a high strength capacity of around three times that of normal steel and their high strength-to-weight ratio means less steel material is required in supporting loads. Self-weight loading in large bridges and roofs can form the majority of the loading to be resisted so reduced structural sections and self-weight can lead to dramatic improvements in overall structural efficiencies and costs. The small sections of cables mean they are attractive in applications where transparency is to be maximized such as in supporting glass facades and where shadowing is to be mini-mized such as in supporting roofs.

Cables have low bending stiffness, but rather than being a limitation, the special nature of cables offers engineers great possibilities for creative design. Many light-weight cable structures are deliberately form-found to ensure cables resist forces efficiently in axial stress. However, cables can be designed to support tension forces, compression forces and also lateral load. These methods of load resistance are described first followed by a description of a number of design solutions using cables.

7.3.2 The ways in which cables resist loads

For structures resisting tension or compression, a rod can be used equally as a cable.

Table 7.1 Ways in which cables resist load

Loading	Diagram	Key points
1. Tension loads		• The most common occurrence is for example guy cables carrying axial forces from end-to-end • At a first approximation the behaviour is little different to that of any tension element: beam or rod, with the performance being characterized by elastic stiffness. However, if the tensile stress is high, then 'tension stiffness' becomes relevant
2. Compression loads		• Cables can only resist compression if prestressed by self-weight or an internal self stress • The net axial load must always be a tension
3. Lateral loads		• Cables have little initial elastic stiffness and resistance to the applied lateral loads • They cannot resist the load in bending, but will move to an equilibrium position to resist the loads via axial stress in the cable • For a uniformly loaded cable the form becomes a catenary. However although this is a well known form, in practice it rarely occurs in such a pure way. Cables are often loaded at discrete points along their length and the equilibrium form becomes facetted

7.3.3 Linear behaviour, non-linear behaviour and large displacements

There can be much confusion about the level of analysis and sophistication required within cable structures. To introduce a discussion of this for cable structures (and for compression structures in Section 7.4), a general summary of the causes of non-linearity of all steel structures is presented in Table 7.2.

Cable systems can exhibit non-linearity from the following sources:

- tension stiffening within individual cable elements loaded axially or laterally
- displacements of the overall system (generally deemed 'large' displacements) as an overall cable system develops equilibrium against forces
- cables dropping out of the overall structural system if they become slack.

Cable *stayed* structures behave basically as linear elastic structures. In structural systems employing 'straight cables' based on applied tension and compression loads to the cable ends (methods 1 and 2 in Table 7.1), the effects of non-linearity can be small and much initial analysis can be carried out on simple linear programs. In general if a structure is 'noded out' and can be resolved by hand or by a computer, then its non-linearity effects will be small.

Structural systems employing cables loaded laterally have little initial elastic resistance and move into their equilibrium shape. These may be generally called cable *net* structures and have to be analysed under loads using software that can handle the geometrically non-linear behaviour. In this instance the structure cannot generally be identified as 'noded out' or statically determinate in the same way as for example a Warren truss. In this instance, although basic hand calculations can inform a solution, non-linear programs are required. These generally move the structure from some initial starting position to a final equilibrium position by solving the structural equations in a step-by-step incremental manner. A well known technique is called 'dynamic relaxation'.

The initial geometry of cable net structures has to be form-found to achieve a figure of equilibrium. This can be done using constant force elements to introduce the prestress forces within a nonlinear program as described above. In this process the prescribed forces and the cable stiffnesses are adjusted until the desired shape is achieved. This becomes the final geometry and the cable lengths and tensions are defined from it. To construct the structure the cables have to be cut to the calculated lengths with compensation for elastic stretch and inelastic or 'construction' stretch. This process is complex and needs to be carried out by experienced engineers.

7.3.4 Structural solutions using cables

Tension structures separate themselves into a number of groups whose form is related to the function of the cables i.e. the way in which the cable is loaded and

Table 7.2   Sources of non-linear structural behaviour

- Material non-linearity arises from:
  - onset of yielding, plasticity and formation of plastic hinges
  - non-linear elastic modulus relationship
  - onset of plasticity in the section influenced by residual stresses

- steel cross-sections will behave linearly, with strain proportional to stress, until yield is reached
- once yield is reached, generally in extreme fibres under bending, the proportionate relationship changes
- once the section exhibits full yield, then the section can maintain load, if the section remains locally stable and within defined ductility limits
- sophisticated analysis packages will be able to model the elasto-plastic stress/strain function for steel

- P-δ non-linearity arises from:
  - force dependent stiffness or geometric stiffness which results in tension stiffening or compression softening

- the increased axial and lateral stiffness of cables/wires/rods under increased tension is well known and termed 'tension stiffening', where the additional stiffness (to the Young's Modulus material stiffness) is a function of a change in stiffness matrix due to the element force
- the corollary of tension stiffening, perhaps to be known as 'compression softening', is also a function of change in stiffness due to force
- the effect of compression flexibility is a gradual vulnerability of an element to buckling effects and is non-linear, being more marked the closer the load is to the Euler Load

- Stability non-linearity arises from:
  - tension elements going slack
  - compression elements having buckled
  - post-buckling or 'post-slack' behaviour where the initial model has changed because some elements cannot participate in resisting force

- at its limit, elements will fail under large compression loads and, depending upon geometry and stress, this might be via buckling or via yield. In either case, non-linear programs will recognize the absence of load carrying ability and shed load into other elements
- non-prestressed tension elements looking to carry compression forces will go slack and a non-linear analysis will look for alternative loadpaths
- A system may be chosen deliberately to have slack cables. In many lightweight structures, cables are configured to give different structural load-paths under downwards and upwards loading cases.

- P-Δ non-linearity arises from:
  - the iterative effect of loads acting on the already deflected shape
  - it is generally important for cable structures having large displacements, and when Bernoulli – simple bending theory – breaks down

- large displacement structures are ones in which the basic simple approximate formula are insufficient and the displacement affects the force distribution
- non-linear programs such as those using kinematic analysis must be used to solve equations iteratively

Table 7.3 Structural solutions employing steel cables

| 1. Cable stayed structures | • the cables are loaded end to end axially
• cable end nodes support generally steel beams or trusses
• a typical application is for cable stayed bridges, e.g. Queen Elizabeth II Bridge at Dartford and the Second Severn Crossing
• applications for roofs are numerous and include Inmos, Newport and National Exhibition Centre, Birmingham. |

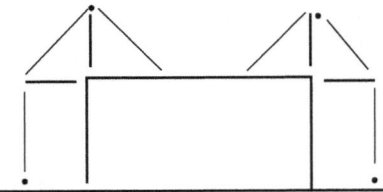

| 2. Suspension structures | • the cable is loaded laterally
• the most typical applications are suspension bridges, such as the Old Severn Crossing, and in these structures a catenary or parabolic cable provides the primary support to the hangers and deck
• in major suspension structures the dip to span ratio is normally greater than 1:12 and the primary cable is stressed by the dead load of the structure. In this case the impact of the cable stretch on the deflection is small. |

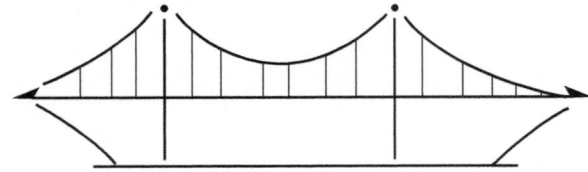

| 3. Surface stressed structures | • the cables are pre-stressed by jacking against the supporting members to induce a state of self stress and are then loaded laterally
• the cable extension is very important for the deflections, and nonlinear calculations have to be used to calculate the deflections and forces
• the structure can consist of individual straight cables or a net of cables basically at right angles to each other
• a cable net may be flat or prestressed against a 3D form
• the initial geometry is governed by the equilibrium of the tension forces under the initial pre-stress forces and hence can only be defined by calculations involving displacements to the equilibrium position
• the final geometry is governed by the equilibrium of the tension forces under the forces resulting from the pre-stress and applied loading, and are similarly only defined by calculations involving displacements to the equilibrium position
• there are numerous recent applications where facades are restrained against lateral wind loading by cables in a single flat plane and generally deflections are large
• cable nets have been used successfully to form Aviary enclosures such as at Munich Zoo (Addis, 2001[2]) and other famous examples include Munich Olympic Stadium, 1972, and Calgary Olympic Saddledome (see Bobrowksi, 1985[3]). |

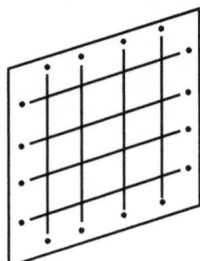
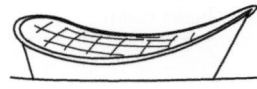

Table 7.3 (*Continued*)

4. 2D cable trusses	• cable trusses can be fully triangulated in which case the individual elements resist load by method (1) of Table 7.1, i.e. by applied tensions at the end nodes • however, commonly, the term cable truss is also used (perhaps incorrectly) to describe systems such as shown below: 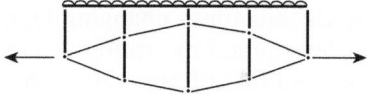 • the system is not fully triangulated and the cable resists load in method (3) of Table 7.2 and they are a particular example of a surface-stressed structure with the geometry chosen to be funicular to the most common loading • there are numerous cases of this system being used to restrain vertical facades against wind loading.
5. 3D cable net	• the term '3D cable net' does not completely define the nature of some complex unusual systems but is used as a generic term for some of those structures which might have cables working in different ways • certain structures have been developed in 3D using straight cables where all cables are 'noded-out', and they resist loads by method (1) of Table 7.1; applications include the BA London Eye (see Wernick, 2000[4]) • other applications, including cables loaded laterally include the Millennium Dome, Greenwich (see Liddell *et al.*, 1999[5]). 

the way in which the cable resists loads. A complex structure may incorporate a number of these functions, but the main groups are in Table 7.3.

7.3.5 Analysis and design issues for cables and cable systems

• Codes of Practice:
  ○ The most relevant Eurocode is EC3, Part 1-11, BS EN 1993-1-11, Design of structures with tension components.[6] There is information on cables in other industry guides developed for bridges and post-tensioned concrete.[7,8]

- Steel cable types:
  - EC3 identifies a number of cable types: spiral strand rope, strand rope, locked coil rope, parallel wire strand.
- Cable stiffness:
  - Cable stiffness is partly a function of the material modulus and partly a function of the strands and ropes changing length due to their winding. Precise values must be determined by testing or data from the manufacturer. (Note: cables are often sized for stiffness rather than strength).
- Working stress or load factor design:
  - Early pioneers of cable designs for buildings operated with working stress design. Cable structures have seen a slow transition to load factor design, partly because some engineers feel that form-found structures adhere best to working stress design. The new Eurocodes adopt a load factor approach.
- Cable strength
  - Cable strength is normally defined by manufacturers as the Minimum Breaking Load (MBL). Cables were, in the past, designed using a working load approach with low utilization factors compared to the breaking strength. The maximum unfactored force is generally limited to 50% MBL. However, conversion of the MBL to design values using either working stress or limit state methods needs care to ensure that different manufacturers are using the same approach for fatigue within their quoted MBL. There is a trend to move to ultimate limit state approach. Also, care is needed to verify that connector designs are stronger than cables.
- Load factors:
  - The general non-linearity of lightweight tension structures means that using load factors needs an unusual amount of care to ensure that their use results in a safe, effective and realistic set of loading conditions. The behaviour of the structure in response to changes of factors needs to be understood from first principles and at a greater level of thought than for other structures. Sometimes an approach using a working load and an ultimate limit approach may be required for different steel elements in the same structure, i.e. for steel cables and steel tubes.
  - Load factors will be generally the same as for other structures. Care needs to be taken when considering prestress. Sometimes prestress loading and its factors are bracketed with dead loading. However, if these are independent (e.g. if the prestress is jacked into the system), then the load factors should be treated as independent variables. At the BA London Eye, a critical design consideration was maximum force in the lower cables and minimum force in the upper cables (to ensure they remained active and did not go slack).

    The influence of four main loading conditions on the wheel cables is shown in Figure 7.3.

The extreme values of forces in the cables were derived using load factors for prestress that were independent from the load factors on dead load as explained below for two generic loadcases:

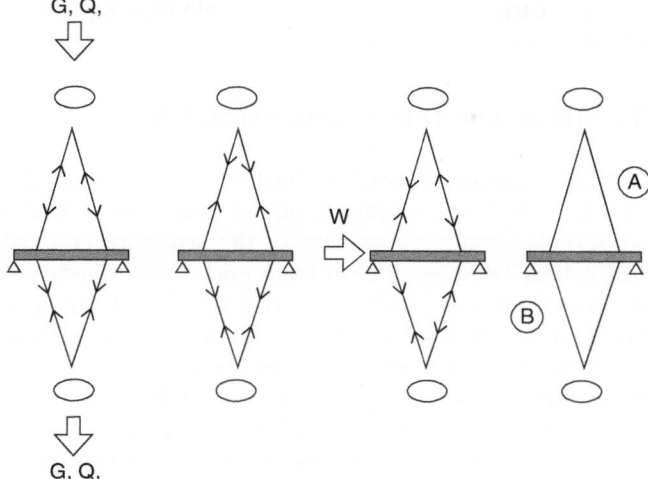

Figure 7.3   Loading conditions from dead load, imposed load, prestress and wind load applied to the BA London Eye

- For derivation of the maximum tension in the lower cables 'B':

$$\gamma_{\mathrm{fmax}}\, G + \gamma_{\mathrm{fmax}}\, Q + \gamma_{\mathrm{fmax}}\, PS + \gamma_{\mathrm{fmax}}\, W$$

- For derivation of the minimum tension in the upper cables 'A':

$$\gamma_{\mathrm{fmax}}\, G + \gamma_{\mathrm{fmax}}\, Q + \gamma_{\mathrm{fmin}}\, PS + \gamma_{\mathrm{fmax}}\, W$$

- Construction stretch and cable compensation:
  - Tolerances on cable length and its impact on the design need to be understood. Cables should be prestretched and adjustable length connectors such as turn-buckles may be installed.
- Cable vibration:
  - Cables need to be checked against the effects of wind induced galloping or vortex shedding and rain-induced vibration (see Caetano, 2007[9]).
- Cable fatigue:
  - Even if the worst effects of aeroelastic instabilities are not present, cables need to be checked against fatigue loading (see Raoof, 1991[10]).
- Cable end connectors:
  - Installation and service movements mean that connections may rotate and displace much more than normal steelwork connections. Fork end connections are normal for cable structures and allow rotation about one axis, but special connections may be required to accommodate larger than normal rotations about more than one axis.
- Cable saddles and diverters:
  - Cable end connectors are expensive and thus it is preferable to take a large diameter cable through a joint, if possible, using a saddle or clamp or diverter. These joints need appraisal of issues including reduction of permissible axial

stress on the cable, friction capability, permissible bearing stress and Poisson's ratio effects affecting installation.

Case study: The Millennium Dome, Greenwich, UK

Built to house the Millennium Exhibition, the Dome was to provide a cover over the whole site under which the elements of the exhibition could be built later (Figures 7.4 and 7.5). The main component of the structure is a radial net of 72 straight tensioned cables These span 25 m from node to node and carry tensioned fabric panels. The nodes are supported by an array of cables hanging from the 100m high masts. Around the perimeter the radial cables are supported by masts and tied down to anchor points where the vertical forces are taken by tension piles and the radial forces by a compression ring beam on the ground.

Since the cables deflect both vertically and horizontally under loads the cables are connected at each node with a detail that allows movement in both directions. This avoids the risk of fatigue at the cable terminations. The radial geometry on the spherical cap minimizes the risk of ponding under snow loading.

Case study: 2012 Olympic Stadium Roof, London, UK

The roof is an example of a complex system exhibiting cables working in different ways. Pairs of radial spoke cables are tensioned between chords of the external

Figure 7.4   Aerial view of The Millennium Dome (courtesy of Buro Happold)

Figure 7.5   External view of The Millennium Dome (courtesy of Buro Happold)

trussed horizontal compression ring and an inner horizontal tension ring (Figure 7.6). The whole cable net is pre-tensioned prior to installing the fabric cladding, and the lower radial spoke cables carry the tensioned fabric cladding (see Crockford et al.[11]).

7.4 Lightweight compression steel structures

7.4.1 Introduction

This section concerns the design of lightweight steel roofs where the structural steel elements are acting primarily in compression in a single layer. Solution variants include arches, barrel vaults, domes and shell structures.

Triangulated grids on a doubly curved surface act like shell structures and can carry non-uniform loads by axial forces in the members which can be pinned at the nodes. To achieve shell action the boundary must be rigidly supported so that the shell is effectively closed. An eggshell is an example of a closed shell and can resist local loads very effectively. If the shell is cut and left open then it is severely

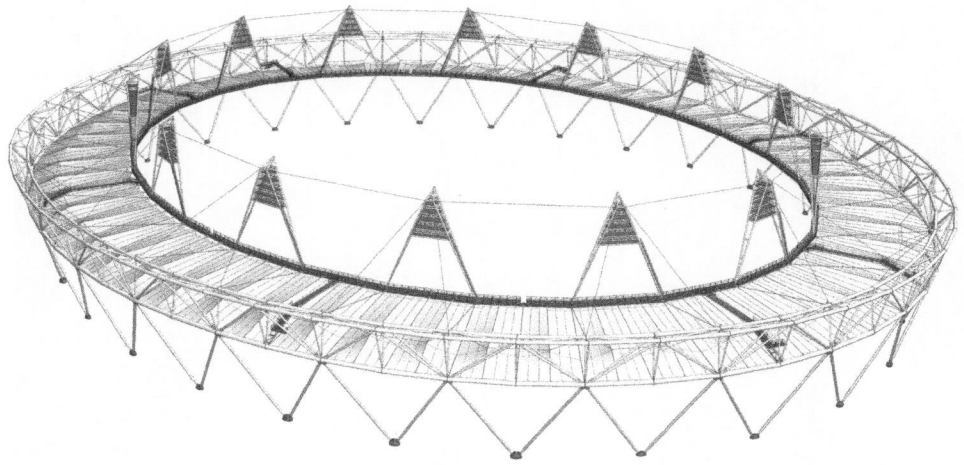

Figure 7.6   Render of steel roof structure and supporting columns of London 2012 Olympic Stadium (courtesy of Populous,™ used with permission)

weakened. In the 1950s and 1960s a number of triangulated spherical cap shells were built, starting with the Dome of Discovery in 1951. A number of such structures were built around the world with spans up to 250 m or so. An interesting early example is Fort Regent leisure centre (see Davies *et al.*[12]).

7.4.2 Funicular geometry of compression structures

Triangulated grid domes can adopt a wide range of shapes so long as a double syn-clastic curvature is maintained. Shells with two way grids are also possible but they have less resistance to random loading. Typically, the structural design of these types of roof will aim to use a funicular geometry to minimize bending moments and carry imposed and self-weight loads via axial compression. Careful form finding using this principle can lead to very efficient load carrying and reduce the size of the structural steel members to a minimum. Additional stiffness is still required to resist large deflections and buckling. This can come from adding diagonal ties and adding bending stiffness of the members for both in-plane and out-of-plane bending.

A funicular geometry for a compression structure is the inverse of a hanging chain geometry. Form-finding techniques used for tensile structures can therefore be applied to also find the optimum shapes for compression structures. These can include physical models, or more commonly computer analysis using specialist 'form-finding' software. It should be noted that a funicular geometry is unique to a particular loading. A given arch geometry can be derived with pure axial compression for loadcases where the applied load is uniformly applied in a downward direction e.g. a uniform snow loading. However, an uneven loadcase, such as wind or patterned live loading, will cause bending moments in the structure. A funicular

Table 7.4 Structural solutions – compression structures

1. Arches; singly curved sheets	• e.g. Chur railway/bus station, Switzerland which is an elegant example of paired tubular steel arch elements with tie restraints to avoid buckling (see Addis, 2001[2])
2. Domes and synclastic shapes	• e.g. British Museum Courtyard roof (see Brown, 2005[13]; Williams, 2000[14])
3. Barrel vaults with shell action	• e.g. Atrium of the Imperial War Museum, London, UK (see Pearce *et al.*, 2002[15]).
4. Anticlastic shapes	• these appear more free form, e.g. Sage Centre, Gateshead, UK (see Cook, 2006[16])

shape is most appropriate then, when the magnitude of known loading (generally dead loading and uniform loading) is relatively large compared to the magnitude of variable (generally environmental) loading.

7.4.3 Structural solutions using steel compression structures

Compression structures may be separated into a number of groups, the main ones are in Table 7.4.

7.4.4 Boundary conditions

A funicular structure acting mainly in compression imparts an in-plane thrust on its supports. The horizontal component of this thrust must be either resisted by the supporting structure or by connecting the ends of the funicular structure with a structural tie.

If the supporting structure is to resist the horizontal thrust then it is not possible to provide a movement joint between it and the roof structure. Differential thermal movement between the roof and supporting structure may result in build–up of forces within both.

7.4.5 Creating a structural grid

Barrel vault roofs may have simple structural framing comprising transverse arches linked by longitudinal purlins. However, domes and more complex roof shapes will result in a more complex structural grid within the roof surface. The structural grid will typically determine the geometry of the roof cladding panels which often may span between individual structural members. Economical fabrication of the structure would be best achieved if the connections between structural members were of repetitive geometry and also if the geometry of each cladding panel was identical. Achieving both of these is rarely possible in anything other than the simplest structures, but careful thought at the design stage in regard to the grid geometry can significantly reduce the construction costs of the roof. Specialist bespoke computer packages have also been developed by leading engineering consultants to generate repetitive grid patterns for the most complex shapes. Previously these were generally triangular grids, but squares and rectangles are more often adopted as they result in fewer elements and connections.

7.4.6 Buckling of compression structures

Introduction to buckling

The non-linear reduction of stiffness by compression forces causing potential for buckling is a major issue for these special compression structures. From a philosophical point of view, there are two approaches to an understanding of buckling matched by two main analysis methods.

The Eurocode proposes simple methods to account for the vulnerability of steel columns to buckling by reducing the allowable stress against compression. The interaction formula can therefore be expressed as:

Unity Factor (UF) $= P/P_c \downarrow + M/M_c$

The arrow in the above formula denotes the conventional Code approach of a *reduction* permitted in axial stress capacity. The formulae for simple columns identify the reduction based on interrogation of effective lengths which is, at heart, based on Euler buckling.

Identification of equivalent Euler buckling effects and identification of equivalent effective lengths within a complex system with many numbers of elements can be done using eigenvalue analysis. Therefore mapping of the simple code techniques for 'element' buckling to 'system' buckling can be done.

However, for all structures, but certainly the more complex ones, it can be easier to think in terms of the vulnerability to buckling manifesting itself as increased moments in the system (amplified moments) with an approach therefore represented by:

UF $= P/P_c + M \uparrow / M_c$

The arrow in the above formula represents the approach of an *increase* in bending stresses caused by displacements of a slender structure.

The approach is quite general and thus can be applied to most structures.

The deflections of a slender structure from its theoretical initial geometry can arise from a number of sources:

- initial imperfection
- first-order deflection (which can be derived from a conventional simple linear analysis)
- non-linear deflections from P-δ and P-Δ and other effects.

Two main methods for buckling analysis of complex structures

There are two main methods used for buckling analysis: eigenvalue techniques and non-linear analysis. It is important that these are both undertaken within a code compliant framework.

1. Eigen-solution analysis:
    - This is a simple but very powerful method and strongly recommended. The analysis is generally quick. It uses matrix methods to develop the equivalent of Euler buckling for a complex structure. At its most basic, it can be used to show buckling mode shapes – the *eigenmodes*- and the critical load factors, λ_{cr} – the *eigenvalues* – for a complex structure in a way which is similar to that for any simple column.
    - The buckling shapes reveal flexibility – the weak areas – and this knowledge can be used to identify how the structural design should be corrected or be optimized.
    - Effective lengths for complicated structures cannot always be read off directly, but useful comparisons for early design can be made.
    - The eigenvalues give the ratio of the load to the buckling load and thus define how close the structure is to buckling and thus how vulnerable it is to effects of additional moments and imperfection. They reveal whether account of non-linearities is required and whether they are likely to be significant.
    - The analysis gives no information on the place of the structure on the force-displacement curve.
    - Advanced techniques can be used to generate UF effects using only eigen-value solutions without a full non-linear analysis. These techniques replicate the types of safety margins used to reduce stresses from those at pure Euler buckling to Code buckling permissible loads for simple columns and apply these types of reductions to the complex structures.
    - The eigenmodes, i.e. mode shapes, are required to determine the geometries required to model initial imperfections in a full non-linear analysis.
2. Full non-linear analysis:
    - Can explicitly account for all of the effects of non-linearity defined earlier in Table 7.2, such as large deflection effects.
    - Can be used to interrogate the sensitivity of the structure by analysing the structure at different factors of the ULS load, in the region the ULS load, and plotting selected load/deflection plots.
    - Is difficult to use to identify 'buckling' behaviour or to compare against 'pure buckling' behaviour because mode shapes are not produced.

- Is generally time-intensive.
- Is required to observe snap-through modes.
- Generally requires imperfections to be modeled into the initial geometry based on the mode shapes.
- Accounts explicitly for the P-δ and P-Δ effects, and thus code compliant design is relatively straightforward with elements being designed for buckling using effective lengths between nodes.

Case study: Queen Elizabeth II Great Court, The British Museum, London, UK

This roof was a landmark in the transition from defined geometries to freeform shapes fabricated by automated industrial methods (Figures 7.7 and 7.8). The geometry is 'dome shaped' but is particularly optimized to fit the constraints of the existing museum buildings from which it derives its support. The annular geometry fits a rectangular courtyard with a circular reading room that is not exactly in the centre.

Figure 7.7   Internal view of the Courtyard of the British Museum (copyright Mandy Reynolds/Buro Happold)

Figure 7.8   Aerial view, at night, of courtyard roof (copyright Mandy Reynolds/Buro Happold)

The surface geometry was generated mathematically using complex toroids to fit the boundary. The surface was discretized to triangles which were distributed over the surface using an algorithm that optimized the lengths of the sides of the triangles. The boundary is a ring beam that rests on the existing stone wall of the surrounding building but can only deliver vertical loads to it. Some arching/shell action occurs at the corners where the ring beam generated tie forces. In the centres of the sides there is little arching action and most of the load is taken in bending of the members.

The individual rectangular members were fabricated so the depth could be varied in response to the local bending moments. Node and member geometries and lengths are non-regular, but a careful design of the nodes allowed accommodation of the varying geometry without varying design principles. The star shaped nodes were cut from plate up to 200 mm thick and the ends of the members were fabricated to fit into the recesses of the nodes. The roof was assembled by welding up sections that could be lifted on to the scaffolding and then completed by welding on site.

7.5 Steel for stadiums

The past two decades have seen an explosion in the construction of stadiums, mostly for football. The drivers behind this are:

- the regulations for all-seat stadiums with covered seating
- the clubs' business plans for a larger number of seats, especially premium seats and facilities to maximize income
- the quadrennial World Cup that results in a requirement for new stadiums in the host country.

Steel is the choice for stadium roof designs and a number of innovative solutions have been developed (see McCormick, 2010[17]). The case study presents a recent elegant design for Emirates Stadium (see Liddell, 2006[18]). It shows a refined effective choice of a bending system with optimum choices of element depths and framing. Previous sections of this chapter have shown form-derived solutions of compression and tension structures; but bending systems formed of refined fabricated truss roofs, generally one-way spanning, continue to have a major place in architectural construction. An historical classic is the Sainsbury Centre at the University of East Anglia in Norwich. More modern examples can adopt complex geometries enabled by CAD-Cam processes such as the London 2012 Olympic Games Aquatics Centre.

Case study: Steel roof for Emirates Stadium for Arsenal FC, London, UK

The club wanted a new stadium to be close to the old one to maintain their supporters and the only site available was squeezed between two railway lines.

The designers responded to these requirements by shoe-horning the seating bowl into the available space (Figures 7.9 and 7.10). They also had to take account of the construction sequence in the design of the structure, since some site areas would not be freed up until later. The roof in particular had tight height restrictions and had to be constructed in halves. The solution was to use three chord steel trusses that would be self-stable during construction. The primary truss spanned 210 m and was temporarily propped in the centre. The secondary trusses spanning between the primaries and the tertiary trusses spanning to the sides were also three chord trusses so required no intermediate propping. The large steel trusses were brought to site as individual elements and assembled.

Figure 7.9   Elevation of Emirates Stadium, Arsenal FC (courtesy of Buro Happold)

Figure 7.10   Aerial view of Emirates Stadium (copyright Sealand Aerial Photography, used with permission)

7.6 Information and process in the current digital age – the development of technology

7.6.1 Introduction

This section describes some of the modern ways in which the structural engineer and the steel fabricator work with analysis and design information within the digital age. The engineer justifies his design by analysis of the steel structure and delivers information (generally drawings, but more recently uniquely digital information) to the fabricator. The fabricator receives data from the structural engineer and converts it to manageable information which the automated machinery and craftsman in the fabrication shop use to deliver the built steel product.

For unusual special structures, sophistication is required to control and manage data, and uses of the most advanced digital techniques can yield great efficiencies in the whole process. A goal for the industry is that the architect, engineer and fabricator work on the same type of digital information, hopefully of the same format and hopefully within the same computer platform generating as little paper output as required.

Recent trends in Building Information Modelling (BIM) are pushing the industry towards integration of architectural, engineering and fabrication tools in a way that a single 3D model caters to the wider needs of the industry. This is enabling not only a new generation of tools such as Autodesk Revit, Digital Projects, Bentley BIM and Tekla Structures but also a whole new generation of thinking which focuses on 3D, parametric modelling, and interoperability. A number of advancements are also being made in modelling-analysis integration and modelling-fabrication integration.

This section describes some of the key issues associated with:

- engineer's tools and methodology of geometry generation (7.6.2)
- contractor's methods of steel fabrication and erection (7.6.3).

7.6.2 Engineer's tools: geometry generation, analysis and data transfer

Initiation of geometry

The modern structural geometries of nodes and elements and surfaces are initiated, either drawn with a CAD package on screen or derived by programming techniques or spreadsheets. The most recent CAD developments include parametric, 4-D and collaborative modelling, used in packages such as Digital Projects and Bentley BIM, where forms can be generated or manipulated using mathematical expressions with a history of operations. This has provided a powerful, immediate, link between a graphic control and a mathematical control of geometry. Historically, it might have been architects who tended to draw form only, and engineers to manipulate form with mathematics, but the dual nature of modern information manipulation means both disciplines must employ people skilled in both drawing and programming, and must collaborate closely.

Forms at this stage contain only geometry, potentially some mathematical intelligence but no embodied properties such as material sizing.

Analysis and design development of drawings

Computer packages used by the structural engineer include some which uniquely analyse, but many have now developed element code checkers to allow faster design iterations of the structure. The code checker is a post-process to the analysis and for steel structures requires care and understanding of effective length factors for bending and compression elements to assure effective design.

Most analysis packages have good render capabilities and others embody the production of sophisticated 2D drawing. Such complete design and drawing packages have been around a few years particularly for composite steel floor structures involving regular geometries and have progressed to working with more complex forms. The most complete platforms are Building Information Modelling software which can include all components and systems of the building (which enables integration of services and avoidance of clashes) and also have structural analytical capability. These can help the engineer to synthesize the data within one model so that the calculations, analysis, geometric data, section sizes and connection forces are contained within a compact efficient set of limited information.

As has been described earlier, analysis for tension and compression structures manipulates geometry initiated by, for example, CAD to derive new form-found structures. After the completion of analysis and design, the engineer has a set of element sizes, connection forces and geometry and it is this information that is delivered to the fabricator.

Delivery of data to the fabricator

Ideally, at the completion of the steel design, the information delivered by the engineer to the fabricator is in a form that is most helpful to the fabricator. This may not be always the most obvious and usual deliverable from the engineers point of view as the requirements of the design development process mean that much data is developed to suit communications with other parties such as clients, architects and other consultants.

As such, reality may mean that bespoke manipulation of data by either engineer or contractor is required. The case study below, on Sidra Trees, describes how the engineer delivered bespoke digital geometric and material information direct to the fabricator eliminating the need for conventional drawings which in such applications would be unnecessary and inefficient.

Special steel structures often require special consideration of erection. The contractor will be required to develop a detailed erection methodology involving analysis. Ideally the engineer and contractor can share common analytical tools so that both parties can understand any influences of the temporary stages of design on the final stages quickly and effectively. In some cases, contractual requirements may lead to reluctance to parties sharing information, but some of the most collaborative processes and projects may involve sharing of analytical models.

7.6.3 Contractor's analysis and modelling tools and methods of steel fabrication and erection

The fabricator will synthesize data so that the information is delivered to the fabrication shop in an organized, programmed, efficient manner. Modern steelwork design and detailing software is developed so that every component part can be extracted with its cutting and fabrication information. These data can be linked directly to automated cutting and drilling machines that can process linear steel members without the need for printed drawings. Plates for fabricated connecting brackets are also cut and drilled by automated machines. Automated machines are also available for precise profile cutting of the ends of tubular members so that efficient joints in tubular steel construction can be made. Steelwork detailing software can also include the geometry of the fixings for the secondary steelwork and cladding which can then be made with the primary steel.

For complex special structures the prepared members are often made up into assemblages that can be transported and bolted together on site. The challenge is to set out the mating parts during fabrication so that they will fit together on site. This is a very difficult task where connected members form curved surfaces. In former times the adjoining assemblages would be match fabricated so they fitted together perfectly on site but modern techniques mean that match fitting can occur less.

Today steel fabricators use the CAD design software to extract the planned assemblage and rotate it to the attitude in which it will be fabricated. The software then gives the setting out dimensions and angles of the mating surfaces relative to datum lines on the shop floor. Laser surveying equipment is then used to ensure the joining surfaces are exactly jigged in the right places. The component members of each assemblage can also be identified, prepared and the whole welded together and shipped to site on a just-in-time basis ready for installation.

The benefits of the modern integrated parametric modelling environment directly extend to the fabrication stage. This enables development of details while keeping consistency with the architectural intent and engineering constraints, automatic clash detection, 4-D construction planning, etc.

Modern surveying techniques are required to set out complex structures on site and to check them after complete assembly and erection. Sometimes for special structures the contractor will be required to measure and verify forces on site in addition to geometry.

Case study: Sidra Trees, Qatar Convention Centre, Doha

An architectural showpiece of Qatar's Education City, the building's conceptual design incorporates a huge organic tree-like structure in the main façade, symbolizing the Sidra Tree (Figures 7.11–7.13).

Design and engineering of this 250 m long and 20 m high signature entrance structure was a challenge on several fronts: geometric rationalization and engineering as well as fabrication. The challenges were associated with free-form shape, size (with up to 7 m trunk diameter), and major forces running in the structure, roof and the base of the tree. The tree is wholly made of steel, and supports a complex steel plate girder and tie-bar system that supports the concrete roof. Within the architectural constraints posed by this impressive and iconic statement, the designers provided a truly optimum solution incorporating an efficient structural core that follows the

Figure 7.11   Detail showing steel core and steel surface of element of Sidra Tree (courtesy of Victor Buyck Steel Construction Ltd., used with permission)

Figure 7.12   Construction of Sidra Trees (courtesy of Victor Buyck Steel Construction Ltd., used with permission)

Figure 7.13   Photo of Sidra Tree project near completion (courtesy of Victor Buyck Steel Construction Ltd., used with permission)

centreline of the tree and an intelligently panelized skin that is supported by it. The goal was resolving the geometry and the structure inside to make sure it will keep its organic form while being structurally efficient and buildable. The engineers developed parametric models for the tree geometry that allowed surface smoothing, form optimization including minimizing expensive double-curved panels, detailed finite element analysis as well as digital fabrication eliminating the need for drawings for thousands of panels and framing (see Liddell *et al.*, 2010[19]).

Case study: Steel roof and cladding supports for Aviva Stadium, Lansdowne Road, Dublin, Ireland

The collaborative process of detailed design development and production of construction documentation between architects and structural engineers was enhanced through the use of a shared parametric model used for development and definition of the steel roof structure, steel cladding supports, cladding envelope and floor extents (Figures 7.14–7.16). The stadium envelope design was highly constrained by

Figure 7.14 Internal view of Aviva Stadium (Donal Murphy Photography, courtesy of Populous™ and Scott Tallon Walker Architects)

Figure 7.15 Aerial view of Aviva Stadium (Peter Barrow Photography, courtesy of LRSDC)

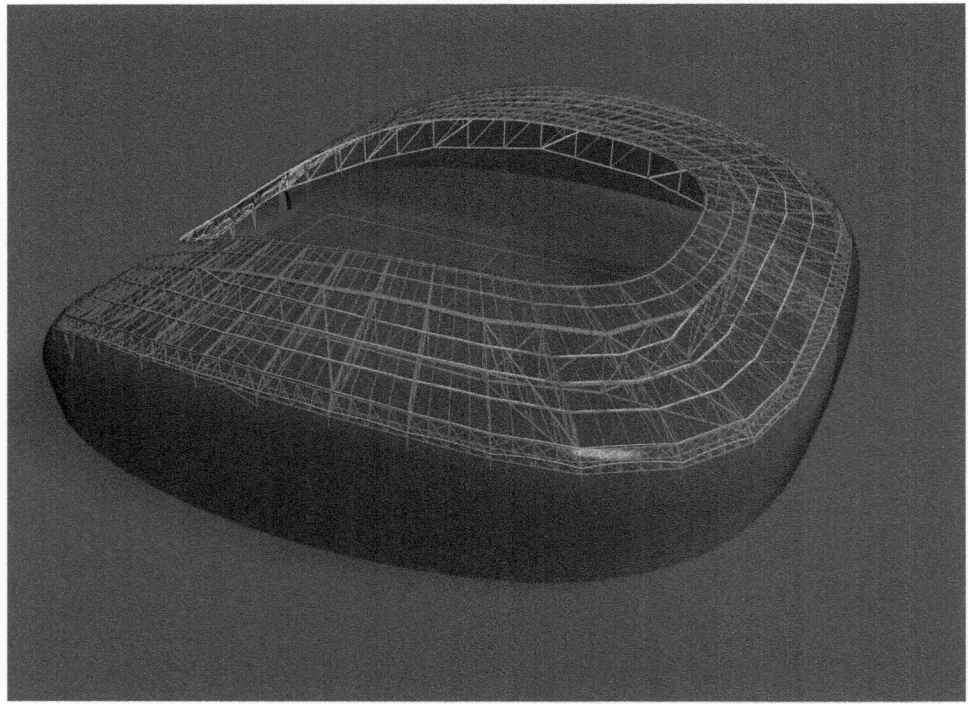

Figure 7.16 Model of steel roof structure of Aviva Stadium (courtesy of Buro Happold)

site boundaries, rights to light, air traffic regulations and planning restrictions (see McCormick *et al.*, 2011[20]). Underlying geometric considerations for the early design were rule-driven. As the project entered the design development phase, the team anticipated changes and adjustments to the constraints and therefore used an integrated technology that could easily accommodate and communicate variation between the architects and engineers. Any changes in external surfaces defined by the architects were issued to the engineers and then the structural geometry and corresponding structural steelwork analysis would automatically update. A parametric approach meant that the geometrical configuration of the stadium could be numerically controlled, removing the need to manually edit and rebuild geometry when a change occurred.

Case study project credits

Eden Project Domes: Engineer: SKM Anthony Hunt, Architect: Grimshaw, Fabricator: Mero; Millennium Dome: E: Buro Happold, A: Richard Rogers Partnership, F: Watson Steel Ltd.; London 2012 Olympic Stadium: E: Buro Happold, A: Populous™, F: Watson Steel Ltd.; British Museum: E: Buro Happold, A: Foster and Partners, F: Wagner Buro; Emirates: E: Buro Happold, A: Populous™, F: Watson Steel Ltd.; Sidra Trees: A: Isozka, E: Buro Happold, F: Victor Buyck Steel Construction Ltd.; Aviva Stadium: E: Buro Happold, A: Populous,™ and Scott Tallon Walker Architects, F: SIAC-Butler and Cimolai.

References to Chapter 7

1. Jones A.C. and Jones M. (2001) Eden Project, Cornwall: design, development and construction, *The Structural Engineer*, Vol. 79, Issue 20, 16th October.
2. Addis W. (2001) *Creativity and Innovation: The Structural Engineer's Contribution to Design*. Oxford, Architectural Press.
3. Bobrowski J. (1985) Calgary's Olympic Saddledome. In: *Space structures*. Elsevier Applied Science Publishers, Barking, England, 1, No. 1, 13–26.
4. Wernick J. (2000) Full circle. The story of the London Eye, *Architecture Today* 108, May.
5. Liddell W.I. and Miller P. (1999) The design and construction of the Millennium Dome, *The Structural Engineer*, Vol 77, Issue 7, 6th April.
6. British Standards Institution (2006) *BS EN 1993-1-11:2006 Eurocode 3 – Design of steel structures – Part 1-11, Design of structures with tension components.* London, BSI.
7. International Federation for Structural Concrete (FIB) *Acceptance of stay cable systems using pre-stressing steels* Bulletin 30, see http://ukfib.concrete.org.uk/ and http://ukfib.concrete.org.uk/bulletins.asp

8. Post Tensioning Institute (2008) *Recommendations for stay cable design, testing and installation*, 5th edn. Phoenix, AZ, PTI.http://post-tensioning.org/product/x_zETPk3lGcmMY2lkPT/Bridges

9. de Sá Caetano E. (2007) *Cable vibrations in cable-stayed bridges*, IABSE, see http://www.iabse.ethz.ch/publications/seddocuments/SED9.php.

10. Raoof M. (1991) Axial fatigue life prediction of structural cables from first principles, *Proceedings of the Institution of Civil Engineers*, Part 2, March, 19–38.

11. Crockford I., Breton M., Westbury P., Johnson P., McCormick F., McLaughlin T. (2011) *The London 2012 Olympic Stadium. Part 1: Concept and Philosophy*. International Association of Steel and Spatial Structures, London. CD-ROM (in publication).

12. Davies W.H., Gray B.A. and West F.E.S. (1973) Fort Regent leisure centre, *The Structural Engineer*, Vol. 51, Issue 11, 1st November.

13. Brown S. (2005) Millennium and Beyond, *The Structural Engineer*, Vol. 83, Issue 20, 18th October.

14. Williams C.J.K. (2000) The definition of curved geometry for widespan enclosures. In: *Widespan roof structures*, Barnes M. and Dickson M. (eds.): 41–49. London, Thomas Telford.

15. Pearce D. and Penton A. (2002) The Imperial War Museum, *Arup Journal*, Issue 2.

16. Cook M., Palmer A. and Sischka J. (2006) SAGE Music Centre, Gateshead – Design and construction of the roof structure, *The Structural Engineer*, Vol. 84, Issue 10, 16th May.

17. McCormick F. (2010) Engineering stadia roof forms. ABSE Conference, Guimaraes, Portugal. CD-ROM.

18. Liddell I. (2006) Pitch perfect, The construction of the new Arsenal Emirates stadium, *Ingenia*, Issue 28, September.

19. Liddell I., McCormick F., Sharma S. (2010) State of the art optimized computer techniques in design and fabrication of unique steel structures – Aviva Stadium, Sidra Trees, The Louvre Museum. In International Association of Steel and Space Structures, Shanghai. CD-ROM.

20. McCormick F., Werran G. (2011) Aviva Stadium, Ireland, *Structural Engineering International*, Vol. 21, No. 1.

Further reading for Chapter 7

Abel R., (2004) *Architecture, Technology and Process*. Oxford, Architectural Press.
Addis W. (1998) *Happold: The Confidence to Build*. London, Taylor and Francis.
Bechtold M., Mordue B., Rentmeister F.-E. (2008) Special bridges – special tension members. Locked coil cables for footbridges. In: *Footbridge 2008, 3rd International Conference*.
Barnes M., Dickson M, (2000) *Widespan Roof Structures*. London, Thomas Telford.

Blanc A., McEvoy E., Plank R. (1993) *Architecture and Construction in Steel.* London, E. and FN. Spon, SCI.

Chilton J. (2000) *Space Grid Structures*, 1st edn. Oxford, Architectural Press.

Chaplin F., Calderbank G., Howes J. London, (1984) *The Technology of Suspended Cable Net Structures*. Longman.

Eggen A., Sandaker B. (1995) *Steel, Structure and Architecture: A Survey of the Material and its Applications*. New York, Watson-Guptill Publications.

Makowski Z.S., (1965) *Steel Space Structures*. London, Michael Joseph.

Manning M. and Dallard P. (1998) Lattice shells, recent experiences, *The Structural Engineer* Vol. 76, Issue 6, 17th March.

Ramaswamy G.S., Eekhout M., Suresh G.R. (2002) *Analysis, Design and Construction of Steel Space Frames*. London, Thomas Telford.

Robbin T. (1996) *Engineering a New Architecture*. New Haven, Yale University Press.

Rice P. (1998) *An Engineer Imagines*. London, Artemis.

Sebestyen G. (2003) *New Architecture and Technology*. Oxford, Architectural Press.

Sharma S., Fisher A. (2010) A SMART integrated optimisation. In: *The New Mathematics of Architecture*, (ed. Jane Burry). London, Thames and Hudson.

Tekin S. (1997) Modern geometric concepts in architectural representation, IAED 501 *Graduate Studio – Commentary Bibliography Series*, p2, December.

Williams, C.J.K. (2001) The analytic and numerical definition of the geometry of the British Museum Great Court Roof, *Mathematics and Design 2001*, Burry, M., Datta, S., Dawson, A., and Rollo, A.J. (eds.), 434–440. Deakin University, Geelong, Victoria 3217, Australia.

American Society of Civil Engineers (1996) *Structural applications of steel cables for buildings, 19–96*. Reston, Virginia, ASCE.

British Standards Institution (2005) *BS EN 1993-1-9, Eurocode 3: Design of steel structures. Part 1-9: Fatigue*. London, BSI.

British Standards Institution (1997) *Eurocode 3, 1993-2:1997. Design of steel structures – Part 2: Steel bridges Annex A – High strength cables*. London, BSI.

Chapter 8
Light steel structures and modular construction

MARTIN HEYWOOD, MARK LAWSON and ANDREW WAY

8.1 Introduction

The use of light gauge cold-formed steel in building construction is becoming increasingly common in the UK across a range of building types and applications.[1] Although similar to hot-rolled steel in terms of many of its physical properties, the thin gauge of most of the cold-formed steel used in construction means that its structural behaviour is significantly different, with the failure mode often dominated by local buckling. The arrangement of the steel members within the building structure is also drastically different from the traditional hot-rolled steel frame, with typical load-bearing walls comprising light steel studs at 400 mm or 600 mm centres, in place of hot-rolled columns at a spacing of 4 m to 6 m. The forms of construction, together with the light nature of the structural members, have consequences for other aspects of the building design such as acoustics, thermal performance and the dynamic response of floors.

This chapter considers the use of light gauge steel in a variety of construction applications. It describes the most common forms of construction for light steel frames, lightweight floors and secondary steelwork and illustrates the use of light gauge steel in these applications. Structural and non-structural design issues are discussed in detail and several practical solutions to these issues are presented. The structural analysis and design of light steel members is considered in greater detail in Chapter 24.

8.1.1 Light gauge steel

The term 'light gauge' steel commonly refers to galvanised cold-formed steel with a gauge ranging from 0.9 mm to 3.2 mm. Gauges from 0.9 mm to 1.6 mm are typical

Steel Designers' Manual, Seventh Edition. Edited by Buick Davison and Graham W. Owens.
© 2012 Steel Construction Institute. Published 2012 by Blackwell Publishing Ltd.

for light steel framing applications, including wall studs, floor joists and roof trusses. Purlins and cladding rails typically range from 1.4 mm to 3.2 mm depending on the section depth (it is necessary to increase gauge as the depth increases to limit the adverse impact of local buckling on the section capacity).

Light gauge steel members are usually cold-formed from hot-dip galvanised strip steel. The steel is supplied to the section manufacturers as pre-galvanised coil, so there is no need to apply any protective coating to the members post manufacture. Specifiers may choose from a range of steel grades and coatings, but in all cases the specified material should conform to BS EN 10346: 2009.[2] This Standard, which gives the technical delivery conditions for continuously hot-dip coated steel flat products, includes requirements for chemical composition, mechanical properties, coatings and surface finish. It supersedes BS EN 10326: 2004, BS EN 10327: 2004, BS EN 10336: 2007 and BS EN 10292: 2007.

For light steel framing applications, the commonly used grades of steel are S350, S390 and S450, while purlin and cladding rail products tend to use S390 and S450. The current trend is for higher steel strengths, with S450 becoming the norm, especially for secondary steelwork. The most common coatings are zinc, zinc–iron, zinc–aluminium, aluminium–zinc and aluminium–silicon. The standard zinc coating for construction products is $275 \, \text{g/m}^2$ (referred to as Z275), which corresponds to a coating thickness of 0.02 mm on each surface. This coating thickness should not be included when calculating section properties for structural purposes. Hence, the steel thickness used for all calculations should be 0.04 mm less than the specified nominal gauge.

8.1.2 Cold-formed sections for construction applications

Unlike hot-rolled steel members, cold-formed sections are not made to a standard range of shapes and sizes. Each manufacturer produces its own range of sections, although for framing applications the industry seems to have settled on a fairly narrow selection of standard sections for studs and joists. The manufacturers of purlins and cladding rails tend to be more imaginative in their section shapes in a drive to maximise the efficiency of the cross-section through the use of lips and stiffeners.

As the name implies, cold-formed products are formed by passing the cold strip steel through a set of rollers to obtain the desired shape (see Figure 8.1).

The cold-forming process is continuous in nature with the steel fed into the rolling machine directly off the coil. Since the standard coils supplied by the steel producers are generally too wide for wall studs, joists and purlins (the coil is typically 1200 mm to 1250 mm wide), it has to be slit to the correct width before starting through the rolling line. The larger roll-forming firms slit the coil themselves, while those without the necessary equipment have to buy it pre-slit from the steel coil supplier. Where manufacturers slit the coils themselves, great care is taken to fine tune the dimensions of the finished sections to ensure that the full width of the parent coil is utilised with no steel going to waste.

Figure 8.1   Roll forms used for cold-formed sections

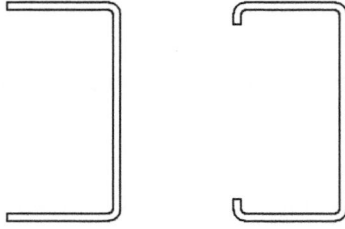

Figure 8.2   Typical C-section shapes used for light steel walls

For light steel framing wall elements, by far the most common section shape is the C-section (either plain or lipped), as shown in Figure 8.2. Depths commonly range from 70 mm to 120 mm. Similar sections are also used for light steel roof trusses.

C-sections may also be used for floor joists, but tend to be deeper due to the higher bending moments encountered, typically 120 mm to 250 mm. To overcome local buckling in the web of the section without increasing the gauge of the steel, it is common practice to roll stiffeners into the web to form a sigma section. Typical section shapes for floor joists are shown in Figure 8.3.

Purlins are typically made from zed or sigma sections as shown in Figure 8.4. Due to the competitive nature of the secondary steelwork market and the use of published load-span data derived from testing, the trend in recent years has been

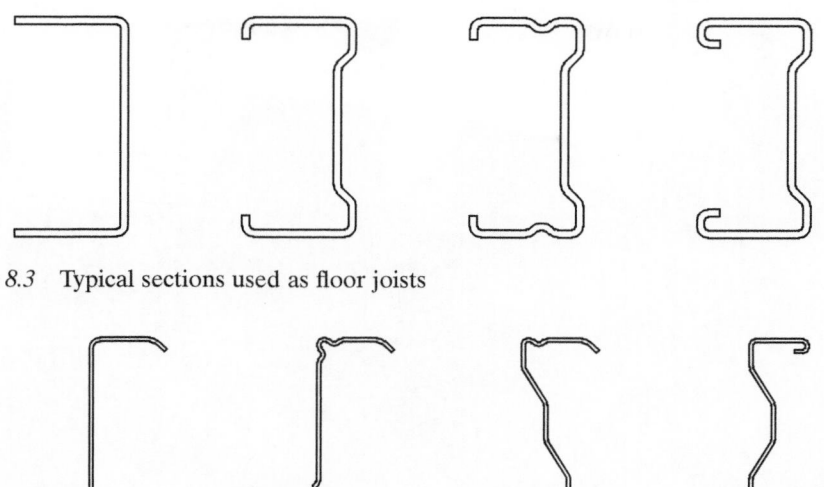

Figure 8.3   Typical sections used as floor joists

Figure 8.4   Typical sections used as purlins

to minimise the weight of steel by the use of complex stiffener arrangements in the flanges and the web. By reducing the effective length for flange and web local buckling, this approach may result in the use of thinner gauges (down to 1.2 mm) than would otherwise be possible. The same sections may also be used for cladding rails, although in this case lipped C-sections are also a viable option.

8.1.3 Forms of construction

Light gauge steel is commonly used in the following building elements:

- load-bearing walls
- non-load-bearing external walls
- internal partitions
- floors
- roofs.

Light gauge steel lends itself to off-site manufacture and the elements listed above may all be supplied with varying degrees of prefabrication as follows:

- individual members (stick build)
- 2-dimensional planar assemblies
- volumetric or modular.

The degree of offsite content will generally depend on the preferences of the steel frame supplier and main contractor (familiarity with the technology is an important

Figure 8.5   Bathroom pods within a light steel frame

issue), the building geometry, the importance of speed of construction and constraints on space for storing or handling materials. Offsite manufactured building elements are generally preferable to stick-build on grounds of speed of installation and quality of build. However, where the geometry is complex leading to problems with installation tolerances or where access for a crane is difficult, stick-build might be the better option. It is not uncommon for buildings to contain a mixture of offsite manufactured and site-assembled elements. A common example of this is the use of bathroom pods within a stick built or planar light steel frame as illustrated in Figure 8.5.

8.2 Building applications

The use of light gauge steel in building structures can generally be divided into two categories:

- light steel framing
- secondary steelwork.

Light steel framing is most commonly associated with residential buildings, especially low-rise apartment blocks.[3] In recent years, there have also been attempts to introduce light steel frames into the single-occupancy housing market, although it has struggled to compete with traditional masonry construction and timber framing. By contrast, light gauge steel has quickly established itself as the material of choice for multi-occupancy social housing and student accommodation due to the need to exceed two or three storeys in height. Light steel framing is also popular in mixed-use developments, in which residential apartments are built above retail outlets. In some cases, a hot-rolled steel or concrete frame is used for the lower storeys, with the light steel residential units being built off a podium. Examples of light steel framing in residential applications are shown in Figure 8.6, Figure 8.7 and Figure 8.8.

Light steel framing is also used in the healthcare and education sectors, often as infill walls within a hot-rolled steel or concrete frame. An example of a secondary school with light steel infill walls is shown in Figure 8.9. In this instance, the infill walls were supplied to site with sheathing board and windows already installed.

Hotels, student accommodation, prisons and army barracks have all seen large-scale use of light steel framing and modular construction in recent years, proving the versatility of the material across a range of building applications.

Figure 8.6   Light steel framed semi-detached house (courtesy of Terrapin Ltd)

Figure 8.7   Light steel framed apartment block (photograph courtesy of Metsec)

Figure 8.8   Social housing using modular construction

Figure 8.9   Light steel infill panels in a secondary school

The other major use of light gauge steel in UK construction is as secondary steelwork on steel portal framed sheds and similar structures. In this case, the light gauge steel fulfils a dual function of supporting the roof and wall cladding (spanning between the primary structural frame members) and providing lateral restraint to the primary steelwork. This form of construction is commonly used for industrial buildings, warehouses, retail outlets, leisure facilities, sports halls and in agricultural buildings. An industrial building using light gauge steel purlins is shown in Figure 8.10. Further design guidance on the use of light gauge secondary steelwork in single-storey buildings is given in Chapter 4.

8.3 Benefits of light steel construction

Light steel construction offers many benefits over traditional forms of construction for the client, contractor and local community. Some of the benefits are associated with the light steel material itself, while others are due to the use of offsite manufacturing (prefabrication of building elements) that is possible with light steel framing. Some of the key benefits are listed below.

Figure 8.10   Light steel purlins on an industrial building

Benefits of light gauge steel:

- reduced dead load compared to traditional construction (this is useful where ground conditions are poor)
- ease of handling on site (for stick-build)
- high strength and stiffness allowing thinner walls and shallower floors
- good long-term durability
- reliable and consistent material quality
- tight rolling and fabrication tolerances, especially where members are cut to length off-site
- steel is readily recyclable
- not susceptible to shrinkage
- not susceptible to fungal or insect attack.

Benefits of off-site manufacturing:

- speed of construction
  - earlier handover to client
  - reduced disruption to local residents
- rapid completion of a dry envelope allowing earlier access for following trades
- less noise and dust
- fewer site vehicles

- fewer commercial vehicle movements (deliveries, plant and waste)
- less waste due to more efficient use of materials and higher rates of recycling
- reduced site management cost
- higher productivity (in terms of site labour)
- reduced storage charges
- suitable for confined sites with limited working and storage space.

As part of a UK Government-sponsored project, the SCI investigated the benefits of Modern Methods of Construction (MMC) in urban locations. The findings of this report are summarised in *Benefits of off-site steel construction in urban locations*,[4] with further details available from SCI report RT1098 *Urban impact case studies – Final project report*.[5] The study considered several construction projects using varying degrees of prefabrication and quantified some of the benefits listed above through the use of case studies. In one example, a secondary school construction programme was reduced from 76 to 54 weeks through the use of offsite manufactured infill wall panels. This represents a 29% time saving. In a second case, which featured a residential building, it was estimated that commercial vehicle movements were reduced by 40% by the use of volumetric modular construction.

The benefits of light steel construction were also highlighted by the recent SmartLIFE initiative,[6] as part of which a series of housing projects near Cambridge were monitored by the Building Research Establishment (BRE). The projects were located on three similar 'greenfield' sites and consisted of 106 two-storey houses. The houses were a mixture of 2-bedroom and 3-bedroom configurations in rectangular and L-shapes. The following forms of construction were used:

- traditional brick-block
- timber framed
- insulated concrete formwork
- light steel framed.

For the 3-bedroom rectangular houses, the construction cost of the light steel system was £51.9k (€66k), which was 3% less than traditional construction. Importantly, the cost of the light steel system was also 18% less than timber framing and 12% less than insulated concrete formwork. For the 3-bedroom L-shaped houses, the construction cost of the light steel system was £53.4k (€68k), which was 8% more than traditional construction on one site and 11% less than traditional construction on the second site. Again, the cost of the light steel system was 10% less than timber framing and 19% less than insulated concrete formwork. For the 2-bedroom houses, the construction cost of the light steel system was £49.8k (€63k), which was only 3% more than the traditional brick-block construction on the same site or 10% less than traditional construction on a second site. Again, the cost of the light steel framing system was 12% less than timber framing and 15% less than insulated concrete.

The total number of man-hours required to build one house when averaged across a number of houses on one site is a good measure of productivity. The light steel framing system took 846 man-hours per house to build (including foundations but excluding site management) in comparison to 1,002 man-hours for timber framing, an average of 1,070 man-hours for traditional construction, and 1,338 man-hours for insulated concrete formwork. The super structure component of the total installation effort in man-hours was only 25% for light steel framing and 60% for traditional construction.

8.4 Light steel building elements

8.4.1 Load-bearing walls

Load-bearing walls consist of a framework of vertical studs, horizontal rails and diagonal bracing members and are designed to carry gravity and wind loading. An example is shown in Figure 8.11. The vertical studs are usually spaced at either 400 mm or 600 mm centres and, in effect, provide continuous support to the walls and floors (or roof) above. They are typically used up to 6 storeys and may be

Figure 8.11   Light steel load-bearing wall (photograph courtesy of Metek)

stick-built, supplied as prefabricated wall panels or as part of a volumetric system. The studs are designed to carry gravity loads in compression and out-of-plane wind loading in bending. In-plane wind loads may either be taken by integral bracing, flat strap bracing or by the racking resistance of any plasterboard or sheathing board attached to the studs. Openings for windows and doors may be provided, but additional steelwork will normally be required either side of the opening (unless it fits within the 600 mm stud spacing).

8.4.2 Non-load-bearing external walls

Non-load-bearing external walls may be used to form the building envelope in steel or concrete framed buildings. They are designed as infill walls (to fill the space bounded by adjacent columns and floors) and may be stick-built or delivered to site as prefabricated panels. Where the latter option is chosen, the panels may either be bare steel, have the sheathing board and insulation already installed but without windows, or be pre-glazed in the factory. The last of these options has a significant advantage in that the panels provide a dry internal environment the moment that they are installed, allowing immediate access by the following trades. A typical example is shown in Figure 8.12.

Figure 8.12 Light steel infill wall

In non-load-bearing walls, the vertical studs are designed purely as wind posts, spanning from floor to floor. To ensure that no gravity loading from the floors or walls above is taken by the studs a suitable deflection head detail must be provided, to allow the rail above to deflect without transferring load to the studs. Failure to do this will result in unwanted vertical loads being carried by the potentially inadequate studs. Due to the susceptibility of light steel sections to lateral-torsional buckling, many designers choose to eliminate this possible failure mode by using the attached board(s) to provide lateral restraint to the flanges of the studs.

8.4.3 Internal partitions

Light gauge steel is commonly used for non-load-bearing internal partition walls with dry lining. In this case, the main function of the steel is to support the plasterboard; it has no other structural function. Consequently, smaller, lighter sections may be used for internal partitions than for external walls. A lower strength of steel may also be used. The partition consists of top and bottom tracks and vertical studs, all made from plain C-sections (which may be slotted). A typical internal partition is shown in Figure 8.13. Care must be taken with the deflection head detail to avoid the transfer of any vertical loads into the partition from the floor above.

Figure 8.13 Light steel internal partition

8.4.4 Floors

Floors may either be constructed from light steel joists or from lattice trusses. For speed of installation, floor joists may be pre-assembled to form floor cassettes. This works well for regular floor plans, but care should be taken when the geometry of the building requires the cassettes to vary in size with location or where non-right angled corners are required.

As noted earlier, lipped C- and sigma sections are normally used for floor joists with depths up to 250 mm (depending on the span and spacing). Restraint may be provided to the top (compression) flange of the joist by the flooring board. C-sections are normally used for lattice trusses. The floors should be designed for the combined effect of dead and imposed load at the Ultimate Limit State. As with any floor beam, Serviceability Limit State checks must be carried out for deflections and walking-induced vibrations. A typical floor truss is shown in Figure 8.14.

8.4.5 Roofs

Light steel is sometimes used to form the roof structure for residential buildings in place of the traditional timber trusses (although the latter are also commonly used

Figure 8.14   Light steel floor truss

Figure 8.15   Light steel framed building with light steel roof (photograph courtesy of Metek)

on light steel framed buildings). With the growing scarcity of available land, planning limits on building heights and the drive to improve the energy efficiency of buildings, habitable roof space is becoming very common in residential buildings. Light steel construction is favoured for this situation due to the smaller sizes of the members (compared to timber) and the use of warm-frame construction. A building with a light steel roof is shown in Figure 8.15.

8.5 Modular construction

The terms 'modular' or 'volumetric' construction refer to a particular type of off-site manufactured building technology in which the floor, walls and ceiling are pre-assembled in a factory to form a prefabricated building unit.[7] These units, commonly referred to as 'modules', are transported to site and stacked together using a crane to form a complete building within a matter of weeks or even days. The modules are usually supplied fully finished internally, virtually eliminating the need for tradesmen on site (the modules still need to be plumbed/wired in and touching up is often required to the finishes, especially at the joints between modules). Unsurprisingly, speed of construction is a significant benefit of this type of building, even compared to other modern methods of construction. Other benefits include a significant reduction in defects (due to improved quality processes within a factory environment), fewer commercial vehicle movements and reduced waste.[8]

Since its introduction to the UK a decade ago, volumetric construction has been successfully used in a range of applications, most notably in the social housing sector and, in recent years, for student accommodation. It has also been used for hotels, prisons and Ministry of Defence accommodation and, to a lesser extent, for modular houses. In addition, volumetric modules are often used in conjunction with other forms of construction, e.g. with planar or stick-built light steel framing in cases where wholly volumetric construction would be impractical. A particular example of this is the use of non-load-bearing kitchen and bathroom modules, often referred to as 'pods', within the load-bearing frame of a residential building.

There are five types of light steel building module in common use in the UK:

- 4-sided modules
- partially open-sided modules
- corner-supported modules
- stair modules
- non-load-bearing modules.

8.5.1 Four-sided modules

As the name suggests, 4-sided modules consist of four walls, comprising vertical studs, horizontal rails and bracing. Floor and ceiling joists span parallel to the short edges of the module. In addition to the four 'external' walls, non-load-bearing partition walls may also be included to divide the enclosed space into rooms of the appropriate size. This type of construction is ideal for applications such as hotels, student accommodation and social housing/key worker accommodation, which are usually characterised by a large number of small rooms. However, it is less ideal for buildings requiring larger open spaces, since the room size is limited by the maximum width of the module, which is in turn limited by transportation constraints. The maximum height for this form of construction is 6–10 storeys, depending on wind loading. A typical 4-sided module is shown in Figure 8.16.

Four-sided modules are assembled from 2-dimensional floor, wall and ceiling panels. Each of the panels is manufactured from light steel sections as described earlier in this chapter. The two longitudinal walls are designed to carry the gravity loads not only from the ceiling joists, but also from all of the modules stacked vertically above. For this reason, the modules are designed to sit directly on top of one another, with the gravity loads being transferred directly through each longitudinal wall to the one below. It follows that the walls of the bottom storey have to be designed to support the entire weight of the building plus the imposed loads from the roof and each floor (excluding the ground floor). The maximum height of the building is, therefore, limited by the compression resistance of the wall studs in the bottom storey. Stability may be another limiting factor; this is discussed later in the chapter.

Another limitation of 4-sided modules is the need to keep the longitudinal load-bearing walls free from windows, doors and other openings. For this reason, the

Figure 8.16   Typical 4-sided module (courtesy of Terrapin Ltd)

doors and windows are always positioned in the ends of this type of module. This clearly has implications for the design of the building and, in particular, the layout of the modules. Where desired, the windows can be set in from the end of the module to form a balcony.

8.5.2 Partially open-sided modules

A variation on the 4-sided module is the partially open-sided module in which doors, windows and other openings are introduced into the longitudinal walls of the module. This provides greater flexibility to the architect in terms of the layout of the modules within the building and also allows larger rooms to be incorporated into the design, by occupying more than one module. A typical example of this is shown in Figure 8.17. The large opening at the left hand end of the module in this photograph marks the centre of an open-plan kitchen/dining/living area.

The provision of openings in the longitudinal sides of the modules is made possible by the introduction of corner and intermediate posts and by the use of a stiff continuous edge beam in the floor cassette. Small section square hollow sections (70 mm × 70 mm to 100 mm × 100 mm) are often used to form the additional intermediate posts with angle sections at the corners. Six to eight storeys are possible

Figure 8.17   Partially open-sided module used at Wyndham Road, London

with this form of construction. The width of the opening is limited by the bending resistance and stiffness of the edge beam in the floor cassette. Larger openings can be accommodated by the inclusion of additional edge beams.

Partially open-sided modules may be used for key worker accommodation, hotels and student residences. They are especially useful where internal corridors or open communal areas are required. This type of module also has applications in the renovation and extension of existing buildings. In this case, the modules are designed to support their own weight, but stability is provided by attachment to the existing building.

8.5.3 Corner-supported modules

Corner-supported modules resemble traditional hot-rolled steel construction in their use of deep longitudinal edge beams (often parallel flange channels) spanning between square hollow section (SHS) corner columns. This is in sharp contrast to the more familiar 4-sides modules with their load-bearing light steel walls. The edge beams are typically 300 to 450 mm deep giving spans (i.e. module lengths) of 5 to 8 m. The width of the modules is generally 3.0 to 3.6 m to allow transportation to site. The key advantage of corner-supported modules is that one or more of the sides can be completely open, allowing the creation of large open plan spaces when modules are placed side by side. This form of construction is ideal for schools and hospitals. The primary steel frame of a typical corner module is shown in Figure 8.18.

Figure 8.18   Primary steel frame for a corner-supported module (courtesy of Terrapin Ltd)

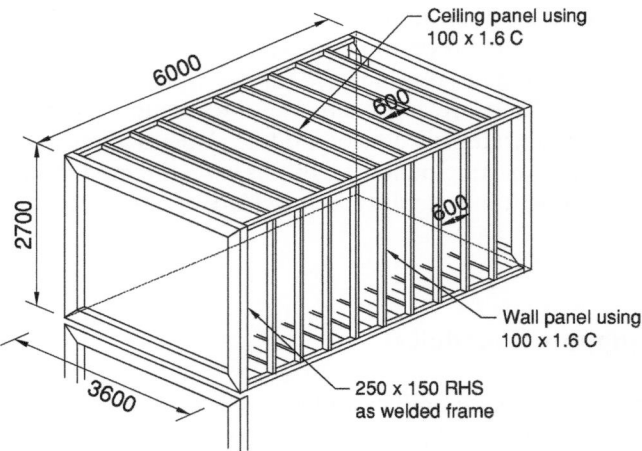

Isometric view of module and welded end frame

Figure 8.19   Open-ended module using a rigid frame

Where a large opening is required at one or both ends of the module only (i.e. with solid sides), a variant of the 4-sided module may be used in which the end walls are replaced by rigid frames as shown in Figure 8.19. The rigid frame removes the need for bracing in this wall, allowing the wall to be fully glazed. The frame also provides attachment points for a balcony.

8.5.4 Stair modules

Stair modules give access to the upper storeys of a modular building by providing stairs (usually two flights with a half landing in between) and a landing in a single prefabricated unit. They are generally used with conventional apartment modules (bedrooms, living room, etc.) to form a modular residential apartment block. The landings and half landings are supported by longitudinal walls with additional steelwork if necessary. This form of construction may generally be used in fully modular buildings up to four storeys in height.

8.5.5 Non-load-bearing modules

Non-load-bearing modules are used to provide an element of volumetric prefabrication within a non-modular building. They are typically used for bathrooms, kitchens and plant rooms, for which there are significant benefits to be gained from installing the services in a factory environment. Kitchens and bathrooms are generally delivered to site fully fitted, including plumbing, electrics, fitted furniture and finishes. As the name implies, non-load-bearing modules possess limited structural strength and must be supported by other structural members, e.g. structural steelwork or a floor slab. They are, however, designed to support their own self-weight and to resist the forces arising from being lifted into position.

8.6 Hybrid construction

There are several applications in which it may be advantageous to combine two or more forms of construction in order to provide the optimum structural solution for the building. For example, where the architectural design requires an irregular shaped building, including perhaps a non-rectangular floor plan for part of the building, it would be quite common to include an element of stick-build within an otherwise prefabricated structure. This compromise would allow advantage to be taken of the benefits of offsite construction (speed of construction, reduced waste etc.) while benefiting from the flexibility of stick-build where the geometry posed more of a challenge.

Two particular forms of hybrid construction are considered in this section:

- mixed modules and panels
- modules on a primary structural frame.

8.6.1 Mixed modules and panels

In this form of construction, modules are stacked to form a core, around which load-bearing walls and floor cassettes are arranged. The modules often form the

Figure 8.20   Mixed modular and panel construction at Lillie Road in Fulham, London

stairwell and provide a central zone for bringing the services into the building. Heavily serviced areas such as kitchens and bathrooms may also be accommodated within the modular part of the building, with the non-modular areas forming the bedrooms and living room. From a structural perspective, in addition to its load-bearing function, the modular core also provides stability to the whole building. This arrangement is typically limited to 4–6 storeys and is ideal for residential applications. A typical example of mixed modular and panel construction is shown in Figure 8.20.

8.6.2 Modules on a primary structural frame

Where the height or loading of a building exceeds the structural limitations of a purely light gauge steel solution, an attractive option is to use non-load-bearing modules within a structural frame. The primary frame is constructed as normal and the modules are installed one storey at a time as the construction proceeds. In this case, the load-bearing and stability functions are provided by the structural frame and the light steel walls of the modules act only as partitions. A variation on this theme is the use of an external steel structure or 'exo-skeleton' within which load-bearing modules are placed. In this case, the primary steelwork's main functions are to carry the façade loads and to provide stability to the building. Both forms of construction are used extensively for medium-rise residential developments including key-worker and student accommodation.

An alternative form of hybrid construction involves the construction of a podium, which acts as the foundation for the modular building above. The podium structure may be constructed from hot-rolled structural steel or reinforced concrete and will usually be one or two storeys high. The supporting columns are normally positioned at a spacing of two or three module widths, with the modules sitting on structural steel beams or a concrete slab. The use of conventional commercial building technologies for the podium structure allows for large open spaces, which are ideal for retail premises or use as a car park. Above podium level, the modular building, which would normally be 4–6 storeys in height, is generally used for residential apartments. A typical mixed-use application is shown in Figure 8.21.

Figure 8.21   Typical mixed-use application with podium structure supporting a residential building

8.7 Structural design issues

8.7.1 Member design

Depending on the form of construction and its function within the frame, each light steel member may need to be designed to withstand axial load, bending or a combination of both. In order to simplify the design and reduce the risk of mistakes during construction, it is common practice to rationalise the design of the frame to use as few section sizes as possible. For example, the load-bearing walls of a building will typically all consist of the same size of stud at the same spacing, even though the wind loading is likely to vary between the walls. Therefore, the member design process becomes one of checking the chosen section size for each type of member (wall stud, floor joist, etc.) against the most onerous loading conditions for that member.

The structural adequacy of the chosen light steel section should be verified according to the rules of the appropriate Codes of Practice. In the UK, and other countries of the European Union, this means designing to the structural Eurocodes. These documents, which replaced national Standards in March 2010, provide the methods, design rules and equations to enable the structural engineer to calculate the loading on the structure (e.g. from wind, snow, imposed loads and structure self-weight) and also the structural resistance of the chosen section. Specific rules for light gauge steel can be found in BS EN 1993-1-3,[9] which in the UK replaces BS 5950-5.

EN 1993-1-3 permits the structural adequacy of a member to be verified by calculation or by testing and presents methods for both approaches. Design by calculation is less expensive than design by testing and is ideal for bespoke structural designs. However, as the calculation rules in BS EN 1993-1-3 are somewhat conservative, this approach can sometimes result in heavier structural members than those verified by testing. Despite this, the majority of light steel frames used in building structures are designed by calculation. Member design by calculation to BS EN 1993-1-3 is described in detail in Chapter 24.

Due to the cost of performing physical tests in an accredited laboratory, the verification by testing route is only economically feasible for standardised systems sold in large quantities. Purlin and rail systems used to support the roof and wall cladding in industrial buildings fit into this category and are routinely 'designed' by testing. More precisely, the manufacturer of the system uses the appropriate test methods to derive the structural capabilities of a range of section sizes and publishes this information in the form of load/span tables. These tables are then used by the structural engineer to select a suitable section for the given load and span.

It is sometimes desirable to adopt a hybrid approach in which the member is designed by calculation with the support of test results. For example, in designing a wall stud against wind loading, a structural engineer might choose to assume that the compression flange is fully restrained by the attached sheathing or plasterboard. This simplifies the design process considerably and results in a lighter structure by removing the need to design for lateral-torsional buckling. In this case, the purpose of the testing would be to verify the assumption of full restraint. As noted above, the cost of testing would prohibit this approach from being followed on an individual design basis. However, manufacturers of complete wall systems (light steel frame and board) are in a position to test their systems and publish advice on issues such as lateral restraint for use by structural engineers.

8.7.2 Frame stability

Checking that each member within a steel frame is capable of withstanding the applied loads is an essential part of the structural design process, but it is not sufficient by itself. The stability of the building as a whole must also be checked. While the methods used to provide stability vary between hot-rolled and cold-formed steel, the principles remain the same:

- a suitable load path must be provided to safely transmit horizontal forces to the foundations
- the system used to provide stability must be sufficiently stiff to avoid excessive lateral deflections
- some account should be taken of the instability arising from frame imperfections
- where appropriate, account should be taken of the *P*-delta or second-order effects.

The four issues outlined above are dealt with by the design rules in BS EN 1993-1-1. These rules cover hot-rolled and cold-formed steel structures and no distinction is made between the two forms of construction. Further guidance on the subject of frame stability is given in Chapter 5.

Stability of light steel frames is usually achieved through one of the following methods:

- integral bracing
- X-bracing
- diaphragm action.

Integral bracing consists of C-section members placed diagonally between the vertical wall studs and within the depth of the studs. The use of a C-section means that the integral bracing members are capable of carrying tension and compression forces. However, careful detailing and connection design are important. An example of integral bracing is shown in Figure 8.22.

X-bracing consists of diagonal crossed flat straps attached to the face of the vertical studs. Unlike integral bracing, the flats usually extend across several studs and are connected to every stud that they cross. Each individual bracing element is only capable of acting in tension (hence the need for the X-arrangement). An example of X-bracing is shown in Figure 8.23.

As an alternative to steel bracing members, the frame designer may choose to rely on the racking resistance of the wall itself. In this case, stability is provided by diaphragm action in the plane of the wall due to the attached board or cladding.

Figure 8.22 Example of integral bracing

Figure 8.23 Example of X-bracing

Board options include:

- plywood
- cement particle board
- OSB
- plasterboard.

The racking resistance of a particular board and frame combination should be determined by testing.

8.7.3 Robustness

The term 'robustness' when used in the context of building design relates to the ability of the structure to withstand accidental actions without the spread of damage or disproportionate collapse. In this sense, robustness is synonymous with structural integrity.

The essential principles of robustness, which apply across all forms of construction and materials, are summarised below:

- Robustness relates to the ability of a structure to withstand events such as explosions, impact or the consequences of human error. The Building Regulations do not require buildings to be designed to withstand acts of terrorism or other deliberate actions against the building.
- The aim is to restrict the spread of localised damage and to prevent collapse of the structure disproportionate to the original cause.
- The primary objective is to ensure the safety of the structure while building occupants make their escape and the emergency services are in attendance.
- The structure does not have to be serviceable. Large deformations and plasticity are permitted. It is anticipated that the structure will need to be repaired before it can be re-occupied. In some cases, it will need to be demolished.

The design steps required to comply with Building Regulations depend on the type and use of the building. To this end, Approved Document A (in England and Wales)[10] introduced a classification system relating to robustness. The same system was also adopted by BS 5950, and BS 5950-5 (for light gauge structures) was amended in 2006.

Under the classification system, Class 1 is the least onerous and applies to houses up to four storeys and also to agricultural buildings. Class 2A buildings include residential apartment blocks, hotels and offices up to four storeys. Many light steel framed buildings therefore belong in this category. This class also includes smaller retail buildings and industrial buildings up to three storeys. Class 2B includes hotels, residential, educational, retail and offices up to 15 storeys and hospitals up to three storeys. The final class, Class 3, will rarely be encountered in light steel design as it is reserved for buildings exceeding the limits given above, grandstands and buildings containing hazardous substances.

Class 1 and 2A buildings are generally deemed to satisfy the regulatory require-ments provided that the structural members (or building elements in the case of planar wall units or volumetric modules) are adequately tied together. Equations for the required tying force are given in BS 5950-5 for light steel structures and also in BS EN 1991-1-7[11] (the Eurocode relating to accidental actions). In the latter case, the UK National Annex permits the use of a smaller tying force for lightweight structures (in line with BS 5950-5).

Class 2B buildings are required to satisfy a more onerous set of tying rules, includ-ing the horizontal tying of edge columns and vertical tying of all columns (at splice locations). There is also a rule that requires the bracing systems to be distributed around the building, although this is not normally an issue for light steel framing. Where the tying rules cannot be satisfied, the building designer must either demon-strate that the notional removal of each column or building element does not lead to disproportionate collapse (defined in BS 5950-5 as the lesser of 15% of the floor area or $70\,m^2$) or must design them as 'key elements'. Where the notional removal route is followed, further checks must be conducted to ensure that falling debris does not lead to collapse of the floors below. In the case of modular buildings, the term 'element' relates to an individual module. Modular buildings should, therefore, be designed to withstand the notional removal of a module without collapse. 'Key element' design aims to avoid disproportionate collapse by ensuring that critical members and building elements remain in place following an accidental event such as an explosion. Key elements are designed to a special accidental load case, as defined in BS 5950-5 and BS EN 1991-1-7.

8.7.4 Frame anchorage

Due to their light weight, light steel framed buildings are at risk of sliding, overturn-ing or lifting off their foundations unless they are adequately held in position by a suitable anchorage. Two types of anchorage are generally employed:

- holding-down bolts connecting the stud and track to the ground slab
- steel straps connecting wall studs to the foundations.

8.8 Non-structural design issues

8.8.1 Acoustics

8.8.1.1 Principles of acoustic insulation

Sound insulation between rooms is achieved by applying the principles below in combination:

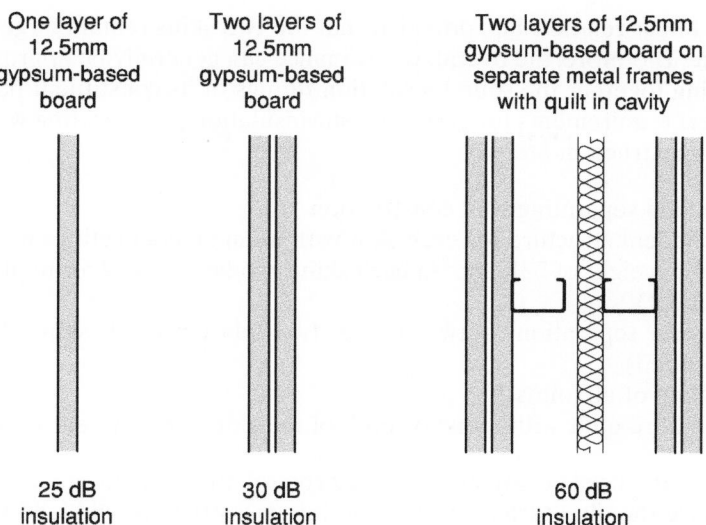

One layer of 12.5mm gypsum-based board

Two layers of 12.5mm gypsum-based board

Two layers of 12.5mm gypsum-based board on separate metal frames with quilt in cavity

25 dB insulation

30 dB insulation

60 dB insulation

Figure 8.24 Sound insulation using mass and isolation of layers

- provision of mass
- isolation of separate layers
- sealing of joints.

The combined principles of mass and isolation of layers are shown in Figure 8.24. It can be seen that increasing the mass without the use of separate layers has a diminishing reward in terms of sound insulation.

The acoustic insulation properties of walls or floors vary with the frequency of the noise. Certain frequencies are likely to be attenuated (reduced) more effectively than others by any given construction. Low pitched sounds are usually attenuated less than high pitched sound. The attenuation of different frequencies is, in part, a function of the cavity width. When using dry lining board, it is important to ensure efficient sealing of air paths that can lead to local sound transfer. Furthermore, flanking at the floor-wall junctions should be minimised by good detailing at these positions. Further information on acoustic insulation in steel construction is provided in SCI Publication P372.[12]

8.8.1.2 Separating walls

In separating or party walls using light steel framing, generally, two walls are constructed alongside one another. Each skin is structurally and physically independent of the other, in order to provide the necessary acoustic insulation. In a double skin wall, the sound insulation of individual components is combined in a simple cumulative linear relationship, provided that the two skins remain largely structurally separate. Therefore, the overall performance can generally be approximated by simply adding together the sound insulation ratings of its constituent parts.

The typical requirements for good acoustic insulation of separating walls in lightweight dry construction are:

- a double skin separating wall construction
- an independent structure for each skin with minimal connections between
- a minimum weight of $22\,\mathrm{kg/m^2}$ in each skin (two layers of 12.5 mm plasterboard, or equivalent)
- wide cavity separation between the two plasterboard skins (200 mm is recommended)
- good sealing of all joints
- a mineral fibre quilt within one or both of the skins or between the skins.

Resilient bars, used to attach the plasterboard to the light steel framing, can further reduce the direct transfer of sound into the structure and lead to enhancement of acoustic insulation.

8.8.1.3 Separating floors

For a separating floor construction between dwellings, both airborne and impact sound transmission must be addressed. High levels of acoustic insulation are achieved in lightweight floors by using a similar approach to that described for walls. It is important to separate (as far as possible) the top surface layer from the ceiling layer. This is usually done by the use of a resilient layer between the top floor finish and the structure below, and by resilient bars used to isolate the ceiling.

Impact sound transmission in lightweight floors is reduced by:

- specifying an appropriate resilient layer with correct dynamic stiffness under imposed loading
- ensuring that the resilient layer has adequate durability and resonance
- isolating the floating floor surface from the surrounding structure at the floor edges; this can be achieved by returning the resilient layer up the edges of the walking surface.

Airborne sound insulation in lightweight floors is achieved by:

- structural separation between layers
- appropriate mass in each layer
- sound absorbent quilt
- minimising flanking transmission at floor-wall junctions.

Further improvements in the design of lightweight floors can be achieved by complete separation of the floor structure from the ceiling structure, in a similar way to the double skin walls described above. The resilient layer beneath the floor finish contributes to insulation against both airborne and impact sound. Generally, mineral fibre with a density between 70 and 100 kg/m^3, provides sufficient stiffness to prevent local deflection but is soft enough to function as a vibration insulator. At the underside of the steel joists, resilient bars partially isolate the dry lining layer from the structure. A mineral wool quilt in the cavity between the steel joists provides sound absorption. The precise specification of each layer needs to be considered to optimise the floor performance. Increasing the mass of the top (floating) layer can have a significant improvement on the airborne sound insulation. Limited evidence suggests that floor joists at 600 mm centres have slightly better sound insulation than joists at 400 mm centres. Thicker plasterboard layers and gypsum fibreboard will have a higher mass, thus reducing sound transmission.

8.8.1.4 Flanking transmission

Flanking transmission occurs when airborne sound travels around the separating element of structure through adjacent building elements. Flanking transmission is difficult to predict, because it depends on the details of the floor and wall junction and the quality of construction on site. It is possible for a building to have separating walls and floors built to a high specification, but for sound to be transmitted through side walls which are continuous across the separating elements. Flanking transmission is dependent on:

- the properties of the surrounding structure, and whether it allows for indirect passage of sound
- the size of the wall or floor and, therefore, the proportionate effect of flanking losses
- the details of the floor/wall connections.

Flanking transmissions can add 3–7 dB to the sound transfer of real constructions in comparison to those tested acoustically in the laboratory. To reduce flanking transmission, it is important to prevent the floor boarding from touching the wall studs by including a resilient strip between the wall and floor boarding. Furthermore, the air space between the wall studs can be filled with mineral wool insulation to a height of 300 mm above the floor level in separating and external walls.

8.8.2 Thermal performance

8.8.2.1 Energy efficiency

The energy efficiency of buildings is becoming increasingly important in terms of regulatory requirements (Part L of the Building Regulations in England and

Wales) and also the expectations of building owners and tenants. In the residential sector, the Code for Sustainable Homes[13] now sits at the heart of the building design process and a similar code is likely to follow for non-domestic buildings. In general, measures to improve the energy efficiency of a building can be divided into 3 categories:

- reduce energy waste associated with building operation (e.g. heating and lighting)
- improve the energy efficiency of building services and electrical appliances
- install renewable energy sources (e.g. wind turbine, solar or photovoltaic).

Of these options, the first should always be the highest priority, with particular emphasis on minimising heat loss through the building envelope. There is little point installing the highest efficiency boiler or generating energy through solar collectors, if the heat generated is allowed to leak from the building due to poor design or poor quality construction.

Heat loss through the envelope is generally due to one or more of the following[14]:

- thermal transmittance through the wall, floor and roof
- air leakage through joints
- thermal bridging.

8.8.2.2 Thermal transmittance

The thermal transmittance of a construction is given by the U-value. This is defined as the heat in watts (W) passing through one square metre of construction per degree temperature difference from inside to outside. The lower the U-value, the better the insulation rating for the wall or roof. Consequently, for a number of years, the Building Regulations have attempted to promote energy efficiency through the prescription of maximum permissible U-values. Although Part L of the Building Regulations for England and Wales no longer relies on elemental U-values to achieve energy efficiency (since 2006 a more holistic approach has been taken to energy performance), the reduction of U-values remains central to efforts to improve the thermal performance of buildings.

Control of the U-value of a particular building element is achieved through the specification of the appropriate thickness of insulation. The less dense the insulation, the greater the thickness will need to be for a given U-value. Therefore, mineral wool insulation tends to be thicker than polyurethane. In light steel framed buildings in the UK, it is common practice to place the majority of the insulation on the external face of the frame to create a so-called 'warm frame'. This technique minimises the risk of condensation forming on the steel frame. Lower U-values may be achieved by placing insulation in between the studs, in addition to that on the outside of the frame. A typical wall build-up is shown in Figure 8.25.

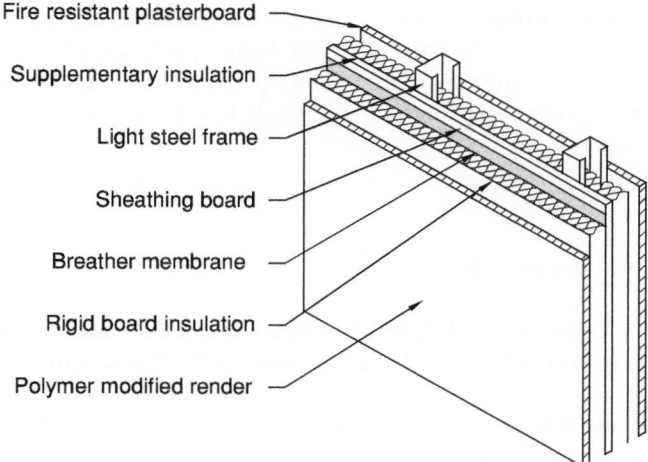

Fire resistant plasterboard

Supplementary insulation

Light steel frame

Sheathing board

Breather membrane

Rigid board insulation

Polymer modified render

Figure 8.25   Insulated render cladding attached to light steel wall

8.8.2.3 Air-tightness

To minimise the amount of heat being lost due to air leakage through joints, designers must ensure that the building envelope is airtight. The importance of air-tightness is recognised in the Building Regulations through the imposition of air permeability limits and the requirement for post-completion air leakage tests on each new building. Indeed, with U-values now reaching the point of diminishing returns, air-tightness has become a major focus of energy savings measures in construction.

The degree of air-tightness that can be achieved in practice will depend on the care (and hence cost) afforded to the building envelope by the contractor. With tight construction tolerances and properly sealed joints, typical walls of the type shown in Figure 8.25 are easily capable of exceeding the minimum requirements of the Building Regulations, thereby giving the building designer the opportunity to relax the U-value requirements while simultaneously improving energy efficiency. However, care must be taken when relying on air-tightness for compliance with Building Regulations, since as-built performance is very much dependent on build quality.

8.8.2.4 Thermal bridging

Thermal bridges are areas or components within the building envelope whose thermal insulation properties are lower (often much lower) than those of the surrounding material, thereby permitting local high heat flows through the building envelope. They can also lead to a reduction in the internal surface temperature of the envelope, causing condensation to form under certain conditions. There are several examples of thermal bridges in typical light steel framed buildings and good detailing is required to minimise their effect. Of particular concern is the thermal bridge between the sole plate and concrete slab, but door and window details can also cause problems.

References to Chapter 8

1. Grubb P.J., Gorgolewski M.T. and Lawson R.M. (2001) *Light Steel Framing in Residential Construction.* SCI Publication 301. Ascot, Steel Construction Institute.
2. British Standards Institution (2009) *BS EN 10346 Continuously hot-dip coated steel flat products – Technical delivery conditions.* London, BSI.
3. Lawson R.M. (2003) *Multi-storey Residential Buildings using Steel.* SCI Publication 329. Ascot, Steel Construction Institute.
4. Heywood M.D. (2007) *Benefits of off-site steel construction in urban locations.* SCI Publication 350. Ascot, Steel Construction Institute.
5. Heywood M.D. (2007) *Urban impact case studies – Final project report.* SCI Report RT1098. Ascot, Steel Construction Institute.
6. Cartwright P., Moulinier E., Saran T., Novakovic O. and Fletcher K. (2008) *Building Research Establishment SmartLIFE – Lessons Learned.* Watford, BRE.
7. Gorgolewski M.T., Grubb P.J. and Lawson R.M. (2001) *Modular Construction using Light Steel Framing.* SCI Publication 302. Ascot, Steel Construction Institute.
8. Lawson R.M. (2007) *Building Design Using Modules.* SCI Publication 348. Ascot, Steel Construction Institute.
9. British Standards Institution (2006) *BS EN 1993-1-3 Eurocode 3: Design of steel structures – Part 1-3: General Rules – Supplementary rules for cold-formed members and sheeting.* BSI, London.
10. Building Regulations 2000 – Approved Document A (2004 Edition) Structure. The Stationery Office.
11. British Standards Institution (2006) *BS EN 1991-1-7 Eurocode 1: Actions on structures – Part 1-7: General actions – Accidental actions.* London, BSI.
12. Way A.G.J. and Couchman G.H. (2008) *Acoustic Detailing for Steel Construction.* SCI Publication 372. Ascot, Steel Construction Institute.
13. Department for Communities and Local Government (2009) *Code for Sustainable Homes – Technical Guide – Version 2.* Communities and Local Government Publications.
14. The Steel Construction Institute (2009) *Code for Sustainable Homes: How to satisfy the code using steel technologies.* Ascot, Steel Construction Institute.

Chapter 9
Secondary steelwork

RICHARD WHITE

9.1 Introduction

In order to define the term 'secondary steelwork' it is necessary to start with a definition of primary structure. The primary structure consists of those elements that are essential to the integrity and robustness of the structural frame. The structural frame is the skeleton that supports all other building components. Steelwork that is supported by the primary structure, but which is not required to modify its strength or stiffness in any way, is known as secondary steelwork.

This chapter is in two sections. The first section describes the issues that should be considered when designing and procuring secondary steelwork whilst the second section provides guidance for the design and procurement of a range of secondary steel components.

9.2 Issues for consideration

9.2.1 Design criteria

9.2.1.1 Strength and stiffness

Unlike primary structure, strength is rarely the dominant requirement in the design of secondary steelwork. This means that secondary steel structures will tend to be lightweight and their design is likely to be governed by effects that would usually, when designing the primary structure, be thought of as minor. These can include the following:

- Footfall vibration – components that are subject to footfall, such as stairs and atrium bridges, will vibrate under loading. Their dynamic response must be such that normal walking does not cause discomfort to the building users. Discomfort

Steel Designers' Manual, Seventh Edition. Edited by Buick Davison and Graham W. Owens.
© 2012 Steel Construction Institute. Published 2012 by Blackwell Publishing Ltd.

is a function of both the physical response of the structure and the subjective response of the user. For this reason the acceptable response factors for secondary components may be higher than those of the floors which they serve.

- Imposed loads – the magnitude of imposed loads can be a function of the shape and size of the area over which they act. This is generally codified; for example wind pressures on cladding panels (especially at corners) and local drifting of snow. The supporting purlins and mullions will be designed for these increased loads. Also, individual stair treads can experience their full design load whenever they are in use simply because of their relatively small area. Frequently the critical load will be particular to the component and may include: vehicle impact, crowd loading, plant, and lift motor equipment. Where necessary guidance should be sought from the relevant sub-contractor.
- Deflection limits – normal code deflection limits are intended to prevent cracking of finishes applied to primary structure and will not normally be relevant when designing secondary components. The deflection limits should ensure that both the functionality of the component and the comfort of the users are maintained. For example, lift guides and their supporting brackets should be stiff enough for the lifts to work and handrails should not deflect alarmingly under crowd loading.
- P-Δ effects – these are the additional forces that are induced by the applied load acting on the displaced form of a structure. In the case of a secondary framework which is lightweight, it is likely to be sensitive to sway under gravity loading. This effect can generate significant additional buckling moments in the columns and other compression elements. It is, therefore, important that compression elements are effectively restrained and that additional moments are adequately assessed.

9.2.1.2 Robustness

Although the building regulations consider robustness in the context of disproportionate collapse of buildings, a similar approach should be adopted in the design of secondary components. The criterion to be considered is life safety. If the consequence of collapse is considered unacceptable, such as might be the case for a handrail, balcony, or walkway, then the component, and its fixings to the supporting structure, should be designed to have an adequate degree of redundancy. This may be achieved by the provision of alternative load paths. Where this is not possible a suitably conservative design approach should be adopted. Particular care should be taken with the design of support fixings, for which BS 6180,[1] the code of practice for barriers in and about buildings, provides this guidance:

- All joints should be designed to provide the full strength of the members being joined.
- Where any uncertainty exists with regard to the strength of any component in the fixing, the design loading should be increased by 50%.
- reliance on the pull-out capacity of a single fixing should be avoided.

These recommendations are intended to ensure that under an extreme load condition, failure will be indicated by excessive deflection and not by total collapse, as would be brought about by a failure of the fixing, attachment or anchorage system.

9.2.1.3 Movement and tolerance

By definition, secondary steelwork is attached to the primary structure. The primary structure will be built to tolerances and it will move under load. The designer of the secondary component will need to understand both the tolerances in the location of the supports, and how they are expected to move under service loads. The secondary component must, in turn, avoid providing any restraint to the movement of the primary structure. The point of reference in most cases will be the movement and tolerance report for the project.

The movement and tolerance report is written by the project structural engineer primarily as a guide for other members of the design and construction team, in order to avoid lack of fit or damage to other elements. Ideally the report will have been prepared in consultation with all members of the design team to ensure that the predicted movements are acceptable to all parties.

The designer of the secondary component should note that the tolerance in the fixing locations will include an allowance for both constructional tolerances and building movement that may occur prior to the installation of the secondary steelwork. Generally this will be the deflection due to the self-weight of the structure.

9.2.1.4 Other criteria

- Fire – the Building Regulations for England and Wales 2000[2] (Part B, Schedule 1) contain the mandatory requirement for a building to have structural fire resistance: 'The building shall be designed and constructed so that, in the event of fire, its stability will be maintained for a reasonable period'. It follows therefore that secondary structures will not generally need to be fire protected. It is, however, essential that the integrity of the fire compartmentation of the building be maintained. All elements, including secondary components, which provide support or restraint to a fire compartment wall, will need to be fire protected to the same rating as the compartment. Additionally, where secondary components attach to primary elements, the integrity of the protection to the primary element must be maintained.
- Thermal – thermal effects will need to be considered if there is a likelihood of a temperature differential arising between the component and its supporting structure. This situation will occur where external components are fixed through the cladding envelope and where a temperature differential can exist internally (e.g. within a glass clad steel stair enclosure). If the forces resulting from thermal restraint are excessive then the supports should be detailed to allow an appropriate amount of thermal movement. The finishes that are applied to these connections will also need to accommodate this movement.

9.2.2 Interface with primary structure

Interfaces are where things tend to go wrong. There is frequently uncertainty about who is responsible for what, what the design requirements are and the timing of the exchange of information.

9.2.2.1 Design interfaces

In the context of secondary steelwork, the connection design is the responsibility of the designer of the element that is to be fixed to the primary structure. The designer of the fixing and the structural engineer must exchange the following information in a timely fashion:

- Loads – the structural engineer will need to review both the location and magnitude of the applied fixing loads to check that their impact on the primary structure is acceptable. This information is provided by the designer of the fixing although frequently it will not be available until after the design of the structure is complete. In this case the engineer will have made assumptions in the design which should be shared with the fixing designer.
- Flexibility – although the primary structure will not rely on the secondary steelwork in any way, the flexibility of the support may have a significant effect on the design of the secondary item; for example lift guide brackets may not perform satisfactorily if the structure to which they are connected is too flexible. Generally, the sub-contractor designing the secondary element will assume the supporting structure is rigid unless advised otherwise by the structural engineer.
- Tolerances – the structure will have been built to tolerances and will have deflected under its self-weight prior to the attachment of the secondary steelwork. The arising potential deviations should be communicated by the structural engineer to the interfacing designers in the movement and tolerance report so that they may design suitably tolerant connections.
- Movement – the anticipated building movements are assessed by the structural engineer and communicated to the interfacing designers, usually via the movement and tolerance report. The critical interface is generally with the cladding contractor. As it is generally more economic to stiffen the structure than to design flexible cladding systems, it is normal practice to limit the imposed load deflections of the perimeter structure to approximately 10 mm.

Generally the structure will be designed in advance of the secondary steelwork using codified assumptions for deflection and loading. These should allow the designers of the secondary elements to develop standard, rather than bespoke, solutions.

Figure 9.1   Example of thermal break

9.2.2.2 Material interfaces

In addition to transferring load and accommodating tolerance and movement, the interface should:

- Have adequate corrosion resistance – this will depend upon the environmental conditions, accessibility requirements for maintenance, and design life. Guidance is given in the National Structural Steelwork Specification.[3]
- Avoid cold bridging – steel is a very good thermal conductor and, if the interface bridges the thermal envelope, a proprietary thermal break should be built into the connection to provide thermal discontinuity and minimise the risk of condensation forming (N.B. these items are quite bulky and expensive so should be identified in the tender documents). Figure 9.1 shows a typical example.

The connection design will also be influenced by the material of the primary structure as follows:

- Steel to concrete – in Figure 9.2 channels are cast into the concrete. This is preferable to using drilled fixings which may clash with reinforcement. It also avoids noise, vibration and dust associated with drilling operations. Adjustment is provided both horizontally and vertically using shims, whilst seating brackets provide temporary support of the steel beam. The capacity of the connection will be limited

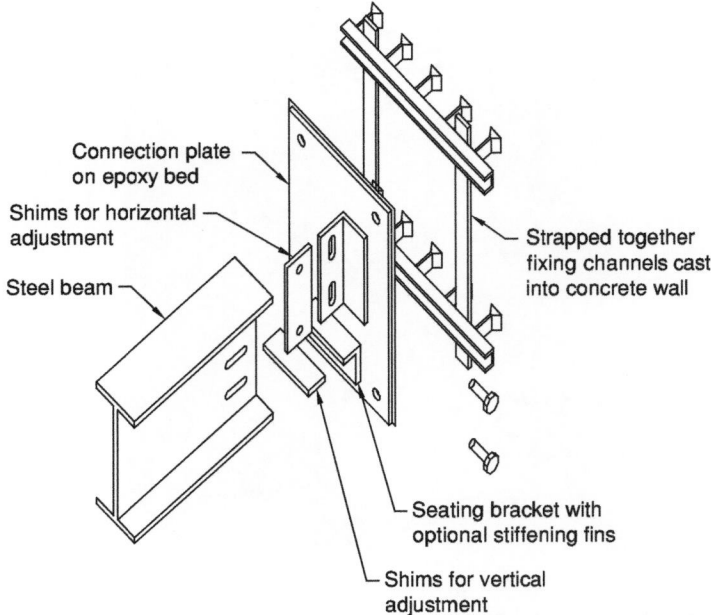

Connection plate
on epoxy bed

Shims for horizontal
adjustment

Steel beam

Strapped together
fixing channels cast
into concrete wall

Seating bracket with
optional stiffening fins

Shims for vertical
adjustment

Figure 9.2   Generic detail – steel to concrete wall

by the pull-out capacity of the cast-in channels and so excessive eccentricities should be avoided. The detail is suitable for both in situ and pre-cast concrete.
- Steel to masonry – masonry is a variable material, unsuitable for the support of concentrated loads. In Figure 9.3 the steel beam is supported on a concrete pad-stone which is sized to limit the stress applied to the supporting wall. The pad-stone incorporates a cast-in slot and is set to line and level on a mortar bed. Fine adjustment is provided with steel shims. The beam is fixed using bolts grouted into the slot.

Further guidance may be found in SCI publication 102,[4] *Connections between steel and other materials*.

9.2.3 Responsibilities

CIRIA guide C556,[5] Managing Project Change, provides guidance on the allocation of design responsibilities for a building project. This section is based on that advice.

Initially the architect will be responsible for the design of all secondary elements. The structural engineer may have an advisory role to ensure that the architects proposals are structurally feasible. The architect is responsible for obtaining this advice, and co-ordinating the details, which are shown on his drawings.

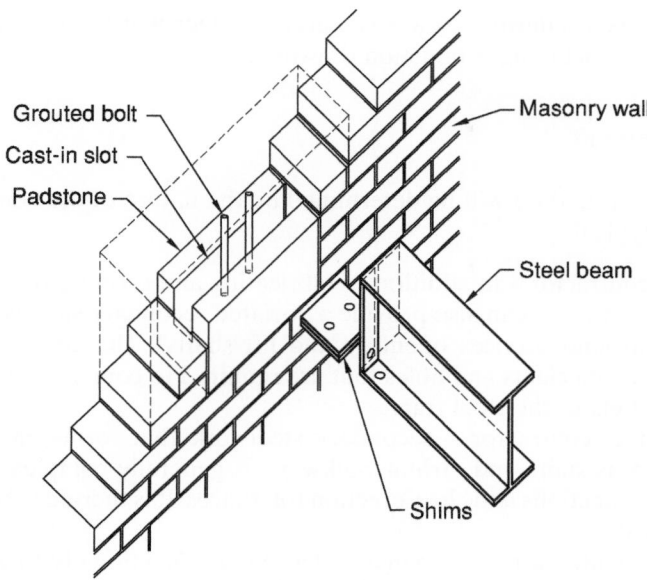

Figure 9.3   Generic detail – steel to masonry wall

Once the Works are tendered, design responsibility will generally be devolved to the relevant trade contractor or specialist supplier, although the architect will retain overall responsibility for co-ordination of the details. Thus, all secondary components that are specific to the cladding, or the lifts, or the building services, will become the responsibility of those particular sub-contractors. There are, however, exceptions where the design team should retain responsibility, as follows:

● Multi-function elements. If a component has more than one function, such as a beam, or framework, that supports both cladding brackets and lift guide brackets the design team should retain design responsibility. This will simplify both detailed design co-ordination and the duties of the trade contractors.
● Programme dependent elements. Where a contractor will not have time to provide design information, responsibility should remain with the design team. For example, if cladding is to be supported on pre-cast concrete wall panels the pre-cast contractor will want details of any cast-in connections long before the cladding contractor has had time to develop them. In this scenario, the cast-in details are developed by the architect and given to the cladding contractor as a constraint on his design.
● Bespoke elements. If an element is a major architectural feature, it will have a strong visual requirement and the design team will wish to retain control of its final design.

In situations where the architect retains responsibility, the structural engineer will continue to provide advice to be incorporated into the architects' drawings. In the case of secondary frameworks, the structural engineer will be responsible for the structural design, including production drawings.

9.2.4 Procurement

The procurement method will be determined by the nature of the component. The following are typical:

- Main steel contractor – in addition to fabricating and erecting the primary steel frame, this contractor can also provide associated secondary steelwork (e.g. trimming steel around services openings and lift shafts). The contractor will also provide the fixing cleats and holes that are required to connect the cladding and roofing brackets to the steel frame.
- Specialist steel contractor – secondary steelwork that forms an architectural feature, such as stairs and atrium walkways, is generally complex with visually expressed connections and close erection tolerances. It is therefore best produced by a specialist.
- Other trade contractor – components that belong functionally to a single trade contract, such as cladding supports, lift guide supports and services supports and walkways, are best provided by that trade contractor (cladding, lifts, or building services in this example).
- Specialist supplier – proprietary components such as windposts, fixings, brackets, cold-rolled purlins, partition frames and chequer-plate flooring will be obtained from a specialist supplier by the relevant trade contractor.

The principal difficulty with procurement of secondary steelwork is that it is seldom adequately defined at the time the works are tendered. It is essential therefore that the design team provide adequate early guidance to allow the appropriate allowances to be built into the tender documents – see Table 9.1.

9.2.4.1 Project example

Tendered at stage D – with no allowances for secondary structure not shown on drawings.
Allowances have been listed for each zone, but the actual distribution between zones may well vary.
Items labelled with an * will be contractor-designed.
Unless noted otherwise, assume the following:

Mild steel

1. Steel grade S275 in Zone A; steel grade S355 in zone B.
2. 100 man-hours workmanship per tonne of steel (cutting, drilling, welding).
3. Each item: 25% steel black; 25% steel galvanised; 50% fire-protected with thin film intumescent paint.

Table 9.1   Example of secondary steel allowances

Item	Zone A	Zone B
* Door trimmers	5T of 200 × 75PFC	8T of 305 × 305 × 97 UC
* Windposts (3.5 m tall)	100no.HalfenBW11606 stainless steel angle windposts	50no.HalfenBW1 1606 stainless steel angle windposts
Wall-restraints to floor beams	1T of 150 × 150 × 10 RSA	1T of 150 × 150 × 10 RSA
* Ledger angles for brickwork	50 m of HalfenHZA 100 × 100 × 10 stainless steel angles	125 m of HalfenHZA 100 × 100 × 10 stainless steel angles
Services openings in slabs	2T of 305 × 102 × 25 UB	2T of 305 × 102 × 25 UB
Rooftop plant support steel	2T of 305 × 65 × 46 UB	2T of 305 × 65 × 46 UB
*Stairs – hangers and trimmers for half landings	5T of 152 × 152 × 30 UC	5T of 152 × 152 × 30 UC
Minor trimmers to steps and recesses in slabs	2T of 100 × 100 × 8 RSA	2T of 100 × 100 × 8 RSA
* Lifting beams – lifts and plant	2T of 203 × 203 × 46 UC	2T of 203 × 203 × 46 UC
Plant screens and enclosures – general	2T of 152 × 152 × 30 UC	Not applicable
* Cold-rolled side and roof rail	2T of 172Z14 Metsec purlins	2T of 172Z14 Metsec purlins
* Cleats, lugs and brackets	3T of 100 × 100 × 8 RSA &1000 mm of 6 mm FW	1T of 100 × 100 × 8 RSA &500 mm of 6 mm FW
Drilled anchor bolts and stainless steel ledge angles	350 no. Hilti HST-R M20/60 and 150 m of 150 × 150 × 10 RSA	350 no. Hilti HST-R M20/60 and 150 m of 150 × 150 × 10 RSA

Additional items of Secondary Steelwork:
The following items are not included in the allowances listed above. All items labelled *will be contractor-designed.
* Built-in supports for M, E & P items
* Bolt-on supports for M, E & P items
* Welded on lugs for M, E & P items
* M, E & P support brackets
* M, E & P support droppers
* M, E & P support trapezes
* Access gantries on/around plant
Access gantries to/from plant rooms
* Services support racks
* Chequer plate infills to service risers
Cast-in Unistrut/Halfen channels
Service trench cover plates
* Mansafe system eyes and attachments
* Handrailing
Theatre equipment support rails
* Theatre lighting support and access gantries
* Signage bracketry and supports

Stainless steel

1. Stainless steel grade S316L.
2. 200 man-hours workmanship per tonne of steel (cutting, drilling, welding).

9.3 Applications

9.3.1 Stairs

The basic components of a staircase are the treads, risers, stringers, landings and their supports. These can be arranged in a variety of ways to create stairs that range from being utilitarian in nature to being major architectural features.

BS 5395,[6] the code of practice for the design, construction and maintenance of straight stairs and winders, provides the following recommendations:

- Geometry – this is determined by building regulation requirements. The relationship between rise, going and pitch must be such that the stair is safe and comfortable to use. This will depend upon the whether the stair is intended for private or public use. Typical dimensions for public stairs are:
  - ○ Rise: 100 – 190 mm
  - ○ Going: 250 – 350 mm
  - ○ Pitch: maximum 38 degrees
  - ○ Clear width: minimum 1000 mm.

  Stairs that are often used by large numbers of people at the same time (assembly stairs in public buildings) should be designed with a large going and a small rise to achieve a maximum pitch of 33 degrees. Stairs that are used as means of escape may require a clear width greater than 1000 mm.

  All stairs are required to have a minimum of three and a maximum of 16 rises per flight and the clear width of all landings should never be less than the stair clear width.
- Loads and robustness – stairs should be designed to allow for accidental loading, especially if they are required to provide a means of escape. In this case the stairs should not collapse in the event that the building becomes damaged by accidental loads. The connections to the primary structure must be suitably robust and be detailed to provide sufficient bearing area and tie resistance.

  If individual treads are used, their design can be governed by the dynamic effect of repeated foot loading and so a conservative design should be adopted.

  Dynamic response can be critical as steel stairs tend to have little inherent damping.
- Other criteria – safety, slip resistance, durability, acoustic requirements and lighting requirements all influence stair design and are addressed in BS 5395.

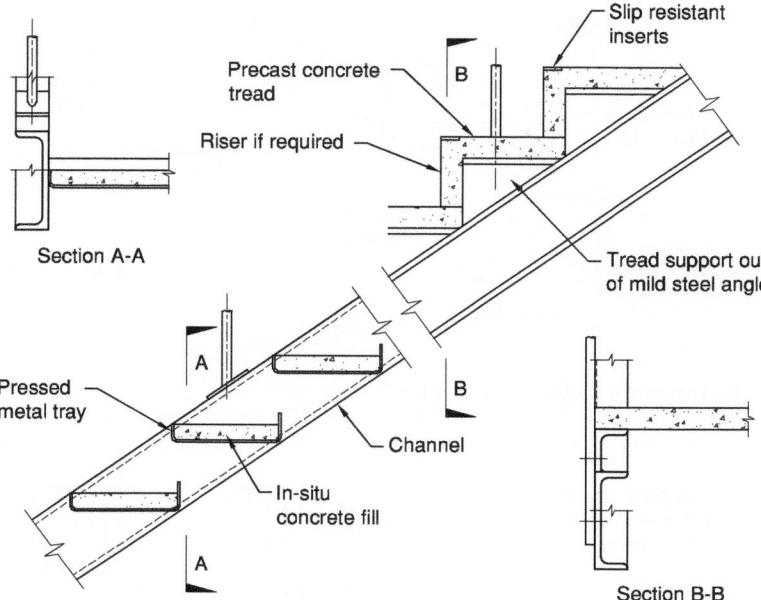

Figure 9.4   Steel strings and alternative tread arrangements

9.3.1.1 *Forms of construction*

The treads and risers are supported by the stringers to form a stair flight. Normal geometry requires that there are two flights per storey height and that these are configured at 180 degrees to one another occupying a footprint no greater than 6 m × 3 m (stairs in assembly buildings may be larger). Each end of the stair flight connects to a landing. The simplest form of stair construction is where the staircase is located internally, within a hole in the primary floor structure. In this case both the floor level and the half level landings may be directly supported by the primary structure, and the stair flight may span directly between landings. The floor level landing may be designed either as part of the staircase or as part of the floor structure. Stair treads may be located either above, or in the plane of, the stringers (see Figure 9.4).

If the treads are located in the plane of the stringers the stringer depth generated by minimum planning dimensions will be structurally adequate. Furthermore, if folded steel plate is used for both the treads and risers the stair flight will inherently have sufficient rigidity to achieve an adequate dynamic response. This form of construction is very efficient (see Figure 9.5).

If a staircase is located at the edge of a floor slab, the support of the landings (especially the half landings) will be of critical importance. A simple generic example is shown below (see Figure 9.6).

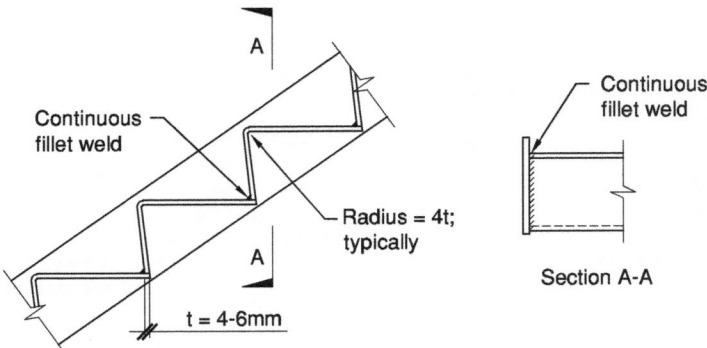

Figure 9.5   Stair flight with folded steel plate

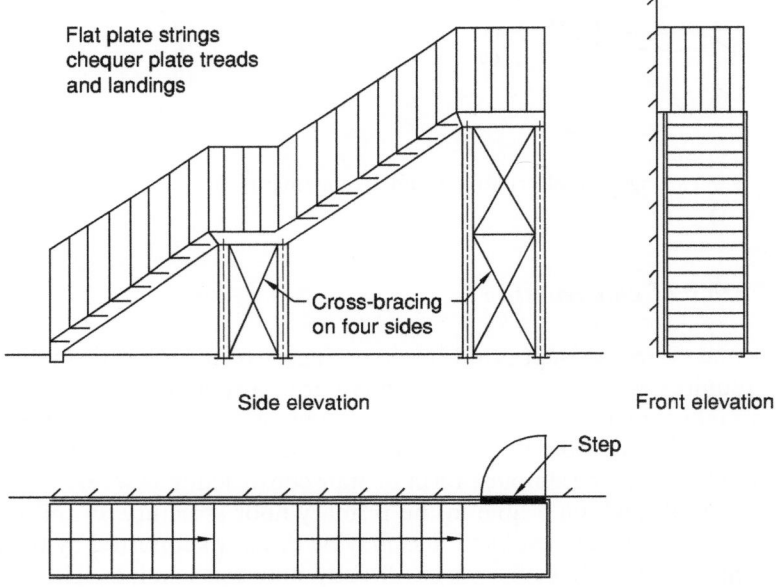

Figure 9.6   Simple external stair

Figure 9.6 shows a single-storey external escape stair. The stairflights span (horizontally as well as vertically) onto the landings which are in turn supported by an arrangement of braced steel columns. A more complicated example of this type of stair is described in Section 9.3.1.3.

9.3.1.2 Design responsibilities and procurement

The architect is responsible for staircase design. The structural engineer will provide the architect with advice which the architect will incorporate into his detailed design drawings to enable a contractor to be appointed. There are many specialist fabricators who have the capacity to design and manufacture staircases so most commonly responsibility for the construction design will pass to the fabricator. This will allow the fabricator to develop the construction details to suit his method of production. The architect will remain responsible for approving the contactors details and for co-ordinating those into the overall design. The structural engineer remains responsible for assessing any impact that the stair design may have on the supporting primary structure.

9.3.1.3 Project example

In this example an 18-storey concrete framed commercial office building is serviced by six perimeter cores. These cores house toilets, service risers, lifts, and stairs. They are glass clad steel framed structures which form a major architectural feature of the building. The principal design considerations for the stairs are:

- the stairs cantilever on plan from the side of the core and so have little inherent stability
- the stair structure supports the enclosing glass cladding
- the prominent and transparent nature of the stairs requires exceptional attention to architectural detail
- the need to accommodate movement and tolerance.

Figure 9.7 indicates the plan of a typical stair core.

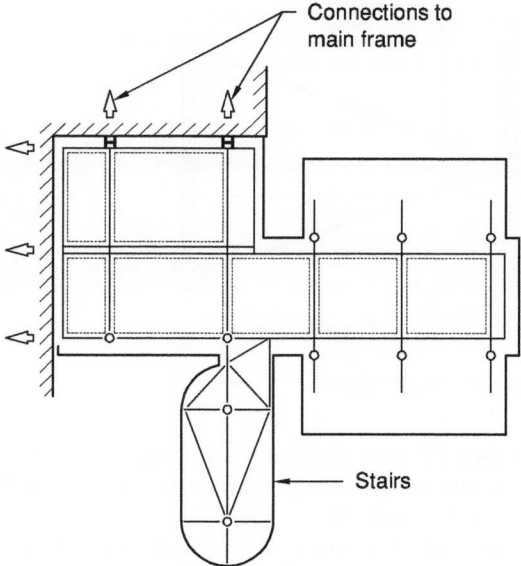

Figure 9.7   Diagrammatic plan of typical stair core

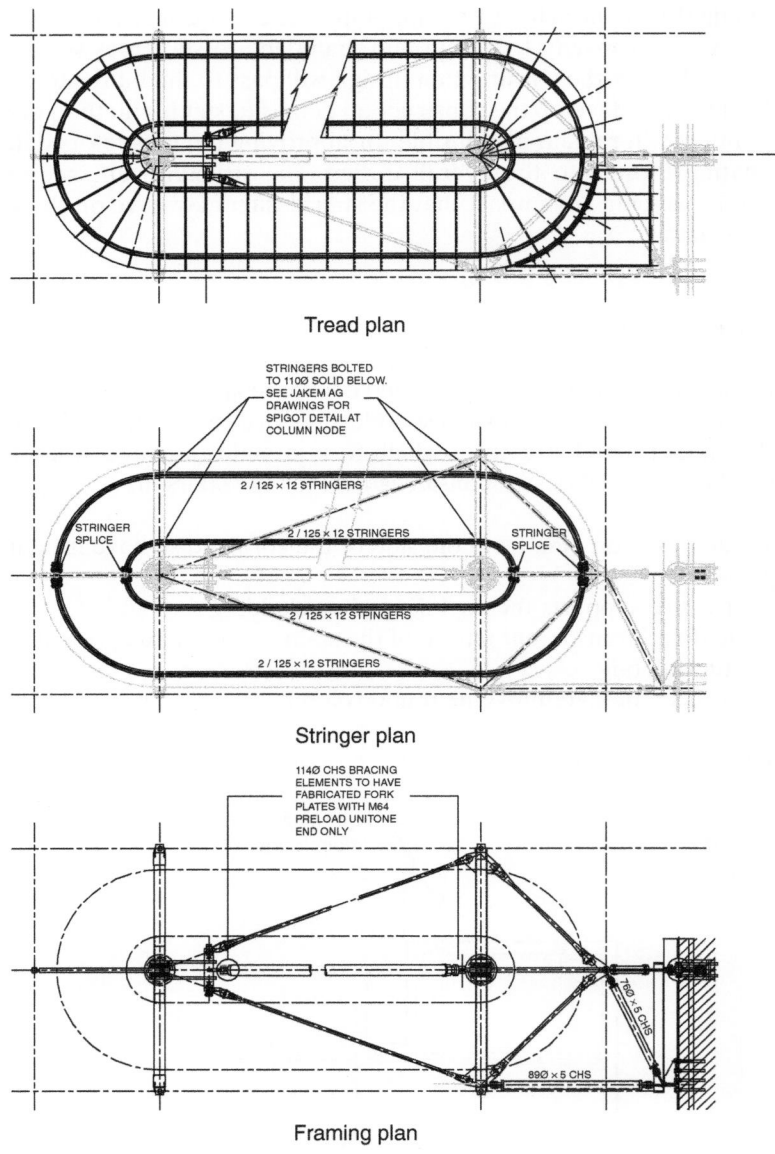

Tread plan

Stringer plan

Framing plan

Figure 9.8   Typical support arrangement

The stair flight consists of open tread units formed of 5 mm folded plate bolted to plate stringers. The flights and landings span onto radial arms that cantilever from two columns, see Figures 9.8 and 9.9.

The radial arms also guide tension rods that act as mullions to support lateral wind loads from the glazing. The rods are stressed against the stair columns to

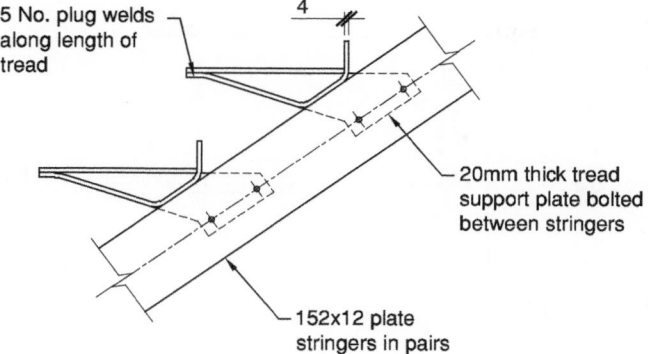

Figure 9.9   Typical tread and stringer

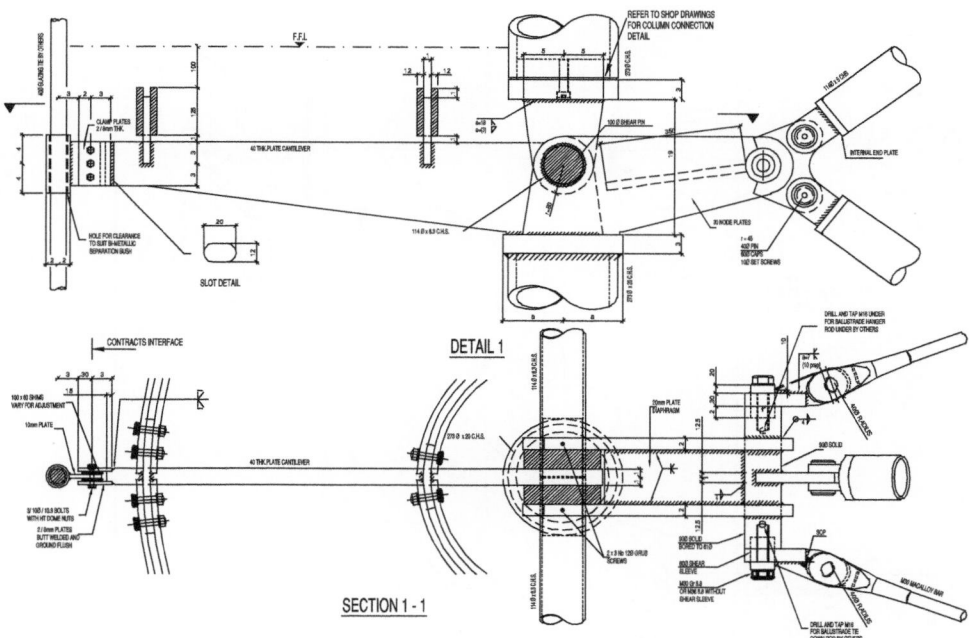

Figure 9.10   Typical column node detail at half landing

provide the stiffness required to resist out-of-plane forces. A frame at the top and bottom of the towers provides anchorages and stressing points for these rods.

As the stair flights have little inherent stiffness, the columns are stabilised on plan by a system of pre-tensioned rod bracing. The bracing runs under the stair flights to the column node at the half landing connecting back to the ends of the radial arms at the inner column. These nodes are then triangulated below the landing back to the core floor plate, see Figures 9.8, 9.10 and 9.11.

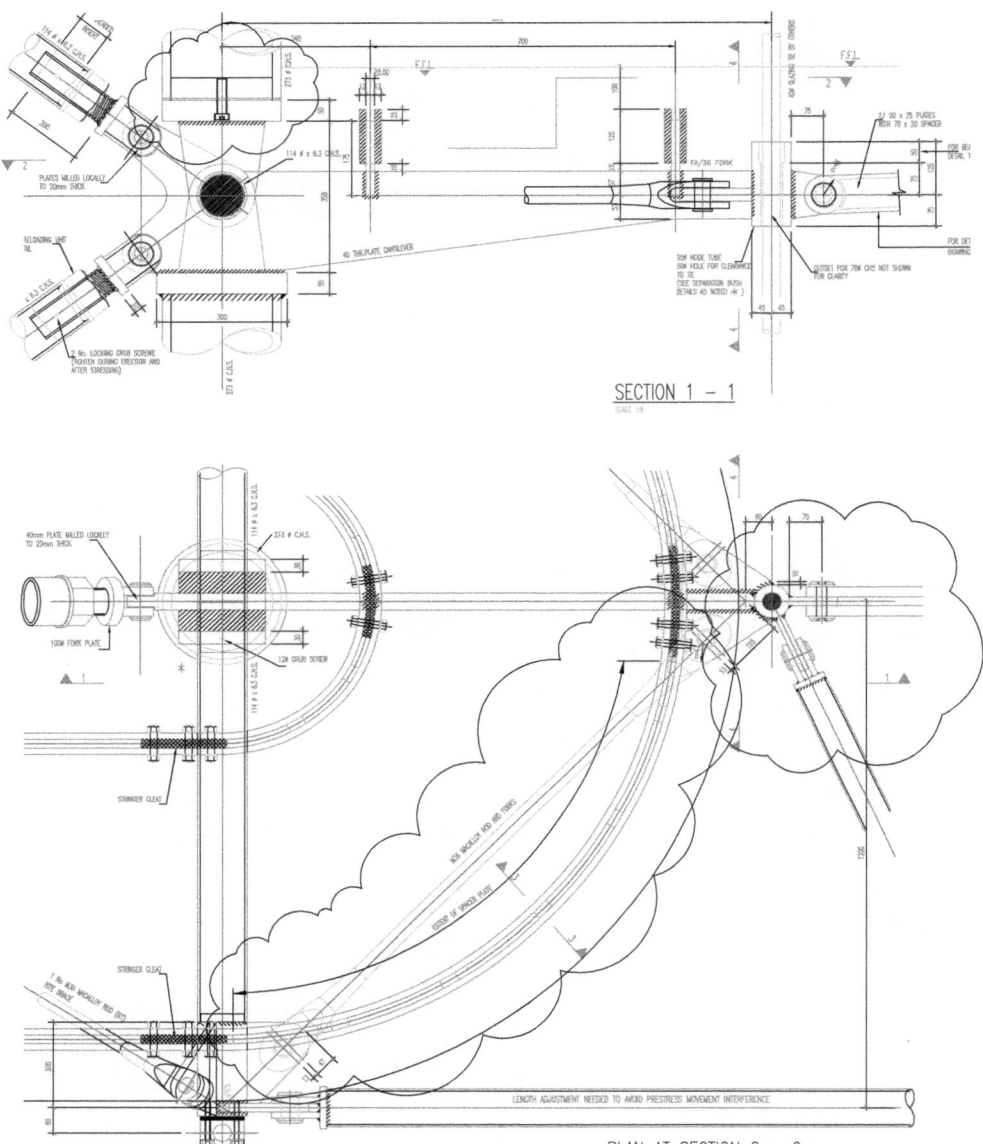

Figure 9.11   Typical detail at floor level

In order to resist out-of-balance forces acting on the column nodes they are linked by a series of braces in elevation.

All bracing elements, including those connecting to the floorplate, are fitted with threaded length adjustment to enable accurate fit-up during erection and stressing. The connection to the floorplate is also articulated to accommodate any

Figure 9.12   Completed staircase

vertical movement arising from prestress during erection and thermal movements in service.

The completed stair is shown below, Figures 9.12 and 9.13.

A (slightly) simpler version of this stair was provided at the rear of the building where architectural constraints were less onerous. Here the flights act as cantilevered vierendeel girders to stabilise the furthest column from the floor plate, thus obviating the need for the system of prestressed tension rods. Also, the mullions are normal structural sections supported at the ends of the radial arms requiring no prestress either.

The design of the stair structures remained the responsibility of the structural engineer due to their complexity. The construction details were developed with input from the specialist steel fabricator and the architect retained responsibility for overall co-ordination with the interfacing components (cladding and handrails).

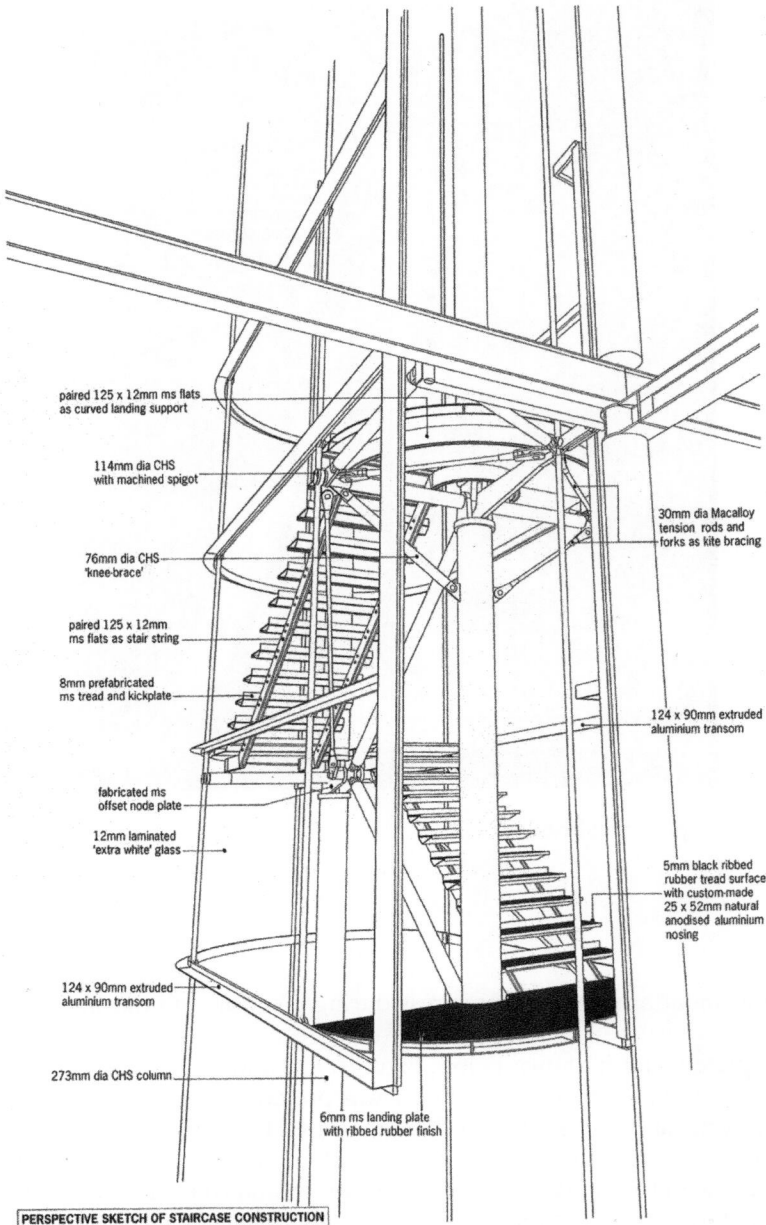

paired 125 x 12mm ms flats
as curved landing support

114mm dia CHS
with machined spigot

76mm dia CHS
'knee-brace'

paired 125 x 12mm
ms flats as stair string

8mm prefabricated
ms tread and kickplate

fabricated ms
offset node plate

12mm laminated
'extra white' glass

124 x 90mm extruded
aluminium transom

273mm dia CHS column

6mm ms landing plate
with ribbed rubber finish

30mm dia Macalloy
tension rods and
forks as kite bracing

124 x 90mm extruded
aluminium transom

5mm black ribbed
rubber tread surface
with custom-made
25 x 52mm natural
anodised aluminium
nosing

PERSPECTIVE SKETCH OF STAIRCASE CONSTRUCTION

Figure 9.13   Completed staircase

9.3.2 Lifts

The principal components of a lift installation are the lift car, the counterweights and the motors. Traditionally the car and counterweights are suspended from pulleys within the lift motor room which is located directly above the lift shaft and is supported on the primary structure. Secondary steel brackets are required to support the lift guides. These brackets are provided by the lift manufacturer. It should, however, be noted that:

- Lift guides are required both for the lift car and counterweight. They must be held rigidly in place for the lift to function properly.
- The lift manufacturer will assume that the structure to which he fixes his brackets will provide a rigid support.
- The lift guides require support at a maximum of 4.5 metre intervals vertically, although intermediate supports are required for high speed lifts.
- The lift guide brackets provide lateral restraint only; vertical loads are transmitted by the rails to the lift pit. BS 5655,[7] the code of practice for the installation of lifts and service lifts, provides guidance for design loads and deflection limits.
- The lift guide brackets must have sufficient horizontal adjustment to enable accurate installation of the lift guides.

The above conditions can be met quite easily if the lift is located within a concrete or masonry shaft.

If the main building frame is steel, it is likely that the lift shaft will be formed from lightweight partitions. SCI publication 103[8] provides guidance for the provision of electric lifts in steel framed buildings and makes the following recommendations:

- Guide brackets will be supported at floor level. They will either be fixed directly to the primary steelwork or to the edge of the concrete floor. Fixings to concrete should be made to cast-in channels to avoid the need for post-drilled fixings. Proprietary edge trims can be provided which incorporate fixing channels in their profile.
- If several lifts are grouped together they will be separated by steel divider beams which are provided to support the guide rails. Divider beams are either bolted to the primary steelwork or to the edge of the concrete floor in the same way as the guide brackets.
- The lift doors will need to be supported on secondary steel frames. These should be supported at each floor level, either suspended or bottom fixed, and should be detailed to accommodate floor deflections.
- The framing that supports the partitioning around the lift shaft will need to be detailed to allow relative vertical movement so that it does not act as a prop to the floors.

If, however, there is no enclosed lift shaft then additional framing will be required to support the lift guide brackets. This framing must provide adequate lateral stiffness and the connections for the brackets must provide sufficient adjustment (horizontally and rotationally) to enable their accurate installation. This secondary framing will normally be designed by the structural engineer. The engineer should obtain the required stiffness criteria from the lift manufacturer since a purely strength-based design is likely to be too flexible.

Finally, lifting beams should be provided above the centre of the liftwell for the purpose of lift installation and plant replacement.

9.3.3 Secondary cores and atrium structures

If overall building stability is achieved without reliance upon the cores then to the structural engineer they are simply an assembly of holes and applied loads. This allows the architect and services engineer freedom to optimise their designs and to provide the required degree of adaptability for future change. The architect may in this case wish to make a feature of the stairs, lifts and/or toilets by locating them off the floorplate; either at the building perimeter or within an atrium. The core components will be supported by a secondary steel framework which will rely on the floorplate for restraint. The key factors in the design of this secondary framework will be:

- Stiffness – to limit footfall response of the stairs and to provide sufficient restraint to the lift guides
- P-Δ effects – the columns will tend to be both slender and only flexibly restrained and so will have limited buckling resistance.
- Robustness – to ensure adequate resistance to collapse in the event of accidental column removal or loss of column restraint.
- Column shortening – columns in secondary cores tend to be closely spaced to one another and to the primary structure. If they are reasonably tall then differential shortening effects will need to be considered, particularly differential thermal effects. In the case of lift shaft columns, the dominant load is from the lift motor room. This is applied at the top of the column and so shortening effects during erection will need to be assessed. If the primary structure is concrete, long term creep and shrinkage effects may require articulation of the connections between the core and the primary frame.

Secondary cores are therefore expensive components despite their limited structural function for which an appropriate allowance should be made in the cost plan.

Design responsibility should remain with the structural engineer due to the multi-functional nature of the structure and the complexity of the design. The steelwork should either be procured as part of the main steel frame contract (if there is one) or through a specialist steel fabricator if required by the architectural design.

9.3.3.1 Project example – the St Botolph Building, London EC3

The St. Botolph Building is a 15 storey commercial office development built by the developer M1 Limited. The primary structure is a steel frame stabilised by concrete cores. There is a centrally located atrium space which houses a set of feature lifts arranged either side of a lift lobby. One end of the lift lobby connects to the main floorplate, whilst there is provision at the other end of the lobbies to connect to bridges spanning across the atrium.

As shown in Figures 9.14 and 9.15, the key features of the design are:

- The lobby floors – these cantilever horizontally from the main floor to stabilise the columns. In order to allow the floor units to be procured independently of the core steelwork, a horizontal truss is formed by bracing between the steel beams and the floor units are considered to be non-structural elements.
- The atrium bridges – the design assumes a minimum of three bridges, spaced at three storey intervals, to be permanent. These are used to prop the lobby

Figure 9.14   St Botolph Building, London EC3: image of core structure

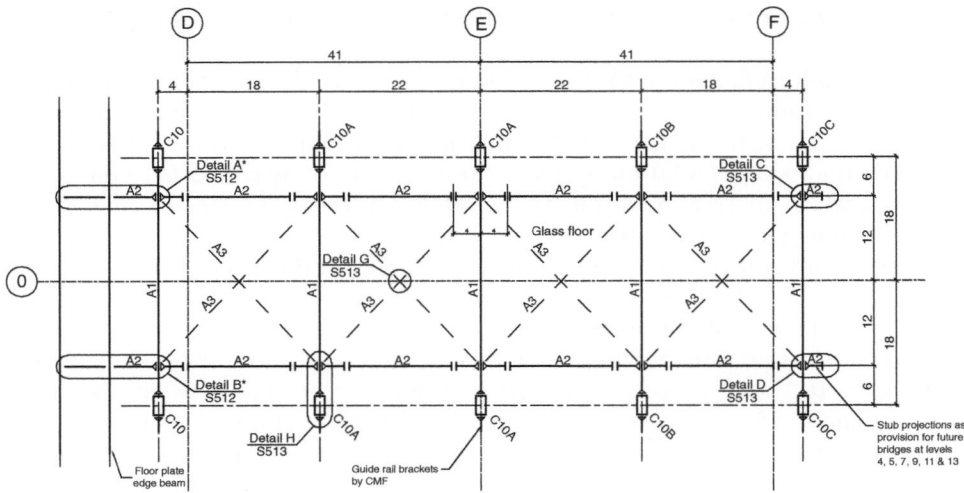

Figure 9.15   Typical lobby plan

floors laterally and so increase the lateral stiffness of the structure. The connection between the bridge and the lobby floor must allow a degree of thermal movement in addition to transmitting the restraining forces. This is achieved with a set of disc springs (Figure 9.16) that compress by a maximum of 10 mm under design load. The bridge design itself is governed by its response to footfall vibration.

- The columns – these are heavily loaded as the design adopts two lift cars per shaft, resulting in a lift motor room that is double the weight of a conventional one. Additionally, the outermost columns support the atrium bridges. Column restraint is provided only by cantilevering floor beams. The available space is restricted and the architectural design required the use of a rolled box section. Due to the non-availability of the required section, fabricated boxes are adopted which are finished to have the same appearance as a standard RHS.

Figure 9.17 shows the connection that had to be adopted to ensure a smooth outer surface.

- Robustness – due to the critical nature of the structure it was checked both for individual column removal and for the loss of restraint between a lobby floor and main frame connection (effectively removing restraint from all columns at a particular level).

The responsibility for the structural design remained with the structural engineer throughout the project, due to its complexity and the need to simplify the interfaces at the connections to the main frame.

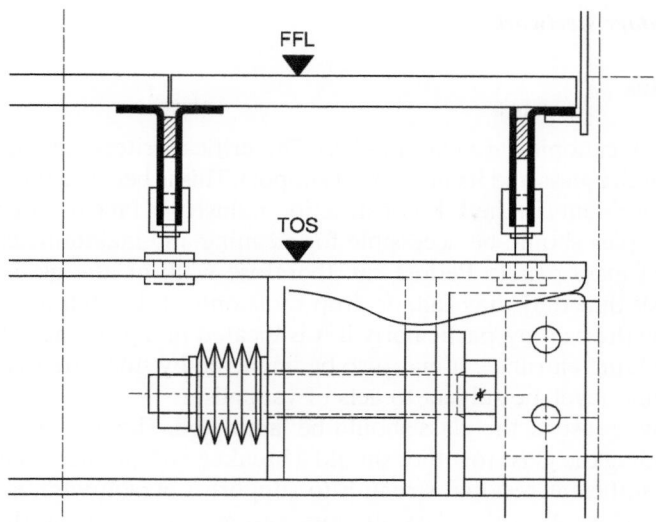

Figure 9.16   Disc spring connection

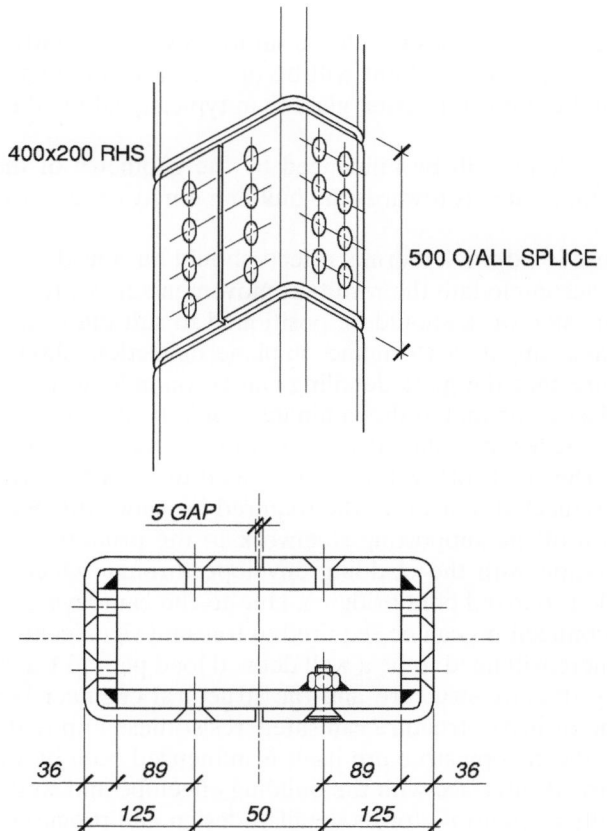

Figure 9.17   Column details

9.3.4 Canopies

Most commonly canopies are clad in glass. The critical criteria in their design are the selection of the glass and its method of support. This is because the biggest single cause of fatal accidents in the UK construction industry is falls through fragile roof materials. Canopies should be accessible for cleaning and maintenance and for the replacement of glass panels. People can therefore walk on the glass and there is therefore a risk that they may fall, or drop tools onto it. In addition many people will pass below the canopy, particularly if it is located in a prominent location such as over the building entrance. It may also be impacted by storm debris or, if reasonably near ground level, be subject to acts of vandalism.

For the above reasons, the glass should be laminated. The laminations provide a method of holding the glass together should a breakage occur and it can be designed so that it has sufficient residual strength to support a person who has fallen onto and broken the glass. Also, the glass retaining system has to prevent the entire pane of laminated glass from falling out. It is therefore preferable to provide continuous support to all four edges of the glazing (i.e. a fully framed system) than to use point fixed supports.

Deflection limits for canopies should be set to prevent unsightly deformation of the leading edge. Typically the limit will be of the order of span/350 which may require stiffer, and therefore heavier, glass than typically adopted elsewhere in the building.

Snow and wind loads will be influenced by the geometry of the canopy – for example if the canopy slopes towards the building it may experience uplift or accumulate snow.

As the structure is external, thermal effects should be considered and the design must be able to accommodate the resultant movements and stresses.

The supporting steelwork should be positioned to suit the available glass panel sizes and when assessing its performance in plane, deflections should be considered in order to ensure that the glass detailing can accommodate the resulting movements. As the glass is laminated the laminate should be assumed to have no shear stiffness and the deflections should be based upon the performance of the steel structure alone. The steelwork will therefore need to be either braced in-plane or to be moment-connected to achieve the required in-plane stiffness.

The connection of the supporting steelwork to the primary frame will require careful co-ordination with the cladding envelope through which it passes and it should be detailed to avoid cold bridging. Due to the cantilevered nature of most canopies, these connections can be required to transmit significant lateral as well as vertical loads. There will need to be a well defined load path to transmit these loads into the primary stability structure and the structural engineer is responsible for checking that the primary structure can safely resist these imposed loads.

The design of the canopy structure itself is influenced both by the cladding that it supports and by its interface with the building envelope and so the cladding contractor is normally responsible for its detailed design and procurement. An exception to this arrangement can occur if the canopy is sufficiently large that it should be fabricated and installed by a steel contractor. If the primary structure is also steel, then logically this would be the main frame contractor; otherwise (and especially if the architectural detailing is important) procurement should be through a specialist steel contractor.

9.3.5 Façade supports

The principal functions of facade supports are to transfer load between the facade and supporting structure and to accommodate movement and tolerance. It is very common on projects of all sizes for problems of fit and in-service movement capacity to exist at the interface between the facade and the supporting structural framework. The design of the facade supports is critical to the resolution of these problems.

The support design is influenced both by the performance of the primary structure and by the nature of the facade construction, which will generally consist of the following components:

- Cladding panel – an element which provides a 'skin' to the building. It may attach directly to the building structure or be supported by a secondary frame of mullions and transoms.
- Mullion – a vertical beam element which spans between floors, or other suitable restraints, and supports a cladding panel. It is usually made from aluminium.
- Transom – a horizontal beam element which spans between mullions.

The two commonest forms of facade are:

- Stick system – individual mullions and transoms (sticks) are assembled on site to form a secondary frame to which the cladding panels are attached.
- Panel system – mullions, transoms and cladding panels are factory-assembled into units often 1.5 metres wide by a storey height before delivery to site.

The facade will be designed, manufactured and installed by the cladding contractor. However, before the cladding contractor is appointed the architect and structural engineer will have had to make several key decisions that will influence the facade performance. These include:

- the weight of the cladding
- the type and locations of cladding supports and hence the local forces that these transmit to the structure
- suitable deflection limits for the structural elements to which the cladding will be attached
- the cladding zone
- out-of-position limits to which the structure will be built.

These decisions should be documented in the movement and tolerance report by the structural engineer to enable the cladding contractor to develop his design. As the facade cost will typically be double that of the structure it will be a false economy to optimise the structural design at the expense (literally) of over constraining the facade design. The perimeter structure should therefore be suitably stiff.

9.3.5.1 Common problems

Facades are supported at the slab edge. Where the slab provides support the cladding bracket should be attached to a proprietary cast-in channel to avoid the need for drilling into the concrete. During the early stages of the building design the structural engineer and architect must agree sufficient space in the floor finishes to accommodate the cladding support brackets (It is not acceptable from a CDM standpoint for brackets to be fixed overhead). This check should be carried out at all locations, not just the typical ones, and provision should be made in the design for any additional steelwork that may be needed in non-typical areas, such as stairwells or areas where there are large slab openings at the building perimeter.

If the primary structure is a steel frame supporting a thin lightweight concrete slab, the slab edge may not be sufficiently strong to support the cladding loads. In this case an edge beam should be provided to which the cladding may be attached. This edge beam should have sufficient torsional stiffness to resist local effects of eccentrically applied load (e.g. wind restraint fixings attached to the bottom flanges of universal beam sections).

When considering movement at the interface between cladding and structure, the corners can present particular problems as the out-of-plane movement in one elevation will equate to the in-plane movement in the other. Ideally the lateral movement of the building will not exceed the racking capability of the cladding (normally storey height/500). If this is not the case (e.g. thermal movement of large roofs) then non-standard, and possibly complex, movement capability will need to be provided by the corner brackets (e.g. by using the façade, rather than the structure to restrain the corner mullion – see Figure 9.18).

Two types of connection are used in both stick and panel systems. The primary connection resists both gravity and lateral loads. The secondary connection provides lateral restraint only and allows relative vertical movement between the cladding and structure. Each cladding element is supported by a primary and a secondary connection. The primary connection may be located at either the top or bottom of the cladding element. As the curtain wall is located in front of the structural frame, gravity loads will induce bending in the connections. Wind pressures acting on the cladding will be governed by corner effects and by the relatively small area of the cladding elements. This will therefore generate significantly greater pressures than those for which the building structure is designed.

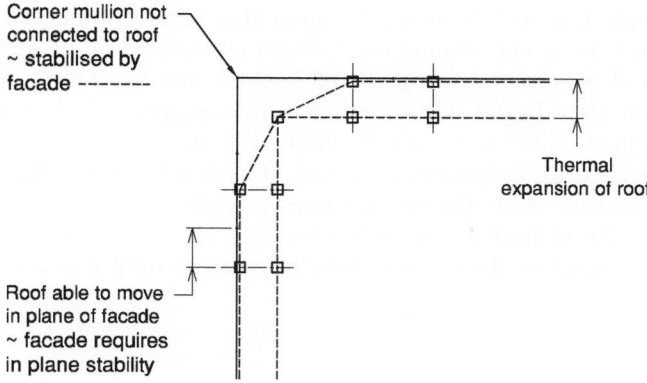

Figure 9.18   Corner detail

These connections must also accommodate all necessary movement and tolerance requirements.

9.3.5.2 Stick systems

The mullions are connected to the support brackets and the rest of the wall is assembled by attaching pieces to the mullions. Mullions are normally bottom supported and can be continuous over two or three stories. Vertical movement is accommodated by sliding spigot connections between mullion sections. Rotational movement, caused by deformation of the structure, induces minor axis bending in the mullion and requires no special provision in the connection. The support brackets can thus be the same at each floor level: see Figure 9.19. They are normally

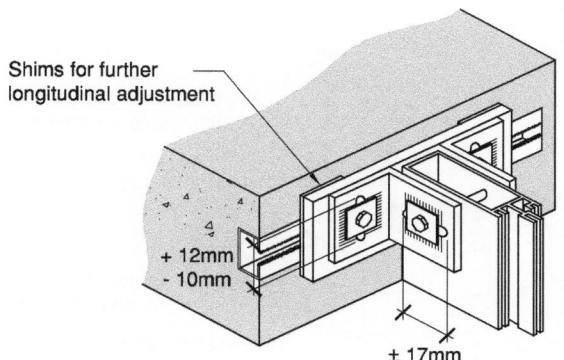

Figure 9.19   Typical stick system fixing bracket

made from simple rolled steel channels and angles, usually stainless steel or occasionally hot-dipped galvanised mild steel. Normally, side-to-side adjustment in the plane of the wall is achieved using slotted holes in the bracket attached directly to the structural slab. In-out adjustment is achieved using slotted holes at right angles to the plane of the wall and/or shims. The slots can be in the mullion or the bracket attached to the mullion. Vertical adjustment can be achieved either by packing pieces placed under the bracket fixed directly to the structural slab, or by providing vertically slotted holes on the bracket. Serrated washers engage with grooves in the brackets to lock the mullion in position once it has been lined and levelled.

9.3.5.3 *Panel systems*

Each panel is prefabricated as a structural element. Generally, their weight is supported by two brackets, conventionally located at either the top or bottom corners of the panel. Restraint to horizontal loads is provided by additional brackets or fixings located on the opposite edge. In plane, racking of the cladding can generate some panel interlocking which redistributes the loads on the support brackets. In extreme cases one vertical support bracket may carry up to three panels.

Panel system brackets are usually more sophisticated and substantial than those used for stick systems. See Figure 9.20 for a typical example. They are fabricated from steel or aluminium plate or are cast aluminium or steel. They are generally fixed to channel sections cast into the floor slab; these run parallel to the slab edge to provide side-to-side adjustment. Serrated washers lock the bracket in position once it has been lined and levelled. The panels are hung off the bracket and up/down adjustment is provided by threaded rods or screws.

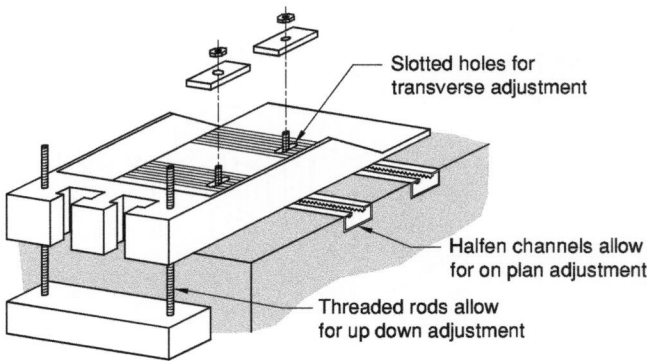

Figure 9.20   Typical panel system fixing bracket

9.3.5.4 Corrosion and fire protection

Fixings tend to be made from galvanised or stainless steel to minimise the risk of oxidation. However, bi-metallic corrosion can occur when dissimilar metals, such as stainless and ferrous steels are in contact in the presence of moisture. In these situations it is essential that the dissimilar metals are separated by gaskets or coatings of PTFE, nylon, or nepoprene.

The gap between the cladding and the slab is usually sealed to prevent smoke and fire from spreading between compartments. Usually, the cladding supports themselves need no fire protection as the cladding is not normally fire rated. If, however, the fixings are attached to a protected member, such as a steel beam, then their design must ensure that the performance of the supporting beam is not compromised. In this case, if fire protection is required it must not prevent the free movement of the cladding at its supports.

9.3.5.5 Responsibilities

Prior to the appointment of the cladding contractor, the architect is responsible for the cladding design and for determining the spatial allowances for the support brackets. If the project is complex, he may be assisted by a façade engineer. The structural engineer is responsible for assessing the predicted building movements and for the preparation of the movement and tolerance report. Once appointed, the cladding contractor becomes responsible for the cladding design, including the design of the support brackets. The architect retains responsibility for co-ordination of the cladding with the overall design, whilst the structural engineer remains responsible for assessing the ability of the structure to resist the loads imposed at the façade support locations.

9.3.6 Balconies

The traditional method for constructing balconies has been to simply extend the primary floor structure outside the building. This is no longer preferred as it is now necessary to prevent cold bridging across the cladding envelope. Balconies have, literally, become bolt-on secondary structures.

SCI publication P332,[9] *Steel in multi storey residential buildings*, identifies three generic balcony types as follows:

- Stacked ground-supported modules – the balconies are supported on a set of columns that extend to ground level. These can be lifted into place as a group. See Figure 9.21.
- Cantilever balconies – the balconies are supported by beams that are moment connected to the primary floor structure. See Figure 9.22.

Figure 9.21   Typical stacked balcony

- Tied balconies – the outer edge of the balcony is supported by a pair of ties. These are either inclined to connect to the primary structure at the floor level above, see Figure 9.23, or are vertical and connect to a supporting structure at roof level.

In the first case, no vertical load is transferred to the primary structure, except for horizontal restraints, providing the connections to the main frame are detailed to accommodate any movements that may arise from differential settlement or thermal effects. In the second case, the size of the balcony is limited by the ability of the primary structure to resist the cantilever bending moments. In the third case, the inclined tie, whilst unobtrusive, will impose lateral load on the primary structure.

In all cases the connections between the balcony and the primary structure should incorporate a proprietary thermal break to minimise the risk of condensation forming as a result of cold bridging.

As balconies tend to be relatively simple structures the responsibility for their design will initially belong with the architect. He will prepare his designs with advice from the structural engineer. The trade contractor will usually adopt responsibility for the detailed design, manufacture and installation of the balcony. The architect will retain responsibility for overall co-ordination and the structural engineer will

Figure 9.22   Typical cantilever balcony

remain responsible for checking that the primary structure can resist the loads that are applied to it.

9.3.7 Barriers

The following advice is taken from BS 6180,[10] the code of practice for the design of barriers in and around buildings. The purpose of a barrier in this context is either to protect a person from falling or to protect a critical structural element from vehicle impact, where vehicle speed is no greater than 16 km/hr.

Barrier height is governed by the building regulations. Barrier spacing can be critical in the case of sports grounds where guidance may be obtained from the *Guide to Safety at Sports Grounds*.[11] Barriers that are intended to protect people

Figure 9.23   Typical tied balcony

should avoid sharp edges, thin sections, open ended tubes, or projecting details that might cause injury. Loading should be obtained from BS 6399-1.[12] Deflection under normal service conditions should not cause alarm and so should be limited to 25 mm.

Barriers designed to resist vehicular impacts may be distorted by such impacts but should remain substantially in place. Loading should be obtained from BS 6180. They should be designed where possible to redirect vehicles after impact so as to minimise damage.

The supporting structure should be able to resist all applied loads without excessive stress or distortion and the support fixings should be designed so that:

- all joints can provide the full strength of the members being joined
- where any uncertainty exists with regard to the strength of any component in the fixing, the design loading should be increased by 50%
- reliance on the pull-out capacity of a single fixing is avoided.

These recommendations are intended to ensure that under an extreme load condition, failure will be indicated by distortion and not by total collapse, as would be brought about by a failure of the fixing, attachment or anchorage system.

Steel should either be stainless, if an attractive appearance or significant durability is required, or be hot-dip galvanised or be sprayed with zinc or aluminium. Alternatively pre-galvanised or pre-coated steel strip is available. Fasteners and fittings should either be stainless steel, or should be hot-dip galvanised.

9.3.8 Industrial stairs, ladders, and walkways

It is necessary to provide access for inspection, cleaning and maintenance purposes to all areas of a building and in particular to roofs and plantrooms. This will require the provision of stairs, ladders and walkways. Their design requirements are described in BS 5395–3.[13] The principal consideration is safety, to protect those using them from slipping or falling. Edge protection is provided to prevent tools or equipment from being inadvertently pushed over an edge.

Treads and walking surfaces are generally formed from metal plate or open mesh using galvanised steel.

These components are generally procured through the relevant trade contract, usually the services or cladding sub-contract, with design and manufacture by a specialist supplier.

References to Chapter 9

1. British Standards Institution (2011) BS 6180:2011 *Barriers in and about buildings – Code of practice*. London, BSI.
2. The Building Regulations for England and Wales 2000: Part B. B1 Means of warning and escape, B2 Internal fire spread (linings), B3 Internal fire spread (Structure), B4 External fire spread, B5 Access and facilities for the fire service. London, The Stationery Office, 2004.
3. National Structural Steelwork Specification (2007) *Specification for Building Construction*, 5th ed., 2007 (BCSA Publication 203/07). London, BCSA.
4. The Steel Construction Institute (2006) *Connections between steel and other materials*, SCI Publication 102. Ascot, SCI.
5. CIRIA (2001) *Managing Project Change – a best practice guide*, Guide C556. London, CIRIA.
6. British Standards Institution (2010) BS5395-1:2010 *Code of practice for the design, construction and maintenance of straight stairs and winders*. London, BSI.
7. British Standards Institution (1998) BS 5655-1-1986 *Code of practice for the installation of lifts and service lifts*. Replaced by BS EN 81-1:1998+A3:2009. London, BSI.

8. The Steel Construction Institute (1994) *Electric lift installations in steel frame buildings*, SCI Publication 103. Ascot, SCI.

9. The Steel Construction Institute (2004) *Steel in multi storey residential buildings* SCI Publication P332. Ascot, SCI.

10. British Standards Institution (2011) BS 6180:2011 *Barriers in and about buildings – Code of practice*. London, BSI.

11. Great Britain. Department of Culture, Media and Sport and The Scottish Office (1997) *Guide to Safety at Sports Grounds*. London, The Stationery Office.

12. British Standards Institution (1996) BS6399-1:1996 *Loading for buildings – Part 1: Code of practice for dead and imposed loads*. Replaced by BS EN 1991-1-1:2002, BS EN 1991-1-7:2006. London, BSI.

13. British Standards Institution (2001) BS 5395-3:1985 *Code of practice for the design of industrial type stairs, permanent ladders and walkways*. Replaced in part by BS EN ISO 14122. London, BSI.

Chapter 10
Applied metallurgy of steel

by RICHARD THACKRAY and MICHAEL BURDEKIN

10.1 Introduction

The versatility of steel for structural applications rests on the fact that it can be readily supplied, at a relatively cheap price, in a wide range of different product forms and with a useful range of material properties. The key to understanding the versatility of steel lies in its basic metallurgical behaviour. Steel is an efficient material for structural purposes because of its good strength-to-weight ratio. A diagram of strength-to-weight ratio against cost per unit weight for various structural materials is shown in Figure 10.1. Steel can be supplied with strength levels from about $250 \, \text{N/mm}^2$ up to about $2000 \, \text{N/mm}^2$ for common structural applications, although the strength requirements may limit the product form. The material is normally ductile with good fracture toughness for most practical applications. Product forms range from thin sheet material, through optimised structural sections and plates, to heavy forgings and castings of intricate shape. Although steel can be made to a wide range of strengths, it generally behaves as an elastic material with a high (and relatively constant) value of the elastic modulus up to the yield or proof strength. It also usually has a high capacity for accepting plastic deformation beyond the yield strength, which is valuable for drawing and forming of different products, as well as for general ductility in structural applications.

Steel derives its mechanical properties from a combination of chemical composition, heat treatment and manufacturing processes. While the major constituent of steel is always iron, the addition of very small quantities of other elements can have a marked effect upon the type and properties of steel. These elements also produce a different response when the material is subjected to heat treatments involving cooling at a prescribed rate from a particular peak temperature. The manufacturing process may involve combinations of heat treatment and mechanical working, which are of critical importance in understanding the subsequent performance of steels and what can be done satisfactorily with the material after the basic manufacturing process.

Steel Designers' Manual, Seventh Edition. Edited by Buick Davison and Graham W. Owens.
© 2012 Steel Construction Institute. Published 2012 by Blackwell Publishing Ltd.

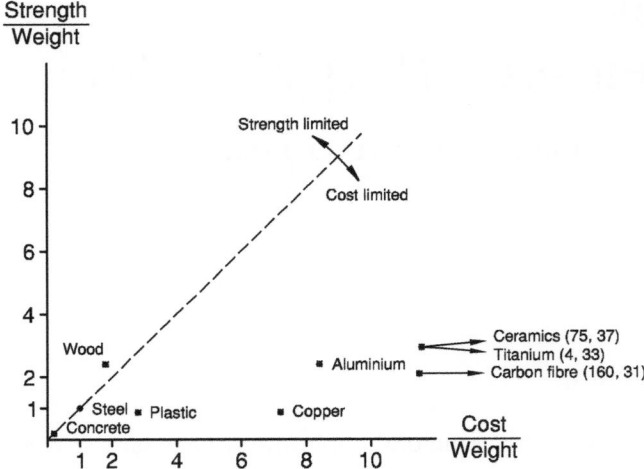

Figure 10.1   Strength/weight and cost/weight ratios for different materials normalised to steel

Although steel is such an attractive material for many different applications, two particular problems which must be given careful attention are those of corrosion behaviour and fire resistance, which are dealt with in detail in Chapters 36 and 35 respectively. Corrosion performance can be significantly changed by choice of a steel of appropriate chemical composition and heat treatment, as well as by corrosion protection measures. Although normal structural steels retain their strength at temperatures up to about 300°C, there is a progressive loss of strength above this temperature so that in an intense fire, bare steel may lose the major part of its structural strength. The hot strength and creep strength of steels at high temperature can be improved by special chemical formulation but it is usually cheaper to either provide fire protection for normal structural steels by protective cladding, or adopt a structural fire engineering approach.

10.2 Chemical composition

10.2.1 General

The key to understanding the effects of chemical composition and heat treatment on the metallurgy and properties of steels is to recognise that the properties depend upon the following factors:

1. microstructure
2. grain size
3. non-metallic inclusions
4. precipitates within grains or at grain boundaries
5. the presence of absorbed or dissolved gases.

Steel is basically iron with the addition of small amounts of carbon up to a maximum of 1.67% by weight and other elements added to provide particular mechanical properties. Above 1.67% carbon, the material generally takes the form of cast iron. As the carbon level is increased, the effect is to raise the strength level but reduce the ductility and make the material more sensitive to heat treatment. The cheapest and simplest form is, therefore, a plain carbon steel commonly supplied for the steel reinforcement in reinforced concrete structures, for wire ropes, for some general engineering applications in the form of bars or rods, and for some sheet/strip applications. However, plain carbon steels at medium to high carbon levels give rise to problems where subsequent fabrication/manufacturing takes place, particularly where welding is involved, and more versatility can be obtained by keeping carbon to a relatively low level and adding other elements in small amounts. When combined with appropriate heat treatments, addition of these other elements produces higher strength while retaining good ductility, fracture toughness and weldability, or the development of improved hot strength, or improved corrosion-resistance. The retention of good fracture toughness with increased strength is particularly important for thick sections, and for service applications at low temperatures where brittle fracture may be a problem. Hot strength is important for service applications at high temperatures such as pressure vessels and piping in the power generation and chemical process plant industries. Corrosion-resistance is important for any structures exposed to the environment, particularly for structures immersed in sea water. Weathering grades of steel are designed to develop a tight adherent oxide layer which slows down and stifles continuing corrosion under normal atmospheric exposure of alternate wet and dry conditions. Stainless steels are designed to have a protective oxide surface layer which reforms if any damage takes place to the surface, and these steels are therefore designed not to corrode under oxidising conditions. Stainless steels find particular application in the chemical industry.

10.2.2 Added elements

The addition of small amounts of carbon to iron increases the strength and the sensitivity to heat treatment (or hardenability, see later). Other elements which also affect strength and hardenability, although to a much lesser extent than carbon, are manganese, chromium, molybdenum, nickel and copper. Their effect is principally on the microstructure of the steel, enabling the required strength to be obtained for given heat treatment/manufacturing conditions, while keeping the carbon level very low. Refinement of the grain structure of steels leads to an increase in yield strength and improved fracture toughness and ductility at the same time, and this is therefore an important route for obtaining enhanced properties in steels. Although heat treatment and, in particular, cooling rate are key factors in obtaining grain refinement, the presence of one or more elements which promote grain refinement by aiding the nucleation of new grains during cooling is also extremely beneficial. Elements

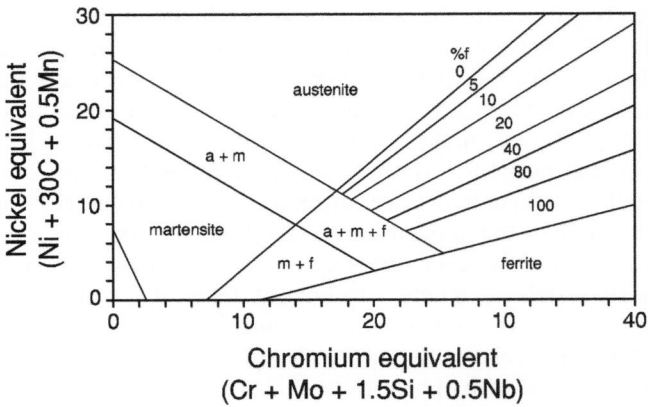

Figure 10.2   Constitution (Schaeffler) diagram for stainless steels

which promote grain refinement, and which may be added in small quantities up to about 0.050%, are niobium, vanadium and aluminium.

The major elements which may be added for hot strength and also for corrosion-resistance are chromium, nickel and molybdenum. Chromium is particularly beneficial in promoting corrosion-resistance as it forms a chromium oxide surface layer on the steel, which is the basis of stainless steel corrosion protection in oxidising environments. When chromium and nickel are added in substantial quantities with chromium levels in the range 12% to 25%, and nickel content up to 20%, different types of stainless steel can be made. As with the basic effect of carbon in the iron matrix, certain other elements can have a similar effect to chromium or nickel but on a lesser scale. From the point of view of the effects of chemical composition, the type of stainless steel formed by different combinations of chromium and nickel can be shown on the Schaeffler diagram in Figure 10.2. The three basic alternative types of stainless steel are ferritic, austenitic and martensitic stainless steels, which have different inherent lattice crystal structures and microstructures, and hence may show significantly different performance characteristics.

10.2.3 Non-metallic inclusions

The presence of non-metallic inclusions has to be carefully controlled for particular applications. Such inclusions arise as a residue from the ore or scrap in the steelmaking process, through unwanted reoxidation during refining, or due to erosion of refractories, and steps must be taken to reduce them to the required minimum level. The cleanness of steel is a measure of the number and size of inclusions in the steel. Even small aggregates of inclusions can lead to problems in components subjected to large stresses or fatigue.

The commonest impurities are sulphur and phosphorus, high levels of which lead to reduced resistance to ductile fracture and the possibility of cracking problems in welded joints. For weldable steels, the sulphur and phosphorus levels must be kept less than 0.050%, and, with modern steel-making practice, should now preferably be less than 0.010%. They are not always harmful, however, and in cases where welding or fracture toughness are not important, deliberate additions of sulphur may be made up to about 0.15% to promote free machining qualities of steel, and small additions of phosphorus may be added to non-weldable weathering grade steels. Other elements which may occur as impurities and may sometimes have serious detrimental effects in steels, are tin, antimony and arsenic, which in certain steels may promote a problem known as temper embrittlement, in which the elements migrate to grain boundaries if the steel is held in a temperature range between about 500°C and 600°C for any length of time. At normal temperature, steels in this condition can have very poor fracture toughness, with failure occurring by inter-granular fracture. It is particularly important to ensure that this group of tramp elements is eliminated from low alloy steels. Steels with high levels of dissolved gases, particularly oxygen and nitrogen, can behave in a brittle manner. The level of dissolved gases can be controlled by addition of small amounts of deoxidising elements such as aluminium, silicon or manganese. These elements have a high affinity particularly for oxygen, and will combine with the gas to either float out of the liquid steel at high temperatures, or remain as a distribution of solid non-metallic inclusions, some of which can float to the top of the liquid steel and be absorbed by the steelmaking slag.

The composition of the inclusion is also relevant. Some inclusions are ductile and will deform when the steel is hot worked but some, such as alumina, will not and must be avoided in certain applications, for example in the production of rod or wire. A steel with no such additions to control oxygen level is known as a rimming steel. Aluminium also helps in controlling the free nitrogen level, which it is important to keep at low levels where the phenomenon of strain ageing embrittlement may be important. Removal of gases can also be carried out by degassing the liquid steel under a vacuum, or by remelting processes such as vacuum arc remelting, or electro-slag refining.

10.3 Heat treatment

10.3.1 Effect on microstructure and grain size

During the manufacture of steel, the required chemical composition is achieved while it is in the liquid state at high temperature. As the steel cools, it solidifies at the melting temperature at about 1350°C, but substantial changes in structure take place during subsequent cooling and may also be affected by further heat treatments. If the steel is cooled slowly, it is able to take up the equilibrium type of lattice crystal structure and microstructure appropriate to the temperature and chemical composition.

These conditions can be summarised on a phase or equilibrium diagram for the particular composition; the equilibrium diagram for the iron–iron carbide system is shown in Figure 10.3. Essentially this is a diagram of temperature against percentage

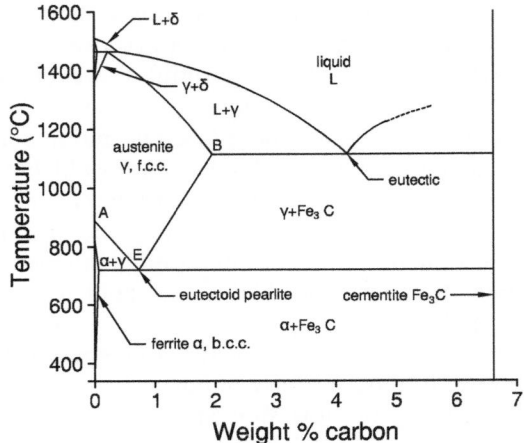

Figure 10.3 Equilibrium phase diagram for iron – iron carbide system (f.c.c., face-centred cubic; b.c.c., body-centred cubic)

of carbon by weight in the iron matrix. At 6.67% carbon, an inter-metallic compound called cementite is formed, which is an extremely hard and brittle material. At the left-hand end of the diagram, with very low carbon contents, the equilibrium structure at room temperature is ferrite. At carbon contents between these limits the equilibrium structure is a mixture of ferrite and cementite in proportion depending on the carbon level. On cooling from the melting temperature, at low carbon levels a phase known as delta ferrite is formed first, which then transforms to a different phase called austenite. At higher carbon levels, the melting temperature drops with increasing carbon level and the initial transformation may be direct to austenite. The austenite phase has a face-centred cubic lattice crystal structure, which is maintained down to the lines AE and BE on Figure 10.3. As cooling proceeds slowly, the austenite then starts to transform to the mixture of ferrite and cementite which results at room temperature. However, point E on the diagram represents a eutectoid at a composition of 0.83% carbon at which ferrite and cementite precipitate alternately in thin laths to form a structure known as pearlite. At compositions less than 0.83% carbon, the type of microstructure formed on slow cooling transformation from austenite is a mixture of ferrite and pearlite. Each type of phase present at its appropriate temperature has its own grain size, and the ferrite/pearlite grains tend to precipitate in a network within and based on the previous austenite grain boundary structure. The lattice crystal structure of the ferrite material which forms the basic matrix is essentially a body-centred cubic structure. Thus in cooling from the liquid condition, complex changes in both lattice crystal structure and microstructure take place dependent on the chemical composition. For the equilibrium diagram conditions to be observed, cooling must be sufficiently slow to allow time for the transformations in crystal structure and for the diffusion/migration of carbon to take place to form the appropriate microstructures.

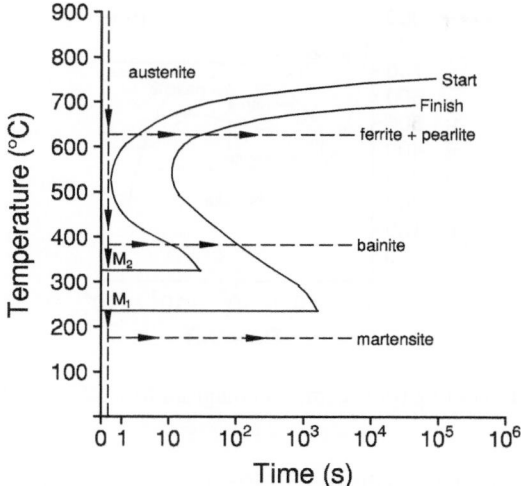

Figure 10.4 Isothermal transformation diagram for 0.2% C, 0.9% Mn steel

If a steel is cooled from a high temperature and held at a lower constant temperature for sufficient time, different conditions may result; these are represented on a diagram known as the isothermal transformation diagram. The form of the diagram depends on the chemical analysis and in particular on the carbon or related element content. In plain carbon steels, the isothermal transformation diagram typically has the shape of two letters 'C', each with a horizontal bottom line as shown in Figure 10.4. The left-hand/upper curve on the diagram of temperature against time represents the start of transformation, and the right-hand/lower curve represents the completion of transformation with time. For steels with a carbon content below the eutectoid composition of 0.83% carbon, holding at a temperature to produce isothermal transformation through the top half of the letter C leads to the formation of a ferrite/pearlite microstructure. If the transformation temperature is lowered to pass through the lower part of the C curves, but above the bottom horizontal lines, a new type of microstructure is obtained, which is called bainite, which is somewhat harder and stronger than pearlite, but also tends to have poorer fracture toughness. If the transformation temperature is dropped further to lie below the two horizontal lines, transformation takes place to a very hard and brittle substance called martensite. In this case the face-centred cubic lattice crystal structure of the austenite is not able to transform to the body-centred cubic crystal structure of the ferrite, and the crystal structure becomes locked into a distorted form known as a body-centred tetragonal lattice. Bainite and martensite do not form on equilibrium cooling but result from quenching to give insufficient time for the equilibrium transformations to take place.

The position and shape of the C curves on the time axis depend on the chemical composition of the steel. Higher carbon contents move the C curve to the right on the time axis, making the formation of martensite possible at slower cooling rates. Alloying elements change the shape of the C curves, and an example for a low-alloy

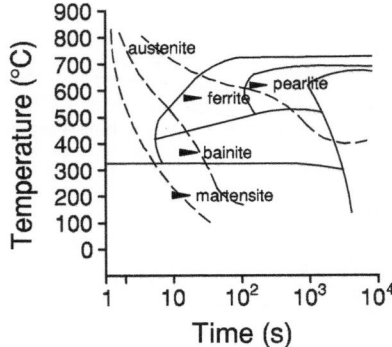

Figure 10.5   Continuous cooling transformation diagram for 0.4% C, 0.8% Mn, 1% Cr, 0.2% Mo steel

steel is shown in Figure 10.5. Additional effects on microstructure, grain size and resultant properties can be obtained by combinations of mechanical work at appropriate temperatures during manufacture of the basic steel.

In addition to the effect of cooling rate on microstructure, the grain size is significantly affected by time spent at high temperatures and subsequent cooling rate. Long periods of time at higher temperatures within a particular phase lead to the merging of the grain boundaries and growth of larger grains. For ferritic crystal structures, grain growth starts at temperatures above about 600°C, and hence long periods in the temperature range 600°C to 850°C with slow cooling will tend to promote coarse grain size ferrite/pearlite microstructures. Faster cooling through the upper part of the C curves will give a finer grain structure but still of ferrite/pearlite microstructure.

The type of microstructure present in a steel can be shown and examined by the preparation of carefully polished and etched samples viewed through a microscope. Etching with particular types of reagent attacks different parts of the microstructure preferentially, and the etched parts are characteristic of the type of microstructure. Examples of some of the more common types of microstructure mentioned above are shown in Figure 10.6. The basic microstructure of the steel is usually shown by examination in the microscope at magnifications of 100 to about 500 times. Where it is necessary to examine the effects of very fine precipitates or grain boundary effects, it may be necessary to go to higher magnifications. With the electron microscope it is possible to reach magnifications of many thousands and, with specialised techniques, to reach the stage of seeing dislocations and imperfections in the crystal lattice itself.

10.3.2 Heat treatment in practice

In practical steelmaking or fabrication procedures, cooling occurs continuously from high temperatures to lower temperatures. The response of the steel to this form of

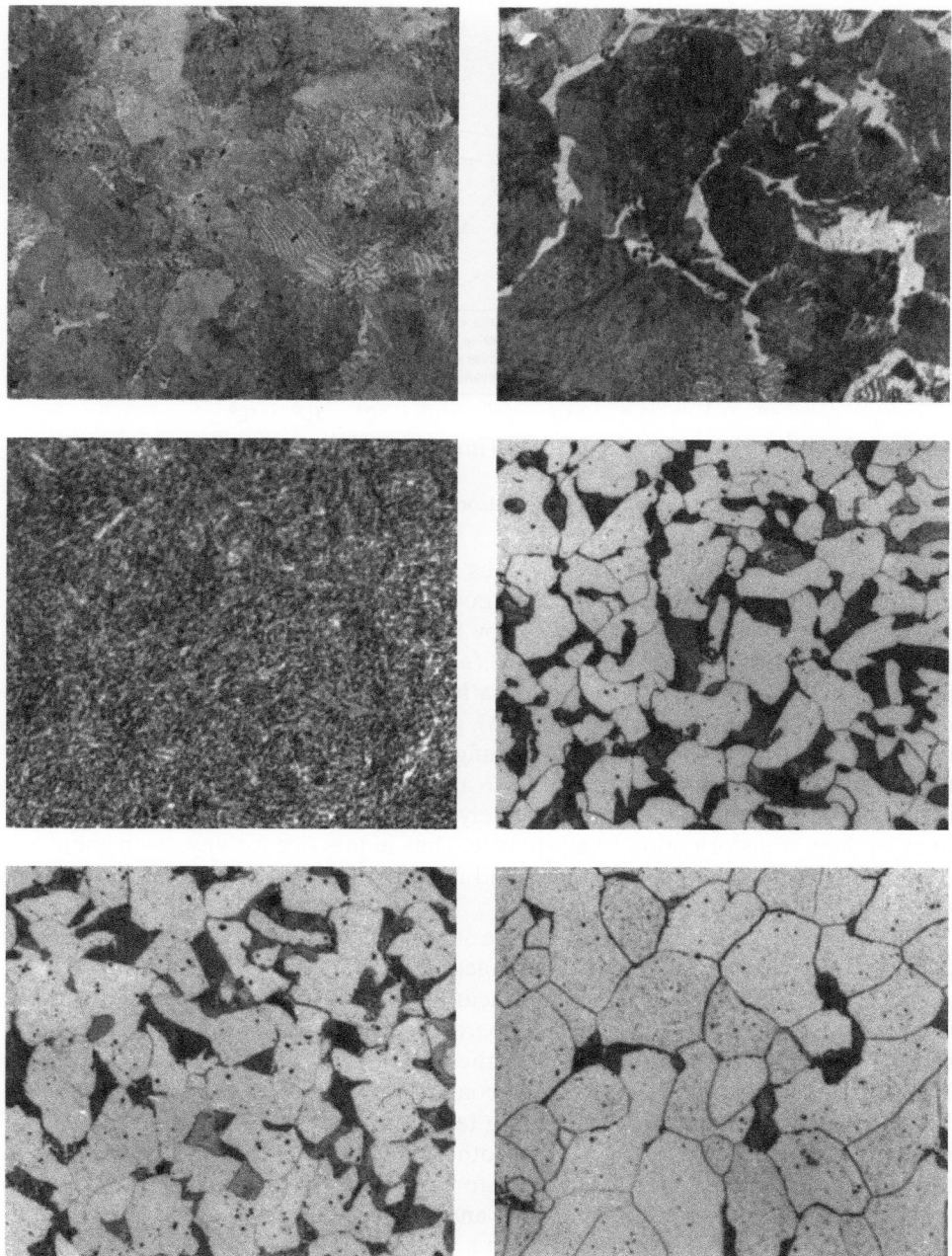

Figure 10.6   Examples of common types of microstructure in steel (magnification × 500) (courtesy of Manchester Materials Science Centre, UMIST)

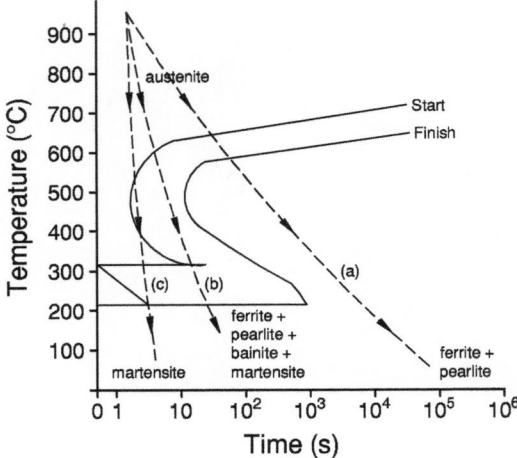

Figure 10.7   Continuous cooling transformation diagram for 02% C, 0.9% Mn steel

cooling can be shown on the continuous cooling transformation diagram (CCT diagram) of Figure 10.7. This resembles the isothermal transformation diagram, but the effect of cooling rate can be shown by lines of different slopes on the diagram. For example, slow cooling, following line (a) on Figure 10.7, passes through the top part of the C curve and leads to the formation of a ferrite/pearlite mixture. Cooling at an intermediate rate, following line (b), passes through pearlite/ferrite transformation at higher temperatures, but changes to bainite transformation at lower temperatures so that a mixture of pearlite and bainite results. Rapid cooling following line (c) misses the C curves completely and passes through the two horizontal lines to show transformation to martensite. Thus in practice for any given composition of steel different microstructures and resultant properties can be produced by varying the cooling rate.

The microstructure and properties of a steel can be changed by carefully chosen heat treatments after the original manufacture of the basic product form. A major group of heat treatments is effected by heating the steel to a temperature such that it transforms back to austenite, this temperature being normally in the range 850°C to 950°C. It is important to ensure that the temperature is sufficient for full transformation to austenite, otherwise a very coarse-grained ferritic structure may result. It is also important that the austenitising temperature is not too high, and that the time at this temperature is not too long, otherwise a coarse-grained austenite structure will form, making subsequent transformation to fine grains more difficult. A heat treatment in which cooling is slow and essentially carried out in a furnace is known as annealing. This tends to lead to a relatively coarse-grained final structure, as predicted by the basic equilibrium phase diagram, and is used to put materials into their softest condition. If the steel is allowed to cool freely in air from the austenitising temperature, the heat treatment is known as normalising, which gives a finer grain size and hence tends to higher yield strength and better toughness for a given composition of steel. Normalising may be combined with rolling of a particular product form over a relatively narrow band of temperatures, followed by natural cooling in air, in which case it is known as controlled rolling. When the steel product form is cooled more rapidly by immersing it directly into oil or water, the heat treatment is known as quenching. Quenching into a water bath is generally more severe than quenching into an oil bath.

A second stage heat treatment to temperatures below the austenitising range is frequently applied, known as tempering. This has the effect of giving more time for the transformation processes which were previously curtailed to develop further, and can permit changes in the precipitation of carbides, allowing them to merge together and develop into larger or spheroidal forms. These thermally activated events are highly dependent on temperature and time for particular compositions. The net effect of tempering is to soften previously hardened structures and make them tougher and more ductile.

Both plain carbon and low alloy steels can be supplied in the quenched and tempered condition for plates and engineering sections to particular specifications. The term 'hardenability' is used to describe the ability of steel to form martensite to greater depths from the surface, or greater section sizes. There are, therefore, practical limits of section thickness or size at which particular properties can be obtained. It should be noted that the term 'hardenability' does not refer to the absolute hardness level which can be achieved, but to the ability to develop uniform hardening throughout the cross-section. Cooling rates vary at different positions in the cross-section as heat is conducted away in a quenching operation from the surface.

It is sometimes necessary to apply heat treatment to components or structures after fabrication, particularly when they have been welded. The aim is mainly to relieve residual stresses but heat treatment may also be required to produce controlled metallurgical changes in the regions where undesirable effects of welding have occurred. Applications at high temperatures may also lead to metallurgical changes taking place in service. It is vitally important that, where any form of heat treatment is applied, the possible metallurgical effects on the particular type of steel are taken into account.

Heat treatments are sometimes applied to produce controlled changes in shape or correction of distortion and again temperatures and times involved in these heat treatments must be carefully chosen and controlled for the particular type of steel being used.

10.4 Manufacture and effect on properties

10.4.1 Steelmaking

There are currently two major steelmaking routes. The traditional blast furnace/ basic oxygen steelmaking (BOS) route is responsible for about 65% of the world's steel production. The second route uses an electric arc furnace to melt and refine scrap steel. In the blast furnace/BOS route, iron ore is used as the source of iron. When combined with coke, the iron ore is sintered, limestone is added and this burden is fed into the top of a blast furnace, which is essentially a cylindrical steel shell lined with refractory bricks. Preheated air is introduced through nozzles near the base of the blast furnace, the oxygen in this air combines with the coke to produce carbon monoxide gas at very high temperatures. This gas then rises up through the burden, heating the ore and causing it to convert to molten metallic iron, which is collected at the base of the furnace. This so called hot metal would, on solidification, essentially form a cast iron, rather than steel and would therefore display inferior strength and toughness properties. The steelmaking process is designed to reduce the amount of carbon, manganese and silicon and eliminate phosphorous and sulphur where possible, whilst retaining the iron. In order to achieve this, the hot metal is poured into a basic oxygen steelmaking vessel, to which scrap steel is added along with some limestone. Oxygen gas is then blown at supersonic speeds on to the metal using a water-cooled lance, and the desired refining reactions take place. Temperature and chemical composition can be monitored during this period. At the end of the oxygen blow, the BOS vessel is tilted and the steel is poured into a ladle for further treatment. At this stage, alloying or deoxidising additions can also be made.

The electric arc furnace route for producing steel is different. Rather than using iron ore as the source of iron, the EAF route usually uses scrap steel as the source. The arc furnace is very flexible with regards to the raw materials it can use, but the main component is usually purchased or works arising scrap. Scrap is carefully selected so that the compositional adjustment needed is small, although some impurities will be present and so some refining will be needed. The scrap material is loaded into the arc furnace, the roof closed, the electrodes lowered, and melting of the charge begins. Again, temperature and composition can be monitored, and oxygen blown into the furnace to assist the refining reactions if necessary. At the end of the melting process, the furnace is tilted and the steel poured into a ladle for further treatment.

The importance of secondary steelmaking has increased greatly over the last 20 years. There are a number of processes available to the steelmaker for refining liquid steel and these have resulted in steel with more reproducible and superior properties. The main aims of secondary steelmaking processes are to remove any unwanted elements, particularly sulphur, hydrogen and nitrogen, and give better control of inclusion type and quantity in the liquid steel. This last aim in particular is important, given the need for clean steel making practice for some critical applications.

After secondary steelmaking, the steel is allowed to solidify. Although ingot casting is still used for special grades, around 95% of the steel produced is continuously cast. Molten steel is teemed from the ladle, through a tundish, where inclusions can be removed, then into the mould via a submerged refractory nozzle. The mould oscillates and is lubricated to prevent sticking and to control the solidification rate. On exit from the water-cooled copper mould, the steel forms a continuous strand and is further cooled by water sprays, before being cut to length with a gas torch when solidification is complete. These semi-finished products take the form of blooms, slabs or billets, depending on the dimensions. Generally blooms are larger than billets and can be square or rectangular in cross-section, billets can also be round. These products are then ready to be further processed into finished form via hot and cold working.

Stainless steel can be produced via the EAF route but is more commonly produced using the argon-oxygen decarburisation (AOD) process. In this process, the arc furnace is used to melt scrap steel and then the molten metal is transferred to another vessel for refining. In order to prevent chromium loss during the refining, the oxidising gas is a mixture of oxygen and argon. In this way, the carbon can be removed without excessive losses of expensive chromium, and stainless steel of the correct composition can be formed.

Generally, the blast furnace/BOS route is suitable for producing large quantities of low carbon bulk steel structural grades, whereas the EAF route is suited to production of special, highly alloyed engineering grades.

10.4.2 Casting and forging

If the final product form is a casting, the liquid steel is poured direct into a mould of the required geometry and shape. Steel castings provide a versatile way of achieving the required finished product, particularly where either many items of the same type are required and/or complex geometries are involved. Special skills are required in the design and manufacture of the moulds in order to ensure that good quality castings are obtained with the required mechanical properties and freedom from significant imperfections or defects. High-integrity castings for structural applications have been successfully supplied for critical components in bridges, such as the major cable saddles for suspension bridges, cast node and tubular sections for offshore structures, and the pump bowl casings for pressurised water reactor systems. The size of component that can be made in cast form has increased over the last 10 years, and it is now possible to produce steel castings of over 350 tonnes. However, castings still only account for a small proportion of total finished steel applications.

Another specialist route to the finished steel product is by forging, in which a bloom is heated to the austenitising temperature range and formed by repeated mechanical pressing in different directions to achieve the required shape. The combination of temperature and mechanical work enables high-quality products with good mechanical properties to be obtained. An example of high-integrity forgings is the production of steel rings to form the shell/barrel of the reactor pressure vessel in a pressurised water reactor system. Again, the proportion of steel production as forgings is a relatively small and specialised part of overall steel production.

10.4.3 Rolling

Rolling accounts for by far the largest tonnage of finished steel products. The principle of rolling is simple, the semi-finished steel bloom, billet or slab is reheated and passed through a series of rolls, which reduce the thickness of the material, whilst at the same time increasing the length. Primary rolling, or roughing, is carried out initially to break down the coarse cast structure, and during primary rolling the steel may pass backwards and forwards through the same set of rolls many times. The steel may also be rotated 90° and rolled widthways to attain the correct width. Finish rolling produces material of the correct final thickness and, in this case, the material may pass though several sets of rolls in sequence. The majority of strip products are cold rolled, after the oxide scale has been removed by pickling in acid.

Strip steel is collected at the end of the rolling process as a coil, which can then be further treated. For example, to improve the corrosion resistance, the steel can be coated with zinc in a galvanising process. It can also be coated in plastic, or patterned in order to improve both the aesthetic and corrosion properties.

Slabs are generally used to produce plate and strip, whereas blooms and billets are used to produce rod or bar and structural sections such as joists, columns and beams, as well as rail sections. Tubes or pipes can be made from either plate or strip material.

In 2008, the construction industry accounted for some 29% of UK steel demand (around 4 million tonnes), 14% of which went into structural steelwork and the remaining 15% into building and civil engineering.[1,2]

For the structural industry steel slabs can be rolled into plates of the required thickness, or into structural sections such as universal columns, universal beams, angle sections, rail sections, etc. Round blooms or ingots can be processed by a seamless tube rolling mill into seamless tubes of different diameters and thicknesses, or solid bar subsequently drawn out into wire. Tubes can be used either for carrying fluids in small-diameter pipelines or as structural hollow sections of circular or rectangular shape. The shape of engineering structural sections is determined by the required properties of the cross-section such as cross-sectional area and moments of inertia about different axes to give an effective distribution of the weight of the material for structural purposes. Rolled structural sections are supplied in a standard range of shapes detailed in BS 4: Part 1: 2005, and a selection of typical shapes and section properties is given in the Appendix (pp. 1151–3).

10.4.4 Defects

In any bulk manufacturing process, such as the manufacture of steel, it is inevitable that a small proportion of the production will have imperfections which may or may not be harmful from the point of view of intended service performance of the product. In general, the appropriate applications standards have clauses which limit any such imperfections to acceptable and harmless levels. In castings, a particular family of imperfections can occur which are dependent on the material and the geometry being manufactured. The most serious types of imperfection are cracks caused by shrinkage stresses during cooling, particularly at sharp changes in cross-section. A network of fine shrinkage cracks or tears can sometimes develop, again particularly at changes in cross-section where the metal is subjected to a range of different cooling rates. The second type of imperfection in castings is solid inclusions, particularly in the form of sand where this medium has been used to form the moulds. Porosity, or gaseous inclusions, is not uncommon in castings to some degree and again tends to occur at changes in cross-section. There is usually appreciable tolerance for minor imperfections such as sand inclusions or porosity provided these do not occur to extreme levels.

In rolled or drawn products, the most common types of defect are either cold laps or rolled-in surface imperfections. A lap is an imperfection which forms when the material has been rolled back on to itself but has not fully fused at the interface. Surface imperfections may occur from the same cause where a tongue of material is rolled down but does not fuse fully to the underlying material. Both of these faults are normally superficial and in any serious cases ought to be eliminated by final inspection at the steel mills. A third form of imperfection which can occur in plates, particularly when produced from the ingot route, is a lamination: the failure of the material to fuse together, usually at the mid-thickness of the plate. Laminations tend to arise from the rolling-out of pipes, or separation on the centreline of an ingot at either top or bottom which formed at the time of casting the ingot. Normal practice is that sufficient of the top and bottom of an ingot is cut off before subsequent processing to prevent laminations being rolled into subsequent products, but nevertheless they do occur from time to time. Fortunately the development of cracks in rolled products is relatively rare, although it may occasionally occur in drawn products or as a result of quenching treatments in heat-treated products.

Since much of the manufacture of steels involves processing at, and subsequent cooling from, high temperatures, it will be appreciated that high thermal stresses can develop during differential cooling and these can lead to residual stresses in the finished product. In many cases these residual stresses are of no significance to the subsequent performance of the product but there are situations where their effect must be taken into account. The two in which residual stresses from the steel manufacture are most likely to be of importance are where close tolerance machining is required, or where compression loading is being applied to slender structural sections. For the machining case it may be necessary to apply a stress relief treatment, or alternatively to carry out the machining in a series of very fine cuts. The effect of the inherent manufacturing residual stresses on the structural behaviour of sections is taken account of in Eurocode 3[3] by the choice of buckling curve and imperfection factor α (see Clause 6.3.1.2(2)).

10.5 Engineering properties and mechanical tests

As part of the normal quality control procedures of the steel manufacturer, and as laid down in the different specifications for manufacture of steel products, tests are carried out on samples representing each batch of steel and the results recorded on a test certificate. At the stage when the chemical analysis of the steel is being adjusted in the steelmaking furnace, samples are taken from the liquid steel melt at different stages to check the analysis results. Samples are also taken from the melt just before the furnace is tapped, and the analysis of these test results is taken to represent the chemical composition of the complete cast. The results of this analysis are given on test certificates for all products which are subsequently made from the same initial cast. The test certificate will normally give analysis results for C, Mn, Si, S and P for all steels, and where the specification requires particular elements to be present in a specific range the results for these elements will also be given. Even when additional elements are not specified, the steel manufacturer will often provide analysis results for residual elements which may have been derived from scrap used or which could affect subsequent fabrication of performance during fabrication particularly welding. Thus steel supplied to BS Standard[4,5] will often have test certificates giving Cr, Ni, Cu, V, Mb and Al, as well as the main basic five elements.

In some specifications, the requirement is given for additional chemical analysis testing on each item of the final product form, and this is presented on the test certificates as product analysis, in addition to the cast analysis. This does, however, incur additional costs. The specifications for weldable structural steels give the opportunity for requiring the steelmaker to supply information on the carbon equivalent to assist the fabricator in deciding about precautions during welding (see later). In low-alloy and stainless steels, the test certificates will, of course, give the percentage of the alloying elements such as chromium, nickel, etc.

The test certificates should also give the results of mechanical tests on samples selected to represent each product range in accordance with the appropriate specification. The mechanical test results provided will normally include tensile tests giving the yield strength, ultimate strength and elongation to failure. In structural steels, where the fracture toughness is important, specifications include requirements for Charpy V-notch impact tests to BS 131: Part 2.[6] The Charpy test is a standard notched bar impact test of 10 mm square cross-section with a 2 mm deep V-notch in one face. A series of specimens is tested under impact loading either at one specification temperature or over a range of temperatures, and the energy required to break the sample is recorded. In the European Standards,[7] these notch ductility requirements are specified by letter grades JR, JO, J2 and K2. Essentially these requirements are that the steel should show a minimum of 27 J energy absorption at a specified testing temperature corresponding to the letter grade.

The specifications normally require the steelmaker to extract specimens with their length parallel to the main rolling direction. In fact, it is unlikely that the steel will be wholly isotropic and significant differences in material properties may occur under different testing directions, which would not be evident from the normal test certificates unless special tests were carried out. It is possible, in some specifications, to have material tests carried out both transverse to the main rolling direction and in the through-thickness direction of rolled products. Testing in the through-thickness direction is particularly important where the material may, in fact, be loaded in this direction in service by welded attachments. Since such tests are additional to the normal routine practice of the steel manufacturer, and cost extra both for the tests themselves and for the disruption to main production, it is not unexpected that steels required to be tested to demonstrate properties in other directions are more expensive than the basic quality of steel tested in one standard direction only. The quality-control system at the steel manufacturing plant normally puts markings in the form of stamped numbers or letters on each length or batch of products so that it can be traced back to its particular cast and manufacturing route. In critical structural applications, it is important that this numbering system is transferred on through fabrication to the finished structure so that each piece can be identified and confirmed as being of the correct grade and quality. The test certificate for each batch of steel is therefore a most important document to the steel manufacturer, to the fabricator and to the subsequent purchaser of the finished component or structure. In addition to the chemical composition and mechanical properties, the test certificate should also record details of the steelmaking route and any heat treatments applied to the material by the steel manufacturer.

It is not uncommon for some semi-finished products to be sold by the steel manufacturer to other product finishers or to stockholders. Unless these parties retain careful records of the supply of the material, it may be difficult to trace specific details of the properties of steel bought from them subsequently, although some stockholders do maintain such records.

Where products are manufactured from semi-finished steel and subsequently given heat treatment for sale to the end user, the intermediate manufacturer should produce his own test certificates detailing both the chemical analysis of the steel and the mechanical properties of the finished product. For example, bolts used for structural connections are manufactured from bar material and are normally stamped with markings indicating the grade and type of bolt. Samples of bolts are taken from manufactured batches after heat treatment and subjected to mechanical tests to give reassurance that the correct strength of steel and heat treatment have been used.

10.6 Fabrication effects and service performance

Basic steel products supplied from the steel manufacturer are rarely used directly without some subsequent fabrication. The various processes involved in fabrication may influence the suitability for service of the steel and, over the years, established procedures of good practice have been developed which are acceptable for particular industries and applications.

10.6.1 Cutting, drilling, forming and drawing

Basic requirements in the fabrication of any steel component are likely to be cutting and drilling. In thin sections, such as sheet material, steel can be cut satisfactorily by guillotine shearing and, although this may form a hardened edge, it is usually of little or no consequence. Thicker material in structural sections up to about 15 mm thickness can also be cut by heavy-duty shears, useful for small part pieces such as gussets, brackets, etc. Heavier section thicknesses will usually have to be cut by cold saw or abrasive wheel or by flame cutting. Cold saw and abrasive wheel cutting produce virtually no detrimental effects and give good clean cuts to accurate dimensional tolerances. Flame cutting is carried out using an oxyacetylene torch to burn the steel away in a narrow slit, and this is widely used for cutting of thicker sections in machine-controlled cutting equipment. The intense heating in flame cutting does subject the edge of the metal to rapid heating and cooling cycles and so produces the possibility of a hardened edge in some steels. This can be controlled by either preheating just ahead of the cutting torch or using slower cutting speeds or, alternatively, if necessary, any hardened edge can be removed by subsequent machining. In recent years, laser cutting has become a valuable additional cutting method for thin material, in that intricate shapes and patterns can be cut out rapidly by steering a laser beam around the required shape.

Drilling of holes presents little problem and there are now available numerical/computer controlled systems which will drill multiple groups of holes to the required size and spacing. For thinner material, hole punching is commonly used and, although this, like shearing, can produce a hardened edge, provided the punch is sharp no serious detrimental effects occur in thinner material.

It is sometimes necessary to bend, form or draw steel into different shapes. Reinforcing steel for reinforced concrete structures commonly has to be bent into the form of hooks and stirrups. The curved sections of tubular members of offshore structures or cylindrical parts of pressure vessels are often rolled from flat plate to the required curvature. In these cases, yielding and plastic strain take place as the material is deformed beyond its elastic limit. This straining moves the material condition along its basic stress–strain curve, and it is therefore important to limit the amount of plastic strain used up in the fabrication process so that availability for subsequent service is not diminished to an unacceptable extent. The important variable in limiting the amount of plastic strain which occurs during cold forming is usually the ratio of the radius of any bend to the thickness or diameter of the material. Provided this ratio is kept high, the amount of strain will be limited. Where the amount of cold work which has been introduced during fabrication is excessive, it may be necessary to carry out a reheat treatment in order to restore the condition of the material to give its required properties.

In the manufacture of wire, the steel is drawn through a series of dies gradually reducing its diameter and increasing the length from the initial rod sample. This cold drawing is equivalent to plastic straining and has the effect of both increasing the strength of the material and reducing its remaining ductility, as the material moves along its stress/strain curve. In certain types of wire manufacture, intermediate heat treatments are necessary in order to remove damaging effects of cold work and enhance and improve the final mechanical properties.

10.6.2 Welding

One of the most important fabrication processes for use with steel is welding. There are many different types of welding and this subject is itself a fascinating multi-disciplinary world involving combined studies in physics, chemistry, electronics, metallurgy and mechanical, electrical and structural engineering. Although there is a huge variety of different welding processes, probably the most common and most important ones for general applications are the group of arc welding processes and the group of resistance welding processes. Among the newer processes are the high energy density beam processes such as electron beam and laser welding.

Arc welding processes involve the supply of an intense heat source from an electric arc which melts the parent material locally, and may provide additional filler metal by the melting of a consumable electrode. These processes are extensively used in the construction industry, and for any welding of material thicknesses above the range of sheets. The resistance group of welding processes involve the generation of heat at the interface between two pieces of material by the passage of very heavy current directly between opposing electrodes on each face. The resistance processes do not involve additional filler metal and can be used to produce local joints as spot welds, or a series of such welds to form a continuous seam. This group of processes is particularly suitable for sheet material and is widely used in the automotive and domestic equipment markets.

It will be appreciated that fusion welding processes, as described above, involve rapid heating and cooling locally at the position where a joint is to be made. The temperature gradients associated with welding are intense, and high thermal stresses and subsequent residual stresses on cooling are produced. The residual stresses associated with welding are generally much more severe than those which result during the basic steel manufacturing process itself as the temperature gradients are more localised and intense. Examples of the residual stress distribution resulting from the manufacture of a butt weld between two plates and a T-butt weld with one member welded on to the surface of a second, are shown in Figure 10.8. As will be

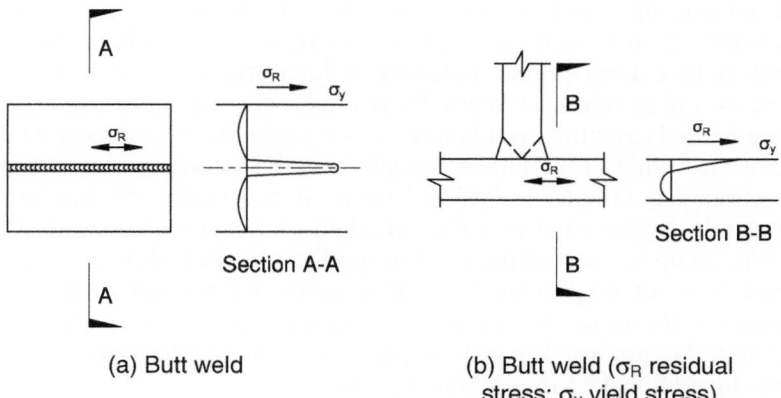

(a) Butt weld

(b) Butt weld (σ_R residual stress; σ_y yield stress)

Figure 10.8 Typical weld residual stress distributions: (a) butt weld, (b) T-butt weld (σ_R residual stress; σ_y yield stress)

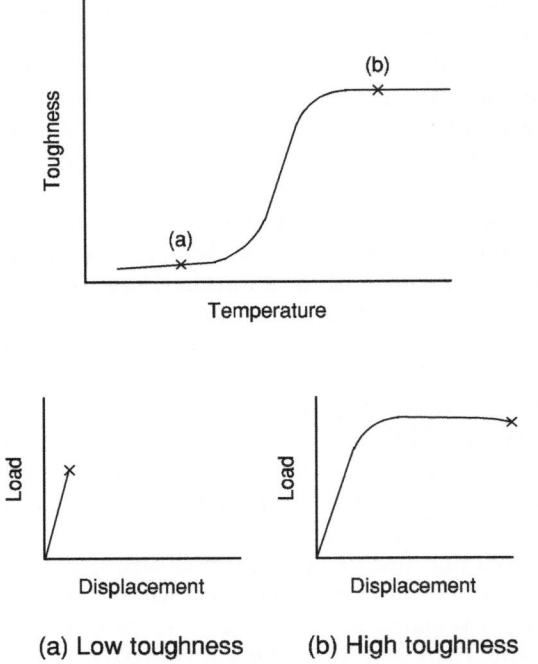

Figure 10.9   Effective toughness and residual stresses on strength in tension

seen from other chapters, residual stresses can be important in the performance of steel structures because of their possible effects on brittle fracture, fatigue and distortion. If a steel material has low fracture toughness and is operating below its transition temperature, residual stresses may be very important in contributing to failure by brittle fracture at low applied stresses. If, on the other hand, the material is tough and yields extensively before failure, residual stresses will be of little importance in the overall structural strength. These effects are summarised in Figure 10.9.

In fatigue-loaded structures, residual stresses from welding are important in altering the mean stress and stress ratio. Although these are secondary factors compared to the stress range in fatigue, the residual stress effect is sufficiently important that it is now commonly assumed that the actual stress range experienced at a weld operates with an upper limit of the yield strength due to locked-in residual stresses at this level. Thus, although laboratory experiments demonstrate different fatigue performance for the same stress range at different applied mean stress in plain unwelded material, the trend in welded joints is for the applied stress ratio effect to be overridden by locked-in residual stresses.

The effect on distortion from welding residual stresses can be significant, both at the time of fabrication and in any subsequent machining which may be required.

The forces associated with shrinkage of welds are enormous and will produce overall shrinkage of components and bending/buckling deformations out of a flat plane. These effects have to be allowed for either by pre-setting in the opposite direction to compensate for any out-of-plane deformations, or by making allowances with components initially over-length to allow for shrinkage.

Just as the basic steel manufacturing process can lead to the presence of imperfections, welding also can lead to imperfections which may be significant. The types of imperfection can be grouped into three main areas: planar discontinuities, nonplanar (volumetric) discontinuities, and profile imperfections. By far the most serious of these are planar discontinuities, as these are sharp and can be of a significant size. There are four main types of weld cracking which can occur in steels as planar discontinuities. These are solidification (hot) cracking, hydrogen-induced (cold) cracking, lamellar tearing, and reheat cracking. Examples of these are shown in Figure 10.10. Hot cracking occurs during the solidification of a weld due to the rejection of excessive impurities to the centreline. Impurities responsible are usually sulphur and phosphorus, and the problem is controlled by keeping them to a low level and avoiding deep narrow weld beads. Cold cracking is due to the combination of a susceptible hardened microstructure and the effects of hydrogen in the steel lattice. The problem is avoided by control of the steel chemistry, arc energy heat input, preheat level, quenching effect of the thickness of joints being welded, and by careful attention to electrode coatings to keep hydrogen potential to very low levels. Guidance on avoiding this type of cracking in the heat affected zones of weldable structural steels is given in BS 5135.

Lamellar tearing is principally due to the presence of excessive non-metallic inclusions in rolled steel products, resulting in the splitting open of these inclusions under the shrinkage forces of welds made on the surface. The non-metallic inclusions usually responsible are either sulphides or silicates; manganese sulphides are probably the most common. The problem is avoided by keeping the impurity content low, particularly the sulphur level to below about 0.010%, also by specifying tensile tests in the through-thickness direction to show a minimum ductility by reduction of area dependent on the amount of weld shrinkage anticipated (i.e. size of welded attachment, values of R of A of 10% to 20% are usually adequate). Reheat cracking is a form of cracking which can develop during stress relief heat treatment or during high temperature service in particular types of steel (usually molybdenum or vanadium bearing) where secondary precipitation of carbides develops before relaxation of residual stresses has taken place.

Other forms of planar defect in welds are the operator or procedure defects of lack of penetration and lock of fusion. The volumetric/non-planar imperfections divide into the groups of solid inclusions and gaseous inclusions. The solid inclusions are usually slag from the electrode/flux coating and the gaseous inclusions result from porosity trapped during the solidification of the weld. In general, the non-planar defects are much less critical than planar defects of the same size and are usually limited in their effect because their size is inherently limited by their nature.

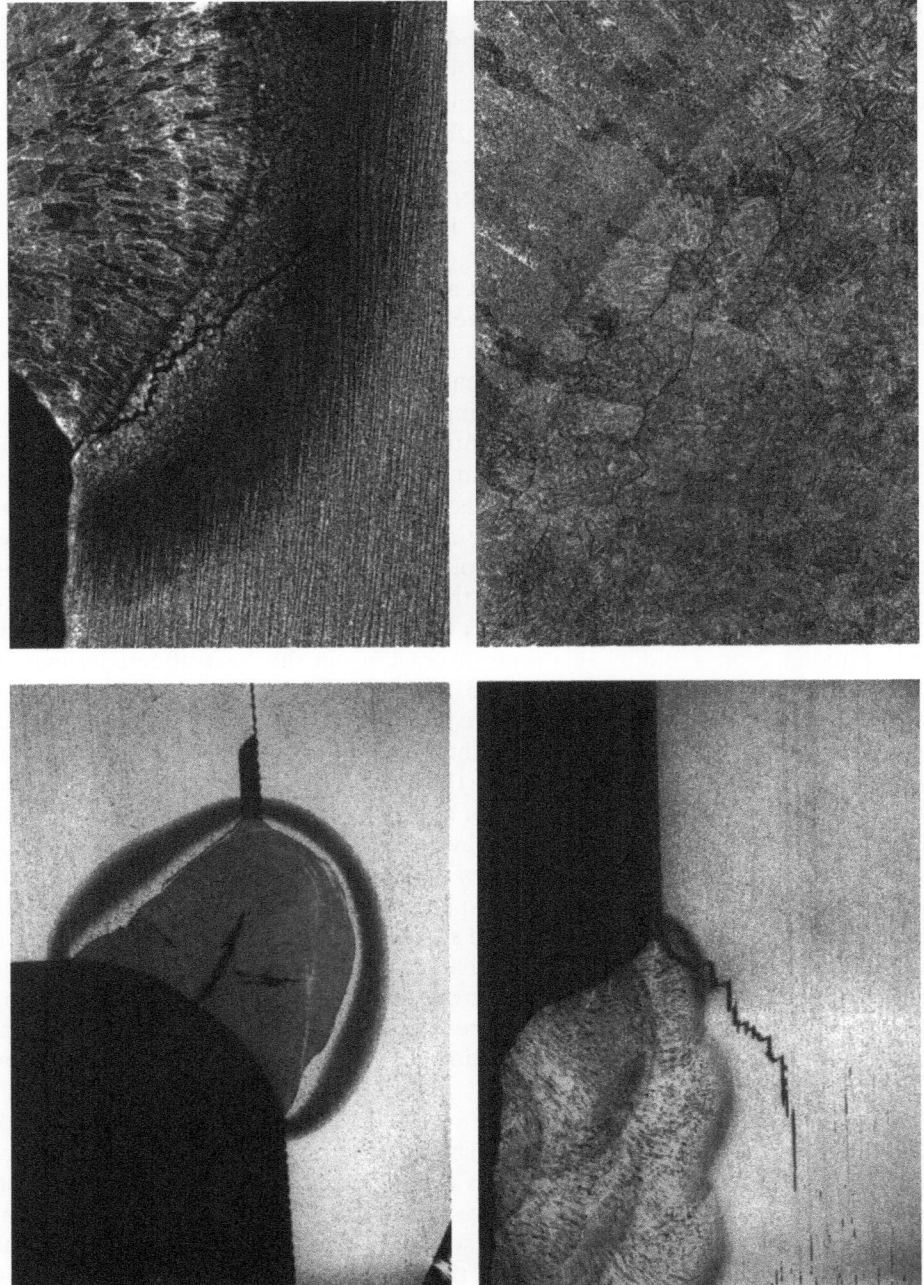

Figure 10.10   Examples of different types of cracking which may occur in welded steel joints (courtesy of The Welding Institute)

10.7 Summary

10.7.1 Criteria influencing choice of steel

The basic requirement in the choice of a particular steel is that it must be fit for the product application and design conditions required. It must be available in the product form and shape required and it should be at the minimum cost for the required application. Clearly, before the generic type of material is chosen as steel, it must be shown to be advantageous to use steel over other contending materials and, therefore, the strength-to-weight ratio and cost ratios must be satisfactory.

The steel must have the required strength, ductility and long-term service life in the required environmental service conditions. For structural applications, the steel must also have adequate fracture toughness, this requirement being implemented by standard Charpy test quality control levels.

Where the steel is to be fabricated into components or structures, its ability to retain its required properties in the fabricated condition must be clearly established. One of the most important factors for a number of industries is the weldability of steel and, in this respect, the chemical composition of the steel must be controlled within tight limits and the welding processes and procedures adopted must be compatible with the material chosen.

The corrosion-resistance and potential fire-resistance/high temperature performance of the steel may be important factors in some applications. A clear decision has to be taken at the design stage as to whether resistance to these effects is to be achieved by external or additional protection measures, or inherently by the chemical composition of the steel itself. Stainless steels with high quantities of chromium and nickel are significantly more expensive than ferritic carbon or carbon manganese steels. Particular application standards generally specify the range of material types which are considered suitable for their particular application.

Increased strength of steels can be obtained by various routes, including increased alloying content, heat treatment, or cold working. In general, as the strength increases so does the cost and there may be little advantage in using high-strength steels in situations where either fatigue or buckling are likely to be ruling modes of failure. It should not be overlooked that, although there is some increase in cost of the basic raw material with increasing strength, there is likely to be a significant increase in fabrication costs, with additional precautions necessary for the more sophisticated types of higher-strength material.

Certain product forms are available only in certain grades of steel. It may not be possible to achieve high strength in some product shapes and retain dimensional requirements through the stage of heat treatment because of distortion problems.

Wherever possible, guidance should be sought on the basis of similar previous experience or prototype trials to ensure that the particular material chosen will be suitable for its required application.

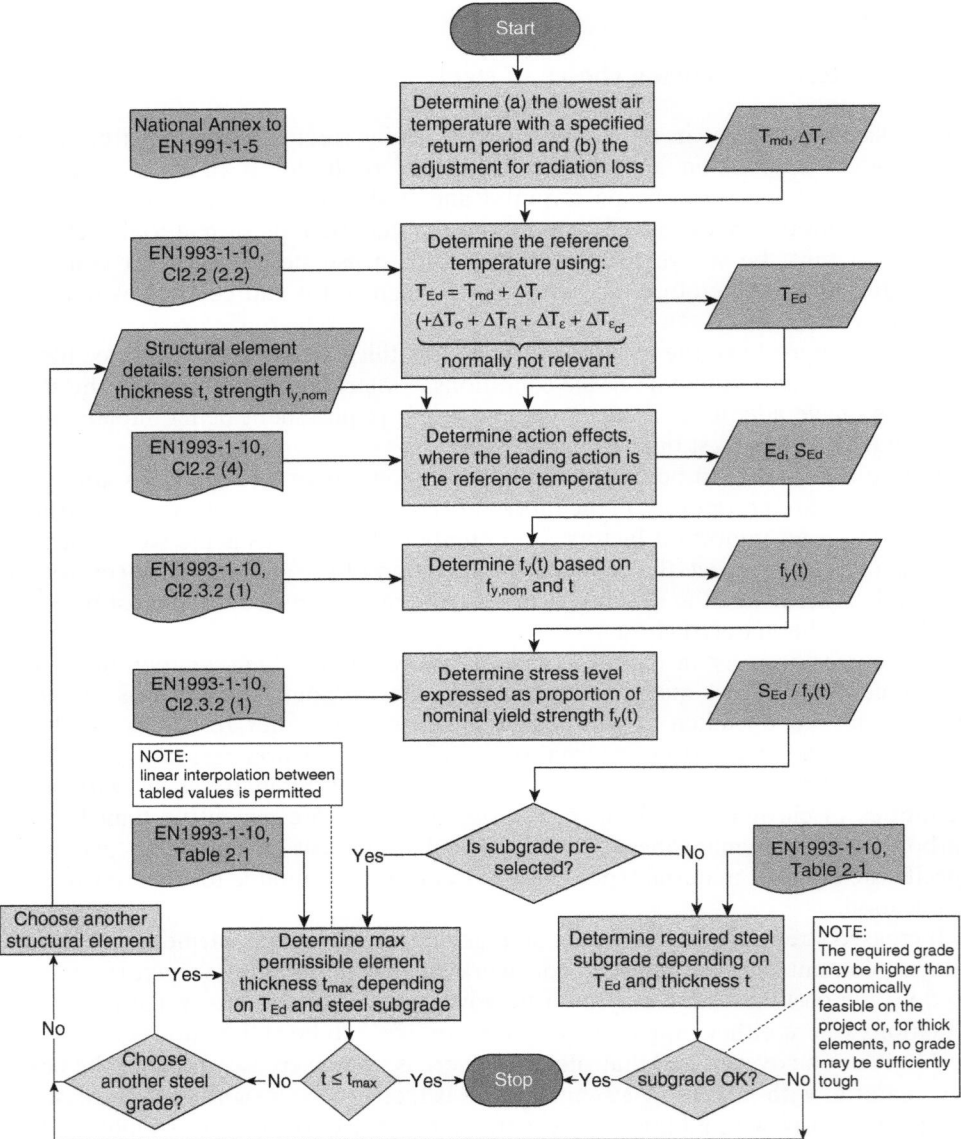

Figure 10.11   Choosing a steel subgrade

10.7.2 Steel specifications and choice of grade

The designation systems used in the European standards are explained in the following section. For example, one of the common structural steels, is designated S355. S means that the steel is structural and 355 indicates the minimum yield strength of the steel in MPa at 16 mm. Further designations, such as S355J2 +AR would indicate (J2) that the steel has longitudinal Charpy V-notch impact values not lower than 27J at −20°C, and that it was supplied in the as-rolled condition (AR). Material supplied in the normalised or normalised rolled condition would have the letter N at the end. Another example would be for flat products of high yield strength structural steels in the quenched and tempered condition (EN 10025: Part 6). Steel designated S460QL would denote a structural steel, with a minimum yield strength (R_{eH}) of 460 MPa at 16 mm and longitudinal Charpy V-notch impact values of not less than 50J at 0°C.

For information on grade designation systems and information regarding the comparison of steel grades in EN 10025 (2004) and equivalent versions in previous standards, publications such as the Corus (now Tata) European Structural Steel Standard Information brochure (2004) are useful.

Structural steelwork, comprising rolled products of plate, sections and hollow sections, is normally of a weldable carbon or carbon manganese structural steel conforming to the standard BS EN 10025. Figure 10.11 (reproduced from Reference 8) is a useful guide to the selection of a steel subgrade and References 9 and 10 are detailed worked examples illustrating the various choices that need to be made.

References to Chapter 10

1. UK Steel, *UK Steel Key Statistics Leaflet*, EEF, 2009, www.uksteel.org.uk
2. Corus *EN10025 Information brochure*, 2004, www.tatasteeleurope.com
3. British Standards Institute (2005) BS EN 1993-1-1:2005 Eurocode 3: *Design of steel structures – Part 1-1: General rules and rules for buildings*. London, BSI.
4. British Standards Institution (2004) BS EN 10025 *Hot rolled products of structural steels. Part 3: Technical delivery conditions for normalised/normalised rolled weldable fine grain structural steels*. London, BSI.
5. British Standards Institution (2006) BS EN 10210 *Hot finished structural hollow sections of non-alloy and fine grain structural steels, Part 1: Technical delivery requirements*. London, BSI.
6. British Standards Institution (1998) BS 131-6:1998 *Notched bar tests. Method for precision determination of Charpy V-notch impact energies for metals*. London, BSI.
7. British Standards Institution (2005), BS EN 1993-1-10:2005 *Eurocode 3: Design of steel structures – Part 1-10: Material toughness and through-thickness properties*. London, BSI.
8. Access-steel *Flowchart: Choosing a steel subgrade*, SF013a-EN-EU. www.access-steel.com
9. Brettle M. E. (2009) *Steel Building Design: Worked Examples – Open Sections In accordance with Eurocodes and the UK National Annexes*. SCI Publication P364. Ascot, The Steel Construction Institute.
10. Example: Choosing a steel sub-grade, SX005a-EN-EU, www.access-steel.com

Further reading for Chapter 10

Baddoo, N.R. and Burgan, B.A. (2001) *Structural Design in Stainless Steel*. Ascot, Steel Construction Institute.

Dieter, G.E. (1988) *Mechanical Metallurgy*, 3rd edn (SI metric edition). New York, McGraw-Hill.

Gaskell, D. (1981) *Introduction to Metallurgical Thermodynamics*, 2nd edn. New York, McGraw-Hill.

Honeycombe, R.W.K. (1995) *Steels: Microstructure and Properties*, 2nd edn. London, Edward Arnold.

Llewellyn, D.T. (1992) *Steels : Metallurgy and Applications*, 2nd Edition. Oxford, Butterworth-Heinemann.

Lancaster, J.F. (1987) *Metallurgy of Welding*, 4th edn. London, Allen and Unwin.

Porter, D.A. and Easterling, K.F. (2001) *Phase Transformations in Metals and Alloys*, 2nd edn. Cheltenham, Nelson Thomas.

Smallman, R.E. (1999) *Modern Physical Metallurgy*, 6th edn. Oxford, Butterworth-Heinemann.

World Steel Association, Steel University online resource, www.steeluniversity.org

Chapter 11
Failure processes

JOHN YATES

Structural steelwork is susceptible to many failure processes, some of which may interact with each other. The principal amongst these are wet corrosion, plastic collapse, fatigue cracking and rapid fracture. Rapid fracture itself encompasses several mechanisms, the best known being brittle fracture and ductile tearing. In this chapter, failure by rapid fracture and fatigue cracking will be discussed.

11.1 Fracture

11.1.1 Introduction

The term brittle fracture is used to describe the fast, unstable fractures that occur with very little energy absorption. In contrast, ductile tearing is a relatively slow process that absorbs a considerable amount of energy, usually through plastic deformation. Some metals, such as copper and aluminium, have a crystalline structure that enables them to resist fast fracture under all loading conditions and at all temperatures. This is not the case for many ferrous alloys, particularly structural steels, which can exhibit brittle behaviour at low temperatures and ductile behaviour at higher temperatures. The consequence of a brittle fracture in a structure may be an unexpected, catastrophic failure. An understanding of the fundamental concepts of this subject is therefore important for all structural engineers.

11.1.2 Ductile and brittle behaviour

Ductile fracture is normally preceded by extensive plastic deformation. Ductile fracture is slow and generally results from the formation and coalescence of voids. These voids are often formed at inclusions due to the large tensile stresses set up

Steel Designers' Manual, Seventh Edition. Edited by Buick Davison and Graham W. Owens.

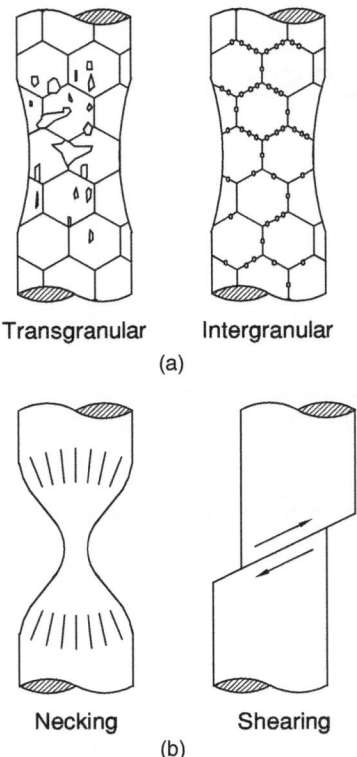

Transgranular   Intergranular

(a)

Necking   Shearing

(b)

Figure 11.1 (a) Plastic deformation by voids growth and coalescence, (b) plastic deformation by necking or shearing

at the interface between the inclusion and the metal, as seen in Figure 11.1(a). Ductile fracture usually goes through the grains and is termed transgranular. If the density of inclusions or of pre-existing holes is higher on grain boundaries than it is within the grains, then the fracture path may follow the boundaries. In cases where inclusions are absent, it has been found that voids are formed in severely deformed regions through localised slip bands and macroscopic instabilities, resulting in either necking or the formation of zones of concentrated shear, as depicted in Figure 11.1(b). The fracture path of a ductile crack is often irregular and the presence of a large number of small voids gives the fracture surface a dull fibrous appearance.

The capacity of many structural steels for plastic deformation and work hardening beyond the elastic limit is extremely valuable as a safeguard against design oversight, accidental overloads or failure by cracking due to fatigue, corrosion or creep.

Brittle fracture is often thought to refer to rapid propagation of cracks without any plastic deformation at a stress level below the yield stress of the material. In

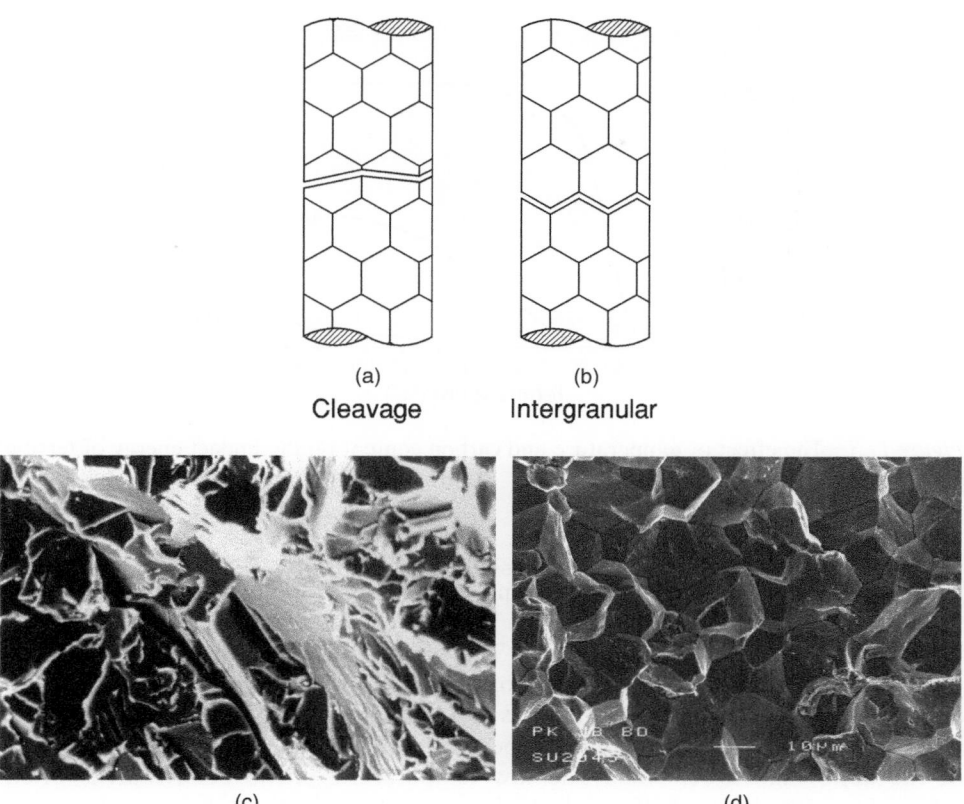

(a)   (b)
Cleavage   Intergranular

(c)   (d)

Figure 11.2   (a) Transgranular brittle fracture; (b) intergranular brittle fracture; (c) photomicrograph of typical fracture surfaces in ferritic steels: transgranular cleavage;
(d) photomicrograph of typical fracture surfaces in ferritic steels: intergranular
cracking

practice, however, most brittle fractures show some, very limited, plastic deformation ahead of the crack tip. Brittle fractures occur by a transgranular, cleavage
mechanism or an intergranular mechanism, as depicted in Figure 11.2(d).

It is also important to know that ferritic steels, which often show ductile behaviour, can behave in a brittle fashion at low temperatures or high loading rates. This
results in fast unstable crack growth and has been clearly demonstrated over the
years by some unfortunate accidents involving ships, bridges, offshore structures, gas
pipelines, pressure vessels and other major constructions. The Liberty ships and the
King Street bridge in Melbourne, Australia, the Sea Gem drilling rig for North Sea
gas and the collapse of the Alexander Kielland oil rig are a few examples of the
casualties of brittle fracture.

An important feature of ferritic steels is the transition temperature between the
ductile and brittle fracture behaviour. Understanding the factors which influence

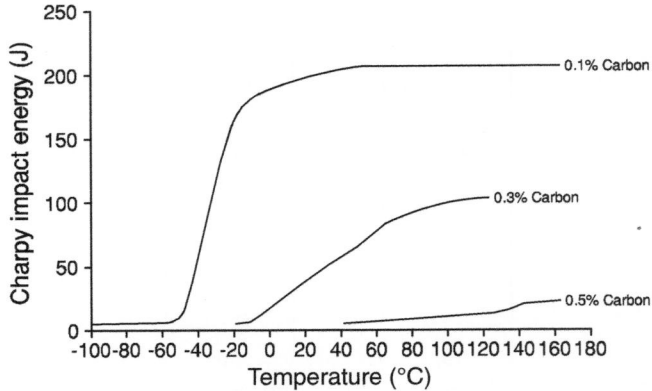

Figure 11.3   The effect of temperature and carbon content on the impact energy of ferritic steels

the transition temperature allows designers to be able to select a material which will be ductile at the required operating temperatures for a given structure. The traditional procedure for assessing the ductile to brittle transition in steels is by impact testing small notched beams.[1] The energy absorbed during the fracture process is a measure of the toughness of the material and varies from a low value at low temperatures to a high value as the temperature is raised. The characteristic shape of the impact energy (temperature graph) has led to the terminology of the *upper* and *lower shelf*. The low temperature, brittle behaviour, is often referred to as the lower shelf and the high temperature, ductile behaviour, the upper shelf.[2]

The Charpy V-notch test[1] is the most popular impact testing technique and is described later in this chapter. The transition in impact toughness values obtained from Charpy tests on carbon steels at different temperatures is shown in Figure 11.3. Table 9 in BS EN 10025-2:2004[3] gives minimum toughness values for a range of structural steel grades at different temperatures and section thicknesses. In terms of limiting steel section sizes, Table 2.1 in BS 1993-1-10:2005 provides guidance on the maximum thickness of steel section, as a function of temperature, which should be used to avoid brittle fracture for a given steel grade.

Impact transition curves are a simple way of defining the effect that variables such as heat treatment, alloying elements and effects of welding, have on the fracture behaviour of steel. Charpy values are useful for quality control but more sophisticated tests[4,5] are required if the full performance of a material is to be exploited.

An understanding of the fracture behaviour of steel is particularly important when considering welded structures. Welding can considerably reduce the toughness of plate in regions close to the fusion line and introduce defects in the weld area. This, coupled with residual tensile stresses that are introduced by the heating and cooling during welding, can lead to cracking and eventual failure of a joint.

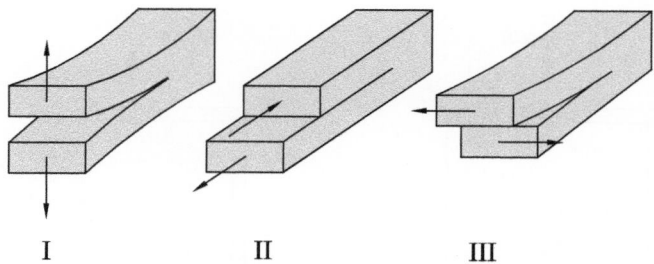

I   II   III

Figure 11.4   Modes of crack opening

11.2 Linear elastic fracture mechanics

The introduction and development of fracture mechanics allows the designer to assess the susceptibility of a steel structure to failure by fracture or fatigue mechanisms. A fracture mechanics assessment is built around the size of known or postulated defects, the measured or intended stresses on the structure, and the resistance of the steel to fracture or fatigue. Fracture mechanics is a particularly useful tool when dealing with welded joints, since the defects found there are invariably larger than those occurring in rolled or wrought products.

The science of dealing with relatively large cracks in essentially elastic bodies is linear elastic fracture mechanics. The assumptions upon which it is based are that the crack is embedded in an isotropic, homogeneous, elastic continuum. In engineering practice, this means that a crack must be much larger than any microstructural feature, such as grain size, but it must be small in relation to the dimensions of the structure. In addition, the stresses present in the structure must be less than about 1/3 of the yield stress. These conditions mean that the local plastic deformation at the defect has a negligible effect on the assessment of failure.

A crack in a solid can be opened in three different modes, as shown in Figure 11.4. The commonest is when normal stresses open the crack. This is termed mode I, or the opening mode. The essential feature of linear elastic fracture mechanics is that all cracks adopt the same parabolic profile when loaded in mode I and that the tensile stresses ahead of the crack tip decay as a function of $1/\sqrt{(\text{distance})}$ from the crack tip.

The absolute values of the opening displacements and the crack tip stresses depend on the load applied and the length of the crack. The scaling factor for the stress field and crack displacements is called the *Stress Intensity Factor*, K_I, and is related to the crack length, a, and the remote stress in the body, σ, by:

$$K_{\mathrm{I}} = Y\sigma\sqrt{\pi a} \tag{11.1}$$

where Y is a geometric correction term to account for the proximity of the boundaries of the structure and the form of loading applied. The subscript I denotes the mode I, or opening mode, of loading.

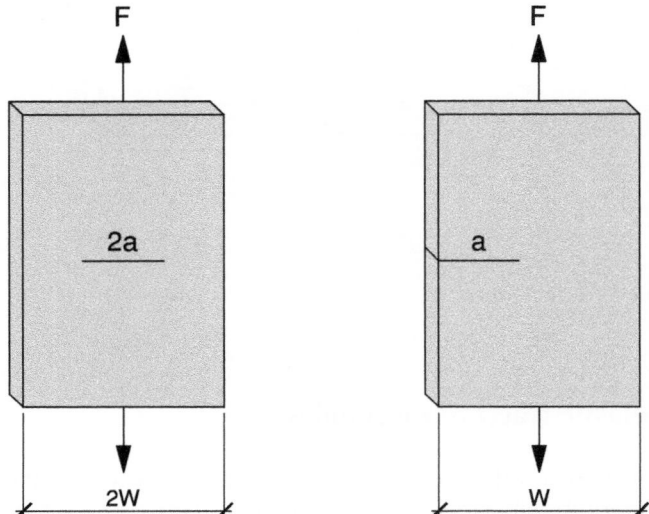

Figure 11.5   Convention for describing the length of embedded and edge cracks

It is important to note that the convention is that an embedded crack that has two tips has a length of $2a$, and an edge crack, which only has one tip, has a length of a, see Figure 11.5. Values of the correction term, Y, have been compiled for many geometries and load cases and are published in References 5 and 6. It is also important to take care over the units used in calculating stress intensity factors. Values may be quoted in $MPa\sqrt{m}$, $MN.m^{-1.5}$ or $N.mm^{-1.5}$. The conversion between them is:

$$1\ MPa\sqrt{m} = 1\ MN.m^{-1.5} = 31.62\ N.mm^{-1.5}$$

The usefulness of the stress intensity factor lies in the concept of similitude. That is, if two cracks, one in a small laboratory specimen and one in a large structure, have the same value of K_I then they have identical opening displacements and identical crack tip stress fields. If the laboratory specimen fails in a catastrophic manner at a critical value of K_I, then the structure will also fail when the stress intensity factor reaches that value. This critical value is a material property, termed the *Fracture Toughness* and, when measured in accordance with a standard such as BS 7448: Part 1:1991,[5] is represented by K_{Ic}.

The practical applications of linear elastic fracture mechanics are in assessing the likelihood that a particular combination of loading and crack size will cause a sudden fracture. Given that a structure is at risk of failing if:

$$Y\sigma\sqrt{\pi a} \geq K_{Ic} \tag{11.2}$$

Then, knowing two of the three parameters of maximum stress, crack size and fracture toughness will allow the limiting value of the third parameter to be determined. For example, if the maximum allowable stress is 160 MPa and the maximum allowable crack size in an edge cracked plate, where $Y = 1.12$ (see for example references 6 and 7), is 100 mm, then the minimum fracture toughness of the material must be 100 MPa√m to avoid fracture. Remember that it is important to keep track of the units and 1 MPa√m = 31.62 N.mm$^{-1.5}$.

It has been found by experiment that the fracture toughness value measured in a laboratory is influenced by the dimensions of the specimen. In particular, the size and thickness of the specimen and the length of the remaining uncracked section, called the ligament, control the level of constraint at the crack tip. Constraint is the development of a triaxial stress state which restricts the deformation of the material. Thick specimens generate high constraint or plane strain conditions and give lower values of the fracture toughness than those found from thin specimens under low constraint or plane stress conditions. The smallest specimen dimensions to give the maximum constraint, and hence the minimum value of toughness are:

$$\text{thickness, width and ligament} \geq 2.5 \left(\frac{K_{Ic}}{\sigma_y} \right)^2 \qquad (11.3)$$

This ensures that the localised plasticity at the crack tip is less than 2% of any of the dimensions of the body and therefore does not disturb the elastic crack tip stress field.

The size requirement of Equation 11.3 gives rise to practical problems in testing. For example, steel with a room temperature yield strength of 275 N.mm^{-2} and a minimum fracture toughness of 100 MPa√m needs to be tested using a specimen at least 160 mm thick to ensure plane strain conditions. In reality, much structural steelwork is made from sections substantially thinner than that needed to ensure plane strain conditions and its fracture toughness will be significantly higher than the K_{Ic} value derived by following BS 7448: Part 1:1991. It is good practice in many engineering disciplines to measure fracture toughness values using specimens of similar thickness to the proposed application to ensure that the most appropriate value is used in structural integrity assessments.

Linear elastic fracture mechanics is of much greater use on the lower shelf of the ductile-brittle behaviour of steels than on the upper shelf. At low temperatures, the yield strength is higher, the fracture toughness much lower and plane strain conditions can be achieved in relatively thin sections. On the upper shelf, linear elastic fracture mechanics tends to be inapplicable and other techniques need to be used.

11.3 Elastic-plastic fracture mechanics

The need to consider fracture resistance of materials outside the limits of validity of plane strain linear elastic fracture mechanics (LEFM) is important for most engineering designs. To obtain valid K_{Ic} results for relatively tough materials, it would be necessary to use a test piece of dimensions so large that they would not be representative of the sections actually in use.

Historically, the approaches to fracture mechanics when significant plasticity has occurred, have been to consider either the crack tip opening displacement or the

J-integral. The current British Standard on Fracture Mechanics Toughness Tests, BS 7448: Part 1:1991, describes how a single approach to linear elastic and general yielding fracture mechanics tests can be adopted.

Significant yielding at a crack tip leads to the physical separation of the surfaces of a crack, and the magnitude of this separation is termed crack tip opening displacement (CTOD), and has been given the symbol δ. The CTOD approach enables critical toughness test measurements to be made in terms of δ_c, and then applied to determine allowable defect sizes for structural components.

The J-integral is a mathematical expression that may be used to characterise the local stress and strain fields around a crack front. Like the CTOD, the J-integral simplifies to be consistent with the stress intensity factor approach when the conditions for linear elastic fracture mechanics prevail. When non-linear conditions dominate, either CTOD or J are useful parameters for characterising the crack tip fields.

The relationships between K_I, δ and J_I under linear elastic conditions are

$$J_I = G_I = \frac{K_1^2}{\sigma_\gamma E} \tag{11.4}$$

$$\delta = \frac{K_1^2}{\sigma_\gamma E} \tag{11.5}$$

where G_I is the strain energy release rate, which was the original, energy-based approach to studying fracture.

In all cases, either linear elastic or general yielding, there is a parameter that describes the loading state of a cracked body. This might be K_I, δ or J_I, as appropriate to the conditions. The limiting case for a particular material is the critical value at which failure occurs. Under elastic conditions this is the sudden fracture event and the material property is K_{Ic}. Under elastic-plastic conditions, failure is not usually rapid and the critical condition, δ_c or J_{Ic}, is usually associated with the onset, or initiation of, the ductile fracture process.

The assessment of flaws in fusion-welded structures is covered by BS 7910: 2005[8] and allows the use of material fracture toughness values in the form of K_{Ic}, δ_c or J_{Ic}. In the absence of genuine fracture mechanics derived toughness data, estimates of K_{Ic} may be made from empirical correlations with Charpy V-notch impact energies. Caution should be exercised when doing so as the degree of fit of the correlations tends to be poor.

The BS 7910: 2005 document prescribes the procedure for assessing the acceptability of a known flaw at a given stress level. There are three levels of sophistication in the analysis, requiring more precise information about the stresses and material properties as the assessment becomes more advanced. The procedure is very powerful as it considers the possibility of either fracture or plastic collapse as alternative failure processes. The first level uses a simple approach with built-in safety factors and conservative estimates of the material fracture toughness and the applied and residual stresses in the structure. The Standard allows for toughness estimates from Charpy impact tests to be used at Level 1. Level 2 is the normal assessment route for steel structures and requires more accurate estimates of the stresses, material properties and defect sizes and shapes. Level 3 is much more sophisticated and can accommodate the tearing behaviour of ductile metals. The details of dealing with multiple flaws, residual stresses and combinations of bending and membrane stresses are all dealt with in the document. In many practical cases, a Level 1 assessment is sufficient.

Annex N in BS 7910: 2005 describes the manual procedure for determining the acceptability of a flaw in structure using the Level 1 procedure. An equivalent flaw parameter, $\bar{a}$, is defined as the half length of a through-thickness flaw in an infinite plate subjected to a remote tension loading. An equivalent tolerable flaw parameter, $\bar{a}_m$, can then be estimated and used to represent a variety of different defect shapes and sizes of equivalent severity.

$$\bar{a}_m = \frac{1}{2\pi}\left(\frac{K_{\text{mat}}}{\sigma_{\text{max}}}\right)^2 \tag{11.6}$$

Where the parameter K_{mat} in Equation 11.6 is a measure of the resistance of the material to fracture, or fracture toughness. This may well not be valid plane strain K_{Ic} value, as determined by BS 7448: Part 1: 1999,[4] but it should be the toughness appropriate to the conditions and section sizes under review. Furthermore, Level 1 allows for the possibility of estimating a fracture toughness value from Charpy impact data as described in Annex J of the Standard. However, these estimates are highly approximate and care should be taken when making use of them.

The possibility of plastic collapse of the cracked section must be checked by calculating the ratio, known as S_r, of a reference stress to the flow stress of the material. The flow stress is taken to be the average of the yield and tensile strengths. The reference stress is related to the applied and residual stresses in the structure and depends on the geometry of the structure and the defect. Details of its calculation are given in Annex P. Provided that the S_r parameter is less than 0.8, there is no risk of failure by collapse and any failure will be as a result of fracture.

Tough structural steels are extremely tolerant of the presence of cracks. If one considers a typical structural steel, with a yield stress of 275 N.mm^{-2} and a minimum fracture toughness of 100 MPa$\sqrt{\text{m}}$, subjected to a maximum allowable stress of 165 N.mm^{-2}, then the maximum allowable flaw from Equation 11.6 is about 60 mm long. (Remember that $\bar{a}_m$ is the half length of a through thickness crack in an infinite plate.) The correction for a finite width plate makes less than 10% difference to the crack size, provided that the width of the plate is more than four times the length of the crack.

Reducing the maximum allowable stress in the structure has a big effect on the allowable flaw size. Halving the maximum stress increases the allowable flaw size to around 240 mm, a four-fold increase. The corollary, of course, is that the use of high strength metals makes a structure more sensitive to the presence of defects. High strength materials mean that higher structural stresses are allowed; high strength steels also tend to be of lower fracture toughness than low strength steels. This combination means that the maximum allowable flaw size can become so small that critical defects are not readily observed during manufacture or commissioning. In these circumstances, the structure is at risk of fracture without any prior indication of cracking and it would be inadvisable to use such as structure without a structural integrity assessment conducted by a suitably qualified and experienced person.

11.4 Materials testing for fracture properties

There are, essentially, two approaches to fracture testing. The traditional impact test is on a small notched bar, following the work of Izod or Charpy, or a fracture toughness test on a pre-cracked beam or compact tension specimen.

The principal advantage of an impact test is that it is relatively quick, simple and cheap to carry out. It provides qualitative information about the relative toughness of different grades of material and is well suited to quality control and material acceptance purposes.

Fracture toughness tests on pre-cracked specimens provide a direct measure of the fracture mechanics toughness parameters K_{Ic}, J_{Ic} or critical crack tip opening displacement. They are, however, more expensive to perform, use larger specimens and require more complex test facilities. The main advantage of fracture toughness tests is that they provide quantitative data for the design and assessment of structures.

11.4.1 Charpy test

Reference 1 specifies the procedure for the Charpy V-notch impact test. The test consists of measuring the energy absorbed in breaking a notched bar specimen by one blow from a pendulum, as shown in Figure 11.6. The test can be carried out at a range of temperatures to determine the transition between ductile and brittle behaviour for the material. The Charpy impact value is usually denoted by the symbol C_v and is measured in Joules.

Many attempts have been made to correlate Charpy impact energies with fracture toughness values. The large scatter found in impact data makes such relationships difficult to describe with any great degree of confidence. No single method is currently able to describe the entire temperature-toughness response of structural ferritic steels. The correlation method that forms the basis of Annex E in BS 7910:2005 and also the latest European guidelines on flaw assessment methods,[9] is known as the 'Master Curve'. The temperature at which a certain specified Charpy impact energy is achieved, usually 27J, is used as a fixed point and the shape of the transition curve is generated from that temperature.

11.4.2 Fracture mechanics testing

The recommended method for determining fracture toughness values of metallic material are described in BS 7448: Part 1: 1991.[5] This covers both linear elastic and

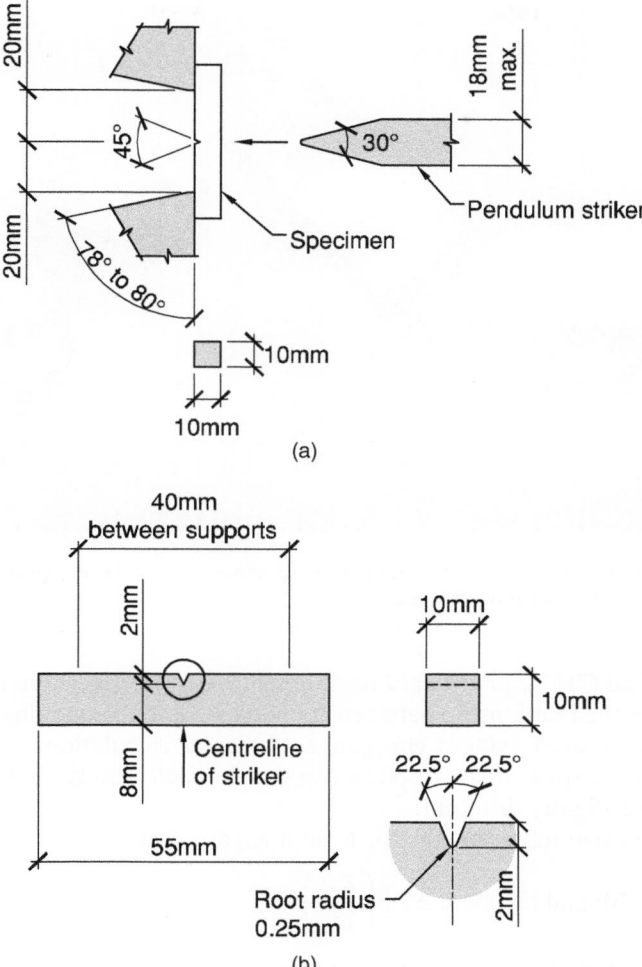

Figure 11.6   Charpy test (a) test arrangement; (b) specimen

elastic-plastic conditions. The advantage of a single test procedure is that the results may be re-analysed to give a critical CTOD or critical *J* value if the test is found not to conform to the requirements for a valid plane strain K_{Ic} result.

The principle behind the tests is that a single edge notched bend or compact tension specimen, see Figure 11.7, is cyclically loaded within prescribed limits until a sharp fatigue crack is formed. The specimen is then subjected to a displacement-controlled monotonic loading until either brittle fracture occurs or a prescribed maximum force is reached. The applied force is plotted against displacement and, provided specific validity criteria are met, a plane strain K_{Ic} may be found by analysis of the data. When the validity criteria are not met, the data may be re-analysed to

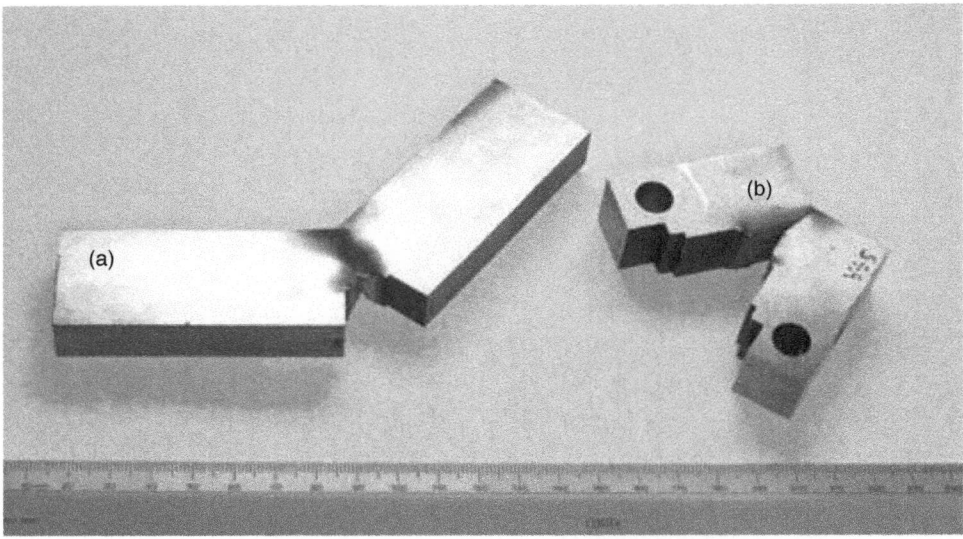

Figure 11.7 Typical fracture toughness test specimens. (a) Three point bend beam. (b) Compact tension specimen

evaluate a critical CTOD or critical *J* for that material. The determination of critical CTOD requires the relationship between applied load and the opening of the mouth of the crack, measured using a clip gauge. Critical *J* calculations need the load-against-load line displacement response, so the detailed arrangements of the two types of test are slightly different.

The specimen size requirements for a valid K_{Ic} are that:

$$\text{thickness, width and ligament} \geq 2.5\left(\frac{K_{Ic}}{\sigma_y}\right)^2 \tag{11.7}$$

The size requirements for the *J* value to be valid are that:

$$\text{thickness, width and ligament} \geq 25\left(\frac{J_{Ic}}{\sigma_y}\right) \tag{11.8}$$

This implies that significantly smaller specimens are required for critical *J* values tests than for K_{Ic} tests.

11.4.3 Other tests

There are other tests that can be carried out such as the wide plate test used for testing welded plate joints which was developed by Wells at the Welding Institute.[10] A large full-thickness plate, typically one metre square, is butt welded using the process and treatments to be used in the production weld. This test has the advantage of representing failure of an actual welded joint without the need for machining prior to testing.

11.4.4 Test specimens

As has already been described, each test procedure requires a sample of a certain size. In addition, the position of a sample in relation to a weld or, in the case of a thick plate, its position through the thickness is important. In modern structural steels toughness of the parent plate is rarely a problem. However, once a weld is deposited the toughness of the plate surrounding the weld, particularly in the heat-affected zone (HAZ), will be reduced. Although lower than for the parent plate, the C_v or fracture toughness values that can be obtained should still provide adequate toughness at all standard operating temperatures. In the case of thicker joints, appropriate post-weld heat treatments should be carried out to reduce the residual stresses that are created by welding.

11.5 Fracture-safe design

The introduction into EN 1090-2:2008[11] of execution classes, EXC1 to 4, distinguishes the increase in quality of both the steel and the welding needed to provide sufficient integrity to avoid failure under increasingly severe operating conditions. The execution grades are determined by the combination of the service category, the production category and the consequence class.

Service category SC1 is for structures that are essentially quasi-statically loaded, or are likely to see very few loading cycles during their life. These are structures, such as buildings, for which the failure mechanism to be avoided is fracture of welds, joints or design details during erection or due to excessive loading in service. Service category SC2 is for structures that experience fluctuating or cyclic loads from, for example, vibrations, wind loading, or simply frequent changes in operating conditions. These are structures in which small defects may grow by fatigue until they become large enough to cause fracture.

The two production classes distinguish between non-welded components, or those made by welding low strength steels, PC1, from those made by welding higher strength steels or components, PC2, whose production involves the risk of introducing large defects. In essence, components classified as PC1 will either have insignificantly small defects or be made of high toughness steel operating at low stresses. Components in PC2 may contain large defects either in the welds or at the edges due to flame cutting, or they are lower toughness steels operating at relatively high stresses. The risk of failure by both fatigue and fracture is higher in PC2 than PC1.

The third consideration is the consequence of any failure, from CC1 to CC3 with increasingly severe consequences in term of human, economic or environmental impact. Table B.3 in EN 1090-2:2008 shows the resulting execution grades from EXC1 for a quasi-statically loaded structure made from non-welded steel component with minor consequences if it were to fail, to EXC4 for a fatigue loaded structure made from a welded high strength steel for which failure would have severe consequences.

The design requirements for steel structures in which brittle fracture is a consideration are given in most structural codes. There are several key factors which need to be considered when determining the risk of brittle fracture in a structure. These are:

- minimum operating temperature
- loading, in particular, rate of loading
- metallurgical features such as parent plate, weld metal or HAZ
- thickness of material to be used.

Each of these factors influences the likelihood of brittle fracture occurring.

As discussed earlier, at normal operating temperatures and slow rates of loading, valid K_{Ic} values are not usually obtained for structural steels and in the offshore and nuclear industries critical CTOD and J tests are widely used. Under these conditions, valid K_{Ic} values from low temperature tests will be conservative. If the structure is acceptable when assessed using such conservative data then there is often no great need to pursue the problem, particularly as all forms of fracture mechanics testing are expensive compared to routine quality control tests such as Charpy testing.

In general, the fracture toughness of structural steel increases with increasing temperature and decreasing loading rates. The effect of temperature on fracture toughness is well known. The effect of loading rate may be equally important, not only in designing new structures, but also in understanding the behaviour of existing ones which may have been built from material with low toughness at their service temperature. The shift in the ductile-brittle transition temperature for structural steels can be considerable when comparing loading rates used in slow bend tests to those in Charpy tests. Results from experimental work have shown that the transition temperature for a BS EN 10025 S355 J steel can change from around 0°C to −60°C with decreasing loading rate.

Materials standards set limits for the transition temperatures of various steel grades based on Charpy tests. When selecting steel for a given structure it must be remembered that the Charpy values noted in the standards apply to the parent plate. Material toughness varies in the weld and heat affected zones of welded joints and these should be checked for adequate toughness, see BS 7910-2005 Annex L. Furthermore, since larger defects may be present in the weld area than the parent plate, appropriate procedures should be adopted to ensure that a welded structure will perform as designed. These could involve non-destructive testing including visual examination. In situations where defects are found, fracture mechanics procedures such as those in References 8 and 9 can be used to assess their significance.

11.6 Fatigue

11.6.1 Introduction

A component or structure which survives a single application of load may fracture if the application is repeated many times. This is classed as fatigue failure. Fatigue life can be defined as the number of cycles and hence the time taken to reach a predefined failure criterion. Fatigue failure is by no means a rigorous science and the idealisations and approximations inherent in it prevent the calculation of an absolute fatigue life for even the simplest structure.

In the analysis of a structure for fatigue there are three main areas of difficulty in prediction:

- the operational environment of a structure and the relationship between the environment and the actual forces on it
- the internal stresses at a critical point in the structure induced by external forces acting on the structure
- the time to failure due to the accumulation of damage at the critical point.

The promotion of cyclically loaded structures from execution class EXC1 to EXC2, EXC2 to EXC3, and EXC3 to EXC4, depending on the production and consequence classes, reflects the increasing quality of steel, design details and fabrication processes needed to ensure that small defects do not grow sufficiently large during the lifetime of the structure to cause a catastrophic failure.

There are three approaches to the assessment of fatigue life of structural components. The traditional method, called the *S-N* approach, was first used in the mid-19th century. This relies on empirically derived relationships between applied elastic stress ranges and fatigue life. A development of the S-N approach is the *strain-life* method in which the plastic strains are considered important. Empirical relationships are derived between strain range and fatigue life. The third method, based on fracture mechanics, considers the growth rate of an existing defect. The concept of *defect tolerance* follows directly from fracture mechanics assessments.

11.6.2 Loadings for fatigue

Fluctuating loads arise from a wide range of sources. Some are intentional, such as road and rail traffic over bridges. Others are unavoidable, such as wave loading on offshore oil rigs, and some are accidental. Occasionally the loads are entirely unforeseen; resonance of a slender tower under gusting wind loads can induce large numbers of small amplitude loads. References 12, 13, 14 and 15 are useful sources of information on fatigue loading.

The designer's objective is to anticipate the sequence of service loading throughout the life of the structure. The magnitude of the peak load, which is vital for

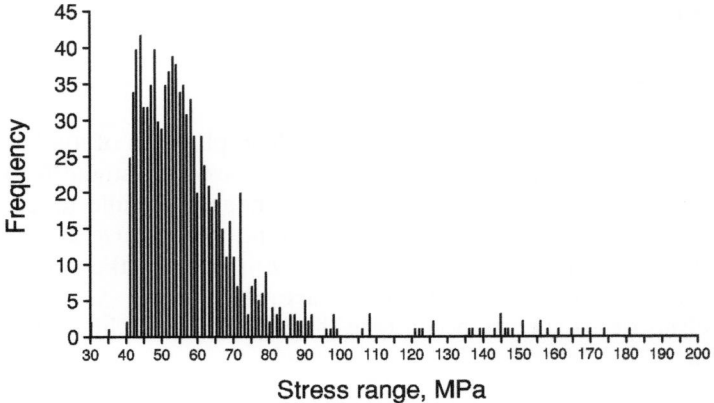

Figure 11.8   Fatigue loading spectrum for a sample of 1000 cycles. Note the large number of small cycles and the small number of large cycles

static design purposes, is generally of little concern as it only represents one cycle in millions. For example, highway bridge girders may experience 100 million significant cycles in their lifetime. The sequence is important because it affects the stress range, particularly if the structure is loaded by more than one independent load system.

For convenience, loadings are usually simplified into a load spectrum, which defines a series of bands of constant load levels and the number of times that each band is experienced, as shown in Figure 11.8.

11.6.3 The nature of fatigue

Materials subject to a cyclically variable stress of a sufficient magnitude change their mechanical properties. In practice, a very high percentage of all engineering failures are due to fatigue. Most of these failures can be attributed to poor design or manufacture.

Fatigue failure is a process of crack propagation due to the highly localised cyclic plasticity that occurs at the tip of a crack or metallurgical flaw. It is not a single mechanism but the result of several mechanisms operating in sequence during the life of a structure: propagation of a defect within the microstructure of the material; slow incremental propagation of a long crack and final unstable fracture.

Crack initiation is a convenient term to cover the early stages of crack growth that are difficult to detect. The reality is that a crack starts to grow from the first loading cycles and continues right through to failure. In welded structures or cast components the initiation phase is bypassed as substantial existing defects are already likely to be present.

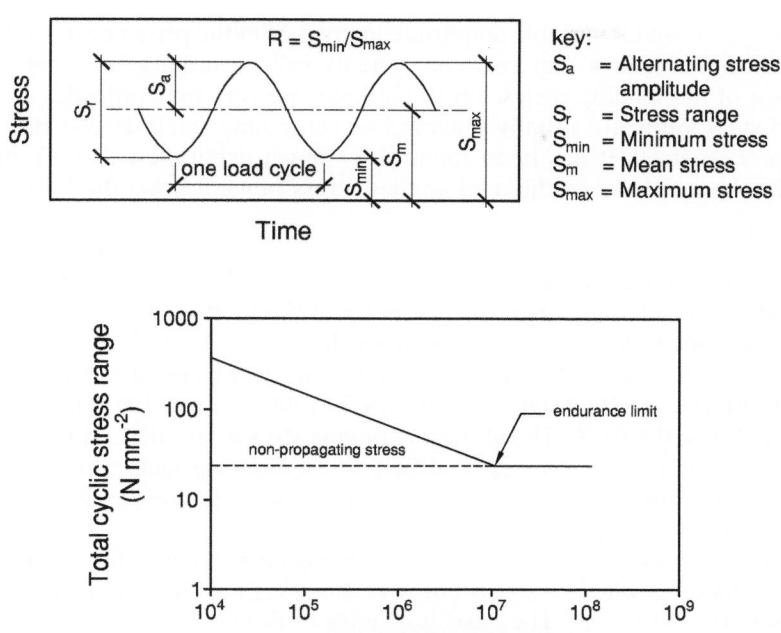

Figure 11.9   Typical S-N curve and nomenclature used in fatigue

11.6.4 S-N curves

The traditional form for presenting fatigue data is the S-N curve, where the total cyclic stress range (S) is plotted against the number of cycles to failure (N). A typical curve is shown in Figure 11.9 with a description of the usual terminology used in fatigue. Logarithmic scales are conventionally used for both axes. However, this is not universal and S-N data may also be presented on linear stress axes instead of logarithmic; stress amplitudes instead of ranges and reversals instead of cycles.

Fatigue endurance data are obtained experimentally. S-N curves can be obtained for a material, using smooth laboratory specimens, for components or for detailed sub-assemblies such as welded joints. In all cases, a series of specimens is subject to cycles of constant load amplitude to failure. A sufficient number of specimens are tested for statistical analysis to be carried out to determine both mean fatigue strength and its standard deviation. Depending on the design philosophy adopted, design strength is taken as mean minus an appropriate number of standard deviations.

Under some circumstances, laboratory tests on steel specimens appear to have an infinite life below a certain stress range. This stress range is variously known as the 'fatigue limit' or 'endurance limit'. In practice, the tests are stopped after two, ten or one hundred million cycles as appropriate, and if it has not broken then the stress range is assumed to be below the fatigue limit. The fatigue, or endurance, limit will tend to disappear under variable amplitude loading or in the presence of a corrosive environment. This means that real components will eventually fail whatever the stress range of the loading cycles, but the life may be very long indeed.

Welded steel joints are usually regarded as containing small defects due to the welding process itself. It has been found after much experimental work that the relationship between fatigue life and applied stress range follows the form:

$$N_f \Delta\sigma^a = b \tag{11.9}$$

where N_f is the number of cycles to failure, $\Delta\sigma$ is the applied stress range and a and b are constants which depend on the geometry of the joint. The value of a is in the range of 3 to 4 for welded joints in ferritic steels.

In BS EN 1993-1-9:2005,[12] fatigue assessments are conducted using different S-N curves for different design details, as described in Tables 8.1 to 8.10 of the Eurocodes Standard (BS EN 1993-1-9). The allowable fatigue stresses are determined from the nominal direct and shear stresses modified by appropriate factors to account for stress concentrations, secondary bending moments, section size, variable amplitude loading and other relevant parameters.

The British Standard BS 7608:1993[13] contains a code of practice for fatigue design of welded structures which is consistent with the assessments for fatigue in welded structures in BS 7910:2005.[8] The basis is a series of design S-N curves for different welded joint configurations and crack locations. Each potential crack location in each type of joint is classified by a letter S, B, C, D, E, F, F2, G and W. Each detail class has a specific design S-N curve, Figure 11.10. This curve indicates the number of cycles at any given stress range that should be achieved with 97.7% probability of surviving. The design S-N curves for each class of joint are based on extensive experimental data and are suitable for ferritic steels. Details of dealing with other metals are given in BS 7910:2005.

The procedure is to identify the worst weld detail in the design and the ranges of the stress cycles experienced by the structure. If there is only one stress range, then the corresponding lifetime can be read off from the appropriate S-N curve. This is

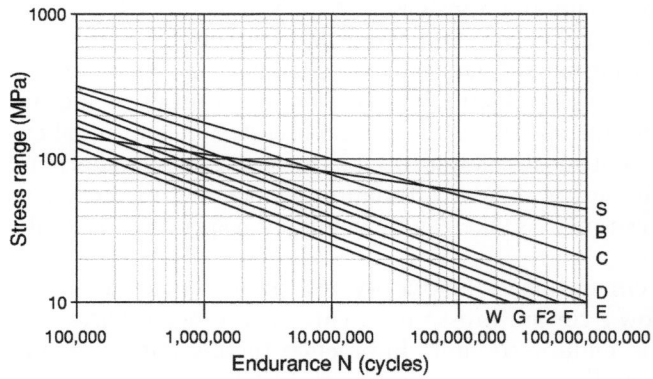

Figure 11.10   Fatigue design curves for welded joints according to BS 7608:1993

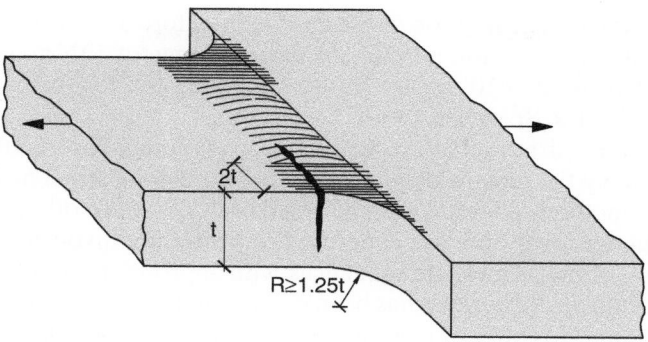

Figure 11.11   An example of an F2 cracked weld

the number of repeated cycles that the structure can endure and have a 97.7% probability of survival.

Consider the weld detail in Figure 11.11. This is classified as type F2. If this is subjected to a cyclic stress with a range of 20 N.mm^{-2} every 10 seconds, then reading off the type F2 curve in Figure 11.10 indicates that the structure should survive 66 million cycles. This corresponds to 4.8×10^8 seconds or about 14 years.

11.6.5 Variable-amplitude loading

For constant-amplitude loading, the permissible stress range can be obtained directly from Figure 11.10 by considering the required design life. In practice, it is more common for structures to be subjected to a loading spectrum of varying amplitudes or random vibrations. In such cases, use is made of Miner's rule.[16]

Miner's rule is a linear summation of the fatigue damage accumulated during the life of the structure. For a joint subjected to a number of repetitions, n_i, each of several stress ranges $\Delta\sigma_i$, the value of n_i corresponding to each $\Delta\sigma_i$ should be determined from stress spectra measured on similar equipment or by making reasonable assumptions as to the expected service history. The permissible number of cycles, N_i, at each stress range, $\Delta\sigma_i$, should then be determined from Figure 11.10 for the relevant joint class and the stress range adjusted so that the linear cumulative damage summation does not exceed unity:

$$\frac{n_1}{N_1} + \frac{n_2}{N_2} + \frac{n_3}{N_3} + \ldots + \frac{n_j}{N_j} = \sum_{i=1}^{i=j} \frac{n_i}{N_i} < 1.0 \tag{11.10}$$

The order in which the variable amplitude stress ranges occur in a structure is not considered in this procedure.

An example would be a class F2 weld that is subjected to 3×10^7 cycles at a stress range of 20 N.mm^{-2} and then the stress range increases to 30 N.mm^{-2} and the remaining life needs to be estimated. The lifetime at 20 N.mm^{-2} is read off from Figure 11.10

and is 6.6×10^7 cycles. The fraction of life used is therefore $3 \times 10^7 / 6.6 \times 10^7 = 0.455$ and the remaining life fraction is 0.545. At a stress range of $30\,\mathrm{N.mm^{-2}}$ the lifetime from Figure 11.10 is 1.9×10^7 cycles, so the remaining life for this welded joint is $0.545 \times 1.9 \times 10^7 = 1 \times 10^7$ cycles.

Various methods exist to sum the spectrum of stress cycles. The 'rainflow' counting method is probably the most widely used for analysing long stress histories using a computer. This method separates out the small cycles that are often superimposed on larger cycles, ensuring both are counted. The procedure involves the simulation of a time history, or use of measure sequences, with appropriate counting algorithms. Once the spectrum of stress cycles has been determined, the load sequence is broken down into a number of constant load range segments. The 'reservoir' method, which is easy to use by hand for short stress histories, is described in Annex A of BS 1993-1-9:2005.[12]

11.6.6 Strain-life

The notion that fatigue is associated with plastic strains led to the strain-life approach to fatigue. The endurance of laboratory specimens is correlated to plastic strain range, or amplitude, in a strain controlled fatigue test. Typical data for a S355 steel[17] and the similar grade BS 4360 50D[18] is shown in Figure 11.12. A common empirical curve fit to the fatigue lifetime data takes the form:

$$\varepsilon_a = \frac{\sigma'_f}{E}(2N_f)^b + \varepsilon'_f(2N_f)^c \qquad (11.11)$$

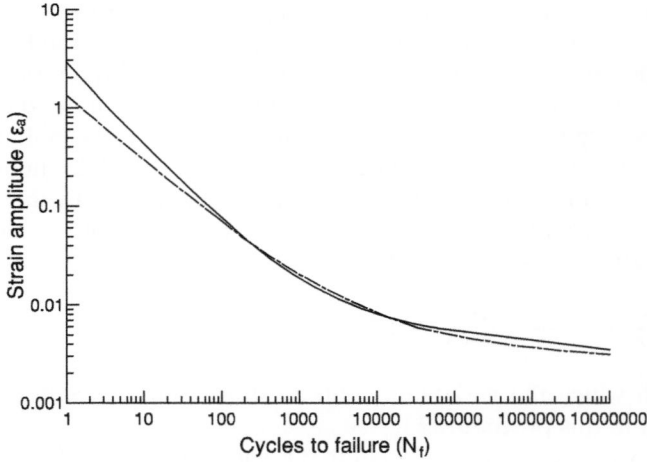

Figure 11.12   Strain-life curve for structural steels. Dotted line is from data in EN S355, solid line is from data in BS4360 50D

where E is Young's modulus and σ'_f, ε'_f, b and c are considered to be material properties. It is worth noting that the first term of the right hand side is an elastic stress term which dominates at low loads and long lifetimes. The second term is a plastic strain term which dominates at high loads and short lifetimes.

The strain-life approach is preferred over the S-N method since it is almost identical to the S-N approach at long lives and elastic stresses, and is more general for problems of short lives, high strains, high temperatures or localised plasticity at notches.

The technique is:

1. Determine the strains in the structure, often by finite element analysis.
2. Identify the maximum local strain range, the 'hot spot' strain.
3. Read off, from the strain-life curve, the lifetime to first appearance of a crack at that strain at that position.

Variable amplitude loads are dealt with in the same way as the S-N method with the local hot spot strains and their associated lifetimes being determined for each block of loading.

The strain-life, or local strain, method has wider use in fatigue assessments in the engineering industry than the S-N method and is available in commercial computer software.

11.6.7 Fracture mechanics analysis

Fatigue life assessment using fracture mechanics is based on the observed relationship between the change in the stress intensity factor, ΔK, and the rate of growth of fatigue cracks, da/dN. If experimental data for crack growth rates are plotted against ΔK on a logarithmic scale, an approximate sigmoidal curve results, as shown in Figure 11.13. Below a threshold stress intensity factor range, ΔK_{th}, no growth occurs. For intermediate values of ΔK, the growth rate is idealised by a straight line. This approach was first formulated by Paris and Erdogan[19] who proposed a power law relation of the form:

$$\frac{da}{dN} = A(\Delta K)^m \tag{11.12}$$

where da/dN is the crack extension per cycle, A, m are crack growth constants, and

$$\Delta K = K_{max} - K_{min}$$

where K_{max} and K_{min} are the maximum and minimum stress intensities respectively in each cycle. Since the crack growth rate is related to ΔK raised to an exponent, it is important that ΔK should be known accurately, if meaningful crack growth predictions are to be made.

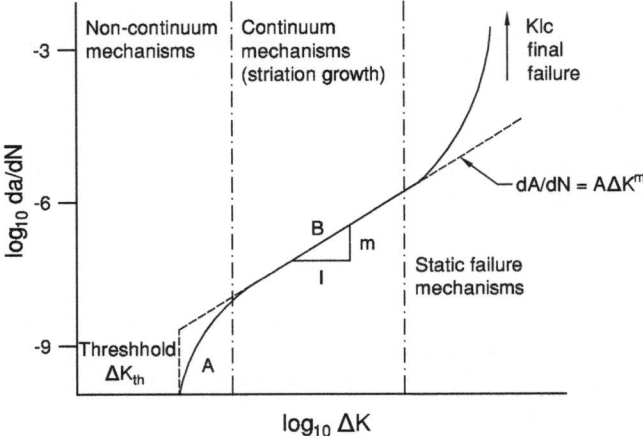

Figure 11.13   Schematic presentation of crack growth

Values of A and m to describe the fatigue crack growth rate can be obtained from specific tests on the materials under consideration. From published data reasonable estimates of fatigue growth rates in ferritic steels are given by:[9]

$m = 3$

$A = 5.21 \times 10^{-13}$ for non aggressive environments at temperatures up to 100°C

$A = 2.3 \times 10^{-12}$ for marine environments at temperatures up to 20°C

for crack growth rates in $mm/cycle$ and stress intensity factors in $N\,mm^{-1.5}$

Some care should be taken with fatigue lifetime estimates in corrosive environments to make sure that the fatigue crack growth data are appropriate to the particular combination of steel grade and environment. Small changes in steel composition or environmental conditions can result in very large changes in crack growth behaviour.

The fracture mechanics convention in the welded steel structures industry is slightly different to that of other practitioners. For a crack at the toe of a welded joint:

$$\Delta K = M_K Y \Delta \sigma \sqrt{\pi a} \tag{11.13}$$

where:
$\Delta \sigma$ = applied stress range
a = crack depth
Y = a correction factor dependent on crack size, shape and loading
M_K = a function which allows for the stress concentration effect of the joint and depends on crack site, plate thickness, joint and loading.

In many industries the magnification factor, M_K, is incorporated in the geometric correction term, Y. This makes Equation 11.12 the fatigue version of Equation 11.1.

Substituting Equation 11.12 in Equation 11.11, rearranging and integrating gives:

$$N_f = \frac{1}{A\pi^{m/2}\Delta\sigma^m} \int_{a_i}^{a_f} \frac{da}{\left[M_K Y \sqrt{a} \right]^m}$$ (11.14)

where a_i is the initial crack depth and a_f is the final crack depth corresponding to failure.

Equation 11.13 forms an extremely powerful tool. If a welded joint contains a crack or crack-like flaw, then Equation 11.13 can be used to predict its fatigue endurance. This makes the reasonable assumption that the life consists of crack growth from a pre-existing crack. The techniques requires knowledge of:

- the crack propagation behaviour described by Equation 11.11
- the initial flaw size
- the final flaw size
- the geometry and loading correction terms Y and M_K.

The integration is straightforward if Y and M_K are independent of crack length, which is not usually the case. Otherwise, some suitable numerical technique must be used. The life is fairly insensitive to the final crack length but highly dependent on the initial flaw size. Engineering judgement is often required when selecting appropriate values of these crack sizes.

If Y and M_K are independent of crack length, and $m \neq 2$ then:

$$N_f = \frac{1}{\left(m/2 - 1\right) A.\pi^{m/2} \left(M_k Y \Delta\sigma\right)^m} \left[a_i^{\left(1-\frac{m}{2}\right)} - a_f^{\left(1-\frac{m}{2}\right)} \right]$$ (11.15)

The initial crack size can be taken as the largest flaw that escapes the detection technique used. This is likely to be several millimetres for visual or ultrasonic inspection of large welded structures.

The final allowable flaw size might be taken from Equation 11.5, or it may be a crack that penetrates the wall of a containment vessel and allows fluid to escape. Other failure conditions could be a crack that allows excessive displacements to occur or a crack that is observable to the naked eye and is therefore unacceptable to the customer.

Guidance on the assessment of fatigue by fracture mechanics methods is provided in Section 8 of BS 7910:2005. This also includes assessing fatigue life from S-N curves by consideration of the quality category of a weld detail ranging from Q1, best, to Q10, worst.

The design S-N curves, described by Equation 11.8, and given in the British Standards BS EN1993-1-9:2005 and BS 7910:2005 are effectively the same as Equation 11.14, the integration of the crack propagation equation.

11.6.8 Improvement techniques

11.6.8.1 Introduction

The fatigue performance of a joint can be enhanced by the use of weld improvement techniques. There is a large amount of data available on the influence of weld improvement techniques on fatigue life but as yet little progress has been made into developing practical design rules. Modern steelmaking has led to the production of structural steels with excellent weldability. The low fatigue strength of a welded connection is generally attributed to a very short period of crack initiation because of the flaws and defects introduced during welding. An extended crack initiation life can be achieved by:

- reducing the stress concentration of the weld
- removing crack-like defects at the weld toe
- reducing tensile welding residual stresses or introducing compressive stresses.

The methods employed fall broadly into two categories as illustrated below:

- *weld geometry improvement*: grinding: weld dressing; profile control
- *residual stress reduction*: peening; thermal stress relief.

Most of the current information relating to weld improvement has been obtained from small-scale specimens. When considering actual structures, one important factor is size. In a large structure, long range residual stresses due to the assembly of the members are present and will influence the fatigue life. In contrast to small joints where peak stress is limited to the weld toe, the peak stress region in a large multi-pass joint may include several weld beads and cracks may initiate anywhere in this highly stressed area.

However, it is good practice not to seek benefit from improvement techniques in the design office.

11.6.8.2 Grinding

The improvement of the weld toe profile and the removal of slag inclusions can be achieved by grinding either with a rotary burr or with a disc. To obtain the maximum benefit from this type of treatment, it is important to extend the grinding to a sufficient depth to remove all small undercuts and inclusions. The degree of improvement achieved increases with the amount of machining carried out and the care taken by the operator to produce a smooth transition.

The performance of toe-ground cruciform specimens is fully investigated in Reference 11. Under freely corroding conditions, the benefit from grinding is minimal. However, in air, the results appear to fall on the safe side of the mean curve: endurance is altered by a factor of 2.2. It is, therefore, recommended that an increase in fatigue life by a factor of 2.2 can be taken if controlled local machining or grinding is carried out.

11.6.8.3 Weld toe remelting

Weld toe remelting by TIG and plasma arc dressing are performed by remelting the toe region with a torch held at an angle of 50° or 90° to the plate (without the addition of filler material). The difference between TIG and plasma dressing is that the latter requires a higher heat input.

Weld toe remelting can result in large increases in fatigue strength due to the effect of providing low contact angle in the transition area between the weld and the plate and by the removal of slag inclusions and undercuts at the toe.

11.6.8.4 Hammer peening

Improved fatigue properties of peened welds are obtained by extensive cold working of the toe region. These improved fatigue properties are due to:

- introduction of high compressive residual stresses
- flattening of crack-like defects at the toe
- improved toe profile.

It can be shown that weld improvement techniques greatly improve the fatigue life of weldments. For weldments subject to bending and axial loading, peening appears to offer the greatest improvement in fatigue life, followed by grinding and TIG dressing.

11.6.9 Fatigue-resistant design

The nature of fatigue is well understood and analytical tools are available to calculate the fatigue life of complex structures. The accuracy of any fatigue life calculation is highly dependent on a good understanding of the expected loading sequence during the whole life of a structure. Once a global pattern has been developed then a more detailed inspection should be carried out of particular areas of a structure, where the effects of loading may be more important, due to the geometries of joints for example.

Data have been gathered for many years on the performance of bridges, towers, cranes and offshore structures where fatigue is a major design consideration. Codes of Practice, such as BS 7608, give details for the estimation of fatigue lives. Where a structure is subjected to fatigue, it is important that welded joints are considered carefully. Fatigue and brittle fractures can be initiated at changes in section, holes, notches and cracks which give rise to locally high stresses. The avoidance of local structural and notch peak stresses by good design is the most effective means of increasing fatigue life.

It is important, during the design process, that consideration is given to how the structure is to be made. Poor access and difficult welding conditions can lead to defects in the weld and it can be difficult to carry out non-destructive testing in the critical locations. For example, acute-angled welds of less than 30° are difficult to fabricate, particularly in tubular structures. Similarly, if several parts join at the same place, access to some of the welds may be restricted. Despite these problems, some structures do contain such features.

The competence of the welders is critical to the durability and integrity of the structure. Poor welding practices can introduce large defects and high residual stresses, which increase the risk of fracture and reduce the fatigue lifetime. It is also vital that any repairs to rectify defects found during commissioning or in service are carried out by competent welders.

Repairs carried out to structures in service are expensive and, in the worst case, may require that a facility is closed down temporarily. Care needs to be taken when specifying secondary attachments to main primary steelwork. These are often not considered in detail during the design process as they themselves are not complex. However, there have been a number of failures in offshore structures due to fatigue crack growth resulting from a welded attachment. The following general suggestions can assist in the development of an appropriate design of a welded structure with respect to fatigue strength:

• Use butt or single and double bevel butt welds in preference to fillet welds.
• Make double-sided fillet welds rather than single-sided fillet welds.
• Aim to place the weld, particularly toe, root and weld ends in regions of low stress.
• Ensure good welding procedures are adopted and adequate non-destructive testing (NDT) undertaken.
• Consider the effects of localised stress concentration factors.
• Consider the potential effects of residual stresses and the possibility of post-weld heat treatment to reduce the magnitude of these internal stresses.

11.7 Final comments

The consistent message in BS EN 1993-1-9:2005, BS EN 1993-1-10:2005 and EN 1090-2:2008 is that the risk of failure can be brought to an acceptable level by good design, that minimises welding and sharp changes of section, together with good fabrication and inspection, that gives confidence that any defects are acceptably small, and the choice of steels with sufficiently high toughness.

The introduction of execution classes to EN 1090-2:2008 highlights the need for good control of the fabrication of steel structures to ensure that the defects that are introduced can be tolerated by the steel under the anticipated loading conditions and design stresses, at an acceptable level of risk for the consequences of failure. Under the most demanding of conditions, high toughness steels should be used with low design stresses and the structures should be manufactured by competent staff.

References to Chapter 11

1. British Standards Institution (2004) BS EN 10045-1 *Charpy impact test on metallic materials*. London, BSI.
2. Bhadeshia H.K.D.H. and Honeycombe, Sir Robert (2006) *Steels Microstructure and Properties*, 3rd edn. Oxford, Elsevier Butterworth-Heinemann.
3. British Standards Institution (2005) BS EN 1993-1-10 *Eurocode 3: Design of steel structures. Part 1-10: Material toughness and through-thickness properties*. London, BSI.
4. American Society for Testing and Materials (1981) *The standard test for J_{Ic} a measure of fracture toughness*. ASTM E813-81.
5. British Standards Institution (1991) BS7448-1:1991 *Fracture mechanics toughness tests. Part 1. Method for determination of KIc, critical CTOD and critical J values of metallic materials*. London, BSI.
6. Murakami Y. (1987) *Stress Intensity Factors Handbook*. Oxford, Pergamon Press.
7. Tada H., Paris P.C. and Irwin G.R. (1985) *The Stress of Cracks Handbook*. Del Research Corporation, Hellertown, PA, USA.
8. British Standards Institution (2005) BS 7910:2005 *Guidance on methods for assessing the acceptability of flaws in fusion welded structures*. London, BSI.
9. SINTAP (1999) *Structural Integrity Assessment Procedures for European Industry*. Project BE95-1426. Final Procedure, British Steel Report, Rotherham, UK.
10. American Society for Testing and Materials (1982) *Design of fatigue and fracture resistant structures*. ASTM STP 1061.
11. British Standards Institution (2008) BS EN 1090-2 *Execution of steel structures and aluminium structures*. London, BSI.
12. British Standards Institution (2005) BS EN 1993-1-9 *Eurocode 3: Design of steel structures. Part 1-9: Fatigue*. London, BSI.
13. British Standards Institution (1993) BS7608:1993 *Fatigue design and assessment of steel structures*. London, BSI.
14. Department of Energy (1990) *Offshore Installations: Guidance Design Construction and Certification*, 4th edn. London, HMSO.
15. British Standards Institution (1983 and 1980) BS 2573 *Rules for the design of cranes. Part 1: Specification for classification, stress calculations and design criteria for structures. Part 2: Specification for classification, stress calculations and design of mechanisms*. London, BSI.

16. Miner M.A. (1945) Cumulative damage in fatigue, *Journal of Applied Mechanics*, 12: A159–A164.
17. Nip K.H., Gardner L., Davies C.M. and Elghazouli A.Y. (2010) Extremely low cycle fatigue tests on structural carbon steel and stainless steel, *Journal of Constructional Steel Research*, 66: 96–110.
18. Divsalar F., Wilson Q. and Mathur S.B. (1988) Low cycle fatigue behaviour of a structural steel (B.S. 4360-50D rolled plate), *International Journal of Pressure Vessels and Piping*, 33: 301–315.
19. Paris P.C. and Erdogan F. (1963) A critical analysis of crack propagation laws, *Journal of Basic Engineering*, 85: 528–534.

Further reading for Chapter 11

Dowling, N.E. (1999) *Mechanical Behavior of Materials: Engineering Methods for Deformation, Fracture and Fatigue*, 2nd edn. Englewood Cliffs, NJ, Prentice-Hall, Inc.
Gray, T.F.G., Spence, J. and North, T.H. (1975) *Rational Welding Design*, 1st edn (2nd edn 1982). London, Newnes-Butterworth.
Gurney, T.R. (1979) *Fatigue of Welded Structures*, 2nd edn. Cambridge, Cambridge University Press.
Pellini, W.S. (1983) *Guidelines for Fracture-Safe and Fatigue-Reliable Design of Steel Structures*. Cambridge, The Welding Institute.
Radaj, D. (1990) *Design and Analysis of Fatigue Resistant Welded Structures*. Cambridge, Woodhead Publishing.

Chapter 12
Analysis

RICHARD DOBSON and ALAN J RATHBONE

12.1 Introduction

This chapter is concerned with analysis – the process of obtaining the forces and moments to which the members in a structure are subject. The traditional hand methods – moment distribution, method of joints, virtual work, to name but a few – are rather, perhaps sadly, a dying art. There are many textbooks that provide details on such methods and so it is not the purpose of this chapter to reiterate such information. Rather it is to place analysis into the context of the modern design office – this inevitably means discussing the subject almost entirely from the point of view of software.

For many years, analysis by software was a separate operation that was carried out on an isolated computer confined to some corner of the office. With the advent of the personal computer and meteoric progress in the price and performance of the processing power, most designers now have direct access to software. The software and its use has also changed – from individual user interfaces running in an MS DOS environment to common Windows interfaces presenting full 3D imaging. Further, the traditional approach of isolating 2D frames from a structure that behaves in 3D was relatively easy to visualise and, for simple rectilinear structures, was safe despite ignoring some of the 3D effects and not explicitly taking account of second-order effects. Guidance produced by the SCI[1] only 15 years ago was skewed towards the use of 2D analysis and much of the general advice therein is still applicable.

However, many multi-storey buildings are not simple – they may be curved on plan, multi-faceted and have severe restrictions on where bracing systems can be placed. For these structures, the interpretation, application and accuracy of a simple treatment of the building as a series of 2D frames can become more uncertain. That is not to say that there is not a place for 2D analysis; 2D analysis has a number of advantages, the most clear of which is reduced complexity, for example nodes in 2D

Steel Designers' Manual, Seventh Edition. Edited by Buick Davison and Graham W. Owens.
© 2012 Steel Construction Institute. Published 2012 by Blackwell Publishing Ltd.

analysis have only three degrees of freedom whereas in 3D, each node has 6 degrees of freedom. The classic case of the application of 2D analysis is in the analysis and design of portal frames (see Section 12.7).

Increasingly sophisticated analysis methods continue to improve the accuracy with which the behaviour of structures can be predicted. These advances are reflected in more complex codes of practice, such as Eurocode 3, which permits the use of a wide range of analysis types but requires the designer to understand when the use of each is appropriate. The aim is to use more accurate mathematical prediction of real world behaviour so that structures become more efficient and have fewer constraints on layout. The use of 3D in analysis and design matches this trend and also supports the burgeoning use of 3D models for visualisation, drawing production, steelwork detailing and the latest trend towards Building Information Modelling (BIM). Here all elements of the building, not just the main structure, can be modelled; sub-sets can be exported to the analysis/design software and the whole model passed down the construction chain and form the basis of the, now common, 3D steelwork detailing models.

Modelling is not just about choosing the appropriate software, it also involves using the appropriate people, training them and above all using experience and engineering judgement to at least know when the answers are reasonable. More information is given in the IStructE Guidelines 2002.[2]

12.2 The basics

For common building structures, analysis is concerned with determining the building displacements together with the external and internal forces and moments that result from the applied loading. In order to achieve this, the analysis process combines the structural stiffness of the analysis model together with the loading mathematically within the 'solver' to determine the results.

12.2.1 A few definitions

A few definitions for clarity:

- A structure has *joints* and *connections* and these occur at analysis *nodes*. A joint can be a number of connections.
- A *member* (a beam, brace, column, floor, wall) in a structure can be made up of one or more analysis *elements*.
- The *solver* is the mathematical solution finder for the analysis model with its loading used to determine the nodal displacements and element end forces.

- The *absolute* displacements and rotations are the displacements of nodes and elements in the model relative to the structural supports.
- *Relative* displacements and rotations are the displacements of the elements relative to their end nodes – not taking into account the absolute end node displacement and rotation.
- *Degrees of freedom* are the translation and rotation freedoms that a node is 'free' to move in. Typically there are 6 degrees of freedom in a 3D analysis model, three translational and three rotational.
- A *first-order analysis* considers the structure geometry to remain in its initial unstressed state.
- A *second-order analysis* takes into account the effect of deformation on the distribution of internal forces and moments and can be rigorous or approximate:
  - A *rigorous second-order analysis* is iterative and considers the geometry and stressed condition resulting from the previous analysis. The secondary effects of how the structure responds to both loading and its internal forces are hence accounted for in each analysis at the next step.
  - An *approximate second-order analysis* is a first-order analysis that takes account of the second-order effects by increasing the horizontal loads to replicate the deflected shape of the swaying structure thereby accounting for some of the second-order effects using a first-order analysis.

12.2.2 Mathematical modelling

The analysis of structures is a mathematical modelling process undertaken to determine the structural response to applied loading. All structural analysis models are idealisations or approximations of the real world, including simplifications of the geometry, material properties, structural supports and loading. The assessment of structural response is therefore the best estimate that can be obtained in light of the simplifying assumptions implicit in the structure. Idealisations necessary to simplify the modelling of a structure include:

- The physical dimensions of the structural components. For example, skeletal structures are represented by a series of stick elements. The nodes between the elements are typically assumed to be of negligible size and the members joining at the nodes are assumed to remain at the same angle to each other at the node (the node does not deform). The imperfections in the member straightness and structure out of plumb are also typically ignored in the model geometry and allowed for by other means (see Sections 12.6 and 12.7).
- Material behaviour is simplified. For example, the stress–strain characteristic might be assumed to be linearly elastic, and then perfectly plastic. No account is taken of the variation of yield stress along or across the member. The influence of residual stresses due to thermal processes (such as hot rolling and flame

cutting), as well as that due to cold working and roller straightening, is not usually included directly (although some research analysis tools do so).

- The local effects of actions are frequently ignored. For example, the development of local plasticity at connections or the possible effects of change of geometry causing local instability are rarely, if ever, accounted for in the analysis.
- The design loads employed in assessing structural response are themselves approximate.

The type of analysis chosen should therefore be adequate for the purpose and should be capable of providing the required solutions at an economical cost.

One of the fundamental responsibilities of the engineer is to appreciate the limitations of the modelling process and to have a full understanding of the idealisations being used together with a full appreciation of the accuracy of results obtained.

12.2.3 The analytical process

The analytical model of the structure is created by defining the idealised geometry, material properties and the structural supports. Next the load model is created defining the location, magnitudes and directions of loading on the structure. These loads are typically grouped into load cases by type e.g. dead, imposed, wind and snow. Finally, combinations of load cases are created which add together the loads in the load cases multiplying them by relevant factors as appropriate to codes of practice (BS EN 1990 Clause 6.10 and Chapter 3).[3]

Should the material properties be non-linear, for example for tension-only elements or compression-only supports, then a 'non linear' analysis will be required. Similarly, should the loads be other than static loads, for example time-dependent loading from a machine or an acceleration spectrum to model an earthquake, then a time history or a response spectrum analysis will be required. This is beyond the scope of this chapter. For more information see ASCE7-05.[4]

Once submitted to the analysis, the structure stiffness and load matrices are created and acted upon to determine the structure deformations, depending upon the type of analysis run. From the structural deformations, element end forces and moments can then be determined. From these and the loading, the element internal forces and moments are calculated along element lengths.

12.2.4 The structural model

Any complex structure can be looked upon as being built up of simpler units or elements. Broadly speaking, these can be classified into three categories:

- Skeletal structures consisting of 1D elements. Elements where the length is much larger than the breadth and height. These elements are variously termed as beam or truss elements and represent complete or partial beams, columns, braces or ties. A variety of structures are modelled by connecting such elements together using pinned or rigid joints. Should all the elements lie in a single plane, the structure is termed a plane frame or 2D structure. Where all elements do not lie in a single plane, the structure is typically called a space frame or 3D structure.

- Structures consisting of planar elements. 2D elements where the length and breadth are of the same order but much greater than the thickness. Such structural elements are further classified as membranes, plates or shells depending upon whether they have stiffness in plane, out of plane or both respectively. Typically these 2D elements are used in combination with 1D elements in a building model. Flat slabs, tented structures and shear walls are examples requiring meshed 2D elements.
- The third category consists of structures composed of members having length, breadth and depth of the same order – 3D elements. The analysis of such structures is complex, even when simplifying assumptions are made. Dams, some raft foundations, steel castings, caissons are all examples of complex structures potentially requiring 3D elements.

For the most part the structural engineer is concerned with skeletal structures consisting of 1D elements which combine with 2D elements to model walls and floors as appropriate to give a 3D model of the structure. Analysis of structures incorporating 3D elements is only rarely carried out for building structures and is beyond the scope of this chapter.

12.2.5 Structural stiffness

Structural stiffness is affected by the modelling of a number of items in the structure. Many of these are discussed in more detail later in this chapter but are listed here:

- geometry of the structural elements
- types of structural element used
- cross-section properties of these elements
- material properties of the elements
- release conditions at each end of the element
- structure supports to ground and other boundaries.

12.2.6 Static equilibrium

Static equilibrium is one of the fundamental principles of structural analysis. From Newton's law of motion, the conditions under which a body remains in static equilibrium can be expressed as follows:

- The sum of the components of all forces acting on a body, resolved along any arbitrary direction, is equal to zero. This condition is completely satisfied if the components of all forces resolved along the x, y, z directions individually add up to zero.
- The sum of the moments of all forces resolved in any arbitrarily chosen plane about any point in that plane is zero. This condition is completely satisfied when all the moments resolved into xy, yz and zx planes all individually add up to zero.

12.2.7 The principle of superposition

The principle of superposition is only applicable when the displacements are linear functions of applied loads – for instance in a first-order analysis.

In first-order analysis, for structures subjected to multiple loading, the total effect of several loads can be computed as the sum of the individual effects calculated by applying the loads separately.

This principle is very useful in computing the combined effects of many loads, as they can be calculated separately during the analysis and then summed after the analysis is complete. In contrast it should be noted that when the displacements are non-linear functions of applied loading, for instance in rigorous second-order analysis or non-linear analysis, then the principle of superposition cannot be applied. This means that the full combination of load cases must be applied to the model during the analysis to obtain the required results.

12.2.8 The results

The primary results from any static analysis are nodal deflections and rotations, and element end forces and moments, from which element internal forces, moments and deflections can be derived. Note that there are two components to deflection – bending deflection and shear deflection. Shear deflection tends to dominate in short stiff elements and bending deflection in long flexible elements.

In the case of a dynamic analysis the primary results are the structure natural frequencies for a number of modes of vibration and the structural modes of vibration.

12.3 Analysis and design

12.3.1 Introduction

Whilst the title of this chapter is primarily concerned with analysis, in modern steelwork design the analysis and design processes are not as clearly separated as they once were. This sub-section of the chapter explores the modern aspects of analysis associated with design.

There is an increasing tendency to build sophisticated 3D models for analysis and design – partly because modern complex building forms require them but also because with modern software for building design it can be more efficient in overall design terms.

For many years analysis of the structure was separate from the design component. The forces and moments in the structure were established by analysis models. For simple structures such as multi-storey buildings using Simple Construction (see Section 12.6) or planar trusses, the analysis could easily be carried out by hand. More complex areas of the structure could be entered into analysis software to establish the forces and moments – almost always in 2D. These forces and moments were then available for design by hand or by simple component software. Analysis software also had built-in 'design checkers' as a post-processor but the model was still built from the point of view of analysis. That is, the model was built from a set of nodes and elements and the model was 'unaware' of what function the elements provided.

Design models on the other hand as adopted in modern software for 3D design, are built from physical items that engineers understand, i.e. slabs, beams, columns, braces, trusses, supports and connections (Figure 12.1, Members in a design model).

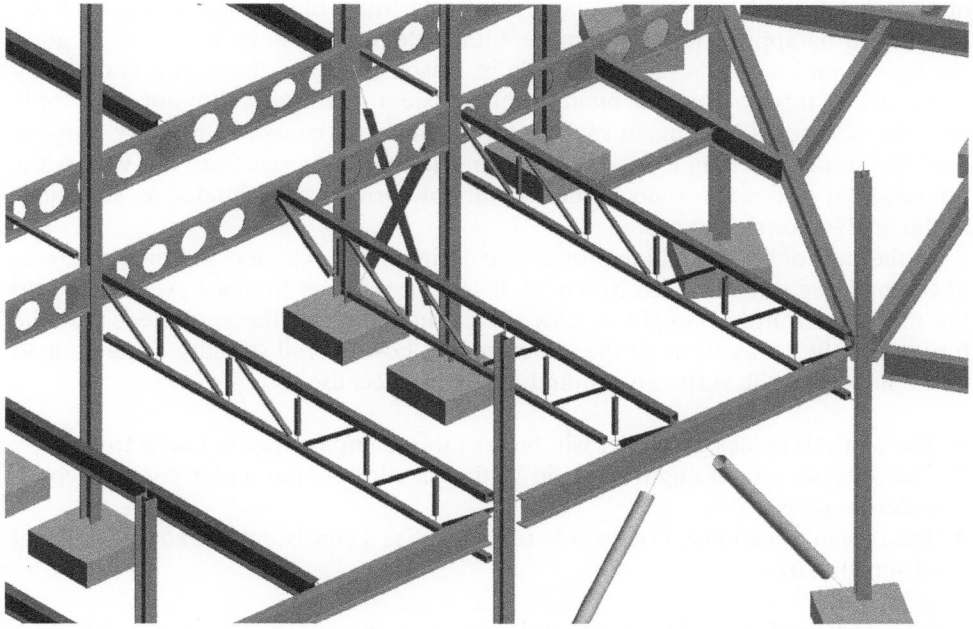

Figure 12.1 Members in a design model (Fastrak Building Designer Courtesy CSC (UK) Ltd)

Each of these has a particular behaviour and more importantly can be built into the model in a similar way to how they are put together on site. In this case, the analysis is subservient to the design and so behind the screen these design members are split appropriately into analysis nodes and elements to create the analysis model e.g. the internal members of a truss in the design model are given pinned ends in the analysis model. As well as representing how the structure would be built, a design model will tend to replicate the processes that the designer would go through traditionally. For example, designers would ensure by hand that all the lateral loads are applied to the braced frames or moment frames ignoring any contribution from the simple columns. In a good design model the software will execute this by pinning the simple columns above floor level so that all the lateral load finds its way to the lateral load resisting system.

12.3.2 Analysis and design models

Irrespective of whether the analysis/design is by hand, uses 2D software or 3D analysis or design modelling software, in all cases the designer is idealising the structure and creating a 'model'.

In the 'real world' a structure contains many items that do not affect (in any significant way) structural behaviour and so can clearly be left out of the design model e.g. false ceilings, windows, large doors, etc. Many of these can be represented as loads to be applied to the structure. Other elements may be less clear. A good example is an internal wall, particularly in masonry. Unless the wall is completely isolated structurally from the main framing system it will in reality not only apply load but also resist loading, in racking for example. Of course its primary purpose could be to take racking loads and provide stability to the structure but where it is not required or desired to do so then it will not normally be included as 'structure' in the analysis or design model.

At the start of the modelling process the designer needs to decide which elements of the building it is intended to model. It is poor practice to place everything into the model with little thought of how the elements are to be connected, how the forces find their way through the structure and how overall stability is maintained in the hope that the software will find a solution because:

- The analysis or design model will be far more complex than it needs to be.
- The analysis model might struggle to find a solution, particularly with second-order analysis.
- The design model may not be able to rationalise a consistent and/or efficient set of section sizes.

All design models (and their subservient analysis models) will have many idealisations – a summary of some of these is given below and more information is provided in later sections. It should be noted that analysis/design idealisations give approximate solutions to the 'real' behaviour. Within engineering accuracy, this is perfectly acceptable and the best that can be achieved even with software especially when considering the accuracy to which some of the loads are known.

- Beams and columns in Simple Construction do not, by definition, attract any moments due to frame action and, hence, it is important that they are modelled in that way.
- Floor diaphragms are often used to transfer lateral loads to the lateral load resisting system. These are usually considered to be (almost) infinitely stiff in their plane with (almost) no stiffness out of plane.
- Unlike the true physical models required for detailing, models for analysis and design are 'centre-line' models, i.e. the properties of the member is assumed to be concentrated on their 'centre-line'. However, for beams in multi-storey buildings, it is usual and acceptable to take the top of steel as the position of the 'line element' in the model.
- These centre-line elements may not always join ('node') at the same point. In the idealisation of such effects, it needs to be decided whether any eccentricity can be safely ignored, can be dealt with at the design stage or must be modelled.
- At the ends of members there will exist connections. These are idealised in their simplest form as 'pinned' or 'fixed'. The default in analysis software (as opposed to design) is for the latter. Careful consideration needs to be given to how these connections are to be fabricated and hence how they should be modelled. Note that in a 3D analysis/design system each connection has six degrees of freedom and may have quite different in-plane and out-of plane characteristics.

12.3.3 The basics of analysis

As well as the modelling idealisations discussed above, there are also analysis idealisations and simplifications. To determine the internal forces and moments in a building frame the following should be taken into account (if significant):

- second-order P-Δ effects – additional forces caused by deformation of the frame
- second-order P-δ effects – additional forces caused by deformation of the member
- global imperfections in the structure – e.g. the 'lean' in out-of-plumb columns
- local imperfections in members – e.g. initial bow of the member
- residual stresses in (steel) members
- flexural, shear and axial deformations
- connection behaviour.

There are probably many more, particularly those that have a tertiary effect on the behaviour e.g. lack of fit in connections, inelastic behaviour, settlement of supports. The 'Holy Grail' for 'structural analysts' in the context of building design is to allow for all of these effects in the global analysis of the structure, so-called

Advanced Analysis. Add in checking a yield criteria and the analysis subsumes the design, i.e. there is no need for member design checks. Such methods, although available for research purposes and special structures, are some way off for application to normal building structures in a practical environment. There is also the concern that such developments could lead to engineers relinquishing their grasp of the behaviour of the structure and relying too heavily on a computer – the 'black box' effect. The current application of reasonably sophisticated analysis, with well tried and tested design rules based on sensible idealisations, gives a pragmatic, safe solution.

BS EN 1993-1-1[5] assumes that a second-order analysis will be carried out and only under special circumstances can use be made of first-order analysis (see Section 12.6). BS EN 1993-1-1 addresses the points listed at the start of this section and provides two alternatives for dealing with them:

1. Where second-order effects in members and relevant member imperfections are totally accounted for in the global analysis, no individual member stability checks are required, i.e. it is only necessary to check strength.
2. Where this is not the case then those effects not included in the analysis must be taken into account in the individual member stability checks.

The first of these requires some form of Advanced Analysis (research software) whilst the second is the approach most likely to be found in modern commercial software.

As well as providing the forces and moments for design at the Ultimate Limit State the analysis can provide the resulting deflections under serviceability loads. It is normally the relative deflections in which the designer is interested and these are compared with given limits under particular loading conditions, or calculated for the total loading, and compared with some absolute maximum value. It has been mentioned earlier that connections are usually idealised as pinned or fixed and so the approximate deflections that result are checked against limits that experience has shown do not present problems in practice; the calculated deflections are unlikely to be an accurate prediction of those that occur in practice.

12.4 Analysis by hand

12.4.1 Some general rules

The calculations required to obtain the shear forces and bending moments in simply-supported beams form the basis of many other calculations required for the analysis of built-in beams, continuous beams and other statically indeterminate structures.

There are a number of general rules which are applicable to all beams that are subject to a first-order analysis (linear static). The following rules relate to the shear force and bending moment diagrams for beams:

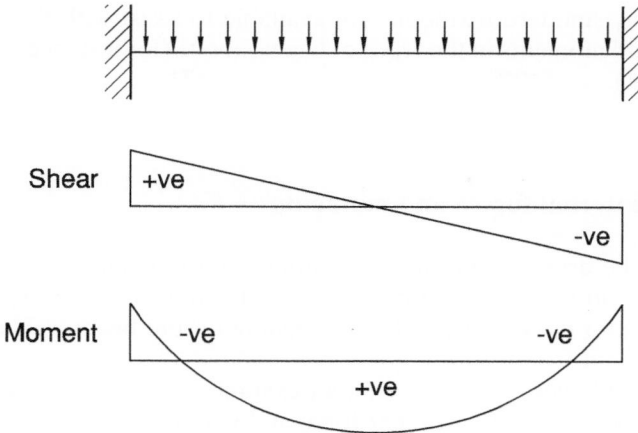

Figure 12.2   Shear and moment sign conventions

1. The maximum bending moment occurs at the point of zero shear along a beam, where such exists.
2. The shear force at any cross-section is the algebraic sum of forces acting to one side of the cross-section, including the element end shear.
3. Shear is usually considered positive when the resultant shear force to the left of the cross-section calculated as above is upwards, see Figure 12.2.
4. The bending moment at any cross-section is the algebraic sum of the moments about that cross-section of all forces to one side of the cross-section including the element end moment.
5. Moments are usually considered positive when the middle of a beam sags with respect to its ends or when tension occurs in the lower fibres of the beam, see Figure 12.2.

12.4.2 Simply supported beams

Appropriate formulae for cantilevers and simple beams under various loading conditions are presented in the Appendix on pages 1115–1124.

In the case of simple beams it is necessary to calculate the support reactions before the bending moments can be evaluated; the procedure is reversed for built-in or continuous beams.

12.4.3 More complex beams

Beams which are not simply supported are more complex because their connectivity induces end moments into the beam. For typical cases refer to the Appendix, pages

1125–1139. Many standard textbooks are available to assist with the hand calculation of forces and moments in the types of beams above (for instance Coates, Coutie
and Kong, 1998[6]).

12.4.4 Beam internal forces and moments

All beams can be assessed internally for forces and moments by considering the
'fixed' element end shears and moments in conjunction with the 'free' shears and
moments, the latter caused purely by the loading on the beam as if it was simply
supported.

This is best explained by considering an example beam in an analytical model.
Figure 12.3 shows the fixed end shear force and bending moment diagrams. From
analysis, the 'fixed' end forces and moments are F_1 and M_1 at end 1 and F_2 and
M_2 at end 2 resulting from the presence of the structure around the member in
question.

Note that the shears (F_1 and F_2) and the moment (M_1 and M_2) are influenced by
both the structural stiffness around the member and the stiffness of the member
itself and the member loading.

The 'free' shear force and bending moment diagram is equivalent to the shear
forces and bending moment due to loading on a simply supported beam (refer to
the Appendix, pp. 1117–1124, for typical simple beams). See Figure 12.4.

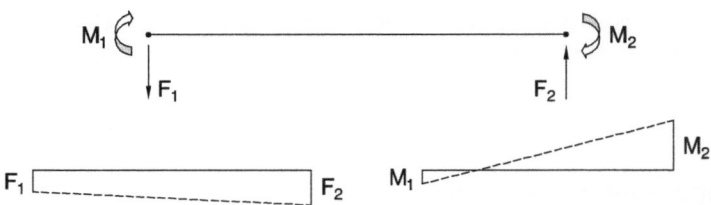

Figure 12.3   Member shears and moments – 'fixed'

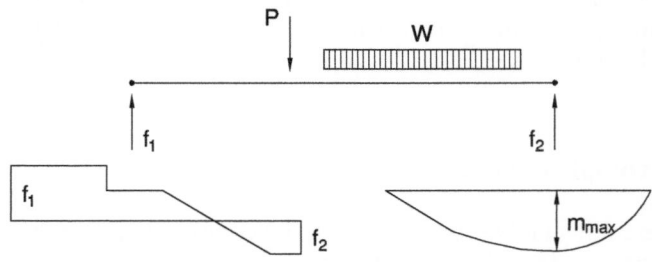

Figure 12.4   Member shears and moments – 'free'

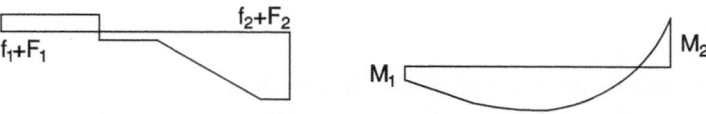

Figure 12.5 Member shears and moments – total

The total shear force and bending moment diagram is the sum of the two above as shown in Figure 12.5.

12.5 Analysis by software

There are many analysis packages currently available on the market. They offer a wide range of elements and analysis capabilities. The purpose of this section is to help unravel some of the features and functions of these pieces of software so that it is possible to have at least a level of understanding of the more complex picture.

It is worth noting that in steelwork building design, generally only the simplest of the items listed below will be required.

12.5.1 Modern 3D analysis

General analysis software packages perform analysis based on a defined structural model. They are applicable to a wide range of structural forms: buildings, bridges, towers, masts, tented structures, etc. Usually, the engineer has to define all the member sizes before the analysis can be run.

The modern trend is moving towards the analysis being a subset of 'design software'. A user builds a 'physical model' in the design software defining the members and connections, the software selects initial sizes for members based on engineering rules of thumb and then creates an analysis model automatically. Typically this process would be hidden from the user but a reasonable degree of control would remain available.

12.5.2 Axes

All analysis packages have at least two axis systems – the global axis system within which the whole structure exists and the local axis system relative to the individual element being considered.

Global axes

Typically referred to as X, Y and Z with Z acting vertically. In general analysis packages, Z is typically +ve upwards which means normal gravity loading has to be applied as a negative Z force.

Local axes

Typically referred to as x, y and z. For linear elements, x is typically along the member length with y and z being the cross-sectional axes (see BS EN 1993-1-1 Cl 1.7 and Figure 1.1). For 2D elements, z is typically normal to the plane of the element with x and y in the plane.

It is usually possible to rotate the element about its local axis system:

- for linear elements around the local x-axis using what is sometimes called the beta or gamma angle
- for 2D elements around the local z-axis. This is particularly useful to ensure that the element local axes align for instance in the plane of a floor.

12.5.3 Analysis element types

Analysis packages have a range of different element types. Some of the more common are as follows.

Inactive elements

An inactive element's contribution to the stiffness and mass matrices is ignored. This member type allows the user to experiment with the model. Usually, deleting an element deletes all of its properties and loads. Making the element inactive preserves the data for future use. The use of inactive elements is particularly useful for cross-bracing allowing one brace to be inactive whilst the other remains active.

Linear elements – applicable to all analysis types

Beam element – a beam element, by default, supports all 6 degrees of freedom at each end. The degrees of freedom align with axial, shear y and z, torsion, bending y and z. Typically, it will be possible to release an element end in any axis or rotation so that the member does not add stiffness to the connected node for that degree of freedom. Similarly, the connected node will not distribute force to the member end in that axis or rotation.

Truss element – the truss element, by definition, supports axial compression and tension and has only one degree of freedom at each node – a translation in the direction along the element x direction.

Linear spring – a linear spring element provides stiffness in one or more degrees of freedom. The user inputs the stiffness value.

Link element or rigid beam element – a 2-noded element which has effectively an infinite stiffness. This type of element is used to refine analysis models to reflect more closely true behaviour. For instance, where the two lifts of a column are offset to align the flanges along one face –a rigid link might be introduced to model the local effects of the lack of alignment in centre-lines. It is worth noting that often in steel structures these effects are either considered small and ignored (e.g. beams in floors with top flanges aligned) or accommodated in member design (simple columns with eccentricity moments).

Non-linear elements – only applicable to non-linear analyses

Tension-only elements – are truss elements which only have stiffness when in tension. During a non-linear analysis, the element will have a zero stiffness if, during the previous analysis iteration, the element was subject to compression.

Compression only elements – are truss elements which only have stiffness when in compression.

Cable element – only has stiffness when in tension. During a non-linear analysis, the element will have a zero stiffness if, during the previous analysis iteration, the element was subject to compression. In addition, a true cable element will include the effects of axial pre-stress as well as large deflection theory, such that the flexural stiffness of the cable will be a function of the axial force in the cable. In other words, for a true cable element the axial force will be applied to the deflected shape of the cable instead of being applied to the initial (undeflected) shape.

Non-linear spring – a non-linear spring element provides stiffness in one or more degrees of freedom. The stiffness is typically defined as a curve relating stiffness to displacement.

Gap element – a 2-noded element provides stiffness in one direction for a degree of freedom but zero stiffness in the other direction.

2D elements

Membrane element – triangular or quadrilateral thin membrane elements have in-plane (membrane) stiffness only and thus can only resist in-plane loads. These elements have two degrees of freedom at each node. With x, y in the plane of the element and z normal to it, the nodal degrees of freedom are translations in the x- and y-directions. Membrane elements are typically used for modelling tented structures.

Plate element – triangular and quadrilateral thick plate elements have out-of-plane (bending) stiffness only and so out-of-plane loads are permitted. These elements have three degrees of freedom at each node. With x, y in the plane of the element and z normal to it, the nodal degrees of freedom are translation in the z-direction, and rotations about the x- and y-directions. Plate elements are typically used to model floors.

Shell element – triangular and quadrilateral shell elements support both in-plane and out-of-plane loading. It has typically five or six degrees of freedom at each node, depending on the element shape and the capabilities of the analysis package; the extra degree of freedom being rotation about the z-axis: this is sometimes known as the drilling degree of freedom. It is most easily visualised when a beam intersects a wall, modelled in shell elements, at right angles. If the beam is subject to a torsional moment, some shell elements, i.e. those with the drilling degree of freedom correctly implemented, will correctly handle the application of the torsion, while some will not, i.e. those with it not implemented. Typically a triangular shell will lack this drilling degree of freedom, whereas a quadrilateral shell element will provide stiffness in this degree of freedom. Shell elements are typically used to model walls.

12.5.4 Types of analysis

12.5.4.1 Setting the scene

It is important to note that in the design of the majority of building structures in the UK, the only analysis types that are likely to be used are first-order analysis (linear static) and second-order analysis (*P*-delta static) – the latter is for structures which are susceptible to second-order effects. However, to aid understanding of the multitude of analysis types that are currently available in commercial software, they can be broken down into a number of 'classes' which are described below:

1. Static analysis – used to determine the nodal displacements, the element deflections together with the element forces, moments and stresses.
2. Dynamic analysis – also called vibration analysis – used to determine the natural frequencies and corresponding mode shapes of vibration.
3. Buckling analysis – used to determine the modes and associated load factors for buckling and assess whether or not the structure is prone to buckling at a higher or lower load than has been applied.
4. Response spectrum analysis – used in earthquake situations to apply an acceleration spectrum to a structure and to determine from this the design shears and moments in the elements.
5. Time history analysis – used to apply time dependent loading to a structure.

12.5.4.2 First-, second-order or non-linear analysis

Engineers today typically use first-order analysis (linear static) to determine design forces and moments resulting from loads acting on a building structure. First-order analysis assumes small deflection behaviour, i.e. the resulting forces and moments take no account of the additional effect due to the deformation of the structure under load, the assumption being that the additional effect is small and therefore can be ignored.

There are two classes of second-order analysis, rigorous and approximate:

1. A rigorous second-order analysis accounts for two *P*-delta effects to reach a solution:
   - Large displacement theory – the resulting forces and moments take account of the effects due to the deformed shape of both the structure and its members.
   - 'Stress stiffening' – the effect of element axial loads on structure stiffness. Tensile loads straighten the geometry of an element thereby stiffening it. Compressive loads accentuate deformation thereby reducing the stiffness of the element.
2. An approximate second-order analysis accounts for only one *P*-delta effect:
   - The second-order approximation is achieved by applying 'pseudo' horizontal loads to the structure to create a horizontal displacement to mimic the sway deflection. The resulting forces and moments take account of the effect due to the deformed shape of the structure.

P-delta effects

As structures become more slender and less resistant to deformation, the need to consider the *P*-delta effect increases. To reflect this, Codes of Practice, the Eurocode included, are referring engineers more and more to the use of second-order analysis to ensure that *P*-delta and 'stress stiffening' effects are accounted for, when appropriate, in design. This is as true in concrete and timber design as it is in the design of steelwork.

Engineers have been aware of the *P*-delta effects for many years. However, it is only relatively recently that the computational techniques and power have become widely available to provide the necessary analytical approximations. In the past, in the absence of more rigorous analysis capabilities, many design codes incorporated empirical checks and 'Good Practice' design guidance to ensure that the magnitude of the *P*-delta effect stays within limits, for which allowance has inherently been made. While this has ensured safe design, it has perhaps obscured a clear understanding of the *P*-delta 'effect' on a structure.

Initially let us consider static analysis problems only, for example a structure under gravity loads. Usually these problems are considered to be small displacement and hence linear. This means that if the load is doubled, all deflections, moments etc. double too, and the principle of superposition is deemed to apply. As the loads increase, eventually failure will occur by either yielding or buckling and the structural response will cease to be linear. In addition, for many structures, displacements are large enough that secondary (*P*-delta) effects become significant. Modern structural design codes recognise that small displacement theory (first-order analysis) may not be appropriate for all structures, and that *P*-delta effects must be accounted for in the design in some way. Such structures are often referred to as 'sway sensitive' (although BS EN 1993-1-1 does not use this term). When structures fall into this category, second-order analysis is necessary (refer to BS EN 1993-1-1 5.2.1(3)).

What are the *P*-delta effects?

P-delta is a non-linear effect that occurs in every structure where elements are subject to axial load. *P*-delta is actually only one of many second-order effects. It is a genuine 'effect' that is associated with the magnitude of the applied axial load (*P*) and a displacement (delta).

There are two *P*-delta effects:

- *P*-'Big' delta (*P*-Δ) – a structure effect, see Figure 12.6
- *P*-'little' delta (*P*-δ) – a member effect, see Figure 12.7.

The magnitude of the *P*-delta effect is related to the magnitude of axial load P, the stiffness of the structure as a whole and the stiffness of individual elements. It is worth noting that at first glance, the theory of static equilibrium may look to be broken for a rigorous *P*-delta analysis but in fact it is not in reality (Figure 12.8).

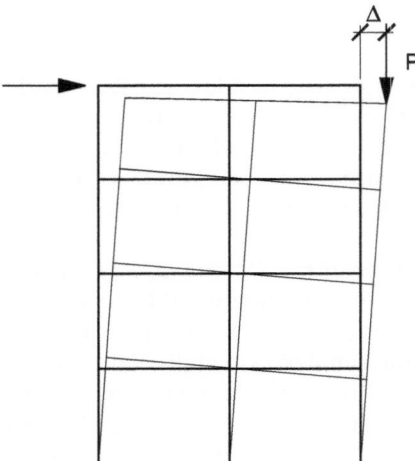

Figure 12.6   The *P*-Δ effect

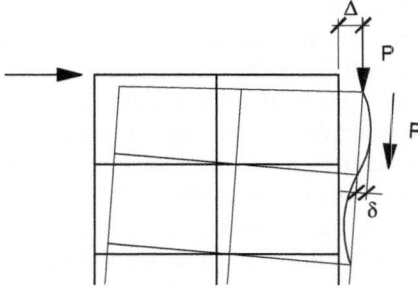

Figure 12.7   The *P*-δ effect

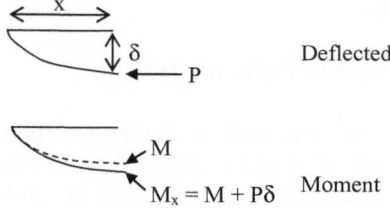

Figure 12.8   Member moment including *P*-delta moment

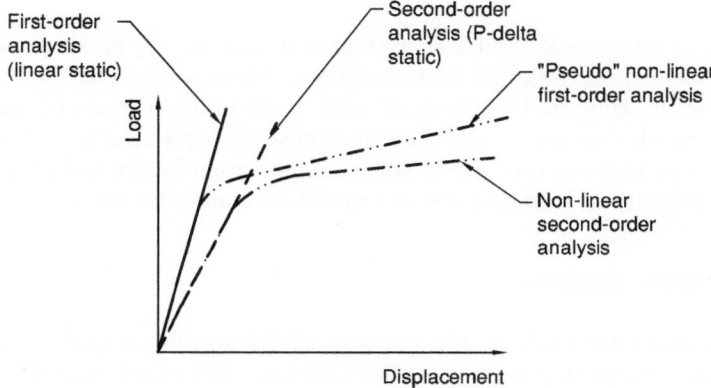

Figure 12.9   First-order and second-order analysis types

Non-linear effects

As soon as any structural non-linearity is defined in a model (for instance tension-only members, compression-only supports or cable elements), it is necessary to use a non-linear analysis. A non-linear analysis will usually, by default, also include P-Δ (geometric non-linearity) effects.

Which type of analysis?

To summarise, it is not the analysis that is linear or non-linear, it is the behaviour of the structure. It is essential to ensure that the analysis type chosen is appropriate to the structure:

- first-order analysis (linear static) – if the increase of internal forces or moments caused by deformation is small enough to be neglected
- second-order analysis (P-delta static) – if second-order effects increase the internal forces and moments significantly or modify significantly the structural behaviour
- non-linear – if any non-linear properties are defined (first- or second-order).

Figure 12.9 shows the different load displacement paths potentially followed by the different analysis types.

12.5.4.3 Static analysis

First-order analysis – a linear static analysis

The purpose of a linear static analysis is to determine the nodal displacements, the element deflections together with the element forces, moments and stresses. It is assumed that all stiffness effects and applied loads are independent of time.

A rigorous second-order analysis – a *P*-delta static analysis

The purpose of a rigorous *P*-delta analysis is to determine the nodal displacements, the element deflections together with the element forces, moments and stresses. In a rigorous *P*-delta analysis, the effects of axial loads on the stiffness of the structure are considered. However, it is assumed that these stiffness effects and the applied loads are independent of time. Load combinations must be applied in the analysis rather than unfactored load cases as superposition cannot be used.

Non-linear static analysis

First- or second-order analysis. The purpose of the non-linear static analysis is to determine the nodal displacements, the element deflections together with the element forces, moments and stresses due to loading conditions that are independent of time. In a non-linear analysis, stiffness effects and applied loads usually depend on deformation. Again load combinations must be applied in the analysis rather than superposition of unfactored load cases.

The non-linear solution

A full non-linear iterative solution allows for several non-linear conditions to be considered simultaneously, including 'stress stiffening' and both the P-Δ and P-δ effects. The solution is carried out in an incremental step-by-step analysis with the total applied loads divided into a number of load steps. A popular method of solution for non-linear equations is the Newton-Raphson method (see Süli and Mayers 2003).[7] When a general 'geometric (stress) stiffness matrix' approach is used in the method, there are no significant limitations on its use or applicability.

12.5.4.4 Buckling analysis

This analysis determines the modes and associated load factors for buckling. Therefore, it is possible to determine whether or not the structure is prone to buckling at a higher or lower load than has been applied. This analysis is particularly useful when investigating errors or warnings that are given during any rigorous *P*-

delta or full non-linear analysis as it can indicate areas of a structure particularly prone to buckling at low loads.

An elastic buckling analysis may be conducted to determine the critical load factors and corresponding buckling mode shapes. The analysis determines the smallest (critical) load factor required to buckle a structure (an individual element, set of elements or the entire structure) for a specified load case or combination. Buckling analyses are particularly useful for understanding the likely behaviour of a structure and tracking down why a linear static or a rigorous *P*-delta analysis fails (the failure may be due to buckling).

12.5.4.5 *Dynamic or vibration analysis*

There are a number of dynamic analysis types, all are focused at determining the natural frequencies for the primary and further modes of vibration for a structure. Dynamic analyses can be either stressed or unstressed. Unstressed vibration analysis will almost certainly be appropriate for structures where *P*-delta effects are not significant. Until a structure is quite highly stressed, it will generally be found that the vibration frequencies extracted from a *P*-delta stressed vibration analysis are similar to those for the unstressed case.

Axial or membrane stresses in elements (whether tension or compression) influence the natural vibration frequencies, just as the tension in guitar strings alters the pitch. Vibration analyses typically assume that there are no time-varying loads, damping effects are ignored and the vibration is simple harmonic. It is also necessary to ensure that the analysis type is appropriate to the structure, i.e. first-order analysis (unstressed vibration) where the effect of structure deformations can be ignored, second-order analysis (*P*-delta stressed vibration) if *P*-delta effects may be significant, and non-linear stressed vibration if any non-linear properties are defined (first- or second-order).

Unstressed vibration analysis

A first-order analysis. The purpose of the free-vibration analysis is to determine the natural frequencies and corresponding mode shapes of vibration. This information is needed for instance for seismic design. An unstressed vibration analysis is a good approximation when the structure is not overly sensitive to sway nor highly stressed and does not have non-linear elements or springs.

P-delta stressed vibration analysis

A second-order analysis. Natural frequencies and corresponding mode shapes of vibration may be determined while taking into consideration the axial loads due to loading conditions that are independent of time. A *P*-delta stressed vibration analysis should be used when the structure is sensitive to sway but is not highly stressed and has no non-linear elements or springs.

Non-linear stressed vibration

First- or second-order analysis. The analysis determines the natural frequencies and corresponding mode shapes of vibration while taking into consideration non-linear effects due to time-independent loads.

A non-linear stressed vibration analysis can be used whether a structure is sensitive to sway or not, whether the structure is highly stressed or not, and whether there are non-linear elements or springs in the structure or not.

12.5.4.6 Response spectrum analysis

Traditionally response spectrum analysis is most widely utilised as an aid to design in seismic regions. This analysis uses an acceleration spectrum to shake the supports of the structure so that it excites a number of modes of vibration. Mass in the structure participates in the modal vibration, a differing amount in each mode. The effects of these accelerating masses are assessed, and by using sophisticated combination techniques the design forces and moments in the structural elements can then be determined.

Response spectrum analysis is typically used to assess structures as they undergo seismic accelerations.

The assumptions are typically that:

- The structure is excited at all support nodes by the same defined spectrum.
- The spectrum's direction and frequency content are known.

As previously described, it is not the analysis that is linear or non-linear, it is the structure. It is therefore necessary to ensure that the analysis type is appropriate to the structure:

- first-order analysis (unstressed response spectrum) – if small displacement theory applies
- second-order analysis (stressed response spectrum) – if *P*-delta effects may be significant
- non-linear stressed response spectrum – if any non-linear properties are defined (first- or second-order).

As for the dynamic analysis above, there are three types of response spectrum analysis – unstressed response spectrum analysis, stressed response spectrum analysis and non-linear stressed response spectrum analysis.

12.5.4.7 Time history analysis

A time history analysis would typically be used to determine the response of a structure to defined time-varying loads such as blast loading, impact loading, loading from vibrating machinery (nodal excitation), or seismic loading (support excitation), more precisely than a response spectrum analysis.

12.5.4.8 Summary

In summary, the analysis of normal building structures will utilise 1D and 2D elements usually in a first-order linear static or a second-order static analysis.

It is worth noting here that:

- EN1993-1-1 starts from the premise of using an analysis that accounts for second-order *P*-delta effects.
- An analysis that accounts for *P*-delta effects does not have to be rigorous. Other types of analysis are typically used for nuclear installations, more complex structures, structures subject to vibration or structures in seismic zones.

12.6 Analysis of multi-storey buildings

Multi-storey buildings are defined as those buildings which have horizontal floors with pre-cast or composite floor slabs, horizontal steel or composite (steel/concrete) beams and vertical columns, although sloping columns are becoming more common. Typically, these structures have two or more floors and a roof. Lateral stability is generally achieved by braced frames, moment frames or shear walls, or a combination of the three. Figure 12.10 offers an interesting example of such a building.

12.6.1 Procedure

The following procedure will assist for braced simple multi-storey framed buildings. Steel frame multi-storey structures in the UK are typically analysed and designed for two types of loading – gravity and lateral. For a structure where floor grids repeat, greater levels of repeatability within the structure, and thus a more economic design, can be achieved by analysing the structure in the following order:

1. Analyse and design for gravity loading design:
    (i)   one typical floor first – ensuring common sizes are used where possible to maximise standardisation
    (ii) using this floor, replicate it up and down the building as many times as possible; design all floors and the columns for gravity loading.

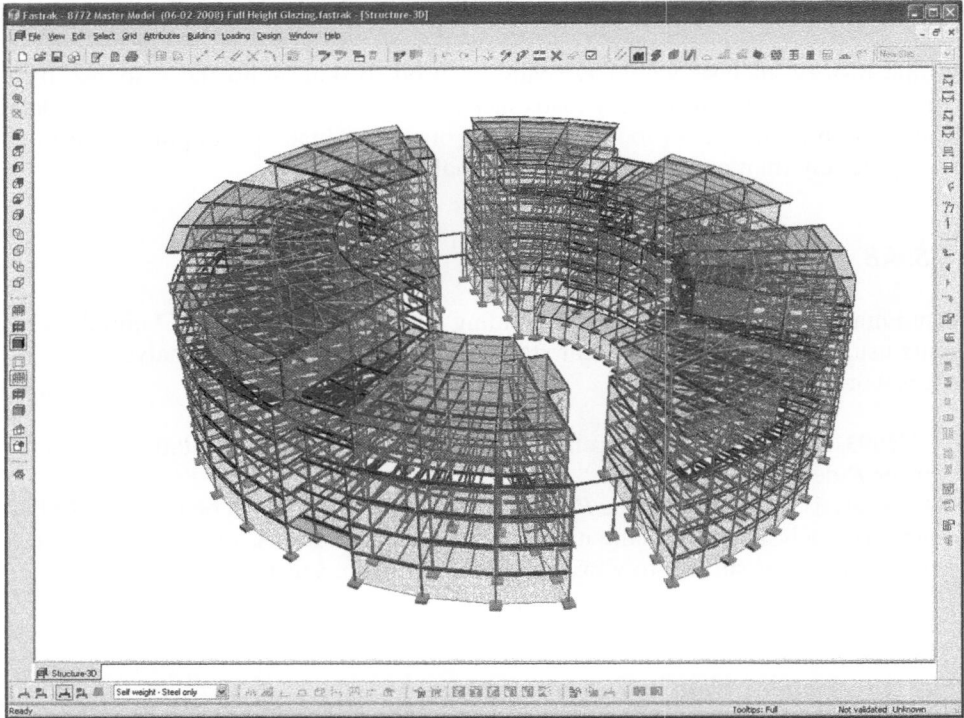

Figure 12.10   Building model (Fastrak Building Designer. Courtesy of CSC (UK) Ltd and Robinson Construction Ltd)

2. Analyse and design for lateral loading and design the lateral load resisting system.

    Other advice to take account of during the design process would be to:

-   Design for the anticipated critical combinations and check the design for the others – this can save significant amounts of time and enhance understanding of structural behaviour.
-   Always run first-order analyses before attempting to run second-order analyses. This ensures that the basic structure is stable before submitting it to an analysis which will fail if there are any instabilities in the structure.
-   If permissible use approximate methods to account for second-order effects, for instance, by increasing the horizontal loads on the structure rather than a rigorous second-order solution.

12.6.1.1 Modelling

Steel framed multi-storey buildings typically consist of:

- A structure to resist gravity loads (self-weight, imposed and snow loads) comprising horizontal floors which are often a composite deck acting either compositely or non-compositely with steel beams and vertical columns.
- A structure to resist lateral loading arising from wind and the effects of vertical loads acting through an assumed initial frame imperfection (BS EN 1993-1-1 Cl 5.3). The lateral load resisting system can consist of one or more of the following:
  - ○ braced frames – with bays containing diagonal braces or cross-bracing which resist the lateral loading in tension and/or compression
  - ○ continuous frames with bays resisting lateral load due to frame action and moment-resisting connections between beams and columns
  - ○ concrete shear walls which are typically planar elements or groups of planar elements which resist the lateral load in shear or shear and bending respectively.

12.6.1.2 Beams and columns

A number of simplifying assumptions are made when modelling the building for analysis:

- Analysis elements are aligned with the tops of steel beams in floors thereby ignoring the small offsets in centre-line between beams of different depth.
- The horizontal offset of edge beams is usually small enough to be ignored.
- All columns are typically modelled as being co-linear along their centre-line.
- Small offsets of columns from grids are typically ignored in design.
- To ensure that all the lateral loading is carried by the braced or moment frames (continuous framing), it is typical to assume that all columns not in braced bays or moment frames are pinned at each floor level, so they do not attract lateral loads.

  If columns are modelled in this manner, it is then essential to ensure that the assumptions inherent in column design in Simple Construction are valid – e.g. eccentricity moments (see Section 12.6.6.4).

12.6.1.3 Braced frames and moment-resisting frames

Braces in braced frames are usually modelled between beam-to-column nodes at each floor although their positions in the final structure are governed more by efficient gusset connection design and fabrication.

Moment-resisting frames typically model beam-to-column joints as rigid and full-strength in the plane of the frame but as pins out of the plane. Moment-resisting connections are typically only designed for the in-plane moments. (Moment connections are usually to a column flange. A moment connection to a column web requires significant fabrication and is best avoided.)

12.6.1.4 Shear walls

Shear walls can be modelled in a number of ways – the two most common are:

- The mid pier model – idealised as a central vertical member of equivalent stiffness and rigid arms at each floor linking to the structure it supports (see Section 12.8.4).
- Meshed using shell elements. The refinement of the mesh within the wall is a matter of engineering judgement – for more details see Arnott, 2005.[8] It is worth noting that some implementations of shell elements cannot deal with the torsional moment of a beam coming in at right angles to the shell (the drilling degree of freedom – see Section 12.5.3).

Special care should be given to modelling walls with openings. This can be achieved effectively by either of the methods above given an appropriate model.

12.6.1.5 Connections

Connections between members, beam-to-beam and beam-to-column are usually:

- simple joints – pinned in all axes
- continuous (full strength, rigid) joints – fixed in the major axis and pinned in the minor axis
- fully fixed – fixed in both axes.

Within analysis, these joint types are achieved by element end releases.
The use of continuous joints and fully fixed joints should be considered carefully to ensure that their behaviour can be achieved in practice (see Section 12.8.5).

12.6.1.6 Supports

Special attention should be exercised in the assessment of structural supports. There are two components to consider at a support:

- the base connection, i.e. the connection between the column and the base using a base plate
- the foundation itself, the pad base, ground beam, pile cap and piles etc.

The design forces for both components are required from the analysis results. Base connections in the UK will typically be pinned, however this depends upon the level of moment transfer required into the foundation, introduced typically to reduce lateral deflections (see Section 12.8.6).

12.6.1.7 *Material properties*

Material properties for steel are clearly defined in BS EN 1993-1-1 Cl 3.2.6:

- Modulus of elasticity $E = 210,000 \, N/mm^2$
- Poisson's ratio $v = 0.3$
- Shear modulus $G = 80,770 \, N/mm^2$ $(G = E/2(1 + v))$
- Coefficient of linear thermal expansion
  - structural steel $\alpha = 12 \times 10^{-6}$ per K
  - composite concrete/steel $\alpha = 10 \times 10^{-6}$ per K (BS EN 1994-1-1).[9]

If the building is to be assessed for second-order effects and it is composed of concrete members in addition to steelwork, then due consideration should be given to whether the concrete elements are cracked or not in bending. This can be accounted for by adjusting their bending section properties. Typical values would be (ACI 318-08 Cl 10.10.4.1):[10]

- beams – $0.35 \times I_{gross}$
- columns – $0.7 \times I_{gross}$
- walls – uncracked – $0.7 \times I_{gross}$
- walls – cracked – $0.35 \times I_{gross}$

Also when modelling concrete members, it is necessary to consider whether creep and shrinkage need to be accounted for. This is usually achieved by using either a short (no account) or a long term value for E_c.

12.6.2 **Loading and combinations**

Gravity loading (or action) will typically be applied to floors. In modern building design software, this can be applied anywhere on the floor with the software distributing the loads around the floor onto the supporting beams and columns automatically. Similarly, roof gravity loading, snow and snow drift loading will need to be applied to the top level of the building.

The principal lateral loading on any building is most likely to be wind. However, as the Eurocode requires that all frames should be considered to have an initial global imperfection, this is most easily accounted for by applying Equivalent

Horizontal Forces (EHFs). These fictitious horizontal forces arise from all load combinations and will vary in magnitude depending upon the vertical load. Lateral wind combination should therefore be combined with the EHF.

According to BS EN 1990 there are many possible combinations of actions. The net result for an analysis is that there are likely to be many load combinations created and analysed with member and connection design performed on the critical design forces (see Section 12.8.8 and Chapter 3).

12.6.3 Initial sizing

In the analysis of a structure, the forces and moments in a member which result from loading are dependent upon the stiffness of both the structure and the member itself. Some analysis packages require sections to be sized manually prior to analysis. Other more design-oriented packages (like CSC's Fastrak Building Designer) will size members automatically without the need for initial sizes. This can be achieved by using typical engineering rules of thumb for the initial size of a member before it goes through analysis which ensures that the first analysis is a reasonable estimate and the design forces and moment do not change significantly on a second pass through the analysis.

12.6.4 Global analysis

According to BS EN 1993-1-1 Cl 5.2.1, the analysis of a structure to determine the internal forces and moments can be determined from:

- a first-order analysis using the initial geometry of the structure
- a second-order analysis taking into account the influence of deformation of the structure.

A first-order analysis can be used up to the point where the second-order effects are too large to ignore. BS EN 1993-1-1 limits the use of first-order analysis to multi-storey structures where the design loads are less than 10% of those that would be needed to cause elastic instability. The code expresses this limit as $\alpha_{cr} \geq 10$ (α_{cr} is defined as the factor by which the design loading would have to be increased to cause elastic instability in a global mode). Helpfully, the code permits an approximate method to be used to calculate α_{cr} on a storey-by-storey basis within the building:

$$\alpha_{cr} = (H_{Ed} / V_{Ed}) \times (h / \delta_{H,Ed}) \qquad \text{BS EN 1993-1-1 Eqn 5.2}$$

where

  H_{Ed} is the total design horizontal reaction at the bottom of the storey taking account of both the horizontal loads and fictitious loads

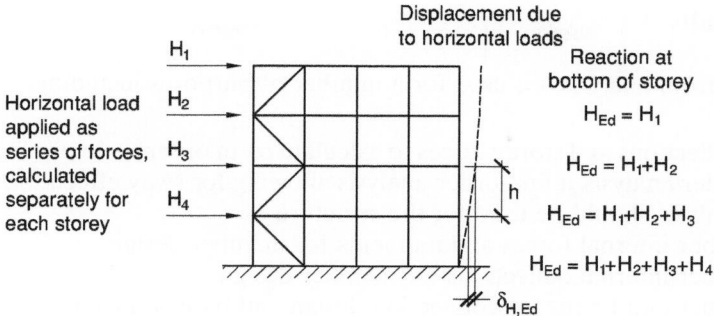

Figure 12.11 Displacement of a multi-storey frame due to horizontal loads

V_{Ed} is the total design vertical load on the structure at the bottom of the storey

H is the storey height

$\delta_{H,Ed}$ is the horizontal displacement at the top of the storey.

See Figure 12.11.

For structures where $\alpha_{cr} < 10$, the second-order effects are deemed too large to ignore, and a second-order analysis must be used. (BS EN 1993-1-1 Cl 5.2.1(3)). A second-order analysis can be either a rigorous *P*-delta analysis or an approximation using an amplification factor with first-order analysis, then:

- If $\alpha_{cr} < 3.0$ – a rigorous second-order analysis must be run – for instance using a rigorous *P*-delta analysis (BS EN 1993-1-1 Cl 5.2.1(5)B).
- If $10 > \alpha_{cr} \geq 3$, the second-order analysis does not have to be a rigorous *P*-delta analysis, but can be approximate. The Eurocode offers an alternative method to account for the second-order sway effects (*P-Δ*) by applying an amplification factor to all horizontal loading in a first-order analysis (BS EN 1993-1-1 Cl 5.2.1(5)B).

The second-order sway effects can be included in a first-order analysis by increasing the horizontal loads H_{Ed} (eg wind) and the EHF loads ($V_{Ed} \times \phi$) by the factor

$1/(1-1/\alpha_{cr})$ provided that $\alpha_{cr} \geq 3$

For details on the EHF loads see Section 12.6.6.1 (BS EN 1993-1-1 Cl 5.2.1(5)B).

For multi-storey buildings, this approach can be taken if all storeys have a similar distribution of vertical loads, horizontal loads and frame stiffness with respect to the applied storey shear forces (BS EN 1993-1-1 Cl 5.2.1(6)B).

12.6.5 Results

The results from analysis are used for a number of purposes including:

- storey deflections and storey forces to calculate α_{cr} in order to determine whether a first-order analysis, a first-order analysis allowing for sway effects, or a second-order analysis should be used for the structure
- the member internal forces and moments for member design
- the member internal deflections for member design
- the member end forces for connection design and base connection design
- the foundation forces for foundation design.

See Section 12.9.1.

12.6.6 Special considerations

12.6.6.1 Stability

BS EN 1993-1-1 requires that the stability of frames should take account of imperfections and second-order effects where they are deemed to be significant. Imperfections include the effect of residual stresses, geometric imperfections such as lack of verticality, lack of straightness, lack of fit and minor joint eccentricities.

The following imperfections should be taken into account, see BS EN 1993-1-1 Cl 3.1(3):

- global imperfections for frame and bracing systems
- local imperfections for individual members.

It is normal for the global analysis to take account of the global imperfections but local imperfections are usually treated implicitly within the design checks for individual members for flexural and lateral torsional buckling. The global imperfections are typically included in analysis as a horizontal force set that would give the same sway deflection as the required imperfection. These are termed Equivalent Horizontal Forces (EHFs). See Figure 12.12.

The EHF can be calculated from (BS EN 1993-1-1 Cl 3.2.3(a)):

$$\phi = \phi_0 \times \alpha_h \times \alpha_m$$

where

$\phi_0 = 1/200$

$\alpha_h = 2/\sqrt{h}$ but $2/3 \leq \alpha_h \leq 1.0$, h is the height of the building[11]

$\alpha_m = \sqrt{[0.5 \times (1 + 1/m)]}$, m is the number of columns acting in the sway mode.

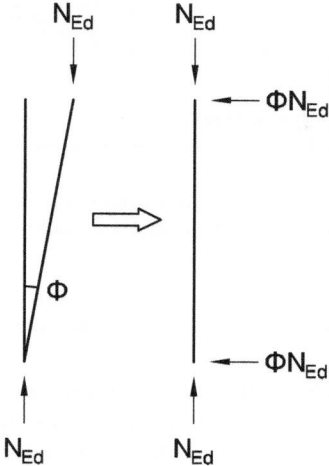

Figure 12.12   Initial sway imperfections

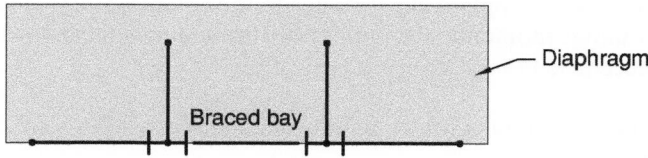

Figure 12.13   Case 1 – all nodes in diaphragm – no axial force in any beam

For heavily compressed members with at least one moment connection at an end, the Eurocode introduces a condition to add local imperfections into the analysis model for these members where:

$$\bar{\lambda} > 0.5\sqrt{[(A \times f_y)/N_{Ed}]}$$ (BS EN 1993-1-1 Cl 5.3.2(6))

which equates to $N_{cr}/N_{Ed} < 4.0$ for the member.

12.6.6.2 Floor diaphragms

Floors can be modelled as diaphragms. Diaphragms are horizontal plates in which all the nodes in the plane move together horizontally but move independently out of the plane of the diaphragm (vertically). It should be noted that all elements with both nodes in the diaphragm carry no axial load as the diaphragm carries this load. Consequently, if it is considered that beams in braced bays should be designed for axial load, then the effect of any diaphragm modelling on the axial load in both those beams and the surrounding beams should be considered carefully e.g. compare Figures 12.13 and 12.14.

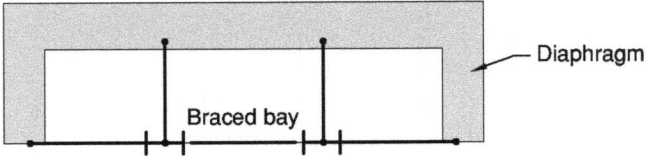

Figure 12.14   Case 2 – braced bay beam end nodes not in diaphragm – axial load in all incoming beams – is this the true axial load in the braced bay beam?

12.6.6.3 Simple columns – eccentricity moments

Beams and columns in 'simple' multi-storey building construction are modelled for analysis to meet at idealised nodes where the beams have pinned ends. Hence no moments are generated in the columns from the beam end reactions. In reality, however, the beams either connect to the column flanges or the column web and so the beam end reaction is applied at an eccentricity to the column centre-line. In order to cater for the potential local moment in the column that results from the beam end reaction being applied at a distance to the column centre-line, eccentricity moments or nominal moments are applied to the column local to the supported beams. These eccentricity moments are typically set to be:

- 0.5D + 100mm for connection to column flange; where 'D' is the depth of the column section
- 0.5t + 100mm for connection to the column web; where 't' is the thickness of the column web.

It should be noted that these are meant to be nominal moments that principally confine their effect local to the connection with the supported beam. They should be treated differently to 'real' moments in the column due to frame action (Access-Steel, 2005).[12,13]

12.6.6.4 Imposed load reductions

It is unlikely that all parts of a building will be simultaneously fully loaded hence imposed loads on beams and columns can be reduced. However, it is not permitted to apply imposed load reductions to beams and columns simultaneously.

The UK National Annex to BS EN 1991-1-1[14] permits an alternative method of reduction using NA 2.6:

- $\alpha_n = 1.1 - n/10$ if $1 \leq n \leq 5$
- $\alpha_n = 0.6$ if $5 \leq n \leq 10$
- $\alpha_n = 0.5$ if $n > 10$

where

n is the number of storeys with loads qualifying for reduction – in BS EN 1991-1-1 Eqn 6.2, n is the number of storeys above from the same 'category of load'. Imposed loads on roofs do not qualify for reduction.

NOTE the imposed load reduction may only be applied when the imposed load is the leading variable action in a load combination. When it is an accompanying action the load reduction α_n must not be applied.

12.6.6.5 X-Braced frames

Braced frames that use X-braces typically treat the X-braces as tension-only members. Although it is possible to consider X-braced frames with a non-linear analysis, which will treat the member correctly if in tension and ignore it if in compression (see Section 12.5.3 tension-only elements), this is not recommended for building structures because in non-linear analyses, X-bracing can be a source of analytical instability. Under gravity loading, the X-braces in a building will all go marginally into compression and thus immediately have zero stiffness thereby often resulting in an analysis that cannot find a solution.

It is preferable to manually make active and inactive the X-braces in tension and compression to arrive at an analytical solution, but this can take time. For complex models, this can also be difficult to achieve. Both +ve and –ve lateral loads should be considered to obtain the correct column axial forces and foundation forces.

There are automatic 'tricks' that can be used to simulate non-linear behaviour in a linear analysis; however, the user should be wary of these solutions as they do not always provide the correct solution for the X-braced members.

12.7 Portal frame buildings

12.7.1 Modelling

Proprietary software dedicated to the analysis of portal frames generally involves an elastic analysis to check frame deflection at the serviceability limit state, and an elastic-plastic analysis to determine the forces and moments in the frame at the ultimate limit state. These methods have largely replaced the rigid-plastic method which is unable to account for the important second-order effects in lightweight portal frames. Nevertheless a description of both methods is provided since the rigid-plastic method continues to be allowed by BS EN 1993-1-1 Cl 5.4.3 (1) providing certain criteria are met – see BS EN 1993-1-1 Cl 5.2.1(3) – that ensure second-order effects can safely be disregarded.

12.7.2 Analysis

BS EN 1993-1-1 provides three alternatives for 'plastic global analysis':

- Elastic-plastic analysis with plastified sections and/or joints as plastic hinges (referred to below as the 'elastic-plastic method').
- Non-linear plastic analysis considering the partial plastification of members in plastic zones. This method is not generally used in commercial portal frame design and so is not discussed below.
- Rigid plastic analysis neglecting the elastic behaviour between hinges. This is included as the 'rigid-plastic method' below for historical and comparison purposes.

The approach to analysis in BS EN 1993-1-1 is to presuppose that a second-order analysis is carried out except under special circumstances when the second-order effects are small enough to be ignored. For ease of explanation, the text below describes first-order elastic-plastic analysis and then makes note of some items that are expected from second-order analysis. The basics of analysis types are dealt with in Section 12.5.4 and the text here limits itself to second-order elastic-plastic analysis – second-order elastic analysis is dealt with more fully in Section 12.5.4.2.

12.7.2.1 The rigid-plastic method

The rigid plastic method is a simplified approach suitable for hand calculation and graphical methods. In this method the frame is assumed not to deform under load (no linear elastic component) until all hinges required for a given mechanism have formed. The frame then collapses. The design process involves comparing a number of predetermined mechanisms to evaluate which one has the lowest load factor and hence represents the maximum load which could be carried by the frame prior to collapse. In each case, the bending moment diagram along the members is constructed to check that the plastic moment is nowhere exceeded. For simple structures such as single-span frames this process is a relatively simple matter since there are a very limited number of possible failure mechanisms. However for more complex frames, e.g. multi-span, steps in eaves height, sprung supports or valley bases, the number of potential failure mechanisms, particularly under complex loading conditions, is vast. Alternative approaches are therefore usually incorporated to quickly establish a close approximation to the critical mechanism without the need to try all possibilities.

12.7.2.2 The elastic-plastic method

The elastic-plastic method, in addition to finding the collapse load, determines the order in which the hinges form, the load factor associated with each hinge formation,

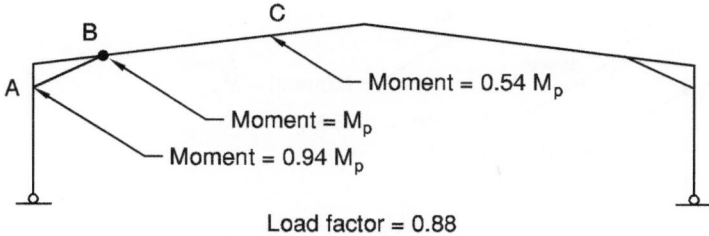

Figure 12.15   Incremental approach – first step

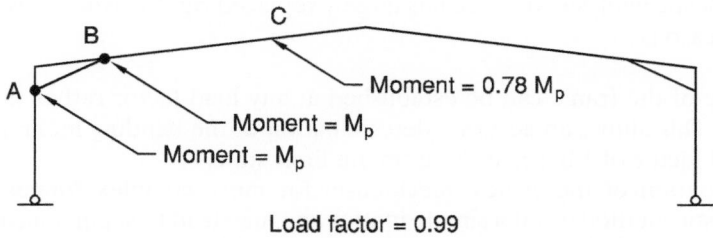

Figure 12.16   Incremental approach – second step

and how the bending moments around the frame vary between each hinge forma-tion. The frame is assumed to behave linearly between each hinge formation.

The incremental approach of the method means that it can determine whether hinges form and later 'un-form', i.e. hinges cease to rotate and begin unloading as a result of the necessary redistribution of moment around the frame. This phenom-enon and the incremental approach is best illustrated by an example. Consider the frame in Figure 12.15. The elastic-plastic analysis indicates that in this particular example, the first hinge would form at the sharp end of the haunch, B, at a load factor of 0.88. This can be confirmed by a linear elastic analysis since the frame remains elastic until the formation of the first hinge. The corresponding moment at the top of the stanchion, A, is less than M_p.

As more load is introduced, the next hinge to form is at the top of the stanchion at a load factor of 0.99 (Figure 12.16). Thus hinges now exist at positions A and B although as applied load is increased still further, the moment at hinge B would begin to reduce because of the continued redistribution of moment around the frame. This is known as hinge reversal.

Finally, the last hinge to form would be in the rafter close to the apex, C, at a collapse load factor of 1.05 (Figure 12.17). It may be noted that at the Ultimate Limit State (load factor = 1.0), the moment at B will be very close to M_p and impor-tantly will have undergone some rotation.

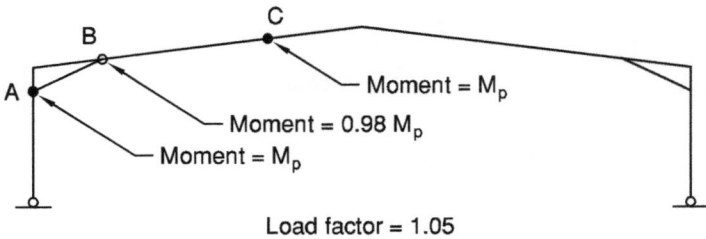

Figure 12.17   Incremental approach – final bending moments

Elastic-plastic analysis software has largely replaced rigid-plastic software for the following reasons:

• The state of the frame can be established at any load factor rather than only at collapse. This allows an accurate determination of the bending moment diagram at a load factor of 1.0, i.e. at the ultimate limit state.
• Determination of the critical mechanism for more complex frames using the rigid-plastic method is not a simple matter and may lead to slight approximations. The elastic-plastic method will always find the critical collapse mechanism.
• The elastic-plastic method has a complete hinge formation history, whereas the rigid-plastic method takes account of only those hinges which exist at collapse. Therefore any hinge which forms, rotates, ceases to rotate and then unloads is not identified by the rigid-plastic method.

12.7.2.3 Walk-through of the elastic-plastic method

It is possible to use an elastic analysis program in a 'step-wise' manner to produce a pseudo elastic-plastic analysis. This is relatively easy in conceptual terms but can be very tedious for anything but the simplest of frames. The process is an aid to understanding the way elastic-plastic analysis operates.

The first step is to carry out an elastic analysis at the full design loading. It is then necessary to investigate the bending moment diagram around the frame and determine the point or node at which the ratio of the applied moment to the plastic moment of resistance of the section (appropriately reduced for axial force) is the greatest. This is the position of the first hinge formation. A new model is then created with a pin at that point, and a pair of equal and opposite moments equal to M_p of the section applied at the pin. This new model is then reanalysed to determine the position of the next hinge formation. A further pin and pair of moments are inserted at that position, the model reconstituted and the process continued.

This was the basic approach of early software for elastic-plastic analysis, although the re-creation of the model was incorporated internally within the software by reconstituting the stiffness matrix at each hinge formation. Computationally this was found to be inefficient and, as with the hand method, did not cope easily with complex features such as hinge reversal.

12.7.2.4 Results

In addition to following the advice given in Section 12.9.1, there are some particular effects that might arise with elastic-plastic analysis. It is important to check the 'hinge history' for each design combination to see if it appears sensible:

- Are there enough hinges to form a collapse mechanism? This does not have to be a complete collapse, i.e. there do not have to be sufficient to cause a collapse of the whole frame, only a local area of the frame may collapse. For example in a multi-span frame the combination of one column and one rafter with sufficient hinges may cause collapse whilst the remainder of the structure is stable.
- Do any hinges form and then 'reverse'? Reversed hinges may need to be stabilised with purlins or side rails depending upon at what stage they form.
- Check for symmetrical hinges. For example, whilst under vertical loading the results may show a hinge at the sharp end of the left hand haunch, due to symmetry this could equally have formed at the right hand side. It is only due to very minor differences in the 'mathematical model' that in this case the hinge decided to form at the left hand side. Hence, the sharp end of the haunch needs to be restrained at both sides of the frame.

When using second-order elastic-plastic analysis, a hinge history will still be present although in this case there is a more significant likelihood that there will appear insufficient hinges to cause collapse. This is because the effect of a hinge formation is similar to inserting a 'pin' at that position. Thus at each hinge formation, the stiffness of the frame reduces and consequently the second-order effects (principally P-Delta effects) increases non-linearly. This can have a 'runaway' effect and can lead to a 'falling branch' in the load-response history. Figure 12.18 shows a

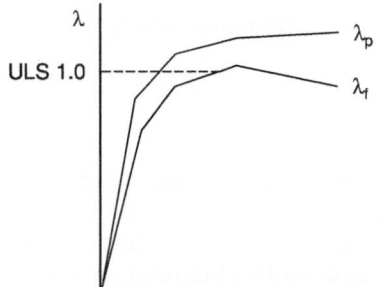

Figure 12.18   Load response history

load response history for a typical portal frame – the ordinate is the load factor against collapse where at Ultimate Limit State the load factor is 1.0 and the abscissa is some measure of deformation/deflection. The upper line shows a classic response with each change in slope of the line indicating a hinge formation. Eventually sufficient hinges form to create a mechanism and the frame collapses, i.e. the line becomes horizontal – infinite deformation for no increase in load. The lower line is a condition in which the second-order effects are sufficient to cause the falling branch effect – the deterioration in stiffness with each hinge formation is sufficient to rapidly increase the second-order effects at the next hinge formation to the extent that less load can be applied (lower load factor) before the next hinge forms.

The forces/reactions on the stanchion bases (base plates) and the supporting foundations will be required – the latter possibly early in the project. There are several important aspects to these results. The first is generally applicable to analysis models of steel structures and the second is specific to elastic-plastic analysis.

1. Care should be exercised in interpreting the sets of results given for either foundations or base plates. Either from the documentation accompanying the analysis software or from running a very simple (understandable) model, it should be established whether the results given are 'forces' or 'reactions' – they might also be termed loads although strictly they are not loads. Also, the sign convention needs to be observed. For example in a standard portal frame, the horizontal reaction at the base would be expected to be reacting in towards the frame to resist the component of thrust in the rafter that is horizontal. In one (typical) sign convention (positive reaction in the global X) the horizontal reaction at the left base would be positive and that at the right base negative.
2. Depending upon the application, the required results might need to be unfactored forces/reactions from loadcases or factored forces/reactions from combinations. Elastic-plastic analysis is essentially non-linear and so the principle of superposition does not hold. In addition the distribution of the forces around the frame due to the formation of hinges will not be the same as that from an elastic analysis (despite the same loads). Thus for the Ultimate Limit State only the elastic-plastic analysis results for combinations are strictly correct. Unfactored loadcase results can only be obtained from an elastic analysis. Note that the unfactored loadcase results adjusted for the load factors in a combination will not sum to those from the elastic-plastic analysis. See Section 12.7.4.

12.7.3 Haunches

Haunches are frequently provided at the eaves and apex connections of a portal frame. Typically analysis software does not have the facility to use a 'tapered element'. In such cases it is acceptable to model tapered members as a series of uniform prismatic elements as described below in the context of portal frame design.

Eaves haunches of normal proportions may satisfactorily be modelled with two 'rafter' elements and one 'column' element, evaluated at the cross-sections shown

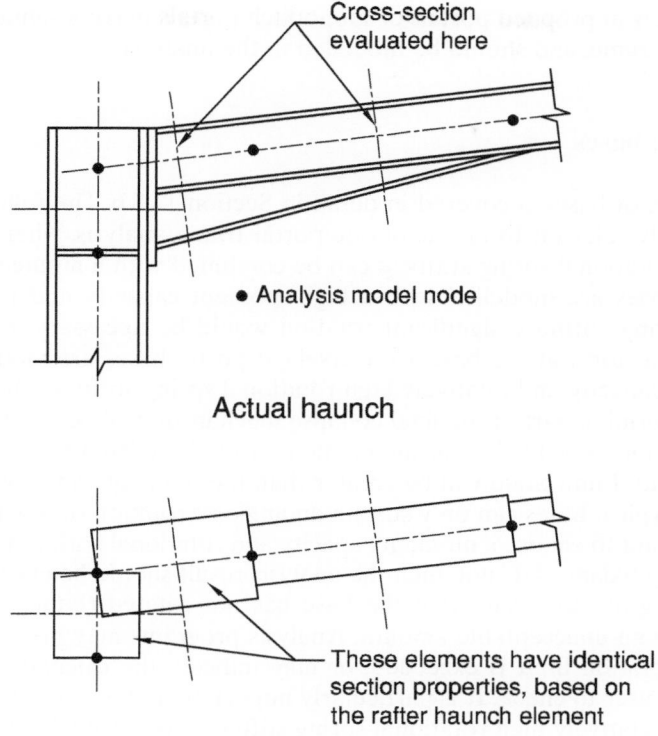

Cross-section
evaluated here

• Analysis model node

Actual haunch

These elements have identical
section properties, based on
the rafter haunch element

Prismatic equivalent haunch

Figure 12.19   Modelling of eaves haunch

in Figure 12.19. The haunched rafter is modelled with average section properties for lengths corresponding to ⅓ and ⅔ of the haunch as shown. The top of the column may be modelled using the section properties of the deeper haunch section. The assumption that the neutral axis remains at the centre-line of the rafter and does not descend towards the haunch is safe, since it tends to overestimate both the compression in the bottom flange, and the shear. Increased refinement is not justified by improved accuracy for most normally proportioned eaves haunches. The equivalent elements should be connected rigidly at their intersections.

Plastic hinges must not be allowed to form in the haunched region during the elastic-plastic analysis. Hence, when defining the properties for the haunch elements, the moment capacity should be set to a large value. (Depending on the software, this may be by direct input of a high moment capacity, or, for example, by input of a high section modulus.)

Apex haunches of normal proportions generally have no significant influence on the frame analysis, and so do not need to be included in the analysis model. So-called 'apex' haunches in propped portals or monopitch portals make a significant contribution to the frame, and should be modelled in the analysis.

12.7.4 Portal bases

The modelling of bases is covered in detail in Section 12.8.6. The following points are particularly relevant to elastic-plastic portal frame analysis, where horizontal, vertical and rotational spring stiffness can be combined with a moment capacity.

If portal bases are modelled with a high moment capacity and relatively low rotational spring stiffness, significant rotation would be necessary in order for a plastic hinge to form at the base. Conversely, if portal bases are modelled with a low moment capacity and relatively high rotational spring stiffness, then it is likely that a hinge forming part of the final collapse mechanism will occur at a base position. Furthermore, it is likely that the moment at the base from the elastic analysis at Serviceability Limit State will be greater than the moment capacity of the base.

In reality, typical bases can only sustain an angle of rotation of less than 10° and so it is important to choose a moment capacity and rotational spring stiffness which are reasonably balanced. If not, then the analysis result should be checked by hand or by the program to ensure that the base has not rotated (either elastically or plastically) by an unacceptable amount. Analysis programs may present a warning if a pre-set rotation limit is exceeded, or may indicate the calculated rotation at nodes for the user to check. It is particularly important, in the case of low moment capacity and relatively high rotational spring stiffness, to judge whether the plastic rotation which is inferred can be accommodated by the base detail, i.e. is the base sufficiently ductile? Embedded holding-down bolts and the welds should not be relied upon to provide ductility.

12.7.5 Valley supports

'Hit and miss' valley frames are common in portal frame buildings (see Figure 12.20) – this is where one or more internal stanchions in a multi-span frame are omitted in every second frame, the 'miss' frame. Valley beams running longitudinally support the rafters at the miss positions and react back onto the columns of the adjacent frames, the 'hit' frames.

In a typical two dimensional elastic-plastic analysis, valley supports are usually modelled by the inclusion of vertical, horizontal and rotational spring stiffness at those positions. Specifying a support moment capacity at a valley beam would imply plastic torsional behaviour of the valley beam, and this option is generally not available in proprietary portal frame analysis software.

The behaviour of the hit frame is influenced by that of the miss frame and vice versa. Hence, an interactive approach to the analysis and design of both is required because the reaction from the valley beam has to be included in the loading on the 'hit' frame but will be unknown until the 'miss' frame has been analysed and

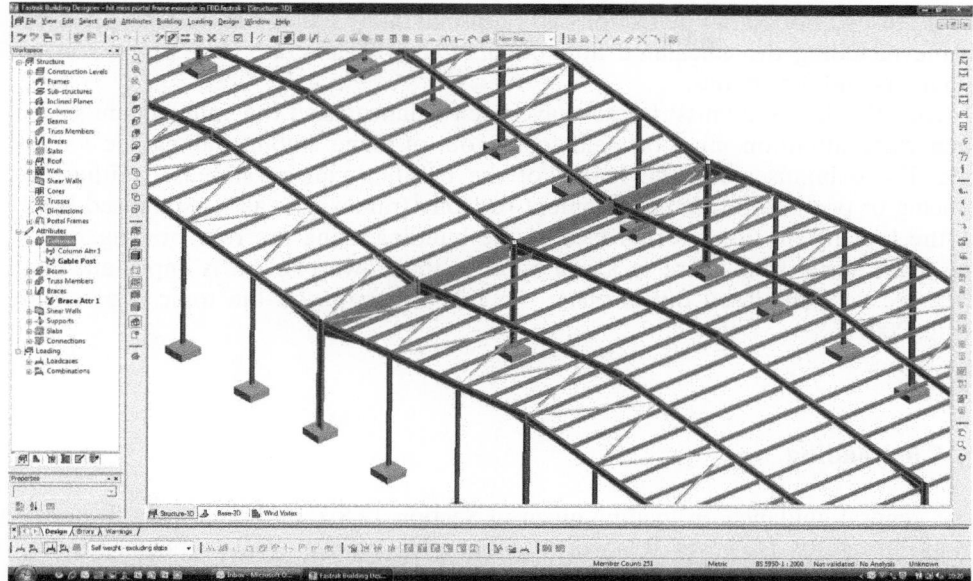

Figure 12.20   Typical hit and miss portal frames (CSC Fastrak Portal Frame Design)

designed; and the spring stiffness of the valley beam in the miss frame will be unknown until the beam has been designed or a section size estimated. The horizontal deflections of both frames need to be similar, since in reality the sheeting, which is very stiff, constrains the two frames to move together.

The vertical spring stiffness is relatively easy to calculate, knowing (or estimating) the section size of the valley beam. The vertical deflection of the beam due to a unit point load can be calculated using Engineer's bending theory, and defined by the ratio of [deflection]/[force]. The spring stiffness is the inverse of this, in appropriate units. The vertical deflection of the valley beam will depend on the degree of fixity assumed at the supports on the 'hit' frames.

A horizontal spring stiffness can be calculated in a similar manner using the weak axis properties of the valley beam. The horizontal supports to the valley beam (the 'hit' frames) are however not rigid, and the horizontal deflection calculation must include the support deflection before determining the equivalent spring stiffness. A horizontal spring stiffness will produce a horizontal load on the valley beam, requiring the valley beam to be designed for biaxial bending. Alternatively, bracing in the plane of the roof may be provided to the valley beam. In both cases the horizontal reactions must be included in the design of the 'hit' frame.

One convenient approach when using bracing to the valley beam, is to apply an assumed horizontal 'support' force at the valley of the 'miss' frame, which should be released horizontally. An equal and opposite force is applied at the valley of the 'hit' frame, and the horizontal deflections compared. This approach is then repeated, until the two horizontal deflections are approximately equal. The analysis of each frame including the calculated horizontal force will generate the correct forces, moments and deflections.

Once the 'hit' and 'miss' frames have been balanced and the valley beam sized, then close attention must be paid to the out-of-plane stability along the valley line. The columns along the length of the building can be stabilised either by bracing or portalised framing (moment resisting frames) or can be stabilised back to the bracing on the external face of the building using the roof bracing. Since the main portal frames are 'working hard' in their own plane it is important that they are well stabilised out-of-plane – a full discussion on the topic is given in King, 2007.[15]

12.7.6 Loading

The general guidance on loading given in Section 12.8.7 applies equally to portal frame structures. However, it is worth noting a few points in the particular context of portal frames with regard to assessment of loads and their application.

Unlike most multi-storey buildings, a portal frame building is dominated by roof loads and wind loads. The assessment of the former needs to take account of drift – in the context of BS EN 1990 drift loads are dealt with as 'accidental' loads and so are combined in different ways to normal dead and imposed loads. This can only be a note in this section and is dealt with more thoroughly elsewhere (ref to Chapter 4). Assessment of wind loads can be complex and whilst many multi-storey type buildings are relatively unaffected by detailed refinement of the wind loading, efficient portal frame structures are best achieved by paying close attention to the assessment of wind loads.

Application of the various loads 'wind', 'snow' and 'roof imposed' are initially through the cladding which is then decomposed onto purlins and side rails. In turn the end reactions from the purlins and side rails are applied to the portal frame. In 3D 'Building Design' software and 3D general analysis software these loads will be transferred as point loads on the rafters and stanchions providing the purlins and side rails are included in the model – see Section 12.3.2. In dedicated 2D portal frame design software and in 2D general analysis software it is most likely that these point loads will be modelled as the equivalent uniformly distributed load (udl). In all cases it is accepted practice to model these loads as udls.

When using dedicated 2D portal frame design software or 2D general analysis software, care needs to be taken to ensure that load effects from other structure not in the plane of the portal frame are taken into account – see also Special Considerations, Section 12.7.7. There are a number of areas where this effect occurs:

- At the gable frame where the gable posts 'collect' wind loads and add reaction out-of-plane of the rafter. Depending upon their head detail they may also apply (upward reaction) to the rafter and provide it with intermediate support.

- Similarly for roof bracing, additional axial forces are produced in the rafters that act as the chords of the 'truss' of which the bracing forms part.
- A typical form of multi-span portal frame building utilises 'hit' and 'miss' frames. The former is a standard set of portal frame spans (two or more) and the latter has one or more internal stanchions omitted. In the latter case the 'valley' of the 'miss' frame is supported on a 'valley girder' that then transfers the reactions back to the 'hit' frame. Since the building as a whole has to move as one, there is an 'art' to balancing the forces and deformations between the 'hit' and 'miss' frames. (See Section 12.7.5.)
- Portal frame buildings often incorporate storage or office floors within them. The interaction between the internal structure and the main portal frames needs careful consideration. The solution will depend upon whether the floors are independently braced, react onto or provide support to the main portal frame, or both. It is not simply a question of transferring the appropriate loading from the floor to the portal frame – depending upon the configuration, the floor may be stabilising or destabilising with respect to the main portal frame structure.

Crane supporting structures, on the other hand, are usually successfully dealt with as additional loads. The corbelling of the support brackets normally produce both a vertical reaction and a moment into the stanchion of the portal frame. Movement of the crane cradle produces horizontal loads due to surge. The vertical loads can also be enhanced due to dynamic effects during lifting. In multi-span frames identifying the worst (but reasonable) effects of the crane loads combined with wind or snow loads can become quite complex.

12.7.7 Special considerations

The main clauses in BS EN 1993-1-1 dealing with analysis are:

- 5.2 Global analysis
- 5.3 Imperfections
- 5.4 Methods of analysis considering material non-linearities
- 5.6 Cross-section requirements for plastic global analysis.

Each of these is considered in the context of the specific requirements of portal frame structures below.

12.7.7.1 Global analysis

Clause 5.2.1(3) is used to determine whether second-order effects are small enough to be ignored. The UK National Annex to BS EN 1993-1-1[16] (UK NA) differs significantly from the base Eurocode in that for plastic global analysis of buildings in general, the limit on the elastic critical buckling load factor, α_{cr}, above which second-order effects can be ignored, is set at 10 (cf. 15 in BS EN 1993-1-1). The UK NA applies a limit of 5 for the plastic analysis of portal frames providing certain criteria are met. This, along with the determination of α_{cr} using 5.2.1(4), is quite likely to allow single-span portal frames of reasonable dimensions to be analysed using first-order plastic analysis.

Where second-order effects must be taken into account, BS EN 1993-1-1 provides two methods:

- amplification of the sway effects by a factor based on α_{cr} – as given in 5.2.2(5)
- use of an appropriate second-order analysis.

The former is the familiar 'amplified forces method' but is only relevant to elastic analysis. The latter can be taken to mean:

- either a modified 'amplified forces method' that makes allowance for the relationship between the plastic collapse factor and the elastic critical buckling load factor (based on 'Merchant Rankine') (see Lim *et al.* 2005[17])
- or a more rigorous second-order analysis that allows for the most important effects directly.

Note that in both cases member buckling checks are still required.

12.7.7.2 Imperfections

Clause 5.3.2 identifies two imperfection effects that need to be allowed for in the global analysis for 'frames sensitive to buckling in the sway mode' – i.e. portal frames. These are initial sway imperfections of the frame and individual bow imperfections of members.

The frame imperfections are given by an initial 'lean', the base value of which is 1 in 200 and this can be equilibrated to a set of 'equivalent horizontal forces'. The sway imperfection is given as:

$$\phi = \phi_0 \, \alpha_h \, \alpha_m$$

Where ϕ_0 is the base value of 1in 200 and α_h and α_m are factors that allow for the height of the column and the number of columns respectively. The formulae are:

$$\alpha_h = 2/\sqrt{h} \qquad \alpha_m = \sqrt{[0.5(1+1/m)]}$$

With an upper limit of 1.0 and a lower limit of 0.67 applied to α_h, typical values for portal frames for ϕ would be:

$\phi = 0.7\ \phi_0$ for a moderately sized single-span portal frame

$\phi = 0.5\ \phi_0$ for a three-span portal frame of eaves height $\geq 9\,$m.

The sway imperfections can be applied in the analysis either as a deformation of the nodes and members from the vertical or (more usually) as a series of small horizontal forces ('equivalent horizontal forces') applied at the nodes. These forces are often calculated automatically by software.

The member imperfections are given as an initial bow e_0 as a proportion of the length of the member, i.e. as e_0/L. BS EN 1993-1-1 indicates that local bow imperfections may be neglected since their effect is covered in the member design checks. However, for members with at least one moment connection at their ends and with a 'limiting slenderness' greater than a given limit, the initial bow imperfection should be built into the analysis model. For stanchions in portal frames this is unlikely to be a requirement but for rafter members such an adjustment to the analysis model prior to obtaining the final design forces might be required.

12.7.7.3 *Methods of analysis considering material non-linearities*

Two methods are offered – elastic analysis and plastic analysis; it is the second that is of interest here in relation to portal frames. The former can be used in all cases whereas the latter has additional requirements.

- The members of the frame must be either doubly symmetric, or if singly symmetric, the plane of symmetry must be in the same plane as the rotation of the plastic hinge.
- Plastic hinges are allowed to occur in the joints or in the member. Without special attention the former is not recommended and the use of haunches in portal frames helps ensure that the hinges form in the member.
- The cross-section must be Class 1 at hinge positions.
- The members should have sufficient rotation capacity to allow moments to be redistributed during the load-response history up to Ultimate Limit State. The plastic hinges should be restrained to ensure stability of the member.
- A bilinear stress-strain relationship is described but a more precise relationship is allowed. The former is normally adequate unless strain hardening is to be taken into account.

Note that a number of the above are 'design' rather than 'analysis' requirements e.g. ensuring restraint at plastic hinges. The analysis will proceed without knowledge of whether such restraint is or will be provided.

12.7.7.4 Cross-section requirements for plastic global analysis

Essentially the cross-section must be Class 1 at hinge locations – more information is given in Chapters 4 and 14.

12.8 Special structural members

12.8.1 Beams with variable inertia

Included here are those beams in which a simple representation of their properties with the conventional second moment of area (inertia) in the two orthogonal directions is not an exact description of their behaviour. Within this category are:

- composite beams
- cellular beams and beams with multiple openings
- haunched or tapered beams.

Haunched beams can be taken to have three flanges and be made from a standard beam with an additional part section (or series of plates) welded to it. A tapered beam has two flanges and the web component tapers. These are almost exclusively used in portal frame structures – in that context, guidance is given in Section 12.7.3 and hence no further discussion is provided here.

12.8.1.1 Composite beams

These are 'special' since the elastic properties required by analysis vary with time due to the long term effects of shrinkage and creep. Also, in sagging bending the gross cracked or uncracked inertia can be used whereas in hogging bending the concrete is in tension and so is discounted – although account could be taken of any significant reinforcement in the concrete flange. This makes the properties variable with sign and possibly magnitude of the bending moment. Finally where the composite beam does not have full shear connection the concrete and the steel beam cannot be considered as monolithic and the behaviour will be somewhere between the 'bare' beam and the fully composite beam.

Thankfully in most cases the above have little or no impact. Most composite beams are simply supported, i.e. have pinned ends. This means that in terms of the global analysis the second moment of area of the beam has no influence on the rest of the structure. Also, the beam is only subject to sagging moments. Hence, an approximate value of second moment of area can be assumed in the analysis somewhere between that of the bare steel beam or the fully composite section. However, the consequence is that the deflections reported in the global analysis will **not** be those that result from the overall Serviceability Limit State design of the individual beam.

Continuous composite beams and composite columns are not often used and it is not possible therefore to give general guidance in this publication. Concrete-filled hollow section columns will, in general, behave fully compositely due to the confining nature of the hollow section surrounding and restraining the concrete.

12.8.1.2 Cellular beams and beams with multiple openings

Large or regular multiple openings influence the deflection of a beam due to Vierendeel effects around the openings. These effects are normally allowed for during the Serviceability Limit State design of the beam but are not normally explicitly taken into account in the global analysis. This would require a special 'beam element' or the individual beam could be modelled using a mesh of 'shell elements'. The authors are not aware of any analysis solver containing the former and the level of sophistication of the latter is unwarranted for normal building structures.

Most beams of this kind are usually simply supported – this means that in terms of the global analysis the second moment of area of the beam has no influence on the rest of the structure. Also, the beam is only subject to sagging moments. Hence, any second moment of area that is 'sensible' can be assumed in the analysis. For cellular beams or other types with regular openings the net section over the centre-line would be a simple approach. For irregular openings of various sizes the net section properties will vary along the beam and the gross section second moment of area or minimum net section could be adopted. In all cases the deflections of the beam in the global analysis will **not** be those that result from the overall Serviceability Limit State design of the individual beam. Depending upon the size, number and disposition of the openings their effect on the deflection can be significant and should not be ignored.

More information on establishing the deflection of beams with openings is contained in several publications from the Steel Construction Institute.[18-20] A matrix solution that could form the basis of a special 'beam element' is given in Section 6.5 of the American Institute of Steel Construction, Design Guide #2, 2003.[21]

12.8.2 Curved members

Most analysis solvers can only handle straight elements, i.e. there is no specific 'curved element'. It is necessary therefore to model curved members as a series of straight elements. The more elements used, the more accurate will be the model but at the risk of making the overall model very large and unwieldy. A reasonable approach is to aim for each straight element to subtend an angle of around 2.5° (see Section 12.9.2).

Support conditions are important for curved members:

- For members curved in elevation the member will tend to arch and the stiffness of the supports considerably influences the amount of arching action. Inaccurate modelling of the support will give significantly inaccurate results for the curved member.

- For members curved on plan the interaction of the torsion induced in the beam and the stiffness of the supports is important. The assumptions made in the analysis model must be achievable in the real detail otherwise the beam will twist more than anticipated.

When a curved member is modelled as a series of straight elements, these elements will intersect at an angle that is a function of the overall radius of the curve and the number of elements defined. In the true curved member these points lie on the tangent to the curve and so no angle exists between any two points. This can raise two issues depending upon the angle subtended by each of the straight elements. First, the forces from the analysis will be in the local coordinate axis system of the straight elements and these are not aligned. Hence, the forces, moments and torsion will need to be resolved to give the actual effects tangential to the curved member. Second, the slight differences in the member end forces can produce anomalies in the expected bending moment diagram (see Section 12.9.2). These are one and the same problem but manifest themselves differently. Of course the forces at the ends of the straight elements will be in equilibrium.

More information on the treatment of curved members is given in the SCI publication P281[22] and covers both design and analysis requirements.

12.8.3 Trusses

12.8.3.1 Modelling of trusses

There are several models that may be used for the analysis of a truss. These include:

1. pin-jointed frames
2. continuous chords and pin-jointed internal (and side) members
3. rigid frames.

The second option is the most common and is preferred since it more usually reflects the behaviour in practice. In this case the chords resist some bending due to loading off 'panel points' but are axial load dominated and behave primarily as an axially loaded continuous beam. The internals are axially loaded only – moments due to self-weight are usually ignored. The analysis model should be set up to reflect this behaviour. This model has the advantage that in most situations there will be no bending moments to be included in the connection design. In tubular trusses using hollow sections, despite the fact that the internals are fully welded to the chords, this type of connection behaves as if it were pinned at the final stage of loading, i.e. at Ultimate Limit State. This is due to the relatively thin walls of hollow sections and the large deformations that such connections can sustain. In reality for all truss types, secondary moments will be present due to the change in geometry as the truss deflects, the actual rigidity at the connection and the stiffness of the members. How this is dealt with in BS EN 1993-1-8[23,24] is given later.

The third option is usually only particularly relevant to trusses that use Vierendeel action. In this case the behaviour of the truss, in the way it resists loads, is through bending action in both the chords and internals. This has the advantage that the diagonal internals are omitted but the efficacy of the hollow section is somewhat lost as the behaviour switches from axial load to moment-dominated. The connections between the hollow section also have to be more robust since they have to be designed to resist significant moments due to the Vierendeel action.

Most trusses are 'plane frames', i.e. their primary behaviour is in 2D. Hence, they can be modelled in this way in analysis. Where trusses are 'taken out of' the structure and analysed separately, care must be taken with the support conditions. One end of the truss must be restrained in the horizontal direction whilst the other end can be either restrained or released in the horizontal direction. The resulting forces, deflections and reactions will be different between the cases. In reality the truss is likely to be supported by columns and the support conditions are then neither released nor restrained in the horizontal direction. A similar situation can occur in the vertical direction when trusses overhang a supporting beam. This situation supports the case for using 3D analysis of the whole structure. Clearly, if the truss is a 'space frame' or a triangular (in cross-section) lattice then the use of 3D analysis becomes a necessity.

12.8.3.2 Code requirements

In most analysis and design situations, it is both convenient and reasonable that the connection design, being subsequent to both analysis and member design, is carried out in a manner consistent with the assumptions previously made. Thus in most situations, the type of connection (nominal pin, moment resisting etc.) follows the assumptions of the analysis.

In truss and lattice construction however, the design of the joints between chord and internal members frequently dominates the member design (BCSA, 2005).[25] Gap or overlap joints are often used to increase joint capacity, or to improve the fabrication details. This introduces eccentricity into the setting out of the elements, and it will generally be necessary to include the effect of this eccentricity in the calculation of member forces and moments. As the eccentricities are not known prior to the choice of member, this is usually done manually, following an initial analysis with nodes at centre-line intersections. If the eccentricities are known, they may be modelled as illustrated in Figure 12.21.

Clause 5.1.5 of BS EN 1993-1-8 describes the code requirements with respect to analysis of 'lattice girders'. The first few sub-clauses of Clause 5.1.5 confirm the general advice given above regarding pin-jointed frameworks and moments between panel points, etc. Moments resulting from eccentricity do not need to be taken into account in the following circumstances:

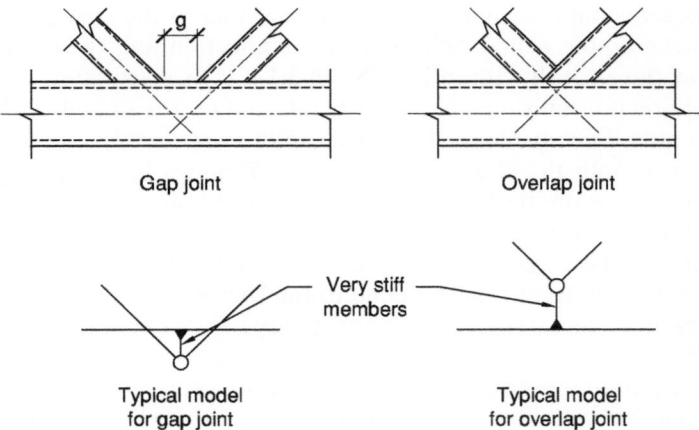

Figure 12.21   Modelling of gap and overlap connections

- tension chord members and internal (brace) members
- the connections between internal (brace) members and chord members providing the set out complies with certain limits on the nodal eccentricity.

This leaves connections in which the eccentricity is outside of the valid limits and compression chords. It is worth noting that:

- the connection design and the member design are often separate operations even when carried out by the same designer
- the member design may be carried out by one organisation (the consultant) and the connection design by another (the steelwork contractor)
- the software application for member design is likely to be different to that for connection design even in an integrated Building Design Model system.

The consequence of the above is that if the connection design requires an eccentricity for either efficiency or practical reasons, there may be no explicit feedback loop to ensure that the eccentricity moments are taken into account in the member design.

Also note that the code requirements, whilst generally applicable to lattice girders, are written from the point of view of those constructed from hollow sections.

12.8.3.3 *Practice and process*

As the member sizes are so influential on the efficiency of the connection design it is best to guess the initial member sizes generously and be prepared for some iteration. An outline of the analysis/design process is summarised below.

1. Determine the truss layout, span, depth, panel lengths, lateral bracing by the usual methods, but keep the number of connections to a minimum, and maintain a minimum angle of 30° between chords and internal members.
2. Determine loads; where possible simplify these to equivalent loads at the nodes.
3. Determine axial forces in all members by assuming that the joints are pinned and that all member centre-lines intersect at nodes.
4. Determine preliminary member sizes and check if secondary stresses due to change in geometry and the simplifying assumption of pinned connections can be ignored. If secondary stresses cannot be ignored, the best option is to re-configure the truss. An alternative is to re-analyse the truss with rigid connections although this is also a simplification in that the connections are unlikely to be 'rigid'.
5. Check the joint geometry and joint capacities. Modify the joint geometry, with particular attention to the eccentricity limits. Consider the fabrication procedure when deciding on a joint layout.
6. Check the effect of primary moments on the design of the chord members. Add the effect of joint eccentricity where required, by manual methods, or by creating a new analysis model representing the actual setting out.

12.8.4 Shear and core walls

Shear and core walls have sufficient 'solidity' that it might be difficult to see how these can be modelled as line elements, as is the case for all the steelwork in buildings. Designers now routinely have access to highly capable 3D analysis software that can deal effectively with 'finite elements' (more particularly 2D 'shell elements'). It is natural, therefore to try to model these solid members – shear and core walls – with solid elements. This is of course possible and practical advice on use of finite elements is given in Rombach, 2004.[26] Such a large and diverse subject cannot be dealt with here. However, some guidance is given on modelling of shear and core walls using line elements and shell elements (for more information on the latter see Arnott, 2005).[8]

Before outlining the modelling techniques it is worth noting their objectives. If deflections and the distribution of forces around the structure are important then all of the aspects of the model need particular consideration if a reasonably accurate prediction is to be made. This means:

*   accurate material properties for each member
*   accurate section properties for each member
*   a good arrangement of members to idealise the overall physical geometry.

BS EN 1992-1-1[27] indicates a potentially broad range of properties for concrete of any given grade. For example, the short term Young's modulus for $30\,N/mm^2$ cylinder strength (37 cube strength) concrete is given in Table 3.1 as $33\,kN/mm^2$. However, Clause 3.1.3 (2) indicates that this figure is for concrete with quartzite aggregate and that for other aggregate types the figure should be modified. The modification varies from minus 10 to 30% for limestone aggregates to plus 20% for basalt aggregate, i.e. for the strength previously quoted a modulus of somewhere between 23 and $40 \, \text{kN}/\text{mm}^2$. This then needs to be adjusted to allow for load duration (and potentially other factors as well). The gross section properties of elements may need to be adjusted to allow for cracking. Therefore there is a good deal of judgement involved in the selection of the section and material properties, this directly affects results and must be borne in mind when debating the intricacies and relative merits of alternative idealised models. Given the variations in the base material properties then irrespective of the type of model (line element or finite element), sensitivity studies may be appropriate.

If deflection is not a concern and the aim is to produce design forces then the traditional approach would be to ensure that the properties are consistent throughout the analysis model and an exact set of properties appropriate to each member type, e.g. Young's modulus is less important. However, BS EN 1992-1-1 infers that the same model is used for all purposes, i.e. determination of deflection and determination of member forces.

Simple shear walls that are continuous and uniform for their full height can be modelled as a series of 'beam elements', i.e. line elements with the properties of the shear wall concentrated at the centre-line. In order to allow beams and floors to be connected to the shear wall a 'mid-pier' idealisation using beam elements can be made as described below and shown in Figure 12.22:

- two horizontal elements at the bottom of the wall running between the two set out points and the mid point
- two horizontal elements at the top of the wall running between the two set out points and the mid point
- further pairs of horizontal elements for any intermediate level that is acting as a floor

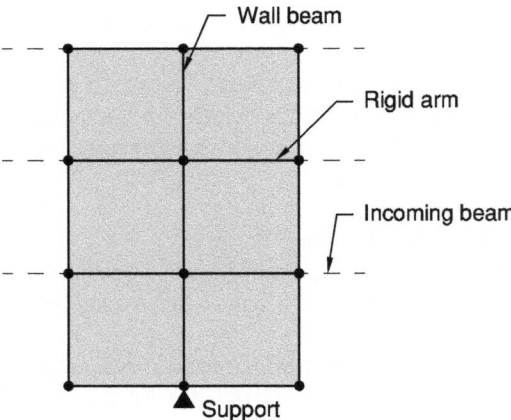

Figure 12.22   Mid-pier model

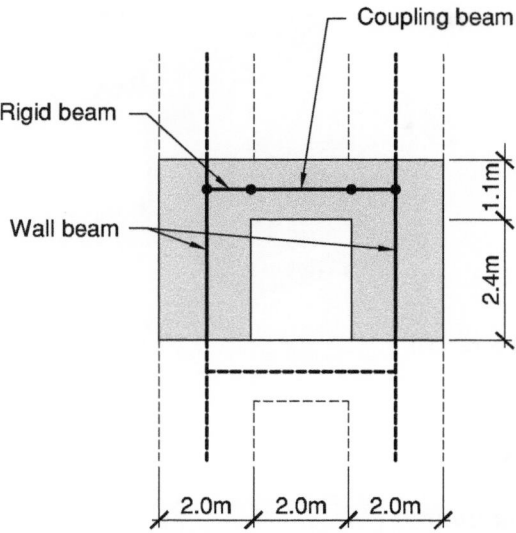

Figure 12.23   Mid-pier model with openings

- vertical elements joining the mid-points at the top and bottom of the wall and any intermediate floor levels
- a fixity of the support at the midpoint of the wall baseline depends upon what is supporting the base of the shear wall e.g. a foundation, one or more columns, another shear wall, or a transfer beam.

If openings have been added to the wall the mid-pier model should be modified accordingly – see Figure 12.23. Additional vertical elements can be introduced to the sides of the opening and a coupling beam introduced above. Addition of openings will reduce the strength, stiffness and self-weight of the wall.

Core walls (a series of shear walls) can be modelled in a similar fashion as indicated in Figures 12.23 and 12.24 – the first shows a facsimile of the 'real' wall and the beams that it supports and the second shows the analysis model idealisation.

Comparisons between meshed models of shear and core walls and the 'mid pier' model (using beam elements) are made in (Arnott, 2005).[8] It is worth noting here some of the words from the conclusions of the paper:

'Regardless of which way you have idealised the structure, if you have doubts about the results the best thing you can do is model it another way and compare.

You should not think that the world of shell elements offers a new level of accuracy – in many cases it might better be regarded as a new way to get the same answers, or perhaps more worryingly as a new way to make some new mistakes?'

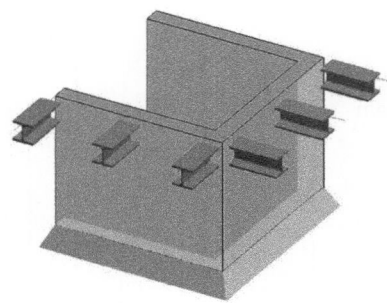

Figure 12.24   Physical core wall

12.8.5 Connections

12.8.5.1 Joint behaviour

Within a frame, joint behaviour affects the distribution of internal forces and moments
and the overall deformation of the structure. In many cases, however, the effect of
modelling a stiff joint as fully rigid, or a simple joint as perfectly pinned, compared
to modelling the real behaviour, is sufficiently small to be neglected. This is allowed
by BS EN 1993-1-8 in its opening clause on classification of joints – 5.1.1(1).

Elastic analysis programs consider only the stiffness of the joint and it is convenient to define three joint types as given in Clause 5.1.1(2) of BS EN 1993-1-8 (see
Section 12.7.4).

Note on terminology – Eurocodes use the term 'joint' to mean what in the UK
we would normally understand as a 'connection'. Both terms are interchanged
within the foregoing text to suit the context of either reference to the Eurocode or
familiarity to UK designers.

- *Simple* – a joint which may be assumed not to transmit bending moments.
  Sometimes referred to as a pinned connection, it must also be sufficiently flexible
  to be regarded as a pin for analysis purposes.
- *Continuous* – a joint which is stiff enough for the effect of its flexibility on the
  frame bending moment diagram to be neglected. Sometimes referred to as 'rigid';
  they are, by definition, moment-resisting.
- *Semi-continuous* – a joint which is too flexible to qualify as continuous, but is
  not a pin. The behaviour of this type of joint must be taken into account in the
  frame analysis.

BS EN 1993-1-8 requires that joints be 'classified' according to their stiffness for
elastic analysis and should also have sufficient strength to transmit the forces and
moments acting at the joint that result from the analysis. For elastic-plastic analysis

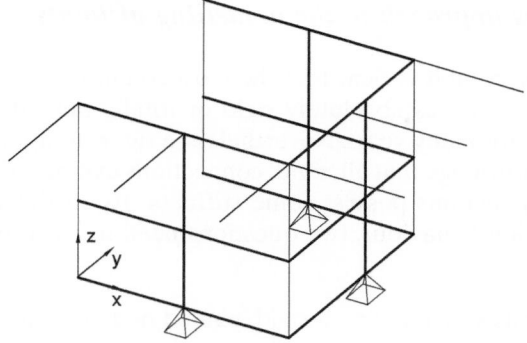

Figure 12.25 Analysis idealisation of core wall

joints should be classified by both stiffness and strength. BS EN 1993-1-8 provides information on how to classify joints with regard to both stiffness and strength. Rules for determining the stiffness of I/H joints is given in Clause 6.3 (see 'Assessment of actual joint stiffness' in 12.8.5.2). However, BS EN 1993-1-8 also allows joints to be classified on the basis of experimental evidence or experience of previous satisfactory performance. For buildings, it is the latter route that is taken by the UK National Annex for simple (pinned) connections, for continuous connections and for semi-continuous connections.

When classifying the stiffness of a joint:

Simple joints – are described as 'nominally pinned' rather than pinned since it is accepted that some moment is transferred but in this definition these moments are insufficient to adversely affect the member design. The UK NA[24] indicates that connections designed in accordance with the principles given in *Joints in Steel Construction, Simple Connections* (BCSA/SCI, 2002)[28] may be classified as nominally pinned.

Continuous joints – are described as 'rigid'. The UK NA refers the designer to the publication *Joints in Steel Construction, Moment Connections* (BCSA/SCI, 1995)[29] in which for many, but importantly not all, cases connections designed for strength alone can be considered as rigid.

Semi-continuous joints – are 'semi-rigid' and are usually also 'partial strength'. These are described as 'ductile connection in UK practice and are used in plastically designed semi-continuous frames. For braced semi-continuous frames the UK NA indicates that these may be designed using the principles given in the publication *Design of Semi-continuous Braced Frames* (SCI, 1997)[30] with connections designed to the principles given in Section 2 of *Joints in Steel Construction – Moment Connections*. In addition, for unbraced semi-continuous frames (known as wind moment frames) the UK NA indicates that these may be designed using the principles given in the publication *Wind Moment Design of Low Rise Frames* (SCI, 1997).[31]

12.8.5.2 *Rigorous approach to the modelling of joints*

From the previous section it is clear that the most common technique of modelling joints in analysis is either as absolutely rigid or totally pinned and this has been successfully applied for many years. Nevertheless, a rigorous approach to modelling of joints would acknowledge that all 'rigid' connections exhibit a degree of flexibility, and all 'pinned' connections possess some stiffness. To do this or to utilise semi-continuous connection behaviour, two questions need to be resolved by the structural engineer:

- What are the limits which define a rigid, pinned or semi-rigid connection?
- How stiff is the particular connection?

These two issues are considered in the following two sub-sections.

Stiffness limits

Figure 12.26 shows a number of moment-rotation curves, representing joints of varying stiffness, and shows the dividing lines between rigid, semi-rigid and pinned joints given in BS EN 1993-1-8. For a joint to be considered as rigid its stiffness must be greater than:

- 8EI/L for braced frames
- 25EI/L for unbraced frames.

For a joint to be classified as (nominally) pinned its stiffness must be less than:

- 0.5EI/L.

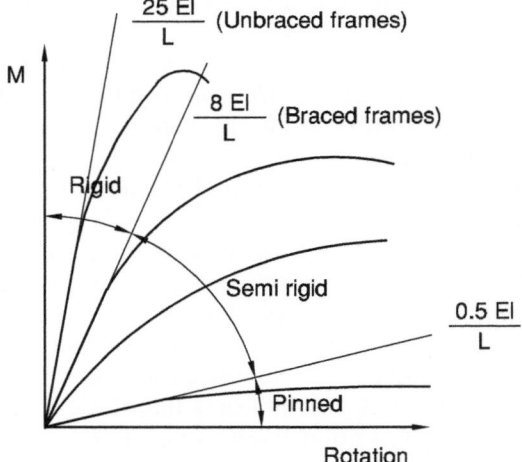

Figure 12.26   Joint moment-rotation curves

Alternatively nominally pinned joints can be classified by their strength. A joint can be assumed to be nominally pinned if its moment capacity is less than 25% of the member bending capacity, providing it has sufficient rotation capacity.

All joints between the stiffness limits are classified as semi-rigid.

Assessment of actual joint stiffness

The only accurate way at the present time to determine the moment-rotation characteristics of a joint is by testing. A method of calculating joint stiffness (for I- and H-sections) is given in Clause 6.3 of BS EN 1993-1-8. However, the UK NA gives the following statement:

'Until experience is gained with the numerical method of calculating rotational stiffness given in BS EN 1993-1-8:2005, 6.3 and the classification by stiffness method given in BS EN 1993-1-8:2005, 5.2.2, semi-continuous elastic design should only be used where either it is supported by test evidence according to BS EN 1993-1-8:2005, 5.2.2.1(2) or where it is based on satisfactory performance in a similar situation.'

Recommendations

Due to the uncertainties described above, it is relatively uncommon to determine joint stiffness prior to, or during analysis. In particular, frame analysis with springs representing joint stiffness is uncommon, although if the joint characteristics are known, or can be calculated, procedures do exist for incorporating the effects of joint flexibility into standard methods of frame analysis (Li, Choo and Nethercot, 1995).[32]

Modelling of semi-rigid joint behaviour is currently not recommended for 'everyday design'. The modelling of semi-rigid joints as a standard technique in the future may prove feasible, although the overall benefit of such an approach may need justification. For the present, the usual practice of defining a joint in the model as rigid or pinned is recommended, with the joint design following the assumptions made in analysis.

12.8.5.3 Bracing connections

The modelling of bracing systems is often not straightforward and can lead to misunderstanding between the structural designer carrying out the analysis and the connection designer. The bracing, columns and floor beams are generally modelled on centre-line intersections, as shown in Figure 12.27.

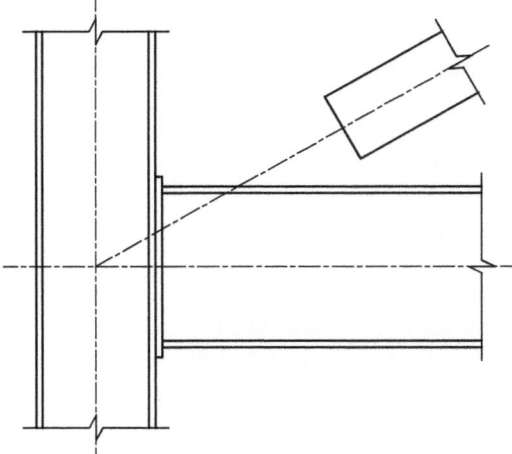

Figure 12.27   Typical brace connection noding on centre-line

The main area of debate concerns the resolved vertical forces from the bracing system. The end reactions of the floor beam in the output from the analysis will only contain (assuming a pinned connection) the shear forces from the applied floor loads, together with the axial force from the bracing system. Because the joint is modelled on centre-line intersection the vertical brace force in the analysis model (correctly) 'disappears' directly into the column. Depending upon the detail of the bracing connection this may not be the case in reality.

Two examples are given in the figures (Figure 12.28). In the Figure (a) the braces clearly apply forces independent of the beam whilst all centre-lines still intersect. However, the physical detail indicates that the load path might be less straightforward. Figure (b) shows a joint with the brace centre-lines intersecting the beam centre-line on the face of the column. In the latter case any eccentricity moments normally calculated based on the end reaction of the beams must include the resolved brace force (which of course may cancel out). In the former case whether the resolved brace forces should be considered to act at an eccentricity (at least from the centre-line of the column if not 100 mm from the face also) is a matter of debate (BCSA, SCI, Corus, Oct 2006).[33]

Such details can be modelled locally with the analysis model and in some cases require the use of small rigid members. This is normally avoided for several reasons:

- The detail of the analysis model adjustments rely on the sizes of the members and connection configurations which are usually unknown at the analysis stage. Also, any changes to section size, etc., would require rebuilding of the model.
- The analysis model size can 'grow' substantially with all the additional nodes and members, although with modern computing power this is unlikely to be a significant issue.

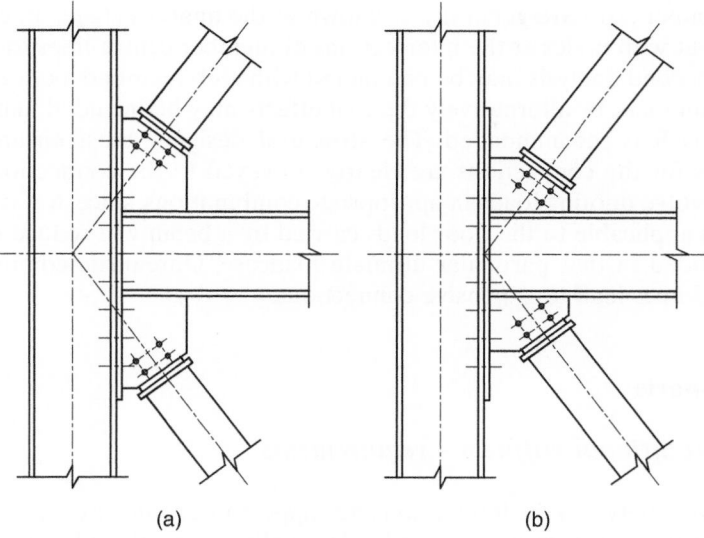

Figure 12.28   Alternative bracing configurations

- The use of small, very stiff members adjacent to other large less stiff members can cause instability in the mathematical solution form that underlies all analysis software, especially when considering second-order analysis.

Hence, it is not normally practical to improve the modelling and such perturbations are handled as part of the design process. The exception is perhaps the use of the brace force at an eccentricity since most building modelling software (as opposed to general analysis software) already addresses eccentricity moments from beam end reactions in the design of columns.

12.8.5.4 Summary

In principle, joints should be modelled for global analysis in a way which appropriately reflects their expected behaviour under the relevant loading. Beware that the default in general analysis software (as opposed to building modelling software) is for all joints to be fully restrained in all directions (six degrees of freedom restrained). Consider too that a moment joint between a beam and column is likely to be stiff enough to be considered as rigid in the plane of the bending but out of plane has little stiffness and is therefore better modelled in that direction as unrestrained.

In general, nodes at centre-line intersections are recommended in the analysis for beam and column structures. Real eccentricities, if present, should be taken into account during member design.

Since member sizes are generally unknown at the analysis stage, bracing should still be set out with nodes at the intersections of member centre-lines for the initial analysis. A second analysis may be completed with stub members between bracing and main members, or alternatively the real effects may be included manually. The latter approach is recommended. The structural designer must ensure that the design loads for the connections are clearly conveyed to the connection designer. This may involve quoting loads in appropriate combinations, since, for example, the load factors applicable to the floor loads carried by a beam will reduce when wind load is included in that particular ultimate loadcase. Unrealistic combinations of connection forces lead to expensive connections.

12.8.6 Supports

12.8.6.1 Rotational stiffness – requirements

The interaction between the foundation and supporting ground is complex. Detailed modelling of the soil-structure interaction is usually too involved for general analysis. BS EN 1993-1-8 has no specific recommendations covering rotational stiffness of base connections, and so these fall into the same categories of nominally pinned, semi rigid and rigid as given in the previous section on 'Joints'. However, an empirical approach is provided in the NCCI document from Access Steel, SN045, 'Column base stiffness for global analysis'.[34]

Nominally pinned bases – no guidance is given in BS EN 1993-1-8 on the configuration of bases that can be considered as nominally pinned. Recourse can be made to Clause 5.2.2.1(2) which allows normal practice to be adopted. Given that the base can be considered as nominally pinned and that the foundation is designed assuming that the base moment is zero then:

- The base should be assumed to be pinned when using elastic global analysis to establish the design forces and moments at the Ultimate Limit State.
- The base may be assumed to have a stiffness equal to 10% of the column stiffness (which can be taken as $4EI/L$) when checking frame stability, i.e. when checking whether the frame is susceptible to second-order effects.
- The base may be assumed to have a stiffness equal to 20% of the column stiffness when calculating deflections at the Serviceability Limit State.

Nominally rigid bases – again no guidance is given in BS EN 1993-1-8 on the configuration of base that can be considered as nominally rigid and recourse should be made to Clause 5.2.2.1(2). Given that the base can be considered as nominally rigid:

- The stiffness of the base should be limited to the stiffness of the column when using elastic global analysis to establish the design forces and moments at the Ultimate Limit State.

- The base may be assumed to be rigid when calculating deflections at the Serviceability Limit State.
- For elastic-plastic global analysis, the assumed stiffness of the base must be consistent with the assumed moment capacity of the base but should not exceed the stiffness of the column. Any base moment capacity between zero and the plastic moment of resistance of the column may be assumed, provided that the foundation and base plate are designed to resist a moment equal to the assumed moment capacity, together with the forces obtained from the analysis.

Semi-rigid bases – the NCCI SN045 suggests, based on previous UK practice, that a nominal base stiffness of up to 20% of the column stiffness may be assumed in elastic global analysis, provided that the foundation is designed for the moments and forces obtained from this analysis.

The determination of column base stiffness by calculation is subject to the same caveat contained in the UK NA as outlined in Section 12.8.5.2.

12.8.6.2 Rotational stiffness – modelling

Despite the apparent clarity of the requirements for stiffness above, it is important to realise that the base stiffness has to be treated as a beam stiffness, not a column stiffness, as shown in Figure 12.29 (for a detailed explanation see SCI, 1991).[35]

In many cases this can be visualised and modelled for analysis by rigidly connecting a beam member to the base. If the dummy beam is given a length and second moment of area identical to the column, with the beam end remote from the column fixed, this will achieve the required base stiffness.

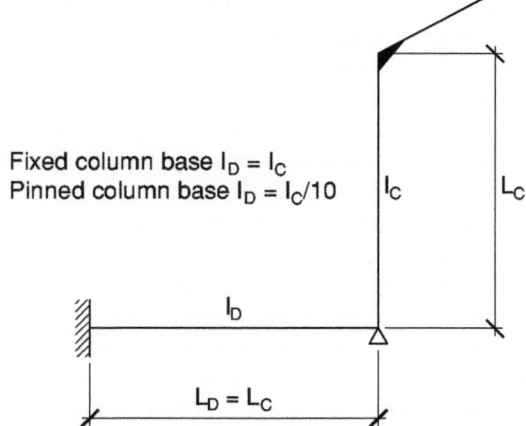

Figure 12.29   Model of base stiffness

420   *Analysis*

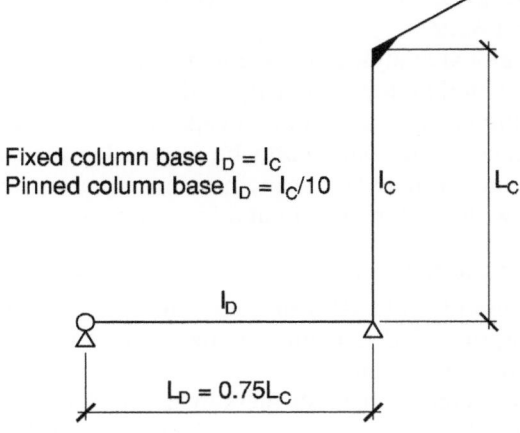

Figure 12.30   Alternative model of base stiffness

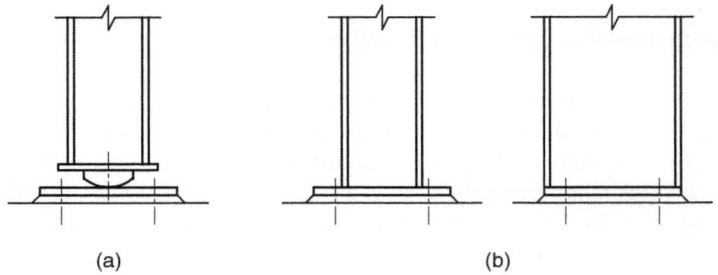

Figure 12.31   Base details: (a) uncommon, (b) typical

In order to reduce the confusion caused by the moment at the end of the dummy member, it is more convenient to pin the remote end of the dummy beam as shown in Figure 12.30, and reduce the length to 0.75 × column length.

In both models, the second moment of area of the dummy member is equal to the column second moment of area in the case of a rigid base, and a value of 10% or 20% (as appropriate) of the column second moment of area in a nominally pinned base.

Many programs permit base stiffness to be input directly as a spring stiffness. In this case a rigid base is input as $4EI_c/L_c$ and a nominally pinned base as $4EL_c/5L_c$ for deflection calculations and $4EL_c/10L_c$ for frame stability checks.

The column base connection to the foundation is an area of uncertainty, both in modelling and reality, and the distinction between pinned and fixed bases can be difficult to define in practical details. Portal frames are usually analysed with pinned bases, since the cost of moment-resisting foundations often exceeds the savings in frame weight achieved by using fixed bases. It is uncommon, however, to see details which are immediately recognised as pinned (Figure 12.31 (a)). More common are details shown in Figure 12.31 (b) which are frequently deemed to be pinned in analysis. Details such as these are preferred for two reasons:

1. The use of four holding-down bolts allows the column to be erected without guying or propping, and permits easier adjustment and plumbing.
2. A moment-resisting base may be required for stability during fire, if the column is situated near a site boundary. The Building Regulations define when a building must incorporate this requirement. The reader is referred to *Single Storey Steel Framed Buildings in Fire Boundary Conditions* (SCI, 2002).[36]

It will be noted, however, that the bases shown in Figure 12.31 (b) could also be classed as moment-resisting. In the case of boundary columns, the base detail must be capable of resisting moment, although current practice is to model them as pinned for the frame analysis.

12.8.6.3 Horizontal and vertical fixity

Rotational base fixity was discussed in the previous section, but most analysis software also allows vertical and horizontal support options of restrained, free and spring stiffness. The reader's attention is drawn to the statement at the start of the previous section that, 'detailed modelling of the soil-structure interaction is usually too involved for general analysis'.

Differential settlement is usually more damaging than overall settlement. Often ignored, settlement of isolated foundations can have a dramatic effect on the bending moments in continuous frames. Figure 12.32 shows a typical bending moment diagram provided as a result of the third foundation being displaced vertically by 30 mm, relative to the remainder.

If foundation details and soil properties are known, spring supports can be introduced to model the behaviour of the soil. There might also be a requirement for 'compression only' springs, 'gap elements' etc. which improve the modelling, add complexity and which require non-linear analysis.

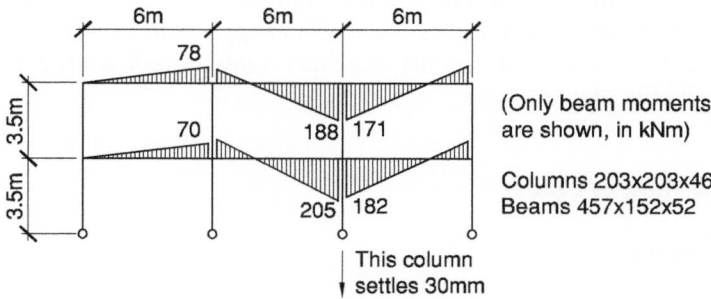

Figure 12.32   Example bending moment due to differential settlement

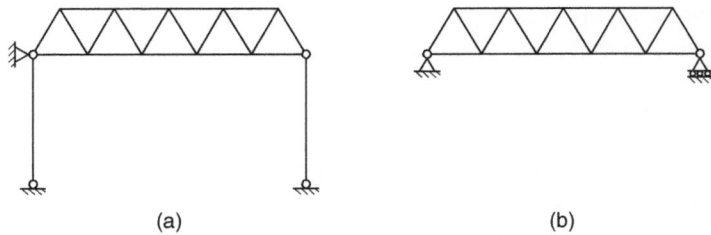

Figure 12.33   Roof truss model: (a) including columns, (b) isolated

Horizontal releases are frequently necessary if the model is to properly reflect reality. Apart from the foundations, any other support is almost certain to allow the structure to 'spread', and one or more supports must be released to reflect this.

To illustrate this point, consider a triangulated roof truss, simply supported by two columns in Figure 12.33. If designing the truss in isolation, the supports must be modelled with a horizontal release, or the analysis will produce compression in some panels of the bottom boom – clearly incorrect in this situation.

12.8.6.4 Summary

If a structure is analysed with pinned bases, but in reality the bases are semi-rigid, the bending moments produced by the analysis are generally conservative with regard to the frame members. Hence, analysis using pinned bases can usually be justified, even if the base details appear capable of resisting moment.

Fixed bases should not be specified without consideration of the effect of the fixity on the foundation costs, which can become prohibitively expensive. It should be noted that the guidance given above allows full fixity to be assumed in the analysis for Serviceability Limit State calculations but precludes full fixity for analysis at the Ultimate Limit State.

The capacity of most nominally pinned bases to resist moment may be used to advantage, particularly in the reduction of sway deflections.

Without detailed investigation of ground conditions and foundation behaviour, it is generally acceptable to assume the foundation supports to be rigid vertically and horizontally, acknowledging the ability of a steel frame to redistribute moment and behave in a ductile manner.

12.8.7 Modelling of loads

12.8.7.1 General

In most structures, the magnitude of loads cannot be determined precisely, and the loads used in analysis represent an estimate of the likely maximum load to which the structure will be exposed. Some actions, such as the self-weight of a structure, may appear easier to estimate than others, such as wind loads. The estimate of imposed loads such as wind and snow can be based on observation of previous conditions and the application of a probabilistic approach to predict maximum effects which might occur within the design life of the structure.

Actions associated with the use of the structure, such as imposed floor loads, can only be estimated based on nature of usage. Insufficient data are available in most cases for a fully statistical approach and notional values are therefore assigned either by the client and/or by standards – see BS EN 1991 series.

In limit state design, characteristic values of loads are used as the basis of all design. They are values which statistically have only a small probability of being exceeded during the life of a structure. To provide a margin of safety, particularly against collapse, partial safety factors are applied to these characteristic values to obtain design loads. The design loads are included together in a series of design combinations. The partial safety factors used for a particular type of load can vary depending upon the other types of load that are contained within the design combination. These partial safety factors can also be modified by 'ψ-factors'. A detailed treatment of the requirements of BS EN 1990 is given elsewhere – see Chapter 3.

12.8.7.2 Modelling of actions

Once the actions to be taken into account have been identified, the application of the loads to the analysis model will depend largely on the degree of simplification present in the analysis model. A three-dimensional model including secondary elements is likely to have a complex application of loads, compared with a plane frame analysis where the characteristic loads will be further simplified. Considering a portal frame, the wind load and roof load will be applied to the main frames via secondary elements such as purlins or sheeting rails, which in turn are loaded via the cladding. If the purlins or rails are included in a three-dimensional model, the load should be applied to these elements, whereas if the cladding is also modelled then area loads can be applied to these and the model will decompose them onto the purlins and sheeting rails. If single two-dimensional frames are modelled, the equivalent point loads can be calculated at purlin positions (if known at that stage), or, more usually, an equivalent uniformly distributed load is applied to the frame.

Simplification in calculating equivalent point loads and distributed loads is recommended. In the floor slab shown in Figure 12.34, the floor beams will usually be designed for a uniformly distributed load as shown in (a), not the distribution shown in (b).

Similarly, multiple point loads on any member may be treated as a distributed load. Five or more equally-spaced identical point loads on a member may generally be considered as a distributed load, without significant loss in the accuracy of the analysis.

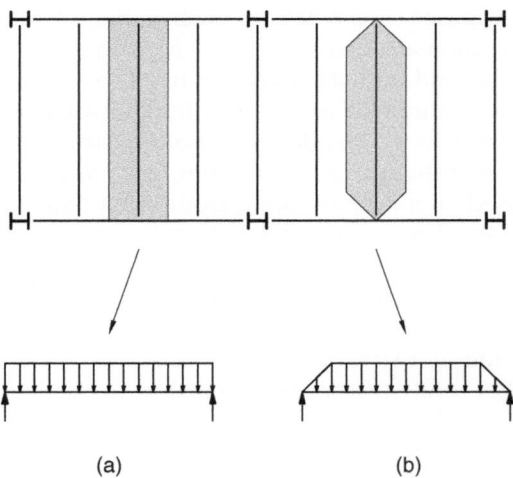

Figure 12.34   Modelling of loads

12.8.7.3 Load types

General analysis programs will have a range of ways in which the model may be loaded. Actions will require definition relative to either global or local (member) axes. The following list is not comprehensive, but indicates the options that are usually available:

- joint loads
- joint displacements
- uniformly distributed loads on elements
- varying distributed loads on elements
- point loads on elements
- self-weight
- temperature change.

Building modelling software often allows more general load types such as global floor loads, area loads, uniform and varying patch loads and line loads. Each of these will be decomposed into the loads required by the analysis solver that is a component part of the building modelling software.

12.8.8 Load combinations

It is usual to specify actions as a series of unfactored loadcases and each of these may have an associated 'type' e.g. 'dead', 'imposed', 'wind', 'snow'. These should be combined into a series of factored design combinations. Generally the software will provide the results in the following two forms:

- as individual loadcase results, for checking deflections for example
- as design combination results for checking strength or resistance at the Ultimate Limit State.

Note that in first-order analysis it is possible to analyse only the unfactored loadcases and then use superposition to obtain the design combination results. However, for rigorous second-order analysis the design combinations must be analysed separately as superposition does not hold.

More information on the requirements of BS EN 1990 are covered in Chapter 3.

12.9 Very important issues

12.9.1 Checking the results

Some basic checks on the results should always be carried out; the purpose of these is to confirm the validity of the model (not to verify the accuracy of the software). The checks are:

- a thorough review of the input
- the sum of the applied forces equals the sum of the reactions for each load case and combination
- structural deflections are of the right order of magnitude and correct direction
- the axial load, shear force and bending moment diagrams are a sensible shape.

It is much easier to perform all these by looking at graphical images of the results than by any other means. Often an expedient way to check the model is to introduce additional 'test' load cases with very simple point loads. If the results look and feel right then they probably are right – if something looks odd then it should be investigated further.

Simple checks on structural stiffness can be made by approximating a building to a single vertical cantilever and undertaking a hand calculation to determine the approximate expected horizontal deflection resulting from a horizontal load. The same load can be applied to the full structural model and the results compared.

Similarly simple hand calculations on the expected foundation load distribution should be made and compared with the analysis results. It should always be possible to explain or correct large discrepancies. Software is a tool and does not (and never will) substitute for the engineer's intuitive feel for structural behaviour. The old adage is true – 'garbage in garbage out'.

It is not uncommon for engineers to verify the results of one software package by running the same model in another package, but all software packages will give slightly different answers. However, in the case of first-order analysis, these differences should be very small – in the 4th or 5th significant figure. Such small discrepancies in results will most likely be because the models are different and not because of a fault in the software. Automatic transfer of an analysis model between analysis packages does not verify that the model is correct; it usually results in an identical model in both packages. To fully verify a model, the model should be built twice and the results compared.

12.9.2 General advice and common mistakes

A number of common mistakes regularly seen in analysis models could be avoided by following the advice below:

- It is ***not necessary to model all the detail***. Often in a building structure, there is much detail that is unaffected by the loading on the primary structure – for instance, door and window framing steel. In order to simplify the analytical model, it is worth spending time assessing what should and should not be modelled. The bigger the model, the longer it will take to create, the longer it will take to check, the longer it will take to locate errors and the more room there will be for mistakes.
- It is essential to ***keep things simple*** where possible. Approximation for design purposes is still an important part of the engineer's role. The more complex an analytical model is, the greater the risk of error and the more time-consuming the inevitable changes will be when having to rework the model.
- Always ***estimate the answers*** expected; look at the results and compare the two.
- Open steel sections are very weak in ***torsion***. It is important that the structure does not rely on the torsional stiffness of its members for its stability. Generally engineers 'design torsion out' so that torsion is not a dominant factor in design. Having run an analysis, check the torsion diagrams to ensure that torsion is not affecting the structure. (Some analysis software permits a global factor to be applied to reduce the torsional stiffness of members to ensure torsion is not a dominant factor in the results.)
- Most analysis software will give ***Warnings and Errors*** if the model is not correctly constructed. These may vary from being helpful to being full of 'analysis speak' and needing translation. All warnings and errors resulting from analysis are there for a reason. More often than not, the reason will affect the integrity of the results. It is important to assess, and if necessary correct, the sources of warnings and errors in order to complete a satisfactory analysis.
  Common reasons for these errors are:
  ○ A ***lack of support*** for the structure.
  ○ ***A part of a model is unstable***. This could be an area of floor or roof or a small part of the building which acts as a mechanism within the context of the whole model.

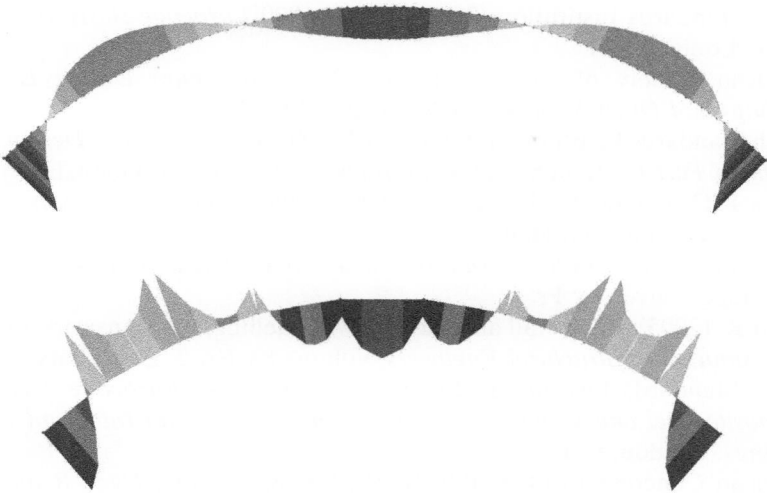

Figure 12.35   Coarse and fine models (pictures from CSC S-Frame – courtesy of CSC (UK) Ltd)

○ **Member properties** that have not been defined and so are taken as zero.
○ Nodes in the model which '**spin**' – all the interconnecting elements end releases result in a node that has no rotational stiffness about a particular axis and can thus spin in space.
○ Elements which meet at nodes but have **bending stiffness that differs** by a factor of 10^6 or greater cause many analysis packages to fail to provide a robust solution.
○ Elements that are co-linear and **released for torsion** at the extreme ends – so the elements themselves can spin about their own axis.
○ **Members that are modelled too coarsely** thereby introducing secondary effects which hide the real design forces. A nice example (shown in Figure 12.35) is a curved member with axial load. The upper structure has been modelled with 90 elements around the curve so the angular difference between adjacent elements is small enough that the secondary effect due to the change in direction of element really is a secondary effect. The lower structure has the same curved beam but modelled using only 9 elements making the secondary effects greater than the primary effects and hence not resulting in the correct design forces.

References to Chapter 12

1. The Steel Construction Institute (1995). *Modelling of steel structures for computer analysis*. SCI Publication P148. Ascot, SCI.
2. The Institution of Structural Engineers (2002) *Guidelines for the use of computers for engineering calculations*. London, IStructE.

3. British Standards Institution (2002) BS EN 1990 *Eurocode Basis of Structural Design*. London, BSI.
4. American Society of Civil Engineers (2005) *Minimum Design Loads for Buildings and Other Structures*. USA, ASCE/SEI 7-05.
5. British Standards Institution (2005) BS EN 1993-1-1 *Eurocode 3 Design of Steel Structures. Part 1-1 General rules and rules for buildings*. London, BSI.
6. Coates R.C., Coutie M.G. and Kong F.K. (1998) *Structural Analysis*, 3rd edn. London, Chapman and Hall.
7. Süli E. and Mayers D. (2003) *An Introduction to Numerical Analysis*. Cambridge, Cambridge University Press.
8. Arnott K. (2005) Shear wall analysis – new modelling, same answers. *Journal of The Institution of Structural Engineers*, Volume 83, No. 3, 1 February.
9. British Standards Institution (2004) BS EN 1994-1-1 *Eurocode 4 Design of Composite steel and Concrete Structures. Part 1-1 General rules and rules for buildings*. London, BSI.
10. American Concrete Institute (2008) *ACI318-08, Building Code Requirements for Structural Concrete (ACI318-08) and Commentary*. Farmington Hills, MI, ACI.
11. The Steel Construction Institute (2009) Advisory Desk AD 333 *New Steel Construction*, Volume 17, No. 4, April. BCSA, SCI, Corus.
12. Access Steel (2005) SN005 NCCI *Determination of moments on columns in simple construction*, www.access-steel.com
13. Access Steel (2005) SN048b-EN-GB, NCCI *Verification of columns in simple construction – a simplified interaction criterion*, www.access-steel.com
14. British Standards Institution (2002) National Annex to BS EN 1991-1-1 *UK National Annex to Eurocode 1: Actions on Structures – Part 1-1 General Actions – Densities, self-weight, imposed loads for buildings*. London, BSI.
15. King C. M. (2007) Overall stability of multi-span portal sheds at right-angles to the portal spans. *New Steel Construction*, May.
16. British Standards Institution (2005) UK National Annex to BS EN 1993-1-1 *UK National Annex to Eurocode 3 Design of Steel Structures. Part 1-1 General rules and rules for buildings*. London, BSI.
17. Lim J., King C.M., Rathbone A.J. *et al.* (2005) Eurocode 3: The in-plane stability of portal frames. *The Structural Engineer*, Volume 83, No. 21, November.
18. Knowles P. R. (1986) *Design of castellated beams. For use with BS 5950 and BS 449*. Ascot, SCI.
19. Ward J.K. (1990) *Design of Composite and Non-composite Cellular Beams*. SCI Publication P100. Ascot, SCI.
20. CIRIA/The Steel Construction Institute (1987) *Design for openings in the webs of composite beams*. SCI Publication P068. Ascot, SCI.
21. Darwin D. (2003) *Steel and Composite Beams with Web Openings*. American Institute of Steel Construction, Steel Design Guide Series No.2. USA, AISC.
22. The Steel Construction Institute (2001) *Design of Curved Steel*. SCI Publication P281. Ascot, SCI.

23. British Standards Institution (2005) BS EN 1993-1-8 *Eurocode 3 Design of Steel Structures. Part 1-8 Design of Joints*. London, BSI.
24. British Standards Institution (2005) UK NA to BS EN 1993-1-8 *UK National Annex to Eurocode 3 Design of Steel Structures. Part 1-8 Design of Joints*. London, BSI.
25. British Constructional Steelwork Association (2005) *Steel Details*. BCSA Publication 41/05, Chapter 3, 10-15. London, BCSA.
26. Rombach G. A. (2004) *Finite Element Design of Concrete Structures*. London, Thomas Telford Publications.
27. British Standards Institution (2005) BS EN 1992-1-1 *Eurocode 2 Design of Concrete Structures. Part 1-1 General rules and rules for buildings*. London, BSI.
28. The British Constructional Steelwork Association and The Steel Construction Institute (2002) *Joints in Steel Construction, Simple Connections*. BCSA/SCI Publication P212. London, Ascot BCSA/SCI.
29. The British Constructional Steelwork Association and The Steel Construction Institute (1995) *Joints in Steel Construction, Moment Connections*. BCSA/SCI Publication P207. London, Ascot BCSA/SCI.
30. The Steel Construction Institute (1997) *Design of Semi-continuous Braced Frames*. SCI Publication P183. Ascot, SCI.
31. The Steel Construction Institute (1999) *Wind Moment Design of Low Rise Frames*. SCI Publication P263. Ascot, SCI.
32. Li T.Q., Choo B.S. and Nethercot D.A. (1995) Connection element method for the analysis of semi-rigid frames. *Journal of Constructional Steel Research*, Volume 32.
33. The Steel Construction Institute (2006) Advisory Desk AD 304. *New Steel Construction*, Volume 14, No. 9. BCSA, SCI, Corus, October.
34. Access Steel, SN045, *Column base stiffness for global analysis*, www.accesssteel.com
35. The Steel Construction Institute (1991) Advisory Desk AD 090. *Steel Construction Today*, Volume 5, No. 6. Ascot, SCI.
36. The Steel Construction Institute (2002) *Single Storey Steel Framed Buildings in Fire Boundary Conditions*. SCI Publication P313. Ascot, SCI.

Chapter 13
Structural vibration

ANDREW SMITH

13.1 Introduction

Structural vibration can be caused by a number of sources from both inside and outside the structure. Typical examples are presented in Table 13.1.

There are three principal effects that may need to be considered, depending on the frequency of occurrence and magnitude of the vibration. These are:

- *Strength* – The structure must be strong enough to resist the peak dynamic forces that arise.
- *Fatigue* – Fatigue cracks can initiate and propagate when large numbers of cycles of vibration that induce significant stress are experienced, leading to reduction in strength and failure.
- *Perception* – Human occupants of a building can perceive very low amplitudes of vibration, and if these amplitudes are too high they can cause discomfort or alarm. Certain items of precision equipment are also extremely sensitive to vibration. Perception will generally be the most onerous dynamic criterion in occupied buildings.

Some of these effects are more common than others, and it is beyond the scope of this book to address them all; references are suggested for more detailed guidance. By far the most common is the perception of vibration within buildings, and the most typical source of vibration is human activity, usually walking. The structural response (i.e. the accelerations) to walking activities is discussed here, but the theory also extends to other inputs and outputs.

13.1.1 Vibration basics

The motion of a vibrating system can be defined in terms of three parameters. The frequency defines how quickly it vibrates, the amplitude how much it vibrates, and the damping how long it vibrates for.

Steel Designers' Manual, Seventh Edition. Edited by Buick Davison and Graham W. Owens.
© 2012 Steel Construction Institute. Published 2012 by Blackwell Publishing Ltd.

Table 13.1   Sources of vibration

Forces generated inside a structure	Machinery Impacts Human activity (walking, dancing, etc.)
External forces	Wind buffeting and other aerodynamic effects Waves (offshore structures) Impacts from vehicles, etc.
Ground motions	Earthquakes Ground-borne vibration due to railways, roads, pile driving, etc.

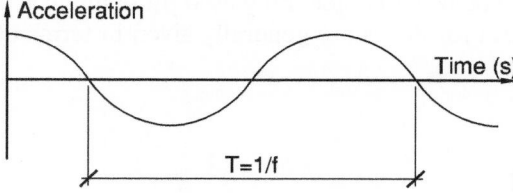

Figure 13.1   Frequency

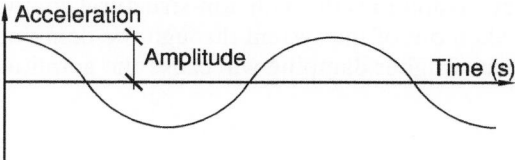

Figure 13.2   Amplitude

13.1.1.1 *Frequency*

As shown in Figure 13.1, the frequency (*f*) is the rate at which an oscillating system vibrates – the inverse of the time (*T*) it takes for a particular point to get from one extreme of motion to the other and return to its initial position. Frequency is given in terms of cycles per second (s^{-1}) or Hertz (Hz), and is proportional to the square root of stiffness divided by mass. Historically, vibration design has been limited to ensuring that the beam frequencies are above a certain limit.

13.1.1.2 *Amplitude*

As shown in Figure 13.2, the amplitude of vibration defines the displacement, velocity or acceleration of the system and is measured from the mean position to the extreme. For floor vibrations, accelerations are generally used to determine acceptability, and the acceleration can be calculated as the applied force divided by the active mass (or modal mass), multiplied by a dynamic magnification factor that takes into account the inertial effects of the response. This magnification factor will compare the natural frequency of the structure (i.e. the frequency at which the structure tends to vibrate without outside interference) with the frequency of the input force. As these two frequencies get closer together, the inertial effects become more significant and so the amplitude becomes larger. When the two frequencies are identical, resonance occurs and the amplitude reaches its maximum.

Amplitudes can be given either in terms of the peak amplitude, which is measured from the mean position to the extreme, or in terms of the root-mean-square (rms) value. This latter case provides an average of the output that will not be dominated by an abnormally high peak, and is generally used for defining acceptability criteria.

Acceptability criteria for floors are generally given in terms of the amplitude of the acceleration.

13.1.1.3 Damping

Damping refers to the loss of mechanical energy in a system. Many aspects of a structure contribute to its damping, including friction at the connections, furniture and fit-out, and energy dissipations through non-structural components such as partitions. As energy is taken out of the system through the damping, the amplitude of the response will reduce. Higher damping will cause the amplitude to reduce more.

13.1.2 Current documentation

BS EN 1990,[1] Annex A1.4.4 says that vibration should be limited to avoid discomfort to users, and to ensure the functionality of the structure or structural members. It states that the acceptability criteria may be given in terms of a frequency limit, or can be determined using 'a more refined analysis of the dynamic response of the structure, including the consideration of damping.' It refers the reader to ISO 10137,[2] and also lists a number of possible sources of vibration, including walking, synchronised movement of people and machinery.

BS EN 1993-1-1[3] states that 'the vibrations of structures on which the public can walk should be limited to avoid significant discomfort to users, and limits should be specified for each project and agreed with the client.' The UK National Annex refers to specialist literature for more detailed advice.

13.2 Causes of vibration

Possible sources of vibration are listed in Table 13.1, and they can be described in terms of different types of vibration: continuous excitation; impulsive excitation; and

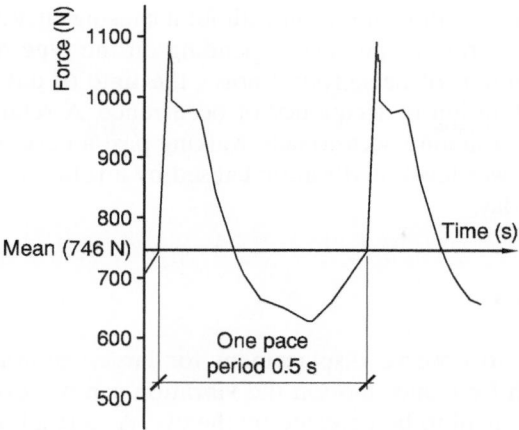

Figure 13.3 Typical force-time plot for walking

a combination of the two. A rotary compressor is an example of continuous excitation, as the compressor works at a steady frequency and so any force transmitted into the structure will act at the same frequency. A bus driving over a pot-hole or speed bump will cause an impulsive excitation, as the structure jolts under the sudden force and gradually settles back to its initial state.

Some sources of vibration, notably walking, comprise elements of both types of excitation. As Figure 13.3 shows, a typical footstep has an impulsive force as the heel hits the ground, as well as a steady cyclical force that represents the rise and fall of the person's centre of mass as they walk.

The cyclical part of the excitation can be analysed as the summation of a series of sinusoids at individual frequencies by using Fourier series. In the case shown, the walking is taking place at 2 Hz (calculated as the reciprocal of the period), and the force breaks down into significant sinusoidal contributions at frequencies of 2 Hz, 4 Hz, 6 Hz and 8 Hz. The first of these will be the largest (generally by a factor of approximately 4), but the subsequent frequencies also transmit enough energy to the floor to potentially cause excessive responses.

13.3 Perception of vibration

13.3.1 Human perception

Just as the deflections of a structure are limited to ensure that the occupants feel safe and secure, it is also important to ensure that the building feels solid in terms of its dynamic performance. If the magnitude of the vibration is not limited, the reaction from the building occupants can vary between irritation and insecurity.

A person can perceive some vibration without it causing an adverse reaction, and their perception of vibration can vary depending on the type of activity they are undertaking, the amount of background noise, the time of day, the source of the vibration and the duration or frequency of occurrence. A relatively high level of vibration caused by someone occasionally walking past a desk may well cause less annoyance than a lower level of vibration caused by a rotary machine that runs in the background all day.

13.3.2 Floor response

It may be possible to observe displacement for large-amplitude, low frequency motion, but for high frequency motion the vibration can be severe even when the displacement is too small to be detected by the eye. As a result of this, human perception of motion is usually related to acceleration levels rather than displacements, and this also makes it easier to measure the floor response through testing using accelerometers. The floor response may be defined as the acceleration of a floor in motion in response to an applied force.

13.3.3 Effect of frequency

A person's ability to hear sound varies with frequency, with the human ear unable to hear very low frequencies (bass notes) or very high frequencies (ultra-sonic sound, such as radar). There is not a step change in hearing – there is no frequency at which noise will suddenly cut in and out – but instead there is a gradual increase in the ability to hear a constant amplitude sound as its frequency rises before it then starts to decrease again. The same effect can be seen with the human perception of motion, except that the body is more sensitive at lower frequencies than the ear.

Figure 13.4 shows the base-curve of human perception from ISO 10137 – the relationship between frequency and acceleration amplitude below which the 'average' person should not be able to perceive vibration. From this it is clear that humans are most sensitive to vibrations in the frequency range 4 Hz to 8 Hz, and that in this range an rms acceleration of 5 mm/s^2 will just about be noticeable. This acceleration value is known as the threshold of human perception.

13.3.4 Multiplication factors

As described above, different situations may require more strict or lenient acceptability criteria. In an operating theatre, it is important that the surgeon cannot feel any vibration in case it causes him to slip at the wrong moment, whereas in his consulting room where the work is not so delicate he may be content occasionally feeling a small amount of vibration. The different acceptability criteria are given in terms of multiplying factors that apply to the base curve shown in Figure 13.4, and

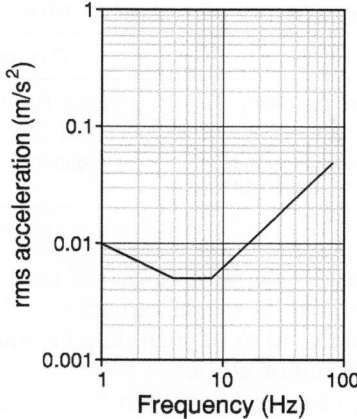

Figure 13.4   Threshold of human perception against frequency from ISO 10137

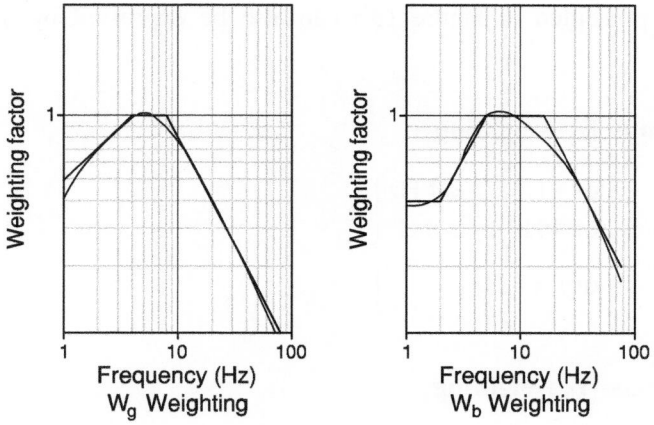

Figure 13.5 W_g and W_b weighting curves

so for an operating theatre a multiplying factor of 1.0 would apply, but for a consulting room a multiplying factor of 8.0 is a more reasonable design criterion (values taken from HTM 08-01[4]).

13.3.5 Weighting factors

Rather than comparing a multi-frequency response to the base curve, weighting curves are used instead to normalise acceleration to the typical human perception. Values of frequency weighting are given by inverting the base curves such as the one shown in Figure 13.4. Two weighting curves are commonly used: W_g for vision and hand control, and W_b for perception. These are illustrated in Figure 13.5, and

Table 13.2 Weighting factors appropriate for floor design

Room type	Category	Weighting curve
Critical working areas (e.g. hospital operating theatres, precision laboratories)	Vision/Hand control	W_g
Residential, offices, wards, general laboratories, consulting rooms, workshop and circulation spaces	Perception	W_b

their use is defined in Table 13.2. In most cases, the aim of vibration analysis is to reduce or remove discomfort, but in special circumstances, such as operating theatres, the level of vibration will need to be such that it cannot be perceived and does not affect the steadiness of hand or vision.

To illustrate the use of the curves, using curve W_b for discomfort, a sine wave of 8 Hz has the same feel as a sine wave at 2.5 Hz or 32 Hz with double the amplitude, so the absolute values of acceleration at these frequencies would be halved for comparison to acceptability criteria.

The curves presented in Figure 13.5 can also be expressed by the following equations.

z-axis vibrations W_g weighting

$$\left. \begin{array}{ll} W = 0.5\sqrt{f} & \text{for } 1\,\text{Hz} < f < 4\,\text{Hz} \\ W = 1.0 & \text{for } 4\,\text{Hz} \leq f \leq 8\,\text{Hz} \\ W = \dfrac{8}{f} & \text{for } f > 8\,\text{Hz} \end{array} \right\} \tag{13.1}$$

z-axis vibrations W_b weighting

$$\left. \begin{array}{ll} W = 0.4 & \text{for } 1\,\text{Hz} < f < 2\,\text{Hz} \\ W = \dfrac{f}{5} & \text{for } 2\,\text{Hz} \leq f < 5\,\text{Hz} \\ W = 1.0 & \text{for } 5\,\text{Hz} \leq f \leq 16\,\text{Hz} \\ W = \dfrac{16}{f} & \text{for } f > 16\,\text{Hz} \end{array} \right\} \tag{13.2}$$

13.4 Types of response

13.4.1 Steady-state response

When a continuous cyclic force is applied to a system, it will start to respond at the same frequency as the input. The inertia of the moving mass will gradually build

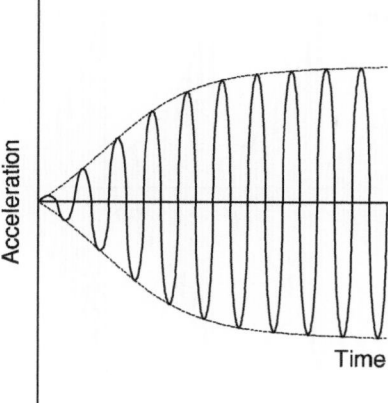

Figure 13.6   Idealised steady-state response

until the response has a constant amplitude – this is the steady-state. An example of this response is shown in Figure 13.6.

The magnitude of the amplitude depends on the amount of mass that is mobilised, the magnitude of the force, the ratio of the forcing frequency and the natural frequency of the system, and the damping present in the system. The largest response occurs when the forcing frequency is the same as the natural frequency of the system, and this is known as resonance. For a typical floor, the response will be in the region of sixteen times larger at resonance that it will be when the frequencies are separated.

13.4.2 Transient response

The transient response is the response of the system to an impulsive force, or a series of impulsive forces. The response is comparable to plucking a guitar string – the response will jump to its highest level and gradually die away as the damping dissipates the energy. A typical example of a transient, or impulsive, response is given in Figure 13.7.

In this case the system will respond at its own natural frequency and the magnitude of the response is governed simply by the magnitude of the impulsive force compared to the amount of mass that is mobilised.

13.5 Determining the modal properties

The modal properties define the vibration characteristics of the structure, and comprise four elements: frequency, modal mass, mode shape and modal damping. For any structure, there are an infinite number of modes for which each of these

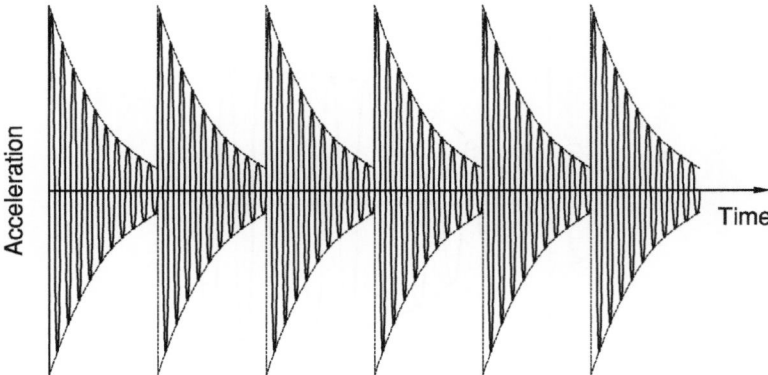

Figure 13.7   Response to a series of impulses

Table 13.3   Critical damping ratios, ζ, for various floor types

ζ	Floor finishes
0.5%	For fully welded steel structures, e.g. staircases.
1.1%	For completely bare floors or floors where only a small amount of furnishings are present.
3.0%	For fully fitted out and furnished floors in normal use.
4.5%	For a floor where the designer is confident that partitions will be appropriately located to interrupt the relevant mode(s) of vibration (i.e. the partition lines are perpendicular to the main vibrating elements of the critical mode shape).

parameters are different, though for analysis purposes only those with the lowest frequencies will be relevant.

For each individual mode: the natural frequency is the rate at which it would naturally tend to vibrate; the mode shape is the deformed shape that it would naturally tend to exhibit at that frequency; the modal mass is a function of this mode shape and the distributed mass, and defines how much mass is participating in the motion; and the modal damping defines the energy dissipation within the mode. The first three can be calculated based on the dimensions, mass and stiffness of the structure, while the last depends on the finishes on the structure, and an appropriate value will generally have to be assumed. Typical critical damping ratios are given in Table 13.3, and in the vast majority of cases, $\zeta = 3\%$ can be used for design.

The mode shape has no physical units, and so is generally scaled in comparison to the modal mass – it is common to either scale the mode shapes to a maximum value of 1.0, or to scale the mode shapes so that the modal mass is 1 kg. The designer needs to ensure that compatible values are used.

As the modal frequencies increase, the complexity of the mode shape will also increase. This can easily be seen for a simply-supported beam, as shown in Figure 13.8, where the lowest frequency will correspond to the single half-sine wave. For this simple system, the frequencies and other modal properties can be easily defined,

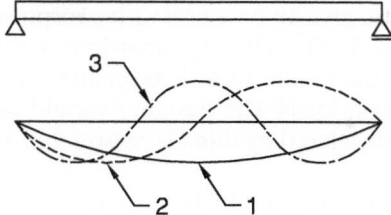

Figure 13.8   First three mode shapes of a simply-supported beam

Table 13.4   Modal properties of a simply-supported beam

Mode	Frequency (Hz)	Mode shape	Modal Mass (kg)
1	$f = \dfrac{\pi}{2}\sqrt{\dfrac{EI}{mL^4}}$	$\mu = \sin\left(\dfrac{\pi x}{L}\right)$	$M = \dfrac{mL}{2}$
2	$f = 2\pi\sqrt{\dfrac{EI}{mL^4}}$	$\mu = \sin\left(\dfrac{2\pi x}{L}\right)$	$M = \dfrac{mL}{2}$
3	$f = \dfrac{9\pi}{2}\sqrt{\dfrac{EI}{mL^4}}$	$\mu = \sin\left(\dfrac{3\pi x}{L}\right)$	$M = \dfrac{mL}{2}$
n	$f = \dfrac{(n\pi)^2}{2\pi}\sqrt{\dfrac{EI}{mL^4}}$	$\mu = \sin\left(\dfrac{n\pi x}{L}\right)$	$M = \displaystyle\int_0^L \mu^2 m\,dx$

as shown in Table 13.4 (L is the span, x is the position along the beam, EI is the stiffness of the beam, m is the distributed mass per unit length of the beam).

In practice, as structures are built up from a number of components, determining the modal properties may not be as straightforward as for a uniform simply-supported beam. The most accurate way to assess the structure is by using computational methods, generally finite element software, but simplified methods can also be used on some simpler structures.

13.5.1 Computational methods

The frequencies of a structure, along with its mode shapes and modal masses, are best calculated using finite element (FE) software. This best allows for the complex interaction of different building elements, geometries, restraints and loadings. Modal analysis of a structure is a fairly straightforward task for an FE package, but it is important to make the correct modelling assumptions to ensure the accuracy of the results.

The mass of the structure should include not just the dead weight of the beams and slab, but also a reasonable allowance for the amount of mass that will be present on the floor in service. Including too much mass in the analysis will yield lower frequencies, but will also give unconservative responses; conversely including too little mass will give higher frequencies which may affect the type of response that the floor will exhibit and also give an unconservative response. It is generally recommended that all of the dead load, including super-imposed dead load from ceilings and services, is included in the model, as well as an allowance of around 10% of the design imposed and partition loads. The designer should take a pragmatic view of this though, and only include what they think is reasonable depending on the specific design situation.

In modelling a structure it can generally be assumed that all connections are fixed, even those that are designed as pinned; deflections and rotations caused by the vibration of the structure are generally so small that they will not overcome the friction at the connections. Core walls can also be assumed to be fully fixed and, in general, cladding can be assumed to prevent deflection of the edge beams of the building, though they should not be restrained from rotating. Care should be taken over the modelling of composite beams and slabs to try and ensure that the orthogonal behaviours are replicated. Note that concrete can be assumed to have a higher modulus of elasticity in dynamic conditions compared with short-term loads (typically $38 \, \mathrm{kN/mm^2}$ for normal weight concrete or $22 \, \mathrm{kN/mm^2}$ for lightweight concrete), as the timescales are so small that creep does not have an effect.

13.5.2 Simplified methods

The natural frequencies of a uniform simply-supported beam subjected to a uniform load can be calculated using the equation given in Table 13.4. A more general version of the equation, which takes account of different end conditions, is:

$$f_n = \frac{\kappa_n}{2\pi} \sqrt{\frac{EI}{mL^4}} \tag{13.3}$$

where:
- EI is dynamic flexural rigidity of the member ($\mathrm{Nm^2}$)
- m is the effective mass (kg/m)
- L is the span of the member (m)
- κ_n is a constant representing the beam support conditions for the nth mode of vibration.

Some standard values of κ_n for elements with different boundary conditions are given in Table 13.5.

A convenient method of determining the fundamental (i.e. the lowest) natural frequency of a beam f_1 (sometimes referred to as f_0), is by using the maximum deflection δ caused by the weight of a uniform mass per unit length m. For a simply-supported element subjected to a uniformly distributed load (for which $\kappa_1 = \pi^2$), this is the familiar expression:

$$\delta = \frac{5mgL^4}{384EI} \tag{13.4}$$

where g is the acceleration due to gravity ($9.81 \, \mathrm{m/s^2}$).

Table 13.5 κ_n coefficients for uniform beams

Support conditions	κ_n for mode n		
	$n = 1$	$n = 2$	$n = 3$
Pinned/pinned ('simply-supported')	π^2	$4\pi^2$	$9\pi^2$
Fixed both ends	22.4	61.7	121
Fixed/free (cantilever)	3.52	22	61.7

Rearranging Equation 13.4, and substituting the value of m and κ_1 into Equation 13.3 gives the following equation, in which δ is expressed in mm:

$$f_1 = \frac{17.8}{\sqrt{\delta}} \approx \frac{18}{\sqrt{\delta}} \tag{13.5}$$

where δ is the maximum deflection due to the self-weight and any other loads that may be considered to be permanent.

It can also be easily shown that a numerator of approximately 18 would again be achieved if the above steps were repeated for a beam with different support conditions, with the appropriate equation for deflection and κ_n inserted within Equation 13.3. Therefore, for design, Equation 13.5 may be used as the generalised expression for determining the natural frequency of individual members, even when they are not simply-supported, providing that the appropriate value of δ is used.

In addition, Dunkerly's approximation (Equation 13.6) demonstrates that Equation 13.5 will give the fundamental natural frequency of a floor system when δ is taken as the sum of the deflections of each of the structural components (for example the primary and secondary beams and the slab).

$$\frac{1}{f_1^2} = \frac{1}{f_s^2} + \frac{1}{f_b^2} + \frac{1}{f_p^2} \tag{13.6}$$

where f_s, f_b and f_p are the component frequencies of the slab, secondary beam and floor beam respectively.

In conventional steel-concrete floor systems, the fundamental frequency may be estimated by using engineering judgement on the likely deflected shape of the floor (the mode shape), and considering how the supports and boundary conditions will affect the behaviour of the individual structural components. For example, on a simple composite floor comprising a slab continuous over a number of secondary beams that are, in turn, supported by stiff primary beams, two possible mode shapes may sensibly be considered:

In the secondary beam mode, the primary beams form nodal lines (i.e. they have zero deflection). This enables consecutive secondary beam spans to move in opposite directions with a small amount of rotation in the primary beams, and so the secondary beam frequency is determined with a simply-supported end condition (see Figure 13.9(a)). Inertia effects in the slab as it is moved by the secondary beams will give a mode shape with the slab behaving as fixed-ended between the secondary beams, so this boundary condition should be used to calculate the slab frequency.

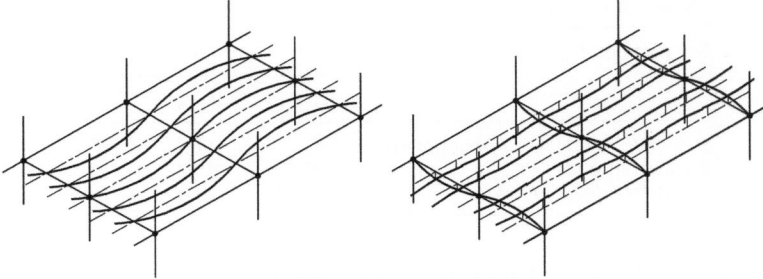

(a) Governed by secondary beam flexibility   (b) Governed by primary beam flexibility

Figure 13.9   Typical mode shapes for steel-concrete floor systems

Table 13.6   Calculation of deflection for different framing arrangements

Framing arrangement	Secondary beam mode of vibration	Primary beam mode of vibration
	$\delta = \dfrac{mgb}{384\,E}\left(\dfrac{5L^4}{I_b} + \dfrac{b^3}{I_s}\right)$	–
	As above	$\delta = \dfrac{mgb}{384\,E}\left(\dfrac{64\,b^3 L}{I_p} + \dfrac{L^4}{I_b} + \dfrac{b^3}{I_s}\right)$
	As above	$\delta = \dfrac{mgb}{384\,E}\left(\dfrac{368\,b^3 L}{I_p} + \dfrac{L^4}{I_b} + \dfrac{b^3}{I_s}\right)$

In the primary beam mode, the primary beams vibrate about the columns as simply-supported members (see Figure 13.9(b)), and the inertia effects cause the secondary beams and slab to behave as if they are fixed-ended.

The natural frequency should be calculated for each mode using Equation 13.5. The fundamental frequency, f_0, is the lower value for the two modes considered. δ should be taken as the total deflection (in millimetres) of the slab, secondary beams and primary beams using the end conditions described above and based on the gross second moment of area of the components, with a load corresponding to the self-weight and other permanent loads, plus a proportion of the imposed load that may be considered permanent.

For cases when the adjacent spans are approximately equal, δ can be calculated from the equations given in Table 13.6.

where:

m is the distributed floor loading (kg/m^2)

g is the acceleration due to gravity (= 9.81m/s^2)

E is the elastic modulus of steel (N/m^2)

I_b is the composite second moment of area of the secondary beam (m^4)

I_s is the second moment of area of the slab per unit width in steel units using the dynamic value of the elastic modulus (m^4/m)

I_p is the composite second moment of area of the primary beam (m^4).

13.6 Calculating vibration response

13.6.1 Computational methods

Once the modal properties have been determined, the predicted accelerations can be calculated by applying the forcing function(s) to each mode in turn and combining the results using modal superposition. The accelerations from the two different kinds of response are calculated using separate equations.

13.6.1.1 Steady-state response

The steady-state response of a system is significant when one or more of the harmonic components of the forcing function are close to one of the natural frequencies of the floor. In these circumstances, it is recommended that all modes of vibration having natural frequencies up to 3 Hz higher than the highest frequency component of the input force should be considered, to account for off-resonant vibration of the highest harmonic of the activity. The weighted rms acceleration response of a single mode to a single frequency force is calculated from:

$$a_{\text{w,rms,e,r}} = \mu_e \mu_r \frac{F}{M\sqrt{2}} DW \tag{13.7}$$

where:

μ_e is the mode shape amplitude at the point on the floor where the excitation force F is applied (non dimensional – see Figure 13.10)

μ_r is the mode shape amplitude at the point where the response is to be calculated (non dimensional – see Figure 13.10)

F is the excitation force

M is the modal mass

D is the dynamic magnification factor for acceleration (see Equation 13.8)

W is the appropriate code-defined weighting factor for human perception of vibrations corresponding to the forcing frequency f_p.

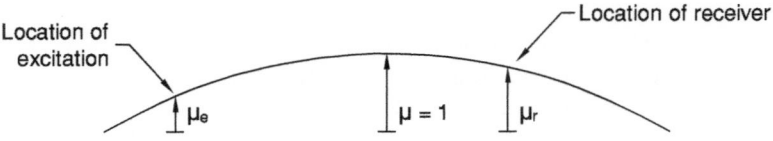

Figure 13.10   Mode shape amplitude

The dynamic magnification factor for acceleration, which is the ratio of the peak amplitude to the static amplitude, is given by the following:

$$D = \frac{\beta^2}{\sqrt{\left(1-\beta^2\right)^2 + \left(2\zeta\beta\right)^2}} \qquad (13.8)$$

where:

β is the frequency ratio (taken as f_p / f)
ζ is the damping ratio
f_p is the frequency corresponding to the first harmonic of the activity
f is the frequency of the mode under consideration.

When the frequency of the input force can vary within a range, such as for walking, the most critical response will be found when the forcing frequency (or a harmonic of it) is the same as the frequency of the mode under consideration. In this case $\beta = 1$, and $D = 1/(2\zeta)$. The mode shape amplitudes take into account the magnitude of the mode shape at the excitation and response points, and take into account the different areas of a structure that may be affected by different mode shapes (for example a response point over a primary beam would only be in motion when the primary beam mode is being excited, and would be stationary when the secondary beam mode is being excited), or can be used to establish the transmission of vibration from one part of a structure to another.

The total response to each harmonic of the activity is found by summing the acceleration response of each mode of vibration of the system at each harmonic of the forcing function. The combined effect of all the harmonics can then be calculated using a square-root-sum-of-squares.

Generally the response will be dominated by a single mode, but with contributions from a number of other modes. Repeating the calculations and summations for different excitation and response points can enable the designer to establish which parts of a structure are more responsive than others, and can assist with locating critical working areas.

13.6.1.2 *Transient response*

The transient response is the response of the system to an impulse or series of impulses. In these circumstances, it is recommended that all modes with natural frequencies up to twice the fundamental (first mode) frequency or 20 Hz, whichever is larger, should be taken into account, as above this the effects of the frequency weighting will make the results insignificant. The weighted peak acceleration response for a single mode of vibration may be obtained from the following:

$$a_{\text{w,peak,e,r}} = 2\pi\, f_{\text{d}}\mu_{\text{e}}\mu_{\text{r}} \frac{F_{\text{I}}}{M} W \tag{13.9}$$

where:

f_{d} is the damped natural frequency of the mode under consideration (equal to $f = \sqrt{1-\zeta^2}$)

f is the natural frequency of the mode under consideration

ζ is the critical damping ratio for the mode under consideration

μ_{e} is the mode shape amplitude at the point on the floor where the impulse force F_{I} is applied (non dimensional – see Figure 13.10)

μ_{r} is the mode shape amplitude at the point where the response is to be calculated (non dimensional – see Figure 13.10)

F_{I} is the impulsive excitation force

M_{n} is the modal mass

W_{n} is the appropriate code-defined weighting factor for human perception of vibrations corresponding to f_{d}

The acceleration to each impulse is found by assuming the response of each mode starts at the same time and decays exponentially, as shown in the following equation:

$$a_{\text{w,e,r}}(t) = 2\pi\, f_{\text{d}}\mu_{\text{e}}\mu_{\text{r}} \frac{F_{\text{I}}}{M} \sin(2\pi\, f_{\text{d}}t)e^{-\zeta 2\pi f_{\text{d}}t} W \tag{13.10}$$

By summing the acceleration responses of each mode and performing a root-mean-square integration, the overall acceleration can be determined.

13.6.2 Simplified methods

Simplified methods of determining the response of floor structures to walking activities are proposed in SCI P354,[5] CCIP-016,[6] HiVOSS[7] and AISC DG11.[8] Each of them tries to produce a method of analysis that can be used in design without needing to make heavy use of computers, and the approaches vary. By nature of the simplification, there can be a degree of conservatism in each method.

The SCI approach uses the dimensions and stiffness properties of the floor to define an effective modal mass. This modal mass reflects the combined effect of several harmonics of the forcing function acting on a number of different modes of vibration, and so is not directly comparable to the modal mass that an FE package would calculate. There are two sets of equations, depending on the flooring system used.

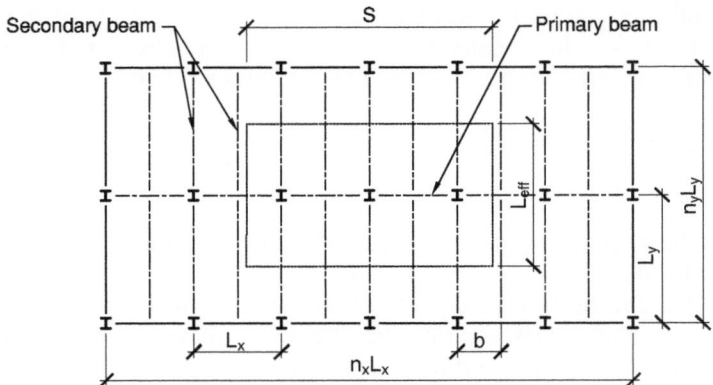

Figure 13.11   Definition of variables used to establish effective modal mass

13.6.2.1 Modal mass for floors using downstand beams with shallow decking

For floors constructed with shallow decking and downstand beams (i.e. the decking is supported on the top flange of the beam), the variables L_{eff} and S should be calculated from the following equations:

$$L_{eff} = 1.09(1.10)^{n_y-1}\left(\frac{EI_b}{mbf_0^2}\right)^{1/4} \quad L_{eff} \leq n_y L_y \tag{13.11}$$

where:

    n_y is the number of bays (but $n_y \leq 4$) in the direction of the secondary beam span

    EI_b is the dynamic flexural rigidity of the composite secondary floor beam (expressed in Nm^2 when m is expressed in kg/m^2)

    b is the floor beam spacing (expressed in m)

    f_0 is the fundamental frequency (as defined in Section 13.5.2)

    L_y is the span of the secondary beams (expressed in m; see Figure 13.11).

The effective floor width, S, should be calculated from the following equation:

$$S = \eta(1.15)^{n_x-1}\left(\frac{EI_s}{mf_0^2}\right)^{1/4} \quad S \leq n_x L_x \tag{13.12}$$

where:

    L_x is the span of primary beam (expressed in m; as shown in Figure 13.11)

    n_x is the number of bays (but $n_x \leq 4$) in the direction of the primary beam span (see Figure 13.11).

    η is a factor that accounts for the influence of floor frequency on the response of the slab (see Table 13.7)

Table 13.7 Frequency factor η

Fundamental frequency, f_0	η
$f_0 < 5\,\text{Hz}$	0.5
$5\,\text{Hz} \leq f_0 \leq 6\,\text{Hz}$	$0.21\,f_0 - 0.55$
$f_0 > 6\,\text{Hz}$	0.71

EI_s is the dynamic flexural rigidity of the slab (expressed in Nm^2/m when m is expressed in kg/m^2)

f_0 is the fundamental frequency (as defined in Section 13.5.2)

L_x is the width of a bay (see Figure 13.11).

13.6.2.2 Modal mass for floors using slim floor beams with deep decking

For floors constructed with deep decking placed on the bottom flange of the support member (i.e. systems such as *Slimdek*), the variables L_{eff} and S should be calculated from the following equations:

$$L_{eff} = 1.09 \left(\frac{EI_b}{mL_x f_0^2} \right)^{1/4} \qquad L_{eff} \leq n_y L_y \qquad (13.13)$$

where:

EI_b is the dynamic flexural rigidity of the composite secondary floor beam (expressed in Nm^2 when m is expressed in kg/m^2).

L_x is floor beam spacing (expressed in m).

n_y is the number of bays (but $n_y \leq 4$) in the direction of the secondary beam span (see Figure 13.12).

f_0 is the fundamental frequency (as defined in Section 13.5.2).

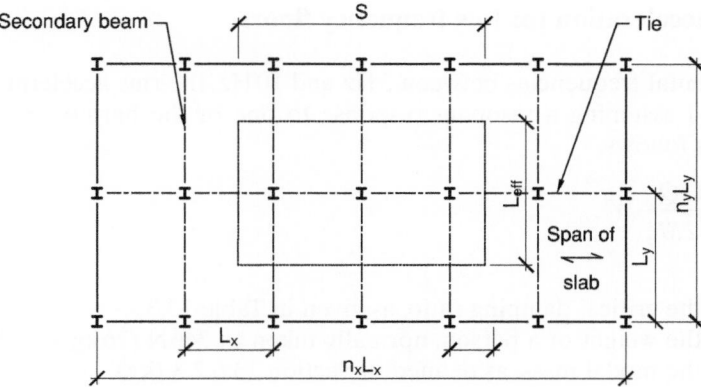

Figure 13.12 Definition of variables used to establish effective modal mass

The effective floor width S should be calculated from the following equation:

$$S = 2.25 \left(\frac{EI_s}{mf_0^2} \right)^{1/4} \qquad S \le n_x L_x \tag{13.14}$$

where:

EI_s is the dynamic flexural rigidity of the slab in the strong direction (expressed in Nm^2/m when m is expressed in kg/m^2)

f_0 is the fundamental frequency (as defined in Section 13.5.2)

L_x is the width of a bay (see Figure 13.12)

n_x is the number of bays (but $n_x \le 4$) in the direction of the primary beam span (see Figure 13.12).

13.6.2.3 Effective modal mass

The effective modal mass, M, is calculated from an effective plan area of the floor participating in the motion as follows:

$$M = mL_{eff}S \tag{13.15}$$

where:

m is the floor mass per unit area including dead load and imposed load that will be present in service (see Section 13.5.1)

L_{eff} is the effective floor length (see Equation 13.11 or 13.13)

S is the effective floor width (see Equation 13.12 or 13.14).

13.6.2.4 Response acceleration

Response acceleration for low frequency floors

For fundamental frequencies between 3 Hz and 10 Hz, the rms acceleration should be calculated assuming a resonant response to one of the harmonics of walking frequency as follows:

$$a_{w,rms} = \frac{0.1Q}{2\sqrt{2}M\zeta} W \tag{13.16}$$

where:

ζ is the critical damping ratio, as given in Table 13.3

Q is the weight of a person, normally taken as 746 N ($76 \, kg \times 9.81 \, m/s^2$)

M is the modal mass, as defined in Section 13.6.2.3 (kg)

W is the appropriate code-defined weighting factor for human perception of vibrations (see Section 13.3.5), based on the fundamental frequency, f_0.

Response acceleration for high frequency floors

If the fundamental frequency is greater than 10 Hz, the rms acceleration should be calculated from the following expression, which assumes that the floor exhibits a transient response:

$$a_{\mathrm{w,rms}} = 2\pi \frac{185}{M f_0^{0.3}} \frac{Q}{700} \frac{1}{\sqrt{2}} W \qquad (13.17)$$

where:

f_0 is the fundamental frequency of the floor (Hz)

M is the modal mass, as defined in Section 13.6.2.3 (kg)

Q is the static force exerted by an 'average person' (normally taken as $76\,\mathrm{kg} \times 9.81\,\mathrm{m/s^2} = 746\,\mathrm{N}$)

W is the appropriate code-defined weighting factor for human perception of vibrations (see Section 13.3.5), based on the fundamental frequency, f_0.

13.6.3 Response factors

Response factors can be calculated from the frequency-weighted rms acceleration simply by dividing by the threshold acceleration of $5\,\mathrm{mm/s^2}$ identified in Section 13.3.3.

13.7 Acceptability criteria

13.7.1 Multiplying factors

The most commonly used acceptability criteria are multiplying factors that are applied to the base curve shown in Figure 13.4. These can be used as an upper limit to the response factor calculated from the weighted rms acceleration above. Multiplying factor limits are given in ISO 10137 (as well as HTM 08-01 for health-care buildings), and correspond to 'a low probability of adverse comment'. The limits from ISO 10137, for both continuous vibration and impulsive excitation with several occurrences per day, are reproduced here in Table 13.8.

In addition to the values presented in Table 13.8, further limits are recommended in SCI P354 for other areas such as shopping malls and staircases. Among these is a recommended limit for offices of 8.0, but this only applies for single person excitation from walking activities, and so should not be used for machinery-generated vibration. This limit reflects the intermittent nature of walking.

13.7.2 Vibration dose values

As shown in Table 13.8, limiting multiplying factors are given for continuous vibration and for intermittent vibration with only a few occurrences. However, there are

Table 13.8 Limiting multiplying factors from ISO 10137

Place	Time	Continuous and intermittent vibration	Impulsive vibration excitation with several occurrences per day
Critical working areas		1	1
Residential	Day	2 to 4	30 to 90
	Night	1.4	1.4 to 20
Quiet office		2	60 to 128
General office		4	60 to 128
Workshops		8	90 to 128

Table 13.9 Limiting VDVs from BS 6472 & ISO 10137

Time	Low probability of adverse comment	Adverse comment possible	Adverse comment probable
Day (16 hours)	0.2 to 0.4	0.4 to 0.8	0.8 to 1.6
Night (8 hours)	0.13	0.26	0.51

For offices these limits can be multiplied by 2, and for workshops these limits can be multiplied by 4.

some activities, walking among them, where the vibration will not fall into either of these categories. To take this into consideration, vibration dose values (VDVs) may be used as an alternative to response factors and multiplying factors. VDVs are the only limiting criterion defined in BS 6472,[9] and are calculated as a function of the root-mean-quad of the acceleration response (similar to the root-mean-square but the fourth power is used instead of the second to make the peaks more significant). However, the VDV may be estimated from the calculated weighted rms acceleration, $a_{w,rms}$, using the following formula:

$$VDV = 1.4\, a_{w,rms} \sqrt{t}$$

where t is the total duration of the vibration over the exposure period, which is usually taken as a day of 16 hours or a night of 8 hours. The calculated VDV can then be compared to the limiting values given in BS 6472, which are reproduced in Table 13.9.

The use of VDVs for design is problematic, as it is difficult to define exactly the duration of the activity. In general, therefore, design is performed by calculating response factors and comparing the result with the limiting multiplying factors, such as those given in ISO 10137 and SCI P354.

13.8 Practical considerations

There are a number of issues that a designer should consider to minimise the likelihood of unsatisfactory dynamic performance.

13.8.1 Cantilevers

Cantilevers combine low frequencies with low modal masses. As a result, they can be very responsive to dynamic excitation and need to be carefully designed to ensure that the levels of response are limited.

13.8.2 Continuity of slab

When using precast units, it is important that the designer ensures that there is suitable continuity of the slab between units and across beams. If this continuity is not provided, the vibration may only affect a single bay or single unit at a time and so significantly less mass will be mobilised and the response will be higher. Structural toppings generally provide this continuity, but careful grouting between units and infill around the beams may also prove to be acceptable.

13.8.3 Architectural layout

Many vibration issues can be solved by careful consideration of the building layout and framing arrangements, rather than by attempting to improve a structure once it is already designed. By considering the likely mode shapes that the structure will adopt, it is possible to ensure that corridors and critical areas are not involved in the same motion, and so there is a natural isolation effect. For example, by splitting a structure into long office and ward spans and short corridor spans, rather than providing two equal spans, the effects of the different stiffnesses and spans will provide a natural isolation and so will reduce the transmitted response.

The location of gymnasia, dance floors and plant rooms in relation to other facilities is also important, as these areas will typically generate significant responses. Ideally these areas should be placed directly onto the ground floor slab, but if this is not possible then the designer must be confident that the vibration caused in these areas will not transmit into other areas of the building where the levels of vibration may cause annoyance.

13.8.4 Steel staircases

As fully welded structures, steel staircases exhibit significantly low damping, with typical critical damping ratios of 0.5%. There is also significantly less mass to mobilise in a staircase compared with a floor, and the forces and frequencies that can be achieved by people travelling up and down the stairs are also significantly higher than walking. As a result of these three factors, the response of staircases to human activity may be very significant, and so the vibration design will need to be carefully considered. Generally a fundamental frequency in excess of 10 to 12 Hz will be required for an acceptable staircase, though even at these frequencies the response should still be assessed.

13.9 Synchronised crowd activities

The UK National Annex to BS EN 1991-1-1[10] states in Clause NA.2.1.2 that:

'Structures with elements subject to dancing and jumping are liable to inadvertent or deliberate synchronised movement of occupants, sometimes accompanied by music with a strong beat, such as occurs at pop concerts and aerobics events. These activities generate dynamic effects that can result in enhanced vertical and horizontal loads. If a natural frequency of a structure matches the frequency of the synchronised movement, or an integer multiple of it, then resonance can occur that greatly amplifies the dynamic response.'

This clause is especially important for the design of stadia that may be used for concerts, for gymnasia or sports halls that may be used for aerobics or dancing, and for dance floors such as found in nightclubs. The loading in these cases affects the Ultimate Limit State, and there are two methods that are generally used to ensure compliance.

The first is to design the structural elements so that the fundamental frequency of the floor system (note that this is the combined system, rather than just the floor beams) is greater than 8.4 Hz, and the horizontal frequency of the structure is greater than 5 Hz. Above these frequencies the forces that can be generated by large groups of people will be lower than the normal design loads.

The second method is to calculate explicitly the forces that will be applied to the structure should resonance occur between the human activity and the structure. In general when using this approach, the loads from light aerobic activities will be within the normal imposed loads considered for design, but the loads from dense groups of people jumping or dancing may be higher. The calculation procedure is given in SCI P354 or BRE Digest 426.[11]

References to Chapter 13

1. British Standards Institution (2002) BS EN 1990: 2002 *Eurocode – Basis of structural design*. London, BSI.
2. International Organization for Standardization (2007) ISO 10137: 2007. *Bases for design of structures – Serviceability of buildings and walkways against vibrations*. Geneva, ISO.
3. British Standards Institution (2005) BS EN 1993-1-1: 2005. *Eurocode 3: Design of steel structures – Part 1-1: General rules and rules for buildings*. London, BSI.

4. Department of Health *Health Technical Memorandum 08-01: Acoustics*. London, The Stationery Office, 2008.
5. Smith A.L., Hicks S.J. and Devine P.J. (2007) *SCI P354 – Design of Floors for Vibration: A New Approach*. Ascot, SCI.
6. The Concrete Centre, Willford M.R. and Young P. (2006) CCIP-016 – *A Design Guide for Footfall Induced Vibration of Structures*. London, TSC.
7. www.stb.rwth-aachen.de/projekte/2007/HIVOSS/download.php
8. Murray T.M., Allen D.E. and Ungar E. E. (1997) *AISC DG11 – Floor Vibrations Due to Human Activity*. Chicago, AISC.
9. British Standards Institution (2008) BS 6472-1: 2008 *Guide to evaluation of human exposure to vibration in buildings. Part 1: Vibration sources other than blasting*. London, BSI.
10. British Standards Institution (2002) BS EN 1991-1-1: 2002 *Eurocode 1: Actions on structures – Part 1-1: General actions – Densities, self-weight, imposed loads for buildings*. London, BSI.
11. Ellis B.R. and Ji T. (2004) BRE Digest 426 – *The response of structures to dynamic crowd loads*. Watford, Building Research Establishment.

Chapter 14
Local buckling and cross-section classification

LEROY GARDNER and DAVID NETHERCOT

14.1 Introduction

The efficient use of material within a steel member requires those structural properties that most influence its load-carrying capacity to be maximised. This, coupled with the need to make connections between members, has led to the majority of structural sections being thin-walled, as illustrated in Figure 14.1. Moreover, apart from circular tubes, structural steel sections (such as universal beams and columns, cold-formed purlins, built-up box columns and plate girders) normally comprise a series of flat plate elements. Simple considerations of minimum material consumption frequently suggest that some plate elements be made extremely thin but limits must be imposed if certain potentially undesirable structural phenomena are to be avoided. The most important of these in everyday steelwork design is local buckling.

Figure 14.2 shows a short UC section after it has been tested as a column. Considerable deformation of the cross-section is evident with the flanges being displaced out of their original flat shape. The web, on the other hand, appears to be comparatively undeformed. The buckling has therefore been confined to certain plate elements and has not resulted in any overall lateral deformation of the member (i.e. its centroidal axis has not deflected). In the particular example of Figure 14.2, local buckling did not develop significantly until well after the column had sustained its yield load, equal to the product of its cross-sectional area and its material strength. Local buckling did not prevent attainment of the yield load because the proportions of the web and flange plates are sufficiently stocky. The fact that the local buckling appeared in the flanges before the web is due to the former being the more slender.

Terms such as stocky (or compact) and slender are used to describe the proportions of the individual plate elements of structural sections, based on their

Steel Designers' Manual, Seventh Edition. Edited by Buick Davison and Graham W. Owens.
© 2012 Steel Construction Institute. Published 2012 by Blackwell Publishing Ltd.

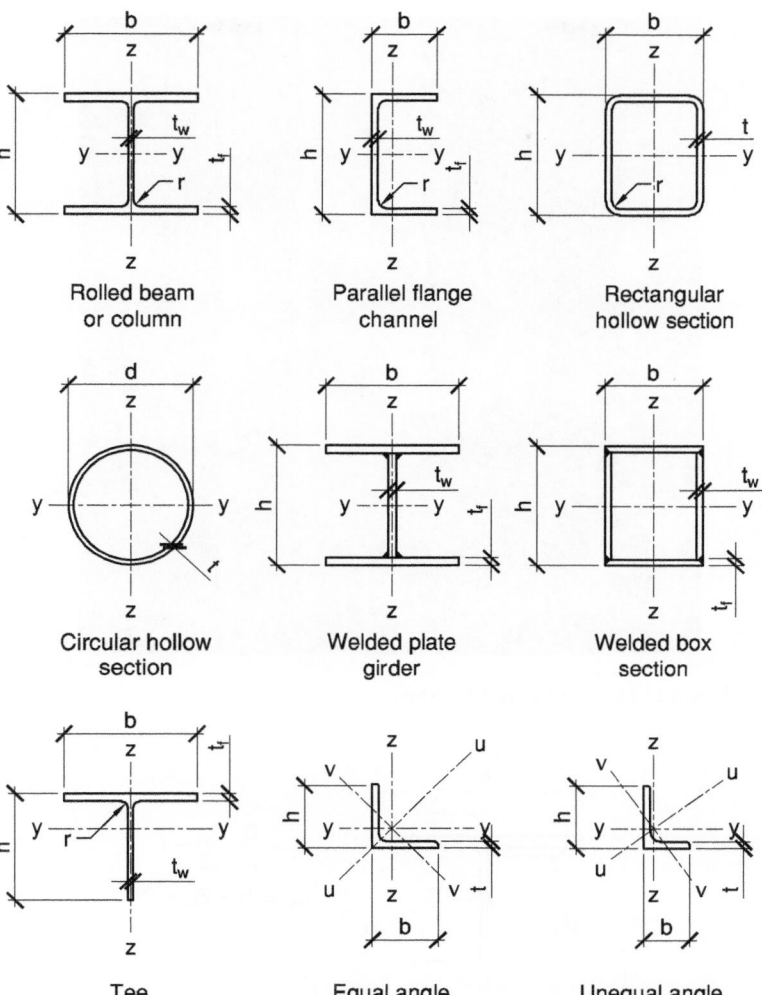

Figure 14.1 Structural cross-sections

susceptibility to local buckling. The most important governing property is the ratio of plate width to plate thickness, λ_p, referred to as the c/t ratio in Eurocode 3 or, more generally, the b/t ratio. Other influential factors are material strength, the stress system to which the plate is subjected and the support conditions provided. Note that the Eurocode width to thickness (c/t) ratios are based on the flat widths only of the plated elements, e.g. the flat width between the ends of the root radii for the web of an I-section, or the flat width from the end of the root radius to the tip of the flange for an outstand element, as illustrated in Figure 14.3.

Figure 14.2   Local buckling of column flange

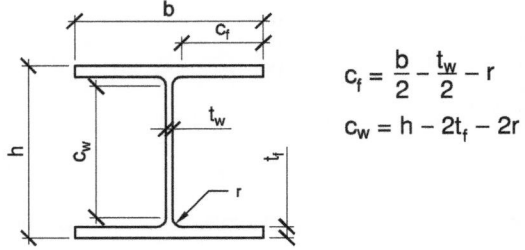

$$c_f = \frac{b}{2} - \frac{t_w}{2} - r$$

$$c_w = h - 2t_f - 2r$$

Figure 14.3   Definition of flat element widths c_f and c_w

Although the rigorous treatment of plate buckling is a complex topic,[1,2] it is possible to design safely and, in most cases, economically with no direct consideration of the subject. For example, the properties of the majority of standard hot-rolled sections have been selected such that local buckling effects are unlikely to affect significantly their load-carrying capacity when used as beams or columns. Greater care is, however, necessary when using fabricated sections, for which the proportions are under the direct control of the designer. Also, cold-formed sections are often proportioned such that local buckling effects must be accounted for, as described in Chapter 24.

14.2 Cross-sectional dimensions and moment-rotation behaviour

Figure 14.4 shows a rectangular box section subject to major axis bending and illustrates the elastic stress distributions in the flanges and webs. The plate slenderness ratios for the flanges and webs are c_f/t_f and c_w/t_w, respectively. If the beam is subject to equal and opposite end moments M, Figure 14.5 shows, in a qualitative manner, different forms of relationship between M and the corresponding rotation θ.

Assuming c_w/t_w to be such that local buckling of the webs does not occur, which of the four forms of response given in Figure 14.5 applies depends on the compression flange slenderness c_f/t_f. The four cases are defined as:

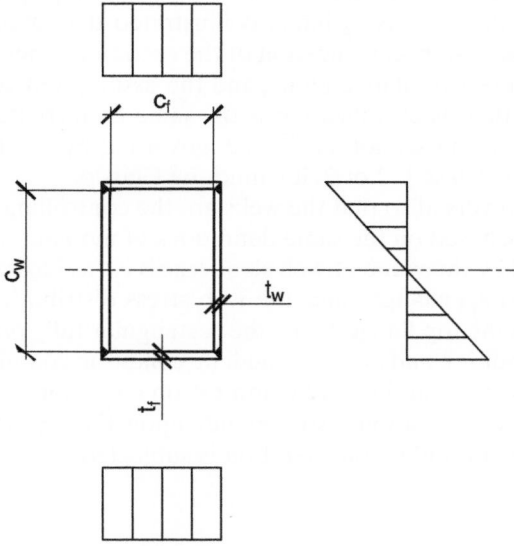

Figure 14.4 Rectangular box section in bending

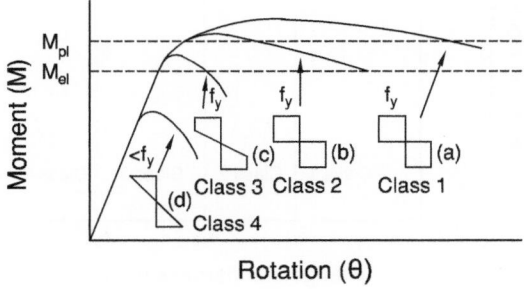

Figure 14.5 Behaviour in bending of different classes of section

1. $c_f/t_f \leq \lambda_{p1}$, the full plastic moment capacity M_{pl} is attained and maintained for large rotations and the member is suitable for plastic design – Class 1 cross-section
2. $\lambda_{p1} < c_f/t_f \leq \lambda_{p2}$, the full plastic moment capacity M_{pl} is attained but is only maintained for small rotations and the member is suitable for elastic design using its full capacity – Class 2 cross-section
3. $\lambda_{p2} < c_f/t_f \leq \lambda_{p3}$, the elastic moment capacity M_{el} (but not M_{pl}) is attained and the member is suitable for elastic design using this limited capacity – Class 3 cross-section
4. $\lambda_{p3} \leq c_f/t_f$, local buckling limits moment capacity to less than M_{el} – Class 4 (slender) cross-section.

The relationship between moment capacity M_c and compression flange slenderness c_f/t_f indicating the various λ_p limits is illustrated diagrammatically in Figure 14.6. In the above discussion, classification of the section has been based on considerations of the compression flange alone, and the assumption concerning the web slenderness c_w/t_w is that its classification is the same as, or better than, that of the flange. For example, if the section is Class 3, governed by the flange proportions, then the web must be Class 1, 2 or 3; it cannot be Class 4.

If the situation is reversed so that the webs are the controlling elements, then the same four categories, based on the same definitions of moment–rotation behaviour, are now determined by the value of web slenderness c_w/t_w. However, the governing values of λ_{p1}, λ_{p2} and λ_{p3} change since the web stress distribution differs from the pure compression in the top flange. Since the rectangular fully plastic condition, the triangular elastic condition and any intermediate condition contain less compression (i.e. they are less severe than the pure compression condition), the values of λ_p are larger. Thus section classification also depends upon the type of stress system to which the plate element under consideration is subjected.

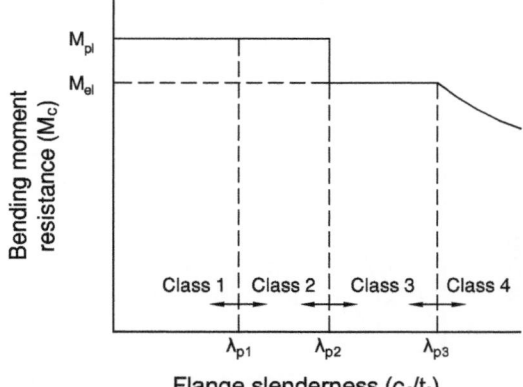

Figure 14.6 Bending moment resistance as a function of flange slenderness

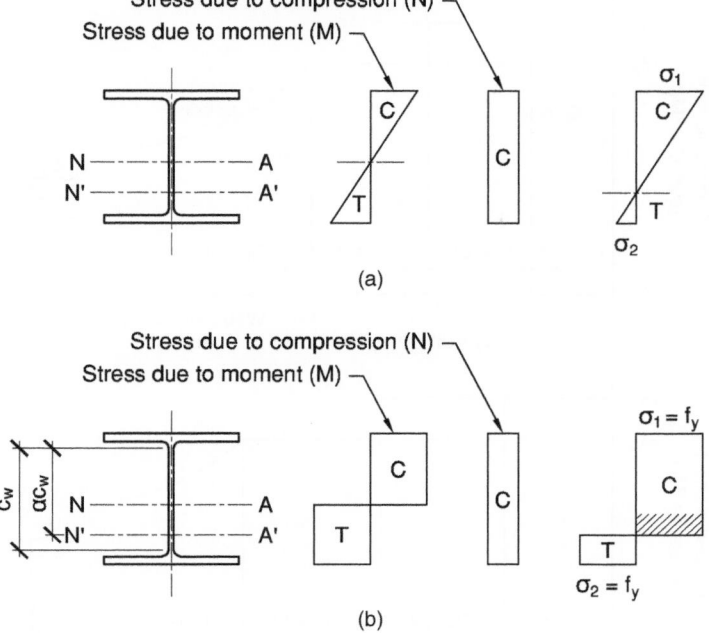

Figure 14.7 Stress distributions in webs of symmetrical sections subject to combined bending and compression. (a) Class 3, elastic stress distribution. (b) Class 1 or 2, plastic stress distribution

If, in addition to the moment M, an axial compression N is applied to the member, then for elastic behaviour the pattern of stress in the web is of the form shown in Figure 14.7(a). The values of the end stresses σ_1 and σ_2 are dependent on the ratio N/M. When the axial load is high and the bending moment is low, $\sigma_2 \approx \sigma_1$, while, when the bending moment is high and the axial load is low, $\sigma_2 \approx -\sigma_1$. As would be expected, the appropriate λ_p limits will be somewhere between the values for pure compression and pure bending, approaching the former when $\sigma_2 \approx \sigma_1$, and the latter when $\sigma_2 \approx -\sigma_1$. A qualitative indication of this, assuming elastic material behaviour, is given in Figure 14.8, which shows M_c as a function of c_w/t_w for three different σ_2/σ_1 ratios including pure compression, $\sigma_2/\sigma_1 = 0$ and pure bending. If the value of c_w/t_w is sufficiently small that the web may be classified as Class 1 or 2, then the stress distribution will adopt the alternative plastic arrangement of Figure 14.7(b), where α is the proportion of the web in compression.

For a plate element in a member which is subject to pure compression, the load-carrying capacity is not affected beyond first yield by the degree of deformation since there is no marked change in the stress distribution, i.e. it remains at f_y. This is not the case for beams, in which increasing deformation brings about increasing plastification of the web and hence greater moment capacity, as illustrated in Figure 14.4. For pure compression, the Class 1 and 2 classifications do not therefore have

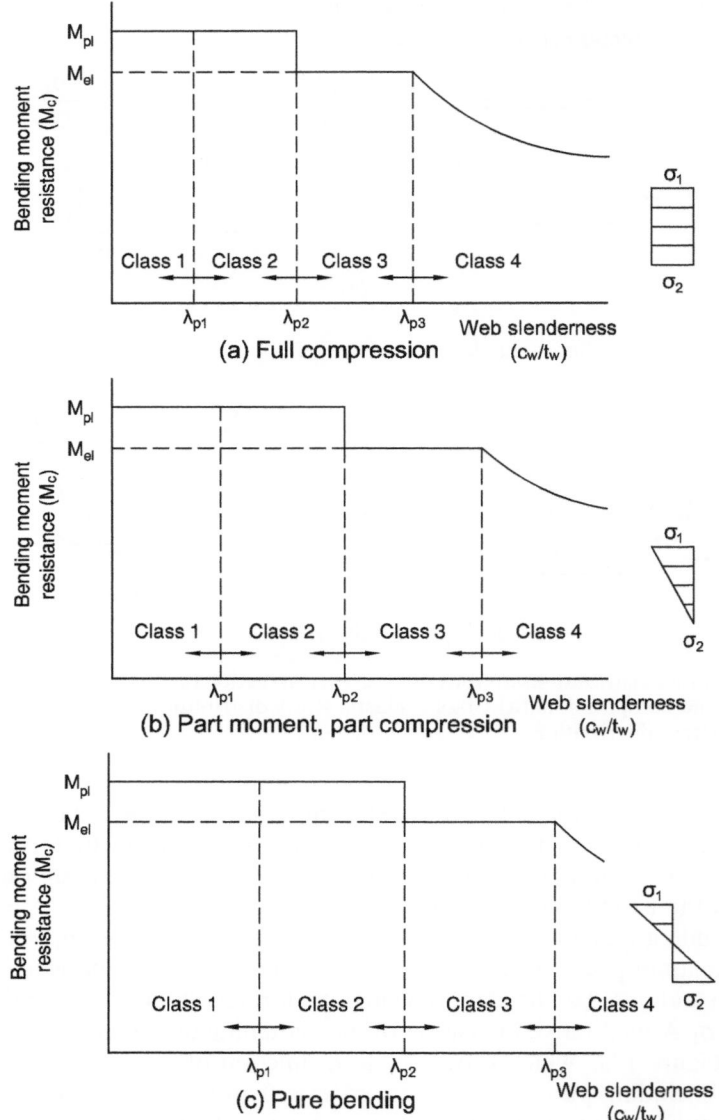

Figure 14.8 Bending moment resistance as a function of c_w/t_w for different web stress patterns:(a) full compression; (b) part moment, part compression; (c) pure bending

any particular significance; the only decision required is whether or not the member is Class 4 (slender).

In the introduction to this chapter, several other factors which affect local buckling are listed. These have a corresponding influence on the λ_p limits. As an example, the flanges of an I-section receive support along one longitudinal edge only, with the result that their local buckling resistance is less than that of the flange of a box section (in which both longitudinal edges are supported), and lower λ_p limits are defined. Note that in Eurocode 3 no distinction in made in cross-section classification between rolled sections and welded sections.

In the case of plate girders, efficient design often leads to deep webs of slender proportions, which may well be susceptible to local buckling, shear buckling or both. Design procedures for such sections are provided in BS EN 1993-1-5 and are discussed in Chapter 18.

14.3 Effect of moment-rotation behaviour on approach to design and analysis

The types of member present in a structure must be compatible with the method employed for analysis and design. This is particularly important in the context of section classification.

Take the most restrictive case first. For a plastically designed structure, in which plastic hinge action in the members is being relied upon as the means to obtain the required load-carrying capacity, only Class 1 sections are admissible. Sections which contain any plate elements that do not meet the required λ_{pl} limit for the stress condition present are therefore unsuitable. This restriction could be relaxed for those members in a plastically-designed structure that are not required to participate in plastic hinge action (i.e. members other than those in which the plastic hinges correspond to the collapse mechanism form) – see Clause 5.6 of BS EN 1993-1-1. However, such an approach could be considered unsound on the basis of the effects of over-strength material, changes in the elastic pattern of moments due to settlement or lack of fit, and so on.

When elastic design is being used, in the sense that an elastically determined set of member forces and moments forms the basis for member selection, any of Class 1, Class 2 or Class 3 sections may be used, provided member resistances are properly determined. This point is discussed more fully in Chapters 16–20, which deal with different types of member. As a simple illustration, however, for members subject to pure bending the available cross-section moment resistance M_c should be taken as M_{pl} for Class 1 and 2 sections or M_{el} for Class 3 sections. If Class 4 (slender) sections are being used, the loss of effectiveness due to local buckling will reduce not just their strength but also their stiffness. Moreover, reductions in stiffness are dependent upon load level, becoming greater as stresses increase sufficiently to cause local buckling effects to become significant. This is of most importance for cold-formed sections, which are covered in Chapter 24.

In practice, the designer, having decided upon the design approach (essentially either elastic or plastic), should check section classification using whatever design aids are available. Since most hot-rolled sections are at least Class 2 in both S275 and S355 steel, this will normally be a relatively trivial task. When using cold-formed sections, which will often be Class 4, sensible use of manufacturers' literature will often eliminate much of the actual calculation. Greater care is required when using

Table 14.1   Extract from table of section classification limits (BS EN 1993-1-1)

Type of element and loading	Class of section		
	Class 1 (λ_{p1})	Class 2 (λ_{p2})	Class 3 (λ_{p3})
Outstand flange in pure compression	$c_f/t_f \leq 9\varepsilon$	$c_f/t_f \leq 10\varepsilon$	$c_f/t_f \leq 14\varepsilon$
Internal element in pure compression	$c_f/t_f \leq 33\varepsilon$	$c_f/t_f \leq 38\varepsilon$	$c_f/t_f \leq 42\varepsilon$
Web with neutral axis at mid-depth	$c_w/t_w \leq 72\varepsilon$	$c_w/t_w \leq 83\varepsilon$	$c_w/t_w \leq 124\varepsilon$
Web, generally	when $\alpha > 0.5$: $c_w/t_w \leq 396\varepsilon/(13\alpha-1)$ when $\alpha \leq 0.5$: $c_w/t_w \leq 36\varepsilon/\alpha$	when $\alpha > 0.5$: $c_w/t_w \leq 456\varepsilon/(13\alpha-1)$ when $\alpha \leq 0.5$: $c_w/t_w \leq 41.5\varepsilon/\alpha$	when $\psi > -1$: $c_w/t_w \leq 42\varepsilon/(0.67+0.33\psi)$ when $\psi \leq -1$: $c_w/t_w \leq 62\varepsilon(1-\psi)(-\psi)^{0.5}$

where $\psi = \sigma_2/\sigma_1$ is the ratio of end stresses in the case of an elastic stress distribution, and α is the compressed portion of the web in the case of a plastic stress distribution.

sections fabricated from plate, for which the freedom to select dimensions and thus c_f/t_f and c_w/t_w ratios means that any class is possible.

14.4 Classification table

Part of a typical classification table, extracted from BS EN 1993-1-1, is given in Table 14.1. Values of λ_{p1}, λ_{p2}, and λ_{p3} for outstand flanges, defined as plates supported along one longitudinal edge, under pure compression, and internal elements or webs, defined as plates supported along both longitudinal edges, under pure compression, pure bending and combined compression and bending are listed. Note that the slenderness limits for combined compression and bending reduce to the pure compression limit in the case where $\alpha = 1$ or $\psi = 1$ and to the pure bending limit in the case where $\alpha = 0.5$ or $\psi = -1$.

14.5 Economic factors

When design is restricted to a choice of suitable standard hot-rolled sections, local buckling is not normally a major consideration. For plastically designed structures, only Class 1 sections are suitable where plastic hinges may occur. Thus the designer's choice is slightly restricted, although no UBs and only 3 UCs in S275 steel and 1 UB and 6 UCs in S355 steel are outside the Class 1 limits of BS EN 1993-1-1 when used in pure bending. Although considerably more sections are unsatisfactory if their webs are subject to high compression, the number of sections barred from use in plastically designed portal frames is, in practice, extremely small. Similarly, for elastic design, no UB is other than Class 3 or better, provided it is not required to carry high compression in the web, while all UCs are at least Class 3 even when carrying their full squash load.

The designer should check the class of any trial section at an early stage. This can be done most efficiently using information of the type given in Reference 3. For webs under combined compression and bending, the first check should be for pure compression as this is the most severe case. Provided the section is satisfactory, no additional checks are required; if it does not meet the required limit, a decision on whether it is likely to do so under the less severe combined load case must be made.

The economic use of cold-formed sections, including profiled sheeting of the type used as decking and cladding, often requires that sections are of slender proportions. The forming process is exploited to provide carefully proportioned shapes, typically containing Class 4 plate elements, and often with intermediate or edge stiffening. Since cold-formed sections are proprietary products, manufacturers normally provide design literature in which member capacities which allow for the presence of slender plate elements are listed. If rigorous calculations are, however, required, then BS EN 1993-1-3 and BS EN 1993-1-5 contain the necessary procedures, which are discussed in Chapter 24.

When using fabricated sections (see Chapter 18) the opportunity exists for the designer to optimise on the use of material. This leads to a choice between three courses of action:

1. Eliminate all considerations of local buckling, by ensuring that the width-to-thickness ratios of every plate element are sufficiently small.
2. Determine section capacities allowing for reductions due to local buckling when employing higher width-to-thickness ratios, where the relevant slenderness limits are exceeded.
3. Use stiffeners to reduce plate proportions, either to the extent that local buckling is eliminated completely, or until the desired resistance is achieved with the less severe reductions for local buckling.

Effectively only the first of these is available if plastic design is being used. For elastic design when the third approach is being employed and the sections are Class 4, then calculations inevitably are more involved as even the determination of basic cross-sectional capacities requires allowances for local buckling effects through the use of concepts such as the effective width technique,[1] which is described in Chapter 24.

References to Chapter 14

1. Bulson P.S. (1970) *The Stability of Flat Plates*. London, Chatto and Windus.
2. Trahair N.S., Bradford M.A., Nethercot D.A. and Gardner L. (2008) *The Behaviour and Design of Steel Structures to EC3*, 4th edn. London, Taylor and Francis.
3. The Steel Construction Institute and the British Constructional Steel Association (2009) *Steelwork Building: Design Data, In Accordance with Eurocodes and the UK National Annexes*. SCI Publication No. P363. Ascot, SCI, and London, BCSA.

Chapter 15
Tension members

DAVID NETHERCOT and LEROY GARDNER

15.1 Introduction

A tension member (or tie) transmits a direct tensile force between two points in a structure and is, theoretically, the simplest and most efficient structural element. In many cases, this efficiency is seriously impaired by the end connections required to join tension members to other members in the structure. In some situations (for example, in cross-braced panels), the load in the member reverses, usually by the action of wind and then the member must also act as a strut. Where the load can reverse, the designer often permits the member to buckle, with the load then being taken up by another member.

15.2 Types of tension member

The main types of tension member, their applications and behaviour, are:

- Open and closed single-rolled sections such as angles, tees, channels and structural hollow sections. These are the main sections used for tension members in light trusses and lattice girders for bracing.
- Compound sections consisting of double angles or channels. At least one axis of symmetry is present and so the eccentricity in the end connection can be minimised. When angles or other shapes are used in this fashion, they should be interconnected at intervals to prevent vibration, especially when moving loads are present.
- Heavy rolled sections and heavy compound sections of built up H- and box sections. The built-up sections are tied together either at intervals (batten plates) or continuously (lacing or perforated cover plates). Batten plates or lacing do not add any load-carrying capacity to the member but they do serve to provide rigidity and to distribute the load among the main elements. Perforated plates can be considered as part of the tension member.

Steel Designers' Manual, Seventh Edition. Edited by Buick Davison and Graham W. Owens.
© 2012 Steel Construction Institute. Published 2012 by Blackwell Publishing Ltd.

- Bars and flats. In the sizes generally used, the stiffness of these members is very low; they may sag under their own weight or that of construction workers. Their small cross-sectional dimensions also mean high slenderness values and, as a consequence, they may tend to flutter under wind loads or vibrate under moving loads.
- Ropes and cables. Further discussion on these types of tension members is included in Section 15.7 and Chapter 7, Section 7.3.

The main types of tension members are shown in Figure 15.1.
Typical uses of tension members are:

- tension chords and internal ties in trusses and lattice girders in buildings and bridges
- bracing members in buildings
- main cables and deck suspension cables in cable-stayed and suspension bridges
- hangers in suspended structures.

Typical uses of tension members in buildings and bridges are shown in Figure 15.2.

15.3 Design for axial tension

Rolled sections behave similarly to tensile test specimens under direct axial tension (Figure 15.3).

For a straight member subject to axial tensile force, N:

Tensile stress $\quad f_t = \dfrac{N}{A}$

Elongation $\quad \delta_L = \dfrac{NL}{EA}$ (in the linear elastic range)

Load at yield $\quad N_{pl} = Af_y =$ load at failure neglecting strain hardening

where:

A is the cross-sectional area of the tension member
L is the member length
E is the Young's modulus
f_y is the material yield strength.

Typical stress-strain curves for structural steel and wire rope are shown in Figure 15.3.

The design of axially loaded tension members is given in Clause 6.2.3 of BS EN 1993-1-1. The design tension resistance, $N_{t,Rd}$, is taken as the lesser of the yielding (plastic) resistance of the gross cross-section, $N_{pl,Rd}$ (to prevent excessive deformation of the member) and the ultimate fracture resistance of the net cross-section, $N_{u,Rd}$, where:

$$N_{pl,Rd} = \frac{Af_y}{\gamma_{M0}}$$

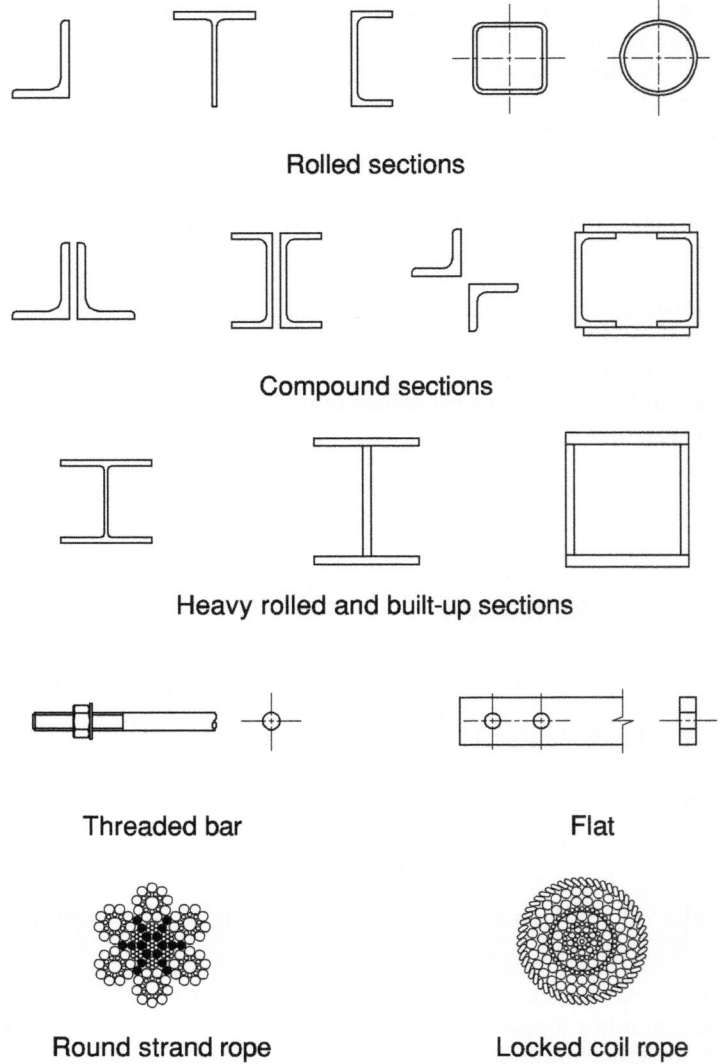

Rolled sections

Compound sections

Heavy rolled and built-up sections

Threaded bar   Flat

Round strand rope   Locked coil rope

Figure 15.1   Tension members

in which γ_{M0} is the partial factor for cross-section resistance, specified as 1.00 for both buildings and bridges, and

$$N_{u,Rd} = \frac{0.9 A_{net} f_u}{\gamma_{M2}}$$

in which A_{net} is net area of the cross-section allowing for bolt holes or other openings (defined in Clause 6.2.2.2 of BS EN 1993-1-1), f_u is the ultimate tensile strength

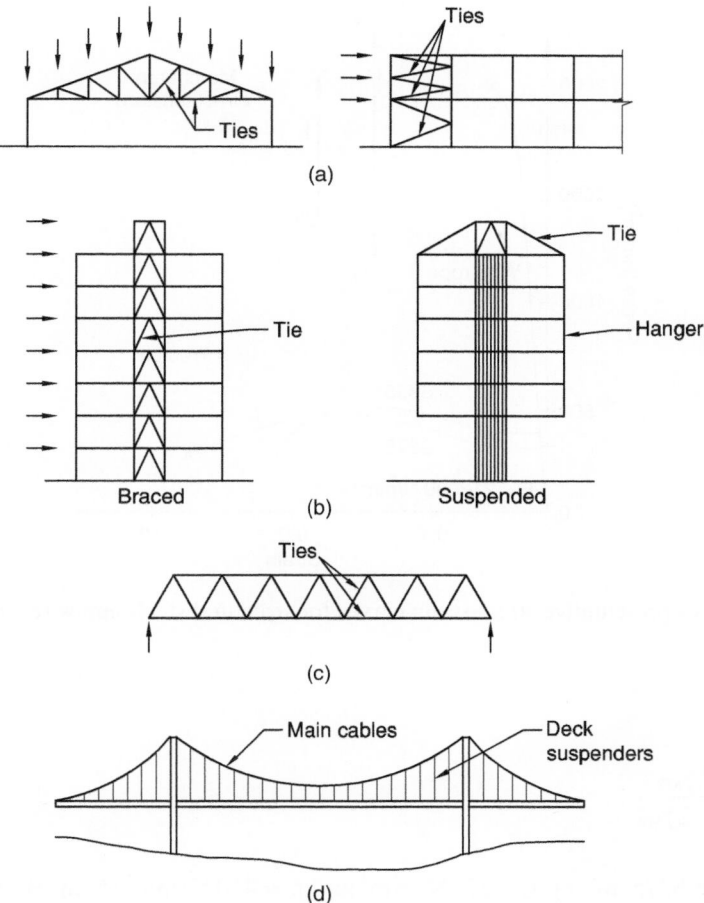

Figure 15.2   Tension members in buildings and bridges: (a) single-storey building – roof and truss bracing, (b) multi-storey building, (c) bridge truss, (d) suspension bridge

of the material and γ_{M2} is the partial factor for fracture resistance. The value of γ_{M2} recommended in both BS EN 1993-1-1 (for buildings) and BS EN 1993-2 (for bridges) is 1.25, but, while the UK National Annex to BS EN 1993-2 states that this value should be retained for bridges, the UK National Annex to BS EN 1993-1-1 allows a more relaxed value of 1.1 for buildings.

Reference should be made to Clause 6.2.2.2(4) of BS EN 1993-1-1 for members with staggered holes. To ensure ductile behaviour of the tension member, which may be required in seismic design scenarios (see BS EN 1998), yielding of the gross cross-section should occur prior to ultimate fracture of the net cross-section (i.e. $N_{pl,Rd} < N_{u,Rd}$). Combining the above two resistance equations reveals that a ductile response will be achieved provided:

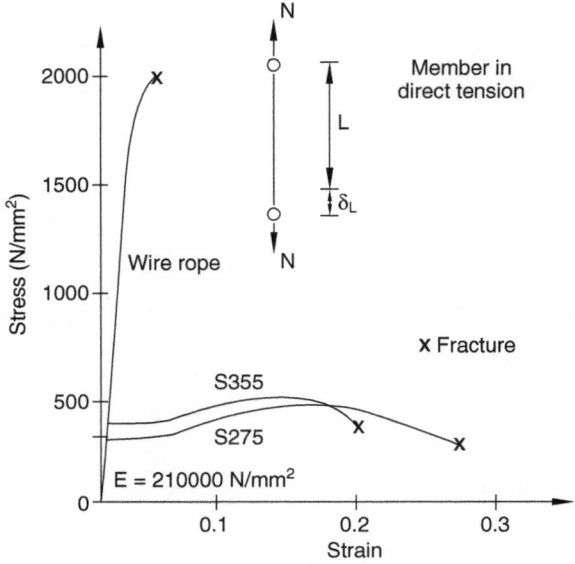

Figure 15.3　Representative stress-strain curves for structural steels and wire rope

$$\frac{A_{\text{net}}}{A} \geq \frac{f_y \gamma_{M2}}{0.9 f_u \gamma_{M0}}$$

For Grade S275, taking $f_y = 275\,\text{N}/\text{mm}^2$ and $f_u = 410\,\text{N}/\text{mm}^2$ (from BS EN 10025-2), and with $\gamma_{M0} = 1.0$ and $\gamma_{M2} = 1.1$, the minimum value of A_{net}/A that satisfies this expression is 0.82.

15.4 Combined bending and tension

Bending in tension members may arise from:

- eccentric connections
- lateral loading on members
- rigid frame action.

When a structural member is subjected to axial tension combined with bending about the y-y and z-z axes, then, assuming elastic behaviour, the maximum stress in the member is the sum of the maximum stresses from the individual actions:

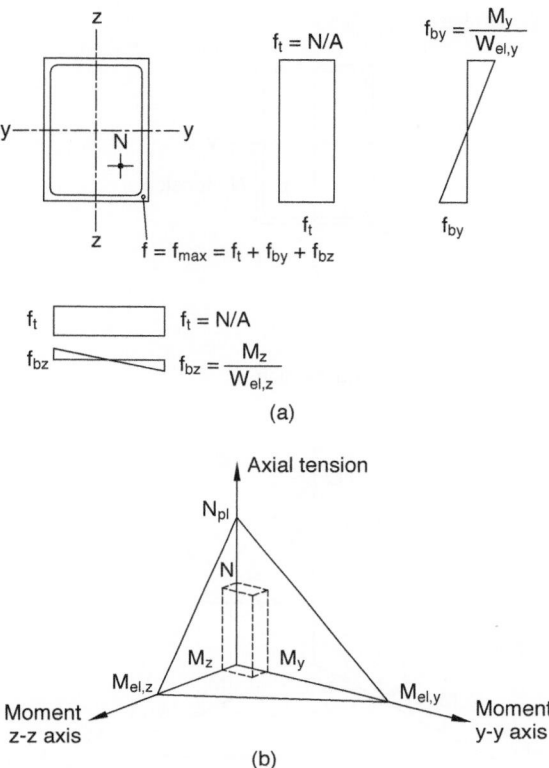

Figure 15.4 Combined bending and tension – elastic analysis: (a) stress diagram, (b) elastic interaction surface

$$f_{\max} = f_t + f_{by} + f_{bz}$$

in which the tension stress $f_t = N/A$, the maximum bending stress about the y-y axis $f_{by} = M_y/W_{el,y}$, and the maximum bending stress about the z-z axis $f_{bz} = M_z/W_{el,z}$

The separate stress diagrams are shown in Figure 15.4(a). The load values causing yield when each of the three actions act alone are:

Tension load at yield	$N_{pl} = Af_y$
Moment at yield (y-y axis)	$M_{el,y} = W_{el,y}\, f_y$
Moment at yield (z-z axis)	$M_{el,z} = W_{el,z}\, f_y$

The values N_{pl}, $M_{el,y}$ and $M_{el,z}$ form part of a three-dimensional interaction surface and any point on this surface gives a combination of N, M_y and M_z for which the maximum stress equals the yield stress. An elastic interaction surface is shown in Figure 15.4(b).

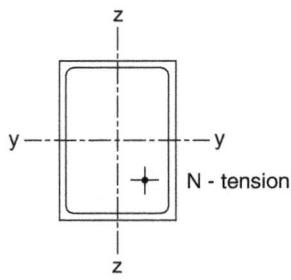

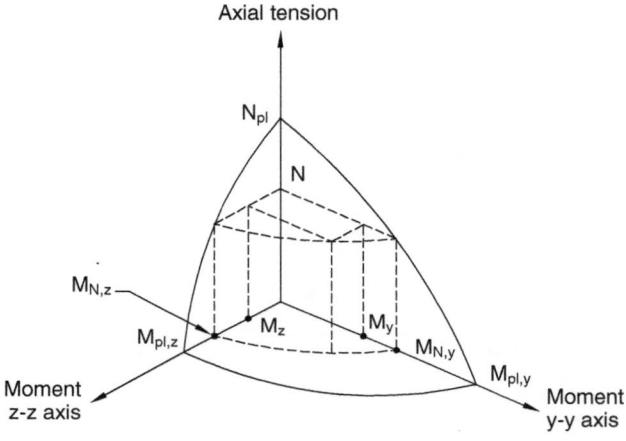

Figure 15.5   Plastic interaction surface

The maximum values for axial tension and plastic moment for the same member are:

Axial tension	$N_{pl} = A f_y$
Plastic moment (y-y axis)	$M_{pl,y} = W_{pl,y}\, f_y$
Plastic moment (z-z axis)	$M_{pl,z} = W_{pl,z}\, f_y$

The difference between the elastic and plastic interaction surfaces represents the additional design strength available if plasticity is taken into account; a convex plastic interaction surface is shown in Figure 15.5.

For the design of tension members with moments, BS EN 1993-1-1 provides a conservative linear interaction formula in Clause 6.2.1(7) and a more accurate approach in Clause 6.2.9. The linear interaction is given as:

$$\frac{N_{Ed}}{N_{Rd}} + \frac{M_{y,Ed}}{M_{y,Rd}} + \frac{M_{z,Ed}}{M_{z,Rd}} \leq 1$$

in which N_{Ed}, $M_{y,Ed}$ and $M_{z,Ed}$ are the applied axial tension, major axis bending moment and minor axis bending moment, respectively. N_{Rd}, $M_{y,Rd}$ and $M_{z,Rd}$ are the corresponding resistances. This formula applies to Class 1, 2 and 3 cross-sections, with the bending resistance of the Class 1 and 2 sections based on the plastic moment capacity ($W_{pl}f_y$) and the bending resistance of Class 3 sections based on the elastic moment capacity ($W_{el}f_y$).

Greater economy may be achieved for Class 1 and 2 cross-sections by following the design rules of Clause 6.2.9 of BS EN 1993-1-1. For bending in a single plane, Clause 6.2.9(4) defines, for doubly symmetric I- and H-sections, axial load levels below which there is no reduction to the full plastic moment capacity. For higher levels of axial load, the bending moment may not exceed a reduced bending moment resistance. Formulae for determining reduced bending moment resistances for I- and H-sections and hollow sections about the major and minor axes (denoted $M_{N,y,Rd}$ and $M_{N,z,Rd}$ respectively, with 'N' signifying the presence of axial load) are provided in Clause 6.2.9(5). For bi-axial bending, the following interaction criterion is given in Clause 6.2.9(6):

$$\left(\frac{M_{y,Ed}}{M_{N,y,Rd}}\right)^{\alpha} + \left(\frac{M_{z,Ed}}{M_{N,z,Rd}}\right)^{\beta} \leq 1$$

where the constants α and β have values of 2 for circular hollow sections (CHS), $\alpha = 2$ and $\beta = 5n$ (but $\beta \geq 1.0$) where $n = N_{Ed}/N_{pl,Rd}$ for I- and H-sections, and $\alpha = \beta = (1.66/1-1.13\,n^2) \leq 6$ for rectangular hollow sections (RHS), and $\alpha = \beta = 1$ for all other sections.

BS EN 1993-1-1 does not provide any specific guidance for assessing the lateral torsional buckling resistance of members in bending with co-existent tension. The member may be conservatively checked under bending alone, ignoring the stabilising effect of the tension.

15.5 Eccentricity of end connections

Simplified design rules are given in BS EN 1993-1-8 for the effects of combined tension and bending caused solely by the eccentric load introduced into the member by the end connection.

The rules, presented in Clause 3.10.3 of BS EN 1993-1-8, are limited in scope and cover only single angles connected by a single row of bolts in one leg, while for other section types, reference should be made to Section 15.4. To allow for the effects of eccentricity in single angles connected by one leg, the members are designed for axial tension only, but with an effective (reduced) net cross-sectional area in place of the actual net cross-sectional area. The design expressions given in Clause 3.10.3(2) apply to equal angles and unequal angles connected by the longer leg. For unequal angles connected by the shorter leg, design is based on a fictitious equal angle, which has both legs equal to the shorter leg length of the real unequal angle. In the case of single angles with welded end connections, the effective area should be taken as the gross area for equal angles and unequal angles connected by their longer leg (see Clause 4.13(2) of BS EN 1993-1-8). For unequal angles connected by their shorter leg, the effective area should be taken as the gross area of a fictitious equal angle having leg lengths equal to the shorter leg of the real unequal angle.

15.6 Other considerations

15.6.1 Serviceability and corrosion

Ropes and bars are not normally used in building construction because they lack stiffness but they have been used in some cases as hangers in suspended buildings. Very light, thin tension members are susceptible to excessive elongation under direct load as well as lateral deflection under self-weight and lateral loads. Special problems may arise where the members are subjected to vibration or conditions leading to failure by fatigue, such as can occur in bridge deck hangers. Damage through corrosion is undesirable, and adequate protective measures must be adopted. All these factors can make the design of tension members a complicated process in some cases.

The light rolled sections used for tension members in trusses and for bracing are easily damaged during transport. It is customary to specify a minimum size for such members to prevent this from happening. For angle ties, a general rule is to make the leg length not less than one-sixtieth of the member length.

Ties subject to load reversal, e.g. under the action of wind, could buckle. Where such buckling is dangerous (e.g. due to lack of alternative load paths) or merely unsightly, tension members should also be checked as a compression member (see 16.3).

15.6.2 Stress concentration factors and fatigue

In cases of geometrical discontinuity, such as a change of cross-section or an aperture, the resulting stress concentrations may be determined either by numerical analysis or by the use of stress concentration factors. Stress concentrations are not usually important in ductile materials but can be the cause of failure due to fatigue or brittle fracture in certain conditions. A hole in a flat member increases the stresses locally on the net section by a factor which depends on the ratio of hole diameter to net plate width.

When designing against fatigue, it is convenient to consider three levels of stress concentration:

- stress concentrations from structural action due to the difference between the actual structural behaviour and the static model chosen
- macroscopic stress concentrations due to large scale geometric interruptions to stress flow
- microscopic and local geometry stress concentrations due to imperfections within the weld or the heat affected zone.

For the fatigue assessment of the design of welded details and connections, reference should be made to BS EN 1993-1-9. However, for non-standard situations, it may be necessary to determine the stress concentration factor directly from a numerical analysis or from an experimental model.

In many cases, the detail under consideration may not fit neatly into one of the classes. On site, the actual stress range for a particular loading occurrence is likely to be strongly influenced by detailed fit of the joint and overall fit of the structure. Therefore, the overall form should be such that load paths are as smooth as possible and unintended load paths should be avoided, particularly where fit could significantly influence behaviour. Discontinuities must be avoided by tapering and appropriate choice of radii.

15.6.3 Fabrication and erection

The behaviour of tension members in service depends on the fabrication tolerances and the erection sequence and procedure. Care must be taken to ensure that no tension member is slack after erection, so that they are all immediately active in resisting service loads.

Screwed ends and turnbuckles can be used to adjust lengths of bars and cables after they are in place. Bracing members fabricated from rolled sections should be installed and properly tightened before other connections and column-base plates are bolted up to bring the structure into line and square. Bracing members are usually specified slightly shorter than the exact length to avoid sagging and allow them to be immediately effective.

Complete or partial trial shop assembly is often specified in heavy industrial trusses and bridge members to ensure that the fabrication is accurate and that erection is free from problems.

15.7 Cables

Cables are not commonly used in buildings, but Chapter 7, Section 7.3 presents some appropriate practical examples.

15.7.1 Composition

A cable may be composed of one or more structural ropes, structural strands, locked coil strands or parallel wire strands. A strand, with the exception of a parallel wire strand, is an assembly of wires formed vertically around a central wire in one or more symmetrical layers. A strand may be used either as an individual load-carrying member, where radius of curvature is not a major requirement, or as a component in the manufacture of structural rope.

A rope is composed of a plurality of strands vertically laid around a core. In contrast to the strand, a rope provides increased curvature capability and is used where curvature of the cable becomes an important consideration. The significant differences between strand and rope are as follows:

- At equal sizes, a rope has lower breaking strength than a strand.
- The modulus of elasticity of a rope is lower than that of a strand.
- A rope has more curvature capability than a strand.
- The wires in a rope are smaller than those in a strand of the same diameter; consequently, a rope for a given size coating is less corrosion-resistant because of the thinner coating on the smaller diameter wires.

15.7.2 Application

Cables used in structural applications fall into the following categories:

- parallel-bar cables
- parallel-wire cables
- stranded cables (see Figure 15.1)
- locked-coil cables (see Figure 15.1).

The final choice depends on the properties required by the designer, such as modulus of elasticity, ultimate tensile strength and durability. Other criteria include economy and structural detailing.

15.7.3 Parallel-bar cables

Parallel-bar cables are formed of steel rods or bars, parallel to each other in metal ducts, kept in position by polyethylene spacers. The process of tensioning the bar or rods individually is simplified by the capability of the bars to slide longitudinally. Cement grout, injected after erection, makes sure that the duct plays its part in resisting the stresses due to live loads.

Transportation in reels is only possible for the smaller diameters while for the larger sizes, delivery is made in straight bars 15.0–20.0 m in length. Continuity of the bars has to be provided by the use of couplers, which considerably reduces the fatigue strength of the stay.

The use of mild steel necessitates larger sections than when using high-strength wires or strands. This leads to a reduction in the stress variation and thus lessens the risk of fatigue failure.

15.7.4 Parallel-wire cables

Parallel wires are used for cable-stayed bridges and pre-stressed concrete. Their fatigue strength is satisfactory, mainly because of their good mechanical properties.

15.7.5 Corrosion protection

Wires in the cables should be protected from corrosion. The most effective protection is obtained by hot galvanising by steeping or immersing the wires in a bath of melted zinc, automatically controlled to void overheating. A wire is described as terminally galvanised or galvanised re-drawn, depending on whether the operation has taken place after drawing or in between two wire drawings prior to the wire being brought to the required diameter. For reinforcing bars and cables, the first method is generally adopted. A quantity of zinc in the range of 250–330 g/m^2 is deposited, providing a protective coating 25–45 μm thick.

15.7.6 Coating

The coating process used currently for locked-coil cables consists of coating the bare wires with an anti-corrosion product with a good bond and long service life. The various substances used generally have a high dropping point so as not to run back towards the lower anchorages. They are usually high viscosity resins or oil-based grease, paraffin or chemical compounds.

15.7.7 Protection of anchorages

The details of the connections between the ducts and the anchorages must prevent any inflow or accumulation of water. The actual details depend on the type of anchorages used, on the protective systems for the cables, and on their slope. There are different arrangements intended to ensure watertightness of vital zones.

15.7.8 Protection against accidents

Cables should be protected against various risks of accident, such as vehicle impact, fire, explosion and vandalism. Measures to be taken may be based on the following:

- protection of the lower part of the stay, over a height of about 2.0 m, by a steel tube fixed into the deck and fixed into the duct; the tube dimensions (thickness and diameter) must be adequate
- strength of the lower anchorage against vehicle impact
- replacement of protective elements is possible without affecting the cables themselves and, as far as possible, without interrupting traffic.

Further reading for Chapter 15

Adams, P.F., Krentz, H.A. and Kulak, G.L. (1973) *Canadian Structural Steel Design*. Ontario, Canadian Institute of Steel Construction.

Dowling, P.J., Knowles, P. and Owens, G.W. (1988) *Structural Steel Design*. London, Butterworths.

Gardner, L. and Nethercot, D.A. (2005) *Designers' Guide to EN 1993-1-1: Eurocode 3: Design of Steel Structures*. London, Thomas Telford Publishing.

Horne, M.R. (1971) *Plastic Theory of Structures*, 1st edn. Walton-on-Thames, Nelson.

Owens, G.W. and Cheal, B.D. (1989) *Structural Steelwork Connections*. London, Butterworths.

Timoshenko, S.P. and Goodier, J.N. (1970) *Theory of Elasticity*. London, McGraw-Hill.

Toy, M. (1995) *Tensile Structures*. London, Academy Editions.

Trahair, N.S., Bradford, M.A., Nethercot, D.A. and Gardner, L. (2008) *The Behaviour and Design of Steel Structures to EC3*, 4th edn. London and New York, Taylor and Francis.

Troitsky, M.S. (1988) *Cable-Stayed Bridges*, 2nd edn. London, BSP Professional Books.

Vandenberg, M. (1988) *Cable Nets*. London, Academy Editions.

Chapter 16
Columns and struts

DAVID NETHERCOT and LEROY GARDNER

16.1 Introduction

Members subject to compression are typically referred to as either columns when they are orientated vertically or struts more generally. Compression members represent one of the basic types of load-carrying component and may be found, for example, as the vertical elements in building frames, in the compression chords of trusses or in any position in a space frame.

In many practical situations, columns or struts are not subject solely to compression but, depending upon the exact nature of the load path through the structure, are also required to resist some degree of bending. For example, a corner column in a building is normally bent about both axes by the action of the beam loads, a strut in a space frame is not necessarily loaded concentrically and the compression chord of a roof truss may also be required to carry some lateral loads. Thus many compression members are actually designed for combined loading as beam-columns. Notwithstanding this, the ability to determine the pure compressive resistance of members is of fundamental importance in design, both for the struts loaded only in compression and as one component in the interaction type of approach normally used for beam-column design.

The most significant factor that must be considered in the design of struts is buckling. Depending on the type of member and the particular application under consideration, this may take several forms. One of these, local buckling of individual plate elements in compression, has already been considered in Chapter 14. Much of the present chapter is devoted to the consideration of the way in which member buckling is handled in strut design.

16.2 Common types of member

Various types of steel section may be used as struts to resist compressive loads; Figure 16.1 illustrates a number of them. Practical considerations such as the

Steel Designers' Manual, Seventh Edition. Edited by Buick Davison and Graham W. Owens.
© 2012 Steel Construction Institute. Published 2012 by Blackwell Publishing Ltd.

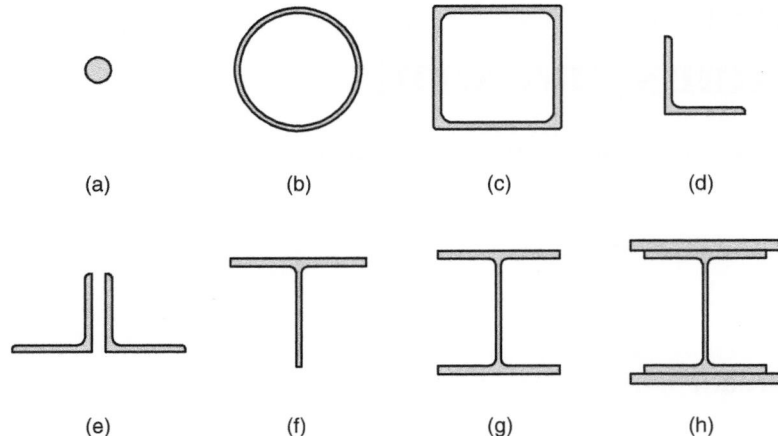

(a)   (b)   (c)   (d)

(e)   (f)   (g)   (h)

Figure 16.1   Typical column cross-sections

methods to be employed for making connections often influence the choice, especially for light members. Although closed sections such as tubes are theoretically the most efficient, it is normally much easier to make simple site connections, using the minimum of skilled labour or special equipment, to open sections. Typical section choices for a range of applications are as follows:

1. light trusses and bracing – angles (including compound angles back to back) and tees
2. larger trusses – circular hollow sections, rectangular hollow sections, compound sections and universal columns
3. frames – universal columns, fabricated sections e.g. reinforced UCs
4. bridges – box columns
5. power stations – stiffened box columns.

16.3 Design considerations

The most important property of a compression member, as far as the determination of its load-carrying capacity is concerned, is its overall member slenderness. In Eurocode 3, member slenderness or non-dimensional slenderness $\bar{\lambda}$ is defined as:

$$\bar{\lambda} = \sqrt{\frac{Af_y}{N_{cr}}}$$

where A is the cross-sectional area, f_y is the material yield strength and N_{cr} is the elastic buckling load. For the most common mode of member buckling in compression – flexural buckling, which is typically characterised by bending of the member about the weaker principal axis – N_{cr} is given by:

$$N_{cr} = \frac{\pi^2 EI}{L_{cr}^2}$$

in which E is the Young's modulus, I is the second moment of area about the axis of buckling and L_{cr} is the effective length of the column – see Section 16.7. Note that Eurocode 3 generally refers to an 'effective length' as a 'buckling length'. Alternatively, member slenderness may be expressed in a normalised form as:

$$\bar{\lambda} = \frac{\lambda}{\lambda_1}$$

where λ is the effective column length L_{cr} divided by the appropriate radius of gyration i, and λ_1 is the slenderness at which the yield load ($A f_y$) and elastic buckling load are coincident, which, for flexural buckling, may be shown to be:

$$\lambda_1 = \pi \sqrt{\frac{E}{f_y}}$$

Codes of practice such as BS 5950 (prior to the 2000 version) used to place upper limits on member slenderness so as to avoid the use of flimsy construction, i.e. to ensure that a member which will ordinarily be subject only to axial load does have some limited resistance to an accidental lateral load, does not rattle, etc. Although not explicitly stated in Eurocode 3, it remains good practice to limit the maximum slenderness of compression members for the reasons outlined above. The maximum recommended values of member slenderness are given in Table 16.1; these are presented in terms of non-dimensional member slenderness $\bar{\lambda}$ and have been derived from BS 5950.

Strut design will normally require that, once a trial member has been selected and its loading and support conditions determined, attention be given to whichever of the following checks are relevant for the particular application.

Table 16.1 Recommended maximum slenderness values $\bar{\lambda}$ for compression members

Condition	Slenderness limit for $f_y = 275\,\text{N/mm}^2$	Slenderness limit for $f_y = 355\,\text{N/mm}^2$
Members in general	2.1[a]	2.4[a]
Members resisting self-weight and wind loads only	2.9[a]	3.3[a]
Members normally in tension but subject to load reversal due to wind	4.0	4.6

[a] Check for self-weight deflection if $\bar{\lambda} > 2.1$, allow for bending effects in design if this deflection exceeds L/1000.

1. Overall flexural buckling – largely controlled by the non-dimensional member slenderness $\bar{\lambda}$ which is a function of member length, cross-sectional shape and the support conditions provided; also influenced by the type of member.
2. Local buckling – controlled by the width-to-thickness ratios of the component plate elements (see Chapter 15); with some care in the original choice of member this need not involve any actual calculation.
3. Buckling of component parts – only relevant for built-up sections such as laced and battened columns; the strength of individual parts must be checked, often by simply limiting distances between points of interconnection.
4. Torsional or torsional-flexural buckling – for cold-formed sections and in extreme cases of unusually shaped heavier open sections, the inherent low torsional stiffness of the member may make this form of buckling more critical than simple flexural buckling.

In principle, local buckling and overall buckling (flexural, torsional or torsional-flexural) should always be checked. In practice, provided cross-sections that at least meet the Class 3 limits for pure compression are used, then no local buckling check is necessary since the cross-section will be fully effective.

16.4 Cross-sectional considerations

Since the maximum attainable load-carrying capacity for any structural member is controlled by its local cross-sectional capacity (factors such as buckling may prevent this being achieved in practice), the first step in strut design must involve consideration of local buckling as it influences axial capacity. Only two classes of section are relevant for purely axially compressed members: either the section is not Class 4 (slender), in which case its full capacity, Af_y, is available, or it is Class 4 (slender) and some allowance in terms of a reduced capacity is required. The distinctions between Class 1, Class 2 and Class 3 sections, as described in Chapter 15, have no relevance when the type of member under consideration is a strut.

The general Eurocode 3 treatment for a member containing slender plate elements is to define an effective cross-section to be used in place of the gross cross-section in design calculations. Essentially, the portions of any Class 4 elements that are rendered ineffective by local buckling are removed from the cross-section, and the section properties of the remaining effective section are determined. The reduction factors ρ for the local buckling of slender plate elements in compression, illustrated in Figure 16.2, are defined in Clause 4.4 of BS EN 1993-1-5, and are discussed further in the context of cold-formed sections in Chapter 24. A slightly simpler, but generally much more conservative approach is to design the member using a reduced yield strength, with the magnitude of the reduction being dependent on the extent to which the Class 3 limits (the boundary between fully effective and slender) are exceeded. With a reduced yield strength, the slenderness limits effectively become more relaxed through the influence of the ε factor; the yield strength is reduced

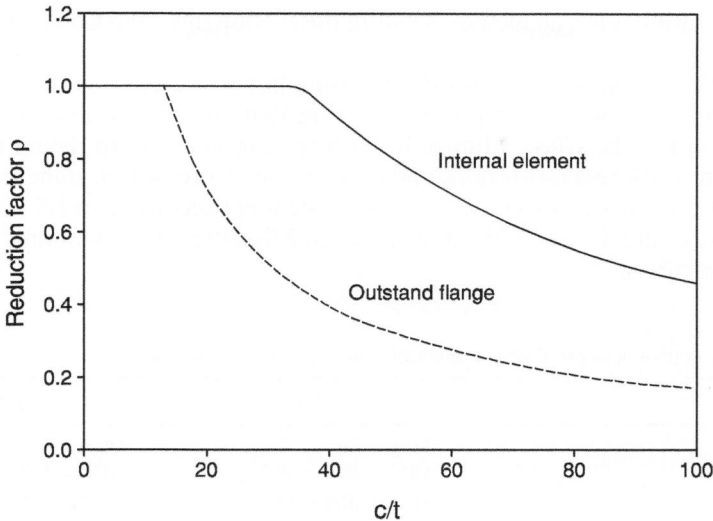

Figure 16.2   Relationship between reduction factor ρ and element width-to-thickness (c/t) ratio for pure compression with $f_y = 275\,\text{N}/\text{mm}^2$

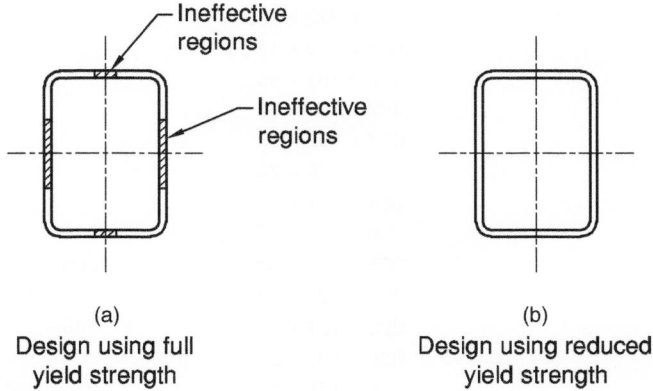

Figure 16.3   Two alternative methods for treating local buckling in Class 4 sections: (a) effective area, full yield strength; (b) full area, uniformly reduced yield strength

until all elements within the cross-section satisfy the Class 3 limits, thus becoming fully effective. Essentially, the member is assumed to be made of a steel having a lower material strength than is actually the case.

The two approaches to the design of members containing Class 4 plate elements are illustrated in Figure 16.3. In Method 1, the full yield strength is employed in conjunction with a reduced cross-sectional area. The effective section is defined as the gross section minus the ineffective portions of slender elements removed according to the reduction factor ρ. In Method 2, the yield strength of the material is reduced uniformly until all elements within the section meet the Class 3 slenderness limits.

It is worth emphasising that for design using hot-rolled sections, the majority of situations may be treated simply by ensuring that the proportions of the cross-section lie within the Class 3 limits. Reference 1, in addition to listing flange and web width-to-thickness ratios (c_f/t_f and c_w/t_w, respectively) for all standard sections, also identifies those sections that are Class 4 (slender) according to BS EN 1993-1-1 when used in either S275 or S355 steel. Table 16.2 lists these for commonly employed hot-rolled sections.

Table 16.2 Sections that are Class 4 (slender) under axial compression

Section type	S275	S355
Universal beam	All *except*	All *except*
	$1016 \times 305 \times 487$	$1016 \times 305 \times 487$
	$1016 \times 305 \times 437$	$1016 \times 305 \times 437$
	$1016 \times 305 \times 393$	
	$914 \times 419 \times 388$	
	$610 \times 305 \times 238$	$610 \times 305 \times 238$
	$610 \times 305 \times 179$	$610 \times 305 \times 179$
	$533 \times 312 \times 150$	
	$533 \times 210 \times 122$	
	$457 \times 191 \times 98$	
	$457 \times 191 \times 89$	
	$457 \times 152 \times 82$	
	$406 \times 178 \times 74$	
	$356 \times 171 \times 67$	
	$356 \times 171 \times 57$	
	$305 \times 165 \times 54$	$305 \times 165 \times 54$
	$305 \times 127 \times 48$	$305 \times 127 \times 48$
	$305 \times 127 \times 42$	$305 \times 127 \times 42$
	$305 \times 127 \times 37$	
	$254 \times 146 \times 43$	$254 \times 146 \times 43$
	$254 \times 146 \times 37$	
	$254 \times 146 \times 31$	
	$254 \times 102 \times 28$	
	$254 \times 102 \times 25$	
	$203 \times 133 \times 30$	$203 \times 133 \times 30$
	$203 \times 133 \times 25$	$203 \times 133 \times 25$
	$203 \times 102 \times 23$	$203 \times 102 \times 23$
	$178 \times 102 \times 19$	$178 \times 102 \times 19$
	$152 \times 89 \times 16$	$152 \times 89 \times 16$
	$127 \times 76 \times 13$	$127 \times 76 \times 13$
Universal column	None	None

Table 16.2 (*Continued*)

Section type	S275	S355
Hot-finished square hollow sections		400 × 400 × 10.0 350 × 350 × 8.0 300 × 300 × 6.3 and 8.0 260 × 260 × 6.3 250 × 250 × 6.3 200 × 200 × 5.0
Hot-finished rectangular hollow sections		500 × 300 × 8.0 to 12.5 500 × 200 × 8.0 to 12.5 450 × 250 × 8.0 and 10.0 400 × 300 × 8.0 and 10.0 400 × 200 × 8.0 and 10.0 400 × 150 × 5.0 to 10.0 400 × 120 × 5.0 to 10.0 350 × 250 × 5.0 to 8.3 300 × 150 × 5.0 to 8.0 300 × 250 × 5.0 to 8.0 300 × 200 × 6.3 and 8.0 300 × 150 × 8.0 300 × 100 × 8.0 260 × 140 × 5.0 and 6.3 250 × 150 × 5.0 and 6.3 200 × 120 × 5.0 200 × 100 × 5.0 160 × 80 × 4.0 150 × 125 × 4.0
Hot-finished circular hollow sections		406.4 × 6.3
Hot-finished elliptical hollow sections		500 × 250 × 10.0 to 16.0 400 × 200 × 8.0 to 12.5 300 × 150 × 8.0 and 10.0 250 × 125 × 6.3 and 8.0 200 × 100 × 5.0 and 6.3 150 × 75 × 4.0 and 5.0
Rolled steel angles	200 × 200 × 16.0 and 18.0 150 × 150 × 10.0 and 12.0 120 × 120 × 8.0 and 10.0 100 × 100 × 8.0 90 × 90 × 7.0 and 8.0 75 × 75 × 6.0 70 × 70 × 6.0 60 × 60 × 5.0 50 × 50 × 4.0 200 × 150 × 12.0 and 15.0 200 × 100 × 10.0 and 12.0 150 × 90 × 10.0 150 × 75 × 10.0 125 × 75 × 8.0 100 × 75 × 8.0 100 × 65 × 7.0 100 × 50 × 6.0 65 × 50 × 5.0	200 × 200 × 16.0 to 20.0 150 × 150 × 10.0 to 15.0 120 × 120 × 8.0 to 12.0 100 × 100 × 8.0 to 10.0 90 × 90 × 7.0 and 8.0 80 × 80 × 8.0 75 × 75 × 6.0 and 8.0 70 × 70 × 6.0 and 7.0 60 × 60 × 5.0 and 6.0 50 × 50 × 4.0 and 5.0 200 × 150 × 12.0 to 18.0 200 × 100 × 10.0 to 15.0 150 × 90 × 10.0 and 12.0 150 × 75 × 10.0 and 12.0 125 × 75 × 8.0 and 10.0 100 × 75 × 8.0 100 × 65 × 7.0 and 8.0 100 × 50 × 6.0 and 8.0 80 × 60 × 7.0 80 × 40 × 6.0 75 × 50 × 6.0 70 × 50 × 6.0 65 × 50 × 5.0 60 × 40.5.0

16.5 Column buckling resistance

The axial load-carrying capacity for a single compression member is a function of its slenderness, its material strength, cross-sectional shape and method of manufacture. Using BS EN 1993-1-1, column buckling resistance $N_{b,Rd}$ is given by Clause 6.3.1.1(3) as:

$$N_{b,Rd} = \frac{\chi A f_y}{\gamma_{M1}} \quad \text{for Class 1, 2 and 3 sections}$$

$$N_{b,Rd} = \frac{\chi A_{eff} f_y}{\gamma_{M1}} \quad \text{for Class 4 sections}$$

in which A is the cross-sectional area of the column, A_{eff} is the effective cross-sectional area required only for Class 4 sections to allow for local buckling, γ_{M1} is the partial factor for member buckling, specified in both BS EN 1993-1-1 and its UK National Annex as 1.0 and χ is the buckling reduction factor defined in Clause 6.3.1.2(1) of BS EN 1993-1-1 as:

$$\chi = \frac{1}{\Phi + \sqrt{\Phi^2 - \bar{\lambda}^2}} \leq 1.0$$

where

$$\Phi = 0.5[1 + \alpha(\bar{\lambda} - 0.2) + \bar{\lambda}^2]$$

in which $\bar{\lambda}$ is the non-dimensional member slenderness discussed in Section 16.3 and α is the imperfection factor, which can take one of five values. The five values of α, set out in Table 6.1 of BS EN 1993-1-1 and repeated in Table 16.3, correspond to the five buckling curves – a_0, a, b, c and d.

The five buckling curves are plotted in Figure 16.4, directly from the expressions given above. The buckling curves are equivalent to those provided in BS 5950-1 (2000) in tabular form in Table 24 but with one additional curve (curve a_0). These curves have resulted from a comprehensive series of full-scale tests, supported by detailed numerical studies, on a representative range of cross-sections.[2] They are often referred to as the European Column Curves.

A number of buckling curves (five in the case of Eurocode 3) are used in recognition of the fact that for the same slenderness certain types of cross-section consistently perform better than others as struts. This is due to the arrangement of the material and the influence of differing levels of geometric imperfections and resid-

Table 16.3 Imperfection factors α

Buckling curve	a_0	a	b	c	d
Imperfection factor α	0.13	0.21	0.34	0.49	0.76

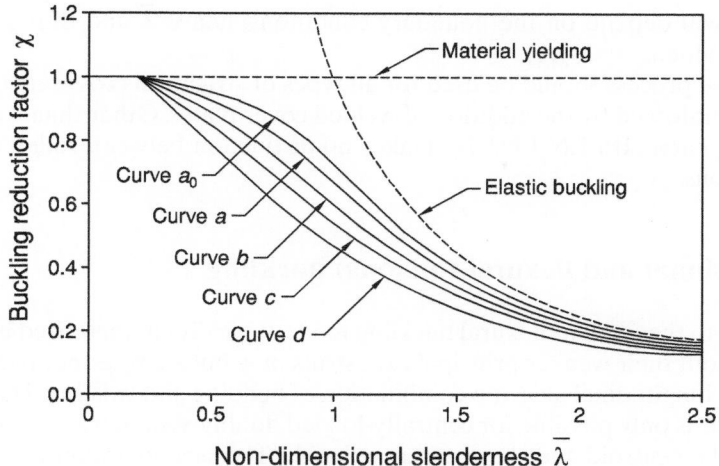

Figure 16.4   Column buckling curves of BS EN 1993-1-1

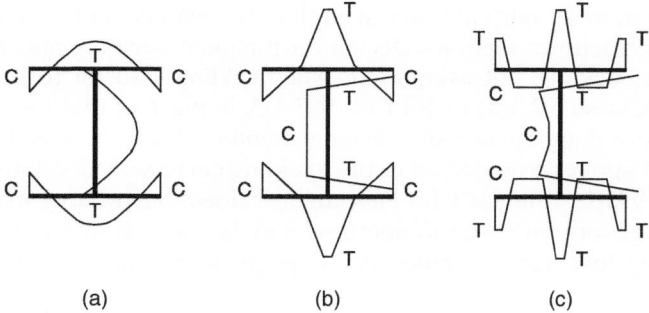

Figure 16.5   Typical residual stress patterns in (a) hot-rolled, (b) welded I-sections with mill cut plates and (c) welded I-sections with oxygen-cut plates (T = tension, C = compression)

ual stresses. Typical residual stress patterns for hot-rolled and welded I-sections are shown in Figure 16.5. The different relative buckling performance is catered for in design by using the buckling curve selection table given as Table 6.2 in BS EN 1993-1-1.

The first step in column design is therefore to consult Table 6.2 of BS EN 1993-1-1 to see which buckling curve (and hence imperfection factor α) is appropriate. For example, if the case being checked is a UC liable to buckle about its minor axis, buckling curve c should be used. Selection of a trial section fixes the radius of gyration i and the cross-sectional area A of the section; the geometrical length of the member L will be defined by the application required, while the effective (buckling) length L_{cr} will depend on the boundary conditions; hence $\bar{\lambda}$ and thus χ and $N_{b,Rd}$ may be obtained.

The above process should be used for all types of structural cross-section, including those reinforced by the addition of welded cover plates. Other than a difference in buckling curve, BS EN 1993-1-1 makes no distinction between rolled and fabricated sections.

16.6 Torsional and flexural-torsional buckling

In addition to the familiar flexural buckling mode, typically characterised by bending of struts about their weaker principal axis, struts may buckle by either pure twisting about their longitudinal axis or a combination of bending and twisting. The first type of behaviour is only possible for centrally-loaded doubly-symmetrical cross-sections for which the centroid and shear centre coincide. The second, rather more general, form of response occurs for centrally-loaded struts such as channels for which the centroid and shear centre do not coincide.

In practice, pure torsional buckling of hot-rolled structural sections is highly unlikely, with the pure flexural mode normally occurring at a lower load, unless the strut is of a somewhat unusual shape such that its torsional and warping stiffnesses are low (e.g. a cruciform section). Design for torsional buckling may be conducted as for flexural buckling, but using the elastic buckling load for torsional buckling $N_{cr,T}$, given in Clause 6.2.3(5) of BS EN 1993-1-3, in place of that for flexural buckling, when determining the non-dimensional member slenderness. A different buckling curve also applies, as specified in the buckling curve selection table – see Table 6.3 of BS EN 1993-1-3. Similarly, for unsymmetrical sections, where torsional-flexural buckling may govern, member slenderness may be calculated on the basis of the elastic buckling load for this mode $N_{cr,TF}$, as given in Clause 6.2.3(7) of BS EN 1993-1-3.

Torsional and torsional-flexural buckling are discussed further, together with distortional buckling, in Chapter 24, since these modes are of greater practical significance in the design and use of cold-formed sections. There are two primary reasons for this:

1. The torsion constant of a section I_t depends on t^3, hence the use of thin material results in the ratio of torsional to flexural stiffness being much reduced as compared with hot-rolled sections.
2. The forming process leads naturally to a preponderance of singly-symmetrical or unsymmetrical open sections as these can be produced from a single sheet.

As noted above, formulae for determining the elastic buckling loads for torsional and torsional-flexural buckling in terms of the basic geometric properties of the member are provided in BS EN 1993-1-3. Figure 16.6 shows elastic buckling curves for the various buckling modes for a typical channel section. In the case shown, the critical mode (i.e. the lowest elastic buckling mode) changes from torsional-flexural

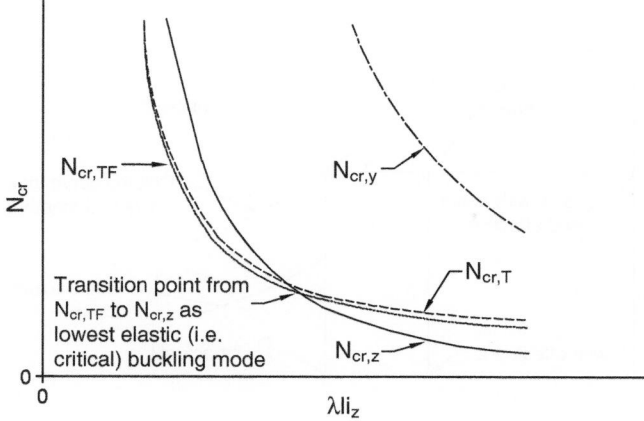

Figure 16.6   Elastic buckling curves in the flexural, torsional and torsional-flexural modes for a typical channel section

buckling, which governs at low slenderness, to minor axis flexural buckling, which governs at higher slendernesses. In general, minor axis flexural buckling will tend to be critical for longer columns, but the transition to this mode will depend on the particular geometry of the section under consideration.

16.7 Effective (buckling) lengths L_{cr}

Basic design information relating column buckling resistance (through the buckling reduction factor χ) to slenderness is normally founded on the concept of a pin-ended member, e.g. Figure 16.4. A pin-ended member is one whose ends are supported such that they cannot translate relative to one another but are able to rotate freely. Compression members in actual structures are provided with a variety of different support conditions which are likely to be less restrictive in terms of translational restraint, with or without more restriction in terms of rotational restraint.

The usual way of treating this topic in design is to use the concept of an effective column length (or buckling length) L_{cr}. An engineering definition of effective length is the length of an equivalent pin-ended column having the same load-carrying capacity as the member under consideration provided with its actual conditions of support. This definition of effective length is illustrated in Figure 16.7, which compares a column buckling curve for a member with some degree of rotational end restraint with the basic curve for the same member when pin-ended. Recent practice, and that implied by Eurocode 3, is to use theoretical effective lengths in design, defined on the basis of an equivalent pin-ended strut with the same elastic buckling load N_{cr}.

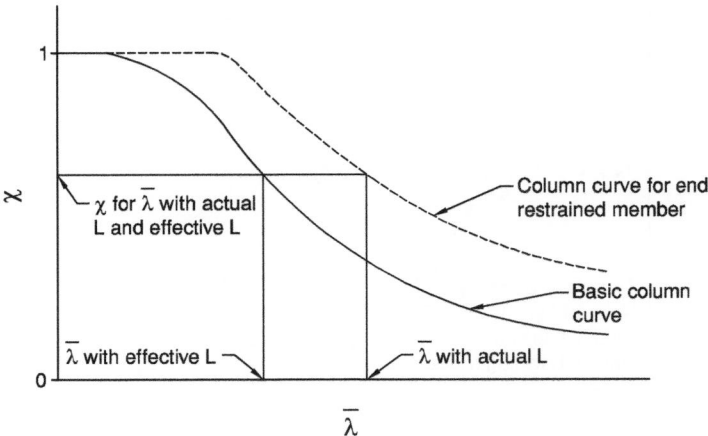

Figure 16.7   Effective length concept

Hence, in determining the non-dimensional slenderness $\bar{\lambda}$ of a column, the geometrical length L is replaced by the effective or buckling length L_{cr}. Values of effective length factors $k = L_{cr}/L$ are not presented in BS EN 1993-1-1, but typical values from existing design practice for a series of standard cases are illustrated in Figure 16.8. When compared with theoretical values determined on the basis of elastic buckling loads,[2] these appear to be high for those cases in which reliance is being placed on externally provided rotational fixity; this is in recognition of the practical difficulties of providing sufficient rotational restraint to approach the condition of full fixity. On the other hand, translational restraints of comparatively modest stiffness are quite capable of preventing lateral displacements. A certain degree of judgement is required of the designer in deciding which of these standard cases most nearly matches the particular arrangement under consideration. In cases of doubt, the safe approach is to use a high approximation, leading to an overestimate of column slenderness and thus an underestimate of column buckling resistance. The idea of an effective column length may also be used as a device to deal with special types of column, such as compound or tapered members, the idea then being to convert the complex problem into one of an equivalent simple column for which the basic design approach of the relationship between buckling reduction factor and member slenderness may be employed.

Of fundamental importance when determining suitable effective lengths is the classification of a column as either a sway case for which translation of one end relative to the other is possible or a non-sway case for which end translation is prevented. For the first case, effective lengths will be at least equal to the geometrical length, tending in theory to infinity for a pin-base column with no restraint at its top, while for the non-sway case, effective lengths will not exceed the geometrical length, decreasing as the degree of rotational fixity increases.

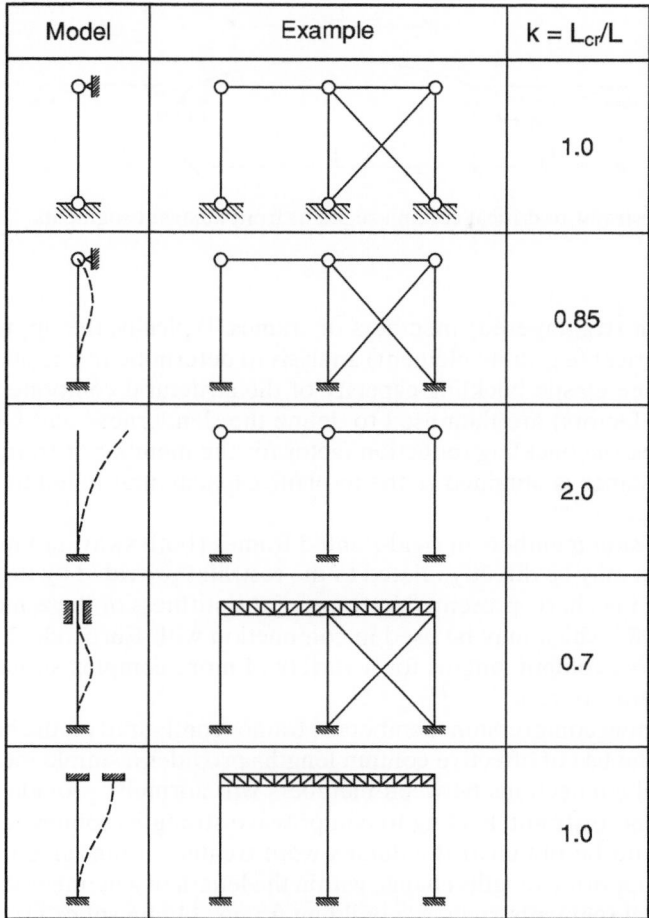

Model	Example	k = L$_{cr}$/L
		1.0
		0.85
		2.0
		0.7
		1.0

Figure 16.8   Typical effective length factors $k = L_{cr}/L$ for use in column design

For non-standard cases, an elastic buckling analysis of the structural system may be performed or reference may be made to published results obtained from elastic stability theory. Provided these relate to cases for which buckling involves the inter-action of a group of members with the less critical restraining the more critical, as illustrated in Figure 16.9, available evidence suggests that the use of effective lengths derived directly from elastic stability theory in conjunction with a column design curve of the type shown in Figure 16.4 will lead to good approximations of the true load-carrying capacity. A 'general method for lateral and lateral-torsional buckling of structural components' is also included in Clause 6.3.4 of BS EN 1993-1-1. This method is aimed at non-standard structural scenarios, such as the design

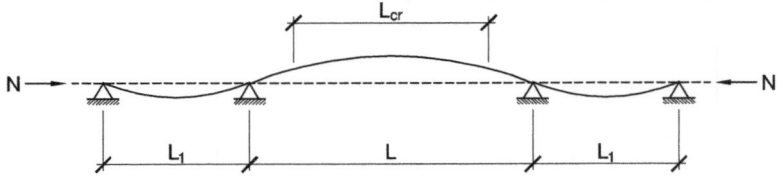

Figure 16.9 Restraint to critical column segment from adjacent segments

of non-uniform (e.g. tapered) members or frames. Typically, the approach initially employs numerical (e.g. finite element) analysis to determine the in-plane resistance and out-of-plane elastic buckling capacity of the structural component. These two loads (or load factors) are then used to define the slenderness, and hence, through buckling curves, the buckling reduction factor for the member or frame. Finally, the structural resistance is obtained as the in-plane capacity multiplied by the buckling reduction factor.

For compression members in rigid-jointed frames (both sway and non-sway), the effective length may be directly related to the restraint provided by the surrounding members by using charts presented in terms of the stiffness of these members given in NCCI SN008,[3] which may be used in conjunction with Eurocode 3. Useful guidance on effective column lengths for a variety of more complex situations is available from several sources.[4,5]

When designing compression members in frames configured on the basis of simple construction, the use of effective column lengths provides a simple means of recognising that real connections between members will normally provide some degree of rotational end restraint, leading to compressive strengths somewhat in excess of those that would be obtained if columns were treated as pin-ended. If axial load levels and unsupported lengths change within the length of a member that is continuous over several segments, such as a building frame column spliced so as to act as a continuous member but carrying decreasing compression with height or a compression chord in a truss, then the less heavily loaded segments will effectively restrain the more critical segments. Even though the distribution of internal member forces has been made on the assumption of pin joints, some allowance for rotational end restraint when designing the compression members is therefore appropriate. Thus, the apparent contradiction of regarding a structure as pin-jointed but using compression member effective lengths that are less than their actual lengths does have a rational basis. Figure 16.10 presents results obtained from elastic stability theory for columns continuous over a number of storeys which show how the effective length of the critical segment will be reduced if more stable segments (shorter unbraced lengths in this case) are present. A practical equivalent for each case in terms of simple braced frames with pinned beam-to-column connections is also shown.

For compression members in rigid-jointed frames the effective length is directly related to the restraint provided by all the surrounding members. Strictly speaking an interaction of all the members in the frame occurs because the real behaviour is

Model	Example	0	0.1	0.2	0.3	0.4	0.5	0.6	0.7	0.8	0.9	1.0	a/L
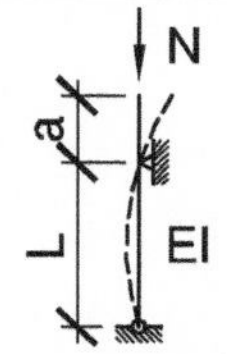	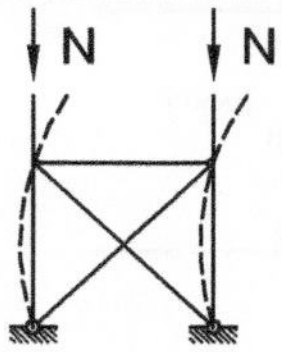	1.10	1.11	1.24	1.40	1.56	1.74	1.93	2.16	2.31	2.50	2.70	L_{cr}/L
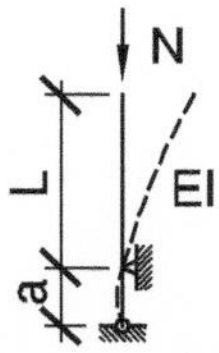	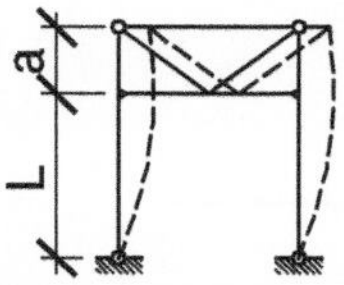	2.0	2.07	2.13	2.20	2.27	2.34	2.41	2.48	2.55	2.62	2.70	L_{cr}/L
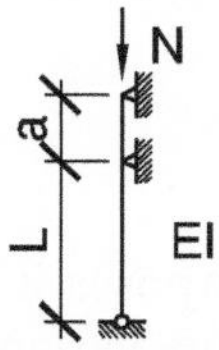	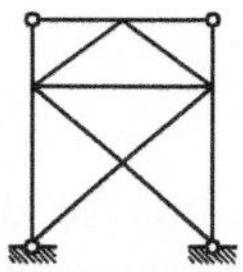	0.70	0.72	0.74	0.77	0.79	0.81	0.84	0.87	0.91	0.95	1.0	L_{cr}/L
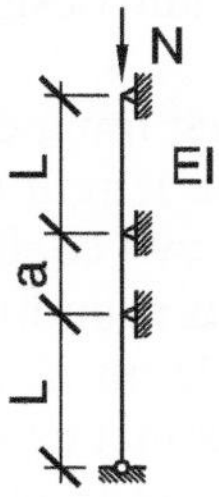	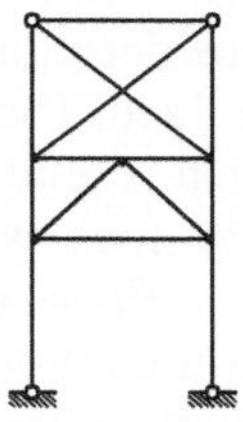	0.70	0.73	0.76	0.79	0.82	0.85	0.88	0.91	0.94	0.97	1.0	L_{cr}/L
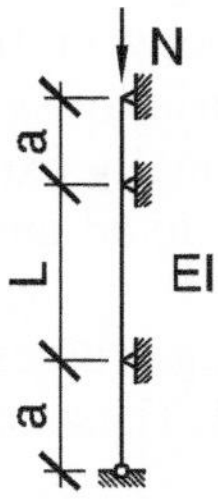	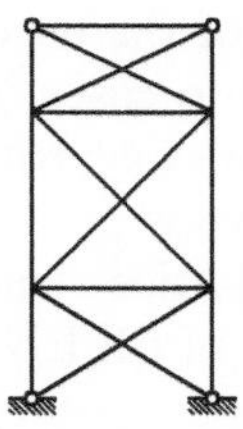	0.50	0.53	0.57	0.61	0.65	0.70	0.75	0.81	0.87	0.93	1.0	L_{cr}/L

Figure 16.10 Effective length factors for continuous columns based on elastic stability theory

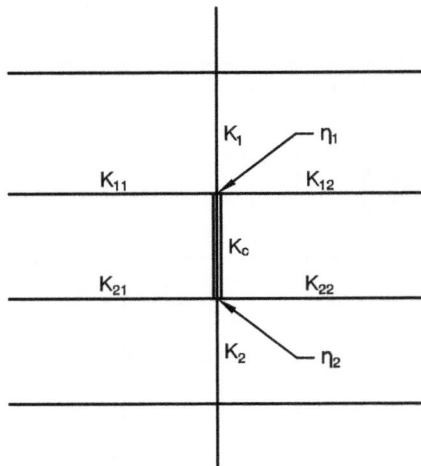

Figure 16.11   Limited frame defined in NCCI SN008

one of frame buckling rather than column buckling, but for design purposes it is often sufficient to consider the behaviour of a limited region of the frame. Variants of the 'limited frame' concept are to be found in several codes of practice and design guides. That used in NCCI SN008[3] is illustrated in Figure 16.11.

The limited frame comprises the column under consideration and each immediately adjacent member treated as if its far end were fixed. The effective length of the critical column is then obtained from a chart which is entered with two coefficients η_1 and η_2, the values of which depend on the stiffnesses of the surrounding members K_1, K_2, K_{11}, K_{12}, K_{21} and K_{22}, relative to the stiffness of the column K_C, a concept similar to the well-known moment distribution method. Two distinct cases are considered: columns in non-sway frames and columns in frames that are free to sway. Figure 16.12(a) and Figure 16.12(b) illustrate both cases as well as giving the associated effective length charts. For the former, the factors will vary between 0.5 and 1.0 depending on the values of η_1 and η_2, while for the latter, the variation will be between 1.0 and ∞. These end points correspond to cases of: rotationally fixed ends with no sway and rotationally free ends with no sway; rotationally fixed ends with free sway and rotationally free ends with free sway.

For the buckling of members in triangulated and lattice structures (e.g. trusses), guidance on effective lengths is given in Clause BB.1 of BS EN 1993-1-1. These are discussed in Chapter 20.

For beams not rigidly connected to the column or for situations in which significant plasticity either at a beam end or at either column end would prevent the restraint being transferred into the column, the K (and thus the η) values must be suitably modified. Similarly at column bases, η_2 values, in keeping with the degree of restraint provided, should be used. Guidance is also provided on K values for beams, distinguishing between both non-sway and sway cases and beams supporting

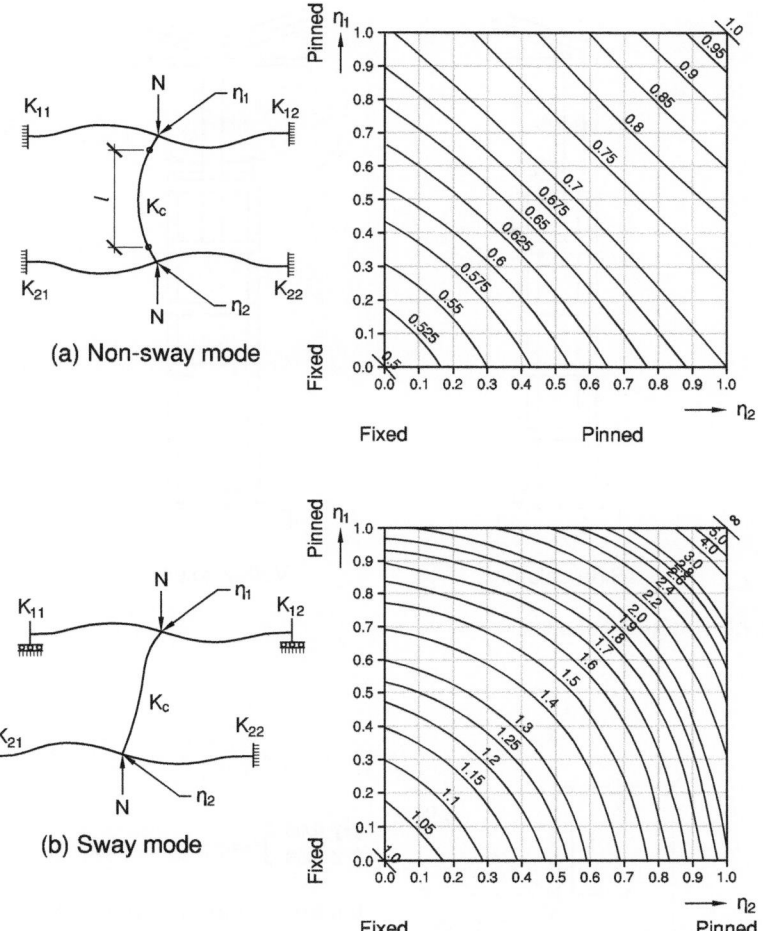

Figure 16.12 Charts from NCCI SN008 for obtaining effective length factors L_{cr}/L for (a) non-sway frames and (b) sway frames

concrete floors and bare steelwork. Full details of the background to this approach to the determination of effective lengths in rigidly jointed frames may be found in the work of Wood.[5]

16.8 Special types of strut

The design of two types of strut requires that certain additional points be considered:

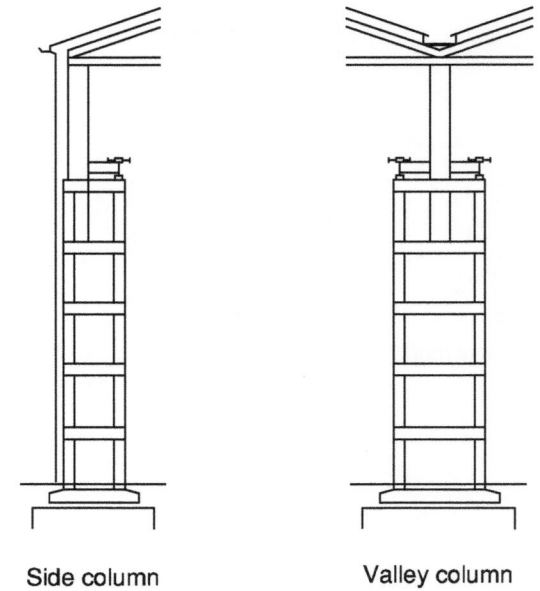

Side column   Valley column

Figure 16.13   Built-up columns

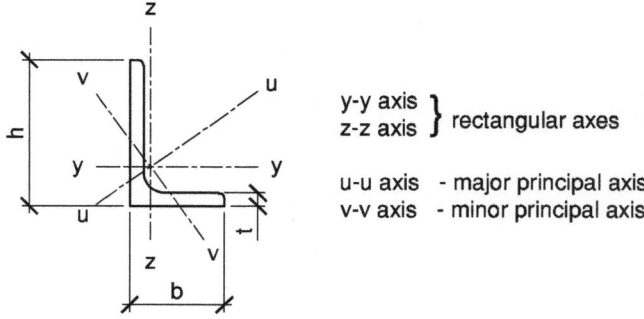

y-y axis ⎫
z-z axis ⎬ rectangular axes

u-u axis - major principal axis
v-v axis - minor principal axis

Figure 16.14   Geometrical properties of an angle section

1. built-up sections or compound struts (see Figure 16.13), for which the behaviour of the individual components must be taken into account
2. angles (see Figure 16.14), for which the eccentricity of loading produced by normal forms of end connection must be acknowledged.

   In both cases, however, it will often be possible to design this more complex type of member as an equivalent single axially-loaded strut.

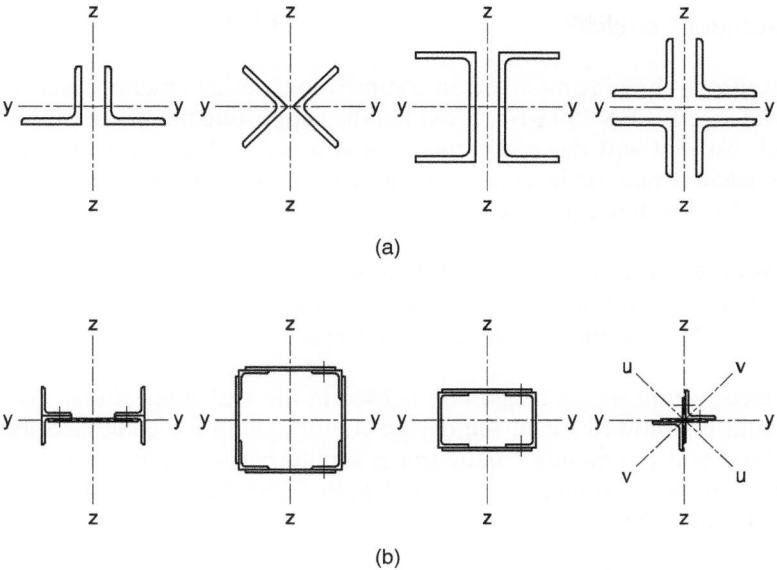

Figure 16.15 Typical arrangements for compound columns: (a) closely spaced; (b) laced or battened

16.8.1 Design of compound struts

Individual members may be combined in a variety of ways to produce a more efficient compound section. Figure 16.15 illustrates the most common arrangements. In each case the concept is one of providing a compound member whose overall slenderness will be such that its load-carrying capacity will significantly exceed the sum of the axial resistances of the component members, i.e. for the case of Figure 16.15(b) the laced strut will have significantly more resistance than the four corner angles treated separately.

Built-up members whose chords are in contact or closely spaced, such as those illustrated in Figure 16.15(a), may be designed as a single integral member, as described in Section 16.5. Conditions to be met in order to qualify for this simple treatment, in which the shear stiffness of the member is assumed to be infinite, are set out in Clause 6.4.4 of BS EN 1993-1-1.

For compound members that do not satisfy these conditions, the shear stiffness S_v of the member must be calculated, as described in Clause 6.4.2 of BS EN 1993-1-1 for laced members and Clause 6.4.3 for battened members. Design is then based on a second-order analysis of an equivalent continuous member with smeared shear stiffness S_v, as set out in Clause 6.4.1 of BS EN 1993-1-1. Design checks on the individual laces or battens are also required.

16.8.2 Design of angles

For angles used as bracing members in compression, design guidance is available in Clause BB.1.2 of BS EN 1993-1-1. Provided the supporting members supply appropriate end restraint and the end connection is made with at least two bolts, then load eccentricities may be ignored and end fixity may be allowed for by using the following effective slendernesses $\bar{\lambda}_{\mathrm{eff}}$:

$$\bar{\lambda}_{\mathrm{eff,v}} = 0.35 + 0.7\bar{\lambda}_{\mathrm{v}} \text{ for buckling about the v-v axis}$$
$$\bar{\lambda}_{\mathrm{eff,y}} = 0.50 + 0.7\bar{\lambda}_{\mathrm{y}} \text{ for buckling about the y-y axis}$$
$$\bar{\lambda}_{\mathrm{eff,z}} = 0.50 + 0.7\bar{\lambda}_{\mathrm{z}} \text{ for buckling about the z-z axis.}$$

In instances where only a single bolt is used in the end connections, load eccentricity should be allowed for in the design (by designing for combined axial load plus bending) and the member slenderness should be determined on the basis of an effective length L_{cr} equal to the system length L. Further discussion on this topic is given in Chapter 20.

16.9 Economic points

Strut design is a relatively straightforward design task involving choice of cross-sectional type, assessment of end restraint and thus effective length, calculation of slenderness, determination of buckling resistance and hence checking that the trial section can withstand the design load. Certain subsidiary checks may also be required part way through this process to ensure that the chosen cross-section is not Class 4 (slender), or make suitable allowances if it is, or to guard against local failure in compound members. Thus only limited opportunities occur for the designer to use judgement and to make choices on the grounds of economy. Essentially these are restricted to control of the effective length, by introducing intermediate restraints where appropriate, and the original choice of cross-section.

However, certain other points relating to columns may well have a bearing on the overall economy of the steel frame or truss. Of particular concern is the need to be able to make connections simply. In a multi-storey frame, the use of heavier UC sections thus may be advantageous, permitting beam-to-column connections to be made without the need for stiffening the flanges or web. Similarly in order to accommodate beams framing into the column web an increase in the size of UC may eliminate the need for special detailing.

While compound angle members were a common feature of early trusses, maintenance costs due both to the surface area requiring painting and to the incidence of corrosion caused by the inherent dirt and moisture traps have caused a change to the much greater use of tubular members. If site joints are kept to the minimum, tubular trusses can be transported and handled on site in long lengths and a more economic as well as a visually more pleasing structure is likely to result.

Worked examples follow which are relevant to Chapter 16. Other worked examples may be found in References 6–11.

References to Chapter 16

1. SCI and BCSA (2009) *Steel Building Design: Design Data – In accordance with Eurocodes and the UK National Annexes*. SCI Publication No. P363. Ascot, The Steel Construction Institute.
2. Ballio G. and Mazzolani F.M. (1983) *Theory and Design of Steel Structures*. London, Taylor and Francis.
3. NCCI SN008 (2006) *Buckling lengths of columns: rigorous approach*. http://www.steel-ncci.co.uk
4. Allen H.G. and Bulson P.S. (1980) *Background to Buckling*. New York, McGraw-Hill.
5. Wood R.H. (1974) Effective lengths of columns in multi-storey buildings. *The Structural Engineer*, 52, Part 1, July, 235–44, Part 2, Aug., 295–302, Part 3, Sept., 341–6.
6. Access-Steel (2007) SX002 *Example: Buckling resistance of a pinned column with intermediate restraints*, www.access-steel.com
7. Access-Steel (2006) SX010 *Example: Continuous column in a multi-storey building using an H-section or RHS*, www.access-steel.com
8. Trahair N.S., Bradford M.A., Nethercot D.A. and Gardner L. (2008) *The Behaviour and Design of Steel Structures to EC3*, 4th edn. London, Taylor & Francis.
9. Gardner L. and Nethercot D.A. (2005) *Designers' Guide to EN 1993-1-1: Eurocode 3: Design of Steel Structures*. London, Thomas Telford Publishing.
10. Martin L. & Purkiss J. (2008) *Structural design of steelwork to EN 1993 and EN 1994*, 3rd edn. Oxford, Butterworth-Heinemann.
11. Brettle M. (2009) *Steel Building Design: Worked Examples – Open Sections – In accordance with Eurocodes and the UK National Annexes*. SCI Publication No. P364. Ascot, The Steel Construction Institute.

Further reading for Chapter 16

Galambos T.V. (Ed.) (1998) *Guide to Stability Design Criteria for Metal Structures*, 5th edn. New York, Wiley.
Gardner L. (2011) *Stability of Beams and Columns*. SCI Publication No. P360. Ascot, The Steel Construction Institute.
Trahair N.S. and Nethercot D.A. (1984) Bracing requirements in thin-walled structures. In: *Developments in Thin-Walled Structures – 2* (ed. by J. Rhodes and A.C. Walker), 92–130. Barking, Elsevier Applied Science Publishers.

	Job No.		Sheet 1 of 2		Rev
The Steel Construction Institute Silwood Park, Ascot, Berks SL5 7QN Telephone: (01344) 623345 Fax: (01344) 622944 **CALCULATION SHEET**	Job Title		Rolled Universal Column Design		
	Subject		Compression Members		
	Client	Made by	LG	Date	2010
		Checked by	DAN	Date	2010

Compression members

Rolled Universal Column design

Problem
Check the ability of a 203 × 203 × 52 UC in grade S275 steel to withstand a design axial compressive load of 1150 kN over an unsupported height of 3.6 m assuming that both ends of the member are pinned. Design to BS EN 1993-1-1.
The problem is as shown in the sketch below:

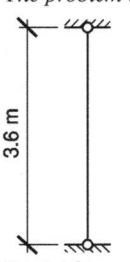

Partial factors:
$\gamma_{M0} = 1.0$; $\gamma_{M1} = 1.0$

Geometric properties:
$A = 66.3\ cm^2 = 6630\ mm^2$; $i_z = 5.18\ cm = 51.8\ mm$;

$t_f = 12.5\ mm$; $c_f/t_f = 7.04$; $c_w/t_w = 20.4$

Material properties:
Yield strength $f_y = 275\ N/mm^2$ since $t_f \leq 16\ mm$

Check cross-section classification under pure compression:
Need only check that section is not Class 4 (slender)
For outstand flange $c_f/t_f \varepsilon \leq 14$

For web $c_w/t_w \varepsilon \leq 42$

$$\varepsilon = \sqrt{\frac{235}{f_y}} = \sqrt{\frac{235}{275}} = 0.92$$

Actual $c_f/t_f \varepsilon = 7.04/0.92 = 7.62$; within limit
Actual $c_w/t_w \varepsilon = 20.4/0.92 = 20.2$; within limit
∴ Section is not Class 4

UK NA to BS EN 1993-1-1

Steel Building Design: Design Data

BS EN 10025-2

BS EN 1993-1-1 Table 5.2

Rolled Universal Column Design	Sheet 2 of 2	Rev
Cross-section compression resistance: $N_{c,Rd} = \dfrac{Af_y}{\gamma_{M0}} = \dfrac{6630 \times 275}{1.0} \times 10^{-3} = 1823\,kN > 1150\,kN = N_{Ed}$ *OK* *Member buckling resistance:* *Take effective length $L_{cr} = 1.0\ L = 1.0 \times 3600 = 3600\,mm$* *On the assumption that minor axis flexural buckling will govern, use buckling curve 'c'.* $\lambda_1 = \pi\sqrt{\dfrac{E}{f_y}} = \pi\sqrt{\dfrac{210000}{275}} = 86.8$ $\bar{\lambda}_z = \dfrac{\lambda_z}{\lambda_1} = \dfrac{L_{cr}/i_z}{\lambda_1} = \dfrac{3600/51.8}{86.8} = 0.80$ $\Phi_z = 0.5[\,1 + \alpha(\bar{\lambda}_z - 0.2) + \bar{\lambda}_z^2\,] = 0.5[\,1 + 0.49(0.80 - 0.2) + 0.80^2\,] = 0.97$ $\chi_z = \dfrac{1}{\Phi_z + \sqrt{\Phi_z^2 - \bar{\lambda}_z^2}} = \dfrac{1}{0.97 + \sqrt{0.97^2 - 0.80^2}} = 0.66 \le 1.0$ $N_{b,z,Rd} = \dfrac{\chi_z Af_y}{\gamma_{M1}} = \dfrac{0.66 \times 6630 \times 275}{1.0} \times 10^{-3} = 1207\,kN > 1150\,kN = N_{Ed}$ *OK* *∴ Use $203 \times 203 \times 52$ UC in grade S275 steel* *It should be noted that the same answer could have been obtained directly by the use of Reference 1.*	*BS EN 1993-1-1 Cl 6.2.4* *BS EN 1993-1-1 Table 6.2* *BS EN 1993-1-1 Cl 6.3.1.2* *BS EN 1993-1-1 Cl 6.3.1.1*	

		Job No.	PUB 809		Sheet	1	of	3	Rev

<table>
<tr><td rowspan="2" colspan="2"><img> The Steel Construction Institute

Silwood Park, Ascot, Berks
SL5 7QN
Telephone: (01344) 623345
Fax: (01344) 622944

CALCULATION SHEET</td><td>Job No.</td><td colspan="2">PUB 809</td><td>Sheet 1 of 3</td><td>Rev</td></tr>
</table>

The Steel Construction Institute Silwood Park, Ascot, Berks SL5 7QN Telephone: (01344) 623345 Fax: (01344) 622944 **CALCULATION SHEET**	Job No.	PUB 809		Sheet 1 of 3		Rev
	Job Title	Pinned Column with Intermediate Lateral Restraints				
	Subject	Compression Members				
	Client		Made by	LG	Date	2010
			Checked by	DAN	Date	2010

Compression members

Pinned column with intermediate lateral restraints

<u>*Problem*</u>

A 254 × 254 × 89 UC in grade S275 steel is to be used as a 12.0 m column with pin ends and intermediate lateral braces provided restraint against minor axis buckling at third points along the column length. Check the adequacy of the column, according to BS EN 1993-1-1, to carry a design axial compressive load of 1250 kN.

The problem is as shown in the sketch below:

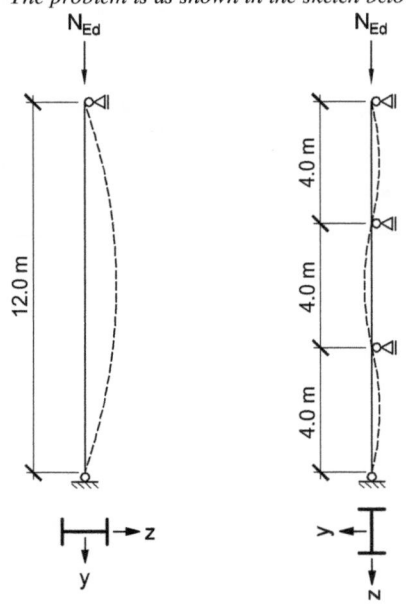

Partial factors:

$\gamma_{M0} = 1.0; \; \gamma_{M1} = 1.0$

UK NA to BS EN 1993-1-1

Pinned Column with Intermediate Lateral Restraints	Sheet 2 of 3	Rev

Geometric properties:

$A = 113\,cm^2 = 11300\,mm^2$; $i_y = 11.2\,cm = 112\,mm$; $i_z = 6.55\,cm = 65.5\,mm$;
$t_f = 17.3\,mm$; $c_f/t_f = 6.38$; $c_w/t_w = 19.4$

Material properties:

Yield strength $f_y = 265\,N/mm^2$ since $16 > t_f \geq 40\,mm$

Check cross-section classification under pure compression:

Need only check that section is not Class 4 (slender)

For outstand flange $c_f/t_f\varepsilon \leq 14$

For web $c_w/t_w\varepsilon \leq 42$

$$\varepsilon = \sqrt{\frac{235}{f_y}} = \sqrt{\frac{235}{265}} = 0.94$$

Actual $c_f/t_f\varepsilon = 7.04/0.94 = 6.77$; within limit

Actual $c_w/t_w\varepsilon = 20.4/0.94 = 20.7$; within limit

∴Section is not Class 4

Cross-section compression resistance:

$$N_{c,Rd} = \frac{Af_y}{\gamma_{M0}} = \frac{11300 \times 265}{1.0} \times 10^{-3} = 2995\,kN > 1250\,kN = N_{Ed} \qquad OK$$

Effective lengths:

$L_{cr,y} = 1.0\,L = 1.0 \times 12000 = 12000\,mm$ *for buckling about the y-y axis*

$L_{cr,z} = 1.0\,L = 1.0 \times 4000 = 4000\,mm$ *for buckling about the z-z axis.*

Non-dimensional slendernesses:

$$\lambda_1 = \pi\sqrt{\frac{E}{f_y}} = \pi\sqrt{\frac{210000}{265}} = 88.4$$

$$\overline{\lambda}_y = \frac{\lambda_y}{\lambda_1} = \frac{L_{cr,y}/i_y}{\lambda_1} = \frac{12000/112}{88.4} = 1.21$$

$$\overline{\lambda}_z = \frac{\lambda_z}{\lambda_1} = \frac{L_{cr,z}/i_z}{\lambda_1} = \frac{4000/65.5}{88.4} = 0.69$$

Buckling curves:

For major axis buckling, use buckling curve 'b'

For minor axis buckling, use buckling curve 'c'.

Buckling reduction factors χ:

Steel Building Design: Design Data

BS EN 10025-2

BS EN 1993-1-1 Table 5.2

BS EN 1993-1-1 Cl 6.2.4

BS EN 1993-1-1 Table 6.2

Pinned Column with Intermediate Lateral Restraints	Sheet 3 of 3	Rev
$\Phi_y = 0.5[1 + \alpha(\bar{\lambda}_y - 0.2) + \bar{\lambda}_y^2] = 0.5[1 + 0.34(1.21 - 0.2) + 1.21^2] = 1.40$		*BS EN 1993-1-1 Cl 6.3.1.2*
$\chi_y = \dfrac{1}{\Phi_y + \sqrt{\Phi_y^2 - \bar{\lambda}_y^2}} = \dfrac{1}{1.40 + \sqrt{1.40^2 - 1.21^2}} = 0.47 \leq 1.0$		
$\Phi_z = 0.5[1 + \alpha(\bar{\lambda}_z - 0.2) + \bar{\lambda}_z^2] = 0.5[1 + 0.49(0.69 - 0.2) + 0.69^2] = 0.86$		*BS EN 1993-1-1 Cl 6.3.1.2*
$\chi_z = \dfrac{1}{\Phi_z + \sqrt{\Phi_z^2 - \bar{\lambda}_z^2}} = \dfrac{1}{0.86 + \sqrt{0.86^2 - 0.69^2}} = 0.73 \leq 1.0$		
Buckling resistances:		
$N_{b,y,Rd} = \dfrac{\chi_y A f_y}{\gamma_{M1}} = \dfrac{0.47 \times 11300 \times 265}{1.0} \times 10^{-3} = 1422 \, kN > 1250 \, kN = N_{Ed} \quad OK$		*BS EN 1993-1-1 Cl 6.3.1.1*
$N_{b,z,Rd} = \dfrac{\chi_z A f_y}{\gamma_{M1}} = \dfrac{0.73 \times 11300 \times 265}{1.0} \times 10^{-3} = 2189 \, kN > 1250 \, kN = N_{Ed} \quad OK$		*BS EN 1993-1-1 Cl 6.3.1.1*
$\therefore$ *Use* $254 \times 254 \times 89$ *UC in grade S275 steel.*		

Chapter 17
Beams

DAVID NETHERCOT and LEROY GARDNER

17.1 Introduction

Beams are possibly the most fundamental type of member present in a civil engineering structure. Their principal function is the transmission of vertical load by means of flexural (bending) action into, for example, the columns in a rectangular building frame or the abutments in a bridge which support them.

17.2 Common types of beam

Table 17.1 provides general guidance on the different structural forms suitable for use as beams in a steel structure; several of these are illustrated in Figure 17.1. For modest spans, including the majority of those found in buildings, the use of standard hot-rolled sections (normally Universal Beams or IPE sections but possibly Universal Columns or HE sections, if minimising floor depth is a prime consideration, or channels if only light loads need to be supported) will be sufficient. The range of spans for which rolled sections may be used can be extended if cover plates are welded to both flanges. Lightly loaded members such as the purlins supporting the roof of a portal-frame building are frequently selected from the range of proprietary cold-formed sections produced from steel sheet only a few millimetres thick, normally already protected against corrosion by galvanising, in a variety of highly efficient shapes, advantage being taken of the roll-forming process to produce sections with properties carefully selected for the task they are required to perform. The design of cold-formed steelwork is the subject of Chapter 24. Beams with web openings – for which proprietary sections with circular web openings formed by splitting and re-welding Universal Beam sections have largely replaced the original castellated sections – are structurally efficient, visually attractive and facilitate the passage of services within the floor zone but require careful design if required to carry significant concentrated loads.

Steel Designers' Manual, Seventh Edition. Edited by Buick Davison and Graham W. Owens.
© 2012 Steel Construction Institute. Published 2012 by Blackwell Publishing Ltd.

Table 17.1 Typical usages of different forms of beam

Beam type	Span range (m)	Notes
Angles	1–6	Used for roof purlins, sheeting rails, etc. where only light loads have to be carried
Cold-formed sections	2–8	Used for roof purlins, sheeting rails, etc. where only light loads have to be carried
Rolled sections: UBs, IPEs, UCs, HEs	1–30	Most frequently used type of section; proportioned to eliminate several possible modes of failure
Open web joists	4–40	Prefabricated using angles or tubes as chords and round bar for the web diagonals, used in place of rolled sections
Cellular beams	6–60	Used for long spans and/or light loads; depth of rolled section increased by 50%; web openings may be used for services etc.
Compound sections	5–30	Used when a single rolled section would not provide sufficient resistance
Plate girders	10–100	Made by welding 3 plates, often automatically, with web depths up to 3–4 m; may need stiffening – see Chapter 18
Trusses	10–100	Fabricated from angles, tubes or, if spanning large distances, rolled sections – see Chapter 20
Box girders	15–200	Fabricated from plate, usually stiffened; used for overhead travelling cranes and bridges due to good torsional and transverse stiffness properties

Alternatively, a beam fabricated entirely by welding plates together may be employed, allowing variations in properties by changes in depth and/or flange thickness. In certain cases where the use of very thin webs is required, stiffening to prevent premature buckling failure is necessary. A full treatment of the specialist aspects of plate girder design is provided in Chapter 18. If spans are so large that a single member cannot economically be provided, then a truss may be a suitable alternative. In addition to the deep truss fabricated from open hot-rolled sections, SHS or both, used to provide long clear spans in sports halls and supermarkets, smaller prefabricated arrangements using RHS or CHS provide an attractive alternative to the use of standard sections for more modest spans. Truss design is discussed in Chapter 20.

Since the principal requirement of a beam is to provide adequate resistance to vertical bending, a very useful indication of the size of section likely to prove suitable may be obtained through the concept of the span-to-depth ratio. This is simply the value of the clear span divided by the overall depth of the beam. An average figure for a properly designed steel beam is between 15 and 20, perhaps more if a particularly slender form of construction is employed or possibly less if very heavy loadings are present.

When designing beams, attention must be given to a series of issues, in addition to simple vertical bending, to ensure satisfactory behaviour. Torsional loading may often be eliminated by careful detailing or its effects reasonably regarded as of

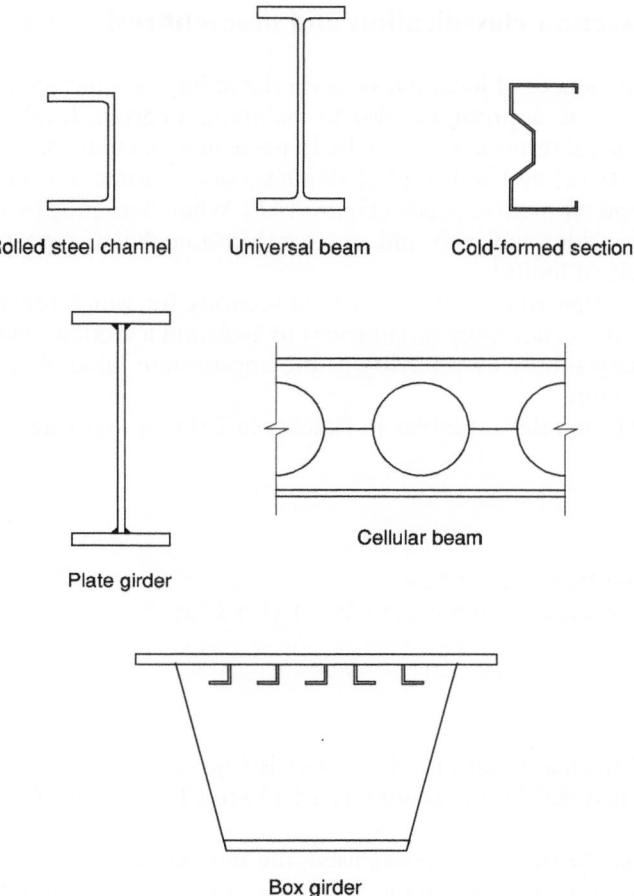

Figure 17.1 Types of beam cross-section

negligible importance by a correct appreciation of how the structure actually behaves; in certain instances it should, however, be considered. Section classification (allowance for possible local buckling effects) for beams is more involved than is the case for struts since different elements in the cross-section are subject to different patterns of stress; the flanges in an I-beam in the elastic range will be in approximately uniform tension or compression while the web will contain a stress gradient. The possibility of members being designed for an elastic or a plastic state, including the use of a full plastic design for the complete structure, also affects section classification. Various forms of instability of the beam as a whole, or of parts subject to locally high stresses, such as the web over a support, also require attention. Finally, certain forms of construction may be prone to unacceptable vibrations; although this is likely to be affected by the choice of beams, this topic is addressed in Chapter 13.

17.3 Cross-section classification and moment resistance $M_{c,Rd}$

The possible influences of local buckling on the ability of a particular cross-section to attain, and where appropriate also to maintain, a certain level of moment are discussed in general terms in Chapter 14. In particular, Section 14.2 covers the influence of flange (c_f/t_f) and web (c_w/t_w) slenderness on moment–rotation behaviour (Figure 14.5) and moment capacity (Figure 14.6). When designing beams it is usually sufficient to consider the web and the compression flange separately using the appropriate sets of limits.[1]

In building design when using hot-rolled sections, for which relevant properties are tabulated,[2] it will normally be sufficient to ascertain a section's classification and moment capacity simply by referring to the appropriate table. Worked example 1 illustrates the point.

Inspection of the relevant tables in Reference 2 shows that when used as beams:

S275 steel

- All Universal Beams are Class 1.
- All but 1 Universal Columns are Class 1 (1 is Class 2).

S355 steel

- All but 1 Universal Beams are Class 1 (1 is Class 2).
- All but 6 Universal Columns are Class 1 (3 are Class 2, 3 are Class 3).

Whichever grade of steel is being used, the moment capacity, $M_{c,Rd}$, for Class 1 and 2 sections will be the maximum attainable value corresponding to the section's full plastic moment capacity, M_{pl}, given by:

$$M_{c,Rd} = \frac{W_{pl}f_y}{\gamma_{M0}} \tag{17.1}$$

where W_{pl} is the plastic section modulus and $\gamma_{M0} = 1.0$. In those cases of Class 3 sections, moment capacity is based on the moment corresponding to first yield, M_{el} given by:

$$M_{c,Rd} = \frac{W_{el}f_y}{\gamma_{M0}} \tag{17.2}$$

where W_{el} is the elastic section modulus.

The full list of non-Class 1 sections in bending is given in Table 17.2.

When using fabricated sections, individual checks on the web and compression flange using the actual dimensions of the trial section must be made. It normally

Table 17.2   Non-Class 1 Universal Beams and Universal Columns in bending

Section type	Class 2	Class 3
S275		
Universal Column	$356 \times 368 \times 129$	
S355		
Universal Beam	$356 \times 171 \times 45$	
Universal Column	$356 \times 368 \times 153$	$356 \times 368 \times 129$
	$254 \times 254 \times 73$	$305 \times 305 \times 97$
	$203 \times 203 \times 46$	$152 \times 152 \times 23$

proves much simpler if proportions are selected so as to ensure that the section is Class 3 or better since the resulting calculations need not then involve the various complications associated with the use of slender (Class 4) sections. When Class 4 sections are used, the amount of calculation increases considerably due to the need to allow for loss of effectiveness of some parts of the cross-section due to local buckling when determining $M_{c,Rd}$. Probably the most frequent use of slender sections involves cold-formed shapes used for example as roof purlins. Owing to their proprietary nature, the manufacturers normally provide design information, much of it based on physical testing, listing such properties as moment capacity. In the absence of design information, reference should be made to Part 1.3 of Eurocode 3 for suitable calculation methods.

Although the part of the web between the flange and the horizontal edge of the opening in a cellular beam frequently exceeds the Class 2 limit for an outstand, sufficient test data exist to show that this does not appear to influence the moment capacity of such sections. Section classification should therefore be made in the same way as for solid web beams, the value of $M_{c,Rd}$ being obtained using the net section modulus value for the section at the centre of a web opening.

17.4 Basic design

One (or more) of a number of distinct limiting conditions may, in theory, control the design of a particular beam as indicated in Table 17.3, but in any particular practical case only a few of them are likely to require full checks. It is therefore convenient to consider the various possibilities in turn, noting the conditions under which each is likely to be important. For convenience the various phenomena are first considered principally within the context of using standard hot-rolled sections, i.e. Universal Beams, Universal Columns, joists and channels; other types of cross-section are covered in the later parts of this chapter.

Probably the most important design decision is whether the beam should be regarded as laterally restrained. The exact way in which this needs to be done is presented in Section 17.5. The remaining parts of this Section (17.4.1 to 17.4.5) cover

Table 17.3 Limiting conditions for beam design

Ultimate limit state	Serviceability limit state
Moment resistance (including influence of local buckling) $M_{c,Rd}$	Deflections due to bending (and shear if appropriate)
Lateral torsional buckling resistance $M_{b,Rd}$	Twist due to torsion
Shear resistance $V_{c,Rd}$	Vibrations
Shear buckling resistance $V_{b,Rd}$	
Moment-shear interaction	
Resistance to transverse load F_{Rd}	
Torsional resistance T_{Rd}	
Bending-torsion interaction	

design checks needed for both the laterally restrained and the laterally unrestrained cases. The additional checks required only for the laterally unrestrained case are presented in Section 17.5.

17.4.1 Moment resistance $M_{c,Rd}$

The most basic design requirement for a beam is the provision of adequate in-plane bending strength. This is provided by ensuring that $M_{c,Rd}$ for the selected section exceeds the maximum moment produced by the factored loading. Determination of $M_{c,Rd}$, which is closely linked to section classification (see Chapter 14), is covered in Section 17.2.

For a statically determinate structural arrangement, simple considerations of statics provide the moment levels produced by the applied loads against which $M_{c,Rd}$ must be checked. For indeterminate arrangements, a suitable method of elastic analysis is required. The justification for using an elastically obtained distribution of moments with, in the case of Class 1 and 2 cross-sections, a plastic cross-sectional resistance has been fully discussed by Johnson and Buckby.[3]

17.4.2 Shear resistance $V_{c,Rd}$

Only in cases of high coincident shear and moment, found for example at the internal supports of continuous beams, is the effect of shear likely to have a significant influence on the design of typical beams.

Shear resistance $V_{c,Rd}$ is normally calculated as the product of the yield stress of steel in shear, which is $1/\sqrt{3}$ of the uniaxial tensile yield stress f_y, and an appropriate shear area A_v. The process approximates the actual distribution of shear stress in a beam web as well as assuming some degree of plasticity. While suitable for rolled sections, it may not therefore be applicable to plate girders. An alternative design approach, more suited to webs containing large holes or having variations in thickness, is to work from first principles and to limit the maximum shear stress to a suitable value, say $0.7 f_y$. In cases where the web slenderness h_w/t_w (h_w being the clear distance between the flanges and t_w the web thickness) exceeds 72ε (taking $\eta = 1.0$ as specified in the UK National Annex to BS EN 1993-1-5), where $\varepsilon = \sqrt{235/f_y}$, shear buckling limits the effectiveness of the web and reference to Chapter 18 should be made for methods of determining the reduced capacity.

In principle the presence of shear in a section reduces its moment resistance. In practice the reduction may be regarded as negligibly small up to quite large fractions of the shear resistance $V_{c,Rd}$. For example, BS EN 1993-1-1 requires a reduction in $M_{c,Rd}$ only when the applied shear exceeds $0.5V_{c,Rd}$. Thus for many practical cases, e.g. the region of maximum moment in a simply supported beam, the need to make explicit allowance for the shear/moment interaction is unlikely. Situations in which high shear and high moment occur at the same location e.g. support regions of continuous beams, may be checked using the interaction approach described in Chapter 18 in Section 18.4.5; Figure 18.6, which illustrates its application, shows how the BS EN 1993-1-1 rule also permits the full shear capacity to be used for applied moments up to the bending resistance of the flanges alone, which is typically about 75% of the full sectional moment resistance. Specific applications and methods for dealing with Class 3 or 4 cross-section and/or sections with webs that are so slender that shear buckling becomes a potential mode of failure (which is not the case for any UKB or UKC) are explained in Sections 18.4.5.1 and 18.4.5.2 respectively. For the great majority of commonly encountered situations, design may be based on the full cross-section shear and moment values.

17.4.3 Deflections

When designing according to limit state principles it is customary to check that deflections at working load levels will not impair the proper function of the structure. For beams, examples of potentially undesirable consequences of excessive serviceability deflections include:

- cracking of plaster ceilings
- allowing crane rails to become misaligned
- causing difficulty in opening large doors.

Although earlier codes of practice specified limits for working load deflections, the tendency with more recent documents[1] is to draw attention to the need for deflection checks and to provide advisory limits to be used only when more specific guidance is not available. Clause NA.2.23 of the UK National Annex to BS EN 1993-1-1 gives 'suggested limits' for certain types of beams and purlins and states that 'circumstances may arise where greater or lesser values would be more appropriate'.

When checking deflections of steel structures under serviceability loading, the central deflection Δ_{max} of a uniformly-loaded simply-supported beam, assuming linear–elastic behaviour, is given by:

$$\Delta_{max} = \frac{5}{384}\frac{WL^3}{EI}\times10^{12} \tag{17.3}$$

in which W = total load (kN)

$\quad\quad E$ = Young's modulus (N/mm²)

$\quad\quad I$ = second moment of area (mm⁴)

$\quad\quad L$ = span (m).

If Δ_{max} is to be limited to a fraction of L, Equation 17.3 may be rearranged to give

$$I_{rqd} = 0.62\times10^{-2}\,\alpha WL^2 \tag{17.4}$$

in which α defines the deflection limit as

$$\Delta_{max} = L/\alpha \tag{17.5}$$

and I_{rqd} is now in cm⁴.

Writing $I_{rqd} = KWL^2$, Table 17.4 gives values of K for a range of values of α.

Since deflection checks are essentially of the 'not greater than' type, some degree of approximation is normally acceptable, particularly if the calculations are reduced as a result. Converting complex load arrangements to a roughly equivalent UDL permits Table 17.4 to be used for a wide range of practical situations. Table 17.5 gives values of the coefficient K by which the actual load arrangement shown should be multiplied in order to obtain an approximately equal maximum deflection.

Tables of deflections for a number of standard cases are provided in the Appendix.

17.4.4 Torsion

Beams subjected to loads which do not act through the point on the cross-section known as the shear centre normally suffer some twisting. Methods for locating the shear centre for a variety of sectional shapes are given in Reference 4. For doubly symmetrical sections such as Universal Beams and Universal Columns, the shear centre coincides with the centroid, while for channels it is situated on the opposite

Table 17.4   K values for uniformly loaded simply-supported beams for various deflection limits

α	200	240	250	325	360	400	500	600	750	1000
K	1.24	1.49	1.55	2.02	2.23	2.48	3.10	3.72	4.65	6.20

Table 17.5 Equivalent UDL coefficients K for beams of span *L*

a/L	K	No. of equal loads	b/L	c/L	K
0.5	1.0	2	0.2	0.6	0.91
0.4	0.86		0.25	0.5	1.10
0.375	0.82		0.333	0.333	1.3
0.333	0.74	3	0.167	0.333	1.05
0.3	0.68		0.2	0.3	1.14
0.25	0.58		0.25	0.25	1.27
0.2	0.47	4	0.125	0.25	1.03
0.1	0.24		0.2	0.2	1.21
		5	0.1	0.2	1.02
			0.167	0.167	1.17
		6	0.083	0.167	1.01
			0.143	0.143	1.15
		7	0.071	0.143	1.01
			0.125	0.125	1.12
		8	0.063	0.125	1.01
			0.111	0.111	1.11

$$K = 1.6\frac{a}{L}\left[3 - 4\left(\frac{a}{L}\right)^2\right]$$

a/L	0.01	0.05	0.1	0.15	0.2	0.25	0.3	0.35	0.4	0.45	0.5
K	0.05	0.24	0.47	0.70	0.91	1.10	1.27	1.41	1.51	1.58	1.60

side of the web from the centroid. For rolled channels its location is included in the tables of Reference 2. Figure 17.2 illustrates its position for a number of rolled sections and provides values for the warping constant I_w.

The effects of torsional loading may often be minimised by careful detailing, particularly when considering how loads are transferred between members. Proper attention to detail can frequently lead to arrangements in which the load transfer is organised in such a way that twisting should not occur.[4] Whenever possible this approach should be followed as the open sections normally used as beams are inherently weak in resisting torsion. In circumstances where beams are required to withstand significant torsional loading, consideration should be given to the use of a torsionally more efficient shape such as a structural hollow section. The treatment of combined bending and torsion as it arises in a number of practical situations is explained in Reference 4.

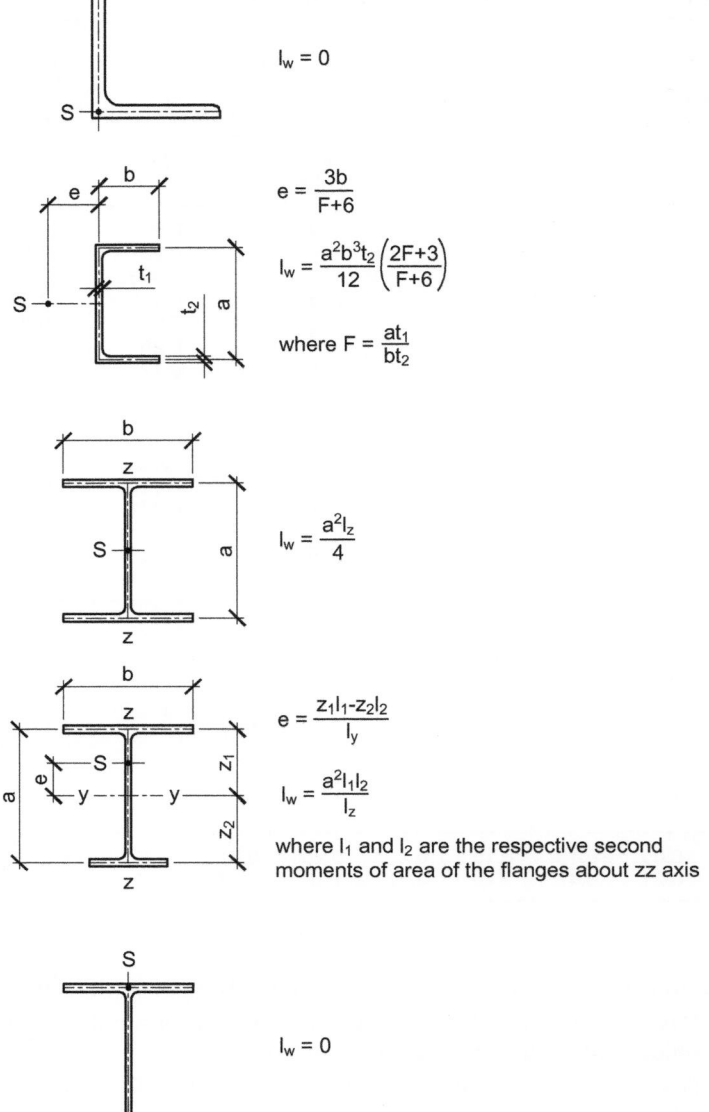

$I_w = 0$

$e = \dfrac{3b}{F+6}$

$I_w = \dfrac{a^2 b^3 t_2}{12}\left(\dfrac{2F+3}{F+6}\right)$

where $F = \dfrac{a t_1}{b t_2}$

$I_w = \dfrac{a^2 I_z}{4}$

$e = \dfrac{z_1 I_1 - z_2 I_2}{I_y}$

$I_w = \dfrac{a^2 I_1 I_2}{I_z}$

where I_1 and I_2 are the respective second moments of area of the flanges about zz axis

$I_w = 0$

Figure 17.2   Location of shear centre (S) for standard rolled sections

17.4.5 Local transverse forces on webs

At points within the length of a beam where vertical loads act, the web is subject to concentrations of stresses, additional to those produced by overall bending. Failure by buckling, rather in the manner of a vertical strut, or by the development of unacceptably high bearing stresses in the relatively thin web material immediately adjacent to the flange, are both possibilities. Methods for assessing the likelihood of both types of failure are given in BS EN 1993-1-5, and tabulated data to assist in the evaluation of the formulae required are provided for rolled sections in Reference 2. The parallel treatment for cold-formed sections is discussed in Chapter 24.

In cases where the web is found to be incapable of resisting the required level of load, additional strength may be provided through the use of stiffeners. Design provisions for web stiffeners are given in BS EN 1993-1-5 and discussed in Chapter 18. Further guidance is given in Reference 5.

17.5 Laterally unrestrained beams

Beams for which none of the conditions listed in Table 17.6 are met are liable to have their load-carrying capacity governed by the type of failure illustrated in Figure 17.3. Thus a first check in any design is to make an assessment of the actual conditions to see if any of the conditions of Table 17.6 may safely be assumed; since design for the laterally restrained case is both simpler and leads to higher load-carrying capacities, it should clearly be adopted wherever possible.

Beams bent about their minor principal axis will respond by deforming in that plane i.e. there is no tendency when loaded in a weaker direction to buckle by

Table 17.6 Types of beam not susceptible to lateral-torsional buckling

---

Loading produces bending about the minor axis only
Beam provided with closely spaced or continuous lateral restraint
Closed sections (provided aspect ratio is not excessive)

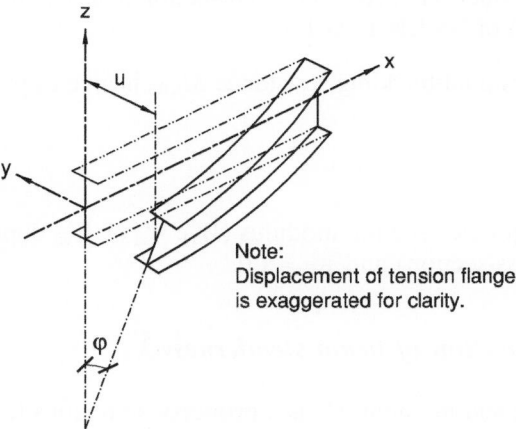

Figure 17.3 Lateral-torsional buckling

deflecting in a stiffer direction. Similarly, if the sort of deflections illustrated in Figure 17.3 are prevented by the form of construction e.g. by attaching the beam's top flange to a laterally very stiff concrete slab, then buckling of this type cannot occur. Treatment of arrangements which partly or fully meet this condition are discussed in Section 17.5.2. and more fully in Reference 6. Finally, if the beam's cross-section is torsionally very stiff, as is the case for all SHS, its resistance to lateral torsional buckling for all practical arrangements will be so great that it will not influence the design.

Lateral–torsional instability is normally associated with beams, subject to vertical loading, buckling out of the plane of the applied loads by deflecting sideways and twisting; behaviour analogous to the flexural buckling of struts. The presence of both lateral and torsional deformations does cause both the governing mathematics and the resulting design treatment to be rather more complex.

17.5.1 Design of laterally unrestrained beams

The design of a beam taking into account lateral–torsional buckling consists essentially of assessing the maximum moment that can safely be carried from knowledge of the section's material and geometrical properties, the support conditions provided and the arrangement of the applied loading. Codes of practice, such as BS EN 1993-1-1, and supporting NCCI material include detailed guidance on the subject. Essentially the basic steps required to check a trial section (using BS EN 1993-1-1 for a Universal Beam as an example) are:

1. Determine beam slenderness either by calculating M_{cr}, the elastic critical moment, as explained under Method 1 of Reference 7, or by using the simpler Method 3 of the same reference and determining $\bar{\lambda}_{LT}$ directly. The simpler method is also set out in NCCI SN002.[8]
2. Calculate the reduction factor on cross-sectional moment capacity χ_{LT} using Clause 6.3.2.3(1) of BS EN 1993-1-1.

Hence, lateral torsional buckling resistance $M_{b,Rd}$ is given by:

$$M_{b,Rd} = \frac{\chi_{LT} W_y f_y}{\gamma_{M1}} \tag{17.6}$$

where W_y is the major axis section modulus (W_{pl}, W_{el} or W_{eff} depending on the classification of the cross-section) and $\gamma_{M1} = 1.0$.

17.5.1.1 Determination of beam slenderness $\bar{\lambda}_{LT}$

A beam's elastic critical moment M_{cr} is a property analogous to the Euler load for a strut, N_{cr}. However unlike the value of N_{cr}, which is readily calculated from the expression:

$$N_{cr} = \frac{\pi^2 EI}{L_{cr}^2} \tag{17.7}$$

determination of M_{cr} for any given arrangement is more difficult. For a doubly symmetrical I-section, loaded at the shear centre, M_{cr} is given by:

$$M_{cr} = C_1 \frac{\pi^2 EI_z}{L_{cr}^2} \sqrt{\frac{I_w}{I_z} + \frac{L_{cr}^2 GI_t}{\pi^2 EI_z}} \tag{17.8}$$

in which E and G are the Young's modulus and shear modulus of steel (210000 N/mm² and 81000 N/mm², respectively), I_z is the second moment of the area about the minor axis, I_t is the torsion constant, I_w is the warping constant, $L_{cr} = k_L$ is the buckling length of the member and C_1 is a factor whose value depends on the pattern of moments; values for C_1 for a series of common cases are provided in Table 17.7.

Knowing the value of M_{cr}, the non-dimensional beam slenderness $\bar{\lambda}_{LT}$ may be obtained from:

Table 17.7 Values of $1/\sqrt{C_1}$ and C_1 for various bending moment distributions when load is not destabilising

End Moment Loading	Ψ	$\frac{1}{\sqrt{C_1}}$	C_1
	+1.00	1.00	1.00
	+0.75	0.92	1.17
	+0.50	0.86	1.36
	+0.25	0.80	1.56
	0.00	0.75	1.77
	−0.25	0.71	2.00
	−0.50	0.67	2.24
	−0.75	0.63	2.49
	−1.00	0.60	2.76

Intermediate Transverse Loading		
	0.94	1.17
	0.62	2.60
	0.86	1.35
	0.77	1.69

$$\bar{\lambda}_{LT} = \sqrt{\frac{W_y f_y}{M_{cr}}} \tag{17.9}$$

In principle the method whereby M_{cr} is calculated explicitly may be used for any arrangement, including more complex cross-sectional shapes, e.g. unequal flanged I-sections, complex patterns of moment and complex arrangements of lateral support. A degree of ingenuity in locating a method to calculate M_{cr} correctly plus some understanding of the physical features of the problem in hand as they influence lateral torsional buckling will, however, be required. A useful tool for the determination of the elastic buckling moment M_{cr} for numerous section types, loading configurations and support conditions is also available in the form of freely downloadable software called LTBeam.[9]

As a simpler alternative to working with M_{cr}, beam slenderness $\bar{\lambda}_{LT}$ may be determined directly from the following relationship, as set out in NCCI SN002[8] and Method 3 of Reference 7:

$$\bar{\lambda}_{LT} = \frac{1}{\sqrt{C_1}} U V \bar{\lambda}_z \sqrt{\beta_w} \tag{17.10}$$

in which:

$\dfrac{1}{\sqrt{C_1}}$ is a parameter dependant on the shape of the bending moment diagram, which is given in Table 17.7 (Table 6.4 of Reference 7) for loads which are not destabilising.

U is a section property (given in section property tables, or may conservatively be taken as 0.9).

V is a parameter related to slenderness, and for symmetric rolled sections where the loads are not destabilising, may be conservatively taken as 1.0 or as:

$$V = \frac{1}{\sqrt[4]{1 + \frac{1}{20}\left(\frac{\lambda_z}{h/t_f}\right)^2}} \tag{17.11}$$

where:

$\lambda_z = \dfrac{L_{cr}}{i_z} = \dfrac{kL}{i_z}$, in which k may conservatively be taken as 1.0 for beams supported and restrained against twist at both ends. With certain additional restraint conditions k may be less than 1.0, as described in Section F.1 of Reference 7. For the value of k for cantilevers, see Section F.3 of Reference 7.

$$\bar{\lambda}_z = \frac{\lambda_z}{\lambda_1} \tag{17.12}$$

L is the distance between points of lateral restraint

$$\lambda_1 = \pi \sqrt{\frac{E}{f_y}} \tag{17.13}$$

$$\beta_w = \frac{W_y}{W_{pl,y}} \tag{17.14}$$

It is conservative to assume that the product $UV = 0.9$ and that $\beta_w = 1.0$.

In its most conservative form, $\bar{\lambda}_{LT} = \dfrac{L/i_z}{96}$ for S275 and $\bar{\lambda}_{LT} = \dfrac{L/i_z}{85}$ for S355.

17.5.1.2 Determination of buckling reduction factor χ_{LT}

The buckling reduction factor χ_{LT} is controlled by the slenderness of the beam $\bar{\lambda}_{LT}$. The method to determine χ_{LT} for beams is analogous to that for columns, and for the 'general case' (Clause 6.3.2.2 of BS EN 1993-1-1) is derived from the same relationship:

$$\chi_{LT} = \frac{1}{\Phi_{LT} + \sqrt{\Phi_{LT}^2 - \bar{\lambda}_{LT}^2}} \text{ but } \chi_{LT} \le 1.0 \tag{17.15}$$

with

$$\Phi_{LT} = 0.5[1 + \alpha_{LT}(\bar{\lambda}_{LT} - 0.2) + \bar{\lambda}_{LT}^2] \tag{17.16}$$

in which α_{LT} is the imperfection factor corresponding to the appropriate buckling curve defined in Table 6.4 of BS EN 1993-1-1 and the 'LT' subscripts indicate lateral torsional buckling. Note that for rolled sections and hot-finished and cold-formed hollow sections, no reduction is required for lateral torsional buckling (i.e. $\chi_{LT} = 1.0$) provided the beam slenderness $\bar{\lambda}_{LT} \le 0.4$.

For rolled or equivalent welded sections, the modified buckling curves (Equations 17.17 and 17.18) given in Clause 6.3.2.3 of BS EN 1993-1-1 may be applied. Reference is now made to Table 6.5 of BS EN 1993-1-1 for the selection of buckling curves for different section types, but note that this table is replaced in the UK National Annex.

$$\chi_{LT} = \frac{1}{\Phi_{LT} + \sqrt{\Phi_{LT}^2 - \beta\bar{\lambda}_{LT}^2}} \text{ but } \begin{cases} \chi_{LT} \le 1.0 \\ \chi_{LT} \le \dfrac{1}{\bar{\lambda}_{LT}^2} \end{cases} \tag{17.17}$$

with

$$\Phi_{LT} = 0.5[1 + \alpha_{LT}(\bar{\lambda}_{LT} - \bar{\lambda}_{LT,0}) + \beta\bar{\lambda}_{LT}^2] \tag{17.18}$$

where, from the UK National Annex, $\bar{\lambda}_{LT,0} = 0.4$ and $\beta = 0.75$ for rolled sections and hot-finished and cold-formed hollow section and $\bar{\lambda}_{LT,0} = 0.2$ and $\beta = 1.00$ for welded sections. Further economy for rolled or equivalent welded sections is available through the f factor, which may be used to derive a modified buckling reduction factor $\chi_{LT,mod}$, as described in Clause 6.3.2.3 of BS EN 1993-1-1.

17.5.2 Lateral bracing

The stability of a beam liable to fail by lateral-torsional buckling may be improved by providing suitable forms of lateral restraint that limit the buckling type deformations. Arrangements may, if desired, be configured so that the beam's full in-plane resistance can be developed i.e. there is no loss of load-carrying capacity due to lateral-torsional buckling effects.

To be effective, lateral restraints should:

- possess sufficient stiffness to limit the buckling deformations
- possess sufficient strength to resist the load transmitted as a result of restricting these buckling deformations.

The former property acts directly to improve member resistance; the latter is a necessary consequence of the former.

Useful guidance on the provision of lateral restraint, including detailed consideration of a number of practical arrangements, is provided in Reference 6. In particular, Section 2 includes a helpful introduction to all the basic concepts, identifies the key physical features and provides direct comment on the use of the Eurocode.

A key design issue for individual lateral restraints is to determine how much force they are required to resist. Although restraints require both strength and stiffness, design codes often specify only minimum strength, with the assumption that practical bracing members that meet this requirement will also possess sufficient stiffness. Clause 5.3.3 of BS EN 1993-1-1, however, allows explicitly for the stiffness of the bracing system (through calculation of its deflection δ_q under the stabilising forces and any other external loads, e.g. wind loads) in determination of the force that it has to resist. Hence, stiffer bracing systems are required to resist lower bracing forces. For typical bracing systems in buildings, deflections will be such that the restraint forces are unlikely to exceed 2.0% of the maximum design force in the compression flange of a beam for any single restraint.[6]

A further important topic is the procedure for deciding on the maximum unbraced length for which the full member resistance may be achieved. Clearly, positioning restraints at these intervals produces an efficient and easily understood design situation i.e. no consideration of lateral-torsional buckling is then necessary. Although the material of Annex BB.3 is presented in the context of columns and rafters in portal frame construction, the provisions may be used more generally – providing the underlying principles are correctly appreciated. These are:

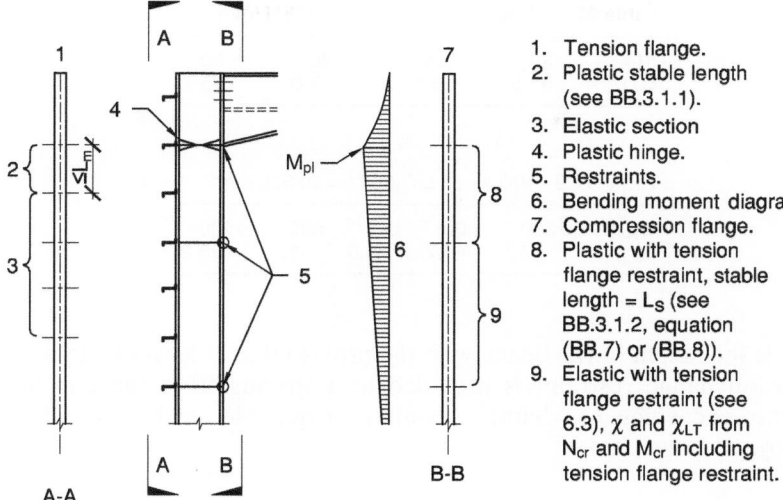

1. Tension flange.
2. Plastic stable length (see BB.3.1.1).
3. Elastic section.
4. Plastic hinge.
5. Restraints.
6. Bending moment diagram.
7. Compression flange.
8. Plastic with tension flange restraint, stable length = L$_S$ (see BB.3.1.2, equation (BB.7) or (BB.8)).
9. Elastic with tension flange restraint (see 6.3), χ and χ_{LT} from N$_{cr}$ and M$_{cr}$ including tension flange restraint.

Figure 17.4 Stable lengths between restraints for members containing plastic hinges

- torsional restraints are defined as arrangements that prevent both lateral deflection and twisting e.g. restraint to both the tension and compression flanges
- lateral restraints are defined as arrangements that only prevent lateral deflection of the compression flange i.e. lateral deflection of the tension flange and twisting are still possible.

Figure 17.4, reproduced from **BS EN 1993-1-1**, illustrates both arrangements, together with several other conceptual features:

- the method covers members containing plastic hinges (for which the rotation capacity essential for the use of plastic design is necessary) but may (conservatively) be used for regions where rotation capacity is not required as well as for wholly elastic regions
- identification of fully braced i.e. torsionally restrained locations is important.

The basis of the approach is the determination of the maximum length L_k (or L_s in the presence of a moment gradient) of a segment of a beam between a pair of fully restrained points for which lateral-torsional buckling effects may be ignored. When such a length is required to develop a plastic hinge at one end i.e. the most severe practical arrangement, this stable length is defined as:

$$L_k = \frac{\left(5.4 + \dfrac{600 f_y}{E}\right)\left(\dfrac{h}{t_f}\right) i_z}{\sqrt{5.4\left(\dfrac{f_y}{E}\right)\left(\dfrac{h}{t_f}\right)^2 - 1}}$$
(17.19)

Table 17.8   Values of L_k/i_z for $f_y = 275\,\text{N}/\text{mm}^2$

h/t_t	20	30	40	50
L_k/i_z	91.5	80.1	77.0	75.7

Table 17.9   Values of L_k/i_z for $f_y = 275\,\text{N}/\text{mm}^2$

$W_{pl,y}^2/AI_t$	200	400	600	800	1000
L_k/i_z	63.1	44.7	36.5	31.6	28.2

in which h is the height of the beam, with the proviso that at least one intermediate lateral tension flange restraint is provided at a spacing from the end torsional restraints not exceeding L_m, defined (assuming negligible axial load and uniform moment) by:

$$L_m = \frac{38 i_z}{\dfrac{1}{27.5}\left(\dfrac{f_y}{235}\right)\sqrt{\dfrac{W_{pl,y}^2}{AI_t}}} \tag{17.20}$$

where A is the cross-sectional area of the member, $W_{pl,y}$ is the plastic section modulus of the member, I_t is the torsion constant of the member and f_y is the yield strength in N/mm^2.

For $f_y = 275\,\text{N}/\text{mm}^2$, Equation 17.19 yields the values of L_k/i_z given by Table 17.8.

If no intermediate lateral restraints are present, the maximum permitted distance between torsional restraints is L_m. Taking some typical sections and assuming uniform moment loading ($C_1 = 1.0$), negligible axial force and $f_y = 275\,\text{N}/\text{mm}^2$, safe values for L_m/i_z are given by Table 17.9.

More accurate values may be obtained by following the full process of Annex BB.3 of BS EN 1993-1-1 noting that its focus is on the columns and rafters of portal frames. It should also be noted that the above use of Annex BB.3 presumes that plastic hinge action is required; for the more usual and less severe case of simply attaining the full moment capacity at the point of maximum moment the results will therefore be conservative.

17.6 Beams with web openings

Advances in fabrication techniques – specifically the ability to split I-section beams using curved or straight line cuts using automatic methods plus the ability to re-weld these sections – has led to ever more imaginative types of beam containing hexagonal, circular and elliptical web openings. These are largely proprietary products, for which design information – often in the form of specialist software – is provided by the supplier. This typically covers all important design checks and may even extend beyond basic considerations to cover features such as design for fire resistance.

A series of worked examples follow which are relevant to Chapter 17. Further worked examples may be found in References 10–13.

References to Chapter 17

1. British Standards Institution (2005) *BS EN 1993-1-1 Eurocode 3: Design of steel structures – Part 1-1: General rules and rules for buildings.* CEN. London, BSI.
2. SCI and BCSA (2009) *Steel Building Design: Design Data – In accordance with Eurocodes and the UK National Annexes.* SCI Publication No. P363. Ascot, SCI.
3. Johnson R.P. and Buckby R.J. (1979) *Composite Structures of Steel and Concrete, Vol. 2: Bridges with a Commentary on BS 5400: Part 5*, 1st edn. (also 2nd edn, 1986). London, Granada.
4. Hughes A.S. and Malik A. (2010) *Steel Building Design: Combined bending and torsion.* SCI Publication 385. Ascot, SCI.
5. Hendy C.R. and Murphy C.J. (2007) *Designers' Guide to EN 1993-2. Eurocode 3: Design of Steel Structures. Part 2: Steel bridges.* London, Thomas Telford Publishing.
6. Gardner L. (2011) *Stability of Beams and Columns.* SCI Publication No. P360. Ascot, SCI.
7. SCI and BCSA. (2009) *Steel Building Design: Concise Eurocodes.* SCI Publication No. P362. Ascot, SCI.
8. NCCI SN002 Access Steel *Determination of non-dimensional slenderness of I-and H-sections.* www.access-steel.com
9. CTICM: LTBeam. *Lateral torsional buckling of beams software.* http://www.steelbizfrance.com/telechargement/desclog.aspx?idrub=1andlng=2
10. Trahair N.S., Bradford M.A., Nethercot D.A. and Gardner L. (2008) *The Behaviour and Design of Steel Structures to EC3*, 4th edn. London and New York, Taylor and Francis.
11. Gardner L. and Nethercot D.A. (2005) *Designers' Guide to EN 1993-1-1: Eurocode 3: Design of Steel Structures.* London, Thomas Telford Publishing.
12. Martin L. and Purkiss J. (2008) *Structural Design of Steelwork to EN 1993 and EN 1994*, 3rd edn. Oxford, Butterworth-Heinemann.
13. Brettle M. (2009) *Steel Building Design: Worked Examples – Open Sections – In accordance with Eurocodes and the UK National Annexes.* SCI Publication No. P364. Ascot, SCI.

		Job No.		Sheet 1 of 2		Rev
The Steel Construction Institute		Job Title				
Silwood Park, Ascot, Berks SL5 7QN		Subject	Beam example 1			
Telephone: (01344) 623345 Fax: (01344) 622944		Client	Made by	DAN	Date	2010
CALCULATION SHEET			Checked by	LG	Date	2010

Beam example 1

Rolled Universal Beam using Design Tables

<u>***Problem***</u>

Determine the cross-section classification and moment resistance for a $533 \times 210 \times 82$ UKB when used as a beam in (1) S275 steel and (2) S355 steel.

(1) Using S275 steel

From Reference 2, page C-66, cross-section is Class 1 and $M_{c,y,Rd} = 566 \, kNm$

Alternatively, from Reference 2, page B-4:
$c_f/t_f = 6.58$, $c_w/t_w = 49.6$, $W_{pl,y} = 2060 \, cm^3$

Yield strength $f_y = 275 \, N/mm^2$ since $t_f = 13.2 \, mm \leq 16 \, mm$ *BS EN 10025-2*

$$\varepsilon = \sqrt{\frac{235}{f_y}} = \sqrt{\frac{235}{275}} = 0.92$$

From Table 5.1 of BS EN 1993-1-1:   *BS EN 1993-1-1 Table 5.1*

For a Class 1 outstand flange in compression: $c_f/t_f \varepsilon \leq 9$

For a Class 1 web in bending: $c_w/t_w \varepsilon \leq 72$

Actual $c_f/t_f \varepsilon = 6.58/0.92 = 7.12$; within limit

Actual $c_w/t_w \varepsilon = 49.6/0.92 = 53.7$; within limit

∴ Cross-section is Class 1

$$\therefore M_{c,y,Rd} = M_{pl,y,Rd} = \frac{W_{pl,y}f_y}{\gamma_{M0}} = \frac{2060 \times 275}{1.0} \times 10^{-3} = 566.5 \, kNm$$ *BS EN 1993-1-1 Cl 6.2.5(2)*

(2) Using S355 steel

From Reference 2, page D-66 cross-section is Class 1 and $M_{c,y,Rd} = 731 \, kNm$

Beam example 1	Sheet 2 of 2	Rev

Alternatively, from Reference 2, page B-4: $c_f/t_f = 6.58$, $c_w/t_w = 49.6$, $W_{pl,y} = 2060\ cm^3$ *Yield strength* $f_y = 355\ N/mm^2$ *since* $t_f = 13.2\ mm \leq 16\ mm$	*BS EN 10025-2*
$\varepsilon = \sqrt{\dfrac{235}{f_y}} = \sqrt{\dfrac{235}{355}} = 0.81$	
From Table 5.1 of BS EN 1993-1-1: *For a Class 1 outstand flange in compression:* $c_f/t_f\varepsilon \leq 9$ *For a Class 1 web in bending:* $c_w/t_w\varepsilon \leq 72$ *Actual* $c_f/t_f\varepsilon = 6.58/0.81 = 8.09$; *within limit* *Actual* $c_w/t_w\varepsilon = 49.6/0.81 = 61.0$; *within limit* $\therefore$ *Cross-section is Class 1*	*BS EN 1993-1-1* *Table 5.1*
$\therefore M_{c,y,Rd} = M_{pl,y,Rd} = \dfrac{W_{pl,y}f_y}{\gamma_{M0}} = \dfrac{2060 \times 355}{1.0} \times 10^{-3} = 731.3\ kNm$	*BS EN 1993-1-1* *Cl 6.2.5(2)*

		Job No.			Sheet 1 of 3	Rev
The Steel Construction Institute		Job Title				
Silwood Park, Ascot, Berks SL5 7QN Telephone: (01344) 623345 Fax: (01344) 622944 **CALCULATION SHEET**		Subject	Beam example 2			
		Client		Made by	DAN	Date	2010
				Checked by	LG	Date	2010

Beam example 2

Laterally restrained Universal Beam

Problem

Select a suitable UKB in S275 steel to function as a simply supported beam carrying a 140 mm thick solid concrete slab together with an imposed load of 7.0 kN/m². The beam span is 7.2 m and beams are spaced at 3.6 m intervals. The slab may be assumed capable of providing continuous lateral restraint to the beam's top flange. Assume a concrete density of 2400 kg/m³ and a deflection limit of span/360.

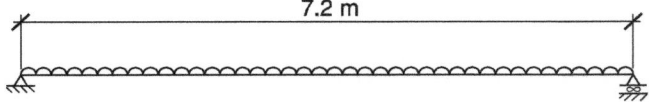

Due to the lateral restraint from the concrete slab, there is no possibility of lateral torsional buckling, so design the beam for:

i)   Cross-section bending resistance

ii)   Shear resistance

iii)   Deflections

Loading

Assume self-weight of beam		*= 1.0 kN/m*
Permanent load (concrete slab)	*= 2400 × 9.81 × 0.14 × 10–3 × 3.6*	*= 11.9 kN/m*
Variable imposed loading	*= 7.0 × 3.6*	*= 25.2 kN/m*
Design combination at ULS	*= (1.35 × [1.0 + 11.9]) + (1.5 × 25.2)*	*= 55.2 kN/m*
Design ultimate moment M_{Ed}	*= 55.2 × 7.22/8*	*= 357.5 kNm*
Design ultimate shear force V_{Ed} = 55.2 × 7.2/2		*= 198.6 kN*

Initial sizing

Adopting S275 steel and assuming no material is greater than 16 mm thick (to be confirmed later), the nominal yield strength f_y is 275 N/mm².

BS EN 10025-2

Beam example 2	Sheet 2 of 3	Rev

Assuming that the cross-section is Class 1 or 2 in bending (to be confirmed when a section is chosen):

Required $W_{pl,y} = 357.5 \times 10^6/275 = 1.30 \times 10^6\,mm^3 = 1300\,cm^3$

From section tables, a 457 × 152 × 67 UKB has a value of $W_{pl,y}$ *of 1450 cm³, and a self-weight less than that assumed.*

Maximum component thickness is flange thickness $t_f = 15.0\,mm \leq 16.0\,mm$
∴ $f_y = 275\,N/mm^2$ *is OK.*

BS EN 10025-2

Check cross-section classification

$$\varepsilon = \sqrt{\frac{235}{f_y}} = \sqrt{\frac{235}{275}} = 0.92$$

From Table 5.1 of BS EN 1993-1-1:

For a Class 1 outstand flange in compression: $c_f/t_f\varepsilon \leq 9$

BS EN 1993-1-1 Table 5.2

For a Class 1 web in bending: $c_w/t_w\varepsilon \leq 72$

Actual $c_f/t_f\varepsilon = 4.15/0.92 = 4.49$; *within limit*

Actual $c_w/t_w\varepsilon = 45.3/0.92 = 49.0$; *within limit*

∴ *Cross-section is Class 1.*

Bending resistance

BS EN 1993-1-1 Cl 6.2.5

$$M_{c,y,Rd} = \frac{W_{pl,y}f_y}{\gamma_{M0}} \text{ for Class 1 or 2 cross-sections}$$

$$M_{c,y,Rd} = \frac{1450 \times 10^3 \times 275}{1.00} \times 10^{-6} = 398.8\,kNm > 357.5\,kNm$$

∴ *Cross-section resistance in bending is OK.*

Shear resistance

BS EN 1993-1-1 Cl 6.2.6

$$V_{pl,Rd} = \frac{A_v(f_y/\sqrt{3})}{\gamma_{M0}}$$

For a rolled I-section, loaded parallel to the web, the shear area A_v *is given by:*

$A_v = A - 2bt_f + (t_w + 2r)t_f$ *(but not less than* $\eta h_w t_w$)

From UK NA to BS EN 1993-1-5, $\eta = 1.0$. *With* $\eta = 1.0$, $A_v > \eta h_w t_w$ *for all UKB and UKC.*

UK NA to BS EN 1993-1-5

Beam example 2	Sheet 3 of 3	Rev

$\therefore A_v = 8560 - (2 \times 153.8 \times 15.0) + (9.0 + [2 \times 10.2]) \times 15.0 = 4387\,mm^2$

$V_{pl,Rd} = \dfrac{4387 \times (275/\sqrt{3})}{1.00} \times 10^{-3} = 696.5kN > 198.6kN$

$\therefore$ *Shear resistance is OK*

Deflections

Check deflection under unfactored variable loads.

Assumed deflection limit is span/360 = 7200/360 = 20.0 mm

Actual deflection:

$\delta = \dfrac{5}{384} \dfrac{wL^4}{EI} = \dfrac{5 \times 25.2 \times 7200^4}{384 \times 210000 \times 28900 \times 10^4} = 14.5\,\text{mm} < 20.0\ \text{mm}$

$\therefore$ *Deflections OK*

$\therefore$ *Use 457 × 152 × 67 UKB in Grade S275 steel.*

UK NA to BS EN 1993-1-1

 **The Steel Construction Institute**	Job No.			Sheet 1 of 2		Rev
	Job Title					
Silwood Park, Ascot, Berks SL5 7QN Telephone: (01344) 623345 Fax: (01344) 622944 **CALCULATION SHEET**	Subject	Beam example 3				
	Client		Made by	DAN	Date	2010
			Checked by	LG	Date	2010

Beam example 3

Laterally unrestrained Universal Beam

Problem

For the same loading and support conditions of example 2, select a suitable UKB in S275 steel assuming that the member must now be designed as laterally unrestrained.

It is not now possible to arrange the calculations in such a way that a direct choice of section can be made (since susceptibility to lateral torsional buckling is not yet known); an iterative design approach is therefore required. In this example, the simplified method set out in NCCI SN002 and referred to as Method 3 in Reference 7 will be employed.

Try $610 \times 229 \times 125$ UKB

Geometric properties:

$h = 612.2\,mm;$ $b = 229.0\,mm;$ $t_w = 11.9\,mm;$ $t_f = 19.6\,mm;$
$c_f/t_f = 4.89;$ $c_w/t_w = 46.0;$ $i_z = 4.97\,cm = 49.7\,mm;$
$W_{pl,y} = 3680\,cm^3 = 3680 \times 10^3\,mm^3$

                                                             Steel Building Design: Design Data

Maximum component thickness is flange thickness $t_f = 19.6\,mm > 16.0\,mm$ $\therefore f_y = 265\,N/mm^2$.   *BS EN 10025-2*

Check cross-section classification:

$$\varepsilon = \sqrt{\frac{235}{f_y}} = \sqrt{\frac{235}{265}} = 0.94$$

From Table 5.1 of BS EN 1993-1-1:

For a Class 1 outstand flange in compression: $c_f/t_f \varepsilon \leq 9$ *BS EN 1993-1-1 Table 5.2*

For a Class 1 web in bending: $c_w/t_w \varepsilon \leq 72$

Actual $c_f/t_f \varepsilon = 4.89/0.94 = 5.19$; within limit

Actual $c_w/t_w \varepsilon = 46.0/0.94 = 48.8$; within limit

$\therefore$ Cross-section is Class 1

Using NCCI SN002 (Method 3 of Reference 7), lateral torsional buckling slenderness is defined as:

Beam example 3	Sheet 2 of 2	Rev

$$\bar{\lambda}_{LT} = \frac{1}{\sqrt{C_1}} UV\bar{\lambda}_z \sqrt{\beta_w}$$

$\beta_w = 1.0$ for a Class 1 or 2 section. Take $UV = 0.9$.

For the present shape of bending moment diagram, $\dfrac{1}{\sqrt{C_1}} = 0.94$

Table 17.7

$$\lambda_1 = \pi\sqrt{\frac{E}{f_y}} = \pi\sqrt{\frac{210000}{265}} = 88.4$$

$$\bar{\lambda}_z = \frac{\lambda_z}{\lambda_1} = \frac{L_{cr}/i_z}{\lambda_1} = \frac{7200/49.7}{88.4} = 1.64$$

$$\therefore \bar{\lambda}_{LT} = \frac{1}{\sqrt{C_1}} UV\bar{\lambda}_z \sqrt{\beta_w} = 0.94 \times 0.9 \times 1.64 = 1.39$$

Buckling curve selection:

$h/b = 612.2/229.0 = 2.67$

For the case of rolled and equivalent welded sections, for I-sections with $2 < h/b \le 3.1$, use buckling curve 'c' ($\alpha = 0.49$). For rolled sections, $\beta = 0.75$ and $\bar{\lambda}_{LT,0} = 0.4$.

UK NA to BS EN 1993-1-1

Buckling reduction factor χ_{LT}:

$$\begin{aligned}\Phi_{LT} &= 0.5[1 + \alpha_{LT}(\bar{\lambda}_{LT} - \bar{\lambda}_{LT,0}) + \beta\bar{\lambda}_{LT}^2] \\ &= 0.5[1 + 0.49(1.39 - 0.4) + (0.75 \times 1.39^2)] = 1.46\end{aligned}$$

BS EN 1993-1-1 Cl 6.3.2.3

$$\chi_{LT} = \frac{1}{\Phi_{LT} + \sqrt{\Phi_{LT}^2 - \beta\bar{\lambda}_{LT}^2}} = \frac{1}{1.46 + \sqrt{1.46^2 - 0.75 \times 1.39^2}} = 0.44$$

Lateral torsional buckling resistance:

$$\begin{aligned}M_{b,Rd} &= \chi_{LT}W_y\frac{f_y}{\gamma_{M1}} = 0.44 \times 3680 \times 10^3 \frac{265}{1.0} \times 10^{-6} \\ &= 424.7\ kNm < 357.5\ kNm = M_{Ed} \qquad\qquad \therefore OK\end{aligned}$$

BS EN 1993-1-1 Cl 6.3.2.1

Since section is larger than before, $V_{c,Rd}$ and δ will also be satisfactory.

$\therefore$ *Use $610 \times 229 \times 125$ UKB in Grade S275 steel.*

The Steel Construction Institute	Job No.			Sheet 1 of 4		Rev
	Job Title					
Silwood Park, Ascot, Berks SL5 7QN Telephone: (01344) 623345 Fax: (01344) 622944 **CALCULATION SHEET**	Subject	Beam example 4				
	Client		Made by	LG	Date	2010
			Checked by	DAN	Date	2010

Beam example 4

Universal beam supporting point loads

Problem

Select a suitable UKB in S275 steel to carry a pair of point loads at the third points transferred by cross-beams as shown in the accompanying sketch.

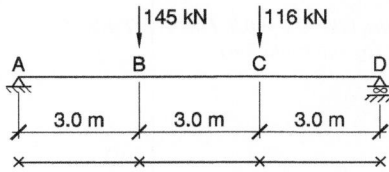

The cross-beams may reasonably be assumed to provide full lateral and torsional restraint at B and C; assume further that ends A and D are similarly restrained. Thus the actual level of transfer of load at B and C (relative to the main beam's shear centre) will have no effect; the general lateral torsional buckling aspects of the design are therefore to consider the segments AB, BC and CD separately. In this example, the simplified method set out in NCCI SN002 and referred to as Method 3 in Reference 7 will be employed.

From statics, the bending moment diagram (BMD) and shear force diagram (SFD) are as follows:

BMD

```
              406 kNm   377 kNm
```

SFD

```
                      126 kN
          10 kN
135 kN
```

For initial trial section, select a UKB with $M_{c,Rd} > 406$ kNm

A $457 \times 152 \times 74$ UKB is Class 1 and provides $M_{c,Rd} = 431$ kNm. Given that each of the unrestrained segments (AB, BC and CD) have the same length ($L_{cr} = 3.0$ m), it is clear from the BMD that segment BC is critical (since it contains the most severe magnitude and distribution of bending moments). Therefore, only segment BC need be considered for lateral torsional buckling.

Steel Building Design: Design Data

Beam example 4	Sheet 2 of 4	Rev

Try 457 × 152 × 74 UKB

Geometric properties:

$h = 462.0\,mm$; $b = 154.4\,mm$; $t_w = 9.6\,mm$; $t_f = 17.0\,mm$;
$c_f/t_f = 3.66$; $c_w/t_w = 42.5$; $i_z = 3.33\ cm = 33.3\,mm$;
$W_{pl,y} = 1630\ cm^3 = 1630 \times 10^3\,mm^3$

Steel Building Design: Design Data

Maximum component thickness is flange thickness $t_f = 19.6\,mm > 16.0\,mm$
$\therefore f_y = 265\ N/mm^2$

BS EN 10025-2

Using NCCI SN002 (Method 3 of Reference 7), lateral torsional buckling slenderness is defined as:

$$\bar{\lambda}_{LT} = \frac{1}{\sqrt{C_1}} UV\bar{\lambda}_z \sqrt{\beta_w}$$

For class 1 or 2 sections, $\beta_w = 1.0$. Conservatively for I-sections, take $UV = 0.9$. For the ratio of end moments $\psi = 377/406 = 0.93$, $1/\sqrt{C_1} = 0.98$ (by interpolation) from Table 16.7 (Table 6.4 of Reference 7).

Table 17.7

$$\lambda_1 = \pi\sqrt{\frac{E}{f_y}} = \pi\sqrt{\frac{210000}{265}} = 88.4$$

$$\bar{\lambda}_z = \frac{\lambda_z}{\lambda_1} = \frac{L_{cr}/i_z}{\lambda_1} = \frac{3000/33.3}{88.4} = 1.02$$

$$\therefore \bar{\lambda}_{LT} = \frac{1}{\sqrt{C_1}} UV\bar{\lambda}_z \sqrt{\beta_w} = 0.98 \times 0.9 \times 1.02 \times 1.0 = 0.90$$

Buckling curve selection:

$h/b = 462.0/154.4 = 2.99$

For the case of rolled and equivalent welded sections, for I-sections with $2 < h/b \leq 3.1$, use buckling curve 'c' ($\alpha = 0.49$).

UK NA to BS EN 1993-1-1

For rolled sections, $\beta = 0.75$ and $\bar{\lambda}_{LT,0} = 0.4$.

Buckling reduction factor χ_{LT}:

$$\Phi_{LT} = 0.5[1 + \alpha_{LT}(\bar{\lambda}_{LT} - \bar{\lambda}_{LT,0}) + \beta\bar{\lambda}_{LT}^2]$$
$$= 0.5[1 + 0.49(0.90 - 0.4) + (0.75 \times 0.90^2)] = 0.92$$

BS EN 1993-1-1 Cl 6.3.2.3

$$\chi_{LT} = \frac{1}{\Phi_{LT} + \sqrt{\Phi_{LT}^2 - \beta\bar{\lambda}_{LT}^2}} = \frac{1}{0.92 + \sqrt{0.92^2 - 0.75 \times 0.90^2}} = 0.70$$

Beam example 4	Sheet 3 of 4	Rev

Lateral torsional buckling resistance:

$$M_{b,Rd} = \chi_{LT} W_y \frac{f_y}{\gamma_{M1}} = 0.70 \times 1630 \times 10^3 \frac{265}{1.0} \times 10^{-6}$$

$$= 303 \, kNm < 406 \, kNm = M_{Ed} \qquad \therefore \textit{Not OK, try larger section}$$

BS EN 1993-1-1 Cl 6.3.2.1

Try 457 × 191 × 82 UKB

Geometric properties:

$h = 460.0 \, mm; \, b = 191.3 \, mm; \, t_w = 9.9 \, mm; \, t_f = 16.0 \, mm;$
$i_z = 4.23 \, cm = 42.3 \, mm; \, W_{pl,y} = 1830 \, cm^3 = 1830 \times 10^3 \, mm^3; \, Section \, is \, Class \, 1$
$A = 104 \, cm^2 = 10400 \, mm^2; \, I_y = 37100 \, cm^4 = 371 \times 10^6 \, mm^4.$

Steel Building Design: Design Data

Maximum component thickness is flange thickness $t_f = 16.0 \, mm \leq 16.0 \, mm$
$\therefore f_y = 275 \, N/mm^2$

BS EN 10025-2

As above, using NCCI SN002 (Method 3 of Reference 7), lateral torsional buckling slenderness is defined as:

$$\bar{\lambda}_{LT} = \frac{1}{\sqrt{C_1}} UV\bar{\lambda}_z \sqrt{\beta_w}$$

For Class 1 or 2 sections, $\beta_w = 1.0$. Conservatively for I-sections, take $UV = 0.9$. As above, ratio of end moments $\psi = 377/406 = 0.93$. $\therefore 1/\sqrt{C_1} = 0.98$.

Table 17.7

$$\lambda_1 = \pi \sqrt{\frac{E}{f_y}} = \pi \sqrt{\frac{210000}{275}} = 86.8$$

$$\bar{\lambda}_z = \frac{\lambda_z}{\lambda_1} = \frac{L_{cr}/i_z}{\lambda_1} = \frac{3000/42.3}{86.8} = 0.82$$

$$\therefore \bar{\lambda}_{LT} = \frac{1}{\sqrt{C_1}} UV\bar{\lambda}_z \sqrt{\beta_w} = 0.98 \times 0.9 \times 0.82 \times 1.0 = 0.72$$

Buckling curve selection:

$h/b = 460.0/191.3 = 2.40$

For the case of rolled and equivalent welded sections, for I-sections with $2 < h/b \leq 3.1$, use buckling curve 'c' ($\alpha = 0.49$). For rolled sections, $\beta = 0.75$ and $\bar{\lambda}_{LT,0} = 0.4$.

UK NA to BS EN 1993-1-1

Buckling reduction factor χ_{LT}:

$$\Phi_{LT} = 0.5[1 + \alpha_{LT}(\bar{\lambda}_{LT} - \bar{\lambda}_{LT,0}) + \beta\bar{\lambda}_{LT}^2]$$

$$= 0.5[1 + 0.49(0.72 - 0.4) + (0.75 \times 0.72^2)] = 0.77$$

BS EN 1993-1-1 Cl 6.3.2.3

$$\chi_{LT} = \frac{1}{\Phi_{LT} + \sqrt{\Phi_{LT}^2 - \beta\bar{\lambda}_{LT}^2}} = \frac{1}{0.77 + \sqrt{0.77^2 - 0.75 \times 0.72^2}} = 0.81$$

Beam example 4	Sheet 4 of 4	Rev

Lateral torsional buckling resistance:

$$M_{b,Rd} = \chi_{LT} W_y \frac{f_y}{\gamma_{M1}} = 0.81 \times 1830 \times 10^3 \frac{275}{1.0} \times 10^{-6}$$

$$= 409 \, kNm > 406 \, kNm = M_{Ed} \quad \therefore OK$$

BS EN 1993-1-1 Cl 6.3.2.1

A 457 × 191 × 82 UKB provides sufficient resistance to lateral torsional buckling for segment BC and, by inspection, also segments AB and CD. The beam is therefore satisfactory in bending.

Shear resistance:

BS EN 1993-1-1 Cl 6.2.6

$$V_{pl,Rd} = \frac{A_v (f_y / \sqrt{3})}{\gamma_{M0}}$$

For a rolled I section, loaded parallel to the web, the shear area A_v is given by:

$$A_v = A - 2bt_f + (t_w + 2r)t_f \quad \text{(but not less than } \eta h_w t_w)$$

From UK NA to BS EN 1993-1-5, $\eta = 1.0$. With $\eta = 1.0$, $A_v > \eta h_w t_w$ for all UKB and UKC.

UK NA to BS EN 1993-1-5

$$\therefore A_v = 10400 - (2 \times 191.3 \times 16.0) + (9.9 + [2 \times 10.2]) \times 16.0 = 4763 \, mm^2$$

$$V_{pl,Rd} = \frac{4763 \times (275 / \sqrt{3})}{1.00} \times 10^{-3} = 756 \, kN > 135 \, kN = V_{Ed}$$

∴ Shear resistance is OK

Deflections:

Check deflection under unfactored variable loads. Assume for initial design that the point loads can be represented as a UDL and that the design loads can be factored down by 1.5 to estimate the serviceability loads.

UK NA to BS EN 1993-1-1

Assumed deflection limit is span/360 = 3000/360 = 25.0 mm

Assumed serviceability UDL $w = \dfrac{145 + 116}{1.5 \times 9} = 19.3 \, kN/m$

Actual deflection:

$$\delta = \frac{5}{384} \frac{wL^4}{EI} = \frac{5 \times 19.3 \times 9000^4}{384 \times 210000 \times 371 \times 10^6} = 21.2 \, mm < 25.0 \, mm \qquad \therefore OK$$

Beam is clearly satisfactory for deflection since these (approximate) calculations have used the full load and not just the imposed (variable) load.

∴ Use 457 × 191 × 82 UKB in Grade S275 steel.

Chapter 18
Plate girders

LEROY GARDNER

18.1 Introduction

Plate girders are fabricated sections employed to support heavy vertical loads over long spans for which the resulting bending moments are larger than the moment resistance of available rolled sections. In its simplest form the plate girder is a built-up beam consisting of two flange plates, fillet welded to a web plate to form an I-section (see Figure 18.1). The primary function of the top and bottom flange plates is to resist the axial compressive and tensile forces caused by the applied bending moments; the main function of the web is to resist the shear. Indeed this partition of structural action is used as the basis for design in some codes of practice.

For a given bending moment, the required flange areas can be reduced by increasing the distance between them. Thus, for an economical design it is advantageous to increase the distance between flanges. To keep the self-weight of the girder to a minimum the web thickness should be reduced as the depth increases, but this leads to web buckling considerations being more significant in plate girders than in rolled beams.

Plate girders are sometimes used in buildings, as transfer beams for example, and are often used in small to medium span bridges. Design provisions are contained in BS EN 1993-1-5 (2006).[1] This chapter explains current practice in designing plate girders, with an emphasis on buildings, and makes reference to the relevant code clauses.

18.2 Advantages and disadvantages

The development of highly automated workshops has reduced the fabrication costs of plate girders considerably; box girders and trusses still have to be largely fabricated manually with consequently high fabrication costs. Optimum use of material can be made with fabricated plate girders compared with rolled sections, since the designer has greater freedom to vary the section to correspond with changes in the

Steel Designers' Manual, Seventh Edition. Edited by Buick Davison and Graham W. Owens.
© 2012 Steel Construction Institute. Published 2012 by Blackwell Publishing Ltd.

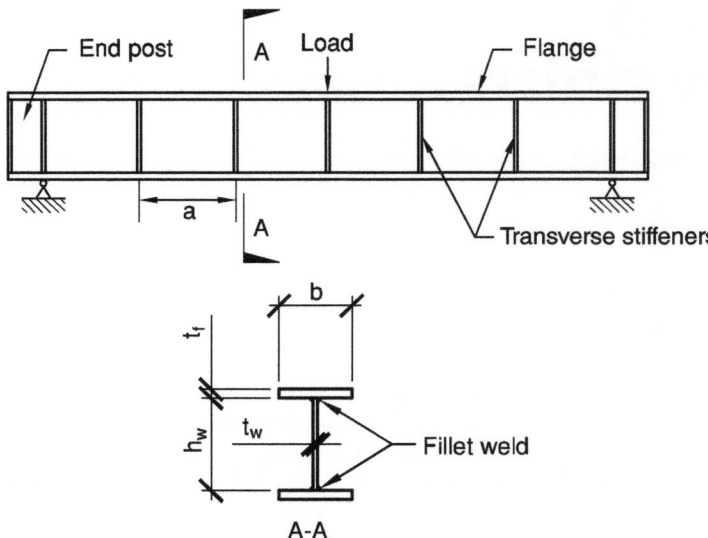

Figure 18.1   Elevation and cross-section of a typical plate girder

applied forces. Thus, variable depth plate girders have been increasingly designed in recent years. Plate girders are aesthetically more pleasing than trusses and are easier to transport and erect than box girders.

There are a few limitations to the use of plate girders. Compared with trusses, they are heavier, more difficult to transport and have larger wind resistance. The provision of openings for services is also more difficult. Plate girders can sometimes pose problems during erection because of concern for the stability of compression flanges.

18.3 Initial choice of cross-section for plate girders

18.3.1 Span-to-depth ratios

Advances in fabrication methods allow the economic manufacture of plate girders of constant or variable depth. Traditionally, constant-depth girders have been more common in buildings; however, this may change as designers become more inclined to modify the steel structure to accommodate services.[2] Recommended span-to-depth ratios are given in Table 18.1.

18.3.2 Recommended plate thickness and proportions

In general, the slenderness of the cross-sections of plate girders used in buildings should not exceed the limits specified for Class 3 cross-sections (presented in Clause 5.5 of BS EN 1993-1-1[3] and discussed in Chapter 14), even though more slender

Table 18.1 Recommended span-to-depth ratios for plate girders used in buildings

Applications	Span-to-depth ratio
(1) Constant-depth beams used in simply-supported composite girders, and for simply-supported non-composite girders with concrete decking	12 to 20
(2) Constant-depth beams used in continuous non-composite girders using concrete decking (N.B. continuous composite girders are uncommon in buildings)	15 to 20
(3) Simply-supported crane girders (non-composite construction is usual)	10 to 15

cross-sections are permitted. The choice of plate thickness is related to buckling of the cross-section. If the plates are too thin they may require stiffening to restore adequate stiffness and strength, and the extra workmanship required is expensive.

In view of the above, the maximum depth-to-thickness ratio of the webs (c_w/t_w) of plate girders in buildings is usually limited to:

$$c_w / t_w < 124\varepsilon = 124\left(\frac{235}{f_{yw}}\right)^{\frac{1}{2}}$$

where f_{yw} is the yield strength of the web plate.

The outstand width-to-thickness ratio of the compression flange (c_f/t_f) is typically limited to:

$$c_f / t_f < 14\varepsilon = 14\left(\frac{235}{f_{yf}}\right)^{\frac{1}{2}}$$

where f_{yf} is the yield strength of the compression flange. Note that c_w and c_f are the flat element widths, measured from the edges of the fillet welds (or root radii for rolled sections), as discussed in Chapter 14. For initial design purposes, when the weld size may be unknown, it is conservative to ignore the weld and take $c_w = h_w$ (the distance between the flanges) for webs and $c_f = b/2 - t_w/2$ for outstand flanges.

Changes in flange size along the girder are not usually worthwhile in buildings. For non-composite girders the flange width is usually within the range 0.3–0.5 times the depth of the section (0.4 is most common). For simply-supported composite girders these guidelines can still be employed for preliminary sizing of compression flanges. The width of tension flanges can be increased by 30%.

18.3.3 Stiffeners

Longitudinal web stiffeners are not usually required for plate girders used in buildings. Transverse (vertical) web stiffeners may be provided to enhance the resistance to shear near the supports or to carry high concentrated transverse forces acting on the flanges in cases where the resistance of the unstiffened web would otherwise be exceeded. The need for intermediate stiffening lessens away from supports due to the reduced shear in these regions.

The provision of intermediate transverse web stiffeners increases both the elastic shear buckling strength τ_{cr} and the ultimate shear buckling strength τ_u (including post-buckling or post-critical strength) of web panels. The elastic shear buckling strength is further increased by a reduction in the web panel aspect ratio a/h_w (width/depth). Ultimate shear buckling strength is increased by enhanced tension field action, whereby diagonal tensile membrane stresses, which develop during the post-buckling phase, are resisted by the boundary members (transverse stiffeners and flanges).

Intermediate transverse stiffeners are usually spaced such that the web panel aspect ratio is between 1.0 and 2.0, since there is little increase in strength for smaller panel aspect ratios. Sometimes pairs of stiffeners are employed at the end supports, to form what is known as a rigid end post, as discussed in Section 18.4.6. The overhang of the girder beyond the support is generally limited to a maximum of one eighth of the depth of the girder.

18.4 Design of plate girders to BS EN 1993-1-5

18.4.1 General

Any cross-section of a plate girder will normally be subjected to a combination of shear force and bending moment, present in varying proportions. Design checks are required for in-plane bending resistance, shear resistance and resistance to combined bending and shear. For laterally unrestrained girders, lateral torsional buckling should also be checked, as described in Chapter 17.

18.4.2 Dimensions of webs and flanges

A minimum web thickness is needed to prevent the compression flange buckling into the web (see Clause 8(1) of BS EN 1993-1-5) and to ensure satisfactory serviceability performance including, for example, avoiding unsightly buckles developing during erection and in service.

The buckling resistance of slender webs can be increased by the provision of web stiffeners. In general, the webs of plate girders used in buildings are either unstiffened or have transverse stiffeners only (see Figure 18.1).

Minimum web thickness values to avoid serviceability problems are not prescribed in BS EN 1993-1-5, but the following values, taken from BS 5950-1, are recommended.

For unstiffened webs, $t_{\mathrm{w}} \geq \dfrac{h_{\mathrm{w}}}{250}$

For transversely stiffened webs,

For $a > h_{\mathrm{w}}$, $t_{\mathrm{w}} \geq \dfrac{h_{\mathrm{w}}}{250}$

For $a \leq h_{\mathrm{w}}$, $t_{\mathrm{w}} \geq \left(\dfrac{h_{\mathrm{w}}}{250} \right) \left(\dfrac{a}{h_{\mathrm{w}}} \right)^{1/2}$

A further serviceability consideration, though one that can also contribute to fatigue failure, is that of web breathing. Web breathing refers to the recurring out-of-plane deflection of the web (due to plate buckling) under variable loads. It is principally associated with bridges due to the repeated nature of the loading; web breathing may be neglected provided slenderness criteria set out in Clause 7.4 of BS EN 1993-2 are met.

The following minimum web thickness values are prescribed in Clause 8(1) of BS EN 1993-1-5 to avoid the compression flange buckling into the web (i.e. flange-induced web buckling).

$$t_{\mathrm{w}} \geq \left(\frac{h_{\mathrm{w}}}{k} \right) \left(\frac{f_{\mathrm{yf}}}{E} \right) \sqrt{\frac{A_{\mathrm{fc}}}{A_{\mathrm{w}}}}$$

where A_{w} is the cross-sectional area of the web, A_{fc} is the effective cross-sectional area of the compression flange, h_{w} is the height of the web, t_{w} is the thickness of the web and k depends on the utilisation of the section in bending: $k = 0.3$ when plastic rotation is utilised (i.e. the formation and rotation of a plastic hinge in the member as part of a collapse mechanism); $k = 0.4$ when the plastic moment resistance is utilised; and $k = 0.55$ when the elastic moment capacity is utilised.

Local buckling of the compression flange may also occur if the flange plate is of slender (Class 4) proportions. In general there is seldom good reason for the $c_{\mathrm{f}} / t_{\mathrm{f}}$ ratio of the compression flanges of plate girders used in buildings to exceed the Class 3 limit of 14ε set out in Clause 5.5 of BS EN 1993-1-1.

18.4.3 Moment resistance

Determination of the moment resistance $M_{\mathrm{c,Rd}}$ of laterally restrained plate girders depends upon the co-existent level of shear force. Provided that the applied shear force is less than 50% of the plastic shear resistance $V_{\mathrm{pl,Rd}}$ (for webs that are not susceptible to shear buckling) or 50% of the shear buckling resistance (for webs that are susceptible to shear buckling), the section is deemed to be in a state of low shear and the full in-plane bending resistance $M_{\mathrm{c,Rd}}$ can be achieved. High shear is covered in Section 18.4.5. Cross-section bending resistance should be determined in accordance with Clause 6.2.5 of BS EN 1993-1-1, and depends on the classification of the cross-section, as discussed in Chapter 14.

For Class 1 and Class 2 cross-sections:

$$M_{c,Rd} = M_{pl,Rd} = \frac{W_{pl}f_y}{\gamma_{M0}}$$

where f_y is the yield strength of the material, W_{pl} is the plastic section modulus and $\gamma_{M0} = 1.0$ is the partial factor for cross-section resistance. Note that, if the yield strength is not constant through the section (e.g. the flange plate material is of higher strength than the web), the plastic section modulus cannot be used and the cross-section resistance should be determined directly from the plastic stress distribution.

For Class 3 cross-sections:

$$M_{c,Rd} = M_{el,Rd} = \frac{W_{el}f_y}{\gamma_{M0}}$$

where W_{el} is the elastic section modulus.

For Class 4 sections, local buckling occurs prior to yielding, resulting in bending resistances below the elastic moment capacity. To account for the loss of effectiveness due to local buckling, an effective section modulus should be determined and used in place of the elastic section modulus. The bending resistance of Class 4 cross-sections is therefore given by:

$$M_{c,Rd} = \frac{W_{eff}f_y}{\gamma_{M0}}$$

where W_{eff} is the effective section modulus of the cross-section.

18.4.4 Shear resistance

18.4.4.1 Web not susceptible to shear buckling

Susceptibility to shear buckling depends on the slenderness (height to thickness ratio) of the web. According to Clause 6.2.6(6) of BS EN 1993-1-1, for unstiffened webs, provided that:

$$\frac{h_w}{t_w} \leq 72\frac{\varepsilon}{\eta}$$

where h_w is the height of the web taken as the clear distance between the flanges (see Figure 18.1), t_w is the thickness of the web, $\varepsilon = (235/f_{yw})^{0.5}$ and $\eta = 1.0$ (specified for all steels in Clause NA.2.4 of the UK National Annex to BS EN 1993-1-5), the web is not susceptible to shear buckling. In such instances, the shear resistance may be taken as the plastic shear resistance $V_{\mathrm{pl,Rd}}$, as given in Clause 6.2.6(2) of BS EN 1993-1-1:

$$V_{\mathrm{pl,Rd}} = \frac{A_{\mathrm{v}} f_{\mathrm{yw}}}{\sqrt{3}\ \gamma_{\mathrm{M0}}}$$

where A_{v} is the shear area, defined in Clause 6.2.6(3) of BS EN 1993-1-1 as:

    $A_{\mathrm{v}} = h_{\mathrm{w}} t_{\mathrm{w}}$ for welded I-sections loaded parallel to the web

and

    $A_{\mathrm{v}} = A - h_{\mathrm{w}} t_{\mathrm{w}}$ for welded I-sections loaded parallel to the flanges.

18.4.4.2 Web susceptible to shear buckling

Webs of slender proportions (i.e. height-to-thickness ratio exceeding specified limiting values) are susceptible to shear buckling. The response of such webs from the onset of loading to collapse may be described in three successively occurring stages, as illustrated in Figure 18.2. In stage 1 (see Figure 18.2(a)), at low load levels, the web panel is in a state of pure shear, with the principal compressive and tensile stresses orientated at 45° to the horizontal. For slender web panels, the limit of stage 1, i.e. the pure shear field, is reached when the applied shear stress τ reaches the elastic shear buckling stress τ_{cr} and buckling occurs along the direction of the compressive diagonal. In stage 2, the post-buckling regime, no further increase in stress

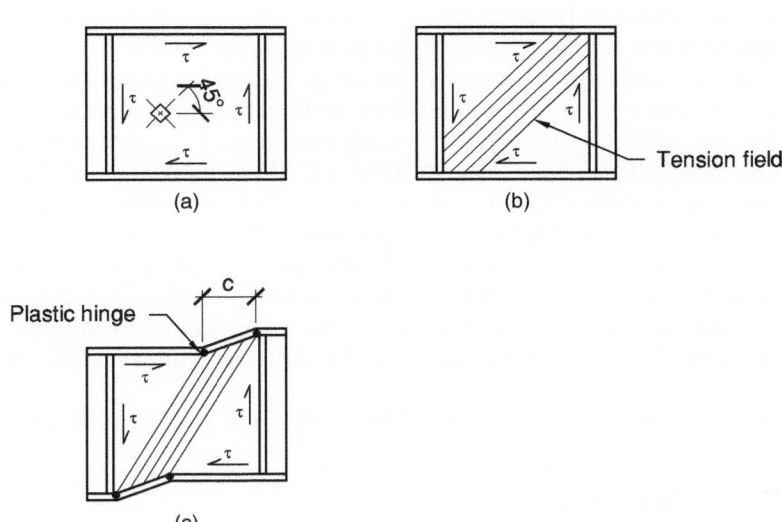

Figure 18.2   Response of slender web panels from the onset of loading to collapse: (a) pure shear, (b) post-buckling and (c) collapse mechanism

can take place along the compressive diagonal, but stresses can continue to increase along the tensile diagonal, resulting in so-called tension field action (see Figure 18.2(b)). Under increasing load, the direction of principal tensile stress rotates towards the horizontal. These tensile stresses are resisted by the horizontal and vertical boundaries of the web panel (i.e. the flanges and the transverse stiffeners) forming a load-carrying mechanism similar to that of a Pratt truss. Ultimate shear capacity is reached in stage 3 (see Figure 18.2(c)), with failure occurring due to yielding of the web panel and the formation of plastic hinges in the beam flanges. Further discussion of the above is presented in References 4 and 5.

According to Clause 5.1(2) of BS EN 1993-1-5, unstiffened webs become susceptible to shear buckling when

$$\frac{h_{\mathrm{w}}}{t_{\mathrm{w}}} > 72\frac{\varepsilon}{\eta},$$

while stiffened webs become susceptible to shear buckling when

$$\frac{h_{\mathrm{w}}}{t_{\mathrm{w}}} > 31\frac{\varepsilon}{\eta}\sqrt{k_{\tau}}$$

where k_{τ} is the shear buckling coefficient, defined in Annex A.3 of BS EN 1993-1-5 for webs without longitudinal stiffeners as:

$$k_{\tau} = 5.34 + 4.00(h_{\mathrm{w}}/a)^2 \text{ for } a/h_{\mathrm{w}} \geq 1$$
$$k_{\tau} = 4.00 + 5.34(h_{\mathrm{w}}/a)^2 \text{ for } a/h_{\mathrm{w}} < 1$$

in which a is the distance between centrelines of transverse stiffeners. The above relationships between k_{τ} and a/h_{w} are illustrated in Figure 18.3.

When a web is susceptible to shear buckling, transverse stiffeners should be specified at the supports to prevent the web from buckling as a column (see Clause 5.1(2) of BS EN 1993-1-5) and shear buckling resistance $V_{\mathrm{b,Rd}}$ should be checked in accordance with Clause 5.2 of BS EN 1993-1-5. Shear buckling resistance comprises a dominant contribution from the web $V_{\mathrm{bw,Rd}}$, and an additional contribution from the flanges $V_{\mathrm{bf,Rd}}$, provided they are not fully utilised in resisting co-existent bending. The shear buckling resistance may not exceed the plastic resistance of the section. The design approach in BS EN 1993-1-5 is based on the rotated stress field method.[6] For slender webs, this method allows the utilisation of considerable post-buckling strength (i.e. load-carrying capacity beyond elastic buckling), as shown in Figure 18.4.

Shear buckling resistance is defined in Clause 5.2(1) of BS EN 1993-1-5 as:

$$V_{\mathrm{b,Rd}} = V_{\mathrm{bw,Rd}} + V_{\mathrm{bf,Rd}} \leq \frac{\eta f_{\mathrm{yw}} h_{\mathrm{w}} t_{\mathrm{w}}}{\sqrt{3}\, \gamma_{\mathrm{M1}}}$$

in which $\gamma_{\mathrm{M1}} = 1.0$ is the partial factor for instability. The contribution of the web $V_{\mathrm{bw,Rd}}$ is given by:

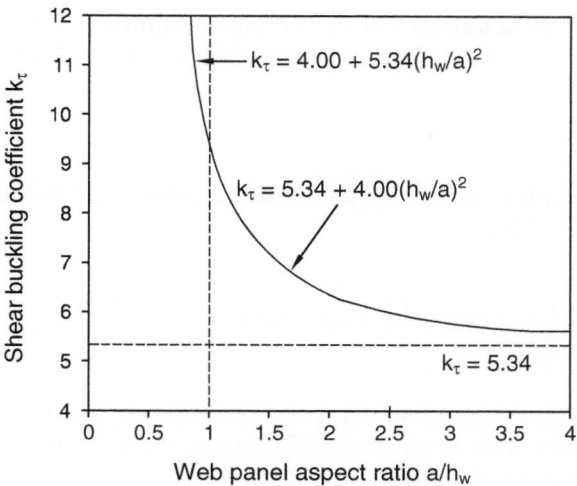

Figure 18.3 Relationship between shear buckling coefficient k_τ and web panel aspect ratio a/h_w

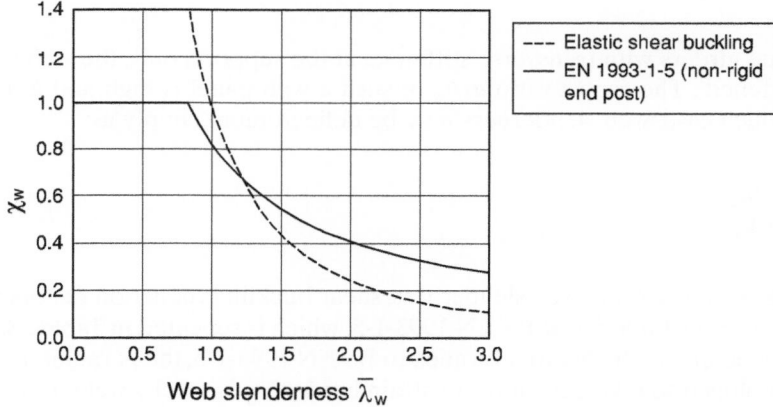

Figure 18.4 Comparison between elastic shear buckling resistance and BS EN 1993-1-5 resistance for web with a non-rigid end post

$$V_{bw,Rd} = \frac{\chi_w f_{yw} h_w t_w}{\sqrt{3}\ \gamma_{M1}}$$

where χ_w is the reduction factor for shear buckling, which depends on the slenderness of the web and the rigidity of the end posts (rigid or non-rigid) at the beam supports (see Section 18.4.6). Web slenderness is defined in general as:

$$\overline{\lambda}_w = \sqrt{\frac{\tau_{yw}}{\tau_{cr}}} = 0.76\sqrt{\frac{f_{yw}}{\tau_{cr}}}$$

in which τ_{yw} is the yield strength of the web material in shear and τ_{cr} is the elastic buckling stress of the web in shear, which may be determined from:

$$\tau_{cr} = k_\tau \sigma_E$$

where k_τ is the shear buckling coefficient described above and σ_E is defined in Annex A.1 of BS EN 1993-1-5 as:

$$\sigma_E = \frac{\pi^2 E}{12(1-v^2)}\left(\frac{t}{b}\right)^2 = 190000\left(\frac{t}{b}\right)^2 \text{ in N/mm}^2.$$

Substituting τ_{cr} into the general expression for web slenderness given above, allows $\bar{\lambda}_w$ for webs with transverse stiffeners at the supports and intermediate transverse stiffeners to be defined directly as:

$$\bar{\lambda}_w = \frac{h_w}{37.4 t_w \varepsilon \sqrt{k_\tau}}$$

For plate girders with transverse stiffeners at the supports only, the web is said to be 'unstiffened'. The aspect ratio a/h_w of such a web panel is high and k_τ tends to 5.34, in which case, web slenderness may be defined more simply as:

$$\bar{\lambda}_w = \frac{h_w}{86.4 t_w \varepsilon}$$

Having established the web slenderness, shear buckling reduction factors may be determined from Table 5.1 of BS EN 1993-1-5, which is repeated in Table 18.2. Note that, as stated in the UK National Annex to BS EN 1993-1-5, the parameter η, which can be employed to take advantage of strain hardening in stocky webs, has been set equal to unity.

For end panels, shear buckling reduction factors have been found to depend on the rigidity of the end posts, with greater rigidity offering greater anchorage to tension field action and hence higher load-carrying capacities. In BS EN 1993-1-5,

Table 18.2 Shear buckling reduction factors for web χ_w (with $\eta = 1.0$)

Slenderness range	Rigid end post	Non-rigid end post
$\bar{\lambda}_w < 0.83$	1.0	1.0
$0.83 \leq \bar{\lambda}_w < 1.08$	$0.83/\bar{\lambda}_w$	$0.83/\bar{\lambda}_w$
$\bar{\lambda}_w \geq 1.08$	$1.37/(0.7+\bar{\lambda}_w)$	$0.83/\bar{\lambda}_w$

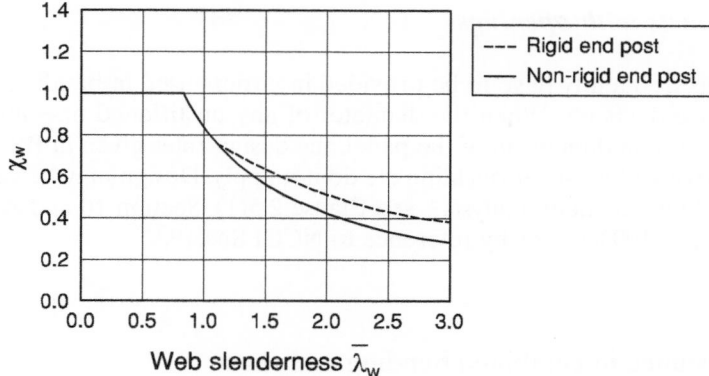

Figure 18.5 Comparison between shear buckling reduction factors for rigid and non-rigid end posts

end posts are classified as either rigid or non-rigid; distinction between the two is discussed in Section 18.4.6, while the reduction factors for the two cases (derived from Table 18.2) are shown in Figure 18.5. For intermediate panels, rigid end post restraint may be assumed to be provided by the adjacent panels.

The flange contribution to shear buckling resistance $V_{bf,Rd}$ is defined in Clause 5.4(1) of BS EN 1993-1-5 as:

$$V_{bf,Rd} = \frac{b_f t_f^2 f_{yf}}{c\, \gamma_{M1}}\left(1-\left(\frac{M_{Ed}}{M_{f,Rd}}\right)^2\right)$$

where b_f is the total flange width, but should be taken as no larger than $15\varepsilon t_f$ either side of the web, $M_{f,Rd}$ is the moment resistance of the flanges alone calculated on the basis of effective section properties if the compression flange is Class 4, and c is the width of the tension band (see Figure 18.2 (c)) given by:

$$c = a\left(0.25 + \frac{1.6 b_f t_f^2 f_{yf}}{t_w h_w^2 f_{yw}}\right)$$

The full flange contribution can only be mobilised when the co-existent bending moment M_{Ed} is 0. With increasing bending moment, the contribution of the flanges reduces until $M_{Ed} = M_{f,Rd}$, at which point the flanges are fully utilised in bending and make no contribution to the shear resistance. Note also that the flange contribution is inversely proportional to the width of the tension band c which itself is proportional to the spacing of the transverse stiffeners a. Thus, the flange contribution reduces as the spacing of the transverse stiffeners increases and, for unstiffened webs (other than at the supports), the flange contribution is negligible.

Resistance of plate girders to combined bending and shear is covered in Section 18.4.5.

18.4.4.3 Panels with openings

Web openings frequently have to be provided in girders used in building construction for service ducts etc. When the diameter of any unstiffened opening exceeds 5% of the minimum dimension of the panel, the design rules given in BS EN 1993-1-5 for effective widths, shear buckling etc do not apply. Design may be carried out by means of finite element analysis – see Clause 2.5(1), Section 10 and Annexes B and C of BS EN 1993-1-5, or by reference to NCCI SN019.[7]

18.4.5 Resistance to combined bending and shear

18.4.5.1 Web not susceptible to shear buckling

For webs not susceptible to shear buckling, resistance to combined bending and shear is covered in Clause 6.2.8 of BS EN 1993-1-1. In this Clause it is stated that, provided the design shear force V_{Ed} is less than 50% of the plastic shear resistance $V_{pl,Rd}$ of the section, the interaction between bending and shear can be ignored and the full bending moment resistance can be attained. Where the design shear force exceeds 50% of the plastic shear resistance, the bending resistance of the cross-section must be reduced; this is achieved by applying a reduced yield strength to the shear area of the section. The reduced yield strength f_{yr} is defined on the basis of the level of applied shear force as follows:

$$f_{yr} = (1-\rho)f_y$$

in which f_y is the basic material yield strength and ρ is determined as:

$$\rho = \left(\frac{2V_{Ed}}{V_{pl,Rd}} - 1 \right)^2$$

On the above basis, the reduced major axis bending moment resistance $M_{y,V,Rd}$ of a Class 1 or 2 I-section with equal flanges and uniform yield strength, in the presence of a shear force greater than 50% of the plastic shear resistance, may be determined directly from:

$$M_{y,V,Rd} = \frac{\left[W_{pl,y} - \dfrac{\rho A_w^2}{4t_w} \right] f_y}{\gamma_{M0}} \le M_{y,c,Rd}$$

where $A_w = h_w t_w$ is the area of the web and $M_{y,c,Rd}$ is the full in-plane bending resistance of the cross-section, as defined in Clause 6.2.5(2) of BS EN 1993-1-1. The above interaction is illustrated for a typical I-section in Figure 18.6. It is unlikely that a

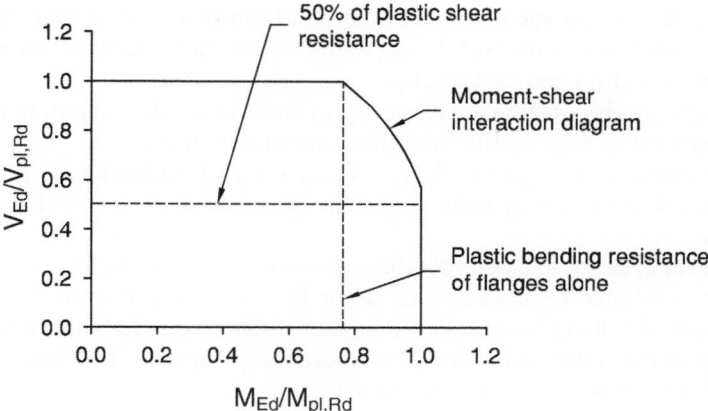

Figure 18.6   Shear-moment interaction for Class 1 and 2 cross-sections where the web is not susceptible to shear buckling

section which is sufficiently slender to be categorised as Class 3 will not be susceptible to shear buckling, but if this situation does arise (e.g. in the case of closely spaced transverse stiffeners), it is recommended that the above interaction (based on the plastic moment resistance) be used, but with the maximum bending resistance limited to the elastic moment capacity of the cross-section. For Class 4 sections, the maximum bending resistance should be limited to the elastic moment capacity of the effective cross-section, allowing for local plate buckling. Further discussion on Class 3 and 4 sections under combined bending and shear is given in Reference 4.

18.4.5.2   Web susceptible to shear buckling

For webs susceptible to shear buckling, the interaction between moment and shear is effectively covered in Clause 5 of BS EN 1993-1-5, when the flanges are not fully utilised in resisting bending ($M_{Ed} < M_{f,Rd}$), and in Clause 7.1 of BS EN 1993-1-5, when the flanges are fully utilised in resisting bending ($M_{Ed} \geq M_{f,Rd}$). The former case is discussed in Section 18.4.4.2 while, for the latter case, the following interaction expression, set out in Clause 7.1(1) of BS EN 1993-1-5, applies:

$$\frac{M_{Ed}}{M_{pl,Rd}} + \left(1 - \frac{M_{f,Rd}}{M_{pl,Rd}}\right)\left(\frac{2V_{Ed}}{V_{bw,Rd}} - 1\right)^2 \leq 1.0$$

The key features of the full bending-shear interaction diagram for Class 1 and 2 cross-sections are as follows:

1. Provided the design shear force V_{Ed} is less than or equal to half of the shear buckling resistance of the web $V_{bw,Rd}$, then the full plastic moment capacity $M_{pl,Rd}$ of the cross-section can be achieved.
2. The shear capacity of the web alone $V_{bw,Rd}$ and the bending moment capacity of the flanges alone $M_{f,Rd}$ can be mobilised simultaneously.
3. The maximum shear capacity ($V_{b,Rd} = V_{bw,Rd} + V_{bf,Rd}$), including the full contribution of the flanges, is only achieved in the presence of zero co-existent bending moment.
4. As the level of the co-existent bending moment increases, the flange contribution to shear resistance reduces by the factor $(1 - M_{Ed}/M_{f,Rd})^2$ until $M_{Ed} = M_{f,Rd}$, at which point the flanges are fully utilised in resisting bending and make no contribution to the shear resistance. This interaction is set out in Clause 5 of BS EN 1993-1-5 and discussed in Section 18.4.4.2.
5. For $M_{Ed} > M_{f,Rd}$, the interaction between bending and shear is set out in Clause 7.1(1) of BS EN 1993-1-5, as described above.

The above points are illustrated on the full interaction diagram shown in Figure 18.7.

For Class 3 and 4 cross-sections, the same interaction formulae apply (i.e. based on plastic moment resistance), but the maximum bending resistance that can be achieved is limited. For Class 3 cross-sections, the bending resistance is limited to the elastic moment capacity, while for Class 4 cross-sections, the effective section modulus is employed, as illustrated in Figure 18.7. For sections with fully effective flanges but a Class 4 web, calculation of effective section properties may be avoided by assuming that the bending is resisted by the flanges alone, while the shear is resisted by the web alone.

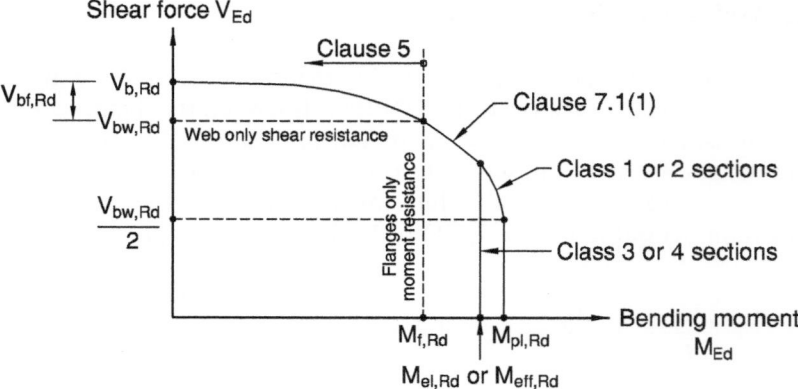

Figure 18.7   Shear-moment interaction for cross-sections where the web is susceptible to shear buckling

18.4.6 End posts and end anchorage

18.4.6.1 General

When a web is susceptible to shear buckling, some form of end anchorage is required at the supports to resist the post-buckling membrane stresses (tension field) that develop in a buckled web. The situation at the end of the plate girder is clearly more onerous than that at internal locations where support is present from adjacent panels. End anchorage need not be provided if the plastic shear resistance $V_{pl,Rd}$ (see Section 18.4.4.1) rather than the shear buckling resistance $V_{b,Rd}$ (see Section 18.4.4.2) is the governing design criterion (i.e. when the web is not susceptible to shear buckling).

End posts are classified in BS EN 1993-1-5 as either rigid or non-rigid. For web slendernesses $\bar{\lambda}_w$ of less than 1.08, shear buckling resistance is unaffected by this distinction (see Table 18.2), but for higher slendernesses, the greater anchorage afforded by a rigid end post results in improved shear buckling resistance. End posts also act as bearing stiffeners resisting the reaction from the supports, as discussed in Section 18.4.7. Rigid end posts should therefore be designed to resist the horizontal component of the tension field together with compressive forces due to the support reactions. Non-rigid end posts need only be designed as bearing stiffeners to resist the compressive forces due to the support reactions.

18.4.6.2 Rigid end posts

A rigid end post, as defined in Clause 9.3.1(2) of BS EN 1993-1-5, should comprise two double-sided transverse stiffeners or a rolled section connected to the end of the web plate – see Figure 18.8. In both cases, the end post acts as a short beam spanning vertically between the flanges and resisting the longitudinal membrane force arising from tension field action.

The minimum section modulus W_{min} for a rigid end post comprising either a rolled section or a pair of double-sided stiffeners is specified in Clause 9.3.1(3) of BS EN 1993-1-5 as:

$$W_{min} = 4h_w t_w^2$$

where t_w is the thickness of the web of the plate girder. For a pair of double-sided flat stiffeners separated by a distance e (which must be at least $0.1h_w$, according to Clause 9.3.1(3) of BS EN 1993-1-5), this corresponds (ignoring the web contribution) to a minimum area A_{min} for each stiffener of:

$$A_{min} = 4h_w t_w^2 / e$$

The magnitude of the longitudinal membrane force N_H to be resisted by a rigid end post is not specified in BS EN 1993-1-5, but has been derived in Reference 4 for a perfect plate and modified for design to give:

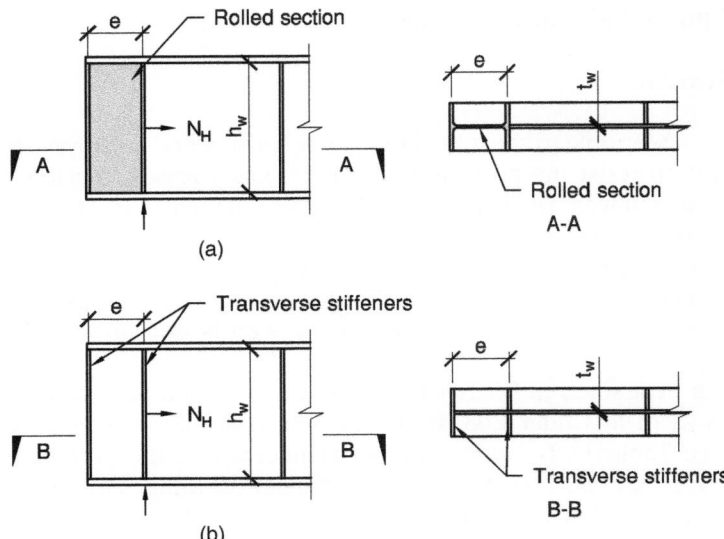

Figure 18.8   Rigid end post: (a) formed by a rolled section, and (b) formed by a pair of double-sided stiffeners

$$N_{\mathrm{H}} = h_{\mathrm{w}} t_{\mathrm{w}} \left(\frac{\tau_{\mathrm{Ed}}^2}{\tau_{\mathrm{cr}}/1.2} - \tau_{\mathrm{cr}}/1.2 \right) \geq 0$$

in which τ_{Ed} may be taken as $V_{\mathrm{Ed}}/h_{\mathrm{w}} t_{\mathrm{w}}$ and τ_{cr} is the elastic shear buckling stress of the web defined in Section 18.4.4.2. This membrane force may be assumed to act on the end post as a uniformly distributed load, which gives a maximum moment at mid-height of $N_{\mathrm{H}} h_{\mathrm{w}}/8$.

The bending moment due to the tensile membrane force is to be resisted in conjunction with the compressive force arising from the support reaction. An additional bending moment will also be present if the support reaction is eccentric to the bearing stiffener, due, for example, to construction tolerances or allowance for thermal expansion and contraction.[4] The capacity of the rigid end post should first be verified under the individual actions, followed by checks under combined axial compression and bending, including both a cross-section check and an overall stiffener buckling check. For the cross-section check, a linear elastic interaction is recommended, in order to avoid plasticity in the stiffener,[4] while, for the overall stiffener buckling check, the interaction formulae set out in Clause 6.2.3 of BS EN 1993-1-1 apply. A simplified interaction formula is recommended in Reference 4, and presented in Section 18.4.7.

Note that the stiffener cross-section, when considering action as a rigid end post to resist the tensile membrane force, differs from that when considering action as a

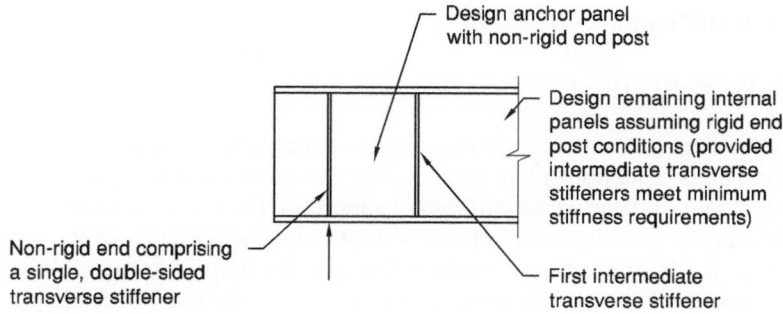

Figure 18.9 Non-rigid end post and anchor panel

bearing stiffener. For the former, the cross-section is simply an I-section comprising either a rolled section or the paired stiffeners acting as the flanges plus the plate girder web acting as the web. When resisting bearing, the effective bearing stiffener section extends along the web to a distance of $15\varepsilon t_w$ either side of the transverse stiffeners (where such material is available), as shown in Clause 9.1(2) of BS EN 1993-1-5 and Section 18.4.7.1.

Designing an end post as rigid may be avoided either by accepting the reduced capacity (when $\bar{\lambda}_w > 1.08$) associated with a non-rigid end post (comprising only one double-sided transverse stiffener at the support) or by positioning the first intermediate transverse stiffener closer to the support stiffener to create an anchor panel – see Figure 18.9. The anchor panel itself should be designed as non-rigid (see Clause 9.3.1(4) of BS EN 1993-1-5), but the reduced capacity may be compensated for by the reduced stiffener spacing which will give a lower web slenderness. Beyond the first intermediate transverse stiffener, the web panel may be designed assuming rigid end post conditions (provided the minimum stiffness criteria set out in Clause 9.3.3(3) of BS EN 1993-1-5 and discussed in Section 18.4.7 are met). The two transverse stiffeners bounding the anchor panel should be designed in a similar manner to the stiffeners of a twin end post.

18.4.6.3 *Non-rigid end posts*

A non-rigid end post comprises only a single double-sided stiffener, as shown in Figure 18.9. The reduced anchorage, compared with a rigid end post, afforded to the adjacent web panel, results in lower shear buckling capacity for slender webs ($\bar{\lambda}_w > 1.08$), as discussed in Section 18.4.7 and illustrated in Figure 18.9.

Non-rigid end posts should be designed as bearing stiffeners, as described in Section 18.4.7.3. They do not need to be designed to carry any tensile anchorage force, since this is assumed to be resisted by the web itself.

18.4.7 Web stiffeners

18.4.7.1 *Types of stiffeners*

Transverse stiffeners are generally required to ensure the satisfactory performance of the web panels of slender plate girders. The two most common types of stiffeners are (1) intermediate transverse stiffeners to increase the shear capacity of the web, and (2) bearing or load-bearing stiffeners to prevent local failure of the web under concentrated loads or reactions applied through the flange. A particular stiffener may serve more than one function, in which case it should be designed for the combined effects: e.g. an intermediate transverse stiffener may also be load-bearing. The effective cross-section of a stiffener (used both for verifying minimum stiffness and for buckling checks) is defined in Clause 9.1(2) as the stiffener(s) plus a width of web equal to $15\varepsilon t_w$, but clearly not greater than the actual material available, as shown in Figure 18.10.

The outstand proportions for flat stiffeners should generally be limited to:

$$\frac{h_s}{t_s} \leq 13.0\varepsilon$$

in order to avoid torsional buckling, where h_s and t_s are the outstand width and thickness of the stiffener, respectively. This requirement may be derived from that set out in Clause 9.2(8) of BS EN 1993-1-5.

18.4.7.2 *Intermediate transverse web stiffeners*

Intermediate transverse web stiffeners are used to increase both the elastic shear buckling resistance and ultimate shear buckling resistance $V_{b,Rd}$, of slender web panels. They are usually placed on one side of the web only, and are designed for minimum stiffness and buckling resistance, as specified in Clause 9.3.3(1) of BS EN 1993-1-5.

Intermediate transverse web stiffeners should have the following minimum second moment of area I_{st}, (see Clause 9.3.3(3) of BS EN 1993-1-5) in order to control the lateral deflection of the web at their locations and to be considered as rigid restraints:

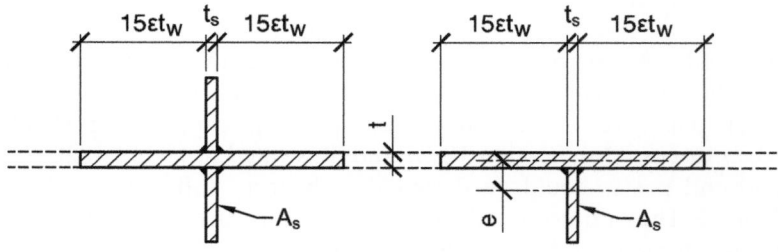

Figure 18.10   Extent of effective stiffener cross-section

$$I_{st} \geq 1.5 h_w^3 t_w^3 / a^2 \quad \text{for} \quad a/h_w < \sqrt{2}$$

$$I_{st} \geq 0.75 h_w t_w^3 \quad \text{for} \quad a/h_w \geq \sqrt{2}$$

The buckling resistance of intermediate transverse web stiffeners should also be checked. The effective stiffener cross-section should be designed to resist a compressive axial force, P_{Ed}, arising from the shear-induced tension field, plus any externally applied loads and moments. P_{Ed} acts at the mid-plane of the web and therefore causes both axial and bending stresses in the effective stiffener sections. P_{Ed} is defined in Clause NA.2.7 of the UK National Annex to BS EN 1993-1-5 as:

$$P_{Ed} = V_{Ed} - 0.8\tau_{cr} h_w t_w \sqrt{1 - \frac{\sigma_{x,Ed}}{0.8\sigma_{cr,x}}} \qquad \text{for } a \geq h_w$$

$$P_{Ed} = \left[V_{Ed} - 0.8\tau_{cr} h_w t_w \sqrt{1 - \frac{\sigma_{x,Ed}}{0.8\sigma_{cr,x}}} \right] \frac{a}{h_w} \quad \text{for } a < h_w$$

in which $\sigma_{cr,x}$ is the elastic plate buckling stress for the web under direct longitudinal compression, τ_{cr} is the elastic shear buckling stress of the web and $\sigma_{x,Ed}$ is longitudinal direct stress in the web, taken as zero for symmetric sections and the algebraic mean of the stresses at the top and bottom of the web in the case of asymmetric sections.

The buckling resistance of the effective stiffener section under axial compression may be evaluated as for a conventional compression member in accordance with Clause 6.3.1 of BS EN 1993-1-1, using buckling curve c. The effective buckling length of the stiffener may be taken as $0.75h_w$ if both ends are assumed to be fixed laterally. If there is less end restraint, the effective length should be taken equal to the actual length. The resistance of the effective stiffener section under combined axial load and bending may be evaluated in accordance with Clause 6.3.3 of BS EN 1993-1-1, though a simplified interaction expression, proposed in Reference 4, that assumes no lateral torsional buckling of the stiffener, is also suitable:

$$\frac{N_{Ed}}{\chi_y A f_y} + \frac{1}{1 - (N_{Ed}/N_{cr,y})} \frac{M_{y,Ed}}{W_{el,y} f_y} + \frac{M_{z,Ed}}{W_{el,z} f_y} \leq 1.0$$

with the axis convention as shown in Figure 18.11. Note that no amplifier is applied to the $M_{z,Ed}$ moment since the web prevents buckling in this plane.

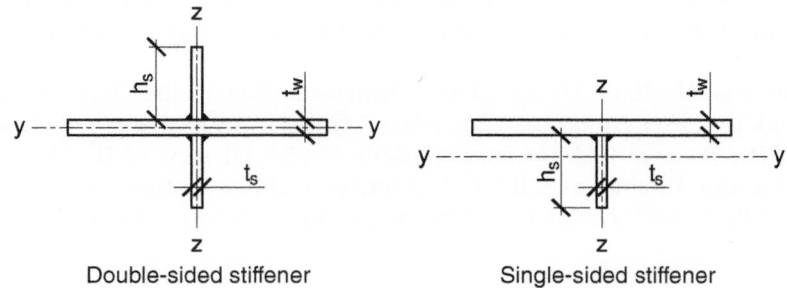

Figure 18.11 Axis convention for effective stiffener cross-section

18.4.7.3 Bearing or load-bearing stiffeners

Bearing or load-bearing stiffeners are employed to prevent local failure of the web under concentrated loads or reactions applied through the flange. Clause 5.1(2) of BS EN 1993-1-5 requires such stiffeners to be provided at the supports of beams in which the web slenderness h_w/t_w exceeds $72\varepsilon/\eta$. Bearing stiffeners should generally be placed symmetrically about the web centre-line, else the resulting eccentricity should be allowed for in their design (see Clause 9.4(3) of BS EN 1993-1-5). Other than at the supports, load-bearing stiffeners are required at the location of any concentrated loads, if the resistance of the unstiffened web, evaluated in accordance with Clause 6 of BS EN 1993-1-5, would otherwise be exceeded (see Clause 9.4(1) of BS EN 1993-1-5). Clause 6 of BS EN 1993-1-5 also applies to patch loading, when concentrated loads act between stiffener locations; this is common, for example, in the case of travelling loads.

Checks for cross-section resistance and buckling resistance under combined loading are required as described above. The simplified interaction expression given in Section 18.4.7.2 may also be applied to load-bearing stiffeners. Note that, for bearing stiffeners that also act as rigid end posts, the additional bending moments associated with anchoring of the tension field must be considered in conjunction with the support reaction and that, as described in Section 18.4.6.2, different cross-sections are defined when resisting the different loading components.

A worked example follows which is relevant to Chapter 18.

References to Chapter 18

1. British Standards Institution (2006) *BS EN 1993-1-5 Eurocode 3 – Design of steel structures – Part 1.5: Plated structural elements.* CBS EN. London, BSI.
2. Owens G.W. (1989) *Design of Fabricated Composite Beams in Buildings.* Ascot, The Steel Construction Institute.
3. British Standards Institution (2005) *BS EN 1993-1-1 Eurocode 3 – Design of steel structures – Part 1.1: General rules and rules for buildings.* CBS EN. London, BSI.
4. Hendy C. R. and Murphy C. J. (2007) *Designers' Guide to BS EN 1993-2 – Eurocode 3: Design of steel structures. Part 2: Steel bridges.* London, Thomas Telford.
5. Subramanian N. (2008) *Design of Steel Structures.* New Delhi, OUP India.
6. Höglund T. (1981) *Design of Thin Plate I Girders in Shear and Bending, with Special Reference to Web Buckling.* Bulletin No. 94, Division of Building Statics and Structural Engineering, Royal Institute of Technology, Sweden.
7. NCCI SN019. (2009) *Design rules for web openings in beams*, www.steel-ncci.co.uk

	Job No.		Sheet 1 of 10	Rev
The Steel Construction Institute	Job Title			
Silwood Park, Ascot, Berks SL5 7QN	Subject	Plate girders		
Telephone: (01344) 623345 Fax: (01344) 622944	Client		Made by	LG	Date
CALCULATION SHEET			Checked by	KAC	Date

Plate girders

Plate girder design example

Design brief

The plate girder shown below is fully laterally restrained along its length. For the design loading specified below, design a transversely stiffened plate girder in S275 steel.

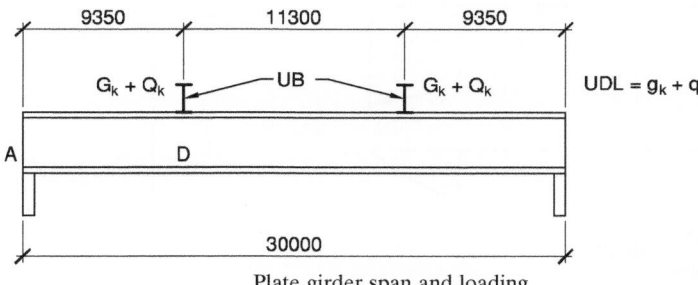

Plate girder span and loading

Actions (loading)

Permanent actions:	*UDL*	g_k	$= 25\ kN/m$
	Point loads	G_k	$= 200\ kN$
Variable actions:	*UDL*	q_k	$= 40\ kN/m$
	Point loads	Q_k	$= 450\ kN$

Partial factors for actions   BS EN 1990

Partial factor for permanent actions $\gamma_G = 1.35$

Partial factor for variable actions $\gamma_Q = 1.50$

Reduction factor for permanent actions $\xi = 0.925$

ULS combination of actions

Design UDL $F_d = (1.35 \times 0.925 \times 25) + (1.5 \times 40) = 91.2\,kN/m$

Design point loads $F_d = (1.35 \times 0.925 \times 200) + (1.5 \times 450) = 925\ kN$

Plate girder design example	Sheet 2 of 10	Rev

Design shear force and bending moment diagrams

The shear force and bending moment diagrams corresponding to the ULS design loads are shown below.

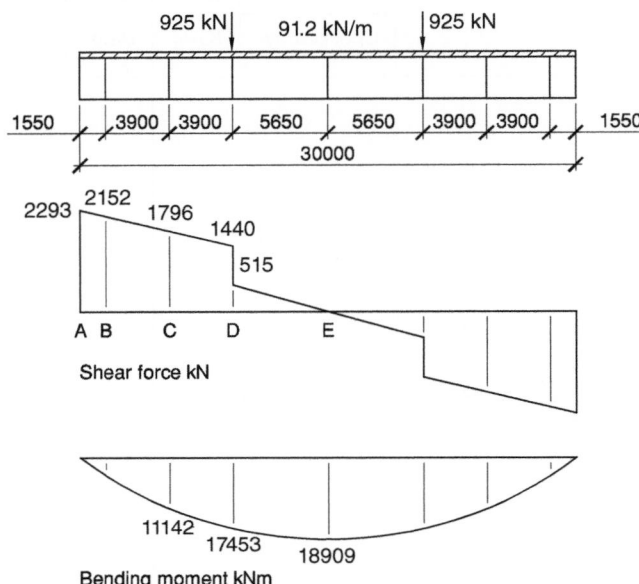

Design shear forces, bending moments and stiffener spacing

Initial sizing of plate girder

The recommended span/depth ratio for simply supported non-composite plate girders ranges between 12 for short span girders and 20 for long span girders. Herein the depth is assumed to be span/15.

$$h_w = \frac{span}{15} = \frac{30000}{15} = 2000 \ mm$$

Estimate flange area assuming $f_{yf} = 255 \ N/mm^2$ (i.e. assuming $40 \ mm < t_f < 63 \ mm$).

$$A_f = \frac{M_{max}}{h_w f_{yf}} = \frac{18909 \times 10^6}{2000 \times 255} = 37076 \ mm^2$$

For non-composite plate girders, the flange width is usually within the range of 0.3 and 0.5 of the depth. Assume a flange $750 \times 50 = 37500 \ mm^2$.

The minimum web thickness for plate girders in buildings is usually $t_w \geq h_w/124\varepsilon$ to ensure a non-slender section. Assume the web thickness $t_w = 15 \ mm$, slightly less than that required for a Class 3 section due to proposed transverse stiffening.

BS EN 10025-2

Plate girder design example	Sheet 3 of 10	Rev

Cross-section classification

Initial sizing proposed a plate girder with $750 \times 50\,mm$ flanges and $2000 \times 15\,mm$ web. Check cross-section classification.

Flange:

For $t_f = 50\,mm$, $f_{yf} = 255\,N/mm^2$

$$\varepsilon = \sqrt{\frac{235}{f_{yf}}} = \sqrt{\frac{235}{255}} = 0.96$$

Ignoring weld size in determination of plate width:

$$c_f = \frac{750 - 15}{2} = 367.5\ mm$$

$$\frac{c_f}{t_f} = \frac{367.5}{50} = 7.35$$

Limiting slenderness for Class 1 flange is $9\varepsilon = 8.64 > 7.35$

∴ Flange is Class 1

Web:

For $t_w = 15\,mm$, $f_{yw} = 275\ N/mm^2$

$$\varepsilon = \sqrt{\frac{235}{f_{yf}}} = \sqrt{\frac{235}{275}} = 0.92$$

Ignoring weld size in determination of plate width:

$c_w = 2000\,mm$

$$\frac{c_w}{t_w} = \frac{2000}{15} = 133$$

Limiting slenderness for Class 3 web in bending is $124\varepsilon = 114.63 < 133$

∴ Web is Class 4.

Since $\dfrac{h_w}{t_w} > \dfrac{72\varepsilon}{\eta} = 72\varepsilon = 66.56$, the web must be checked for shear buckling.

Reference column:

- BS EN 10025-2
- BS EN 1993-1-1 Table 5.2
- BS EN 10025-2
- BS EN 1993-1-1 Table 5.2
- BS EN 1993-1-1 Cl 6.2.6(6)

Plate girder design example	Sheet 4 of 10	Rev

Dimensions of web and flanges

Assume stiffener spacing $a > h_w$

Minimum web thickness to avoid serviceability problems:

$$t_w = 15 \geq \frac{h_w}{250} = \frac{2000}{250} = 8.0 \text{ mm} \qquad OK$$

To avoid the flanges buckling into the web:

$$t_w = 15 \geq \left(\frac{h_w}{k}\right)\left(\frac{f_{yf}}{E}\right)\sqrt{\frac{A_{fc}}{A_w}} = \left(\frac{2000}{0.55}\right)\left(\frac{255}{210000}\right)\sqrt{\frac{750 \times 50}{2000 \times 15}} = 4.94 \text{ mm} \quad OK$$

BS EN 1993-1-5 Cl 8(1)

Bending moment resistance

For sections with class 1-3 flanges but a class 4 web, it may be assumed that the bending moment is resisted by the flanges alone while the web is designed to carry the shear only.

The bending resistance of the flanges alone $M_{f,Rd} = A_f f_{yf} \times (h_w + t_f)$

$M_{f,Rd} = (750 \times 50 \times 255) \times (2000 + 50)/10^6 = 19603 \text{ kNm}$

$M_{f,Rd} = 19603 > M_{max} = 18909 \text{ kNm} \qquad OK$

Shear buckling resistance of end (anchor) panel AB

The shear resistance of panel AB will be calculated assuming a non-rigid end post, with the reduced capacity compensated for by closer stiffener spacing. The contribution of the flanges will be ignored, as described above.

$h_w = 2000 \text{ mm}$; $a = 1550 \text{ mm}$; $f_{yw} = 275 \text{ N/mm}^2$; $V_{Ed} = 2293 \text{ kN}$

Web aspect ratio $a/h_w = 1550/2000 = 0.78$

Buckling coefficient (for $a/h_w < 1$):
$k_\tau = 4.00 + 5.34(h_w/a)^2 = 4.00 + 5.34(2000/1550)^2 = 12.89$

BS EN 1993-1-5 Cl A.3(1)

Web slenderness:

BS EN 1993-1-5 Cl 5.3(3)

$$\bar{\lambda}_w = 0.76\sqrt{\frac{f_{yw}}{\tau_{cr}}} = \frac{h_w}{37.4 t_w \varepsilon \sqrt{k_\tau}} = \frac{2000}{37.4 \times 15 \times 0.92\sqrt{12.89}} = 1.07$$

Web buckling reduction factor (for non-rigid end post):

BS EN 1993-1-5 Cl 5.3(1)

$$\bar{\lambda}_w < 1.08 \quad \therefore \chi_w = \frac{0.83}{\bar{\lambda}_w} = \frac{0.83}{1.07} = 0.77$$

Shear resistance of web $V_{bw,Rd}$:

BS EN 1993-1-5 Cl 5.2(1)

$$V_{bw,Rd} = \frac{\chi_w f_{yw} h_w t_w}{\sqrt{3}\,\gamma_{M1}} = \frac{0.77 \times 275 \times 2000 \times 15}{\sqrt{3} \times 1.0 \times 10^3} = 3681 \text{ kN} > 2293 \text{ kN} = V_{Ed} \quad OK$$

Plate girder design example	Sheet 5 of 10	Rev

Shear buckling resistance of panel BC

The shear panel resistance of the end panel BC is calculated assuming a rigid end post but ignoring the contribution of the flange.

$h_w = 2000 \, mm; \, a = 3900 \, mm; \, f_{yw} = 275 \, N/mm^2; \, V_{Ed} = 2152 \, kN$

Web aspect ratio $a/h_w = 3900/2000 = 1.95$

Buckling coefficient (for $a/h_w \geq 1$):
$k_\tau = 5.34 + 4.00(h_w/a)^2 = 5.34 + 4.00(2000/3900)^2 = 6.39$

BS EN
1993-1-5
Cl A.3(1)

Web slenderness:

$$\bar{\lambda}_w = 0.76\sqrt{\frac{f_{yw}}{\tau_{cr}}} = \frac{h_w}{37.4 t_w \varepsilon \sqrt{k_\tau}} = \frac{2000}{37.4 \times 15 \times 0.92\sqrt{6.39}} = 1.53$$

BS EN
1993-1-5
Cl 5.3(3)

Web buckling reduction factor (for rigid end post):

$$\bar{\lambda}_w \geq 1.08 \quad \therefore \chi_w = \frac{1.37}{(0.7 + \bar{\lambda}_w)} = \frac{1.37}{(0.7 + 1.53)} = 0.62$$

BS EN
1993-1-5
Cl 5.3(1)

Shear resistance of web $V_{bw,Rd}$:

$$V_{bw,Rd} = \frac{\chi_w f_{yw} h_w t_w}{\sqrt{3} \, \gamma_{M1}} = \frac{0.62 \times 275 \times 2000 \times 15}{\sqrt{3} \times 1.0 \times 10^3} = 2932 \, kN > 2152 \, kN = V_{Ed} \quad OK$$

BS EN
1993-1-5
Cl 5.2(1)

Shear buckling resistance of panel DE

The shear resistance of panel DE is calculated assuming a rigid end post but ignoring the contribution of the flange.

$h_w = 2000 \, mm; \, a = 3900 \, mm; \, f_{yw} = 275 \, N/mm^2; \, V_{Ed} = 515 \, kN$

Web aspect ratio $a/h_w = 5650/2000 = 2.83$

Buckling coefficient (for $a/h_w \geq 1$):
$k_\tau = 5.34 + 4.00(h_w/a)^2 = 5.34 + 4.00(2000/5650)^2 = 5.84$

BS EN
1993-1-5
Cl A.3(1)

Web slenderness:

$$\bar{\lambda}_w = 0.76\sqrt{\frac{f_{yw}}{\tau_{cr}}} = \frac{h_w}{37.4 t_w \varepsilon \sqrt{k_\tau}} = \frac{2000}{37.4 \times 15 \times 0.92\sqrt{5.84}} = 1.60$$

BS EN
1993-1-5
Cl 5.3(3)

Web buckling reduction factor (for rigid end post):

$$\bar{\lambda}_w \geq 1.08 \quad \therefore \chi_w = \frac{1.37}{(0.7 + \bar{\lambda}_w)} = \frac{1.37}{(0.7 + 1.60)} = 0.60$$

BS EN
1993-1-5
Cl 5.3(1)

Shear resistance of web $V_{bw,Rd}$:

$$V_{bw,Rd} = \frac{\chi_w f_{yw} h_w t_w}{\sqrt{3} \, \gamma_{M1}} = \frac{0.60 \times 275 \times 2000 \times 15}{\sqrt{3} \times 1.0 \times 10^3} = 2843 \, kN > 515 \, kN = V_{Ed} \; OK$$

BS EN
1993-1-5
Cl 5.2(1)

Plate girder design example	Sheet 6 of 10	Rev

Design of bearing stiffener at A

The bearing stiffener at A should be designed for the compressive force due to the support reaction equal to 2293 kN. The single (double-sided) stiffener constitutes a non-rigid end post and does not therefore need to be designed to resist any tensile anchorage force.

Try double-sided stiffening consisting of two flats $280 \times 24\,mm$ (i.e. $h_s = 280\,mm$; $t_s = 24\,mm$)

Check outstands:

For $t_s = 24\,mm$, $f_y = 265\ N/mm^2$

BS EN 10025-2

$$\varepsilon = \sqrt{\frac{235}{f_{yf}}} = \sqrt{\frac{235}{265}} = 0.94$$

$h_s = 280 \leq 13\varepsilon t_s = 13 \times 0.94 \times 24 = 294\,mm$

Since $h_s / t_s \leq 13.0\varepsilon$, torsional buckling is avoided, as is local buckling since $h_s / t_s \leq 14.0\varepsilon$, which is the class 3 limit for a compressed outstand.

BS EN 1993-1-5 Cl 9.2.1(8)

The effective stiffener section comprises the area of the stiffeners themselves, plus an effective web width equal to $15\varepsilon t_w = 15 \times 0.92 \times 15 = 208\,mm$ (where ε relates to the web material) either side of the stiffeners, where such material is available. At location A, at the end of the plate girder, web material is available on one side of the stiffeners only. The effective stiffener section is shown below:

Effective stiffener section at A

It is assumed that the support reaction acts at the centroid of the effective stiffener section, such that there is no bending moment induced. The effective stiffener section will therefore be designed to resist an axial compression equal to the support reaction $N_{Ed} = 2293\ kN$. It is assumed that both ends of the stiffeners are fixed laterally such that their effective length may be taken as $0.75h_w$.

Effective stiffener properties:

$A_s = (2 \times 280 \times 24) + ([208 + 24] \times 15) = 16920\,mm^2$

$I_{st} = (24 \times [15 + 2 \times 280]^3 / 12) + (208 \times 15^3 / 12) = 380.28 \times 10^6\,mm^4$

Plate girder design example	Sheet 7 of 10	Rev

$$i_s = \sqrt{\frac{I_{st}}{A_s}} = \sqrt{\frac{380.28 \times 10^6}{16920}} = 149.9 \, mm$$

Stiffener cross-section resistance:

$$N_{c,Rd} = \frac{A f_y}{\gamma_{M0}} = \frac{16920 \times 265}{1.0} \times 10^{-3} = 4484 \, kN > 2293 \, kN = N_{Ed} \qquad OK$$

Stiffener buckling resistance:

$$\lambda_1 = \pi \sqrt{\frac{E}{f_y}} = \pi \sqrt{\frac{210000}{265}} = 88.4$$

$$L_{cr} = 0.75 h_w = 0.75 \times 2000 = 1500 \, mm$$

BS EN 1993-1-5 Cl 9.4(2)

$$\bar{\lambda} = \frac{\lambda}{\lambda_1} = \frac{L_{cr}/i_s}{\lambda_1} = \frac{1500/149.9}{88.4} = 0.11$$

Since $\bar{\lambda} < 0.2$, buckling effects may be ignored and only cross-section checks apply.

BS EN 1993-1-1 Cl 6.3.1.2(4)

Design of intermediate transverse stiffeners at B and C

The stiffeners at B and C should be designed to have a minimum stiffness and sufficient buckling resistance to withstand a compressive axial force, P_{Ed}, arising from the tension field.

For panels BC and CD, a = 3900 mm.

Try double-sided stiffening consisting of two flats 80 × 15 mm (i.e. h_s 80 mm; $t_s = 15$ mm)

Check outstands:

For $t_s = 15$ mm, $f_y = 275 \, N/mm^2$

BS EN 10025-2

$$\varepsilon = \sqrt{\frac{235}{f_{yf}}} = \sqrt{\frac{235}{275}} = 0.92$$

$$h_s = 80 \leq 13 \varepsilon t_s = 13 \times 0.92 \times 15 = 180 \, mm$$

Since $h_s/t_s \leq 13.0\varepsilon$, torsional buckling is avoided, as is local buckling since $h_s/t_s \leq 14.0\varepsilon$, which is the Class 3 limit for a compressed outstand.

BS EN 1993-1-5 Cl 9.2.1(8)

The effective stiffener section comprises the area of the stiffeners themselves plus an effective web width equal to $15\varepsilon t_w = 15 \times 0.92 \times 15 = 208 \, mm$ (where ε relates to the web material) either side of the stiffeners, where such material is available. The effective stiffener section for locations B and C is shown below:

Plate girder design example	Sheet 8 of 10	Rev

Effective stiffener section at B and C

Check minimum stiffness:

$$\frac{a}{h_w} = \frac{3900}{2000} = 1.95 \geq \sqrt{2}$$

Minimum $I_{st} = 0.75 h_w t_w^3 = 0.75 \times 2000 \times 15^3 = 5.06 \times 10^6 \ mm^4$

BS EN 1993-1-5 Cl 9.3.3(3)

Actual
$I_{st} = (15 \times [15 + 2 \times 80]^3 / 12) + (2 \times 208 \times 15^3 / 12) = 6.82 \times 10^6 \, mm^4$

Actual $I_{st} = 6.82 \times 10^6 > $ *Required* $I_{st} = 5.06 \times 10^6 \, mm^4$

Double-sided intermediate transverse stiffeners are designed to resist a compressive axial force P_{Ed}, where, for $a \geq h_w$:

$$P_{Ed} = V_{Ed} - 0.8 \tau_{cr} h_w t_w \sqrt{1 - \frac{\sigma_{x,Ed}}{0.8 \sigma_{cr,x}}}$$

BS EN 1993-1-5 Cl NA.2.7

$\sigma_{x,Ed} = 0$ for *a symmetric section;* $V_{Ed} = 2152 \, kN$

$\tau_{cr} = k_\tau \sigma_E$

$$\sigma_E = \frac{\pi^2 E}{12(1-v^2)} \left(\frac{t}{b}\right)^2 = 190000 \left(\frac{t}{b}\right)^2 = 190000 \left(\frac{15}{2000}\right)^2 = 10.7 \ N/mm^2$$

For $a/h_w \geq 1$,
$k_\tau = 5.34 + 4.00(h_w/a)^2 = 5.34 + 4.00(200/3900)^2 = 6.39$

BS EN 1993-1-5 Annex A.3

$\tau_{cr} = k_\tau \sigma_E = 6.39 \times 10.7 = 68.2 \, N/mm^2$

$$P_{Ed} = V_{Ed} - 0.8 \tau_{cr} h_w t_w \sqrt{1 - \frac{\sigma_{x,Ed}}{0.8 \sigma_{cr,x}}}$$
$$= 2152 - (0.8 \times 68.2 \times 2000 \times 15 \times 10^{-3}) = 514 \, kN$$

Plate girder design example	Sheet 9 of 10	Rev

Effective stiffener properties:

$A_s = (2 \times 80 \times 15) + ([2 \times 208 + 15] \times 15) = 8865\,mm^2$

$I_{st} = (15 \times [15 + 2 \times 80]^3 / 12) + (2 \times 208 \times 15^3 / 12) = 6.82 \times 10^6\,mm^4$

$i_s = \sqrt{\dfrac{I_s}{A_s}} = \sqrt{\dfrac{6.82 \times 10^6}{8865}} = 27.7\,mm$

Stiffener cross-section resistance:

$N_{c,Rd} = \dfrac{A f_y}{\gamma_{M0}} = \dfrac{8865 \times 275}{1.0} \times 10^{-3} = 2438\,kN > 514\,kN = P_{Ed}$ *OK*

Stiffener buckling resistance:

$\lambda_1 = \pi \sqrt{\dfrac{E}{f_y}} = \pi \sqrt{\dfrac{210000}{275}} = 86.8$

$L_{cr} = 0.75 h_w = 0.75 \times 2000 = 1500\,mm$ *BS EN 1993-1-5 Cl 9.4(2)*

$\bar{\lambda} = \dfrac{\lambda}{\lambda_1} = \dfrac{L_{cr}/i_s}{\lambda_1} = \dfrac{1500/27.7}{86.8} = 0.62$

For buckling of stiffeners, use buckling curve 'c', which has an imperfection factor $\alpha = 0.49$.   *BS EN 1993-1-5 Cl 9.4(2)*

$\Phi = 0.5[1 + \alpha(\bar{\lambda} - 0.2) + \bar{\lambda}^2] = 0.5[1 + 0.49(0.62 - 0.2) + 0.62^2] = 0.80$ *BS EN 1993-1-1 Cl 6.3.1.2*

$\chi = \dfrac{1}{\Phi + \sqrt{\Phi^2 - \bar{\lambda}^2}} = \dfrac{1}{0.80 + \sqrt{0.80^2 - 0.62^2}} = 0.77 \leq 1.0$

$N_{b,Rd} = \dfrac{A f_y}{\gamma_{M1}} = \dfrac{0.77 \times 8865 \times 275}{1.0} \times 10^{-3} = 1881\,kN > 514\,kN = P_{Ed}$ *OK*   *BS EN 1993-1-1 Cl 6.3.1.1*

Design of intermediate load-bearing stiffener at D

The stiffener at D should be designed to have a minimum stiffness and sufficient buckling resistance to withstand the externally applied load at D of 92 kN plus a compressive axial force P_{Ed} arising from the tension field.

Try double-sided stiffening consisting of two flats $80 \times 15\,mm$ (i.e. $h_s = 80\,mm$; $t_s = 15\,mm$), as employed at B and C – see above Figure. The minimum stiffness requirements are satisfied as before.

Plate girder design example	Sheet 10 of 10	Rev

For panels DE, a = 5650 mm. For a ≥ h_w, the compressive axial force P_{Ed} is given by:

BS EN 1993-1-5 Cl NA.2.7

$$P_{Ed} = V_{Ed} - 0.8\tau_{cr}h_w t_w \sqrt{1 - \frac{\sigma_{x,Ed}}{0.8\sigma_{cr,x}}}$$

$\sigma_{x,Ed} = 0$ *for a symmetric section;* $V_{Ed} = 1440\ kN$

$\tau_{cr} = k_\tau \sigma_E$

$$\sigma_E = \frac{\pi^2 E}{12(1-v^2)}\left(\frac{t}{b}\right)^2 = 190000\left(\frac{t}{b}\right)^2 = 190000\left(\frac{15}{2000}\right)^2 = 10.7\ N/mm^2$$

For $a/h_w \geq 1$, $k_\tau = 5.34 + 4.00(h_w/a)^2 = 5.34 + 4.00(200/5650)^2 = 5.84$

BS EN 1993-1-5 Annex A.3

$\tau_{cr} = k_\tau \sigma_E = 5.84 \times 10.7 = 62.4\ N/mm^2$

$$P_{Ed} = V_{Ed} - 0.8\tau_{cr}h_w t_w \sqrt{1 - \frac{\sigma_{x,Ed}}{0.8\sigma_{cr,x}}}$$
$$= 1440 - (0.8 \times 62.4 \times 2000 \times 15 \times 10^{-3}) = -96.7\ kN \quad \therefore Take\ P_{Ed} = 0.$$

The total compressive force to be resisted is therefore 925 + 0 = 925 kN.

Buckling resistance of the stiffener (as above) $N_{b,Rd} = 1881\ kN > 925\ kN$ *OK*

Final girder dimensions and details

Based on the above calculations, the final plate girder dimensions and details are as below:

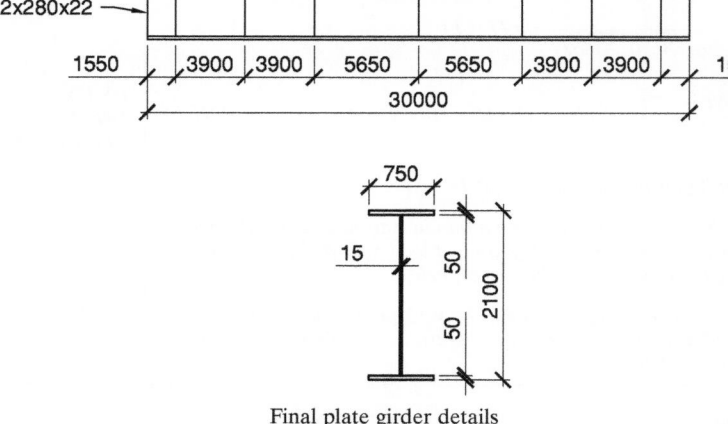

Final plate girder details

Chapter 19
Members with compression and moments

DAVID NETHERCOT and LEROY GARDNER

19.1 Occurrence of combined loading

Chapters 16 and 17 deal respectively with the design of members subject to compression and bending when these loadings act in isolation. However, in practice, a combination of the two effects is frequently present. Figure 19.1 illustrates a number of common examples.

The balance between compression and bending, which may be induced about one or both principal axes, depends on a number of factors, the most important being the type of structure, the form of the applied loading, the member's location in the structure and the way in which the connections between the members function.

For building frames designed according to the principles of simple construction, it is customary to regard column moments as being produced only by beam reactions acting through notional eccentricities as illustrated in Figure 19.2. Thus, column axial load is accumulated down the building but column moments are only ever generated by the floor levels under consideration, with the result that they typically contain increasing ratios of compression to moment. Many columns are therefore designed for high axial loads but rather low moments. Corner columns suffer bending about both axes, but may well carry less axial load; edge columns are subject to bending about at least one axis; and internal columns may, if both the beam framing arrangements and the loading are symmetrical, be designed for axial load only.

Conversely, the columns of portal frames are required to carry high moments in the plane of the frame but relatively low axial loads, unless directly supporting cranes. Rafters also attract some small axial load. Portals employ rigid connections between members permitting the transfer of moments around the frame. Similarly, multi-storey frames designed on the basis of rigid beam-to-column connections are likely to contain columns with large moments.

Members required to carry combined compression and moments are not restricted to rectangular building frames. Although trusses are often designed on the basis that member centrelines intersect at the nodes, this is not always possible and the

Steel Designers' Manual, Seventh Edition. Edited by Buick Davison and Graham W. Owens.
© 2012 Steel Construction Institute. Published 2012 by Blackwell Publishing Ltd.

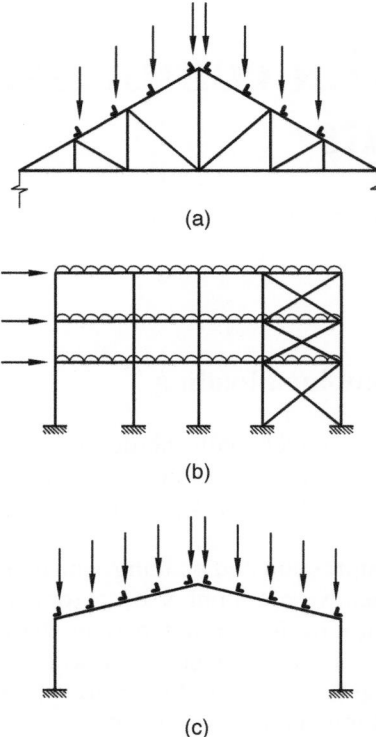

Figure 19.1 Occurrence of beam-columns in different types of steel frame. (a) Roof truss
– top chord members subject to bending from purlin loads and compression
due to overall bending. (b) Simple framing – columns subject to bending from
eccentric beam reactions and compression due to gravity loading. (c) Portal
frame – rafters and columns subject to bending and compression due to frame
action

resulting eccentricities induce some moments in the predominantly axially loaded
members. Sometimes the transfer of loads from secondary members into the main
booms of trusses is arranged so that they do not coincide with the nodal points,
leading to beam action between nodes being superimposed on the compression
produced by overall bending.

19.2 Types of response – interaction

Members subject to combined compression and bending are often referred to as
beam-columns. This term is helpful in appreciating member response because the
combination of loads produces a combination of effects incorporating aspects of the

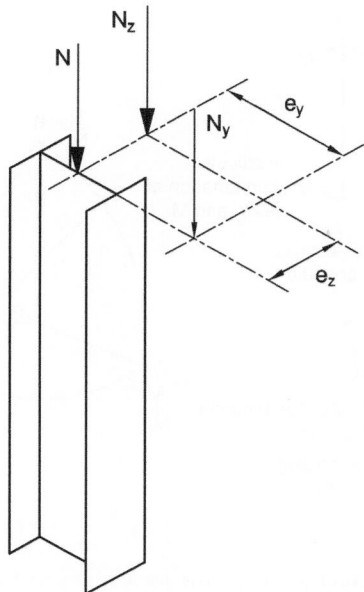

Figure 19.2   Loading on a beam-column in a 'simply-designed' frame

behaviour of the two extreme examples: a beam acting only in bending and a column carrying only compressive load. This then leads naturally to the concept of interaction as the basis for design, an approach in which the proportions of the member's resistance to all types of loading are combined using diagrams or formulae. Figure 19.3 illustrates the concept in general terms for cases of two and three separate load components. Any combination is represented by a point on the diagram, and an increasing set of loads with fixed ratios between the components corresponds to a straight line starting from the origin. Points that fall inside the boundary given by the design resistance are safe, those that fall on the boundary just meet the design resistance and those that lie outside the boundary represent an unsafe load combination. For the two-load component case if one load type is fixed and the maximum safe value of the other is required, the vertical co-ordinate corresponding to the specified load is first located; projecting horizontally to meet the design boundary, the horizontal co-ordinate can be read off.

The exact version of Figure 19.3 appropriate for a design basis depends upon several factors, which include the form of the applied loading, the type of response that is possible, the member slenderness and the cross-sectional shape. Design methods for beam-columns must therefore seek to balance the conflicting requirements of rigour, which would try to adjust the form of the design boundary of Figure 19.3 to reflect the influence of each of these factors, and simplicity. However, some appreciation of the role of each factor is necessary if even the simplest design approach is to be properly understood and applied.

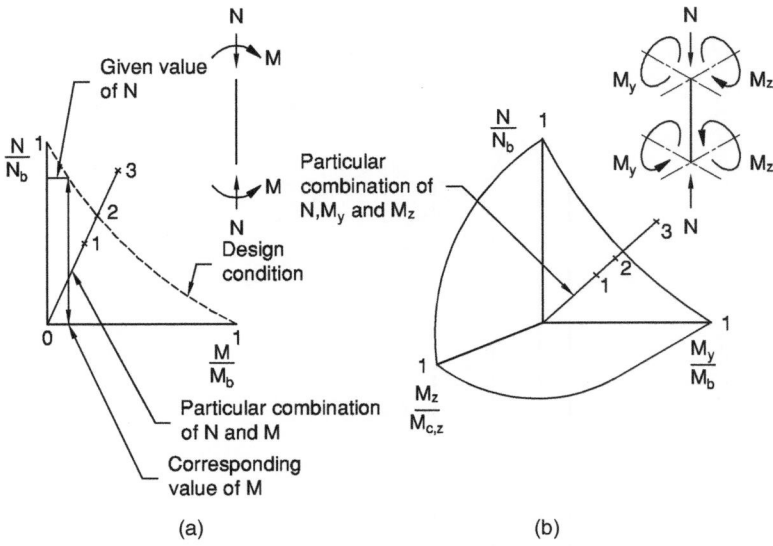

Figure 19.3 Concept of interaction diagrams for combined loading: (a) two-dimensional, (b) three-dimensional

The importance of member slenderness may be appreciated readily with reference to the two-dimensional example illustrated in Figure 19.4. The member is loaded by compression plus equal and opposite end moments and is assumed to respond simply by deflecting in the plane of the applied loading. Bending occurs under the action of the applied moments leading to a lateral deflection, v. The moment at any point within the length comprises two components: a constant primary moment M due to the applied end moments plus a secondary moment Nv due to the axial load N acting through the lateral deflection, v. Summing the effects of compression and bending gives

$$\frac{N}{N_b} + \frac{M_{max}}{M_c} = 1.0 \qquad (19.1)$$

in which N_b and M_c are the resistance as a strut and a beam respectively and M_{max} is the total moment.

Analysis of beam-column problems shows that M_{max} may be closely approximated by:

$$M_{max} = \frac{M}{(1 - N/N_{cr})} \qquad (19.2)$$

in which $N_{cr} = \pi^2 EI / L^2$ is the elastic critical load.

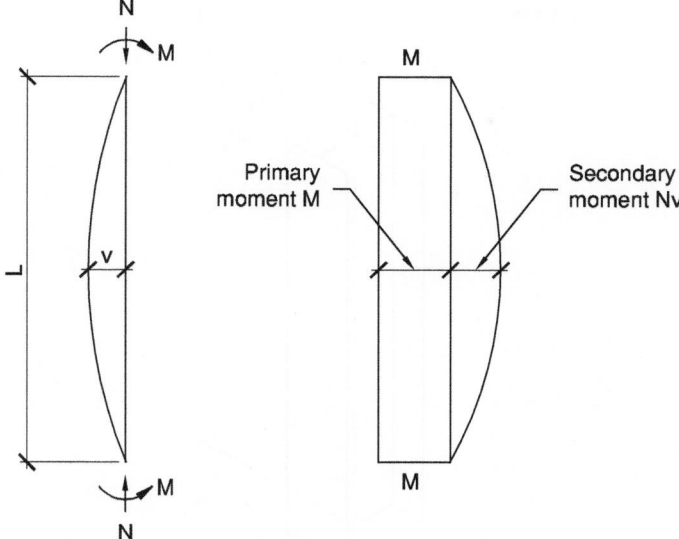

Figure 19.4   In-plane behaviour of beam-columns

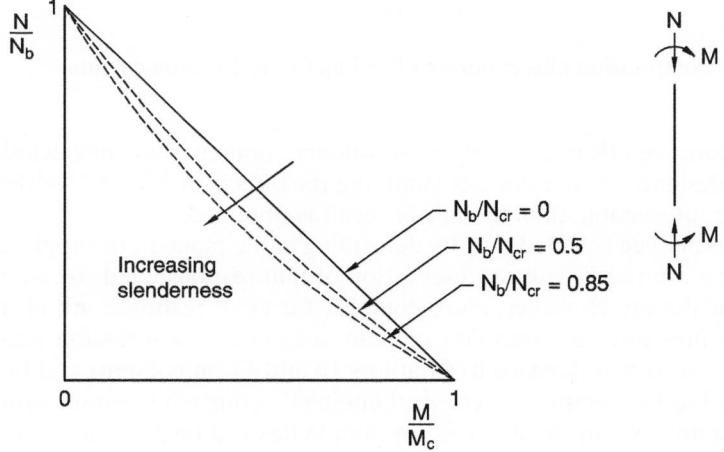

Figure 19.5   Effect of slenderness on form of interaction according to Equation 19.3

Combining these two expressions gives:

$$\frac{N}{N_b} + \frac{M}{M_c(1 - N/N_{cr})} = 1.0 \qquad (19.3)$$

Figure 19.5 shows how this expression plots in an increasingly concave fashion as member slenderness increases and the amplification of the primary moments *M*

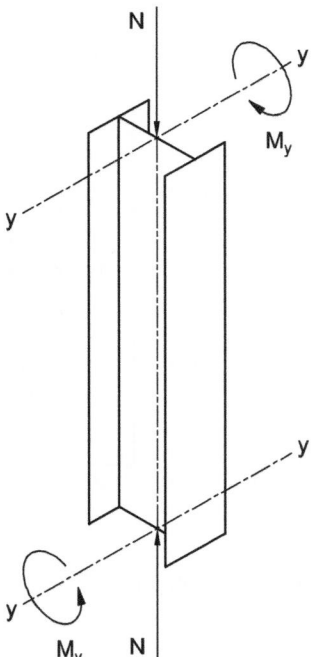

Figure 19.6   Compression plus major axis bending for an I-section column

becomes more significant. Clearly, if secondary moments are neglected, and the member is designed on the basis of summing the effects of N and M without allowing for their interaction, then an unsafe result is obtained.

If the member can respond only by deforming in the plane of the applied bending, the foregoing is an adequate representation of that response and so can be used as the basis for design. However, more complex forms of response are also possible. Figure 19.6 illustrates an I-section column subject to compression and bending about its major axis. Reference to Chapters 16 and 17 on columns and beams indicates that either the compression or the bending if acting alone would induce failure about the minor axis, in the first case by simple flexural buckling at a load $N_{b,z}$ and in the second by lateral–torsional buckling at a bending moment M_b. Both tests and rigorous analysis confirm that the combination of loads will also produce a minor axis failure. Noting the presence of the amplification effect of the axial load acting through the bending deflections in the plane of the web, and not the out-of-plane deformations associated with the eventual failure mode, leads therefore to a modified form of Equation 19.3 as an interaction equation that might form a suitable basis for design:

$$\frac{N}{N_{b,z}} + \frac{M_y}{M_b(1 - N/N_{cr,y})} = 1.0 \tag{19.4}$$

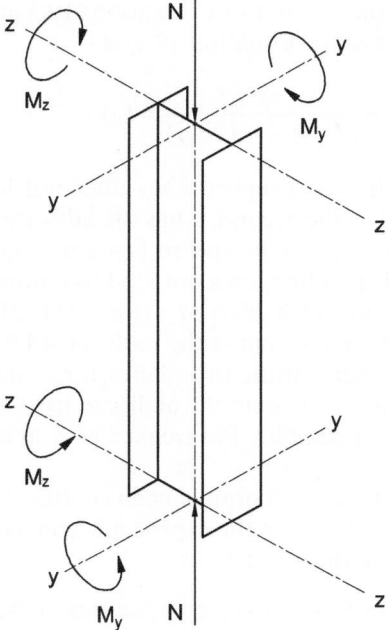

Figure 19.7   Compression plus major and minor axis bending for an I-section column

Note that both resistances $N_{b,z}$ and M_b relate to out-of-plane failure but that the amplification depends upon the in-plane Euler load. Consideration of the term $N/N_{cr,y}$, given that N is limited to $N_{b,z}$ which is less than $N_{b,y}$ and which is, in turn, less than $N_{cr,y}$, suggests that amplification effects are less significant than for the purely in-plane case.

Applied moments are not necessarily restricted to a single plane. For the most general case illustrated in Figure 19.7, in which compression is accompanied by moments about both principal axes, the member's response is a three-dimensional one involving bending about both axes combined with twisting. This leads to a complex analytical problem which cannot really be solved in such a way that it provides a direct indication of the type of interaction formulae that might be used as a basis for design. A practical view of the problem, however, suggests that some form of combination of the two previous cases might be suitable, provided any proposal were properly checked against data obtained from tests and reliable analyses. This leads to two possibilities: combining the acceptable moments M_y and M_z that can safely be combined separately with the axial load N obtained by solving Equations 19.3 and 19.4 to give

$$\frac{M_y}{M_{ay}} + \frac{M_z}{M_{az}} = 1.0 \tag{19.5}$$

in which M_{ay} and M_{az} are the solutions of Equations 19.3 and 19.4; or simply adding the minor axis bending effect to Equation 19.4 as:

$$\frac{N}{N_{\mathrm{b,z}}} + \frac{M_{\mathrm{y}}}{M_{\mathrm{b}}(1 - N/N_{\mathrm{cr,y}})} + \frac{M_{\mathrm{z}}}{M_{\mathrm{c,z}}(1 - N/N_{\mathrm{cr,z}})} = 1.0 \qquad (19.6)$$

Although the first of these two approaches does not lead to such a seemingly straightforward end result as the second, it has the advantage that interaction about both axes may be treated separately, and so leads to a more logical treatment of cases for which major axis bending does not lead to a minor axis failure, as for the case of a rectangular tube in which $M_{\mathrm{b}} = M_{\mathrm{c,y}}$ (for practical situations), and $N_{\mathrm{b,y}}$ and $N_{\mathrm{b,z}}$ are likely to be much closer than is the case for a UB. Similarly for members with different effective lengths for the two planes, for example, due to intermediate bracing acting in the weaker plane only, the ability to treat in-plane and out-of-plane responses separately and to combine the weaker with minor axis bending leads to a more rational result.

The foregoing discussion has deliberately been conducted in rather general terms, the main intention being to illustrate those principles on which beam-column design should be based. Collecting them together:

1. Interaction between different load components must be recognised; merely summing the separate components can lead to unsafe results.
2. Interaction tends to be more pronounced as member slenderness increases.
3. Different forms of response are possible, depending on the form of the applied loading.

Having identified these three principles it becomes easier to recognise their inclusion in the design procedures of BS EN 1993-1-1, which are discussed in Section 19.5.

19.3 Effect of moment gradient loading

Returning to the comparatively simple in-plane case, Figure 19.8 illustrates the patterns of primary (first-order) and secondary (second-order) moments in a pair of members subject to unequal end moments that produce either single- or double-curvature bending. For the first case, the point of maximum combined moment occurs near the mid-height where secondary bending effects are greatest. On the other hand, for double-curvature bending the two individual maxima occur at quite different locations, and for the case illustrated, in which the secondary moments have deliberately been shown as small, the point of absolute maximum moment is at the top. Had larger secondary moments been shown, as is the case in Figure 19.9, then the point of maximum moment moves down slightly but is still far from that of the single-curvature case.

Theoretical and experimental studies of steel beam-columns constrained to respond in-plane and subject to different moment gradients, as represented by the

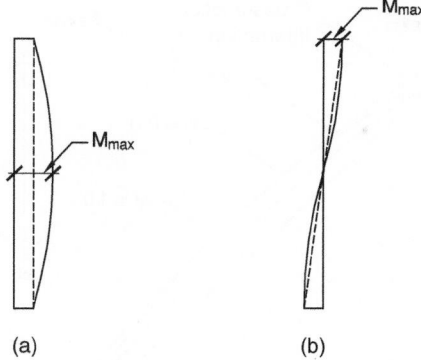

Figure 19.8   Primary and secondary moments 1: (a) single curvature, (b) double curvature

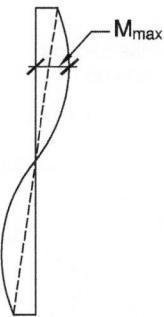

Figure 19.9   Primary and secondary moments 2

factor of the ratio of the numerically smaller end moment to the numerically larger end moment ψ, show clearly that, when all other parameters are held constant, failure loads tend to increase as ψ is varied from +1 (uniform single-curvature bending) to –1 (uniform double-curvature bending). Figure 19.10 illustrates the point in the form of a set of interaction curves. Clearly, if all beam-column designs were to be based upon the $\psi = +1$ case, safe but rather conservative designs would result.

Figure 19.10 also shows how, for high moments, the curves for $\psi \neq +1$ tend to merge into a single line corresponding to the condition in which the more heavily stressed end of the member controls design. Reference to Figure 19.8 and Figure 19.9 illustrates the point. The left-hand side (i.e. relatively low bending moments) and lower curves (i.e. ψ closer to unity) of Figure 19.10 correspond to situations in which Figure 19.9 controls, while the right-hand side and upper curves represent failure at one end. The two cases are sometimes referred to as 'stability' and 'strength' failure (or 'buckling' and 'cross-section' failure) respectively. While this may offend

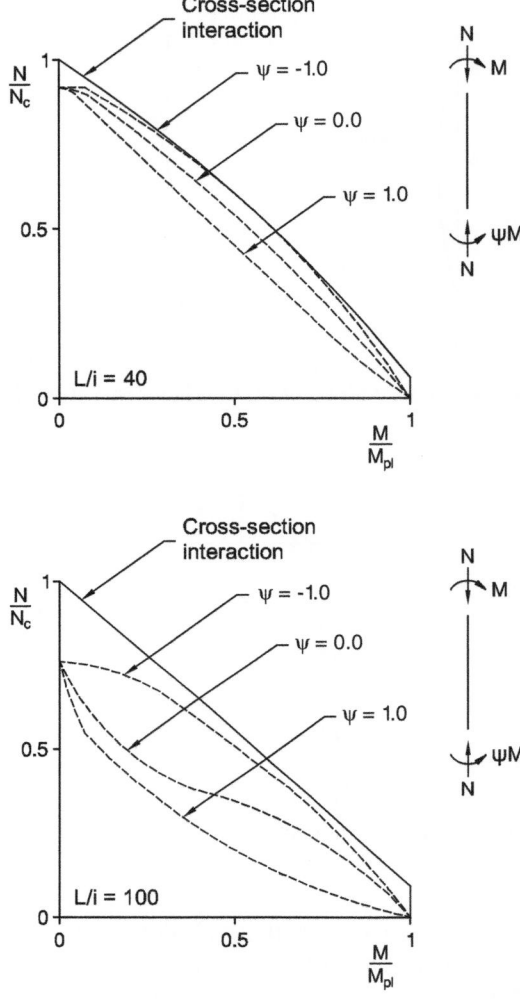

Figure 19.10   Effect of moment gradient on interaction

the purists, for in both cases the limiting condition is one of exhausting the cross-sectional capacity, though at different locations within the member length, it does, nonetheless, serve to draw attention to the principal difference in behaviour. Also shown in Figure 19.10 is a line corresponding to the cross-sectional interaction: the combinations of *N* and *M* corresponding to the full strength of the cross-section. This is the 'strength' limit, representing the case where the primary moment acting in conjunction with the axial load accounts for all the cross-section's capacity.

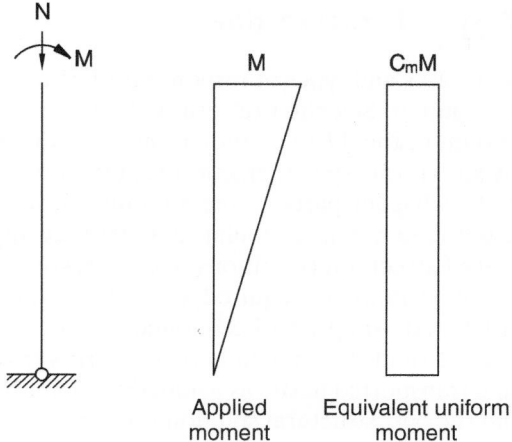

Figure 19.11   Concept of equivalent uniform moment applied to primary moment on a beam-column

The substance of Figure 19.10 can be incorporated within the type of interaction formula of Section 19.2 through the concept of equivalent uniform moment, presented in Chapter 17 in the context of the lateral–torsional buckling of beams; its meaning and use for beam-columns are virtually identical. Hence, for beam-columns under moment gradients, by introducing an equivalent uniform moment factor C_m, stability is checked under combined axial load N and an equivalent uniform moment $C_m M$, as shown in Figure 19.11.

The situation corresponding to the upper boundary or strength failure of Figure 19.10 must be checked separately using an appropriate means of determining cross-sectional capacity under N and M. The strength check is superfluous for $\psi = +1$ as it can never control, while as $\psi \to -1$ and $M \to M_c$ it becomes increasingly likely that the strength check will govern. For in-plane behaviour, the procedure is:

1. Check stability using an interaction formula in terms of buckling resistance N_b and moment capacity M_c with axial load N and equivalent moment $C_m M$.
2. Check strength using an interaction procedure in terms of axial capacity N_c and moment capacity M_c with coincident values of axial load N and maximum applied end moment M_1 (this check is unnecessary if the equivalent uniform moment factor $C_m = 1.0$; $C_m M = M_1$ is used in the stability check).

Consideration of other cases involving out-of-plane failure or moments about both axes shows that the equivalent uniform moment concept may also be applied. For simplicity the same C_m values are normally used in design, although minor variations for the different cases can be justified. For biaxial bending, two different values of C_m for bending about the two principal axes may be appropriate.

19.4 Selection of type of cross-section

Several different design cases and types of response for beam-columns are outlined in Section 19.2 of this chapter. Selection of a suitable member for use as a beam-column must take account of the differing requirements of these various factors. In addition to the purely structural aspects, practical requirements such as the need to connect the member to adjacent parts of the structure in a simple and efficient fashion must also be borne in mind. A tubular member may appear to be the best solution for a given set of structural conditions of compressive load, end moments, length, etc., but if site connections are required, very careful thought is necessary to ensure that they can be made simply and economically. On the other hand, if the member is one of a set of similar web members for a truss that can be fabricated entirely in the shop and transported to site as a unit, then simple welded connections should be possible and the best structural solution is probably the best overall solution too.

Generally speaking when site connections, which will normally be bolted, are required, open sections which facilitate the ready use of, for example, cleats or endplates are preferred. UCs are designed principally to resist axial load but are also capable of carrying significant moments about both axes. Although buckling in the plane of the flanges, rather than the plane of the web, always controls the pure axial load case, the comparatively wide flanges ensure that the strong-axis moment capacity $M_{c,y}$ is not reduced very much by lateral-torsional buckling effects for most practical arrangements. Indeed the condition $M_b = M_{c,y}$ will often be satisfied.

In building frames designed according to the principles of simple construction, the columns are unlikely to be required to carry large moments. This arises from the design process by which compressive loads are accumulated down the building but the moments affecting the design of a particular column length are only those from the floors at the top and bottom of the storey height under consideration. In almost all cases, the most economic column will be a UC. Preliminary member selection may conveniently be made by adding a small percentage to the actual axial load to allow for the presence of the relatively small moments and then choosing an appropriate trial size from the tables of compressive resistance given in Reference 1. For moments about both axes, as in corner columns, a larger percentage to allow for biaxial bending is normally appropriate, while for internal columns in a regular grid with no consideration of pattern loading, the design condition may actually be one of pure axial load.

The natural and most economic way to resist moments in columns is to frame the major beams into the column flanges since, even for UCs, $M_{c,y}$ will always be comfortably larger than $M_{c,z}$. For structures designed as a series of two-dimensional frames in which the columns are required to carry quite high moments about one axis but relatively low compressive loads, UBs may well be an appropriate choice of member. The example of this arrangement usually quoted is the single-storey portal building, although here the presence of cranes, producing much higher axial loads, the height, leading to large column slenderness, or a combination of the two, may result in UCs being a more suitable choice. UBs used as columns also suffer from the fact that the c_w/t_w values for the webs of many sections are slender (Class 4) when the applied loading leads to a set of web stresses that have a mean compressive component of more than about $70–100\,\text{N}/\text{mm}^2$.

19.5 Basic design procedure to Eurocode 3

When the distribution of moments and forces throughout the structure has been determined, for example, from a frame analysis in the case of continuous construction or by statics for simple construction, the design of a member subject to compression and bending consists of checking that a trial member satisfies the design conditions being used by ensuring that it falls within the design boundary defined by the type of diagram shown as Figure 19.3. BS EN 1993-1-1 provides sets of interaction formulae that approximate this boundary. Use of these rely on the procedures already explained in Chapters 16 and 17 for buckling resistance under axial compression and major axis bending respectively to define the end points. If allowance is made for the pattern of moment through a C_m factor of less than unity, then a separate check on cross-section resistance at the most heavily loaded location (which might not be obvious and which might, therefore, involve checking several possible locations) will also be necessary.

19.5.1 Cross-section resistance

Cross-section resistance should be checked first using Clause 6.2 of BS EN 1993-1-1. The simplest, but often rather conservative, approach, valid for Class 1, 2 and 3 cross-sections, uses the linear interaction of Clause 6.2.1(7):

$$\frac{N_{Ed}}{N_{Rd}} + \frac{M_{y,Ed}}{M_{y,Rd}} + \frac{M_{z,Ed}}{M_{z,Rd}} \leq 1 \tag{19.7}$$

in which N_{Ed}, $M_{y,Ed}$ and $M_{z,Ed}$ are the applied loads and moments, and N_{Rd}, $M_{y,Rd}$ and $M_{z,Rd}$ are the axial and bending resistances.

For Class 1 and 2 cross-sections, a more economic solution is to use Clause 6.2.9.1(2) of BS EN 1993-1-1, which states:

$$M_{Ed} \leq M_{N,Rd} \tag{19.8}$$

in which $M_{N,Rd}$ is the design plastic moment of resistance allowing for the presence of the axial force N_{Ed}.

For doubly symmetric I- and H-sections bent about the y-y axis no reduction in moment capacity is necessary providing both:

$$N_{Ed} \leq 0.25 N_{pl,Rd} \tag{19.9}$$

and

$$N_{Ed} \leq \frac{0.5 h_w t_w f_y}{\gamma_{M0}} \tag{19.10}$$

are satisfied. Similarly for bending about the z-z axis, no reduction in moment capacity is necessary when:

$$N_{Ed} \leq \frac{h_w t_w f_y}{\gamma_{M0}} \tag{19.11}$$

is satisfied. For larger values of N_{Ed}, use may be made of the following expressions for standard rolled I- and H-sections.

$$M_{N,y,Rd} = M_{pl,y,Rd}(1-n)/(1-0.5a) \quad \text{but} \quad M_{N,y,Rd} \leq M_{pl,y,Rd} \tag{19.12}$$

$$M_{N,z,Rd} = M_{pl,z,Rd} \quad \text{for} \quad n \leq a \tag{19.13}$$

$$M_{N,z,Rd} = M_{pl,z,Rd}\left[1 - \left(\frac{n-a}{1-a}\right)^2\right] \quad \text{for} \quad n > a \tag{19.14}$$

in which $n = N_{Ed}/N_{pl,Rd}$ and $a = (A - 2bt_f)/A$, but $a \leq 0.5$.

For RHS, reduced moment capacities to allow for the presence of axial load may be obtained from:

$$M_{N,y,Rd} = M_{pl,y,Rd}(1-n)/(1-0.5a_w) \quad \text{but} \quad M_{N,y,Rd} \leq M_{pl,y,Rd} \tag{19.15}$$

$$M_{N,z,Rd} = M_{pl,z,Rd}(1-n)/(1-0.5a_f) \quad \text{but} \quad M_{N,z,Rd} \leq M_{pl,z,Rd} \tag{19.16}$$

where $a_w = (A - 2bt)/A$, but $a_w \leq 0.5$ and $a_f = (A - 2bt)/A$, but $a_f \leq 0.5$.

For Class 1 and 2 cross-sections subject to biaxial bending, the following interaction expression applies:

$$\left[\frac{M_{y,Ed}}{M_{N,y,Rd}}\right]^\alpha + \left[\frac{M_{z,Ed}}{M_{N,z,Rd}}\right]^\beta \leq 1 \tag{19.17}$$

in which, for I- and H-sections, $\alpha = 2$ and $\beta = 5n$, but $\beta \geq 1$.

19.5.2 Member buckling resistance

For member buckling resistance, Equations 6.61 and 6.62 of BS EN 1993-1-1 for Class 1,2 and 3 cross-sections may be simplified to:

$$\frac{N_{Ed}}{N_{b,y,Rd}} + k_{yy}\frac{M_{y,Ed}}{M_{b,Rd}} + k_{yz}\frac{M_{z,Ed}}{M_{c,z,Rd}} \leq 1 \tag{19.18}$$

and

$$\frac{N_{Ed}}{N_{b,z,Rd}} + k_{zy}\frac{M_{y,Ed}}{M_{b,Rd}} + k_{zz}\frac{M_{z,Ed}}{M_{c,z,Rd}} \leq 1 \tag{19.19}$$

in which $N_{b,y,Rd}$ and $N_{b,z,Rd}$ are the design buckling resistances about the major and minor axes, respectively, $M_{b,Rd}$ is the design lateral torsional buckling resistance moment, $M_{c,z,Rd}$ is the minor axis bending resistance and k_{yy}, k_{yz}, k_{zy} and k_{zz} are interaction factors determined from either Annex A or Annex B of BS EN 1993-1-1.

Two alternative approaches (Annex A or Annex B) for the calculation of the interaction factors are provided. Both are permitted by the UK National Annex. Although neither is particularly simple, the Annex B method involves fewer calculations. Figure 19.12 provides a graphical means of deriving the interaction factors from Annex B for Class 1 and 2 I-sections – k_{yy} may be found from Figure 19.12(a), k_{zy} from Figure 19.12(b) conservatively assuming $C_{mLT} = 1.0$, k_{zz} from Figure 19.12(c) and k_{yz} may be taken as $0.6k_{zz}$; similar graphs for further scenarios are presented in Appendix D of Reference 2.

Maximum (conservative) values for these interaction factors are set out in Reference 2 and repeated in Table 19.1, in which expressions for the equivalent uniform moment factors C_{my} and C_{mz} may be obtained from Annex B of BS EN 1993-1-1. For the case of end moment loading with a linear variation from M to ψM, C_m may be calculated from:

$$C_m = 0.6 + 0.4\,\psi \geq 0.4 \tag{19.20}$$

A special procedure is also available for Class 1, 2 and 3 hot-rolled I- and H-section and RHS columns in structures designed according to the principles of 'simple construction'. Simple construction applies to rectangular multi-storey frames braced against side-sway such that the beams and columns need be designed to resist gravity loads only, with 'simple' beam to column connections that transmit limited moments to the columns. The simplified design procedure, set out in the NCCI SN048,[3] effectively reduces the pair of interaction formulae to a single formula, and provides conservative values for the interaction factors:

$$\frac{N_{Ed}}{N_{b,z,Rd}} + \frac{M_{y,Ed}}{M_{b,Rd}} + 1.5\frac{M_{z,Ed}}{M_{c,z,Rd}} \leq 1.0 \tag{19.21}$$

The method assumes that the first term (i.e. the axial component) is the dominant term in the interaction, and that failure occurs about the weaker axis.

19.6 Special design methods for members in portal frames

19.6.1 Design requirements

Both the columns and the rafters in a typical pitched roof portal frame represent particular examples of members subject to combined bending and compression.

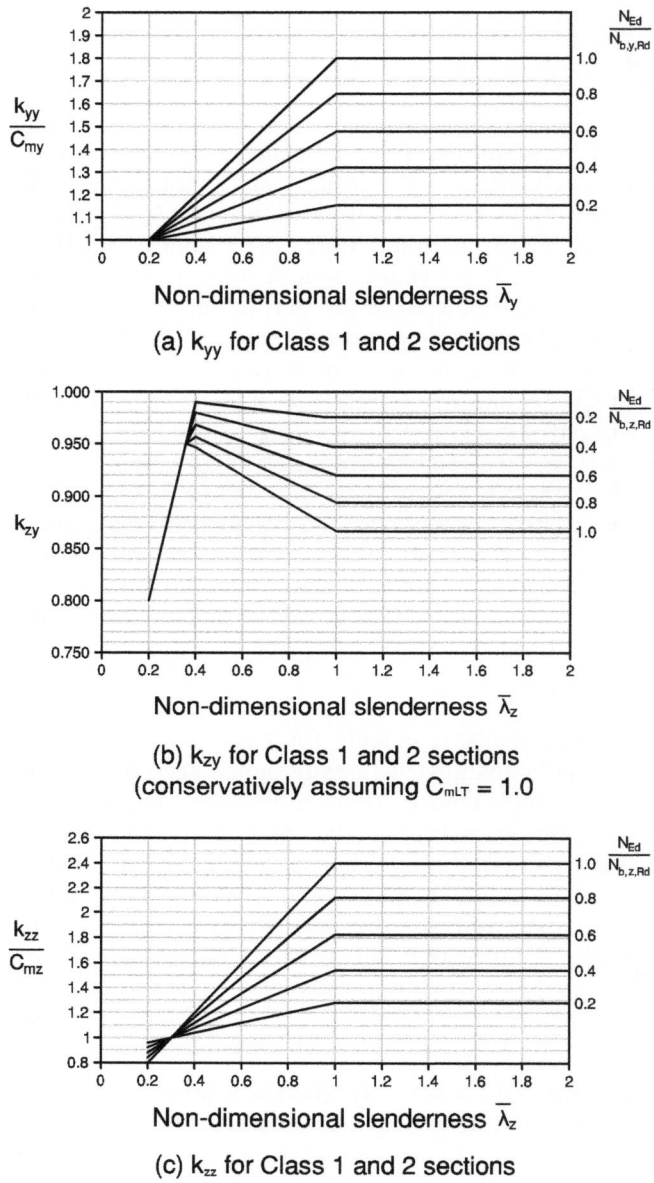

Figure 19.12   Graphical determination of interaction factors for Class 1 and 2 I-sections

Table 19.1 Conservative determination of interaction factors

Interaction factor	Class 1 and 2	Class 3
k_{yy}	$1.8C_{my}$	$1.6C_{my}$
k_{yz}	$0.6k_{zz}$	k_{zz}
k_{zy}	1.0	1.0
k_{zz}	$2.4C_{mz}$	$1.6C_{mz}$

Provided such frames are designed elastically, the methods already described for assessing local cross-sectional capacity and overall buckling resistance may be employed. However, these general approaches fail to take account of some of the special features present in normal portal frame construction, some of which can, when properly allowed for, be shown to enhance buckling resistance significantly.

When plastic design is being employed, the requirements for member stability change somewhat. It is no longer sufficient simply to ensure that members can safely resist the applied moments and thrust; rather for members required to participate in plastic hinge action, the ability to sustain the required moment in the presence of compression during the large rotations necessary for the development of the frame's collapse mechanism is essential. This requirement is essentially the same as that for a Class 1 cross-section discussed in Chapter 14. The performance requirement for those members in a plastically designed frame actually required to take part in plastic hinge action is therefore equivalent to the most onerous type of response shown in Figure 14.5. If they cannot achieve this level of performance, for example because of premature unloading caused by local buckling, then they will prevent the formation of the plastic collapse mechanism assumed as the basis for the design, with the result that the desired load factor will not be attained. Put simply, the requirement for member stability in plastically-designed structures is to impose limits on slenderness and axial load level, to ensure stable behaviour while the member is carrying a moment equal to its plastic moment capacity suitably reduced so as to allow for the presence of axial load. For portal frames, advantage may be taken of the special forms of restraint inherent in that form of construction by, for example, purlins and sheeting rails attached to the outside flanges of the rafters and columns respectively.

Figure 19.13 illustrates a typical collapse bending moment diagram for a single-bay pinned-base portal frame subject to gravity load only (dead load + imposed load), this often being the governing load combination in the UK. The frame is assumed to be typical of UK practice with columns of somewhat heavier section than the rafters, and haunches of approximately 10% of the clear span and twice the rafter depth at the eaves. It is further assumed that the purlins and siderails which support the cladding and are attached to the outer flanges of the columns, and rafters provide positional restraint to the frame, i.e. prevent lateral movement of the flange, at these points. Four regions in which member stability must be ensured may be identified:

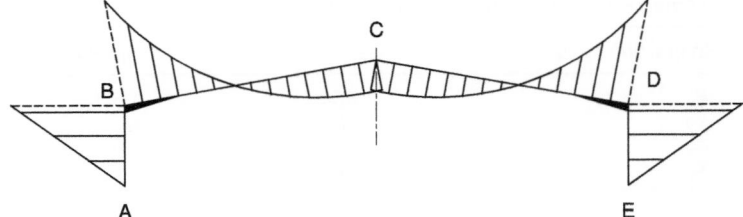

Figure 19.13 Typical bending moment distribution in a portal frame under dead plus imposed load combination

1. full column height AB
2. haunch, which should remain elastic throughout its length
3. eaves region of rafter for which the lower unbraced flange is in compression due to the moments, from end of haunch
4. apex region of the rafter between top compression flange restraints.

19.6.2 Column stability

Figure 19.14 provides a more detailed view of the column AB, including both the bracing provided by the siderails and the distribution of moment over the column height. Assuming the presence of a plastic hinge immediately below the haunch, the design requirement is to ensure stability up to the formation of the collapse mechanism.

According to Clause 6.3.5.2(4)B of BS EN 1993-1-1, effective torsional restraint must be provided no more than $h/2$ (where h is the overall column depth, measured along the column axis) from the underside of the haunch. This may conveniently be achieved by means of the knee brace arrangement of Figure 19.15, where effective torsional restraint is achieved by providing lateral restraint to both flanges. The simplest means of ensuring adequate stability for the region adjacent to this braced point is to provide another torsional restraint within a distance of not more than the stable length L_{stable}; the most straightforward expression for L_{stable} is given in Clause 6.3.5.3(1)B of BS EN 1993-1-1 as:

$$L_{\text{stable}} = 35\varepsilon i_z \quad \text{for} \quad 0.625 \leq \psi \leq 1 \tag{19.22}$$

$$L_{\text{stable}} = (60 - 40\psi)\varepsilon i_z \quad \text{for} \quad -1 \leq \psi \leq 0.625 \tag{19.23}$$

Depending on the grade of steel, ε will be slightly less than unity and noting from Figure 19.14 that ψ for the column segment immediately below the rafter haunch will be in a little less than unity gives an approximate limit of $35 i_z$ for use when deciding on the approximate positioning of the second side rail and its associated knee brace.

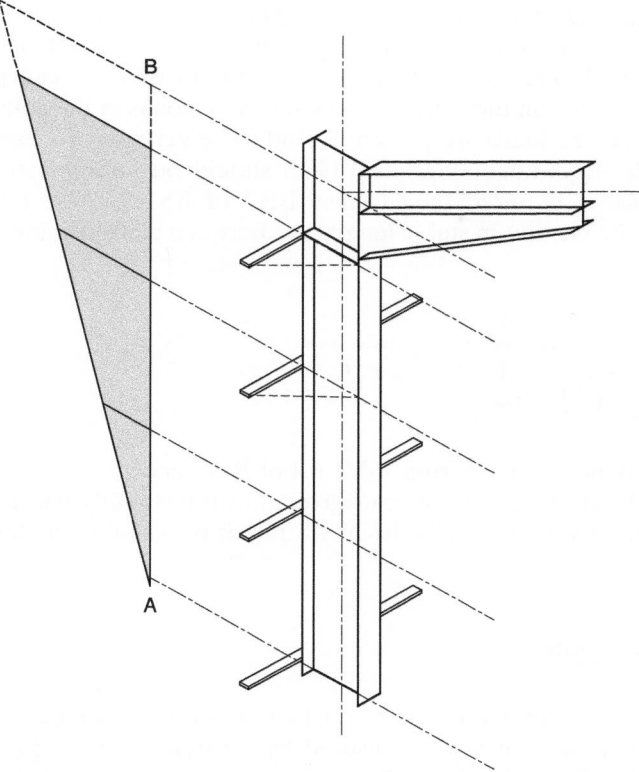

Figure 19.14 Member stability – column

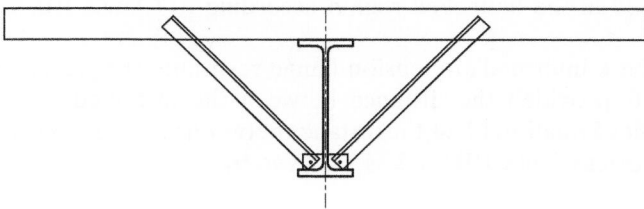

Figure 19.15 Effective torsional restraints

Below this region, the distribution of moment in the column normally ensures that the remainder of the length is elastic. Its stability may therefore be checked using the procedures of Section 19.5. Frequently no additional intermediate restraints are necessary, the elastic stability condition being much less onerous than the plastic one.

Strictly speaking, Equations 19.22 and 19.23 only apply to uniform I-sections with $h/t_f \leq 40\varepsilon$, a linear variation of moment and when there is no significant axial load. The majority of UKB sections will meet the geometrical limit – exceptions tend to be the lightest weights in the larger serial sizes. Axial loads in the columns of portal frames – unless crane loads are present – tend to be very low. So, used carefully as an initial guide, Equations 19.22 and 19.23 should be appropriate. When more precise guidance is required then Clause BB.3 of BS EN 1993-1-1 provides an expression for the maximum stable length, L_m, between a plastic hinge and adjacent restraints as:

$$L_m = \frac{38i_z}{\sqrt{\dfrac{1}{57.4}\left(\dfrac{N_{Ed}}{A}\right)+\dfrac{1}{756C_1^2}\left(\dfrac{W_{pl,y}^2}{AI_t}\right)\left(\dfrac{f_y}{235}\right)^2}} \tag{19.24}$$

in which C_1 may be obtained from Table 6.4 of Reference 3.

Further guidance on the use of background of the stability checks for partially restrained members in plastically designed frames is available in Reference 4 and Chapter 4.

19.6.3 Rafter stability

Stability of the eaves region of the rafter (where the inner flanges are typically in compression) may most easily be ensured by satisfying the conditions of Clause BB.3 of BS EN 1993-1-1. If tension flange restraint is not present between points of compression flange restraint, i.e. widely spaced purlins and a short unbraced length requirement, this simply requires the use of Equation 19.24 to check that the distance between compression flange restraints does not exceed L_m. Note that L_m may be modified in the haunched region according to Clause BB.3.2.1 of BS EN 1993-1-1.

For cases where intermediate tension flange restraints are present, as illustrated in Figure 19.16, provided the distance between the intermediate tension flange restraints satisfies Equation 19.24, the distance between compression flange restraints may be taken from Clause BB.3.1.2 as L_s, given by:

$$L_s = L_k = \frac{\left(5.4+\dfrac{600f_y}{E}\right)\left(\dfrac{h}{t_f}\right)i_z}{\sqrt{5.4\left(\dfrac{f_y}{E}\right)\left(\dfrac{h}{t_f}\right)^2-1}} \quad \text{for uniform moment} \tag{19.25}$$

$$L_s = \sqrt{C_m}L_k\left(\frac{M_{pl,y,Rk}}{M_{N,y,Rk}+aN_{Ed}}\right) \quad \text{for linear moment gradients} \tag{19.26}$$

$$L_s = \sqrt{C_n}L_k \quad \text{for nonlinear moment gradients} \tag{19.27}$$

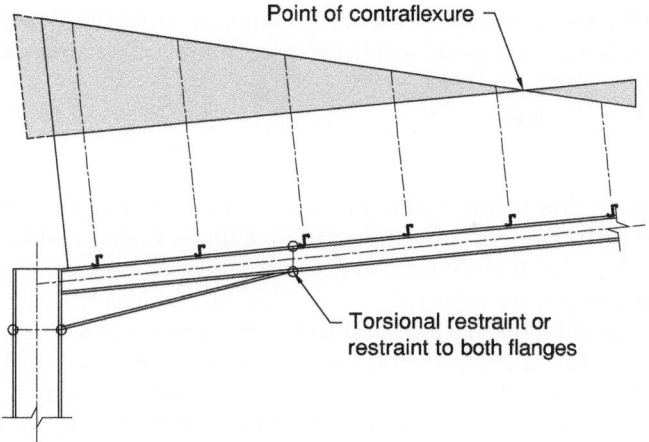

Figure 19.16   Member stability in portal frame rafters

where C_m is a modification factor for linear moment gradients, C_n is a modification factor for nonlinear moment gradients, a is the distance between the centroid of the member with the plastic hinge and the centroid of the restraint members, $M_{pl,y,Rk}$ is the characteristic major axis plastic moment resistance, which is equal to $M_{pl,y,Rd}$ when $\gamma_{M0} = 1.0$ (as is the case in the UK), and $M_{N,y,Rk}$ is the reduced plastic moment capacity in the presence of an axial force N_{Ed}. The above expressions are modified to account for the taper in haunched regions, as set out in Clause BB.3.2.2 of BS EN 1993-1-1.

19.6.4 Bracing

The general requirements of lateral bracing systems have already been referred to in Chapter 17 – Sections 17.3, 17.4 and 17.5 in particular. When purlins or siderails are attached directly to a rafter or column compression flange it is usual to assume that adequate bracing stiffness and strength are available without conducting specific calculations. BS EN 1993-1-1 does, however, provide specific design rules for restraints to members with plastic hinges within their length in Clause 6.3.5.2. It states:

- At each plastic hinge location, the lateral restraint must be capable of withstanding a force equal to 2.5% of the maximum force in the compression flange $N_{f,Ed}$ of the braced member at the plastic hinge location, without any combination with other loads.
- For the design of the bracing system, in addition to the checks for restraint forces due to member imperfections (see Clause 5.3.3 of BS EN 1993-1-1), the bracing

system should also be able to resist, in combination with other loads, local forces Q_m applied at the plastic hinge locations of each braced member, where:

$$Q_m = 1.5\alpha_m \frac{N_{f,Ed}}{100} \tag{19.28}$$

where α_m is a reduction factor, defined in Clause 5.3.3(1), based on the number of members being restrained. In order to ensure adequate restraint stiffness, it is also recommended that the non-dimensional slenderness of each restraint acting in compression be not greater than 1.2.

When purlins or siderails are attached to the main member's tension flange, any positional restraint to the compression flange must be transferred through both the bracing to main member interconnection and the webs of the main member. When full torsional restraint is required so that interbrace buckling may be assumed, the arrangement of Figure 19.14 is often used. The stays may be angles, tubes (provided simple end connections can be arranged) or flats (which are much less effective in compression than in tension). In theory a single member of sufficient size would be adequate, but practical considerations such as hole clearance[4,5] normally dictate the use of pairs of stays. It should also be noted that for angles to the horizontal of more than 45° the effectiveness of the stay is significantly reduced.

19.6.5 Effect of wind on rafter stability

The foregoing discussion relates to portal frames subject to maximum gravity load. Under dead load and maximum wind load, some portal frames may go into overall uplift. Where this occurs, it will be necessary to carry out additional stability checks on the mid span section of the rafter, where the lower flange is in compression. The methods in Section 19.6.3 should be followed.

A series of worked examples follows which are relevant to Chapter 19. Further worked examples may be found in References 6, 7, 8 and 9.

References to Chapter 19

1. SCI and BCSA (2009) *Steel Building Design: Design Data – In accordance with Eurocodes and the UK National Annexes*. SCI Publication No. P363. Ascot, SCI.
2. SCI and BCSA (2009) *Steel Building Design: Concise Eurocodes*. SCI Publication No. P362. Ascot, SCI.
3. NCCI SN048. *Verification of columns in simple construction – a simplified inter-action criterion*. http://www.steel-ncci.co.uk
4. Gardner L. (2011) *Stability of Steel Beams and Columns*. SCI Publication No. P360. Ascot, SCI.

5. Morris, L.J. (1981 and 1983) A commentary on portal frame design. *The Structural Engineer*, 59A, No. 12, 394–404 and 61A, No. 6 181-9.
6. NCCI SN002. *Determination of non-dimensional slenderness of I and H sections.* http://www.steel-ncci.co.uk
7. Trahair N.S., Bradford M.A., Nethercot D.A. and Gardner L. (2008) *The Behaviour and Design of Steel Structures to EC3*, 4th edn. London and New York, Taylor and Francis.
8. Gardner L. and Nethercot D.A. (2005) *Designers' Guide to EN 1993-1-1: Eurocode 3: Design of Steel Structures*. London, Thomas Telford Publishing.
9. Martin L and Purkiss J. (2008) *Structural Design of Steelwork to EN 1993 and EN 1994*, 3rd edn. Oxford, Butterworth-Heinemann.
10. Brettle M. (2009) *Steel Building Design: Worked Examples – Open Sections – In accordance with Eurocodes and the UK National Annexes*. SCI Publication No. P364. Ascot, SCI.

Further reading for Chapter 19

Chen W.F. and Atsuta T. (1977) *Theory of Beam-Columns, Vols 1 and 2*. New York, McGraw-Hill.
Davies J.M. and Brown B.A. (1996) *Plastic Design to BS 5950*. Oxford, Blackwell Science.
Galambos T.V. (1998) *Guide to Stability Design Criteria for Metal Structures*, 5th edn. New York, Wiley.
Horne M.R. (1979) *Plastic Theory of Structures*, 2nd edn. Oxford, Pergamon.
Horne M.R., Shakir-Khalil H. and Akhtar S. (1967) The stability of tapered and haunched beams. *Proc. Instn Civ. Engrs*, 67, No. 9, 677–94.
Morris L.J. and Nakane K. (1983) Experimental behaviour of haunched members. In: *Instability and Plastic Collapse of Steel Structures* (ed. by L.J. Morris), pp. 547–59. London, Granada.

			Sheet 1 of 3	Rev
The Steel Construction Institute	**Job No.**			
	Job Title			
Silwood Park, Ascot, Berks SL5 7QN	**Subject**	Beam-column example 1		
Telephone: (01344) 623345 Fax: (01344) 622944	**Client**		**Made by** DAN	**Date** 2010
CALCULATION SHEET			**Checked by** LG	**Date** 2010

Beam-column example 1

Rolled Universal Column

Problem

Select a suitable UKC in S275 steel to carry safely a combination of 840 kN in direct compression and a uniform bending moment about the minor axis of 12 kNm, over an unsupported height of 3.6 m.

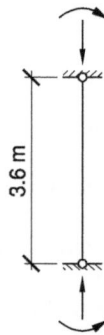

3.6 m

Try 203 × 203 × 60 UKC in S275 steel – member capacity tables suggest a minor axis buckling resistance $N_{b,z,Rd}$ of approximately 1400 kN will provide correct sort of margin to carry the moment.

Steel Building Design: Design Data

Partial factors:

$\gamma_{M0} = 1.0; \ \gamma_{M1} = 1.0$

UK NA to BS EN 1993-1-1

Geometric properties:

$h = 209.6\,mm;$ $b = 205.8\,mm;$ $t_w = 9.4\,mm;$ $t_f = 14.2\,mm;$
$c_f/t_f = 6.20;$ $c_w/t_w = 17.1;$ $A = 76.4\ cm^2 = 7640\,mm^2;$
$i_y = 8.96\ cm = 89.6\,mm;\ i_z = 5.20\ cm = 52.0\,mm;$
$W_{pl,y} = 656\ cm^3 = 656 \times 10^3\,mm^3;$ $W_{pl,z} = 305\ cm^3 = 305 \times 10^3\,mm^3;$

Steel Building Design: Design Data

Material properties:

Yield strength $f_y = 275\ N/mm^2$ since $t_f < 16\,mm$

BS EN 10025-2

Check cross-section classification:

For a Class 1 outstand flange in compression $c_f/t_f \varepsilon \leq 9$

BS EN 1993-1-1 Table 5.2

Beam-column example 1	Sheet 2 of 3	Rev

For a Class 1 web in compression $c_w/t_w \varepsilon \leq 33$

$$\varepsilon = \sqrt{\frac{235}{f_y}} = \sqrt{\frac{235}{275}} = 0.92$$

Actual $c_f/t_f \varepsilon = 6.20/0.92 = 6.70$; within limit

Actual $c_w/t_w \varepsilon = 17.1/0.92 = 18.5$; within limit

Cross-section is Class 1 under pure compression, so will also be Class 1 under the more favourable stress distribution arising from compression plus bending.

Major and minor axis column buckling resistances

Effective lengths:

$L_{cr,y} = 1.0\ L = 1.0 \times 3600 = 3600\ mm$ for buckling about the y-y axis

$L_{cr,z} = 1.0\ L = 1.0 \times 3600 = 3600\ mm$ for buckling about the z-z axis

Non-dimensional column slendernesses:

$$\lambda_1 = \pi\sqrt{\frac{E}{f_y}} = \pi\sqrt{\frac{210000}{275}} = 86.8$$

$$\overline{\lambda}_y = \frac{\lambda_y}{\lambda_1} = \frac{L_{cr,y}/i_y}{\lambda_1} = \frac{3600/89.6}{86.8} = 0.46$$

$$\overline{\lambda}_z = \frac{\lambda_z}{\lambda_1} = \frac{L_{cr,z}/i_z}{\lambda_1} = \frac{3600/52.0}{86.8} = 0.80$$

Buckling curves:

$h/b = 209.6/205.8 = 1.02 < 1.2$.
For major axis buckling, use buckling curve 'b' ($\alpha = 0.34$)
For minor axis buckling, use buckling curve 'c' ($\alpha = 0.49$)

BS EN 1993-1-1 Table 6.2

Buckling reduction factors χ:

$$\Phi_y = 0.5[1 + \alpha(\overline{\lambda}_y - 0.2) + \overline{\lambda}_y^2] = 0.5[1 + 0.34(0.46 - 0.2) + 0.46^2] = 0.65$$

$$\chi_y = \frac{1}{\Phi_y + \sqrt{\Phi_y^2 - \overline{\lambda}_y^2}} = \frac{1}{0.65 + \sqrt{0.65^2 - 0.46^2}} = 0.90 \leq 1.0$$

BS EN 1993-1-1 Cl 6.3.1.2

$$\Phi_z = 0.5[1 + \alpha(\overline{\lambda}_z - 0.2) + \overline{\lambda}_z^2] = 0.5[1 + 0.49(0.80 - 0.2) + 0.80^2] = 0.96$$

BS EN 1993-1-1 Cl 6.3.1.2

$$\chi_z = \frac{1}{\Phi_z + \sqrt{\Phi_z^2 - \overline{\lambda}_z^2}} = \frac{1}{0.96 + \sqrt{0.96^2 - 0.80^2}} = 0.66 \leq 1.0$$

Beam-column example 1	Sheet 3 of 3	Rev

Column buckling resistances:

$$N_{b,y,Rd} = \frac{\chi_y A f_y}{\gamma_{M1}} = \frac{0.90 \times 7640 \times 275}{1.0} \times 10^{-3} = 1892\ kN > 840\ kN = N_{Ed} \qquad OK$$

BS EN
1993-1-1
Cl 6.3.1.1

$$N_{b,z,Rd} = \frac{\chi_z A f_y}{\gamma_{M1}} = \frac{0.66 \times 7640 \times 275}{1.0} \times 10^{-3} = 1395\ kN > 840\ kN = N_{Ed} \qquad OK$$

BS EN
1993-1-1
Cl 6.3.1.1

Minor axis bending resistance

$$M_{c,z,Rd} = \frac{W_{pl,z} f_y}{\gamma_{M0}} = \frac{305 \times 10^3 \times 275}{1.0} \times 10^{-6} = 83.9\ kNm > 12\ kNm = M_{Ed}$$

Combined axial load plus bending:

To verify resistance under combined axial loading plus bending, both Equations 6.61 and 6.62 of BS EN 1993-1-1 must be satisfied. The major axis bending term is absent in this example since $M_{y,Ed} = 0$.

Maximum (conservative) values of k_{yz} may be taken from Table 18.1, as $k_{yz} = 0.6k_{zz}$. $k_{zz} = 2.4C_{mz}$. For $\psi = 1$, $C_{mz} = 1.0$ from Table B.3 of BS EN 1993-1-1.

BS EN
1993-1-1
Table B.3

$$\therefore k_{zz} = 1.4C_{mz} = 2.4 \times 1.0 = 2.4;\ k_{yz} = 0.6k_{zz} = 0.6 \times 2.4 = 1.44$$

$$\therefore \frac{N_{Ed}}{N_{b,y,Rd}} + k_{yz}\frac{M_{z,Ed}}{M_{c,z,Rd}} = \frac{840}{1892} + 1.44\frac{12}{83.9} = 0.44 + 0.21 = 0.65 < 1.0 \quad \therefore OK$$

BS EN
1993-1-1
Equation
6.61

$$\therefore \frac{N_{Ed}}{N_{b,z,Rd}} + k_{zz}\frac{M_{z,Ed}}{M_{c,z,Rd}} = \frac{840}{1395} + 2.4\frac{12}{83.9} = 0.60 + 0.34 = 0.95 < 1.0 \quad \therefore OK$$

BS EN
1993-1-1
Equation
6.62

$\therefore$ *Adopt $203 \times 203 \times 60$ UKC*

	Job No.			Sheet 1 of 8	Rev
The Steel Construction Institute	Job Title				
Silwood Park, Ascot, Berks SL5 7QN Telephone: (01344) 623345 Fax: (01344) 622944 **CALCULATION SHEET**	Subject	Beam-column example 2			
	Client		Made by	DAN	Date	2010
			Checked by	LG	Date	2010

Beam-column example 2

Rolled Universal Beam

Problem

Check the suitability of a $533 \times 210 \times 82$ UKB in S355 steel for use as a column in a portal frame of clear height 5.6 m if the axial compression is 160 kN, the moment at the top of the column is 530 kNm and the base is pinned. Bending is about the major axis. The ends of the columns are adequately restrained against lateral displacement (i.e. out of plane) and rotation. Use Annex B of BS EN 1993-1-1 to determine the beam-column interaction factors.

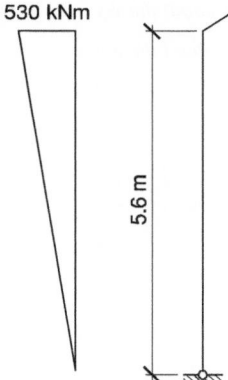

530 kNm

5.6 m

Check initially over the full column height.

Partial factors:

$\gamma_{M0} = 1.0; \; \gamma_{M1} = 1.0$

UK NA to BS EN 1993-1-1

Geometric properties:

$h = 528.3\,mm;$ $b = 208.8\,mm;$ $t_w = 9.6\,mm;$ $t_f = 13.2\,mm;$

$c_f/t_f = 6.55;$ $c_w/t_w = 49.6;$ $A = 105\;cm^2 = 10500\,mm^2;$

$i_y = 21.3\;cm = 213\,mm;\; i_z = 4.38\;cm = 43.8\,mm;$

$W_{pl,y} = 2060\;cm^3 = 2060 \times 10^3\,mm^3;$ $W_{pl,z} = 300\;cm^3 = 300 \times 10^3\,mm^3;$

Steel Building Design: Design Data

Material properties:

Yield strength $f_y = 355\,N/mm^2$ since $t_f < 16\,mm$

BS EN 10025-2

| Beam-column example 2 | Sheet 2 of 8 | Rev |

Check cross-section classification:

For a Class 1 outstand flange in compression $c_f/t_f\varepsilon \leq 9$

For a Class 1 web in bending $c_w/t_w\varepsilon \leq 72$

$$\varepsilon = \sqrt{\frac{235}{f_y}} = \sqrt{\frac{235}{355}} = 0.81$$

Actual $c_f/t_f\varepsilon = 6.55/0.81 = 8.05$; within limit

Actual $c_w/t_w\varepsilon = 49.6/0.81 = 61.0$; within limit for pure bending

Assume since compression of 160 kN is very low compared with axial resistance of cross-section ($Af_y = 10500 \times 355/10^3 = 3727\,kN$) that section is Class 1 under combined loading.

Major and minor axis column buckling resistances

Effective lengths:

$L_{cr,y} = 1.0\ L = 1.0 \times 5600 = 5600\,mm$ for buckling about the y-y axis

$L_{cr,z} = 1.0\ L = 1.0 \times 5600 = 5600\,mm$ for buckling about the z-z axis

Non-dimensional column slendernesses:

$$\lambda_1 = \pi\sqrt{\frac{E}{f_y}} = \pi\sqrt{\frac{210000}{355}} = 76.4$$

$$\bar{\lambda}_y = \frac{\lambda_y}{\lambda_1} = \frac{L_{cr,y}/i_y}{\lambda_1} = \frac{5600/213}{76.4} = 0.34$$

$$\bar{\lambda}_z = \frac{\lambda_z}{\lambda_1} = \frac{L_{cr,z}/i_z}{\lambda_1} = \frac{5600/43.8}{76.4} = 1.67$$

Buckling curves:

$h/b = 528.3/208.8 = 2.53 > 1.2$.
For major axis buckling, use buckling curve 'a' ($\alpha = 0.21$)
For minor axis buckling, use buckling curve 'b' ($\alpha = 0.34$)

Buckling reduction factors χ:

$$\Phi_y = 0.5[1 + \alpha(\bar{\lambda}_y - 0.2) + \bar{\lambda}_y^2] = 0.5[1 + 0.21(0.34 - 0.2) + 0.34^2] = 0.57$$

$$\chi_y = \frac{1}{\Phi_y + \sqrt{\Phi_y^2 - \bar{\lambda}_y^2}} = \frac{1}{0.57 + \sqrt{0.57^2 - 0.34^2}} = 0.97 \leq 1.0$$

$$\Phi_z = 0.5[1 + \alpha(\bar{\lambda}_z - 0.2) + \bar{\lambda}_z^2] = 0.5[1 + 0.34(1.67 - 0.2) + 1.67^2] = 2.15$$

BS EN 1993-1-1 Table 5.2
BS EN 1993-1-1 Table 6.2
BS EN 1993-1-1 Cl 6.3.1.2
BS EN 1993-1-1 Cl 6.3.1.2

Beam-column example 2	Sheet 3 of 8	Rev

$$\chi_z = \frac{1}{\Phi_z + \sqrt{\Phi_z^2 - \overline{\lambda}_z^2}} = \frac{1}{2.15 + \sqrt{2.15^2 - 1.67^2}} = 0.29 \le 1.0$$

Column buckling resistances:

$$N_{b,y,Rd} = \frac{\chi_y A f_y}{\gamma_{M1}} = \frac{0.97 \times 10500 \times 355}{1.0} \times 10^{-3} = 3604\,kN > 160\,kN = N_{Ed} \quad OK$$

BS EN 1993-1-1 Cl 6.3.1.1

$$N_{b,z,Rd} = \frac{\chi_z A f_y}{\gamma_{M1}} = \frac{0.29 \times 10500 \times 355}{1.0} \times 10^{-3} = 1065\,kN > 160\,kN = N_{Ed} \quad OK$$

BS EN 1993-1-1 Cl 6.3.1.1

Lateral torsional buckling resistance

Non-dimensional beam slendernesses is determined using the simplified approach given in NCCI SN002[6] and Reference 2 as follows:

$$\overline{\lambda}_{LT} = \frac{1}{\sqrt{C_1}} UV \overline{\lambda}_z \sqrt{\beta_w}$$

NCCI SN002

For Class 1 or 2 sections, $\beta_w = 1.0$. Conservatively for I-sections, take $UV = 0.9$. For the ratio of end moments $\psi = 0$, $1/\sqrt{C_1} = 0.75$ from Table 6.4 of Reference 2 – see also Chapter 17.

$$\overline{\lambda}_{LT} = \frac{1}{\sqrt{C_1}} UV \overline{\lambda}_z \sqrt{\beta_w} = 0.75 \times 0.9 \times \frac{5600/43.8}{76.4} \times 1.0 = 1.13$$

For the case of rolled and equivalent welded sections, for I-sections with $2 < h/b \le 3.1$, use buckling curve 'c' ($\alpha = 0.49$). For rolled sections, $\beta = 0.75$ and $\overline{\lambda}_{LT,0} = 0.4$.

UK NA to BS EN 1993-1-1

Buckling reduction factor χ_{LT}:

$$\Phi_{LT} = 0.5[1 + \alpha_{LT}(\overline{\lambda}_{LT} - \overline{\lambda}_{LT,0}) + \beta \overline{\lambda}_{LT}^2]$$
$$= 0.5[1 + 0.49(1.13 - 0.4) + (0.75 \times 1.13^2)] = 1.16$$

BS EN 1993-1-1 Cl 6.3.2.3

$$\chi_{LT} = \frac{1}{\Phi_{LT} + \sqrt{\Phi_{LT}^2 - \beta \overline{\lambda}_{LT}^2}} = \frac{1}{1.16 + \sqrt{1.16^2 - 0.75 \times 1.13^2}} = 0.56$$

Lateral torsional buckling resistance:

$$M_{b,Rd} = \chi_{LT} W_y \frac{f_y}{\gamma_{M1}} = 0.56 \times 2060 \times 10^3 \frac{355}{1.0} \times 10^{-6}$$
$$= 412\,kNm < 530\,kNm = M_{Ed} \quad \therefore Member\ fails$$

BS EN 1993-1-1 Cl 6.3.2.1

Clearly the lateral torsional buckling resistance is insufficient. This may be improved by adding bracing to reduce the minor axis buckling length and hence $\overline{\lambda}_{LT}$.

Beam-column example 2	Sheet 4 of 8	Rev

Add bracing to reduce minor axis buckling length

Estimate suitable bracing location as 1.6 m from top of column.

For minor z-z axis buckling and lateral torsional buckling (LTB), $L_{cr} = 1.6$ m for the upper part of the member and 4.0 m for the lower part, while for major y-y axis buckling, $L_{cr} = 5.6$ m. Major axis buckling resistance is therefore unaltered, while minor axis buckling resistance and LTB resistance are increased. For the upper part of the member, the bending moment varies linearly from 530 kNm to 379 kNm ($= 530 \times 4/5.6$), while for the lower part of the member, the bending moment varies linearly from 379 kNm to zero.

For the upper part of the member:

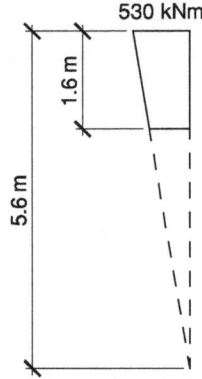

Minor axis column buckling resistance

Effective length:

$L_{cr,z} = 1.0\ L = 1.0 \times 1600 = 1600$ mm for buckling about the z-z axis

Non-dimensional column slenderness:

$$\lambda_1 = \pi\sqrt{\frac{E}{f_y}} = \pi\sqrt{\frac{210000}{355}} = 76.4$$

$$\bar{\lambda}_z = \frac{\lambda_z}{\lambda_1} = \frac{L_{cr,z}/i_z}{\lambda_1} = \frac{1600/43.8}{76.4} = 0.48$$

Buckling curve:

$h/b = 528.3/208.8 = 2.53 > 1.2$.
For minor axis buckling, use buckling curve 'b' ($\alpha = 0.34$)

BS EN 1993-1-1 Table 6.2

Buckling reduction factor χ:

$\Phi_z = 0.5[1 + \alpha(\bar{\lambda}_z - 0.2) + \bar{\lambda}_z^2] = 0.5[1 + 0.34(0.48 - 0.2) + 0.48^2] = 0.66$

BS EN 1993-1-1 Cl 6.3.1.2

Beam-column example 2	Sheet 5 of 8	Rev

$$\chi_z = \frac{1}{\Phi_z + \sqrt{\Phi_z^2 - \overline{\lambda}_z^2}} = \frac{1}{0.66 + \sqrt{0.66^2 - 0.48^2}} = 0.89 \le 1.0$$

Column buckling resistance:

$$N_{b,z,Rd} = \frac{\chi_z A f_y}{\gamma_{M1}} = \frac{0.89 \times 10500 \times 355}{1.0} \times 10^{-3} = 3332\,kN > 160\,kN = N_{Ed} \qquad OK$$

BS EN 1993-1-1 Cl 6.3.1.1

Lateral torsional buckling resistance

Non-dimensional beam slendernesses is determined using the simplified approach given in NCCI SN002[6] and Reference 2 as follows:

$$\overline{\lambda}_{LT} = \frac{1}{\sqrt{C_1}} UV \overline{\lambda}_z \sqrt{\beta_w}$$

NCCI SN002

For Class 1 or 2 sections, $\beta_w = 1.0$. Conservatively for I-sections, take $UV = 0.9$. For the ratio of end moments $\psi = 379/530 = 0.71$, $1/\sqrt{C_1} = 0.91$ (by interpolation) from Table 6.4 of Reference 2 – see also Chapter 17.

$$\overline{\lambda}_{LT} = \frac{1}{\sqrt{C_1}} UV \overline{\lambda}_z \sqrt{\beta_w} = 0.91 \times 0.9 \times \frac{1600/43.8}{76.4} \times 1.0 = 0.39$$

Since $\overline{\lambda}_{LT} < 0.4$, there is no reduction for lateral torsion buckling and $M_{b,Rd} = M_{c,y,Rd}$.

$$M_{c,y,Rd} = W_{pl,y} \frac{f_y}{\gamma_{M0}} = 2060 \times 10^3 \frac{355}{1.0} \times 10^{-6} = 731\,kNm > 530\,kNm = M_{Ed} \; \therefore OK$$

Combined axial load plus bending:

To verify resistance under combined axial loading plus bending, both Equations 6.61 and 6.62 of BS EN 1993-1-1 must be satisfied. The minor axis bending term is absent in this example since $M_{z,Ed} = 0$.

Values of k_{yy} and k_{zy} are determined graphically from Figure 19.12.

For $\psi = 0.71$, $C_{my} = 0.6 + 0.4 \times 0.71 = 0.89$.

BS EN 1993-1-1 Table B.3

For $\dfrac{N_{Ed}}{N_{b,y,Rd}} = \dfrac{160}{3604} = 0.04$ and $\overline{\lambda}_y = 0.34$, $\dfrac{k_{yy}}{C_{my}} \approx 1.01$.

$\Rightarrow k_{yy} = 1.01 \times 0.89 = 0.89$

Figure 18.12(a)

For $\dfrac{N_{Ed}}{N_{b,z,Rd}} = \dfrac{160}{3332} = 0.05$ and $\overline{\lambda}_z = 0.48$, $k_{zy} \approx 1.00$.

Figure 18.12(b)

Beam-column example 2	Sheet 6 of 8	Rev

Applying the interaction equations (Equations 6.61 and 6.62) of BS EN 1993-1-1:

$$\frac{N_{Ed}}{N_{b,y,Rd}} + k_{yy}\frac{M_{y,Ed}}{M_{b,Rd}} = \frac{160}{3604} + 0.89\frac{530}{731} = 0.04 + 0.65 = 0.69 < 1.0 \therefore OK$$

BS EN 1993-1-1 Equation 6.61

$$\frac{N_{Ed}}{N_{b,z,Rd}} + k_{zy}\frac{M_{y,Ed}}{M_{b,Rd}} \le 1 = \frac{160}{3332} + 1.0\frac{530}{731} = 0.05 + 0.73 = 0.78 < 1.0 \therefore OK$$

BS EN 1993-1-1 Equation 6.62

For the lower part of the member:

Minor axis column buckling resistance

Effective length:

$L_{cr,z} = 1.0\ L = 1.0 \times 4000 = 4000\,mm$ *for buckling about the z-z axis*

Non-dimensional column slenderness:

$$\lambda_1 = \pi\sqrt{\frac{E}{f_y}} = \pi\sqrt{\frac{210000}{355}} = 76.4$$

$$\bar{\lambda}_z = \frac{\lambda_z}{\lambda_1} = \frac{L_{cr,z}/i_z}{\lambda_1} = \frac{4000/43.8}{76.4} = 1.20$$

Buckling curve:

$h/b = 528.3/208.8 = 2.53 > 1.2.$
For minor axis buckling, use buckling curve 'b' ($\alpha = 0.34$)

BS EN 1993-1-1 Table 6.2

Buckling reduction factor χ:

$$\Phi_z = 0.5[1 + \alpha(\bar{\lambda}_z - 0.2) + \bar{\lambda}_z^2] = 0.5[1 + 0.34(1.20 - 0.2) + 1.20^2] = 1.38$$

BS EN 1993-1-1 Cl 6.3.1.2

$$\chi_z = \frac{1}{\Phi_z + \sqrt{\Phi_z^2 - \bar{\lambda}_z^2}} = \frac{1}{1.38 + \sqrt{1.38^2 - 1.20^2}} = 0.48 \le 1.0$$

Column buckling resistance:

$$N_{b,z,Rd} = \frac{\chi_z A f_y}{\gamma_{M1}} = \frac{0.48 \times 10500 \times 355}{1.0} \times 10^{-3} = 1792\,kN > 160\,kN = N_{Ed} \quad OK$$

BS EN 1993-1-1 Cl 6.3.1.1

Beam-column example 2	Sheet 7 of 8	Rev

Lateral torsional buckling resistance

Non-dimensional beam slendernesses is determined using the simplified approach given in NCCI SN002[6] and Reference 2 as follows:

$$\overline{\lambda}_{LT} = \frac{1}{\sqrt{C_1}} UV\overline{\lambda}_z\sqrt{\beta_w}$$

NCCI SN002

For Class 1 or 2 sections, $\beta_w = 1.0$. Conservatively for I-sections, take $UV = 0.9$. For the ratio of end moments $\psi = 0$, $1/\sqrt{C_1} = 0.75$ from Table 6.4 of Reference 2 – see also Chapter 17.

$$\overline{\lambda}_{LT} = \frac{1}{\sqrt{C_1}} UV\overline{\lambda}_z\sqrt{\beta_w} = 0.75 \times 0.9 \times \frac{4000/43.8}{76.4} \times 1.0 = 0.81$$

For the case of rolled and equivalent welded sections, for I-sections with $2 < h/b \le 3.1$, use buckling curve 'c' ($\alpha = 0.49$). For rolled sections, $\beta = 0.75$ and $\overline{\lambda}_{LT,0} = 0.4$.

UK NA to BS EN 1993-1-1

Buckling reduction factor χ_{LT}:

$$\Phi_{LT} = 0.5[1 + \alpha_{LT}(\overline{\lambda}_{LT} - \overline{\lambda}_{LT,0}) + \beta\overline{\lambda}_{LT}^2]$$
$$= 0.5[1 + 0.49(0.81 - 0.4) + (0.75 \times 0.81^2)] = 0.84$$

BS EN 1993-1-1 Cl 6.3.2.3

$$\chi_{LT} = \frac{1}{\Phi_{LT} + \sqrt{\Phi_{LT}^2 - \beta\overline{\lambda}_{LT}^2}} = \frac{1}{0.84 + \sqrt{0.84^2 - 0.75 \times 0.81^2}} = 0.76$$

Lateral torsional buckling resistance:

$$M_{b,Rd} = \chi_{LT} W_y \frac{f_y}{\gamma_{M1}} = 0.76 \times 2060 \times 10^3 \frac{355}{1.0} \times 10^{-6}$$
$$= 555\,kNm < 379\,kNm = M_{y,Ed} = 530 \times (4.0/5.6) \quad \therefore OK$$

BS EN 1993-1-1 Cl 6.3.2.1

Combined axial load plus bending:

To verify resistance under combined axial loading plus bending, both Equations 6.61 and 6.62 of BS EN 1993-1-1 must be satisfied. The minor axis bending term is absent in this example since $M_{z,Ed} = 0$.

Values of k_{yy} and k_{zy} are determined graphically from Figure 19.12.

For $\psi = 0$, $C_{my} = 0.6$

BS EN 1993-1-1 Table B.3 Figure 18.12(a)

For $\dfrac{N_{Ed}}{N_{b,y,Rd}} = \dfrac{160}{3604} = 0.04$ and $\overline{\lambda}_y = 0.34$, $\dfrac{k_{yy}}{C_{my}} \approx 1.01$, as for the upper part.

$$\Rightarrow k_{yy} = 1.01 \times 0.60 = 0.60$$

For $\dfrac{N_{Ed}}{N_{b,z,Rd}} = \dfrac{160}{1792} = 0.09$ and $\overline{\lambda}_z = 1.20$, $k_{zy} \approx 0.99$

Figure 18.12(b)

Beam-column example 2	Sheet 8 of 8	Rev

Applying the interaction equations (Equations 6.61 and 6.62) of BS EN 1993-1-1:

$$\frac{N_{Ed}}{N_{b,y,Rd}} + k_{yy}\frac{M_{y,Ed}}{M_{b,Rd}} = \frac{160}{3604} + 0.60\frac{379}{555} = 0.04 + 0.41 = 0.46 < 1.0 \quad \therefore OK$$

BS EN 1993-1-1 Equation 6.61

$$\frac{N_{Ed}}{N_{b,z,Rd}} + k_{zy}\frac{M_{y,Ed}}{M_{b,Rd}} \leq 1 = \frac{160}{1792} + 0.99\frac{379}{555} = 0.09 + 0.67 = 0.76 < 1.0 \quad \therefore OK$$

BS EN 1993-1-1 Equation 6.62

$\therefore$ *Adopt 533 × 210 × 82 UKB*

Note that if the maximum values of the interaction factors given in Table 19.1 were to be used, the member would have failed. This is mainly due to the k_{yy} factor, which has a maximum value of $1.8C_{my}$, but for low axial loads and low major axis non-dimensional column slenderness, the value of k_{yy} approaches $1.0C_{my}$, as may be seen in Figure 19.12(a).

The Steel Construction Institute Silwood Park, Ascot, Berks SL5 7QN Telephone: (01344) 623345 Fax: (01344) 622944 **CALCULATION SHEET**	Job No.		Sheet 1 of 3		Rev	
	Job Title					
	Subject	Beam-column example 3				
	Client		Made by	LG	Date	2010
			Checked by	DAN	Date	2010

Beam-column example 3

Rolled Universal Column in Simple Construction

<u>Problem</u>

A 254 × 254 × 73 UKC is to be assessed for use as an internal column in a simple frame (i.e. designed on the assumptions of simple construction). The column length is 5.0 m and the steel grade is S275. Connection eccentricity causes a design major axis bending moment $M_{y,Ed}$ of 9.4 kNm and a design minor axis bending moment $M_{z,Ed}$ of 2.3 kNm. The design axial load in the column N_{Ed} is 1253 kN. Check the adequacy of this section to carry the applied loads.

The simplified interaction expression from columns in simple construction is as follows:

$$\frac{N_{Ed}}{N_{b,z,Rd}} + \frac{M_{y,Ed}}{M_{b,Rd}} + 1.5\frac{M_{z,Ed}}{M_{c,z,Rd}} \leq 1.0$$

Partial factors:

$\gamma_{M0} = 1.0$; $\gamma_{M1} = 1.0$

UK NA to BS EN 1993-1-1

Geometric properties:

$h = 254.1\,mm$; $b = 254.6\,mm$; $t_w = 8.6\,mm$; $t_f = 14.2\,mm$;

$c_f/t_f = 7.77$; $c_w/t_w = 23.3$; $A = 93.1\ cm^2 = 9310\,mm^2$;

$i_y = 11.1\ cm = 111\,mm$; $i_z = 6.48\ cm = 64.8\,mm$;

$W_{pl,y} = 992\ cm^3 = 992 \times 10^3\,mm^3$; $W_{pl,z} = 465\ cm^3 = 465 \times 10^3\,mm^3$;

Steel Building Design: Design Data

Material properties:

Yield strength $f_y = 275\,N/mm^2$ *since* $t_f < 16\,mm$

BS EN 10025-2

Check cross-section classification:

For a Class 1 outstand flange in compression $c_f/t_f\varepsilon \leq 9$

BS EN 1993-1-1 Table 5.2

For a Class 1 web in compression $c_w/t_w\varepsilon \leq 33$

$$\varepsilon = \sqrt{\frac{235}{f_y}} = \sqrt{\frac{235}{275}} = 0.92$$

Beam-column example 3	Sheet 2 of 3	Rev

Actual $c_f/t_f\varepsilon = 7.77/0.92 = 8.40$; within limit

Actual $c_w/t_w\varepsilon = 23.3/0.92 = 25.2$; within limit

Cross-section is Class 1 under pure compression, so will also be class 1 under the more favourable stress distribution arising from compression plus bending.

Minor axis column buckling resistance

Effective length:

$L_{cr,z} = 1.0\ L = 1.0 \times 5000 = 5000\ mm$ *for buckling about the z-z axis*

Non-dimensional column slenderness:

$$\lambda_1 = \pi\sqrt{\frac{E}{f_y}} = \pi\sqrt{\frac{210000}{275}} = 86.8$$

$$\bar{\lambda}_z = \frac{\lambda_z}{\lambda_1} = \frac{L_{cr,z}/i_z}{\lambda_1} = \frac{5000/64.8}{86.8} = 0.89$$

Buckling curve:

$h/b = 254.1/254.6 = 1.0 \le 1.2$.
For minor axis buckling, use buckling curve 'c' ($\alpha = 0.49$)

Buckling reduction factor χ:

$$\Phi_z = 0.5[1 + \alpha(\bar{\lambda}_z - 0.2) + \bar{\lambda}_z^2] = 0.5[1 + 0.49(0.89 - 0.2) + 0.89^2] = 1.06$$

$$\chi_z = \frac{1}{\Phi_z + \sqrt{\Phi_z^2 - \bar{\lambda}_z^2}} = \frac{1}{1.06 + \sqrt{1.06^2 - 0.89^2}} = 0.61 \le 1.0$$

Column buckling resistance:

$$N_{b,z,Rd} = \frac{\chi_z A f_y}{\gamma_{M1}} = \frac{0.61 \times 9310 \times 275}{1.0} \times 10^{-3} = 1553\,kN > 1253\,kN = N_{Ed} \qquad OK$$

Lateral torsional buckling resistance

Non-dimensional beam slendernesses is determined using the simplified approach given in NCCI SN002[6] and Reference 2 as follows:

$$\bar{\lambda}_{LT} = \frac{1}{\sqrt{C_1}} UV\bar{\lambda}_z\sqrt{\beta_w}$$

For Class 1 or 2 sections, $\beta_w = 1.0$. Conservatively for I-sections, take $UV = 0.9$ and $C_1 = 1.0$.

$\therefore \bar{\lambda}_{LT} = 0.9\bar{\lambda}_z = 0.9 \times 0.89 = 0.80$

Right column references:

BS EN 1993-1-1 Table 6.2

BS EN 1993-1-1 Cl 6.3.1.2

BS EN 1993-1-1 Cl 6.3.1.1

NCCI SN002

Beam-column example 3	Sheet 3 of 3	Rev

For the case of rolled and equivalent welded sections, for I-sections with h/b ≤ 2.0, use buckling curve 'b' (α = 0.34). For rolled sections, β = 0.75 and $\bar{\lambda}_{LT,0} = 0.4$.

UK NA to BS EN 1993-1-1

Buckling reduction factor χ_{LT}:

$$\Phi_{LT} = 0.5[1 + \alpha_{LT}(\bar{\lambda}_{LT} - \bar{\lambda}_{LT,0}) + \beta\bar{\lambda}_{LT}^2]$$
$$= 0.5[1 + 0.34(0.80 - 0.4) + (0.75 \times 0.80^2)] = 0.81$$

BS EN 1993-1-1 Cl 6.3.2.3

$$\chi_{LT} = \frac{1}{\Phi_{LT} + \sqrt{\Phi_{LT}^2 - \beta\bar{\lambda}_{LT}^2}} = \frac{1}{0.81 + \sqrt{0.81^2 - 0.75 \times 0.80^2}} = 0.82$$

Lateral torsional buckling resistance:

$$M_{b,Rd} = \chi_{LT}W_y\frac{f_y}{\gamma_{M1}} = 0.82 \times 992 \times 10^3 \frac{275}{1.0} \times 10^{-6}$$
$$= 223 \ kNm > 9.4 \ kNm = M_{y,Ed} \qquad \therefore OK$$

BS EN 1993-1-1 Cl 6.3.2.1

Minor axis bending resistance

$$M_{c,z,Rd} = \frac{W_{pl,z}f_y}{\gamma_{M0}} = \frac{465 \times 10^3 \times 275}{1.0} \times 10^{-6} = 128 \ kNm > 2.3 \ kNm = M_{z,Ed} \therefore OK$$

BS EN 1993-1-1 Cl 6.2.5(2)

Combined axial load plus bending

For a column in simple construction, the following simplified interaction check may be performed:

$$\frac{N_{Ed}}{N_{b,z,Rd}} + \frac{M_{y,Ed}}{M_{b,Rd}} + 1.5\frac{M_{z,Ed}}{M_{c,z,Rd}} = \frac{1253}{1553} + \frac{9.4}{223} + 1.5\frac{2.3}{128}$$
$$= 0.81 + 0.04 + 0.03 = 0.88 \leq 1.0 \quad \therefore OK$$

∴ Adopt 254 × 254 × 73 UKC

Note how the first term (axial load) dominates for this arrangement, illustrating why great precision is not required with the two bending terms and justifying the use of conservative interaction factors.

Chapter 20
Trusses

LEROY GARDNER and KATHERINE CASHELL

20.1 Introduction

A truss is a structural framework in which loads are resisted primarily by axial forces in the individual members. Trusses are commonly used in buildings to support roofs, floors and other internal components such as services and suspended ceilings. They are generally more economical than standard steel beams over relatively large spans. Trusses are also employed for lateral bracing as is discussed in Section 20.7 of this chapter and, in greater detail, in Chapters 4, 5 and 17 in the context of portal frames, multi-storey buildings and beams, respectively. Lattice girders are a form of truss, most often used in bridge construction, and comprise parallel top and bottom chords which are connected by intersecting diagonal members.

20.2 Types of truss

There are many types and forms of truss; some of the most widely used are illustrated in Figure 20.1. The type of truss adopted in design is usually governed by architectural and client requirements, in addition to geometric and economic factors.

The Pratt truss, shown in Figure 20.1 (a) and (e), has diagonals in tension under normal vertical loading, and consequently, the shorter vertical web members are in compression. This advantage is partially offset by the fact that the compression chord is more heavily loaded than the tension chord at mid-span under normal vertical loading. It should be noted that for a light-pitched Pratt roof truss, wind loads may cause a reversal of load thus inducing compressive forces in the longer web members.

The converse of the Pratt truss is the Howe truss (or English truss), which is shown in Figure 20.1(b). The Howe truss can be advantageous for very lightly loaded roofs in which reversal of load due to wind will occur. In addition, the tension chord is more heavily loaded than the compression chord at mid-span under normal vertical loading. The Fink truss, depicted in Figure 20.1(c), offers greater economy in terms of steel weight for long-span high-pitched roofs as the members are subdivided into

Steel Designers' Manual, Seventh Edition. Edited by Buick Davison and Graham W. Owens.
© 2012 Steel Construction Institute. Published 2012 by Blackwell Publishing Ltd.

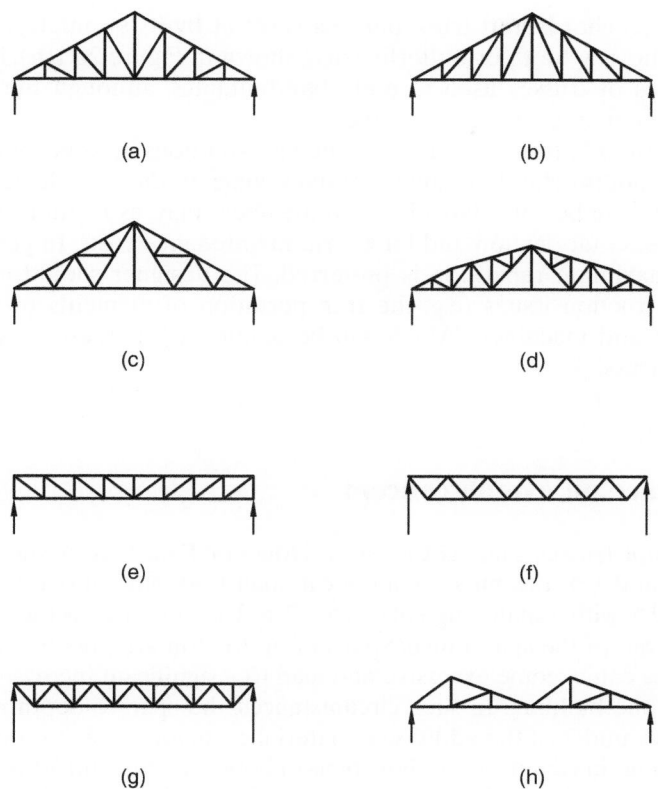

Figure 20.1 Common types of roof truss: (a) Pratt-pitched; (b) Howe; (c) Fink; (d) mansard; (e) Pratt-flat; (f) Warren; (g) modified Warren; and (h) saw-tooth

shorter elements. There are many ways of arranging and subdividing the chords and web members, according to the designer's requirements.

The mansard truss, shown in Figure 20.1(d), is a variation of the Fink truss and offers the advantage of reducing the unusable roof space thereby lowering the running costs of the building. On the other hand, the main disadvantage of the mansard truss is that the forces in the top and bottom chords are increased due to the smaller span-to-depth ratio.

The Warren truss, presented in Figure 20.1(f), has diagonal web members only which are of equal length, thus reducing fabrication costs. Unlike the Pratt truss, the diagonal web members at the centre of the Warren truss are in compression under gravity loading. For longer spans, the modified Warren truss may be adopted in which vertical members are included to provide support to the chords at closer intervals, as shown in Figure 20.1(g). This reduces both the effective buckling length of the compression chord members and the secondary stresses arising from local bending (see Section 20.5.2.). The modified Warren truss requires more material than the parallel-chord Pratt truss, but this is offset by its symmetry and pleasing appearance. The saw-tooth or butterfly truss, shown in Figure 20.1(h), is just one of many examples of trusses used in multi-bay buildings, although the other types described above may be equally suitable.

Trusses can offer extremely efficient structural solutions in terms of material use. However, it is noteworthy that any potential savings to the overall steel weight by using a greater number of relatively small members may, as is often the case, substantially increase fabrication and long-term maintenance costs. In general, simple design with maximum repetition is preferred. The designer should also consider practical construction issues (e.g. the transportation of elements to site and the erection plant and machinery which will be employed) as these may dictate the design of the truss.

20.3 Guidance on overall concept

For pitched-roof trusses such as the Pratt, Howe or Fink trusses shown in Figure 20.1 (a), (b) and (c), the most economical span-to-depth ratio (at the apex) is between 4 and 5, with a span range of 6 m to 12 m. The Fink truss is the most efficient at the higher end of the span range. Spans of up to 15 m are possible but the unusable roof space can become excessive and lead to a significant increase in the operating costs of the building. In such circumstances, the span-to-depth ratio may be increased to around 7 as the additional material costs are offset to some extent by long-term savings in running costs. For spans of between 15 m and 30 m, the mansard truss shown in Figure 20.1(d) reduces the unusable roof space whilst retaining the pitched appearance. It offers an economically-efficient structural solution for span-to-depth ratios of about 7 or 8.

Parallel chord trusses (i.e. lattice girders) have an economic span range of between 6 m and 50 m, with a span-to-depth ratio of 15–25 depending on the intensity of the applied loads. For the top end of the span range, the bay width should be such that the web members are inclined at approximately 50° or slightly steeper. For long, deep trusses, the bay widths become too large and are often subdivided with secondary web members.

The most economical spacing between roof trusses is a function of the overall span and load intensity as well as, to a lesser extent, the span and spacing of the secondary transverse roof members (i.e. purlins). However, as a general rule, the spacing should be between 1/4 and 1/5 of the span which results in a spacing of between 4 m and 10 m for the most cost-effective span range. For short-span roof trusses (6–15 m), the minimum spacing should be limited to between 3 m and 4 m.

Buildings with light pitched roofs may experience load reversal as a result of wind suction and internal pressure (shown in Figure 20.2). This can have significant consequences for the structural response as the light sections which are normally required to act in tension (under dead and imposed loads) may buckle when sub-

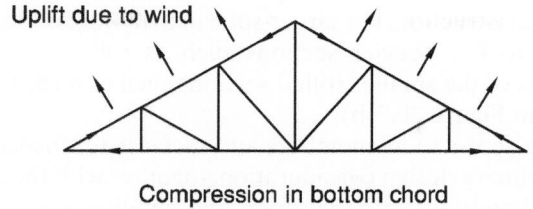

Figure 20.2   Example of load reversal due to wind forces

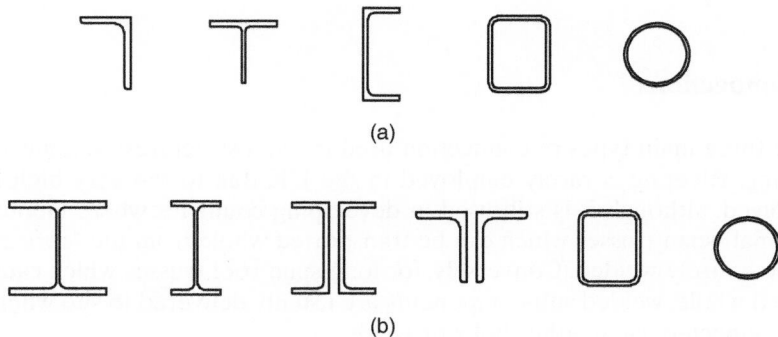

Figure 20.3   Typical element cross-sections: (a) light building trusses; and (b) heavy building trusses

jected to compressive forces. Under wind load, the bottom chord is likely to be in compression and checking for out-of-plane buckling, in particular, is important. It may be necessary to include a line of ties (i.e. bracing) between the bottom chords of adjacent trusses to provide lateral restraint. For heavy pitched or flat roofs, load reversal is rarely a problem as the dead load typically exceeds the wind uplift forces.

20.4 Selection of elements and connections

20.4.1 Elements

The selection of members depends on the location, use, span, type of connection to be used and appearance required. For the individual members in light roof trusses, the most cost-effective sections are usually angles, channels and tee-sections; these are illustrated in Figure 20.3(a). Structural hollow sections are also becoming more popular due to their efficiency in compression and their neat and pleasing appearance. However, these sections have higher fabrication costs and are only suitable for welded construction. For larger-span heavily-loaded roof trusses, it often becomes necessary to use heavier sections such as rolled universal beams and columns, or multiples of the smaller rolled sections such as back-to-back angles and channels, as shown in Figure 20.3(b).

Providing suitable access to all members and surfaces for inspection and maintenance should be a primary design consideration, together with the associated details. This is particularly important in harsh operating conditions (e.g. highly corrosive environments) where regular maintenance is required. For that reason, welded closed-box or circular hollow sections with welded connections are often used in such circumstances.

20.4.2 Connections

There are three main types of connection used in truss structures: welding, bolting and riveting. Riveting is rarely employed in the UK due to the very high labour costs involved, although it is still used in developing countries where labour costs are low. Small-span trusses which can be transported whole from the fabricators to site can be entirely welded. Conversely, for long-span roof trusses which cannot be transported whole, welded sub-components are usually delivered to site where they are then connected using either bolts or welds.

Welded trusses are generally lighter than those using bolts and are also easier to paint and maintain. However, bolted site splices can generally be made more rapidly than welded connections and therefore, are more cost-effective and preferred by contractors. Furthermore, the use of bolts may permit on-site erection of the individual elements without the need for heavy craneage. A possible disadvantage of bolted connections is that they often necessitate the use of gusset plates which can be cumbersome and obtrusive in appearance. The size of the gusset plate is dependent on the dimensions of the incoming members as well as the space available for bolting. Ideally, gussets should be designed such that adjacent edges are at right angles and the number of edges is kept to a minimum. If correctly designed and assembled, gusset plates can be very useful as the incoming members can be positioned such that their centroidal axes meet at a single point, thus avoiding load eccentricities (see Section 20.5.2).

Some typical joint details are illustrated in Figure 20.4.

20.5 Analysis of trusses

20.5.1 General

Loads are generally assumed to be applied at the intersection points of the members, so that they are subjected principally to direct axial stresses. To simplify the analysis, the self-weight of the truss is usually apportioned to the nodal points along the

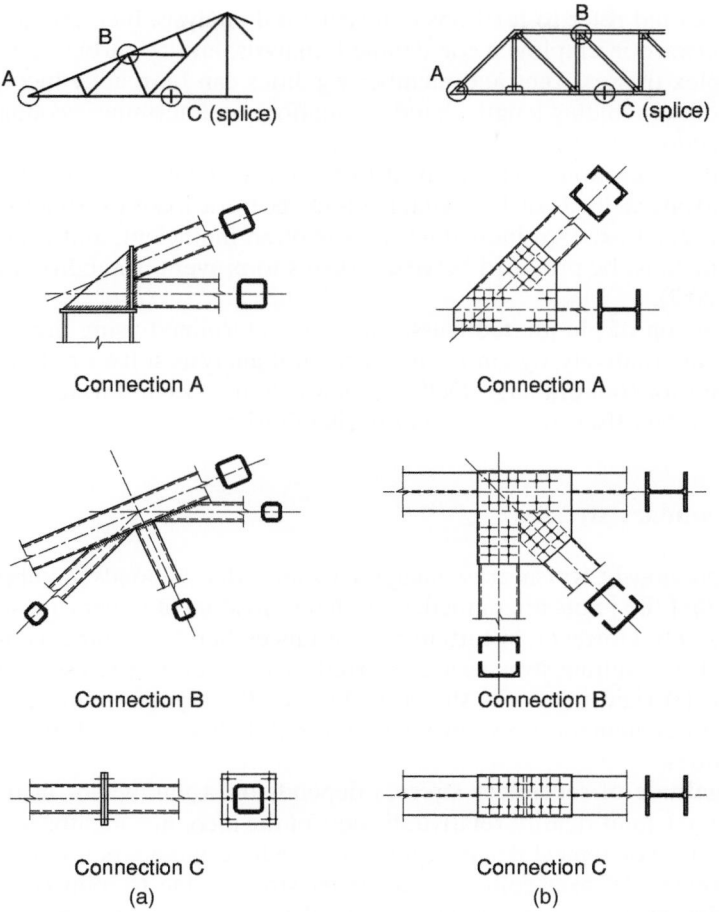

Figure 20.4 Typical joints in trusses: (a) welded RHS building roof truss; and (b) heavy bolted truss

chords. Traditionally, it is also assumed that the individual members are pin-ended, although this is often not the case in practice as will be discussed in the following section.

Manual methods of analysis may be used to determine member forces for simple trusses. For statically determinate trusses, methods of analysis include joint resolution, graphical analysis (Bow's notation or Maxwell diagram) and the method of sections. The last of these is particularly useful if the forces in only a few critical members are required.

Statically indeterminate trusses are more laborious to analyse manually; methods available include virtual work, least work, the reciprocal theorem with influence lines, and the stiffness method. For a full discussion on these methods of analysis, the reader should refer to textbooks on structural analysis. Increasingly, structural analysis software is employed for detailed analysis, and is particularly useful for more complex trusses. Joint and member rigidities can be readily incorporated in the models, thus avoiding lengthy hand calculations to determine secondary stresses (refer to Section 20.5.2.).

Careful consideration must be given to the out-of-plane stability of a truss and resistance to lateral loads such as wind loads or eccentric loads causing torsion about their longitudinal axis. An individual truss is often inefficient, and generally sufficient bracing must be provided between trusses to prevent instability (as discussed in Section 20.7).

The deflection of pin-jointed trusses may be determined using the virtual work method or, alternatively, by employing structural analysis software. This is particularly relevant for roof drainage. Deflections which may occur during the service life of a truss can be offset to some extent during fabrication, by pre-cambering.

20.5.2 Secondary stresses

As stated previously, typical truss analysis assumes that all loads are applied at the nodes and that the joints are pinned, such that individual members are subjected to axial forces only. However, in certain circumstances, bending moments may arise in a member. The resulting stresses are referred to as 'secondary stresses' and may be induced by: (i) rigid joints (as shown in Figure 20.5(a)); (ii) joint eccentricity or misalignment of elements (as shown in Figure 20.5(b)); or (iii) loads that are applied between nodes.

The magnitude of secondary stresses depends on a number of factors such as member layout, joint rigidity, relative stiffness of the incoming members at the joints and lack of fit. For typical design applications, where members are arranged such that their centroidal axes coincide, secondary stresses due to joint eccentricity do not arise. For eccentric connections, the members should be designed to resist the resulting bending moments caused by the eccentricities as well as the axial forces. Similarly, bending moments due to loads applied between nodal points should be accounted for. Despite the assumption of pinned-joints, a certain degree of fixity is likely to be present at the connections, thus inducing secondary moments. These can usually be neglected although, with experience, the designer should be able to recognise when the response of the truss is likely to be significantly influenced by joint rigidity (e.g. when the elements are short and stocky). In this case, a detailed structural analysis should be conducted.

BS EN 1993-1-8[1] provides particular guidance for lattice girders with hollow section joints. It states that bending moments resulting from joint rigidity cannot be ignored in the design if the member length-to-depth ratio, in the plane of the girder, is less than 6. Furthermore, moments induced by secondary stresses arising from joint eccentricity should be distributed between the compression chord members on either side of the connection. This distribution is based on the relative stiffness coefficient I/L of both members, in which I is the second moment of area of the

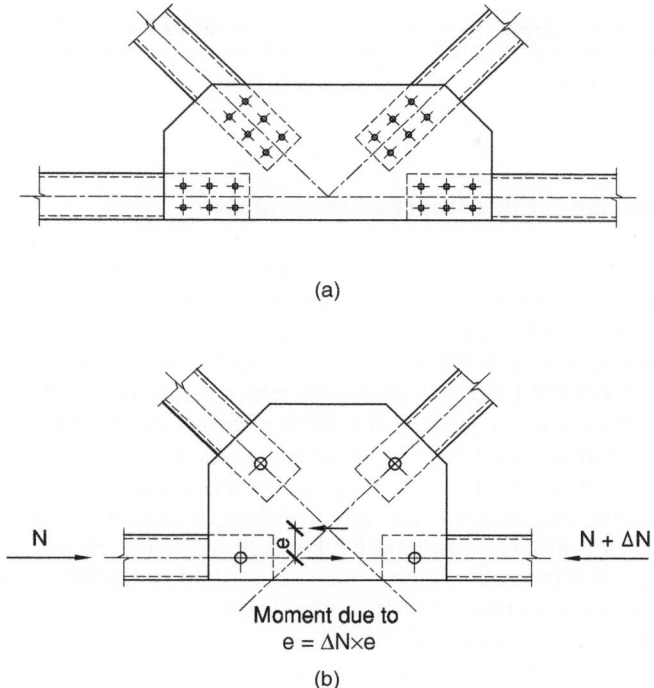

(a)

(b)

Figure 20.5 Secondary moments induced by (a) rigid joints; and (b) eccentricity at joint of incoming members

element in the plane of the truss. Web members are thus subjected to axial forces only. Moments induced by loading between the joints need only be considered in the design of the elements to which they are applied. As a corollary to this, the web members may still be considered as pin-ended whereas the chords should be treated as continuous.

20.6 Detailed design considerations for elements

20.6.1 General

Design actions (loads) are set out in BS EN 1991. Dead and imposed loads are given in BS EN 1991-1-1,[2] snow loads are contained in BS EN 1991-1-3[3] and wind loads are specified in BS EN 1991-1-4.[4] BS EN 1990[5] provides expressions for the determination of load combinations. With experience, the designer should be able to identify the critical load combinations for each element. Further guidance on Eurocode actions and load combinations is given in Reference 6 and Chapter 3.

For guidance on the detailed design of axially loaded members the reader should refer to Chapters 15 and 16, and to Chapter 19 for members subject to combined axial load and bending.

20.6.2 Effective length of compression members

The effective (buckling) length L_{cr} of web members in trusses for both in- and out-of-plane buckling may be conservatively taken as the system length L. This also applies for in-plane buckling of the chord members. For out-of-plane buckling of chords, the effective length is taken as the distance between lateral restraints.

Annex BB.1 of BS EN 1993-1-1[7] allows for lower values of effective length to be used in certain situations to account for fixity of joints or the rigidity of adjacent members. For example, when two or more bolts are used for joints, as shown in Figure 20.6, a reduced effective buckling length (L_{cr}) may be used for the web member in buckling calculations. The effective length factors ($k = L_{cr}/L$) given in the Eurocode for in- and out-of-plane buckling of webs and chords, are presented in Table 20.1, for various section types. In all cases, smaller buckling lengths may be used during design calculations if they can be justified either through experimental results or a more rigorous analysis.

For angles used as web members in compression, provided that there are at least two bolts at the connection and the chords provide adequate end restraint, Annex BB.1.2 of BS EN 1993-1-1 allows the eccentricities to be ignored and accounts for end fixities. The effective non-dimensional slenderness of the member $\overline{\lambda}_{\mathrm{eff}}$ may be obtained from:

$\overline{\lambda}_{\mathrm{eff,v}} = 0.35 + 0.7\overline{\lambda}_{\mathrm{v}}$ for buckling about the v-v axis
$\overline{\lambda}_{\mathrm{eff,y}} = 0.5 + 0.7\overline{\lambda}_{\mathrm{y}}$ for buckling about the y-y axis
$\overline{\lambda}_{\mathrm{eff,z}} = 0.5 + 0.7\overline{\lambda}_{\mathrm{z}}$ for buckling about the z-z axis

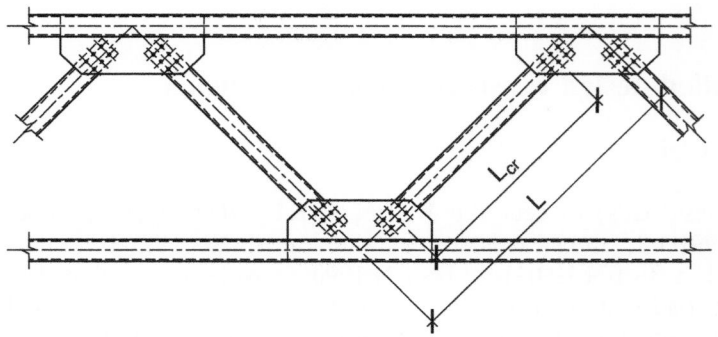

Figure 20.6 System length L and buckling length L_{cr} of web members in trusses.

Table 20.1 Effective length factors according to BS EN 1993-1-1 Annex BB.1

	Effective buckling length factors $k = L_{cr}/L$	
	In-plane buckling	Out-of-plane buckling
1. General Rules		
– chord members	1.0	1.0
– web members	0.9[1]	1.0
2. I- and H-sections		
– chord members	0.9	1.0
– web members	0.9[1]	1.0
3. Hollow sections		
– chord members	0.9	0.9
– web members	1.0	1.0
4. Angles		
– chord members	1.0	1.0
– web members	referred to in text	referred to in text

1 provided that adequate end restraint is provided and the end connections supply appropriate fixity (i.e. at least 2 bolts in bolted connections).

where $\overline{\lambda}$ is determined about the relevant axis based on the system length L. If only one bolt is used for the end connections of angle web members, the eccentricities must be accounted for in design and the effective non-dimensionless slenderness defined above does not apply. Instead, the slenderness should be determined from Clause 6.3.1.3 on the basis of the system length L.

For lattice girders in which the ratio of web width (or diameter) to chord width (or diameter) β is less than 0.6, BS EN 1993-1-1 Annex BB.1.3(3) states that L_{cr} for a hollow section web member which is welded around its perimeter to the chords, may generally be taken as $0.75L$ for in- and out-of-plane buckling. Additional guidance can also be found in NCCI SN031a[8] which gives detailed formulations relating L_{cr} to β for both circular and rectangular hollow sections.

20.7 Bracing

A common use for trusses in buildings is to provide stability to the structure in the form of triangulated bracing and to transmit horizontal loads (such as wind) or crane surges to foundation level. Some examples are shown in Figure 20.7 (a) and (b) for single- and multi-storey construction, respectively.

Triangulated bracing can generally be accommodated within the walls and roof of a building. To avoid the use of heavy compression bracing members, the elements are usually arranged so that a tensile load path always exists and compressive resistance is neglected (e.g. through the use of cross-bracing). Vertical bracing is often required to ensure overall stability, particularly if the structure has simple connections. For single-storey buildings, such as that shown in Figure 20.7(a), plan bracing should be also provided in the roof plane to transfer horizontal loads to the vertical

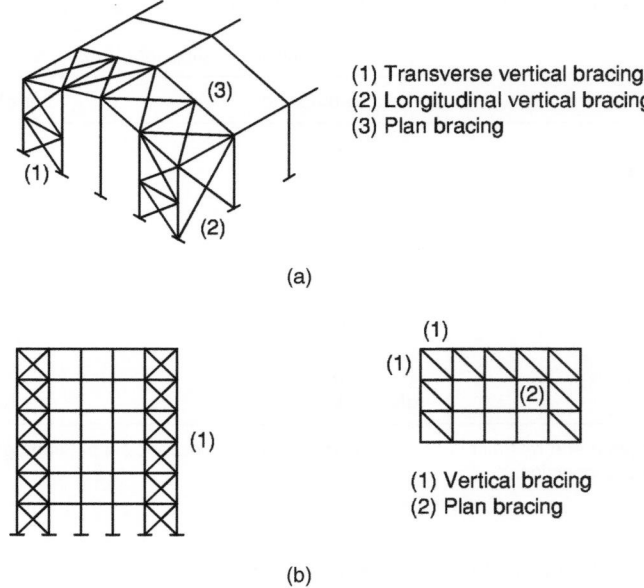

(1) Transverse vertical bracing
(2) Longitudinal vertical bracing
(3) Plan bracing

(a)

(1) Vertical bracing
(2) Plan bracing

(b)

Figure 20.7   Trusses used for bracing of: (a) single-storey building; and (b) multi-storey building

members. On the other hand, the floor slabs in multi-storey structures usually act inherently as horizontal bracing. However, if the steel frame of a building is erected before the floors are constructed, temporary horizontal bracing must be employed until the floors are in place. Furthermore, permanent bracing might be required at floor level if the slabs are discontinuous.

20.8 Rigid-jointed Vierendeel girders

20.8.1 Use of Vierendeel girders

Trusses can be designed without the need for diagonal elements by having rigid joints between the vertical web members and the top and bottom chords. Such trusses are known as Vierendeel girders and typically have chords that are parallel or near-parallel; some typical forms are shown in Figure 20.8(a).

Unlike conventional triangulated trusses where the members are primarily designed for axial loads, the elements in Vierendeel girders are subjected to bending moments and shear forces as well as direct tension or compression. Vierendeel girders are usually more expensive than conventional trusses and their use is limited to instances where diagonal web members are either obtrusive or undesirable. In

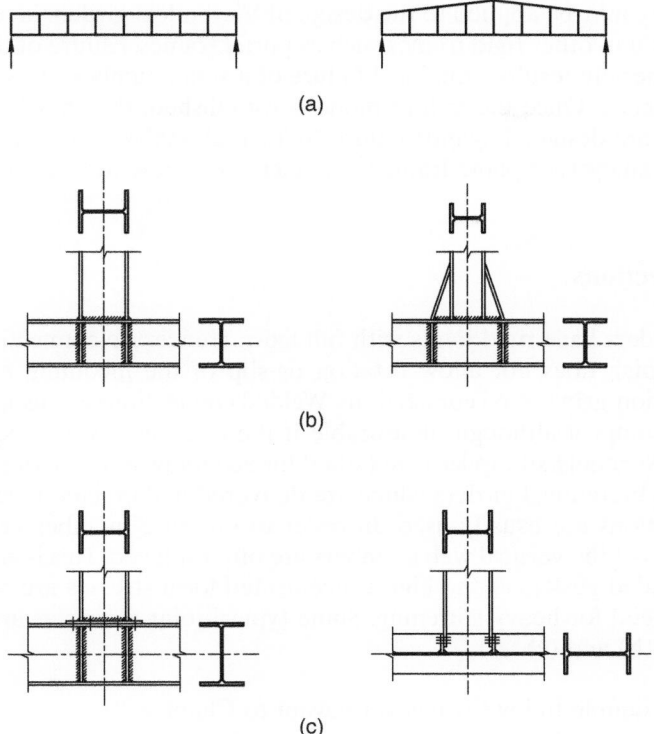

(a)

(b)

(c)

Figure 20.8   Typical details of Vierendeel girders: (a) typical forms; (b) welded connections; and (c) bolted connections

this context, these trusses are most commonly employed in buildings when openings are required within the depth of the truss for doors, windows, corridors or service ducts.

The economic proportions and span lengths are similar to those of the parallel chord trusses already discussed in Section 20.2.

20.8.2 Analysis

Vierendeel girders are statically indeterminate structures, but various manual methods of analysis have been developed. The statically determinate method assumes pin joints at the mid-points of the vertical web members and horizontal chords, for each panel. However, this approach is only suitable for girders with parallel chords of constant stiffness and when the loads are applied at the node points. Nowadays, structural analysis software usually offers the most accurate and efficient means of analysing Vierendeel girders.

Plastic theory may be applied to the design of Vierendeel girders in a similar way to its application to other rigid frames such as portal frames. Failure of the structure, as a whole, generally results from local failure of a small number of its members to form a mechanism. Once the failure mode is established, the chords and vertical web members are designed against failure. Structural analysis software is available for the plastic analysis of plane frameworks including Vierendeel arrangements.

20.8.3 Connections

Vierendeel girders have rigid joints with full fixity. Hence, the connections must be of the type which does not allow rotation or slip of the incoming members, i.e. welded or friction-grip bolted connections. Welded connections are usually the most efficient and compact although undesirable if the connections are required to be made on site. Normally, site splices are bolted for economy and ease of construction. For very large Vierendeel girders which are delivered and erected piecemeal, fully bolted connections are usually used. In order to optimise member and joint efficiency, the ends of the vertical web members are often splayed. This is advantageous for heavily-loaded girders as the high concentrated local stresses are reduced thus avoiding the need for heavy stiffening. Some typical joint examples are illustrated in Figure 20.8 (b) and (c).

A worked example follows which is relevant to Chapter 20.

References to Chapter 20

1. British Standards Institution BS EN 1993-1-8 (2005) *Eurocode 3: Design of steel structures – Part 1-8: Design of joints*. CEN. London, BSI.
2. British Standards Institution BS EN 1991-1-1 (2002) *Eurocode 1 – Actions on structures – Part 1-1: General actions – Densities, self-weight, imposed loads for buildings*. CEN. London, BSI.
3. British Standards Institution BS EN 1991-1-3 (2003) *Eurocode 1 – Actions on structures – Part 1-3: General actions – Snow loads*. CEN. London, BSI.
4. British Standards Institution BS EN 1991-1-4 (2005) *Eurocode 1 – Actions on structures – Part 1-4: General actions – Wind actions*. CEN. London, BSI.
5. British Standards Institution BS EN 1990 (2002) *Eurocode – Basis of structural design*. CEN. London, BSI.
6. Gardner L. and Grubb P.J. (2010) *Eurocode load combinations for steel structures*. London, BCSA Publication No. 53/10. BCSA.
7. British Standards Institution BS EN 1993-1-1 (2005) *Eurocode 3: Design of steel structures – Part 1-1: General rules and rules for buildings*. CEN. London, BSI.
8. NCCI SN031a (2009) *Effective lengths of columns and truss elements in truss portal frame construction*. http://www.steel-ncci.co.uk

		Job No.		Sheet 1 of 10		Rev
The Steel Construction Institute		Job Title				
		Subject	Design of Roof Truss			
Silwood Park, Ascot, Berks SL5 7QN Telephone: (01344) 623345 Fax: (01344) 622944		Client		Made by	KAC	Date
CALCULATION SHEET				Checked by	LG	Date

Roof Truss

Roof truss design example

Design brief

Design the roof trusses for an industrial building which is 120 m long and 25 m wide. The roofing is insulated metal sheeting with purlins at each of the nodal positions on the trusses.

Structural form

For a span of 25 m, a single pitched roof is uneconomical as the height at the apex would be in the order of 5.5 m.

Ideal solutions are to either use a mansard truss or a parallel chord Pratt, Howe or Warren truss.

For the purpose of this example, a mansard truss will be adopted.

The most economical span-to-depth ratio is between 7 and 8.

For 3.5 m depth: $\dfrac{span}{depth} = \dfrac{25\,m}{3.5\,m} = 7.14 \rightarrow acceptable$

Truss spacing should ideally be between 1/4th to 1/5th of the span.

For trusses at 6 m centres: $\dfrac{spacing}{span} = \dfrac{6\,m}{25\,m} = \dfrac{1}{4.17} \rightarrow acceptable$

A truss spacing of 6 m is convenient and suitable for a building which is 120 m long. There will be 20 equal bays.

The connections will be generally shop-welded, with site-splices used at appropriate locations for ease of transportation and to avoid the need for heavy cranage.

Example Concept Design	Sheet 2 of 10	Rev

Truss dimensions

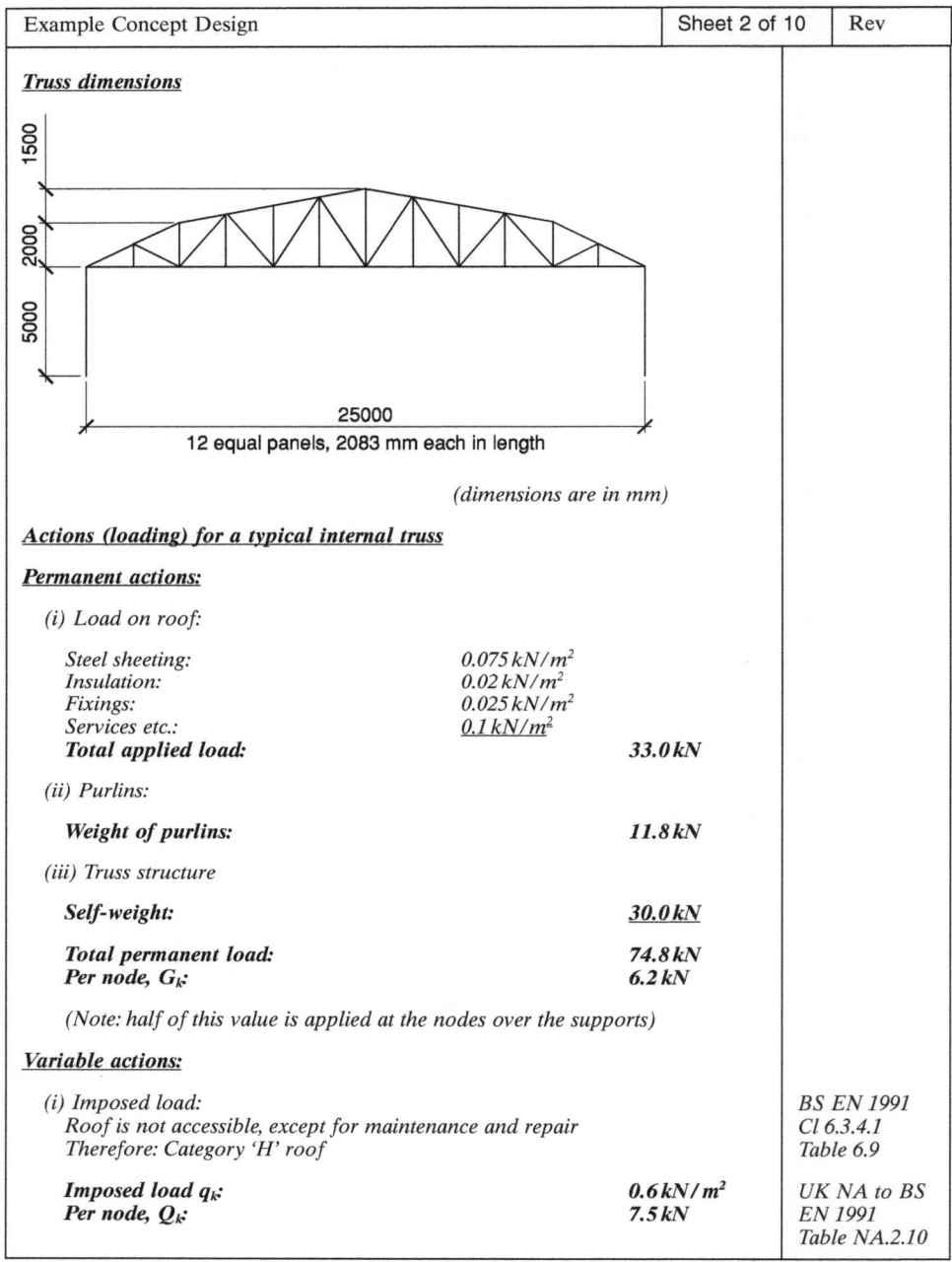

25000

12 equal panels, 2083 mm each in length

(dimensions are in mm)

Actions (loading) for a typical internal truss

Permanent actions:

(i) Load on roof:

Steel sheeting:	$0.075\,kN/m^2$
Insulation:	$0.02\,kN/m^2$
Fixings:	$0.025\,kN/m^2$
Services etc.:	<u>$0.1\,kN/m^2$</u>
Total applied load:	**33.0 kN**

(ii) Purlins:

Weight of purlins: **11.8 kN**

(iii) Truss structure

Self-weight: <u>**30.0 kN**</u>

Total permanent load: **74.8 kN**
Per node, G_k: **6.2 kN**

(Note: half of this value is applied at the nodes over the supports)

Variable actions:

(i) Imposed load:
Roof is not accessible, except for maintenance and repair
Therefore: Category 'H' roof

Imposed load q_k: **$0.6\,kN/m^2$**
Per node, Q_k: **7.5 kN**

BS EN 1991
Cl 6.3.4.1
Table 6.9

UK NA to BS
EN 1991
Table NA.2.10

Example Concept Design	Sheet 3 of 10	Rev

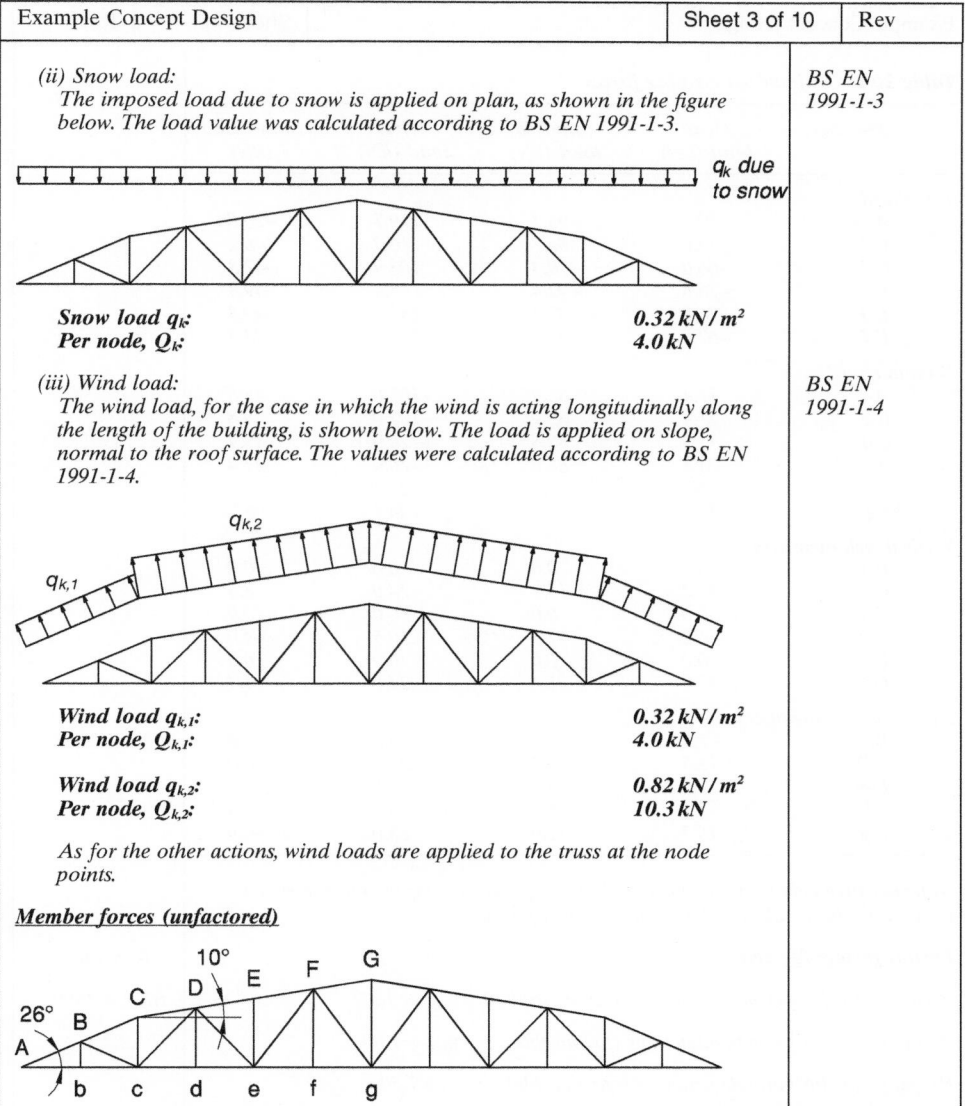

(ii) Snow load:
The imposed load due to snow is applied on plan, as shown in the figure below. The load value was calculated according to BS EN 1991-1-3.

BS EN 1991-1-3

Snow load q_k: **0.32 kN/m²**
Per node, Q_k: **4.0 kN**

(iii) Wind load:
The wind load, for the case in which the wind is acting longitudinally along the length of the building, is shown below. The load is applied on slope, normal to the roof surface. The values were calculated according to BS EN 1991-1-4.

BS EN 1991-1-4

Wind load $q_{k,1}$:   **0.32 kN/m²**
Per node, $Q_{k,1}$:   **4.0 kN**

Wind load $q_{k,2}$:   **0.82 kN/m²**
Per node, $Q_{k,2}$:   **10.3 kN**

As for the other actions, wind loads are applied to the truss at the node points.

Member forces (unfactored)

Example Concept Design			Sheet 4 of 10	Rev

Table 20.2   *Unfactored member forces*

Member	Dead load (kN)	Imposed load (kN)	Wind load (kN)	Snow load (kN)
Top chord				
A-B	−79.2	−95.3	119.7	−50.8
B-C	−72.0	−86.7	116.3	−46.2
C-D	−66.0	−79.4	108.4	−42.3
D-E	−76.8	−92.4	135.4	−49.3
E-F	−76.8	−92.4	137.2	−49.3
F-G	−67.9	−81.6	127.9	−43.5
Bottom chord				
A-b	71.4	85.9	−107.0	45.8
b-c	71.4	85.9	−107.0	45.8
c-d	73.8	88.8	−120.9	47.4
d-e	73.8	88.8	−120.9	47.4
e-f	72.7	87.5	−124.1	46.7
f-g	72.7	87.5	−124.1	46.7
Vertical web members				
B-b	0.0	0.0	0.0	0.0
C-c	13.2	15.9	−24.0	8.5
D-d	0.0	0.0	0.0	0.0
E-e	−6.2	−7.5	10.5	−4.0
F-f	0.0	0.0	0.0	0.0
G-g	17.8	21.4	−24.9	11.4
Diagonal web members				
B-c	−7.2	−8.7	5.6	−4.6
c-D	−13.5	−16.2	28.7	−8.6
D-e	2.6	3.2	−8.8	1.7
e-F	5.1	6.1	−4.7	3.3
F-g	−10.7	−12.9	15.0	−6.9

Note: negative values in the above table correspond to compression forces. The values have been calculated assuming pinned joints.

__Partial factors for actions__

Partial factor for permanent actions (unfavourable) $\gamma_{G,sup} = 1.35$

Partial factor for permanent actions (favourable) $\gamma_{G,inf} = 1.00$

Partial factor for variable actions (unfavourable) $\gamma_Q = 1.50$

Reduction factor for permanent actions $\xi = 0.925$

__ULS combination of actions__

The most critical load combinations have been identified as:

Load combination 1 $(1.35 \times 0.925 \times Dead\ Load) + (1.5 \times Imposed\ Load)$

Load combination 2 $(1.0 \times Dead\ Load) + (1.5 \times Wind\ Load) - uplift\ case$

UK NA to

BS EN 1990

NA.2.2.3.2

BS EN 1990

Cl 6.4.3.2 (3)
Eq. 6.10b

Example Concept Design	Sheet 5 of 10	Rev

Member forces (factored)

Table 20.3   Factored member forces

Member	Load combination 1 (kN)	Load combination 2 (kN)
Top chord		
A-B	−242.0	100.4
B-C	−220.0	102.4
C-D	−201.5	96.6
D-E	−234.5	126.3
E-F	−234.5	129.0
F-G	−207.3	124.0
Bottom chord		
A-b	218.2	−89.0
b-c	218.2	−89.0
c-d	225.5	−107.5
d-e	225.5	−107.5
e-f	222.2	−113.4
f-g	222.2	−113.4
Vertical web members		
B-b	0.0	0.0
C-c	40.5	−22.8
D-d	0.0	0.0
E-e	−19.0	9.6
F-f	0.0	0.0
G-g	54.4	−19.6
Diagonal web members		
B-c	−22.0	1.2
c-D	−41.2	29.6
D-e	8.1	−10.6
e-F	15.6	−1.9
F-g	−32.7	11.8

Member design

Hollow sections will be used for all members.

1. Design of top chord

The most heavily loaded top chord member in compression is A-B. This is also the longest top chord member with a system length L of 2311 mm.

∴ *Maximum compressive force* N_{Ed} *242.0 kN*

Try section size $80 \times 80 \times 3.6$ *SHS (S355)*

c = 69.2 mm; i = 31.1 mm; I = 1.05 × 10⁶ mm⁴; A = 1090 mm²; fy = 355 N/mm²;
$c = 69.2\,mm; i = 31.1\,mm; I = 1.05 \times 10^6\,mm^4; A = 1090\,mm^2; f_y = 355\,N/mm^2;$
$E = 210 \times 10^3\,N/mm^2$

(i) Section classification

$\varepsilon = \sqrt{235/f_y} = 0.81$

$c/t = 69.2/3.6 = 19.2$

BS EN 1993-1-1 Cl 5.5 Table 5.2

Example Concept Design	Sheet 6 of 10	Rev

Limit for Class 1 $= 33\varepsilon = 26.8$

$26.8 > 19.2$ $\therefore$ *section is Class 1*

(ii) Cross-section compression resistance

$N_{c,Rd} = \dfrac{A f_y}{\gamma_{M0}}$ *for Class 1, 2 or 3 cross-sections*

$N_{c,Rd} = \dfrac{1090 \times 355}{1.0} \times 10^{-3} = 386.9 \text{ kN} > 242.0 \text{ kN} = N_{Ed}$

$\therefore$ *OK in compression*

(iii) Member flexural buckling resistance

The buckling length L_{cr} is taken as $0.9 \times L$ for both in- and out-of-plane buckling of the chord. L is considered as the distance between nodes for in-plane buckling and the distance between lateral restraints (i.e. the distance between nodes in this case) for out-of-plane buckling.

$\therefore$ *Buckling length* L_{cr} $0.9 \times L = 2080 \text{ mm}$

Section is symmetrical with identical buckling lengths in both planes; therefore the buckling resistance is the same in both planes.

$N_{b,Rd} = \dfrac{\chi A f_y}{\gamma_{M1}}$ *for Class 1, 2 or 3 cross-sections*

$\chi = \dfrac{1}{\Phi + \sqrt{\Phi^2 - \bar{\lambda}^2}}$ *but $\chi \leq 1.0$*

where

$\Phi = 0.5 \left[1 + \alpha \left(\bar{\lambda} - 0.2 \right) + \bar{\lambda}^2 \right]$

and

non-dimensional slenderness $\bar{\lambda} = \sqrt{\dfrac{A f_y}{N_{cr}}}$ *for Class 1, 2 or 3 cross-sections*

Elastic critical force and non-dimensional slenderness for flexural buckling

$N_{cr} = \dfrac{\pi^2 EI}{L_{cr}^2} = \dfrac{\pi^2 \times 210000 \times 1050000}{2080^2} \times 10^{-3} = 503.1 \text{ kN}$

$\bar{\lambda} = \sqrt{\dfrac{1090 \times 355}{503.1 \times 10^3}} = 0.88$

Reference column:

BS EN 1993-1-1 Cl 6.2.4

BS EN 1993-1-1 Cl BB.1

BS EN 1993-1-1 Cl 6.3.1.1

Example Concept Design	Sheet 7 of 10	Rev

Selection of buckling curve and imperfection factor α.

For a hot-rolled square hollow section, use buckling curve a

For buckling curve a, α = 0.21

Buckling curves

$\Phi = 0.5[1 + 0.21(0.88 - 0.2) + 0.88^2] = 0.96$

$\chi = \dfrac{1}{0.96 + \sqrt{0.96^2 - 0.88^2}} = 0.75$

$\therefore N_{b,Rd} = \dfrac{0.75 \times 1090 \times 355}{1.0} \times 10^{-3} = 289.8 \ kN > 242.0 \ kN = N_{Ed}$

$\therefore OK$ *for member buckling*

2. Design of bottom chord

The maximum compressive force in the bottom chord is in member's c-d and d-e, whilst the most heavily loaded members in tension are e-f and f-g. All elements are of equal length with L = 2083 mm.

$\therefore$ *Maximum compressive force N_{Ed} 113.4 kN*

   Maximum tensile force N_{Ed} 225.5 kN

Try same section size as top chord   80 × 80 × 3.6 SHS (S355)

   c = 69.2 mm; i = 31.1 mm; I = 1.05 × 10^6 mm^4; A = 1090 mm^2; f_y = 355 N/mm^2; f_u = 510 N/mm^2; E = 210 × 10^3 N/mm^2

(i) Section classification

As before, section is Class 1.

(ii) Cross-section compression resistance

As before, $N_{c,Rd} = 386.9 \ kN > 113.4 \ kN = N_{Ed}$ $\therefore$ OK in compression

(iii) Member flexural buckling resistance

As with the upper chord, the buckling length L_{cr} for the bottom chord is taken as 0.9 × L for both in- and out-of-plane buckling where L is considered as the distance between nodes for in-plane buckling and the distance between lateral restraints for out-of-plane buckling. Longitudinal ties are positioned at every second nodal point along the bottom chord, thus resulting in an out-of-plane system length of 4167 mm.

$\therefore$ *In-plane buckling length L_{cr} 0.9 × L = 1875 mm*
   Out-of-plane buckling length L_{cr} 0.9 × L = 3750 mm

In-plane buckling

   As before, $N_{b,Rd} = 289.8 \ kN > 113.4 \ kN = N_{Ed}$
   $\therefore$ *OK for in-plane member buckling*

Right column references:

BS EN 1993-1-1 Cl 6.3.1.2(2) Table 6.2

BS EN 1993-1-1 Cl 5.5 Table 5.2

BS EN 1993-1-1 Cl 6.2.4

BS EN 1993-1-1 Cl BB.1

Example Concept Design	Sheet 8 of 10	Rev

Out-of-plane buckling

  Elastic critical force and non-dimensional slenderness

$$N_{cr} = \frac{\pi^2 EI}{L_{cr}^2} = \frac{\pi^2 \times 210000 \times 1050000}{3750^2} \times 10^{-3} = 154.8\,kN$$

$$\bar{\lambda} = \sqrt{\frac{1090 \times 355}{154.8 \times 10^3}} = 1.58$$

  Selection of buckling curve and imperfection factor α

  For a hot-rolled rectangular hollow section, use buckling curve a

  For buckling curve a, α = 0.21

  Buckling curves

  $\Phi = 0.5[1 + 0.21(1.58 - 0.2) + 1.58^2] = 1.90$

$$\chi = \frac{1}{1.90 + \sqrt{1.90^2 - 1.58^2}} = 0.34$$

$$\therefore N_{b,Rd} = \frac{0.34 \times 1090 \times 355}{1.0} \times 10^{-3} = 131.6\,kN > 113.4\,kN = N_{Ed}$$

∴ *OK for out-of-plane member buckling*

BS EN 1993-1-1 Cl 6.3.1.2(2) Table 6.2

(iv) Cross-section tension resistance

$N_{t,Rd}$ *is the lesser of the design plastic resistance of the gross cross-section* $N_{pl,Rd}$ *and the design ultimate resistance of the net cross-section* $N_{u,Rd}$, *determined as:*

BS EN 1993-1-1 Cl 6.2.3

$$N_{pl,Rd} = \frac{Af_y}{\gamma_{M0}} = \frac{1090 \times 355}{1.0} \times 10^{-3} = 386.9\,kN$$

$$N_{u,Rd} = \frac{0.9 A_{net} f_u}{\gamma_{M2}} = \frac{0.9 \times 1090 \times 510}{1.1} \times 10^{-3} = 454.8\,kN$$

∴ $N_{t,Rd} = 386.9\,kN > 225.5\,kN = N_{Ed}$ ∴ *OK in tension*

3. Design of web members

The maximum tensile force in the web members is in element G-g. The maximum compressive force is in element c-D.

Maximum compressive force N_{Ed} *41.2 kN*

Maximum tensile force N_{Ed} *54.4 kN*

Try section size   *80 × 60 × 3 RHS (S355)*

  $c_w/t = 21.7$; $c_f/t = 15$; $i_y = 30\,mm$; $i_z = 24\,mm$; $I_y = 0.7 \times 10^6\,mm^4$;
  $I_z = 0.449 \times 10^6\,mm^4$; $A = 781\,mm^2$; $f_y = 355\,N/mm^2$; $f_u = 510\,N/mm^2$;
  $E = 210 \times 10^3\,N/mm^2$

Example Concept Design	Sheet 9 of 10	Rev

(i) Section classification

$\varepsilon = \sqrt{235/f_y} = 0.81$

$c_w/t = 21.7; \; c_f/t = 15$

Limit for Class 1 = 33ε = 26.8 ∴ section is Class 1

(ii) Cross-section compression resistance

$N_{c,Rd} = \dfrac{A f_y}{\gamma_{M0}}$ *for Class 1, 2 or 3 cross-sections*

$N_{c,Rd} = \dfrac{781 \times 355}{1.0} \times 10^{-3} = 277.3 \, kN > 41.2 \, kN = N_{Ed}$

∴ OK in compression

(iii) Member flexural buckling resistance

The buckling length L_{cr} is taken as 1.0 × L for both in- and out-of-plane buckling of web members, where L is considered as the distance between nodes. L for the most heavily loaded element in compression c-D is 3159 mm. However, the buckling resistance is also related (inversely) to the buckling length of the member and therefore consideration should also be given to longer elements. In this example, each of the other web members has been checked and c-D was found to be the most critical for buckling; hence, only the calculations relevant to c-D are included herein.

∴ Buckling length L_{cr} 1.0 × L = 3159 mm

The section is non-symmetrical; it is attached such that the weak direction is in the plane of the truss. As the buckling lengths are identical in both directions, only the weak direction will be checked.

Elastic critical force and non-dimensional slenderness for flexural buckling

$N_{cr} = \dfrac{\pi^2 EI}{L_{cr}^2} = \dfrac{\pi^2 \times 210000 \times 449000}{3159^2} \times 10^{-3} = 93.2 \, kN$

$\overline{\lambda} = \sqrt{\dfrac{781 \times 355}{93.2 \times 10^3}} = 1.72$

Selection of buckling curve and imperfection factor α

For a hot-rolled rectangular hollow section, use buckling curve a

For buckling curve a, α = 0.21

Buckling curves

$\Phi = 0.5[1 + 0.21(1.72 - 0.2) + 1.72^2] = 2.15$

Right column references:

BS EN 1993-1-1 Cl 5.5 Table 5.2

BS EN 1993-1-1 Cl 6.2.4

BS EN 1993-1-1 Cl BB.1

BS EN 1993-1-1 Cl 6.3.1.2(2) Table 6.2

Example Concept Design	Sheet 10 of 10	Rev

$$\chi = \frac{1}{2.15 + \sqrt{2.15^2 - 1.72^2}} = 0.29$$

$$\therefore N_{b,Rd} = \frac{0.29 \times 781 \times 355}{1.0} \times 10^{-3} = 80.9 \ kN > 41.2 \ kN = N_{Ed}$$

$\therefore$ *OK for member buckling*

(iv) Cross-section tension resistance

$N_{t,Rd}$ *is the lesser of the design plastic resistance of the gross cross-section* $N_{pl,Rd}$ *and the design ultimate resistance of the net cross-section* $N_{u,Rd}$, *determined as:*

BS EN
1993-1-1
Cl 6.2.3(2)

$$N_{pl,Rd} = \frac{Af_y}{\gamma_{M0}} = \frac{781 \times 355}{1.0} \times 10^{-3} = 277.3 \ kN$$

$$N_{u,Rd} = \frac{0.9A_{net}f_u}{\gamma_{M2}} = \frac{0.9 \times 781 \times 510}{1.1} \times 10^{-3} = 325.9 \ kN$$

$\therefore N_{t,Rd} = 277.3 \ kN > 54.4 \ kN = N_{Ed} \ \therefore$ *OK in tension*

BS EN
1993-1-1
Cl 6.3.1(1)

Final truss design

All shop connections to be welded

Bolted site splice

80x80x3.6 SHS

80x60x3 RHS

80x60x3 RHS

80x80x3.6 SHS

2000

3500

Positions of longitudinal ties

12500

Chapter 21
Composite slabs

by MARK LAWSON

21.1 Definition

Composite slabs comprise profiled steel decking that acts as permanent formwork to support an in-situ concrete slab during construction. Later, when the concrete has gained sufficient strength, the decking acts compositely with the concrete to resist imposed loading. The decking supports the loads during construction and is usually designed to be unpropped at this stage. Mesh reinforcement is placed in the concrete to act as 'fire reinforcement', to distribute local loads and to control crack widths at the internal supports.

Composite slabs are also generally used in conjunction with composite beams, and some aspects of the slab design also influence the effective composite action of the slab with the beams, for example the use of welded shear connectors (refer to Chapter 22). End anchorage of the decking by through-deck welded shear connectors may also be taken into account in the composite slab design. A typical cross-section through a composite slab at an edge beam is shown in Figure 21.1, showing also the use of an edge trim to form the slab edge.

In principle, any shape of profiled decking may be used as permanent formwork if the slab is designed to act non-compositely with the decking and subsequent imposed loads are resisted by provision of reinforcing bars in the ribs of the decking. However, most modern composite decks achieve a suitable degree of shear connection with the concrete by means of mechanical interlock through the embossments or indentations or re-entrant portions rolled into the deck profile. BS EN 1994-1-1 Eurocode 4[1] and formerly BS 5950-4[2] follow the same principles for composite slab design, although the test regimes differ slightly.

21.2 General description

21.2.1 Deck profiles

Deck profiles used in composite slabs are in the range of 45 to 80 mm height and 150 to 333 mm rib spacing. Many deck profiles contain multiple stiffeners and

Steel Designers' Manual, Seventh Edition. Edited by Buick Davison and Graham W. Owens.
© 2012 Steel Construction Institute. Published 2012 by Blackwell Publishing Ltd.

Figure 21.1   Composite slab at an edge beam (courtesy Kingspan)

re-entrant stiffeners to facilitate attachments of service hangers. There are two well known types: the trapezoidal profile with various types of indentations or re-entrant portions and re-entrant (or dovetail) profile. In recent years, a number of new deck profiles have been developed, as shown in Figure 21.2, which are aimed at achieving unpropped spans of 4 to 4.5 m. Guidance on the practical use of composite slabs is presented in SCI Publication 300.[3]

A deep deck profile, known as SD225, and illustrated in Figure 21.2(d) is used in 'Slimdek' construction in which the decking sits on the bottom flange of the beam or on an extended flange plate. The slab and beam occupy the same depth to create a slim floor of approximately 300 mm depth. Deep decking may be used in conjunction with composite slim floor beams when shear connectors are welded to the top flange of the beam.

21.2.2 Steel grades and thicknesses

The galvanised steel used for composite decking applications is now specified to BS EN 10327,[4] and is typically 0.9 to 1.2 mm thick. However, steel thicknesses as low

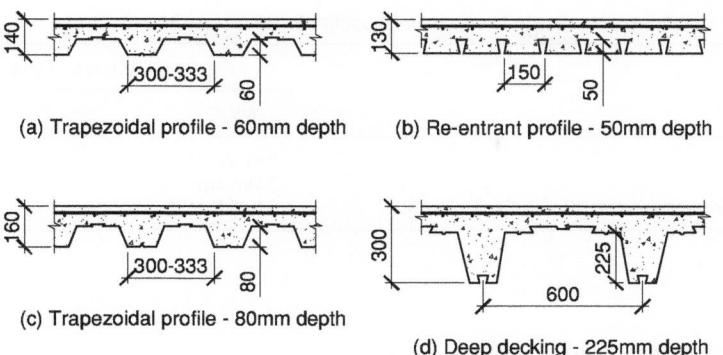

(a) Trapezoidal profile - 60mm depth

(b) Re-entrant profile - 50mm depth

(c) Trapezoidal profile - 80mm depth

(d) Deep decking - 225mm depth

Figure 21.2   Typical re-entrant and trapezoidal deck profiles and slab depths used in composite slabs (other profile shapes are available)

as 0.7 mm are now permitted in BS EN 1994-1-1 (but not in BS 5950-4). A nominal allowance of 0.04 mm is made for the thickness of zinc coating in the total thickness. Steel yield strengths of 280 or 350 N/mm^2 are generally specified, the higher strength steel often being used for longer span deeper profiles.

21.2.3 Concrete type and grade

Normal weight (NWC) and lightweight (LWC) concrete can both be used in composite slabs, although LWC tends only to be used on large projects where minimising of self-weight is important. The modern method of concrete placement is by pumping. In the UK, lightweight concrete is generally in the form of 'Lytag' with sand aggregate and is of 1800 to 1900 kg/m^3 dry density. The wet density is used when determining the loads on the decking in the construction stage and is typically 100 kg/m^3 greater than the dry density. BS EN 1991-1-6[5] recommends that the wet weight of normal weight concrete is taken as 25 kN/m^3 plus the weight of the reinforcement (mesh may be taken as a nominal 50 kg/m^3) which is higher than the 2400 kg/m^3 wet density in BS 5950-4. The equivalent figure for lightweight concrete is 20 kN/m^3 – see Table 21.1.

The concrete grade is specified in terms of its cylinder and cube strength (cylinder/cube strengths in N/mm^2). C25/30 or C30/37 are the common concrete grades for use in composite construction. The concrete type slightly affects the elastic stiffness of the section, but has little effect on shear-bond strength for the same concrete grade.

Table 21.1   Densities of concrete to be used in design to BS EN 1991-1-6

	Concrete Type	
Condition	NWC	LWC
Wet density	25 kN/m³	20 kN/m³
Dry density	24 kN/m³	19 kN/m³
Dry density inc. nominal reinforcement	24.5 kN/m³	19.5 kN/m³

21.2.4 Slab spans and depths

The minimum slab depth for composite design is 90 mm to BS EN 1994-1-1 Cl 9.2.1, and the minimum depth of concrete above the deck profile is 50 mm when the composite slab is used in conjunction with composite beams (or 40 mm if not). In practice, slab depths largely depend on fire insulation requirements and are typically between 120 and 170 mm. Although guidance on serviceability performance is not given in BS EN 1994-1-1, for most designs the slab span-to-depth ratio should not exceed the limits given in Section 21.7.2.

In BS EN1994-1-1, the minimum bearing length of the slab at its supports is 75 mm (compared with 70 mm in BS 5950-4), or 100 mm where the decking is continuous at an internal support. The minimum bearing of the decking on its supports is 50 mm.

The most efficient use of composite slabs is for spans between 3 and 3.6 m, but deeper profiles can achieve spans of up to 4.5 m without requiring propping during construction. The maximum span-to-depth ratio for the decking will normally be in the range of 50 to 60, as influenced by the post-construction deflection of the deck soffit.

21.3 Design for the construction condition

21.3.1 Construction loading for decking design

The decking supports the weight of the concrete in the finished slab. An appropriate allowance has to be made for the excess concrete arising from the deflection of the decking (known as 'ponding'), the temporary weight of the operatives and any impact loads. The construction load, taken to act in addition to the self-weight of the slab and beam, is not specified in BS EN 1994-1-1 but cross-refers to BS EN 1991-1-6.[5] The following loads should be used in the design of the steel decking during construction:

- a uniform load of 1.5 kN/m² acting over a working area of 3 × 3 m
- outside this working area, a reduced uniform load of 0.75 kN/m².

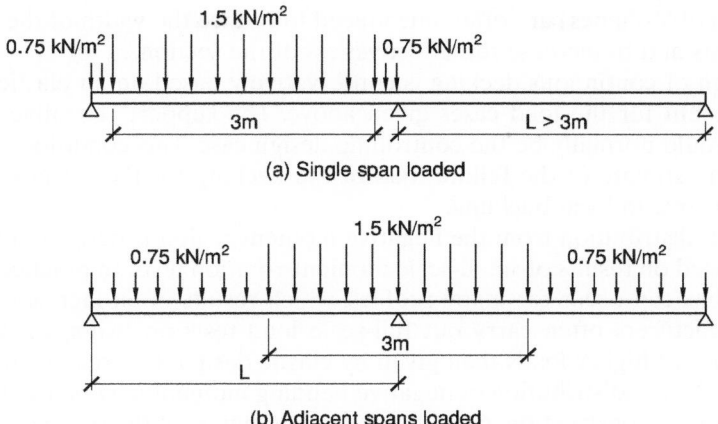

Figure 21.3 Construction load cases applied to the steel decking

For multiple deck spans, the applied loads take account of the sequential nature of the concreting operation on the decking. Therefore, design cases to be considered are:

(a) single span loaded to $1.5\,\text{kN}/\text{m}^2$ over a 3 m length and $0.75\,\text{kN}/\text{m}^2$ elsewhere plus self-weight; adjacent spans not loaded, or
(b) adjacent spans loaded to $0.75\,\text{kN}/\text{m}^2$ plus self-weight throughout plus a localised additional load of $0.75\,\text{kN}/\text{m}^2$ over a 3 m length.

These load cases are defined in Figure 21.3. The self-weight of the slab and beam are also included in combination with the construction load (see densities in Table 21.1). Case (a) corresponds to the maximum elastic moment at mid-span. Case (b) corresponds to the maximum elastic moment at the supports.

According to BS EN 1991-1-6, partial factors in the construction stage are taken as 1.5 for imposed construction loads and also 1.5 for self-weight of the concrete (as concrete is considered as a variable load for this state), which is higher than to BS 5950-4. However, the partial factor for the self-weight of the decking and reinforcement may be taken as 1.35. Loads exceeding these factored loads should not be applied until the slab has gained adequate strength for composite action.

21.3.2 Design of decking

The design of steel decking is covered by BS EN 1993-1-3,[6] which relates to the design of thin steel sections, decking and roof sheeting. It is similar in approach to BS 5950-6.[7] The elastic moment resistance of the section is established taking account of the effective breadth of the thin steel elements in compression. Stiffeners (in the form of V-shapes) are often introduced to reduce the width of the compression elements and to increase the effectiveness of the section.

The design of continuous decking is conservatively based on an elastic distribution of moment for the load cases given above. The support (negative moment) condition would normally be the controlling design case. This condition represents a safe under-estimate of the failure load of the decking for these Class 4 profiles that are sensitive to local buckling.

Moment redistribution from the negative moment region is permitted in BS EN 1993-1-3, based on results of small-scale moment-rotation tests. In practice, moment redistributions can occur for many profiles, which results in an increase in failure load. Manufacturers often carry out full-scale load tests on two-span decking to justify the use of higher loads than given by elastic design. It is recommended that a maximum 30% redistribution of negative bending moment may be used in design of the decking in construction, provided the performance of the decking is justified by testing. This leads to a design moment of approximately $w_u L^2/10$ in both the negative and positive bending regions when considering pattern loads in construction.

21.3.3 Post-construction deflection limits

No deflection limits are specified in BS EN 1994-1-1 for the deflection of the decking after concreting, but the effect of 'ponding' of concrete in terms of its additional weight, should be taken into account when the calculated deflection of the decking exceeds 10% of the slab depth (typically 15 mm).

The additional weight of concrete is equivalent to a uniform load given by 0.7δ of additional concrete depth, where δ is the deflection due to the nominal weight of concrete. However, the UK National Annex to BS EN1994-1-1 permits the same deflection limits as in BS 5950-4 to be used, as follows:

$\delta_{s,max} = L/180$ but less than 20 mm, where the effects of 'ponding' of concrete are not included
$\delta_{s,max} = L/130$ but less than 30 mm, where the effects of 'ponding' of concrete are included.

It may also be necessary to further limit the deflection of the slab if it is visually exposed.

21.4 Design of composite slabs

Composite slabs are usually designed as simply supported members and failure normally occurs due to loss of shear-bond strength between the decking and the

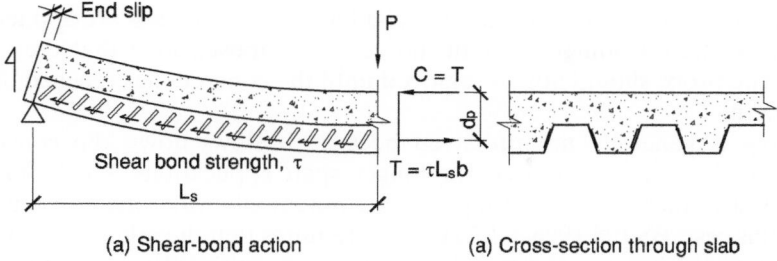

Figure 21.4   Typical shear-bond failure of a composite slab

concrete. This failure mode occurs before the plastic moment resistance of the composite section is reached. It is known as 'partial shear connection'.

Modern deck profiles are rolled with embossments and re-entrant portions, and their shear-bond resistance is good. This means that composite slabs possess imposed load resistances in excess of those required in most buildings. The design of unpropped composite slabs is generally controlled by the construction condition. Propped slabs also resist the self-weight of the slab on removal of the temporary props.

21.4.1 Modes of failure

The moment resistance of composite slabs is determined by the breakdown of chemical bond and mechanical interlock between the decking and the concrete, known generically as 'shear-bond' failure. This occurs at slips (relative displacements) of 2–3 mm at the ends of the span, but for deck profiles with limited composite action, slips are relatively sudden at failure. The shear-bond mode of failure and its associated internal forces for an idealised composite slab are illustrated in Figure 21.4.

For good performance, the load resistance of composite slabs should be considerably greater than that corresponding to initial slip, due to the embossments or indentations in the deck profile, which cause the concrete to 'ride-over' these points. Re-entrant profiles restrict separation of the two materials, giving improved shear-bond performance by mechanical interlock.

For long-span slabs, failure can occur in pure bending. This is a ductile mode of failure and occurs when the shear-bond strength is sufficient to cause tensile yielding in the decking. Pure shear failure rarely occurs except for cases of punching shear close to the supports.

21.4.2 Propped or unpropped slabs

If the slab is unpropped during construction, then the decking resists the self-weight loads, and subsequent loads are applied to the composite slab. If the slab is propped during construction, then factored self-weight and imposed loads are applied to the composite section, leading to a reduction in the imposed load that the slab can support. The props should not be removed until the concrete has reached its specified strength.

Propping is generally not preferred because it slows down the construction process, but it may be required for longer span applications. Furthermore, for propped slabs, cracks at internal supports are potentially wider than for unpropped slabs. In this case, BS EN 1994-1-1 Cl 9.8.1(2) requires that the minimum percentage of reinforcement is increased to 0.4% of the cross-sectional area in the negative moment region of the slab in order to achieve reasonable crack control. This amount of reinforcement may have to be increased even further in order to control crack widths to precise limits for durability. The comparable minimum reinforcement for unpropped slabs is 0.2%, which also provides resistance to local forces on the slab and also acts as transverse reinforcement in the region of the shear connectors.

21.4.3 Alternative methods of design

Two methods of design of composite slabs are permitted by BS EN 1994-1-1 Cl 9.7.2. Both use test information on the 'shear-bond' resistance of composite slabs as a means of interpolating to other spans and depths.

The preferred method in the UK is the so-called 'm and k' method, which has been traditionally used since the introduction of BS 5950-4 in 1982. It is discussed in the following section.

An alternative method based on the principles of partial shear connection is presented in Cl 9.7.2(8). This second approach uses a shear-bond strength, τ, established from tests, which is applied over the plan area corresponding to the shear span L_s of the slab from which the bending resistance of the composite slab can be established directly (see Section 21.4.5).

BS EN 1994-1-1 also permits the design of continuous composite slabs by the provision of reinforcement in the negative moment region. In this case, Cl 9.7.2(6) states that the effective span length for consideration of shear-bond action is given by 0.8L for end spans and 0.7L for internal spans, where L is the actual continuous span.

A method is presented in Cl 9.7.4 for provision of end anchorage by welded shear connectors and for reinforcement in the positive moment region, acting in combination with the composite action of the decking – see Section 21.4.6.

21.4.4 Composite design by 'm and k' method

The performance of any particular deck profile used in a composite slab can only readily be assessed by testing – see Section 21.6. According to BS EN 1994-1-1 and

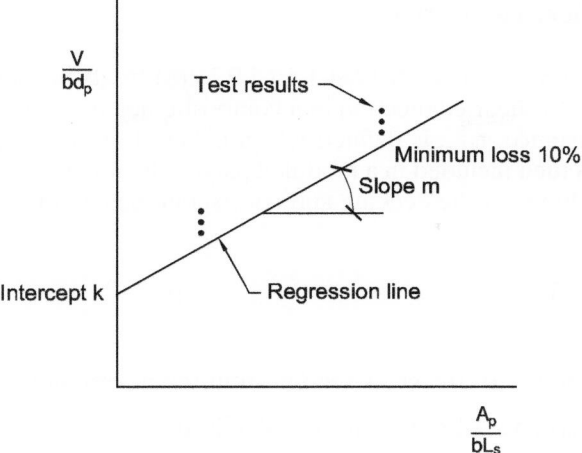

Figure 21.5   Derivation of *m* and *k* values from tests

BS 5950-4, a minimum of 6 tests are required, covering the key design parameters. The test results are presented in terms of the end shear which includes the self-weight of the slab. The characteristic failure load is taken as the minimum in the group reduced by 10%.

The characteristic failure loads of the two groups of tests are then presented in terms of a straight-line relationship, which leads to the empirical coefficients (*m* and *k* in units of N/mm^2) as the slope and y-axis intercept of the line. These coefficients broadly define the mechanical interlock and chemical bond components of the shear-bond resistance.

This relationship between the groups of tests used to derive *m* and *k* is illustrated in Figure 21.5. The design shear resistance of all slab configurations within the limits of the test data may be determined according to Cl 9.7.2(4), as follows:

$$V_{Rd} = \frac{bd_p}{\gamma_{vs}}\left(\frac{mA_p}{bL_s}+k\right) \tag{21.1}$$

Where
A_p is the cross-sectional area of the deck profile
d_p is the effective depth of the slab to the centroid of the decking
and
L_s is the shear span (taken as $L/4$ for uniformly distributed loading).

The characteristic resistance is divided by a partial safety factor of $\gamma_{vs} = 1.25$ for shear-bond action. Because of the empirical nature of the design of composite slabs, manufacturers normally present direct load-span design tables and may not give *m* and *k* values in their literature. Also, because of differences in their method of interpretation from the tests, the *m* and *k* values in BS EN 1994-1-1 and BS 5950-4 are not transferable.

21.4.5 Partial shear connection

The alternative method in BS EN 1994-1-1 Cl 9.7.2(8) treats the shear-bond resistance as analogous to shear connection in a composite beam. A characteristic longitudinal shear resistance, $\tau_{u,Rd}$, is defined which is based on tests. This longitudinal shear resistance is then included in a modified partial shear connection analysis. The tensile force developed in the decking and corresponding compression force in the slab is given by:

$$N_c = \tau_{u,Rd}\, bL_x \le A_p f_y \tag{21.2}$$

where

L_x is the distance of the cross-section from the nearer support

The bending resistance of the composite slab is given by:

$$M_{Rd} = N_c (d_p - 0.5\, x_{pl}) + M_{Rd,p} \tag{21.3}$$

where

$$\begin{aligned}
x_{pl} &= N_c / (bf_{cd}) \\
f_{cd} &= \text{design strength of concrete} = 0.85\, f_{ck}/\gamma_c
\end{aligned}$$

and

$M_{Rd,p}$ is the bending resistance of the decking, which is small in relation to M_{Rd}.

The variation of bending resistance and degree of shear connection is illustrated in Figure 21.6. Both the 'm and k' and partial shear connection methods give similar results within the range of test parameters used. (The two methods are compared in this figure.) It should be noted that, because of slight differences in the design approaches used to calculate m, k and τ, the load resistances may not be exactly the same for both methods.

21.4.6 End anchorage

End anchorage in composite slabs may be provided by welded shear connectors or some other means, and is used to enhance its longitudinal shear resistance. The end anchorage force develops a tensile force in the decking which increases the composite bending resistance. The anchorage resistance per shear connector is given in BS EN 1994-1-1 Cl 9.7.4, as follows:

$$P_{pb,Rd} = k_\phi d_{do}\, t\, f_{yp,d} \tag{21.4}$$

where:

$$\begin{aligned}
k_\phi &= 1 + a/d_{do} \le 6.0
\end{aligned}$$
d_{do} is the diameter of the weld collar taken as $1.1 \times$ stud diameter

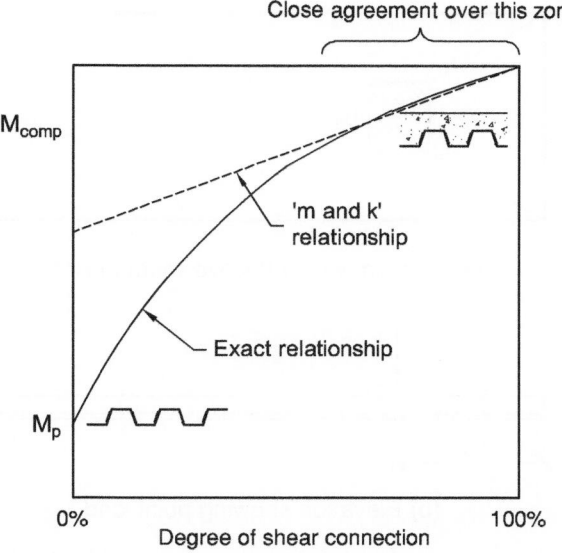

Figure 21.6   Comparison of Eurocode 4 and BS 5950-4 methods in terms of partial shear connection

    a   is the distance of the centre of the stud from the edge of the sheet (not less than $1.5d_{do} \approx 32\,\text{mm}$)

    t   is the sheet thickness.

Typically, $P_{pb,Rd}$ per shear connector is as high as $30\,\text{kN/mm}$ sheet thickness. The same end anchorage resistance is used in checks on transverse reinforcement of composite beams (refer to Chapter 22).

21.4.7 Plastic bending resistance

For longer span slabs, it is possible to develop the plastic bending resistance of the composite slab acting effectively as a reinforced concrete slab. The upper bound to the value $\tau_{u,Rd}\,bL_x$ is given by $A_p f_{yp}$, in Equation 21.2, where f_{yp} is the strength of the steel used in the decking.

21.5 Design for shear and concentrated loads

At concentrated loads or point loads, the composite slab should be checked as in BS EN 1994-1-1 Cl 9.7.6. The shear resistance is calculated assuming the shear force

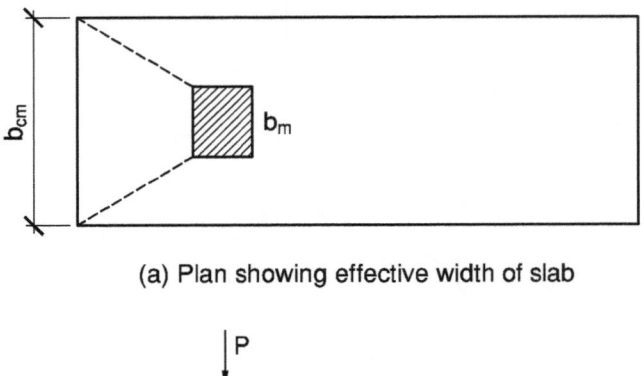

(a) Plan showing effective width of slab

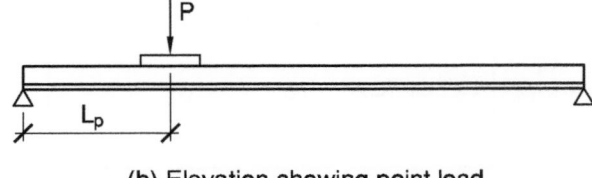

(b) Elevation showing point load

Figure 21.7   Load spread for shear-bond action at concentrated load positions

acts over an effective width given by $(b_m + 2h_c)$, where b_m is the applied load width perpendicular to the span.

For design in bending and longitudinal shear in single-span slabs, the effective slab width considered when the load is applied close to the supports is given by:

$$b_{cm} = b_m + 2L_p(1 - L_p / L) \qquad (21.5)$$

where:
> b_m is the loaded width perpendicular to the span

and
> L_p is the distance of the point load from the nearer support.

For pure shear, the formula for the effective slab width becomes:

$$b_{cm} = b_m + L_p(1 - L_p / L) \qquad (21.6)$$

This approach is illustrated in Figure 21.7, which is the same as in BS 5950-4. For line loads parallel to the span, the loads should be divided into discrete portions to calculate b_{cm} for each portion, and the resulting shear forces added.

The minimum percentage of reinforcement in the slab should be increased at heavy point loads/positions or where required for crack control (see Section 21.7).

For punching shear checks at concentrated loads, an effective perimeter around the point load is considered (refer to Figure 9.8 of BS EN 1994-1-1). The effective depth for calculation of the punching shear resistance is taken as the minimum concrete depth parallel to the ribs and the effective depth perpendicular to the ribs.

21.6 Tests on composite slabs

The design methods presented in Section 21.4 are based on test parameters, which are determined as in BS EN 1994-1-1 Annex B3, which is an Informative Annex.

Composite slabs are cast flat so that they are effectively fully propped when tested. Tests are carried out under 2 point loads with line loads applied at L/4 and 3L/4 of the span as shown in Figure 21.8. The minimum width of the line load is 100 mm to avoid punching failure through the slab. Crack inducers are required at the load application points to eliminate the tensile strength of concrete at these positions. The crack inducers need only be included within the deck depth, whereas in BS 5950-4, they are used across the full depth of the slab.

Two groups of 3 tests are performed:

- longer span, shallower slabs that are close to the limit of failure in longitudinal shear rather than pure bending
- shorter span, deeper slabs that will fail in longitudinal shear rather than pure shear.

One of the slabs in the group is first tested to failure to establish its working load. The subsequent tests should be subject to a cyclic load between $0.2w_t$ and $0.6w_t$, where w_t is the failure load of the initial static test. Loading should be applied over 5000 cycles in a time of not less than 3 hours. On completion of the cyclic loading test, loading is increased statically to failure. The failure load includes the slab self-weight.

Within a group of tests, the characteristic value is taken as the minimum failure load in the group of tests reduced by 10%, provided the deviation of any of the tests from the mean is less than 10%. (This is different to BS 5950-4, which uses the mean of the 3 tests less 15%). Therefore, the two approaches are the same when the minimum test value is approximately 5% lower than the mean of a group.

A straight line is plotted between the characteristic values to determine the slope, m, and the intercept k on a graph of effective shear stress, $V_t/(bd_p)$ against the inverse of shear span $A_p/b(L_s)$, as in Figure 21.5. The design shear resistance is then established from Equation (21.1) and should only be used within the range of test

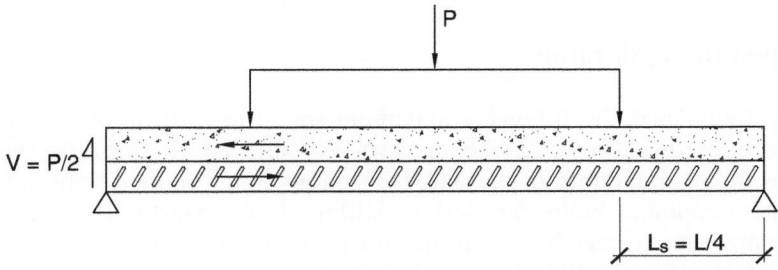

Figure 21.8 Two point load test on a composite slab (showing the load points)

parameters (i.e. span and depth). Some longer span slabs may fail in pure bending in which case the value of *m* is under-estimated, as the shear-bond resistance is not fully mobilised.

An alternative approach in BS EN 1994-1-1 Annex B3.6 is used to determine the longitudinal shear-bond strength τ_a that is established by back-analysis of the composite slab, based on the principles of partial shear connection. In this approach, it is necessary to establish the bending resistance of the steel decking, although small in relation to the composite resistance. The characteristic value of τ_{Rd} is again the minimum of the test results reduced by 10%, and divided by a partial safety factor γ_{vs}.

21.7 Serviceability limits and crack control

21.7.1 Serviceability design

Calculations of deflections of reinforced concrete slabs are conservative, and designers often use simple rules to ensure that the serviceability performance is acceptable. The same approach may be adopted for composite slabs as defined in BS EN 1994-1-1 Cl 9.8.2(4), which refer conservatively to the span:depth ratios of BS EN 1992-1 Cl 7.4.

For continuous slabs, (with nominal mesh reinforcement) deflections may be calculated using the following approximations:

- The second moment of area of the composite section may be taken as the average of the cracked and uncracked sections.
- The average of the modular ratios for short term and long term loads may be used in determining these properties.
- For end spans, the end slip in tests should be less than 0.5 mm at a load of 1.2 times the design service load, or otherwise end anchorage should be provided. This limitation minimises the effects of slip on deflections.

21.7.2 Span-to-depth ratios

For general guidance, the following maximum span-to-depth ratios of composite slabs are proposed in Table 21.2 when using NWC and LWC. In this ratio, the depth is taken as the overall depth of the slab. Deflections of unpropped slabs are generally within acceptable limits for designs within these span-to-depth ratios. For propped slabs, the effects of creep in the concrete should be included.

The limits in BS EN 1992-1-1[8] for lightly stressed concrete construction are also presented and are more conservative. These limits are not considered to be

Table 21.2 Maximum span:depth ratios of composite slabs for adequate serviceability performance

Design Case	UK Practice for composite slabs		Reinforced Concrete slabs
	NWC	LWC	BS EN 1992-1-1(NWC)
Simply supported slabs with bar reinforcement	28	26	20
Continuous slabs with mesh reinforcement – end bay	36	33	26
– internal bay	38	35	30

LWC has dry density of 19 kN/m³

appropriate for composite slabs because of the significant tensile area of the steel decking.

21.7.3 Crack control

It is necessary to control cracking of concrete in cases where the proper functioning of the structure or its appearance would otherwise be impaired. This would be the case in industrial buildings or car parks, where the slab is subject to traffic. Internally within buildings, durability is not affected by cracking. Similarly when raised floors are used, cracking is not visually important.

Where it is necessary to control cracking, the amount of reinforcement should exceed a minimum value in order to distribute cracks uniformly in the negative moment region. This minimum percentage, ρ, is given in BS EN 1992-1-1 by:

$$\rho = \frac{A_s}{A_c} \times 100\% = k_s k_c k \frac{f_{ct}}{\sigma_s} \times 100\% \tag{21.7}$$

where:

- k_s is a coefficient which allows for the effect of the reduction in normal force on the slab due to initial cracking and slip in the shear connectors and may be taken as 0.9
- k_c is a coefficient due to the bending stress distribution in the section with a value between 0.4 and 0.9
- k is a coefficient accounting for the decrease in tensile strength ($k = 0.8$)
- f_{ct} is the effective tensile strength of concrete. A value of 3N/mm^2 is the minimum adopted
- σ_s is the stress in the reinforcement.

BS EN 1994-1-1 Cl 9.8.1(2) recommends a minimum of 0.2% reinforcement of the cross-sectional area of the slab for unpropped slabs and 0.4% for propped slabs.

Table 21.3 Maximum bar spacing for various design crack widths

Steel stress σ_s (N/mm²)		≤160	200	240	280	320	360	400
Maximum bar spacing (mm)	$w_k = 0.2$ mm	200	150	100	50	–	–	–
	$w_k = 0.3$ mm	300	250	200	150	100	50	–
	$w_k = 0.4$ mm	300	300	250	200	150	100	80

where w_k = design crack width

A typical value of ρ for crack control is 0.4 to 0.6%. This is well in excess of the minimum of 0.2% necessary for shrinkage control and transverse load distribution in unpropped slabs. In propped slabs, a higher amount of reinforcement is required to control cracking when the props are removed. These bars or mesh need only be placed in the negative moment region of the beams or slabs. This reinforcement may also act as fire reinforcement and as transverse reinforcement.

An additional criterion is that the bars should be of small diameter and should be spaced relatively close together in order to be more effective in crack control. Maximum bar spacings for a given steel stress are given in Table 21.3. Otherwise checks on crack widths are made as in BS EN 1992-1-1.[8]

21.8 Shrinkage and creep

Shrinkage of the concrete may be an issue in simply supported slabs, where it affects long term deflection, or in long span continuous slabs, where it affects the risk of cracking. BS EN 1994-1-1 Annex C states that in dry environments within buildings the free shrinkage strain may be taken as:

$\varepsilon_s = 325 \times 10^{-6}$ for normal weight concrete (NWC)
$\varepsilon_s = 500 \times 10^{-6}$ for lightweight concrete (LWC)

Shrinkage-induced deflections of composite slabs may be calculated from a simplified formula:

$$\delta_{sh} = \frac{\varepsilon_s h_c (y_e - 0.5 h_c).L^2}{8 I_c} \tag{21.8}$$

where:
   h_c is the depth of concrete topping
   y_e is the depth of the elastic neutral axis of the composite slab (as for I_c)
   I_c is the second moment of area of the composite slab using n_L (in concrete stiffness units).

n_L is the modular ratio for long term loads given in BS EN 1992-1-1 Clause 5.4.2.2 as:

$n_L = n_o (1 + \phi_L \phi_t)$

$\phi_L = 1.1$ for permanent loads and 0.55 for shrinkage calculations

$n_o = E_a / E_{cm}$

For design purposes, $n_L = 2.5 n_o$ under sustained loading, e.g. loading by the slab weight in propped construction.

ϕ_t = creep coefficient to BS EN 1992-1-1 Clause 3.1.4, depending on the age of concrete and the age at first loading.

For C30/37 concrete, $E_{cm} = 33 \, \text{kN/mm}^2$ and so $n_o = 6.3$.

Typically the effective depth of a composite slab is taken as twice its actual depth d_p, (200 to 300 mm) for consideration of the creep factor and shrinkage over time, as moisture movement only occurs upwards.

Shrinkage is often ignored in continuous slabs with spans up to 3.5 m, but must be considered in simply supported slabs where it affects the serviceability performance.

21.9 Fire resistance

The fire resistance of composite slabs is covered by BS EN 1994-1-2.[9] In principle, this follows the same method as in BS 5950-8.[10] The minimum slab depth is controlled by the fire insulation requirements and the amount of reinforcement is determined from the load to be supported at the fire limit state.

However, BS EN1994-1-2 presents the minimum slab depths in terms of an 'average' slab depth, which leads to a slight reduction in slab depth in comparison with the requirements of BS 5950-8, which is based on the slab depth over the deck profile (see Table 21.4).

In BS 5950-8, a further 10% reduction in slab depth is permitted for lightweight concrete slabs because of the better insulating properties of the aggregate. However, in all cases, the minimum slab depth over the steel decking is 50 mm, which will control for 30 minutes fire resistance.

Table 21.4   Minimum slab depths (mm) and reinforcement in fire resistant design of composite slabs

Code	Parameter	Fire resistance (minutes)			
		R30	R60	R90	R120
BS EN 1994-1-2	Min. slab depth (mm)	110	110	130	150
BS 5950-8		120	130	140	150
BS 5950-8	Min. Mesh	A142	A142	A193	A252

Data for trapezoidal profiles of 60 mm depth. Mesh area in mm²/m

The use of Simplified Design Tables is permitted for practice in the UK. Calculation procedures are presented in BS EN 1994-1-2 for design cases outside the limits of test data upon which the Simplified Tables were prepared.

For scheme design, it may be assumed that a 130 mm deep composite slab is adequate for up to 90 minutes fire resistance. Standard A142 or A193 mesh (142 and 193 mm^2/m respectively) reinforcement may be used in the slab, depending on the span and loading configuration (see Table 21.4), but should not be less than the minimum slab percentage of reinforcement required for unpropped or propped construction.

A worked example follows, which is relevant to Chapter 21.

References for Chapter 21

1. British Standards Institution (2004) BS EN 1994-1-1 *Eurocode 4 : Design of composite steel and concrete structures. Part 1. General rules and rules for buildings*. London, BSI.
2. British Standards Institution (1992) BS 5950-4 *Structural use of steelwork in buildings, Part 4: Code of practice for design of composite slabs using profiled steel sheeting*. London, BSI.
3. Rackham J.W., Couchman G.H., and Hicks S.J. (2009) *Composite slabs and beams using steel decking: Best practice for design and construction* (Revised Edition) The Steel Construction Institute Publication 300, (with the MCRMA). Ascot, SCI.
4. British Standards Institution (2004) BS EN 10327 *Continuously hot dip coated strip and steel of low carbon steels for cold forming. Technical Delivery Conditions*. London, BSI.
5. British Standards Institution (2005) BS EN 1991-1-6 *Actions on structures. General Actions, Actions during Execution*. London, BSI.
6. British Standards Institution (2006) BS EN 1993-1-3 *Eurocode 3: Design of steel structures Part 1.3. General rules. Supplementary rules for cold formed steel members and sheeting*. London, BSI.
7. British Standards Institution (1995) BS 5950-6 *Structural use of steelwork in buildings. Part 6 Code of practice for design of light steel profiled steel sheeting*. London, BSI.
8. British Standards Institution (2004) BS EN 1992-1-1: *Eurocode 2: Design of concrete structures Part 1.1 General rules and rules for buildings*. London, BSI.
9. British Standards Institution (2005) BS EN 1994-1-2 *Design of composite steel and concrete structures, Part 1-2 Structural fire design*. London, BSI.
10. British Standards Institution (2003) BS 5950-8 *Structural use of steelwork in buildings. Part 8: Code of practice for fire resistant design*. London, BSI.

	Job No		Sheet 1 of 6		Rev	
The Steel Construction Institute	Job Title	Steel Designers Manual Chapter 21				
	Subject	Design Example for propped composite slab to BS EN 1994-1				
Silwood Park, Ascot, Berks SL5 7QN Telephone: (01344) 623345 Fax: (01344) 622944	Client		Made by	RML	Date	Nov 2009
CALCULATION SHEET			Checked by	GWO	Date	Feb 2010

Design Example of 5 m Span Propped Composite Slab to BS EN 1994-1-1 and BS EN 1994-1-2

Consider the design of a simply supported composite slab that is propped during construction by a single line of props at mid-span. Although propped slabs are less often used than unpropped slabs, this form of construction is often used in residential buildings requiring spans of 4.5 to 6 m. The calculation procedure demonstrates the magnitude of the deflections on de-propping of the slab and also long term creep and shrinkage effects.

Data:

Span, L = 5 m
Slab depth, $h_c + h_p$ = 180 mm in normal weight concrete
Deck profile, h_p = 80 mm
Deck area, A_p = 1850 mm^2
Deck rib spacing = 300 mm
Steel thickness, t = 1.2 mm
Steel grade = S350 ($f_y = 350\,N/mm^2$)
Concrete grade = C25/30 ($f_{cu} = 30\,N/mm^2$)
Reinforcement = Single T12 bar per rib
Diameter of bar φ = 12 mm
Mesh reinforcement = A252(252 mm^2/m)

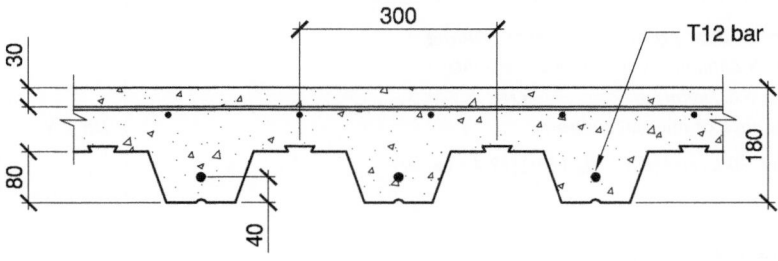

Cross section through the slab

Loading

Self-weight of slab = 3.35 kN/m^2 (using dry concrete density = 24 kN/m^3)
Self-weight of slab (wet) = 3.5 kN/m^2 (using wet concrete density = 25 kN/m^3)
Self-weight of decking = 0.15 kN/m^2
Construction loading, q_c = 1.5 kN/m^2 acting over 3 m span length
and 0.75 kN/m^2 acting over (L-3) m span length
Imposed loading, q_i = 2.5 kN/m^2(for a residential building)
Partitions = 0.5 kN/m^2
Other dead loads = 0.5 kN/m^2

BS EN 1991-1-6

Example Design of propped composite slab to BS EN 1994-1-1	Sheet 2 of 6	Rev

Partial factors

For the construction stage, the self-weight of concrete is treated as a variable load and hence the partial factor is taken as that for an imposed load. Partial factors are used as follows:

γ_{fi} = 1.5 for imposed and construction loads

γ_{fd} = 0.925 × 1.35 = 1.25 for self-weight loads

ξ *The 0.925 factor is adopted for the self-weight of the finished slab when the imposed load is the predominant loading.*

Concrete properties

Concrete strength f_{cd}	= 0.85 × 25/1.5 = 14.2 N/mm²	*BS EN 1992-1-1*
Initial elastic modulus E_{cm}	= 31 kN/m² for C25/30 concrete	*Table 3.1*
Long term elastic modulus E_c	= 10 kN/m² (creep factor ≈ 2)	*BS EN 1994-1-1*
Modular ratio n_s	≈ 7 short term loading	*Annex C*
or n_L	≈ 21 long term loading	
Free shrinkage strain ε_s	= 325 × 10⁻⁶	

Design for Construction Stage

Factored construction load, q_c = 1.5 × 1.5 + 1.5 × 3.65 = 7.7 kN/m²

Factored construction loading $(L > 3 m) = 1.5 × 0.75 + 1.5 × 3.65 = 6.6 kN/m²$

Bending moment on simply supported decking

$M_{Ed} = 7.7 × 5^2/8 - (7.7 - 6.6) × 1.5^2/2 = 22.8 kNm/m$

Bending resistance of deck profile in positive bending

$M_{Rd} = 12.0 kNm/m$ – *taken from manufacturer's tables. This shows that the decking is inadequate without temporary propping.*

For single line of props, bending moment;

$M_{Ed} = q_c(0.5L)^2/8 = 7.7x\ 2.5^2/8 = 6.0 kNm/m$

Prop force, $N_{Ed} = 1.25 × q_c × (0.5L) = 1.25 × 7.7 × 2.5 = 24.0 kN/m$

Bending resistance of deck profile in negative bending

From manufacturer's data, negative bending resistance

$M_{Rd} = 10.5 kNm/m > 6.0 kNm/m$ OK

Web crushing resistance of internal support:

$$R_{W,Rd} = \alpha t^2 \sqrt{f_{yb}E} \left(1 - 0.1\sqrt{r/t}\right)\left(0.5 + \sqrt{l_a/(50t)}\right)\left(2.4 + (\phi/90)^2\right)$$

Where:

l_a = bearing length = 100 mm at prop position

$\phi = 70°$ to horizontal, $\alpha = 0.15$ for sheeting profiles, and $r = 2t$

$R_{W,Rd} = 0.15 × 1.16^2\ (350 × 210 × 10^3)^{0.5}\ (1 - 0.1 × 0.71)\ (0.5 + (100/(50 × 1.16))^{0.5})$
    $(2.4 + (70/90)^2) × 10^{-3}$

    $= 1.73 × 0.93 × 1.8 × 3.0 = 8.7 kN/web = 2 × 8.7/0.3 = 58 kN/m$

Combining bending moment and web crushing:

$$\frac{M_{Ed}}{M_{Rd}} + \frac{F_{Ed}}{R_{W,Rd}} \leq 1.25$$

$$\frac{6.0}{10.5} + \frac{23.9}{58} = 0.57 + 0.41 = 1.08 < 1.25$$

Therefore a single line of props with 100 mm minimum width is adequate during construction.

BS EN 1993-1-3 Cl 6.1.7.3(2)

BS EN 1993-1-3 Cl 6.1.11

Example Design of propped composite slab to BS EN 1994-1-1	Sheet 3 of 6	Rev

Design of Composite Slab at Ultimate Limit State

Factored loading, $q_{Ed} = 1.5 \times (2.5 + 0.5) + 1.25 \times (3.35 + 0.15 + 0.5)$
$= 9.5\,kN/m^2$

Shear force, $V_{Ed} = 9.5 \times 5.0/2 = 23.8\,kN/m$

Bending moment, $M_{Ed} = 9.5 \times 5.0^2/8 = 29.7\,kNm/m$

BS EN 1994-1-1
Cl 9.7.3(8)
and Annex B3.6

a. Use the effective longitudinal shear resistance method:

From composite tests, $\tau_{Rd} = 0.45/1.25 = 0.36\,N/mm^2$

Tensile force in decking over shear span, $N = \tau_{Rd} L_v$,*where* $L_v = L/4 = 1250\,mm$

$N = 0.36 \times 10^3 \times 5/4 = 450\,kN$

Depth of concrete in compression

$y_c = N/(f_{cd}b) = 450 \times 10^3/(14.2 \times 10^3) = 32\,mm$

Effective depth of composite slab to centroid of decking, $d_p = 180 - 40 = 140\,mm$

Bending resistance of composite slab:

$M_{Rd} = N(d_p - 0.5y_c) = 450 \times (140 - 0.5 \times 32) \times 10^{-3} = 55.8\,kNm/m$

This exceeds the applied moment of 29.7 kNm/m (Unity Factor= 0.53).

b. Use the 'm and k' shear force method:

From tests ; $m = 0.204$ *and* $k = 0.156$ *using test parameters;*
Shear resistance, $V_{Rd} = bd_p (m A_p/L_v + k)/1.25$:

$V_{Rd} = 10^3 \times 140 \times (0.204 \times 1850/1250 + 0.156)/1.25 \times 10^{-3} = 51.3\,kN/m$

This exceeds the applied shear force of 23.8 kN/m (Unity Factor = 0.46).

BS EN 1994-1-1
Annex B3.2

Composite Slab-Elastic Properties

For deflection calculations, the average of the uncracked (gross) and the cracked stiffnesses of the composite slab may be used. The area of the bar reinforcement may be included, in this case at a cover of 40 mm to the soffit of the deck rib.

9.8.2(5)

Uncracked inertia of composite slab

Gross inertia, I_g *(in concrete units per unit slab width):*

$I_g = bd_p^3/12 + n (A_p + A_r)(d_p - y_g)^2 + bd_p (y_g - 0.5 d_p)^2 + n I_p$

Where the elastic neutral axis depth is given by:

$y_g = d_p (n (A_p + A_r)+ 0.5 bd_p)/(n (A_p + A_r) + bd_p)$

and:

$A_r =$ *cross-sectional area of rebar per unit width* $= 113/0.3 = 376\,mm^2/m$

$A_p + A_r = 2226\,mm^2/m$

$I_p =$ *second moment of area of decking per unit width (from manufacturer's data)* $= 2.37 \times 10^6\,mm^4/m$

For the short term deflection using a modular ratio, $n_s = 7$:

$y_g = 140 \times (7 \times 2226 + 0.5 \times 140 \times 10^3)/(7 \times 2226 + 140 \times 10^3) = 77\,mm$

$I_g = 10^3 \times 140^3/12 + 7 \times 2226 \times (140 - 77)^2 + 140 \times 10^3 \times (77 - 70)^2 + 7 \times 2.37 \times 10^6$

$= (228 + 62 + 7 + 16) \times 10^6 = 313 \times 10^6\,mm^4/m$

$E_c I_g = 30 \times 313 \times 10^6 = 9.39 \times 10^9\,kN/mm^2/m$

Example Design of propped composite slab to BS EN 1994-1-1	Sheet 4 of 6	Rev

For the long term deflection using a modular ratio, $n_L = 21$;

$y_g = 87 \, mm$

$I_g = (228 + 131 + 40 + 50) \times 10^6 = 449 \times 10^6 \, mm^4/m$

$E_c I_g = 10 \times 449 \times 10^6 = 4.49 \times 10^9 \, kN/mm^2/m$

(a 52% reduction relative to the short term stiffness).

Cracked inertia of composite slab

Cracked inertia, I_{cr} (in concrete units per unit slab width):

$I_{cr} = b y_e^3/3 + n (A_p + A_r)(d_p - y_e)^2$

Where the elastic neutral axis depth of the cracked section from the top of the slab is given by:

$y_e = d_p \left[-nr + \sqrt{(2nr + (nr)^2)} \right]$ *and* $r = (A_p + A_r)/(bd_p)$

For the short term deflection using a modular ratio, $n_s = 7$:

$r = 2226/(140 \times 10^3) = 0.016$, *and* $nr = 7 \times 0.016 = 0.11$

$y_e = 140 \times \left[-0.11 + \sqrt{2 \times 0.11 + 0.11^2} \right] = 52 \, mm$

$I_{cr} = 52^3 \times 10^3/3 + 7 \times 2226 \times (140 - 52)^2 + 7 \times 2.37 \times 10^6$

$\quad = (47 + 120 + 16) \times 10^6 = 183 \times 10^6 \, mm^4/m$ *(58% of uncracked inertia)*

$EI_{cr} = 30 \times 154 \times 10^6 = 4.62 \times 10^9 \, mm^4/m$

For the long term deflection using a modular ratio, $n_L = 21$;

$n_L r = 0.016 \times 21 = 0.33$

$y_e = 140 \times \left[-0.33 + \sqrt{2 \times 0.33 + 0.33^2} \right] = 76 \, mm$

$I_{cr} = 76^3 \times 10^3/3 + 21 \times 2226 \times (140 - 76)^2 + 21 \times 2.37 \times 10^6$

$\quad = (146 + 191 + 50) \times 10^6 = 387 \times 10^6 \, mm^4/m$ *(86% of uncracked inertia)*

$EI_{cr} = 10 \times 3.87 \times 10^6 = 3.87 \times 10^9 \, mm^4/m$

For a propped slab, check the deflection under self-weight of the slab as a long term load due to re-application of a point load due to the removed prop force:

Prop force = $1.25 \times 3.65 \times 5/2 = 11.4 \, kN/m$

Effective inertia of composite section

Consider an effective stiffness of the average of the cracked and uncracked long term stiffness: **9.8.2(5)**

$I_{eff} = (I_g + I_{cr})/2 = 0.5 \times (449 + 387) \times 10^6 = 418 \times 10^6 \, mm^4/m$

$\delta_{de} = \dfrac{11.4 \times 5.0^3 \times 10^9}{48 \times 10 \times 418 \times 10^6} = 7.1 \, mm \; (span/705)$ *OK*

Check deflection under imposed load considered as a short term load:

For the deflection due to prop removal, consider an effective stiffness of the average of the cracked and uncracked stiffness:

$I_{eff} = (313 + 183) \times 10^6/2 = 248 \times 10^6 \, mm^4/m$

$\delta_i = \dfrac{5 \times 2.5 \times 5.0^4 \times 10^9}{384 \times 30 \times 248 \times 10^6} = 2.7 \, mm \; (span/1850)$ *very stiff*

Check deflection under other dead loads considered as a long term load:

Example Design of propped composite slab to BS EN 1994-1-1	Sheet 5 of 6	Rev

$\delta_d = \dfrac{5 \times 1.0 \times 5.0^4 \times 10^9}{384 \times 10 \times 418 \times 10^6} = 2.0 \, mm$

Shrinkage induced deflection

Shrinkage-induced deflection (long term condition using the cracked slab stiffness) for $\varepsilon_s = 325 \times 10^{-6}$ strain. For shrinkage calculations, the long term cracked stiffness may be used conservatively.

$\delta_s = \dfrac{\varepsilon_s b h_c \, (y_e - 0.5 h_c) L^2}{8 \, I_{cr}}$ *where h_c = depth of concrete topping*

$= \dfrac{325 \times 10^{-6} \times 10^3 \times 100 \times (76 - 50) \times 5.0^2 \times 10^6}{8 \times 387 \times 10^6}$

Eqn (21.8)

$= 6.8 \, mm \ (= span/735)$

Total deflection = 7.1 + 2.7 + 2.0 + 6.8 = 18.6 mm (span/270)

This is less than the total deflection limit of span/250 = 20 mm and is acceptable.

Vibration sensitivity- simplified design :

Check natural frequency of floor:

$f = 18/\sqrt{\delta_d} \ > 5 \, Hz$,

where δ_d is the instantaneous deflection due to re-application of the self-weight of the slab, permanent loads and 10% imposed load (= 4.65 kN/m^2)

$\delta_d = 2.7 \times (4.65/2.5) = 5.3 \, mm$

$f = 18/\sqrt{5.3} \ = 7.8 \, Hz > 5 \, Hz$

This shows that the 180 mm deep floor is not sensitive to vibrations for a 5 m span.

Design of Composite Slab At Fire Limit State

The composite slab is designed to BS EN 1994-1-2 at the fire limit state. Consider a fire resistance of 90 minutes. The steel decking is assumed to be ineffective in fire and the reinforced slab resists the moment at the fire limit state.

Check the minimum slab depth to Table D.6 of BS EN 1994-1-2:
Min. effective thickness = 100 mm
Average slab depth, $h_{eff} = h_c + 0.5 h_p = 140 \, mm > 100 \, mm$ OK

BS EN 1994-1-2 Table D.6

Load at fire limit state

$q_{fi} = \phi q_i + q_d$ *where ϕ = partial factor on variable loads in fire = 0.6*
$= 0.6 \times 2.5 + 3.5 + 1.0 = 6.0 \, kN/m^2$

Moment at fire limit state

$M_{fi} = 6.0 \times 5.0^2/8 = 18.8 \, kNm/m$

Use single T12 reinforcing bar per rib at 40 mm bottom cover.

Table D.5

From Table D5 of BS EN 1994-1-2, temperature after 90 mins = 428°C.

Example Design of propped composite slab to BS EN 1994-1-1	Sheet 6 of 6	Rev

For cold-worked reinforcing bar, strength reduction factor is obtained from Table 3.4 of BS EN 1994-1-2, and is $k_{y,\phi} = 0.86$.

Tensile resistance of rebar, $N_{Rd} = 0.86 \times 113 \times 460 \times 10^{-3} = 44.7\,kN$
Compression strength of concrete in fire, $f_{cd,fi} = 0.85 \times 25 = 21\,N/mm^2$
Depth of concrete in compression, $y_c = 44.7 \times 10^3/(21 \times 300) = 7\,mm$

Reduced bending resistance in fire:

$M_{Rd,fi} = 44.7 \times (140 - 0.5 \times 7) \times 10^{-3}/0.3 = 20.3\,kNm/m$

This exceeds the applied moment of 18.8 kNm/m and shows that a single T12 bar per rib is able to provide 90 minutes fire resistance for a 180 mm deep slab.

Mesh Reinforcement

The mesh reinforcement is taken as A252 generally, which corresponds to 0.25% reinforcement. Because the composite slab is designed as simply supported, it is not necessary to provide reinforcement for crack control. However, if the slab had been continuous over a beam, it would have been necessary to use 2 layers of A252 mesh to satisfy the 0.4% reinforcement requirement.

Conclusions

These calculations show that a 180 mm deep composite slab is acceptable for a 5 m span. The limiting design criterion is control of total deflections to a maximum of span/250, when the effects of shrinkage and creep induced deflections are considered. Single T12 bars per deck rib provide 90 minutes fire resistance. It may be shown that the maximum effective span for this design case may be extended to 5.2 m (equivalent to span: depth ratio = 29).

Rev column:
BS EN 1994-1-2 Table 3.4

Chapter 22
Composite beams

MARK LAWSON and KWOK-FAI CHUNG

22.1 Introduction

BS EN 1994-1-1 Eurocode 4: *Design of composite steel and concrete structures: General rules and rules for buildings*[1] deals with the design of composite beams, slabs and columns. The term 'composite' in this context means the structural action between a concrete slab and a steel section in which the concrete slab resists compression and the steel section is largely in tension. Composite action increases the bending resistance and stiffness of the beam and leads to a reduction in steel weight of 30 to 50% relative to non-composite design. The former design standards were BS 5950-3[2] and BS 5950-4.[3]

22.1.1 Scope

This chapter concentrates on the design of composite beams as used in modern building construction. The SCI publication *Design of composite slabs and beams with steel decking*[4] dealt with composite design to BS 5950-3.[2] This chapter reviews the design principles and application rules of BS EN 1994-1-1 and its UK National Annex for composite beams using composite slabs.

It covers the common case of design of simply supported composite beams in braced frames for which a typical application is shown in Figure 22.1. Simply supported steel beams effectively satisfy the requirements for Class 1 sections to BS EN 1993-1-1[5] (or 'plastic' to BS 5950-1[6]) by their attachment to the concrete or composite slab. This means that plastic design is permitted for all composite beams using hot-rolled steel sections.

Continuous composite beams and partially encased steel beams are covered by BS EN 1994-1, but are treated only in outline in this chapter. Guidance on composite beam design using precast concrete slabs is presented in a recent SCI publication[7] and is not covered here.

The design of composite slabs is presented in Chapter 21. The design of composite columns is presented in Chapter 23.

Steel Designers' Manual, Seventh Edition. Edited by Buick Davison and Graham W. Owens.
© 2012 Steel Construction Institute. Published 2012 by Blackwell Publishing Ltd.

Figure 22.1   Composite decking on steel beams before concreting

22.1.2 National annexes

The National Annex to BS EN 1994-1-1: 2004 presents nationally determined parameters, which include partial factors, as well as other issues such as shear connector performance and some material and application limits.

The partial safety factors used for design of composite beams and slabs at the ultimate and the serviceability limit states are consistent with the Eurocodes for steel and concrete. The partial factors for loads are common to all materials and are specified for EN 1990-1-1[8] (see Table 22.1). The partial factor for self-weight loads may be reduced when they are not the predominant loads. The nationally determined value for ξ is taken as 0.925, and the partial factor for self-weight loads becomes $0.925 \times 1.35 = 1.25$.

In BS EN 1993-1-1 and 4-1-1, the partial factor for structural steel, γ_a, is taken as 1.0. The statistical variation of steel strengths and geometric properties of hot-rolled sections ensures a sufficient margin of safety so that the characteristic strength may be taken as the guaranteed minimum strength of steel, f_y. This partial factor is also a 'nationally determined value'. (A partial factor of 1.05 was formerly adopted in the BS ENV version of BS EN 1993-1-1).

Table 22.1 Partial safety factors for loads and materials

Design parameter		Limit state		
		Ultimate	Serviceability	Fire
Loads, γ_Q	Imposed (variable) load	1.5	1.0	0.5
	Dead (permanent) load	1.25	1.0	1.0
Materials, γ_G	Structural steel, γ_a	1.0	1.0	1.0
	Concrete, γ_c	1.5	1.3	1.0
	Shear connectors, γ_{vs}	1.25	1.0	1.0
	Shear-bond. γ_{sb}	1.25	1.0	–
	Reinforcement, γ_s	1.15	1.0	1.0

Other partial factors are specified in BS EN 1993-1-1[5] and BS EN 1992-1-1.[9] It follows that applied loads are multiplied by γ_f and material strengths are reduced by γ_a, γ_c, etc. For serviceability performance and structural fire design, the partial factors for materials are reduced. Also the load levels for serviceability and in fire are reduced by a ψ factor reflecting the statistical variation of loads at these limit states.

A more detailed clause by clause review of BS EN 1994-1-1 is given in the publication by Johnson and Anderson,[10] which should be referred to for background information.

22.2 Material properties

22.2.1 Steel properties

Various grades of steel can be used in composite beam design. S275 and S355 are commonly specified in the UK, the higher grade often being preferred for composite construction except for serviceability controlled design. The design rules do not cover the use of steel of higher grade than S460.

Grades of steel for profiled steel decking are specified in BS EN 10 326[11] (which has replaced BS EN 10147). S280 and S350 are the common grades for strip steel. The minimum steel thickness for use in composite decking is 0.7 mm although, in practice, thicknesses of 0.9 to 1.2 mm are generally used.

22.2.2 Concrete

The concrete grade to BS EN 1992-1-1[9] is specified in terms of the cylinder strength, f_{ck} in addition to the cube strength f_{cu} (which is used in BS 8110[12] and BS 5950-3). The approximate conversion is:

$$f_{ck} \approx 0.8\,f_{cu}$$

Table 22.2 Concrete properties in BS EN 1992-1-1 Table 3.1

Properties of concrete	Strength class of concrete						
	C20/25	C25/30	C30/37	C35/45	C40/50	C50/60	C60/75
f_{ck}	20	25	30	35	40	50	60
f_{cu}	25	30	37	45	50	60	75
f_{ct}	2.2	2.6	2.9	3.2	3.5	4.1	4.4
E_c	29	30.5	32	33.5	35	37	39

Note: All values in N/mm^2, except elastic modulus E_c which is in kN/mm^2.

Hence C30/37 concrete, based on cylinder strength is approximately $37 N/mm^2$ cube strength. Relevant data on the mean tensile strength, f_{ct}, and elastic moduli of concrete, E_c, are presented in Table 22.2. Clause 3.1 states that the application rules cover the strength grades of normal weight concrete between C20/25 and C60/75.

In BS EN 1992-1-1, the design strength of concrete, f_{cd}, is taken as equal to 0.85 f_{ck}/γ_c for plastic design, and is justified by the bending model that is adopted for composite design. This is equivalent to a concrete compression strength of 0.45 f_{cu} in bending, which is compatible with BS 5950-3.

The use of lightweight concrete is permitted by BS EN 1994-1-1 for strength grades LC20/22 to LC60/66. The elastic modulus of lightweight concrete is assumed to vary as $(\rho/2400)$,[2] where ρ is the dry density in kg/m^3.

Data on the free shrinkage strain of concrete are also given in Annex C of BS EN 1994-1-1, but specific calculations may only be required for this effect in exceptional circumstances, such as long span beams supporting columns.

22.2.3 Reinforcement

Reinforcement grades are given in BS EN 10 080[13] and properties are given by reference to BS EN 1992-1-1. The commonly used reinforcement grade is S500B, which has a tensile strength of $500 N/mm^2$ and is recommended for composite design because of its higher ductility than S500A.

The minimum percentage of reinforcement in composite slabs is 0.2% of the concrete topping for distribution of local loads and to act as transverse reinforcement. Additional bar reinforcement may be required to control cracking at supports of composite beams or in an exposed slab.

22.2.4 Shear connectors

The properties and proportions of headed stud shear connectors are defined in BS EN 1994-1-1 Clause 6.6 and in EN ISO 13918.[14] As shown in Figure 22.2, one or

Figure 22.2   Pairs of headed shear connectors attached to the beam by through deck welding

more shear connectors are used per trough, depending on the magnitude of shear to be transferred. Normally, the steel used in headed stud shear connectors is of $450 \, \text{N} / \text{mm}^2$ ultimate tensile strength, but BS EN 1994-1-1 Clause 6.6.3.1 permits shear connectors of up to $500 \, \text{N} / \text{mm}^2$ ultimate strength to be used in solid slabs. The dimensions of the heads of studs are important in preventing separation of the beam and slab. When specifying stud shear connectors, the designation SD is used. For example SD 19×100 means $19 \, \text{mm}$ dia $\times 100 \, \text{mm}$ height.

Other forms of shear connectors are permitted provided that they achieve adequate deformation capacity as justified by tests. The Hilti HVB shear connector, which is fixed by powder actuated pins, may be used where the forces to be transferred are relatively modest (see Figure 22.3) or where site welding is not permitted.

22.3 Composite beams

The structural system of a composite beam is essentially one of a series of parallel T-beams with thin wide flanges. The concrete flange acts in compression and the

Figure 22.3   Powder actuated HVB shear connectors arranged in pairs

steel section is largely in tension. The longitudinal forces between the two materials are transferred by shear connectors. The benefit of composite action is the increased bending resistance and stiffness of the composite section, leading to economy in the size of the steel section used.

22.3.1 Construction condition

In unpropped construction, the steel beam is sized first to support the self-weight of the concrete slab and other construction loads before the concrete has gained adequate strength for composite action. No specific guidance is given in BS EN 1994-1-1 regarding the magnitude of this construction load used in the design of the steel beam, but it refers to BS EN 1991-1-6.[15] A construction load of $0.75\,kN/m^2$ may be assumed to be applied to the entire supported area of steel decking in addition to the self-weight of the slab. However, this construction load and also the self-weight of the floor slab are both treated as a variable load with a partial factor of 1.5. (The corresponding construction load is $0.5\,kN/m^2$ in BS 5950-3).

The steel beam is then designed in accordance with BS EN 1993-1-1. It is assumed that beams are laterally restrained by the steel decking in cases where the decking spans perpendicular to the beams and is directly attached to them. Restrained beams can develop their full bending resistances. In cases where the decking spans parallel to the beam, lateral restraints are provided only by the beam-to-beam connections, and the buckling resistance of the steel beam is based on the effective length between these points of lateral restraint.

22.3.2 Effective width of slab

In a T-beam, the contribution of the concrete flange is limited by the influence of 'shear lag' associated with in-plane strains across the slab. The effective width of the slab takes this effect into account, and is the notional width of the slab acting at the compressive strength of the concrete. The effective width depends on the form of loading and position in the span and a representative value is used in most Codes. The typical distribution of longitudinal forces in the slab is illustrated in Figure 22.4.

BS EN 1994-1-1 Clause 5.4.1.2 states the effective slab width is span/8 on each side of the beam (see Figure 22.5) for designs at both the ultimate and serviceability limit states. This results in an effective width of span/4 for simply supported internal

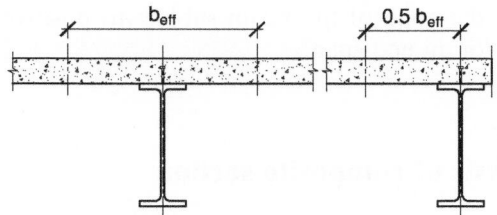

(a) Cross-section showing effective slab width

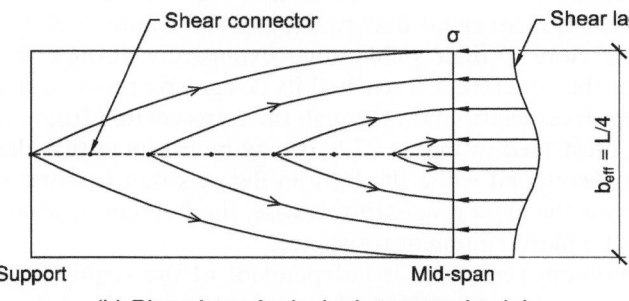

(b) Plan view of principal stresses in slab

Figure 22.4 Compression forces developed in the concrete flange of a composite beam

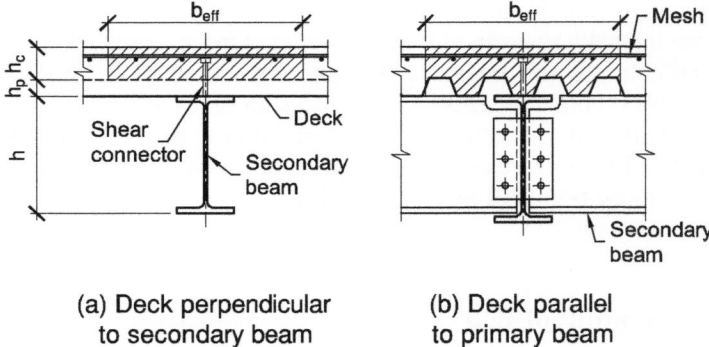

(a) Deck perpendicular
to secondary beam

(b) Deck parallel
to primary beam

Figure 22.5 Effective width of slab used to determine the properties of the composite
section

beams, but not exceeding the actual slab width acting with each beam. (This is the
same as in BS 5950-3.) The depth of concrete in the ribs can be included for primary
beams, as in Figure 22.5(b). No allowance is made for co-existing stresses in the slab
due to its flexural action and its compression action as a composite beam, as these
effects are relatively small.

The treatment of continuous beams is similar in BS EN 1994-1-1 and BS 5950-3
(see Section 22.8). For the end span of a continuous beam, the effective width of
the slab is based on the zone of the beam subject to positive (sagging) bending,
which is $L_e = 0.85\ L$ for an end span.

22.4 Plastic analysis of composite section

22.4.1 Plastic stress blocks

The moment resistance of a composite section may be determined using plastic
analysis principles. It is assumed that strains are sufficiently high so that the steel
stresses are at or close to their yield values extensively through the depth of the
section, and that the concrete has reached its design compressive strength.

The change in stress distribution through the cross-section from elastic to plastic
stress blocks is illustrated in Figure 22.6. It may be assumed that the plastic stress
blocks are fully developed when the bottom flange strain reaches $5\varepsilon_y$, where ε_y is
the yield strain for the steel grade. In this case, the bending resistance is approximately 95% of the plastic moment resistance.

The plastic moment resistance is independent of the sequence of loading (i.e.
propped or unpropped construction). The moment resistance of the section is then
compared to the total factored moment applied to the beam.

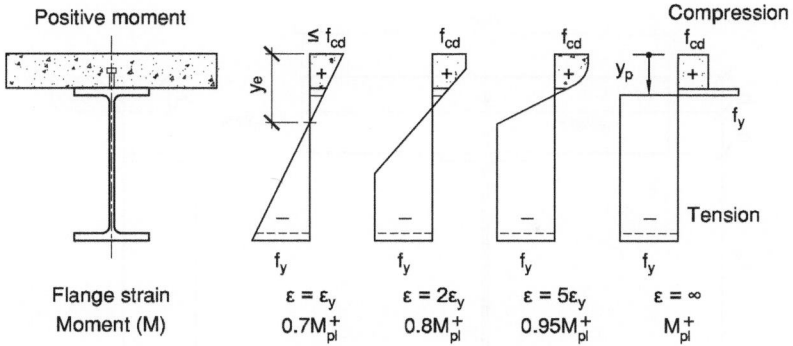

Figure 22.6   Elastic, elastic-plastic and plastic stress distributions in a composite section

The compressive resistance of the concrete slab is:

$$R_c = 0.56 \, f_{ck} \, b_{eff} \, h_c = 0.45 \, f_{cu} \, b_{eff} \, h_c \tag{22.1}$$

where:

h_c is the depth of the concrete slab above the profiled decking
b_{eff} is the effective width of the slab (see 22.3.2)

Account may be taken of the concrete contained within the ribs of the profile in cases where the ribs run parallel to the beam (this benefit is usually neglected in practice where the deck dimensions may not be known precisely at the early design stages).

The tensile resistance of the steel section is:

$$R_s = f_y \, A_a \tag{22.2}$$

The bending resistance of the cross-section may be evaluated by equating compression and tension forces across the section, the concrete being assumed to be ineffective in tension. There are three possible cases of the plastic neutral axis position depending on this force equilibrium, which are in the slab, top flange and web. No direct formulæ are given in BS EN 1994-1-1, but the following formulæ are appropriate for symmetric steel I-sections subject to positive (sagging) moment. For asymmetric sections, the formulæ should be calculated from first principles, based on the three cases of plastic neutral axis position.

22.4.2 Plastic neutral axis in the concrete slab

Where $R_c > R_s$, the plastic neutral axis lies within the concrete slab. This case is illustrated in Figure 22.7. The compressive force in the slab is reduced to the value

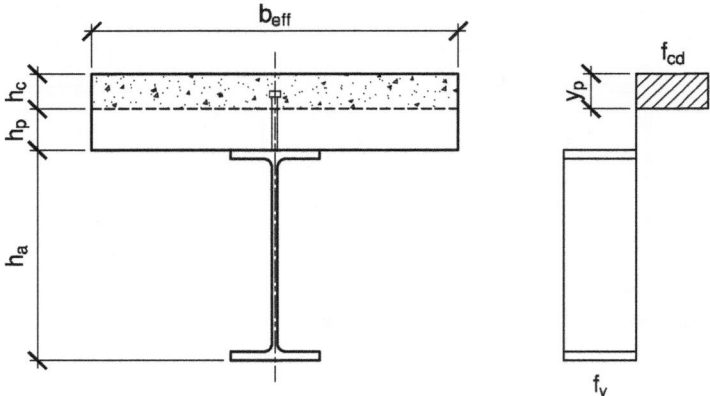

Figure 22.7   Plastic analysis for the case where the plastic neutral axis lies within the concrete
slab

of the tensile resistance of the steel section. The plastic neutral axis depth in the
slab is given by:

$$y_p = \frac{R_s}{R_c} h_c \tag{22.3}$$

The bending resistance is established by taking moments about the centre of
compression in the slab, as follows:

$$M_{pl,Rd} = R_s \left[\frac{h_a}{2} + h_p + h_c - \frac{R_s}{R_c} \frac{h_c}{2} \right] \tag{22.4}$$

where:

h_a is the depth of the steel section
h_p is the depth of the profiled decking.

22.4.3 Plastic neutral axis in the top flange of the steel section

Where $R_c \le R_s$, the full compressive resistance of the slab is developed, and the
plastic neutral axis falls into the steel section. This case is illustrated in Figure 22.8.
The plastic neutral axis lies within the top flange when $R_c \ge R_w$, where R_w is the
tensile resistance of the web of the section. In this case, taking moments around the
centre of the top flange of the section gives an approximate formula for the bending
resistance of the composite section, given by:

$$M_{pl,Rd} = R_s \frac{h_a}{2} + R_c \left(\frac{h_c}{2} + h_p \right) \tag{22.5}$$

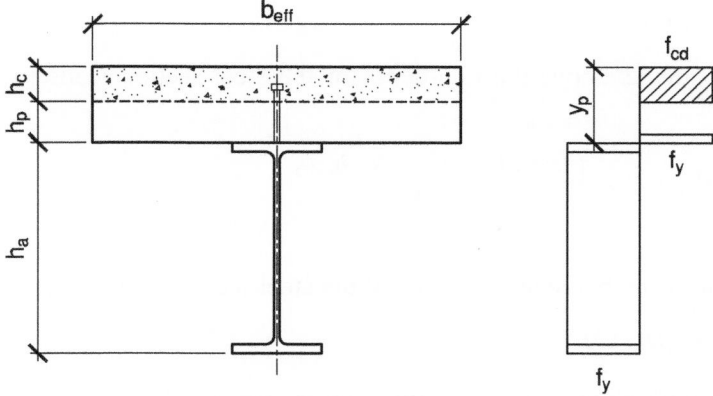

Figure 22.8   Plastic analysis for the case where the plastic neutral axis lies in the top flange of the section

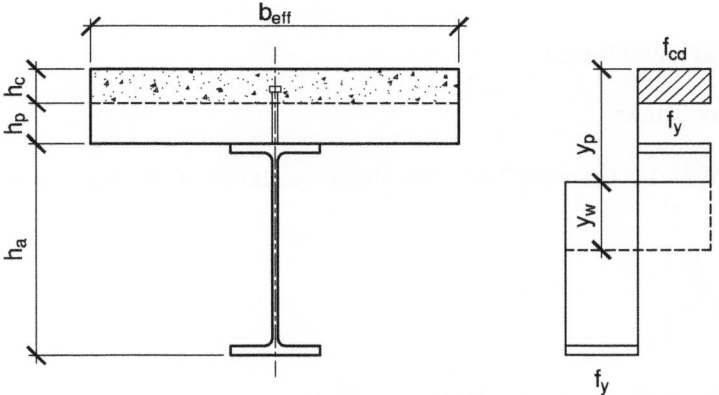

Figure 22.9   Plastic analysis for the case where the plastic neutral axis lies in the web of the section

22.4.4 Plastic neutral axis in the web of the steel section

Where $R_c < R_w$, the plastic neutral axis lies in the web of the section, as illustrated in Figure 22.9. The plastic neutral axis depth is given by:

$$y_p = 0.5\,h_a + h_c + h_p - y_w$$

where y_w is the distance of the plastic neutral axis from the centre-line of the steel section, given by:

$$y_w = R_c / (2t_w f_y)$$

Taking moments about the mid-depth of the steel section leads to a bending resistance of:

$$M_{pl,Rd} = M_{a,pl,Rd} + R_c \left(\frac{h_c + 2h_p + h_a}{2} \right) - \frac{R_c^2}{R_w} \frac{h_a}{4} \tag{22.6}$$

where:

$M_{a,pl,Rd}$ is the bending resistance of the steel section alone, and;

$$R_w = f_y t_w (h_a - 2t_f)$$

t_w and t_f are the web and the flange thicknesses respectively.

As defined in BS EN 1993-1-1 Clause 6.2.2.4(6), the depth of the web in compression should not exceed $40 \, t_w \varepsilon$ in order for it to be treated as 'Class 2', where $\varepsilon = \sqrt{(235/f_y)}$. This case is only likely to occur in highly asymmetric plate girders.

22.5 Shear resistance

22.5.1 Pure shear

In BS EN 1993-1-1 Clause 6.2.6(2), the shear resistance of the web of a steel section is taken as:

$$V_{pl,Rd} = \frac{f_y}{\sqrt{3}\,\gamma_a} \cdot A_v = 0.58 \, f_y \, A_v \tag{22.7}$$

where:

A_v is the shear area of the steel section.

The shear strength of steel compares to $0.6 f_y$ in BS 5950-1. The definition of A_v may be taken as $A_v = h_a t_w$ for hot-rolled sections, as in BS 5950-1. When $d/t > 76\varepsilon$, the shear buckling strength of the web should replace the pure shear strength. In all cases, the applied shear force, $V_{Ed} \leq V_{pl,Rd}$.

22.5.2 Combined bending and shear

Where high shear and moment co-exist at the same point in the span (i.e. the beam is subject to point loads), the influence of vertical shear can cause a reduction in the bending resistance of the beam. The interaction equation in BS EN 1993-1-1 Clause 6.2.8 (3) is as follows:

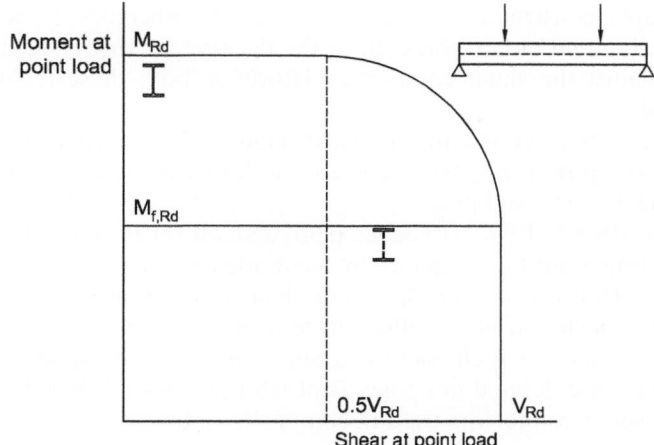

Figure 22.10   Interaction curve for co-existing moment and shear in a composite beam

$$M_{Ed} \leq M_{f,Rd} + (M_{Rd} - M_{f,Rd})(1 - (2V_{Ed}/V_{pl,Rd} - 1)^2) \tag{22.8}$$

where:

$M_{f,Rd}$ is the bending resistance of the composite section ignoring the web of the section

M_{Ed} and V_{Ed} are the applied moment and the applied shear force respectively at the same point in the span considered.

This relationship is presented in Figure 22.10. It follows that, if $V_{Ed} \leq 0.5\ V_{pl,Rd}$, no reduction to the bending resistance is made. This reduction does not apply for uniformly loaded beams, as the maximum moment and shear are not coincident.

22.6 Shear connection

22.6.1 Forms of shear connector

The modern form of shear connector is the headed stud shear connector of 19 mm diameter and 100 or 125 mm height. Studs are generally welded through the trough of the profiled decking using a hand tool connected via a control unit to a power generator (see Figure 22.2).

There are, however, some limitations to through-deck welding: Firstly, the top of the steel flange should not be painted or, alternatively, the paint should be removed from the zone where the shear connectors are to be welded; Secondly, the galvanised steel decking should not be more than 1.25 mm thick and be clean and free from moisture.

As an alternative construction procedure, the shear connectors can be pre-welded to the steel section and either holes cut in the decking, or single sheet spans used that butt up against the shear connectors. However, both these techniques have buildability limitations.

The 'shot-fired' shear connector shown in Figure 22.3 is often used in smaller projects where site power may be a problem. In this case, the shear connectors are attached by powder-actuated pins.

As defined in BS EN 1994-1-1 Clause 6.6.1.1(8), all shear connectors should be capable of resisting uplift forces caused by the tendency of the slab to separate from the steel section. Hence headed rather than plain studs are used.

Shear connectors should have sufficient deformation capacity for use in plastic design, which is defined by a characteristic slip of 6 mm. Tests for shear connectors used in solid slabs are defined in Annex B of BS EN 1994-1-1, but this test should be adapted for shear connectors used in composite slabs.

22.6.2 Resistance of stud shear connectors in solid slabs

The resistance of headed stud shear connectors in a solid slab is defined in BS EN 1994-1-1 Clause 6.6.3.1 by two design equations; the first corresponding to failure of the concrete, and the second to shear failure of the stud (at its weld collar). The smaller of the two values is used in design, as follows:

$$P_{Rd} = 0.29\,\alpha\, d^2 \sqrt{f_{ck} E_{cm}}\,/\gamma_v \qquad (22.9)$$

$$P_{Rd} = 0.8\, f_u \frac{\pi d^2}{4\gamma_v} \qquad (22.10)$$

α takes into account the height of the stud and $\alpha = 0.2\,(h_{sc}/d + 1) \leq 1,0$
where:

h_{sc} is the overall height of the stud
d is the diameter of the stud in the range 16 mm $\leq d \leq$ 25 mm

Concrete properties f_{ck} and E_{cm} are defined in Table 22.2. For the case of light-weight concrete, its density should not be less than 1750 kg/m^3. The value of ultimate tensile strength of the steel, f_u, used in the through-deck welded stud shear connectors should not exceed 450 N/mm^2 for design purposes.

The partial safety factor γ_v is taken as 1.25 at the ultimate limit state. This is the inverse of the 0.8 factor used to modify the basic resistances of shear connectors in BS 5950-3. However, the partial factor is modified when using the reduction factors taking account of the shape of the profiled decking (see Section 22.6.3).

The design resistances of the common range of stud shear connectors are presented in Table 22.3 These resistances are up to 10% lower than the equivalent values in BS 5950-3 Table 5. The cut-off in resistance in Equation 22.10 defines pure shear failure in the shear connectors and applies when concrete strengths exceed C35/45.

Table 22.3 Design resistances (kN) of common stud shear connectors to BS EN 1994-1-1

Stud diameter and height (mm)	Concrete strength (N/mm²)			
	C20/25	C25/30	C30/37	C35/45
19 mm dia × 100 mm	63	73	81	81
22 mm dia × 100 mm	85	98	108	108
16 mm dia × 75 mm	45	52	57	57

Design resistance = 0.8 × Characteristic resistance

For lightweight concrete, BS EN 1994-1-1 assumes that the design resistance in Equation 22.9, varies as $\rho/2400$ due to influence of the lower E_{cm}, where ρ is the dry density of the concrete in kg/m^3. This approach is considerably more conservative than BS 5950-3, where a strength reduction of 10% is permitted for concrete densities of 1800 to 2000 kg/m^3. For design in the UK, this resistance of shear connectors in lightweight concrete (of this density range) may be taken as 90% of the resistance of shear connectors for the equivalent grade of normal weight concrete.

22.6.3 Influence of deck shape on shear connection

The efficiency of shear connection between the composite slab and the steel beam may be reduced as a result of the shape of the deck profile and the number of shear connectors placed in each rib. The resistance of the shear connectors is highly dependent on the area of concrete around them and the embedment of the heads of the shear connectors into the slab topping. In most standards for composite design, a simple reduction factor formula is used to take account of the shape of the profiled decking and the number of shear connectors in a group in a deck trough.

The possible failure modes of shear connectors in a composite slab are illustrated in Figure 22.11. For relatively wide ribs in the composite slab, failure occurs by shear and separation due to development of a concrete cone over the stud as in Figure 22.11(a). For narrow ribs, rotational failure can occur, as in Figure 22.11(b). The modes of failure in Figure 22.11(c) and Figure 22.11(d) are prevented by introducing various geometrical limits in design.

The case where the profiled decking is orientated perpendicular to the beam (i.e. for secondary beams) is considered first.

22.6.3.1 Strength reduction factor for decking perpendicular to beam

According to BS EN 1994-1-1, the strength reduction factor, k_t, for shear connectors (relative to a solid slab) is determined from the empirical formula given in Clause 6.6.4.2, as follows:

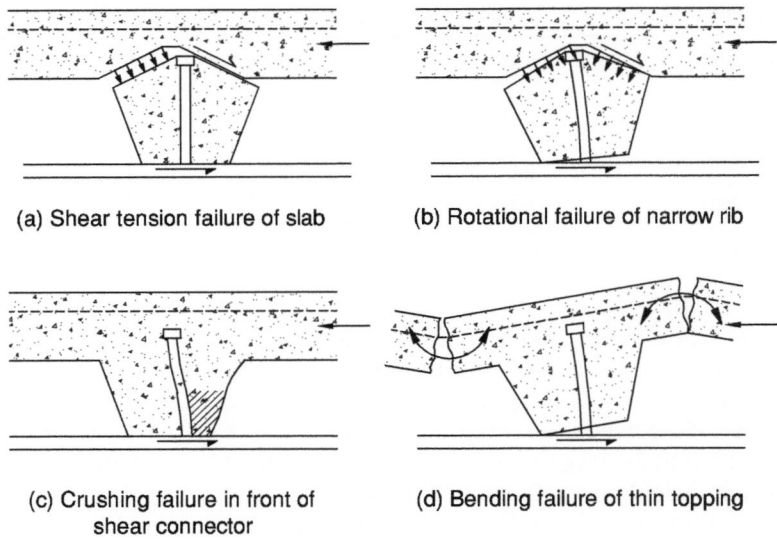

(a) Shear tension failure of slab   (b) Rotational failure of narrow rib

(c) Crushing failure in front of   (d) Bending failure of thin topping
shear connector

Figure 22.11   Shear connector failure in composite slabs

$$k_t = \frac{0.7}{\sqrt{n_r}} \frac{b_o}{h_p} \frac{(h_{sc} - h_p)}{h_p} \qquad\qquad (22.11)$$

where:

    b_o is the average rib width (or minimum width for re-entrant profiles)
    h_{sc} is the stud height
    n_r is the number of studs per rib ($n_r \leq 2$)

BS EN 1994-1-1 Clause 6.6.5.8 states that the shear connectors should project at least 2 × stud diameters above the top of the profiled decking. For 19 mm diameter shear connectors, the 38 mm limit is stricter than the 35 mm projection required by BS 5950-3 and requires use of shear connectors of 125 mm, or even 150 mm height for deeper profiled decking.

The coefficient of 0.7 in Equation 22.11 has been established statistically on the basis of recent test evidence. It is a reduction from the coefficient of 0.85 used in BS 5950-3. It is recognised that the formula is conservative for single shear connectors per rib but may be unconservative for shear connectors in pairs. Therefore, the upper limits on k_t are given in Table 22.4, which are a function of the thickness of the steel sheet, as it is recognised that the decking plays a beneficial role in transferring shear into the composite slab.

In the UK National Annex to BS EN 1994-1-1, the partial factor, $\gamma_v = 1.25$. The National Annex also defines the height of the deck profile as to the top of the web, and stiffeners in the top of the profile can be ignored, provided they are less than 15 mm deep and 55 mm wide.

Table 22.4 Upper limits on reduction factor, k_t in BS EN 1994-1-1 Table 6.2

Number of stud connectors per rib	Thickness of steel used in profiled decking (mm)	Studs not exceeding 20 mm in diameter and welded through profiled decking	Profiled decking with holes and studs 19 mm or 22 mm in diameter
$n_r = 1$	≤1.0	0.85	0.75
	>1.0	1.0	0.75
$n_r = 2$	≤1.0	0.70	0.60
	>1.0	0.80	0.60

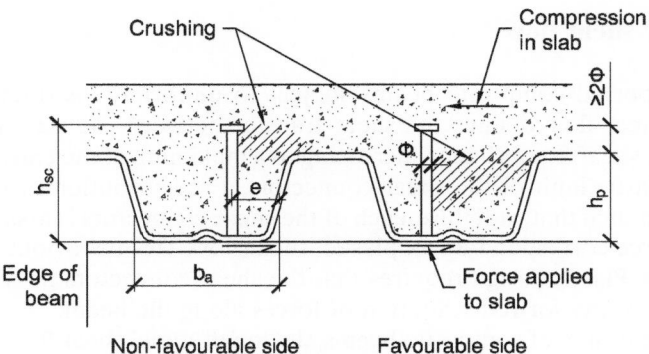

Figure 22.12   Influence of off-centre shear connectors in the troughs of the profiled decking

22.6.3.2 *Strength reduction factor in off-centre placing of shear connectors*

Many modern deck profiles have a central stiffening fold in the rib which requires the shear connector to be located off-centre. The preferred position of attachment is where the shear connectors are placed on the side of the rib closest to the end of the nearest support (see Figure 22.12). This requires a change in orientation of the shear connectors at mid-span. If this arrangement cannot be assured on site, then a conservative view of the strength reduction factor should be taken.

No guidance on design using off-centre shear connectors is given in BS EN 1994-1-1, but BS 5950-3 may be used in this case. The important dimension is e, which is the distance from the centre of the shear connector to the mid-height of the adjacent deck (see Figure 22.16). In using the above equation b_o should be taken as $2e$ where the shear connectors are welded in the non-favourable location. This only applies to cases where the profiled decking spans perpendicular to the beams (i.e. for secondary beams), as illustrated.

Alternatively, manufacturers' test information, as obtained from standard 'push out' tests may be used. This test result should be reduced statistically depending on the number of tests and divided by the appropriate partial factor of 1.25.

22.6.3.3 Strength reduction factor for decking parallel to beam

For the case where the profiled decking is orientated parallel to the beams, Clause 6.6.4.1(2) gives the same equation as in 22.11 but the constant is reduced from 0.7 to 0.6 (as in BS 5950-3). However, no reduction is made for the number of shear connectors in a row, n_r for this case.

22.7 Full and partial shear connection

22.7.1 Plastic shear flow

In simply supported composite beams subject to uniformly distributed load, the longitudinal shear flow defines the shear transfer between the slab and the steel section. Elastic shear flow is linear, increasing to a maximum at the ends of the beam. Beyond the elastic limit of the shear connectors, a redistribution of forces occurs along the beam such that, at failure, each of the shear connectors is assumed to resist equal shear force corresponding to 'plastic' shear flow. This behaviour is illustrated in Figure 22.13. Plastic design requires that the shear connectors possess adequate deformation capacity for redistribution of forces along the beam.

In the plastic design of composite beams, the longitudinal shear force to be transferred between the points of zero and maximum moment should be the smaller of R_c or R_s (see Section 22.4.1). If so, full shear connection is achieved.

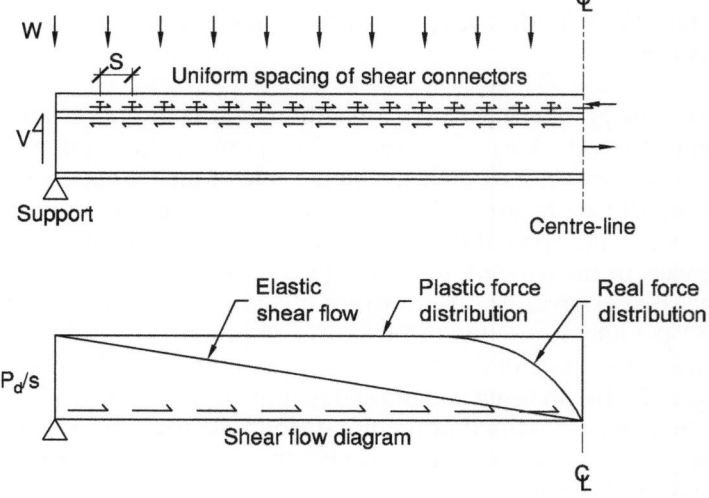

Figure 22.13   Re-distribution of longitudinal shear forces in a composite beam subject to uniform loading

22.7.2 Degree of shear connection

In cases where fewer shear connectors than the number required for full shear connection are provided, it is not possible to develop the full plastic moment resistance of the composite section. This is known as 'partial' shear connection.

The degree of shear connection may be defined as:

$$\eta = \frac{n}{n_f} = \frac{R_q}{R_s} \quad \text{for} \quad R_s < R_c \tag{22.12}$$

or

$$\eta = \frac{n}{n_f} = \frac{R_q}{R_c} \quad \text{for} \quad R_c < R_s \tag{22.13}$$

where:

R_q is the total shear force transferred by the shear connectors between the points of zero and maximum moment

n_f is the number of shear connectors required for full shear connection

n is the number of shear connectors provided over the part of the span between the points of zero and maximum moment.

22.7.3 Linear interaction method

There are two methods of determining the moment resistance of a composite section with partial shear connection. The simplest method is the so-called 'linear-interaction' approach, which is the one directly covered by BS EN 1994-1-1 Clause 6.2.1.2. The reduced bending resistance is given by:

$$M_{Rd} = M_{a,pl,Rd} + \eta(M_{pl,Rd} - M_{a,pl,Rd}) \tag{22.14}$$

where:

$M_{pl,Rd}$ is the moment resistance of the composite section for *full* shear connection

$M_{a,pl,Rd}$ is the moment resistance of the steel section.

For adequate design, $M_{Ed} \le M_{Rd}$, where M_{Ed} is the ultimate moment applied to the beam. The check may be repeated at point load positions by redefining n as the number of shear connectors from the support to the point considered.

The linear interaction method is conservative with respect to the stress block method (see following section), as illustrated in Figure 22.14.

666   *Composite beams*

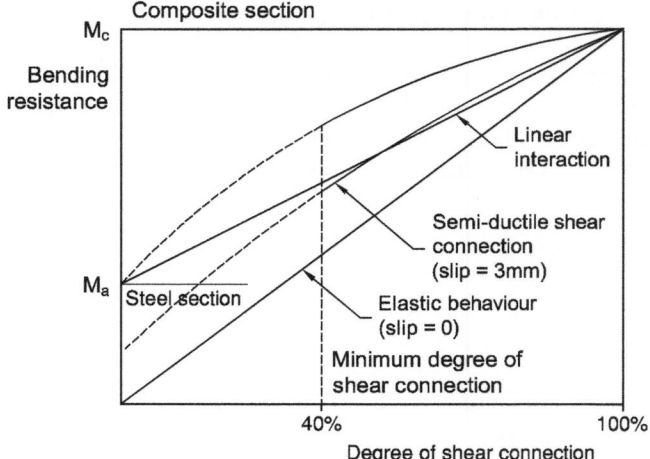

Figure 22.14 Interaction between moment resistance and degree of shear connection in composite beams

22.7.4 'Stress block' method

The 'stress block' approach is presented in BS EN 1994-1-1 Clause 6.2.1.3. It is an 'exact' method in that the equilibrium of the section is achieved by equating the compression force in the concrete slab to the longitudinal shear force transferred by the shear connectors, R_q. No design formulæ are given in BS EN 1994-1-1, but the formulæ in Appendix B of BS 5950-3 are based on the same principles.

The 'stress block' method may be used by replacing the term R_c by R_q, the longitudinal shear force in Equations 22.5 and 22.6 for the plastic resistance of the composite beam. The case where the plastic neutral axis lies within the slab does not apply for partial shear connection because the compression force in the slab is limited by the longitudinal shear force. Equation 22.5 corresponds to the case where the plastic neutral axis lies in the top flange of the steel section, which is satisfied when R_q exceeds R_w, where R_w is the tensile resistance of the steel web.

Equation 22.6 corresponds to the case where the plastic neutral axis lies in the web of the steel section, which is satisfied when R_w exceeds R_q. In this case, it is also necessary to check that the web remains Class 2 by checking that the depth of web in compression exceeds $40t\varepsilon$. This condition may not be satisfied for asymmetric sections where the plastic neutral axis is closer to the bottom flange than in rolled symmetric sections.

These two methods lead to a variation of bending resistance with degree of shear connection, as shown in Figure 22.14. The 'stress block' method leads to significantly higher moment resistance when compared to the linear interaction method for degrees of shear connection between 0.4 and 0.7, which is confirmed by tests on composite beams with this level of shear connection.

22.7.5 Minimum degree of shear connection

In using the above methods, a minimum degree of shear connection is specified in BS EN 1994-1-1 Clause 6.6.1.2 and is based on research by Johnson and Molenstra.[16] The minimum limit is introduced in order to ensure adequate deformation capacity of the shear connectors as defined by a characteristic slip of 6 mm. In principle, the use of the stress-block method requires larger deformations on the shear connectors at failure than the linear interaction method. Therefore, these limits are conservative for the linear interaction method.

For symmetric steel sections, the general limit on the degree of shear connection is defined as:

$$L \leq 25\,\text{m:} \quad \eta_\text{f} \geq 1 - \left(\frac{355}{f_\text{y}}\right)(0.75 - 0.03L) \tag{22.15}$$

$$L > 25\,\text{m:} \quad \eta_\text{f} \geq 1.0$$

where:

   L is the beam span in metres.

The influence of the steel strength f_y is introduced because of the higher deformation capacity implied by plastic design using higher steel strengths.

For asymmetric steel sections in which the bottom flange area does not exceed three times the top flange area, the minimum degree of shear connection is defined by:

$$L \leq 20\,\text{m:} \quad \eta_\text{f} \geq 1 - \left(\frac{355}{f_\text{y}}\right)(0.30 - 0.015L) \tag{22.16}$$

$$L > 20\,\text{m:} \quad \eta_\text{f} \geq 1.0$$

A relaxation of the degree of shear connection is permitted in Clause 6.6.1.2(3) when all the following conditions are met:

- through-deck welding of headed stud shear connectors of 19 mm diameter is used
- there is one stud per deck rib
- the rib is of proportions $b_\text{o}/h_\text{p} \geq 2$ and $h_\text{p} \leq 60$ mm
- the linear interaction method, described in Section 22.7.3, is used.

In this case of single shear connectors per rib:

$$L \leq 25\,\text{m:} \quad \eta_\text{f} \geq 1 - \left(\frac{355}{f_\text{y}}\right)(1 - 0.04L) \tag{22.17}$$

$$L > 25\,\text{m:} \quad \eta_\text{f} \geq 1.0$$

The relevant limits on the minimum degree of shear connection in BS 5950-3 are given by:

$$L \geq 10\,\text{m:} \quad \eta_\text{f} \geq (L - 6)/10 \tag{22.18}$$

$$L \leq 16\,\text{m:} \quad \eta_\text{f} \leq 1.0$$

The lower limit of 0.4 is adopted for composite beams of less than 10 m span. In BS 5950-3.1, no distinction is made between steel grades or for asymmetric or

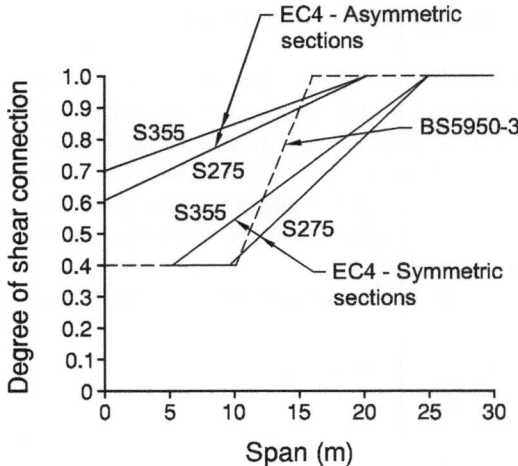

Figure 22.15   Variation of minimum degree of shear connection with the span of a composite beam

symmetric steel sections. The approach in BS 5950 was based on beam tests existing prior to 1990 which generally used higher steel strengths, and was not calibrated numerically against the characteristic slip of 6 mm as in BS EN 1994-1-1.

These limits are compared in Figure 22.15. Equation 22.14 is similar to the requirements of BS 5950-3, but gives a reduction in shear connection for spans longer than 13 m. It is clear that the BS EN 1994-1-1 limits for asymmetric sections are more restrictive than the limits in BS 5950-3.

22.7.6 Influence of proportionate loading

In design to BS EN 1994-1-1, the composite beam is assumed to be loaded to its plastic resistance. However, in practice, most composite beams are designed for serviceability limits of deflection and natural frequency, for which the Unity Factor (UF) in bending is much less than 1.0 (typically 0.7 to 0.9).

It is proposed that when using BS EN 1994-1-1, the ratio $(355/f_y)$ in Equations 22.15 to 22.17 may be taken as the ratio $(355/\sigma)$, where σ is the stress in the bottom flange necessary to resist the applied moment acting on the beam. For the general case of a symmetric composite beam, it follows that:

$$\eta_f \geq 1 - (1.0 - 0.04L)/UF \quad \text{for S275 steel} \tag{22.19}$$

$$\eta_f \geq 1 - (0.75 - 0.03L)/UF \quad \text{for S355 steel} \tag{22.20}$$

In all cases, $\eta_f \geq 0.4$, which is the minimum value independent of UF, in order to avoid the possibility of failure in elastic conditions. This is proposed as a way forward for composite beams designed for less than their plastic resistance which, although not explicitly stated, is within the principles of BS EN 1994-1-1 Clause 6.6.1.2.

22.7.7 Spacing of shear connectors

The limits on spacing of the shear connectors may also influence the maximum or minimum degree of shear connection that may be achieved in practice. The following limits refer to the relevant clauses in BS EN 1994-1-1:

The minimum spacing of the shear connectors is defined in Clause 6.6:
$5d$ longitudinally and $4d$ laterally.
Clause 6.6.5.5(2) requires that the spacing of the shear connectors should be such that Class 3 flanges are treated as Class 2. In this case, the spacing should not exceed $15\, t_f\, (235/f_y)^{0.5}$, where t_f is the flange thickness.
Clause 6.6.5.5(3) states that the maximum spacing of the shear connectors is the smaller of:
$6 \times$ slab depth or 800 mm.
Clause 6.6.5.6(2) states that the edge distance (defined as from the edge of the shear connector to the tip of the flange) should not be less than 20 mm.
Clause 6.6.5.6(2) states that t_f should not be less than $0.4 \times$ stud diameter for adequate welding.

22.7.8 Other checks

A further requirement is that the moment resistance of composite beams subject to point loads should be adequate at all locations along the beam. It may be necessary to check the shear connection provided at intermediate points using the linear interaction method in Section 22.7.3, or alternatively, to distribute the total number of shear connectors in proportion to the shear force diagram along the beam. This second approach is more conservative because it neglects the contribution of the steel section.

22.8 Transverse reinforcement

The requirement for transverse reinforcement ensures an effective transfer of force from the shear connectors into the concrete slab without splitting the concrete. Both BS 5950-3 and the ENV version of Eurocode 4 used an approach based on potential shear planes through the concrete slab on either side of the shear connectors. The BS EN 1994-1-1 method is different and is based on the method for concrete T-beams to BS EN 1992-1-1 and generally leads to less transverse reinforcement than to BS 5950-3.

22.8.1 Stress block model

The treatment of transverse reinforcement to resist splitting along the line of the shear connectors refers directly to the method in Clause 6.2.4 of BS EN 1992-1-1 for concrete T-beams. It is based on a simplified 'stress block' model in which the longitudinal shear force transferred from the shear connectors, F_{sc} causes a compression force F_c in the slab, and the outward component of this force F_t is resisted by the reinforcement transverse to the beam span. This action is illustrated in Figure 22.16(a).

The angle to the axis of the beam at which this compression force acts can vary between 26.5° and 45°, depending on the tensile resistance of the reinforcement. Generally, the minimum amount of transverse reinforcement is obtained when $\theta = 26.5°$ (or a 1 in 2 slope of the compression force to the beam span). Equilibrium of the horizontal and longitudinal forces is established from:

$$F_L = F_l \tan \theta \qquad (22.21)$$

where:
$$F_t \leq A_s f_{yp} s \text{ and } F_L = n_r P_{Rd}/2$$

and

F_L is the longitudinal shear force per shear plane, which is the force transferred by the shear connectors divided by 2 for internal beams

n_r is the number of shear connectors in a rib of the profiled decking

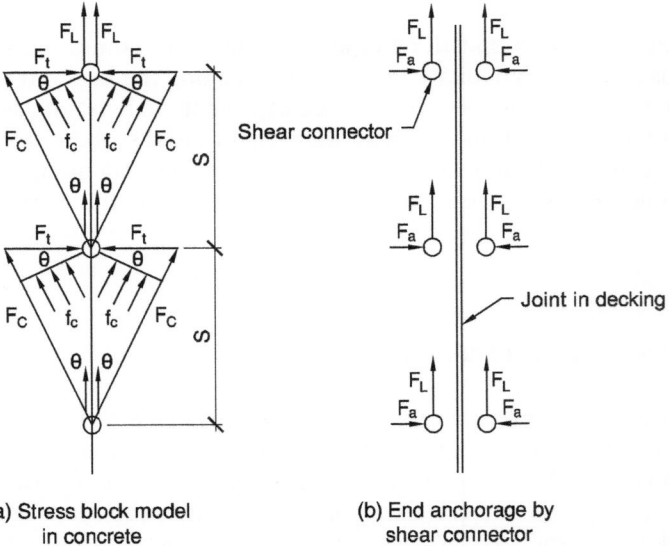

(a) Stress block model
in concrete

(b) End anchorage by
shear connector

Figure 22.16 Forces transferred from shear connectors and action of transverse reinforcement

    s is the longitudinal spacing of the ribs or of the group of shear connectors

    A_s is the cross-sectional area of the transverse reinforcement per unit length of the beam.

The compression force acting on the slab in front of the shear connectors is given by:

$$F_C = F_L / \cos \theta = F_t / \sin \theta \qquad (22.22)$$

The local compression or bearing resistance of the slab is given by:

$$F_C \leq f_c h_c s \sin \theta \qquad (22.23)$$

The compressive strength of the concrete, f_c, is given by:

$$f_c = 0.6(1 - f_{ck}/250)\, f_{ck}/\gamma_c \qquad (22.24)$$

Rearranging these equations and setting $\theta = 26.5°$, the maximum percentage of transverse reinforcement which may be used before compression in the concrete controls is given by:

$$\frac{A_S}{h_c s} \leq 0.29 \frac{f_{ck}}{f_y} \qquad (22.25)$$

For grade C30/37 concrete and S500 reinforcement, this is equivalent to a maximum of 1.6% of the concrete topping area, or approximately $1100\,\text{mm}^2/\text{m}$ of reinforcement for $h_c = 70\,\text{mm}$. It also follows that the minimum spacing of shear connectors in pairs is approximately 130 mm for $h_c = 70\,\text{mm}$, when limited by the local compression resistance of the concrete slab.

22.8.2   End anchorage by profiled decking

An additional component arising from the end anchorage of the decking, F_a, may be included in this longitudinal shear resistance. The full tensile strength of the profiled decking can be used when it crosses the beams (i.e. secondary beams) and is continuous. Where the profiled decking is discontinuous, the end anchorage resistance, P_{bp}, may be included in the calculation of F_t, provided both ends of the decking are properly attached (see Figure 22.16(b)). The anchorage force per shear connector is given in BS EN 1994-1-1 Clause 9.7.4(3) and is presented in Chapter 21.4.6.

For shear connectors placed in pairs per rib, each shear connector anchors one edge of the decking, whereas for single shear connectors, the shear connector positions should be 'staggered' along their length to anchor the decking effectively at its joints. The tensile forces developed in the profiled decking are assumed to be distributed uniformly by shear-bond action into the concrete.

Hence, the minimum requirement for transverse reinforcement is given by:

$$A_s f_y s \geq n_r (P_d \tan \theta - P_{pb})/2 \qquad (22.26)$$

For $\theta = 26.5°$, it follows that the minimum amount of transverse reinforcement per unit length of the beam is given by:

$$A_s \geq n_r (P_d - 2P_{pb})/(4 f_y s) \qquad (22.27)$$

Typically for $n_r = 2$, $s = 300\,\text{mm}$ and $P_d = 73\,\text{kN}$, with an end anchorage resistance of $P_{pb} = 20\,\text{kN}$; the minimum amount of transverse reinforcement is $A_s \geq 111\,\text{mm}^2/\text{m}$. This can be achieved by A193 mesh reinforcement.

The contribution of the profiled decking should be *neglected* where it is not properly anchored – see below for primary beams. It is generally found that this approach leads to a smaller amount of reinforcement and higher compression forces in the slab than in the longitudinal shear method in BS 5950-3.

22.9 Primary beams and edge beams

22.9.1 Primary beams

For primary beams, more shear connectors should be placed in higher shear zones, as illustrated in Figure 22.17. The total number of shear connectors may be distrib-

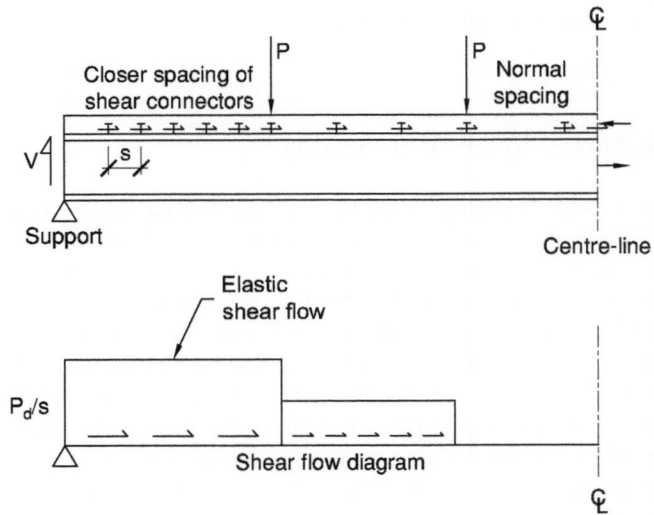

Figure 22.17   Longitudinal shear flow in primary beams

uted in proportion to the shear force in each zone. However, this may lead to an impractical shear connector arrangement, in which case, the bending resistance should be calculated at the point load positions according to the actual distribution of shear connectors.

It is also necessary to increase the amount of transverse reinforcement. Furthermore, discontinuities in the profiled decking placed parallel to the primary beams often mean that the term in P_{pb} should be ignored. Generally, in the high shear zones, a second layer of mesh reinforcement or additional bars are provided, extending over the effective width of the concrete slab.

22.9.2 Edge beams

For edge beams where shear connectors are placed close to the slab edge, the effect of end anchorage is generally neglected, while U-bars are provided and looped around the shear connectors, as required by BS EN 1994-1-1 Clause 6.6.5.3(2). If U-bars are not provided, then composite action may be ignored conservatively but included only in stiffness calculations.

22.10 Continuous composite beams

In BS EN 1994-1-1, the design of continuous composite beams may be based on two approaches to determine the design bending moments in negative (hogging) and positive (sagging) bending:

- Clause 5.4.4 states that linear elastic analysis may be used for composite beams with all section classifications using maximum permitted moment redistributions.
- Clause 5.4.5 states that rigid plastic analysis may be used for Class 1 sections (based on the proportions of the bottom flange of the section).

This general approach is similar to that presented in BS 5950-3.

22.10.1 Elastic analysis

The elastic analysis method depends on whether the composite cross-section is considered to be uncracked, or alternatively, cracked in negative bending. In the first case, the stiffness of the beam is treated as being constant along its length. In the second case, the stiffness of the beam is reduced in the negative (hogging) moment region and hence lower percentage redistributions of moment are permitted in comparison to the uncracked case.

Table 22.5 Maximum moment redistributions for elastic global analysis of continuous composite beams

Analysis method for composite section	Section class to BS EN 1993-1-1			
	1	2	3	4
Uncracked section	40%	30%	20%	10%
Cracked section	25%	15%	10%	0%

The section class determines the degree of moment redistribution that is permitted in order that local buckling is prevented. Plastic section properties may be used for Class 1 or 2 sections. For steel grades higher than S355, only Class 1 or 2 cross-sections may be used for elastic redistribution of moments.

The maximum values of redistribution of negative (hogging) bending moment are presented in BS EN 1991-1-1 Table 5.1 (see Table 22.5). The redistributed moment is transferred to the positive (sagging) moment regions in order to maintain equilibrium.

22.10.2 Plastic analysis

For rigid-plastic analysis, the following requirements apply, as stated in BS EN 1994-1-1 Clause 5.4.5(4):

- The beams are subject to uniformly distributed loading.
- Adjacent spans do not differ by more than 50% of the shorter span.
- End spans do not exceed 115% of the length of the adjacent internal span.

Flanges of the steel section should be Class 1 at the support locations, but webs that are Class 3 may be re-classified as Class 1 by considering their effective width in compression.

22.10.3 Bending resistances

The effective width of concrete slab in the negative (hogging) bending region is given by $L_e = 0.5L$ and corresponds to an effective width of $L/8$, where L is the clear span (rather than $L/4$ for the positive (sagging) moment in a simply supported beam). This means that the bar reinforcement is concentrated in a relatively narrow width over the internal supports.

The negative (hogging) bending resistance of the composite section is calculated using plastic stress blocks for Class 1 or 2 sections. In this case, the same formulae as in Section 22.4 are used but the tensile resistance of the reinforcement R_r replaces the compression resistance of the slab R_c.

If the depth of the web of the steel section in compression exceeds $40t\varepsilon$, then the Class 3 web is reduced to Class 2 by considering the effective portion of the web area, which is taken as $d_{eff} = 40t\varepsilon$ in plastic design.

The positive (hogging) bending resistance of the composite section is calculated using the plastic stress blocks presented in Section 22.4, but with a modified effective slab width based on an effective span of $L_e = 0.85L$.

22.11 Serviceability limit states

22.11.1 General criteria

The serviceability requirements for composite beams concern the control of deflections, cracking of concrete and vibration response, as stated in BS EN 1994-1-1 Clause 7.1 to 7.3. Control of deflections is important in order to prevent cracking or deformation of partitions and cladding, or to avoid noticeable deviations of floors or ceilings. Floor vibrations may be important in long span applications, but these calculations are outside the scope of BS EN 1993-1-1 and BS EN 1994-1-1. Reference is made to BS EN 1990 A1.4.4.[8]

Assessments at the serviceability limit state are based on elastic behaviour (with certain modifications for creep and cracking). To avoid consideration of post-elastic effects, the stresses existing in the steel section at the serviceability limit state should be less than the steel yield strength. However, no stress limits at the serviceability limit state are required by BS EN 1994-1-1, because it is concluded that:

- Slight local yielding in the positive moment region has a limited effect on deflections.
- Beneficial effects of continuity due to the connections on deflections are ignored.

Deflection limits are not specified in BS EN 1994-1-1, and reference is made to national requirements for limits due to permanent and variable loads. Many designers feel that 'total' deflections are less important than imposed load deflections, for example, where a raised floor or suspended ceiling is used. It may be justified to use a total deflection limit of span/200 (limited to maximum of 60 mm for adjustment of raised floors), or to consider pre-cambering or propping of long span beams.

In practice, actual deflections will be less than calculated deflections due to continuity provided by the beam to column connections as well as the slab reinforcement.

22.11.2 Second moment of area

Deflections are calculated using the second moment of area of the composite section based on elastic properties. When subject to positive (sagging) moment, the concrete

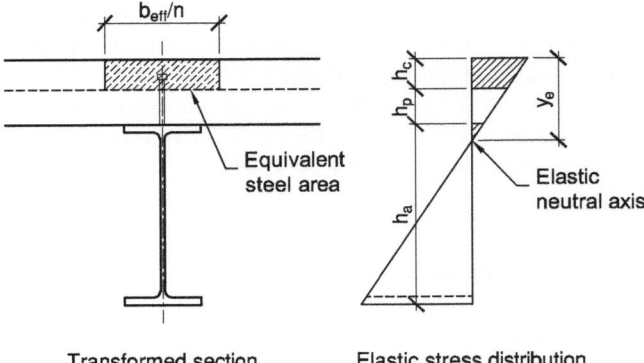

Transformed section   Elastic stress distribution

Figure 22.18   Transformed section for elastic properties of composite beams

may be assumed to be uncracked. To avoid the inconvenience of working in two materials, the concrete is transformed into an equivalent area of steel, as shown in Figure 22.18. The second moment of area of the transformed steel section is:

$$I_c = \frac{A_a \left(h_c + h_p + h\right)^2}{4(1+nr)} + \frac{b_{\text{eff}} h_c^3}{12n} + I_{\text{ay}} \tag{22.28}$$

where:

- n is the ratio of the elastic moduli of steel to concrete (see Section 22.11.2), taking into account the creep of the concrete
- r is the ratio of the cross-sectional area of the steel section relative to the concrete section
- I_{ay} is the second moment of area of the steel section.

Note: other terms as defined previously.

The ratio I_c/I_{ay} in Equation 22.28 therefore defines the improvement in the stiffness of the composite section relative to the steel section. Typically, I_c/I_{ay} is in the range of 2.5 to 4.5, indicating that one of the main benefits of composite action is to reduce imposed load deflections and vibration response.

22.11.3 Modular ratio

The values of elastic modulus of concrete under short term loads are given in BS EN 1992-1-1 Table 3.1. The elastic modulus of concrete under long term loads is affected by creep, which causes a reduction in the stiffness of the concrete and is defined in BS EN 1994-1-1 Clause 5.4.2.2(2). The modular ratio, n, is the ratio of the elastic modulus of steel to the time-dependent modulus of concrete. The modular ratio that may be used for normal weight concrete is $n_s = 6.5$ for short term (variable) loading. For longer term loads, BS EN 1994-1-1 suggests a representative modular ratio corresponding to an elastic modulus of $E_{cm}/2$, or approximately $n_L = 13$. A modular ratio of 20 may be used for beams subject to permanent loads (e.g. propped beams) using a creep coefficient determined in BS EN 1992-1-1 Clause 3.1.4. A higher modular ratio should be used for lightweight concrete.

For buildings of normal usage, surveys have shown that the proportions of variable and permanent imposed loads usually exceed 3:1. Although separate deflection calculations may be made for the variable and permanent deflections, a representative modular ratio of 10 is usually appropriate for imposed load deflection calculations as was recommended in BS 5950-3.

22.11.4 Influence of partial shear connection

In general, all shear connectors are not rigid but deformable under load. Therefore slip or relative movement occurs between the concrete slab and the top flange of the steel section. However, at the serviceability limit state, slip effects are relatively small and hence deflections due to slip are small. According to BS EN 1994-1-1 Clause 7.3.1(4), these additional deflections may be ignored when:

- the degree of shear connection exceeds 50%, and
- the elastic forces on the shear connectors do not exceed P_{Rd}
- the height of the decking does not exceed 80 mm.

In comparison, for cases of partial shear connection less than 50%, deflections are increased according to Clause 7.3.1(7), as follows:

$$\frac{\delta}{\delta_c} = 1 + C(1-\eta)\left(\frac{\delta_a}{\delta_c} - 1\right) \tag{22.29}$$

where:
η is the degree of shear connection at the ultimate limit state
δ_c is the deflection of the composite beam with full shear connection
δ_a is the deflection of the steel beam
C is 0.3 for unpropped construction and 0.5 for propped construction.

It is recommended that additional deflections due to the effects of slip are included in propped composite construction, where the forces acting on the shear connectors are higher than those in unpropped beams.

22.11.5 Shrinkage induced deflections

BS EN 1994-1-1 states that shrinkage deflections should be calculated for simply supported beams when the span-to-depth ratio of the beam exceeds 20, and when the free shrinkage strain of the concrete exceeds 400×10^{-6}. In practice, these deflections will only be significant for spans longer than 15 m in exceptionally warm dry environments.

The curvature, K_s, due to a free shrinkage strain, ε_s, is:

$$K_s = \frac{\varepsilon_s (h_c + 2h_p + h_a) A_a}{2(1+nr) I_c} \tag{22.30}$$

n is the modular ratio appropriate for shrinkage calculations ($n \approx 20$ for normal weight concrete).

The deflection due to shrinkage induced curvature is calculated from:

$$\delta_s = 0.125 \, K_s L^2 \tag{22.31}$$

This deflection formula ignores continuity effects at the beam supports and therefore over-estimates shrinkage deflections.

22.11.6 Stress checks

Checks on serviceability stress limits are not required in BS EN 1994-1-1, which represents a reduction in design effort relative to BS 5950-3. However, serviceability stress checks are prudent in order to avoid inelastic deflections. Examples might be beams supporting heavy cladding or heavily loaded columns.

22.11.7 Vibration sensitivity

This section is included here because a check on the vibration sensitivity and response may be necessary for long span beams designed for light imposed loads. BS EN 1994-1-1 contains no specific guidance on vibration sensitivity, other than recognising its importance in many building types.

A simple measure of the natural frequency of a beam, f, is:

$$f = \frac{18}{\sqrt{\delta_{sw}}} \; \text{cycles/sec} \tag{22.32}$$

where:

   δ_{sw} is the instantaneous deflection (in mm) caused by re-application of the self-weight of the floor and other permanent loads to the composite beam. The permanent loads are taken as 10% of the specified imposed loads plus services but excluding partitions.

In UK practice, the minimum limit on f is taken as 4 cycles/sec for most building applications. A lower value of 3 cycles/sec is appropriate for car parks, but a higher limit of 6 cycles/sec may be used for hospitals or other buildings where low response is required.

A response factor approach should be used for hospitals or other sensitive buildings, as given in SCI publication 354.[17]

22.11.8 Span-to-depth ratios

Adequate serviceability performance may be generally assumed when the span:depth ratio of composite beams is less than a certain value. This precise limit depends on the form of loading and steel grade. The following limits may be used for choosing composite beam sizes at the Scheme Design stage:

Uniform loading: Span-to-depth ratio = 18 to 20
Two point loading: Span-to-depth ratio = 15 to 18

where the 'depth' is the combined slab and beam depth. Spans longer than these limits will generally be controlled by serviceability criteria, as covered in this section.

22.11.9 Crack control

It is necessary to control cracking of concrete in cases where the proper functioning of the structure or its appearance would be impaired. This may be the case in industrial buildings or car parks, where the floor is subject to traffic. Internally within buildings, durability is not affected by concrete cracking. Similarly when raised floors are used, cracking is not visually important.

Where it is necessary to control cracking, the amount of reinforcement should exceed a minimum value in order to distribute cracks uniformly. This minimum percentage, ρ, is given in BS EN 1994-1-1 Clause 7.4.2 by:

$$\rho = \frac{A_s}{A_c} \times 100\% = k_s k_c k \frac{f_{ct}}{\sigma_s} \times 100\% \tag{22.33}$$

where:

$\quad k_s$ is a coefficient which allows for the effect of the reduction in normal force on the slab due to initial cracking and slip in the shear connectors, and may be taken as 0.9

$\quad k_c$ is a coefficient due to the bending stress distribution in the section with a value between 0.4 and 0.9

$\quad k$ is a coefficient accounting for a decrease in the concrete tensile strength ($k = 0.8$)

$\quad \sigma_s$ is the stress in the reinforcement

$\quad f_{ct}$ is the effective tensile strength of concrete. A value of $3\,N/mm^2$ is the minimum adopted.

Table 22.6 Maximum bar spacing for high bond bars (Table 7.2)

Steel stress σ_s (N/mm^2)		≤160	200	240	280	320	360	400
Maximum bar spacing (mm)	$w_k = 0.2$ mm	200	150	100	50	–	–	–
	$w_k = 0.3$ mm	300	250	200	150	100	50	–
	$w_k = 0.4$ mm	300	300	250	200	150	100	80
where w_k = design crack width								

A typical value of ρ for crack control is 0.4% to 0.6%, which is in excess of the minimum of 0.2% necessary for transverse load distribution in unpropped slabs. In propped slabs, it may be necessary to increase the amount of reinforcement to control cracking. These bars or mesh need only be placed in the negative moment regions of the beams or slabs. This reinforcement may also act as fire reinforcement and as transverse reinforcement.

An additional criterion is that the bars should be of small diameters and should be spaced relatively close together in order to be more effective in crack control. Maximum bar spacings for a given steel stress are given in Table 22.6. Otherwise, checks on crack widths should be made as in BS EN 1992-1-1.

22.12 Design tables for composite beams

'Direct' design tables may be used at the Scheme Design stage and require no additional calculations (but they should not be substituted for final design).

22.12.1 General information

The selection of the size of steel sections to be used in a composite beam depends upon a number of variables: span, loading, beam spacing, slab depth, concrete type and grade, steel grade, shape of profiled decking etc. Some of these variables have been fixed for the purposes of preparing indicative Design Tables:

- The steel beams considered are Universal Beam (UKB).
- The typical height of profiled decking is 50 to 60 mm for beam spacings of 3 to 3.75 m, and 80 mm for 4 m beam spacings.
- The depth of the composite slab is determined by the fire resistance requirements. For 90 minutes fire resistance, a typical slab depth is 130 mm, increasing to 150 mm for deeper profiled decking.

- The concrete grade is C25/30 in normal weight concrete. The concrete grade has little effect on the section properties or chosen beam size.
- The steel grade is usually chosen as S355 for primary beams and S275 for secondary beams, where design is controlled by deflection limits.
- Welded shear connectors are placed in each deck rib. Shear connectors are normally welded singly or in pairs for wide-rib decking.
- The span of primary beams is a function of the number and the spacing of the secondary beams.
- The steel beams are unpropped during construction and therefore support the self-weight of the concrete slab.
- The imposed loading is $3.5 \, kN/m^2$ plus $1 \, kN/m^2$ for partitions and service loads. The self-weight of the concrete slab is established from its depth.

22.12.2 Design tables

Two design cases are considered: secondary beams loaded directly by the composite slabs, and primary beams (which are subject to point loads from the secondary beams). For point-loaded beams, the secondary beams are connected at spacings of 3 to 3.75 m.

A design table for secondary beams of 3 m and 4 m spacing and subject to uniformly distributed loading is presented in Table 22.7.

A design table for primary beams and secondary beams used in various standard column grids is presented in Table 22.8. The optimum grid tends to be where the

Table 22.7 Scheme design table for composite secondary beams to BS EN 1994-1-1

Beam Span (m)	Beam Spacing = 3 m	Beam Spacing = 4 m
	Slab depth =130 mm Deck height = 50 to 60 mm	Slab depth =150 mm Deck height = 80 mm
6.0	254 × 146 × 31 UKB	254 × 146 × 37 UKB
7.0	254 × 146 × 37 UKB	305 × 127 × 42 UKB
8.0	305 × 127 × 48 UKB	356 × 171 × 45 UKB
9.0	356 × 171 × 57 UKB	406 × 178 × 60 UKB
10.0	406 × 178 × 54 UKB	406 × 178 × 67 UKB
11.0	406 × 178 × 67 UKB	457 × 191 × 74 UKB
12.0	457 × 191 × 74 UKB	457 × 191 × 98 UKB
13.0	457 × 191 × 98 UKB	533 × 210 × 101 UKB
14.0	533 × 210 × 92 UKB	610 × 229 × 101 UKB
15.0	610 × 210 × 101 UKB	686 × 254 × 125 UKB
16.0	610 × 210 × 125 UKB	686 × 254 × 140 UKB

Data for use of S275 steel in unpropped construction. Imposed loading of (3.5 + 1.0) kN/m^2 plus self-weight of the NWC composite slab of the depths shown.

Table 22.8 Typical primary and secondary beam sizes for various floor grids

Span of primary beam	Span of secondary beam	Size of primary beam	Size of secondary beam
6 m	6 m	305 × 127 × 37 UKB	245 × 126 × 31 UKB
Single point load	7.5 m	305 × 165 × 46 UKB	305 × 127 × 37 UKB
	9 m	356 × 171 × 51 UKB	356 × 171 × 51 UKB
	10.5 m	406 × 178 × 54 UKB	406 × 178 × 60 UKB
	12 m	457 × 191 × 67 UKB	457 × 191 × 74 UKB
	13.5 m	457 × 191 × 74 UKB	533 × 210 × 82 UKB
	15 m	457 × 191 × 89 UKB	610 × 210 × 101 UKB
7.5 m	7.5 m	356 × 171 × 51 UKB	305 × 165 × 40 UKB
Single point load	9 m	406 × 178 × 54 UKB	356 × 171 × 57 UKB
	10.5 m	457 × 191 × 67 UKB	406 × 178 × 74 UKB
	12 m	457 × 191 × 89 UKB	457 × 191 × 89 UKB
	13.5 m	533 × 210 × 92 UKB	533 × 210 × 101 UKB
	15 m	533 × 210 × 101 UKB	610 × 210 × 125 UKB
9 m	7.5 m	406 × 198 × 74 UKB	305 × 127 × 37 UKB
Two point loads	9 m	457 × 191 × 82 UKB	356 × 121 × 51 UKB
	10.5 m	533 × 210 × 92 UKB	406 × 178 × 60 UKB
	12 m	533 × 210 × 101 UKB	457 × 191 × 74 UKB

Data for use of S355 steel as primary beams and S275 steel as secondary beams. Imposed loading of $(3.5 + 1.0)$ kN/m^2 plus self-weight of 130 mm deep NWC composite slab.

span of the primary beams is 50 to 67% of the span of the secondary beams so that both beams are of similar size (i.e. in a column grid of 6 m × 9 m).

A worked example follows.

References to Chapter 22

1. British Standards Institution (2004) BS EN 1994-1-1 *Eurocode 4. Design of composite steel and concrete structures Part 1-1: General rules and rules for buildings.* London, BSI.
2. British Standards Institution (1990) BS 5950-3 *Structural use of steelwork in buildings Part 3: Design in composite construction.* London, BSI.
3. British Standards Institution (2004) BS 5950-4 *Structural use of steelwork in buildings Part 4: Code of practice for design of floors with profiled steel sheeting.* London, BSI.
4. Lawson R.M. (1989) *Design of composite slabs and beams with steel decking.* The Steel Construction Institute P55. Ascot, SCI.
5. British Standards Institution (2005) BS EN 1993-1-1 *Eurocode 3: Design of steel structures Part 1.1 General rules and rules for buildings.* London, BSI.
6. British Standards Institution (2000) BS 5950 *Structural use of steelwork in building: Code of practice for design in simple and continuous construction Part 1: Hot rolled sections.* London, BSI.

7. Lawson R.M. and Chung K.F. (1993) *Composite beam design to Eurocode 4.* The Steel Construction Institute P121. Ascot, SCI.

8. British Standards Institution (2002) BS EN 1990-1-1 *Eurocode – Basis of structural design.* London, BSI.

9. British Standards Institution (2004) BS EN 1992-1-1 *Design of concrete structures Part 1-1: General rules and rules for building.* London, BSI.

10. Johnson R P and Anderson D (2005), *Designers Guide to EN 1994-1-1 Eurocode 4: Design of composite steel at concrete structures. Part 1.1 General Rules and Rules for Buildings.* London, Thomas Telford.

11. British Standards Institution (2004) BS EN 10 326 *Continuously hot-dip coated strip steel of structural steel: Technical delivery conditions.* London, BSI.

12. British Standards Institution (1997) BS 8110 *Structural use of concrete Part 1: Code of practice for design and construction.* London, BSI.

13. British Standards Institution (2005) BS EN 10 080 *Steel for reinforcement of concrete. Weldable reinforcing steels. General.* London, BSI.

14. British Standards Institution (2008) BS EN ISO 13918 *Welding studs and ceramic ferrules for arc stud welding.* London, BSI.

15. British Standards Institution (2005) BS EN 1991-1-6 *Eurocode 1 – Actions on structures Part 1.6 – Actions during execution.* London, BSI.

16. Johnson R.P. and Molenstra N. (1991) Partial shear connection in composite beams for buildings, *Proc Inst Civil Engineers* Vol 91 No4 Dec, 679–704.

17. Hicks S.J. and Smith A. (2009) *Design of floors for vibrations – A new approach.* The Steel Construction Institute P354. Ascot, SCI.

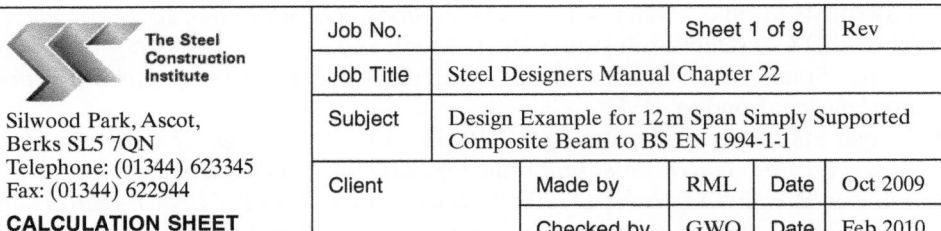

The Steel Construction Institute	Job No.			Sheet 1 of 9	Rev
	Job Title	Steel Designers Manual Chapter 22			
Silwood Park, Ascot, Berks SL5 7QN Telephone: (01344) 623345 Fax: (01344) 622944	Subject	Design Example for 12 m Span Simply Supported Composite Beam to BS EN 1994-1-1			
	Client	Made by	RML	Date	Oct 2009
CALCULATION SHEET		Checked by	GWO	Date	Feb 2010

Design Example of 12 m Span Simply Supported Composite Beam to BS EN 1994-1-1

Consider an internal secondary composite beam A-A of 12 m span between columns and subject to uniform loading. Choose a 457 × 191 × 74 kg/m UKB in S 355 steel.

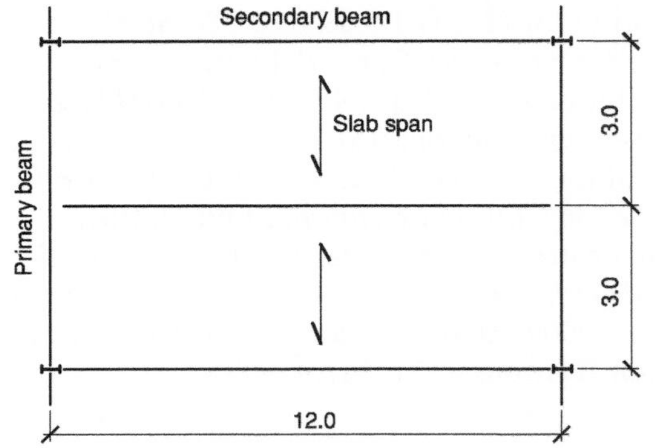
Secondary beam
Slab span
Primary beam
3.0
3.0
12.0

(a) Plan on floor

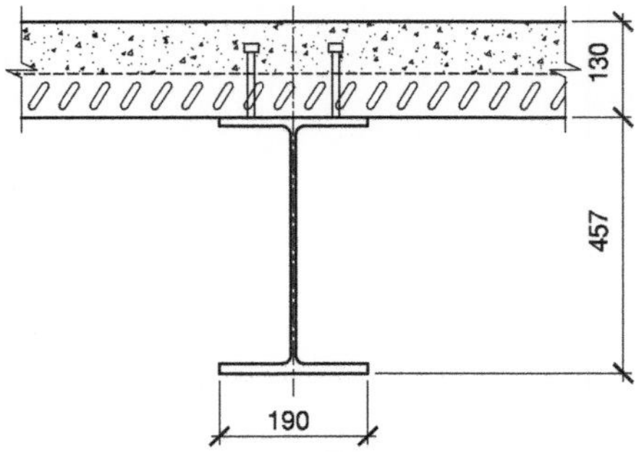
130
457
190

(b) Cross-section through beam

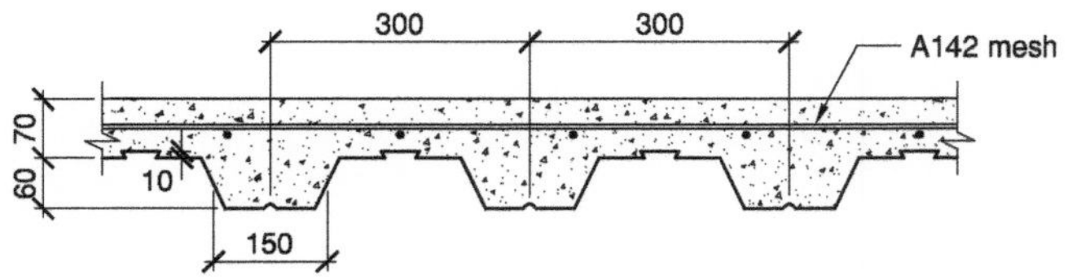
300
300
A142 mesh
70
60
10
150

(c) Cross-section through slab

Example of 12 m span composite beam to BS EN 1994-1-1	Sheet 2 of 9	Rev

Design criteria

Slab designed for 90 minutes fire resistance
Imposed load $5\,kN/m^2$ ($4\,kN/m^2$ occupancy load and $1\,kN/m^2$ partitions)

Floor dimensions

Span L	$= 12.0\,m$	
Beam spacing b	$= 3.0\,m$	
Slab depth	$= 130\,mm$	
Depth above profile h_c	$= 60\,mm$	
Deck profile height, h_p	$= 60\,mm$ (allow a further 10 mm for re-entrant stiffener)	

The beam and slab are both unpropped during construction.

Materials

Steel: Grade S 355. nominal value of yield strength, $f_y = 355\,N/mm^2$
  Partial safety factor $\gamma_a = 1.0$

Design strength $f_d = \dfrac{f_y}{\gamma_a} = 355\,N/mm^2$

Cl 3.3.2
Table 3.3

Concrete: Normal weight concrete strength class C25/30
Density $= 24\,kN/m^3$
Concrete cylinder strength, f_{ck} $= 25\,N/mm^2$
Elastic modulus, E_{cm} $= 30.5\,kN/mm^2$

Cl 3.1.2

Decking: Steel thickness $= 0.9\,mm$
  Steel grade $= S\,350$ ($f_y = 350\,N/mm^2$)

Reinforcement: A 142 mesh
  Steel grade $= S\,460$ ($f_y = 460\,N/mm^2$)

Shear connectors

19 mm diameter stud and 100 mm overall height
95 mm length after welding

Loading

Construction stage loading

Concrete slab

Weight $= [130 - 60/2]24/10^3 = 2.4\,kN/m^2$

kN/m

Concrete slab	$= 2.40$
Steel decking (allow)	$= 0.12$
Reinforcement (allow)	$= 0.04$
Steel beam (allow)	$= \underline{0.25}$
	$2.81\,kN/m^2$

Construction load $= 0.75\,kN/m^2$ acting on the supported area of floor
(note this is different from the deck design which uses a construction load
$1.5\,kN/m^2$ over 3 m length and $0.75\,kN/m^2$ elsewhere on the decking)

BS EN 1991-1-6

Example of 12 m span composite beam to BS EN 1994-1-1	Sheet 3 of 9	Rev

Imposed loads and other loads

$Occupancy = 4.0 \, kN/m^2$
$Partitions \quad = \underline{1.0} \, kN/m^2$
$Total \qquad\quad 5.0 \, kN/m^2$

The design uses BS EN 1991 for actions (loading). According to BS 6399, imposed loads may be reduced with respect to the total area supported by the beam. For the purposes of the design example, this reduction is omitted.

BS EN 1991 National Annex

$Ceiling \, and \, Services = 0.50 \, kN/m^2$

Initial selection of beam size

A suitable section for a secondary beam subject to an imposed load of $5.0 \, kN/m^2$ is a $457 \times 191 \times 74 \, UKB$ – Grade S 355 steel.

Section properties and dimensions:

$h \quad = 457 \, mm$	$d \quad = 407 \, mm$
$b \quad = 190 \, mm$	$A_a \quad = 9460 \, mm^2$
$t_w \quad = 9.0 \, mm$	$I_{ay} \quad = 33320 \, cm^4$
$t_f \quad = 14.5 \, mm$	$W_{pl} = 1650 \, cm^3$
$c \quad = 190/2 = 95 \, mm$	

$\varepsilon = \sqrt{(235/f_y)} = 0.81$

Section classification

BS EN 1993-1-1

$c/t_f = 6.6 < 9 \, \varepsilon = 7.3$

Table 4.1

$d/t_w = 45.3 < 72 \, \varepsilon = 58.3$

Table 4.2

The cross-section is Class 1 to BS EN 1993-1-1 for the construction stage.

Construction stage design

Dead load factor	γ_G	$= 0.925 \times 1.35 = 1.25$
Imposed load factor	γ_Q	$= 1.5$
Slab and beam	$= 2.81 \times 1.25$	$= 3.51 \, kN/m^2$
Construction load	$= 0.75 \times 1.5$	$= \underline{1.13} \, kN/m^2$
Total factored load		$4.64 \, kN/m^2$

$Design \, moment \quad = M_{Ed} = \dfrac{4.64 \times 3 \times 12^2}{8} = 250 \, kNm$

The beam is laterally restrained as the decking spans perpendicular to the beam and is directly attached to it.

Example of 12 m span composite beam to BS EN 1994-1-1	Sheet 4 of 9	Rev

Moment resistance of steel beam = $M_{a,pl,Rd}$, where:

*BS EN
1993-1-1*

$M_{a,pl,Rd} = W_{pl} \times f_d$ $\quad = 1650 \times 355 \times 10^{-3} kNm$
$\qquad\qquad\qquad = 585\,kNm > M_{Ed} = 250\,kNm$

The beam is satisfactory in the construction stage.

Composite stage design

Slab and beam $= 2.81 \times 1.25$ $\quad = 3.51\,kN/m^2$
Serviced ceiling $= 0.50 \times 1.25$ $\quad = 0.63\,kN/m^2$
Imposed load $\;= 5.0 \times 1.5$ $\qquad = \underline{7.50}\,kN/m^2$
$\qquad\qquad\qquad\qquad\qquad\quad 11.6\,kN/m^2$

Shear force, V_{Ed} $\quad = 11.6 \times 3 \times 12/2 = 209\,kN$

Design moment, $M_{Ed} = \dfrac{11.6 \times 3 \times 12^2}{8} = 626\,kNm$

Effective width of compression flange, b_{eff}

$b_{eff} = \dfrac{2 \times l_0}{8}$ *(for a simply supported beam, l_0 = span)*

$\qquad = \dfrac{2 \times 12}{8} = 3.0\,m = 3\,m$ *(beam centres)*

Compressive resistance of slab, R_c

$R_c = \dfrac{0.85\,f_{ck}}{\gamma_c} \times b_{eff} \times h_c$

where $\quad \gamma_c$ $\;= partial\ safety\ factor\ for\ concrete = 1.5$

$\qquad\quad f_{ck}$ $\;= characteristic\ cylinder\ strength = 25\,N/mm^2$

$\qquad\quad h_c$ $\;= 130 - 60 - 10 = 60\,mm$ *(allowing for re-entrant stiffener)*

$\qquad\quad R_c$ $\;= 0.85 \times (25/1.5) \times 3000 \times 60/10^3 = 2550\,kN$

Tensile resistance of steel section, R_c

$R_s = f_{yd} \times A_a$

$\quad = 355 \times 9460/10^3 = 3358\,kN > 2550\,kN$

Moment resistance for full shear connection

Since $R_s > R_c$, the plastic neutral axis (PNA) lies in the steel flange. Therefore, the moment resistance of the composite beam is given by taking moments about the centre of the top flange of the beam, as follows:

*Cl
5.9.1(5)*

Example of 12 m span composite beam to BS EN 1994-1-1	Sheet 5 of 9	Rev

$$M_{a,pl,Rd} = R_s \frac{h}{2} + R_c \left[\frac{h_c}{2} + h_p \right]$$

$$= 3358 \times 0.457/2 + 2550 \times (30 + 70) \times 10^{-3}$$

$$= 767 + 255 = 1022\,kNm > M_{Ed}$$

Unity Factor $= 626/1022 = 0.61$

Shear resistance of the steel section

$$V_{pl,Rd} = \frac{457 \times 9.0 \times 355}{\sqrt{3} \times 10^3}$$

$$= 843\,kN > V_{Ed} = 179\,kN$$

NB: For uniformly distributed load, shear at the supports does not influence the moment resistance of the section.

Shear connector resistance

The design shear resistance of a shear connector is:

$$P_{Rd} = 0.29\,\alpha \times d^2 \sqrt{(f_{ck}E_{cm})}/\gamma_v \ or$$

$$P_{Rd} = 0.8\,f_u\,(\pi\,d^2/4)\,\gamma_v$$

whichever is smaller.

For $d = 19\,mm$, $h_{sc} = 95\,mm$, $f_u = 450\,N/mm^2$, $\gamma_v = 1.25$, $f_{ck} = 25\,N/mm^2$ *Cl 6.3.2.1*

and $E_{cm} = 30.5\,kN/mm^2$

$h/d = 95/19 > 4 \ \therefore \ \alpha = 1.0$

$$P_{Rd} = 0.29 \times 1.0 \times 19^2 \left(\sqrt{25 \times 30.5/10^3} \right)/1.25$$

$$= 73\,kN$$

$$P_{Rd} = 0.8 \times 450 \times (\pi \times 19^2/4) \times 10^{-3}/1.25 = 81.7\,kN > 73\,kN$$

$$P_{Rd} = 73\,kN \ for \ a \ solid \ slab$$

Influence of deck shape – Deck perpendicular to the beam   *Cl 6.3.3.2*

One stud per rib, i.e. $n_r = 1$:

$$k_t = \frac{0.7}{\sqrt{n_r}}\,(b_0/h_p)\,[(h_{sc}/h_p) - 1] \le 1.0$$

where $k_t = $ *reduction factor due to deck shape*

$$= 0.7 \times \frac{150}{60}[(95/60) - 1] = 1.02 > 1.0$$

The upper limit on k_t *is dependent on the number of shear connectors and the thickness of the profiled decking, and is given by* $k_t = 0.85$ *in Table 22.4.*   *Table 22.4*

Example of 12 m span composite beam to BS EN 1994-1-1	Sheet 6 of 9	Rev

$P_{Rd} = 0.85 \times 73 = 62\,kN$

Two studs per rib, i.e. $n_r = 2$:

$$k_t = \frac{0.7}{\sqrt{n_r}}\ (b_0/h_p)\ [(h/h_p) - 1] \le 0.8$$

$$= \frac{0.7}{\sqrt{2}}\ (150/60)\ [(95/60) - 1] = 0.72$$

The upper limit on k_t for $n_r = 2$ and $t_s = 0.9\,mm$ is 0.75. Therefore, $k_b = 0.72$.

$P_{Rd} = 0.72 \times 73 = 52\,kN$ *for pairs of shear connectors per rib*

Shear connector layout

A total of 19 deck ribs is available for the positioning of the stud shear connectors from the support to mid-span.

Longitudinal shear force transfer, R_q

R_q *(1 stud/rib)* $= 19 \times P_{Rd}\quad = 19 \times 62\,kN\quad = 1178\,kN$

R_q *(2 studs/rib)* $= 19 \times 2 \times P_{Rd} = 19 \times 2 \times 52\,kN = 1976\,kN$

Degree of shear connection, n/n_f (one stud per rib)

$$\eta = \frac{R_q}{R_c} = \frac{1178}{2550} = 0.46$$

Minimum degree of shear connection (one stud per rib)

Cl 6.6.1.2 (3)

$$\eta \ge 1 - \left(\frac{355}{f_y}\right)(1 - 0.04L) \ge 0.4$$

For $f_y = 355\,N/mm^2$ and $L = 12\,m$, $\eta > 0.48$. It follows that single shear connectors are not adequate.

Degree of shear connection, n/n_f (two studs per rib):

$$\eta = \frac{R_q}{R_c} = \frac{1976}{2550} = 0.70$$

Minimum degree of shear connection (two studs per rib)

$$\eta \ge 1 - \left(\frac{355}{f_y}\right)(0.75 - 0.03L) = 0.61\ (> 0.4)$$

The actual degree of shear connection of $\eta = 0.70$ exceeds this limit of 0.61.

Moment resistance for partial shear connection

Consider pairs of shear connectors per rib with a degree of shear connection, $\eta = 0.70$.

Example of 12 m span composite beam to BS EN 1994-1-1	Sheet 7 of 9	Rev

Use the simpler linear interaction method:

$$M_{Rd} = M_{a,pl,Rd} + \eta(M_{pl,Rd} - M_{a,pl,Rd})$$
$$= 585 + 0.70 \times (1022 - 585)$$
$$= 891\,kNm > 626\,kNm$$

Unity Factor = 626/891 = 0.70

Transverse reinforcement checks

Check the resistance of the concrete flange to longitudinal splitting. Use A142 mesh reinforcement in the slab.

Shear resistance per shear surface, V_{Rd}

Tensile resistance of reinforcement perpendicular to axis of beam

$F_t \quad = 142 \times 460 \times 10^{-3} = 65\,kN/m$

Longitudinal force per shear connector per plane (at 300 mm centres)

$F_L \quad = 2 \times 52/(0.3 \times 2) = 173\,kN/m$

$tan\theta = F_t/F_L = 65/173 = 0.37 < 0.5$ *not OK*

It follows that the amount of transverse reinforcement is not acceptable without considering the beneficial influence of the profiled decking. For pairs of shear connectors per rib, the effect of end anchorage is included with a minimum anchorage distance as follows:

BS EN 1994-1-1 Cl 9.7.4

$P_{pb,Rd} = 2.5 \times 19 \times 0.9 \times 350 \times 10^{-3} = 15\,kN$ *per stud or 50 kN/m*

$F_t \quad = 65 + 50 = 115\,kN/m$

$tan\theta \quad = F_t/F_L = 115/173 = 0.66 > 0.5 \qquad OK$

$\theta \quad = 26°$ *to the axis of the beam*

Compression force along the slab

$F_C \quad = 0.5F_L/cos\theta = 0.5 \times 173/cos26° = 96\,kN/m$

Compressive strength of concrete

$f_c \quad = 0.6\,(1 - f_{ck}/250)f_{ck}/\gamma_c$

$\quad = 0.6\,(1 - 25/250)25/1.5 = 9\,N/mm^2$

Compression resistance of slab

$F_C \quad = f_c h_c\,s\,sin\theta$

$\quad = 9 \times 60 \times sin26° \times 10^3 \times 10^{-3} = 236\,kN/m > 96\,kN/m \qquad OK$

It follows that A142 mesh is acceptable for this case. For the case where the profiled decking is discontinuous, A193 mesh should be used.

Serviceability Limit States

No stress checks are required in BS EN 1994 for normal conditions.

Example of 12 m span composite beam to BS EN 1994-1-1	Sheet 8 of 9	Rev

Non-composite stage deflection, δ_d

Self-weight of slab and beam $q_d = 2.81\,kN/m^2$

From properties in sheet 2, the deflection of the steel beam due to the self-weight of the slab and beam is:

$$\delta_d = \frac{5q_d b L^4}{384 E_a I_{ay}} = \frac{5 \times 2.81 \times 3.0 \times \left(12 \times 10^3\right)^4 \times 10^{-3}}{384 \times 210 \times 33320 \times 10^4}$$

$$= 32.5\,mm$$

Composite stage deflection

The second moment of area of the composite section, based on elastic properties (uncracked section), I_c is given by:

$$I_c = \frac{A_a \left(h + 2h_p + h_c\right)^2}{4(1 + nr)} + \frac{b_{eff} \times h_c^3}{12n} + I_{ay}$$

$$r = \frac{A_a}{b_{eff} \times h_c} = \frac{9460}{3000 \times 60} = 0.052$$

n = modular ratio = 10 for normal weight concrete subject to variable loads.

$$I_c = \frac{9460(457 + 2 \times 70 + 60)^2}{4(1 + 10 \times 0.052)} + \frac{3000 \times 60^3}{12 \times 10} + 33320 \times 10^4$$

$$= (6.71 + 0.05 + 3.33) \times 10^8 = 10.09 \times 10^8\,mm^4$$
(203% increase compared to the steel beam stiffness).

Imposed load deflection for full shear connection

Imposed load and services load = $5.5\,kN/m^2$

$$\delta_i = \frac{5q_i b L^4}{384 E_a I_c} = \frac{5 \times 5.5 \times 3 \times \left(12 \times 10^3\right)^4 \times 10^{-3}}{384 \times 210 \times 10.09 \times 10^8}$$

$$= 21.0\,mm$$

The imposed load part of this deflection is 19.1 mm.

For pairs of shear connectors per rib, the degree of shear connection is 70%, in which case, the effects of slip do not have to be considered.

No deflection limits are given in BS EN 1993-1-1 and the designer should refer to national requirements. The imposed load deflection is less than the deflection limit of L/360 in BS 5950-1, and indicates that the beam is acceptably stiff.

Total deflection

Construction stage = 32.5 mm
Imposed load   = 19.1 mm (no slip effect)
Ceiling and services = <u>1.9</u> mm
Total   53.5 mm [= L/224]

Example of 12 m span composite beam to BS EN 1994-1-1	Sheet 9 of 9	Rev

Normally, in UK practice, the limit on the maximum total deflection for a composite

beam is $\dfrac{L}{200}(=60\,mm)$, which is satisfactory and is within the range of adjustment on

the raised floor and suspended ceiling.

To BS EN 1993-1-1, the suggested total deflection limit is L/250= 48 mm, which is not satisfied using this beam. However, pre-cambering would not normally be considered for a beam with a span of 12 m. In this case, it may be appropriate to consider nominal partial fixity of the connections to reduce deflections.

Sensitivity to vibration

Slab and beam　　　　$= 2.81\,kN/m^2$

Ceiling and services　$= 0.50\,kN/m^2$

10% of imposed load $= \underline{0.50}\,kN/m^2$

Total　　　　　　$3.81\,kN/m^2$

In this simplified approach , it is appropriate to increase the inertia, I_o of the composite beams by 10% to allow for the increased dynamic stiffness of the composite beam, I_{cl}

$I_{cl} = 10.09 \times 10^8 \times 1.1 = 11.1 \times 10^8\,mm^4$

Simplified natural frequency

Instantaneous deflection caused by the self-weight of the composite slab and the beam plus 10% of the imposed load re-applied to the composite beam;

$$\delta_d = \frac{5q_d bL^4}{384\,E_a I_{cl}}$$

$$\delta_d = \frac{5 \times 3.81 \times 3.0 \times (12 \times 10^3)^4 \times 10^{-3}}{384 \times 210 \times 11.10 \times 10^8}$$

$$= 13.2\,mm$$

Natural frequency, $f = \dfrac{18}{\sqrt{\delta_{sw}}} = \dfrac{18}{\sqrt{13.2}} = 4.9\,Hz > 4\,Hz$　OK

This simplified check of natural frequency shows that the composite beam is satisfactory.

Conclusions

The beam size $457 \times 191 \times 74\,kg/m$ UKB in S355 steel is satisfactory for a 12 m span secondary beam. The Unity Factor on bending resistance is 0.61, when using S355 steel. Shear connectors may be placed in pairs per deck rib at 300 mm centres in which case, the degree of shear connection is 70%. The design is strongly influenced by the requirements for limitation of total deflections, rather than the bending resistance or other serviceability criteria. It would have been possible to use S275 steel for this secondary beam design.

The Response Factor method in Chapter 13 may be used to determine the acceptability regarding floor vibrations.

The Steel Construction Institute	Job No.		Sheet 1 of 8	Rev
	Job Title	Steel Designers Manual Chapter 22		
Silwood Park, Ascot, Berks SL5 7QN Telephone: (01344) 623345 Fax: (01344) 622944	Subject	Design Example for 15 m Span Continuous Composite Beam to BS EN 1994-1-1		
	Client		Made by	RML	Date	Nov 2009
CALCULATION SHEET			Checked by		Date	

	BS EN 1994-1-1 Clauses
Design Example of 15 m Span Continuous Composite Beam to BS EN 1994-1-1	
Consider a continuous secondary composite beam of 15 m span and subject to uniform loading. Choose a 457 × 191 × 74 kg/m UKB in S355 steel, based on the 12 m span simply supported beam worked example. The same beam size has been chosen to simplify this Worked Example, but it may be necessary to increase the beam size to satisfy vibration response. The design of continuous beams introduces new design checks for lateral torsional buckling in the construction stage and at the ultimate limit state. Construction stage checks are carried out to BS EN 1993-1-1.	
Design Criteria	
Use the same loading and design criteria as for the simply supported beam example.	
Basis of Design	
The bending moments acting on the continuous beam and the resulting design checks are established as follows:	
• In construction, the negative (hogging) bending moment at the internal support is determined elastically for the case of both spans loaded by the self-weight of the beam and slab and a construction load of 0.75 kN/m². Partial factors of 1.5 are used for all these variable loads during construction, as required by BS EN 1991-1-6.	
• For consideration of stability of the beam under negative bending, the critical case is that when one span is loaded, and the bottom flange of the other span is unrestrained over its length.	
• Under factored loading acting on the composite beam, the negative bending resistance is determined from the reinforcement placed across the effective slab width in the support region. The plastic collapse load of the semi-continuous beam is based on the combined negative and positive bending resistances and is compared to the 'free' bending moment acting on the beam. This is consistent with a 40% redistribution of negative moment from the supports to mid-span, which is permitted for Class 1 sections.	5.4.5 Table 5.1
The effective span for consideration of the effective width of the slab in positive (sagging) bending is taken as 0.8× span, which means that the properties of the composite section are the same as for the 12 m span simply supported beam.	Cl 5.4.2.1(5)
Construction stage design	
<u>***Ultimate limit state loading:***</u>	
Dead load factor γ_G *= 0.925 × 1.35 = 1.25*	Table 2.2
Imposed load factor $\gamma_Q = 1.5$	
Slab and beam = 2.81 × 1.25 = 3.51 kN/m²	

Example of 15 m span continuous composite beam to BS EN 1994-1-1	Sheet 2 of 8	Rev

Construction load $= 0.75 \times 1.5$ $= \underline{1.13}\,kN/m^2$

Total factored load $4.64\,kN/m^2$

Consider the case where one span is loaded during construction

Design moment $= M_{Ed} = \dfrac{4.64 \times 3 \times 15^2}{16} = 196\,kNm$

Check bending resistance of steel beam which is Class 1:

$M_{a,pl,Rd} = 404\,kNm > 196\,kNm$

For the purpose of this design example, the beam is stable when both spans are loaded.

BS EN 1993-1-1 5.4.5.2

See Simply Supported Worked Example

Lateral torsional buckling (LTB) check during construction

Consider the stability of the bottom flange of the beam when one span is loaded with no intermediate lateral restraint.

Slenderness ratio for LTB:

$$\bar{\lambda}_{LT} = C_1^{-0.5}\, uv\, \bar{\lambda}_z \sqrt{\beta_w}$$

BS EN 1993-1-1

Assume the bending moment diagram in the unloaded span decreases linearly from the internal to zero at the external support. Therefore the factor for the variation of moment is:

$$C_1^{-0.5} = 1.33^{-1} = 0.75$$

Table 6.6

Data for $457 \times 191 \times 74\,kg/m$ UKB section:

u $= 0.9$ *for hot-rolled steel beams*

$h_a/t_f = 457/14.5$ $= 31.5$

v $= (1 + 0.05\ (h_s/t)^2)^{-0.25}$

         $= (1 + 0.05\ (31.5)^2)^{-0.25}$ $= 0.37$

λ_1 $= \pi\,(205000/355)^{0.5}$ $= 75.4$

λ_z $= 15000/42$ $= 357$

$\bar{\lambda}_z$ $= 357/75.4$ $= 4.73$

$\bar{\lambda}_{LT}$ $= 0.75 \times 0.9 \times 0.37 \times 4.73 = 1.18$

The bending strength reduction factor due to LTB is given by:

$$\chi_{LT} = \left[\phi_{LT} + \left(\phi_{LT}^2 - \bar{\lambda}_{LT}^2 \right)^{0.5} \right]^{-1}$$

where: $\varphi_{LT} = 0.5\ [1 + \alpha_{LT}\ (\bar{\lambda}_{LT} - 0.2) + \bar{\lambda}_{LT}^2]$

BS EN 1993-1-1 Cl 6.3.2.3

Example of 15 m span continuous composite beam to BS EN 1994-1-1	Sheet 3 of 8	Rev

and $\alpha_{LT} = 0.34$ for buckling curve b

$\varphi_{LT} = 0.5 \ [1 + 0.34 \ (1.18 - 0.2) + 1.18^2] = 1.36$

$\chi_{LT} = [1.36 + (1.36^2 - 1.18^2)^{0.5}]^{-1} = 0.49$

This just exceeds the design moment $M_{Ed} = 196 \, kN/m$

BS EN 1993-1-1

Bending resistance $M_{b,Rd} = 0.49 \times 404 = 198 \, kNm > 196 \, kNm$

Composite stage design

<u>Factored loading</u>

The composite design stage includes the factored imposed and self-weight loads and other permanent loads, as follows:

Slab and beam	*= 2.81 × 1.25*	*= 3.51 kN/m²*
Services and ceiling	*= 0.50 × 1.25*	*= 0.63 kN/m²*
Imposed load	*= 5.0 × 1.5*	*= <u>7.50 kN/m²</u>*
		11.6 kN/m²

Shear force $= V_{Ed} = 11.6 \times 3 \times 15/2 \ = 261 \, kN$

Design moment $= M_{Ed} = \dfrac{11.6 \times 3 \times 15^2}{8} \ = 979 \, kNm$

<u>Effective width of concrete flange, b_{eff}</u>

Effective span $l_e = 0.8l$ in positive bending

5.4.1.2(5)

$b_{eff} = \dfrac{2 \times 0.8 l_e}{8} \ $ (for a simply supported beam, $l_0 =$ span)

$= \dfrac{2 \times 0.8 \times 15}{8} = 3m$

Effective span $l_e = 0.5 l_0$ in negative bending

$b_{eff} = 2 \times (0.5 \times 15)/8 = 1.87 \, m$

The negative moment reinforcement in the slab is distributed over this effective width of 1.85 m.

Moment resistance for full shear connection

$M_{aplRd} = 1022 \, kNm > 979 \, kNm$

However, consider partial shear connection as follows:

Example of 15 m span continuous composite beam to BS EN 1994-1-1	Sheet 4 of 8	Rev

Shear connector resistance

$P_{Rd} = 0.72 \times 73 = 52\,kN$ for 2 shear connectors per rib.

A total of 19 deck ribs are available for the positioning of the shear stud connectors from the support to 0.8× span (over 12 m).

R_q (2 studs/rib) $= 19 \times 2 \times P_{Rd} = 19 \times 2 \times 52\,kN = 1976\,kN$

Degree of shear connection, n / n_f

Degree of shear connection n/n_f (two studs per rib):

$$\eta = \frac{R_q}{R_c} = \frac{1976}{2550} = 0.77$$

Minimum degree of shear connection (two studs per rib)

$$\eta \geq 1 - \left(\frac{355}{f_y}\right)(0.75 - 0.03L) = 0.61 > 0.4 \text{ for } L = 12\,m \text{ as simply supported beam}$$

Bending resistance for partial shear connection
$M_{pl,Rd} = 404 + 0.77 \times (1022 - 404)$
$\qquad = 880\,kNm$

Shear resistance, $V_{pl,Rd}$

$$V_{pl,Rd} = A_v \frac{f_{yd}}{\sqrt{3}}$$

where $A_v = h\,t_w$ (as a simplification)

$$V_{pl,Rd} = \frac{457 \times 9.0 \times 355}{\sqrt{3} \times 10^3}$$

$$\qquad = 843\,kN > V_{Ed} = 261\,kN$$

BS EN 1993-1-1 Cl 6.2.8

NB: For a continuous beam, shear at the internal support may influence the negative (hogging) moment resistance of the section but, in this case, the unity factor is less than 0.5 and so shear effects may be ignored.

Moment resistance in negative (hogging) bending

Use 20 mm dia. reinforcing bars (in S460 steel) at 200 mm centres which are placed over an effective slab width of 1875 mm. This corresponds to 10 bars with a cross-sectional area of: $10 \times 3.14 \times (20^2/4) = 3140\,mm^2$.

Tensile resistance of reinforcement, $R_r = 3140 \times 460 \times 10^{-3} = 1444\,kN$

Depth of web in compression, $y_w = d/2 + R_r/(2t_w f_y)$

$y_w = 0.5 \times 407 + 1444 \times 10^3/(2 \times 9.0 \times 355) = 429\,mm > d = 407\,mm$

It follows that the whole of the web is in compression, and the plastic neutral axis lies in the top flange. However, it is necessary to take into account local buckling of the web.

Example of 15 m span continuous composite beam to BS EN 1994-1-1	Sheet 5 of 8	Rev

The plastic bending resistance in negative bending is given as follows:

Limiting depth of Class 3 web in compression
$= 40 t_w \ \varepsilon = 40 \times 9.0 \times 0.81 = 291\,mm < 407\,mm$

Reduced compression resistance of web

$R_{w,red} = 291 \times 9.0 \times 355 \times 10^{-3} = 931\,kN$

Plastic bending resistance of reduced section (PNA in top flange)

$M_{pl,Rd} = 0.5(h_c + h_p)R_r + h_s(R_f + 0.5R_{w,ned})$

where $R_f = 0.5 \times (3358 - 9.0 \times 407 \times 355 \times 10^{-3}) = 1029\,kN$

$M_{pl,Rd} = 100 \times 1444 \times 10^{-3} + 457 \times (1029 + 0.5 \times 931) \times 10^{-3}$

$\qquad = 144.4 + 682.7 = 827\,kNm$

> *BS EN 1993-1-1 Cl 6.2.2.4*

Plastic failure load of continuous composite beam

The plastic failure load of the semi-continuous beam may be determined from the formula for the end span case:

$$M_{pl,Rd} = M_{pl,Rd,p} + 0.5 M_{pl,Rd,n}\left(1 - \frac{M_{pl,Rd,n}}{q_u L^2}\right) \ge q_u L^2 / 8$$

or: $M_{pl,Rd} = 880 + 0.5 \times 827 \times (1 - 827/(11.8 \times 3 \times 15^2))$

$\qquad = 1250\,kNm \ge 979\,kNm \qquad\qquad OK$

> *Worked Example 1 for $M_{pl,\,Rd}$*

Shear connection in negative bending region

Tensile force in reinforcement is developed over a length of :
$2M_{pl,Rd,n}/(qbL) = 3.1\,m$ *(or 20% of span)*

Number of rib positions = 10 for shear connectors in pairs

Shear connector resistance $= 10 \times 2 \times 52 = 1040\,kN < 1444\,kN$

Partial shear connection exists in the negative moment region. (Degree of shear connection = 0.72).

Reduced negative bending resistance, $M_{pl,Red,Rd} = 144.4 \times 0.72 + 682.7 = 786\,kNm$

Modified plastic bending resistance of semi-continuous beam = 1230 kNm > 979 kNm

> *Above Equation*

Lateral torsional buckling check for composite beam

Consider distortional buckling of the bottom flange and web of the section with no deformation of the slab.

The slenderness ratio for LTB of the bottom flange of the composite section as a strut is reduced by the factor:

$$\bar{\lambda}_{LT,mod} = \bar{\lambda}_{LT}\left[1 + \frac{k_s}{EI_f}\left(\frac{L}{\pi}\right)^4\right]^{-0.5}$$

> *6.4.1(6)*

Example of 15 m span continuous composite beam to BS EN 1994-1-1	Sheet 6 of 8	Rev

where k_s is the bending stiffness of the web

$$k_s = \frac{Et_w^3}{4\,h_s^3}$$

The reduction factor on the slenderness of the bottom flange is now given by:

$$\bar{\lambda}_{LT,mod} = \bar{\lambda}_{LT}\left[1 + 0.03\left(\frac{t_w}{t_f}\right)^3 \left(\frac{t_f}{b}\right)^2 \left(\frac{h_s}{b}\right)\left(\frac{L}{h_s}\right)^4\right]^{-0.5}$$

Inserting the parameters for the UKB:

$$\bar{\lambda}_{LT,mod} = \bar{\lambda}_{LT}[1 + 0.03 \times 0.62^3 \times 13.1^{-2} \times 2.4 \times 32.8^4]^{-0.5}$$

$$= 0.092\,\bar{\lambda}_{LT}$$

$\bar{\lambda}_{LT} = C_1^{-0.5}\,uv\,\bar{\lambda}_z$ *(v = 1.0 for distortional buckling)*

$$= 0.75 \times 0.9 \times (15000/42)/75.4 \qquad = 3.19$$

$$\bar{\lambda}_{LT,mod} = 0.092 \times 3.19 = 0.29$$

$\varphi_{LT} \quad = 0.5\,[1 + 0.34\,(0.29 - 0.2) + 0.29^2] \ = 0.56$

$\chi_{LT} \quad = [0.56 + (0.56^2 - 0.29^2)^{0.5}] = 1.04 > 1.00$

Buckling resistance moment, $M_{pl,Red,Rd}$ $\quad = 827\,kNm$	*Sheet 5*
Applied negative bending moment for one span loaded	*Sheet 5*
$M_{Ed} = 0.5 \times 979 = 490 < 827\,kN/m$ $\qquad$ OK	
Unity factor in LTB = 0.59	*Sheet 3*

Serviceability Limit State

No stress checks are required in BS EN 1994-1-1 for normal conditions.

<u>**Non-composite stage deflection, δ_d**</u>

Self-weight of slab and beam, $w_d = 2.81\,kN/m^2$

From properties in sheet 2, the deflection of the continuous steel beam due to the self-weight of the slab and beam is:

$$\delta_d = \frac{2.1w_d b L^4}{384EI} = \frac{2.1 \times 2.81 \times 3.0 \times (15 \times 10^3)^4 \times 10^{-3}}{384 \times 210 \times 33320 \times 10^4}$$

$$= 33.3\,mm$$

Example of 15 m span continuous composite beam to BS EN 1994-1-1	Sheet 7 of 8	Rev

Composite stage deflections

Imposed load $w_i = 5.0 \, kN/m^2$

$I_c = 10.09 \times 10^8 \, mm^4$ (203% increase on steel beam)

Imposed load deflection for full shear connection

Imposed load and services load $= 5.5 \, kN/m^2$

$$\delta_d = \frac{2.1 w b_i L^4}{384 E_a I_c} = \frac{2.1 \times 5.5 \times 3 \times \left(15 \times 10^3\right)^4 \times 10^{-3}}{384 \times 210 \times 10.09 \times 10^8}$$

$$= 21.5 \, mm$$

The imposed load part of this deflection is 19.6 mm.

The degree of shear connection is 77% for pairs of shear connectors per rib, in which case, the effects of slip do not have to be considered.

No deflection limits are given in BS EN 1993-1-1 and the designer should refer to national requirements, in this case, BS 5950-1. This imposed load deflection is much less than the deflection limit of L/360 in BS 5950-1, and indicates that the beam is acceptably stiff.

Total deflection

Construction stage $= 33.2 \, mm$
Imposed load $= 19.5 \, mm$ (no slip effect)
Ceiling and services $= \underline{2.0} \, mm$
Total $54.7 \, mm \; [= L/274] < L/250$ OK

Vibration sensitivity : Simplified approach

Permanent loading

Slab and beam $= 2.81 \, kN/m^2$

Ceiling and services $= 0.50 \, kN/m^2$

10% of imposed load $= \underline{0.50} \, kN/m^2$

Total $3.86 \, kN/m^2$

Increase the inertia, I_o by 10% to allow for the increased dynamic stiffness of the composite beam, I_{cl}

$I_{cl} = 10.09 \times 10^8 \times 1.1 = 11.10 \times 10^8 \, mm^4$

Example of 15 m span continuous composite beam to BS EN 1994-1-1	Sheet 8 of 8	Rev

Simplified natural frequency

Instantaneous deflection caused by the self-weight of the floor and the beam and 10% of the imposed load reapplied to the composite beam. However, for a continuous beam, anti-symmetric movement of adjacent spans is considered in dynamic cases, in which case, the deflection formula is:

$$\delta_d = \frac{5_{wd} bL^4}{384\, E_a I_{cl}}$$

$$\delta_d = \frac{5 \times 3.86 \times 3.0 \times (15 \times 10^3)^4 \times 10^{-3}}{384 \times 210 \times 11.10 \times 10^8}$$

$$= 32.7\, mm$$

Natural frequency, $f = \dfrac{18}{\sqrt{\delta_{sw}}} = \dfrac{18}{\sqrt{32.7}} = 3.1\, Hz < 4\, Hz\ not\ OK$

This simplified natural frequency check shows that the composite beam is not satisfactory according to the limit. However, the natural frequency exceeds the minimum system frequency of 3 Hz specified in the Response Factor approach in SCI publication 354 and in the method presented in Chapter 13. of this publication.

Conclusions

The beam size $457 \times 191 \times 74\, kg/m$ UKB in S 355 steel is satisfactory for 15 m span continuous secondary beam, except for the simplified natural frequency check. The Unity Factor on bending resistance is 78%. The beam is stable during construction and in the composite stage and requires no further lateral restraints. Shear connectors are placed in pairs per deck rib at 300 mm centres in which case, the degree of shear connection of 77% is acceptable.

The design is strongly influenced by the requirements for control of vibrations rather than the bending resistance or other serviceability criteria. The use of the response factor method is recommended to demonstrate that vibration sensitivity is acceptable. Alternatively, a heavier beam in this serial size may be chosen.

Chapter 23
Composite columns

KWOK-FAI CHUNG and MARK LAWSON

23.1 Introduction

Steel-concrete composite columns are compression members in the form of concrete encased H-sections or concrete-filled hollow sections. A typical cross-section of a composite column with a steel H-section that is fully encased in concrete is shown in Figure 23.1(a). A typical cross-section of a composite column with a concrete partially encased H-section is shown in Figure 23.1(b). A typical cross-section of a composite column with a partially encased cruciform I-section is also shown in Figure 23.1(c). Figure 23.2 shows typical cross-sections of composite columns with concrete-filled hollow sections.

The early development of composite columns was based on the need to provide effective fire protection by encasing steel stanchions in concrete. Increases in strength and stiffness due to concrete encasement were ignored, although it was recognised that the buckling resistances of encased columns were increased. An example of this is the so-called 'encased strut' method in BS 449. By the early 1960s, research studies showed that concrete encasement increased the axial resistances of steel columns significantly, which led to the development of composite columns of all types.

The advantages of composite columns are:

- increased resistance for a given member size, leading to economy in the use of the steel sections
- increased stiffness, leading to reduced slenderness and increased buckling resistance
- improved connection behaviour with attachments made via the steel sections
- good fire resistance
- excellent corrosion protection of encased columns
- enhanced seismic resistance.

The first Code in the U.K. to provide design guidance for composite columns was BS 5400: Part 5.[1] BS EN 1994-1-1 Eurocode 4[2] presents the latest design

Steel Designers' Manual, Seventh Edition. Edited by Buick Davison and Graham W. Owens.
© 2012 Steel Construction Institute. Published 2012 by Blackwell Publishing Ltd.

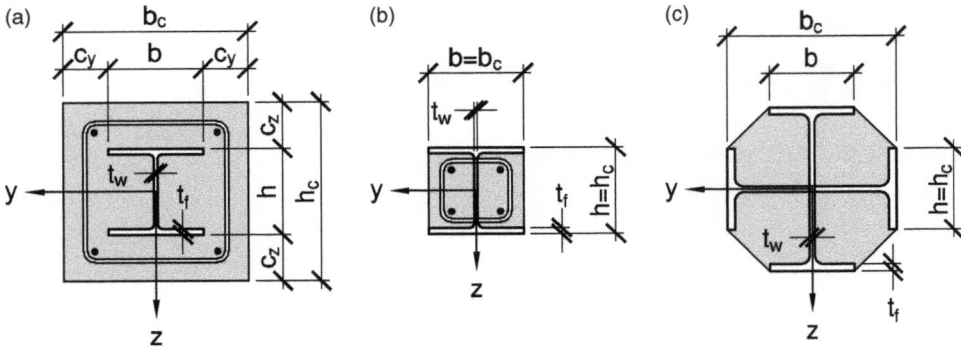

Figure 23.1 Typical cross-sections of composite columns with fully or partially concrete-encased H-sections

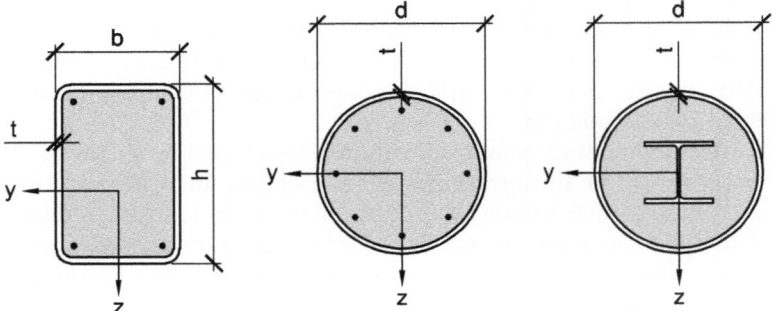

Figure 23.2 Typical cross-sections of composite columns with concrete-filled hollow sections

recommendations for all types of composite columns. This chapter reviews the Eurocode method for composite columns of concrete encased H-sections and concrete filled hollow sections. A previous SCI publication[3] presented guidance to the EN version of BS EN 1994-1-1.

For further information, refer to *Designers' Guide to EN 1994-1-1*[4] by Johnson and Anderson. It should also be noted that the UK National Annex to Eurocode 4[5] was published in 2008. A detailed two-part investigation into the design methodology on composite columns was reported by Leon *et al.*[6] in 2007 and Leon and Hajjar[7] in 2008.

23.2 Design of composite columns

Clause 6.7 of BS EN1994-1-1 presents the design of composite columns and composite compression members with concrete fully and partially encased H-sections, and concrete filled rectangular and circular hollow sections. It is applicable to columns and compression members with steel grades S235 to S460 and normal weight concrete of strength classes C20/25 to C50/60.[8]

In general, a composite column should be checked at the ultimate limit state for:

- geometric limits of various elements of the steel sections against local buckling under compression
- resistances of cross-sections and members to internal forces and moments
- buckling resistance of the members, depending on their effective slenderness
- local resistances to interfacial shear forces between the steel sections and the concrete
- local resistances of the cross-sections at load introduction points.

23.2.1 Design methods

Two methods of design for isolated composite columns in braced or non-sway frames are given in BS EN 1994-1-1, Clause 6.7:

- general design method for composite columns applicable to both prismatic and non-prismatic members with either symmetrical or non-symmetrical cross-sections
- simplified design method specifically developed for prismatic composite columns with doubly symmetrical cross-sections.

The use of the simplified design method is presented in detail in this Chapter. It should be noted that when the limits of applicability of this method are not satisfied, the general design method should be used.

23.2.2 Fire resistance

In general, composite columns possess much higher fire resistances than the parent steel columns. Composite columns are usually designed in the normal (cool) state and then checked under fire conditions. Additional reinforcement may be provided to achieve the required fire resistance of the columns.

Design guidance on the fire resistances of composite columns may be found in BS EN1994-1-2[9] and BS 5950-8.[10] General rules on the structural performance of composite columns under fire conditions are summarised as follows:

- The fire resistance of composite columns with full encasement of H-sections may be treated in the same way as reinforced concrete columns. The steel sections are insulated by the concrete cover, and light reinforcement is also required in order to maintain integrity of the concrete cover. In such cases, a fire resistance period of 120 minutes may be achieved with a minimum concrete cover of 40 mm.

- For composite columns with partial encasement of H-sections, the structural performance of the columns under fire conditions is different from that of reinforced concrete columns, as the flanges of the steel sections are exposed and a smaller amount of concrete acts as a 'heat sink'. In general, a fire resistance of up to 60 minutes may be achieved provided that the strength of the concrete is neglected in normal design. For composite design, additional reinforcement is required to achieve a fire resistance period longer than 60 minutes. Normally, the reinforcing bars are fixed by shear links welded onto the web of the steel section.

- For concrete-filled hollow sections under fire conditions, the steel sections are exposed to direct heating while the concrete core behaves as a 'heat sink'. In general, sufficient redistribution of internal stress occurs between the hot steel sections and the relatively cool concrete core so that a fire resistance period of 60 minutes may usually be achieved. For longer periods of fire resistance, additional reinforcement is required, which should be neglected in normal design. Steel fibre reinforcement is also effective in improving the fire resistance of composite concrete-filled hollow sections.

23.3 Simplified design method

The simplified method has been developed specifically for prismatic composite columns with doubly symmetrical cross-sections with hot-rolled, cold-formed or welded steel sections.

To prevent local buckling, the width-to-thickness ratio of various elements in the steel sections in compression must satisfy the following limits, given in Table 6.3 of BS EN 1994-1-1, as follows:

- $\dfrac{d}{t} \le 90\,\varepsilon^2$ for concrete-filled circular hollow sections   (23.1a)

- $\dfrac{h}{t} \le 52\,\varepsilon$ for concrete-filled rectangular hollow sections   (23.1b)

- $\dfrac{b}{t_{\mathrm{f}}} \le 44\,\varepsilon$ for partially concrete-encased H-sections   (23.1c)

where:

$$\varepsilon = \sqrt{\frac{235}{f_y}}$$ and f_y is the yield strength of the steel section in N/mm^2.

For fully encased sections, local buckling in the steel sections is not possible, and hence, verification for local buckling is not necessary.

23.3.1 Compression resistances of cross-sections

The plastic resistance of a composite cross-section in compression represents the maximum load that can be applied to a short column which does not exhibit member buckling. It should be noted that, in concrete-filled hollow sections, a higher compression resistance is achieved in the concrete owing to the confinement provided by the hollow section. Moreover, further strength enhancement is achieved in concrete-filled circular hollow sections owing to the development of circumferential action in the circular hollow section.

23.3.1.1 Concrete encased H-sections and concrete-filled rectangular hollow sections

The plastic resistance of a concrete-encased H-section or a concrete-filled rectangular or square hollow section in compression is given by the sum of the resistances of the components in compression, defined in BS EN 1994-1-1, Clause 6.7.3.2, as follows:

$$N_{pl,Rd} = A_a f_{yd} + \alpha_c A_c f_{cd} + A_s f_{sd} \tag{23.2}$$

where:

A_a, A_c and A_s are the cross-sectional areas of the steel section, the concrete and the reinforcing bars respectively

f_{yd}, f_{cd} and f_{sd} are the design strength of the steel section, the design compressive strength of the concrete, and the design strength of the reinforcement bars respectively which are given by:

$$f_{yd} = \frac{f_y}{\gamma_a}; \quad f_{cd} = \frac{f_{ck}}{\gamma_c}; \quad f_{sd} = \frac{f_{sk}}{\gamma_s}$$

f_y, f_{ck} and f_{sk} are the yield strength of the steel section, the characteristic compressive strength of the concrete and the yield strength of the reinforcing respectively

γ_a, γ_c and γ_s are the material factors of the steel section, the concrete and the reinforcing steel respectively

α_c is the strength coefficient for concrete, which is equal to 1.0 for concrete-filled rectangular or square hollow sections, and 0.85 for fully or partially encased H-sections.

Figure 23.3 shows the idealised stress distribution on which Equation 23.2 is based.

An important design parameter is the steel contribution ratio, δ, which is defined as follows:

$$\delta = \frac{A_a f_{yd}}{N_{pl.Rd}} \tag{23.3}$$

It is important to note that δ should lie within 0.2 and 0.9.

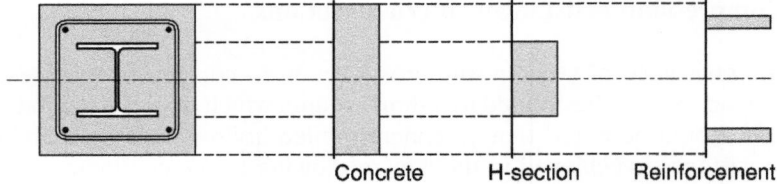

Concrete   H-section   Reinforcement

Figure 23.3   Stress distribution of the plastic resistance of a fully concrete-encased H-section

23.3.1.2 Concrete-filled circular hollow sections

For composite columns with concrete-filled circular hollow sections, the increased compression resistance of the concrete due to the confinement provided by the circular hollow section should be included. It should be noted that these enhancement effects depend on the slenderness of the columns, and they are only significant in stocky columns. In addition, the eccentricity, e, of the applied load should not exceed $0.1d$ where d is the outer dimension of the circular hollow section.

The plastic resistance of concrete-filled circular hollow sections in compression to BS EN 1994-1-1, Clause 6.7.3.2(6) is given by:

$$N_{pl,Rd} = \eta_a A_a \, f_{yd} + \left[1 + \eta_c \frac{t}{d} \frac{f_y}{f_{ck}}\right] A_c f_{cd} + A_s f_{sd} \tag{23.4}$$

where:

$$\eta_a = \eta_{a0} + (1 - \eta_{a0}) \frac{10\,e}{d} \tag{23.5a}$$

$$\eta_c = \eta_{c0} \left[1 - \frac{10\,e}{d}\right] \tag{23.5b}$$

The basic values η_{a0} and η_{c0} depend on the relative slenderness $\bar{\lambda}$ and are given by:

$$\eta_{a0} = 0.25(3 + 2\bar{\lambda}) \text{ but } \leq 1{,}0 \tag{23.6a}$$

$$\eta_{c0} = 4.9 - 18.5\,\bar{\lambda} + 17\bar{\lambda}^2 \text{ but } \geq 0 \tag{23.6b}$$

The relative slenderness $\bar{\lambda}$ is defined in Section 23.3.5.2. If the eccentricity e exceeds the value $0.1d$, or if $\bar{\lambda}$ exceeds 0.5 then $\eta_c = 0$ and $\eta_a = 1.0$ Table 23.1 presents the values of η_{a0} and η_{c0} for different values of $\bar{\lambda}$.

Table 23.1 Values of η_{ao} and η_{co} to allow for confinement in concrete-filled circular hollow sections

Parameter	$\bar{\lambda} = 0$	$\bar{\lambda} = 0.1$	$\bar{\lambda} = 0.2$	$\bar{\lambda} = 0.3$	$\bar{\lambda} = 0.4$	$\bar{\lambda} = 0.5$
η_{ao}	0.75	0.80	0.85	0.90	0.95	1.00
η_{co}	4.90	3.22	1.88	0.88	0.22	0.00

23.3.2 Bending resistances of cross-sections

The plastic resistance of a composite cross-section in bending is given by:

$$M_{pl,\mathrm{Rd}} = f_y(W_p - W_{pn}) + 0.5\alpha_c\, f_{cd}(W_{pc} - W_{pcn}) + f_{sd}(W_{ps} - W_{psn}) \tag{23.7}$$

where:

α_c $\qquad\qquad$ = 0.85 for fully or partially concrete-encased H-sections
$\qquad\qquad\qquad$ = 1.0 for concrete-filled rectangular or square hollow sections

W_p, W_{pc}, W_{ps} $\quad$ are the plastic section moduli for the steel section, the concrete, and the reinforcing bars of the composite cross-section respectively (for the calculation of W_{pc}, the concrete is assumed to be uncracked)

W_{pn}, W_{pcn}, W_{psn} $\quad$ are the plastic section moduli of the corresponding components within the region of $2h_n$ from the middle line of the composite cross-section

h_n $\qquad\qquad$ is the depth of the plastic neutral axis from the centre line of the cross-section.

23.3.3 Shear resistance of cross-sections

In general, the applied shear force, V_{Ed}, may be conservatively assumed to be resisted entirely by the steel section. No reduction to the resistances of the cross-section in compression and in bending are needed when V_{Ed} is smaller than half of the shear resistance of the steel section, i.e. $0.5V_{a,\mathrm{Rd}}$.

However, when $V_{\mathrm{Ed}} > 0.5V_{a,\mathrm{Rd}}$, the resistances of the cross-section in compression and in bending should be evaluated according to a reduced design strength $(1 - \rho)$ f_{yd} in the shear area, A_v of the steel section in accordance with Clause 6.2.2.4(2), and ρ is given by:

$$\rho = \left(2\frac{V_{\mathrm{Ed}}}{V_{a,\mathrm{Rd}}} - 1\right)^2 \tag{23.8}$$

When $V_{\mathrm{Ed}} > V_{a,\mathrm{Rd}}$, it is possible to allocate a proportion of V_{Ed} to be resisted by the concrete; refer to Clause 6.7.3.2 (4) for details.

23.3.4 Resistances of cross-sections in combined compression and bending

The resistance of composite cross-sections to combined compression, N, and uni-axial bending, M, and the corresponding non-linear N-M interaction curve are evaluated according to rectangular stress blocks in various components of the cross-sections.

It should be noted that in a typical N-M interaction curve of a steel section, its moment resistance undergoes an almost linear reduction with increasing axial load, as shown in Figure 23.4(a). However, as shown in Figure 23.4(b), a composite cross-section may exhibit significant increases in its moment resistance in the presence of axial load. This is because, under some favourable conditions, the compressive axial load will prevent concrete cracking, and enable the composite cross-section to be more effective in resisting moments.

Such a non-linear N-M interaction curve for a composite cross-section may be readily simplified into a multi-linear interaction curve with 3 to 5 key points, as shown in Figure 23.5.

The coordinates of these key points of the multi-linear interaction curve are determined from the internal forces and moments, based on the rectangular stress

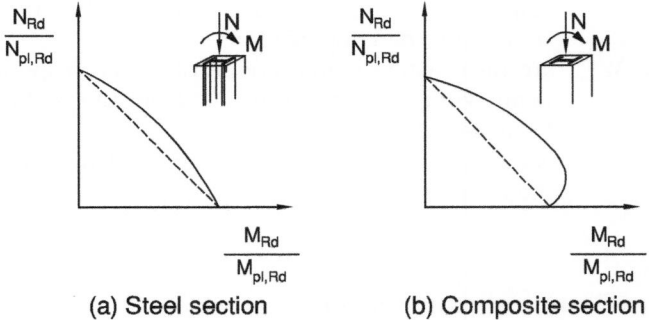

(a) Steel section   (b) Composite section

Figure 23.4   Typical N-M interaction curves for combined compression and uni-axial bending

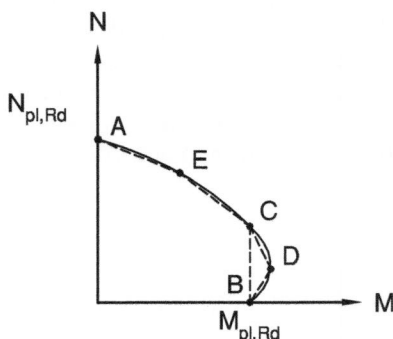

Figure 23.5   Typical multi-linear N-M interaction curve

blocks of the various elements of the composite cross-section with different positions of the neutral axis of the cross-section, h_n, shown in Figure 23.6.

- Point A defines the plastic compression resistance of the cross-section:

$$N_A = N_{pl} \tag{23.9}$$

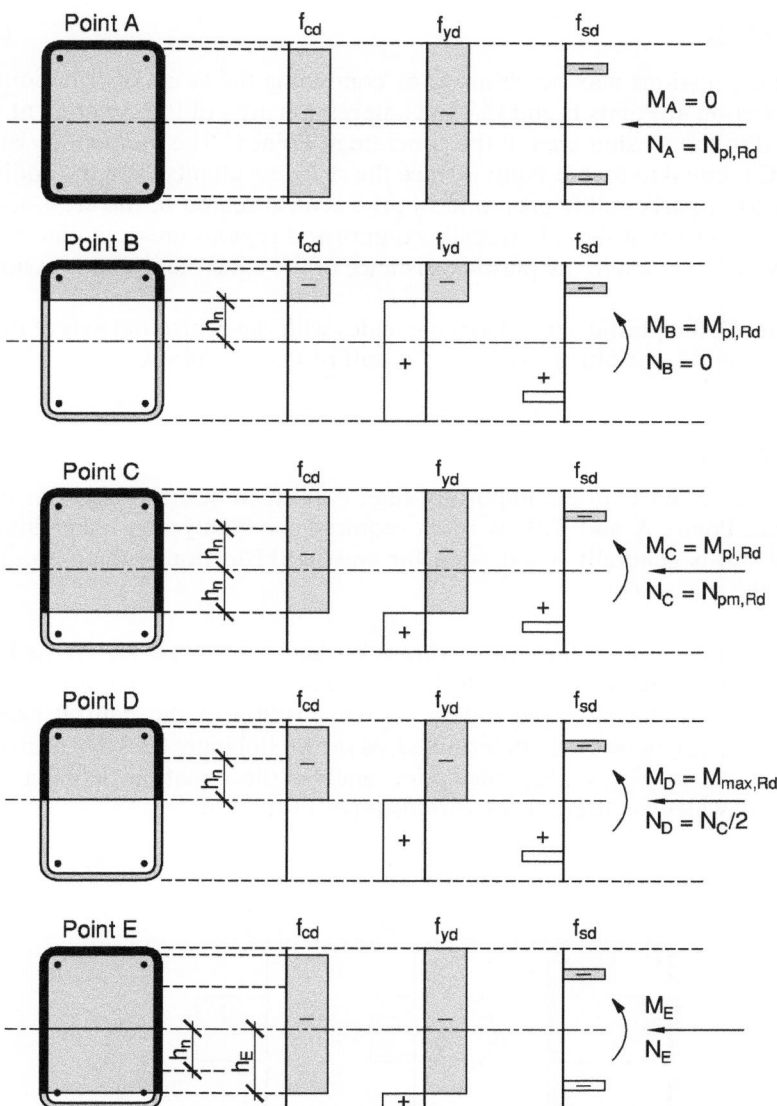

Figure 23.6   Rectangular stress blocks for various key points of the N-M multi-linear interaction curve for a concrete-filled rectangular hollow section

- Point B corresponds to the plastic bending resistance of the cross-section:

$$M_B = M_{pl,Rd} \tag{23.10}$$

- At Point C, the plastic resistances of the cross-section in compression and in bending are given respectively as follows:

$$N_C = N_{pm,Rd}(\text{or } N_{c,Rd}) = A_c f_{cd} \tag{23.11a}$$

$$M_C = M_{pl,Rd} \tag{23.11b}$$

The expressions may be obtained by combining the stress distributions of the cross-section at Points B and C. The compression area of the concrete at Point B is equal to the tension area of the concrete at Point C. The moment resistance at Point C is equal to that at Point B since the stress resultants from the additionally compressed parts nullify each other in the central region of the cross-section. It may be shown that the additionally compressed regions create an internal axial force which is equal to the plastic resistance of the concrete in compression, $N_{pl,Rd}$ or $N_{c,Rd}$.

- At Point D, the plastic neutral axis coincides with the centroidal axis of the cross-section, and the resulting axial force is half of that at Point C.

$$N_D = M_{pm\,Rd}/2 \tag{23.12a}$$

$$M_D = M_{max\,Rd} \tag{23.12b}$$

- In general, Point D is less important than Point C in design. Point E is mid-way between Points A and C. It is often required for highly non-linear interaction curves, but is generally not needed for encased H-sections subject to moments about the major axis.

It is important to note that the positions of the neutral axes for Points B and C, h_n, can be determined from the difference in stresses at Points B and C. The resulting axial forces, which are dependent on the position of the neutral axis of the cross-section, h_n, can be easily determined, as shown in Figure 23.7. The sum of these forces is equal to $N_{pm,Rd}$. This calculation enables the equation defining h_n to be determined, which is different for various types of sections.

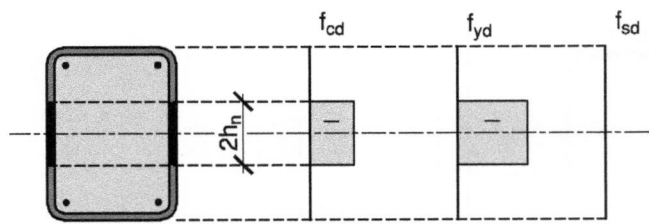

Figure 23.7 Variation in the position of the plastic neutral axis around the centre-line of the column

23.3.5 Buckling resistance of columns

Composite columns may fail in buckling due to the second-order or '*P-Δ*' effects, and it is possible to evaluate their member resistances through the conventional use of buckling curves. In this approach, the conventional column buckling concept as given in BS EN 1993-1-1 Eurocode 3[11] is adopted, and the member resistance of a composite column depends on its relative slenderness and use of an appropriate column buckling curve. Hence, the buckling resistance of the composite column is determined as a reduction to the plastic resistance of the composite cross-section, according to its slenderness.

23.3.5.1 Critical buckling load

It is important to evaluate the elastic critical buckling load, N_{cr}, of the composite column which is defined in BS EN 1994-1-1, Clause 6.7.3.3(3), as follows:

$$N_{cr} = \frac{\pi^2 (EI)_{eff}}{l^2} \tag{23.13}$$

where:

l	is the buckling length of the column
$(EI)_{eff}$	is the characteristic value of the effective flexural stiffness of the composite column, and it is obtained by combining the flexural stiffness of various components of the cross-section:

$$(EI)_{eff} = E_a I_a + 0{,}6 E_{cm} I_c + E_s I_s \tag{23.14}$$

I_a, I_c and I_s	are the second moments of area of the steel section, the concrete (assumed uncracked) and the reinforcement about the axis of bending considered respectively
E_a and E_s	are the moduli of elasticity of the steel section and the steel reinforcement respectively
E_{cm}	is the secant modulus of the concrete according to BS EN 1992-1-1 (see Table 22.2).

In general, the buckling length, l of an isolated non-sway composite column may conservatively be taken as its system length, L. Alternatively, the buckling length may be determined using Annex E of BS EN 1993-1-1.

Moreover, for slender composite columns under long-term loading, creep and shrinkage of concrete may cause a reduction in the effective flexural stiffness of the composite columns, thereby reducing its buckling resistance. In such cases, E_{cm} should be reduced by multiplying by the following factor:

$$\frac{1}{1 + \left(\dfrac{N_{G,Ed}}{N_{Ed}}\right)\varphi_t} \tag{23.15}$$

where φ_t is the creep coefficient according to Clause 5.4.2.2(2), N_{Ed} is the total design compression force, and $N_{G,Ed}$ is the part of N_{Ed} that is permanent. As a simple rule, the long-term effects should be considered in a composite column if its buckling length-to-depth ratio exceeds 15.

23.3.5.2 *Relative slenderness*

The buckling resistance of a composite column is expressed as a proportion χ of the plastic resistance of the cross-section in compression, $N_{pl,Rd}$, and it depends on the relative slenderness of the column, $\bar{\lambda}$, which is given by:

$$\bar{\lambda} = \sqrt{\frac{N_{pl.Rk}}{N_{cr}}} \tag{23.16}$$

where:

$N_{pl,Rk}$ is the characteristic value of the plastic resistance at the cross-section in compression based on the characteristic values of the material strengths

N_{cr} is the elastic critical force of the column for the relevant buckling mode.

23.3.5.3 *Column buckling curves*

For both principal axes of the column, it is necessary to verify that:

$$N_{Ed} \le \chi N_{pl,Rd} \tag{23.17}$$

where:

$N_{pl,Rd}$ is the plastic resistance of the cross-section in compression, and

χ is the reduction factor due to column buckling, which is determined according to the relative slenderness of the composite column, and an appropriate column buckling curve.

Three column buckling curves, as defined in BS EN 1993-1-1[11], are adopted. These curves may be expressed mathematically by:

$$\chi = \frac{1}{\phi + \sqrt{\phi^2 - \bar{\lambda}^2}} \le 1.0 \tag{23.18}$$

where:

$$\phi = 0.5\left[1 + \alpha(\bar{\lambda} - 0.2) + \bar{\lambda}^2\right] \tag{23.19}$$

The factor α is used to allow for different levels of both geometrical and mechanical imperfections in the columns, and the values of α are 0.21, 0.34 and 0.49 for buckling curves a, b and c respectively, as shown in Figure 23.8.

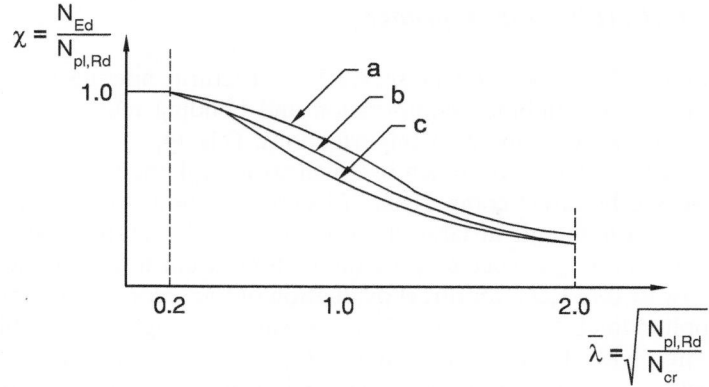

Figure 23.8   Column buckling curves according to Eurocode 3

Table 23.2   Selection of appropriate column buckling curves

Cross-section		Axis of buckling	Buckling curve	Member imperfection
Concrete-encased H-section	Fully encased H-section	y-y	b	L / 200
		z-z	c	L / 150
	Partially encased H-section	y-y	b	L / 200
		z-z	c	L / 150
	Partially encased crossed I-sections	y-y or z-z	b	L / 200
Concrete-filled hollow section	Circular and rectangular hollow steel sections	y-y or z-z	a ($\rho_t \leq 3\%$)	L / 300
		y-y or z-z	b ($3 \leq \rho_t \leq 6\%$)	L / 200
	Circular and rectangular hollow sections with additional H-sections	y-y	b	L / 200
		z-z	b	L / 200

ρ_t is the percentage of reinforcement.

It should be noted that the choice of the value α is made according to the type of the composite columns and the axis of buckling as given in Table 6.5 of BS EN 1994-1-1, which is presented in Table 23.2.

23.3.6 Resistance of members in combined compression and bending

Under combined compression and bending, slender composite columns may fail primarily in lateral buckling under the second order, or 'P-Δ' effects. Hence, it is necessary to evaluate the internal forces and moments of slender columns accurately.

23.3.6.1 Direct evaluation approach

According to BS EN 1994-1-1 Clause 6.7.3.5, structural adequacy of a slender composite column to combined compression and bending may be verified using second-order analysis with member imperfections. This approach may be referred to as the Direct Evaluation Approach in which structural adequacy of a composite column is assessed by direct comparison of its cross-section resistances against the applied forces and moments at large deformations. In this method, an initial geometrical imperfection, i_o, is specified for the column according to Table 6.5 of BS EN 1994-1-1, or to Table 23.2 for direct evaluation of the second-order moment, δM under the applied load, N_{Ed}, i.e. $\delta M_i = N_{Ed}\, i_o$, as shown in Figure 23.9. This moment δM_i is combined with the applied moment M_{Ed} for direct comparison against the bending resistance of the composite cross-section in the presence of the applied load, $M_{pl,N,Rd}$, by using the N-M interaction curve of the composite cross-section.

In general, for typical structural forms with well established structural behaviour, only conventional linear elastic analysis is needed, and the internal forces and moments of a column so obtained are usually referred to as 'first-order' forces and moments. Depending on the slenderness of the column, amplification factors, k, may be applied to these 'first-order' forces and moments to account for any second-order effect as necessary.

However, for structures with irregular member configurations and very slender columns and beams, accurate internal forces and moments should be obtained through the use of appropriate advanced structural analysis software including both geometrical and material non-linearities. In such cases, the important considerations are the initial geometrical and mechanical imperfections of the members, interfacial

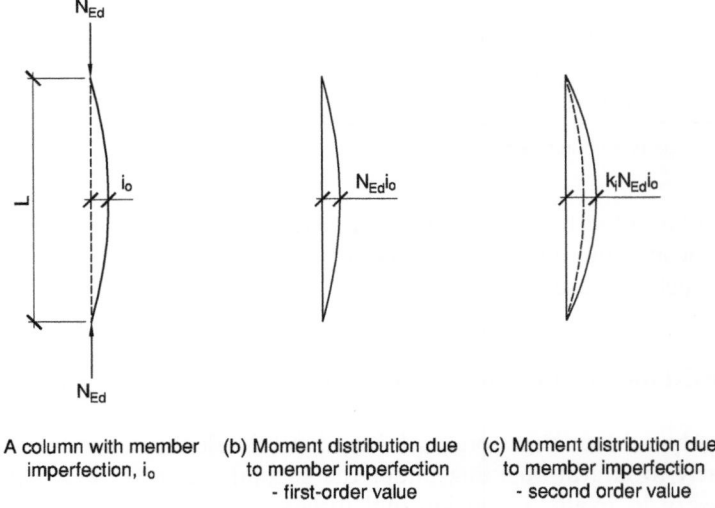

(a) A column with member imperfection, i_o

(b) Moment distribution due to member imperfection - first-order value

(c) Moment distribution due to member imperfection - second order value

Figure 23.9   Additional moments for a column under compression

shear behaviour between steel sections and concrete, crushing and cracking of concrete, flexural rigidities of the beam-column joints and column splices, and the behaviour of the concrete under long term loads.

23.3.6.2 Amplification factor, k

For slender composite columns with the relative slenderness $\bar{\lambda}$ smaller than 2.0, the second-order effect may be allowed for by multiplying the largest first-order design moment $M_{Ed,max}$, by a correction factor k_m, which is defined in BS EN 1994-1-1, Clause 6.7.3.4 (5), as follows:

$$k_m = \frac{\beta}{1 - \dfrac{N_{Ed}}{N_{cr,II}}} \geq 1.0 \tag{23.20}$$

where

       N_{Ed} is the design applied load
       $N_{cr,II}$ is the effective elastic critical force of the composite column based on the system length, L, and is given by:

$$N_{cr,II} = \frac{\pi^2 (EI)_{eff,II}}{L^2} \tag{23.21}$$

It should be noted that for the determination of the internal forces, the design value of the effective flexural stiffness $(EI)_{eff,II}$ is given by:

$$(EI)_{eff,II} = 0.9(E_a I_a + 0.5 E_{cm} I_c + E_s I_s) \tag{23.22}$$

Both values, 0.9 and 0.5, were specified after calibration against column test data.

       β is an equivalent moment factor which is defined according to Table 6.4 of BS EN 1994-1-1, as follows:

$$\beta = 0.66 + 0.44\, r \text{ but } \geq 0.44 \tag{23.23}$$

where r is the ratio of the smaller to the larger end moments.

For columns with transverse loading within the column lengths, $\beta = 1.0$. Moreover, for the determination of the second-order moment in a slender column due to member imperfection, β_i may be taken as 1.0 conservatively.

Figure 23.10 illustrates the typical first-order and the second-order moments in a column bending in single curvature subject to end moments.

23.3.6.3 Check for structural adequacy

As shown in Figure 23.11, BS EN 1994-1-1, Clause 6.7.3.6 considers that the design is adequate when the following condition is satisfied:

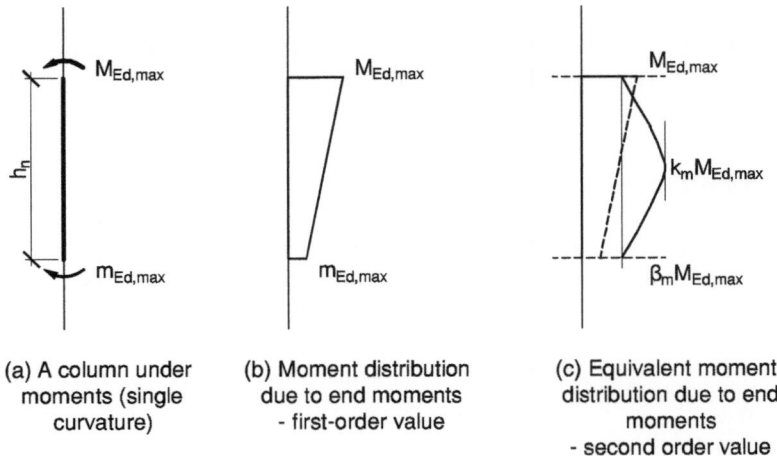

(a) A column under
moments (single
curvature)

(b) Moment distribution
due to end moments
- first-order value

(c) Equivalent moment
distribution due to end
moments
- second order value

Figure 23.10   Design moments in a column bending in a single curvature under end moments
with r ≥ 0

$$\frac{M_{Ed}}{M_{pl,N,Rd}} = \frac{M_{Ed}}{\mu_d M_{pl,Rd}} \leq \alpha_M \tag{23.24}$$

where:

M_{Ed} is the design bending moment, which may be increased to allow for
second-order effects, if necessary

$M_{pl,N,Rd}$ is the plastic resistance of the composite cross-section in bending in
the presence of axial force

μ_d is the moment resistance ratio obtained from the N-M interaction
curve

$M_{pl,Rd}$ is the plastic bending resistance of the composite cross-section

α_M = 0.9 for S235 to S355 steel and 0.8 for S420 and S460 steel.

For simplicity,

$$\mu_d = 1 - \frac{(\chi_d - \chi_{pm})}{(1 - \chi_{pm})} \text{ when } \chi_d > \chi_{pm} \tag{23.25a}$$

$$= 1 \qquad\qquad \text{when } \chi_d \leq \chi_{pm} \tag{23.25b}$$

χ_{pm} is the axial resistance ratio due to the concrete, which is given by $\dfrac{N_{pm,Rd}}{N_{pl,Rd}}$

χ_d is the design axial resistance ratio, which is given by $\dfrac{N_{Ed}}{N_{pl,Rd}}$

The expressions are obtained from geometry consideration of the multi-linear
interaction curve, as illustrated in Figure 23.11.

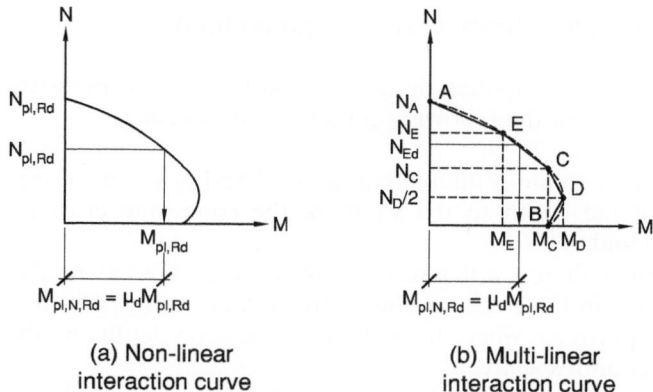

(a) Non-linear
interaction curve

(b) Multi-linear
interaction curve

Figure 23.11   Typical multi-linear N-M interaction curves of a composite cross-section under compression and uni-axial bending

23.3.7 Resistances of members in combined compression and bi-axial bending

For the design of a composite column under combined compression and bi-axial bending, the value of μ_d should be calculated separately for each axis according to Section 23.4.2. Imperfections should be considered only in the plane in which failure is expected to occur. If it is not evident which plane is the more critical, checks should be made for both planes.

After the evaluation of the moment resistance ratios μ_{dy} and μ_{dz} for both axes, the interaction of the moments should also be checked according to Clause 6.7.3.7(2) at various positions along the member length against the resistance of the cross-section in bending in the presence of axial load, as follows:

$$\frac{M_{y,Ed}}{\mu_{dy}\,M_{pl,y,Rd}} \leq \alpha_{M,y} \tag{23.26}$$

$$\frac{M_{z,Ed}}{\mu_{dz}\,M_{pl,z,Rd}} \leq \alpha_{M,z} \quad \text{and} \tag{23.27}$$

$$\frac{M_{y,Ed}}{\mu_{dy}\,M_{pl,y,Rd}} + \frac{M_{z,Ed}}{\mu_{dz}\,M_{pl,z,Rd}} \leq 1{,}0 \tag{23.28}$$

where:

$M_{y,Ed}, M_{z,Ed}$ are the design bending moments which may be factored to allow for second-order effects, if necessary

$M_{pl,y\ Rd}, M_{pl,z\ Rd}$ are the plastic resistances of the cross-section in bending

μ_{dy}, μ_{dz} are the moment resistance ratios obtained from the N-M interaction curves

$\alpha_{M,y}, \alpha_{Mz}$ = 0.9 for S235 to S355 steel and 0.8 for S420 and S460 steel.

23.3.8 Limit of applicability of the design method

In order to apply the simplified design method to design composite columns, it is necessary to ensure all the following conditions are satisfied:

- The composite column is doubly symmetrical and prismatic along its length.
- The ratio of the depth to the width of the composite cross-section, h/b_c, is between 0.2 and 5.0.
- For composite columns with concrete fully encased H-sections, the values of the concrete cover in the z- and in the y-directions, c_z and c_y, are smaller than $0.3\,h$ and $0.4\,b$ respectively, where h and b are the section depth and the flange width of the H-section respectively.
- The steel contribution ratio of the composite column, δ, is between 0.2 and 0.9.
- The cross-sectional area of the steel reinforcement, A_r, does not exceed $0.06\,A_c$ where A_c is the cross-sectional area of the concrete.
- The relative slenderness of the composite column, $\bar{\lambda}$, is smaller than 2.0.

23.4 Illustrative examples of design of composite columns

The following examples are provided to illustrate the results for composite columns designed to the simplified design method. Figure 23.12 shows the N-M interaction curves for concrete-filled 500×300 RHS of various thicknesses for bending about the major and the minor axes.

Figure 23.13 presents the variations of the member resistances of a composite column with different member lengths, L, for buckling about both the major and the minor axes. It should be noted that the column is designed for a concentric compression force, N_{Ed}. Moreover, both the member resistances obtained from the *Buckling Curve Approach (BCA)* using the column buckling curves and the *Direct Evaluation Approach (DEA)* are plotted on the same graphs for comparison.

It is shown that the results obtained from the *DEA* are marginally lower than those obtained from the *BCA*, owing to the use of the large member imperfection, i_o, given in Table 23.2.

Figure 23.14 presents the variations of the member resistances of a composite column of different member lengths for buckling about both the major and the minor axes. It should be noted that the column is subject to a compression force, N_{Ed}, with the eccentricities e at the top of the column, and $e \times r$ at the bottom of the column, where r is the end moment ratio in the range from −1.0 to 1.0.

In general, the effects of the end moment ratio in this design lead to a 15 to 25% reduction in the compression resistance of the composite column.

For details of the design procedure, refer to the worked example on a composite column with a concrete-filled rectangular hollow section.

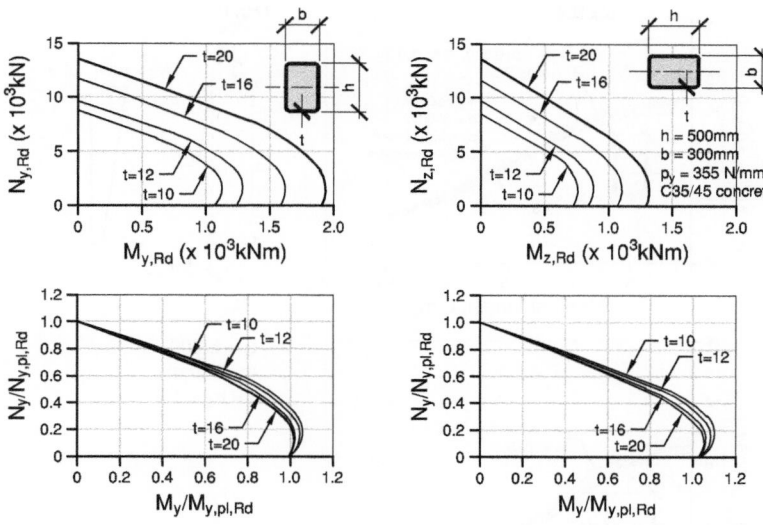

Figure 23.12   Non-linear N-M interaction curves for a composite cross-section with concrete-filled rectangular hollow sections

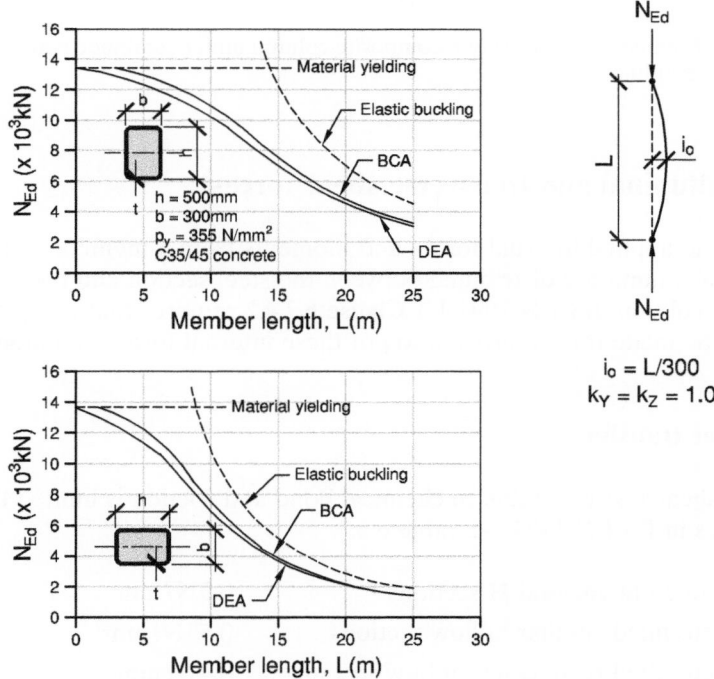

Figure 23.13   Member resistances of a composite column under compression

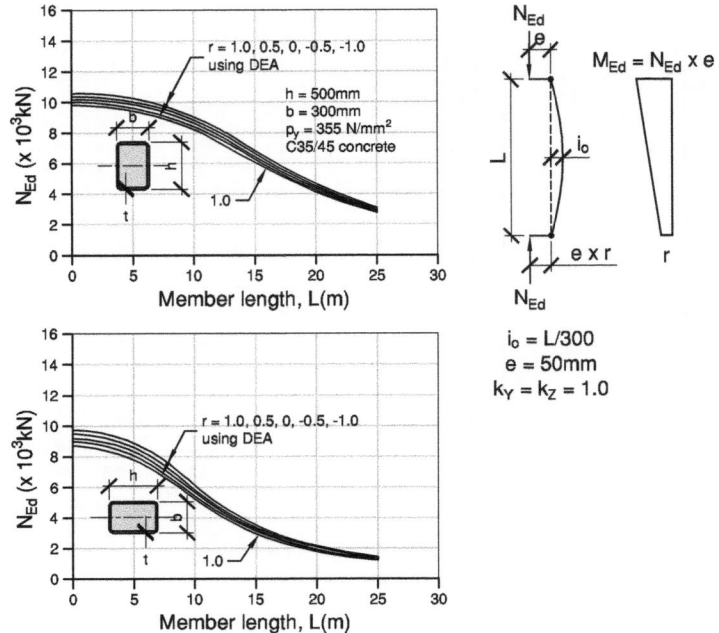

Figure 23.14   Member resistances of a composite column under combined compression and bending

23.5 Longitudinal and transverse shear forces

In general, the applied internal forces and moments from a member connected to the ends of a column are distributed between the steel section and the concrete of a composite column. BS EN 1994-1-1 Clause 6.7.4.2 requires that adequate provision should be made for the distribution of these internal forces and moments.

23.5.1 Shear transfer

The design shear resistance due to chemical bond and friction is limited to the following values in BS EN 1994-1-1 Table 6.6:

- for fully concrete-encased H-sections $0.3 \, \text{N}/\text{mm}^2$
- for concrete-filled circular hollow sections $0.55 \, \text{N}/\text{mm}^2$
- for concrete-filled rectangular hollow sections $0.4 \, \text{N}/\text{mm}^2$
- for flanges in partially encased H-sections $0.2 \, \text{N}/\text{mm}^2$
- for webs in partially encased H-sections   zero.

For axially loaded columns, it is usually found that this interface shear resistance is sufficient to develop the combined strengths of both materials at the critical cross-section (mid-column height). For columns with significant end moments, development of longitudinal shear forces between the concrete and the steel section is required. For simplicity, the design transverse shear forces may be assumed to act on the steel section alone.

23.5.2 Regions of load introduction

Where a load is applied to a composite column, it must be ensured that within a specified introduction length, the individual components of the composite cross-section are loaded to below their resistances. For this purpose, a division of the loads between the steel section and the concrete must be made in a manner similar to that described in BS EN 1994-1-1 Clause 6.7.3.2 (4).

In order to estimate the distribution of the applied loads and moments, the stress distributions at the beginning and the end of the region of introduction must be known. From the differences in these stresses, the loads which are transferred to the cross-section components may be determined. The length of the region of load introduction should be less than $2d$ and $L/3$, where d is the cross-section dimension normal to the bending axis, and L is the system length of the column.

If the load is applied through a connection to the steel section, the elements of the load introduction, e.g. the shear connectors, must be designed to transmit that part of the loading that is to be resisted by the concrete section.

For single-storey columns, head plates are generally used as the elements for load introduction. Special detailing is necessary for continuous columns. For these cases, headed studs have proved to be economic when used with open cross-sections, as shown in Figure 23.15.

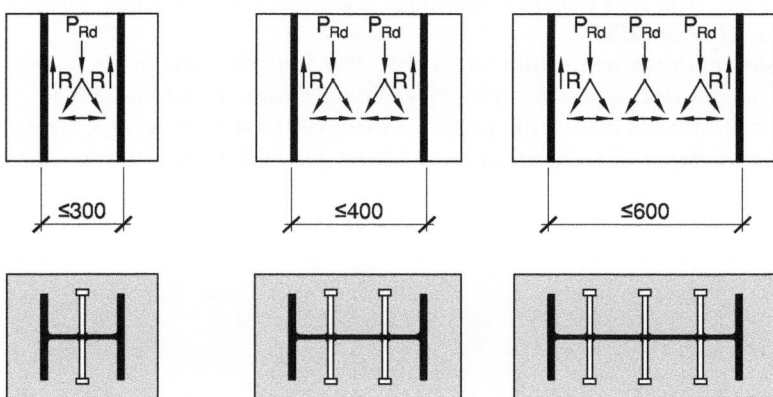

Figure 23.15   Shear resistances of headed stud connectors used to create direct load transfer into the concrete

A worked example for a concrete-filled composite column follows on page 723.

References to Chapter 23

1. British Standards Institution (2005) BS 5400-5 *Steel, concrete and composite bridges. Part 5: Code of practice for the design of composite bridges*. London, BSI.
2. British Standards Institution (2005) BS EN 1994-1-1 *Design of composite steel and concrete structures. Eurocode 4 Part 1-1: General rules and rules for buildings*. London, BSI.
3. Chung K.F. and Narayanan R. (1994) *Composite column design to Eurocode 4*. The Steel Construction Institute, Ascot, SCI.
4. Johnson R.P. and Anderson D. (2004) *Designers' Guide to EN 1994-1-1. Eurocode 4: Design of composite steel and concrete structures. Part 1.1: General rules and rules for buildings*. London, Thomas Telford Services Ltd.
5. British Standard Institution (2008) *UK National Annex to EN 1944: Design of composite steel and concrete structures. Part 1-1: General rules and rules for buildings*. London, BSI.
6. Leon R.T., Kim D.K. and Hajjar J.F. (2007) Limit State Response of Composite Columns and Beam-Columns. Part 1: Formulation of Design Provisions for the 2005 AISC Specification, *Engineering Journal*, AISC, 4th Quarter, 341–358.
7. Leon R.T. and Hajjar J.F. (2008) Limit State Response of Composite Columns and Beam-Columns. Part II: Application of Design Provisions for the 2005 AISC Specification, *Engineering Journal*, AISC, 1st Quarter, 21–46.
8. British Standards Institution (2004) BS EN 1992-1-1 *Eurocode 2: Design of concrete structures. Part 1-1: General rules and rules for buildings*. London, BSI.
9. British Standards Institution (2005) BS EN 1994-1-2 *Eurocode 4: Design of composite steel and concrete structures. Part 1-1: General rules – Structural fire design*. London, BSI.
10. British Standards Institution (2003) BS 5950 *Structural use of steelwork in building: Part 8: Code of practice for fire resistant design*. London, BSI.
11. British Standards Institution (2005) BS EN 1993-1-1 *Eurocode 3: Design of steel structures. Part 1-1: General rules and rules for buildings*. London, BSI.

![The Steel Construction Institute]	Job No.			Sheet 1 of 10	Rev
The Steel Construction Institute	Job Title	Composite Column Design to BS EN 1994-1-1			
Silwood Park, Ascot, Berks SL5 7QN	Subject	Composite Column with Concrete-filled Rectangular Hollow Section			
Telephone: (01344) 623345 Fax: (01344) 622944	Client		Made by	KFC	Date	Dec 2009
CALCULATION SHEET			Checked by	RML	Date	Jan 2010

Design of composite column using concrete-filled rectangular hollow section

This Worked Example presents the design procedure and results for a composite concrete-filled column subject to compression and bending.

BS EN 1994-1-1

Composite column details

Column length: 4.8 m

Buckling coefficient: $k_y = 1.0$ *(major axis direction)*
             $k_z = 0.85$ *(minor axis direction)*

Type: $500 \times 300 \times 20$ mm thick RHS

Steel: Grade S355

Concrete: C35/45 *(cylinder/cube strength)*

Reinforcement: None

Design values of actions

Design axial compressive force N_{Ed} $= 11000\,kN$

Maximum design bending moment about y-y (major) axis $M_{y,max,Ed}$ $= 215\,kNm$

Maximum design bending moment about z-z (minor) axis $M_{z,max,Ed}$ $= 0\,kNm$

End moment ratio *(see below)* r $= -0.5$

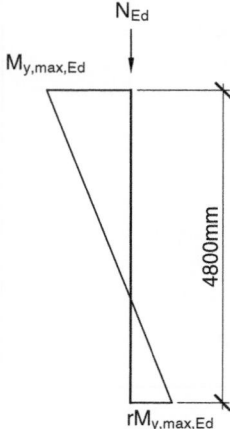

Design of Composite Columns	Sheet 2 of 10	Rev

Material properties

Structural steel

Nominal steel grade	*S355*		*Cl 3.3.2*
Nominal yield strength	f_y	$= 355\ N/mm^2$	
Modulus of elasticity	E_a	$= 210\ kN/mm^2$	

Concrete (normal weight concrete)

Concrete strength class	*C35/45*		*Cl 3.1.2*
Characteristic strength	f_{ck}	$= 35\ N/mm^2$ *(cylinder strength)*	
Secant modulus of elasticity	E_{cm}	$= 33.5\ kN/mm^2$	

Partial safety factors

γ_a = *1.0 steel*

γ_c = *1.5 concrete* — *BS EN 1994-1-1 UK NA*

γ_s = *1.15 reinforcement*

Design strengths

$$f_{yd} = \frac{f_y}{\gamma_a} = \frac{355}{1.0} = 355\ N/mm^2$$

$$f_{cd} = \frac{f_{ck}}{\gamma_c} = \frac{35}{1.5} = 23.3\ N/mm^2$$

Limits of the applicability of the simplified method

The scope of the simplified method in BS EN 1994-1-1 is limited, as follows: — *Cl 6.7.3.1*

(a) The column is doubly symmetrical and uniform cross-section over its length.

(b) For a fully encased steel section, the cover satisfies the following limits:

in the z direction, max $c_z = 0.3h$
in the y direction, max $c_y = 0.4b$

(c) $0.2 \le \delta$ *(steel contribution ratio)* ≤ 0.9.

(d) Slenderness ratio $\bar{\lambda} \le 2.0$.

(e) The cross-sectional area of bar reinforcement does not exceed $0.06\ A_c$.

(f) The aspect ratio, $0.2 \le \frac{h}{b} \le 5$*, where* h = *column depth and* b = *column width.*

Design of Composite Columns	Sheet 3 of 10	Rev

Cross-section geometry and properties of the RHS section

Cross-section dimensions and section properties of 500 × 300 × 20 RHS

Eurocode Notation	Values
b	$300\,mm$
h	$500\,mm$
t_a	$20\,mm$
A_a	$30.4 \times 10^3\,mm^2$
I_{ay}	$1016 \times 10^6\,mm^4$
I_{az}	$451 \times 10^6\,mm^4$

Cross-section properties of the concrete infill

$A_c = 260 \times 460 = 119.6 \times 10^3\,mm^2$

$$I_{cy} = \frac{260 \times 460^3}{12} = 2109 \times 10^6\ mm^4$$

$$I_{cz} = \frac{460 \times 260^3}{12} = 674 \times 10^6\ mm^4$$

$$W_{pc,y} = \frac{260 \times 460^2}{4} = 13.7 \times 10^6\ mm^3$$

Design of Composite Columns	Sheet 4 of 10	Rev

Design checks at ultimate limit state

Plastic resistance of the composite cross-section in compression

Plastic resistance, $N_{pl,Rd}$ of the composite cross-section is obtained by summing the plastic resistances of the components:

Cl 6.7.3.2(1)

$N_{pl,Rd} = A_a f_{yd} + 1.0 A_c f_{cd} + A_s f_{sd}$

$= (30.4 \times 355 + 1.0 \times 119.6 \times 23.3 + 0) \times 10^3 \times 10^{-3}$

$= 10792 + 2786 = 13578 \, kN$

Note: The factor of 1.0 applied to f_{cd} is permitted for concrete-filled RHS

Effective flexural stiffnesses of the composite cross-section

About the major axis:

Cl 6.7.3.3(3)

$(EI)_{eff,y} = E_a I_{ay} + K_e E_{cm} I_{cy} + E_s I_{sy}$

$E_a I_{ay} = 210 \times 1016 \times 10^6 \times 10^{-6} = 213360 \, kNm^2$

$K_e E_{cm} I_{cy} = 0.6 \times 33.5 \times 2109 \times 10^6 \times 10^{-6} = 42390 \, kNm^2$

$(EI)_{eff,y} = 213360 + 42390 = 255750 \, kNm^2$

About the minor axis:

Cl 6.7.3.3(3)

$(EI)_{eff,z} = E_a I_{az} + E_s I_{sz} + K_e E_{cm} I_{cz}$

$E_a I_{az} = 210 \times 451 \times 10^6 \times 10^{-6} = 94710 \, kNm^2$

$K_e E_{cm} I_{cz} = 0.6 \times 33.5 \times 674 \times 10^6 \times 10^{-6} = 13547 \, kNm^2$

$(EI)_{eff,z} = 94710 + 13547 = 108257 \, kNm^2$

Relative slenderness $\bar{\lambda} = \sqrt{\dfrac{N_{pl,Rk}}{N_{cr}}}$

Cl 6.7.3.3(2)

$N_{pl,Rk} = A_a f_y + 1.0 A_c f_{ck} + A_s f_{sk}$ (partial factors set to unity)

Cl 6.7.3.2(1)

$= (30.4 \times 355 + 1.0 \times 119.6 \times 35 + 0) \times 10^3 \times 10^{-3}$

$= 10792 + 4186 = 14978 \, kN$

$l_{ey} = k_y L = 1.0 \times 4.8 = 4.8 \, m$

$N_{cr,y} = \dfrac{\pi^2 (EI)_{eff,y}}{l_{ey}^2} = \dfrac{\pi^2 \times 255750}{4.8^2} = 109560 \, kN$

Major axis slenderness ratio:

$\bar{\lambda}_y = \sqrt{\dfrac{N_{pl,Rk}}{N_{cr,y}}} = \sqrt{\dfrac{14978}{109560}} = 0.37$

$l_{ez} = k_z L = 0.85 \times 4.8 = 4.08 \, m$

Design of Composite Columns	Sheet 5 of 10	Rev

Minor axis slenderness ratio:

$$N_{cr,z} = \frac{\pi^2 (EI)_{eff,z}}{I_{ez}^2} = \frac{\pi^2 \times 108257}{4.08^2} = 64165 \, kN$$

$$\bar{\lambda}_z = \sqrt{\frac{N_{pl,Rk}}{N_{cr,z}}} = \sqrt{\frac{14978}{64165}} = 0.48 > 0.37$$

Check on limits of simplified method:

The steel contribution ratio is obtained as follows:

$$\delta = \frac{A_a f_{yd}}{N_{p\ell,Rd}} = \frac{30.4 \times 10^3 \times 355 \times 10^{-3}}{13578} = 0.795$$

Cl 6.7.1(4)

Hence $0.2 \leq \delta \leq 0.9$ is satisfied.

$$\bar{\lambda}_y = 0.37; \bar{\lambda}_z = 0.48$$

Cl 6.7.3.1(1)

Hence $\bar{\lambda} \leq 2$ is satisfied.

$$\frac{h}{b} = \frac{500}{300} = 1.67; \text{ hence } 0.2 \leq \frac{h}{b} \leq 5 \text{ is satisfied.}$$

Cl 6.7.3.1(4)

Buckling resistance of the composite column in axial compression

$N_{Ed} \quad \leq \chi \, N_{pb\,Rd}$

Cl 6.7.3.5(2)

$N_{pb\,Rd} = 13578 \, kN$

χ is the reduction factor for column buckling, obtained as follows:

Using buckling curve 'a' for concrete-filled RHS:

$$\chi = \frac{1}{\phi + \sqrt{\phi^2 - \bar{\lambda}^2}} \text{ but } \chi \leq 1.0$$

Where $\phi = 0.5\left[1 + \alpha(\bar{\lambda} - 0.2) + \bar{\lambda}^2\right]$ and $\alpha = 0.21$

$\phi \quad = 0.5[1 + 0.21 \times (0.48 - 0.2) + 0.48^2] = 0.646$

$$\chi = \frac{1}{0.646 + \sqrt{0.646^2 - 0.48^2}} = 0.93$$

$\chi N_{pb\,Rd} = 0.93 \times 13578$

$\qquad = 12628 \, kN > N_{Ed} = 11000 \, kN$

Unity factor for buckling in pure compression = 0.87

Design of Composite Columns	Sheet 6 of 10	Rev

Bending resistance of the composite column

The plastic bending resistance $M_{pb\,Rd}$ is obtained by summing the relevant contributions of the steel section, reinforcement and concrete, which are obtained as follows for a rectangular section:

$$W_{pa} = \frac{b \times h^2}{4} - W_{pc,a}$$

Sheet 3

$$= \frac{300 \times 500^2}{4} - 13.8 \times 10^6$$

$$= 4950 \times 10^3\,mm^3$$

Depth of plastic neutral axis from the outside of flange:

$$h_n = \frac{A_c f_{cd} - A_{sn}(2f_{sd} - f_{cd})}{2bf_{cd} + 4t(2f_{yd} - f_{cd})}$$

Sheet 2

$$= \frac{119.6 \times 10^3 \times 23.3 - 0}{2 \times 300 \times 23.3 + 4 \times 20 \times (2 \times 355 - 23.3)}$$

$$= 40.4\,mm$$

W_{pcn} $= (b - 2t) \times h_n^2 - W_{psn}$

$= (300 - 2 \times 20) \times 40.4^2 - 0 = 424 \times 10^3\,mm^3$

W_{pan} $= b\,h_n^2 - W_{pcn} - W_{psn}$

$= 300 \times 40.4^2 - 424 \times 10^3 = 65.6 \times 10^3\,mm^3$

Note: $W_{psn} = 0$, as there is no reinforcement within the region of $2h_n$ from the centre-line of the cross-section.

The plastic moment resistance of the composite cross-section is obtained, as follows:

$M_{pb\,Rd}$ $= f_{yd}\,(W_{pa} - W_{pan}) + 0.5\,f_{cd}\,(W_{pc} - W_{pcn}) + f_{sd}\,(W_{ps} - W_{psn})$

$M_{pb\,Rd}$ $= [355 \times (4950 \times 10^3 - 65.6 \times 10^3) + 0.5 \times 23.3 \times (13.7 \times 10^6 - 424 \times 10^3)] \times 10^{-6}$

$= 1734 + 155$

$= 1889\,kNm$

$N_{pm,Rd}$ $= f_{cd}\,A_c$

$= 23.3 \times 119600 \times 10^{-3} = 2786\,kN$

$M_{max,Rd}$ $= f_{yd}\,W_{pa} + 0.5\,f_{cd}\,W_{pc,a}$

$= [355 \times 4950 \times 10^3 + 0.5 \times 23.3 \times 13.7 \times 10^6] \times 10^{-6}$

$= 1917\,kNm$

Establish the key points on the axial force-moment interaction diagram:

M_E $= M_{max,Rd} - \Delta M_E$

Design of Composite Columns	Sheet 7 of 10	Rev

where $\Delta M_E = (bh_E^2 - W_{cE})f_{yd} + 0.5\ W_{cE}\ f_{cd}$

$h_E \quad = 0.25h + 0.5h_n$

$\quad\quad\ = 0.25 \times 500 + 0.5 \times 40.4 = 145.2\ mm$

$W_{aE} \quad = bh_E^2 - W_{cE}$

$\quad\quad\ = 300 \times 145.2^2 - 5481 \times 10^3$

$\quad\quad\ = 844 \times 10^3\ mm^3$

where:

$W_{cE} \quad = (b - 2t)\ h_E^2$

$\quad\quad\ = (300 - 2 \times 20)\ \times 145.2^2$

$\quad\quad\ = 5481 \times 10^3\ mm^3$

$\Delta M_E \quad = (bh_E^2 - W_{cE})f_{yd} + 0.5\ W_{cE}\ f_{cd}$

$\quad\quad\ = [(300 \times 145^2 - 5481 \times 10^3) \times 355$
$\quad\quad\quad + 0.5 \times 5481 \times 10^3 \times 23.3] \times 10^{-6} = 299.6 + 63.9 = 363.5\ kNm$

$M_E \quad = 1917 - 363.5 = 1553.5\ kNm$

$N_E \quad = 4h_E\ t\ f_{yd} + [0.5A_c + (b - 2t)h_E]\ \times f_{cd}$

$\quad\quad\ = \{4 \times 145.2 \times 20 \times 355 +$
$\quad\quad\quad [0.5 \times 119600 + (300 - 2 \times 20) \times 145.2]\ \times 23.3\} \times 10^{-3}$

$\quad\quad\ = 4124 + 2273 = 6397\ kN$

The simplified M-N interaction diagram is presented below based on points A to E and overlayed by the complete interaction curve.

Simplified interaction curve of the composite cross-section subject to combined compression and uniaxial bending

Design of Composite Columns	Sheet 8 of 10	Rev

The equation for line AE is given by:

$$\frac{N - N_{pl,Rd}}{M} = \frac{N_{Ed} - N_{pl,Rd}}{M_{Ed}}$$

Therefore, N = 13578 − 4.622M, where M = applied moment.

Effective flexural stiffness for second-order linear elastic analysis

About the major axis:

$(EI)_{eff,y}$ $= K_o (E_a I_{ay} + K_{e,II} E_{cm} I_{cy} + E_s I_{sy})$ *Cl 6.7.3.4(2)*

where: $K_{e,II} = 0.5$ *and* $K_o = 0.9$

$E_a I_{ay}$ $= 210 \times 1016 \times 10^6 \times 10^{-6} = 213360 \, kNm^2$

$K_{e,II} E_{cm} I_{cy}$ $= 0.5 \times 33.5 \times 2109 \times 10^6 \times 10^{-6} = 35325 \, kNm^2$

$(EI)_{eff,y}$ $= 0.9 \times (213360 + 35325 + 0) = 223816 \, kNm^2$

$N_{cr,y,eff}$ $= \dfrac{\pi^2 (EI)_{y,eff,II}}{l_{ey}^2} = \dfrac{\pi^2 \times 223816}{4.8^2} = 95876 \, kN$

$\dfrac{N_{Ed}}{N_{cr,y,eff}}$ $= \dfrac{11000}{95876} = 0.11 \geq 0.1$ *Cl 5.2.1(3)*

It follows that second-order effects should be considered in this calculation, although in practice the limit of 0.1 is approximate.

Compression resistance of composite column based on second-order linear elastic analysis

$e_{o,y} = \dfrac{L}{300} = \dfrac{4800}{300} = 16 \, mm$ *Table 6.5*

M_i $= N_{Ed} e_{o,y} = 11000 \times 16 \times 10^{-3} = 176 \, kNm$

Determine the amplication factor, k_i ($\beta_i = 1$)   *Table 6.4*

$k_i = \dfrac{\beta_i}{1 - \dfrac{N_{Ed}}{N_{cr,y,eff}}} = \dfrac{1}{1 - \dfrac{11000}{95876}} = 1.13$ *Cl 6.7.3.4(5)*

$M_{Ed,11,1} = 1.14 \times 176 = 198.9 \, kNm$

The bending moment distribution along the column is presented below:

Design of Composite Columns	Sheet 9 of 10	Rev

M from initial bow (kNm)

$k_i = 1.13$

$M_{Ed,II,i} = 198.9$

$M_i = 176$

--- *First order*

$k_i M_i$

x *(m)*

2.4   4.8

Second-order bending moment due to member imperfection

According to the simplified interaction curve, for 0.9 $M_{Ed,II,i} = 179.0\,kNm$;

$N_{Rd} = 13578 - 4.622 \times 179.0$ *Cl 6.7.3.5(1)*

$= 12750\,kN$ *(but $\chi\,N_{pb\,Rd} = 12628\,kN$)*
$> N_{Ed} = 11000\,kN$ *(according to the column buckling curve)*

The design is satisfactory for axial compression with member imperfections.

Combined compression and bending resistance of composite column based on second-order linear elastic analysis

Determine the amplication factor, k_m (for $r = -0.5$, $\beta = 0.66 + 0.44\,r = 0.44$)   *Table 6.4*

Cl 6.7.3.4(5)

$$k_m = \frac{\beta_m}{1 - \dfrac{N_{Ed}}{N_{cr,y,eff}}} = \frac{0.44}{1 - \dfrac{11000}{95876}} = 0.497$$

$M_{Ed,II,m} = k_m\,M_{Ed,max} = 0.497 \times 215 = 106.9\,kNm$

The influence of the end moments on the column is presented below:

$M_{y,Ed}$ (kNm)

$k_m = 0.497$

$M_{y,max,Ed} = 215$

--- First order

$M_{Ed,II,i} = 106.9$

$\beta M_{Ed} = 94.6$

$k_m M_{Ed}$

x (m)

2.4   4.8

$r M_{Ed} = -107.5$

Second-order bending moments due to end moments

Design of Composite Columns	Sheet 10 of 10	Rev

Total moment at mid-height = 198.9 + 106.9

= 305.8 kNm > $M_{y,Ed}$ = 215 kNm

Therefore, $M_{y,max,Ed,II}$ = 305.8 kNm

According to the simplified interaction curve, for N_{Ed} = 11000 kN;

$$M_{y,Rd} = \frac{13578 - 11000}{4.622} = 558 \ kNm$$

As $M_{y,Rd}$ lies on line AC, so the additional verification to Clause 6.7.1 (7) will not affect the result. It follows that:

$$\frac{M_{y,Ed,max}}{\mu_{dy} M_{y,pl,Rd}} = \frac{305.8}{1.0 \times 558} = 0.55 < 0.9$$

Cl 6.7.3.6(1)

Conclusion: The composite 500 × 300 RHS column is acceptable for combined bending and compression when subject to buckling over its 4.8 m length.

Unity factor for combined bending and compression= 0.55/0.9 = 0.61

Chapter 24
Design of light gauge steel elements

MARTIN HEYWOOD and ANDREW WAY

24.1 Introduction

Light gauge cold-formed steel members are commonly used in a range of building types as secondary steelwork (e.g. purlins and cladding rails in industrial buildings) and in the primary load-bearing elements in light steel frames (e.g. in residential buildings). They may be used as individual structural members (e.g. floor joists) or as part of a structural frame. Light steel members are often prefabricated off-site to form wall panels, floor cassettes or volumetric modular units, but are equally suited to stick build applications (see Chapter 8 for further information).

This chapter focuses on the design of light gauge steel members as used in structural applications. It considers members in compression and members in bending following the design rules presented in BS EN 1993-1-3.[1] Since light steel members are especially prone to local buckling, this issue is dealt with in depth, including the calculation of effective areas. Interaction rules for members subjected to simultaneous axial compression and bending are briefly discussed along with simplified guidance on the connection of light steel members. Worked examples are included at the end of the chapter to illustrate the application of the design rules to practical building applications.

24.1.1 Light gauge steel

For the purposes of this chapter, the term 'light gauge' steel refers to galvanised cold-formed steel with a maximum thickness of 4 mm, although gauges from 1.2 mm to 2.0 mm are the most common for light steel framing applications. For purlins and cladding rails, the thickness generally lies in the range 1.4 mm to 3.2 mm. Light gauge steel members are usually cold-formed from hot-dip galvanised strip steel, which is supplied to the section manufacturers as pre-galvanised coil conforming to BS EN

Steel Designers' Manual, Seventh Edition. Edited by Buick Davison and Graham W. Owens.
© 2012 Steel Construction Institute. Published 2012 by Blackwell Publishing Ltd.

10346: 2009.[2] For light steel framing applications, the commonly used grades of steel are S350, S390 and S450, while purlin and cladding rail products tend to use S390 and S450.

A lipped C-section is the most common section shape for light steel framing applications, including wall studs and floor joists.[3] The C-section shape is simple to roll and widely manufactured, while the lip provides additional stiffness to the flange and increases the stress at which local buckling occurs in the flange. Depths commonly range from 70 mm to 120 mm for wall studs and from 120 mm to 250 mm for floor joists. Purlins are typically made from Zed or Sigma sections with depths ranging from 140 mm to 300 mm.

A typical light steel frame is shown in Figure 24.1. The photograph shows a prefabricated volumetric module, but similar framing arrangements are also used in panellised and stick-built construction.

24.1.2 Design to BS EN 1993-1-3

Like other structural members used in construction applications, light gauge steel members are generally designed following the recommendations of a recognised

Figure 24.1   Load bearing light steel frame in a volumetric module

code of practice. The appropriate document for light gauge steel design in Europe (including the UK) is BS EN 1993-1-3,[1] which replaces BS 5950-5. Like its predecessor, BS EN 1993-1-3 has been specifically written for light gauge steel and makes special allowances for the structural behaviour commonly encountered with this material. Codes of practice intended for heavy gauge hot-rolled structural steelwork (e.g. BS EN 1993-1-1) should not be used for light gauge steel, unless referenced by BS EN 1993-1-3 or an equivalent light gauge code of practice.

BS EN 1993-1-3 permits two alternative design routes:

- design by calculation
- design by testing.

As the name suggests, with the former route the structural designer follows an analytical procedure laid down in BS EN 1993-1-3 to arrive at a calculated value of the section's resistance (to compression, bending etc.). The method is fairly complex due to the need to take account of local buckling through the use of effective widths, but can be attempted by hand for relatively simple sections such as lipped Cees. One disadvantage of this approach is that it tends to be conservative due to the assumptions and simplifications on which the method is based. For this reason, this approach is rarely used by purlin and cladding rail manufacturers, as it would place their products at a commercial disadvantage. Furthermore, while the scope of BS EN 1993-1-3 includes a range of section shapes, it is less well suited to some of the more complex sections, especially those with multiple stiffeners and curved webs, flanges or lips. However, despite the apparent limitations of this approach, design by calculation remains the preferred option for designers of light steel frames and floor joists and is the primary focus of this chapter.

Design by testing overcomes the limitations of the calculation approach by permitting design resistances to be obtained accurately for almost any shape of section. However, the benefits must be weighed against the costs of undertaking a programme of tests followed by the statistical analysis required by BS EN 1993-1-3 to convert raw test data into usable design values. Where testing is undertaken, such as for purlins and cladding rails, the design data are usually tabulated in the form of load span tables. These tables are published by the product manufacturer and are used by potential specifiers to choose the most suitable section for the given span and load.

The testing route can be sub-divided into two options:

- design data derived directly from test data
- design data derived from a numerical model.

In the former, an appropriate number of tests is undertaken on a range of test specimens and the results are analysed statistically to obtain a characteristic resistance for each section size within the range. The statistical analysis is laid out in BS EN 1993-1-3 (or alternatively the method in Annex D of BS EN 1990 may be used) and involves subtracting a prescribed multiple of the standard deviation from the mean test result. The multiple is dependent on the number of tests undertaken. This option is the simpler of the two, but has the disadvantage that tests must be performed on multiple samples of every section size within the product range. It is, therefore, not suitable for products with a wide range of dimensional variations, such as purlins and cladding rails, which are normally sold in a wide range of depths and gauges. It is, however, a useful method where the product range is limited, such as purlin cleats or tie wires.

In the latter option, the testing is only conducted on a limited number of specimens from within the full range. The test results are then normalised (by comparison with the equivalent theoretical value) and used to derive a numerical model that aims to accurately predict a safe design resistance across the full product range. This approach avoids the need to test the full range of section sizes and even allows for new sections to be introduced at a later date without the need for additional testing. Furthermore, where test results are normalised against a theoretical value, it is often possible to consider results from different section sizes as belonging to the same 'family' of data, thereby reducing the multiple of standard deviations to be subtracted. The accuracy of the final resistances will depend on the complexity of the numerical model and the number of tests undertaken. For example, a simple model might be used to calculate a realistic bending resistance of a section, while maintaining the theoretical assumption of a simply supported beam. By comparison, a more complex model might take account of the stiffness in the end connections of the beam, using data from a separate set of tests, in order to increase the beam's safe working load.

A typical test on a light gauge steel purlin is shown in Figure 24.2. In this instance, a point load is being applied to a cleat connecting two lengths of purlin at the midpoint of a simply supported span. The purpose of this test was to assess the moment rotation behaviour of the joint in order to model accurately the behaviour of a complete purlin system. The data from this series of tests were combined with results from two other types of test (gravity and uplift loading on a pair of purlins with sheeting attached) to create a numerical model, which was then used to produce load span data for the full range of purlin sizes.

24.2 Section properties

Before the resistance of a light gauge member to bending, compression or other type of loading can be calculated, it is necessary to determine the dimensional properties of the section under consideration. To those unfamiliar with light gauge steel, this first step might appear to be a trivial exercise of calculating the cross-sectional area and moments of inertia, but in reality it is a complex process that lies at the heart of the Eurocode design procedure for light gauge steel.

It is important to distinguish between the following types of section property:

- gross section properties
- effective section properties.

Figure 24.2   Physical testing of purlin specimen

The term 'effective section properties' refers to the properties of a fictitious cross-section that has been reduced in area to take account of the impact of local buckling (see Section 24.3). Further reductions may also be necessary to allow for distortional buckling (see Section 24.4). The bending and compression resistances of light steel members are always calculated using the effective properties of the section. The calculation of the effective section properties is dealt with in detail later in this chapter; the remainder of this section focuses on the determination of the gross section properties.

As the name suggests, the term gross section properties refers to the whole cross-section without any reduction for local buckling. The process of calculating the gross section properties is relatively straightforward for most common section shapes as it involves little more than the summation of elemental areas and first and second moments of area (for flanges, web, stiffeners etc.), the calculation of the position of the major and minor centroidal axes and, from these values, the second moment of area for the whole section. A similar process can be repeated for other properties as required. There are, however, three important issues that need to be addressed when considering light gauge steel sections:

- core steel thickness
- use of mid-line theory
- impact of corner radii.

24.2.1 Core steel thickness

The cold-rolled strip steel of the type used in light gauge steel construction is normally delivered pre-coated. Therefore, when specifying the thickness of the steel it is common practice to include the thickness of the coating in the specified value. However, BS EN 1993-1-3 (Cl 3.2.4) requires that all section properties are based on the core thickness of the steel, excluding the coating. The standard zinc coating for construction products is $275\,\mathrm{g/m^2}$ (referred to as Z275), which corresponds to a coating thickness of 0.02 mm on each surface. Hence, the nominal (specified) steel thickness should be reduced by 0.04 mm for design purposes.

24.2.2 Mid-line theory

When calculating the section properties of light gauge sections, it is standard practice to measure all dimensions along the mid-lines of the individual elements. At this point, corner radii are ignored (see 24.2.3), resulting in an idealised section consisting of a series of thin rectangular elements. In calculating the lengths of the individual elements, an allowance must be made for the intersection between adjacent elements, to avoid double-counting the over-lapping corner regions. Using the mid-line theory, this is simply achieved by measuring each element length between the points of intersection of the mid-lines. This results in a reduction in the element length below its nominal value of either $t/2$ or t, depending on the number of corners. The mid-line dimensions for a lipped C-section are shown in Figure 24.3.

24.2.3 Corner radii

The use of mid-line theory described in 24.2.2 results in an idealised section that is easy to analyse. However, without modification, the impact of the rounded corners on the section properties, which could be significant, is ignored. The problem is illustrated by Figure 24.4.

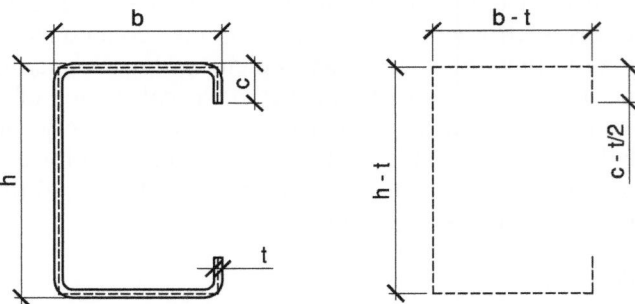

Figure 24.3   Mid-line dimensions for a lipped C-section

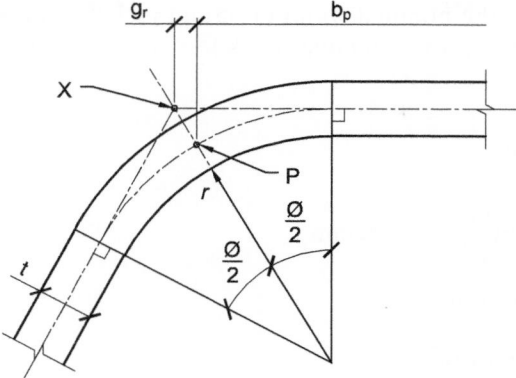

Figure 24.4   Rounded corner of a light steel section

According to mid-line theory, the web and flange intersect at the point X (the intersection of the two mid-lines). The true intersection point is P, a distance g_r from X, given by:

$$g_r = r_m \left(\tan\left(\frac{\phi}{2}\right) - \sin\left(\frac{\phi}{2}\right) \right)$$ (24.1)

where

$$r_m = r + \frac{t}{2}$$

It is apparent that any section properties derived from mid-line theory will contain a degree of error. The important question for designers is whether or not this error is significant. BS EN 1993-1-3 gives some guidance in this respect (in Cl .1) stating that the influence of rounded corners on cross-section resistance may be neglected provided that both of the following conditions are satisfied:

$r \le 5t$

$r \le 0.1b_p$

where b_p is the width of the element measured between the midpoints of the corners (see Figure 24.4).

Consider the following example:

A typical lipped C-section has a nominal width of 65 mm, corner radii of 3.0 mm and a nominal thickness of 1.5 mm.
Allowing for the standard $275 \, g/m^2$ zinc coating, the core thickness $t = 1.46$ mm.
The flange width measured between mid-lines = $65 - 1.5 = 63.5$ mm.

(It is assumed that the nominal width of 65 mm includes the galvanising, so it is appropriate to subtract the nominal thickness when calculating the mid-line dimension.)

$$r_m = r + \frac{t}{2} = 3.73 \text{ mm}$$

$$g_r = r_m \left(\tan\left(\frac{\phi}{2}\right) - \sin\left(\frac{\phi}{2}\right) \right) = 1.09 \text{ mm}$$

$$b_p = 63.5 - 2g_r = 61.32 \text{ mm}$$

Checking the corner radii:

$5t = 7.3$ mm $r = 3.0$ mm, therefore $r \leq 5t$
$0.1 b_p = 6.13$ mm, $r = 3.0$ mm, therefore $r \leq 0.1 b_p$

Therefore, in this instance, the influence of the rounded corners may be neglected when calculating the cross-section resistance.

Note: The influence of rounded corners should always be taken into account when calculating cross-section stiffness properties.

Where the influence of rounded corners needs to be accounted for, this is achieved by first calculating the section properties assuming sharp corners (i.e. ignoring the corner radii) and then applying reduction factors as follows:

For area, $A_g \approx A_{g,sh}(1 - \delta)$ (24.2)

For second moment of area, $I_g \approx I_{g,sh}(1 - 2\delta)$ (24.3)

For warping, $I_w \approx I_{w,sh}(1 - 4\delta)$ (24.4)

In these expressions, the subscript 'sh' denotes the section property based on sharp corners and δ is a reduction factor given by:

$$\delta = 0.43 \frac{\displaystyle\sum_{j=1}^{n} r_j \frac{\phi_j}{90°}}{\displaystyle\sum_{i=1}^{m} b_{p,i}} \tag{24.5}$$

where:
- r_j is the internal radius of curved element j
- n is the number of curved elements (number of corners)
- ϕ_j is the angle between two plane elements
- $b_{p,i}$ is the notional flat width of plane element i
- m is the number of plane elements.

The same reduction factors may also be applied to the effective section properties (A_{eff}, $I_{y,eff}$, $I_{z,eff}$ and $I_{w,eff}$) provided that the notional flat widths of the plane elements are measured to the points of intersection of their midlines.

Where $r > 0.04tE/f_y$ then the resistance of the cross-section should be determined by physical testing. This situation is unlikely to arise for any of the standard sections used in light steel framing, but designers need to be aware of this limit when dealing with some of the more unusual section shapes that are introduced into the light gauge market from time to time.

24.3 Local buckling

Light gauge steel sections of the types used in construction are extremely efficient in terms of their use of material. However, the associated penalty from a designer's point of view is the need to consider local buckling and its impact on the structural resistance of the cross-section.

Designers of hot-rolled structural steelwork will be familiar with the concept of section classification in which a section's susceptibility to local buckling is determined by comparing dimensional ratios (e.g. the depth-to-thickness ratio of the web) with a set of specified limits. This process results in each section being assigned a 'class' (which in some cases is dependent on the magnitude of any compression in the member). BS EN 1993-1-1[4] describes four such classes and prescribes design rules for each class that reflect the impact of local buckling on the resistance of the section. The classifications range from 'Class 1', which is defined as being a section capable of sustaining its full plastic moment while accommodating rotation of a plastic hinge, to 'Class 4', defined as a section whose moment capacity is limited by local buckling to a value below the 'moment at first yield' (that corresponding to the bending stress at the furthest point from the neutral axis reaching the yield strength of the material).

The approach used in light gauge steel design is quite different. Rather than classifying the cross-section, there is an implied assumption that the section is class 4 (although this term is not used in BS EN 1993-1-3). Having made this assumption, the design process focuses on the calculation of effective section properties, following the same procedures as for class 4 sections to BS EN 1993-1-1. The use of effective section properties stems from the need to simplify the complex stress distributions associated with local buckling, in order to minimise the required computational effort, without being over-conservative in terms of the cross-section resistance.

24.3.1 Effective width concept

When dealing with local buckling of slender plate elements, the behaviour of the plate may be analysed approximately using the effective width method, as illustrated by Figure 24.5.

In this approach, the actual stress distribution acting over element width b is replaced by simplified equivalent stresses acting over two equal widths of $b_{eff}/2$. The

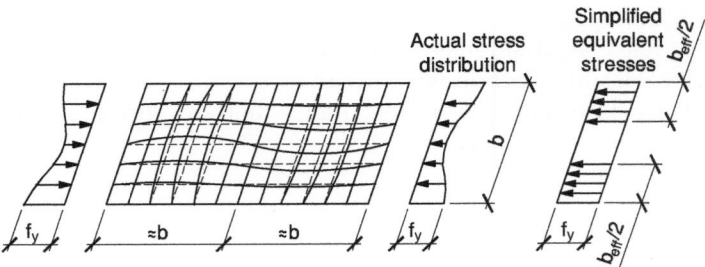

Figure 24.5   The effective width concept applied to a plate

central portion of the plate, the region most affected by local buckling, is assumed to have no stress and is ignored completely. The result is a simple model in which a uniform stress equal to the yield strength of the steel is assumed to act over a reduced width of plate.

The method adopted by BS EN 1993-1-3 takes the effective width concept illustrated above and applies it to the cross-section of a light gauge steel member. The cross-section is divided into elements (flanges, web, lips etc.), with each element being treated like the flat plate in Figure 24.5. An effective width b_{eff} is calculated for all elements that are subjected to compressive stress (either due to applied axial compression or bending). The effective area of the element A_{eff} is then obtained by multiplying b_{eff} by the section thickness t. Elements not subjected to compressive stress are not susceptible to local buckling, so the full element width b may be used in the calculation of the effective section properties.

Having obtained the elemental effective widths and areas, the properties of the effective cross-section are determined in the usual manner by calculating the position of the neutral axis followed by the first and second moments of area about this axis. The resulting set of 'effective properties' should be used when calculating the resistance of the cross-section to bending or compression as appropriate.

Since the distribution of compressive stress across the section differs between sections subjected to pure axial compression and those subjected to bending, it follows that the effective section properties will differ between these two cases. Furthermore, asymmetric sections subjected to bending may have one set of effective properties when sagging and another when hogging. It is important to use the relevant effective properties for the case under consideration.

For sections at the heavier end of the light gauge range, especially those with relatively stocky webs and flanges, it might not be appropriate to reduce the cross-sectional properties to allow for local buckling. Since BS EN 1993-1-3 does not permit the classification of the section in the manner familiar to designers of hot-rolled structural steel, the procedure outlined above must be followed even for stocky sections. However, in this case, the calculation procedure will automatically yield effective widths b_{eff} equal to the full widths b, resulting in effective section properties equal to the gross section properties.

24.3.2 Eurocode calculation procedure for unstiffened plane elements

This section focuses on the calculation of b_{eff} for an element subjected to compression. Once values of b_{eff} have been obtained for all compression elements of a cross-section, the effective properties should be determined using the method outlined in Section 24.3.1.

The calculation of b_{eff} for plane elements without stiffeners is introduced in Cl 5.5.2 of BS EN 1993-1-3. However, the detail of the method, including the relevant equations, can be found in BS EN 1993-1-5.[5]

For each element, the effective width is given by:

$$b_{\text{eff}} = \rho b \tag{24.6}$$

where:
     b is the width of the element
     ρ is the reduction factor to allow for local buckling.

The reduction factor ρ takes account of the slenderness of the element, whether it is an internal or an outstand element, and the stress distribution within the element.

For an internal element ρ is given by:

$$\rho = \frac{\overline{\lambda}_{\text{p}} - 0.055(3 + \psi)}{\overline{\lambda}_{\text{p}}^2} \leq 1.0 \tag{24.7}$$

For an outstand element ρ is given by:

$$\rho = \frac{\overline{\lambda}_{\text{p}} - 0.188}{\overline{\lambda}_{\text{p}}^2} \leq 1.0 \tag{24.8}$$

where ψ is the stress ratio between the ends of the element and $\overline{\lambda}_p$ is the slenderness of the element given by:

$$\overline{\lambda}_{\text{p}} = \sqrt{\frac{f_{\text{y}}}{\sigma_{\text{cr}}}} = \frac{\overline{b}/t}{28.4\varepsilon\sqrt{k_\sigma}} \tag{24.9}$$

where:
     f_{y} is the design strength
     σ_{cr} is the elastic critical plate buckling stress
     $\overline{b}$ is the appropriate width of the compression element
     t is the steel core thickness (i.e. minus the coating)
     k_σ is the buckling factor corresponding to the stress ratio ψ and the boundary conditions. Values of k_σ should be obtained from Tables 4.1 and 4.2 of BS EN 1993-1-5 for internal and outstand elements respectively.
     $\varepsilon = \sqrt{235/f_{\text{y}}}$

From the discussion in 24.3.1, it follows that there must be limits of slenderness below which local buckling does not influence the resistance of the section. These limits correspond to $\rho = 1$ in Equations 24.7 and 24.8 and, according to BS EN 1993-1-5, have the following values:

Internal compression elements:

$$\bar{\lambda}_p \leq 0.673$$

Outstand compression elements:

$$\bar{\lambda}_p \leq 0.748$$

Where $\bar{\lambda}_p$ is lower than the appropriate limit, ρ should be taken as 1.0 in the calculation of the effective width of that element. This does not necessarily mean that the section is fully effective, since there may be other elements for which $\rho < 1.0$

24.4 Distortional buckling

In the discussion on local buckling, it was assumed that the corners of the section remained fixed in position, so that the buckling deformation takes place within the length of the element. This case is represented on the left hand side of Figure 24.6. By contrast, the right hand side of Figure 24.6 shows a situation in which the right hand corners of the flanges are not fixed in position, allowing the flanges to rotate. This is known as distortional buckling.

The susceptibility of a section to distortional buckling depends on the ability of the stiffeners to prevent displacement of the adjacent flange corners. This is dependent on the geometry of the stiffeners relative to the flanges and, in particular, their relative stiffness. BS EN 1993-1-3 presents a detailed procedure for the design of stiffened sections based on a simple spring model. This approach is described in Section 24.4.1 with a summary of the procedure given in 24.4.2.

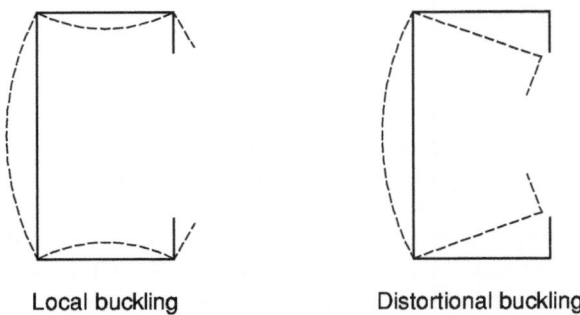

Local buckling   Distortional buckling

Figure 24.6   Local and distortional buckling

24.4.1 Design of stiffened sections

Due to the susceptibility of light gauge steel sections to local and distortional buckling, it is common practice for manufacturers to roll stiffeners into the sections. It
is most common to stiffen the free edge of the flanges by the provision of a lip, or
in some cases a double lip, but intermediate flange stiffeners may also be used in
order to permit the use of thinner gauge steel. It is less common to stiffen the web
in sections used for light steel framing applications, although deeper purlins and
floor joists may benefit from this technique. Flange and web stiffeners are often used
in trapezoidal deck profiles of the types used for roofing and flooring applications
due to the very thin gauge steel used in these products.

BS EN 1993-1-3 provides guidance to cater for all of the options discussed above.
In all cases, the underlying assumption is that the stiffener behaves like a compression member with a continuous partial restraint. This is a reasonable assumption
since, whether the member is subjected to pure axial compression or bending, one
flange and its stiffener at least will be subjected to a longitudinal compressive stress.
In the design model, the stiffener is represented by a linear spring of stiffness K, as
shown in Figure 24.7.

The spring stiffness depends on the boundary conditions and the flexural stiffness
of the adjacent plane elements. The spring is assumed to act at the centroid of the
effective stiffener section. Figure 24.7 shows two varieties of edge stiffener and one
intermediate stiffener. In each case, the 'effective stiffener section', which is depicted
as a dark solid line, comprises the stiffener itself plus an adjacent length (or lengths)
of flange.

In order that the stiffener provides sufficient stiffness and to avoid buckling of
the stiffener, BS EN 1993-1-3 gives limits for the geometry of the stiffener in relation to the adjacent flange as follows:

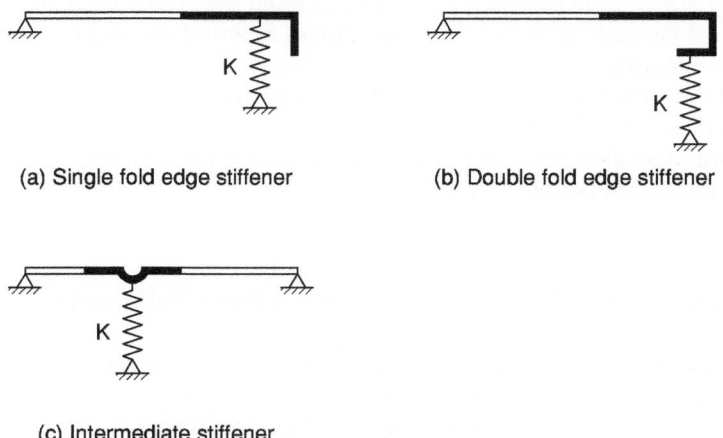

(a) Single fold edge stiffener   (b) Double fold edge stiffener

(c) Intermediate stiffener

Figure 24.7   Linear spring model

- For a single- or double-lipped section, the length of the lip c measured perpendicular to the flange width b should lie in the range $0.2 \leq c/b \leq 0.6$.
- For a double-lipped section, the return length of the lip d measured parallel to the flange should lie in the range $0.1 \leq d/b \leq 0.3$.

For the case of the edge stiffeners of lipped Cee and Zed sections, the spring stiffness K_1 for flange 1 may be obtained from the following equation:

$$K_1 = \frac{Et^3}{4(1-v^2)} \cdot \frac{1}{b_1^2 h_w + b_1^3 + 0.5 b_1 b_2 h_w k_f} \tag{24.10}$$

where:

$\quad b_1$ and b_2 are the distances from the web-to-flange junction to the centroid of the effective stiffener section for flanges 1 and 2 respectively

$\quad h_w$ is the depth of the web

$\quad k_f$ is the ratio of the effective areas of the two edge stiffeners (including the effective portion of the flange).

All other symbols have their usual meanings.
The term k_f can have the following values:

- for sections subjected to axial compression, so that flange 1 and flange 2 are both in compression: $k_f = A_{s2}/A_{s1}$ (where A_{s1} and A_{s2} are the effective areas of the edge stiffeners)
- for sections subjected to bending about the major axis so that flange 1 is in compression and flange 2 is in tension: $k_f = 0$
- for symmetric sections in compression: $k_f = 1$

Having calculated the equivalent spring stiffness of the stiffener, the method described in BS EN 1993-1-3 proceeds to determine the elastic critical stress for the stiffener $\sigma_{cr,s}$ followed by the relative slenderness $\overline{\lambda}_d$ and the reduction factor for the distortional buckling resistance χ_d. Finally, χ_d is used to determine the reduced effective area of the stiffener, which is generally represented as a reduced thickness when determining the effective properties of the section.

A detailed algorithm for determining χ_d and hence the reduced thickness of the stiffener is described below.

24.4.2 Eurocode calculation procedure for stiffened elements

Detailed procedures for the design of plane elements with edge or intermediate stiffeners are presented in Cl 5.5.3 of BS EN 1993-1-3. The procedures combine the calculation of b_{eff} with the calculation of a reduced thickness for the stiffener. The former takes account of local buckling within the length of the element, while the latter makes an allowance for the impact of distortional buckling. The procedure described below is for a flange with an edge stiffener. It is divided into three steps, the last of which involves an optional iteration in order to refine the value of the reduction factor χ_d. BS EN 1993-1-3 also presents procedures for flanges with inter-mediate stiffeners, stiffened webs and trapezoidal decking profiles.

Step 1:

The procedure begins with the calculation of the effective width of the flange b_{eff} following the method described in 24.3.2. At this point in the procedure, it is assumed that the edge stiffener is infinitely stiff and, therefore, provides full restraint to the free end of the flange. This corresponds to the left hand side of Figure 24.6, in which the corners of the section are fixed in position and failure is due to local buckling. It is also assumed that the maximum compressive stress in the flange is equal to the design strength of the material, i.e.

$$\sigma_{\text{com,Ed}} = f_{\text{yb}}/\gamma_{\text{M0}} \qquad (24.11)$$

Step 2:

In the second step, the edge stiffener is considered in isolation in order to calcu-late the reduction factor χ_d for distortional buckling. At this point, the infinitely stiff spring used in Step 1 is replaced by a spring of stiffness K, as illustrated in Figure 24.7. The spring stiffness K may be calculated from Equation 24.10, using the initial effective cross-section of the stiffener determined in Step 1. Once K is known, the elastic critical buckling stress $\sigma_{\text{cr,s}}$ of the stiffener may be calculated from:

$$\sigma_{\text{cr,s}} = \frac{2\sqrt{KEI_s}}{A_s} \qquad (24.12)$$

I_s and A_s are the effective second moment of area and the effective cross-sectional area respectively of the stiffener.

The relative slenderness of the stiffener for distortional buckling $\bar{\lambda}_d$ is given by:

$$\bar{\lambda}_d = \sqrt{\frac{f_{\text{yb}}}{\sigma_{\text{cr,s}}}} \qquad (24.13)$$

where f_{yb} is the basic design strength of the steel.

The reduction factor for distortional buckling χ_d is dependent on the slenderness $\bar{\lambda}_d$ as follows:

For $\bar{\lambda}_d \leq 0.65$, $\chi_d = 1.0$ (24.14a)

For $0.65 < \bar{\lambda}_d < 1.38$, $\chi_d = 1.47 - 0.723\bar{\lambda}_d$ (24.14b)

For $\bar{\lambda}_d \geq 1.38$, $\chi_d = \dfrac{0.66}{\bar{\lambda}_d}$ (24.14c)

Step 3:

The value of χ_d may be refined iteratively by returning to Step 1 and calculating a modified effective flange width b_{eff} based on a revised compressive stress $\sigma_{\text{com,Ed}}$.

This is achieved by calculating a modified value of ρ (see 24.3.2) using a reduced $\bar{\lambda}_p$ given by:

$$\bar{\lambda}_{p,red} = \bar{\lambda}_p \sqrt{\chi_d} \qquad (24.15)$$

Step 2 may then be repeated for the modified effective section to obtain a new value of χ_d. Steps 1 and 2 may be repeated until the desired degree of convergence on the value of χ_d has been achieved.

Step 3 is entirely optional and it is perfectly acceptable to use the initial value of χ_d for the calculation of the reduced area of the stiffener. Where the designer chooses to iterate to obtain an improved value of χ_d, one or two iterations should suffice. Once χ_d has been calculated to the desired degree of refinement, the reduced effective area of the stiffener $A_{s,red}$ may be calculated using:

$$A_{s,red} = \chi_d A_s \frac{f_{yb}/\gamma_{M0}}{\sigma_{com,Ed}} \qquad (24.16)$$

where $\sigma_{com,Ed}$ is the compressive stress at the centreline of the stiffener based on the effective cross-section.

It is usually more convenient to work in terms of a reduced thickness such that:

$$t_{red} = tA_{s,red}/A_s \qquad (24.17)$$

This may be calculated directly using:

$$t_{red} = t\chi_d \qquad (24.18)$$

Worked examples 1 and 2 at the end of this chapter illustrate the use of this procedure for a lipped channel section in bending and under pure axial compression respectively.

24.5 Design of compression members

24.5.1 Design issues

Light gauge steel members are often required to carry axial compression loads, such as in the case of studs in a load-bearing wall. As with their hot-rolled counterparts, the failure of light steel compression members is likely to be governed by buckling rather than yielding of the cross-section, resulting in a member resistance significantly lower than the squash load of the section. The design method for such a member is, therefore, focused on the calculation of its buckling resistance and is similar in many respects to the design of hot-rolled steel columns. However, the behaviour of light steel wall studs differs in a number of respects from that of hot-rolled columns and these differences must be accounted for in the design procedure.

Unlike columns, which act as independent members within a structural frame, light steel wall studs are used in conjunction with plasterboard, and often some form of sheathing board, to form a load-bearing panel. The presence of the boards will provide a certain degree of lateral restraint in the minor axis of the studs, which may be utilised when calculating the buckling resistance. However, any restraint must be verified by testing, using studs of a representative slenderness range and a similar build-up of boards to that used in practice.

While flexural buckling usually governs the behaviour of hot-rolled steel columns, many light steel sections are also susceptible to torsional-flexural buckling. If torsional-flexural buckling occurs at a lower magnitude of load than flexural buckling, this mode of failure will naturally govern the resistance of the member. This is reflected in the Eurocode design rules, in which the elastic critical buckling load used for design is taken to be the smallest of the elastic critical buckling loads for flexural buckling, torsional buckling and torsional-flexural buckling.

Finally, as noted in 24.3 and 24.4, light steel sections are susceptible to local and distortional buckling, both of which can have an adverse impact on the compression resistance of a member. This should be accounted for by using the effective cross-sectional area instead of the area of the gross cross-section when calculating the compression resistance.

24.5.2 Eurocode calculation procedures

The design procedures for light gauge steel compression members are given in Cl 6.2 of BS EN 1993-1-3. However, due to the similarities with the design of hot-rolled columns, designers are referred to Cl 6.3 of BS EN 1993-1-1 for much of the detail, including the buckling curves.

The design buckling resistance of a member subjected to axial compression is given by:

$$N_{b,Rd} = \frac{\chi \, A_{eff} \, f_y}{\gamma_{M1}} \tag{24.19}$$

where

- χ is the reduction factor for flexural buckling
- A_{eff} is the area of the effective cross-section (see 24.3 and 24.4)
- f_y is the design strength
- γ_{M1} is the partial safety factor for buckling.

Since f_y and γ_{M1} are known values ($\gamma_{M1} = 1.0$ in the UK) and the calculation of A_{eff} has been dealt with earlier in this chapter, the procedure described below focuses on the calculation of χ.

The reduction factor χ is used to quantify the reduction in resistance below the squash load of the section due to buckling. It may be obtained from BS EN 1993-1-1 using the appropriate buckling curve and the value of slenderness $\bar{\lambda}$ corresponding

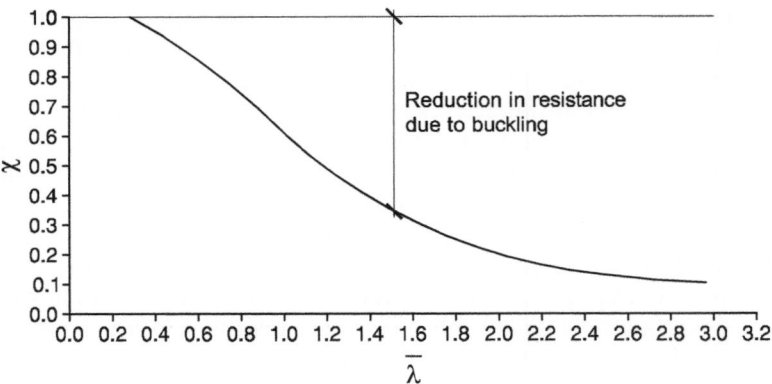

Figure 24.8   Typical buckling curve

to the critical mode of failure. BS EN 1993-1-1 offers a choice of 5 buckling curves, but this is restricted to 3 curves for light gauge steel according to Cl 6.2.2 of BS EN 1993-1-3. The appropriate choice of curve for various types of cross-section is given in Table 6.3 of BS EN 1993-1-3.

The relationship between χ and slenderness $\bar{\lambda}$ for buckling curve b is shown in Figure 24.8. The squash load of the section corresponds to $\chi = 1.0$.

The slenderness $\bar{\lambda}$ is given by:

$$\bar{\lambda} = \sqrt{\frac{A_{eff} f_y}{N_{cr}}} \tag{24.20}$$

N_{cr} is the elastic critical buckling load, which for flexural buckling is equal to the Euler load and is given by:

$$N_{cr} = \frac{\pi^2 EI}{L_{cr}^2} \tag{24.21}$$

where:
   E is Young's modulus for the material.
   I is the appropriate second moment of area (for the gross cross-section).
   L_{cr} is the effective length between points of restraint.

Alternatively, $\bar{\lambda}$ may be obtained from:

$$\bar{\lambda} = \frac{L_{cr}}{i} \frac{\sqrt{A_{eff}/A}}{\lambda_1} \tag{24.22}$$

where i is the radius of gyration and λ_1 is given by:

$$\lambda_1 = \pi \sqrt{\frac{E}{f_y}} \tag{24.23}$$

If the effective lengths differ between the major and minor axes, for example where a mid-height noggin provides restraint to the minor axis of a wall stud, values of $\bar{\lambda}$ should be obtained for both axes (since major axis flexural buckling might govern in this case). In cases where the critical mode of failure is either torsional buckling or torsional-flexural buckling, $\bar{\lambda}$ should be obtained from Equation 24.20 using the value of N_{cr} corresponding to the critical mode of failure (i.e. the elastic critical buckling load for torsional or torsional-flexural buckling). Equations for N_{cr} for both modes of failure are given in BS EN 1993-1-3.

The reduction factor χ may be obtained directly from the buckling curves printed in BS EN 1993-1-1 or from the following equations:

$$\chi = \frac{1}{\Phi + \sqrt{\Phi^2 - \bar{\lambda}^2}} \text{ but } \chi \leq 1.0 \tag{24.24}$$

$$\Phi = 0.5\left[1 + \alpha\left(\bar{\lambda} - 0.2\right) + \bar{\lambda}^2\right] \tag{24.25}$$

α is the imperfection factor corresponding to the chosen buckling curve. Values of α are given in Table 6.1 of BS EN 1993-1-1.

A worked example for the calculation of $N_{b,Rd}$ is included at the end of this chapter.

24.6 Design of members in bending

Several construction applications require light gauge steel members to carry loads in bending. Examples include floor joists, roof purlins and wall studs (when subjected to wind loading). As with hot-rolled steel beams, it is essential that a distinction is made between members that are laterally restrained and those that are susceptible to lateral-torsional buckling, since the bending resistance will differ significantly between the two cases. While many joists, purlins and beams have some form of attachment providing restraint, designers should beware of the potential lack of restraint during construction and the risk of load reversal. In addition to checking the member resistance, it may also be necessary to check serviceability limits such as deflections and dynamic response.

24.6.1 Laterally restrained members

Beams and similar structural members may be considered to be laterally restrained when their compression flange is held in position to the extent that lateral-torsional buckling is prevented. This is often the case with light steel framing members due to the attachment of sheathing or plasterboard (for walls) or flooring products (e.g. wooden floor boards or concrete slabs). Purlins and cladding rails are also often laterally restrained due to the attachment of cladding. However, as the cladding is only attached to one side of the member, which may be subjected to sagging or hogging bending depending on the direction of the loading, purlins and cladding rails must usually be designed for lateral-torsional buckling for at least one load case.

The design of laterally restrained light gauge steel beams is similar to the design of the equivalent hot-rolled members. As such, the following issues need to be considered:

- bending moment resistance
- shear
- local web failure
- deflections.

As noted previously, the key difference between light gauge and hot-rolled steel is the susceptibility to local and distortional buckling, both of which are dealt with by the use of effective section properties. However, the use of thin-gauge material has other consequences, such as the increased risk of shear buckling and of crushing, crippling or buckling of the web under local transverse forces. A typical failure of a light gauge steel member subjected to bending is shown in Figure 24.9. The member shown is a zed section purlin, but similar failure modes can be observed in lipped channels of the type used in framing applications.

Figure 24.9   Bending failure of a light gauge steel purlin

The resistance of a light steel cross-section to bending is considered in Cl 6.1.4 of BS EN 1993-1-3, which states that the moment resistance is given by:

$$M_{c,Rd} = \frac{W_{eff} f_{yb}}{\gamma_{M0}} \tag{24.26}$$

W_{eff} is the elastic modulus of the effective cross-section as discussed in 24.3 and 24.4 (see worked example 1).

Equation 24.26 assumes that failure is due to yielding of the compression flange. Where yielding occurs first in the tension flange, plastic reserves in the tension zone may be utilised as explained in Cl 6.1.4.2 of BS EN 1993-1-3. In this case, the bending moment will be limited by the maximum compressive stress $\sigma_{com,Ed}$ reaching f_{yb}/γ_{M0}.

The resistance to shear is considered in Cl 6.1.5 of BS EN 1993-1-3, which states that the design shear resistance is given by:

$$V_{b,Rd} = \frac{\dfrac{h_w}{\sin \phi} t f_{bv}}{\gamma_{M0}} \tag{24.27}$$

where:

    f_{bv} is the shear strength allowing for buckling
    h_w is the web height between the mid-lines of the flanges
    ϕ is the slope of the web relative to the flanges.

It is apparent from the use of the term $V_{b,Rd}$ that this is a buckling resistance rather than a cross-section resistance. This is due to the susceptibility of some light gauge sections to shear buckling. The risk of failure due to shear buckling is dependent on the slenderness of the web, so deep sections made from very thin gauge steel are most at risk. Shear buckling is accounted for in the design procedure by the use of f_{bv}, which is a function of the basic yield strength f_{yb} and the relative web slenderness $\bar{\lambda}_w$. Values of f_{bv} may be obtained from Table 6.1 of BS EN 1993-1-3.

Local transverse forces are considered in Cl 6.1.7 of BS EN 1993-1-3. Several equations are presented for the local resistance of the web $R_{w,Rd}$, taking into account the location of the applied load (e.g. whether close to the end of the member), the number of webs in the cross-section and the inclusion or absence of stiffeners.

24.6.2 Lateral-torsional buckling

Ideally, light gauge steel members should be incorporated into systems that provide lateral and torsional restraint and, therefore, prevent lateral-torsional buckling. However, there are occasions when this is not possible and the member has to be designed as an unrestrained beam. This situation is considered in Cl 6.2.4 of BS EN 1993-1-3. Alternatively, for purlins and cladding rails, the cladding may provide full lateral restraint when the flange to which it is attached is in compression, but only partial restraint when it is in tension. Design rules for this situation are provided in Cl 10.1.4 of BS EN 1993-1-3.

The behaviour of an unrestrained beam loaded in bending about its major axis is analogous to the behaviour of a column under axial load and the buckling curve depicted in Figure 24.8 is equally applicable to lateral-torsional buckling. The essential feature of this mode of failure is that the compression flange becomes unstable and, as it is not restrained, attempts to buckle laterally. However, since it is attached to the tension flange via the web of the beam, it cannot move freely and must pull the tension flange over as it deforms. The tension flange resists, resulting in the classical combination of lateral and torsional deformation, commonly known as lateral-torsional buckling. This type of failure only occurs when a member is loaded about its major axis; members loaded about their minor axis will always fail by minor axis bending and never by lateral-torsional buckling.

The impact of lateral-torsional buckling on the bending resistance of a beam is dependent on a number of factors. Principal among these are the geometry of the cross-section and the slenderness of the member. In terms of geometry, sections that have a low minor axis flexural rigidity relative to the flexural rigidity of the major axis are most likely to fail by lateral-torsional buckling. The ease with which a section is able to twist and warp is also important. Light gauge steel sections (e.g. channels and zeds) tend to be poor on both counts. By contrast, square hollow sections cannot suffer lateral-torsional buckling.

The relationship between slenderness and bending resistance can be represented by a buckling curve of the type shown in Figure 24.8. For very stocky members, the bending resistance will be limited by the resistance of the cross-section (given by Equation 24.26), but as the slenderness increases, the influence of lateral-torsional buckling also increases, resulting in a significant reduction in the bending resistance. As with the case of axial compression, it is convenient to quantify this reduction in terms of a reduction factor, expressed as a decimal proportion of the section capacity. For unrestrained beams, the Eurocode symbol for this reduction factor is χ_{LT}.

Due to the similarities between the design of light gauge steel unrestrained beams and the equivalent hot-rolled steel members, designers are referred to Cl 6.3 of BS EN 1993-1-1 for the detailed design procedures and the buckling curves. However, important guidance is given in Cl 6.2.4 of BS EN 1993-1-3 regarding the choice of design method and buckling curve.

The design buckling resistance of a member subjected to bending is given by:

$$M_{b,Rd} = \frac{\chi_{LT} \, W_{eff,y} \, f_y}{\gamma_{M1}} \tag{24.28}$$

where
  χ_{LT} is the reduction factor for lateral-torsional buckling
  $W_{eff,y}$ is the elastic modulus of the effective cross-section (major axis)
  f_y is the design strength
  γ_{M1} is the partial safety factor for buckling.

BS EN 1993-1-1 gives two alternative methods for the calculation of χ_{LT}. However, only the 'General case' given in Cl 6.3.2.2 is permitted for light steel members. This method resembles that used for column buckling and uses the same buckling curves. As such, the equations resemble those given in 24.5.2 with the addition of the subscript 'LT'.

The reduction factor χ_{LT} is given by:

$$\chi_{LT} = \frac{1}{\Phi_{LT} + \sqrt{\Phi_{LT}^2 - \bar{\lambda}_{LT}^2}} \text{ but } \chi_{LT} \leq 1.0 \tag{24.29}$$

and

$$\Phi_{LT} = 0.5\left[1 + \alpha_{LT}\left(\bar{\lambda}_{LT} - 0.2\right) + \bar{\lambda}_{LT}^2\right] \tag{24.30}$$

α_{LT} is the imperfection factor corresponding to the chosen buckling curve and is given by Table 6.3 of BS EN 1993-1-1. However, since Cl 6.2.4 of BS EN 1993-1-3 states that buckling curve *b* should always be used for light gauge steel, α_{LT} will always have the value 0.34:

$\bar{\lambda}_{LT}$ is the slenderness for lateral torsional buckling and is given by:

$$\bar{\lambda}_{LT} = \sqrt{\frac{W_{\text{eff},y} f_y}{M_{cr}}} \tag{24.31}$$

where M_{cr} is the elastic critical moment for lateral-torsional buckling based on gross cross-sectional properties.

24.6.3 Serviceability

In addition to checking the resistance of the member to the applied loading and associated bending moments and shear forces, designers should also check that the member is adequate at the serviceability limit state (SLS). For normal building applications, this involves checking the imposed load deflections against specified limits. Occasionally, the dynamic response of light steel floors will also need to be checked. Specialist guidance is available on this subject, for example SCI publication P354, *Design of floors for vibration: A new approach*.[6]

For the purpose of calculating deflections, the second moment of area should be calculated using the following equation:

$$I_{\text{fic}} = I_{gr} - \frac{\sigma_{gr}}{\sigma}\left(I_{gr} - I(\sigma)_{\text{eff}}\right)$$

where:

    I_{gr} is the second moment of area of the gross cross-section

    σ_{gr} is the maximum compressive bending stress at SLS based on gross cross-section properties

$I(\sigma)_{\text{eff}}$ is the second moment of area of the effective cross-section calculated for a maximum stress σ, where $\sigma \geq \sigma_{\text{gr}}$.

A worked example follows.

References to Chapter 24

1. British Standards Institution (2006) BS EN 1993-1-3 *Eurocode 3: Design of steel structures – Part 1-3: General Rules – Supplementary rules for cold-formed members and sheeting.* London, BSI.
2. British Standards Institution (2009) BS EN 10346 *Continuously hot-dip coated steel flat products – Technical delivery conditions.* London, BSI.
3. Grubb P.J., Gorgolewski M.T. and Lawson R.M. (2001) *Light Steel Framing in Residential Construction.* SCI Publication 301. Ascot, Steel Construction Institute.
4. British Standards Institution (2005) BS EN 1993-1-1 *Eurocode 3: Design of steel structures – Part 1-1: General rules and rules for buildings.* London, BSI.
5. British Standards Institution (2006) BS EN 1993-1-5 *Eurocode 3: Design of steel structures – Part 1-5: Plated structural elements.* London, BSI.
6. Smith A.L., Hicks S.J. and Devine P.J. (2009) *Design of floors for vibration: A new approach, Revised Edition.* SCI Publication 354. Ascot, Steel Construction Institute.

		Job No.		Sheet 1 of 5	*Rev*
![SCI logo] **The Steel Construction Institute**		Job Title			
Silwood Park, Ascot, Berks SL5 7QN Telephone: (01344) 623345 Fax: (01344) 622944		Subject	*Worked Example 1: Effective properties of a lipped channel in bending*		
		Client			
CALCULATION SHEET			Made by	AW	Date	
			Checked by	MH	Date	

Calculation of the effective section properties for a cold-formed lipped channel section in bending

This worked example presents the design procedure for the calculation of the effective section properties of a cold-formed lipped channel in bending.

Section dimensions and material properties

Section depth	h	$= 200\,mm$
Flange width	b_1	$= b_2 = 65\,mm$
Stiffener depth	c	$= 25\,mm$
Corner radius	r	$= 3\,mm$
Nominal thickness	t_{nom}	$= 2\,mm$
Core thickness	t	$= 1.96\,mm$
Design strength	f_y	$= 350\,N/mm^2$
Young's modulus	E	$= 210000\,N/mm^2$
Poisson's ratio	v	$= 0.3$
Partial safety factor	γ_{M0}	$= 1.00$

<u>Mid-line dimensions</u>
Web depth $\quad\quad h_p = h - t_{nom} = 200 - 2 = 198\,mm$
Flange width $\quad b_{p1} = b_{p2} = b_1 - t_{nom} = 65 - 2 = 63\,mm$
Stiffener depth $\quad b_{p,c} = c_p = c - t_{nom}/2 = 25 - 2/2 = 24\,mm$

<u>Geometry checks</u>

Checks on the geometry of the cross-section to ensure that the dimensions are within the scope of BS EN 1993-1-3.

$b_1/t = 65/1.96 = 33.16 < 60 \quad\quad OK$
$c/t = 25/1.96 = 12.76 < 50 \quad\quad OK$
$h/t = 200/1.96 = 102.04 < 500 \quad OK$

Check on the dimensions of the stiffener.
$c/b_1 = 25/65 = 0.38 \quad 0.2 < 0.38 < 0.6\ OK$

Check to see whether the rounding of the corners may be neglected.
$r/t = 3/1.96 = 1.53 < 5\ OK$
$r/b_{p1} = 3/63 = 0.05 < 0.10\ OK$

Gross section properties

$A = t(2c_p + b_{p1} + b_{p2} + h_p) = 1.96 \times (2 \times 24 + 63 + 63 + 198) = 729\,mm^2$

As the section is symmetrical about the major axis, the neutral axis is located at the mid-height of the web.
$z_{b1} = 99.0\,mm$

BS EN 1993-1-3 Cl 5.2

BS EN 1993-1-3 Cl 5.1(3)

Example 1 Effective properties – bending	Sheet 2 of 5	Rev

Effective section properties

The effective cross-section for a lipped C in bending is shown below. Note the ineffective portions of the flange and web and the reduced thickness $t\chi_d$ of the stiffener and adjacent section of flange.

BS EN
1993-1-3
Cl 5.5

The effective properties of the flange and web are determined separately as shown below, after which the effective properties of the whole cross-section may be calculated.

Effective properties of the compression flange and lip

*BS EN
1993-1-3
Cl 5.5.3.2*

Step 1

Effective width of the compression flange

For a stress ratio $\psi = 1$ (uniform compression), $k_\sigma = 4$

$\varepsilon = \sqrt{235/f_y}$

$\bar{\lambda}_{p,b} = \dfrac{b_{p1}/t}{28.4\,\varepsilon\sqrt{k_\sigma}} = \dfrac{63/1.96}{28.4 \times \sqrt{235/350} \times \sqrt{4}} = 0.691$

$\rho = \dfrac{\bar{\lambda}_{p,b} - 0.055(3+\psi)}{\bar{\lambda}_{p,b}^2} = \dfrac{0.691 - 0.055 \times (3+1)}{0.691^2} = 0.987 \le 1.0$

$b_{eff} = \rho b_{p1} = 0.987 \times 63 = 62.2\,mm$
$b_{e1} = b_{e2} = 0.5 b_{eff} = 0.5 = 62.2 = 31.1\,mm$

Effective width of the edge stiffener

The buckling factor is given by:
if $b_{p,c}/b_{p1} \le 0.35$: $k_\sigma = 0.5$

if $0.35 < b_{p,c}/b_{p1} \le 0.6$: $k_\sigma = 0.5 + 0.83\sqrt[3]{(b_{p,c}/b_{p1} - 0.35)^2}$

$b_{p,c}/b_{p1} = 24/63 = 0.38$ *so* $k_\sigma = 0.5 + 0.83\sqrt[3]{(0.38-0.35)^2} = 0.58$

*BS EN
1993-1-3
Cl 5.5.2
and
BS EN
1993-1-5
Cl 4.4*

*BS EN
1993-1-5
Cl 4.4*

*BS EN
1993-1-3
Cl
5.5.3.2(5a)*

Example 1 Effective properties – bending	Sheet 3 of 5	Rev

$$\bar{\lambda}_{p,c} = \frac{c_p/t}{28.4\,\varepsilon\sqrt{k_\sigma}} = \frac{24/1.96}{28.4 \times \sqrt{235/350} \times \sqrt{0.58}} = 0.690$$

BS EN 1993-1-5 Cl 4.4

$$\rho = \frac{\bar{\lambda}_{p,c} - 0.188}{\bar{\lambda}_{p,c}^2} = \frac{0.690 - 0.188}{0.690^2} = 1.05$$

but $\rho \leq 1$ so $\rho = 1$

BS EN 1993-1-5 Cl 4.4

The effective width is given by:
$c_{eff} = \rho c_p = 1 \times 24 = 24\,mm$

The effective area of the edge stiffener is:
$A_s = t(b_{e2} + c_{eff}) = 1.96 \times (31.1 + 24) = 108.0\,mm^2$

BS EN 1993-1-3 Cl 5.5.3.2(5a) Cl 5.5.3.2(6)

<u>Step 2</u>

The elastic critical buckling stress for the edge stiffener is given by:

$$\sigma_{cr,s} = \frac{2\sqrt{K\,E\,I_s}}{A_s}$$

BS EN 1993-1-3 Cl 5.5.3.2(7)

where K is the spring stiffness per unit length and I_s is the effective second moment of area of the stiffener.

$$K = \frac{E\,t^3}{4(1 - v^2)} \cdot \frac{1}{b_1^2\,h_p + b_1^3 + 0.5\,b_1\,b_2\,h_p\,k_f}$$

BS EN 1993-1-3 Cl 5.5.3.1(5)

$$b_1 = b_{p1} - \frac{b_{e2}tb_{e2}/2}{(b_{e2} + c_{eff})t} = 63 - \frac{31.1 \times 1.96 \times 31.1/2}{(31.1 + 24) \times 1.96} = 54.23\,mm$$

$k_f = 0$ *(for major axis bending)*
$K = 0.586\,N/mm$

$$I_s = \frac{b_{e2}\,t^3}{12} + \frac{c_{eff}^3\,t}{12} + b_{e2}\,t\left[\frac{c_{eff}^2}{2(b_{e2} + c_{eff})}\right]^2 + c_{eff}\,t\left[\frac{c_{eff}}{2} - \frac{c_{eff}^2}{2(b_{e2} + c_{eff})}\right]^2 = 6100\,mm^4$$

$$\sigma_{cr,s} = \frac{2 \times \sqrt{0.586 \times 210000 \times 6100}}{108.0} = 507.4\,N/mm^2$$

BS EN 1993-1-3 Cl 5.5.3.2(7)

$$\bar{\lambda}_d = \sqrt{f_y/\sigma_{cr,s}} = \sqrt{350/507.4} = 0.831$$

Since $0.65 < \bar{\lambda}_d < 1.38$, $\chi_d = 1.47 - 0.723\,\bar{\lambda}_d$

$\chi_d = 1.47 - 0.723 \times 0.831 = 0.870$

BS EN 1993-1-3 Cl 5.5.3.1(7)

<u>Step 3</u>

EN 1993-1-3 permits the optional iteration to refine the value of χ_d. This iteration has not been undertaken for this example, so the initial value of χ_d and the associated effective properties must be used. The next and, therefore, final step for the flange is the calculation of the reduced thickness for the stiffener.

$t_{red} = t\chi_d = 1.96 \times 0.870 = 1.70\,mm$

BS EN 1993-1-3 Cl 5.5.3.2(12)

Example 1 Effective properties – bending	Sheet 4 of 5	Rev

Effective properties of the web

The position of the neutral axis with regard to the flange in compression is:

$$h_c = \frac{c_p(h_p - c_p/2) + b_{p2} h_p + h_p^2/2 + c_{eff}^2 \chi_d/2}{c_p + b_{p2} + h_p + b_{e1} + (b_{e2} + c_{eff})\chi_d} \qquad h_c = 101.1\,mm$$

The stress ratio is given by:

$$\psi = \frac{h_c - h_p}{h_c} = \frac{101.1 - 198}{101.1} = -0.959$$

Referring to EN 1993-1-5, the buckling factor for the web is given by:
$k_\sigma = 7.81 - 6.29\psi + 9.78\psi^2 \qquad k_\sigma = 22.83$

$$\bar{\lambda}_{p,h} = \frac{h_p/t}{28.4\,\varepsilon\sqrt{k_\sigma}} = \frac{198/1.96}{28.4 \times \sqrt{235/350} \times \sqrt{22.83}} = 0.908$$

$$\rho = \frac{\bar{\lambda}_{p,h} - 0.055(3 + \psi)}{\bar{\lambda}_{p,h}^2} = \frac{0.908 - 0.055 \times (3 - 0.959)}{0.908^2} = 0.965$$

$h_{eff} = \rho h_c = 0.965 \times 101.1 = 97.5\,mm$

$h_{e1} = 0.4 h_{eff} = 0.4 \times 97.5 = 39.0\,mm$

$h_{e2} = 0.6 h_{eff} = 0.6 \times 97.5 = 58.5\,mm$

The effective width of the web is divided into two portions as follows:
$h_1 = h_{e1} = 39.0\,mm$

$h_2 = h_p - (h_c - h_{e2}) = 198 - (101.1 - 58.5) = 155.4\,mm$

Effective properties of the whole cross-section

$A_{eff} = t[c_p + b_{p2} + h_1 + h_2 + b_{e1} + (b_{e2} + c_{eff})\chi_d]$

$A_{eff} = 1.96 \times [24 + 63 + 39 + 155.4 + 31.08 + (31.08 + 24) \times 0.870]$

$A_{eff} = 706.4\,mm^2$

The position of the neutral axis with regard to the compression flange is given by:

$$z_c = \frac{t[c_p(h_p - c_p/2) + b_{p2}h_p + h_2(h_p - h_2/2) + h_1^2/2 + c_{eff}^2 \chi_d/2]}{A_{eff}} = 101.7\ mm$$

The position of the neutral axis with regard to the tension flange is given by:
$z_t = h_p - z_c = 198 - 101.7 = 96.3\,mm$

Rev column note:

BS EN 1993-1-5 Cl 4.4

Example 1 Effective properties – bending	Sheet 5 of 5	Rev

$$I_{eff,y} = \frac{h_1^3 t}{12} + \frac{h_2^3 t}{12} + \frac{b_{p2} t^3}{12} + \frac{c_p^3 t}{12} + \frac{b_{e1} t^3}{12} + \frac{b_{e2}(\chi_d t)^3}{12} + \frac{c_{eff}^3(\chi_d t)}{12} +$$
$$+ c_p t(z_t - c_p/2)^2 + b_{p2} t z_t^2 + h_2 t(z_t - h_2/2)^2 + h_1 t(z_c - h_1/2)^2 +$$
$$+ b_{e1} t z_c^2 + b_{e2}(\chi_d t) z_c^2 + c_{eff}(\chi_d t)(z_c - c_{eff}/2)^2$$

$$I_{eff,y} = 4235508 \, mm^4$$

$$W_{eff,y,c} = \frac{I_{eff,y}}{z_c} = \frac{4235508}{101.7} = 41657 \, mm^3$$

$$W_{eff,y,t} = \frac{I_{eff,y}}{z_t} = \frac{4235508}{96.3} = 43971 \, mm^3$$

			Job No.		Sheet 1 of 4		*Rev*

The Steel Construction Institute

Silwood Park, Ascot, Berks SL5 7QN
Telephone: (01344) 623345
Fax: (01344) 622944

CALCULATION SHEET

Job No.		Sheet 1 of 4	*Rev*
Job Title			
Subject	*Worked Example 2: Effective properties of a lipped channel in compression*		
Client	Made by	AW	Date
	Checked by	MH	Date

Calculation of the effective section properties for a cold-formed lipped channel section in compression

This worked example presents the design procedure for the calculation of the effective section properties of a cold-formed lipped channel in compression. The chosen section is identical to that considered in Example 1, so some of the calculations and checks relating to the gross cross-section have been omitted.

Section dimensions and material properties

Section depth	h	$= 200\,mm$
Flange width	b_1	$= b_2 = 65\,mm$
Stiffener depth	c	$= 25\,mm$
Corner radius	r	$= 3\,mm$
Nominal thickness	t_{nom}	$= 2\,mm$
Core thickness	t	$= 1.96\,mm$
Design strength	f_y	$= 350\,N/mm^2$
Young's modulus	E	$= 210000\,N/mm^2$
Poisson's ratio	v	$= 0.3$
Partial safety factor	γ_{M0}	$= 1.00$

<u>Mid-line dimensions</u>

Web depth	h_p	$= h - t_{nom} = 200 - 2 = 198\,mm$
Flange width	$b_{p1} = b_{p2}$	$= b_1 - t_{nom} = 65 - 2 = 63\,mm$
Stiffener depth	c_p	$= c - t_{nom}/2 = 25 - 2/2 = 24\,mm$

Gross section properties

$$A = t(2c_p + b_{p1} + b_{p2} + h_p) = 1.96 \times (2 \times 24 + 63 + 63 + 198) = 729\,mm^2$$

The neutral axis is located at the mid-height of the web.
$z_{b1} = 99.0\,mm$

Example 2 Effective properties – compression	Sheet 2 of 4	Rev

Effective properties of the flanges and lips	BS EN 1993-1-3 Cl 5.5.3.2
Step 1	
Effective width of the flanges	BS EN 1993-1-3 Cl 5.5.2 *and* BS EN 1993-1-5 Cl 4.4
For a stress ratio $\psi = 1$ (uniform compression), $k_\sigma = 4$	
$\varepsilon = \sqrt{235/f_{yb}}$	
$\bar{\lambda}_{p,b} = \dfrac{b_{p1}/t}{28.4\,\varepsilon\sqrt{k_\sigma}} = \dfrac{63/1.96}{28.4 \times \sqrt{235/350} \times \sqrt{4}} = 0.691$	
$\rho = \dfrac{\bar{\lambda}_{p,b} - 0.055(3+\psi)}{\bar{\lambda}_{p,b}^2} = \dfrac{0.691 - 0.055 \times (3+1)}{0.691^2} = 0.987 \le 1.0$	
$b_{eff} = \rho b_{p1} = 0.987 \times 63 = 62.2\,mm$ $b_{e1} = b_{e2} = 0.5 b_{eff} = 0.5 \times 62.2 = 31.1\,mm$	
Effective width of the edge stiffener	BS EN 1993-1-3 Cl 5.5.3.2(5a)
The buckling factor is given by:	
if $b_{p,c}/b_{p1} \le 0.35$: $k_\sigma = 0.5$	
if $0.35 < b_{p,c}/b_{p1} \le 0.6$: $k_\sigma = 0.5 + 0.83 \sqrt[3]{\left(b_{p,c}/b_{p1} - 0.35\right)^2}$	
$b_{p,c}/b_{p1} = 24/63 = 0.38$ *so* $k_\sigma = 0.5 + 0.83\sqrt[3]{(0.38-0.35)^2} = 0.58$	
$\bar{\lambda}_{p,c} = \dfrac{c_p/t}{28.4\,\varepsilon\sqrt{k_\sigma}} = \dfrac{24/1.96}{28.4 \times \sqrt{235/350} \times \sqrt{0.58}} = 0.690$	BS EN 1993-1-5 Cl 4.4
$\rho = \dfrac{\bar{\lambda}_{p,c} - 0.188}{\bar{\lambda}_{p,c}^2} = \dfrac{0.690 - 0.188}{0.690^2} = 1.05$	
but $\rho \le 1$ so $\rho = 1$	
The effective width is given by: $c_{eff} = \rho c_p = 1 \times 24 = 24\,mm$ *The effective area of the edge stiffener is:* $A_s = t(b_{e2} + c_{eff}) = 1.96 \times (31.1 + 24) = 108.0\,mm^2$	BS EN 1993-1-3 Cl 5.5.3.2(5a) Cl 5.5.3.2(6)
Step 2	
The elastic critical buckling stress for the edge stiffener is given by:	BS EN 1993-1-3 Cl 5.5.3.2(7)
$\sigma_{cr,s} = \dfrac{2\sqrt{K\,E\,I_s}}{A_s}$	
where K is the spring stiffness per unit length and I_s is the effective second moment of area of the stiffener.	

Example 2 Effective properties – compression	Sheet 3 of 4	Rev

$$K = \frac{E\,t^3}{4(1-v^2)} \cdot \frac{1}{b_1^2\,h_p + b_1^3 + 0.5\,b_1\,b_2\,h_p\,k_f}$$

BS EN
1993-1-3
Cl
5.5.3.1(5)

$$b_1 = b_{p1} - \frac{b_{e2}tb_{e2}/2}{(b_{e2}+c_{eff})t} = 63 - \frac{31.1 \times 1.96 \times 31.1/2}{(31.1+24) \times 1.96} = 54.23\ mm$$

$b_2 = b_1 = 54.23\,mm$ *(for a section with equal flanges)*

$$k_f = \frac{A_{s2}}{A_{s1}} = \frac{108}{108} = 1.0\ \text{ for a member in axial compression}$$

$K = 0.421\,N/mm^2$

$$I_s = \frac{b_{e2}\,t^3}{12} + \frac{c_{eff}^3\,t}{12} + b_{e2}\,t\left[\frac{c_{eff}^2}{2(b_{e2}+c_{eff})}\right]^2 + c_{eff}\,t\left[\frac{c_{eff}}{2} - \left[\frac{c_{eff}^2}{2(b_{e2}+c_{eff})}\right]\right]^2 = 6100\ mm^4$$

As the section has equal flanges, the spring stiffness K and second moment of area I_s are applicable to both edge stiffeners. Had the section been asymmetric, it would have been necessary to repeat the process shown above for the upper and lower edge stiffeners.

$$\sigma_{cr,s} = \frac{2 \times \sqrt{0.421 \times 210000 \times 6100}}{108.0} = 430\ N/mm^2$$

$\bar{\lambda}_d = \sqrt{f_y/\sigma_{cr,s}} = \sqrt{350/430} = 0.902$

BS EN
1993-1-3
Cl
5.5.3.1(7)

Since $0.65 < \bar{\lambda}_d < 1.38,\ \chi_d = 1.47 - 0.723\,\bar{\lambda}_d$

$\chi_d = 1.47 - 0.723 \times 0.902 = 0.818$

<u>Step 3</u>

EN 1993-1-3 permits the optional iteration to refine the value of χ_d. This iteration has not been undertaken for this example, so the initial value must be used.

$t_{red} = t\chi_d = 1.96 \times 0.818 = 1.60\,mm$

BS EN
1993-1-3
Cl
5.5.3.2(12)

Effective properties of the web

For uniform compression, the stress ratio $\psi = 1$ and the buckling factor $k_\sigma = 4$ (for an internal compression element).

BS EN
1993-1-5
Cl 4.4

$$\bar{\lambda}_{p,h} = \frac{h_p/t}{28.4\,\varepsilon\sqrt{k_\sigma}} = \frac{198/1.96}{28.4 \times \sqrt{235/350} \times \sqrt{4}} = 2.171$$

$$\rho = \frac{\bar{\lambda}_{p,h} - 0.055(3+\psi)}{\bar{\lambda}_{p,h}^2} = \frac{2.171 - 0.055 \times (3+1)}{2.171^2} = 0.414$$

Example 2 Effective properties – compression	Sheet 4 of 4	Rev

$h_{eff} = \rho h_p = 0.414 \times 198 = 82.0 \, mm$

$h_{e1} = h_{e2} = 0.5 h_{eff} = 0.5 \times 82.0 = 41.0 \, mm$

Effective properties of the whole cross-section

$A_{eff} = t[2b_{e1} + h_{e1} + h_{e2} + 2(b_{e2} + c_{eff})\chi_d]$

$A_{eff} = 459.1 \, mm^2$

Since the section is symmetrical and subjected to pure axial compression, the position of the centroidal axis remains unchanged from that of the gross cross-section, i.e. 99.0 mm from either flange.

The Steel Construction Institute	Job No.			Sheet 1 of 3	*Rev*
	Job Title				
Silwood Park, Ascot, Berks SL5 7QN	Subject	*Worked Example 3: Compression resistance of a cold-formed lipped channel*			
Telephone: (01344) 623345 Fax: (01344) 622944 **CALCULATION SHEET**	Client		Made by	AW	Date
			Checked by	MH	Date

Calculation of the compression resistance of a cold-formed lipped channel

This worked example presents the design procedure for the calculation of the compression resistance $N_{b,Rd}$ of a cold-formed lipped channel. It is assumed that failure will be due to flexural buckling rather than torsional-flexural buckling. However, due to the presence of a mid-height restraint in the minor axis, which halves the effective length in this direction, it is necessary to calculate $N_{b,Rd}$ for major and minor axis buckling. This example does not include the calculation of section properties as these procedures have already been demonstrated in Examples 1 and 2.

Member dimensions and properties

Length of member between restraints:

$L_y = 3.00\,m$

$L_z = 1.50\,m$

Effective lengths (assuming that the member is pin-ended):

$L_{cr,y} = 3.00\,m$

$L_{cr,z} = 1.50\,m$

Section depth	h	$= 200\,mm$
Flange width	b	$= 65\,mm$
Stiffener depth	c	$= 25\,mm$
Corner radius	r	$= 3\,mm$
Nominal thickness	t_{nom}	$= 2\,mm$
Core thickness	t	$= 1.96\,mm$
Design strength	f_y	$= 350\,N/mm^2$
Young's modulus	E	$= 210000\,N/mm^2$
Partial safety factor	γ_{M0}	$= 1.00$

Gross section properties

Area of gross cross-section	A	$= 729\,mm^2$
Second moment of area (major axis)	I_y	$= 440.5\,cm^4$
Second moment of area (minor axis)	I_z	$= 44.26\,cm^4$

Effective section properties

Effective area of the cross-section in compression: $A_{eff} = 459.1\,mm^2$

Example 3 Compression resistance	Sheet 2 of 3	Rev

Resistance to flexural buckling

BS EN 1993-1-3 Cl 6.2.2

$$N_{b,Rd} = \frac{\chi \, A_{eff} \, f_y}{\gamma_{M1}}$$

BS EN 1993-1-1 Cl 6.3.1.1

$$\chi = \frac{1}{\Phi + \sqrt{\Phi^2 - \bar{\lambda}^2}} \quad but \ \chi \le 1.0$$

BS EN 1993-1-1 Cl 6.3.1.2

$$\Phi = 0.5\left[1 + \alpha\left(\bar{\lambda} - 0.2\right) + \bar{\lambda}^2\right]$$

<u>Major axis</u>

$L_{cr,y} = 3.00\,m$

$$N_{cr,y} = \frac{\pi^2 E I_y}{L_{cr,y}^2} = \frac{\pi^2 \times 210000 \times 4405000}{3000^2} = 1014431\,N$$

$$\bar{\lambda}_y = \sqrt{\frac{A_{eff} f_y}{N_{cr,y}}} = \sqrt{\frac{459.1 \times 350}{1014431}} = 0.398$$

BS EN 1993-1-1 Cl 6.3.1.3

Use buckling curve b

BS EN 1993-1-3 Table 6.3

Imperfection factor α for curve b = 0.34

BS EN 1993-1-1 Table 6.1

$$\Phi = 0.5\left[1 + \alpha\left(\bar{\lambda}_y - 0.2\right) + \bar{\lambda}_y^2\right] = 0.5\left[1 + 0.34(0.398 - 0.2) + 0.398^2\right] = 0.613$$

$$\chi_y = \frac{1}{\Phi + \sqrt{\Phi^2 - \bar{\lambda}_y^2}} = \frac{1}{0.613 + \sqrt{0.613^2 - 0.398^2}} = 0.927$$

Flexural buckling resistance

$$N_{b,y,Rd} = \frac{\chi_y \, A_{eff} \, f_y}{\gamma_{M1}} = \frac{0.927 \times 459.1 \times 350}{1.0} = 149\,kN$$

<u>Minor axis</u>

$L_{cr,z} = 1.50\,m$

$$N_{cr,z} = \frac{\pi^2 E I_z}{L_{cr,z}^2} = \frac{\pi^2 \times 210000 \times 442600}{1500^2} = 407707\,N$$

$$\bar{\lambda}_z = \sqrt{\frac{A_{eff} f_y}{N_{cr,z}}} = \sqrt{\frac{459.1 \times 350}{407707}} = 0.628$$

BS EN 1993-1-1 Cl 6.3.1.3

Example 3 Compression resistance	Sheet 3 of 3	Rev
Use buckling curve b		*BS EN 1993-1-3 Table 6.3*
Imperfection factor α for curve b = 0.34		*BS EN 1993-1-1 Table 6.1*
$\Phi = 0.5\left[1 + \alpha\left(\bar{\lambda}_z - 0.2\right) + \bar{\lambda}_z^2\right] = 0.5\left[1 + 0.34\left(0.628 - 0.2\right) + 0.628^2\right] = 0.770$		
$\chi_z = \dfrac{1}{\Phi + \sqrt{\Phi^2 - \bar{\lambda}_z^2}} = \dfrac{1}{0.770 + \sqrt{0.770^2 - 0.628^2}} = 0.823$		
Flexural buckling resistance		
$N_{b,z,Rd} = \dfrac{\chi_z\, A_{eff}\, f_y}{\gamma_{M1}} = \dfrac{0.823 \times 459.1 \times 350}{1.0} = 132\ kN$		
<u>*Governing flexural buckling resistance*</u>		
$N_{b,Rd}$ *is the lesser of* $N_{b,y,Rd}$ *and* $N_{b,z,Rd}$		
Therefore, $\boldsymbol{N_{b,Rd} = 132\,kN}$		

Chapter 25
Bolting assemblies

MARK TIDDY and BUICK DAVISON

25.1 Types of structural bolting assembly

25.1.1 Non-preloaded structural bolting assemblies

The most frequently used bolting assemblies in structural connections are non-preloaded bolts of property classes 4.6 and 8.8 used in 2 mm clearance holes. These assemblies are termed ordinary non-preloaded bolting assemblies conforming to the requirements of BS EN 15048[1] as shown in Table 25.1.[2]

The property class designation system is in accordance with BS EN ISO 898-1.[3] It consists of two figures: the first is one-hundredth of the nominal tensile strength in megapascals (MPa), and the second is 10 times the ratio between the nominal yield strength and the nominal tensile strength. Multiplication of these two figures will give the yield strength in megapascals. A grade 4.6 bolt has a nominal tensile strength of 400 MPa and a yield strength of 240 MPa (0.6 × 400).

Nuts are designated by a number to indicate the maximum appropriate property class of bolts with which they may be mated. Hence a property class 8 nut would normally be used with a property class 8.8 bolt. It is permissible, however, to use a nut of higher property class than the bolt to minimise the risk of thread stripping. Where bolts are galvanised or sherardised, nuts of a higher property class must be used (see footnotes 4 and 5 in Table 25.1). Property class 8.8 bolts conforming to BS EN 15048 are commonly available and are recommended for all main structural connections, with the standard bolt being 20 mm diameter. Property class 4.6 bolts are generally used only for fixing lighter components such as purlins or sheeting rails, when 12 mm or 16 mm bolts may be adopted.

There may be situations, for example, a column splice subjected to large load reversals in a braced bay, where the engineer feels that joint slip is unacceptable. In these cases, high-strength structural bolting assemblies for preloading in accordance with BS EN 14399-3[4] may be used. High-strength structural bolting assemblies for preloading may also be used as non-preloaded in the same way as ordinary non-preloaded bolting assemblies.

Steel Designers' Manual, Seventh Edition. Edited by Buick Davison and Graham W. Owens.
© 2012 Steel Construction Institute. Published 2012 by Blackwell Publishing Ltd.

Table 25.1 Matching Ordinary Assemblies[2]

Grade	Bolt	Nut [1]	Washer
Incorporating full threaded length bolts			
4.6	BS EN ISO 4018	BS EN ISO 4034 (Class 4) [3]	BS EN ISO 7091 (100HV)
8.8	BS EN ISO 4017 [2]	BS EN ISO 4032 [2] (Class 8) [4]	BS EN ISO 37091 (100HV)
10.9	BS EN ISO 4017 [2]	BS EN ISO 4032 [2] (Class 10) [5]	BS EN ISO 7091 (100HV)
Incorporating part threaded length bolts			
4.6	BS EN ISO 4016	BS EN ISO 4034 (Class 4) [3]	BS EN ISO 7091 (100HV)
8.8	BS EN ISO 4014 [2]	BS EN ISO 4032 [2] (Class 8) [4]	BS EN ISO 7091 (100HV)
10.9	BS EN ISO 4014 [2]	BS EN ISO 4032 [2] (Class 10) [5]	BS EN ISO 7091 (100HV)

[1] Nuts of a higher class may also be used.
[2] Grade 8.8 and 10.9 bolts to the strength grades of BS EN ISO 4014 or BS EN ISO 4017 (dimensions and tolerances of BS EN ISO 4016 or BS EN ISO 4018) may also be used, with matching nuts to the strength classes of BS EN ISO 4032 (dimensions and tolerances of BS EN ISO 4034).
[3] Class 5 nuts for size M16 and smaller.
[4] Nuts for galvanised or sherardised 8.8 bolts shall be class 10.
[5] Nuts for galvanised or sherardised 10.9 bolts shall be class 12 to BS EN ISO 4033.

25.1.2 High-strength structural bolting assemblies for preloading

High-strength structural bolting assemblies for preloading are manufactured to the requirements of BS EN 14399-3. This British Standard covers two different property classes 8.8 and 10.9. Property Class 8.8 is the most commonly used type in general structural steelwork. Property Class 10.9 is used infrequently and therefore may require to be manufactured to order. The use of both 8.8 and 10.9 property class preloaded bolting assemblies is governed by BS EN 1090-2[5] (see Table 25.2).

25.1.3 Fully threaded bolts

Common practice in the past has been to use bolts with a short thread length, i.e. 1.5*d*, and to specify them in 5 mm length increments. This can result in an enormous number of different bolts, which is both costly to administer and can prevent rapid erection. It is recommended that fully threaded bolts (technically known as screws) be used as the industry standard. They can be provided longer than necessary for a particular connection and can therefore dramatically reduce the range of bolt lengths specified.

Research has demonstrated that the very marginal increase in deformation with fully threaded bolts in bearing has no significant effect on the performance of a typical joint. In the specific instances where this additional deformation might be of

Table 25.2   Matching preloaded assemblies[2]

	Bolt / nut / washer assembly System HR		Bolt / nut / washer assembly System HRC
General Requirements	**BS EN 14399-1**		
Bolt / nut assembly	BS EN 14399-3 Hexagon Bolt	BS EN 14399-7 Countersunk Bolt	BS EN 14399-10 Tension Control Bolt
Bolt marking	HR	HR	HRC
Nut marking	HR	HR	HR or HRD
Property classes	8.8 / 8 or 8.8 / 10 10.9 / 10	8.8 / 8 or 8.8 / 10 10.9 / 10	10.9 / 10
Washers	BS EN 14399-5 or BS EN 14399-6		
Washer marking	H		
Direct tension indicator, nut face washers and bolt face washers	BS EN 14399-9		At user's discretion
Direct tension indicator marking	H8 or H10		
Nut face washer marking	HN		
Bolt face washer marking	HB	Not applicable	
Suitability test for preloading	BS EN 14399-2 and, if any, additional testing specified in the product standard		

Bolt lengths shall be selected to ensure that a minimum of four full threads (in addition to the thread run-out) remain clear between the bearing surface of the nut and the unthreaded part of the shank.

concern, it is normal and recommended practice to use preloaded bolts. These can be used, for example, in tension and compression splices where the bolts are in shear/bearing or in column splices where the column ends are not in bearing. A paper by Owens[6] gives the background to the use of fully threaded bolts in both tension and shear conditions.

25.2 Methods of tightening and their application

BS EN 1090-2 permits three methods of tightening for preloaded bolts – torque control, torque control followed by part-turn of the nut, and direct tension indicators. Reference 7 briefly describes these methods as follows:

In the **torque control** method, the torque is applied in two steps. The first step, after bedding of the joint, is to apply a torque of up to 75% of the required torque value to all the bolts. The second step is to apply an additional torque to each bolt such that the total applied to the bolt is up to 110% of the required nominal torque value. The extra 10% is to offset the subsequent torsional relaxation of preload in the connection when the tightening wrench is removed.

The **combined method** is a combination of torque control and the traditional 'part-turn' method. After the joint is bedded, the preloading takes places in two steps. The first step is to apply a torque of up to 75% of the required torque value to all bolts. The second step is to apply to each bolt a predetermined rotation or 'part-turn' to a specified angle, depending on the bolt length.

The **direct tension indicator** method is the most popular in the UK and relies on protrusions on direct tension indicators previously known as load indicating washers. These protrusions create a gap prior to preloading in the installed assembly. After the joint is bedded down, the DTI is initially tightened until the protrusions start to deform, at this stage approximately 50% of the preload has been applied. When the gap is closed to the specified value, the bolt force will not be less than the specified preload.

25.3 Geometric considerations

25.3.1 Hole sizes

Ordinary bolts should be used in holes having a suitable clearance in order to facilitate insertion. For bolts up to and including diameters of 24 mm, the clearance should be 2 mm, and above 24 mm should be 3 mm. Table 3.3 in BS EN 1993-1-8[8] gives standard dimensions of holes for use with non-preloaded bolts. When using oversize or slotted holes, care should be taken that the washers used are sufficiently large and thick to span the hole. Large diameter washers to BS 4320[9] may be required.

Normal clearance holes, as given for ordinary bolts, are usually used for preloaded bolting assemblies but it is permissible to use oversize, short or long slotted holes, provided standard hardened washers are used over the holes in the outer plies and not just under the turned part.

Oversize and short slotted holes may be used in all plies but long slotted holes may only be used in one single ply in any connection. If a long slotted hole occurs in an outer ply, it should be covered with a washer plate longer than the slot and at least 8 mm thick.

The assessment of the slip resistance is affected when oversize or slotted holes are used. The constant k_s (Table 3.6 BS EN 1993-1-8), which is 1.0 for bolts in clearance holes, is reduced to 0.85–0.63 depending on the length of the slotted hole and its orientation relative to the direction of load transfer.

25.3.2 Spacing of fasteners, end and edge distances

Spacing is covered fully in Section 3.5 of BS EN 1993-1-8, and summarised as follows:

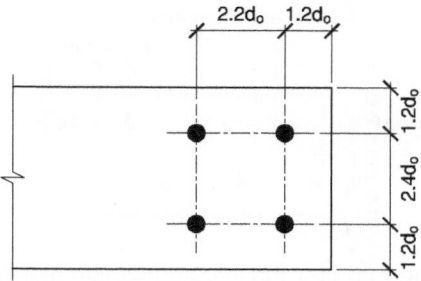

d$_0$ = hole diameter

Figure 25.1   Minimum dimensions

Minimum requirements (see Figure 25.1)

Centres of fasteners   2.2 d_0
Edge or end distances   1.2 d_0

The full bearing value of a bolt through a connected part cannot be developed if the end distance is less than 3 d_0 – see 25.5.3. Bearing.

Maximum requirements (see Figure 25.2)

Centres of fasteners in the direction of stress – the smaller of 14t or 200 mm where t is the thickness of the thinner element.

25.3.2.1 Edge distances

Distance to the nearest line of fasteners from the edge of an unstiffened part is 4t + 40 mm where t is the thickness of the thinner outside ply.

25.3.3 Back marks and cross centres

The back mark is the distance from the back of an angle or channel web to the centre of a hole through the leg or flange. This dimension is determined so as to allow the tightening of a bolt with a standard podger spanner, to be as near as possible to the centroidal axis and to allow the required edge distance. Recommended

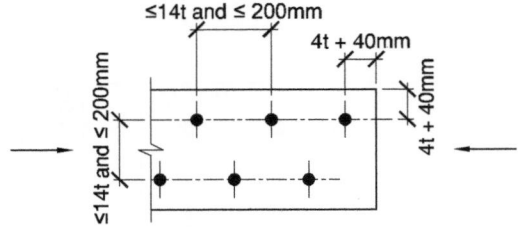

Staggered spacing in compression members

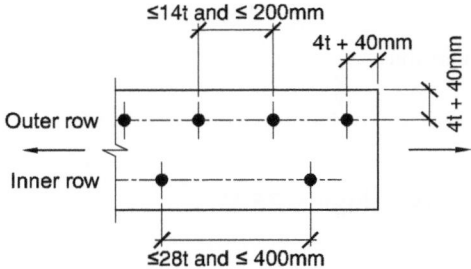

Staggered spacing in tension members

Figure 25.2   Maximum dimensions

back marks and diameters are given in the tables for channels and for angles (Appendix p. 1257).

The distances between centres of holes (cross-centres) in the flanges of joists, universal beams and universal columns are similarly determined after consideration of accessibility and edge distances.

Recommended cross-centres and diameters are given in the Appendix (p. 1258).

25.4 Methods of analysis of bolt groups

25.4.1 Introduction

Any group of bolts may be required to resist an applied load acting through the centroid of the group either in or out of plane producing shear or tension respectively. The load may also be applied eccentrically producing additionally torsional shear or bending tension. Examples are given in Figure 25.3.

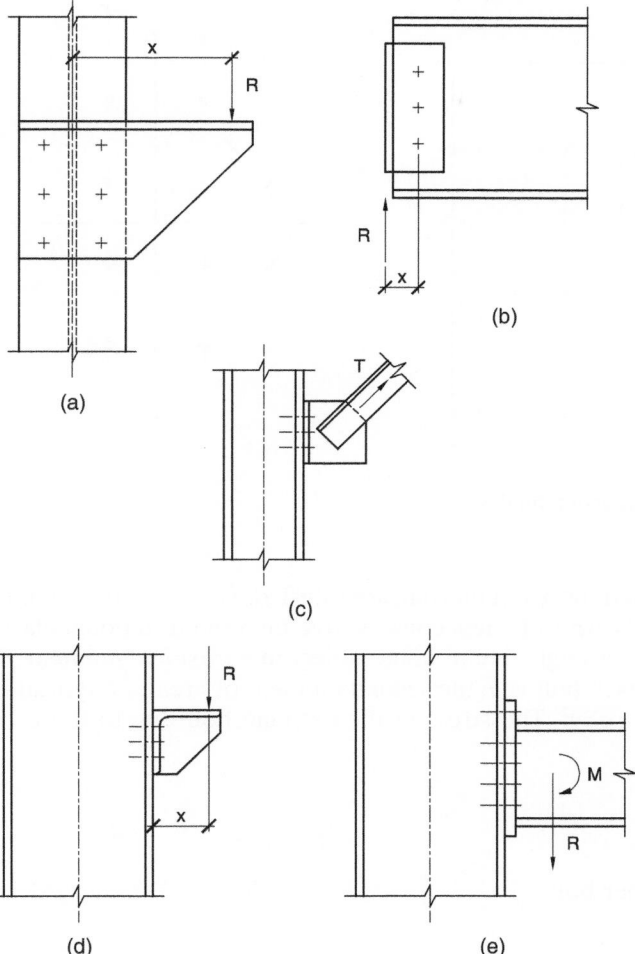

Figure 25.3   Bolt groups

25.4.2 Bolt groups loaded in shear

British and Australian practice is to distribute the torsional shear due to eccentricity elastically in proportion to the distance of each bolt from the centroid of the group. This is referred to as the *polar inertia method*.

(In some countries, notably Canada and in some cases the USA, the instantaneous centre method is used. This is a redistribution system, developed by Crawford and Kulak,[10] in which the assumed centre of rotation is continually adjusted until the

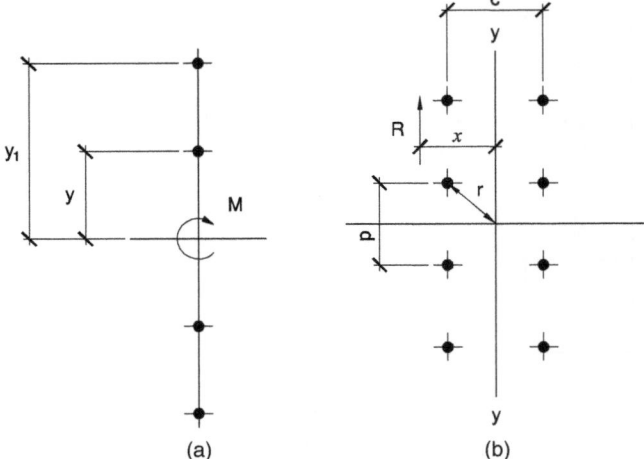

Figure 25.4   Bolt group analysis

three basic equations of equilibrium are satisfied. The method is a limit-state concept and has been shown to be less conservative than the traditional elastic methods.)

Consider first a single line of bolts subject to a torsional moment: Figure 25.4(a). If the area of each bolt is a, the second moment of area of a typical bolt is ay^2 and the total $\Sigma ay^2 = a\Sigma y^2$. The stress in the extreme bolt due to the eccentricity then becomes:

$$\frac{My_1}{I} = \frac{My_1}{a\Sigma y^2}$$

and the force per bolt

$$\frac{May_1}{a\Sigma y^2} = \frac{My_1}{\Sigma y^2}$$

The polar inertia about the centroid for any single line group containing n bolts with constant pitch p is:

$$I_o = \sum_{J=0}^{J=(n-1)} \left[\frac{(n-1-2J)}{2} p \right]^2$$

Consider next a double line of bolts subject to a load R with eccentricity x (Figure 25.4(b)).

The radius to the nearest bolt is given by:

$$r = \sqrt{\left[\left(\frac{p}{2} \right)^2 + \left(\frac{c}{2} \right)^2 \right]}$$

$$r^2 = \left(\frac{p}{2}\right)^2 + \left(\frac{c}{2}\right)^2$$

I_0 for a typical bolt is then ar^2 and it follows that I_0 for the whole group becomes $I_{xx} + I_{yy}$ which, if there are m vertical rows and n horizontal rows, is:

$$I_{00} = m \sum_{J=0}^{J=(n-1)} \left[\frac{(n-1-2J)}{2}p\right]^2 + n \sum_{J=0}^{J=(m-1)} \left[\frac{(m-1-2J)}{2}c\right]^2$$

where c is the cross-centre between the vertical lines. The distance to the extreme bolt is:

$$r = \sqrt{\left\{\left[\frac{(n-1)}{2}p\right]^2 + \left[\frac{(m-1)}{2}c\right]^2\right\}}$$

The force in the extreme bolt due to the moment is:

$$f_m = \frac{Rxr}{I_{00}}$$

The force in each bolt due to the shear (assumed equally divided between all bolts) is:

$$f_v = R/mn$$

The combined force per bolt is the resultant of these two: see Figure 25.5.

The resultant bolt force is then checked against the bolt strength in single shear, double shear or bearing as is appropriate. In the case of bearing, however, it should be remembered that the full strength cannot be achieved if the end distance

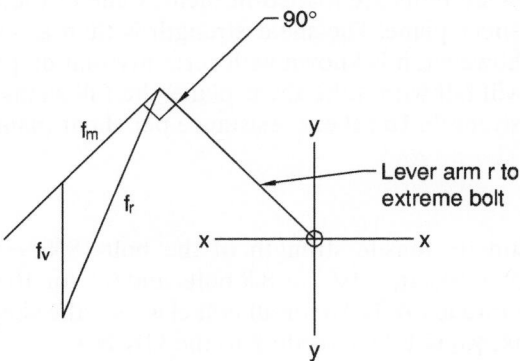

Figure 25.5   Resultant force

measured along the line of the resultant is less than twice the diameter of the bolt. In this case, the bearing strength is reduced in proportion.

25.5 Design strengths

25.5.1 General

Non-preloaded structural bolting assemblies have to resist forces in shear and bearing or tension, or combinations of these. High strength structural bolting assemblies for preloading resist shear by developing friction between the plies; they may also resist external tension. The load capacity of a joint may be affected if it is excessively long, has a large grip length or thickness of packing, or is subjected to prying action.

BS EN 1993-1-8 assigns bolted connections to one of five categories:

Category A: Bearing type – no preloading is required and the resistance is the lesser of the design shear or bearing resistance.
Category B: Slip resistant at serviceability limit state – preloading is sufficient to ensure that slip does not occur under serviceability loading but at the ultimate limit state the bolt acts as a Category A bearing type.
Category C: Slip resistant at ultimate limit state – the design slip resistance should be greater than the design ultimate shear load.
Category D: non-preloaded bolts loaded in tension, which are not suitable in connections where the tensile loading fluctuates.
Category E: preloaded bolts loaded in tension, which require controlled tightening.

25.5.2 Shear

When partially threaded bolts are loaded in shear, some of the threads of the bolt may lie within the shear plane. The shear strength is then assessed on the tensile area of the bolt. If, however, it is known with certainty that no part of the threaded section of the bolt will fall within the shear plane, the full shank area may be used in determining the strength. The shear resistance per shear plane, $F_{v,Rd}$, is given by:

$$F_{v,Rd} = \frac{\alpha_v f_{ub} A}{\gamma_{M2}}$$

where f_{ub} is the ultimate tensile strength of the bolt ($800\,N/mm^2$ for 8.8 bolts, $1000\,N/mm^2$ for 10.9 bolts); α_v is 0.6 for 8.8 bolts and 0.5 for 10.9 bolts if the shear plane is through the threads, α_v is 0.6 for all bolt classes if the shear plane is through an unthreaded shank; γ_{M2} is 1.25 according to the UK NA.[11]

In a preloaded bolted shear connection, shear force is resisted by friction until slip occurs. The slip resistance, $F_{s,Rd}$, is given by:

Table 25.3 Slip factors (reproduced from BS EN 1090-2 Table 18)

Surface Treatment	Class	Slip Factor μ
Surfaces blasted with shot or grit with loose rust removed, not pitted.	A	0.50
Surfaces blasted with short or grit: a) Spray-metallised with aluminium or zinc-based product b) With alkali-zinc silicate paint with a thickness of 50 μm to 80 μm	B	0.40
Surfaces cleaned by wire-brushing or flame cleaning, with loose rust removed	C	0.30
Surfaces as rolled	D	0.20

$$F_{s,Rd} = \frac{k_s n \mu}{\gamma_{M3}} F_{p,C}$$

where:

k_s = 1.0 for clearance holes

= 0.85 for oversized holes, short slotted holes loaded perpendicular to the slot direction

= 0.7 for long slotted holes loaded perpendicular to the slot direction

= 0.76 for short slotted holes loaded parallel to the slot direction

= 0.63 for long slotted holes loaded parallel to the slot direction

μ = slip factor which may be obtained from tests conducted in accordance with BS EN 1090-2 or from Table 18 of that standard (reproduced here as Table 25.3). Depending on the condition of the faying surfaces μ varies from 0.2 to 0.5

$F_{p,C}$ = the preloading force, which for class 8.8 and 10.9 bolts with controlled tightening, may be taken as $0.7 f_{ub} A_s$

n = the number of friction surfaces.

For category B preloaded bolts designed not to slip at the serviceability limit state, γ_{m3} is taken as $\gamma_{m3,ser}$ and Table 2.1 of BS EN 1993-1-8 recommends a value of 1.1 (which is the also the value adopted in the UK NA). For category C preloaded bolts designed not to slip at the ultimate limit state, γ_{M3} is taken as 1.1 in both the standard and the UK NA. For both category B and C preloaded bolts, the design shear resistance, $F_{v,Rd}$, and the design bearing resistance, $F_{b,Rd}$, must be greater than the design *ultimate* shear load. In addition, for category C preloaded bolting assemblies used in a connection in tension, the design plastic resistance of the net cross-section at the bolt holes should be checked.

25.5.3 Bearing

Bearing resistance may be controlled by either deformation of the bolt or, more usually, the bearing resistance of the plates or sections through which the bolts pass, and is a function of the position of the bolt holes, i.e. end, edge and pitch distances.

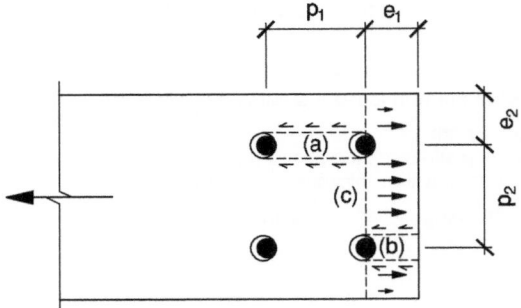

Figure 25.6   Bearing resistance failure modes

The bearing resistance of a bolt, $F_{b,Rd}$ with an ultimate tensile strength f_{ub} less than that of the connected parts, f_u, is given by:

$$F_{b,Rd} = f_{ub}dt / \gamma_{M2}$$

where γ_{M2} is 1.25 in the UK (the same as the recommended value in EC3-1-8 Table 2.1), and d and t are the bolt diameter and plate thickness respectively.

As bolt materials are usually stronger than plates or sections (i.e. $f_{ub}/f_u > 1$), bearing resistance is more commonly limited by the bearing failure of the contact surface of the plate against which the bolt bears. Initially the bolt contacts on a very small area, this yields locally and increases the area of contact to approximately dt. The plate bearing resistance, $F_{b,Rd}$, is determined from:

$$F_{b,Rd} = k \alpha_d f_u dt / \gamma_{M2}$$

The factor α_d is to account for plate tear-out by shearing in the direction of the loading (as shown in areas (a) and (b) Figure 25.6) and takes values of $e_1/3d_0$ (but not greater than 1) or $p_1/3d_0$ - 0.25 (but not greater than 1) for end and internal bolts respectively – d_0 is the bolt hole diameter. For maximum bearing resistance, the minimum end distance (e_1) is $3d_0$ and minimum pitch (p_1) $3.75d_0$. k_1 is concerned with tension fracture of the plate perpendicular to the direction of loading (line (c) in Figure 25.6) and is again related to bolt positioning. For bolts near the edge of a plate, k_1 is $2.8e_2/d_0$ -1.7 and for inner bolts $1.4p_2/d_0$ -1.7; in both cases k_1 is limited to a maximum value of 2.5, thus maximum bearing capacity requires $e_2 > 1.5d_0$ and $p_2 > 3d_0$. The footnotes to EC3-1-8 Table 3.4 gives reductions for the bearing resistance of bolts in oversize, slotted and countersunk holes.

25.5.4 Tension

Connections which put the bolts into tension may use non-preloaded (category D) or preloaded bolts (category E). If non-preloaded, bolt classes from 4.6 up to 10.9

may be used but for preloaded bolts only 8.8 and 10.9 are permitted and these must be tightened in a controlled manner. Non-preloaded bolts should not be used where the connection is subjected to fluctuations in the tensile loading (except where such variations in loading arise from normal wind loading).

The design tension resistance, $F_{t,Rd}$, of a bolt is given in EC3-1-8 Table 3.4 as:

$$F_{t,Rd} = k_2 f_{ub} A_s / \gamma_{M2}$$

where γ_{M2} is 1.25 (the recommended value in the standard and in the UK NA) and k_2 is 0.9, except where the bolt is countersunk, in which case, it is reduced further to 0.63. The k_2 factor accounts for the limited ductility of bolts in tension.

When bolts are loaded in tension, additional axial forces are sometimes induced due to prying action. When designing T-stubs in accordance with EC3-1-8 6.2.4, prying effects are implicitly taken into account. In cases where the prying force is calculated directly, the total applied tension in the bolt should be compared with the design tension resistance, $F_{t,Rd}$. EC3-1-8 provides no guidance on the calculation of prying forces but References 12, 13 and 14 provide some details of how they may be calculated.

25.5.5 Combined shear and tension

Non-preloaded bolts which are subject to both tension and shear should satisfy the following relationship:

$$\frac{F_{v,Ed}}{F_{v,Rd}} + \frac{F_{t,Ed}}{1.4\,F_{t,Rd}} \le 1.0$$

This expression allows a bolt fully loaded in tension to also resist shear forces up to approximately 30% of the design shear resistance. The expression is a conservative approximation to the real tension:shear interaction, as shown in Figure 25.7.

Preloaded bolts in friction grip connections that are also subject to externally applied tension should satisfy:

- for a category B connection (slip-resistant at SLS)

$$F_{s,Rd,ser} = \frac{k_s\, n\, \mu\, (F_{p,C} - 0.8\, F_{t,Ed,ser})}{\gamma_{M3,ser}}$$

- for a category C connection (slip-resistant at the ULS)

$$F_{s,Rd} = \frac{k_s\, n\, \mu\, (F_{p,C} - 0.8\, F_{t,Ed})}{\gamma_{M3}}$$

where the terms are as described in Section 25.5.2. The expressions account for the reduction in clamping force and therefore reduced friction, as a result of an applied tension force, $F_{t,Ed}$.

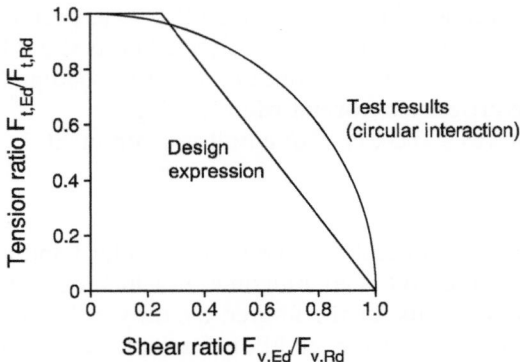

Figure 25.7   Bearing bolts in shear and tension[15]

25.5.6 Long joints and packing

When the joint length in a splice or end connection, L_j, defined as the distance between the first and last bolt on either side of the joint is greater than 500 mm (see, for example, Figure 25.8), the strength of the joint is reduced by the factor β_{LF}.

$$\beta_{Lf} = 1 - \frac{L_{j} - 15d}{200d}$$

but $0.75 \le \beta_{LF} \le 1.0$.

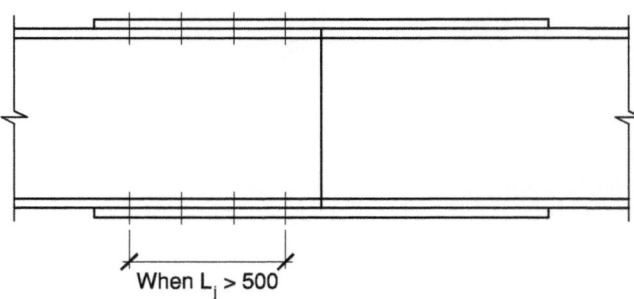

Figure 25.8   Long joint

Where bolts transmitting load in shear and bearing pass through packing of total thickness t_p greater than one-third of the nominal diameter d, (see Figure 25.9) the design shear resistance, $F_{v,Rd}$, should be multiplied by a reduction factor, β_p, given by:

$$\beta_p = \frac{9d}{8d + 3t_p} \quad \text{but } \beta_p \le 1$$

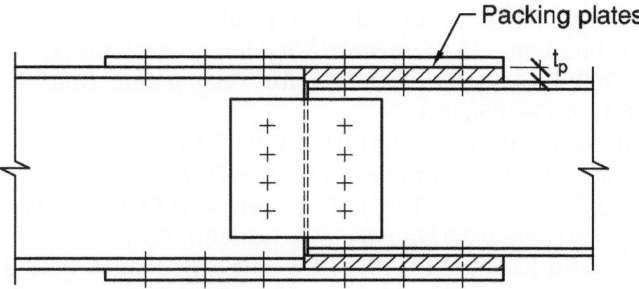

Figure 25.9 Fasteners through packings

The total thickness of steel packing is not limited in BS EN 1998-1-8 but it is recommended that the thickness, t_p, should not exceed $4d/3$, where d is the nominal bolt diameter.

In cases when more than one of the above conditions apply, it is only necessary to apply the factor producing the greater reduction. For preloaded slip resistant connections, the above reduction factors apply to plate bearing and bolt shear checks after slip has occurred but not the design slip resistance calculations.

25.6 Tables of resistance

Tables of bolt resistances are given in the Appendix.

References to Chapter 25

1. British Standards Institution (2007) BS EN 15048 Parts 1 and 2 *Non-preloaded structural bolting assemblies – Part 1: General requirements and Part 2: Suitability Test – specification.* London, BSI.
2. British Constructional Steelwork Association (2010) *National Structural Steelwork Specification for Building Construction*, 5th Edn. (CE Marking Version) London, BCSA.
3. British Standards Institution (2009) BS EN ISO 898–1 *Mechanical properties of fasteners made of carbon steel and alloy steel. Bolts, screws and studs with specified property classes – Coarse thread and fine pitch thread.* London, BSI.
4. British Standards Institution (2005) BS EN 14399-3 *High-strength structural bolting assemblies for preloading – System HR – Hexagon bolt and nut assemblies.* London, BSI.
5. British Standards Institution (2008) BS EN 1090-2 *Execution of steel structures and aluminium structures Technical requirements for the execution of steel structures.* London, BSI.

784 *Bolting assemblies*

6. Owens G.W. (1992) Use of fully threaded bolts for connections in structural steelwork for buildings. *The Structural Engineer*, 1 September, 297–300.
7. BCSA/SCI (2008) *European Standard for Preloadable Bolts*, Steel Industry Guidance Note SN26, 06/2008.
8. British Standards Institution (2005) *BS EN 1993-1-8, Eurocode 3: Design of steel structures – Part 1:8: Design of joints*. London, BSI.
9. British Standards Institution (1968) Metric series BS4320 *Specification for metal washers for general engineering purposes*. London, BSI.
10. Crawsford S.F. and Kulak G.L. (1971) Eccentrically loaded bolted connections. *Journal of the Structural Division, ASCE*, 97, No. ST3, March, 765–83.
11. British Standards Institution (2008) BS EN 1993-1-8 UK National Annex to *Eurocode 3: Design of steel structures – Part 1:8: Design of joints*. London, BSI.
12. Zoetemeijer, P. (1974) A design method for the tension side of statically loaded beam-to-column connections, *Heron*, 20, No. 1, 1–59.
13. Owens G.W. and Cheal B.D. (1989) *Structural Steelwork Connections*. London, Butterworths.
14. Swanson, J.A. (2002) Ultimate strength prying models for bolted T-stub connections, *Engineering Journal*, 39(3), September, 136–147.
15. Trahair, N.S., Bradford, M.A., Nethercot, D.A. and Gardner, L. (2008) *The behaviour and design of steel structures to EC3*, 4th edn. Abingdon, Taylor and Francis.

Further reading for Chapter 25

British Constructional Steelwork Association/The Steel Construction Institute (2010) *Joints in Steel Construction. Simple Connections*. London, BCSA, SCI.
British Constructional Steelwork Association (2010) *National Structural Steelwork Specification for Building Construction*, 5th edn. (CE Marking Version), Publication No. 52/10. London, BCSA.
Kulak G.L., Fisher J.W. and Struik J.H.A. (1987) *Design Criteria for Bolted and Riveted Joints*, 2nd edn. Chicago, John Wiley and Sons.

Chapter 26
Welds and design for welding

JEFF GARNER and RALPH B. G. YEO

Welding is essential in the fabrication of steel structures. Good design leads to cost-effective fabrications that can be made to required standards by the use of coordinated specifications, which provide means for quantitative control of weld quality. This chapter discusses the advantages of welding, the means to control weld quality and design recommendations.

26.1 Advantages of welding

Welding offers many advantages over other joining methods:

- freedom of design, and the opportunity to develop innovative structures
- easy introduction of stiffening elements
- less weight than in bolted joints because fewer plates are required
- welded joints allow increased usable space in a structure
- protection against the effects of fire and corrosion are easier and more effective.

Probably the main benefit of welded construction is the freedom of design, compared with bolted joints. Some important types of structure, such as Vierendeel trusses, tubular frames, tapered beams, and even most T-joints, could not be made as easily by any other method. Hollow sections, both rectangular and circular, can sensibly be joined only by welding, even if the final connection is by bolting.

Welded joints allow more freedom in the use of rolled shapes, high strength steels, and corrosion-resistant steels. Provided that the joints are made using appropriate materials and practices, designers should feel free to develop structures that are aesthetic as well as functional.

Welded construction allows a designer to introduce stiffness and strength where required, in the most discreet and/or the most structurally efficient manner. The transfer of load and the stiffness can, with welding, be introduced in a gradual continuous manner, instead of in step changes through bolted pieces.

Steel Designers' Manual, Seventh Edition. Edited by Buick Davison and Graham W. Owens.
© 2012 Steel Construction Institute. Published 2012 by Blackwell Publishing Ltd.

Total lifecycle costing, including the costs of maintenance, is important in all structures. One key aspect of this is durability, which for steel relates primarily to its corrosion resistance. The corrosion of structural steelwork can be retarded by three means – controlling the surrounding environment, effective coating systems, and the use of steels (and weld metals) with improved inherent corrosion resistance. Controlling the surrounding environment is normally impractical and, as such, the use of effective coating systems is generally the preferred method. Whilst these coating systems are effective on plate surfaces, it is difficult to prevent corrosion in all the crevices of bolted joints. The clean lines of welded structures, however, allow the full effectiveness of modern coatings to be demonstrated. The use of corrosion-resistant (weathering) steels (from BS EN 10025-5[1]) imparts long life, and whilst they can be used in bolted connections, they also perform poorly in crevices and work far better in welded solutions.

26.2 Ensuring weld quality and properties by the use of standards

Because the quality and properties of a weld cannot be readily verified on completion, welding requires continuous control and the use of specified procedures to ensure success. A comprehensive series of harmonised European standards have been developed which detail how to control and verify each aspect of the welding process. The use of the European design standard BS EN 1993[2] automatically initiates these harmonised standards to ensure that the performance of the welded joints will satisfy design requirements. Once specified, the designer needs to have no further input until the steelwork contractor submits all relevant information to demonstrate how the proposed manufacturing methods satisfy these harmonised standards. The flow of information required for the successful design, fabrication, and inspection of welded structures is shown schematically in Figure 26.1.

Perhaps the most important of the harmonised standards is BS EN 1090-2[3] since this brings together all of the relevant quality-related welding standards. It sets out the requirements for the manufacture of structures and components in structural steelwork based on four Execution Classes (EXC1 to EXC4) where the stringency of quality requirements increases from EXC1 to EXC4. Determination of the appropriate EXC is the responsibility of the designer and should be based on the

Figure 26.1   Information required to ensure weld quality and performance

service requirements of the structure and consequence of failure. However, for the steelwork contractor, the specified EXC determines the requirements for development and implementation of a Welding Quality Management System (WQMS) in accordance with the relevant part of BS EN ISO 3834.[4]

In addition to the welding process control aspects, the EXC also determines the requirements for those personnel involved in welding activities and introduces the concept of the 'Responsible Welding Coordinator' (RWC). Welding coordination is required for all except EXC1. A steelwork contractor should nominate at least one RWC who is competent, in terms of both technical knowledge and experience, to supervise its welding operations. Whilst designers are responsible for selecting the joint type, weld size, weld properties and required quality, the RWC is responsible for establishing and monitoring welding activities in accordance with the specified standards. Figure 26.2 shows the range of standards the RWC might typically use and how these interrelate to ensure acceptable quality and properties in welds for structural steelwork.

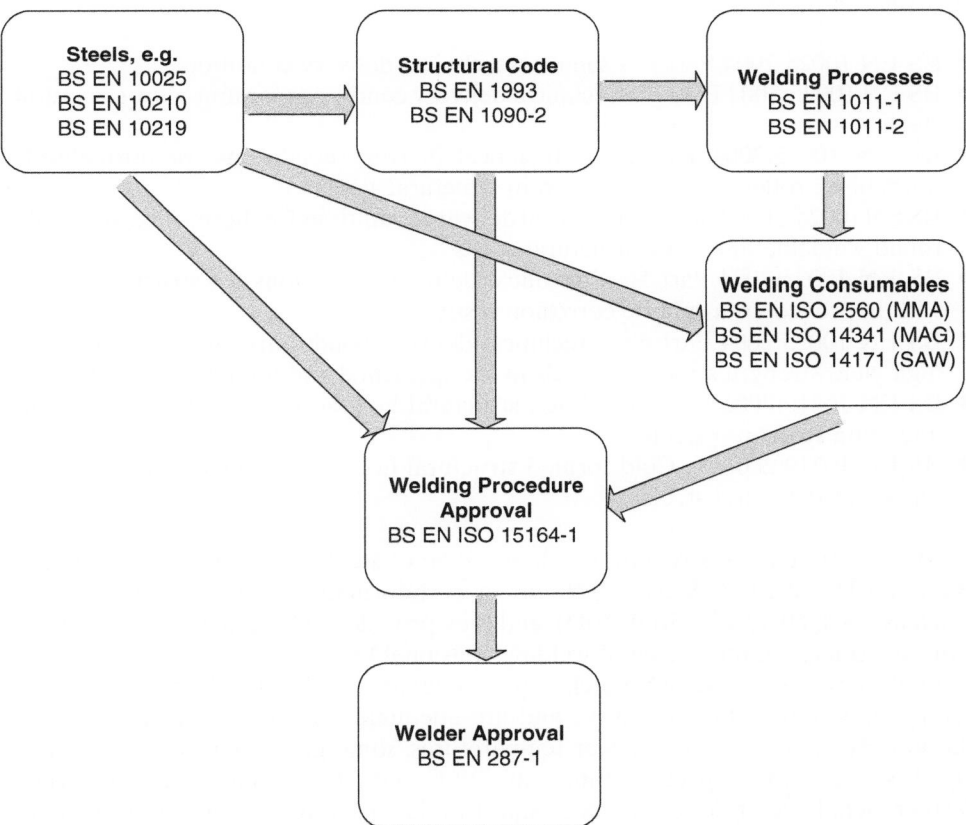

Figure 26.2   Standards that ensure acceptable quality and properties in welds

26.2.1 Standards – joint type, weld type, welding symbols and edge preparation

Designers should identify joint line, weld types, and throat dimensions using symbols shown in BS EN 22553.[5] Steelwork contractors use recommended edge preparations that are included in BS EN ISO 9692-1[6] and BS EN 1011-2.[7] Joint preparations should be appropriate for the welding process to be used, and the designer should not impose a preparation that may not be suitable for all steelwork contractors.

26.2.2 Standards – steel grade, steel selection

Designers should specify weldable structural steels, with grades and quality level suffixes for yield strengths in the range 185–460 N/mm^2, shown in European standards such as:

- BS EN 10025:2004 Part 1[8] – General technical delivery conditions
- BS EN 10025:2004 Part 2[9] – Technical delivery conditions for non-alloy structural steels
- BS EN 10025:2004 Part 3[10] – Technical delivery conditions for normalised/normalised rolled weldable fine grain structural steels
- BS EN 10025:2004 Part 4[11] – Technical delivery conditions for thermo-mechanically rolled weldable fine grain structural steels
- BS EN 10025:2004 Part 5[1] – Technical delivery conditions for structural steels with improved atmospheric corrosion resistance
- BS EN 10025:2004 Part 6[12] – Technical delivery conditions for flat products of high yield strength structural steels in the quenched and tempered condition
- BS EN 10210:2006[13] – Hot finished structural hollow sections of non-alloy and fine grain structural steels
- BS EN 10219:2006[14] – Cold formed structural hollow sections of non-alloy and fine grain structural steels.

The designations of strength, toughness, type of steel, and supply condition, e.g. BS EN 10025-2:2004[9], S355J2C+N, are more informative than previous systems (such as BS 4360:1990[15], Grade 50D), and they provide better guidance to the steelwork contractor in the choice of welding consumables.

Steels have their Charpy V-notch impact toughness (at least 27 J) tested at either room temperature, 0°C, or −20°C, and are adequate for most structural work in the UK. For applications at lower temperatures, some grades covered by BS EN 10025-3[10] have 40 J impact toughness at −20°C and 27 J toughness at −50°C. (For further details refer to Chapters 9 and 10.) In order to prevent cold cracking and lamellar tearing, the recommendations of BS EN 1011-2: 2001[7] should be followed.

The use of consistent product forms (i.e. plates, open sections, rectangular and circular hollow sections, etc.), standard sizes, and available grades in structural steels can result in considerable savings in cutting and welding costs, but not all shapes can be produced in all grades. With this in mind, designers should always check the availability and cost of a product before specifying.

26.2.3 Substitutions – thickness, yield strength, impact toughness, weldability, quality

Steelwork contractors may request, for reasons of availability and cost, the substitution of the specified grade and thickness by another grade and thickness. Both the steelwork contractor and engineer should be familiar with the importance of several factors: thickness, yield strength, ratio of yield to ultimate strengths, impact toughness, weldability, and quality. Although it may appear that a substitute steel with higher yield strength and/or thicker sections may be beneficial because it will provide more strength, its impact energy may not comply with the code requirements and weldability might be adversely affected.

Thicker steels impose several penalties. BS EN 1993-1-10[16] Cl 2.3 and PD 6695-1-10[17] referred to in its National Annex show that moving to a thicker substitute and/or higher yield strength may require higher impact toughness. BS EN 1011-2[7] shows that higher weld preheat temperatures may be required to prevent cracking.

Steels with inferior impact toughness should not be accepted. For instance, a steel with a designation for 27J at 0°C should not be used to replace a steel designated to have 27J at –20°C. Moving in the direction of lower test temperatures, however, is acceptable.

Whilst steelwork contractors will reasonably seek to reduce costs, cheaper steels typically have inferior quality and this generally leads to more expensive welds. Inferior rolling and levelling practices during manufacture of the steel may cause distortion problems when the steel is cut and welded. In some highly restrained welded connections there might also be a risk of lamellar tearing. Non-destructive testing before welding does not reveal the potential for tearing and, as such, steel should have good through-thickness ductility so as to minimise the risk. This is usually achieved by having a good quality steel with a low level of impurities and comes with the added advantage of improving impact toughness. In general, moving to cheaper steels rarely saves on overall cost.

26.2.4 Standards – welding processes and practices

BS EN 1011-2[7] provides guidance for welding practices used in the various codes for buildings and bridges, especially in the avoidance of cold cracks, hot cracks, and other unacceptable discontinuities. Significantly, it provides the steelwork contractor with the means to estimate preheat temperatures required to avoid cold cracking caused by hydrogen. When making recommendations to avoid cold cracking, BS EN 1011 incorporates the principles of diffusible hydrogen content of the weld metal, the carbon equivalent value (CEV) of the parent metal, the combined thickness of the members of the joint being welded, the heat input (determined from the energy input to the weld), and the preheat temperature. The recommendations of BS EN 1011 are incorporated into a Welding Procedure Specification (WPS) to be followed during welding. This ensures that weld properties and soundness, for example the absence of cold cracking, are achieved.

26.2.5 Welding standards – welding consumables

A harmonised set of European standards provides common designations for the yield strength and impact energy of weld metal deposited by the consumables used by the various processes, and additional information specific to the processes (shielding gas, flux type, etc.):

- BS EN ISO 14341:2011[18] Wire electrodes and deposits for gas shielded metal arc welding of non alloy and fine grain steels. Classification.
- BS EN ISO 14175:2009[19] Shielding gases for arc welding and cutting.
- BS EN ISO 2560:2009[20] Covered electrodes for manual arc welding of non-alloy and fine grain steels. Classification.
- BS EN ISO 17632:2008[21] Tubular cored electrodes for metal arc welding with and without a gas shield of non-alloy and fine grain steels. Classification.
- BS EN ISO 14171:2010[22] Welding consumables. Solid wire electrodes, tubular cored electrodes, and electrode / flux combinations for submerged arc welding of non-alloy and fine grain steels. Classification.
- BS EN 760:1996[23] Fluxes for submerged arc welding. Classification.

The standards classify the properties deposited from consumables in test welds under standardised welding conditions. From these test welds, the consumable manufacturer uses undiluted weld metal to demonstrate the strength and toughness of a particular consumable type. The term 'undiluted' relates to the absence of any dilution between the consumable and parent material to which it is being deposited and, as such, test samples are taken clear of the fusion line (see Figure 26.3).

The yield strength and impact toughness of the weld metal, and a series of operating characteristics, such as capability to weld in flat, horizontal, vertical and overhead positions, are used to give classification designations to the consumable.

Whilst the strength and toughness of a production weld may differ from the test values, as a result of dilution with the parent material and heat input, the designations guide the selection of consumables to produce weld metal properties that will match those of the parent plate.

Details of the consumables and the welding conditions to be used are entered into the WPS.

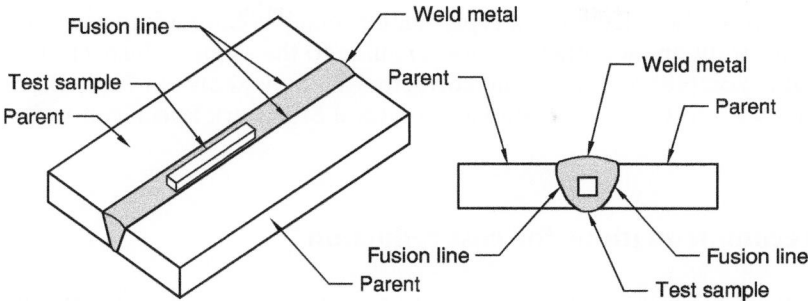

Figure 26.3   Diagram showing typical location of test samples to ensure undiluted weld metal

26.2.6 Standards – welding procedures

In the UK, to comply with the requirements of BS EN 1090-2[3] all welding procedures are typically qualified, using BS EN ISO 15614-1.[24] This standard shows the requirements and methods for the qualification of Welding Procedure Specifications (WPS). All new welding procedure qualifications are to be in accordance with this standard. An approved WPS provides all the information required by a qualified welder to make a joint with the desired properties.

26.2.7 Standards – welder approval

BS EN 1090-2[3] requires the use of suitably qualified welders. The use of a qualified WPS is only part of the quality assurance required in the production of satisfactory welds. The other essential ingredient is welder skill, which is guaranteed by ensuring that welders are qualified in accordance with the requirements of BS EN 287-1.[25] All the various criteria (process, welding position, steel thickness, etc.) specified in this standard are necessary to identify the ability of the welder to make specific welds. In the production of sound welds, the welder must also be capable of following the written instructions given in the WPS.

26.2.8 Standards – inspection and weld quality

All structural materials contain imperfections, and standards define the methods of inspection and the extent to which the imperfections can be accepted. The significance of an imperfection depends mainly on its size, shape and position, and the local applied stresses and temperature. Imperfections that might cause failure should obviously be rejected and repaired, but many common imperfections, such as minor porosity, are acceptable. A useful guide to the scope of inspection and the weld quality acceptance criteria and corrective action is shown in Table B, Annex B and Tables C1 and C2 of the National Structural Steelwork Specification.[26]

26.3 Recommendations for cost reduction

Small changes in design details can have significant effects on welding productivity and costs without adversely affecting the analysis or design of a structure. This section is directed at design improvements that are qualitative in nature but which reduce costs by paying attention to a series of simple principles which are summarised in Section 26.3.5.

26.3.1 Overall principles

Computer aided design is very effective in the analysis of structures to gain maximum efficiency in the use of materials, but the associated fabrication costs are not generally included in the software. Improved efficiency in the use of steel, especially for stiffening, usually leads to increased complication, the need for many short welds with difficult access, and a general increase in fabrication costs that will exceed the cost savings associated with the use of less steel. The use of standard rolled sections can save significant cutting and welding costs. Design priorities vary from one structure to another, but unless weight saving is crucial and is given highest priority, designers should aim for the minimum total cost which will, among other things, be a function of the number of welds to be made. When comparing design alternatives, the fabrication costs will be reduced if the number of individual pieces is minimised. For many structures the use of standard 6 mm leg length fillet welds is convenient but the associated heat input may be insufficient to prevent cold cracking of thick sections. Welding engineers should be consulted for advice when in doubt.

The sizes of welds should be no larger than required to transmit the design forces. The effective throat size a of a fillet weld should be taken as the perpendicular distance from the root of the weld to a straight line joining the fusion faces that lies just within the cross-section of the weld; it should not be taken as greater than 0.7 times the effective leg length, which is what is measured during inspection. Unnecessary weld metal raises costs, and it increases distortion to no good effect. Designers should therefore indicate the throat size required for fillet and butt welds, and give steelwork contractors the responsibility to produce and confirm consistent production of that size.

Care is required in the design of welds for regions where members are closely spaced, and where joints have to be made inside assemblies. All welded joints should be designed for easy access, and the welder or machine operator must be able to see where the weld is to be made, and be able to apply a Manual Metal Arc (MMA) electrode or Metal-arc Active Gas (MAG) gun so that the arc is directed to the bottom of the joint and at the correct angle to ensure root penetration. A MAG gun is essentially a large pistol with a 20 mm barrel, 100 mm long, at 60° to the handle, attached to a heavy cable. MMA welds need space for the manipulation of electrodes that are either 350 mm or 450 mm long, with a flux coating diameter that often exceeds 6 mm, and held in tongs or a holder attached to the welding cable. Where access might be restricted it is recommended that designers should consult a welding engineer to confirm feasibility.

Welding position, shown in Figure 26.4, is one of the important variables that influence the ease of fabrication, the costs of fabrication, and the mechanical properties of the weld.

Welding position influences many important factors:

26.3.1.1 Deposition rate

The PA and PB welding positions allow the highest deposition rates. High deposition rates cannot be used when making welds in the vertical (PF and PG) and especially the overhead (PD and PE) positions.

26.3.1.2 Welder qualification and availability of qualified welders

Welds are more difficult to make in the overhead (PD and PE) positions than in the downhand (PA and PB) positions. Consequently it is more difficult for a welder to gain qualification, and fewer qualified welders are available for the more difficult positions.

26.3.1.3 Weld quality

Welds made in the more difficult PD, PE, and PF positions are more likely to contain defects than similar welds made in easier positions. Aim for the maximum number of welds to be made in the PA or PB positions with minimum re-positioning of the components.

26.3.1.4 Weld metal properties

Weld metal toughness, strength, and hardness are influenced not only by the choice of consumables but also by the heat input into the weld. Heat input is primarily a

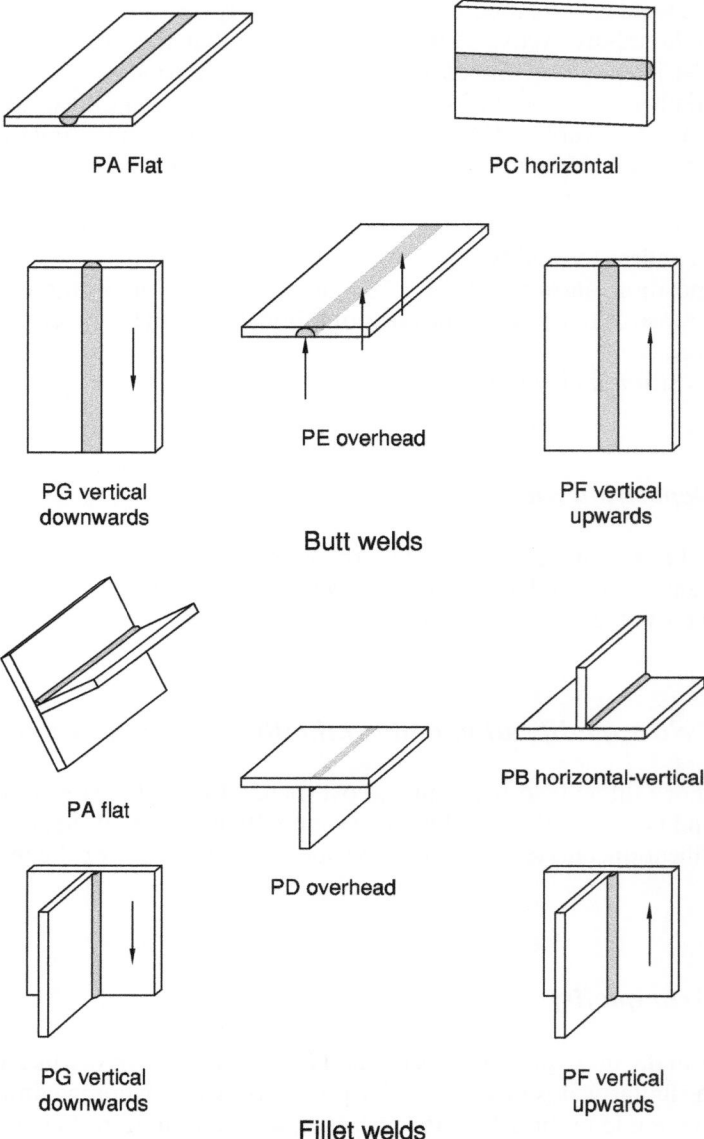

Figure 26.4   Designations of welding positions

function of the thermal efficiency factor of the welding process and the current, voltage and travel speed used to deposit the weld. When impact properties and hardness limits are specified, weld procedures must be qualified to cover the low and high ends of the heat input range. However, if all welds can be made in the PA and PB positions only one set of tests need be made.

The components of a structure will require joints to be made from one or more of five configurations. The most common are butt (in-line), T-, corner, and lap joints. Each type of joint may be connected by several types of weld. The weld types (not to be confused with joint types) recognised by BS 499-1[27] are fillet, butt, compound welds (consisting of both fillet and butt), plug welds, and edge welds. The choice of joint and weld is a major factor in welding costs. They may have similar costs of materials, but significantly different fabrication costs. About 80% of structural joints have a T-configuration, which might require either fillet welds or butt welds. Except where in-line butt joints are necessary, designers should always attempt to choose joints that can be made with fillet welds. Significant costs can be saved by using fillet welds wherever possible; where a butt weld is essential a partial penetration weld should be selected if possible, bearing in mind strength, fatigue and corrosion limitations.

26.3.2 Fillet welds

Fillet welds have triangular cross-section and they are commonly used to make T-joints, corner joints with several variations, and lap joints, shown in Figure 26.5. As stated previously, wherever possible, fillet welds should be made in the PA or PB position. In this welding position, 8 mm is the maximum leg length that can be made in a single-pass weld. Single-sided fillet welds should not be used where tension resulting from the welding operation would introduce a bending stress that would open the root. Bending to close the root is admissible, and fillet welds on both sides of a member will prevent root opening.

26.3.3 Butt welds

Whereas fillet welds join the surfaces of adjacent members, butt welds join all or part of their cross-section, and are consequently called full or partial penetration

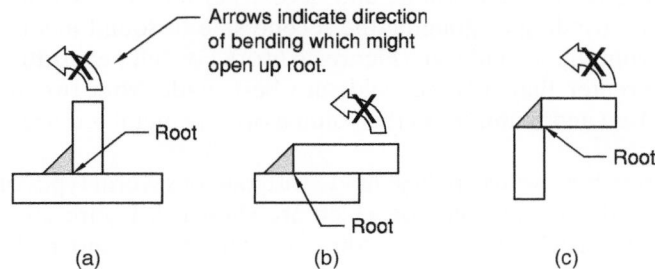

Figure 26.5 Typical fillet weld configurations for: (a) T-joints, (b) lap joints, (c) corner joints. Arrows indicate direction of bending which might open up the root

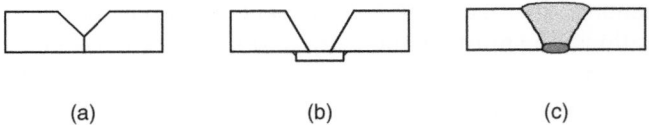

(a)   (b)   (c)

Figure 26.6   Typical butt weld configurations

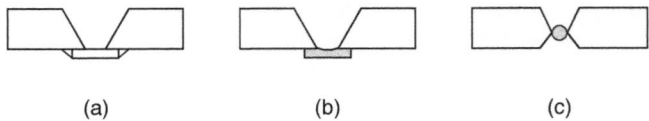

(a)   (b)   (c)

Figure 26.7   Typical joint configurations for full penetration butt welds with temporary or permanent backing: (a) permanent steel strip, (b) copper backing bar, (c) ceramic tiles

welds. Even where the joint configuration requires the use of a butt weld, partial penetration is often sufficient, instead of full penetration. Designers should state the weld throat dimension required instead of routinely demanding full penetration with its extra difficulties and costs. Current practice is to use suitable weld preparations, qualified procedures, and qualified welders to ensure the required depth of fusion for the throat size. When partial penetration is adequate the designer should not state, 'All butt welds must have full penetration'. To a welding engineer, this means full penetration through the sections, when all that is intended is for the steelwork contractor to show proof of achieving the required throat depth.

Figure 26.6 shows different approaches to butt welds. The partial penetration weld (Figure 26.6(a)) is the most economical to prepare and weld. Throat size can be ensured by use of high-current MAG, flux-cored arc, or submerged arc welding, and the steelwork contractor can be asked to provide proof of consistent penetration. For materials up to approximately 12 mm thick, where full penetration is justified, welds should be made from one side with permanent backing (as Figure 26.6(b)).

Where backing is not permissible, and access permits, welds can be made by welding one side, grinding or gouging the second side to sound metal and welding to completion from the second side (Figure 26.6(c)). For full penetration butt welds in thicknesses greater than 12 mm, welds are best made from two sides so as to minimise distortion and to minimise the volume of weld metal required to complete the weld.

Full-penetration butt welds are best made with one of several types of permanent or temporary weld backing, some of which are shown in Figure 26.7. Permanent backing is typically provided by tack welding steel strip to the underside of the joint. Since this becomes an integral part of the finished weld it should be of a similar grade of steel to that being joined. Permanent backing can sometimes be provided by an adjacent member, especially in corner welds. Temporary backing is typically provided by using ceramic tiles or alternatively a copper backing bar. These can be held in place by either the use of heat resistant tape or magnetic supports/fixtures.

The backing allows the welder to use sufficient heat to ensure penetration and full fusion through the full thickness of the members being joined. Where temporary backing is used, this also allows control of the finished penetration profile, minimising the amount of rework required on the underside of the joint.

26.3.4 Consultation with the steelwork contractor

Even though steelwork contractors often hold designers responsible for welding problems, welding staff are reluctant (and often not qualified professionally) to recommend improvements. However, designers cannot be expected to be familiar with all welding innovations, nor the capabilities of individual welding shops and their personnel. Therefore the best way to ensure efficient welding details is for designers to initiate genuine dialogue with steelwork contractors and to consider suggestions for alternative approaches to welding.

26.3.5 Summary of recommendations

The items discussed above can be summarised as follows:

- minimum number of welds and fewest different weld types
- minimum cost welds (i.e. fillets rather than butts)
- full penetration welds only where the following show them to be essential:
  - (a) for a one-sided weld the loading might open the root of the weld
  - (b) the loading is cyclic and fatigue cracks could be initiated in the unfused part of the joint
  - (c) corrosive attack might occur in the non-welded crevice
- smallest size of welds (while recognising that certain sizes are preferred)
- easy welds in terms of access and welding position
- root backing for easy production of full-penetration butt welds
- balanced welding (on both sides of neutral axes) for minimum distortion and least sensitivity to dimensional discrepancies
- easy inspection.

26.4 Welding processes

26.4.1 Introduction

All welding processes have their inherent advantages and disadvantages. The choice of welding process is generally the responsibility of the steelwork contractor, and it depends on the availability of equipment and welders. The objective of this section is to outline some of the features and characteristics of the popular arc welding processes.

The important processes used in steel fabrication are differentiated technically by making reference to the type of consumable electrode and the method of protecting the arc. They are defined and numbered in BS EN ISO 4063[28] as follows:

- 111 Metal-arc welding with covered electrode
- 114 Self-shielded tubular cored arc welding
- 121 Submerged arc welding with solid wire electrode
- 135 Metal-arc active gas welding (MAG) welding with solid wire electrode
- 136 Metal-arc active gas welding (MAG) welding with flux cored electrode
- 138 Metal-arc active gas welding (MAG) welding with metal cored electrode.

National statistics provide an idea of their relative popularity. In the UK market, MAG welding accounts for about 75%, cored wire welding (with and without gas shielding) has a further 7%, and MMA welding now accounts for about 9% of welding consumables consumption. Submerged arc welding has accounted for 7–8% for many years. Popularity depends mainly on productivity, which has the largest influence on welding costs.

26.4.2 Manual metal arc (MMA) welding

MMA welding dominated the structural steelwork industry for many years because of its versatility and suitability for both shop and site application, and the widespread availability of welders qualified for its use. MMA welding is colloquially known as 'stick' welding. MMA equipment is widely available and inexpensive in comparison with other processes.

A typical MMA electrode, which usually has a length in the range of 230–460 mm, is held firmly in tongs or holders that are connected by flexible cables to the power source, which may be a transformer, rectifier, inverter, or engine-driven generator, as shown in Figure 26.8. The electrode coating melts in the arc and (1) protects the weld metal from the surrounding air, (2) forms a slag that protects and supports the weld metal, and (3) usually adds alloying elements to the weld metal.

26.4.3 MAG welding

MAG welding wire is generally copper-coated to provide good electrical contact with the tip in the welding gun. The majority of MAG wires contain sufficiently high levels of silicon and manganese to prevent porosity that would be caused by oxygen that enters the weld metal from the active shielding gases used. A limited range of low-alloy wires is available for high strength and toughness applications.

Figure 26.8   MMA repair welding of a crash barrier (photograph courtesy of the Lincoln Electric Company)

The essential components of equipment for MAG welding are the hand-held gun (or mechanised welding head) through which the continuous wire is fed into the arc, the wire feeder, and a constant-voltage welding power source, as shown in Figure 26.9. The continuous nature of the process allows it to be used for semi-automatic, mechanised, automatic, and robotic welding. The arc conditions (voltage and current) are controlled mainly by the equipment, giving rise to the use of the term semi-automatic welding.

The MAG welding gun feeds the wire, conducts the current and delivers a gas shield. The wire leaves the gun from a copper contact tube, which is surrounded by a concentric gas nozzle. All components of the gun should have adequate current-carrying capacity and be tightly connected to provide good electrical and thermal conductivity.

The wire feeder, fitted with V-grooved rolls for solid MAG wire or knurled rolls for cored and submerged arc wires, pulls the wire from a spool or drum and pushes it through the cable (or welding head on a mechanised machine) to the gun.

Figure 26.9   MAG welding of a thin gauge T-joint in a workshop (photograph courtesy of the Lincoln Electric Company)

Cored wire welding

Cored wire welding is a modification of MAG welding in which the electrode wires are hollow steel tubes filled with powders. Cored wires have some specific benefits, especially productivity and weld quality, compared with solid wire MAG or MMA welding. Most cored wires require the use of a gas shield, but some are self-shielding, thereby making them ideal for site welding where winds would blow the gas shield away and cause weld metal embrittlement.

26.4.4 Submerged arc welding

In the submerged arc process shown in Figure 26.10 the weld is protected by a flux instead of a gas. The granulated flux powder is continually fed around the wire electrode, primarily to protect the arc and molten weld metal during the welding operation. Consequently this process is virtually confined to welds that can be made in the flat (PA) and horizontal vertical (PB) positions. Submerged arc welding is the preferred process for fabricating plate girders because it gives a combination of productivity and high quality that cannot be matched by other processes. The process

Figure 26.10 Automatic submerged arc welding machine fitted with two welding heads feeding flux from hoppers to cover the wire and arc (photograph courtesy of the Lincoln Electric Company)

is generally used for mechanised welding with machines for making accurately placed passes in straight lines in the longitudinal or circumferential direction on girders, drums, and pipes.

26.4.5 Welding productivity

Major improvements in productivity have been made possible by the use of processes that use continuous welding wires to provide the filler metal. The MMA process, which was the first to be used for widespread fabrication, uses individual electrodes made from cut lengths of steel rod. Each rod runs for only about one minute before the stub must be discarded and a new electrode fitted into the holder. This intermittent process has been replaced by the use of the continuous processes (MAG and cored wire) where the welder can continue until lack of access or reach requires the welder to stop welding and move to a new position. If feasible, the weld could continue until the spool (generally about 15 kg) is empty. Stops and starts are the main sites of defects in MMA welding. The continuous wire processes have generally better productivity and fewer stops and starts. Consequently, reduced welding times and improved quality are not incompatible.

26.4.6 Weld quality

The cost of inspection, repair, and re-inspection of a defective weld is about ten times the cost of getting it right first time. Efforts should always be made to achieve acceptable quality. Quality is measured in terms of weld shape, soundness, and mechanical properties.

Weld quality is assessed by a variety of non-destructive tests with the objective of finding, identifying, and measuring the size of imperfections.

Mechanical properties are measured in pre-production procedure tests, and may be confirmed by test welds made under production conditions.

It is impossible to avoid imperfections in commercially produced materials, but not all imperfections will adversely affect the performance of a structure. The acceptability of a weld, therefore, relies on a 'fitness for purpose' assessment. Acceptance criteria are specified in national standards, including the National Structural Steelwork Specification (NSSS).[26]

Imperfections are classified as:

● lack of fusion
● cracks
● porosity
● inclusions
● spatter
● weld geometry and profile.

Of these, two types of defect are generally unacceptable – lack of fusion and cracks. Both of these defects reduce the cross-sectional area of a weld, lowering their load-carrying capacity, and they significantly reduce resistance to fatigue failure in cyclic loading.

26.4.7 Distortion

Even if a weld is sound and acceptable mechanically, it can be rejected if the resulting structure has been distorted to an unacceptable shape. Distortion is caused by the shrinkage of the weld metal and Heat Affected Zone (HAZ) as it cools. Shrinkage cannot be eliminated, but the resultant distortion can be minimised. The most common means of reducing distortion are:

1. selecting the joint type that requires the least amount of weld filler metal
2. using edge preparations that reduce the amount of filler metal required
3. balancing the welding of individual members on either side of the neutral axis
4. balancing the welding on either side of butt welds in thick plate by use of double-V instead of single-V preparations

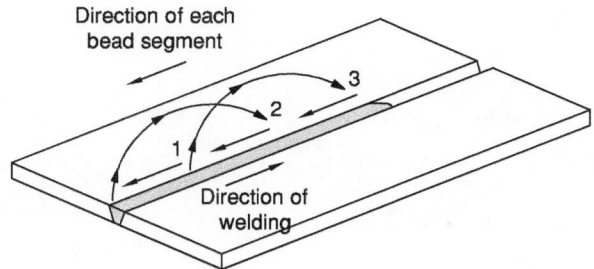

Figure 26.11   Back-step welding sequence to restrain movement

5. making welds at fast travel speeds (by mechanisation) to minimise the heat input into the adjacent metal
6. using tack welds and jigs to restrain movement
7. using a back-step sequence to restrain movement, as shown in Figure 26.11.

26.5 Geometric considerations

26.5.1 Effective throats

The throat thickness of fillet welds is given in Table 26.1. The factors given in the table are approximately equal to the cosine of the half angle between the fusion faces for welds of equal leg length, and when multiplied by the leg length will give the perpendicular distance between the root and a line joining the intersections of the weld with the fusion faces. This by definition is the throat thickness. For welds with an angle less than 90° between the fusion faces, for which the defined distance would be greater than that for a right angle weld, there is an upper limit to the factor of 0.7. Similarly in the case of welds of unequal leg lengths the assumed throat thickness is not to exceed 0.7 multiplied by the shorter of the two legs (Figure 26.12).

Where deep penetration welds are produced by submerged arc welding, the effective throat thickness may be measured to the minimum depth of fusion (see BS EN 1993-1-8:2008[29] Cl 4.5.2).

Table 26.1   Throat thickness of fillet welds

Angle between fusion faces (degrees)	Factor (to be applied to leg lengths)
60 to 90	0.7
91 to 100	0.65
101 to 106	0.6
107 to 113	0.55
114 to 120	0.5

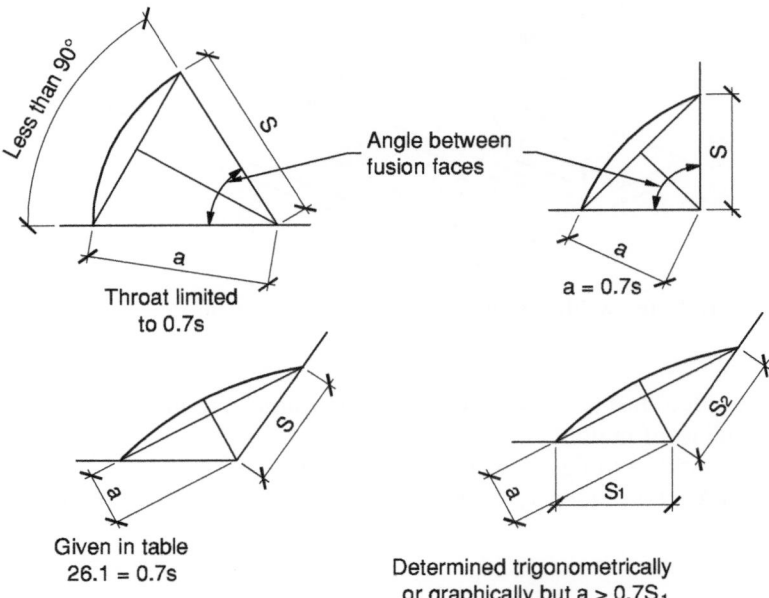

Figure 26.12   Throat thickness of fillet welds

26.5.2 Effective lengths

The effective length of a fillet weld is the actual length less twice the throat thickness to allow for the starting and stopping of the weld. It should not be less than 30 mm or less than six times the throat thickness. When a fillet weld terminates at the end or edge of a plate it should be returned continuously round the corner for a distance of twice the leg length.

Intermittent fillet welds are laid in short lengths with gaps between as specified in BS EN 1993-1-8[29] Figure 4.1. They should not be used in fatigue situations or where capillary action could lead to the formation of rust pockets. The effective length of each run within a length is calculated in accordance with the general requirements for fillet welds.

26.6 Methods of analysis of weld groups

26.6.1 Introduction

Any weld group may be required to resist an applied load acting through the centroid of the group either in or out of plane, producing shear or tension respectively.

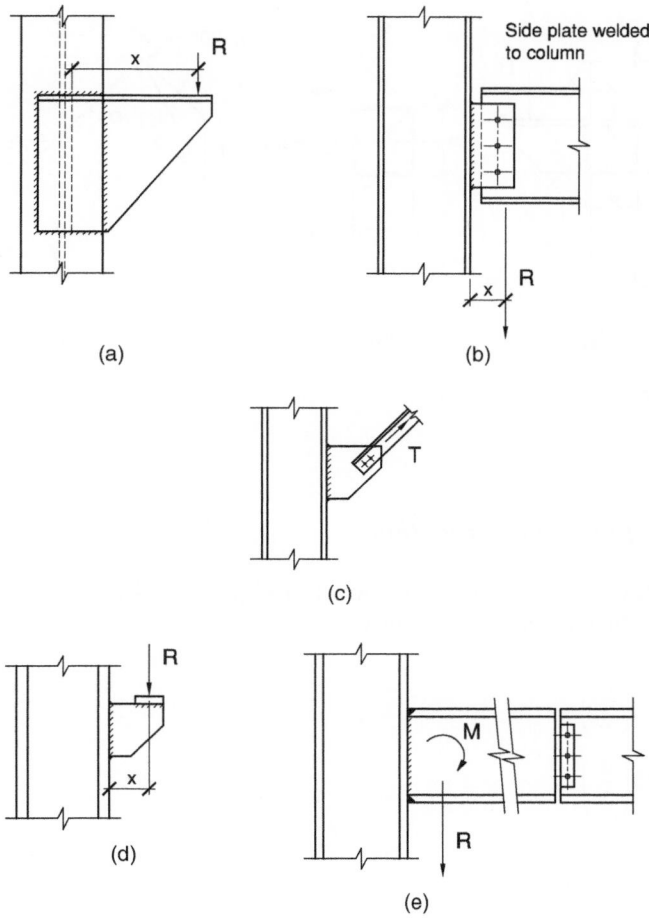

Figure 26.13   Weld groups loaded eccentrically

The load may also be applied eccentrically producing in addition bending tension or torsional shear. Examples are given in Figure 26.13.

26.6.2 Weld groups loaded in shear

British and Australian practice is to distribute the torsional shear due to eccentricity elastically in proportion to the distance of each element of the weld from the centroid of the group. This is referred to as the polar inertia method. In some countries, notably Canada and in some cases the USA, the instantaneous centre method, referred to in Chapter 23 for bolt groups, is also used for weld groups.

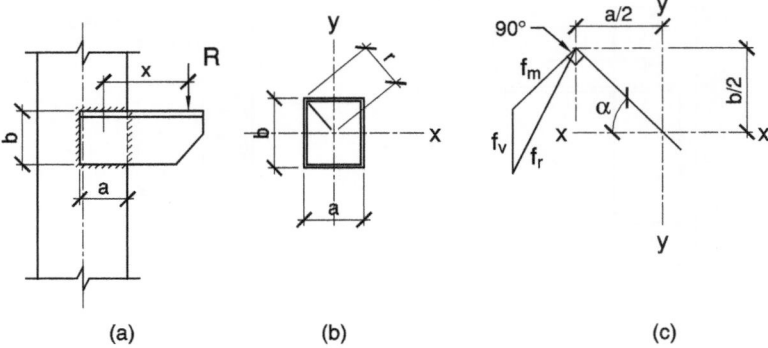

Figure 26.14   Weld groups in shear

26.6.2.1 The polar inertia method

Consider the four-sided weld group shown in Figure 26.14(a).
  Assume the throat thickness is unity.

$$I_{xx} = \frac{2b^3}{12} + 2a\left(\frac{b}{2}\right)^2$$

$$I_{yy} = \frac{2a^3}{12} + 2b\left(\frac{a}{2}\right)^2$$

$$I_{00} = I_{xx} + I_{yy}$$

$$I_{00} = \frac{b^3 + 3ab^2 + 3ba^2 + a^3}{6}$$

Distance r to extreme fibre (Figure 26.14):

$$r = \frac{1}{2}\sqrt{(a^2 + b^2)}$$

$$Z_{00} = \frac{b^3 + 3ab^2 + 3ba^2 + a^3}{3\sqrt{(a^2 + b^2)}}$$

By similar reasoning to that given in Chapter 23 for bolts, f_m, the force vector per unit length from the moment, is:

$$f_m = \frac{R_{xr}}{Z_{00}}$$

f_v, the shear, is assumed to be uniformly distributed around the weld group:

$$f_v = \frac{R}{2a+2b}$$

The resultant force vector per unit length, f_r, may be determined, as shown in Figure 26.14(c), either graphically or trigonometrically:

$$f_r = \sqrt{\left[(f_{00}\cos\alpha + f_v)^2 + (f_{00}\sin\alpha)^2\right]}$$

The value f_r can then be compared with the strength of the weld proposed from Table 8.4, page 1306 in the Appendix as appropriate.

26.7 Design strengths

26.7.1 General

In fillet-welded joints which are subject to compression forces as shown in Figure 26.15, it should not be assumed, unless provision is made to ensure it, that the parent metal surfaces are in bearing contact. In such cases, the fillet weld should be designed to carry the whole of the load. Single-sided fillet welds should not be used in cases where there is a moment about the longitudinal axis: see Figure 26.16. Ideally they should not be used to transmit tension.

In S275 steel a full strength T-connection (as shown in Figure 26.17) can be made using a pair of symmetric fillet welds in which the sum of the throat thicknesses is

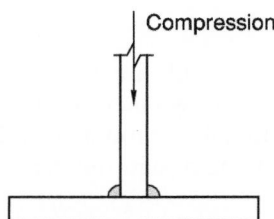

Figure 26.15   Weld in compression

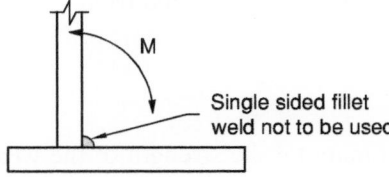

Figure 26.16   Weld with moment

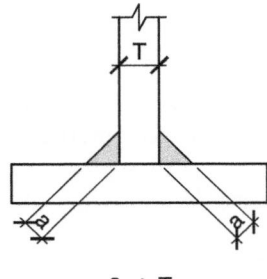

$$2a \geq T$$

Figure 26.17   Full strength T-welds may be formed with symmetrical fillet welds

Table 26.2   Design shear strength of fillet welds, $f_{vw,d}$

Steel Grade	*** (N/mm²)	Thickness of Weaker Jointed Part	**** (N/mm²)
S275	410	3mm ≤ t_p ≤ 100 mm	223
S355	470	3mm ≤ t_p ≤ 100 mm	241

at least equal to the thickness of the connected part. (Note: This rule does not apply to S355 steel.) This simple rule for S275 is permitted because even though the design strength of the weld is less than the parent metal, the higher transverse strength of fillet welds is sufficient to compensate. For higher grades, this does not apply.

Strength

The design shear strengths of fillet welds are given in Table 26.2.

BS EN1993-1-8 provides details of two methods for calculating the design resistance of a fillet weld – the directional method in 4.5.3.2 and a simplified method in 4.5.3.3. In the simplified method, the design resistance of the weld per unit length, $F_{w,Rd}$, is taken as

$$F_{w,Rd} = f_{vw,d}a$$

where $f_{vw,d}$ is the design shear strength of the weld and a is the throat thickness.

The design shear strength, $f_{vw,d}$ is determined from:

$$f_{vw,d} = \frac{f_u\sqrt{3}}{\beta_w\gamma_{M2}}$$

where f_u is the nominal ultimate tensile strength of the weaker part joined; β_w is a correlation factor (from Table 4.1) related to the type and grade of steel used (0.85 for S275 and 0.9 for S355); γ_{M2} takes the recommended value of 1.25 according to the UK National Annex.

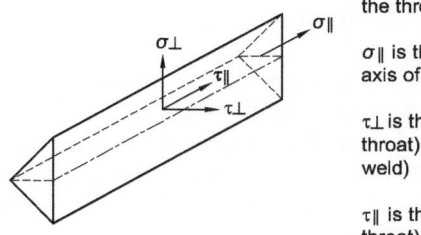

$\sigma_\perp$ is the normal stress perpendicular to the throat

$\sigma_\parallel$ is the normal stress parallel to the axis of the weld

$\tau_\perp$ is the shear stress (in the plane of the throat) perpendicular to the axis of the weld)

$\tau_\parallel$ is the shear stress (in the plane of the throat) parallel to the axis of the weld

Figure 26.18   Stresses on the throat of a fillet weld

In the directional method, the forces transmitted by a unit length of weld are resolved into components parallel and transverse to the longitudinal axis of the weld and normal and transverse to the plane of its throat, as shown in Figure 26.18. This method accounts for the greater resistance of fillet welds in the transverse direction compared with that when loaded along the longitudinal axis. The normal stress parallel to the axis is not considered but the remaining stresses should be combined as shown below and must be less than the design strength of the weld:

$$\sigma_\perp^2 + 3\left(\tau_\perp^2 + \tau_\parallel^2\right)^{0.5} \le f_u/(\beta_w \gamma_{M2}) \quad \text{and} \quad \sigma_\perp \le 0.9\, f_u/\gamma_{M2}$$

where f_u and β_w are as defined above,

$\sigma_\perp$ is the normal stress perpendicular to the throat
$\sigma_\parallel$ is the normal stress parallel to the axis of the weld
$\tau_\perp$ is the shear stress (in the plane of the throat) perpendicular to the axis of the weld
$\tau_\parallel$ is the shear stress (in the plane of the throat) parallel to the axis of the weld.

Where different materials strength grades are joined, the weld resistance should be based on the properties of the material with the lower strength grade.

Pages 1271 and 1286 in the Appendix show fillet weld design resistances calculated using appropriate design shear strengths for S275 and S355 steel.

26.8 Concluding remarks

Good designs lead to cost-effective fabrications that can be made to the required standards by the use of co-ordinated specifications, which provide the means for quantitative control of weld quality. Since the quality and properties of a weld cannot be readily verified on completion, welding requires continuous control and the use of specified procedures to ensure success.

BS EN 1090-2 sets out the requirements for the manufacture of structures and components in structural steelwork based on four Execution Classes (EXC1 to EXC4) where the stringency of requirements increases from EXC1 to EXC4.

Determination of the appropriate EXC is the responsibility of the designer and should be based on the service requirements of the structure and the consequence of failure.

The choice of joint and weld type is a major factor affecting welding costs. Wherever possible the designer should specify fillet welds rather than butt welds and the sizes of welds should be no larger than required to transmit the design forces. To ensure efficient and cost-effective welding details, designers should always seek genuine dialogue with the Steelwork Contractor at the earliest opportunity.

A Steelwork Contractor should nominate at least one Responsible Welding Coordinator who is competent, in terms of both technical knowledge and experience, to supervise its welding operations. BS EN 1090-2 requires the use of suitably qualified Welding Procedure Specifications (WPS) and welders. The Steelwork Contractor is responsible for ensuring that the WPS and welder qualifications are appropriate for the work being undertaken. The National Structural Steelwork Specification (NSSS) provides guidance on a suitable inspection regime and quality acceptance criteria for welds in structural steelwork.

References to Chapter 26

1. British Standards Institution (2004) BS EN 10025-5 *Technical delivery conditions for structural steels with improved atmospheric corrosion resistance.* London, BSI.
2. British Standards Institution (Various) BS EN 1993-All parts. *BSI Eurocode 3. Design of steel structures.* London, BSI.
3. British Standards Institution (2011) BS EN 1090-2 *Execution of steel structures and aluminium structures. Technical requirements for the execution of steel structures.* London, BSI.
4. British Standards Institution (2005) BS EN ISO 3834 *Quality requirements for fusion welding of metallic materials.* London, BSI.
5. British Standards Institution (1995) BS EN 22553 *Welded, brazed and soldered joints. Symbolic representation on drawings.* London, BSI.
6. British Standards Institution (2003) BS EN ISO 9692-1 *Welding and allied processes – Recommendations for joint preparation – Part 1: Manual metal-arc welding, gas-shielded metal-arc welding, gas welding, TIG welding and beam welding of steels.* London, BSI.
7. British Standards Institution (2001) BS EN 1011-2 *Welding. Recommendations for welding of metallic materials. Arc welding of ferritic steels.* London, BSI.
8. British Standards Institution (2004) BS EN 10025-1 *General technical delivery conditions.* London, BSI.
9. British Standards Institution (2004) BS EN 10025-2 *Technical delivery conditions for non-alloy structural steels.* London, BSI.
10. British Standards Institution (2004) BS EN 10025-3 *Technical delivery conditions for normalised/normalised rolled weldable fine grain structural steels.* London, BSI.

11. British Standards Institution (2004) BS EN 10025-4 *Technical delivery conditions for thermo mechanically rolled weldable fine grain structural steels.* London, BSI.

12. British Standards Institution (2004) BS EN 10025-6 *Technical delivery conditions for flat products of high yield strength structural steels in the quenched and tempered condition.* London, BSI.

13. British Standards Institution (2006) BS EN 10210 *Hot finished structural hollow sections of non-alloy and fine grain structural steels.* London, BSI.

14. British Standards Institution (2006) BS EN 10219 *Cold formed structural hollow sections of non-alloy and fine grain structural steels.* London, BSI.

15. British Standards Institution (1990) BS 4360 *Specification for weldable structural steels.* London, BSI.

16. British Standards Institution (2005) BS EN 1993-1-10 *Eurocode 3. Design of steel structures. Material toughness and through-thickness properties.* London, BSI.

17. British Standards Institution (2009) PD 6695-1-10 *Recommendations for the design of structures to BS EN 1993-1-10.* London, BSI.

18. British Standards Institution (2011) BS EN ISO 14341 *Wire electrodes and deposits for gas shielded metal arc welding of non alloy and fine grain steels. Classification.* London, BSI.

19. British Standards Institution (2008) BS EN ISO 14175 *Shielding gases for arc welding and cutting.* London, BSI.

20. British Standards Institution (2009) BS EN ISO 2560 *Covered electrodes for manual arc welding of non-alloy and fine grain steels. Classification.* London, BSI.

21. British Standards Institution (2008) BS EN ISO 17632 *Tubular cored electrodes for metal arc welding with and without a gas shield of non-alloy and fine grain steels. Classification.* London, BSI.

22. British Standards Institution (2010) BS EN ISO 14171 *Welding consumables. Solid wire electrodes, tubular cored electrodes, and electrode/flux combinations for submerged arc welding of non-alloy and fine grain steels. Classification.* London, BSI.

23. British Standards Institution (1996) BS EN 760 *Fluxes for submerged arc welding. Classification.* London, BSI.

24. British Standards Institution (2008) BS EN ISO 15614-1 *Specification and qualification of welding procedures for metallic materials. Welding procedure test.* London, BSI.

25. British Standards Institution (2004) BS EN 287-1 *Qualification test of welders. Fusion welding.* London, BSI.

26. The British Constructional Steelwork Association (2007) *National Structural Steelwork Specification*, 5th edn. London, BCSA/SCI.

27. British Standards Institution (2009) BS 499-1 *Welding terms and symbols. Glossary for welding, brazing and thermal cutting.* London, BSI.

28. British Standards Institution (2010) BS EN ISO 4063 *Welding and allied processes. Nomenclature of processes and reference numbers.* London, BSI.

29. British Standards Institution (2008) BS EN 1993-1-8 *Eurocode 3. Design of steel structures. Design of joints.* London, BSI.

Chapter 27
Joint design and simple connections

DAVID MOORE and BUICK DAVISON

27.1 Introduction

In general the cost of the design, fabrication and erection of the structural frame in a steel framed building is approximately 30% of the total cost of construction. Of these three items, fabrication and erection account for approximately 67%. Any savings in the fabrication and erection costs can significantly reduce the overall cost of construction. The majority of the fabrication costs are absorbed by the connections, and the choice of connection also has a significant influence on the speed, ease, and, therefore, the cost of erection. It is evident that the potential for reducing the cost of steel construction lies in the suitable choice of the beam-to-column and beam-to-beam connections. Indeed, because of the repetitive nature of connections, even small material and labour savings in one connection can have an important effect on the overall economy of the building.

In view of the significance of design and detailing it is surprising that it is often regarded as being of secondary importance in the design process. Eurocode 3 pays much greater attention to connection design than has hitherto been the case in codes of practice, and BS EN 1993-1-8[1] is an entire part devoted to the topic. Consideration of the local effects at connections is usually left to the designer and this has led to a diversity of both connection types and design methods. The traditional split of responsibilities, where the consultant designs the members of the frame and the connections are designed and detailed by the steelwork contractor, has further compounded this problem. Section 5 of BS EN 1993-1-8 outlines in some detail the inter-relationship between analysis of frame behaviour, the type of joint response and how this may be modelled. The code states 'The effects of the behaviour of the joints on the distribution of internal forces and moments within a structure, and on the overall deformations of the structure, should generally be taken into account, but where these effects are sufficiently small they may be neglected.' However, joint behaviour need not be taken into account if the joints are simple i.e. they do not transmit significant bending moments, or continuous, in which case the frame members may be assumed to be perfectly pin-connected.

Steel Designers' Manual, Seventh Edition. Edited by Buick Davison and Graham W. Owens.
© 2012 Steel Construction Institute. Published 2012 by Blackwell Publishing Ltd.

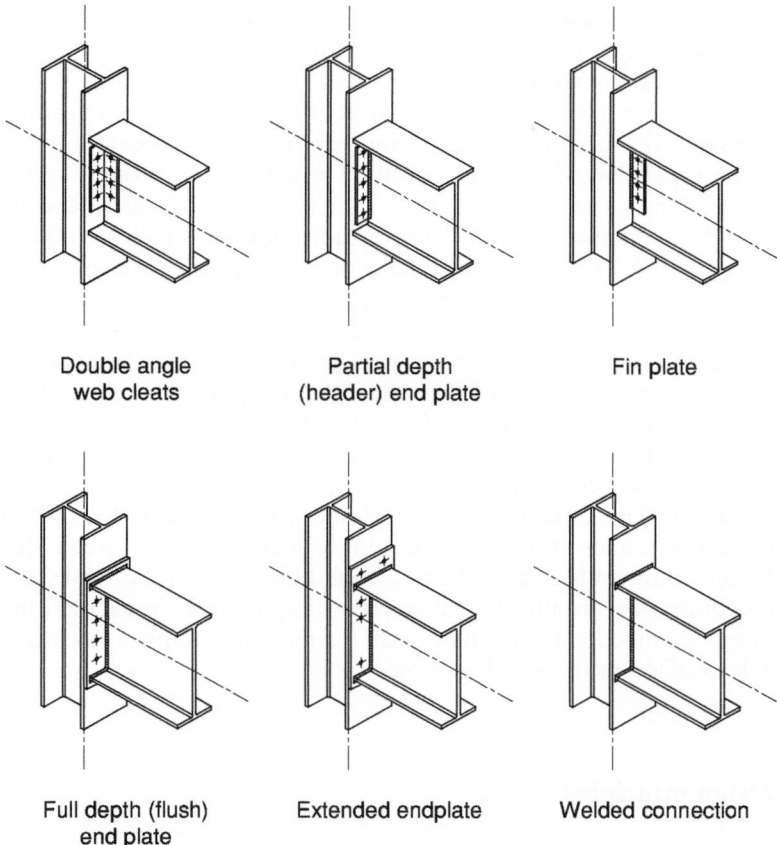

Double angle
web cleats

Partial depth
(header) end plate

Fin plate

Full depth (flush)
end plate

Extended endplate

Welded connection

Figure 27.1   Typical beam-to-column connections

Some of the many types of beam-to-column connections used in multi-storey steel frame construction are shown in Figure 27.1. Choice of connection type is usually based on simplicity, duplication and ease of erection – all for economic reasons. Welded joints provide full moment continuity but are expensive due to the on-site welding involved. Bolted connections have the advantages of requiring less site supervision than welded joints, having a shorter assembly time and support the load as soon as the bolts are in position. They also have a geometry that is easy to comprehend and can accommodate minor discrepancies in the dimensions of the beams and columns. However, when large forces are involved, bolted connections can become quite deep, which may conflict with the architectural need for a 'clean line'.

With such a large number of connection types and the variety within each type (each having different characteristics) design engineers are presented with a vast range of suitable connection types and choosing the appropriate design method can be confusing. Key to making the correct choice is to understand the characteristics of particular joint types (strength, stiffness and ductility) and the relationship between analysis/design methods and connection performance.

In 1987 the Steel Construction Institute (SCI) and the British Constructional Steelwork Association (BCSA) formed the SCI/BCSA connection group with the aim of improving the steel construction industry's effectiveness and efficiency by increasing the repetition of elements within a structure and promoting the use of standard connections. The group aimed to define a range of standard connections and standard design methods, and to develop a framework which would lead to the widespread adoption of rationalised connections using standardised components. Two publications were produced, *Joints in Steel Construction: Simple Connections*[2] and *Joints in Steel Construction: Moment Connections*[3] which have become industry standards presenting standard design methods for the most commonly used connections. Following the publication of Eurocode 3, the first publication has been revised[4] and is now based on a combination of the design principles and the classification systems for connections used in Eurocode 3 and practical aspects associated with current fabrication and erection techniques to produce realistic estimates of a connection's strength. The Green Book for Moment Connections has not yet been revised. However, readers will notice that the principles in the Green Book align with the EC3 recommendations and apart from differences in the symbols used, the design procedures are broadly the same. The design checks presented in Section 27.2 for simple connections and Chapter 28 for moment connections are based on the methods detailed in these publications.

27.1.1 Design principles

Connection design must be consistent with the engineer's assumptions concerning the structural behaviour of the steel frame. Therefore, when choosing and proportioning connections the engineer should always bear in mind the basic requirements such as the stiffness (or flexibility) of the connection, strength and the required rotational capacity. Care should also be taken to ensure that the assumptions made for the design of the various elements of the connection are compatible. For example, sharing the load between ordinary bolts (in shear and tension) and a fillet weld is not acceptable because the deformation characteristics of ordinary bolts and fillet welds are incompatible.

It is also essential to consider economy at the design stage. As a general rule, if fabrication and erection costs are kept to a minimum then the overall cost will also tend to be a minimum. However, it should be realised that the costs are steelwork contractor-dependent and rather than impose a particular connection type on a steelwork contractor it is more cost-effective to state a range of standard connection types that satisfy the design assumptions. The steelwork contractor will then choose that connection which can be fabricated economically. Finally, connections should provide good access for any welding operations and/or the placing and tightening of bolts.

27.1.2 Classification of connections

Connection design depends upon the designer's decision regarding the method by which the structure is analysed. BS EN 1993-1-1[5] Cl 5.1.2(2) (which also refers the reader to BS EN 1993-1-8 Cl 5.1.1) gives three possible joint models which align with distinctly different frame analysis approaches; these are:

1. simple, in which the joint may be assumed not to transmit bending moments
2. continuous, in which the behaviour of the joint may be assumed to have no effect on the analysis
3. semi-continuous, in which the behaviour of the joint needs to be taken into account in the analysis.

Elastic and plastic methods of global analysis may be used. Table 27.1 based on Table 5.1 in 1993-1-8, shows how the joint classification, the type of framing and the method of global analysis are related.

Simple framing is based on the assumption that the beams are simply supported and implies that beam-to-column connections must be sufficiently flexible to avoid the development of significant end moment. Any horizontal forces must be resisted by bracing or shear walls, etc. When using this approach the connections are required to function as nominally pinned no matter what method of global analysis is used.[6] However, research has shown that most connections – even so-called simple connections – are capable of developing some moment capacity.

If the continuous approach is adopted, the type of connection used will depend on the method of global analysis. When elastic analysis is used, joint stiffness is important and connections classified as rigid would usually be used; the appropriate joint model is called 'continuous'. When plastic analysis is used the connections' strength is the key factor and classification according to strength (moment capacity) must be used. The term 'full-strength' relates the strength of the connection to that of the connected beam. If the moment capacity of the connection is higher than that of the connected beam then the connection is termed full-strength. The purpose of

Table 27.1 Relationship between global analysis model, joint classification and type of framing

Method of Global Analysis	Classification of Joint		
Elastic	Nominally pinned	Rigid	Semi-rigid
Rigid-plastic	Nominally pinned	Full-strength	Partial-strength and ductile
Elastic-plastic	Nominally pinned	Rigid and full-strength	Semi-rigid and partial strength Semi-rigid and full strength Rigid and partial strength
Type of framing i.e. joint model	Simple	Continuous	Semi-continuous

this comparison is to determine whether the joint or the connected member will limit the resistance of the structure. If the elastic–plastic method of global analysis is used then the connections are classified according to both their stiffness and strength, and rigid, full-strength connections must be used. These connections must be capable of carrying the design bending moment, shear force and axial load while maintaining the original angle between the connected members. While continuous design can produce economies in beam size with respect to simple design, most of these savings are offset by the need to supply joints with adequate strength or stiffness (or both).

While the simple method ignores stiffness and the continuous method only allows full-strength connections, the semi-continuous method recognises that most practical connections are capable of providing some degree of stiffness and that their moment capacity may be limited. Once again the type of connection used will depend on the method of global analysis. When elastic analysis is used the connections are classified according to their stiffness, and semi-rigid connections can be used. (It should be noted that the term 'semi-rigid' is a general classification and can be used to encompass all connections with pinned and rigid connections at the extremes). If plastic global analysis is used the connections are classified according to their strength. Connections that have a lower moment capacity than the connected member are termed **partial-strength**. In this case the connection will reach its capacity before the connected member and must therefore possess sufficient ductility to allow plastic hinges to form in other parts of the structures. Where the elastic–plastic method of global analysis is used (arguably the most practical accurate representation of real joint behaviour in frame analysis) the connections are classified according to both their stiffness and strength, and semi-rigid, partial strength connections are used.

From the above discussion it is clear that a connection has three fundamental properties:

1. Moment resistance: connection may be either full strength, partial strength or nominally pinned (i.e. not moment-resisting).
2. Rotational stiffness: connection may be rigid, semi-rigid or nominally pinned (i.e. no rotational stiffness).
3. Rotational capacity: connections may need to be ductile. This criterion is probably less familiar to many designers but simply acknowledges that a connection may need to rotate plastically at some stage of the loading cycle without failure. Pinned connections have to perform in this way but the principle also applies to partial strength moment-resisting connections in plastically designed frames.

These three properties are used to classify connections, and Figure 27.2 illustrates the different ways in which a connection may be classified. BS EN 1993-1-8 Cl 5.2.2.1(2) states 'A joint may be classified on the basis of experimental evidence, experience of previous satisfactory performance in similar cases or by calculations based on test evidence.' For UK practice, the National Annex (NA2.6) gives the following important additional information:

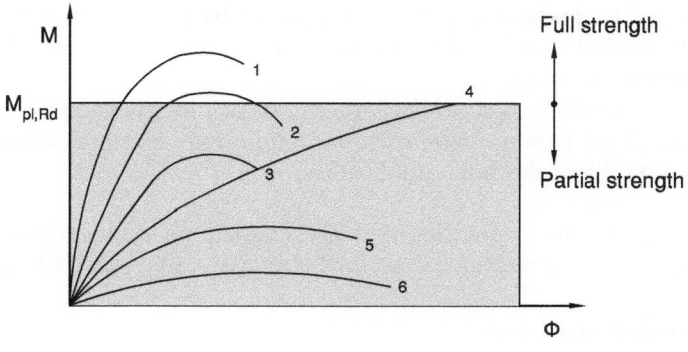

(a) Classification by strength

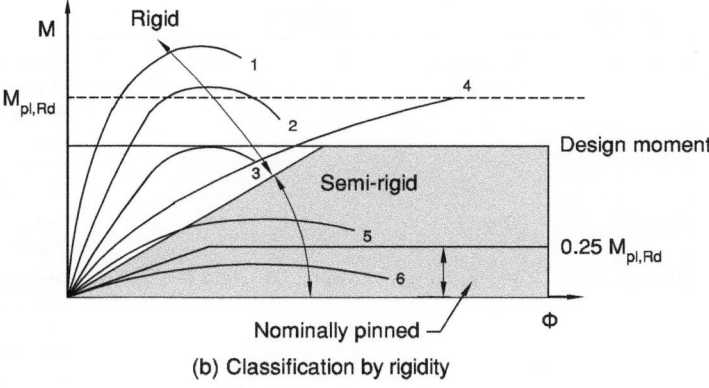

(b) Classification by rigidity

Figure 27.2   Classification of connections

- Nominally pinned joints are described as 'Simple Connections' in UK practice. Connections designed in accordance with the principles given in the publication *Joints in Steel Construction – Simple Connections* may be classified as nominally pinned joints.
- Ductile, partial strength joints are described as 'Ductile Connections' in UK practice. They are used in plastically designed semi-continuous frames. Braced semi-continuous frames may be designed using the principles given in the publication *Semi-continuous Design of Braced Frames* with connections designed to the principles given in Section 2 of *Joints in Steel Construction – Moment Connections*. Unbraced semi-continuous frames (known as wind moment frames) may be designed using the principles given in the publication *Wind-moment Design of Low Rise Frames*.
- Until experience is gained with the numerical method of calculating rotational stiffness given in BS EN 1993-1-8:2005, Cl 6.3 and the classification by stiffness method given in BS EN 1993-1-8:2005, Cl 5.2.2, semi-continuous elastic design

should only be used where either it is supported by test evidence according to BS EN 1993-1-8:2005: Cl 5.2.2.1(2) or where it is based on satisfactory performance in a similar situation.

- Connections designed in accordance with the principles given in the publication *Joints in Steel Construction – Moment Connections* may be classified on the basis given in Section 2.5 of the same publication.

Some guidance on the properties that are needed for connections in frames designed using one of the methods described above are given in Table 27.2.

Table 27.2 Methods of frame design

Design		Connections			Notes
Type of Framing	Global analysis	Properties	Fig 27.2 Example	Section	
Simple	Pin joints	Nominally pinned	6	27.2	Economic method of braced multi-storey frames.
					Connection design is made for shear strength only.
Continuous	Elastic	Rigid	1,2,3,4	28.1	Conventional elastic analysis.
	Plastic	Full strength	1,2,4	28.1	Plastic hinges form in the adjacent member, not in the connections.
					Popular for portal frame designs.
	Elastic-plastic	Full strength and rigid		28.1	
Semi-continuous	Elastic	Semi-rigid	1,2,4	Not covered	Connections are modelled as rotational springs.
					Prediction of connection stiffness presents difficulties.
	Plastic	Partial strength and ductile	5,6	Not covered	Wind-moment design is a variant of this method.
	Elastic-plastic	Partial strength and/ or semi-rigid	Any	Not covered	Full connection properties are modelled in the analysis. A research tool rather than a practical design method at the present time.

27.1.3 Definitions

Eurocode 3 is very precise in its use of terms to describe types of joints and associated behaviour. UK practice has tended to be more liberal in the use of such terms, for example 'rigid joint' has been widely interpreted to mean full-strength, and 'moment-resisting' has probably been taken to be synonymous with rigid for most designers. The greater precision in Eurocode 3 should result in a clearer understanding of the relationship between the choice of global analysis method and the required joint performance. To assist designers with some of these unfamiliar terms, definitions used in this chapter are given below:

1. **Full strength connections** have a moment of resistance at least equal to that of the member.
2. **Partial strength connections** have a moment of resistance which is less than that of the member.
3. A **rigid connection** is stiff enough for the effect of its flexibility on the distribution of the bending moments in the frame to be neglected.
4. A **semi-rigid connection** is too flexible to qualify as rigid, but is not a pin.
5. **Nominally pinned connections** are sufficiently flexible to be regarded as pins for analysis. These connections are by definition flexible, not moment connections, although partial strength connections able to resist less than 25% of the plastic moment capacity of the beam may be regarded as nominally pinned.
6. A **ductile connection** has sufficient rotation capacity to act as a plastic hinge. Connection ductility should not be confused with material ductility.
7. In the **simple design** method of frame design the connections are assumed not to develop moments that adversely affect either the members of the structure or the structure as a whole.
8. The **continuous design** method of frame design does not model the connection properties in the frame analysis. This covers either elastic analysis, where the connections are rigid, or plastic analysis, where the connections are full strength.
9. For **semi-continuous design** of frames the connection properties have to be modelled in the analysis. This covers elastic analysis where semi-rigid connections are modelled as rotational springs, or plastic analysis where partial strength connections are modelled as plastic hinges.
10. A distinction is made between the words **joint** and **connection**. A connection is the 'Location at which two or more elements meet. For design purposes it is the assembly of the basic components required to represent the behaviour during the transfer of the relevant internal forces and moments at the connection.' A joint is the 'Zone where two or more members are interconnected. For design purposes it is the assembly of all the basic components required to represent the behaviour during the transfer of the relevant internal forces and moments between the connected members. A beam-to-column joint consists of a web panel and either one connection (single-sided joint configuration) or two connections (double-sided joint configuration).' The distinction between the

two terms is subtle and joint and connection are likely to be used as synonyms for some time.

27.2 Simple connections

27.2.1 Design philosophy

Simple connections are defined as those connections that transmit end shear only and have negligible resistance to rotation and therefore do not transfer significant moments at the ultimate limit state. This definition underlies the design of the overall structure[7] in which the beams are designed as simply-supported and the columns are designed for axial load and the small moments induced by the end reactions from the beams. In practice, however, the connections do have a degree of fixity, which although not taken into account in the design is often sufficient to allow erection to take place without the need for temporary bracing.

The following three principal forms of simple connection are considered in this section:

1. double-angle web cleats
2. flexible end-plates
3. fin plates.

To comply with the design assumptions, simple connections must allow adequate end rotation of the beam as it takes up its simply-supported deflected profile and practical lack of fit. At the same time this rotation must not impair the shear and tying capacities of the connection (for structural integrity see below). In theory, a 457 mm deep, simply supported beam spanning 6.0 m will develop an end rotation of 0.022 radians (1.26°) when carrying its maximum factored load. In practice this rotation will be considerably smaller because of the restraining action of the connection. When the beam rotates it is desirable to avoid the bottom flange of the beam bearing against the column as this can induce large forces in the connection. The usual way of achieving this is to ensure that the connection extends at least 10 mm beyond the end of the beam. Alternatively, a thin endplate may be used.

27.2.2 Structural integrity

The partial collapse of Ronan Point in 1968 alerted the construction industry to the problem of progressive collapse arising from a lack of positive attachment between principal elements in a structure. This resulted in amendments to both the Building Regulations Approved Document A – Structure, and the UK's steel design code.

Essentially, these changes take cognisance of this failure and require structures to have a minimum level of robustness to resist accidental loading.

The requirement for robustness and structural integrity is reflected in the Eurocodes and is additional to the ultimate and serviceability limit state requirements. Structural integrity/robustness is the ability of a structure to withstand an event without being damaged to an extent disproportionate to the original cause. The events referred to include explosions like the one at Ronan Pont, impact and the consequences of human error. To ensure that the damage is not disproportionate, BS EN 1990 requires the designer to choose one or more of the following measures:

- avoiding, eliminating or reducing the hazard to which the structure can be subjected
- selecting a structural form which has a low sensitivity to the hazards considered
- selecting a structural form and design that will survive adequately the removal of an individual element or limited part of the structure, or the occurrence of acceptable localised damage
- avoiding as far as possible a structural system that can collapse without warning
- tying the structural members together.

BS EN 1991-1-7 gives recommendations for buildings which include a categorisation of building types into consequence classes which is very similar to the classification system used in Approved Document A – Structure. Based on the consequence class of the building BS EN 1991-1-7 recommends a strategy for achieving an acceptable level of robustness which is very similar to the recommendations for tying and/or the notional removal of supporting members given in Approved Document A.

For framed structures horizontal ties should be provided around the perimeter of each floor and roof level, and internally at right-angles to tie the columns and wall elements to the structure of the building. This is most effectively done using members approximately at right angles to each other or by steel reinforcement in concrete floor slabs and profiled steel sheeting in composite steel/concrete flooring systems. These ties and their connections should be able to resist a design tensile force of:

for internal ties $T_i = 0.8(g_k + \Psi q_k)sL$ or 75kN whichever is the greater

for perimeter ties $T_i = 0.4(g_k + \Psi q_k)sL$ or 75kN whichever is greater

where:

s is the spacing of the ties

L is the span of the tie

Ψ is the relevant factor in the expression for combination of action effects for the accidental design situation (see BS EN 1990)

g_k is the characteristic permanent (dead) load

q_k is the characteristic variable (imposed) load.

27.2.3 Design procedures

The design of these simple connections is based on BS EN 1993-1-8. The capacities and design strengths of the fasteners and fittings are based on the rules given in Clause 3.6. The spacing of the fasteners and their distances comply with Clause 3.5 (Table 3.3) and follow the recommendations presented in the Green Book.[2,4]

When fabricating and erecting these connections it is general practice to use the following components:

- untorqued bolts in clearance holes, usually M20 property class 8.8 bolts
- cleats, end-plates, fin-plates and other fittings made from grade S275 steel
- 6mm fillet welds
- punched holes on fittings using semi-automatic equipment.

Wherever possible, general industrial practice should be followed, and guidance is given in the National Structural Steelwork Specification.[8] ECCS publication No.126[6] also provides useful guidance on the design of simple connections to EC3.

27.2.4 Double-angle web cleats

A typical bolted double-angle cleat connection is shown in Figure 27.3. These types of connections provide for minor site adjustments when using untorqued bolts in 2mm clearance holes. Normally the cleats are used in pairs. Any simple equilibrium analysis is suitable for the design of this type of connection. The one recommended in this publication assumes that the line of action of shear transfer between the beam and the column is at the face of the column. Using this model the bolt group connecting the cleats to the beam web must be designed for the shear force and the moment produced by the product of the end shear and the eccentricity of the bolt group from the face of the column. The bolts connecting the cleats to the face of the column should be designed for the applied shear only. In practice the cleats to the column are rarely critical and the bolts bearing on to the web of the beam almost always govern the design. The rotational capacity of this connection is governed largely by the deformation capacity of the angles and the slip between the connected parts. Most of the rotation of the connections comes from the deformation of the angles while fastener deformation is very small. To minimise rotational resistance (and increase rotational capacity) the thickness of the angle should be kept to a minimum and the bolt cross-centres should be as large as is practically possible.

When connecting to the minor axis of a column it may be necessary to trim the flanges of the beam but this does not change the shear capacity of the beam. During erection the beam with the cleats attached is lowered down the column between the column flanges.

The essential detailing requirements for this connection are shown in Figure 27.3, and Table 27.3 shows the detailed design checks. The design procedure given in

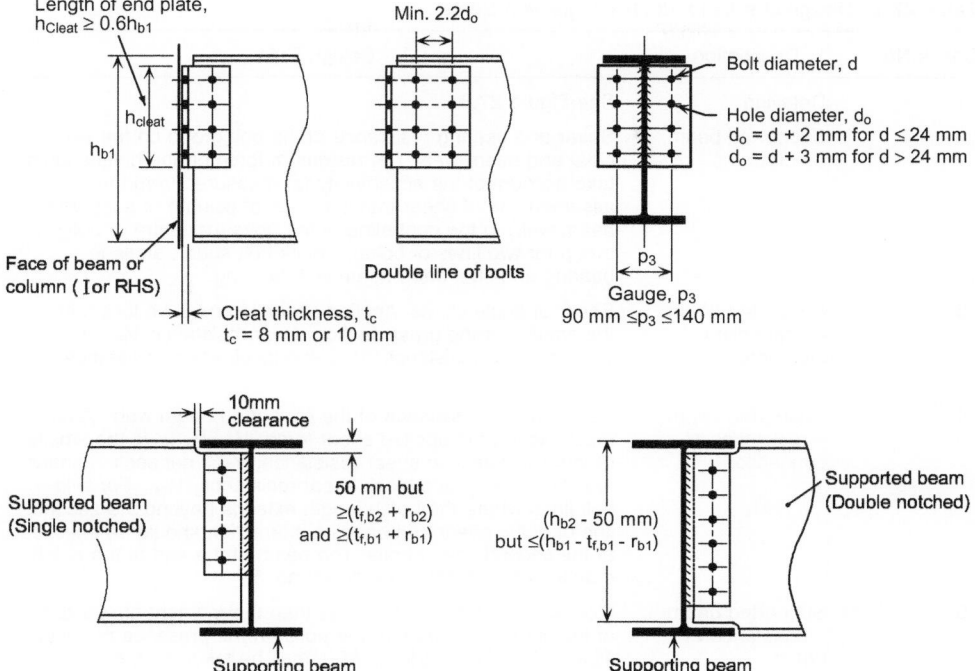

Figure 27.3   Double-angle web cleats

Table 27.3 applies to beams connected to either the column flange or the column web.

27.2.5 Single-angle web cleats

Single-angle web cleats are normally only used for small connections or where access precludes the use of double-angle or end-plate connections.

This type of connection is not desirable from an erector's point of view because of the tendency of the beam to twist during erection. Care should be taken when using this type of connection in areas where axial tension is high (e.g. where the axial tension due to structural integrity is high). The design checks for this type of connection are similar to those shown in Table 27.3. In addition to these checks, the bolts connecting the cleat to the column must also be checked for the moment produced by the product of the end shear force and the distance between the bolts and the centreline of the beam.

Figure 27.3 shows typical beam-to-beam double-angle web cleat connections with single-notched and double-notched beams respectively. Where the top flanges of the connected beams are at the same level, the flange of the supported beam is notched

Table 27.3   Design checks for double-angle web cleats

Check No.	Description	Design Rule
1	Detailing	See Figure 27.3
2	Supported beam – bolt group	Shear and bearing resistance of the bolt group on the web cleat and beam web. The maximum force on the bolts should take account of the eccentricity (z) measured from the assumed line of shear transfer (face of column or supporting beam web) to the centreline of the bolts (or centre of bolt group for two lines of bolts). Check bolt shear, angle in bearing and supported beam web bearing.
3	Supported beam – connecting elements	Shear of angle cleats. Applied shear V_{Ed} must be less than the smaller of the gross section shear resistance $V_{Rd,g}$, net section shear resistance, $V_{Rd,n}$ and block shear resistance, $V_{Rd,b}$.
4	Supported beam – resistance at connection	Design shear resistance of the supported beam web, $V_{Rd,min}$ must exceed the applied shear force, V_{Ed}. $V_{Rd,min}$ is the smaller of the gross section shear resistance, $V_{Rd,g}$, net section shear resistance, $V_{Rd,n}$ and block shear resistance, $V_{Rd,b}$. For twin bolt lines where the notch length extends beyond the second line of bolts, shear and bending interaction should be checked at the second line of bolts. The beam at the end of the notch may also be critical – see check no. 5.
5	Supported beam – resistance at notch	Moment at notch, must be less than the moment of resistance of the supported beam at the notch in the presence of shear $M_{v,Rd}$, When $V_{Ed} \leq 0.5 V_{pl,Rd}$, $M_{v,Rd}$ may be taken as the yield strength of the beam multiplied by the elastic modulus at the notch. When $V_{Ed} > 0.5 V_{pl,Rd}$, $M_{v,Rd}$ must be reduced to account for the presence of high shear (see EC3-1-1 Clause 6.2.8(3)).
6	Supported beam – local stability of notched beam	Satisfied if the beam is restrained against lateral torsional buckling and for a beam notched at the top to a depth not exceeding half the beam depth, or for a beam notched top and bottom, with either notch not exceeding 20% of the beam depth: $l_n \leq h_{b1}$ for $h_{b1}/t_{w,b1} \leq 54.3$ (S275) or ≤ 48.0 (S355) $l_n \leq 160000\, h_{b1}/(h_{b1}/t_{w,b1})^3$ for $h_{b1}/t_{w,b1} > 54.3$ (S275) $l_n \leq 110000\, h_{b1}/(h_{b1}/t_{w,b1})^3$ for $h_{b1}/t_{w,b1} > 48.0$ (S355).
7	Overall stability of notched beam	When a notched beam is unrestrained against lateral torsional buckling, the overall stability of the beam should be checked using EC3 6.3.2. The effective length may be calculated using the method shown.[4]
8	Supporting beam/column – bolt group	Shear and bearing resistance of the bolt group F_{Rd} must be greater than the applied shear V_{Ed}.
9	Supporting beam/column – connecting elements	Shear resistance of the web cleats $V_{Rd,min}$ (smaller of the gross section shear resistance $V_{Rd,g}$, net section shear resistance $V_{Rd,n}$ and block tearing resistance $V_{Rd,b}$) must exceed the applied shear V_{Ed}.
10	Supporting beam/column – Local resistance	Check local shear and punching resistance of the supporting beam web, or column web or wall of RHS or CHS. See check 10 in Reference 2.

Table 27.3   (*continued*)

Check No.	Description	Design Rule
11	Structural integrity – connecting elements	The tension capacity of double-angle web cleats may be conservatively estimated as: $0.6 \, L_e t_{cleat} \, f_y$ for S275 web cleat angles $0.5 \, L_e t_{cleat} \, f_y$ for S355 web cleat angles and L_e is the effective net length of the cleats. (Only applicable for bolt cross-centres into supporting beam or column ≤ 140 mm and $t_{cleat} \geq 8$ mm. A more rigorous calculation is provided in Appendix B of BCSA Simple Connections[2].)
12	Structural integrity – supported beam	Tension and bearing resistance of the web must exceed the tie force. Tension resistance of the beam web is the smallest of the block tearing tension resistance, $F_{Rd,b}$, and the net section tension resistance, $F_{Rd,n}$. Bearing resistance of the bolt group is dependent on the edge and end distance of the bolt holes in the beam web.
13	Structural integrity – tension bolt group	The tensile resistance of the bolt group must be greater than the tying force. To account for extreme prying, the design tensile strength of a property class 8.8 bolt should be taken as not more than 300 N/mm^2 (see Appendix D Reference 2).
14	Structural integrity – supporting column web (UC or UB)	This check is only required for either single-sided connections to a column web or unequally loaded double-sided connections to a column web. Web panel resistance to out of plane bending must exceed the applied tie force. Resistance calculated assuming an 'envelope' hinge collapse mechanism in the web panel.
15	Structural integrity – supporting column wall (RHS)	The tying resistance of an RHS wall in the presence of axial compression in the column must exceed the applied tie force. The check assumes the same collapse mechanism as check 14 but replaces the column web with the RHS wall.
16	NA	

and the web must be checked, allowing for the effect of the notch. The top of the web of the notch, which is in compression, must be checked for local buckling of the unrestrained web. Provided that the supported beam is laterally restrained by a floor slab to ensure that the web at the top of the notch does not buckle, it is recommended that the length of the notch should not exceed the limits shown in check no. 6 of Table 27.3. For supported beams which are not laterally restrained, a more detailed investigation is required on the overall stability of the beam with notched ends against lateral torsional buckling.

Beams with only one flange notched should be checked for lateral torsional buckling using the method given below, but noting that:

1. This check is only applicable for beams with one flange notched (Figure 27.4). Guidance on double-notched beams is given in Section 5.12 of Reference 9.
2. If the notch length l_n and/or notch depth d_{nt} are different at each end, then the larger value for l_n and d_{nt} should be used.

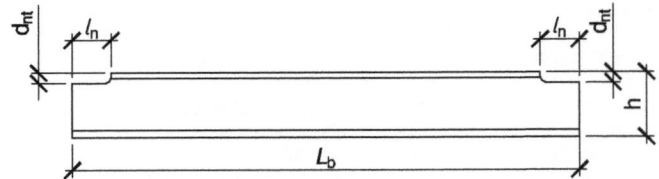

Basic requirement

$$L_E = L_b\left(1+\frac{2l_n}{L_b}\left(K^2+2K\right)\right)^{1/2} \qquad K = K_0/\lambda_b \qquad \lambda_b = \frac{UVL_b}{i_z}$$

where X,U,V and i_z are for the un-notched I beam section and are defined in section data tables (conservatively U = 0.9 and V = 1.0).

For $\lambda_b < 30$ $K_0 = 1.1g_0 X$ but $\leq 1.1K_{max}$
For $\lambda_b \geq 30$ $K_0 = g_o X$ but $\leq K_{max}$

g_0 and K_{max} are as follows:

$\dfrac{l_n}{L_b}$	g_0	K_{max}	
		UB section	UC section
≤0.025	5.56	260	70
0.050	5.88	280	80
0.075	6.19	290	90
0.100	6.50	300	95
0.125	6.81	305	95
0.150	7.13	315	100

Figure 27.4   Unrestrained beam with notched flanges

3. Beams should be checked for lateral torsional buckling to BS EN 1993-1-1, Clause 6.3.2
4. The solution below gives the modified effective length (L_E) based on References 10, 11 and 12. It is only valid for $l_n/L_b < 0.15$ and $d_{nt}/h < 0.2$ (beams with notches outside these limits should be checked as tee sections, or stiffened).

The web angle cleat can become cumbersome when used to connect unequal sized beams. In this case it is necessary to notch the bottom flange of the smaller beam to prevent fouling of the bolts. Alternatively, the cleat of the larger beam could be extended and the bolts placed below the bottom of the smaller beam.

27.2.6 Flexible end-plates

A typical flexible end-plate connection is shown in Figure 27.5. These connections consist of a single plate fillet welded to the end of the beam and site bolted to a

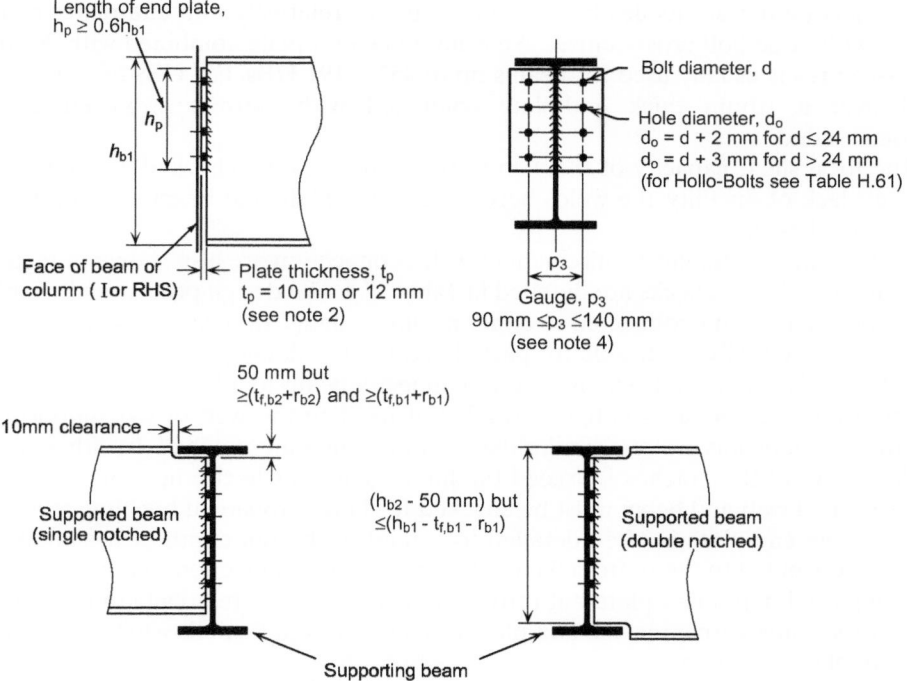

Length of end plate,
$h_p \geq 0.6h_{b1}$

h_p

h_{b1}

Face of beam or
column (I or RHS)

Plate thickness, t_p
$t_p = 10$ mm or 12 mm
(see note 2)

Bolt diameter, d

Hole diameter, d_o
$d_o = d + 2$ mm for $d \leq 24$ mm
$d_o = d + 3$ mm for $d > 24$ mm
(for Hollo-Bolts see Table H.61)

p_3
Gauge, p_3
90 mm $\leq p_3 \leq$ 140 mm
(see note 4)

50 mm but
$\geq (t_{f,b2}+r_{b2})$ and $\geq (t_{f,b1}+r_{b1})$

10mm clearance

Supported beam
(single notched)

$(h_{b2} - 50$ mm) but
$\leq (h_{b1} - t_{f,b1} - r_{b1})$

Supported beam
(double notched)

Supporting beam

Notes
1. The end plate is generally positioned close to the top flange of the beam to provide adequate positional restraint. Plate length of at least 0.6h is usually adopted to give "nominal torsional restraint".
2. Although it may be possible to satisfy the design requirements with $t_p < 8$mm, it is not recommended in practice because of the likelihood of weld distortion during fabrication and damage during transportation. If structural integrity checks are critical the end plate thickness may be increased to 10mm or 12mm.
3. The plate thickness and gauge limitations apply equally to partial depth and to full depth end plates.

Figure 27.5   Typical flexible end plate connection

supporting column. This connection is relatively inexpensive but has the disadvantage that there is no room for site adjustment. Overall beam lengths need to be fabricated within tight limits, although packs can be used to compensate for fabrication and erection tolerances. The end-plate is often detailed to extend to the full depth of the beam but there is no need to weld the end-plate to the flanges of the beam, although many steelwork contractors choose to weld to the flanges to improve the tying capacity.

Sometimes the end-plate is welded to the beam flanges to improve the stability of the frame during erection and avoid the need for temporary bracing. This type of connection derives its flexibility from the use of relatively thin end-plates combined with large bolt cross-centres. An 8 mm thick end-plate combined with 90 mm cross-centres is usually used for beams up to 457×191 UBs. For UBs of 533×210 and over a 10 mm thick end-plate combined with 140 mm cross-centres is recommended.

The local shear capacity of the web of the beam must be checked and, because of their lack of ductility, the welds between the end-plate and beam web must not be the weakest link.

The essential detailing requirements for this connection are also shown in Figure 27.5 and the design checks are detailed in Table 27.4. The design procedure in Table 27.4 applies to beams connected to either a column flange or a column web and the procedure is equally applicable for partial and full depth endplates.

Where this connection type is used to connect a beam to a beam, the top flange of the supported beam is notched to allow it to fit to the web of the supporting beam. If both beams are of a similar depth both flanges are notched. In either case, if the length of the notches l_n exceed the limits given in Check 6 in Table 27.4, the unrestrained web and beam must be checked for lateral torsional buckling.[10,11,12] In practice the end-plate is often detailed to extend to the full depth of the notched beam and welded to the bottom flange. This makes the connection relatively stiffer than a partial depth end-plate but provided the end-plate is relatively thin and the bolt cross-centres are large, the end-plate retains sufficient flexibility to be classified as a simple connection.

If the supporting beam is free to twist there will be adequate rotational capacity even with a thick end-plate. In the cases where the supporting beam is not free to twist, for example in a double-sided connection, the rotational capacity must be provided by the connection itself. In such cases thick, full depth end-plates may lead to overstressing of the bolts and welds. Both partial and full depth end-plates derive their flexibility from the use of relatively thin end-plates combined with large bolt cross-centres. Normally end-plates no more than 8 mm or 10 mm thick should be used.

The essential detailing requirements shown in Figure 27.5 and the design checks in Table 27.4 are applicable to beam-to-beam connections as well as beam-to-column. The procedure in Table 27.4 can be used for both partial depth and full depth end-plates.

27.2.7 Fin plates

Fin plate connections are widely used in both Australian and American practice. This type of connection is primarily used to transfer beam end reactions and is economical to fabricate and simple to erect. There is clearance between the ends of the supported beam and the supporting column, thus ensuring an easy fit. Figure 27.6 shows a typical bolted fin plate connection to a column and a beam. The connection comprises a single plate with either pre-punched or pre-drilled holes that is shop welded to the supporting column flange or web.

Table 27.4 Design checks for flexible end-plate connections

Check No.	Description	Design rule
1	Detailing	See Figure 27.5.
2	Supported beam – welds	Satisfied if the effective weld throat thickness $a \geq 0.45 t_{w,b1}$ for S275 beam or $a \geq 0.53 t_{w,b1}$ for S355 beam.
3	N/A	
4	Supported beam – shear resistance at connection	Design shear resistance of the supported beam at the connection $V_{c,Rd}$, based on a shear area $0.9 h_p t_{w,b1}$, must exceed the applied shear force, V_{Ed}.
5	Supported beam – resistance at notch	Moment at notch, $V_{Ed} (t_p + l_n)$, must be less than the moment of resistance of the supported beam at the notch in the presence of shear $M_{v,Rd}$. When $V_{Ed} \leq 0.5 V_{pl,Rd}$, $M_{v,Rd}$ may be taken as the yield strength of the beam multiplied by the elastic modulus at the notch. When $V_{Ed} > 0.5 V_{pl,Rd}$, $M_{v,Rd}$ must be reduced to account for the presence of high shear (see EC3-1-1 Clause 6.2.8(3).
6	Supported beam – local stability of notched beam	Satisfied if the beam is restrained against lateral torsional buckling and for a beam notched at the top to a depth not exceeding half the beam depth, or for a beam notched top and bottom, with either notch not exceeding 20% of the beam depth: $l_n \leq h_{b1}$ for $h_{b1}/t_{w,b1} \leq 54.3$ (S275) or ≤ 48.0 (S355) $l_n \leq 160000\, h_{b1}/(h_{b1}/t_{w,b1})^3$ for $h_{b1}/t_{w,b1} > 54.3$ (S275) $l_n \leq 110000\, h_{b1}/(h_{b1}/t_{w,b1})^3$ for $h_{b1}/t_{w,b1} > 48.0$ (S355).
7	Overall stability of notched beam	When a notched beam is unrestrained against lateral torsional buckling, the overall stability of the beam should be checked. Details of the check are presented in section 27.2.5.
8	Supporting beam / column – Bolt group	Shear and bearing resistance of the bolt group F_{Rd} must be greater than the applied shear V_{Ed}.
9	Supporting beam/ column – Connecting elements	Shear resistance of the end plate $V_{Rd,min}$ (smaller of the gross section shear resistance $V_{Rd,g}$, net section shear resistance $V_{Rd,n}$ and block tearing resistance $V_{Rd,b}$) must exceed the applied shear V_{Ed}.
10	Supporting beam/ column – Local resistance	Local shear and bearing resistance of the supporting beam web or column web or RHS wall must exceed the total applied shear V_{Ed} arising from the supported beam(s).
11	Structural integrity – connecting elements	The minimum tension resistance of the endplate from Mode 1 (complete yielding of the end plate), Mode 2 (bolt failure with yielding of the endplate) or Mode 3 (bolt failure) must exceed the applied tie force F_{Ed}.
12	Structural integrity – supported beam	The tension resistance of the connected beam, taken as the ultimate tensile strength of the web over the depth of the connected endplate web, must exceed the applied tie force F_{Ed}.
13	Structural integrity – welds	Satisfied if the effective weld throat thickness $a \geq 0.45 t_{w,b1}$ for an S275 supported beam or $a \geq 0.53 t_{w,b1}$ for a S355 supported beam.
14	Structural integrity – supporting column web (UC or UB)	This check is only required for either single-sided connections to a column web or unequally loaded double-sided connections to a column web. Web panel resistance to out-of-plane bending must exceed the applied tie force. Resistance calculated assuming an 'envelope' hinge collapse mechanism in the web panel.
15	Structural integrity – supporting column wall (RHS)	The tying resistance of an RHS wall in the presence of axial compression in the column must exceed the applied tie force. The check assumes the same collapse mechanism as check 14 but replaces the column web with the RHS wall.
16	N/A	

In the design model for a fin plate (Figure 27.6), it is important to identify the appropriate line of action for the shear. There are two possibilities: either the shear acts at the face of the column or it acts along the centre of the bolt group connecting the fin plate to the beam web. For this reason all critical sections should be checked for a minimum moment taken as the product of the vertical shear and the distance between the face of the column (or beam web) and the centre of the bolt group. The critical sections are then checked for the resulting moment combined with the vertical shear. Fin plates with long projections (dimension 'z_p' in Figure 27.6) have a tendency to twist and fail by lateral torsional buckling. Where the projection of the fin plate is small i.e. z_p (the distance between the weld line and the centre of the bolt group) $\leq t_p/0.15$, the buckling resistance of the fin plate is taken as the elastic bending moment resistance of the plate, $W_{el}f_{yp}/\gamma_{M0}$ and $W_{el} = t_p h_p^2/6$ and $\gamma_{M0} = 1.0$ in UK NA. In cases where $z_p \geq t_p/0.15$ the buckling resistance of the fin plate may be based[6,4] on a check derived from the BS 5950 version Green Book[2] and be taken as $W_{el}f_{pLT}/0.6\gamma_{M1} \leq W_{el}f_{yp}/\gamma_{M0}$ where f_{pLT} is the lateral torsional buckling strength of the plate obtained from Table 17 of BS 5950-1 for $\lambda_{LT} = 2.8(z_p h_p/1.5t_p^2)^{0.5}$ and $\gamma_{M1} = 1.0$ in the UK NA.

Fin plate connections derive their in-plane rotational capacity from the bolt deformation in shear, from the distortion of the bolt holes in bearing and from the out-of-plane bending of the fin plate.

The essential detailing requirements for this connection are shown in Figure 27.6 and the detailed design checks are given in Table 27.5. The design procedure given in Table 27.5 applies to beams connected to either the column flange, column web or beam web in a beam-to-beam connection. A beam-to-beam fin plate connection requires either a long fin plate as shown in Figure 27.7(a) or a notched beam as shown in Figure 27.7(b). The designer must therefore choose between the reduced capacity of a long fin plate and the reduced capacity of a notched beam. Another minor consideration is the torsion induced when fin plates are attached to one side of the supported beam web. However, tests have shown that in these cases the torsional moments are small and can be neglected.

27.2.8 Column splices

This section presents design requirements for column splices in braced multi-storey buildings. In this type of building, column splices are required to provide continuity of both strength and stiffness about both axes of the columns. In general they are subject to both axial compression and moments resulting from the end reactions of the beams. If a splice is positioned near to a point of lateral restraint (i.e. within say 500 mm above the floor level), and the column is designed as pinned at that point, the splice may simply be designed for the axial load and any applied moments. If, however, the splice is positioned away from a point of lateral restraint (i.e. more than 500 mm above the level of the floor), or end fixity or continuity has been assumed when calculating the effective length of the column, the additional moment

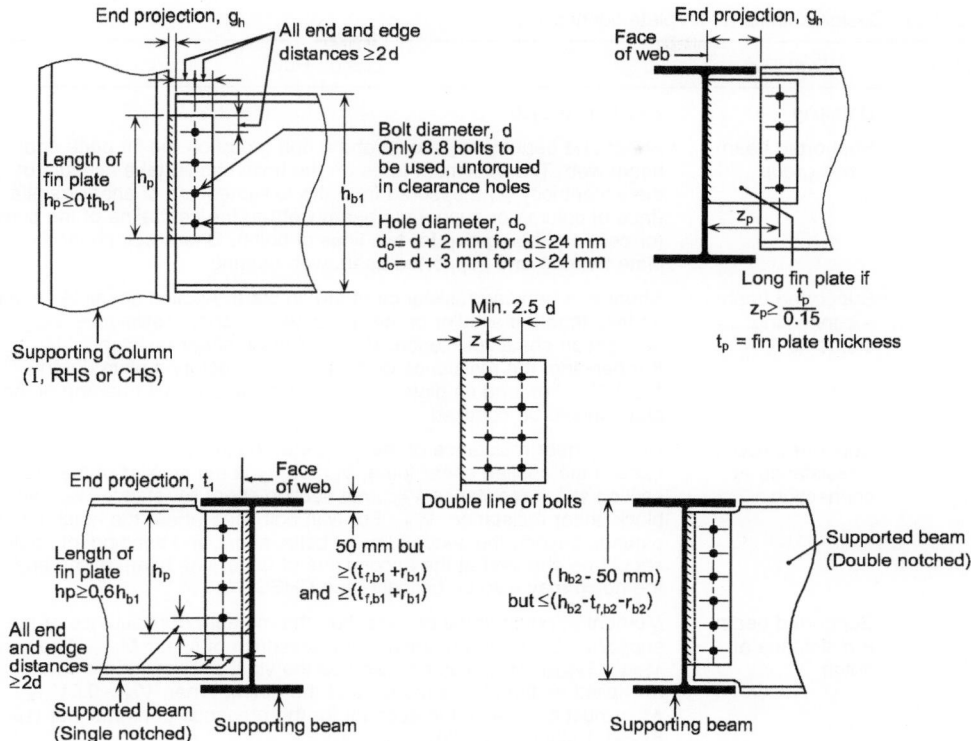

Notes
1. The fin plate is generally poistioned close to the top flange of the beam to provide adequate positional restraint. Its length should be at least 0.6 h_p to give adequate "nominal torsional restraint".
2. For supported beams exceeding a depth of 610mm, the design method given here may only be used when the following three conditions are all met:
   - Supported beam Span/Depth ≤ 20
   - End projection $g_h \geq 20$
   - Vertical distance between extreme bolts $(n-1)\,p \leq 530$mm
3. Bolt spacing and edge distances should comply with the recommendations of BS EN 1993-1-8.
4. Detailing is similar for long fin plates (i.e. the fin plate thickness t_p is less than 0.15z) except the end projection g_h will be considerably greater.

Figure 27.6   Typical fin-plate connection

Table 27.5 Design checks for fin plate connections

Check No.	Detailing	Design Rule
1	Detailing	See Figure 27.6
2	Supported beam – bolt group	Shear and bearing resistance of the bolt group on the fin plate and beam web. The maximum force on the bolts should take account of the eccentricity (z) measured from the assumed line of shear transfer (face of column or supporting beam web) to the centreline of the bolts (or centre of bolt group for two lines of bolts). Check bolt shear, fin plate bearing and supported beam web bearing.
3	Supported beam – connecting elements	Shear and bending resistance of the fin plate. Applied shear V_{Ed} must be less than the smaller of the gross section shear resistance $V_{Rd,g}$, net section shear resistance, $V_{Rd,n}$ and block shear resistance, $V_{Rd,b}$. For bending, the resistance of the plate satisfactory if plate depth $h_p \geq 2.73z$. For long fin plates ($z > 6t_p$), lateral torsional buckling of the plate should be checked.
4	Supported beam – resistance at connection	Design shear resistance of the supported beam web, $V_{Rd,min}$ must exceed the applied shear force, V_{Ed}. $V_{Rd,min}$ is the smaller of the gross section shear resistance $V_{Rd,g}$, net section shear resistance, $V_{Rd,n}$ and block shear resistance, $V_{Rd,b}$. For twin bolt lines where the notch length extends beyond the second line of bolts, shear and bending interaction should be checked at the second line of bolts. The beam at the end of the notch may also be critical – see CHECK 5.
5	Supported beam – resistance at notch	Moment at notch, must be less than the moment of resistance of the supported beam at the notch in the presence of shear $M_{v,Rd}$. When $V_{Ed} \leq 0.5 V_{pl,Rd}$, $M_{v,Rd}$ may be taken as the yield strength of the beam multiplied by the elastic modulus at the notch. When $V_{Ed} > 0.5 V_{pl,Rd}$, $M_{v,Rd}$ must be reduced to account for the presence of high shear (see EC3-1-1 clause 6.2.8(3)).
6	Supported beam – local stability of notched beam	Satisfied if the beam is restrained against lateral torsional buckling and for a beam notched at the top to a depth not exceeding half the beam depth, or for a beam notched top and bottom, with either notch not exceeding 20% of the beam depth: $l_n \leq h_{b1}$ for $h_{b1} / t_{w,b1} \leq 54.3$ (S275) or ≤ 48.0 (S355) $l_n \leq 160000\, h_{b1} /(h_{b1} / t_{w,b1})^3$ for $h_{b1} / t_{w,b1} > 54.3$ (S275) $l_n \leq 110000\, h_{b1} /(h_{b1} / t_{w,b1})^3$ for $h_{b1} / t_{w,b1} > 48.0$ (S355)
7	Overall stability of notched beam	When a notched beam is unrestrained against lateral torsional buckling, the overall stability of the beam should be checked. Details of the check are presented in section 27.2.5.
8	Supporting beam/ column – welds	Strength of the weld connecting the fin plate to the supporting beam or column not critical if $a \geq 0.5t_p$ for a S275 fin plate or $a \geq 0.6t_p$ for a S355 fin plate.
9	NA	
10	Supporting beam/ column – Local resistance	Check local shear and punching resistance of the supporting beam web, or column web or wall of RHS or CHS. V_{Ed} must be less than the local shear resistance $F_{Rd} = h_p t_2 f_{y,2}/3^{0.5}\, \gamma_{M0}$, where t_2 and $f_{y,2}$ refer to the supporting member. Yielding of the fin plate will occur before punching shear provided that $t_p < t_2 f_{u,2}/f_{y,p}$.
11	Structural integrity – connecting elements	The tension and bearing resistance of the fin plate must exceed the applied tie force F_{Ed}. Tension resistance is the smaller of the block tearing tension resistance, $F_{Rd,b}$, and the net section tension resistance, $F_{Rd,n}$; horizontal shear resistance is limited to the ultimate shear strength of the bolts; fin plate bearing is dependent on the edge and end distance of the bolt holes.

Table 27.5 *(continued)*

Check No.	Detailing	Design Rule
12	Structural integrity – supported beam	Tension and bearing resistance of the web must exceed the tie force. Tension resistance of the beam web is the smallest of the block tearing tension resistance, $F_{Rd,b}$, and the net section tension resistance, $F_{Rd,n}$. Bearing resistance of the bolt group is dependent on the edge and end distance of the bolt holes in the beam web.
13	NA	
14	Structural integrity – supporting column web (UC or UB)	This check is only required for either single-sided connections to a column web or unequally loaded double-sided connections to a column web. Web panel resistance to out of plane bending must exceed the applied tie force. Resistance calculated assuming an 'envelope' hinge collapse mechanism in the web panel.
15	Structural integrity – supporting column wall (RHS)	The tying resistance of an RHS wall in the presence of axial compression in the column must exceed the applied tie force. The check assumes the same collapse mechanism as check 14 but replaces the column web with the RHS wall.
16	Structural integrity – supporting column wall (CHS)	Tie force $\leq F_{Rd}$, where $F_{Rd} = 5f_{u,c}t^2(1+0.25h_p/d)*0.67/\gamma_{Mu}$ and $\gamma_{Mu} = 1.1$; $f_{u,c}$, t and d relate to the CHS.

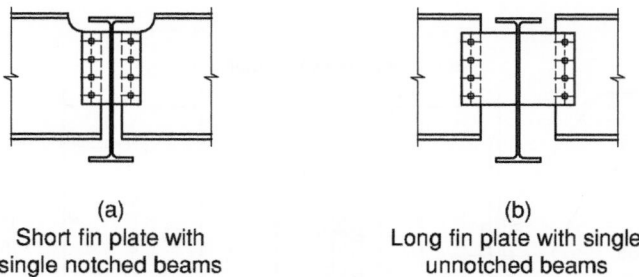

(a)	(b)
Short fin plate with single notched beams	Long fin plate with single unnotched beams

Figure 27.7   Beam to beam fin plates

that can be induced by strut action must be taken in to account. A procedure for calculating these additional moments is not given in EC3 but Reference 13 addresses the issue.

Two types of splices are considered in this section, those where the ends of the members are prepared for contact in bearing, and those where ends of the members are not prepared for contact in bearing.

In both cases the column splices should hold the connected members in line and wherever practicable the members should be arranged so that the centroidal axis of the splice material coincides with the centroidal axes of the column sections above and below the splice.

27.2.8.1 Ends prepared for contact in bearing

Typical details for this type of column splice are shown in Figures 27.8–27.10. In each case the splice is constructed using web and flange cover plates, and packs are used to make up any differences in the thicknesses of the web and the flanges. The flange cover plates may be placed on either the outside or the inside of the column. Placing the cover plates on the inside has the advantage of reducing the overall depth of the column. Each column splice must be designed to carry axial compressive forces, the tension (if any) resulting from the presence of bending moments and any horizontal shear forces.

Axial compressive forces

The ends of the columns are usually prepared for full contact, in which case compressive forces may be transmitted in bearing. However, it is not necessary to achieve an absolutely perfect fit over the entire area of the column. Columns with saw cut ends are adequately smooth and flat for bearing and no machining is required. This is because after erection the ends of the column bed down as successive dead loads are applied to the structure.

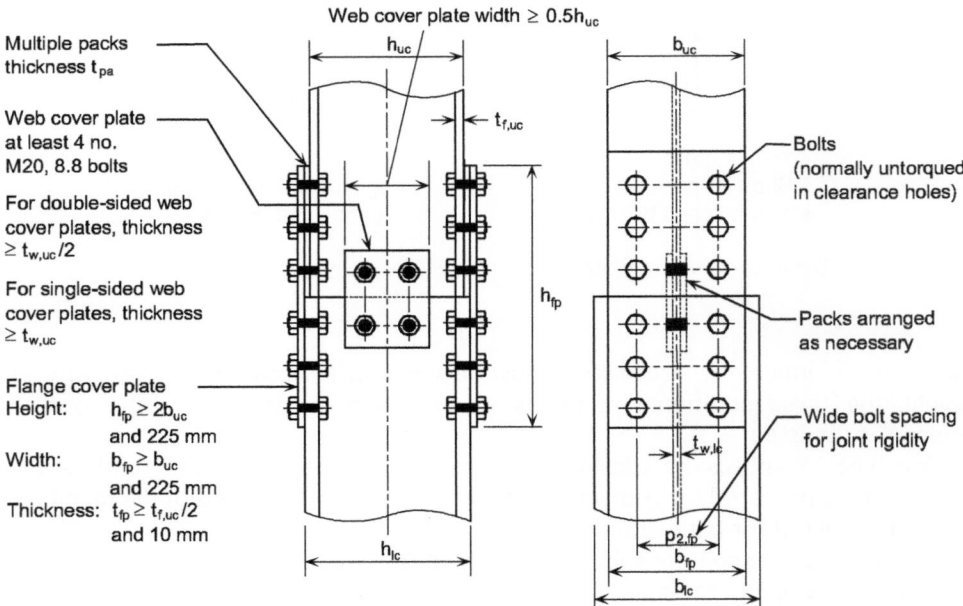

Figure 27.8   Bearing splice with external cover plates

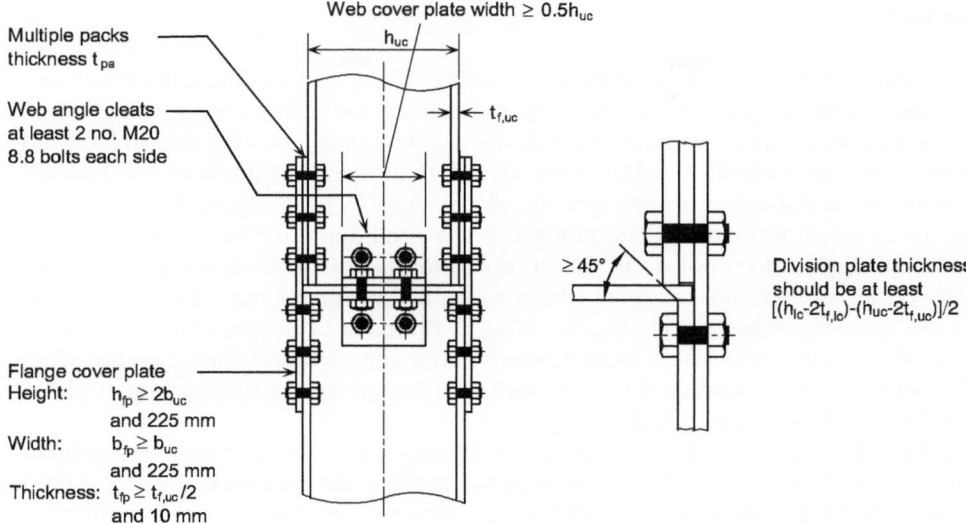

Web cover plate width ≥ 0.5h_{uc}

h_{uc}

Multiple packs thickness t_{pa}

Web angle cleats at least 2 no. M20 8.8 bolts each side

$t_{f,uc}$

≥ 45°

Division plate thickness should be at least $[(h_{lc}-2t_{f,lc})-(h_{uc}-2t_{f,uo})]/2$

Flange cover plate
Height: $h_{fp} \geq 2b_{uc}$
                and 225 mm
Width: $b_{fp} \geq b_{uc}$
                and 225 mm
Thickness: $t_{fp} \geq t_{f,uc}/2$
                and 10 mm

Figure 27.9   Bearing splice with external cover plates and a division plate

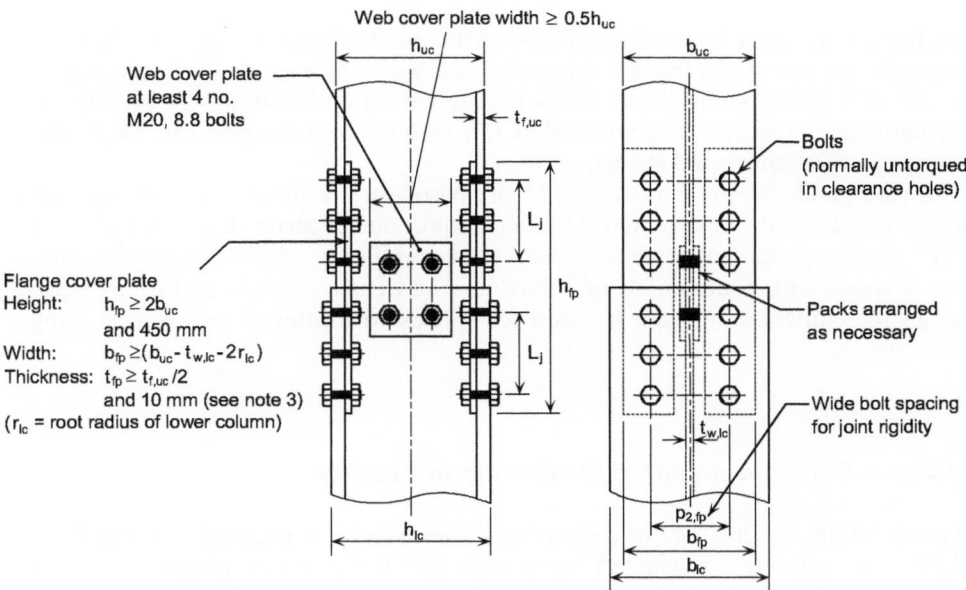

Web cover plate width ≥ 0.5h_{uc}

h_{uc} b_{uc}

Web cover plate at least 4 no. M20, 8.8 bolts

$t_{f,uc}$

L_j

h_{fp}

Bolts (normally untorqued in clearance holes)

Flange cover plate
Height: $h_{fp} \geq 2b_{uc}$
                and 450 mm
Width: $b_{fp} \geq (b_{uc} - t_{w,lc} - 2r_{lc})$
Thickness: $t_{fp} \geq t_{f,uc}/2$
                and 10 mm (see note 3)
(r_{lc} = root radius of lower column)

L_j

Packs arranged as necessary

Wide bolt spacing for joint rigidity

h_{lc}

$t_{w,lc}$

$p_{2,fp}$
b_{fp}
b_{lc}

Figure 27.10   Bearing splice with internal cover plates

Tension

The cover plates provide continuity of stiffness and are designed to resist any tension where the presence of bending moments is sufficiently high to overcome the compressive forces in the column. If the nominal moment, M_{Ed}, due to factored permanent and variable loads (i.e. column design moment) at the floor level immediately below the splice is less than $N_{Ed,G}h/2$ (where $N_{Ed,G}$ is the axial compression due to factored permanent load only and h is overall depth of the smaller column), then net tension does not occur. If this condition is not satisfied, then net tension does occur and the flange cover plates and their fasteners must be checked for tensile force F_{Ed}, where $F_{Ed} = M_{Ed}/h - N_{Ed,G}/2$. Preloaded slip-resistant bolts should be used when the net tension induces stress in the upper column flange greater than 10% of the design strength of that column. The design checks for the flanges plates are described in Section 27.2.8.2.

When checking the splice for structural integrity, the tie force may be assumed to be resisted by the two flange cover plates and the design checks for F_{Ed} should be repeated for a force of $N_{tie}/2$. Preloaded slip-resistant bolts are not required if significant tension only arises as a result of the structural integrity check.

Shear forces

The horizontal shear forces that arise from the moment gradient in the column are normally resisted by the friction across the bearing surfaces of the two columns and by the web cover plates. Wind forces on the external elevations of buildings are normally taken directly in the floor slab. It is rare for column splices in simple construction to transmit wind shears.

The design checks given in Table 27.6 are based on the above assumptions. These design checks and the essential detailing requirements given in Figures 27.9 and 27.10 give empirical rules which preserve the continuity of the finished structure, give a splice with a minimum of robustness and ensure erection stiffness. These design recommendations can be used for splices with internal or external flange cover plates.

27.2.8.2 Ends not prepared for contact in bearing

Typical details for this type of column splice are shown in Figure 27.11 and Figure 27.12. The splice is constructed using web and flange cover plates and, where required, packs are used to make up the differences in the web and flange thicknesses. Where columns of different serial size are to be connected, multiple packs are necessary to take up the dimensional variations.

For this type of splice the cover plates carry all the forces and moments and no load is transferred through direct bearing. The axial load in the column is normally

Table 27.6 Design checks for column splices – ends prepared for contact in bearing

Check No.	Description	Design rule
1	Detailing	See Figure 27.7,Figure 27.8 and Figure 27.9, as appropriate
2	Flange cover plates – presence of net tension	If $M_{Ed} \le N_{Ed,G}h/2$ net tension does not occur where $N_{Ed,Gh}$ = axial compression due to factored permanent load only and h = overall depth of the smaller column. If net tension does occur, flange cover plates and their fasteners must be checked for tensile force F_{Ed}, where $F_{Ed} = M_{Ed}/h - N_{Ed,G}/2$. Preloaded slip-resistant bolts should be used when the net tension induces stress in the upper column flange greater than 10% of the design strength of that column.
3	Flange cover plate – tensile resistance of cover plate	Maximum tensile force developed in the cover plate must not exceed the minimum of the plate tensile resistance ($N_{pl,Rd}$), the ultimate resistance of the net section ($N_{u,Rd}$) or block tearing resistance $N_{bt,Rd}$ (27.2.7.2 for details).
4	Flange cover plate – bolt group	$N_{Ed} < V_{Rd,fp}$ $N_{Ed} = N_{Ed,t}$ $V_{Rd,fp}$ is the design resistance of flange cover plate bolt group: $V_{Rd,fp} = \Sigma F_{b,Rd}$ if $(F_{b,Rd})_{max} \le F_{v,Rd}$ $V_{Rd,fp} = n_{fp}(F_{b,Rd})_{min}$ if $(F_{b,Rd})_{min} \le F_{v,Rd} \le (F_{b,Rd})_{max}$ $V_{Rd,fp} = n_{fp}F_{v,Rd}$ if $F_{v,Rd} \le (F_{b,Rd})_{min}$ where $F_{v,Rd}$ and $F_{b,Rd}$ are the shear and bearing resistances of a single bolt respectively (see Table 3.4 BS EN 1993-1-8). For preloaded bolts used in connections designed to be non-slip under factored loads, the force in the cover plate, N_{Ed}, must be less than the design slip resistance of the bolt group $F_{s,Rd}$ (BS EN 1993-1- 3.9.1).
5	Structural integrity (bearing type)	Checks 3 and 4 should be carried out for a tensile force of $N_{tie}/2$ where N_{tie} is the tensile force from BS EN 1997-1-7, Clause A.6.

shared between the flange and web cover plates in proportion to their areas, while the flange cover plates alone carry any bending moments.

The maximum compressive and tensile forces in the flange cover plate is given by the following two expressions:

$$N_{Ed,0} = \frac{M_{Ed}}{h} + N_{Ed}\left(\frac{A_{f,1}}{A_1}\right)$$

$$N_{Ed,1} = \frac{M_{Ed}}{h} + N_{Ed,G}\left(\frac{A_{f,1}}{A_1}\right)$$

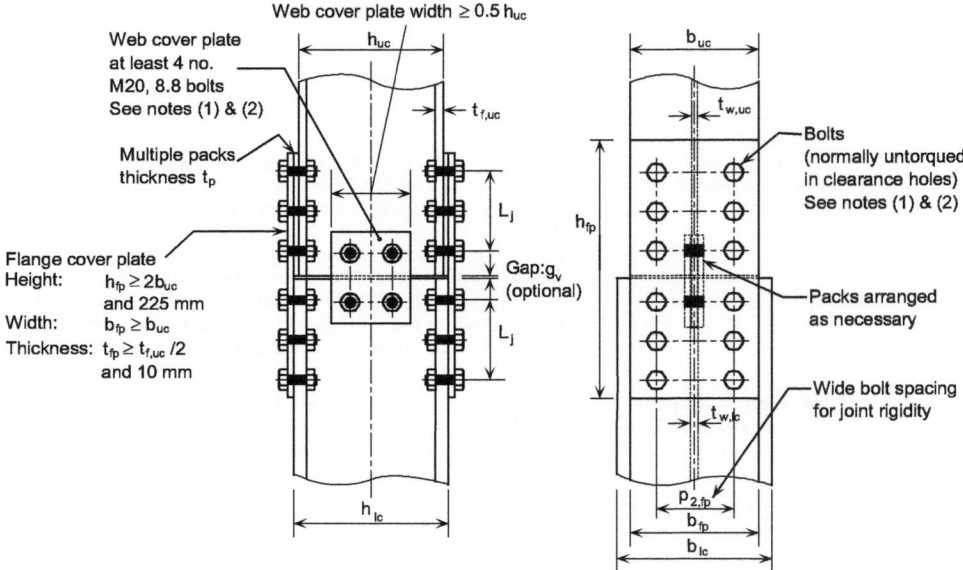

Figure 27.11   Non-bearing column splice with external cover plates

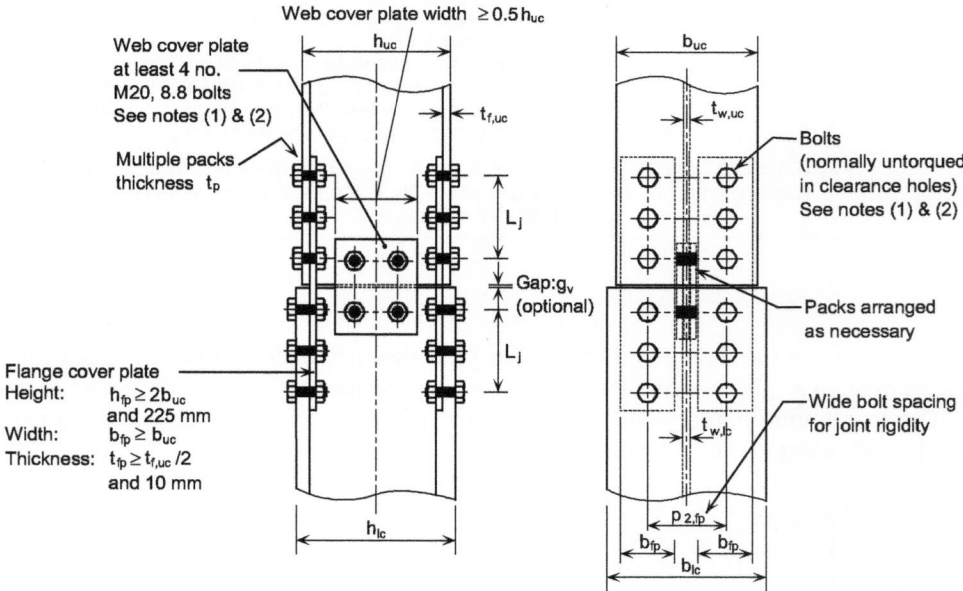

Figure 27.12   Non-bearing column splice with internal cover plates

where:

M_{Ed} is the nominal moment due to factored permanent and variable loads (i.e. column design moment) at the floor level immediately below the splice

N_{Ed} is the axial compression due to factored permanent and variable loads

$N_{Ed,G}$ is the axial compression due to factored permanent load only

h is the overall depth of the smaller column (for external flange cover plates) or the centreline-to-centreline distance between internal flange cover plates

$A_{f,1}$ is the area of one flange of the smaller column and $A1$ is the total area.

The compressive resistance of the flange cover plates attached to one flange is given by:

$$N_{c,Rd} = \chi A_{fp} f_{yp} / \gamma_{M0}$$

where the reduction factor for buckling is from **BS EN1993-1-1** 6.3 using curve *c* and assuming a buckling length equal to the vertical pitch of the bolts and a plate width of $b_{fp}/2$; $\gamma_{M0} = 1.0$ in the UK NA. If the maximum bolt pitch divided by the cover plate thickness is less than 9ε in all cases, then buckling will not occur and χ may be taken as 1.

In tension, the maximum tensile force, $N_{Ed,t}$ must less than the minimum of the flange plate tensile resistance ($N_{pl,Rd}$), the ultimate resistance of the net section ($N_{u,Rd}$) or block tearing resistance $N_{bt,Rd}$.

$$N_{pl,Rd} = \frac{A_{fp} f_{yp}}{\gamma_{M0}}$$

$$N_{u,Rd} = \frac{0.9 A_{fp,net} f_{u,fp}}{\gamma_{M2}}$$

For a concentrically loaded bolt group:

$$N_{bt,Rd} = V_{eff,1,Rd}$$

$$V_{eff,1,Rd} = \frac{f_{u,fp} A_{fp,nt}}{\gamma_{M2}} + \frac{f_{y,fp} A_{fp,nv}}{\sqrt{3}\gamma_{M0}}$$

For an eccentrically loaded bolt group:

$$N_{bt,Rd} = V_{eff,2,Rd}$$

$$V_{eff,2,Rd} = \frac{0.5 f_{u,fp} A_{fp,nt}}{\gamma_{M2}} + \frac{f_{y,fp} A_{fp,nv}}{\sqrt{3}\gamma_{M0}}$$

Where the terms in the above equations are described in Figure 27.13. Table 27.7 summarises the design checks for a non-bearing column splice.

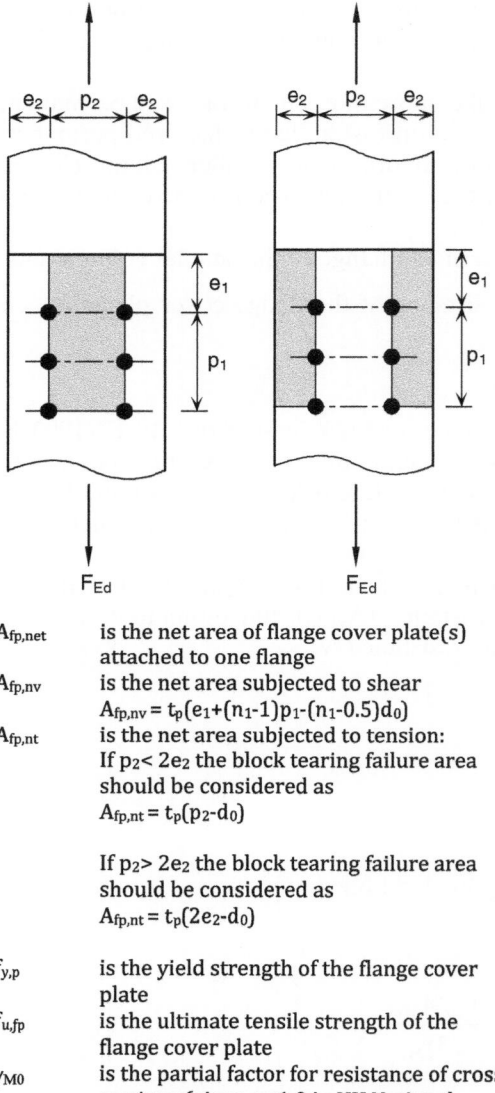

$A_{fp,net}$ is the net area of flange cover plate(s) attached to one flange

$A_{fp,nv}$ is the net area subjected to shear
$A_{fp,nv} = t_p(e_1+(n_1-1)p_1-(n_1-0.5)d_0)$

$A_{fp,nt}$ is the net area subjected to tension:
If $p_2 < 2e_2$ the block tearing failure area should be considered as
$A_{fp,nt} = t_p(p_2-d_0)$

If $p_2 > 2e_2$ the block tearing failure area should be considered as
$A_{fp,nt} = t_p(2e_2-d_0)$

$f_{y,p}$ is the yield strength of the flange cover plate

$f_{u,fp}$ is the ultimate tensile strength of the flange cover plate

γ_{M0} is the partial factor for resistance of cross sections (given as 1.0 in UK National Annex)

γ_{M2} is the partial factor for tension resistance of net sections (given as 1.1 in UK National Annex)

Figure 27.13   Block tearing resistance of a flange cover plate

Table 27.7 Design checks for a non-bearing column splice (based on Green Book)

Check No.	Description	Design rule
1	Detailing	See Figure 27.11 or Figure 27.12 as appropriate
2	Flange cover plates resistance	Compression Maximum compression in the cover plate, $N_{Ed,c}$ (see text for derivation) must be less than the buckling resistance of the cover plate, $N_{c,Rd}$. Tension Maximum tensile force developed in the cover plate must not exceed the minimum of the plate tensile resistance ($N_{pl,Rd}$), the ultimate resistance of the net section ($N_{u,Rd}$) or block tearing resistance $N_{bt,Rd}$ (see text for details).
3	Flange cover plate – bolt group	$N_{Ed} \leq F_{Rd,fp}$ $N_{Ed} = \max(N_{Ed,c}; N_{Ed,t})$ $V_{Rd,fp}$ is the design resistance of the flange cover plate bolt group. $V_{Rd,fp} = \Sigma F_{b,Rd}$ if $(F_{b,Rd})_{max} \leq F_{v,Rd}$ $V_{Rd,fp} = n_{fp}(F_{b,Rd})_{min}$ if $(F_{b,Rd})_{min} \leq F_{v,Rd} \leq (F_{b,Rd})_{max}$ $V_{Rd,fp} = n_{fp}F_{v,Rd}$ if $F_{v,Rd} \leq (F_{b,Rd})_{min}$ Where $F_{v,Rd}$ and $F_{b,Rd}$ are the shear and bearing resistances of a single bolt respectively (see Table 3.4 BS EN 1993-1-8) For preloaded bolts used in connections designed to be non-slip under factored loads, the force in the cover plate, N_{Ed}, must be less than the design slip resistance of the bolt group $F_{s,Rd}$ (BS EN 1993-1- 3.9.1).
4	Web cover plates resistance	Force in the column web, $F_{Ed,c,web}$ (conservatively taken as $F_{Ed,c}A_{w,1}/A_1$) must be less than the sum of the plastic resistance of the cover plates, $\Sigma A_{wp}f_{y,wp}/\gamma_{M0}$ ($\gamma_{M0} = 1.0$ in UK NA)
5	Web cover plates bolt group	Force in the column web cover plates ($F_{Ed,c,web}$) must be less than the design resistance of the bolt group calculated in the same way as check no. 3. If the thickness of the column web is less than the combined thickness of the web cover plates, then the bearing resistance of the column web should also be checked.
6	Structural integrity	Checks 2 and 3 should be carried out for a tensile force of $N_{tie}/2$ (where N_{tie} is the tensile force from BS EN 1997-1-7, Clause A.6) based on the conservative assumption that the flange cover plates only resist the tie force. For non-bearing splices, this check will not be necessary.

Summary

EC3 highlights the inter-relationship between joint behaviour and frame response. Although joints may be assumed to behave in one of the analytically convenient extremes, i.e. as pins or as fully continuous, the code carefully addresses the fact that real joint behaviour will deviate from these assumptions and requires that the designer consider whether the connection details used adequately reflect the assumptions made in the global analysis and design of the frame. In the case of simple connections, it is recognised that the types of details which have been used for many decades meet the necessary conditions for classification as a simple joint and therefore the well established approaches to the design, as outlined in the SCI/ BCSA Simple Connections Green Book, may continue to be used.

A series of worked examples are presented in the following pages. For the benefit of those readers familiar with BS5950, these have been based on the examples in the sixth edition of the *Steel Designers' Manual* and the similarity (and key differences) in approach should be readily apparent. Further detailed examples are available on www.access-steel[14,15,16] along with flowcharts.[17,18,19] Guidance on initial sizing is provided in a number of NCCIs.[20,21,22]

References to Chapter 27

1. British Standards Institution (2005) *BS EN 1993-1-8 Eurocode 3: Design of steel structures – Part 1-8: Design of joints*. London, BSI.
2. The Steel Construction Institute/The British Constructional Steelwork Association LTD (2002) *Joints in Steel Construction: Simple Connections*, Publication No. 212. London, SCI, BCSA.
3. The Steel Construction Institute/The British Constructional Steelwork Association LTD (1995) *Joints in Steel Construction: Moment Connections*, Publication No. 207. Ascot, SCI, BCSA.
4. The Steel Construction Institute/The British Constructional Steelwork Association (2010) *Joints in Steel Construction: Simple Connections*. Ascot, SCI, BCSA.
5. British Standards Institution (2005) *BS EN 1993-1-1 Eurocode 3: Design of steel structures – Part 1-1: General rules and rules for buildings*. London, BSI.
6. Jaspart J., Demonceau J.F., Renkin S. and Guillaume M.L. (2009) *European Recommendations for the Design of Simple Joints in Steel Structures*, ECCS Publication No. 126, European Convention for Constructional Steelwork, Portugal.
7. *NCCI: 'Simple Construction' – concept and typical frame arrangements*, SN020a-EN-EU. www.steel-ncci.co.uk
8. The British Constructional Steelwork Association and the Steel Construction Institute (2010) *National Structural Steelwork Specification for Building Construction*, 5th edn (CE Marking Version), Publication No. 52/10. London, BCSA, SCI.

9. Hogan, T.J. and Thomas, I.R. (1988) *Design of Structural Connections*, 3rd edn, Australian Institute of Steel Construction, Milsons Point, Australia.

10. Cheng J.J.R. and Yura J.A. (1988) Lateral buckling tests on coped steel beams, *Journal of Structural Engineering, ASCE*, 114, No. 1, 1–15.

11. Gupta A.K. (1984) Buckling of coped beams, *Journal of Structural Engineering, ASCE*, 110, No. 9, 1977–87.

12. Cheng J.J.R., Yura J.A. and Johnson C.P. (1988) Lateral buckling of coped steel beams, *Journal of Structural Engineering, ASCE*, 114, No. 1, 16–30.

13. Steel Construction Institute (nd) *Advisory Notes: AD 314: Column splices and internal moments; AD 243: Splices within unrestrained lengths; AD 244: Second order moments.* Ascot, SCI.

14. *Example: Column splice – non-bearing splice*, SX018a-EN-EU, www.access-steel.com

15. *Example: End plate beam-to-column flange simple connection*, SX012a-EN-GB, www.access-steel.com

16. *Example: Fin plate beam-to-column flange connection*, SX13a-EN-GB, www.access-steel.com

17. *Flow Chart: Design model for non-bearing column splices*, SF018a-EN-EU, www.access-steel.com

18. *Flow Chart: Simple end plate connection*, SF008a-EN-EU, www.access-steel.com

19. *Flow Chart: Fin plate connection*, SF009a-EN-EU,www.access-steel.com

20. *NCCI: Initial sizing of non-bearing column splices*, SN024a-EN-EU, www.steel-ncci.co.uk

21. *NCCI: Initial sizing of simple end plate connections*, SN013a-EN-EU, www.steel-ncci.co.uk

22. *NCCI: Initial sizing of fin plate connections*, SN016a-EN-EU, www.steel-ncci.co.uk

	Steel Designers' Manual 7th Edition			**BS EN 1993-1-8**
	Column Splice (UKC bearing splice)			
	Made by DBM	Checked by JBD	Sheet No. 1 of 3	

The splice is between a 305×305×97UKC upper column and a 356×3686×153UKC lower column. Both columns are grade S275 steel.

The factored forces and moments acting on the splice are:-

Axial compression N_{Ed} $= 600kN$ $(N_{Ed,G}=275kN$ $N_{Ed,Q}=325kN)$
Bending moment M_{Ed} $=100kNm$

Check 1 – Detailing satisfies requirements shown in Figures. 27.7 and 27.8

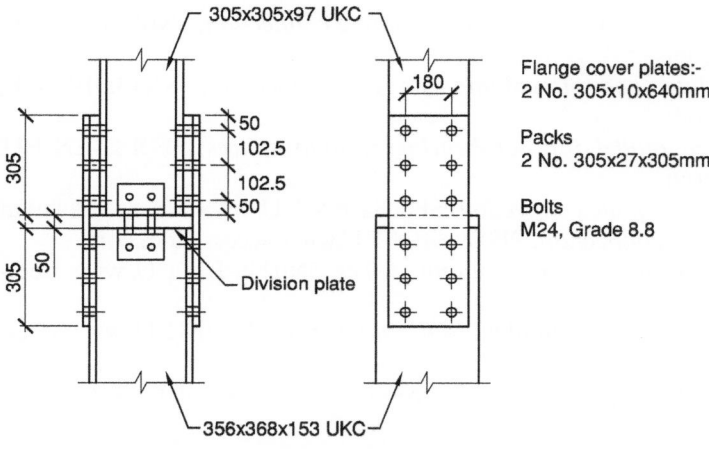

Check 2 – The presence of tension due to axial load and moment

Check for the presence of tension in the splice due to axial load and moment.

If $M_{Ed} \leq N_{Ed,G}\,h/2$

tension does not occur and the splice need only be detailed to transmit axial compression in direct bearing.

If $M_{Ed} > N_{Ed,G}\,h/2$

net tension does occur and the flange cover plates and fasteners should be designed for the tension.
$N_{Ed,G} = 600kN$
$M_{Ed} = 100kNm$
$h = 307.8mm$ *(conservatively taken as the depth of the smaller column)*

$$N_{Ed,G}\frac{h}{2} = \frac{275}{1000} \times \frac{307.8}{2} = 42.3kNm$$

   100 > 42.3

Therefore the splice is subject to net tension.

	Steel Designers' Manual 7th Edition			**BS EN 1993-1-8**
	Column Splice (UKC bearing splice)			
	Made by DBM	Checked by JBD	Sheet No. 2 of 3	

Net Tension

The net tension is given by the following expression:

$$F_{Ed} = \frac{M_{Ed}}{h} - \frac{N_{Ed,G}}{2} = \frac{100}{0.308} - \frac{275}{2} = 187.2\ kN$$

Tension is present in the splice and the flange cover plates and their fasteners should be designed to resist the net tension.

Check 3 – Tensile resistance of the flange cover plate

The maximum tensile force, F_{Ed}, must be less than the minimum of the plate tensile resistance ($N_{pl,Rd}$), the ultimate resistance of the net section ($N_{u,Rd}$) or the block tearing resistance ($N_{bt,Rd}$).

Flange cover plate tensile resistance, $N_{pl,Rd}$

$$N_{pl,Rd} = A_{fp}\, f_{y,fp}\, /\, \gamma_{M0}$$

where

> $f_{y,fp}$ *is the yield strength of the flange cover plate*

> A_{fp} *is the (gross) cross-sectional area of the cover plate*

$$N_{pl,Rd} = \frac{A_{fp} f_{y,fp}}{\gamma_{M0}} = \frac{305 \times 10 \times 275}{1000 \times 1.0} = 838.8\ kN$$

Ultimate resistance of the net section, $N_{u,Rd}$

$$N_{u,Rd} = 0.9 A_{fp,net}\, f_{u,fp}\, /\gamma_{M2}$$

where

> $f_{u,fp}$ *is the ultimate tensile strength of the flange cover plate*

> $A_{fp,net}$ *is the net cross-sectional area of the cover plate*

$$N_{u,Rd} = \frac{0.9 A_{fp,net} f_{u,fp}}{\gamma_{M2}} = \frac{0.9 \times (305 \times 10 - 2 \times 26 \times 10) \times 430}{1000 \times 1.25} = 783.3\ kN$$

Block tearing resistance, $N_{bt,Rd}$

$$N_{bt,Rd} = \frac{f_{u,fp} A_{fp,nt}}{\gamma_{M2}} + \frac{f_{y,fp} A_{fn,nv}}{\sqrt{3}\gamma_{M0}}$$

As $p_2 > 2e_2$ (180 > 2 × 62.5)

then $A_{fp,nt} = t_p\,(2e_2 - d_0) = 10(2 \times 62.5 - 26) = 990mm^2$

$A_{fn,nv} = t_p(e_1 + (n_1 - 1)p_1 - (n_1 - 0.5)d_0)$
   $= 10(50 + (3 - 1)102.5 - (3 - 0.5)26) = 1900\ mm^2$

$$N_{bt,Rd} = \frac{f_{u,fp} A_{fp,nt}}{\gamma_{M2}} + \frac{f_{y,fp} A_{fn,nv}}{\sqrt{3}\gamma_{M0}} = \frac{430 \times 990}{1000 \times 1.25} + \frac{275 \times 1900}{1000 \times \sqrt{3} \times 1.0} = 642.3kN$$

The tensile resistance of the flange cover plates is $N_{bt,Rd}$,
   <u>*642.3kN > 187.2kN*</u> *satisfactory*

	Steel Designers' Manual 7th Edition			**BS EN 1993-1-8**
	Column Splice (UKC bearing splice)			
	Made by DBM	Checked by JBD	Sheet No. 3 of 3	

Check 4 – Design resistance of bolt group connecting flange cover plate to column flange

Design resistance of the bolt group, $V_{Rd,fp}$, must be greater than the applied tensile force, $F_{Ed,}$

$V_{Rd,fp} = \Sigma F_{b,Rd}$ *if* $(F_{b,Rd})_{max} \le F_{v,Rd}$

$V_{Rd,fp} = n_{fp}(F_{b,Rd})_{min}$ *if* $(F_{b,Rd})_{min} \le F_{v,Rd} \le (F_{b,Rd})_{max}$

$V_{Rd,fp} = n_{fp}F_{v,Rd}$ *if* $F_{v,Rd} \le (F_{b,Rd})_{min}$

where $F_{v,Rd}$ and $F_{b,Rd}$ are the shear and bearing resistances of a single bolt respectively   *Table 3.4*

$F_{b,Rd}$ will be maximum for the inner bolts

$F_{b,Rd} = k_1 \alpha_b f_u dt / \gamma_{M2}$

k_1 is the smallest of $1.4 p_2 / d_o - 1.7$ *or* 2.5

           $1.4 \times 180 / 26 - 1.7 = 8.0 > 2.5$ $\therefore k_1 = 2.5$

α_b is the smallest of $\alpha_d = p_1 / 3d_o - 0.25; f_{ub} / f_u; 1.0$

             $\alpha_d = 102.5 / 3 \times 26 - 0.25 = 1.06$

             $f_{ub} / f_u = 800 / 430 = 1.86$ $\therefore \alpha_b = 1.0$

$F_{b,Rd} = k_1 \mathbf{a}b f_u dt / \mathbf{g}M2$

     $= 2.5 \times 1.0 \times 430 \times 24 \times 10 / 1.25 = 206.4kN$ $(F_{b,Rd})_{max} = 206.4kN$

$F_{b,Rd}$ will be minimum for the end bolts

k_1 is the smallest of $2.8 e_2 / d_o - 1.7$ *or* 2.5

           $2.8 \times 62.5 / 26 - 1.7 = 5.0 > 2.5$ $\therefore k_1 = 2.5$

α_b is the smallest of $\alpha_d = e_1 / 3d_o; f_{ub} / f_u; 1.0$

             $\alpha_d = 50 / 3 \times 26 = 0.64$

             $f_{ub} / f_u = 800 / 430 = 1.86$ $\therefore \alpha_b = 0.64$

$F_{b,Rd} = k_1 \alpha_b f_u dt / \gamma_{M2}$

     $= 2.5 \times 0.64 \times 430 \times 24 \times 10 / 1.25 = 132.1kN$ $(F_{b,Rd})_{min} = 132.1kN$

Shear resistance of the bolt, $F_{v,Rd}$

$F_{v,Rd} = \alpha_v f_{ub} A / \gamma_{M2}$

if shear plane passes through threaded area, $\alpha_v = 0.6$ *Table 3.4*
for an M24 bolt threaded area $= 353\ mm^2$

$F_{v,Rd} = 0.6 \times 800 \times 353 / 1.25 = 135.6kN$

<u>*Packing*</u>

The reduction factor for packing is an empirical factor which allows for the effects of bending in the bolt due to thick packing   *3.6.1(12)*

Reduction factor $\beta_p = \dfrac{9d}{8d + 3t_p}$ $\beta_p \le 1$

               $= \dfrac{9 \times 24}{8 \times 24 + 3 \times 27} = 0.79$

$\therefore F_{v,Rd} = 0.79 \times 135.6 = 107.1kN$

Since $F_{v,Rd} \le (F_{b,Rd})_{min}, V_{Rd,fp} = n_{fp}F_{v,Rd}$

$= 6 \times 107.1 = 642.6kN > 33.3kN$

Therefore design resistance of bolt group is OK.

	Steel Designers' Manual 7ᵗʰ Edition			**BS EN 1993-1-8**
	Fin plate connection			
	Made by DBM	Checked by JBD	Sheet No. 1 of 11	

The connection is between a 610×229×101UKB (S275) supporting beam and a 457×152×52UKB (S275) supported beam.

The end reaction of the simply supported beam due to factored loads is 110kN.

Refer to Table 27.5 for design checks.

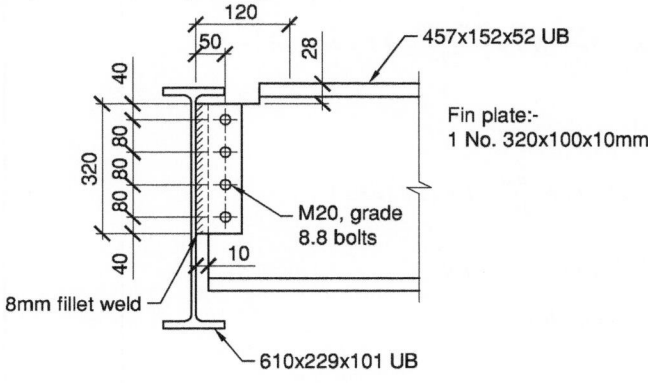

457x152x52 UB

Fin plate:-
1 No. 320x100x10mm

M20, grade
8.8 bolts

8mm fillet weld

610x229x101 UB

Check 1 Recommended detailing practice – *satisfies requirements of Figure 27.5*

Check 2 – Capacity of bolt group connecting fin plate to web of supported beam
(i) bolt shear
basic requirement $V_{Ed} \leq V_{Rd}$

$$V_{Rd} = \frac{nF_{v,Rd}}{\sqrt{(1+\alpha n)^2 + (\beta n)^2}}$$

For a single line of bolts:
$\alpha = 0$, $n = n_1$

$$\beta = \frac{6z}{n_1(n_1+1)p_1} = \frac{6 \times 50}{4(4+1)80} = 0.1875$$

$F_{v,Rd} = \alpha_v f_{ub} A / \gamma_{M2}$
For M20 8.8 bolts: $F_{v,Rd} = 0.6 \times 800 \times 245/(1.25 \times 1000) = 94kN$

$$\therefore V_{Rd} = \frac{4 \times 94}{\sqrt{1 + (0.1875 \times 4)^2}} = 300.8kN$$

$V_{Ed} = 110kN < 300.8kN \therefore OK$

(ii) Fin plate in bearing
Basic requirement $V_{Ed} \leq V_{Rd}$

$$V_{Rd} = \frac{n}{\sqrt{\left(\frac{1+\alpha n}{F_{b,Rd,ver}}\right)^2 + \left(\frac{\beta n}{F_{b,Rd,hor}}\right)^2}}$$

$\alpha = 0$ *and* $\beta = 0.19$ *(as above)*

NCCI: Shear resistance of a fin plate connection SN017a-EN-EU

Table 3.4

	Steel Designers' Manual 7th Edition			**BS EN 1993-1-8**
	Fin plate connection			
	Made by DBM	Checked by JBD	Sheet No. 2 of 11	

The vertical bearing resistance of a single bolt:

$$F_{b,Rd,ver} = \frac{k_1 \alpha_b f_{u,p} d t_p}{\gamma_{M2}}$$

where:

$$k_1 = min\left(2.8\frac{e_2}{d_0} - 1.7; 2.5 \right) = min\left(2.8\frac{e_2}{d_0} - 1.7 = \frac{2.8 \times 50}{22} - 1.7 = 4.66; 2.5 \right) = 2.5$$

$$\alpha_b = min\left(\frac{e_1}{3d_0}; \frac{p_1}{3d_0} - \frac{1}{4}; \frac{f_{ub}}{f_{u,p}}; 1.0 \right)$$

$$= min\left(\frac{40}{3 \times 22} = 0.61; \frac{80}{3 \times 22} - 0.25 = 0.96; \frac{800}{430} = 1.86; 1.0 \right) = 0.61$$

$$\therefore F_{b,Rd,ver} = \frac{2.5 \times 0.61 \times 430 \times 20 \times 10}{1.25} \times 10^{-3} = 104.9 \, kN$$

The horizontal bearing resistance of a single bolt:

$$F_{b,Rd,hor} = \frac{k_1 \alpha_b f_{u,p} d t_p}{\gamma_{M2}}$$

where:

$$k_1 = min\left(2.8\frac{e_1}{d_0} - 1.7; 1.4\frac{p_1}{d_0} - 1.7; 2.5 \right)$$

$$k_1 = min\left(2.8\frac{40}{22} - 1.7 = 3.39; 1.4\frac{80}{22} - 1.7 = 3.39; 2.5 \right) \quad k_1 = 2.5$$

$$\alpha_b = min\left(\frac{e_2}{3d_0}; \frac{f_{ub}}{f_{u,p}}; 1.0 \right)$$

$$\alpha_b = min\left(\frac{50}{3 \times 22} = 0.76; \frac{800}{430} = 1.86; 1.0 \right) \quad \alpha_b = 0.76$$

$$\therefore F_{b,Rd,hor} = \frac{2.5 \times 0.76 \times 430 \times 20 \times 10}{1.25} \times 10^{-3} = 130.7 kN$$

$$V_{Rd} = \frac{n}{\sqrt{\left(\frac{1 + \alpha n}{F_{b,Rd,ver}} \right)^2 + \left(\frac{\beta n}{F_{b,Rd,hor}} \right)^2}} = \frac{4}{\sqrt{\left(\frac{1 + 0 \times 4}{104.9} \right)^2 + \left(\frac{0.19 \times 4}{130.7} \right)^2}} = 358.2 kN$$

$V_{Ed} = 110 kN < 358.2 kN \therefore OK$

(iii) Beam web in bearing
Basic requirement $V_{Ed} < V_{Rd}$

$$V_{Rd} = \frac{n}{\sqrt{\left(\frac{1 + \alpha n}{F_{b,Rd,ver}} \right)^2 + \left(\frac{\beta n}{F_{b,Rd,hor}} \right)^2}}$$

$\alpha = 0$ and $\beta = 0.19$ (as above)

Table 3.4

	Steel Designers' Manual 7th Edition			**BS EN**
	Fin plate connection			**1993-1-8**
	Made by DBM	Checked by JBD	Sheet No. 3 of 11	

The vertical bearing resistance of a single bolt on the supported beam web: | *Table 3.4*

$$F_{b,Rd,ver} = \frac{k_1 \alpha_d f_{u,b1} d t_{w,b1}}{\gamma_{M2}}$$

where:

$$k_1 = min\left(2.8\frac{e_{2,b}}{d_0} - 1.7; 2.5\right) = min\left(2.8\frac{e_{2,b}}{d_0} - 1.7 = \frac{2.8 \times 40}{22} - 1.7 = 3.39; 2.5\right) = 2.5$$

$$\alpha_b = min\left(\frac{e_{1,b}}{3d_0}; \frac{p_1}{3d_0} - \frac{1}{4}; \frac{f_{ub}}{f_{u,b1}}; 1.0\right)$$

$$= min\left(\frac{40}{3 \times 22} = 0.61; \frac{80}{3 \times 22} - 0.25 = 0.96; \frac{800}{430} = 1.86; 1.0\right) \quad \alpha_b = 0.61$$

$$\therefore F_{b,Rd,ver} = \frac{2.5 \times 0.61 \times 430 \times 20 \times 7.6}{1.25} \times 10^{-3} = 79.7 \, kN$$

The horizontal bearing resistance of a single bolt on the supported beam web:

$$F_{b,Rd,hor} = \frac{k_1 \alpha_b f_{u,b1} d t_{w,b1}}{\gamma_{M2}}$$

Table 3.4

where:

$$k_1 = min\left(2.8\frac{e_{1,b}}{d_0} - 1.7; 1.4\frac{p_1}{d_0} - 1.7; 2.5\right)$$

$$= min\left(2.8\frac{40}{22} - 1.7 = 3.39; 1.4\frac{80}{22} - 1.7 = 3.39; 2.5\right) \quad k_1 = 2.5$$

$$\alpha_b = min\left(\frac{e_2}{3d_0}; \frac{f_{ub}}{f_{u,p}}; 1.0\right)$$

$$= min\left(\frac{40}{3 \times 22} = 0.61; \frac{800}{430} = 1.86; 1.0\right) \quad \alpha_b = 0.61$$

$$\therefore F_{b,Rd,hor} = \frac{2.5 \times 0.61 \times 430 \times 20 \times 7.6}{1.25} \times 10^{-3} = 79.7 \, kN$$

$$V_{Rd} = \frac{n}{\sqrt{\left(\frac{1+\alpha n}{F_{b,Rd,ver}}\right)^2 + \left(\frac{\beta n}{F_{b,Rd,hor}}\right)^2}} = \frac{4}{\sqrt{\left(\frac{1+0 \times 4}{79.7}\right)^2 + \left(\frac{0.19 \times 4}{79.7}\right)^2}} = 253.8kN$$

$V_{Ed} = 110kN < 253.8kN \therefore OK$

Check 3 – Shear and bending capacity of fin plate connected to supported beam

(i) For shear

The basic requirement is that the shear capacity of the fin plate must be greater than the shear force, $V_{Rd,min} > V_{Ed}$
For $V_{Rd,min}$ the following checks must be made:
* *shear at the gross section, $V_{Rd,g}$*
* *shear at the net section, $V_{Rd,n}$*
* *block shear, $V_{Rd,b}$*

	Steel Designers' Manual 7th Edition			BS EN 1993-1-8
	Fin plate connection			
	Made by DBM	Checked by JBD	Sheet No. 4 of 11	

Shear (gross section)

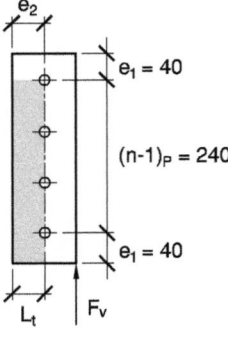

e_2

$e_1 = 40$

$(n-1)_P = 240$

$e_1 = 40$

L_t F_v

$$V_{Rd,g} = \frac{h_p t_p}{1.27} \frac{f_{y,p}}{\sqrt{3}\ \gamma_{M0}}$$

(Note: the coefficient 1.27 is to take into account the reduction of the shear resistance due to the presence of moment).

$$V_{Rd,g} = \frac{320 \times 10}{1.27} \frac{275}{\sqrt{3} \times 1.0} = 400.0 kN$$

Shear (net section)

$$V_{Rd,n} = A_{v,net} \frac{f_{u,p}}{\sqrt{3}\ \gamma_{M2}}$$

where:
Net area , $A_{v,net} = t_p(h_p - n_1 d_0)$

$A_{v,net} = 10(320 - 4 \times 22) = 2320 mm^2$

$$\therefore V_{Rd,n} = 2320 \frac{410}{\sqrt{3} \times 1.1} \times 10^{-3} = 499.3 kN$$

Block shear
The block shear capacity is given by the following expression:

$$V_{Rd,b} = \frac{0.5 f_{u,p} A_{nt}}{\gamma_{M2}} + \frac{1}{\sqrt{3}} f_{y,p} \frac{A_{nv}}{\gamma_{M0}}$$

where:
A_{nt} *is the net area subjected to tension and for a single vertical line of bolts (i.e.* $n_2 = 1$*) is given by:*

$$A_{nt} = t_p \left(e_2 - \frac{d_0}{2} \right)$$

A_{nv} *is the net area subjected to shear*
$= t_p (h_p - e_1 - (n_1 - 0.5)\ d_0)$
$A_{nt} = 10(50 - 22/2) = 390 mm^2$

	Steel Designers' Manual 7th Edition			**BS EN**
	Fin plate connection			**1993-1-8**
	Made by DBM	Checked by JBD	Sheet No. 5 of 11	

$A_{nv} = 10(320 - 40 - (4 - 0.5)22) = 2030mm^2$

$\therefore V_{Rd,b} = \left(\dfrac{0.5 \times 410 \times 390}{1.1} + \dfrac{275 \times 2030}{\sqrt{3} \times 1.0} \right) \times 10^{-3} = 395.0kN$

The shear capacity is given by the block shear capacity <u>395 kN > 110kN</u>
Therefore the shear capacity of the fin plate is OK

(ii) For bending

The basic requirement is that the moment capacity of the fin plate must be greater than applied moment. This is assured if $h_p \geq 2.73\ z$

$h_p = 320 > 2.73 \times 50\ (= 136.5)$

Therefore the fin plate is OK in bending.

Check also lateral torsional buckling of the fin plate

If $z < t_p / 0.15$, then $V_{Rd} = \dfrac{W_{el,p}\ f_{yp}}{z\ \gamma_{M0}}$

Where $W_{el,p} = \dfrac{t_p h_p^2}{6}$

$50 < 10 / 0.15 \quad V_{Rd} = \dfrac{10 \times 320^2}{6 \times 50}\dfrac{275}{1.0} \times 10^{-3} = 939kN\ > 110kN,\ OK\ for\ LTB$

Check 4 – Shear capacity of the supported beam

The basic requirement is that the shear capacity of the beam must be greater than the applied shear, $V_{Rd,min} > V_{Ed}$

For $V_{Rd,min}$ there are three checks to consider:
* *shear at the gross section, $V_{Rd,g}$*
* *shear at the net section, $V_{Rd,n}$*
* *block shear, $V_{Rd,b}$*

Shear (gross section)

$V_{Rd,g} = A_{v,wb}\ \dfrac{f_{y,b1}}{\sqrt{3}\ \gamma_{M0}}$

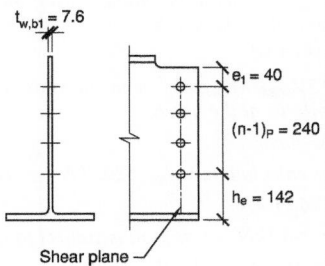

Gross area:
$A_{v,wb} = A_{Tee} - bt_{f,b1} + (t_{w,b1} + 2r)t_{f,b1}/2$
$A_{Tee} = (422–10.9)7.6 + 152.4 \times 10.9 = 4785mm^2$

$A_{v,wb} = 4785 - 152.4 \times 10.9 + (7.6 + 2 \times 10.2)10.9/2$
$\quad\quad = 3276mm^2$

$V_{Rd,g} = 3276\dfrac{275}{\sqrt{3} \times 1.0} \times 10^{-3} = 520.2kN$

	Steel Designers' Manual 7th Edition		**BS EN 1993-1-8**
	Fin plate connection		
	Made by DBM	Checked by JBD	Sheet No. 6 of 11

Shear (net section)

$$V_{Rd,g} = A_{v,b,net} \frac{f_{u,b1}}{\sqrt{3}\gamma_{M2}}$$

where

$$A_{v,wb,net} = A_{v,wb} - n_1 d_0 t_{w,b1}$$
$$= 3276 - 4 \times 22 \times 7.6 = 2607\,mm^2$$

$$\therefore V_{Rd,n} = 2607 \frac{410}{\sqrt{3} \times 1.0} \times 10^{-3} = 617.1kN$$

Block shear (block tearing)

$$V_{Rd,b} = \frac{0.5 f_{u,b1} A_{nt}}{\gamma_{M2}} + \frac{f_{y,b1} A_{nv}}{\sqrt{3}\gamma_{M0}}$$

Net area subject to tension:

$$A_{nt} = t_{w,b1}(e_{2,b} - 0.5 d_0)$$
$$= 7.6(40 - 0.5 \times 22) = 220\,mm^2$$

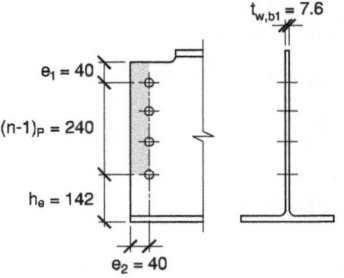

Net area subject to shear:

$$A_{nv} = t_{w,b1}(e_{1,b} + (n_1 - 1)p_1 - (n_1 - 0.5)d_0)$$
$$= 7.6(40 + 3 \times 80 - 2.5 \times 22) = 1710\,mm^2$$

$$V_{Rd,b} = \left(\frac{0.5 \times 410 \times 220}{1.1} + \frac{275 \times 1710}{\sqrt{3} \times 1} \right) \times 10^{-3} = 312.5kN$$

Block tearing (block shear) gives the minimum shear capacity,
$V_{Rd,b}$ *312.5kN > 100kN* $\quad \therefore$ *shear capacity at notch OK*

Check 5 Shear and bending interaction at the notch

The basic requirement is that the moment capacity at the notch in the presence of shear, $M_{v,Rd}$, must be greater than the moment from the product of the end reaction and the distance to the end of the notch,
$M_{v,Rd} > V_{Ed} \times (g_h + l_n)$

When $V_{Ed} \leq 0.5 V_{pl,Rd}$, $M_{v,Rd}$ may be taken as the yield strength of the beam multiplied by the elastic modulus at the notch.

Check for low shear:

from previous calculation, $V_{pl,Rd} = 520.2kN$ (see check 4(i))

$\therefore 0.5 V_{pl,Rd} = 260.1\ kN$

$\quad$ *110kN < 296.1kN* $\quad \therefore$ *section is subject to low shear*

$\quad$ *and therefore $M_{v,Rd} = f_{y,b1} W_{el,notch}/\gamma_{M0}$*

$W_{el,notch}$ = *elastic section modulus of the residual tee section at the notch* = $373\,cm^3$

$\therefore M_{v,Rd} = (275 \times 373 / 1.0) \times 10^{-3} = 102.6kNm$

Applied moment = $110 \times (10 + 110)/10^3\ kNm = 13.2kNm$

$\quad$ *102.6kNm > 13.2kNm*

Therefore the moment capacity of the notched beam is OK.

	Steel Designers' Manual 7th Edition		**BS EN 1993-1-8**
	Fin plate connection		
	Made by DBM	Checked by JBD	Sheet No. 7 of 11

Check 6 – Supported beam – local stability of notched beam

The beam is restrained against lateral torsional buckling, therefore no account need be taken of notch stability provided (for one flange notched in S275 steel and depth of notch not exceeding half the beam depth):

$l_n \leq h_{b1}$ $\qquad$ *for* $h_{b1}/t_{w,b1} \leq 54.3$ (S275)

$l_n \leq 160000\, h_{b1}/(h_{b1}/t_{w,b1})^3$ $\quad$ *for* $h_{b1}/t_{w,b1} > 54.3$ (S275)

$l_n = 120mm;\ h_{b1} = 449.8mm;\ t_{w,b1} = 7.6mm$

$h_{b1}/t_{w,b1} = 449.8/7.6 = 59.2 > 54.3$

$\therefore l_n \leq 160000\, h_{b1}/(h_{b1}/t_{w,b1})^3$

$$= \frac{160000 \times 449.8}{(449.8/7.6)^3} = 347.2\,mm$$

notch length 120mm < 347.2 mm , local stability of notch OK

Check 7 – Overall stability of notched beam

The notched beam is restrained therefore the overall stability of the beam need not be checked.

Check 8 – Supporting beam welds

The basic requirement is that the throat thickness of the fillet weld $\geq 0.5 t_p$ for a S275 fin plate.
Leg length = 8mm, throat thickness = 0.7×8 = 5.6mm > $0.5 \times 10mm$
Therefore 8mm fillet weld is OK.

Check 9 – not applicable

Check 10: Supporting beam – local resistance
The basic requirement is that the local shear capacity of the supporting beam web should be greater that the end reaction.

The following two modes of failure should be considered:-

• *Local shear failure of the supporting beam web*
• *Punching shear capacity*

Local shear failure (beam web)
Basic requirement is $V_{Ed}/2 \leq F_{Rd}$ where

$$F_{Rd} = A_v = \frac{f_y}{\sqrt{3}\gamma_{M0}}$$

$A_v = h_p t_w$
h_p *is the depth of fin plate = 320mm*
t_w *is web thickness of supporting beam = 10.6mm*

$A_v = 320 \times 10.6\ mm^2$

$\quad = 3392\ mm^2$

Steel Designers' Manual 7th Edition			BS EN 1993-1-8
Fin plate connection			
Made by DBM	Checked by JBD	Sheet No. 8 of 11	

$$\therefore F_{Rd} = 3392 \frac{275}{\sqrt{3} \, 1.0} \times 10^{-3}$$

$$= 538.6 \ kN$$

$V_{Ed}/2 = 110/2 = 55kN$
    $538.6kN > 55kN$ *therefore the local shear resistance is OK*

Punching shear capacity

The punching shear capacity of the supporting web is satisfactory provided that

$$t_p \le t_{b,w} \times \frac{f_{u,b}}{f_{y,p}}$$

where

t_p *is the fin plate thickness*
$t_{b,w}$ *is the thickness of the web of the supporting beam*
$f_{u,b}$ *is the ultimate tensile strength of the supporting beam = $410N/mm^2$*
$f_{y,p}$ *is the design strength of the fin plate*
$t_f = 10 \ mm$ *and* $t_{b,w} = 10.6 \ mm$

$$10 \le 10.6 \times \frac{410}{275} = 15.8$$

10mm < 15.8mm therefore the punching shear capacity is OK

Check 11 – Structural integrity – connecting elements

The basic requirement is that the tension and bearing resistance of the fin plate must be greater than the tie force. The minimum tying force is 75 kN but in many cases the tie force will be greater and it will be necessary to check for a tying force equal to the end reaction of the supported beam.

Tension capacity of fin plate

The tension resistance is the smaller of the block tearing resistance, $F_{Rd,b}$, and the net section tension resistance, $F_{Rd,n}$

Block tearing

$$F_{Rd,b} = \frac{f_{u,p} A_{nt}}{\gamma_{Mu}} + \frac{f_{y,p} A_{nv}}{\sqrt{3} \gamma_{M0}}$$

$$A_{nt} = t_p((n_1 - 1)p_1 - (n_1 - 1)d_0)$$
$$= 10 \times (3 \times 80 - (4-1) \times 22) = 1740 mm^2$$

$$A_{nv} = 2t_p(e_2 - d_0/2)$$
$$= 2 \times 10 \times (50 - 22/2) = 780 mm^2$$

$$\therefore F_{Rd,b} = \left(\frac{410 \times 1740}{1.1} + \frac{275 \times 780}{\sqrt{3} \times 1.0} \right) \times 10^{-3}$$

$$= 772.4kN$$

$50 \ \ e_2 = 50$

$e_1 = 40$
$p_1 = 80$
$p_1 = 80$ h_p $F_{Ed} = 75$ kN Tie force
$p_1 = 80$
$e_1 = 40$

	Steel Designers' Manual 7th Edition			**BS EN 1993-1-8**
	Fin plate connection			
	Made by DBM	Checked by JBD	Sheet No. 9 of 11	

Net tension

$$F_{Rd,n} = \frac{0.9\,A_{net,p}\,f_{u,p}}{\gamma_{Mu}}$$

$$A_{net,p} = t_p(h_p - d_0 n_1)$$
$$= 10 \times (320 - 22 \times 4) = 2320\,mm^2$$

$$\therefore F_{Rd,n} = \frac{0.9 \times 2320 \times 410}{1.1} \times 10^{-3} = 778.3\,kN$$

The tension resistance of the fin plate is given by min($F_{Rd,b}$; $F_{Rd,n}$)
772.4kN > 75kN therefore the tension resistance of the fin plate is OK

Horizontal shear resistance
Basic requirement: $F_{Ed} \leq F_{Rd}$ *where* $F_{Rd} = nF_{v,u}$

and $F_{v,u} = \dfrac{\alpha_v f_{ub} A}{\gamma_{Mu}}$

$\alpha_v = 0.6$; $f_{ub} = 800 N/mm^2$; $A = 245 mm^2$

$$F_{v,u} = \frac{0.6 \times 800 \times 245}{1.1} \times 10^{-3} = 107\,kN$$

$F_{Rd} = 4 \times 107 = 428kN$
$\therefore F_{Ed} = 75kN < 428kN$ *bolt shear resistance OK*

Fin plate bearing
Basic requirement: $F_{Ed} \leq F_{Rd}$ *where* $F_{Rd} = \Sigma F_{b,hor,u,Rd}$

and $F_{b,Rd} = \dfrac{k_1 \alpha_b f_u d\, t}{\gamma_{M,u}}$

where:

$$\alpha_b = min\left(\frac{e_2}{3d_o} ; \frac{f_{ub}}{f_{u,p}} ; \; 1.0 \right) = min\left(\frac{50}{3 \times 22} ; \frac{800}{410} ; 1.0 \right) = 0.76$$

For edge bolts,

$$k_1 = min\left(2.8\frac{e_1}{d_o} - 1.7 ; 2.5 \right) = min\left(2.8 \times \frac{40}{22} - 1.7 ; 2.5 \right) = 2.5$$

For inner bolts,

$$k_1 = min\left(1.4\frac{p_1}{d_o} - 1.7 ; 2.5 \right) = min\left(1.4 \times \frac{80}{22} - 1.7 ; 2.5 \right) = 2.5$$

$\gamma_{M,u} = 1,1$ *for tying resistance*
Since for $k_1 = 2.5$ *in both cases inner and edge bolt resistances are equal,*

$$F_{b,hor,u,Rd} = \frac{2.5 \times 0.76 \times 410 \times 20 \times 10}{1.1} \times 10^{-3} = 142\,kN$$

$\therefore F_{Rd} = 4 \times 142 = 568kN > 75kN$ *and the bearing capacity of the fin plate is OK.*

	Steel Designers' Manual 7th Edition			**BS EN 1993-1-8**
	Fin plate connection			
	Made by DBM	Checked by JBD	Sheet No. 10 of 11	

Check 12 Structural integrity – supported beam

The tension and bearing resistance of the supported beam web must exceed the tie force.

Tension
The smaller of the block tearing tension resistance, $F_{Rd,b}$, and the net section tension resistance, $F_{Rd,n}$, must exceed the tie force.

Block tearing

$$F_{Rd,b} = \frac{A_{nt}f_{u,b1}}{\gamma_{M,u}} + \frac{A_{nv}f_{y,b1}}{\sqrt{3}\gamma_{M0}}$$

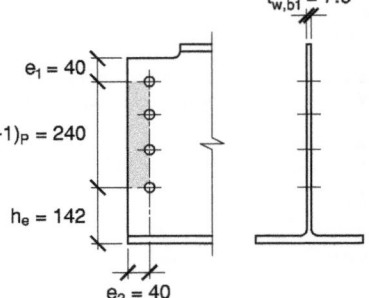

where:

    A_{nt} is the net area subjected to tension
             $= t_{w,b1}\left((n_1\text{-}1)p_1 - (n_1\text{--}1)d_0\right)$
    A_{nv} is the net area subjected to shear
    for single vertical line of bolts

$$A_{nv} = 2t_{w,b1}\left(e_{2,b} - \frac{d_0}{2}\right)$$

$\gamma_{M,u}$ = 1.1 for tying resistance
A_{nt} = 7.6 × (3 × 80 − 3 × 22) = 1322mm²
A_{nv} = 2 × 7.6 (40 − 11) = 441mm²

$$F_{Rd,b} = \left(\frac{1322 \times 410}{1.1} + \frac{441 \times 275}{\sqrt{3} \times 1.0}\right) \times 10^{-3}$$
$$= 562.8kN$$

Net tension

$$F_{Rd,n} = \frac{0.9 A_{net,wb} f_{u,b1}}{\gamma_{Mu}}$$

$A_{net,b1}$ = $t_{w,b1}\,(h_{w,b1} - d_0\,n_1)$
$h_{w,b1}$ = h_p (conservatively)
$A_{net,wb}$ = 7.6(320 − 22 × 4) = 1763mm²

$$\therefore F_{Rd,n} = \frac{0.9 \times 1763 \times 410}{1.1} \times 10^{-3} = 591.4kN$$

The tension resistance of the beam web is given by $min(F_{Rd,b}; F_{Rd,n})$
562.8 <u>kN > 75kN</u> therefore the tension resistance of the fin plate is OK

	Steel Designers' Manual 7ᵗʰ Edition			**BS EN 1993-1-8**
	Fin plate connection			
	Made by DBM	Checked by JBD	Sheet No. 11 of 11	

Bearing

Bearing resistance of the bolt group must be greater than the tying force.

$$F_{Rd} = n \, F_{b,hor,u,Rd}$$

$$F_{b,hor,u,Rd} = \frac{k_1 \alpha_b f_{u,b1} d \, t_{w,b1}}{\gamma_{M,u}}$$

$$\alpha_b = min\left(\frac{e_{2,b}}{3d_o}; \frac{f_{ub}}{f_{u,b1}}; 1.0\right) = \left(\frac{40}{3 \times 22}; \frac{800}{410}; 1.0\right) = (0.62; 1.95; 1.0) = 0.61$$

$$k_1 = min\left(1.4\frac{p_1}{d_o} - 1.7; 2.5\right) = \left(1.4\frac{80}{22} - 1.7; 2.5\right) = (3.39; 2.5) = 2.5$$

$\gamma_{M,u} = 1,1$ *for tying resistance*

$$F_{b,hor,u,Rd} = \frac{2.5 \times 0.61 \times 410 \times 20 \times 9.5}{1.1} \times 10^{-3} = 108kN$$

$F_{Rd} = 4 \times 108 = 432kN > 75kN$ *therefore the bearing capacity of the web is OK.*

Checks 13–16 not applicable

	Job: Steel Designers' Manual 7ᵗʰ Edition			**BS EN 1993-1-8**
	Web cleat connection			
	Made by DBM	Checked by BD	Sheet No. 1 of 10	

The connection is between a 356×171×45 UKB (S275) beam and a 254×254×73 UKC (S275) column.

The end reaction of the simply supported beam due to factored loads is 185kN.

Refer to Table 27.3 for design checks.

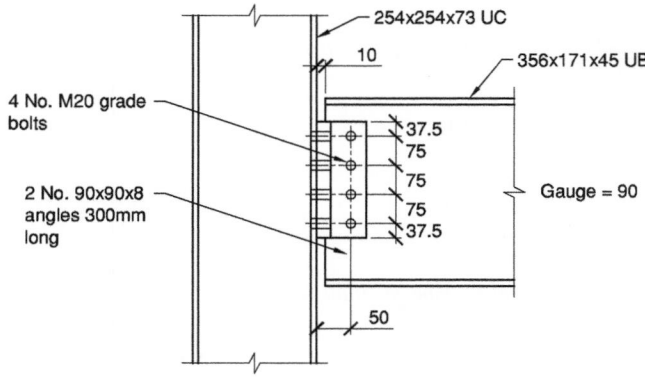

Check 1 Recommended detailing practice – *satisfies requirements of Figure 27.3*

Check 2 Shear and bearing resistance of bolt group connecting web cleats to web of supported beam

(i) bolt shear
basic requirement $V_{Ed} \leq V_{Rd}$

$$V_{Rd} = \frac{nF_{v,Rd}}{\sqrt{(1+\alpha n)^2 + (\beta n)^2}}$$

The shear resistance of a single bolt, $F_{v,Rd}$ is given by: Table 3.4

$$F_{v,Rd} = \frac{\alpha_v f_{ub} A}{\gamma_{M2}}$$

where:

$\gamma_{M2} = 1.25$ for shear resistance

$\alpha_v = 0.6$ for class 8.8 bolts

$A = A_s = 245mm^2$

$\therefore F_{v,Rd} = \frac{0.6 \times 800 \times 245}{1.25} \times 10^{-3} = 94.1kN$ per shear plane

	Job: Steel Designers' Manual 7th Edition			**BS EN 1993-1-8**
	Web cleat connection			
	Made by DBM	Checked by BD	Sheet No. 2 of 10	

For a single vertical line of bolts, $\alpha = 0$ *and*

$$\beta = \frac{6z}{n(n+1)p_1} = \frac{6 \times 50}{4 \times 5 \times 75} = 0.20$$

$$V_{Rd} = \frac{4 \times (94.1 \times 2)}{\sqrt{(1 + 0 \times 4)^2 + (0.2 \times 4)^2}} = 587.8kN$$

$V_{Rd} > V_{Ed}\ (587.8 > 185)$ ∴*bolt shear OK*

(ii) Angle cleats in bearing
basic requirement $V_{Ed}/2 \leq V_{Rd}$

$$V_{Rd} = \frac{n}{\sqrt{\left(\dfrac{1 + \alpha\,n}{F_{b,Rd,ver}}\right)^2 + \left(\dfrac{\beta\,n}{F_{b,Rd,hor}}\right)^2}}$$

For a single vertical line of bolts $\alpha = 0\ \beta = 0.2$ *(as above)*

The bearing resistance of a single bolt, $F_{b,Rd}$ *is given by:*   *Table 3.4*

$$F_{b,Rd} = \frac{k_1 \alpha_b f_u d\,t}{\gamma_{M2}}$$

The vertical bearing resistance of a single bolt on the web cleat is

$$F_{b,Rd,ver} = \frac{k_1 \alpha_b f_{u,cleat} d\,t_{cleat}}{\gamma_{M2}}$$

$$k_1 = min\left(2.8\frac{e_2}{d_o} - 1.7; 2.5\right) = min\left(2.8\frac{50}{22} - 1.7; 2.5\right)$$
$$= min(4.66; 2.5)\quad k_1 = 2.5$$

$$\alpha_b = min\left(\frac{e_1}{3d_0}; \frac{p_1}{3d_0} - \frac{1}{4}; \frac{f_{ub}}{f_{u,cleat}}; 1.0\right)$$
$$= min\left(\frac{37.5}{3 \times 22}; \frac{75}{3 \times 22} - 0.25; \frac{800}{410}; 1.0\right)$$
$$= min(0.57; 0.89; 1.95; 1.0)\quad \alpha_b = 0.57$$

$$\therefore F_{b,Rd,ver} = \frac{2.5 \times 0.57 \times 410 \times 20 \times 8}{1.25} \times 10^{-3} = 74.8kN$$

Horizontal bearing resistance of a single bolt on a web cleat is

$$F_{b,Rd,hor} = \frac{k_1 \alpha_b f_{u,cleat} d\,t_{cleat}}{\gamma_{M2}}$$

	Job: Steel Designers' Manual 7th Edition			BS EN 1993-1-8
	Web cleat connection			
	Made by DBM	Checked by BD	Sheet No. 3 of 10	

$k_1 = min\left(2.8\dfrac{e_1}{d_o} - 1.7; 1.4\dfrac{p_1}{d_o} - 1.7; 2.5 \right) = min\left(2.8\dfrac{37.5}{22} - 1.7; 1.4\dfrac{75}{22} - 1.7; 2.5 \right)$

$= min(3.07; 3.07; 2.5) \quad k_1 = 2.5$

$\alpha_b = min\left(\dfrac{e_2}{3d_o}; \dfrac{f_{ub}}{f_{u,cleat}}; 1.0 \right) = min\left(\dfrac{50}{3 \times 22}; \dfrac{800}{410}; 1.0 \right)$

$= min(0.76; 1.95; 1.0) \quad \alpha_b = 0.76$

$\therefore F_{b,Rd,hor} = \dfrac{2.5 \times 0.76 \times 410 \times 20 \times 8}{1.25} \times 10^{-3} = 99.7kN$

$V_{Rd} = \dfrac{n}{\sqrt{\left(\dfrac{1 + \alpha\, n}{F_{b,Rd,ver}} \right)^2 + \left(\dfrac{\beta\, n}{F_{b,Rd,hor}} \right)^2}} = \dfrac{4}{\sqrt{\left(\dfrac{1 + 0 \times 4}{74.8} \right)^2 + \left(\dfrac{0.2 \times 4}{99.7} \right)^2}} = 256.5kN$

$V_{Ed}/2 = 185/2 < V_{Rd}\ (256.5) \qquad \therefore bearing\ on\ cleats\ OK$

(iii) Beam web in bearing

$V_{Rd} = \dfrac{n}{\sqrt{\left(\dfrac{1 + \alpha\, n}{F_{b,Rd,ver}} \right)^2 + \left(\dfrac{\beta\, n}{F_{b,Rd,hor}} \right)^2}}$

For a single vertical line of bolts $\alpha = 0\ \beta = 0.2$ *(as above)*

Vertical bearing resistance of a single bolt on beam web, $F_{b,Rd,ver}$ *is:*

$F_{b,Rd,ver} = \dfrac{k_1 \alpha_b f_{u,b1} d\, t_{w,b1}}{\gamma_{M2}}$

$k_1 = min\left(2.8\dfrac{e_{2,b}}{d_0} - 1.7; 2.5 \right) = min\left(2.8\dfrac{40}{22} - 1.7; 2.5 \right)$

$= min(3.39; 2.5) \quad k_1 = 2.5$

$\alpha_b = min\left(\dfrac{p_1}{3d_0} - \dfrac{1}{4}; \dfrac{f_{ub}}{f_{u,b1}}; 1.0 \right) = min\left(\dfrac{75}{3 \times 22} - 0.25; \dfrac{800}{410}; 1.0 \right)$

$= min(0.89; 1.95; 1.0) \quad \alpha_b = 0.87$

$F_{b,Rd,ver} = \dfrac{2.5 \times 0.89 \times 410 \times 20 \times 7.0}{1.25} \times 10^{-3} = 102.2kN$

Horizontal bearing resistance of a single bolt on beam web, $F_{b,Rd,hor}$ *is:*

$F_{b,Rd,hor} = \dfrac{k_1 \alpha_b f_{u,b1} d\, t_{w,b1}}{\gamma_{M2}}$

Job: Steel Designers' Manual 7th Edition			BS EN 1993-1-8
Web cleat connection			
Made by DBM	Checked by BD	Sheet No. 4 of 10	

$$k_1 = min\left(1.4\frac{p_1}{d_o} - 1.7; 2.5\right) = min\left(1.4\frac{75}{22} - 1.7; 2.5\right)$$

$$= min(3.07; 2.5) \quad k_1 = 2.5$$

$$\alpha_b = min\left(\frac{e_{2,b}}{3d_o}; \frac{f_{ub}}{f_{u,b1}}; 1.0\right) = min\left(\frac{40}{3 \times 22}; \frac{800}{410}; 1.0\right)$$

$$= min(0.61; 1.95; 1.0) \quad \alpha_b = 0.61$$

$$F_{b,Rd,hor} = \frac{2.5 \times 0.61 \times 410 \times 20 \times 7.0}{1.25} \times 10^{-3} = 70.0kN$$

$$V_{Rd} = \frac{n}{\sqrt{\left(\frac{1+\alpha\,n}{F_{b,Rd,ver}}\right)^2 + \left(\frac{\beta\,n}{F_{b,Rd,hor}}\right)^2}} = \frac{4}{\sqrt{\left(\frac{1+0\times4}{102.2}\right)^2 + \left(\frac{0.20\times4}{70.0}\right)^2}} = 265.9kN$$

$V_{Ed} = 185 < V_{Rd} (265.9) \quad \therefore supported\ beam\ web\ OK\ in\ bearing$

Check 3 – Shear and bending resistance of web cleats connected to supported beam

(i) For shear

The basic requirement is that the shear capacity of one web cleat must be greater than half the shear force, $V_{Rd,min} > V_{Ed}/2$

For $V_{Rd,min}$ the following checks must be made:
* *shear at the gross section, $V_{Rd,g}$*
* *shear at the net section, $V_{Rd,n}$*
* *block shear, $V_{Rd,b}$*

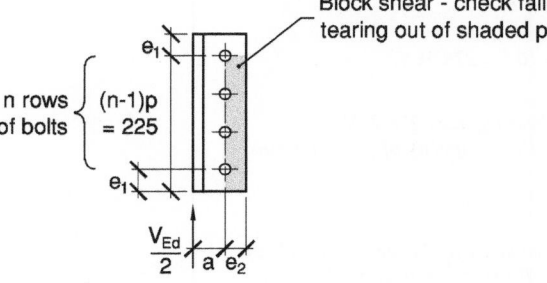

Block shear - check failure by tearing out of shaded portion

n rows of bolts $\begin{cases} (n-1)p \\ = 225 \end{cases}$

e_1

e_1

$\frac{V_{Ed}}{2}$ a e_2

Shear (gross section)

$$V_{Rd,g} = \frac{h_{cleat} t_{cleat}}{1.27} \frac{f_{y,cleat}}{\sqrt{3}\ \gamma_{M0}}$$

(Note: the coefficient 1.27 is to take into account the reduction of the shear resistance due to the presence of moment).

$$V_{Rd,g} = \frac{300 \times 8}{1.27} \frac{275}{\sqrt{3} \times 1.0} = 300.0kN$$

Job: Steel Designers' Manual 7th Edition			**BS EN 1993-1-8**
Web cleat connection			
Made by DBM	Checked by BD	Sheet No. 5 of 10	

Shear (net section)

$$V_{Rd,n} = A_{v,net}\frac{f_{u,cleat}}{\sqrt{3}\,\gamma_{M2}}$$

where:

Net area, $A_{v,net} = t_{cleat}(h_{cleat} - n_1 d_0)$

$A_{v,net} = 8 \times (300 - 4 \times 22) = 1696 mm^2$

$\therefore V_{Rd,n} = 1696\dfrac{410}{\sqrt{3}\times 1.1}\times 10^{-3} = 365.0 kN$

Block shear

The block shear capacity is given by the following expression:

$$V_{Rd,b} = \frac{0.5 f_{u,p} A_{nt}}{\gamma_{M2}} + \frac{1}{\sqrt{3}} f_{y,p}\frac{A_{nv}}{\gamma_{M0}}$$

where:
A_{nt} is the net area subjected to tension and for a single vertical line of bolts is given by:

$$A_{nt} = t_p\left(e_2 - \frac{d_0}{2}\right)$$

A_{nv} is the net area subjected to shear
   $= t_{cleat}(h_{cleat} - e_1 - (n_1 - 0.5)\, d_0)$
$A_{nt} = 8 \times (40 - 22/2) = 232 mm^2$
$A_{nv} = 8 \times (300 - 37.5 - (4 - 0.5) \times 22) = 1484\ mm^2$

$$\therefore V_{Rd,b} = \left(\frac{0.5 \times 410 \times 232}{1.1} + \frac{275 \times 1484}{\sqrt{3}\times 1.0}\right)\times 10^{-3} = 278.9 kN$$

The shear capacity is given by the block shear capacity 278.9kN
V_{Rd} (192.6kN) $> V_{Ed}/2$ (92.5kN) $\therefore$ the shear capacity of the web cleats plate is OK

(ii) For bending

The basic requirement is that the moment capacity of the leg of the web cleat must be greater than applied moment. This is assured if $h_{cleat} \geq 2.73\ z$

$h_{cleat} = 300 > 2.73 \times 50\ (=136.5)$ Therefore the web cleat is OK in bending.

Check 4 – Shear capacity of the supported beam

The basic requirement is that the shear capacity of the beam must be greater than the applied shear, $V_{Rd,min} > V_{Ed}$
For $V_{Rd,min}$ there are three checks to consider:
• shear at the gross section, $V_{Rd,g}$
• shear at the net section, $V_{Rd,n}$
• block shear, $V_{Rd,b}$

	Job: Steel Designers' Manual 7th Edition			BS EN 1993-1-8
	Web cleat connection			
	Made by DBM	Checked by BD	Sheet No. 6 of 10	

Shear (gross section)

$$V_{Rd,g} = A_{v,wb} \frac{f_{y,b1}}{\sqrt{3} \, \gamma_{M0}}$$

$$A_{v,wb} = A - 2bt_{f,b1} + (t_{w,b1} + 2r)t_{f,b1}$$

$$A_{v,wb} = 5730 - 2 \times 171.1 \times 9.7$$
$$+ (7.0 + 2 \times 10.2) \times 9.7$$
$$= 2676 \; mm^2$$

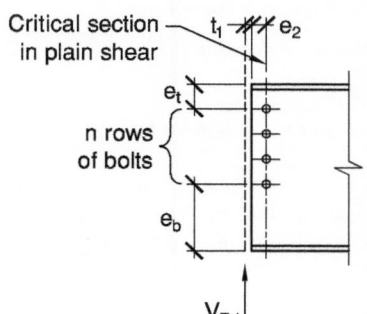

Critical section in plain shear

n rows of bolts

$$V_{Rd,g} = 2676 \frac{275}{\sqrt{3} \times 1.0} \times 10^{-3} = 424.9 kN$$

Shear (net section)

$$V_{Rd,n} = A_{v,b,net} \frac{f_{u,b1}}{\sqrt{3} \, \gamma_{M2}}$$

where $A_{v,wb,net} = A_{v,wb} - n_1 d_0 t_{w,b1}$
$$= 2676 - 4 \times 22 \times 7.0 = 2060 \; mm^2$$

$$\therefore V_{Rd,n} = 2060 \frac{410}{\sqrt{3} \times 1.0} \times 10^{-3} = 487.6 kN$$

Block shear (block tearing)

$$V_{Rd,b} = \frac{0.5 f_{u,b1} A_{nt}}{\gamma_{M2}} + \frac{f_{y,b1} A_{nv}}{\sqrt{3} \gamma_{M0}}$$

Net area subject to tension:

$$A_{nt} = 2t_{w,b1}(e_{2,b} - 2 \times 0.5 d_0)$$
$$= 2 \times 7.0 \times (40 - 0.5 \times 22) = 406 \; mm^2$$

Net area subject to shear:

$$A_{nv} = t_{w,b1}((n_1 - 1)p_1 - (n_1 - 0.5)d_0)$$
$$= 7.0 \times (3 \times 75 - 2.5 \times 22) = 1190 \; mm^2$$

$$V_{Rd,b} = \left(\frac{0.5 \times 410 \times 406}{1.1} + \frac{275 \times 1190}{\sqrt{3} \times 1} \right) \times 10^{-3} = 264.6 kN$$

Block tearing (block shear) gives the minimum shear capacity,
$V_{Rd,b}$ *264.6kN > 185kN* ∴ *shear capacity at notch OK*

	Job: Steel Designers' Manual 7th Edition			**BS EN 1993-1-8**
	Web cleat connection			
	Made by DBM	Checked by BD	Sheet No. 7 of 10	

Checks 5 -7 Not applicable (beam is not notched)

Check 8 Supporting column – bolt group

Basic requirement: $V_{Ed} \le V_{Rd}$

The resistance of the bolt group, F_{Rd}, is given by:

If $(F_{b,Rd})_{max} \le F_{v,Rd}$ *then* $F_{Rd} = F_{b,Rd}$

If $(F_{b,Rd})_{min} \le F_{v,Rd} < (F_{b,Rd})_{max}$ *then* $F_{Rd} = n_{cleat}(F_{b,Rd})_{min}$

If $F_{v,Rd} < (F_{b,Rd})_{min}$ *then* $F_{Rd} = 0.8 n_{cleat} F_{v,Rd}$

Note: The reduction factor 0.8 allows for the presence of tension in the bolts – see Jaspart et al (2009) European Recommendations for the Design of Simple Joints in Steel Structures, ECCS No.126

Shear resistance of a single bolt:

$$F_{v,Rd} = \frac{\alpha_v f_{ub} A}{\gamma_{M2}}$$

For M20 8.8 bolts

$$F_{v,Rd} = \frac{0.6 \times 800 \times 245}{1.25} \times 10^{-3} = 94 kN$$

Bearing resistance of a single bolt:

$F_{b,Rd} = min(F_{b,Rd,cleat}; F_{b,Rd,c})$

Since column flange thickness > cleat thickness, $F_{b,Rd,cleat}$ will be critical

$$F_{b,Rd,cleat} = \frac{k_1 \alpha_b f_u d t_{cleat}}{\gamma_{M2}}$$

Edge bolts –

$$k_1 = min\left(2.8 \frac{e_2}{do} - 1.7; 2.5\right)$$

$$= min\left(2.8 \times \frac{40}{22} - 1.7; 2.5\right) = min(3.39; 2.5) \quad k_1 = 2.5$$

End bolts –

$$\alpha_b = min\left(\frac{e_1}{3d_0}; \frac{f_{ub}}{f_{u,cleat} t}; 1.0\right)$$

$$= m\left(\frac{37.5}{3 \times 22}; \frac{800}{410}; 1.0\right) = m(0.57; 1.95; 1.0) \quad \alpha_b = 0.57$$

Job: Steel Designers' Manual 7th Edition			**BS EN 1993-1-8**
Web cleat connection			
Made by DBM	Checked by BD	Sheet No. 8 of 10	

Inner bolts –

$$\alpha_b = min\left(\frac{p_1}{3d_0} - 0.25; \frac{f_{ub}}{f_{u,cleat}}; 1.0\right)$$

$$= min\left(\frac{75}{3 \times 22} - 0.25; \frac{800}{410}; 1.0\right) = min(0.89; 1.95; 1.0) \quad \alpha_b = 0.87$$

End bolts

$$F_{b,Rd,cleat} = \frac{k_1 \alpha_b f_u d t_{cleat}}{\gamma_{M2}} = \frac{2.5 \times 0.57 \times 410 \times 20 \times 8}{1.25} = 74.8kN$$

Inner bolts

$$F_{b,Rd,cleat} = \frac{k_1 \alpha_b f_u d t_{cleat}}{\gamma_{M2}} = \frac{2.5 \times 0.89 \times 410 \times 20 \times 8}{1.25} = 116.8kN$$

Since the web cleat thickness is smaller than that of the column and the edge and end distances and bolt spacings in the column flange are equal or bigger than those in the cleats, the bearing resistance will be governed by the web cleats.

$(F_{b,Rd})_{min} \leq F_{v,Rd} < (F_{b,Rd})_{max}$ *i.e* $74.8 < 94.0 < 116.8$

$\therefore F_{Rd} = n_{cleat}(F_{b,Rd})_{min}$

Hence, $F_{Rd} = 8 \times 74.8 = 598.4kN > 185kN$ $\therefore$ *bolt group to column flange adequate*

Check 9 Supporting column – connecting elements

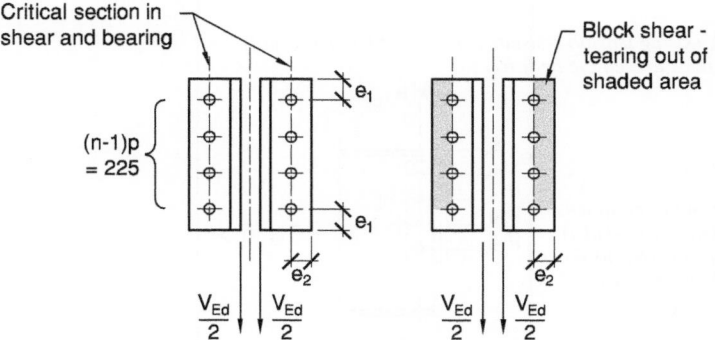

Critical section in shear and bearing

Block shear - tearing out of shaded area

$(n-1)p = 225$

e_1

e_1

e_2

e_2

$\frac{V_{Ed}}{2}$ $\frac{V_{Ed}}{2}$ $\frac{V_{Ed}}{2}$ $\frac{V_{Ed}}{2}$

The basic requirement is that the shear capacity of one web cleat must be greater than half the shear force, $V_{Rd,min} > V_{Ed}/2$

For $V_{Rd,min}$ *the following checks must be made:*
- *shear at the gross section,* $V_{Rd,g}$
- *shear at the net section,* $V_{Rd,n}$
- *block shear,* $V_{Rd,b}$

Job: Steel Designers' Manual 7th Edition			**BS EN 1993-1-8**
Web cleat connection			
Made by DBM	Checked by BD	Sheet No. 9 of 10	

As the web cleats are equal leg angles (90×90×8) and the bolt arrangement is symmetrical (i.e. e_1 is the same at the top and bottom), the checks are identical to those in Check 3 and are adequate.

Check 10 – Not applicable

Check 11 Structural integrity – connecting elements

The basic check is that the tying capacity of double-angle web cleats must be greater that the tie force.

   Tying capacity ≥ tie force

The minimum tie force is 75 kN for either an internal or an edge tie. In many cases the tie force will equal the end reaction.

The tying capacity of a double-angle web cleat is given by the following expression:

Tying capacity = 0.6 L_e t_{cleat} f_y (for S275 steel)

where

   $L_e = 2e_1 + (n-1)p_e - nd_0$

   e_1 *is end distance*

   $p_e = p$ *but* $\leq 2e_2$ *where e_2 is the edge distance*

   d_0 *is the diameter of the bolt hole*

   t_{cleat} *is the thickness of the cleat*

   $\therefore L_e = 2 \times 37.5 + (4 - 1) \times 75 - 4 \times 22 = 212mm$

   $\therefore$ *Tying capacity = $0.6 \times 212 \times 8 \times 275/10^3$ kN*

       $= 279.8kN$

          $279.8kN > 75kN$

The tying capacity of the angle cleats is OK.

Check 12 Structural integrity – supported beam

The tension and bearing resistance of the supported beam web must exceed the tie force.
Tension
The smaller of the block tearing tension resistance, $F_{Rd,b}$, and the net section tension resistance, $F_{Rd,n}$, must exceed the tie force.

Block tearing

$$F_{Rd,b} = \frac{A_{nt}f_{u,b1}}{\gamma_{M,u}} + \frac{A_{nv}f_{y,b1}}{\sqrt{3}\gamma_{M0}}$$

where:

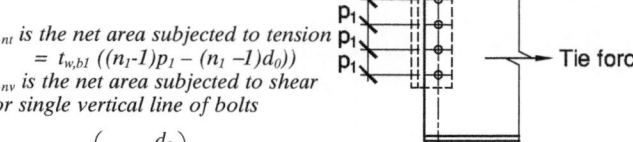

   A_{nt} *is the net area subjected to tension*
       $= t_{w,b1}((n_1-1)p_1 - (n_1 -1)d_0))$
   A_{nv} *is the net area subjected to shear*
   for single vertical line of bolts

   $A_{nv} = 2t_{w,b1}\left(e_{2,b} - \frac{d_0}{2}\right)$

   $\gamma_{M,u} = 1.1$ *for tying resistance*
   $A_{nt} = 7.0 \times (3 \times 75 - 3 \times 22) = 1113mm^2$
   $A_{nv} = 2 \times 7.0 \times (40 - 11) = 406mm^2$

   $F_{Rd,b} = \left(\frac{1113 \times 410}{1.1} + \frac{406 \times 275}{\sqrt{3} \times 1.0}\right) \times 10^{-3}$

       $= 479.3 \, kN$

	Job: Steel Designers' Manual 7th Edition			**BS EN 1993-1-8**
	Web cleat connection			
	Made by DBM	Checked by BD	Sheet No. 10 of 10	

Net tension

$$F_{Rd,n} = \frac{0.9 A_{net,wb} f_{u,b1}}{\gamma_{Mu}}$$

$A_{net,wb} = t_{w,b1} \ (h_{w,b1} - d_0 \ n_1)$

$A_{net,wb} = 7.0 \times (351.4 - 22 \times 4) = 1844 mm^2$

$$\therefore F_{Rd,n} = \frac{0.9 \times 1844 \times 410}{1.1} \times 10^{-3} = 618.6 kN$$

The tension resistance of the beam web is given by $min(F_{Rd,b}; F_{Rd,n})$
479.3kN > 75kN therefore the tension resistance of the beam web is OK

Bearing

Bearing resistance of the bolt group must be greater than the tying force.
$F_{Rd} = n \ F_{b,hor,u,Rd}$

$$F_{b,hor,u,Rd} = \frac{k_1 \alpha_b f_{u,b1} d \ t_{w,b1}}{\gamma_{M,u}}$$

$$\alpha_b = min\left(\frac{e_{2,b}}{3d_o}; \frac{f_{ub}}{f_{u,b1}}; 1.0\right) = \left(\frac{40}{3 \times 22}; \frac{800}{410}; 1.0\right) = (0.61; 1.95; 1.0) = 0.61$$

$$k_1 = min\left(1.4\frac{p_1}{d_o} - 1.7; 2.5\right) = \left(1.4\frac{75}{22} - 1.7; 2.5\right) = (3.07; 2.5) = 2.5$$

$\gamma_{M,u} = 1,1 \ for \ tying \ resistance$

$$F_{b,hor,u,Rd} = \frac{2.5 \times 0.61 \times 410 \times 20 \times 7}{1.1} \times 10^{-3} = 79.6kN$$

$F_{Rd} = 4 \times 79.6 = 318kN > 75kN$ *therefore the bearing capacity of the web is OK.*

Check 13 Structural integrity – tension resistance of bolt group
Tension resistance of the bolts must be greater than the tying force. Due to the gross deformation of the angles and the extreme prying forces likely to be induced, the design tensile strength of the bolts should be limited to $300N/mm^2$ for property class 8.8.
Hence, the tension resistance of the bolt group

$F_{t,Rd} = 2n_1 A_t f_{b,t}^*$

where

        n_1 *is the number of bolt rows*
        A_t *is the tensile stress area of the bolt*
        $f_{b,t}^*$ *is the reduced design tensile strength of the bolt in the presence of extreme prying*

$$F_{t,Rd} = \frac{2 \times 4 \times 245 \times 300}{10^3} = 588 \ kN$$

       Tension capacity of the bolt group is satisfactory.

Checks 14–16 – not applicable

Chapter 28
Design of moment connections

DAVID MOORE and BUICK DAVISON

28.1 Introduction

In multi-storey frames, if moment connections are to be used they are most likely to be flush end-plate connections and extended end-plate connections. If a larger lever arm is required, a haunched connection may be used, but these add extra fabrication and depth to the construction and are best avoided, if possible. For portal frame structures, haunched moment resisting connections at the eaves and apex of a frame are almost always used. Figure 28.1 shows a range of moment resisting connections. Choice of connection type is usually based on simplicity, duplication and ease of fabrication, all for economic reasons. Site welded joints are used extensively in the USA and Japan for the construction of buildings in seismic areas. Such connections can provide full moment continuity but can be expensive to produce. This type of connection is rarely used in the UK but with careful planning there is no reason why they cannot be used for a number of framing systems.

UK practice for the design of both site-welded and shop-welded beam-to-column connections is presented in the SCI/BCSA design guide *Joints in Steel Construction – Moment Connections*.[1] Although this reference has not yet been revised following the publication of BS EN 1993-1-8,[2] the UK NA[3] cites it as a suitable source of guidance and it is likely that designs to this publication will be considered to satisfy the requirements of EC3 until a revised version is prepared. The UK NA also permits the use of 'wind moment frames' (as a specific example of an unbraced semi-continuous frame). In this design method the connections are assumed to behave as pins under gravity loads and the beams they support are designed assuming simple supports. However, under lateral wind loads the connections are assumed to behave rigidly. Connections which behave in this manner are called *wind-moment* connections. These connections generally consist of flush or extended end-plates and have little or no stiffening in the columns. They are easy to fabricate and provide a cost-effective solution for low-rise unbraced buildings. The main requirement for

Steel Designers' Manual, Seventh Edition. Edited by Buick Davison and Graham W. Owens.
© 2012 Steel Construction Institute. Published 2012 by Blackwell Publishing Ltd.

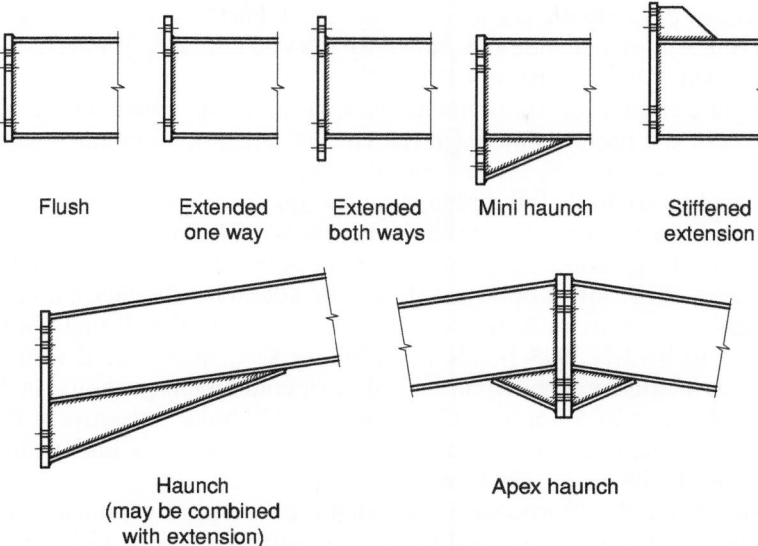

Flush   Extended   Extended   Mini haunch   Stiffened
          one way   both ways   extension

Haunch   Apex haunch
(may be combined
with extension)

Figure 28.1   Typical moment connections

wind-moment connections is that they should be ductile. That is, they must be able to rotate as plastic hinges under gravity loading and still have sufficient strength to resist the moments from wind loads. To ensure this type of behaviour, it is important to design the connection in such a way that the end-plate is thin enough to deform sufficiently so that neither the bolts nor the welds fracture, yet have sufficient strength to resist the moments from the wind loads. Detailed practical guidance is given in Reference 4 as cited in the UK NA.

This chapter will explain the design philosophy contained in BS EN 1993-1-8 for moment-resisting joints and the basis of the component-method for assessing moment resistance. Although emphasis will be placed on the design of bolted end-plate beam-to-column connections, the reader should be aware that the Eurocode also covers in detail column bases, welded connections and joints in tubular steelwork.

28.2 Design philosophy

BS EN 1993-1-8 contains a wealth of information on the design of connections including detailed information on the design of moment connections. However, the code itself is not particularly user-friendly and the majority of joint designs are likely to be conducted with the aid of handbooks, such as the SCI/BCSA Joints in Steel Construction Moment Connections. Although the Moment Connections book has not yet been revised into a Eurocode version, the methods used are based on a combination of capacity checks given in BS 5950: Part 1[5] and the design models given in Annex J of Amendment No. 1 to ENV Eurocode 3: Part 1.1/A1[6], the prestandard which became BS EN 1993-1-8.[3]

An end-plate connection transmits moment by coupling tension in the bolts with compression in the bottom flange (in the case of hogging moment) – see Section 28.7. No assumption need be made about the distribution of bolt forces if each bolt row is allowed to attain its full design strength (on the basis of the strength of the column flange or end-plate, whichever is the lowest). This approach relies on adequate ductility of the connecting part in the uppermost bolt rows to develop the design strength of the lower bolt rows. To ensure adequate ductility, a limit is set on the thickness of the column flange or end-plate relative to the strength of the bolts (see UK NA to BS EN 1993-1-8[3] NA2.7). Where S275 steel is used with property class 8.8 bolts, the thickness of either end-plate or column flange should be less than 18.3 mm, 21.9 mm or 27.5 mm for M20, M24 and M30 bolts respectively. If this criterion is not satisfied then the force in the lower bolt rows is limited to a value resulting from the linear distribution.

The design method in Eurocode 3 uses what is called the *component approach*.[7,8] In this approach the potential resistance (and stiffness and rotation capacity if required) of each component is calculated. Table 28.1 summarises the components. The design moment resistance of a particular joint can be derived by considering the resistance of the individual components and assembling these into a joint model, as shown in Figure 28.2.

A typical end-plate connection is shown in Figure 28.3. For a satisfactory design, 15 principal checks must be made on the beam, the column, the bolts and the welds. These checks can conveniently be split into four zones – the tension zone, the compression zone, horizontal shear and vertical shear. The checks associated with these zones are outlined in the sections given below and relate to the components in Table 28.1.

28.3 Tension zone

The resistance of the tension zone is determined from a consideration of the following modes of failure:

- bolt tension (component 10 in Table 28.1)
- end-plate in bending (component 5 in Table 28.1)
- column flange in bending (component 4 in Table 28.1)
- beam web in tension (component 8 in Table 28.1)
- column web in transverse tension (component 3 in Table 28.1).

In addition to the above, the designer should also check the adequacy of the flange to end-plate welds and the web to end-plate welds in the tension region (component 19 in BS EN 1993-1-8 Table 6.1).

Table 28.1 Basic components and corresponding code clause to determine the design resistance (based on Table 6.1 BS EN 1993-1-8)

Component		Component	
1. Column web panel in shear (6.2.6.1)	V_{Ed} → ← V_{Ed}	7. Flange and web in compression (6.2.6.7)	$F_{c,Ed}$ →
2. Column web panel in transverse compression (6.2.6.2)	→ ← $F_{r,Ed}$	8. Beam web in tension (6.2.6.8)	← → $F_{1,Ed}$
3. Column web in transverse tension (6.2.6.3)	← → $F_{r,Ed}$	9. Plate in tension or compression (EN1993-1-1)	$F_{t,Ed}$ ← ○ → $F_{t,Ed}$ $F_{c,Ed}$ → ← $F_{c,Ed}$
4. Column flange in bending (6.2.6.4)	← → $F_{r,Ed}$	10. Bolts in tension (6.2.6.4-6)	← → $F_{t,Ed}$
5. End-plate in bending (6.2.6.5)	$F_{r,Ed}$	11. Bolts in shear (3.6)	$F_{v,Ed}$
6. Flange cleat in bending (6.2.6.6)	$F_{r,Ed}$ ←	12. Bolts in bearing (3.6)	↑ $F_{b,Ed}$ ↓ $F_{b,Ed}$

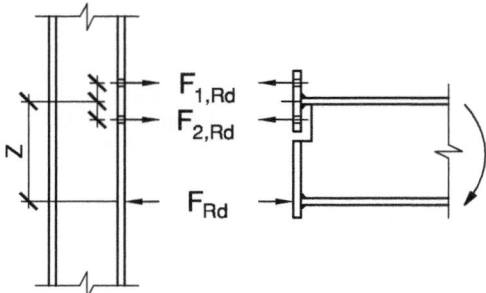

Figure 28.2   Component model for a bolted end-plate

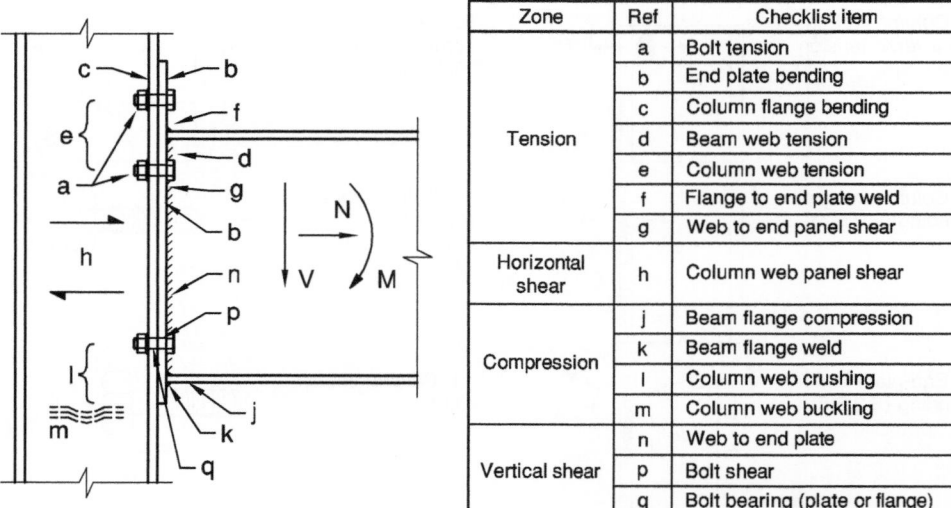

Zone	Ref	Checklist item
Tension	a	Bolt tension
	b	End plate bending
	c	Column flange bending
	d	Beam web tension
	e	Column web tension
	f	Flange to end plate weld
	g	Web to end panel shear
Horizontal shear	h	Column web panel shear
Compression	j	Beam flange compression
	k	Beam flange weld
	l	Column web crushing
	m	Column web buckling
Vertical shear	n	Web to end plate
	p	Bolt shear
	q	Bolt bearing (plate or flange)

Figure 28.3   Design checks for a bolted moment connection[1]

28.3.1 End plate and column flange in bending

EC3 uses the concept of an equivalent T-stub in tension to calculate the design resistance of the column flange and the end-plate in bending. This approach, based on original work by Zoetemeijer[9] and developed over many years for inclusion in EC3, replaces the actual end-plate or column flange with a simpler equivalent T-stub of effective length l_{eff}. Figure 28.4 illustrates the concept. A full description of the effective lengths used for different bolt configurations is beyond the scope of this chapter but full details are provided in BS EN1993-1-8 6.2.4, 6.2.6.4 and 6.2.6.5 and

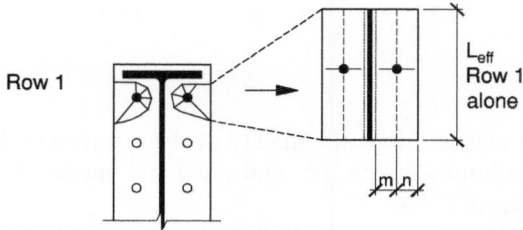

Row 1

L_{eff}
Row 1
alone

Figure 28.4   Concept of an equivalent T-stub

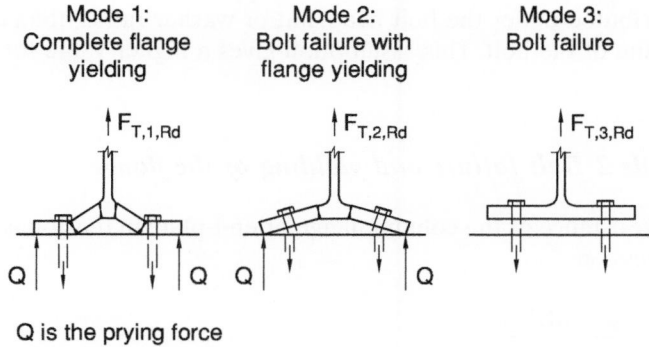

| Mode 1: Complete flange yielding | Mode 2: Bolt failure with flange yielding | Mode 3: Bolt failure |

$F_{T,1,Rd}$ $F_{T,2,Rd}$ $F_{T,3,Rd}$

Q is the prying force

Figure 28.5   T-stub failure modes

also in the SCI/BCSA design guide on Joints in Steel Construction – Moment Connections. It is important to note that the effective length of a T-stub is a notional length and does not necessarily correspond to the actual length of the component it represents.

In a T-stub, failure can occur by one of three mechanisms depending on the relative stiffness of the flange or end-plate and the bolts. These mechanisms are usually referred to as *complete yielding of the flange* (Mode 1), *simultaneous bolt failure and yielding of the flange* (Mode 2) and *bolt failure* (Mode 3). These modes are shown diagrammatically in Figure 28.5. The equations for calculating the potential resistance for each of these modes of failure are given in BS EN 1993-1-8 Table 6.2 and described below.

28.3.1.1 Mode 1 Complete flange yielding

The potential resistance of either the column flange or end-plate, $F_{T,1,Rd}$, can be determined from the following expression:

$$F_{\mathrm{T,1,Rd}} = \frac{4M_{\mathrm{pl,1,Rd}}}{m}$$

where

$M_{\mathrm{pl,1,Rd}}$ is the plastic moment capacity of the equivalent T-stub representing the column flange or end-plate in mode 1 and is equal to $0.25\Sigma l_{\mathrm{eff,1}}t_{\mathrm{f}}^{2}f_{\mathrm{y}}/\gamma_{\mathrm{M0}}$

m is the distance from the bolt centre to a line located 20% into either the column root or end-plate weld.

An alternative, more complicated, equation is also given in BS EN 1993-1-8 Table 6.2 for mode 1. This formula assumes the force applied to the T-stub by a bolt to be uniformly distributed under the bolt head, nut or washer rather than concentrated at the centre-line of the bolt. This assumption gives a higher value for mode 1.

28.3.1.2 Mode 2 Bolt failure and yielding of the flange

The potential resistance of the column flange or end-plate in tension is given by the following expression:

$$F_{\mathrm{T,2,Rd}} = \frac{2M_{\mathrm{pl,2,Rd}} + n\Sigma F_{\mathrm{t,Rd}}}{m+n}$$

where:

$\Sigma F_{\mathrm{t,Rd}}$ is the total tension capacity of all the bolts in the group, $M_{\mathrm{pl,2,Rd}}$ is the plastic moment capacity of the equivalent T-stub representing the column flange or end-plate in mode 2 and is equal to $0.25\Sigma l_{\mathrm{eff,2}}t_{\mathrm{f}}^{2}f_{\mathrm{y}}/\gamma_{\mathrm{M0}}$

m is the distance from the bolt centre to a line located 20% into either the column root or end-plate weld

n is the minimum edge distance e_{min} but $\leq 1.25m$.

28.3.1.3 Mode 3 Bolt failure

The potential resistance of the bolts in the tension zone is give by the following expression:

$$F_{\mathrm{T,3,Rd}} = \Sigma F_{\mathrm{t,Rd}}$$

Where $F_{\mathrm{t,Rd}}$ is the design resistance of a bolt, which EN1993-1-8 Table 3.4 gives as $0.9\,f_{\mathrm{ub}}A_{\mathrm{s}}/1.25$.

BS EN 1993-1-8 recommends that prying forces be assumed to develop in bolted beam-to-column joints but these effects are taken into account when determining the design resistance using the formulae above as prying action is implicit in the expressions for the calculation of the effective length l_{eff}.

Where prying forces may develop, the design tension resistance of a T-stub is taken as the smallest of the three possible failure modes 1, 2 and 3.

In a background publication, Zoetemeijer[9] developed three expressions for the equivalent effective length of an unstiffened column flange taking into account different levels of prying action:

for prying force = 0.0 $l_{\text{eff}} = (p + 5.5m + 4n)$
for maximum prying force $l_{\text{eff}} = (p + 4m)$
for an intermediate value $l_{\text{eff}} = (p + 4m + 1.25n)$

where p is the bolt pitch and m and n are as defined in 28.3.1.2.

Zoetemeijer considered the first expression to have an inadequate margin of safety against bolt failure while the margin of safety in the second was too high. He therefore suggested using the third equation, which allows for approximately 33% prying action. This approach simplifies the calculations by omitting complicated expressions for determining prying action.

28.3.2 Column web in tension

The design resistance of an unstiffened column web in tension, $F_{\text{t,wc,Rd}}$, is given by:

$$F_{\text{t,wc,Rd}} = \frac{\omega b_{\textit{eff,t,wc}} t_{wc} f_{y,wc}}{\gamma_{M0}}$$

where $b_{\text{eff,t,wc}}$ for a bolted connection is the effective width of the column web in tension taken as the length of the equivalent T-stub representing the column flange web; t_{wc} is the thickness of the column web; $f_{\text{y,wc}}$ is the design strength of the steel in the column or beam; ω is a reduction factor to allow for the interaction of shear in the column web panel. ω is found from expressions in BS EN1993-1-8 Table 6.3 and is a function of β defined as a transformation parameter (see 5.3(7) and Table 5.4), which depends upon the relative size and direction of the moments in the beams on either side of the connection. For equal and opposite moments, the web panel is not in shear and β is zero, giving ω a value of 1. For equal moments but in the same direction, β is at a maximum of 2, and ω takes a value less than one.

28.3.3 Beam web in tension

For a bolted end-plate, the design tension resistance of the beam web is:

$$F_{\text{t,wb,Rd}} = b_{\text{eff,t,wb}} t_{\text{wb}} f_{\text{y,wb}} / \gamma_{M0}$$

the effective width $b_{eff,t,wb}$ of the beam web is taken as the effective length of the equivalent T-stub representing the end-plate in bending.

28.4 Compression zone

The checks in the compression zone are similar to those traditionally adopted for web bearing and buckling and include the following:

- column web in bearing (component 2 in Table 28.1)
- column web buckling (component 2 in Table 28.1.)
- beam flange in compression (component 7 in Table 28.1).

In many designs it is common for the column web to be loaded to such an extent that it governs the design of the connection. However, this can be avoided either by choosing a heavier column or by strengthening the web with one of the compression stiffeners shown in Figure 28.6.

The resistance of an unstiffened column web subject to compressive forces, $F_{c,wc,Rd}$, is given by the smaller of the expressions for column web bearing and column web buckling.

28.4.1 Column web bearing

The resistance of the column web to bearing is based on an area of web calculated by assuming the compression force from the beam's flange is dispersed over a length $b_{eff,c,wc}$ shown in Figure 28.6. From this the resistance of the column web to crushing is given by the following expression:

$$F_{c,wc,Rd} = \frac{\omega k_{wc} b_{eff,c,wc} t_{wc} f_{y,wc}}{\gamma_{M0}}$$

where for a rolled column section and bolted end-plate:

$$b_{eff,c,wc} = t_{fb} + 2\sqrt{2}a_p + 5(t_{fc} + s) + s_p$$

which assumes the load disperses at 45° dispersion through the end-plate from the edge of the welds and then at an angle of 1:2.5 through the column flange and root radius (a_p is the weld throat thickness, s is the root radius of the column section, s_p is the length obtained by dispersion at 45° through the end-plate – usually $2t_p$). k_{wc} is a reduction factor to allow for the effect of longitudinal compressive stress in the column when this exceeds 70% of column yield and ω is a reduction factor to allow

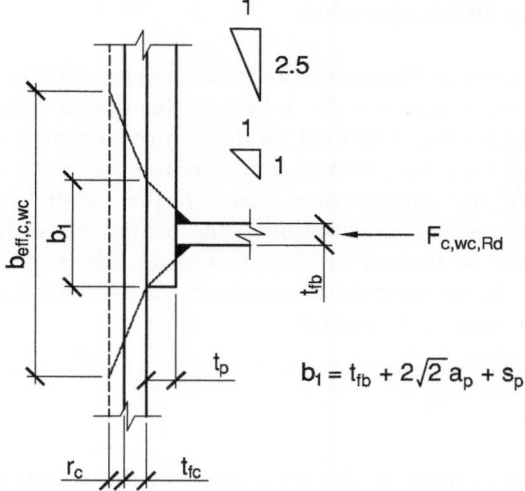

Figure 28.6 Distribution of compressive force

for the interaction of shear in the column web panel, as explained above for the column web in tension.

28.4.2 Column web buckling

The resistance of the column web to buckling uses a similar expression to that for bearing buts adds a further factor, ρ, to allow for the effect of web panel buckling:

$$F_{c,wc,Rd} \le \frac{\omega k_{wc} \rho b_{eff,c,wc} t_{wc} f_{y,wc}}{\gamma_{M1}}$$

the value of ρ depends on the plate slenderness $\bar{\lambda}_p$ calculated as:

$$\bar{\lambda}_p = 0.932 \sqrt{\frac{b_{eff,c,wc} d_{wc} f_{f,wc}}{E t_{wc}^2}}$$

where d_{wc} is the depth of the web measured between the root radius of a rolled section:

- if $\bar{\lambda}_p \le 0.72$: $\rho = 1,0$
- if $\bar{\lambda}_p > 0.72$: $\rho = (\bar{\lambda}_p - 0.2)/\bar{\lambda}_p^2$

28.4.3 Beam flange in compression

The design compression resistance of the beam flange and adjacent web (component 7 in Table 28.1), $F_{c,fb,Rd}$, may be taken as the design moment resistance of the beam cross-section ($M_{c,Rd}$) divided by the centre-to-centre depth between the flanges. For a bolted end-plate connection, this force may be assumed to act at the mid-thickness of the compression flange. If the beam height, including the haunch, exceeds 600mm, the contribution of the beam web to the design compression resistance should be limited to 20%. To ensure this is the case, the values of $F_{c,fb,Rd}$ based on the design moment of resistance of the complete section and that of the flanges alone must be calculated.

28.5 Shear zone

The column web panel must be designed to resist the resulting horizontal shear forces (component 1 in Table 28.1). To calculate these resultant forces the designer must take account of any connection to the opposite column flange. In a single-sided connection with no axial force the resultant shear force will be equal to the compressive force at the beam flange level (i.e. the sum of the bolt forces due to the moment). For a symmetrical two-sided column connection with balanced moments, the resultant shear force will be zero. However, in the case of a two-sided connection subject to moments acting in the same sense, the resultant shears will be additive. For any connection the resulting shear force can be obtained from the following expression:

$$V_{wp,Rd} = (M_{b1,Ed} - M_{b2,Ed})/z - (V_{c1,Ed} - V_{c2,Ed})/2$$

where

$M_{b1,Ed}$ and $M_{b2,Ed}$ are the moments in connections 1 and 2 (hogging positive)

z is the lever arm (usually the distances between the centroids of the beam flanges)

$V_{c1,Ed}$ and $V_{c2,Ed}$ are the horizontal shears in the columns below and above the joint respectively.

The design plastic shear resistance $V_{wp,Rd}$ of an unstiffened column web panel in shear is given by the following expression:

$$V_{wp,Rd} = \frac{0.9 f_{y,wc} A_{vc}}{\sqrt{3}\gamma_{M0}}$$

where A_{vc} is the shear area of the column (i.e. $A - 2b_{tf} + (t_w + 2r)t_f$ for a rolled H-section).

If transverse stiffeners have been used to increase the resistance of the compression and tension zones, the design plastic resistance of the column web panel will be increased to:

$$V_{\mathrm{wp,add,Rd}} = \frac{4M_{pl,fc,Rd}}{d_s} \text{ but } V_{\mathrm{wp,add,Rd}} \leq \frac{2M_{pl,fc,Rd} + 2M_{pl,st,Rd}}{d_s}$$

where:

    d_s is the distance between the centrelines of the stiffeners

    $M_{pl,fc,Rd}$ is the design plastic moment resistance of a column flange

    $M_{pl,st,Rd}$ is the design plastic moment resistance of a stiffener.

Webs of most UC sections will fail in panel shear before they fail in either bearing or buckling and therefore most single-sided connections are likely to fail in shear. The strength of a column web can be increased either by choosing a heavier column section or by using a shear stiffener (see Section 28.6).

28.6 Stiffeners

Most stiffening can be avoided through careful selection of the members during the design process. This will usually lead to a more cost-effective solution. However, where stiffening is unavoidable, one or more of the stiffener types shown in Figure 28.7 may be used.

Horizontal stiffeners such as those shown in Figure 28.7(a) and (b) are used where the concentrated loads from the beam flanges overstress the column web. There is often a high shear stress in the column web, particularly in single-sided connections, and stiffening is required. Diagonal or supplementary web plates can be used (see Figure 28.7(c)). Wherever possible, the angle of diagonal stiffeners should be between 30° and 60°. However, if the depth of the column is considerably less than the depth of the beam 'K' stiffening may be used. In general the type of strengthening must be chosen so that it does not clash with other components at the connections.

28.7 Design moment of resistance of end-plate joints

The design moment resistance of a bolted end-plate beam-to-column joint, $M_{j,Rd}$, may be found by summing the moments created by the effective design tension resistance of each bolt row ($F_{tr,Rd}$) operating at a lever arm (h_r) measured from the assumed centre of compression. This is expressed in BS EN 1993-1-8 eqn 6.25 as:

$$M_{j,Rd} = \Sigma h_r, F_{tr,Rd}$$

Clause 6.2.7.2 outlines the detailed rules for the application of this equation. The effective design tension resistance of an individual bolt row is taken as the smallest value of the basic tension components i.e. the column web in tension, the column

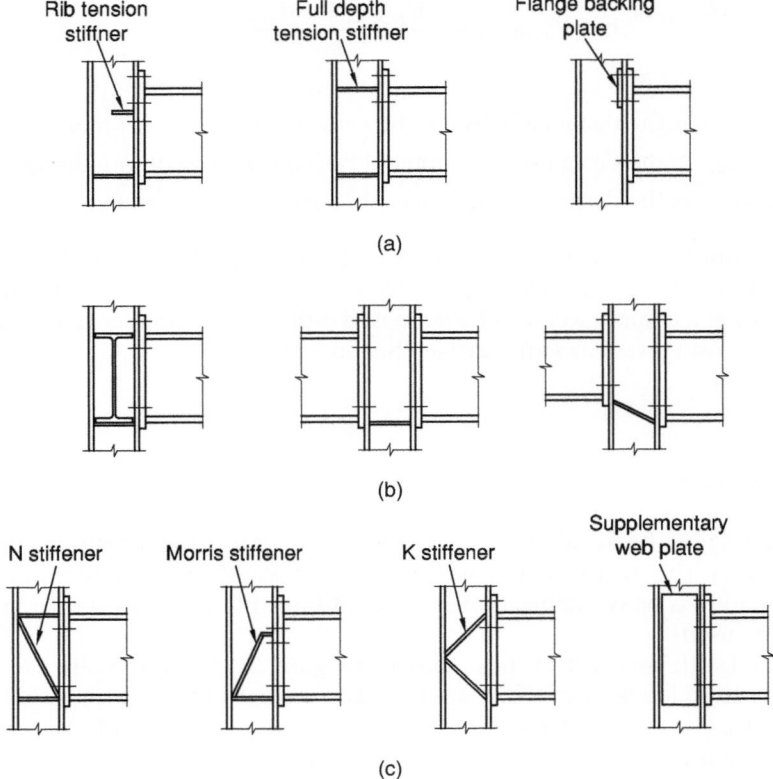

Figure 28.7   Column stiffeners: (a) tension, (b) compression, (c) shear

flange in bending, the end-plate in bending or the beam web in tension (components 3, 4, 5 and 8 in Table 28.1). If the connection fails on the compression side (e.g. column web buckling), the capacity of the tension zone must be reduced to preserve horizontal equilibrium, as explained in Cl 6.2.7.2(7). The resistance of the bolt row nearest the compression would be reduced first. The method is well explained in SCI/BCSA Moment Connections[1] from which the following is an extract (modified for EC3 terminology):

The procedure is to first calculate the potential resistance of each row $F_{tr,Rd}$ as shown in Figure 28.8.

The values of $F_{tr1,Rd}$, $F_{tr2,Rd}$, $F_{tr3,Rd}$ etc. are calculated in turn starting at the top row 1 and working down, ignoring the bolts below the current row. Each row is checked first in isolation and then in combination with successive rows above it (because the resistance of bolt rows acting in a combined pattern of failure may be less than the bolt rows operating in isolation). Hence:

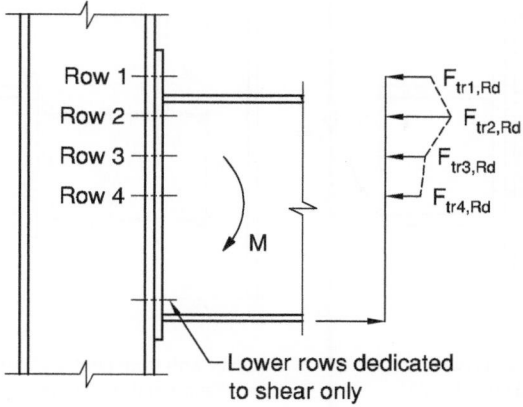

Figure 28.8   Potential resistance of bolt rows

$F_{\text{tr1,Rd}}$ = capacity of row 1 alone
$F_{\text{tr2,Rd}}$ = minimum of (capacity of row 2 alone; combined capacity of rows 2 and 1
    $- F_{\text{tr1,Rd}}$)
$F_{\text{tr3,Rd}}$ = minimum of (capacity of row 3 alone; combined capacity of rows 3 and 2
    $- F_{\text{tr2,Rd}}$; combined capacity of rows 3,2,1 $- F_{\text{tr2,Rd}} - F_{\text{tr2,Rd}}$)

and in a similar manner for subsequent rows.

The calculation method assumes sufficient plastic deformation will be available at each bolt row to permit all rows to attain their full resistance. Connections with relatively thick end-plates and column flanges have relatively little deformation capacity and there is a danger that the upper bolts may fail before the full resistance is developed in the lower rows. This is the problem addressed in BS EN 1993-1-8 6.2.7.2(9) which refers to the National Annex. The UK NA recommends limiting the end-plate thickness $t_p \le (d/1.9)(f_u/f_{y,p})^{0.5}$ or the column flange thickness $t_{fc} \le (d/1.9)$ $(f_u/f_{y,fc})^{0.5}$ where d is the bolt diameter.

BS EN 1993-1-8 6.2.7.1 identifies a number of situations in which a conservative simplification of the above general method is justified. For example, a flush end-plate with only one bolt-row in tension (or where only one bolt row is considered), the design moment resistance may be determined as shown in Figure 28.9(a). For an extended end-plate joint with only two rows of bolts in tension, the design moment of resistance may be conservatively calculated treating the whole tension region as one component operating at a lever arm measured from the centre of the compression flange to the midway point between the two bolt rows (see Figure 28.9(b)). As long as the bolt-rows are approximately equidistant either side of the beam flange, $F_{2,\text{Rd}}$ may be taken to be equal to the calculated value for the upper bolt row, $F_{1,\text{Rd}}$, and so the total design resistance F_{rd} can be taken as $2F_{\text{r1,Rd}}$ (but not greater than 3.8 times the design tension resistance of a single bolt).

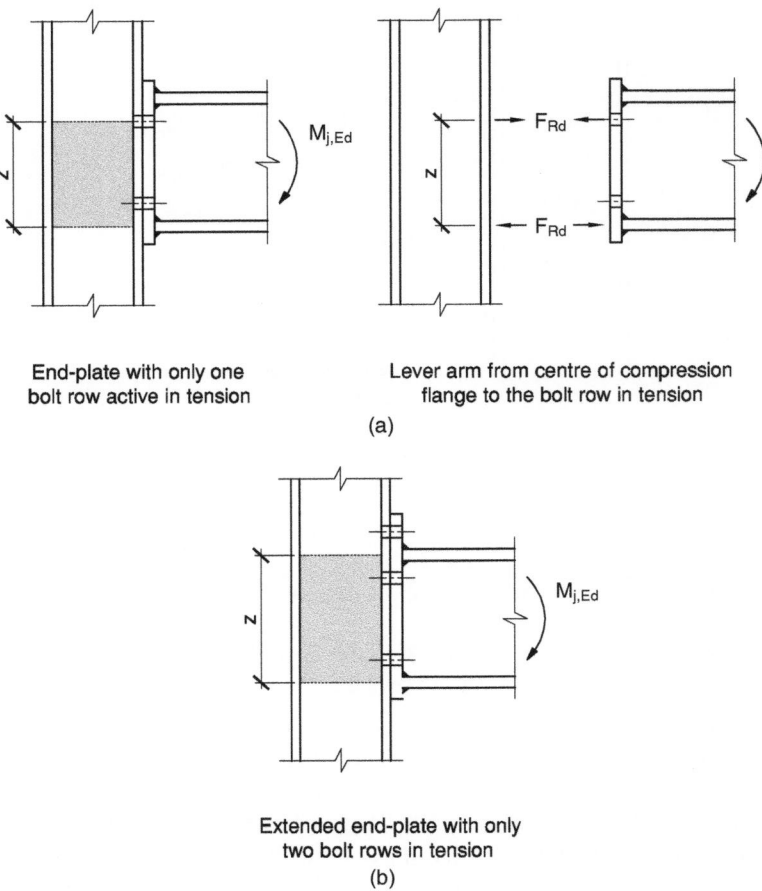

End-plate with only one
bolt row active in tension

Lever arm from centre of compression
flange to the bolt row in tension

(a)

Extended end-plate with only
two bolt rows in tension

(b)

Figure 28.9   Simplified models for flush and extended end-plates

28.8 Rotational stiffness and rotation capacity

The rotational stiffness of a joint may be determined from the flexibilities of the
basic components, which are given in BS EN 1993-1-8 Table 6.11. However, the UK
NA (NA2.6) states 'until experience is gained with the numerical method of calculating rotational stiffness given in BS EN 1993-1-8:2005, 6.3 . . . semi-continuous
elastic design should only be used where either it is supported by test evidence . . . or
based on satisfactory performance in a similar situation'. This comment effectively
prohibits the use of calculated values of rotational stiffness in the UK.

Where rigid plastic global analysis has been used, rotational capacity is clearly
important. If the design moment of resistance of the joint, $M_{j,Rd}$, is at least 20%
greater than the design plastic moment of resistance, $M_{pl,Rd}$, of the connected

member, plastic hinges may be assumed to form in the connected members rather than in the joint and the rotation capacity of the joint need not be checked. In bolted beam-to-column joints in which $M_{\mathrm{j,Rd}}$ is controlled by the design resistance of the column web panel in shear, the joint may be assumed to have adequate rotation capacity when $d/t_{\mathrm{w}} \leq 69\varepsilon$. A bolted end-plate joint may also be assumed to have sufficient rotational capacity if the design moment resistance is governed by the column flange or end-plate in bending *and* the thickness of either the column flange or the beam end-plate $\leq 0.36\, d(f_{\mathrm{ub}}/f_{\mathrm{y}})^{0.5}$ where d is the bolt diameter, f_{ub} is the ultimate tensile strength of the bolt, and f_{y} is the yield strength of the column flange or end-plate.

28.9 Summary

The successful performance of every structural steel frame (in terms of both in-service behaviour and economy of fabrication and erection) is dependent as much on its connections as on the size of its structural members. Bolted connections, and in particular moment connections, are complex in their behaviour and the distribution of the stresses and forces within the connection depends on both the capacity of the welds, bolts etc and on the relative ductility of the connected parts. It is therefore necessary for the design of connections to be consistent with the designer's assumptions regarding the structural behaviour of the steel frame. When choosing and proportioning connections, the engineer should always consider the basic requirements, such as the stiffness/flexibility of the connection, moment and shear resistance, and the required rotational capacity. The design philosophy presented in this chapter, together with the detailed design checks, provide the engineer with a basic set of tools that can be used to design connections that are able to meet the design assumptions. Although BS EN 1993-1-8 contains very detailed information on the design of fasteners, basic components and end-plate connections, application of these rules in routine design would be very time consuming without the use of complementary design aids. Reference 1 remains a very valuable tool for the design of moment connections and it is unlikely that designers would choose to undertake the design of a moment connection by hand. However, Access-steel contains a very detailed worked example for a portal eaves moment connection[10] as well as flow-charts for both eaves[11] and apex connections.[12] NCCIs are also available for the design of portal frame connections.[13,14]

References to Chapter 28

1. The Steel Construction Institute/The British Constructional Steelwork Association LTD (1995) *Joints in Steel Construction: Moment Connections*, Publication No. 207. Ascot, SCI, BCSA.

2. British Standards Institution (2005) *BS EN 1993-1-8:2005, Eurocode 3: Design of steel structures Part 1.8 Design of joints.* London, BSI.
3. British Standards Institution (2008) *NA to BS EN 1993-1-8:2005, UK National Annex to Eurocode 3: Design of steel structures Part 1.8 Design of joints.* London, BSI.
4. Salter P. R., Couchman G. H. and Anderson A. (1999) *Wind-moment design of Low Rise Frames*, Publication No. 263. Ascot, The Steel Construction Institute.
5. British Standards Institution (2000) *BS 5950: Structural use of steelwork in building Part 1: Code of practice for design – Rolled and welded sections.* London, BSI.
6. British Standards Institution (1992) *ENV1993-1-1/A1: 1992 Eurocode 3: Design of steel structures Part 1.1 General rules and rules for buildings.* London, BSI.
7. Faella C., Piluso V. and Rizzano G. (1999) *Structural steel semi rigid connections: theory, design and software.* Florida, CRC Press LLC.
8. Jaspart, J-P. (2000) General report: session on connections. *Journal of Constructional Steel Research*, 55: 69–89.
9. Zoetemeijer P. (1974) *A design method for the tension side of statically loaded bolted beam-to-column connections, Heron 20*, No. 1, Delft University, Delft, The Netherlands.
10. Example: Portal frame – eaves moment connection, SX031a-EN-EU, www.access-steel.com
11. Flow chart: Portal frame eaves connection, SF025a-EN-EU, www.access-steel.com
12. Flow chart: Portal frame apex connection, SF026a-EN-EU, www.access-steel.com
13. NCCI: Design of portal frame eaves connections, SN041a-EN-EU, www.steel-ncci.co.uk
14. NCCI: Design of portal frame apex connections, SN042a-EN-EU, www.steel-ncci.co.uk

Chapter 29
Foundations and holding-down systems

Compiled by GRAHAM OWENS

29.1 Types of foundation

Pad foundations are commonly used to support the major structural elements in
both sheds and multi-storey buildings. The pad foundations for major elements may
be either mass concrete or reinforced concrete, the latter when either heavy loads
or very poor ground conditions are present. In the context of cladding, they may be
used to support intermediate posts carrying sheeting rails, in which case the load is
almost all from wind forces and is horizontal.

Strip foundations are used in steel-framed buildings to support external masonry
or brickwork cladding and masonry internal partitions. In some cases the ground
floor is thickened at these locations to provide a foundation but care should be
taken with respect to the appropriate depth for clay, which may change in volume
with changes in moisture content or frost heave; the compatibility between such
foundations and those of the main frame also needs to be considered.

Piled foundations, either driven, bored or cast in place, are used on sites where
ground conditions are poor or for buildings or structures in which differential settlement is critical. They may also be required in circumstances where heavy concentrations of load occur. In general when piled foundations are used the whole of the
construction should be supported on piles. The ground floor slab, ground floor cladding and internal partitions should be carried by ground beams between the pile
cap locations. If it is necessary for reasons of economy to support the ground floor
independently, provision should be made for differential settlement by the inclusion
of suitable movement joints.

Ground improvement techniques are appropriate for some types of poor ground.
The most usual techniques are vibro-compaction or vibro-replacement but dynamic
compaction can also be useful for improvement of large isolated sites. Ground
improvement specialists or specialist consultants should be approached as economy
will be the most important factor in the decision.

Typical foundation layouts are shown in Figure 29.1.

Steel Designers' Manual, Seventh Edition. Edited by Buick Davison and Graham W. Owens.
© 2012 Steel Construction Institute. Published 2012 by Blackwell Publishing Ltd.

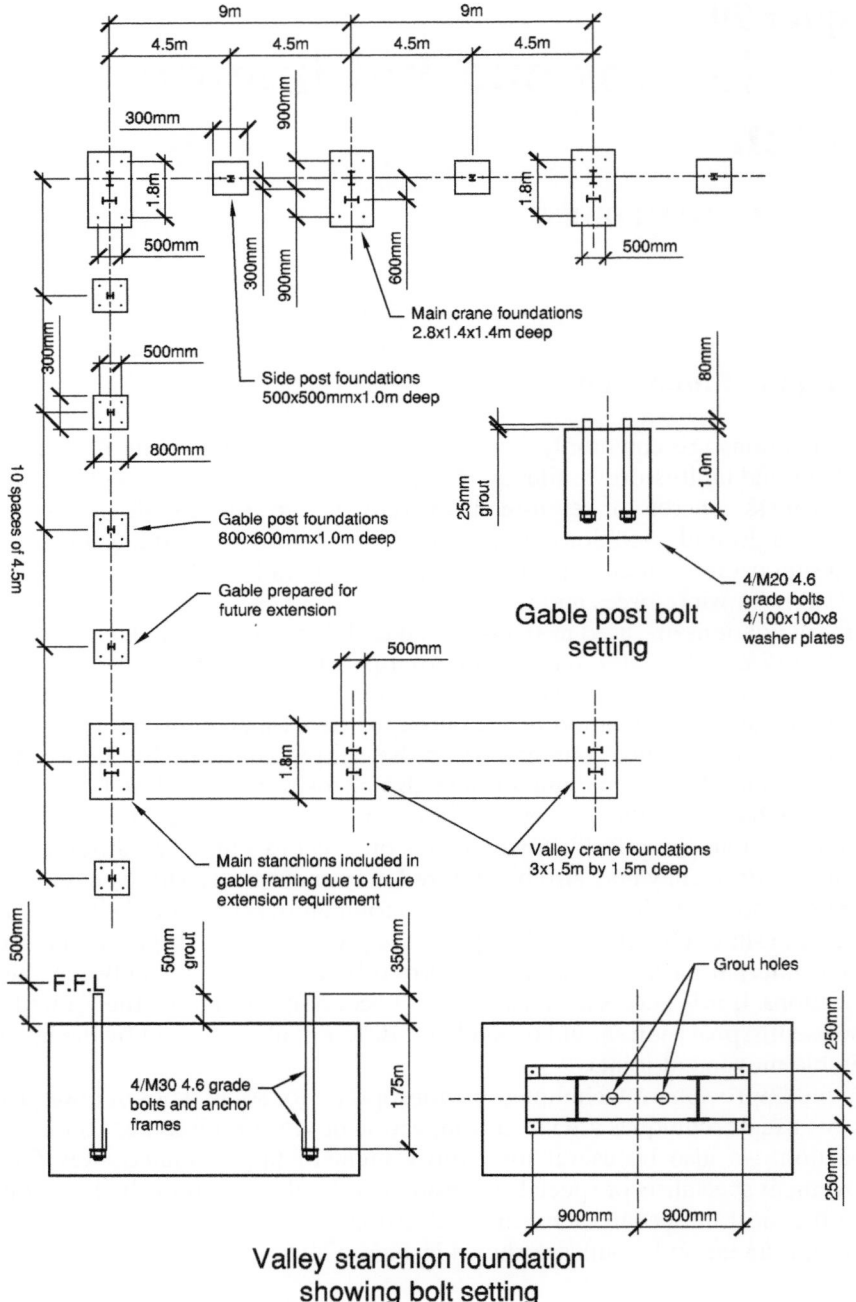

Figure 29.1   Part plan of typical two-bay crane shed

29.2 Design of foundations

In order to assess the distribution of pressure under a foundation it is necessary to make a reasonable estimate of the weight of the foundation. In addition to distributing the forces to the ground, the foundation block is also required to provide stability in cases where overturning moments are present.

Referring to Figure 29.2, loads N_{Ed}, H_{Ed} and M_{Ed} are factored as appropriate while W, the foundation mass, is factored by 1.0, being a restoring moment. Moments about A give:

$$M_{Ed} + H_{Ed}D - N_{Ed}K - \frac{WL}{2} \leq 0$$

From this a minimum value of W for stability is produced.

The minimum value for D for a mass concrete foundation is established by 45° dispersal from the edge of the baseplate shown in Figure 29.3. Shallower foundations can be used if they are suitably reinforced.

The distribution of pressure under the foundation is then assessed as follows.

Case 1

See Figure 29.4(a):

$$\sigma_g = \frac{N_{Ed} + W}{LB} \pm \frac{(M_{Ed} + H_{Ed}.D)6}{BL^2}$$

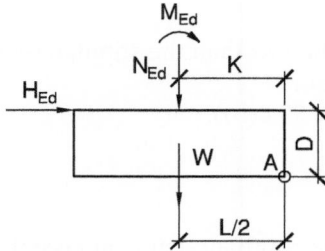

Figure 29.2   Stability of foundation

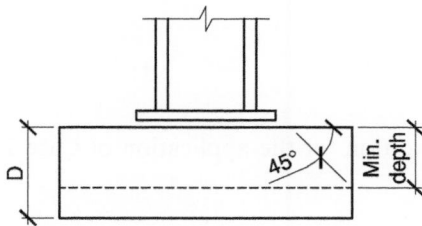

Figure 29.3   Thickness of foundation

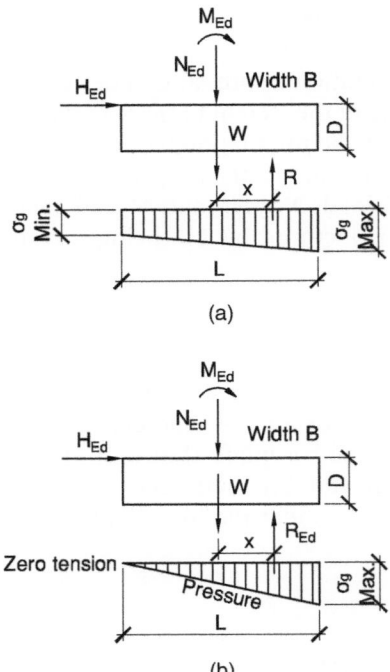

Figure 29.4   Ground pressure – Case 1

It is necessary for σ_{gmax} to be less than the stipulated ground bearing capacity for the foundation to be satisfactory.

For σ_{gmin} to be zero (Figure 29.4(b)):

$$\frac{N_{Ed}+W}{LB} - \frac{(M_{Ed}+H_{Ed}.D)6}{BL^2} = 0$$

Replacing the forces by the resultant acting at eccentricity x:

$$\frac{R_{Ed}}{LB} - \frac{6R_{Ed}x}{BL^2} = 0$$

from which:

$$x = \frac{L}{6}$$

This is the limiting condition for the application of Case 1.

Case 2

This occurs when σ_{gmin} is negative. As no tension can exist between the soil and the underside of the concrete base, a compressive stress wedge is formed at the com-

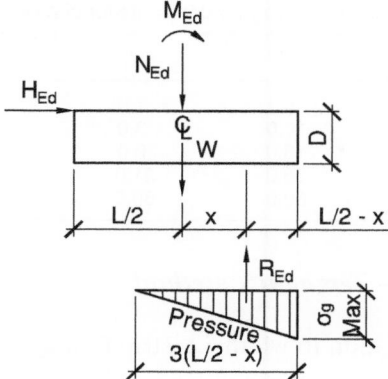

Figure 29.5 Ground pressure – Case 2

pression side of the foundation. The summation of the stress under the block must equal the resultant of the applied loads. When $x > L/6$, the length of the triangular stress wedge is three times the edge distance $(L/2 - x)$ in order that the resultant acts at the centroid of the wedge. The theory proposes that $3(L/2 - x)$ is the length of surface contact between the foundation and the ground (Figure 29.5).

$$\frac{\sigma_{gmax}}{2}\left(\frac{L}{2} - x\right)3B = N_{Ed} + W$$

$$\sigma_{gmax} = \frac{2(N_{Ed} + W)}{3B\left(\dfrac{L}{2} - x\right)}$$

When σ_{gmax} exceeds the stipulated ground bearing pressure, dimensions B or L or both may be increased within the limits of economy, after which piling or ground improvement techniques can be investigated.

29.2.1 Sub-soil bearing pressure

Foundation bearing pressure should always be determined on the basis of experimental results and field assessments taken during the soil investigation. Almost all sites have considerable variation of strength and quality of the sub-strata and it is therefore necessary to undertake a comprehensive investigation producing a large number of test results in order to be satisfied that reasonable average values are obtained for the various parameters. A factor of safety of 3 is usually applied to the bearing strength to obtain the safe foundation bearing pressure, c.

Table 29.1   Terzaghi's constants for clayey soils

ϕ	N_c	N_q	N_r
0	5.7	1.0	0
10	10.0	3.0	2.0
20	18.0	8.0	6.5
30	36.0	21.0	20.0
35	60.0	50.0	45.0

29.2.1.1 *Clayey soils – Terzaghi's method*

1. Foundation long in relation to width, i.e. strip footing.

$$q = cN_c + \gamma z N_q + 0.5\gamma B N \gamma$$

where
  q = bearing capacity (ultimate) (kN/m²)
  c = cohesion (kN/m²)
  γ = bulk density of the soil (kN/m³)
  z = depth of foundation (m)
  B = breadth of foundation (m)
  and N_c, N_q and N_γ are constants dependent upon the angle of cohesion. These constants are given in graph form in soil engineering references; Table 29.1 gives approximate values as guidance only.

2. Square and circular foundations to isolated piers or columns.

The bearing capacity of foundations that are rectangular, square or circular is higher than that of strip footings. Terzaghi's expression is adjusted as follows:

$$q = 1.3cN_c + \gamma z N_q + 0.3\gamma B N \gamma$$

while Skempton applies a factor $(1 + 0.2B/L)$ to the Terzaghi strip footing calculation where B is the breadth and L the length. In the case of a square the enhancement factor is then 1.2.

29.2.1.2 *Sandy or cohesionless soils*

The appropriate site test for cohesionless soils is the standard penetration test (SPT) in which the number of blows of a standard weight is recorded for unit penetration of a standard cylindrical implement. According to Meyerhof the ultimate bearing capacity is given, in kN/m², by:

$$q = 10.7\, NB\,[1 + z/B]$$

where N is the number of blows per metre and q, z and B are as before.

Cohesionless soils subject to flooding will suffer a reduction of capacity at water table level:

Capacity when flooded = $K \times$ unflooded capacity

where $K = (\gamma - 9.8)/\gamma$ and γ is the soil bulk density given for convenience in kN/m^3.

29.3 Fixed and pinned column bases

The function of a column baseplate is to distribute the column forces to the concrete foundation. In general a plain or slab base is used for pinned conditions or when there is very little tension between the baseplate and the concrete. More popular in the past, a gusseted base may still be used occasionally to spread very heavy loads. For fixed bases, plain or slab bases are also likely to be used. However, in conditions of large moment in relation to the vertical applied loads, gusseted bases may be adopted. The principal function of the gusset is to allow the holding-down bolt lever arm to be increased to give maximum efficiency while keeping the baseplate thickness to an acceptable minimum. Gusseted or built-up bases give an ideal solution for compound or twin crane stanchions in industrial shed buildings.

Fixed bases are used primarily in low-rise construction either in portal buildings specifically designed as 'fixed base' or in industrial sheds in which the main columns cantilever from the foundations. They are also used, though less frequently, in multi-storey rigid-frame construction. In each of these cases it is assumed by definition that no angular rotation takes place, and although this is unlikely to be achieved it is generally accepted that sufficient rigidity can be obtained to justify the assumption.

Pinned bases are those in which it is assumed that there is no restraint against angular rotation. Although this is also difficult to achieve it is accepted that sufficient flexibility can be introduced by minimising the size of the foundation and similarly reducing the anchorage system. Pinned bases are used in portal and in multi-storey construction.

Typical pinned and fixed bases using plain or slab bases are shown in the following sections on design.

29.4 Pinned column bases – axially loaded I-section columns

Simple bases for I-section columns transmit an axial compressive force and a shear force to the foundation (i.e. a 'pinned' column base). The rectangular base plate is welded to the column section in a symmetrical position so that it has projections beyond the column flange outer edges on all sides (see Figure 29.6).

Four anchor bolts are required in order to ensure the stability of the column during erection. Anchor bolts provide resistance to any uplift forces which arise in

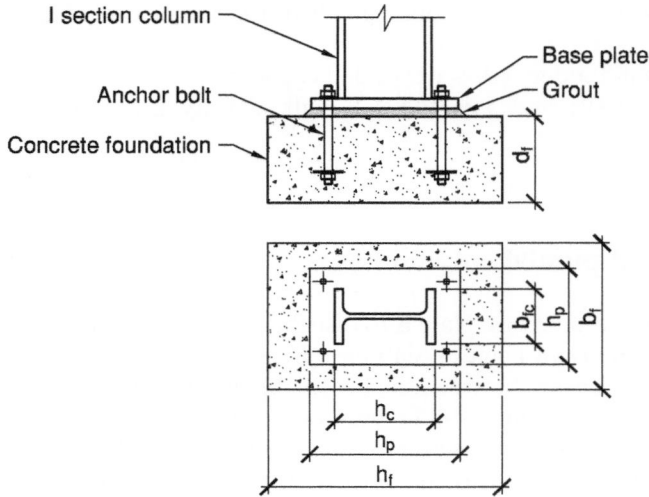

Figure 29.6   Typical simple column bases

the column and also, but only under certain conditions, may be used to provide resistance to shear at the column base, see Sections 29.4.3 and 29.4.4.

In practice, the column section and the axial design force are usually known from the overall building design. Design requires the determination of the dimensions of the base plate.

29.4.1 Design model

The design model for the axial compression force is based on Cl 6.2.5 and Cl 6.2.8.2(1) of BS EN 1993-1-8.[1] The basic design approach is to ensure that the bearing stresses under the base plate neither exceed the design bearing strength of the foundation joint material nor lead to excessive bending of the base plate.

The design model assumes that the bearing resistance of a column base on its foundation is provided by three non-overlapping T-stubs in compression, one for each column flange and one for the column web, as shown in Figure 29.7. For each T-stub, the design bearing resistance is determined by multiplying its bearing area (length by width) by the strength of the foundation joint material.

The length and width of each T-stub depend on the dimensions of the relevant flange or web and on an 'additional' bearing width, cantilevered from the T-stub stem as shown in Figure 29.8. The theoretical value of the 'additional' bearing width depends on the elastic bending resistance of the base plate and on the design strength of the foundation joint material. As discussed below, these theoretical T-stubs may need to be modified if their use leads to overlapping of the individual areas.

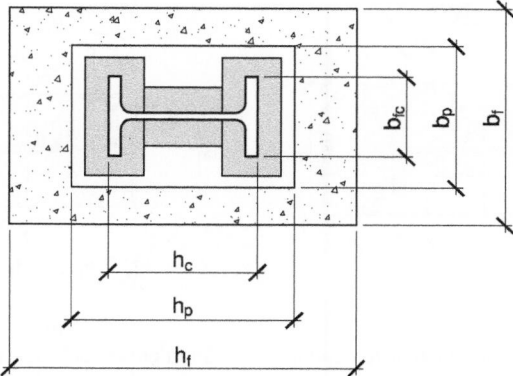

1. T-stub bearing area for left column flange
2. T-stub bearing area for right column flange
3. T-stub bearing area for column web

Figure 29.7   Column base and non overlapping T-stub bearing areas (see Figure 6.19 of BS EN 1993-1-8)

There are two basic types of base plate identified in BS EN 1993-1-8[1], 'large projection' base plates and 'short projection' base plates.

For the 'large projection' base plate, the projection of the base plate beyond the column section perimeter is such that the design bearing width on each side of all three T-stubs is usually equal to the value of the 'additional' width (c). A large projection base plate is illustrated in Figure 29.8(a).

For the 'short projection' base plate, the projection beyond both column flanges towards the base plate edges, while being less than the value of the 'additional' width (c), is adequate to allow fillet welding of the flanges to the base plate. Usually, for the latter purpose, a width approximately equal to the column flange thickness is provided. A 'short projection' base plate is illustrated in Figure 29.8(b).

As noted above, when some H-section columns are used with thick base plates, the flange T-stubs of 'additional' bearing width c on the web side would overlap in the central area between the flanges as shown in Figure 29.8(c) and Figure 29.8(d). In such cases, since there would be no bearing area left for a web T-stub, the effective bearing area would be reduced to a simple rectangular area as follows:

- 'Short' projection base plate:

$$A_{\text{eff.bearing}} = A_{c0} = l_{\text{eff}}\,b_{\text{eff}} = h_p b_p$$

- 'Large' projection base plate:

$$A_{\text{eff.bearing}} = A_{c0} = l_{\text{eff}}\,b_{\text{eff}} = (h_c + c)(b_{\text{fc}} + c) \le h_p b_p$$

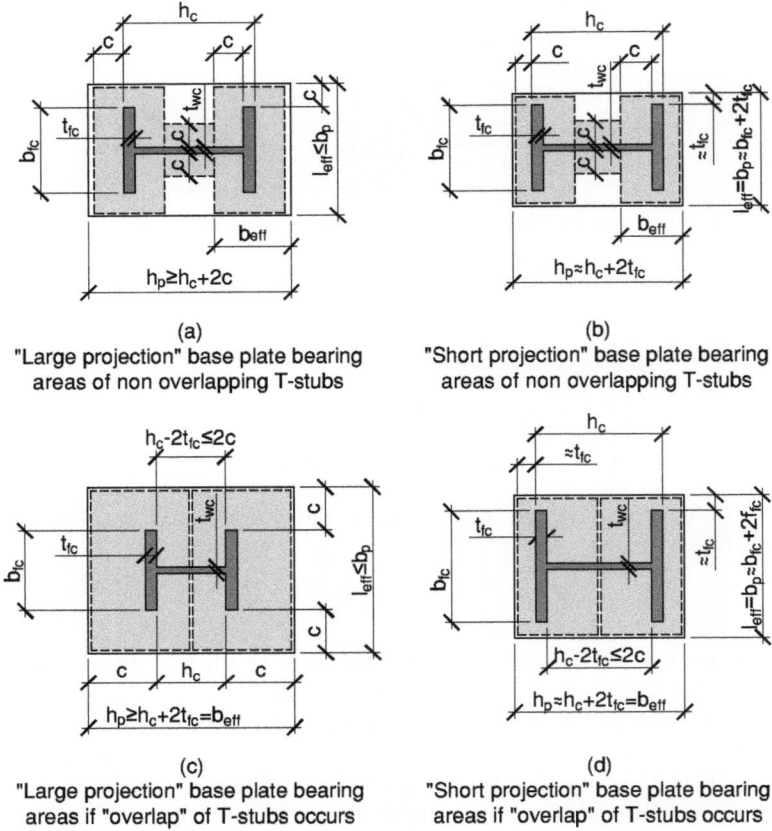

Figure 29.8   Area/dimensions of equivalent T-stubs in compression

29.4.2 Design approach

29.4.2.1 Step 1: Choose the design strengths of the materials

Base plate steel strength

A design value for the yield strength f_{yp} of the base plate steel is adopted, usually S275.

Bearing strength of the foundation joint material (grout)

In most practical cases, the value of the design bearing strength of the joint material can conservatively be taken as equal to that of the design concrete strength in com-

Table 29.2   Bearing strength for typical foundation concrete and foundation joint material

Concrete class f_{ck}	20	25	30	35	40	45
Bearing strength f_{jd} (N/mm^2)	13.3	16.7	20	23.3	26.7	30

pression, i.e. $f_{jd} = f_{cd}$. Table 29.2 provides typical design bearing strengths for typical concrete grades and foundation joint materials.

Annex A of Reference 2 provides guidance on a more precise way of determining f_{jd}, based on Reference 3.

29.4.2.2 Step 2: Make a preliminary estimate of the base plate area

A first estimate of the required base plate area is given by the larger of the following two values:

$$A_{c0} = \frac{1}{h_c b_{fc}}\left[\frac{N_{j,Ed}}{f_{cd}}\right]^2$$

$$A_{c0} = \frac{N_{j,Ed}}{f_{cd}}$$

29.4.2.3 Step 3: Choose the type of base plate

The choice of the base plate type is recommended to be as follows:

$A_{c0} \geq 0.95\, h_c b_{fc}$ adopt a 'large projection' base plate
$A_{c0} < 0.95\, h_c b_{fc}$ adopt a 'short projection' base plate.

Note: A large projection base plate may be adopted in all cases.

29.4.2.4 Step 4: Determine the additional bearing width

The value of the 'additional' bearing width, c, is obtained by satisfying the relevant design bearing resistance condition as follows (see Figure 29.8):

Design bearing resistance of a 'short' projection base plate

Assuming the projections beyond the column flange edges to be equal to the column flange thickness t_{fc}, the design bearing resistance is as follows:

$$N_{j,Rd} = f_{jd}[2(b_{fc} + 2t_{fc})(c + 2t_{fc}) + (h_c - 2c - 2t_{fc})(2c + t_{wc})]$$

Table 29.3 Expressions of the parameters of the quadratic equation for c

Constant	'Short projection' base	'Large projection' base	
	Non overlapping T-stubs	Non overlapping T-stubs	T-stubs overlap predicted
A	2	2	2
B	$-(b_{fc} - t_{wc} + h_c)$	$+(2\,b_{fc} - t_{wc} + h_c)$	$+(b_{fc} + h_c)$
C	$+(N_{j,Ed}/2f_{jd}) - (2b_{fc}t_{fc} + 4t_{fc}^2 + 0{,}5h_c t_{wc} - t_{fc}t_{wc})$	$+(b_{fc}t_{fc} + 0{,}5h_c t_{wc} - t_{fc}t_{wc}) - (N_{j,Ed}/2f_{jd})$	$+(b_{fc}h_c)/2 - (N_{j,Ed}/2f_{jd})$

Design bearing resistance of a 'large' projection base:

Assuming the bearing width about the column perimeter to be equal to the additional bearing width c, the design bearing resistance is as follows:

$$N_{j,Rd} = f_{jd}[2(b_{fc} + 2c)(2c + t_{fc}) + (h_c - 2c - 2t_{fc})(2c + t_{wc})]$$

Replacing $N_{j,Rd}$ by $N_{j,Ed}$ in the above expressions, the solution to the resulting quadratic equations for the unknown c takes the standard form:

$$c = \frac{-B \pm \sqrt{B^2 - 4AC}}{2A}$$

– for which positive solutions only are of interest.

Table 29.3 gives the expressions for the constants A, B and C, under the relevant 'non overlapping T-stub' column.

Check for 'overlapping' T-stubs

The value obtained above for the 'additional' width c sometimes exceeds half the height of the column web, which is unacceptable as it implies having overlapping T-stub bearing areas.

'Short projection' base plate: change to a 'large projection' base plate and recalculate c.

'Large projection' base plate: recalculate c based on having the entire area between the column flanges in bearing in the design expression. The design condition for the 'large projection' base plate then becomes:

$$N_{j,Ed} \leq N_{j,Rd} = f_{jd}[(b_{fc} + 2c)(h_c + 2c)]$$

The corresponding expressions for A, B and C to be used in the solution for c are given in the last column of Table 29.3.

29.4.2.5 Step 5: Determine the required minimum plan dimensions of the base plate

The final plan dimensions of the base plate are based on the following:

'Short projection' base plate:
$b_p \geq (b_{fc} + 2t_{fc})$
$h_p \geq (h_c + 2t_{fc})$
'Large projection' base plate:
$b_p \geq (b_{fc} + 2c)$
$h_p \geq (h_c + 2c)$

29.4.2.6 Step 6: Determine the minimum required base plate thickness

The minimum required thickness of the base plate is obtained from the condition that the plate, assumed to act as a cantilever off the column perimeter, is not subject to more than its elastic design bending resistance under a uniform bearing pressure equal to f_{jd} acting over the 'additional' width c. The value for the minimum required thickness is given by:

$$t_p \geq \frac{c}{\left[\dfrac{f_{yp}}{(3f_{jd}\gamma_{M0})}\right]^{0.5}}$$

29.4.3 Shear resistance of the base plate joint

The design shear resistance is based on the friction resistance developed by the compressive load applied by the base plate on the joint material. It is given in BS EN 1993-1-8[1] Cl 6.2.2(6) as:

$$F_{v,Rd} = F_{f,Rd}$$

where:

$\quad F_{f,Rd} \quad = C_{f,d}\ N_{c,Ed}$

$\quad N_{c,Ed}$ is the column design compressive load and

$\quad C_{f,d}$ is the coefficient of friction between the base plate and the grout layer. A value of 0.2 is specified for sand-cement mortar. Otherwise tests in accordance with BS EN 1990 Annex D are required to determine the coefficient value for any other type of grout.

The design check is: $V_{c,Ed} \leq F_{v,Rd}$

29.4.4 Use of shear nibs to enhance shear capacity

The shear resistance developed by friction between the column base plate in compression and the joint material (grout), as calculated above, is often adequate for most typical simple base plate joints.

However, if there is axial tension on the column in some load cases, shear resistance by friction cannot be developed. Even without net uplift, shear resistance by friction alone may not suffice when high shear is combined with low axial compression.

In these situations, other means are required to transfer the shear force. Options are:

- Shear / bearing of the anchor bolts (see Cl 6.2.2(7) of BS EN 1993-1-8[1]).
- Setting the column end with its base plate within a pocket in the foundation pad. The pocket depth is usually 300 mm or more and is filled with non-shrink concrete once the column is in place. This type is suitable for fixed column base plate joints. The shear force is transferred by lateral bearing of the embedded column part on the pocket infill concrete. The concrete surround of the pocket may require reinforcement in accordance with BS EN 1992-1 to transfer the column end forces and moments.
- Providing a tie from the column end into an adjacent ground floor slab. This may require ensuring that there is appropriate reinforcement in the slab to anchor the horizontal tie force.
- Providing a shear nib (key) welded to the underside of the base plate which is accommodated in a foundation pocket of sufficient depth and size. The pocket is filled with non-shrink concrete after the column and the anchor bolts are positioned.

If anchor bolts are loaded in shear, one must ensure that the shear force transfer to the foundation through the anchor bolts is possible without causing excessive lateral movement at the column base (see Cl 6.2.2(5) of BS EN 1993-1-8[1]). If anchor bolts are grouted in sleeves they may not be dependable in shear/bearing. In addition, oversized holes are often used in base plates in order to account for the usual tolerances in the positioning of anchor bolts set in concrete. In the latter case, the plate washers used under the anchor bolt nuts would need to be welded to the base plates so as to allow transfer of the shear force to the anchor bolts. It is recommended that hole sizes in these plate washers may be reduced to a minimum, for instance $d + 1.5$ mm (where d is the nominal anchor bolt diameter). With these precautions, the design resistance of anchor bolts in shear/bearing, which is given in Cl 6.2.2(7) of BS EN 1993-1-8[1], can be added to the friction resistance when relevant.

Neither the design of foundation pockets (but see comment below for the 'shallow pocket' type) for fixed base plate joints nor that of ties to the floor slab is considered in this chapter; its scope is limited to the design of a shear nib under the base plate for transferring shear forces to the foundation.

A shear nib (or shear key) typically consists of a short length of steel section welded to the underside of the base plate. Once the concrete is poured into the reserve hole for the anchor bolts and the column grouted in its final position, the nib is embedded in the foundation. The shear force acting on the column base can be transmitted to the foundation by the nib acting horizontally leading to compression over the vertical surface of the nib against the concrete foundation.

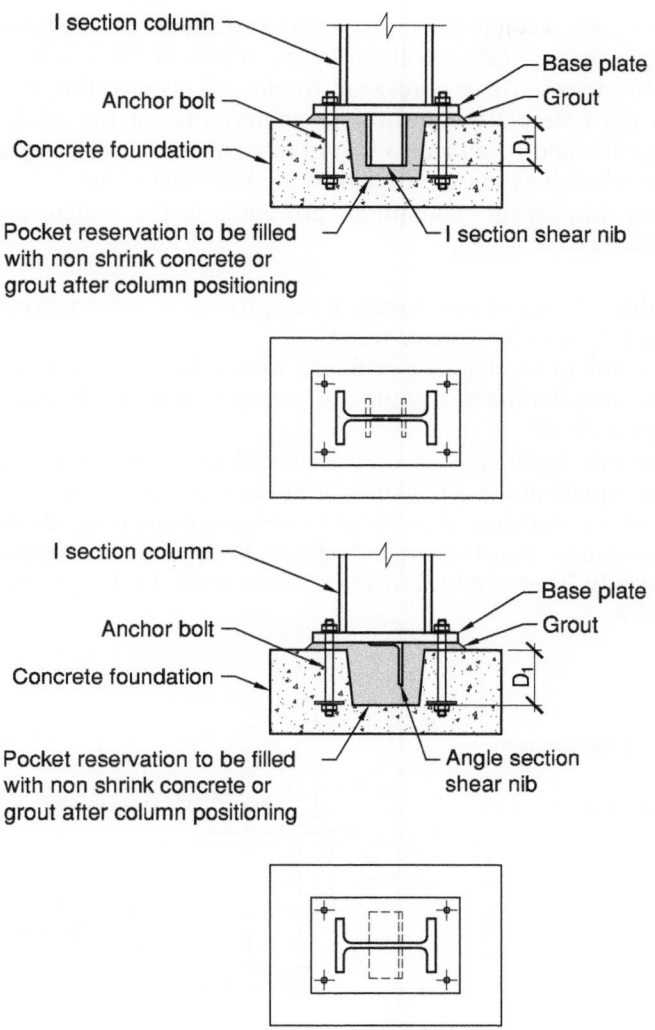

Figure 29.9   Typical column bases with shear nibs

Figure 29.9 shows two types of shear nib in common use, one being a short length of angle capable of resisting relatively modest shear forces and the other a short length of I-section used if the shear forces to be transmitted are relatively high.

The mechanical model adopted for the nib is shown schematically in Figure 29.10. The column base shear force is resisted by pressure developed over the vertical face (or faces) of the nib embedded in sound foundation concrete. The eccentricity between the horizontal reaction on the nib and the applied column base shear causes a secondary moment creating a couple of additional vertical forces ($N_{\text{sec,Ed}}$)

at the base plate joint, a compressive force and a tensile force. The tensile force may be resisted either by the anchor bolts or by the nib itself. Here it is conservatively assumed that the tensile force is resisted by the nib. The additional compression force between the base plate and the joint material (grout) is often neglected in design, although it could be added to that in the column flange compressive T-stub when doing the final check on the design of the base plate joint.

The following simplifying assumptions are made in the design model which is taken from Reference 3:

- Both embedded flanges of an I-section nib provide equal horizontal resistance to the applied column base shear force.
- For the full width of an angle leg or flange within the concrete foundation, there is a triangular distribution of compressive stresses over the effective depth of the nib (see Figure 29.10).
- The effective nib depth, $d_{eff,n}$, is taken as equal to the full height of the nib, d_n, below the base plate minus a thickness at the top surface to allow for the possible inadequacy of the packing of the joint material (grout) beneath the base plate. It is usual to assume that the latter thickness is equal to that of the grout layer, which is typically 30 mm and rarely over 50 mm thick. In the following it is taken as 30 mm thick.

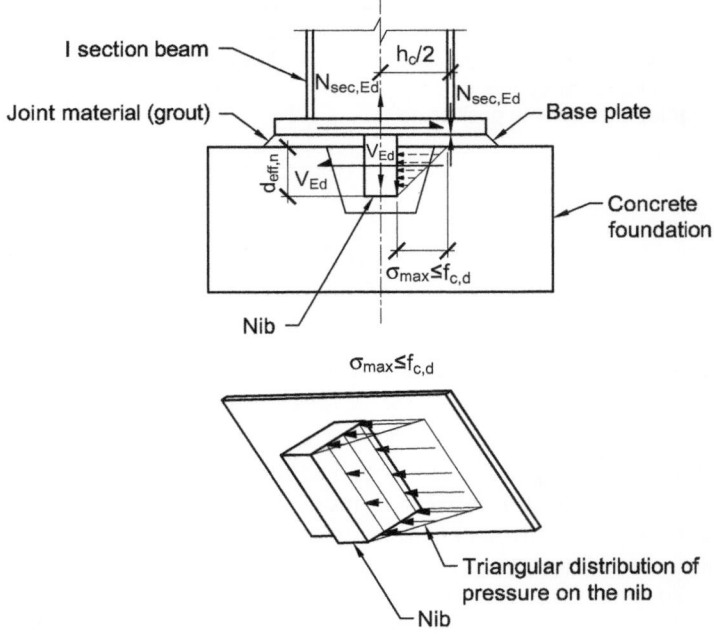

Figure 29.10   Shear nib model showing the forces and stresses induced: distribution of compressive stresses over shear nib and secondary forces

- The secondary moment is considered to be resisted by a couple of forces acting on the column base, one a normal tension force in the base plate over the shear nib and one a compressive force between the base plate and the grout which is centred under one of the column flanges. Assuming the shear nib to be centred at the column centroid and a grout layer thickness of 30mm, one obtains the following axial tension design forces:
  - I-section nib: axial tension in a nib flange, N_{Ed}, is given by

$$N_{Ed} = V_{Ed}\left(\frac{d_{eff,n}}{3}+30\right)\left(\frac{1}{h_n - t_{fn}}\right) + V_{Ed}\left(\frac{d_{eff,n}}{3}+30\right)\left(\frac{2}{h_c}\right)\left(\frac{1}{2}\right)$$

$$= V_{Ed}\left(\frac{d_{eff,n}}{3}+30\right)\left(\frac{1}{h_n - t_{fn}} + \frac{1}{h_c}\right)$$

  - angle nib: axial tension in the vertical leg: $N_{Ed} = V_{Ed}\left(\frac{d_{eff,n}}{3}+30\right)\dfrac{2}{h_c}$

- In order to ensure against pull-out of the nib from the concrete foundation and to have an efficient shear nib, the following limits are placed on the nib dimensions:
  - height of an I-section nib section: $h_n \leq 0.4\,h_c$
  - effective depth in the foundation of an I-section nib: $60\,mm \leq d_{eff,n} \leq 1.5\,h_n$
  - effective depth in the foundation of an angle nib: $60\,mm \leq d_{eff,n} \leq 1.5\,b_n$
    In the case of a simple base plate, the limits on the nib dimension are recommended so as to avoid creating a fixed column base condition.
- Being embedded in the concrete, angle legs or I-section flanges are considered to be subjected to negligible local bending. To support this assumption, the following maximum slenderness criteria are imposed:
  - I-section nib: maximum flange slenderness: $(b_{fn}/t_{fn}) \leq 20$
    (A criterion satisfied by most rolled sections)
  - angle nib: maximum leg slenderness: $(d_n/t_{an}) \leq 10$
    (Not all standard hot-rolled angle sections meet this requirement.)
- For an I-section shear nib, the shear force is transferred from the base through the web. The moment at the underside of the base plate level is resisted by a force couple in the flanges. Rather than assume the anchor bolts to be active, the secondary normal tensile force is considered to be shared between the two flange sections. The column web opposite the flange also resists the total shear force thus obtained.
- For the leg of an angle section shear nib, both the shear force and the secondary normal force are taken by the vertical leg section. Bending at the top of the vertical angle leg is neglected.

The basic design approach is to ensure that the compressive stresses over the vertical surface of the nib in contact with the foundation neither exceed the design compressive strength of the concrete nor lead to excessive stresses in the nib member (leg, flange or web).

Supplementary design checks are required, as follows:

- The column web is checked for the concentrated force corresponding to the secondary tensile force in a nib angle leg or nib flange.
- The base plate to nib fillet weld resistances are checked for both the horizontal shear and for the secondary tensile forces.

More detailed guidance on the design of shear nibs is given in Reference 4.

29.5 Design of fixed column bases

Fixed bases of I-section columns transmit a normal force, a shear force and a moment. The rectangular base plate is welded to the column section in a symmetrical position so that it has projections beyond the column flange outer edges on all sides (see Figure 29.11). Anchor bolts rows, normal to the column axis, are symmetrically placed about the column minor axis. The base plate may be located eccentrically on the concrete foundation.

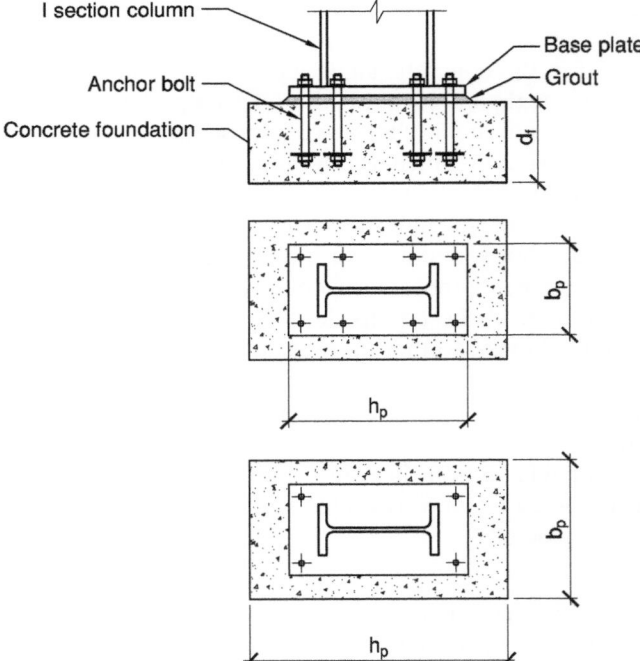

Figure 29.11   Typical fixed column bases

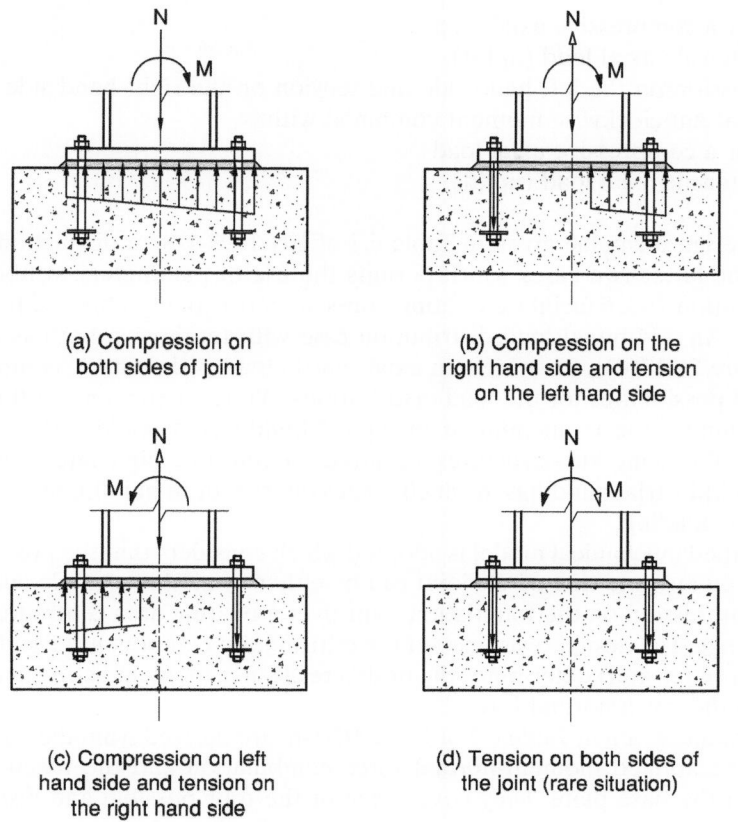

(a) Compression on
both sides of joint

(b) Compression on the
right hand side and tension
on the left hand side

(c) Compression on left
hand side and tension on
the right hand side

(d) Tension on both sides of
the joint (rare situation)

Figure 29.12   Load distribution

Design model

The design model for a fixed column base plate joint for a combined normal force plus a moment about the major column axis is given in Cl 6.2.8 of BS EN 1993-1-8.[1]

The possible load distributions in a fixed column base joint, shown in Figure 29.12(a), (b), (c) and (d), are as follows:

- Compression on both sides of the joint due to a dominant axial compression load combined with:
  - either a clockwise moment
  - or an anticlockwise moment.
- Tension on the left hand side and compression on the right hand side due to a dominant clockwise moment combined with:

  — either a compressive axial load
  — or a tensile axial load (uplift).
- Compression on the left hand side and tension on the right hand side due to a dominant anticlockwise moment combined with:
  — either a compressive axial load
  — or a tensile axial load (uplift).

In the design formulae given in Table 6.7 of EN 1993-1-8,[1] a distinction is made between the latter two cases which permits the use of parameters, symbols and a sign convention which facilitate treating non-symmetric joints subjected to multiple load cases. An additional load distribution case with tension on both sides of the joint (Figure 29.12(d)), for which an axial tensile load is dominant, completes the theoretical possibilities for the load distributions. While having tension throughout a fixed column base is uncommon in typical buildings, it could arise in vertical members of bracing sub-structures required to transmit high lateral loads, for instance in industrial buildings in which cranes operate or in buildings under significant seismic loading.

A simplified mechanical model is adopted which considers that the possible reaction force on any one side of the joint can be either tension in a single anchor bolt row or compression on the foundation joint over a bearing area centred under the column flange. The design resistance of the critical joint component (T-stub in compression or in tension) determines the design resistance moment acting in concomitance with the given normal force.

The formulae given in Table 6.7 of EN 1993-1-8[1] are derived from the equilibrium between the applied moment-normal force combination and the reaction forces induced on the base plate. They cover each of the four possible and distinct load distribution scenarios for the basic configuration of column base plate joint shown in Figure 29.12.

Resistance in bearing

For the compression side of a joint the design approach is to ensure that the bearing stresses under the base plate neither exceed the design bearing strength of the foundation joint material nor lead to excessive bending of the base plate.

The design model assumes that the bearing resistance is provided by one or both of the column flange T-stubs in compression, depending on whether compression reigns over part or all of the column base plate respectively as shown in Figure 29.12. For a flange T-stub in compression the bearing stresses are assumed to be uniformly distributed over the T-stub area centred beneath the flange as shown in Figure 29.13. In the simplified approach given in EN 1993-1-8[1] for the design of column base joints transferring moment, no direct account is taken for any compression force that may be transferred through a column web T-stub in compression.

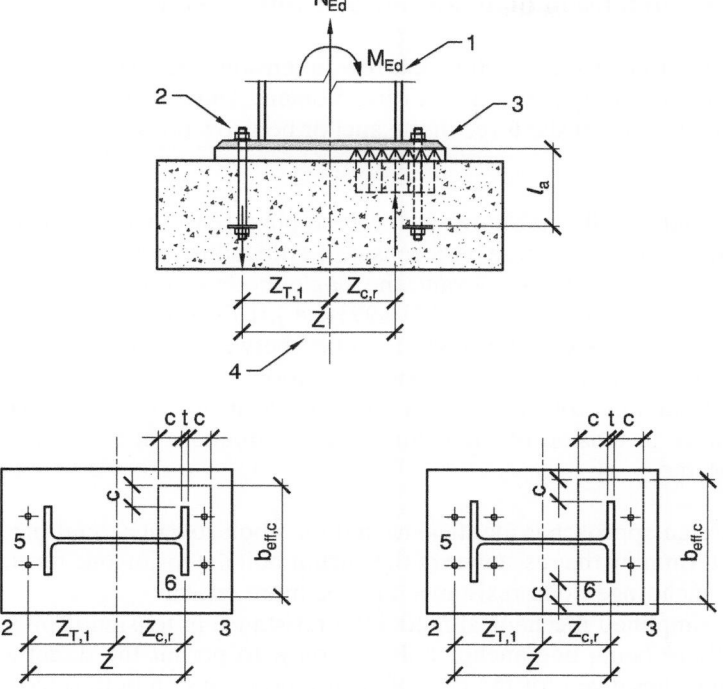

Key:
1. Both the normal force and the moment applied by the column to the column base plate joint are shown acting in the positive sense as defined by EN 1993-1-8, i.e. tensile axial forces are positive and positive moments act clockwise.
2. Left-side of the base plate joint when the anchor bolts are in tension: the tensile force is resisted by the T-stub formed by the base plate and the anchor bolt row.
3. Right side of the base plate joint when in compression: foundation joint offers bearing resistance on the underside of base plate T-stub which is acting in bending off the column flange.
4. Lever arm between the tension force in the anchor bolts and the compression force under the base plate.
5. Anchor bolts.
6. Compression T-stub area.

Figure 29.13 Compression and anchor bolt tension induced by the normal force and moment

Resistance in tension of an anchor bolt row

The design model for an anchor bolt row in tension is similar to that for a bolt row of an end-plate connection transmitting moment. Therefore, the design approach is to ensure that the tensile force in the anchor bolt row does not exceed either of the following:

- The design tensile resistance of the base plate tension T-stub. This involves the consideration of the three basic tension T-stub failure modes as identified in Table 6.2 of EN 1993-1-8[1]. If relevant, the single mode replacing modes 1 and 2 shall be considered (see Table 6.2 EN 1993-1-8[1]). This mode is possible if the prying effect disappears with the loss of contact between the base plate edge and the foundation because of anchor bolt elongation.
- If necessary, i.e. for anchor bolt rows between the column flanges, the design resistance in tension of the column web component of the T-stub needs to be considered.

The design approach is identical to that for a bolt row of an end-plate except that when determining the resistance of the anchor bolt in tension one must also consider that the anchorage bond resistance may be more critical.

In the simplified mechanical model the resistance in tension is presented for the case of there being one anchor bolt row only. To permit the direct application of the design rules given for the case of anchor bolt rows on both sides of the column flange, it is recommended to use an equivalent single row having a total tensile resistance of the two rows acting together at the centroid. It is not recommended to consider that rows other than those adjacent to the column flanges contribute to the resistance of a fixed column base subjected to a moment combined with an axial load.

Further guidance

Further guidance on the design of fixed column bases is given in Reference 5. Its Annex provides detailed data on the design resistance of anchor bolts.

29.6 Holding-down systems

29.6.1 Holding-down bolts

The most generally used holding-down bolts are of grade 4.6, although 8.8 grade are also available. They are usually supplied square head, square shoulder, round shank and hexagon nut. Each bolt must be provided with an anchor washer (square hole to match the shoulder) or an appropriate anchor frame to embed in the concrete in circumstances of high uplift forces. In such cases the anchor frame may be

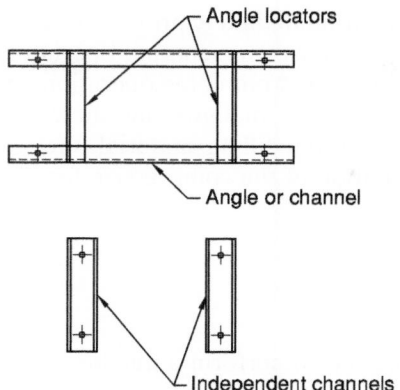

Figure 29.14 Typical anchorages for holding-down bolts

composed of angles or channels. Typical frames are shown in Figure 29.14. When long anchors are required, a rod threaded at both ends may be used and, in exceptional circumstances when prestressing is required, a high tensile rod (usually Macalloy bar) is adopted. In both these cases provision should be made to prevent rotation of the rod during tightening which may result in the embedded nut being slackened.

Corrosion of holding-down bolts to a significant extent has been reported in some instances. This usually occurs between the level of the concrete block and the underside of the steel baseplate, in aggressive chemical environments or at sites where moisture ingress to this level is recurrent. Fine concrete grout, well mixed, well placed and well compacted will provide the best protection against corrosion, but in cases where this is not adequate for the prevailing conditions, an allowance may be made in the sizing of the bolts or by specifying a higher grade bolt which provides a larger factor of safety against tensile failure in the event that some corrosion does occur.

29.6.2 Grouting

The casting-in of the holding-down bolts with adequate provision for adjustment requires that they are positioned in the concrete surrounded by a tube, conical or cylindrical, or a polystyrene former. Ideally, they should be 'waggled' at an appropriate time as the concrete sets, to ensure they may readily be moved during steelwork erection to accommodate the final column position. Alternatively, they may be bent after the concrete is set and prior to steelwork erection to suit the final column centreline. In cases where open tubes are used, they should be provided with a cap or cover to prevent the ingress of water, rubbish and mud. After erection, lining, levelling and plumbing of the frame the grout voids around the bolts should be cleaned out by compressed air immediately prior to grouting. The bolt grouting and baseplate filling should be done as two separate operations to allow shrinkage to take place. During the levelling and plumbing operations, wedges and packings are driven into the grout space. Before final grouting these should be removed, otherwise, after shrinkage of the grout filling material, they will become hard spots, preventing the even distribution of the compressive forces to the concrete base.

29.6.3 Bedding

Bedding materials are required to perform a number of functions, one of which is the provision of the corrosion protection referred to earlier in Section 29.6.1. As noted in Section 29.4.2.1, steel base plates are generally designed for a compression under the plate of $0.67f_{ck}$, where f_{ck} is the characteristic cylinder strength of the concrete base or the bedding material, whichever is less. The bedding material therefore transmits high vertical stresses including those resulting from the applied moment. The third function is to transmit the horizontal forces or shears resulting from wind or crane surge. It is clear therefore that the bedding material is a structural medium and should be specified, controlled and supervised accordingly.

For heavily-loaded columns or those carrying large moments resulting in high compressive forces the bedding should be fine concrete using a maximum aggregate of 10 mm size. The usual mix is 1:1¼:2 with a water-cement ratio of between 0.4 and 0.45 (this is not suitable for filling the bolt tubes as it is too stiff; a pure cement water mix has suitable flow properties and is usually used). It also has high shrinkage properties and should be allowed to set fully before continuing with the bedding. A cement mortar mix is often used for moderately-loaded columns. A suitable mix would be 1:2½. Weaker filling than this should only be used for lightly-loaded columns where the erection packs are left in position and may effectively transfer all the load to the foundation.

In order to facilitate the integrity of the bedding material, holes are cut in the baseplate of the order of 50 mm diameter or more, near to the centre of the plate, in order to allow the escape of air pockets and to ensure that the bedding reaches the centre.

A series of worked examples follows which are relevant to Chapter 29.

References to Chapter 29

1. British Standards Institution (2005) BS EN 1993-1-8:2005 *BS EN 1993-1-8:2005 Eurocode 3. Design of steel structures. Design of joints.* London, BSI.
2. Access-steel *Design model for simple column bases–axially loaded I section columns.* www.access-steel.com

3. Lesconarc'h, Y (1982) *Pinned column bases*. Paris, CTICM.
4. Access-steel *Design of simple column bases with shear nibs,* www.access-steel.com.
5. Access-steel *Design of fixed column base joints*, www.access-steel.com.

Further reading for Chapter 29

British Constructional Steelwork Association, the Concrete Society and Constructional Steel Research and Development Council (1980) *Holding-Down Systems for Steel Stanchions*. London, BCSA.

British Standards Institution (1997) *BS 8110 Structural use of concrete. Part 1: Code of practice for design and construction*. London, BSI.

British Standards Institution (2000) *BS 5950 Structural use of steelwork in building. Part 1: Code of practice for design – Rolled and welded sections*. London, BSI.

Capper, P.L. and Cassie, W.F. (1976) *The Mechanics of Engineering Soils*, 6th edn. London, E. and F.N. Spon.

Capper, P.L., Cassie, W.F. and Geddes, J.W. (1980) *Problems in Engineering Soils*, 3rd edn. London, E. and F.N. Spon.

Lothers, J.E. (1972) *Design in Structural Steel*, 3rd edn. Engleword Cliffs, NJ, Prentice Hall.

Pounder, C.C. (1940) *The Design of Flat Plates*. Association of Engineering and Shipbuilding Draughtsmen.

Skempton, A.W. and McDonald, D.H. (1956) The allowable settlement of buildings. *Proc. Instn Civ. Engrs*, 5, Part 3, 727–68, 5 Dec.

Skempton, A.W. and Bjerrum, L. (1957) A contribution to the settlement analysis of foundations on clay. *Géotechnique*, 7, No. 4, 168–78.

Terzaghi, K., Peck, R.B. and Nesri, G. (1996) *Soil Mechanics in Engineering Practice*, 3rd edn. New York, Wiley.

The Steel Construction Institute/British Constructional Steelwork Association (2002) *Joints in Steel Construction. Simple Connections*. Ascot, SCI/BCSA.

Tomlinson, M.J. (2001) *Foundation Design and Construction*, 7th edn. Harlow, Prentice Hall.

		Job No.		Sheet 1 of 1	Rev
The Steel Construction Institute		Job Title	Example 1: Foundations and Holding-down Systems		
Silwood Park, Ascot, Berks SL5 7QN Telephone: (01344) 623345 Fax: (01344) 622944		Subject	Chapter 29		
		Client	Made by	GWO	Date
CALCULATION SHEET			Checked by	DGB	Date

Problem

Design a simple base plate for a 254 × 254 × 73 UC to carry a factored axial load of 1000 kN

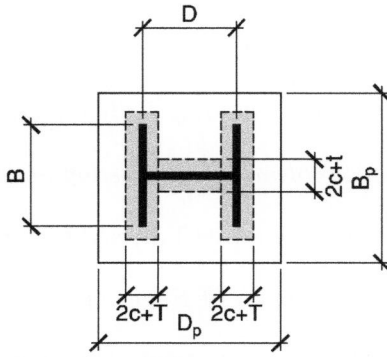

Design to Clause 6.2.8.2

BS EN 1993-1-8: 2005

Bearing strength of concrete/bedding (f_{jd}) for concrete class 40 is 26.7 N/mm²

$$Area\ required = \frac{1000 \times 10^3}{26.7} = 37453\ mm^2$$

Bearing area = hatched area

$= 4c^2 + (column\ perimeter) \times c + column\ area$

Therefore, $4c^2 + (254.6 \times 4 + 2 \times (254.1 - 28.4)) \times c + 9310 = 37453$

$4c^2 + 1470c - 28143 = 0$

$$C = \frac{-1470 \pm \sqrt{1470^2 + 4 \times 4 \times 28143}}{2.4} = 18.2\ mm$$

$$t_p = \left[\frac{6 \times f_{jd} \times c^2}{2 \times f_y}\right]^{0.5} = \left[\frac{6 \times 26.7 \times 18.2^2}{2 \times 275}\right]^{0.5} = 9.8\ mm$$

Use 300 × 300 × 15 grade S275 base plate

[A 10 mm plate would just suffice but leaves no allowance for corrosion and is not very robust for erection.]

	Job No.			Sheet 1 of 1	Rev
 **The Steel Construction Institute**	Job Title	Example 2: Foundations & Holding-down Systems			
Silwood Park, Ascot, Berks SL5 7QN Telephone: (01344) 623345 Fax: (01344) 622944 **CALCULATION SHEET**	Subject	Chapter 29			
	Client		Made by	GWO	Date
			Checked by	DGB	Date

Problem

Design a simple base plate for a 219 × 6.3 CHS to carry a factored axial load of 1010 kN

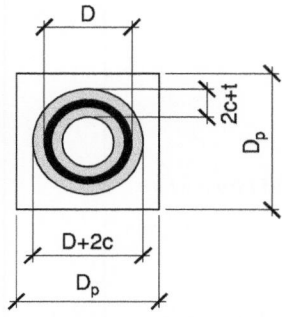

Assumed bedding material $f_{ck} = 40$, bearing strength $f_{jd} = 26.7 \, N/mm^2$

Area required $= \dfrac{1010 \times 10^3}{26.7} = 37828 \, mm^2$

Area of shaded annulus $= (2c + t) \, (D - t) \, \pi = 37828$

$(2c + 6.3)(219 - 6.3) = \dfrac{37828}{0} = 12041 \, mm^2$

$c = 25.1 \, mm$

$t_p = \left[\dfrac{6 \times f_{jd} \times c^2}{2 \times f_y} \right]^{0.5} = \left[\dfrac{6 \times 26.7 \times 25 - 1^2}{2 \times 275} \right]^{0.5} = 13.5 \, mm$

Use 280 × 280 × 15 plate, Grade S275

		Job No.		Sheet 1 of 1		Rev
The Steel Construction Institute		Job Title	Example 3: Foundations & Holding-down Systems			
Silwood Park, Ascot, Berks SL5 7QN Telephone: (01344) 623345 Fax: (01344) 622944 **CALCULATION SHEET**		Subject	Chapter 29			
		Client		Made by	GWO	Date
				Checked by	DGB	Date

Problem

Design a simple base plate for a 273 × 25 CHS S355 to carry a factored axial load of 6340 kN

Strength of bedding material, $f_{ck} = 40$

Bearing strength $= 26.7 \, kN/m^2$

Area required $\dfrac{6340 \times 10^3}{26.7} = 237455 \, mm^2$

Attempt to use simple effective area method $(2c + 25)(273 - 25)\,\pi = 237453$ from which $c = 139 \, mm^2$

Since $C > \dfrac{D - 2t}{2}$, model fails

Therefore, consider effective area shown by hatching, the bearing area is a circle of $(D + 2c)$ diameter

Then $(D + 2c)2 \, \pi/4 = 237453 \, mm^2$

From which $c = 139$

$$t_p = \left[\frac{6 \times f_{jd} \times c^2}{2 \times f_y} \right]^{0.5} = \left[\frac{6 \times 26.7 \times 139^2}{2 \times 255} \right]^{0.5} = 77.9 \, mm$$

Use $600 \times 600 \times 80$ plate Grade S275

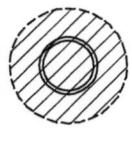

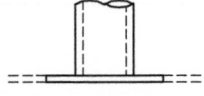

	Job No.		Sheet 1 of 3	Rev
The Steel Construction Institute	Job Title	Example 4: Foundations & Holding-down Systems		
Silwood Park, Ascot, Berks SL5 7QN Telephone: (01344) 623345 Fax: (01344) 622944 **CALCULATION SHEET**	Subject	Chapter 29		
	Client		Made by	GWO	Date
			Checked by	DGB	Date

Problem

Design a built-up base for the valley stanchion of a double bay crane shed, as shown below. The stanchion comprises twin 406 × 178 UB. Each UB is subject to a factored axial load of 239 kN and there is an overall moment of 707 kNm.

Base concrete is C25/30, for which $f_{jd} = 16.7\,N/mm^2$

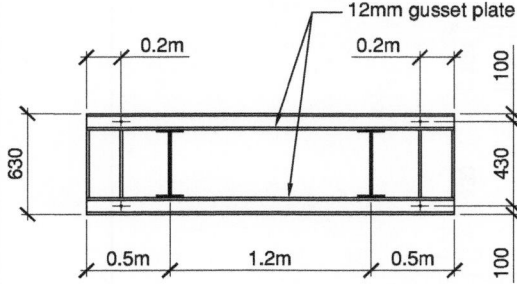

Loading

The overall applied forces resolve to:

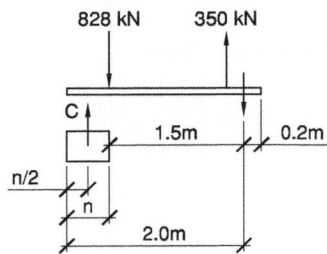

Different models may be postulated

Model 1: assume neutral axis is 0.4 m from compressive end of base plate

Taking moments about tensile bolt

$$828 \times 1.5 - 350 \times 0.3 = C\,(2.0 - 0.2)$$

$$C = \frac{828 \times 1.5 - 350 \times 0.3}{1.8} = 632\,kN$$

$$T = 828 - 350 - 632 = 154\,kN$$

Example Foundations & Holding-down Systems	Sheet 2 of 3	Rev

This gives a mean bearing stress over the entire base plate of $\dfrac{632 \times 10^3}{400 \times 630} = 2.5\,N/mm^2$, well within f_{jd} of $13.3\,N/mm^2$

(In practice, full width of base plate is unlikely to be effective to resist bearing compression).

Model 2: Take n as 0.2 m and taking moments about tensile bolts,
$828 \times 1.5 - 350 \times 0.3 = C\,(2.0 - 0.1)$

$C = 598\,kN$

$T = 598 + 350 - 828 = 120\,kN$

By inspection, model 1 gives a conservative value of compressive bearing that may be resisted by effective area acting off the stiffening shown.

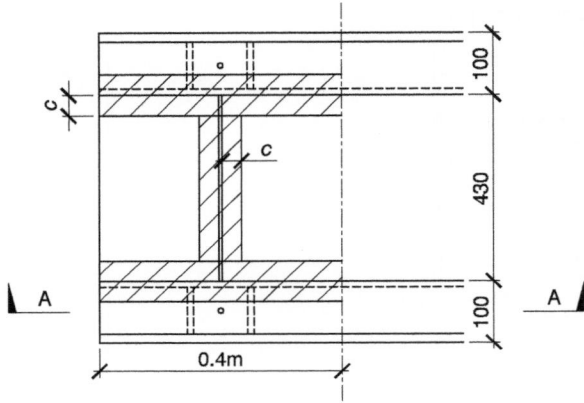

For simplicity, ignore bearing contribution from stiffeners and ignore overlap of bearing areas. Take channel effective t_w and gusset thickness as 10 mm

$$C(400 \times 4 + 430 \times 2) + 10(400 \times 4 + 430) = \frac{632000}{13.3}$$

$2460c + 12300 = 37844$

$c = 10.3\,mm$

Minimum $t = \sqrt{\dfrac{6.13.3.14.32}{2.275}} = 5.4\,mm$

Impractical – adopt 15 mm as practical minimum

Design of Channels and Gusset

$M = 632 \times 0.3 = 189.6\,kNm$

Use 2 no. $230 \times 90 \times 32$ PFC (M_{cx} 98 kNm)

These are satisfactory by inspection because they act compositely with base plate

Example Foundations & Holding-down Systems	Sheet 3 of 3	Rev

Design of Holding-down Bolts and Stiffening

Load/bolt = 77 kN

Use M24 Gr 4.6 bolts

Section A-A   **Section B-B**

Assume channel flange spans between stiffeners, but ignoring hole due to support from channel web

Channel flange capacity to resist point load w is given by:

$$\frac{W\ell^2}{8} = \frac{bd^2}{4} \times f_y$$

$$\frac{W \times 100^2}{8} = \frac{82.5 \times 14^2}{4} \times 0.275$$

W = 44 kN

Use additional cap washer 120 × 90 × 15 to ensure satisfactory strength and stiffness for tensile load path

Chapter 30
Steel piles and steel basements

ERICA WILCOX

30.1 Introduction

Steel piles are widely used in the construction industry and are versatile products which can be used economically for both temporary and permanent purposes. Steel piles are frequently used for quay walls and highways structures. However, this chapter concentrates on the use of steel piles in the context of buildings, specifically focussing on basement retaining walls. Their suitability for this purpose has been substantially improved with recent advances in piling plant and techniques, greatly reducing noise and vibration associated with sheet pile installation.

The use of steel piles for permanent basement retaining wall construction has increased since the 1990s, led to some extent by their use for underground car parks, see Figure 30.1, though they are not frequently used for pure bearing capacity in building foundations. This chapter describes the use of steel basement foundations. The subject is also covered in more depth in the SCI Publication *Steel Intensive Basements*.[1]

30.2 Types of steel piles

Steel bearing piles sections are generally H- or I-sections. These differ from other rolled sections by being of uniform thickness in order to maximise robustness for driving and to minimise reduction from corrosion, see Figure 30.2. Circular hollow section (CHS) may also be used.

As shown in Figure 30.3, standard sheet piles, suitable for basement wall construction, are available in two basic types, U- and Z-sections. These may be combined to create bearing piles, by welding or connector pieces, to form box piles, see Figure 30.4. The sheet piles can also be used as secondary elements in conjunction with steel bearing piles for increased stiffness or to provide improved bearing capacity.

Steel Designers' Manual, Seventh Edition. Edited by Buick Davison and Graham W. Owens.
© 2012 Steel Construction Institute. Published 2012 by Blackwell Publishing Ltd.

Figure 30.1   Permanent sheet pile basement car park in Bristol

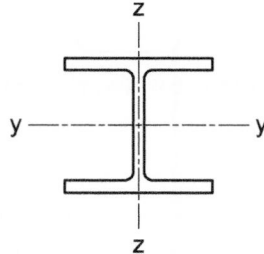

Figure 30.2   Universal bearing pile

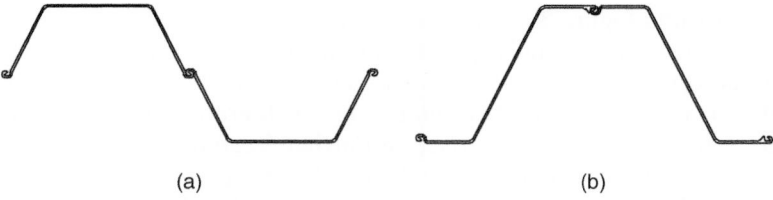

(a)   (b)

Figure 30.3   Sheet piles: (a) U-profile sheet pile, (b) Z-profile sheet pile

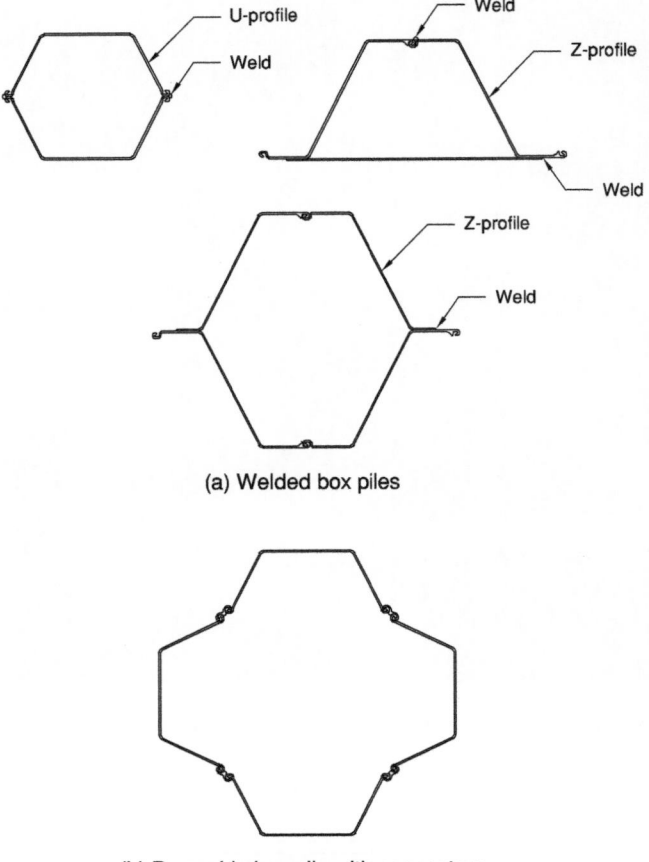

(a) Welded box piles

(b) Pressable box pile with connectors

Figure 30.4   Box piles

These primary or 'king pile' elements are commonly referred to as a combined wall, as shown in Figure 30.5. Some proprietary I-sections are now available which can be interlocked with Z-profile piles, using a specifically designed hot-rolled connector to produce a combined wall. 'High Modulus' walls are created by using elements of the same geometry adjacent to each other for example, I-sections used with Z-profile piles as shown in Figure 30.6 or tube-to-tube sections.

The steel piles most usually specified in the UK are hot-rolled sections, manufactured in accordance with EN 10248. The sections widely available are described in the *Piling Handbook*[2] and other manufacturers' literature; key data may be downloaded from the web, as summarised in Further Reading. These sections are available in strength grades ranging from S240GP to S430GP.

Cold-rolled sheet pile sections (manufactured to EN 10249) are becoming more prevalent in general sheet pile usage. Typically, the sections are slightly lighter and

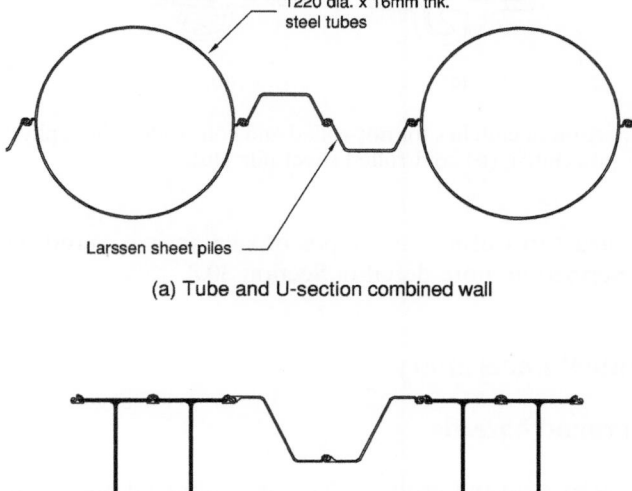

1220 dia. x 16mm thk.
steel tubes

Larssen sheet piles

(a) Tube and U-section combined wall

(b) HZ wall system

Figure 30.5   Combined wall systems

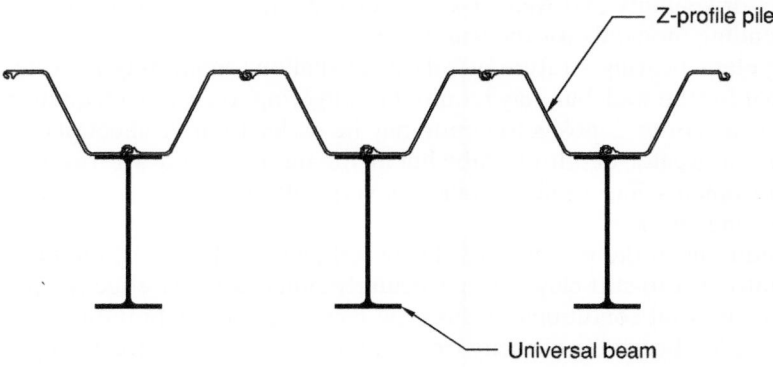

Z-profile pile

Universal beam

Figure 30.6   High modulus walls

thinner than for the equivalent hot-rolled section modulus, which can be an advantage in some respects, but may adversely impact durability. In addition, the method by which the clutches are formed by the cold-rolling process results in a more flexible connection, as shown in Figure 30.7 below. This influences the overall stiffness of the wall, watertightness and the dimensional tolerance during installation. They are therefore of limited suitability for basement construction.

(a)   (b)

Figure 30.7   Comparison of clutches for hot-rolled and cold-rolled sheet piles: (a) hot-rolled
sheet pile clutch, (b) cold-rolled sheet pile clutch

The design of structures using both types of section is covered by Eurocode 3
Part 5,[3] and is described in more detail in Section 30.8.

30.3 Geotechnical uncertainty

30.3.1 Typical ground hazards

Many hazards may exist in the ground. They are discussed here in the context of
their impact on a steel basement project. Most of the issues are discussed further in
later sections.

Soil stratigraphy

Soil strength has an impact at both ends of its range. Broadly speaking, stronger
soils affect driveability and weaker soils create greater earth pressures and hence
greater bending moments for the wall to resist.

A competent bearing stratum at relatively shallow depth may provide a good
embedment for the wall, but may require pre-augering to ensure adequate penetration. However, softer deposits to depth may be easier to drive sheet piles into, but
will require a greater depth of embedment for the piles and will result in larger
ground movements. Sheet piles installed in very soft deposits may require an additional buckling check.

Soil conditions in the presence of clayey soils, where the toe of the piles can be
founded into firm-to-stiff clays, are particularly suited for cost effective steel sheet
pile basements. Soil conditions of this type facilitate silent vibration-free pressing
and provides load-carrying capacity from skin friction on the embedded part of the
pile below formation level.

There are some soil types which are not generally suitable for steel sheet piling,
without particular measures. Chalk and clay-with-boulders are two particular ground
conditions that may be considered unsuitable. Chalk structure will tend to break
down when a sheet pile is driven through it and form a layer of fine soil immediately
adjacent to the piles, resulting in reduced friction. Boulder clay, being a glacial soil
which contains large lumps of rock within a clay matrix, is an example of clay-with-
boulders though there are others with similar conditions. The boulders can cause
deflection and hence potential declutching of piles or early refusal if the pile reaches
a boulder during installation which can't be driven through or pushed to one side.

If peat is identified on site, the soil aggressivity in relation to corrosion should be considered, as humic acid can cause low pH values. Impact on driveability should also be considered as the fibrous nature of the soil can affect the suitability of certain methods.

Groundwater

If the basement is below groundwater level, some level of water resistance will generally be required, depending on the quality of basement space specified. The direction of groundwater flow may need to be considered, if the basement is likely to act as a dam and cause an obstruction to the local groundwater flow regime. The impact of groundwater changes should also be considered; local pumping, or cessation of pumping on adjacent sites may have an impact. Groundwater levels may also fluctuate seasonally or in response to rainfall events, and the potential for emergency conditions, such as flooding, should be considered.

Contamination / chemical environment

Chemically aggressive soil and/or groundwater, from contaminated sites and some naturally occurring conditions may cause corrosion, with an impact on steel pile durability. The presence of ground gas or soil vapours may also have an impact on design of any underground space. Piling through contaminated ground may require specific risk assessments to be undertaken because of the risk of transport of contamination through the ground.

Site history

Previous site development often leaves obstructions in the ground, such as old foundations, sewers and other services, which affect driveability and may cause excessive noise and vibration during driving. Even if successfully driven-through, they may cause damage or declutching to piles during installation.

Sites in areas subject to bombing during the war, or with other reasons for risk of buried ordinance may require specialist studies and investigations to mitigate the risk of unexploded ordnance during ground investigation or driving of steel piles.

Location constraints

Sites in built-up areas may have restrictions on noise, which will influence installation methods. Proximity of adjacent buildings or sensitive infrastructure, including tunnels and buried utilities may limit allowable ground movements or vibrations. Areas subject to seismic activity may require special consideration.

30.3.2 Mitigating geotechnical risk

It is clear from the previous list of hazards that any design and construction on or within the ground involves a higher level of uncertainty compared with the relatively predictable properties of a factory- or site-manufactured material, such as steel or concrete. Even if no specific hazards are present, ground conditions are rarely uniform and variations are possible over small distances.

Unforeseen ground conditions can present a significant risk to construction programme and cost, as well as contributing to failure of constructed walls and foundations, therefore adequate investment in mitigating this risk is essential for any project. Proceeding with overly conservative assumptions to deal with uncertainty without appropriate investigation can also lead to expensive and inefficient design and even buildability problems; therefore an appropriate level of ground risk mitigation is required. This process of ground risk mitigation should be carried out by a suitably experienced geotechnical specialist.

A three stage approach is always recommended:

- desk study
- site reconnaissance
- ground investigation.

These studies may also be augmented by site trials.

30.3.3 Desk studies and site reconnaissance

Desk studies and site reconnaissance visits are essential to establishing the likely ground hazards which need further investigation. Typically a report is compiled at the start of any project, based on a desk-based study of available information, from a variety of sources. These may include historical maps, plans and aerial photographs, utilities and buried services information, geological data, other public records and data collected for adjacent sites. The availability of information will vary from site to site; typically developed urban areas will have a wealth of information available, whereas rural sites may not. This desk-based study should be augmented by a site reconnaissance visit to identify particular hazards which may have visible signs or indicators.

30.3.4 Ground investigations

A ground investigation should be tailored to provide suitable information to mitigate the likely risks identified during the desk study. The ground investigation also provides appropriate information for the type, depth and location of the basement/

foundation elements. This chapter does not explore the various methods and techniques available for ground investigation. A useful summary of the importance of appropriate ground investigation is available in BRE Digest 472 'Optimising Ground Investigation'.[4] The Code of Practice covering Ground Investigation is BS 5930[5], which has been partially updated by BS EN 14688.[6]

It should be recognised that, during any investigation, only a small proportion of the site can be investigated. Therefore, whilst appropriate investigation is essential, it provides no absolute certainty of the ground conditions on site. This fact needs to be recognised in the choice of design parameters and safety factors. It also means that it is essential to observe and record ground conditions encountered during construction, when a much larger percentage of the ground is explored.

30.3.5 Site trials

Further mitigation of geotechnical uncertainty is possible through site trial drives, load testing and verification. Site trials can be particularly useful for:

- proving that a particular section and driving method are appropriate
- measuring noise and vibration during installation, particularly to mitigate impact on neighbours or owners of sensitive infrastructure
- to gather information to inform the driving contract
- load testing.

However, if load testing of trial driven piles is to be undertaken, it should be recognised that there will be a significant difference between the performance of a set of sheet piles at the time of test, prior to excavation, compared with the in-service configuration of a basement wall.

30.4 Choosing a steel basement

30.4.1 Holistic basement design

Utilising underground space is often crucial for optimising land-usage, providing car parking, storage or other space where natural lighting is not required. Basements can be integral to the building above or be situated on their own under open spaces or other infrastructure. Shallow basements tend to comprise a single underground storey, whereas deep basements may have a number of levels.

The intended use and layout of the basement will influence the selection of sheet piles. Overall depth of basement, number and height of basement storeys, ability to use floor plates as permanent props, watertightness requirements and aesthetics are all factors. Steel is more likely to be suitable for shallow basements, however multi-level basements in steel are becoming more common.

A basement design comprises the coordination of various elements. Generally these consist of a retaining wall, floors and building foundations and may also include building cores, drainage and architectural detailing. Whilst this chapter focuses primarily on the use of steel retaining walls, the consideration of the other elements is essential for a holistic design.

30.4.2 General factors

There are many considerations in choosing the appropriate design and construction material for a basement wall. Whatever the construction materials, embedded walls allow the basement footprint to be maximised within a site, compared with concrete gravity walls constructed within battered excavation or even within a temporary sheet pile wall.

The benefits of the use of steel sheet piling include:

- faster construction than any form of reinforced concrete walls
- a thinner wall for the equivalent load capacity of a reinforced concrete wall, saving space and maximising footprint
- installation close to the boundary of the site, maximising usable building space
- suitable for all soil types
- embedded wall means no requirement to excavate for wall foundations
- the steel components are factory quality as opposed to site quality
- high ductility that can reduce bending stresses and soil reactions
- can easily be made aesthetically pleasing
- can be placed in advance of other works
- immediate load-carrying capacity in many cases, allowing basement excavation to proceed without delay once wall is installed
- water-resistance if welded or other water excluding measures implemented, which can be tested and proven to client prior to handover.

However, the choice should be made by considering all relevant factors. These include site specific constraints, design and construction issues, and sustainability.

Table 30.1 shows a comparison between scheme designs for a basement car park wall, illustrating the potential benefits of steel over concrete predicted by the designer, provided that the greater deflections of the steel scheme were acceptable.

Table 30.1 Comparison of scheme design for two-storey basement constructed using top-down sequence, in soft clay over gravel and mudstone

Section	PU32 equivalent	800 mm thick reinforced concrete D-wall
Stiffness	$0.15 \times 10^6\,\mathrm{kNm^2/m}$	$1.5 \times 10^6\,\mathrm{kNm^2/m}$
Section thickness	452 mm	800 mm
Max deflection	56 mm	35 mm
Max Bending Moment	800 kNm	1200 kNm
Estimate construction programme for wall installation	2.5 months	4.5 months

CIRIA C580[7] summarises guidance on the typical applications of different types of embedded retaining walls and indicates that the typical sheet pile wall height range is up to 5 m in cantilever and 4–20 m for a propped wall. These dimensions will vary depending on the various site-specific factors as discussed in the following sections.

In terms of relative costs, CIRIA C580 reports that, for wall thicknesses less than 650 mm, a sheet pile wall is generally the cheapest of the permanently watertight options available, being slightly cheaper than a hard/firm secant pile wall and approximately half the cost, per m^2, of an equivalent diaphragm wall.

30.4.3 Site-specific factors

In the choice of steel over reinforced concrete as a construction material in a building, many of the factors are often very site-specific.

A number of issues govern the suitability of a particular site for the use of steel sheet piling for basement construction; these are the same issues which are considered at the preliminary stages of a project, in the processes described in Section 30.3.3 above. These site factors have various impacts on the construction, and service/performance of a steel basement, such as:

- driveability
- ground movements
- watertightness
- noise and nuisance
- durability/corrosion.

30.4.4 Construction sequence

Where a permanent sheet pile basement is being used, the installation of the perimeter wall will generally be one of the first activities on site, following any demolition or site clearance works required. Dewatering may then commence if excavation is to proceed below the water table.

Following the wall installation, there are three broad sequences possible, cantilever, 'bottom-up' or 'top-down' construction, as follows.

Cantilever

The full depth of excavation is completed with the wall acting as a cantilever. The permanent basement structure is then installed inside the open excavation.

This provides a clear excavation with no obstructions, but may lead to unacceptable deflections, or an uneconomic design for the permanent works and hence is generally only practical for shallow basements.

Bottom-up

The excavation proceeds with temporary supports being installed to provide support to the excavation. Horizontal struts, anchorages, soil berms or raking props may be used to provide the temporary support. Once excavation is complete, the permanent structure is constructed within the propped excavation.

This type of sequence allows economies of sheet pile size and limits deflections; however, large spans may require substantial temporary props, which can result in a congested construction site.

Top-down

A small excavation is made and then the permanent ground level is cast, leaving access holes within the slab. Excavation of the basement proceeds, using the permanent slab as a prop. Further permanent slab levels may be installed as required and the final level of construction is the base slab.

This type of sequence allows economies of sheet pile size and limits deflections, and can be used in conjunction with plunge columns to speed up overall construction programme, as it permits superstructure and basement construction to proceed concurrently. However, excavation underneath permanent slabs can be slow and awkward, and the health and safety aspects of working conditions, including provision of adequate ventilation and emergency egress, need careful consideration.

Combinations of these sequences can also be used to suit the particular project constraints and opportunities. The impact of the construction sequence selected is discussed further in Section 30.5.1.

30.4.5 Watertightness and basement grade

Steel is inherently impervious and in order to achieve water tightness, only the interlocks and base slab connection require any additional measures.

BS 8102[8] defines the requirements in terms of water and vapour ingress permitted for different grades of basement. CIRIA 139[9] is a useful design guide for basement water tightness and the ICE have produced a document 'Reducing the risk of leaking substructure: A Clients guide'[10] which provides useful guidance to both clients and designers what the different grades of basement mean in practical terms and the methods, risks and processes involved in achieving them.

BS 8102 has been revised in 2009, reducing the defined categories from four to three. Table 30.2 compares the two classifications, as the design guides refer to the previous version of the standard.

Table 30.2   Summary of basement grades

Grade	BS8102:1990 Description and usage	BS8102:1990 Performance level	BS8102:2009 Typical usage	BS8102:2009 Performance level
Grade 1	'Basic Utility' Car parking, non-electrical plant-rooms, workshops	Some seepage and damp patches tolerable	Car parking, non-electrical plant-rooms, workshops	Some seepage and damp areas tolerable, dependent on intended use Local drainage may be required to deal with seepage
Grade 2	'Better Utility' Workshops and plant requiring drier environment, retail storage areas	No water penetration but moisture vapour tolerable	Workshops and plantrooms requiring drier environment, retail storage areas	No water penetration acceptable. Damp areas tolerable; ventilation may be required
Grade 3	'Habitable' Ventilated residential and working areas, leisure centres	Dry environment	Ventilated residential and commercial areas, including offices, restaurants etc.; leisure centres	No water penetration acceptable. Ventilation, dehumidification or air conditioning necessary, appropriate to intended use
Grade 4	'Special' Archives and stores requiring special environment	Totally dry environment	Not used	

The guide defines the basic types of protection available for providing protection from water and vapour ingress, as follows:

- Type A Tanked protection
- Type B Structurally integral
- Type C Drained protection.

Welded steel sheet piles can provide Type B protection, and can be used with additional measures, such as a drained cavity, to provide Type C protection.

30.4.6 Types of support

As discussed in 30.4.4, permanent structure may be used to support sheet pile walls during construction, or temporary support may be required, depending on the construction sequence employed. The following types of support can be used:

- floor slab (permanent)
- prestressed anchorage, generally grouted tendon (temporary or permanent);
- non-prestressed anchorage, including grouted, 'dead man' or screwed anchor (temporary or permanent)

- strut across excavation (temporary)
- raking strut (temporary)
- berm (temporary).

When using the floor slab as a prop, provisions for service risers and access stairs/lifts are generally unavoidable, therefore care needs to be taken regarding the positioning of openings within the slab, so that the propping action is not overly compromised. Localised walings or transfer beams may be required where these openings occur at the perimeter of the floorplate.

Anchorages are ways of providing support whilst maintaining a clear excavation. They require space outside of the footprint of the basement, which may not be feasible in congested locations. The method of providing resistance to pull-out can vary, including grouted sockets, 'dead man' walls and other proprietary systems. More flexible than slabs or struts, the acceptability of deflections in relation to the serviceability limit state needs to be considered.

Struts across excavations are generally tubular steel and can provide a much stiffer prop than anchorages, though they may be quite substantial in size. Generally, properly designed struts are the most efficient way of providing temporary support, as they utilise the earth pressure behind the opposite wall to provide a reaction. Once installed, they do not necessarily need to be moved as excavation progresses, if the construction sequence has been carefully planned. Temperature effects need to be considered, especially where props are long. Sway may need to be considered if there is substantial difference in level retained behind opposite walls.

Where sufficient resistance can be obtained from propping off the excavated ground level, raking props can be used. These have the advantage of leaving clear space in the centre of the excavation for other works to proceed and may be required if the basement is very wide. However, they are less efficient than horizontal props and require a competent foundation. They may also need to be re-sited as excavation progresses, though they can be used in conjunction with berms to optimise their use.

Berms provide support by leaving a wedge of soil around the perimeter of the basement as excavation progresses in the middle. They provide temporary support, and can be a useful measure until other temporary or permanent measures can be installed. The use of berms depends on the ground conditions, and they are generally best suited to stiff clays, which may have good short term strength. Stability calculations will be required to establish suitability and required geometry.

Further details and references for support design are included in Sections 30.7 and 30.8.

30.4.7 Fire resistance

The requirements for fire resistance should be considered at the planning stage. The principles of fire engineering apply to a steel basement in the same way as for any other part of a building, including fire loading, geometry, thermal properties of the structure and ventilation.

Fire load, the measure of combustible material within a building space, may be relatively high for basements, given their frequent use for storage or vehicle parking.

Active fire protection systems, such as automatic sprinklers and other fire-fighting measures, can be used prevent the fire spreading.

Passive protection can also be used, which are measures to reduce the temperature affecting structural elements. As for building columns this can include intumescent and fire retardant coatings to the face of the sheet piles. Soil behind a steel sheet pile retaining wall will also have a beneficial effect, reducing the overall temperature and slowing the rate of temperature increase.

30.5 Detailed basement design: Introduction

30.5.1 Overall design approach

The loading conditions on a basement retaining wall will generally be very different during construction from those acting on the completed building. It can therefore sometimes be difficult to determine the critical case, and analysis of various design cases will be required. There are three general approaches and the applicability of each tends to be governed by the client's procurement strategy.

For a multi-level basement, designing for the permanent condition will generally provide a leaner, more efficient sheet-pile design, however, it may require onerous and expensive temporary works by the contractor, to ensure that the loading conditions on the wall during construction are no worse than the service condition.

Optimising the design for the construction phase reduces the need for extensive propping or other temporary works, which may cause site congestion; this will speed up construction. This means that the wall is sized for more onerous conditions during construction, which may lead to a wall which is oversized for the permanent condition and therefore is more costly in terms of the length and section size of sheet pile installed.

A third option can be taken, where a design is based on 'probable' soil parameters, instead of more conservative 'characteristic' parameters. The performance of the wall is then monitored during construction, and reviewed against pre-determined acceptable behaviour limits. Contingency plans are in place for implementation if review of monitoring indicates that the wall performance is likely to exceed acceptable limits. This is known as the 'Observational Method'. More information can be found in the CIRIA Report R185 'Observational method in ground engineering: principles and applications'[11] and in Section 2.7 of Eurocode 7 Part 1.[12] Whilst this option can provide significant materials and construction cost savings, it does require close construction control and may present a risk to programme and cost certainty, if contingency measures are eventually required.

The adoption of the second and third of the three strategies presented above generally requires early contractor involvement of some description, either in an advisory capacity at design stage, or through a Design & Build contract. Alternatively, there needs to be recognition that the contractor may request design changes at construction stage, if the designers assumptions over construction sequencing do not match the contractor's preferred method of working.

30.5.2 Design to Eurocodes

Until the advent of the Eurocodes, geotechnical design of retaining walls was carried out to BS 8002, with additional reference to CIRIA 104 'Design of retaining walls embedded in stiff clay'.[13] An updated CIRIA Guide C580 'Embedded retaining walls – guidance for economic design' was published in 2003[7]. The design approach methodology contained in CIRIA C580 was based on the draft Eurocode 7 in development at the time. Whilst the concepts are similar, the actual design approaches set out in the final Eurocode 7, published in 2004, differ in the detail.

Structural design of retaining walls is carried out in accordance with Eurocode 3 Design of Steel Structures Part 5 Piling,[3] replacing BS 5950 'Structural Use of Steelwork in Building'.

In the foregoing, geotechnical design means the analysis of the interaction of the wall with the ground and hence the determination of loads, moments and deflections on the wall system and definition of stable geometry, including embedment depth of the wall. Structural design means the sizing and specification of the wall elements, i.e. wall section, props and connections. These processes are necessarily interactive and iterative.

30.5.3 Geotechnical limit states

In Eurocode 7 Part 1,[12] Section 9 sets out the design of retaining structures. The Eurocode describes a limit state approach and requires, as a minimum, the consideration of failure in the following limit states:

- overall stability
- structural elements, or connections between structural elements
- combined failure of ground and structural element
- hydraulic heave and piping
- movement-induced damage or collapse of adjacent structures or services
- unacceptable leakage
- unacceptable transport of soil material
- unacceptable change in groundwater regime.

Some of these are illustrated in Figure 30.8.

1) Overall stability	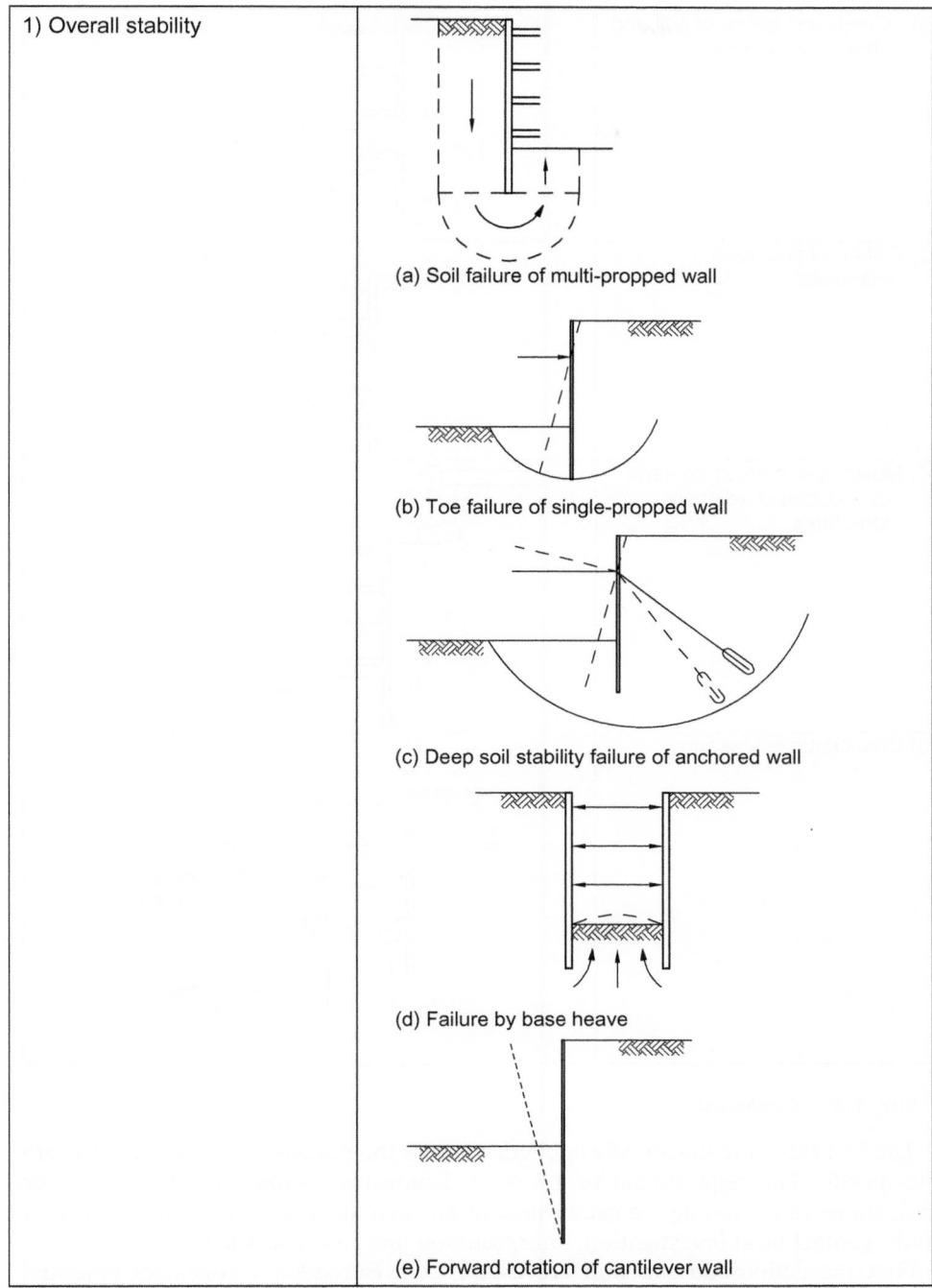

Figure 30.8   Possible failure modes in basements

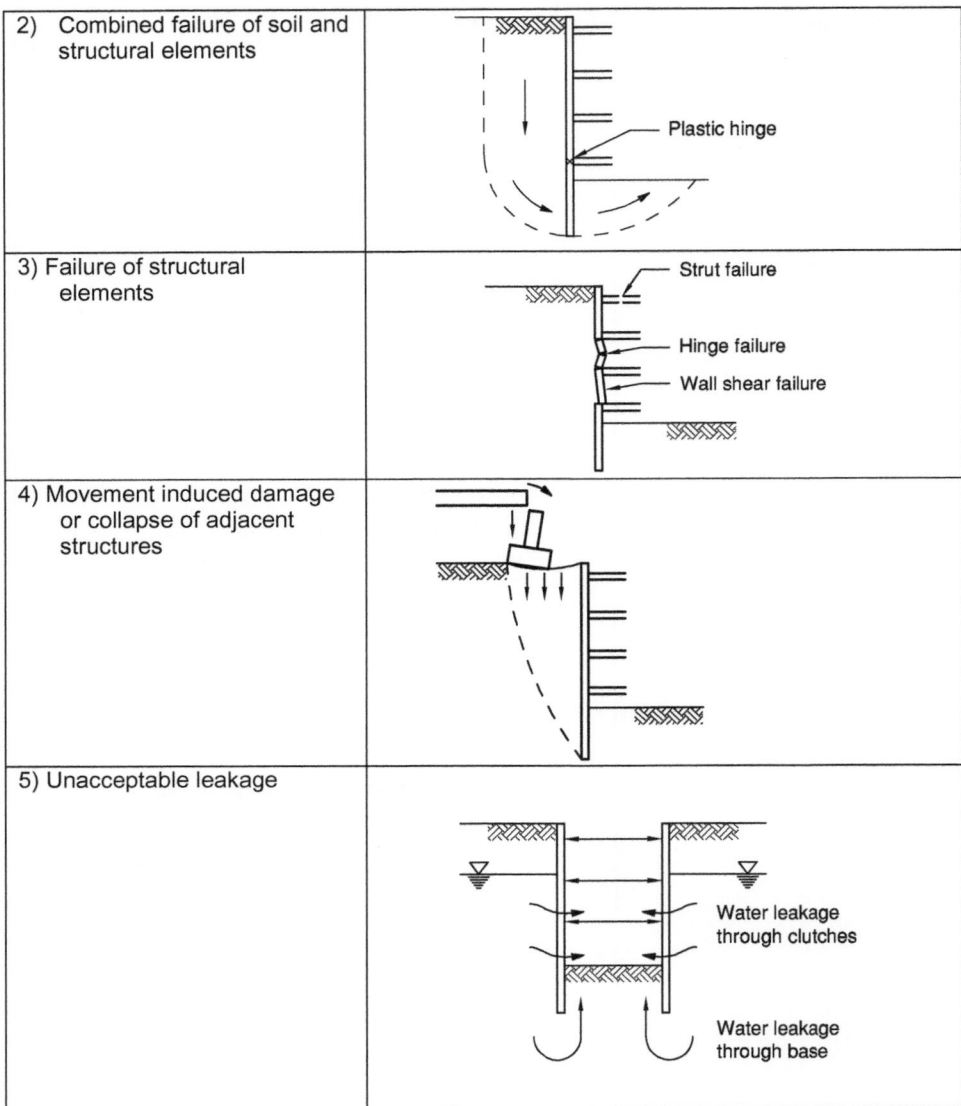

Figure 30.8   (Continued)

The first three are universally applied, whereas the remaining five are much more site-specific. The requirement to assess or demonstrate sufficiency of these latter limit states can generally be established at an early stage, based on the early desk study, geotechnical investigation, interpretation and risk assessment.

First the actions on the wall are established. Eurocode 7 considers potential actions for many types of retaining wall, however, those most relevant to a steel basement wall are as follows:

- weight of backfill material
- ground surcharge behind the wall
- weight of water
- seepage forces.

Vertical loading from building elements, such as walls, columns or floor slabs also require consideration.

Eurocode 7[12] then requires designers to consider the geometry of the situation, including ground levels and water levels. Here the levels of slabs and props should be defined.

Finally, the 'design situations' are considered. All of these may have an element of variation with space, as conditions vary across the site; or with time, as changes occur during construction on a daily basis or as longer term conditions develop. The following have relevance to steel basement retaining walls:

- variability of the soil profile and properties
- variability of the groundwater
- variation in the structural form
- variation in loading actions, both surcharge and structural loading
- impact of ground movement due to other sources
- loss of section due to corrosion.

30.5.4 Analysis tools

Although simple structural forms, such as cantilever walls or single-propped walls, can be analysed by hand using limiting equilibrium methods, computer software programmes are usually used.

For global stability of the wall, it is necessary to ensure that there is sufficient embedment. Simple programmes using fixed method (for cantilever walls) or free-earth support method (for propped walls) may be adopted.

To determine the bending moments in the wall and forces in the props, and also to estimate deflections of the wall, soil-structure interaction software is used. This generally uses either a subgrade reaction or a spring model, or a quasi-finite element approach.

Soil–structure interaction analysis methods predict the earth pressure distribution acting on the design configuration of the wall. As the relative stiffnesses of the wall and the soil are modelled, the earth pressure profiles predicted using these methods are much more realistic than the simplified methods used for global stability and compare favourably with actual earth pressures.

Figure 30.9 shows a typical earth pressure profile for an anchored wall obtained from a soil–structure interaction analysis.

The benefits of soil-structure interaction modelling over simplistic limiting equilibrium methods are as follows:

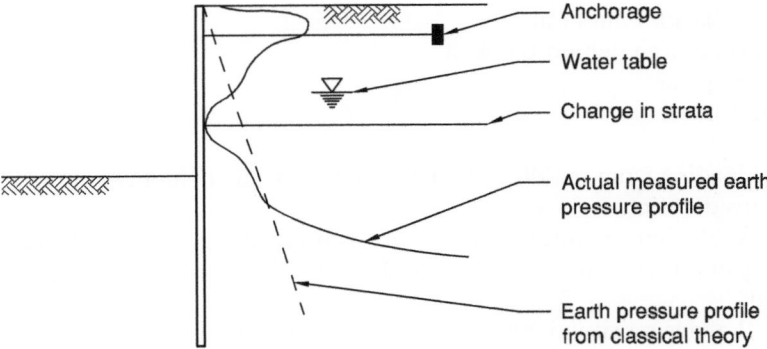

Figure 30.9 Actual horizontal earth pressure distribution for a flexible sheet pile abutment

- soil movements can be estimated
- the effects of construction sequencing can be modelled
- wall flexibility and soil stiffness effects can be modelled.

For more complex problems, for example where three-dimensional effects are influential, or where ground movement assessment of adjacent sensitive structures is critical, full 2D or 3D finite element analysis may be carried out.

Some popular computer programmes used for embedded retaining wall analysis are:

- Geocentrix ReWaRD
- Oasys STAWAL
- Oasys FREW
- Geosolve WALLAP
- Plaxis (Finite Element).

30.6 Detailed basement designs: Selection of soil parameters

30.6.1 General

As described in Section 30.3, there are no absolute intrinsic values that can be attributed to a particular soil parameter for retaining wall design. Variability in ground conditions requires that the potential range of values for any soil parameter be considered. Eurocode 7[12] states:

> *The characteristic value of a geotechnical parameter shall be selected as a cautious estimate of the value affecting the occurrence of the limit state.*

Selection of the appropriate soil parameters should also take account of the type of analysis being undertaken. Different parameter sets may be applicable for the different limit states and for designs using the Observational Method. For cohesive soils, the short term/drained or long term/undrained stages of soil-structure interaction analyses will require different parameters. Also, different strain ranges may also influence parameter selection.

The determination of soil parameters generally requires a combination of methods to determine appropriate values, including:

- direct measurement in situ or in the laboratory
- indirect measurement by correlation with in situ tests or laboratory tests
- local experience
- published data
- back analysis.

Within the spread of data which is likely to be obtained from the above approach, Eurocode 7 requires that a 'characteristic' value be used. This is generally a value less optimistic than an average or most probable value. The characteristic value is explained in more detail in Section 2.4.5.2 of the code.

The following sections cover particular soil properties which are generally used in retaining wall design.

30.6.2 Angle of shearing resistance

Selection of the appropriate effective angle of shearing resistance (φ') values for the soil in different circumstances and design cases is required.

Three different values of φ' may be identified for a particular soil, in order of magnitude: peak, critical state and residual. The more usual choice is between peak angle of shearing resistance φ'_{peak} and the critical state value of shearing resistance, φ'_{crit}. Generally the use of the much lower residual value, φ'_r is unnecessarily conservative, unless the presence of pre-existing slip planes in the soil around the wall needs to be considered as an ultimate limit state failure mechanism.

Determination of angle of shearing resistance can be undertaken by means of laboratory testing. However it is more common to use established empirical correlations with in situ testing, such as the Standard Penetration Test (SPT), and other laboratory testing, including grading and index testing. Local knowledge and experience and back analysis are also used frequently to determine appropriate values.

BS 8002[14] provides conservative correlations for estimating φ'_{crit} for clay soils (based on plasticity) for granular soils (based on angularity and grading) and for weak, fragmented rock (based on rock-type description). The standard also includes an estimation of φ'_{peak} for granular soils by adding a contribution based on the SPT 'N' value correlations. More information on correlations with SPT can be found in CIRIA Report 143.[15]

30.6.3 Soil stiffness

Soil stiffness is an important component in soil-structure interaction analyses and is strain dependent. Typically, a retaining wall operates in what is referred to as the 'small strain' range, which results in greater stiffness than for the 'large strain' range. Determination of soil stiffness can be carried out directly by some in situ methods, for example using the self-boring pressuremeter and by laboratory testing, as well as by correlation with other tests such as SPTs and Cone Penetration Tests. Various correlations exist, some of which are summarised in Section 4.6 of SCI Publication 187[16] and again in CIRIA Report 143.[15] Some correlations have a wide range and therefore appropriate values should be selected carefully based on the anticipated strain conditions. Stiffness can also vary with depth in a particular stratum.

30.6.4 Weight density

Weight Density, γ, (also known as unit weight) is used in calculating the overburden pressure and hence the earth pressures against a wall. Although this property can be tested in the laboratory, the in situ values can be hard to replicate once the sample is disturbed, therefore obtaining a representative result on granular material is unlikely.

Published values are the more usual method of determining the appropriate weight density. These suggest appropriate values based on grading and qualitative assessment of in situ density, which can generally be determined from borehole descriptions and in situ testing.

BS 8002 recommends values for various soils; extracted values are included in Table 30.3. For granular soils, the difference between saturated weight density and bulk weight density needs to be considered for soils above and below the water

Table 30.3 Weight density of soils

Material	Bulk weight density (kN/m³)	Saturated weight density (kN/m³)
Loose gravel	16	20
Dense gravel	18	21
Loose, well graded sand and gravel	19	21.5
Dense, well graded sand and gravel	21	23
Loose, coarse or medium sand	16.5	20
Dense, coarse or medium sand	18.5	21.5
Loose fine or silty sand	17	20
Dense fine or silty sand	19	21.5
Soft clay	17	17
Firm clay	18	18
Stiff clay	19	19
Hard clay	20	20

table respectively, as water replaces air in the voids between the soil particles. For cohesive soils the weight density change is negligible.

30.7 Detailed basement design: Geotechnical analysis

30.7.1 Effective stress analysis

For long-term design, drained effective stress analysis is used. Active soil pressure, σ'_a, and passive soil resistance, σ'_p, at a given depth are both functions of the effective vertical pressure at that depth and the strength of the stratum being considered. Limiting values of effective horizontal active and passive earth pressure at any particular depth are given by the following equations:

$$\sigma'_a = k_a \left(\int_0^z y\,dz - u + q \right) - 2c'\sqrt{k_a}$$

$$\sigma'_p = k_p \left(\int_0^z y\,dz - u + q \right) + 2c'\sqrt{k_p}$$

where

$\sigma_a{}'$	is the effective active pressure acting at a depth in the soil
$\sigma_p{}'$	is the effective passive pressure acting at a depth in the soil
γ	is the bulk weight density (saturated weight density if below water level)
z	is the depth below ground surface
u	is the pore water pressure
q	is any uniform surcharge at ground surface
c'	is the effective shear strength of the soil

k_a and k_p are earth pressure coefficients.

30.7.2 Total stress analysis

The total horizontal active and passive earth pressures acting against the wall are given by:

$$\sigma_a = \sigma_a{}' + u \text{ and } \sigma_p = \sigma_p{}' + u$$

where u is the pore water pressure.

This 'undrained' behaviour is considered only for clay soils in the short term situation, such as construction stages. It may be appropriate for temporary works to be designed solely for short term undrained conditions. However, permanent sheet pile walls should consider both short and long term conditions.

30.7.3 Earth pressure coefficients

Earth pressure coefficients k_a and k_p are functions of the effective angle of shearing resistance of the soil, φ', the wall/soil interface friction, δ, and the angle of the ground surface.

Annex C of Eurocode 7[12] includes charts for determining values of earth pressure coefficient and a method for obtaining the values numerically.

30.7.4 'At rest' earth pressures

Where no movement of the wall takes place, 'at rest' earth pressures are applicable, which utilises the earth pressure coefficient, K_0, defined for level ground in normally consolidated soils as:

$$K_0 = 1 - \sin\varphi'$$

For over-consolidated soils, K_0 can be in excess of 1.0 and may vary with depth. K_0 profiles may be obtained from published case histories for particular strata, or derived from in situ measurements.

Steel sheet piles are generally too flexible for the use of at rest earth pressures to be a typical design load condition, however, there may be specific cases where the at rest value is appropriate. The value is used for initialisation stages of soil-structure interaction analyses, before the wall is installed.

30.7.5 Wall / soil interface friction

The maximum value of the wall/soil interface friction for steel sheet piling in granular soils is generally taken as being $\delta = 2/3\varphi'$, however consideration should be given to lowering this value if low friction conditions exist or friction-reducing coatings are to be applied.

For walls carrying substantial vertical load, movement of the wall relative to the soil may eliminate the beneficial effects of wall friction and $\delta = 0$ should be assumed when calculating earth pressure coefficients in this case. Similarly, where driving assistance measures have been used to deliberately reduce soil friction in order to facilitate installation, $\delta = 0$ should be assumed.

For sheet piles driven in clays, the interface friction and adhesion for undrained analysis should not be relied upon in the very short term as these are disrupted by the driving process and only develop over time. This may affect the values used to model construction stages.

30.7.6 Water pressures

Hydrostatic water pressures, u, are derived using:

$$u = \gamma_w z_w$$

where

z_w is the depth below water level

γ_w is the density of water.

Non-hydrostatic water profiles may need to be taken into account where seepage or under-drainage of the soils behind the wall are possible considerations. In particular construction stages, during which the excavation is being dewatered, may result in non-hydrostatic profiles.

Different groundwater profiles may be applicable to different limit states. Extreme events such as flooding or burst water mains which could conceivably happen during the lifetime of a structure may result in a much higher than usual water table and such an occurrence may be appropriate to consider in an ultimate limit case.

Serviceability limit states should consider the most unfavourable water levels that would occur in normal circumstances, allowing for seasonal variations.

It is recommended that site-specific groundwater data are collected over a number of seasons, but in the absence of site-specific data, BS 8102[8] recommends the following:

- For basements not exceeding 4 m deep a design head of ground water, three-quarters of the full depth below ground (subject to a minimum of 1 m), is usually adequate.
- For basements deeper than 4 m the water table should be taken as being 1 m below ground level.

Where basement slabs are to be installed below the general surrounding water table, uplift will need to be considered. As well as designing the basement slab and connections to withstand this, consideration should be given to how this phenomenon should be represented in the soil-structure interaction model.

30.7.7 Compaction pressures

Where backfilling operations occur on the retained side of a wall, compaction pressures need to be considered. However, compaction immediately behind structures should be carried out with light plant and the depths of fill involved are generally small when considering embedded sheet pile walls. Therefore the magnitude of the contribution of compaction pressures is generally less than the live loading surcharges considered in the construction phase of design and they rarely cause a critical design case.

30.7.8 Representation of structural elements

In order to analyse a retaining wall using soil-structure interaction models, estimates of the properties of the structural elements are required. As the analysis is generally being used to determine the size of sections required, this is often an iterative process, as the relative stiffness of the wall and any prop elements will affect the deflections, forces and moments obtained.

30.7.9 Flexural stiffness of the wall

Most soil-structure interaction models require the wall's flexural stiffness (or its component values) to be defined using EI, where

> E = Young's Modulus for steel
> I = second moment of area.

For sheet piles, where the manual calculation of I is complex, values may generally be obtained from the sheet pile manufacturers' data sheets. However, the published values may need to be modified as Section 6.4 of Eurocode 3 Part 5 defines the effective flexural stiffness for piles as follows:

> $(EI)_{\text{eff}} = \beta_D(EI)$

where β_D is a factor taking account of lack of transmission of shear force across pile interlocks. Z-profile piles are unaffected by this phenomenon and so β_D can be taken as 1.0. The values of β_D for U-profile piles are defined in Table NA.2 of the UK National Annex to Eurocode 3 Part 5[3] for different situations, considering the number of levels of support/restraint, whether the piles are installed singly or in joined pairs and a rating that takes account of the ground conditions.

Values of β_D range as low as 0.3 for unsupported cantilever walls driven as single piles in highly unfavourable ground conditions. As this can severely impact the viability of using sheet piles, measures to improve the applicable value of β_D should be considered. A significant improvement in the value of β_D for the same restraint and ground conditions can be achieved by driving the piles as crimped or welded pairs. The National Annex refers the reader to the Steel Piling Group[17] website for further guidance on selection of values of β_D.

As the stiffness is dependent on the section and the section required is defined by the analysis, for an initial stiffness where no other guidance exists, the minimum section should be determined by considering both the minimum driveable section (see Section 30.8.8) and the need to accommodate section loss due to corrosion (see Section 30.8.9).

30.7.10 Stiffness of supports

Support may be derived from permanent floor slabs; temporary or permanent discrete props (these will generally be steel and may be horizontal or raking); ground anchors or berms. Different ways of modelling the support provided by a berm of soil left in situ are given in Section 7.2 of CIRIA C580.[7]

The axial stiffness, k, of any structural support system is required for soil-structure interaction models. The general form of this is:

$$k = AE\cos^2 \alpha / LS$$

where

A = cross sectional area of the support
E = Young's Modulus of the support
L = effective length of the support
S = spacing
α = angle of inclination of the support from the horizontal.

For support provided by a continuous horizontal slab, this simplifies to:

$$k = AE/L$$

although allowance should be made for any significant openings in the slab.

The effective length of a prop is generally half of the width of the excavation the prop is spanning, but for anchors or raking props the appropriate effective length will depend on the anchor/prop arrangement. It is often prudent to perform a sensitivity analysis with larger and smaller prop stiffnesses to check the overall impact on the results.

30.7.11 Design approaches

Eurocode 7 defines three different design approaches in relation to combinations of partial factors on actions, materials and resistance. The National Annexes defines which Design Approach to use locally and for the UK that is Design Approach 1. For other locations, refer to the corresponding National Annex.

Design Approach 1 requires the consideration of two combinations (for elements other than axial loaded piles or anchors) as follows:

Combination 1: A1 + M1 + R1
Combination 2: A2 + M2 + R1

where A, M and R1 are parameter sets as defined in the National Annex to Eurocode 7 Part 1. The sets for A and M are summarised in Table 30.4 and Table 30.5. For set R1 applied to embedded retaining walls, all values are = 1.0

Table 30.4 Partial factor sets to UK National Annex to Eurocode 7 Part 1 – Actions

Parameter	Set A1	Set A2
Action, Permanent, Unfavourable	1.35	1.0
Action, Permanent, Favourable	1.0	1.0
Action, Variable, Unfavourable	1.5	1.3

Table 30.5 Partial factor sets to UK National Annex to Eurocode 7 Part 1- Materials

Parameter	Set M1	Set M2
Soil: Tan φ'	1.0	1.25
Soil: c' (effective cohesion)	1.0	1.25
Soil: c_u (undrained shear strength)	1.0	1.4
Soil: γ (weight density)	1.0	1.0

As noted in Section 30.5.2, these differ slightly to the Design Approaches A-C defined in CIRIA C580[7], both in the parameter sets and the partial factors used.

The 'actions' for the purposes of geotechnical analysis generally relate to surcharge loads (e.g. live loads, adjacent foundation loads) and imposed loads from other parts of the structure. The resulting moments and prop forces from the geotechnical analysis then form 'actions' for the structural design of the wall components as described in 30.8.6.

For axially loaded piles and anchors the Design Approach 1 Combinations are:

Combination 1: A1 + M1 + R1
Combination 2: A2 + (M1 or M2) + R4.

In Combination 2, set M1 is used for calculating resistances of piles or anchors and set M2 for calculating unfavourable actions on piles such as negative skin friction or transverse loading. Values for R4 in various design situations are given in tables in Annex A of the National Annex to Eurocode 7 Part 1.[12]

30.7.12 Overdig

For ultimate limit state design where the ground in front of the sheet pile wall is contributing to stability, a depth of overdig, Δa, needs to be considered. Generally this is to be taken as 10% of retained height (or for propped walls, height between lowest support and excavated level), limited to a maximum of 0.5 m.

For basements with a permanent slab, it may be argued that the risk of unplanned overdig is negligible and $\Delta a = 0$, however, the temporary construction stages before the base slab is installed should include an appropriate overdig allowance.

30.7.13 Sequence of analysis

As discussed above, the sequence involved in design is generally iterative, but the broad sequence is as follows:

- Determine:
  - geometry
  - soil stratigraphy
  - ground water conditions
  - section loss due to corrosion
  - any other loads.
- Define design cases, dependent on:
  - limit states
  - variation of geometry, including overdig allowance
  - variation of soil stratigraphy
  - normal and extreme groundwater conditions
  - variation in loading conditions
  - Design Approach factor combinations in accordance with Eurocode 7.
- Determine depth of embedment of wall required for the ultimate limit using limiting equilibrium methods (NB some soil-structure interaction programmes will also do this).
- Check depth of pile for vertical load carrying capacity (if required).
- Check length of pile to achieve embedment against pile length limitations (see Sections 30.8.8 and 30.9.1).
- Define assumed construction sequence.
- Analyse by soil-structure interaction model, using parameters defined above.
- Determine critical actions (including bending moments and prop forces) based on output and appropriate partial factors.
- Determine structural sections required.
- Repeat sequence as necessary for optimised structural elements.

30.8 Detailed basement design: Structural design

30.8.1 Use of Eurocode 3 and other guidance

General rules for structural design are covered in the Eurocode: Basis of Structural Design; rules for designing with steel are given in Eurocode 3 Part 1, with additional guidance specific to steel piles covered in Eurocode 3 Part 5.

This section limits its treatment of structural design to issues specific to sheet piles. Reference should be made to Chapters 14 to 19 as appropriate for more general design.

Data sheets in relation to properties of particular sheet pile sections can be found in the *Piling Handbook*[2] and other manufacturers' literature.

30.8.2 Definitions and conventions

Interlocks are the portion of a steel sheet pile or other sheeting that connects adjacent elements by means of a thumb and finger or similar configuration to make a continuous wall. Interlocks may be described as:

- Free: threaded interlocks that are neither crimped nor welded
- Crimped: interlocks of threaded single piles that have been mechanically connected by crimped points
- Welded: interlocks of threaded single piles that have been mechanically connected by continuous or intermittent welding.

Section 1.9 of Eurocode 3 Part 5[3] defines the axis convention used and notes that it differs from Eurocode 3 Part 1.1; care should be taken during cross-referencing. In Part 5 the y-y axis is parallel to the wall rather than related to the section. However in the majority of cases there is only one possible moment axis and no axis is identified.

30.8.3 Steel strength

Tables 30.6 and 30.7 give the nominal values of yield strength for hot-rolled and cold-formed sheet piles, as stated in Eurocode 3 Part 5.[3]

30.8.4 Structural limit states

Eurocode 3 Part 5[3] reiterates the first three ultimate states described in Section 30.5.3 above and then lists the following structural failure modes as applicable to sheet pile retaining walls:

Table 30.6 Nominal values of yield strength and ultimate tensile strength for hot-rolled steel sheet piles according to EN 10248-1

Steel name to EN 10027	Yield strength f_y (N/mm^2)	Ultimate tensile strength f_y (N/mm^2)
S240 GP	240	340
S270 GP	270	410
S320 GP	320	440
S355 GP	355	480
S390 GP	390	490
S430 GP	430	510

Table 30.7 Nominal values of yield strength and ultimate tensile strength for cold-formed steel sheet piles according to EN 10249

Steel name to EN 10027	Yield strength f_y (N/mm^2)	Ultimate tensile strength f_y (N/mm^2)
S235 JRC	235	340
S275 JRC	275	410
S355 JOC	355	490

- failure due to bending and/or axial force
- failure due to overall flexural buckling, taking account of the restraint provided by the soil
- local buckling due to overall bending
- local failure at points of load application (e.g. web crippling)
- fatigue (NB effects of impact or vibration during installation may be neglected).

Serviceability limit state criteria for retaining walls are dependent on the project-specific limitations on wall movement and surrounding ground movement.

Eurocode 3 Part 5 requires that analysis of the wall for serviceability state should be based on a soil-structure interaction model, using a linear elastic model of the structure. No plastic deformations are permitted.

30.8.5 Classification of cross-sections and calculation of design moment resistance

Like Eurocode 3 Part 1-1, Part 5 defines four classes of sheet pile cross-section. These affect the type of structural analysis that may be undertaken and the calculation of design moment resistance. Table 5-1 of Eurocode 3 Part 5[3] defines Class 1 to Class 3 pile sections based on width/thickness ratios and strength of the section. The section properties tables provided in manufacturers' product information generally define the section class for different steel strengths.

The majority of hot-rolled sheet pile sections fall into Classes 2 or 3, though some of the smaller U-sections are Class 4. Cold-rolled sheet pile sections are generally Class 4. It should be noted that section class may change over time due to corrosion (see 30.8.9).

Table 30.8 summarises the design requirements for the different classes of pile section.

As with the factor β_D on flexural stiffness (see Section 30.7.9), β_B is applicable mainly to U-profile piles and should be taken as 1.0 for Z-profile piles. The values of β_B for U-profile piles are defined in table NA.2 of the UK National Annex to Eurocode 3 Part 5[3] for different situations, considering the number of levels of support/restraint, whether installed singly or in joined pairs and how favourable the ground conditions are. The National Annex refers the reader to the Steel Piling Group website for further guidance on selection of values of β_B.

Values of W_{el} and W_{pl} should be obtained from the manufacturers' literature, taking account of any allowance for loss of thickness due to corrosion, see Section 30.8.9 below.

30.8.6 Non-axially loaded sections

For non-axially loaded sections, Eurocode 3 Part 5 requires that:

$$M_{Ed} \leq M_{c,Rd}$$

Table 30.8 Analysis requirements and design moment resistance based on pile class

Class of pile	Type of analysis required	Design moment resistance $M_{c,Rd}$
Class 1	Cross-sections for which a plastic analysis involving moment redistribution may be carried out, provided that they have sufficient rotation capacity (see Annex C)	$\beta_B\ Wpl\ fy/\gamma M0$
Class 2	Cross-sections for which elastic global analysis is necessary, but advantage can be taken of the plastic resistance of the cross-section	$\beta_B\ Wp\ fy/\gamma M0$
Class 3	Cross-sections which should be designed using an elastic global analysis and an elastic distribution of stresses over the cross-section, allowing yielding at the extreme fibres	$\beta_B\ Wel\ fy/\gamma M0$
Class 4	Cross-sections for which local buckling affects the cross-sectional resistance	Design in accordance with Annex A

β_B is a factor that takes account of a possible lack of shear force transmission in the interlocks
Wpl is the plastic section modulus determined for a continuous wall
Wel is the elastic section modulus determined for a continuous wall
fy is the yield strength as given in Reference 3
γ_{M0} partial safety factor = 1.0 in accordance with the UK National Annex to Eurocode 3 Part 1

where
$\quad M_{c,Rd}$ is the design resistance as described in 30.8.5
$\quad M_{Ed}$ is the design bending moment effect, derived from a calculation according to the relevant case of Eurocode 7 Part 1[12], as described in Section 30.7.11.

30.8.7 Axially-loaded sections

Where there will be axial loading of the sheet pile, buckling may need to be considered, as described in Section 5.2.3 of Eurocode 3 Part 5.[3]

In addition, the increase in moment as a result of any eccentricity due to permissible tolerances, and any expected deflections, should be considered.

The capacity of bearing piles is covered in Section 5.3 of Eurocode 3 Part 5.[3]

30.8.8 Driveability

The selection of an appropriate sheet pile section includes consideration of its driveability. The Piling Handbook[2] provides guidance, in the form of a nomogram, on the minimum section size appropriate for different piling lengths and installation methods, under easy, normal and hard driving conditions. In the same section a number of tables define 'easy', 'normal' and 'hard' in relation to SPT values of granular soils and undrained shear strength of cohesive soils.

Driving assistance measures may improve driveability, as discussed further in Section 30.10.7 and thus increases the achievable driven depth. As these techniques have an influence over the design soil parameters used, the potential use of jetting or pre-boring must be considered at design stage.

As improvements are continually being made in the rigs and techniques, it is always worth discussing options with a specialist contractor.

Where there is any uncertainty on the ability to reach design depth with the proposed piling method and chosen section, trial drives may be undertaken to confirm the buildability.

30.8.9 Durability / corrosion

Section 4 of Eurocode 3 Part 5[3] covers the design for durability of steel piles. Two basic approaches exist for designing to counteract corrosion. The first of these is to accept that it occurs and include a sacrificial thickness in the calculation of required section thickness during design. The second is to include some measure to protect the pile in some way, with a coating (paint or galvanising) or encasement (concrete or grout protection), or cathodic protection. For steel basements, including a sacrificial thickness and painting are the most likely protection measures.

Designing to include for a sacrificial thickness requires an estimate of section loss over the design life of the project. Piles with a design life of less than four years do not need to take account of corrosion, but for most basement design, it will need to be considered.

Corrosion rates depend on whether the pile is in contact with soil, air, fresh or salt water, and whether the soil is aggressive or not. Different zones of the pile will be under different conditions and so all relevant combinations and locations need to be considered relative to the moment demand on the sheet pile.

Table 30.9 summarises the section loss per side of sheet pile as recommended in Eurocode 3 Part 5[3] for different conditions adjacent to the wall. Though the marine and freshwater conditions are more applicable to river and quay walls, there may be unusual situations where they are applicable to building retaining walls and hence they are included for completeness.

Example:

For a basement intended for a 50 year design life, on a site with a potentially aggressive natural soil stratum, the total design loss of thickness considered for a section of wall with air on one side and the potentially aggressive natural soil on the other side would be: 0.5 mm + 1.75 mm = 2.25 mm.

Manufacturers' product data typically include charts which can be used to determine the revised elastic section modulus for the calculated total thickness loss for different section types.

Table 30.9 Loss of section (in mm) due to corrosion as recommended in the NA to Eurocode 3 Part 5

Condition	5 years	25 years	50 years	75 years	100 years	125 years
Air – normal atmosphere (0.01 mm/year)	0.05	0.25	0.5	0.75	1.0	1.25
Air – marine atmosphere (0.02 mm/year)	0.1	0.5	1.0	1.5	2.0	2.50
Fresh water	0.15	0.55	0.9	1.15	1.4	1.65
Brackish or very polluted fresh water	0.30	1.3	2.3	3.3	4.3	5.30
Sea water[1], low water and splash zone[2]	0.55	1.90	3.75	5.6	7.5	Protection required
Sea water[1], intertidal and fully immersed[2]	0.25	0.90	1.75	2.60	3.50	4.40
Undisturbed natural soils	0.00	0.30	0.60	0.90	1.20	1.50
Polluted natural soils and industrial sites	0.15	0.75	1.50	2.25	3.00	3.75
Aggressive natural soils	0.20	1.00	1.75	2.50	3.25	4.00
Non-aggressive fills (uncompacted)	0.18	0.70	1.20	1.70	2.20	2.70
Non-aggressive fills (compacted)	0.06	0.35	0.60	0.85	1.10	1.35
Aggressive fills (uncompacted)	0.50	2.00	3.25	4.50	5.75	7.00
Aggressive fills (compacted)	0.25	1.0	1.66	2.25	2.88	3.50

Table notes:
1. Refers specifically to temperate climate.
2. Highest corrosion usually occurs in the splash or low tide zone, however, the critical design case may be at the location of highest bending stresses, which may occur in a permanently immersed zone.

As the determination of section class involves the ratio of the width to thickness, loss of section due to corrosion will increase this ratio and may change the class. The change in class due to thickness loss is not always included in the product data, but relevant dimensions should be available from manufacturers.

30.8.10 Design of high modulus and combined walls

Eurocode 3 Part 5[3] Section 5.5 covers the design of combined walls. The function of the component parts of the wall needs to be recognised during design. The primary element performs the retaining function and the secondary elements fill in the gaps between the primary elements and transmit the loads to them.

30.8.11 Design of anchorages

Design of anchorages is covered by Section 8 of Eurocode 7[12] and design and installation are also covered by EN 1537: 1999.[18]

30.9 Other design details

30.9.1 Pile length

Practical limits exist for the installation of driven steel piles. These are influenced by the ground conditions and also other limitations on noise and vibration, which will in turn govern the appropriate installation technique and driving assistance measures as discussed in Section 30.10. The benefits of driving assistance in increasing depth of installation need to balanced with their effect on design, and adjacent structures.

Typically available leader rigs accommodate a maximum pile length of 15 m–25 m, though some multi-purpose rigs are available with capacity in excess of 30 m. Pressing rigs for sheet piles may be limited to lengths up to 16 m depending on pile section and equipment, however, up to 20 m has been achieved with paired Z-piles. Substantially greater depths have been achieved by pressing rigs in Japan and the range of capability of commonly available rigs continues to improve.

Sheet piles can be supplied up to 31 m in length, although the practicalities of delivery of this length of load should be considered in relation to the site location and access.

Shorter pile lengths of pile can combined by welding; controls on site welding are covered in EN 12063.[19]

30.9.2 Closure

The majority of a basement structure can be completed using standard sections, including hot-rolled corner pieces or other standard junction pieces. However non-standard junctions and corners will most likely need to be bespoke to fit the geometry required. Fabrication of these is covered by EN 12063.[19]

30.9.3 Watertightness

Where watertightness is required for steel sheet piles below the water table, there are a number of options for achieving this:

- welding of interlocks
- interlock sealants.

Welding of sheet piles is generally the most effective method of forming a seal to provide water resistance and is the only method to ensure better than Grade 1 basement where a high water table exists. Welding of piles with small gaps between the interlocks can be achieved with a simple fillet weld. Wider gaps which cannot be dealt with by a fillet weld may be dealt with by the introduction of a small diameter bar or a plate, welded either side to the adjacent piles. The latter can also be used in situations where running water through the interlocks would otherwise affect the quality of welding.

Interlock sealants may be applied in the factory or on site. The former is normally preferred, as conditions are better for the operation, and some of the materials used may be considered hazardous before they are cured and become inert. Two basic types of sealant exist, compression and displacement. The compression sealant is firmer and is 'squashed' to form a compression seal when the piles are interlocked. Displacement sealants are softer, like a paste or gel, and the material is pushed out to fill any gaps between the piles in the interlock. As these reduce friction within the interlock, it may improve driveability, but contribute to lower values of β_D and β_B. The omission of sealants allows an improvement of 0.05 in values of β_D and β_B.

Hydrophilic sealants can also be used, which swell only on contact with water, but should be installed only on the trailing interlock (i.e. not on the one in the ground with an empty interlock) as the hydrophilic properties could be triggered early, making installation of the following pile extremely difficult.

On their own, sealants will only typically provide Grade 1 basement water tightness, however they can be useful to provide temporary water exclusion to allow welding to be carried out in drier conditions.

30.9.4 Detailing base slab connections

Preventing water from entering through the junction between the sheet pile wall and any basement slab may be just as important as preventing leakage through sheet pile interlocks. Suitable detailing is required, as simply pouring the slab against the sheet piles will not guarantee water tightness, due to the potential for shrinkage.

Welding the slab reinforcement to the sheet piles, or including shear studs or a horizontal welded plate are methods to structurally tie the two elements together. Other measures, such as PVC waterbars, waterstops, membranes and other proprietary products can also be placed or included within the slab before the concrete is poured. Some products can be post-applied by surface application, impregnation or grouting through tubes pre-installed within the slab, referred to as 'active' systems, Figure 30.10 illustrates a number of different measures. Further details can be found in SCI P275[1] and SCI P308[20] and the ICE Client Guide[10] provides further illustrations.

30.10 Constructing a steel basement: Pile installation techniques

30.10.1 General

There are two basic driving methods, 'pitch and drive' and 'panel driving'. There are three driving systems that are applicable to both methods:

- impact driving
- vibrodriving
- pressing.

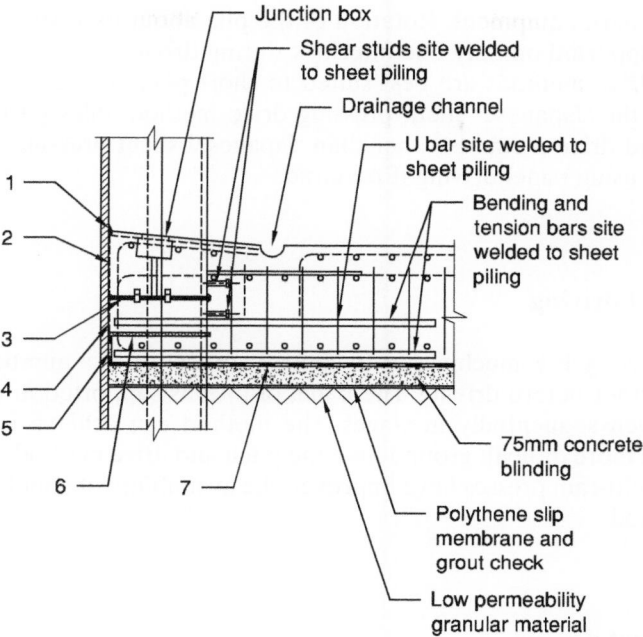

1. 10mm wide x 10mm deep chase formed in top of slab with pourable sealant

2. 100mm wide adhesive waterproofing tape membrane with permanent mechanical bond to concrete

3. Hose injection waterproofing system clipped to face of sheet piles

4. Double sided self adhesive rubber/bitumen waterproofing membrane securely bonded to sheet pile, water bar and hydrophilic strip

5. P.V.C. waterbar returned 125mm

6. 20mm x 5mm hydrophilic waterstop bonded to sheet piling

7. P.V.C. waterbar with co-extruded hydrophilic elements at construction joints

Figure 30.10   Waterproofing with an active injection system

The methods and systems are briefly described in the following sections. More information can be found in the *Piling Handbook*[2] and SCI P308.[20]

30.10.2 Pitch and drive

This method installs piles one by one. This can lead to forward lean and out of tolerance piling, unless verticality is strictly controlled. Better control of this is available with more modern equipment. Rotation of the pile about its vertical axis is also a risk, as it is supported on only one interlock during driving.

Pitch and drive methods are best suited to short piles, and is the only method possible with the 'Japanese' silent pressing drive method. Piles partially installed using pitch and drive methods, other than 'Japanese' silent pressing, can generally be completed using panel driving if required.

30.10.3 Panel driving

With panel driving, it is much easier to control verticality, as a number of piles are threaded together before driving. The panel of piles is supported in a guide frame and then driven sequentially in stages. The method can achieve installations of longer piles in more difficult ground than the pitch and drive method. Recent developments in multi-ram presses have improved the availability of panel driving by the pressing method.

30.10.4 Impact driving

The most common form of impact driving is the drop hammer, which uses a falling weight to create the impact, spread to the top of the pile by a driving cap. The most common form of drop hammer in current use is the hydraulic hammer. Historically, air hammers and diesel hammers were used, which utilise an explosive force to drive the hammer, however, as the newer hydraulic hammers operate at significantly higher efficiencies and are far less noisy than older diesel hammers, the latter are now less frequently used.

30.10.5 Vibrodriving

An oscillating driver is clamped to the top of the pile, to induce vibrations in the pile and reduce friction along the sides of the pile, thus allowing the pile to be inserted into the ground with little extra application of force.

30.10.6 Pressing

Pressing methods operate by jacking the piles into the ground, using the adjacent piles for reaction. This is a low noise and low vibration method, which makes it good for sensitive sites. There are two generic types of pressing rig, the 'Japanese' rigs, such as those by Giken and Tosa, and panel driving rigs. Units have also been developed to adapt leader rigs to use pressing methods.

The Japanese method uses a rig which progresses along the line of piles without needing to be lifted onto each pile individually by crane, which means that access requirements are reduced. The machines are often specific to a particular generic section size, therefore it is important to match pile section to driving method.

The panel driving/pressing rigs are suited mainly to installation in heavy clays and require a crane to move the rams from pile to pile. With older multi-ram presses it was also necessary to bolt plates to each pile, however recent advances have eliminated this requirement.

30.10.7 Driving assistance methods

Driving assistance methods can significantly improve the constructability of a sheet pile wall. Jetting and pre-auguring are the main methods.

Jetting involves delivering a water jet to the soil at the toe of the sheet pile, reducing friction.

Pre-auguring refers to the use of a continuous flight auger to penetrate the ground along the pile line in advance of the sheet pile installation. Soil should only be loosened along the line and not removed when using this technique.

Both methods change the in situ soil properties around the sheet piles and the impact of their use needs to be considered during design. In particular, the definitions of unfavourable ground for determination of β_D and β_B consider different driving assistance methods as they influence the skin friction and interlock friction of the installed sheet piles. The acceptability of these methods for other reasons, including ground movement and creation of flow paths for contamination will also need to be taken into account.

30.10.8 Selection of method

Some ground conditions, particularly layered ground where granular deposits overlie clay deposits (or vice versa) may be best dealt with by a combination of methods. Some specialist plant is available which may provide more than one method, though generally different rigs will be required.

Table 30.10 summarises the methods in relation to the conditions for which they are suited and unsuited.

30.11 Specification and site control

30.11.1 Specification of piling works

The ICE Specification for Piling and Retaining Walls[21] includes a section for sheet piling and covers other generic requirements for embedded retaining walls. This should be used as the baseline specification for steel sheet pile walls in the UK.

Table 30.10 Summary of suitability of different pile installation methods

Installation method	Ideal for	Less suitable for
Impact driving	• Hard and difficult ground • Panel driving	• Noise sensitive areas • Restricted access sites
Vibrodriving	• Loose to medium dense granular soils, mixed soils, soft cohesive soils, saturated granular soils • Bearing piles and non-sheet sections • Extracting piles	• Stiff clay soils • Granular soils with SPT > 50
Pressing – 'Japanese'	• Installation adjacent to sensitive structures and sites with difficult access • Cohesive soils and fine grained soils • Single pile driving • Use in combination with water jetting	• Driving piles already installed by other means • Hard driving conditions • Potential obstructions • Combi wall with mixed sections
Pressing – 'Panel driving'	• Heavy clay soils (e.g. London clay) • Completing drives into clay started through granular strata by other means	• Granular soils • Use with waterjetting • Combi wall with mixed sections

EN 12063[19] covers the handling, preparation and installation of sheet pile and combined walls, including requirements and guidance for welding and is a useful reference when designing and specifying steel sheet pile walls.

30.11.2 Installation tolerances

The required installation tolerances for sheet piles are defined in Table 2 of EN 12063.[19] For piles installed on land (as opposed to over water) these are:

• plan position to be ≤75 mm from design position of the top of the pile, in plan perpendicular to the wall and
• deviation from vertical to be ≤ 1% over the top 1 m of the pile, in any direction.

The latter verticality requirement should be extended to the whole exposed length of the pile when used in basement construction.

Dispensations are given in EN 12063[19] for hard driving conditions to permit some declutching, provided that no strict criteria are required. However, declutching is unlikely to be tolerable for most permanent retaining walls as it impacts performance and watertightness and this relaxation should not generally be applied.

In basement construction, tighter tolerances may be required, for example where column loads are to be supported by the wall or if other site constraints dictate, such as clearance to other structures or building elements. It should be noted that installation to tighter tolerances will carry some cost and time implications, and should be used only where required.

30.11.3 Ground movement

Where adjacent buildings and/or infrastructure would be sensitive to ground movement resulting from sheet pile installation and basement excavation, limits on permissible ground movement during construction can be specified and controlled by monitoring during construction.

30.11.4 Noise and vibration

Historically, sheet piling has been discounted as too noisy for use in urban situations, however, modern hydraulic hammers and the development of non-impact driving methods have improved this.

BS 5228: 2009 Parts 1 and 2[22] cover the control of noise and vibration respectively on construction sites. Annex A of BS 5228-1 covers the legislative controls in the UK for noise. Annex C of the same document presents indicative noise levels of current plant and construction activities including steel piling. Section 8.5 of BS 5228-2 discusses measures to reduce vibration from piling operations on site.

The vibration level at which people feel discomfort or alarm is generally at a lower threshold to that which actually causes damage to buildings or infrastructure. BS 7385-2[23] covers damage to buildings from ground-borne vibrations and BS 6472-1:2008[24] gives guidance on human response to vibration in buildings.

If sensitive receptors are identified, then noise or vibration monitoring may be required during construction.

30.12 Movement and monitoring

Where the wall has been designed using the Observational Method, monitoring is essential. On other projects, at least some monitoring measures should be implemented to ensure that the works are performing as intended during the design. These should be set out in the specification.

Typical installations for monitoring include:

- targets for manual monitoring of position and level
- electrolevels for continuous movement monitoring
- wall inclinometers, installed in tubes mounted on the sheet piles

- soil inclinometers, installed within the soil behind the wall
- load cells on struts and props
- strain gauges mounted on the sheet piles.

If using monitoring as a site control measure, with or without the Observational Method, it is important to review data in a timely manner, set intervention/trigger values and plan the action required if those values are met. Collection of large amounts of data, unless for research purposes, is generally counterproductive as during construction works, there will be little time or inclination to review the data and important signs of problems could be missed.

Remote, real time monitoring with web access and SMS alerts is now readily available, which can ensure alerts are not missed, however, care must be taken to set the triggers at a realistic level to avoid unnecessary alarms or premature implementation of emergency procedures.

Monitoring during installation may be required to control ground movements resulting from measures such as pre-auguring. Noise and vibration monitoring may also be required; as discussed in the previous section it is important that monitoring measures are installed and baselines established before construction commences.

References to Chapter 30

1. Yandzio, E. and Biddle, A.R. (2001) *A R SCI P-275 – Steel intensive basements*. Ascot, SCI.
2. *ArcelorMittal* (2005) *Piling Handbook*, 8th edn. London, ArcelorMittal.
3. British Standards Institution (2007) *BS EN 1993-5:2007 Eurocode 3. Design of steel structures. Piling*. London, BSI.
4. Building Research Establishment (2002) *BRE Digest 472 – Optimising ground investigation*. Watford, BRE.
5. British Standards Institution (1999) *BS 5930:1999 Code of practice for site investigations*. BSI 1999. London, BSI.
6. British Standards Institution (2006) *BS EN ISO 14688 Geotechnical investigation and testing. Identification and classification of soil. Part 1 Identification and description CEN 2002, Part 2 Principles for a classification*. London, BSI.
7. CIRIA (2003) *C 580 – Embedded retaining walls: Guidance for economic design*. London, CIRIA.
8. British Standards Institution (2009) *BS 8102:2009 Code of practice for protection of below ground structures against water from the ground*. London, BSI.
9. CIRIA (1995) *Water-resisting basements Report 139*. London, CIRIA.
10. Maloney M., Skinner H., Vaziri M. and Jan Windle for The Institution of Structural Engineers (2009) *Reducing the Risk of Leaking Substructure. A Clients' Guide*. London, ICE.
11. CIRIA (1999) *The Observational Method in ground engineering – Principles and applications Report 185*. London, CIRIA.

12. British Standards Institution (2007) *BS EN 1997 Eurocode 7. Geotechnical design. Part 1 General rules* BSI 2004 and Part 2 Ground investigation and testing. London, BSI.
13. CIRIA (1984) *Design of retaining walls embedded in stiff clays Report 104.* London, CIRIA.
14. British Standards Institution (2007) *BS EN 1997 Eurocode 7. Geotechnical design.* Part 1 General rules BSI 2004 and Part 2 Ground investigation and testing. London, BSI.
15. CIRIA (1995) *The standard penetration test (SPT): Methods and Use Report 143.* London, CIRIA.
16. Yandzio E. (1998) *Design Guide for Steel Sheet Pile Bridge Abutments.* SCI P-187. Ascot, SCI.
17. The Steel Piling Group, www.steelpilinggroup.org
18. British Standards Institution (2000) *BS EN 1537:2000 Execution of special geotechnical work. Ground anchors.* London, BSI.
19. British Standards Institution (1999) *BS EN 12063:1999 Execution of special geotechnical work. Sheet pile walls.* London, BSI.
20. Yandzio E. and Biddle A. R. (2002) *Specifiers' guide to steel piling.* SCI P-308. Ascot, SCI.
21. The Institution of Structural Engineers (1996) *Specification for Piling and Embedded Retaining Walls - Specification, Contract Document and Measurement, Guidance Notes.* London, ICE.
22. British Standards Institution (2008) *BS 5228 Code of practice for noise and vibration control on construction and open sites. Part 1 Noise and Part 2 Vibration.* London, BSI.
23. British Standards Institution (1993) *BS 7385-2:1993 Evaluation and measurement for vibration in buildings. Guide to damage levels from ground borne vibration.* London, BSI.
24. British Standards Institution (2008) *BS 6472-1:2008 Guide to evaluation of human exposure to vibration in buildings. Vibration sources other than blasting.* London, BSI.

Further reading for Chapter 30

The Steel Construction Institute *H-Pile Design Guide.* SCI Publication P335. Ascot, SCI.

North American Steel Sheet Piling Association (2008) *Steel Sheet Piling Installation Guide: Best Practices.* Alexandria, VA, NASSPA.

Bond A.J. and Harris A.J. (2008), *Decoding Eurocode 7.* London, Taylor and Francis.

Frank R., Bauduin C., Kavvadas M., Krebs Ovesen N., Orr T., and Schuppener B. (2004) *Designers' Guide to EN 1997-1: Eurocode 7: Geotechnical Design – General Rules.* London, Thomas Telford.

North American Steel Sheet Piling Association http://www.nasspa.org/

Arcelor Mittal http://www.arcelormittal.com/sheetpiling/
Corus http://www.corusconstruction.com/en/products/foundations/
Nippon Steel http://www.nsc.co.jp/en/product/construction/catalog.html
JFE Steel corporation http://www.jfe-steel.co.jp/en/products/list.html#Shapes
ThyssenKrupp Steelcom http://www.steelcom.com.au/sheet-pile.htm
Evraz Vitkovice Steel http://www.vitkovicesteel.com/en/seznam-produktu/produkty/
    sheet-piles-8/
Gerdau Ameristeel http://www.sheet-piling.com/main
Hoesch Spundwand und Profil http://www.spundwand.com/e/
WALL-PROFILE (specifically connectors) http://www.wallprofile.com/
Nucor-Yamato Steel http://www.nucoryamato.com/
Independent directory of Geotechnical Software: http://www.ggsd.com/
Oasys (Stawal and Frew) http://www.oasys-software.com/
GeoCentrix (ReWaRd) http://www.geocentrix.co.uk/
GeoSolve (WALLAP) http://www.geosolve.co.uk/
Plaxis (FEA specifically for geotechnical purposes) http://www.plaxis.com/

Chapter 31
Design for movement in structures

Compiled by GRAHAM OWENS

31.1 Introduction

31.1.1 Movement

All structures move to some extent. Movement may be permanent and irreversible or short-term and possibly reversible. The effects can be significant in terms of the behaviour of the structure, its performance during its lifetime and the continued integrity of the materials from which it is built.

Movement can arise from a variety of sources:

- temperature changes and thermal expansion
- differential settlement of the foundations
- creep and shrinkage during drying of the concrete
- vibrations.

The effects of these phenomena are easy to understand in principle; for example, Figure 31.1 illustrates the general effect of providing bracing at both ends of a long building. However, it is difficult to quantify their effects.

31.1.2 Design philosophies

For smaller buildings, and general construction, movements may frequently be ignored. For larger scale construction, or special circumstances, one or more of the following features to accommodate relative movement between different parts of the structure should be adopted:

Steel Designers' Manual, Seventh Edition. Edited by Buick Davison and Graham W. Owens.
© 2012 Steel Construction Institute. Published 2012 by Blackwell Publishing Ltd.

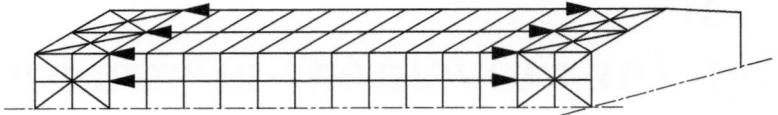

Arrows indicate compression forces when expansion is constrained

Figure 31.1   Effects of restrained expansion in a long building

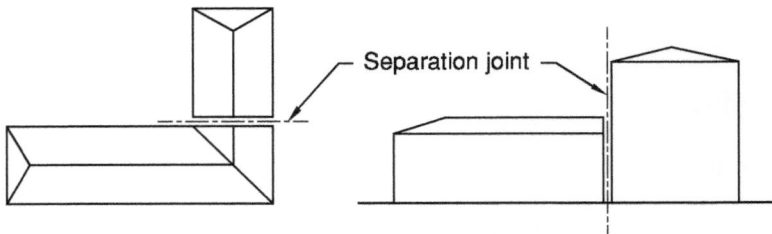

Separation joint

Figure 31.2   Separation of a single building into separate blocks

- **Expansion joints**: these permit displacement to limit thermally-induced forces in long buildings. Their specification depends on temperature range and the thermal expansion coefficient of the materials (see Section 31.2).
- **Construction joints**: these control drying shrinkage of concrete floors and ground slabs.
- **Separation joints**: these ensure separate behaviour of parts of the building that are of different height or structural orientation, see Figure 31.2.
- **Compacting joints**: these are specialist devices that mitigate the effects of the differential settlement that may arise from variations in substrata below the building.

Expansion joints and construction joints are the most common type of movement joint and are discussed in more detail below. Other types of movement joint generally require specialist design and are beyond the scope of this manual.

Irrespective of the nature of the relative movement, one of four methods can be adopted to address it:

1. Ignore the effects, relying on successful past satisfactory performances. This is frequently adopted for small scale construction.
2. Design the structure to withstand all the forces developed by restraint of movement. This is possible with smaller structures (small-span bridges) or structures which are comparatively flexible (portal frames out-of-plane of the frame). The method will avoid joints but may require the use of additional material in construction.

3. Subdivide the structure into small, structurally stable, units, each of which then becomes essentially a structure in its own right, able to move independently of the surrounding units. This principle is ideal for controlling those factors such as thermal movement, which are related to the size of the overall structure. In many cases, the need for bearings can be eliminated. The disadvantage lies in the need to provide joints between the various units of the structure capable of accommodating all the anticipated relative movements between the units, while at the same time fulfilling all the other requirements, i.e. visual, practical, etc. It is, however, generally possible to achieve a balance by subdividing the structure so that the movements at the joints between units are kept relatively small, permitting the joints to be simple and economical (possibly at the expense of larger numbers of joints).

4. Subdivide the structure into fewer but larger sections, and make provision for a smaller number of joints, each with larger movement capacity and thus possibly more complex than those that would be used in (3). Examples are to be found in bridges where use of the least number of road deck joints is preferable both in terms of riding quality and also in the minimisation of long-term maintenance requirements.

The overall design of the buildings must take account of the positioning of joints, in particular, their influences on overall structural behaviour and analysis.

Individual joints must be specified to accommodate the predicted magnitude of the horizontal and/or vertical displacements.

The positioning of vertical and horizontal bracing and their design must be compatible with joint positions. The bracing positions must not inhibit the movements for which the joints have been provided. Each separate part of the building must be adequately braced.

All other components of the building and its equipment (for example a conveyor) must take account of joint positions and their predicted displacements.

31.2 Effects of temperature variation

BS EN 1991-1-5 gives principles and rules for calculating thermal actions on buildings, bridges, other structures and their structural elements.

Values of the maximum shade air temperature T_{max} and minimum shade air temperature T_{min} may be specified by the National Annex to BS EN 1991-1-5.[1]

In steel structures, with a coefficient of linear thermal expansion $a = 12 \times 10^{-6}$ per °C (as given in BS EN 1993-1-1 Cl 3.2.6[2]), the effects of the variations of the temperature can be significant.

In assessing temperature variation, it is important to distinguish between internal and external steelwork. The latter is likely to be subject to much greater variation than the former.

External frames may be exposed to a temperature range from −23°C to +35°C, relative to the temperature at which they are built. The free expansion/contraction

Table 31.1 Maximum spacing of expansion joints

Type of structure	Situation	Spacing
Steel frames – industrial buildings	Generally	150 m [1]
	Buildings subject to high internal temperatures due to plant	125 m [1]
Steel frames – commercial buildings	Simple construction	100 m [1]
	Continuous construction	50 m [2]
Roof sheeting	Down the slope	20 m [3]
	Along the slope	no limit
Brick or block walls	Clay bricks	15 m
	Calcium silicate bricks	9 m
	Concrete masonry	6 m

Notes:
1. Where the stress due to constraint of thermal expansion can be catered for by the members, no limit is necessary in simple construction.
2. Larger spacings are possible where the stresses due to constraint of thermal expansion can be catered for by the members.
3. Longer lengths are possible where provision for expansion is made.

at these temperatures is −3 mm to +0.4 mm per metre length of building. In practice, all expansion is partly constrained and actual movements will be significantly less.

Internal steelwork will be subject to much smaller temperature variations, especially in a heated or air conditioned environment.

Thermal movements can lead to:

- damage at supports, including cracking or even instability of walls supporting long beams or trusses
- connection failure
- significant internal forces in statically indeterminate structures.

31.3 Spacing of expansion joints

The advice circulating on the provision and spacing of expansion joints is variable and conflicting. Table 31.1[3] provides a summary of the best advice that is currently available.

31.4 Design for movement in typical single-storey industrial steel buildings

31.4.1 General

In typical industrial steel buildings, stability in the transverse direction is achieved by portal frame action and longitudinal direction by vertical bracing.

Two design cases need to be considered:

- for portal frames, the expansion in the plane of the frame should be considered by calculation
- for the vertical bracing in the longitudinal direction, the interaction between the expansion and the vertical bracing design needs to be considered.

A part of the elongation of the structural components in the longitudinal direction can generally be absorbed by the slip in connections.

Nevertheless, expansion joints should be provided when the temperature differential becomes important (external structures, or uninsulated construction), or the slips in connections become insufficient to absorb the full thermal expansion. The building length above which expansion joints are used in practice varies between countries. For example central regions of France, which have a relatively continental climate, expansion joints are recommended for expansion lengths above 50 m, i.e. a building length of 100 m with mid-length bracing. In the UK, with a more temperate climate and different construction traditions, expansion joints are only recommended for buildings over 150 m in length. Even above this length, industry advice acknowledges that expansion joints may be omitted if large individual members such as eaves and beams and crane girders are designed to resist stresses due to restraint of expansion.

31.4.1.1 *Position of vertical bracings*

It is not recommended to set out vertical bracing systems at both ends of the building unless there is an expansion joint in between. This arrangement would inhibit the expansion of the longitudinal members and could induce high forces in the structural components of the long sides and in their connections; see Figure 31.3.

For long buildings it is recommended to set out only one vertical bracing at the mid point of the long sides, thus allowing expansion towards the ends in both directions, see Figure 31.4.

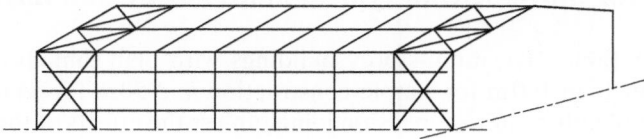

Figure 31.3   Bracing layouts that are NOT recommended for lengths above 75 m

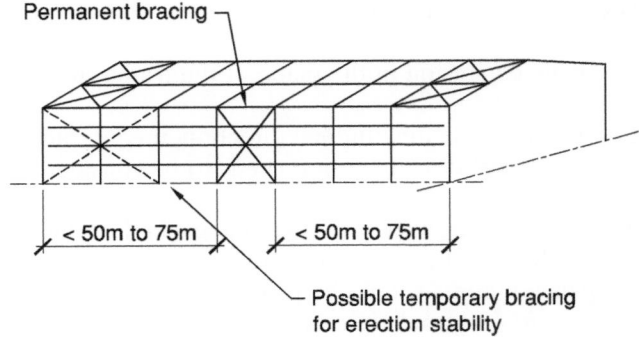

Permanent bracing

< 50m to 75m   < 50m to 75m

Possible temporary bracing
for erection stability

Note:
Where the building erection is required to start at one end of the building,
it will be necessary to provide temporary bracing to stabilise the first two
frames to be erected. This temporary bracing should be removed.

Figure 31.4   Recommended bracing arrangements

31.4.2 Particular cases

● Built-up members
   Components of built-up members can sometimes have very different temperatures, for example when the built-up member comprises a mixture of chords
located outside of the building and chords inside.
   The forces generated in the lacings or battens, due to these local temperature
differences, should be taken into account during their design.
● Erection stage
   In the same way, if the frame is erected in exceptionally hot or cold weather,
adjustment of the components should be carried out in order to allow the construction to return to its null position when the temperature is back to normal.
● Cases of fire
   It may also be necessary to ensure the free expansion of the steel structure in
the event of fire, in order to provide better stability to the components of the
structure.

31.5 Design for movement in typical multi-storey buildings

As indicated in Table 31.1, multi-storey buildings with plan dimensions in either
direction greater than 100 m for simple construction and 50 m for continuous construction, will probably require expansion joints, unless the effects of thermal expansion are considered directly in the frame analysis. The latter approach may enable
these limits to be increased significantly if the building is air conditioned.

31.6 Treatment of movement joints

The primary function of movement joints is to absorb the effects of the thermal expansion during the design working life. However, if necessary they can also act as other types of joint:

- construction joints
- separation joints
- compacting joints.

Design of movement joints has to take account of:

- building architecture
- local and overall geometry
- any forces or reactions transferred across the joint
- specified displacements and notations in one or more directions.

In most steel structures, the movement joint cuts the building into two blocks. Different approaches may be taken at the joint position, as discussed below.

31.6.1 Double frame at expansion joint location

The portal frame (in a single-storey building) or the beams or columns in a multi-storey building are repeated on both sides of the expansion joint, as shown in Figure 31.5.

As shown in Figure 31.6, the purlins are provided with cantilevers with sufficient clearance to accommodate the specified expansion.

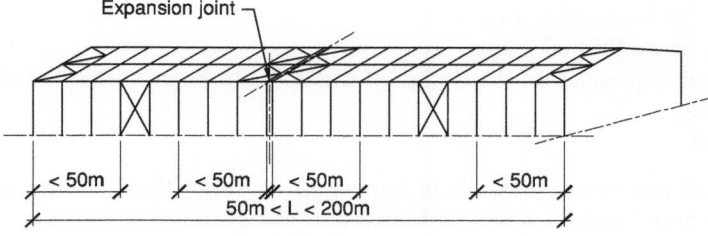

Note:
The 50m expansion length is appropriate in continental climates;
75m may be achievable in the more temperate UK climate.

Figure 31.5 Typical positioning of bracings in a long building

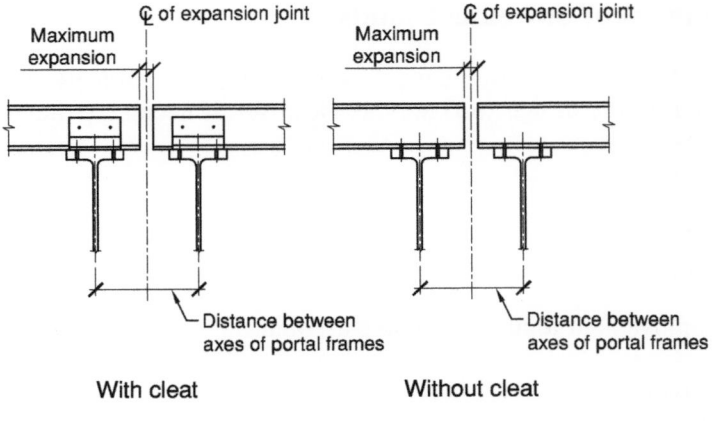

With cleat   Without cleat

Note:
Cleats are preferred with light gauge, cold formed purlins. They may
be omitted if heavier, hot rolled purlins are adopted.

Figure 31.6   Double portal frames at expansion joint

Advantages

- Possibility to absorb substantial horizontal and vertical displacements.
- Use of conventional connections and joints between the elements of the structure.
- Possibility to separate both parts of the building for the fire limit state. A fire wall may readily be built adjacent to the expansion joint.
- Solutions recommended in seismic regions (in this case, the joint must satisfy seismic design rules concerning the gap between blocks).

Disadvantages

- Modification of the grid of the building.
- Doubling of foundation works.
- Requires an additional frame.
- Serious consequences on the design of the joints to be used for cladding, roofing and sealing.
- High costs.

As with all expansion joints, it is important to detail the cladding and roofing carefully, to avoid water ingress and maximise airtightness.

31.6.2 Connection with slotted holes

This form of connection, shown in Figure 31.7, is only suitable for single-storey construction.

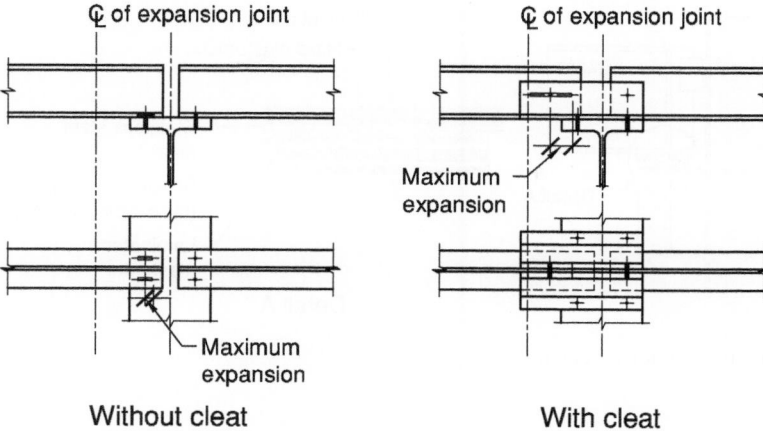

Figure 31.7   Connections with slotted holes

Advantages

- Economy of material.
- Simple fabrication.
- Low cost.
- Possibility of inserting a stainless plate between two sheets of PTFE (for example Teflon), and between two components of the structure to ensure better slip.

Disadvantages

- Very small displacements possible.
- Delicate adjustment on site of the initial position of the bolt in the slotted hole.
- Long term performance may not be achieved due to ingress of dirt and corrosion at sliding surface.

As with all expansion joints, it is important to detail the cladding and roofing carefully, to avoid water ingress and maximise air-tightness.

31.7 Use of special bearings

If high loads have to be transmitted across a movement joint, several types of special structural bearings can be used.

These are the subject of specific standards gathered under the number of European standard BS EN 1337.[4]

Two common types of bearings are presented below.

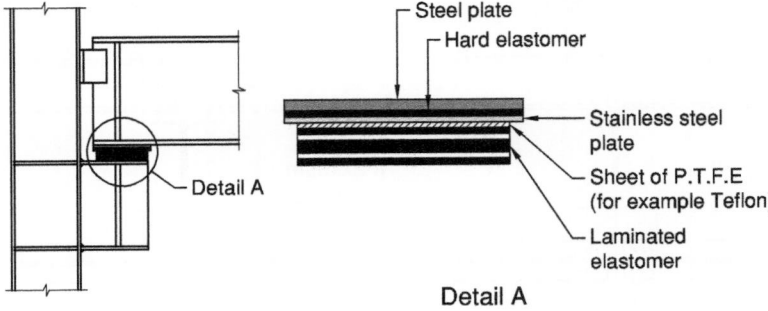

Figure 31.8   Elastomeric bearing

31.7.1 Elastomeric bearings

As shown in Figure 31.8, these bearing systems are made of a thick laminated elastomeric (steel reinforcing plates bonded between layers of elastomer). They allow horizontal displacements by deformation of the elastomeric layer into a parallelogram.

The thickness of elastomer is calculated according to the vertical load and the requirement for rotation and horizontal displacements.

When horizontal displacements are important, a bearing plate on PTFE (for example Teflon) and a stainless steel plate can be added to ensure better slip.

Advantages

- Possibility of absorbing both rotation and small vertical displacements (differential settlement of columns) at the beam support.

Disadvantages

- Expensive detailing of the supporting column.
- Difficult to design and install.

31.7.2 Pot bearings

This form of bearing, shown in Figure 31.9, is primarily used in bridge construction. However, they may reasonably be used in buildings for exceptional loading and movement. In addition to permitting thermal expansion, they can also damp oscillations and vibrations within the structure. A pot bearing may allow one-way or multi-directional slip, as well as rotation at the support. Depending on the design,

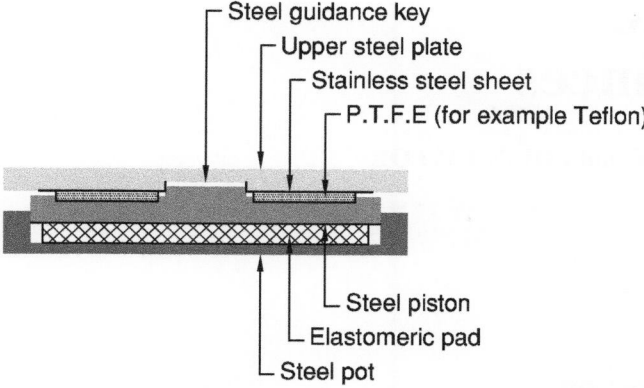

Figure 31.9 Pot bearing

pot bearings comprise a supporting base, a cushioning shock absorber, a piston (with guidance if movement in one direction is prevented), and a slip plate.

Advantages

● Developed for bridges and building structures supporting very high loads.

Disadvantages

● High cost.

References to Chapter 31

1. British Standards Institution (1991) BS EN 1991-1-5: *Eurocode 1: Actions on structures – Part 1.5: General actions – Thermal actions*. London, BSI.
2. British Standards Institution (1993) BS EN 1993-1-1: *Eurocode 3: Design of steel structures – Part 1.1: General rules and rules for buildings*. London, BSI.
3. Steel Construction Institute (2000) *Steelwork Design Guide to BS 5950: Volume 4: Essential Data for Designers*. Ascot, SCI.
4. British Standards Institution (2004) BS EN 1337-6:2004 *Structural bearings (in 11 Parts)*. London, BSI.

Chapter 32
Tolerances

ROGER POPE and COLIN TAYLOR

32.1 Introduction

32.1.1 Why set tolerances?

Compared to other structural materials, steel (and aluminium) structures can be made economically to much closer tolerances. Compared to mechanical parts, however, it is neither economic nor necessary to achieve extreme accuracy.

There are a number of distinct reasons why tolerances may need to be considered. It is important to be quite clear which actually apply in any given case, particularly when deciding the values to be specified, or when deciding the actions to be taken in cases of non-compliance.

The various reasons for specifying tolerances are outlined in Table 32.1. In all cases no closer tolerances than are actually needed should normally be specified, because while additional accuracy may be achievable, it generally increases the costs disproportionately.

32.1.2 Terminology

'Tolerance' as a general term means a permitted range of values. Other terms which need definition are given in Table 32.2.

32.1.3 Classes of tolerance

Table 32.3 defines the classes of tolerances which are recognised in Eurocode 3 and BS EN 1090-2,[1] the technical specification for execution (i.e. fabrication and erection) that supports Eurocode 3.

Table 32.1 Reasons for specifying tolerances

Structural safety	Dimensions (particularly of cross-sections, straightness, etc.) associated with structural resistance and safety of the structure.
Assembly requirements	Tolerances necessary to enable fabricated parts to be put together.
Fit-up	Requirements for fixing non-structural components, such as cladding panels, to the structure.
Interference	Tolerances to ensure that the structure does not foul with walls, door or window openings or service runs, etc.
Clearances	Clearances necessary between structures and moving parts, such as overhead travelling cranes, elevators, etc. or for rail tracks and also between the structure and fixed or moving plant items.
Site boundaries	Boundaries of sites to be respected for legal reasons. Besides plan position, this can include limits on the inclination of outer faces or tall buildings.
Serviceability	Floors must be sufficiently flat and even, and crane gantry tracks, etc. must be accurately aligned, to enable the structure to fulfil its function
Appearance	The appearance of a building may impose limits on verticality, straightness, flatness and alignment, though generally the tolerance limits required for other reasons will already be sufficient.

Table 32.2 Definitions – deviations and tolerances

Deviation	The difference between a specified value and the actual measured value, expressed vectorially (i.e. as a positive or negative value).
Permitted deviation	The vectorial limit specified for a particular deviation.
Tolerance range	The sum of the absolute values of the permitted deviations each side of a specified value.
Tolerance limits	The permitted deviations each side of a specified value, e.g. (a) ±3.5 mm or (b) +5 mm −0 mm.

Table 32.3 Classes of tolerances

Normal tolerances	Those which are generally necessary for all buildings. They include those normally required for structural safety, together with normal structural assembly tolerances.
Particular tolerances	Tolerances which are closer than normal tolerances, but which apply only to certain components or only to *certain* dimensions. They may be necessary in specific cases for reasons of fit-up or interference or in order to respect clearances or boundaries.
Special tolerances	Tolerances which are closer than normal tolerances and which apply to a *complete* structure or project. They may be necessary in specific cases for reasons of serviceability or appearance, or possibly for special structural reasons (such as dynamic or cyclic loading or critical design criteria) or for special assembly requirements (such as interchangeability or speed of assembly).

It is important to draw attention to any particular or special tolerances when calling for tenders, as they usually have cost implications. Where nothing is stated, fabricators will automatically assume that only normal tolerances are required.

32.1.4 Types of tolerances

For structural steel there are three types of dimensional tolerance:

1. **manufacturing tolerances**, such as plate thickness and dimensions of sections
2. **fabrication tolerances**, applicable in the workshops
3. **erection tolerances**, relevant to work on site.

Manufacturing tolerances are specified in standards such as BS 4-1[2], BS EN 10024,[3] BS EN 10029,[4] BS EN 10034[5] and BS EN 10210-2.[6] Only fabrication and erection tolerances will be covered here.

32.2 Standards

32.2.1 Relevant documents

The standards covering tolerances applicable to building steelwork are:

1. BS EN 1090-2 Execution of steel structures and aluminium structures: Part 2: Technical requirements for the execution of steel structures.
2. National structural steelwork specification for building construction NSSS, 5th edition.[7]
3. ISO 10721-2: 1999 Steel structures: Part 2: Fabrication and erection.[8]
4. BS 5606[9] Guide to accuracy in building.

32.2.2 BS EN 1090-2 Execution of steel structures and aluminium structures

This specification of tolerances for steelwork was first introduced into British Standards in 2008. Its scope includes steelwork for buildings, bridges and most other non-marine structures. It is a supporting technical specification for design according to the Eurocode 3. BS EN 1090-2 specifies the 'essential tolerances' that must be to meet the Eurocode requirements for mechanical resistance and stability.

Part 1 of the standard, BS EN 1090-1,[10] gives *Requirements for conformity assessment of structural components* and deals with certification of the manufacturer's factory production control which is necessary for CE Marking of components.

32.2.3 National structural steelwork specification (NSSS)

The *National structural steelwork specification for building construction* was developed to support the application of modern quality management techniques in the steelwork industry. This required a more extensive range of tolerances to be defined, principally for the purposes of manufacturing process control. This is an industry standard based on established sound practice. The widely accepted document, promoted by the British Constructional Steelwork Association (BCSA), is now in its 5th edition. The BCSA has developed a CE Marking version of the 5th edition[7] which aligns with the requirements of both Parts of BS EN 1090, as applied to general building steelwork.

32.2.4 ISO 10721-2 Steel structures: Part 2: Fabrication and erection

This is similar to BS EN 1090-2. It is unlikely to be issued as a BSI standard but may be revised for use in the global procurement market using BS EN 1090-2 as a reference standard.

32.2.5 BS 5606 Guide to accuracy in building

BS 5606 is concerned with buildings generally and is not specific to steelwork. The 1990 version has been rewritten as a guide, following difficulties due to incorrect application of the previous (1978) version, which was in the form of a code.

BS 5606 is not intended as a document to be simply called up in a contract specification. It is primarily addressed to designers to explain the need for them to include means for adjustment, rather than to call for unattainable accuracy of construction. Provided that this advice is heeded, its tables of 'normal' accuracy can then be included in specifications, except where they conflict with overriding structural requirements. This can in fact happen, so it is important to remember that the requirements of BS EN 1090-2 must take precedence over BS 5606.

BS 5606 introduces the idea of *characteristic accuracy*, the concept that any construction process will inevitably lead to deviations from the target dimensions, and its objective is to advise designers on how to avoid resulting problems on site by appropriate detailing. The emphasis in BS 5606 is on the practical tolerances which will normally be achieved by good workmanship and proper site supervision. This can only be improved upon by adopting intrinsically more accurate techniques, which are likely to incur greater costs. These affect the fit-up, the boundary dimensions, the finishes and the interference problems. Data are given on the normal tolerances (to be expected and catered for in detailed design) under two headings:

1. site construction (Table 1 of BS 5606)
2. manufacture (Table 2 of BS 5606).

Unfortunately many of the values for site construction of steelwork are only estimated. No specific consideration is given in BS 5606 to dimensional tolerances necessary to comply with the assumptions inherent in structural design procedures, which may in fact be more stringent. It does, however, recognise that special accuracy may be necessary for particular details, joints and interfaces.

Another important point mentioned in BS 5606 is the need to specify methods of monitoring compliance, including methods of measurement. It has to be recognised that methods of measurement are also subject to deviations; given the methods necessary for monitoring site dimensions, measurement deviations may in fact be quite significant compared to the permitted deviations of the structure itself.

The guidance in BS 5606 was used as a reference during the development of the NSSS and BS EN 1090-2. Hence, BS EN 1090-2 includes a series of 'functional tolerances' that are defined as geometrical tolerances which might be required to meet a function other than mechanical resistance and stability, e.g. appearance or fit-up. The values specified in BS EN 1090-2 for building steelwork are based on experience with those specified in the NSSS, and the values specified in BS EN 1090-2 for the normal (default) functional tolerance class 1 are generally the same as those in the NSSS.

As recommended by BS 5606, BS EN 1090-2 specifies the requirements for the accuracy of measuring instrumentation, e.g. surveying equipment, and the system and methods of measurement to be used for control by reference to ISO 4463-1 *Building setting out and measurement: Part 1: Methods of measuring, planning and organisation and acceptance criteria*. This ISO standard is identical to BS 5964-1,[11] which is the reference standard specified in the NSSS.

32.3 Implications of tolerances

32.3.1 Member sizes

32.3.1.1 Encasement

The tolerances on cross-sectional dimensions have to be allowed for when encasing steel columns or other members, whether for appearance, fire resistance or structural reasons. It should not be forgotten that the permitted deviations represent a further variation over and above the difference between the serial size and the nominal size.

For example, a 356 × 406 × 235UC has a nominal size of 381 mm deep by 395 mm wide, but with tolerances to BS 4 may actually measure 401 mm wide by 387 mm deep one side and have a depth of 381 mm the other side. The same is true of continental sections. A 400 × 400 × 237HD also has a nominal size of 381 mm deep by 395 mm wide, but with tolerances to BS EN 10034 may actually measure 398 mm wide by 389 mm deep one side and have a depth of 380 mm the other side.

32.3.1.2 *Fabrication*

Variations of cross-sectional dimensions (with permitted deviations) may also need to be allowed for, either in detailing the workmanship drawings or in the fabrication process itself, if problems are to be avoided during erection on site.

The most obvious case is a splice between two components of the same nominal size, where packs may be needed before the flange splice plates fit properly, unless the components are carefully matched. Similarly variations in the depths of adjacent crane girders or runway beams may necessitate the provision of packs, unless the members are carefully matched.

Less obviously, if the sizes of columns vary, the lengths of beams connected between them will need some form of adjustment, even if the columns are accurately located and the beams are exactly to length.

32.3.2 Attachment of non-structural components

It is good practice to ensure that all other items attached to the steel frame have adequate provision for adjustment in their fixings to cater for the effects of all steelwork tolerances, plus an allowance for deviations in their own dimensions. Where necessary, further allowances may be needed to cater for structural movements under load and for differential expansion due to temperature changes.

Where possible, the number of fixing points should be limited to three or four, only one of which should be positive with all the others having slotted holes or other means of adjustment.

32.3.3 Building envelope

It must be appreciated that erection tolerances, including variation in the position of the site grid lines, will affect the exact location of the external building envelope relative to other buildings or to site boundaries, and there may be legal constraints to be respected which will have to be taken into account at the planning and preliminary stages of design.

These effects also need to be taken into account where a building is intended to have provision for future extension or where the project is an extension of an existing building, in which case deviations in the actual dimensions have to be catered for at the interface.

In the case of tall multi-storey buildings, the building envelope deviates increasingly with height compared to the location at ground level, even though permitted deviations for column lean generally reduce with height. Unless there are step-backs or other features with a similar effect, it may be necessary to impose particular tolerance limits on the outward deviations of the columns.

32.3.4 Lift shafts for elevators

The deviations from verticality that can be tolerated in the construction of guides for lifts or elevators are commonly more stringent than those for the construction of the building in which they operate. In low-rise buildings, sufficient adjustment can be provided in association with the clearances, but in tall buildings it becomes necessary either to impose 'special' tolerances on column verticality or else to impose 'particular' tolerances on those columns bounding the lift shaft.

In agreeing the limits to be observed with the lift supplier, it should not be overlooked that the horizontal deflections of the building due to wind load also have implications for the verticality of the lift shafts.

32.4 Fabrication tolerances

32.4.1 Scope of fabrication tolerances

The description 'fabrication tolerances' is used here to include tolerances for all normal workshop operations except welding. It thus covers tolerances for:

1. cross sections, other than rolled sections
2. member length, straightness and squareness
3. webs, stiffened plates and stiffeners
4. holes, edges and notches
5. bolted joints and splices
6. column baseplates and cap plates.

However, tolerances for cross-sections of rolled sections and for thicknesses of plates and flats are treated as manufacturing tolerances. Welding tolerances (including tolerances on weld preparations and fit-up and sizes of permitted weld defects) are treated elsewhere.

32.4.2 Relation to erection tolerances

An overriding requirement for accuracy of fabrication must always be to ensure that it is possible to erect the steelwork within the specified erection tolerances.

Due to the wide variety of steel structures and the even wider variety of their components, any recommended tolerances must always be specified in a very general way. Even if it were possible to specify fabrication tolerances in such a way that their cumulative effect would always permit the specified erection tolerances to be satisfied, the resulting permitted deviations would be so small as to be unreasonably expensive, if not impossible to achieve.

Fortunately in most cases it is possible to rely on the inherent improbability of all unfavourable extreme deviations occurring together. Also the usually accepted values for fabrication tolerances do make some limited allowances for the need to avoid cumulative effects developing on site. They are tolerances that have been shown by experience to be workable, provided that simple means of adjustment are incorporated where the effects of a number of deviations could otherwise become cumulative. For example, beams with bolted end cleats usually have sufficient adjustment available due to hole clearances, but where a line of beams all have end-plate connections, provision for packing at intervals may be advisable, unless other measures are taken to ensure that the beams are not all systematically over-length or under-length by the normal permitted deviation. Other possible means for adjustment include threaded rods and slotted holes.

Where it can be seen from the drawings that the fabrication tolerances could easily accumulate in such a way as to create a serious problem in erection, either closer tolerances or means of adjustment should be considered; however, the coincident occurrence of all extreme deviations is highly improbable, and judgement should be exercised both on the need for providing means of adjustment and on the range of adjustment to be incorporated.

32.4.3 Full contact bearing

32.4.3.1 Application

The requirements for contact surfaces in joints which are required to transmit compression by 'full contact bearing' probably cause more trouble than any other item in a fabrication specification, largely due to misapprehension of what is actually intended to be achieved.

First it is necessary to be clear about the kind of joint to which the requirements for full contact bearing should be applied. Figure 32.1(a) shows the normal case, where the profile of a member is required to be in full contact bearing on a baseplate or cap plate or division plate. The stress on the contact area equals the stress in the member: thus full contact is needed to transmit this stress from the member into the plate. Only that part of the plate in contact with the member need satisfy the full contact bearing criteria, though it may be easier to prepare the whole plate.

Figure 32.1(b) shows two end-plates in simple bearing. The potential contact area is substantially larger than the cross-sectional area of the member: thus full contact bearing is not necessary. All that is needed is for the end-plates to be square to the axis of the member. Another common case of simple bearing is shown in Figure 32.1(c).

By contrast, the case shown in Figure 32.1(d) is one where, if full contact bearing is needed, it is also necessary to take special measures to ensure that the profiles of the two members align accurately, otherwise the area in contact may be significantly less than the area required to transmit the load. Particular tolerances should be specified in such cases, based on the maximum local reduction of area that can be accepted according to the design calculations. Alternatively, a division plate could

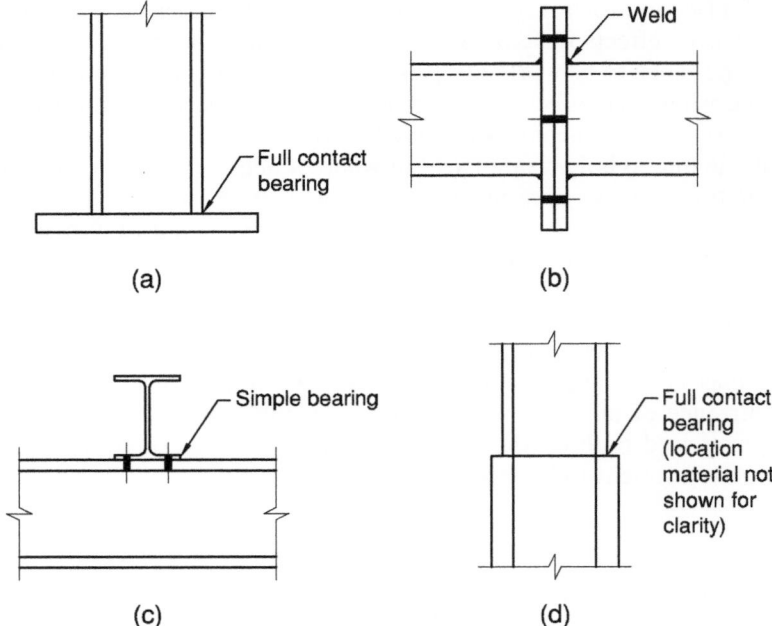

Figure 32.1   Types of member-to-member bearing: (a) profile to plate, (b) plate to plate, (c) flange to flange, (d) profile to profile (accurate alignment necessary)

be introduced; if the stresses are high this may well prove to be the most practical solution.

32.4.3.2 *Requirements*

Where full contact bearing is required, there are, in fact, three different criteria involved:

1. squareness
2. flatness
3. smoothness.

32.4.3.3 *Squareness*

If the ends of a length of column are not square to its axis, then after erection either the column will not be vertical or else there may be tapered gaps at the joints, depending on the extent to which surrounding parts of the structure prevent the column from tilting. Under load any such gap will try to close, exerting extra forces on the surrounding members. In addition, either a gap or a tilt will induce a local eccentricity in the column.

A practical erection criterion is that the column should not lean more than 1 in x (where x is 600 in NSSS and 500 in BS EN 1090-2). This slope is measured relative to a line joining the centres of each end of the column length, referred to as the *overall centreline*. The column is also allowed a *lack of straightness* tolerance of ($L/1000$ in NSSS and $L/750$ in BS EN 1090-2), which corresponds to end slopes of about 1/300 (see Figure 32.2(a)). It is thus necessary to specify end squareness criteria relative to the overall centreline, rather than to the local centreline adjacent to the end (see Figure 32.2(b)).

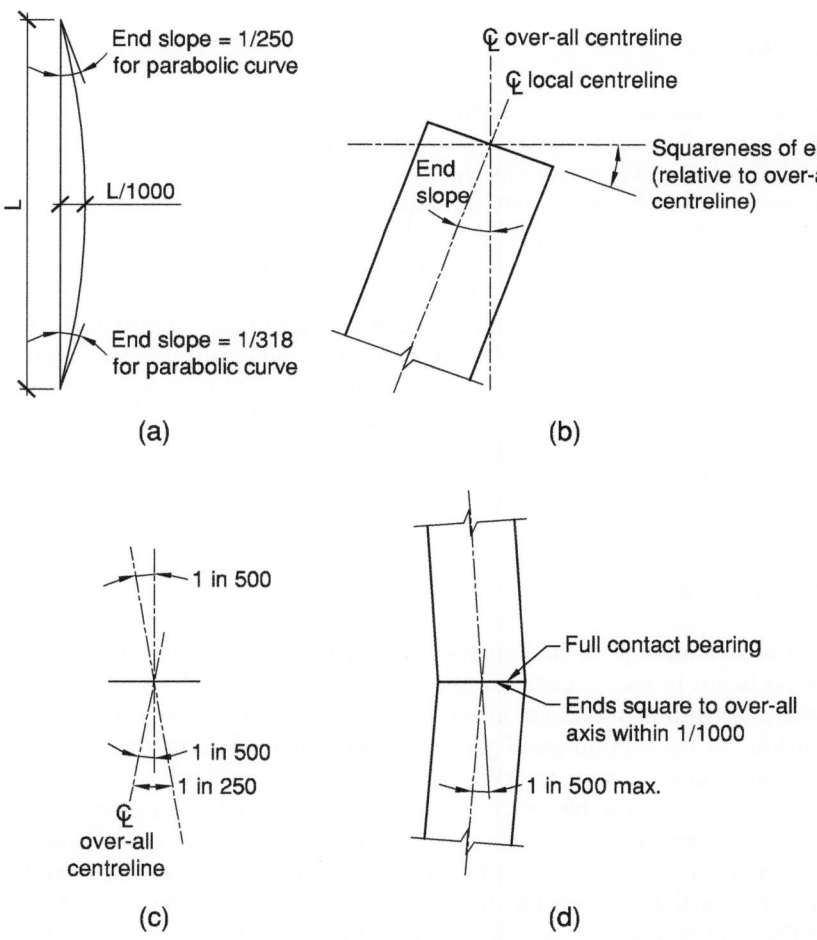

(a)

(b)

(c)

(d)

Figure 32.2 Squareness of column ends: (a) bow of L/1000 giving end slopes of about L/300, (b) squareness of end measured relative to overall centreline, (c) change of direction at a braced joint, (d) end squareness at full contact bearing splice

There is generally a design assumption that the line of action of the force in the column does not change direction at a braced joint by more than 1/250, requiring an end squareness in a simple bearing connection (relative to the overall axis of the member) of 1/500 (see Figure 32.2(c)). However, full contact bearing generally arises at column splices which are not at braced points, so an end squareness tolerance of 1/1000 is usually specified, producing a maximum change of slope of 1/500 (see Figure 32.2(d)).

Once a column has been erected, it is more practical to measure the remaining gaps in a joint. These gaps are affected not only by the squareness of the ends but also by the second criterion, flatness.

32.4.3.4 *Flatness*

Ends have to be reasonably flat (as distinct from curved or grossly uneven) to enable the load to be transferred properly. Following a history of arguments over appropriate specifications, the American Institute of Steel Construction (AISC) commissioned some tests, which are the basis for their current specifications.

It was found that a surprisingly high tolerance was quite acceptable, and that beyond its limit (or to compensate for end squareness deviations) the use of localised packs or shims was acceptable. Basically similar rules are now beginning to appear in other specifications including the EN standard (see Section 32.5.6 in relation to erection tolerances). This is an essentially simple and effective method of correcting excessive gaps on site (see also Section 32.5.6). However, inserting shims into column joints is not a matter to be undertaken lightly. It is normally more economic to avoid the need for shimming by working to close fabrication tolerances in joints where full contact bearing is required.

32.4.3.5 *Smoothness*

In the light of the findings of the flatness tests, it can be appreciated that if absolute local flatness is not in fact needed, absolute smoothness is irrelevant also.

The best description of the smoothness that is needed is the smoothness of a surface produced by a good-quality modern saw in proper working order. This degree of smoothness is indeed very good.

Where sawing is not possible, ending machines (i.e. special end-milling machines) can be used for correcting the squareness (or flatness) of ends of built-up (fabricated) columns, such as box columns or other welded-up constructions. Where baseplates are not flat and are too thick to be pressed flat, either they are milled locally in the contact zone or else planing machines are used.

However, it cannot be overemphasised that the normal preparation for a rolled section column required to transmit compression by full contact in bearing is by saw cutting square to the axis of the member.

It is, of course, unnecessary to flatten the undersides of baseplates supported on concrete foundations.

32.4.4 Other compression joints

Compression joints, transferring compression through end-plates in simple bearing, also need to have their ends square to the axis. If, after the members have been firmly drawn together, a gap remains which would introduce eccentricity into the joint, it should be shimmed.

32.4.5 Lap joints

Steel packs should be used where necessary to limit the maximum step between adjacent surfaces in a lap joint (see Figure 32.3) to 2 mm with ordinary bolts or 1 mm (before tightening the bolts) where preloaded (or HSFG as previously termed) bolts are used.

32.4.6 Beam end-plates

Where the length of a beam with end-plates is too short to fit between the supporting columns, or other supporting members, packs should be supplied to make up the difference.

Gaps arising from distortion caused by welding, as shown in Figure 32.4, need not be packed if the members can be firmly drawn together. However, they may need to be filled or sealed to avoid corrosion where the steelwork is external or is exposed to an aggressive internal environment.

32.4.7 Values for fabrication tolerances

The values for fabrication tolerances currently given in the NSSS are reproduced for convenience in Table 32.4. Each of the specified criteria should be considered

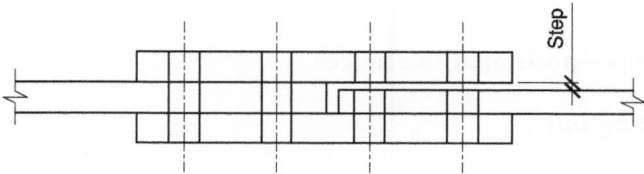

Figure 32.3   Maximum step between adjacent surfaces

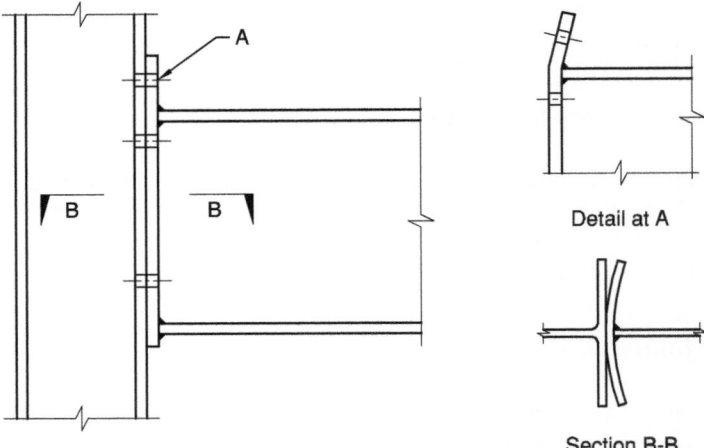

Detail at A

Section B-B

Figure 32.4　End-plate with welding (exaggerated)

and satisfied separately. The cumulative effect of several permitted deviations should not be considered as overriding the specific criteria.

These values represent current practice and are taken from the fourth edition of the NSSS.

The clause members referred to in Table 32.4 are clause numbers in the NSSS, which should be referred to for further information.

32.5 Erection tolerances

32.5.1 Importance of erection tolerances

Erection tolerances potentially have a significant effect on structural behaviour. There are four matters to be considered:

1. overall position
2. fixing bolts
3. internal accuracy
4. external envelope.

32.5.2 Erection – positional tolerances

32.5.2.1 Setting out

The position in plan, level and orientation can only be defined relative to some fixed references, such as the National Grid and the Ordnance datum level. From the

Table 32.4   Extract from National Structural Steelwork Specification 5th Edition)

SECTION 7

WORKMANSHIP - ACCURACY OF FABRICATION

7.1   PERMITTED DEVIATIONS

Permitted deviations in cross section, length, straightness, flatness, cutting, holing and position of fittings shall be as specified in 7.2 to 7.5 below.

7.2   PERMITTED DEVIATIONS FOR ROLLED COMPONENTS AFTER FABRICATION (Δ)
(including structural hollow sections)

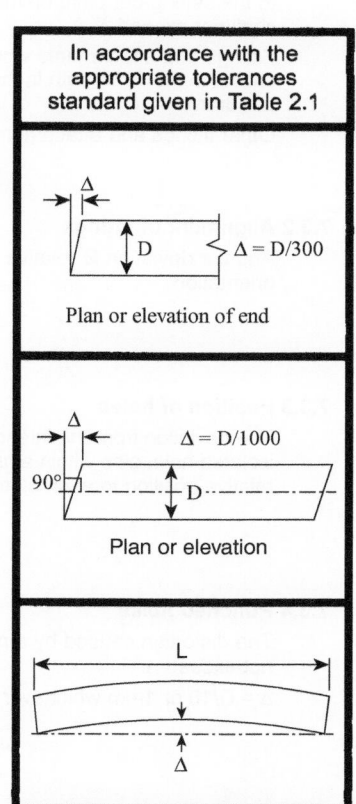

7.2.1 Cross section after fabrication

In accordance with the appropriate tolerances standard given in Table 2.1

7.2.2 Squareness of ends not prepared for bearing
Note: See also 4.3.3

$\Delta = D/300$

Plan or elevation of end

7.2.3 Squareness of ends prepared for bearing

Prepare ends with respect to the longitudinal axis of the member.

Note: See also 4.3.3

$\Delta = D/1000$

Plan or elevation

7.2.4 Straightness on both axes

Generally $\Delta = L/1000$ or 3mm whichever is greater.

For components fabricated from structural hollow sections $\Delta = L/500$ or 3mm whichever is greater.

Table 32.4 (*Continued*)

7.2.5 Length

Length after cutting, measured on the centre line of the section or on the corner of angles.

7.2.6 Curved or cambered

Deviation from intended curve or camber at mid-length of curved portion when measured with web horizontal.

Deviation = L/1000 or 6mm whichever is greater.

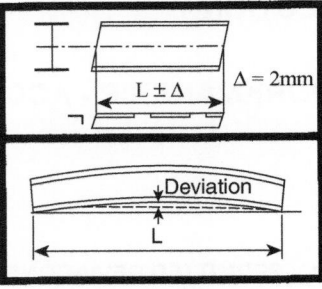

7.3   PERMITTED DEVIATIONS FOR ELEMENTS OF FABRICATED COMPONENTS (Δ)

7.3.1 Position of fittings

The deviation from the intended position relative to the setting-out point on the primary member shall not exceed Δ.

Fittings and attachments whose location is critical to the force path in the structure:
Δ = 3mm

Other fittings and attachments: Δ = 5mm

7.3.2 Alignment of fittings

Angular deviation Ø relative to intended local orientation.

7.3.3 Position of holes

The deviation from the intended position of an isolated hole, also within a group of holes, the relative position to each other shall not exceed Δ.

7.3.4 Punched holes

The distortion caused by a punched hole shall not exceed Δ.

Δ = D/10 or 1mm whichever is greater.

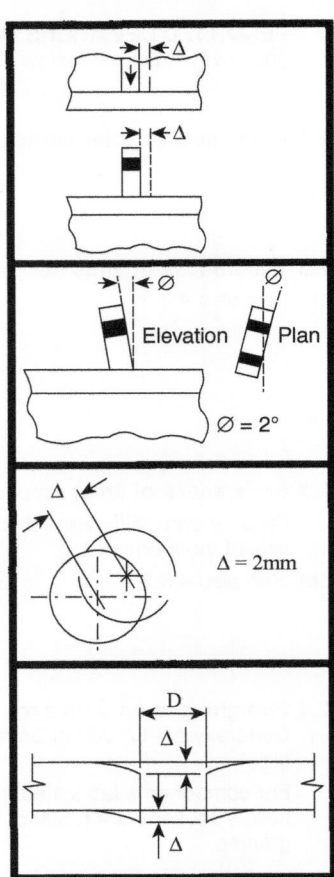

Table 32.4 (*Continued*)

7.3.5 Sheared or cropped edges of plates or angles

The deviation from a 90° edge shall not exceed Δ.

Δ = t/10 up to a maximum of 3mm.

7.3.6 Flatness

Where full contact bearing is specified, the flatness shall be such that when measured against a straight edge not exceeding one metre long, which is laid against the full bearing surface in any direction, the gap does not exceed Δ.

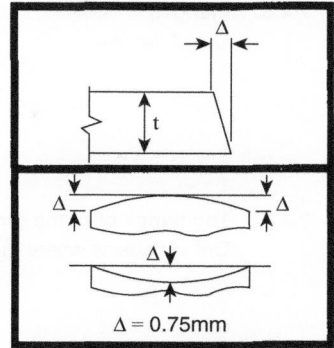

Δ = 0.75mm

7.4   PERMITTED DEVIATIONS FOR PLATE GIRDER SECTIONS (Δ)

7.4.1 Depth

Depth on centre line.

$D \pm \Delta$

Δ = 4mm

7.4.2 Flange width

Width of B_w or B_n.

$B_w \pm \Delta$

B_w or B_n < 300mm
Δ = 3mm

B_w or B_n ≥ 300mm
Δ = 5mm

$B_n \pm \Delta$

7.4.3 Squareness of section

Out of squareness of flanges.

Δ = B/100 or 3mm whichever is greater.

B Flange width

Δ = B/100 or 3mm whichever is greater

7.4.4 Web eccentricity

Position of web from edge of flange.

b
±Δ

b Nominal dimension
Δ = 5mm

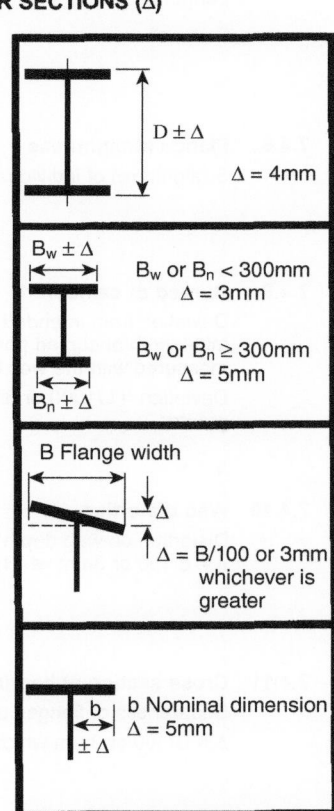

Table 32.4   (*Continued*)

7.4.5   Flanges

Out of flatness.

7.4.6   Top flange of crane girder

Out of flatness where the rail seats.

7.4.7   Length

Length on centre line.

7.4.8   Flange straightness

Straightness of individual flanges.

7.4.9   Curved or cambered

Deviation from intended curve or camber at mid-length of curved portion when measured with the web horizontal.

Deviation = L/1000 or 6mm whichever is greater.

7.4.10   Web distortion

Distortion on web depth or gauge length.
Δ = d/150 or 3mm whichever is greater.

7.4.11   Cross section at bearings

Squareness of flanges to web.
Δ = D/300 or 3mm whichever is greater.

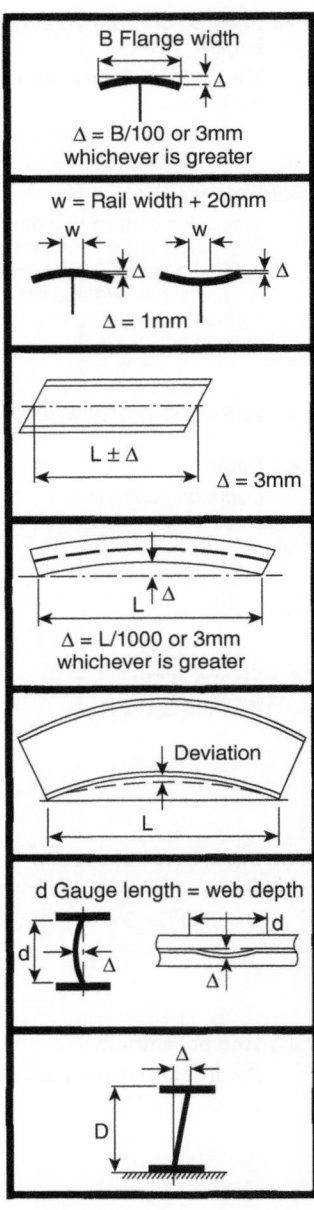

Table 32.4 (*Continued*)

7.4.12 Web stiffeners
Straightness of stiffener out of plane with web after welding.

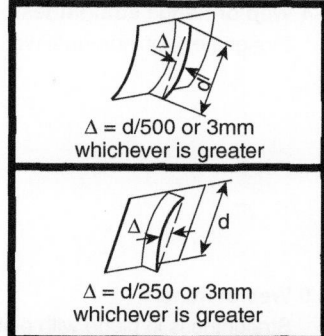

$\Delta = d/500$ or 3mm whichever is greater

7.4.13 Web stiffeners
Straightness of stiffener in plane with web after welding.

$\Delta = d/250$ or 3mm whichever is greater

7.5 PERMITTED DEVIATIONS FOR BOX SECTIONS (Δ)

7.5.1 Plate widths
Width of B_f or B_w.

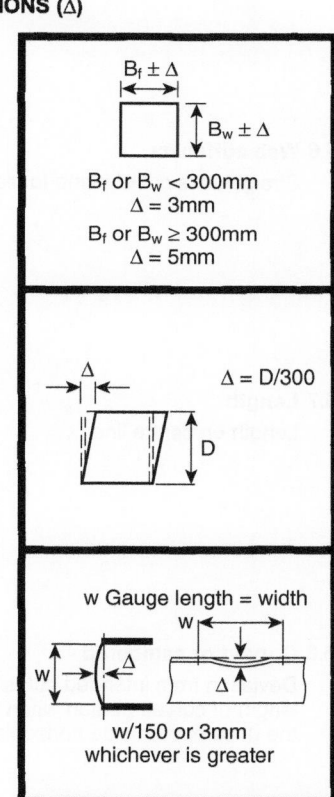

B_f or $B_w < 300$mm
$\Delta = 3$mm
B_f or $B_w \geq 300$mm
$\Delta = 5$mm

7.5.2 Squareness
Squareness at diaphragm positions.

$\Delta = D/300$

7.5.3 Plate distortion
Distortion on width or gauge length.

w Gauge length = width

w/150 or 3mm whichever is greater

988   *Tolerances*

Table 32.4 (*Continued*)

7.5.4 Web or flange straightness

Straightness of individual web or flanges.

7.5.5 Web stiffeners

Straightness in plane with plate after welding.

7.5.6 Web stiffeners

Straightness out of plane to plate after welding.

7.5.7 Length

Length on centre line.

7.5.8 Curved or cambered

Deviation from intended curve or camber at mid-length of curved portion when measured with the uncambered side horizontal.

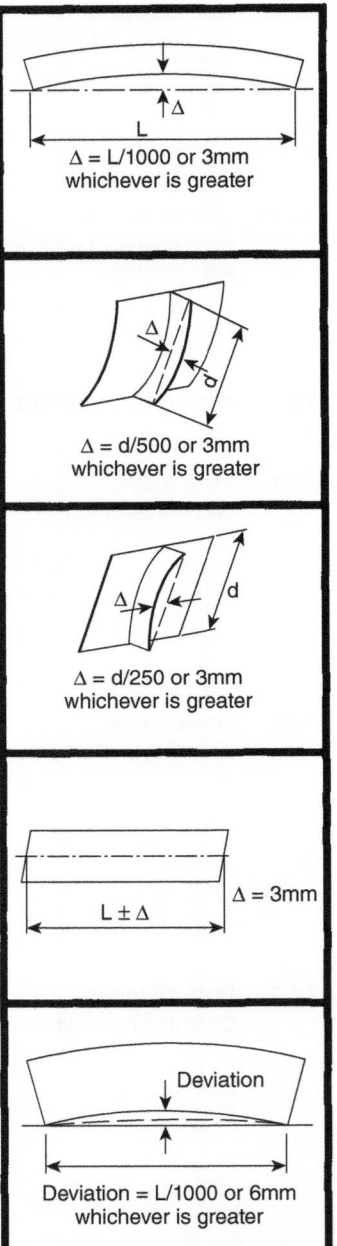

Δ = L/1000 or 3mm
whichever is greater

Δ = d/500 or 3mm
whichever is greater

Δ = d/250 or 3mm
whichever is greater

Δ = 3mm

Deviation = L/1000 or 6mm
whichever is greater

national system, it is usual to set subsidiary site datum points, and often a site datum level, and then refer the accuracy of the structure to these.

For any site the use of a grid of established column lines together with an established site level is strongly recommended. For a large site it is virtually indispensable. To help appreciate this, consider what happens on the site of a steel structure.

32.5.2.2 Site practice

Normal site practice is for the supporting concrete foundations, and other supporting structures, to be prepared in advance of steel erection, generally by an organisation separate from the steel erector. Depending on the system of holding-down bolts or other fixings to be used, this may involve casting-in of holding-down bolts, preparation of pockets in the concrete, and preparation of surfaces to receive fixings to the steelwork.

Even with care, the standard of accuracy achievable is limited, and the concrete requires time to harden to a sufficient strength for steel erection to proceed. Once all the foundations etc. are available for steel erection (or at least a sufficient proportion of them on a large site), it is prudent to survey them to review their accuracy.

32.5.2.3 Established column lines and established site level

From this survey it is convenient to introduce a grid of established column lines (ECL) and an established site level (ESL) of the foundations and other supporting structures in such a way that the positions and levels of steel columns etc. can readily be related to the site grid and site level.

The established column lines are defined as that grid of site grid lines that best represents the actual mean positions of the installed foundations and fixings. Similarly the established site level is defined as that level which best represents the actual mean level of the installed foundations. Of course it should also be verified that the deviation of the ECL grid and the ESL from those specified are within the relevant permitted deviations.

When setting out and measurement are undertaken in accordance with ISO 4463-1 (i.e. BS 5964-1), the ECL and ESL concepts are used to define 'position points' which mark the intended position in plan and level for the erection of individual columns (noting that in practice the actual physical reference mark on site will generally be offset from the position point itself). The term used for the system of established column lines in ISO 4463-1 is the 'secondary net'. BS EN 1090-2 requires that the secondary net survey should be documented and used as the reference system for setting out the steelwork and establishing the deviations of supports – including both foundations for column bases and other fixing positions to embedded supports, anchors or bearings.

32.5.3 Erection – fixing bolts

32.5.3.1 Types of fixing bolts

Fixing bolts include both holding-down bolts for columns and various types of fixing bolts used to locate or to support other members, such as beams or brackets carried by walls or concrete members.

Holding-down bolts and other fixing bolts are either:

1. fixed in position, or
2. adjustable, in sleeves or pockets.

32.5.3.2 Fixed bolts

Fixed bolts used to be solidly cast in, an operation requiring care and the use of jigs or templates to achieve accurately. However, they are now also commonly produced by placing resin-grouted bolts in holes drilled in the concrete after casting. It may also be possible to use expanding bolts.

In whatever way fixed bolts are achieved, they need to be positioned accurately, as the only adjustment possible is in the steelwork, so relatively close tolerances are normally specified.

32.5.3.3 Adjustable bolts

Adjustable bolts are placed in tubes or in tapered trapezoidal or conical holes cast in the concrete, so that a degree of movement of the threaded end of the bolt is possible, while the other end is held in place by a steel washer or other anchoring device embedded in the concrete.

This alternative permits the use of more easily achieved tolerances for the bolts, while using relatively simple details for the steelwork. Adjustment of the bolt necessitates its axis deviating from the vertical to some extent, and the holes in the steelwork need to be large enough to allow for this, particularly if the baseplate is thick. The use of loose plate washers is recommended to span oversize holes if necessary. If required they can be welded in place after the bolts are tightened, but this should not normally be necessary. 'Particular' tolerances need to be worked out for each case, depending on the details, including the length of the bolts, because this affects their slope.

32.5.3.4 Length of bolts

It is also important to ensure that the top of a holding-down bolt is at the correct level so that the nuts can be fitted properly after erection. To provide the necessary tolerances for the fixing of the bolts they should be longer than theoretically required, long threaded lengths should be provided, and the nominal level for the top should be above the theoretical position.

Similar considerations apply to the lengths of fixing bolts located horizontally. BS EN 1090-2 and the NSSS both specify requirements for fixing bolts in foundations and at other support positions.

32.5.4 Erection – internal accuracy

In terms of structural performance, the main erection tolerance is verticality of columns; positions of beams etc. on brackets may also be important. Levels of beams, particularly of one end relative to the other end and of one beam relative to the next one, are important in terms of serviceability.

Otherwise the internal accuracy of one part of the structure relative to another is largely a matter of assembly tolerances, provided that these do not cause any problem of fit-up, interference or clearances. Where the structural accuracy resulting from the assembly tolerances is liable to infringe any of these limits, 'particular' tolerances should be specified.

The necessary tolerances are specified in relation to readily identifiable points and levels. For columns and other vertical members, the reference points are conveniently defined as the actual centre of the member at each end of the fabricated piece. For beams and other horizontal members the reference points are more conveniently defined by the actual centre of the top surface at each end. Either the column system or the beam system should be used for any other cases, and the relevant system should be indicated on the erection drawings. The tolerances are then defined by the permitted deviations of these reference points from the position points established with respect to the secondary documented net survey of supports explained in Section 32.5.2.3.

Further up the building the established floor level EFL is defined as that level which best represents the actual mean level of the as-built floor levels. The EFL must not deviate from the specified floor level (relative to the ESL) by more than the permitted deviation for height of columns.

The reference points for each beam must then be within the permitted deviation from the EFL. In addition the difference in level of each end of a beam and the difference in level between adjacent beams must also be within their respective limits.

In the case of columns, the permitted deviations at each level form an 'envelope' within which the column must lie at all levels. In addition, the permitted inclination of each column within a storey height is limited, but except where columns are fabricated as individual storey-height pieces, the overall envelope normally governs.

32.5.5 Erection – external envelope

Generally the same erection tolerances for verticality apply to external columns as to internal columns. When the envelope of extreme permitted deviations is plotted from the extreme position of the base (allowing for the permitted deviation of the position point from the theoretical position as well as the permitted deviation of the column base from the position point), it may be found that this is unacceptable in terms of site boundaries or building lines, especially for a tall multi-storey building. If so, 'particular' tolerances should be specified.

Fit-up problems with cladding could also occur if alternate adjacent columns at the periphery were allowed too large a deviation alternately in and out from the theoretical line of the building face. Even if the fit-up problems could be overcome, the visual appearance might be affected. Again, 'particular' tolerances should be specified if necessary.

32.5.6 Shimming full contact bearing splices

As mentioned in Section 32.4.3.4 in relation to fabrication tolerances, tests commissioned by AISC, and used as the basis for several modern standards, showed that shims can be used to reduce gaps in full contact bearings to within the specified tolerances. Shimmed gaps up to 6.35 mm were tested, so it is not prudent to permit shimming for gaps exceeding 6 mm; gaps larger than this should be corrected by other means.

Gaps which would otherwise remain over the specified tolerance when the members are in their final alignment should be shimmed. As the tests were on flat shims, it is acceptable to use flat shims in practice. In the tests the shims were of mild steel, and this is permitted in the AISC specification and BS EN 1090-2.

The shims should be inserted such that no remaining gap exceeds the specified permitted deviation. Short lengths of shim are appropriate in a variety of thicknesses in steps not exceeding the permitted deviation. No more than three layers of shims should be used at any point, and preferably only one or two. The shims (and the lengths of columns) may be held in place by means of a partial penetration butt weld extending over the shims (see Figure 32.5(a)).

In bolted compression splices, bolted 'finger' shims (shaped as indicated in Figure 32.5(b)) can be used.

In some cases shims can be driven in, but if so they need to be fairly robust (usually over 2 mm thick), so shims of various thicknesses are needed throughout the joint. Driven shims are best limited to vertical joints e.g. between a beam end-plate and a

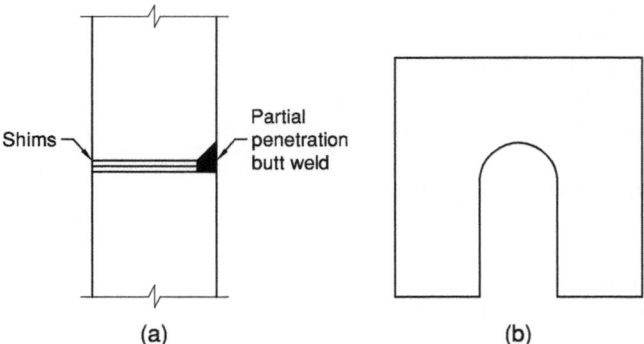

Shims

Partial penetration butt weld

(a)   (b)

Figure 32.5   Shims for full contact bearing: (a) shims with partial penetration butt weld, (b) finger shim

column. More commonly the joint must be jacked or wedged open (or else the upper portion lifted by a crane) so that the shims can be inserted. Tapered shims are particularly difficult to insert; as they are not necessary they are best avoided.

32.5.7 Values for erection tolerances

The values for erection tolerances are given in Table 32.5. Each of the specified criteria should be considered and satisfied separately. The permitted deviations

Table 32.5   Extract from National Structural Steelwork Specification 5th Edition

SECTION 9

WORKMANSHIP - ACCURACY OF ERECTED STEELWORK

9.1   PERMITTED DEVIATIONS FOR FOUNDATIONS, WALLS AND FOUNDATION BOLTS (Δ)

Note: The permitted deviations in 9.1.1 to 9.1.5 are consistent with the National Structural Concrete Specification.

9.1.1 Foundation level

Deviation from specified level.

9.1.2 Vertical wall

Deviation from specified position at steelwork support point.

Δ = ± 15mm up to 4m height

Δ = ± 20mm above 4m plus 1mm for every metre above 8m height to ± 50mm maximum

9.1.3 Pre-set foundation bolt or bolt groups if prepared for adjustment

Deviation from specified position.

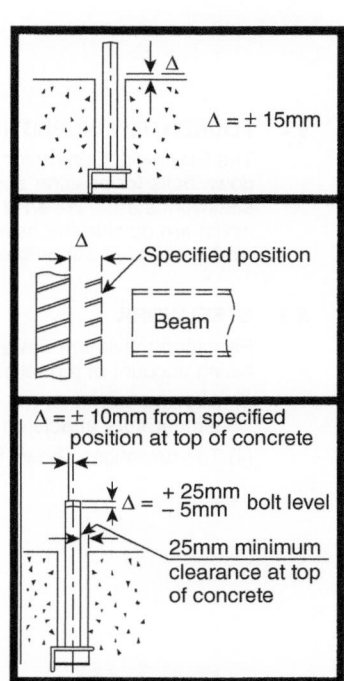

Δ = ± 15mm

Specified position

Beam

Δ = ± 10mm from specified position at top of concrete

Δ = $^{+ 25mm}_{- 5mm}$ bolt level

25mm minimum clearance at top of concrete

Table 32.5 (*Continued*)

9.1.4 Pre-set foundation bolt or bolt groups if not prepared for adjustment

Deviation from specified position.

9.1.5 Pre-set wall bolt or bolt groups if not prepared for adjustment

Deviation from specified position.

Note: This is measured locally relative to the achieved verticality of the wall as specified in 9.1.2.

9.1.6 Embedded cast-in fixing plates

Deviation of centre lines from specified positions.

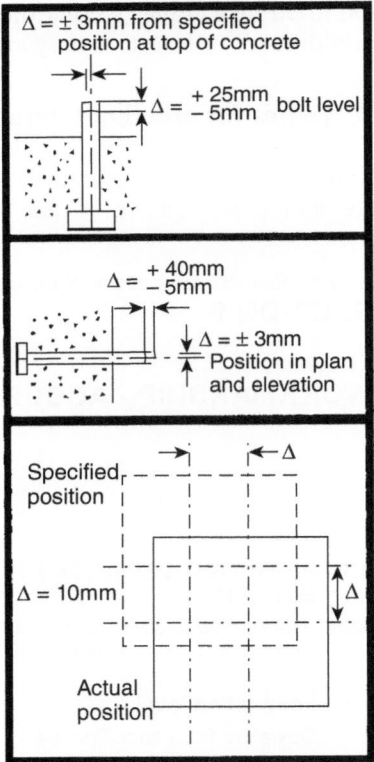

9.2 FOUNDATION INSPECTION

The Steelwork Contractor shall inspect the prepared foundations and holding down bolts for position and level not less than seven days before erection of steelwork starts. He shall then inform the Employer if he finds any discrepancies which are outside the deviations specified in clause 9.1 requesting that remedial work be carried out before erection commences.

9.3 STEELWORK

Permitted maximum deviations in erected steelwork shall be as specified in 9.6 taking account of the following:

(i) All measurements to be taken in calm weather, and due note is to be taken of temperature effects on the structure (see 8.6.2).

(ii) The deviations shown for open sections apply also to box and tubular sections.

Table 32.5 (*Continued*)

9.4 DEVIATIONS

The Steelwork Contractor shall as soon as possible inform the Engineer of any deviation position of erected steelwork which is greater than the permitted deviation in 9.6 so that the effect can be evaluated and a decision reached on whether remedial work is needed.

Note: The survey and assessment of deviations of erected steel frames is described in Annex A of the Commentary.

9.5 INFORMATION FOR OTHER CONTRACTORS

The Engineer shall advise contractors engaged in operations following steel erection of the deviations acceptable in this document in fabrication and erection, so that they can provide the necessary clearances and adjustments.

9.6 PERMITTED DEVIATIONS OF ERECTED COMPONENTS (Δ)

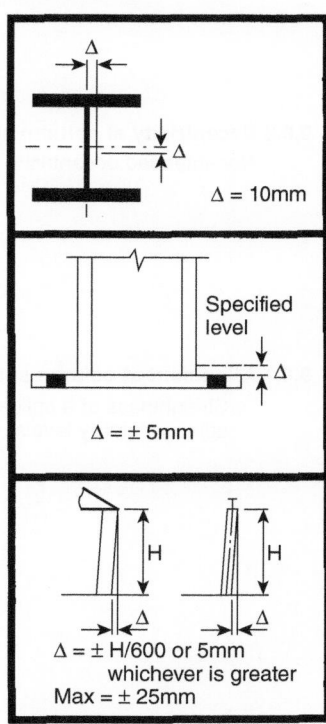

9.6.1 Position of columns at base

Deviation of section centre line from the specified position.

Δ = 10mm

9.6.2 Level of columns at base

Deviation of the top of the base plate from the specified level.

Specified level

Δ = ± 5mm

9.6.3 Single storey columns plumb

Deviation of top relative to base, excluding portal frame columns, on main axes.

Note: See clause 1.2A(xvii) and 3.4.4(iii) regarding pre-setting portal frames.

H

Δ = ± H/600 or 5mm
whichever is greater
Max = ± 25mm

Table 32.5 *(Continued)*

9.6.4 Multi-storey columns plumb

Deviation in each storey and maximum deviation relative to base for up to 10 storeys.

Note: Permitted deviations for columns over 10 storeys to be agreed with the Engineer.

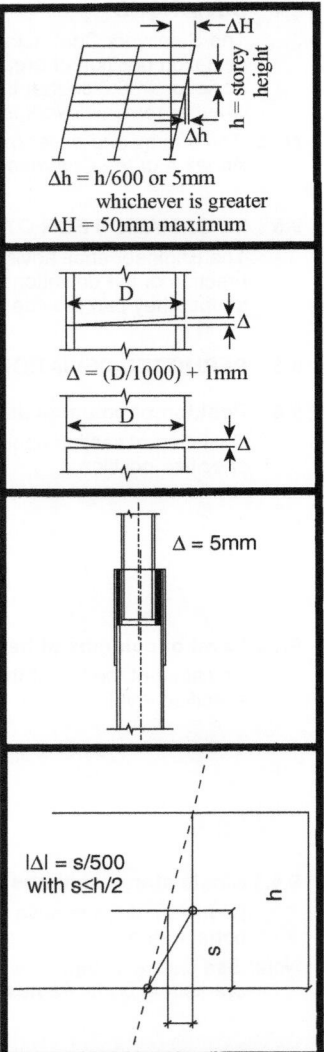

Δh = h/600 or 5mm
whichever is greater
ΔH = 50mm maximum

9.6.5 Gap between bearing surfaces

Note: See also clauses 4.3.3, 6.2.1 and 7.2.3

$\Delta = (D/1000) + 1mm$

9.6.6 Eccentricity at column splice

Non-intended eccentricity about either axis.

$\Delta = 5mm$

9.6.7 Alignment at column splice

Straightness of a spliced column between adjacent storey levels.

$|\Delta| = s/500$
with $s \leq h/2$

Table 32.5 (*Continued*)

9.6.8   Alignment of adjacent perimeter columns

Deviation relative to next column on a line parallel to the grid line when measured at base or splice level.

9.6.9   Beam level

Deviation from specified level at supporting column.

9.6.10   Level at each end of same beam

Relative deviation in level at ends.

9.6.11   Level of adjacent beams within a distance of 5 metres

Deviation from relative horizontal levels (measured on centre line of top flange).

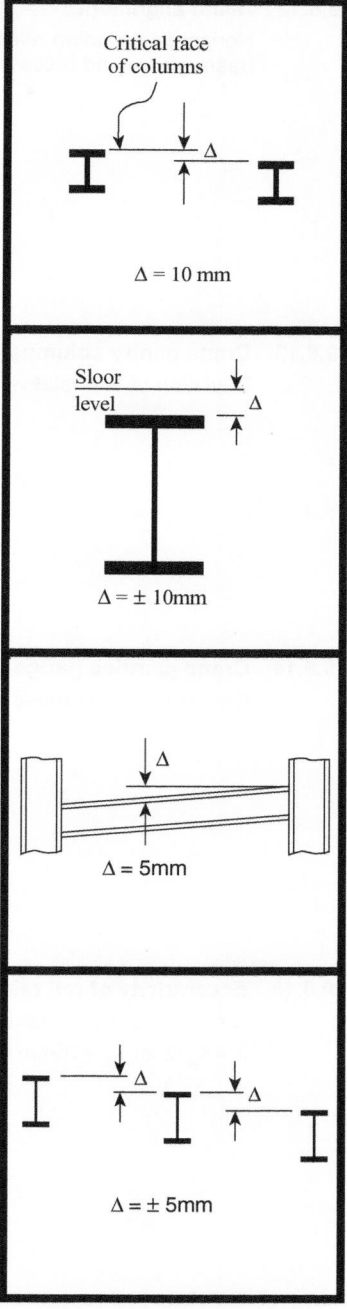

Table 32.5 (*Continued*)

9.6.12 Beam alignment

Horizontal deviation relative to an adjacent beam above and below.

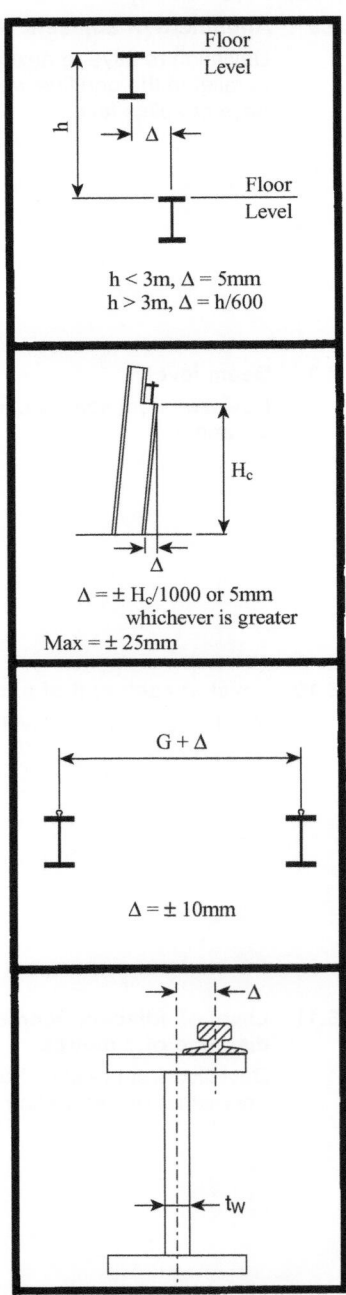

$h < 3m, \Delta = 5mm$
$h > 3m, \Delta = h/600$

9.6.13 Crane gantry columns plumb

Deviation of cap relative to base.

$\Delta = \pm H_c/1000$ or $5mm$
whichever is greater
$Max = \pm 25mm$

9.6.14 Crane gantries gauge of rail tracks

Deviation from nominal gauge.

$G + \Delta$

$\Delta = \pm 10mm$

9.6.15 Eccentricity of rail relative to web

$\Delta = 5mm$ for $t_w < 10mm$
$\Delta = t_w/2$ for $t_w > 10mm$

Table 32.5 (*Continued*)

9.6.16 Rail surface at joints in gantry crane rails Deviation in level at rail joint.	 $\Delta = 0.5\text{mm}$
9.6.17 Rail edge at joints in gantry crane rails Deviation in line at rail joint.	 $\Delta = 1\text{mm}$
9.6.18 Profiled steel floor decking Deviation of dimension between decking edge trim prior to concrete placement and perimeter beam. *Note: The deviation (as shown) between actual beam centre line and intended beam centre line relative to local grid arises from other permitted tolerances (e.g. 9.6.4).*	 $\Delta = \pm\,10\text{mm}$

should not be considered as cumulative, except to the extent that they are specified relative to points or lines that also have permitted deviation. These values represent current practice and are taken from the fifth edition of the NSSS.

The clause numbers referred to in Table 32.5 are clause numbers in the NSSS, which should be referred to for further information.

Acknowledgement

Extracts from the *National Structural Steelwork Specification* 5th edition are reproduced with the kind permission of the British Constructional Steelwork Association.

References to Chapter 32

1. British Standards Institution (2008) BS EN 1090-2: 2008 *Execution of steel structures and aluminium structures: Part 2: Technical requirements for the execution of steel structures*. London, BSI.
2. British Standards Institution (2005) BS 4-1: 2005 *Structural steel sections: Specification for hot-rolled sections*. London, BSI.
3. British Standards Institution (1995) BS EN 10024: 1995 *Hot rolled taper flange I sections: Tolerances on shape and dimensions*. London, BSI.
4. British Standards Institution (1991) BS EN 10029: 1991 *Specification for tolerances and dimensions, shape and mass for hot rolled steel plates 3mm thick and above*. London, BSI.
5. British Standards Institution (1993) BS EN 10034: 1993 *Structural steel I and H sections: Tolerances on shape and dimensions*. London, BSI.
6. British Standards Institution (2006) BS EN 10210-2: 2006 *Hot finished structural hollow sections of non-alloy and fine grain steels: Part 2: Tolerances, dimensions and sectional properties*. London, BSI.
7. The British Constructional Steelwork Association (2010) *National Structural Steelwork Specification*, 5th edn (CE Marking Version). London, BCSA/SCI.
8. International Organization for Standardization (1999) *Steel structures: Part 2: Fabrication and erection*. ISO 10721-2: 1999. Geneva, ISO.
9. British Standards Institution (1990) *BS 5606: 1990 Guide to accuracy in building*. London, BSI.
10. British Standards Institution (2009) BS EN 1090-1: 2009 *Execution of steel structures and aluminium structures: Part 1: Requirements for conformity assessment of structural components*. London, BSI.
11. British Standards Institution (1990) BS 5964-1 (ISO 4463-1) *Building setting out and measurement: Part 1: Methods of measuring, planning and organization and acceptance criteria*. London, BSI.

Further reading for Chapter 32

British Standards Institution (2005) BS 4-1:2005 *Structural steel sections: Specification for hot-rolled sections*. London, BSI.
British Standards Institution (1990) BS 5606:1990 *Guide to accuracy in building*. London, BSI.
British Standards Institution (1990) BS 5964-1 (ISO 4463-1) *Building setting out and measurement: Part 1: Methods of measuring, planning and organisation and acceptance criteria*. London, BSI.
British Standards Institution (2009) BS EN 1090-1:2009 *Execution of steel structures and aluminium structures: Part 1: Requirements for conformity assessment of structural components*. London, BSI.

British Standards Institution (2008) BS EN 1090-2:2008 *Execution of steel structures and aluminium structures: Part 2: Technical requirements for the execution of steel structures*. London, BSI.

British Standards Institution (1995) BS EN 10024:1995 *Hot rolled taper flange I sections: Tolerances on shape and dimensions*. London, BSI.

British Standards Institution (1991) BS EN 10029:1991 *Specification for tolerances and dimensions, shape and mass for hot rolled steel plates 3mm thick and above*. London, BSI.

British Standards Institution (1993) BS EN 10034:1993 *Structural steel I and H sections: Tolerances on shape and dimensions*. London, BSI.

British Standards Institution (2006) BS EN 10210-2:2006 *Hot finished structural hollow sections of non-alloy and fine grain steels: Part 2: Tolerances, dimensions and sectional properties*. London, BSI.

The British Constructional Steelwork Association (2007) *National structural steelwork specification*, 5th edn. London, BCSA/SCI.

International Organization for Standardization (1999) *ISO 10721-2:1999 Steel structures: Part 2: Fabrication and erection*. Geneva, ISO.

Chapter 33
Fabrication

DAVID DIBB-FULLER

33.1 Introduction

The steel-framed building derives most of its competitive advantage from the virtues of prefabricated components which can be assembled speedily on site. Additional economies can be significant provided the designer seeks through the design to minimise the value-added costs of fabrication. This is proper 'value engineering' of the product and is applied to perhaps the most influential sector of the delivered-to-site cost of structural steelwork.

This chapter explains the processes of fabrication and links them with design decisions. It is increasingly important for the designer to understand the skills and techniques available from different fabricators so that the design can be tailored to keep overall costs down. The choice of fabricator can then be made on the basis of ability to conform in production engineering terms to design assumptions made much earlier. It is unacceptable to allow the fabricator or designer to undertake the production engineering element of design in perfect isolation; the dialogue between designers and fabricators must be a continuous one. The effects that fabrication and assembly have on design assumptions and vice versa and, in particular, the achievable fit-up of components and permissible limits of tolerance must be addressed.

Design must be viewed as a complete process, covering strength and stiffness as well as production engineering, to achieve the most economical structures.

33.2 Economy of fabrication

Structural form has a significant effect on the delivered-to-site cost of steelwork. This is due to a number of factors additional to the cost of raw steel from steel suppliers. Some forms will prove to be more costly from some fabricators than others;

Steel Designers' Manual, Seventh Edition. Edited by Buick Davison and Graham W. Owens.
© 2012 Steel Construction Institute. Published 2012 by Blackwell Publishing Ltd.

they tend to attract work by aligning their production facilities to specific market sectors. For example, the industrial building market was largely taken over by the introduction of the portal-framed structure. The fabricators in this sector adopted high-volume, low-cost production and concentrated primarily on this area of the market. By the use of pre-engineered standards they were able to maximise repetition and minimise input from both design and drawing activities: a classic example of a combination of design and production engineering. Other steelwork contractors specialise in tubular structures, lightweight sections and heavy sections.

33.2.1 Fabrication as a cost consideration

Figures 33.1 to 33.4 give an indication of the proportional costs associated with the fabrication and erection of structural steelwork. Actual costs in monetary terms have not been included as they will change with the demand level of the market. The proportions of cost will vary a little from fabricator to fabricator; those shown represent a reasonable average.

The cost headings are:

- raw steel
- fabrication
- painting
- transport
- erection
- site painting.

Raw steel cost covers the average cost of rolled steel in S275 delivered to the steelwork contractor. No allowance has been made for extra costs arising from the use of stockholders' steel or the extremes of section variation.

Fabrication cost covers a number of different sized jobs and incorporates cleaning of the raw steel by shot blasting, preparation of small parts (cleats and plates and connection components), and assembly of the components into complete structural members ready for shop-applied paint treatments. It also includes an allowance for consumables.

Paint cost covers the shop application of $75\,\mu m$ of primer by spray immediately after fabrication. No allowance has been made for blast cleaning of areas affected by welding. A coverage allowance of $28\,m^2$/tonne at the rate of $3\,m^2$/litre has been made, which represents the likely consumption for rolled section beams and columns. This allowance has been adjusted for various specific work types as described in the accompanying text.

Transport cost is based on 20 tonne loads per trailer which delivers finished products to sites within a 50 mile radius of the fabrication shop. Transport costs will rise for loads of less than 20 tonnes or when oversize components need special police escort or permission.

The erection cost is average for the work type and includes preliminaries for high-rise multi-storey work.

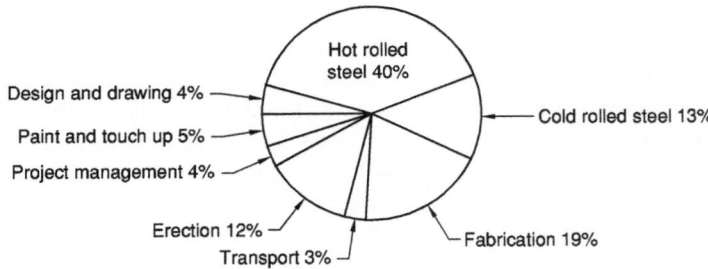

Figure 33.1   Cost breakdown: portal-framed industrial buildings

The cost of site painting has been included but it only covers the average cost for touching-up damage to the primer coat. Other site-applied protection systems vary enormously in cost and so have not been considered.

33.2.1.1 Portal-framed industrial buildings

The breakdown shown in Figure 33.1 follows the assumptions given above. There is no adjustment in either the shop painting or the transport cost as this type of work fits the basic parameters well. It can be seen that design economies come principally from the weight of the structure combined with efficient fabrication processes.

33.2.1.2 Simple beam and column structures

The breakdown shown in Figure 33.2 incorporates the following limitations:

1. The maximum height of the building is three storeys.
2. Erection is carried out using mobile cranes on the ground floor slab.

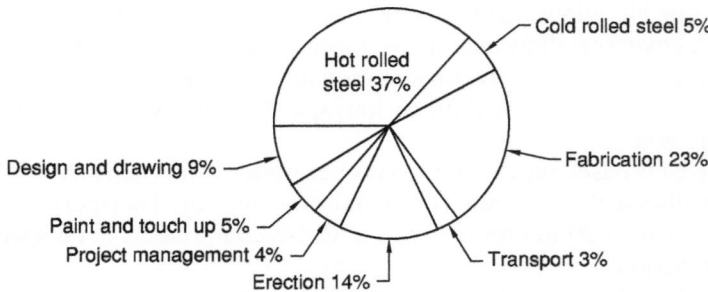

Figure 33.2   Cost breakdown: simple beam and column structures

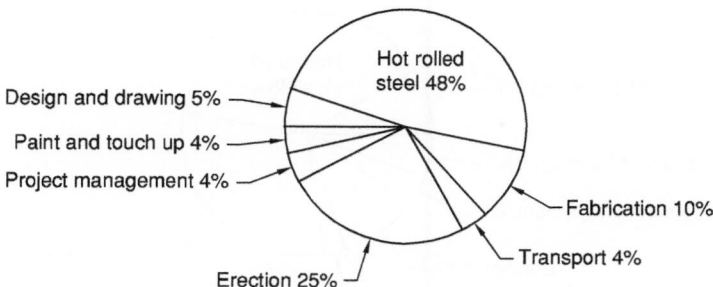

Figure 33.3 Cost breakdown: high-rise multi-storey

It can be seen that again tonnage and fabrication efficiencies are the dominant criteria; 83% of the costs arise from these elements, with a slightly greater emphasis on tonnage than was the case with portal frames.

33.2.1.3 High-rise multi-storey

The breakdown shown in Figure 33.3 incorporates the following adjustments:

1. The steel grade has been taken as S355 with an allowance for cambering.
2. The paint system generally is shot blast and 75 μm primer, 10% of steel coated with 100 μm primer.
3. No allowance has been made for any concrete-encased beams or stanchions.
4. Transport includes for off-site stockpiling, bundling and out-of-hours delivery to site (city centre sites often incur these costs).

This sector of the market has a very different cost profile to those already shown. Raw steel still dominates but erection charges have now overtaken the fabrication element. This type of steelwork lends itself particularly well to automated fabrication techniques featuring drilling lines.

33.2.1.4 Lattice structures

This is perhaps the most difficult of the sectors on which to carry out an analysis as lattice structures vary enormously in size, complexity and element make-up. The costs shown in Figure 33.4 are indicative and incorporate the following:

1. angle booms and lacings
2. welded joints without gusset plates
3. transportable lengths and widths with full-depth splices only.

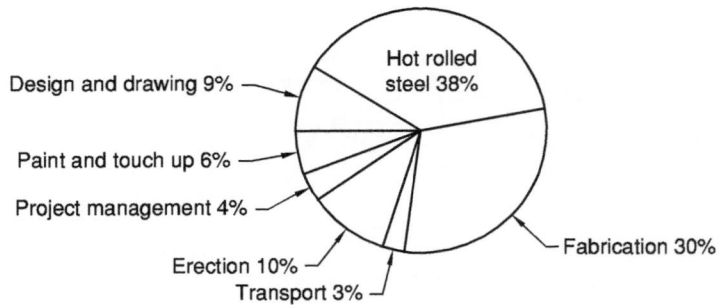

Figure 33.4   Cost breakdown: lattice structures

Here fabrication is virtually the same cost as the raw steel. The designer should work very closely with the steelwork contractor to ensure simplicity of assembly for this form of structure, in particular checking joint capacities at the design stage.

33.2.2 Design for production

The detailed design of any steelwork construction will have a substantial impact on its cost of fabrication. It is therefore very important that the designer has a basic understanding of the implications of his design on construction in practice. Key points need to be addressed if the design is to be fabricated economically and efficiently. This important topic is the subject of Reference 1, from which the following key points are extracted.

33.2.2.1 Fabrication processes

- Modern computer numerically controlled (CNC) fabrication equipment is more effective with:
  - (a) single end cuts, arranged square to the member length
  - (b) one hole diameter on any one piece, avoids drill bit changes
  - (c) alignment of holes on an axis square to the member length, holes in webs and flanges aligned not staggered to reduce piece moves between drill times
  - (d) web holes having adequate side clearance to the flanges.
- To allow efficient production of fittings:
  - (a) rationalise on the range of fittings sizes – use a limited range of flats and angles
  - (b) allow punching and cropping wherever possible.
- If possible select connections which avoid mixing welding and drilling in any one piece. This avoids double handling of the member during fabrication.

33.2.2.2 *Materials grade and section selection*

- The designer should rationalise the range of sections and grades used in any one structure. This will lead to benefits in purchasing and handling during all fabrication, transportation and erection phases of manufacture.
- Make maximum use of S355 material for main sections. This is typically 8% more expensive but up to 30% stronger than S275 steel. The exception is where deflection governs section selection.
- The specification of small quantities of 'special' grade material should be avoided, particularly if the proposed material has poorer welding qualities.
- Choice of fittings material grade should be left with the fabricator wherever possible.
- Limitations on mill lengths should also be remembered.

33.2.2.3 *Connection design considerations*

- Connections directly influence 40–60% of the total frame cost. They must therefore be taken into account during the frame design.
- Least-weight design solutions are rarely the cheapest. Increasing member thickness to eliminate stiffening at connections will often be an economic solution.
- The cost benefits from an integrated approach to frame and connection design will only be realised if the fabricator is given a full package of information at tender stage. Connection styles and design philosophy must be clearly marked on drawings.

33.2.2.4 *Bolts and bolting*

- Non-preloaded bolting is the preferred method for site connections. Preloaded (friction grip) bolts should only be used where joint slip is unacceptable or where there is a danger of fatigue.
- The use of different grade bolts of the same diameter on any one contract should be avoided.
- Threads should be permitted in the shear plane and in bearing.
- Direct and indirect cost savings can accrue by using only a small range of 'standard' bolts.
  Recommended standards are:
  - M20 grade 8.8 for shear connections
  - M24 grade 8.8 for moment connections
  - Mechanical properties to BS 3692, dimensions to BS 4190
  - Fully threaded for shanks up to 70mm long.
- The use of fully threaded bolts generally means additional thread protrusion is visible; specifiers should be aware of this and state at tender stage where this is *not* acceptable.

- Washers are *not* required for strength when using non-preloaded bolts in normal clearance holes; they may still be specified to provide a degree of protection to surface finishes.
- When used with corrosion-protected steelwork, bolts, nuts and washers should be supplied with a coating which does not require further protection applications.

33.2.2.5 Welding and inspection

- The welding content of a fabrication has a significant influence on the total cost of fabrication.
- In designing welded connections consideration should be given to the weldability of materials, access for welding and inspection, and the effects of distortion. Access is of primary importance – good welds cannot be formed without adequate access.
- Fillet welds up to 12 mm leg length are preferred to the equivalent-strength butt weld. Generally two fillet welds whose combined throat thicknesses equal the thickness of the plate to be connected are considered as equivalent in strength to a full penetration butt weld.
- Weld defect inspection and defect acceptance criteria should be defined; the use of the National Structural Steelwork Specification criteria is strongly recommended.

33.2.2.6 Corrosion protection

- In selecting a corrosion protection system the designer must consider the environment in which the steelwork will be placed and the design life of the corrosion protection system.
  If the environment does not require a corrosion protection system don't specify one.
- If a protection system is required, significant advantages are gained by use of a single-coat protection system applied during fabrication. These should be specified where possible.
- Wherever possible avoid using 'named' product specifications; allow the fabricator to use his preferred supplier or even alternative preferred coating system of equal capability.
  Specification of surface conditions should relate to the condition immediately prior to painting, not bound by any time-limit from shot blasting operations.

33.2.2.7 Trusses and lattice girders

- Lattice girders and trusses are effective for medium to long spans where deflection is a major criterion and are able to accommodate services within their depth, but always consider the use of a plain rolled section beam first.
- Most lattice frames are joint critical. Never select a section for the chords or internals without first checking whether it can be effectively joined – preferably without recourse to stiffening.
- Always check the limits on transport before starting the design.
- Be aware that SHS are only available in limited standard lengths, normally from stockists. Long lengths may therefore need additional butt welding.
- For internal members try to detail single-bevel end cuts; for angles square-cut ends are better to allow use of an automatic cropping process.
- In tubular construction use of RHS chords leads to simpler end preparation for internals than that required if CHS chords are used.
- Think about access provisions for welding of internals to chords.
- Access for painting is difficult for double-angle or double-channel members; use of SHS reduces paint area and provides fewer locations for corrosion traps to be formed.

33.2.2.8 Transportation

- Police notification with associated programme and cost penalties will occur for road transport loads greater than 18.3 m long, or 2.9 m wide or 3.175 m high.

33.3 Welding

Fusion welding processes are used to join structural steel components together. These processes can be carried out either in the workshop or on site, though it is generally accepted that site welding should not be used as a primary source of fabrication. Welding should be undertaken only by welders who are certificated to the appropriate level required by British Standards or other recognised authority. Welding is covered in detail in Chapter 26.

33.4 Bolting

Shop bolting may form part of the fabrication process. There are implications arising from the use of shop bolting which need to be appreciated by the designer in order to ensure that a cost penalty does not occur.

At one time it was common practice to assemble components in the workshop using bolts or rivets. With the increased implementation of welding, this practice has declined due to the costs associated with bolted fabrications. Instead of a simple run of fillet weld, holes need to be drilled and bolts introduced, increasing total labour hours and cost. In many respects the ease with which welding can be undertaken has diverted designers' attention from the use of shop bolting. Today, steelwork contractors are looking at increased automation to keep costs down, and machines have been developed which considerably speed up hole drilling.

33.4.1 Shop bolting

There is still a demand for structural members to be bolted arising from a requirement to avoid welding because of the service conditions of the member under consideration. These may be low temperature criteria, the need to avoid welding stresses or the requirement for the component to be taken apart during service (e.g. bolted-on crane rails). For lattice structures, the designer should specify the bolting, bearing in mind the effect of hole clearances around bolt shanks. HSFG bolts will not give problems but other bolts in clearance holes will allow a 'shake-out' which can cause significant additional displacement at joints. Typically, a truss with bolted connections may deflect due to the take-up of lack of fit in clearance holes to such an extent that it loses its theoretical camber. The use of HSFG assemblies avoids this risk.

Large and complex assemblies which are to be bolted together on site may be trial assembled in the fabrication shop. This increases fabrication costs but may pay for itself many times over by ensuring that the steel delivered to site will fit. Restricting trial assembly to highly repetitive items or items critical to the site programme is to be recommended.

33.4.2 Types of bolt

There are three basic types of bolts – structural bolts, friction-grip bolts and close-tolerance bolts. The choice of which type of bolt to use may not necessarily be made on the basis of strength alone but may be influenced by the actual situation in which the bolt is used, e.g. in non-slip connections.

33.4.2.1 Structural bolts

Bolts with low material strength and wide manufacturing tolerance were until recently known as 'black bolts' because of their appearance. Now they are called **structural bolts**, normally have a protective coating which gives them a bright appearance, and are available in a range of tensile strengths.

33.4.2.2 *Preloaded (friction-grip) bolts*

Preloaded (friction-grip) bolts are normally supplied with a protective coating and are differentiated by material grade. Preloaded bolts are used in connections which resist shear by clamping action, in contrast to structural bolts which resist shear and bearing. When considering the use of friction-grip bolts during the fabrication process, adequate means of access must be provided so that the bolt can be properly tensioned.

33.4.2.3 *Close-tolerance bolts*

This type of bolt differs from a structural bolt in that it is manufactured to closer tolerances. To gain the full benefit of close-tolerance bolts, they should be used in close-fitting holes produced by reaming, which adds considerably to the expense of fabrication. Where a limited slip connection is required, close-tolerance bolts can be used in holes of the same nominal diameter as the bolt but not reamed; this gives a connection which is subject to far less slip than would be the case for structural bolts in clearance holes.

33.4.3 Hole forming

To gain the best output, computer numerically controlled (CNC) machines are incorporated in conveyor lines. There is an infeed line which takes the unprocessed raw steel and an outfeed line which distributes the finished product either to the despatch area or further along the fabrication cycle and into an assembly area. The infeed conveyor for punching and drilling machines that handle angles, flats, small channels and joists will normally be configured to handle 12 m long bars. There will be a marking unit which carries a set of marking dyes that stamp an identification mark on to the steel if required. There may be a number of hydraulic punch presses each suitable for accepting up to three punching tools, and it is normal that the tool holders are quick-change units. Typically, hydraulic presses have 1000 kN capacity. The punch presses can form differently-shaped holes so that angles can be produced with slots. Some machines produce the slot by a series of circular punching operations, others have a single slot-shaped dye.

Once the material has passed through the punch presses, it is then positioned for shearing. The hydraulic shear has a capacity of between 3000 kN and 5000 kN. Machines which incorporate drilling facilities can have these positioned next to the punch presses prior to the hydraulic shear. The output of the machine can be as high as a thousand holes per hour. Obviously, these machines can be obtained with different capacities so that if the work of the fabricator is at the light end of the section range, it is not necessary to purchase machines with large capacity.

For larger sections, where drilling and cutting are required, there are two basic machines available. One is a drilling and plate-cutting system which is used for heavy plates; the other machine is of a larger capacity and is used for drilling rolled sections of all sizes. Modern drills normally have three-axis numeric control and air-cooled drills. These machines have sensors which can detect the position of the web to ensure that hole patterns are symmetrical about the web centreline.

Where components only need a small number of holes, mobile drills are used. These are normally magnetic limpet drills hand operated by the fabricator. The holes are formed with rotary-broach drill bits, which have a central guide drill surrounded by a cylindrical cutter.

The fabricator will always carry out hole-forming operations prior to any further fabrication as the presence of any stiffeners or cleats on a bar would severely disrupt the input of NC machines.

33.5 Cutting

The fabrication process of cutting has become highly automated. Universal beams, columns and the larger angles, tees and channels are normally cut to length by saws. Hollow sections are also treated in this way. Small sections of joist, channel, angle and flat are cut to length by shearing, either as a separate operation or as part of a punching and cropping operation which is computer controlled. Large plate sections may be sheared but this involves specialist plant and equipment which is not available to all fabricators.

33.5.1 Cutting and shaping techniques

33.5.1.1 Flame cutting or burning

This technique produces a cut by the use of a cutting torch. The process may be either manual or machine controlled. Manual cutting produces a rough edge profile, which can be very jagged and may need further treatment to improve its appearance. Edges of plates cut manually require greater edge distances to holes for this reason. Notches or holes with square corners should not be hand cut unless the corner is first radiused by a drilled hole, Figure 33.5.

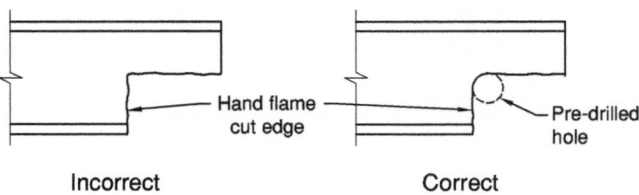

Figure 33.5 Hand flame cutting

When flame cutting is carried out under machine control, the cut edge is smooth and therefore the restrictions on edge distances to holes are relaxed. Notches which are machine cut need radiusing either by a predrilled hole or by the control of the cutting machine. Unradiused notches are to be avoided due to their stress-raising characteristics. Notching or coping of beam ends can now be carried out by computer-controlled machines which burn the webs first and create the radiused corner then cut the flanges with a cut which coincides with the web cut. Numerically controlled coping machines can cut top and bottom notches simultaneously, and the notches can be of different dimensions.

Castellated beams were traditionally formed by cutting a multi-linear, castellated pattern along a UB or UC section followed by realignment and rewelding. The web cutting is always machine controlled to ensure accurate fit-up and an edge which is suitable for subsequent welding.

More recently, a two-cut process has been developed that creates beams with circular openings, frequently called **Cellform** beams. These beams have proved to be very popular both for long span roof structures where the steelwork is exposed and as primary and secondary beams in composite floors. In the latter case, further economy may be achieved by cutting the top tee sections from a lighter rolling than the bottom.

33.5.1.2 Arc plasma cutting

Plates may be cut by arc plasma techniques. In this case the cutting energy is produced electrically by heating a gas in an electric arc produced between the tungsten electrode and the workpiece which ionises the gas, enabling it to conduct an electric current. The high-velocity plasma jet melts the metal of the workpiece and blows it away. The cut so produced is very clean, and its quality can be improved by using a water injection arc plasma torch. Plasma cutting can be used on thicknesses up to about 150 mm but the process is then substantially reduced in speed.

33.5.1.3 Shearing and cropping

Sections can be cut to length or width by cropping or shearing using hydraulic shears. Heavy sections or long plates can be shaped and cut to length by specialist plate shears. These are large and very expensive machines normally to be found in the workshops of those fabricators who specialise in plate girder work for bridges, power stations and other heavy steelwork fabrications. For the more commonplace range of smaller plates and sections, there is a range of equipment available which is suitable for cutting to length or shaping operations. These machines feature a range of shearing knives which can accept the differing section shapes. Shearing can be adjusted so that angled cuts may be made across a section. This is particularly useful for lacings of latticed structures. One version of the shear is a 'notcher' which can cut shaped notches. Special dies are made to suit the notch dimensions, and it is possible to obtain dies to cut the ends of hollow sections in preparation for welding together.

33.5.1.4 Cold sawing

When, because of either specification or size, a section cannot be cut to length by cropping or shearing, then it is normally sawn. All saws for structural applications are mechanical and feature some degree of computer control. Sawing is normally carried out after steel is shot blasted as the saw can be easily incorporated within the conveyor systems associated with shot blast plants.

There are three forms of mechanical saw: circular, band and hack. The circular saw has the blade rotating in a vertical plane which can cut either downwards or upwards, though the former is the more common. The blade is a large milling wheel approximately 5 mm thick. The diameter of the saw blade determines its capacity in terms of the maximum size of section which can be cut. Normally, steelwork contractors will have saws capable of dealing with the largest sections produced. For increased productivity, sections may be nested or stacked together and cut simultaneously. Some circular saws allow the blade to move transversely across the workpiece, which is useful for wide plates. Most circular saws can make raking cuts across a section of any angle, though some saws are restricted to single-side movement. The preferred axis of cut is across the Y–Y axis of the section. In the case of beams, channels and columns, this is with the web horizontal and the flange toes upwards. Depending on the control exercised, a circular saw can make a cut within the flatness and squareness tolerances necessary for end bearing of members.

Band saws have less capacity in terms of section size which can be cut. The saw blade is a continuous metal band edged with cutting teeth which is driven by an electric motor. The speed of cut is adjustable to suit the workpiece. Band saws can make mitre cuts and can cut through stacked sections. Cutting accuracy is dependent on machine set-up but produces results similar to circular saws.

Hack saws are as the name implies mechanically driven reciprocating saws. They have normal format blades carried in a heavy duty hack saw frame. Hack saws have more limited cutting capacity than band saws and have the capability to produce mitre cuts.

All saws feature computer-controlled positioning carriages for accurate length set-up; most also have computer sensing for angle cuts.

33.5.1.5 Gouging

The gouging process is the removal of metal at the underside of butt welds. Gouging techniques are also used for the removal of defective material or welds. The various forms of gouging are flame gouging, air-arc gouging, oxygen-arc gouging and metal-arc gouging.

Flame gouging is an oxy fuel gas cutting process and uses the same torch but with a different nozzle. It is important that a proper gouging nozzle is adopted so that the gouging profile is correct.

Air-arc gouging uses the same equipment as manual metal arc welding. The process differs in that the electrode is made of a bonded mixture of carbon and graphite encased in a layer of copper, and jets of compressed air are emitted from the specially designed electrode holder. These air jets blow away the parent metal from beneath the arc.

A special electrode holder is used in the oxygen-arc gouging process. A special tubular coated steel electrode controls the release of a supply of oxygen. Oxygen is released once the arc is established, and when the gouge is being made the oxygen flow is increased to a maximum.

Standard manual metal arc welding equipment is used for metal-arc gouging. The electrodes, however, are specially designed for cutting or gouging. This process relies on the metal being forced out of the cut by the arc and not blown away as in the other processes.

33.5.2 Surface preparation

Structural sections from the rolling mills may require surface cleaning prior to fabrication and painting. Hand preparation, such as wire brushing, does not normally conform to the requirements of modern paint or surface protection systems.

Blast cleaning is the accepted way of carrying out surface preparation. It involves blasting dry steelwork with either shot or grit at high velocity to remove rust, oil, paint, mill scale and any other surface contaminants. The most productive form of blast cleaning plant has special equipment comprising infeed conveyor, drying oven, blast chamber, spray chamber and outfeed conveyor.

Steel is loaded on to the infeed conveyor either as separate bars or side by side depending on the blast chamber passage opening. It then travels through the drying oven, which ensures that any surface moisture is removed prior to blasting. The blast chamber receives the steel and passes it over racks and between the blasting turbines, which are impellers fed centrally with either shot or grit. The material is thrown out from the edge of the turbine and impacts against the steelwork. The speed of the turbine, its location and aperture determine the velocity and direction of the blasting medium.

The blasting medium is retrieved from within the blast chamber and recycled until it is exhausted. Shot provides a good medium for steel which is to be painted; grit can cause excessive wear of the blast chamber and results in a surface which is more pitted, giving a reduced paint thickness over high points.

The steel leaves the blast chamber and immediately enters the spray chamber. Depending upon the requirements of the specification, the steel is then sprayed with a primer paint which protects it from flash rusting. The use of prefabrication primers should be discussed with the fabricator and paint system supplier as requirements may vary. For most structural applications in building work, there is little need to specify a prefabrication primer. Once the steel has passed through the spray chamber, it is fed on to the outfeed conveyor and continues to its next process centre.

The other form of blast cleaning is called *vacu blasting*, a manual method where the blasting medium is blown out of a hand-held nozzle under the pressure of compressed air. This process is normally performed in a special sealed cabinet with the operator wearing protective clothing and breathing equipment. It relies heavily on the skill of the operator and does not give production rates approaching those of mechanical blasting plants.

Local areas of damage caused by welding can be cleaned by vacu blasting or by needle gunning. The needle gun has a set of hardened needles which are propelled and withdrawn rapidly. The action of the needles on the steel effectively removes weld slag and provides a good surface for painting. Needle gunning is not suitable for large areas or for heavily contaminated surfaces.

Surface preparation is intended to provide a uniform substrate for even paint application. It is important to consider the edges of plates, where paint thickness is often low.

33.5.3 Cambering, straightening and bending

Each of these operations has a different purpose. Cambering is done to compensate for anticipated deflections of beams or trusses under permanent loads, dead loads or superimposed dead loads such as finishes. Straightening is part of the fabrication process aimed at bringing sections back within straightness tolerances. Bending is to form the section to a shape which is outside normal cambering limits.

Rolled sections are normally bought in by the fabricator and then sent to specialist firms for cambering. The steel section is cold-cambered by being passed between rolls. The rolls are adjusted on each pass until the required camber is achieved. The following cambers can be produced:

- circular profile (specify radius)
- parabolic profile (specify equation)
- specified offsets (tabulate co-ordinates)
- reverse cambers (circular, parabolic, offset or composite).

Bending is also carried out by specialists, the process being identical to that for cambering. Curves can be formed on either the X–X or Y–Y axes.

Straightening can be carried out by the fabricator or by the cambering specialists. Sections are often straightened by the fabricator by applying heat local to the flange or web. This is a skilled operation which achieves excellent results.

33.6 Handling and routeing of steel

The fabrication process involves the movement of steel around the workshop from one process activity to the next. Clearly, all these movements should be planned so that the maximum throughput of work is achieved and costs are minimised. Even with the highly automated facilities available today, planning is essential as work centres may become overloaded and cause delays to other activities. In this respect, the work flow has to be balanced between work centres.

The routing of steel through a fabrication workshop is planned on the basis of the work content required on the workpiece. The workpiece will consist of one or more main components which may have smaller items attached (cleats, tabs and brackets). The main components are normally obtained from the steel mills though some fabricators obtain steel from stockholders. The steel is either stored in a stockyard local to the workshop or preferably delivered from the stockist just in time for immediate fabrication. The control of steel stock is an important management function which can balance the cash flow of the fabricator and provide clear identification of the steel for traceability purposes. Stock control systems are commonly computerised and should allow the fabricator to identify:

1. steel on order but not yet received
2. steel which is in the stockyard, and its location
3. allocation of steel to particular contracts
4. levels of unallocated steel (free stock).

Prior to the fabrication cycle, an estimate is made of the work content of a workpiece. This assessment may be crudely based upon tonnage or more commonly based upon time study data of previous similar work. The work content assessment is a production control function which will identify:

1. weight of finished product
2. content of workpiece (parts list)
3. work centre activities required (sawing, drilling, welding, etc.)
4. labour content
5. machine centre utilisation.

Production control systems are now often computerised and can produce a detailed map of the movement of the workpiece and its component parts within the fabrication workshop. Computer systems can identify any potential overloadings of work centres so that the production controller can specify an alternative path.

Generally, the cycle of work centre activity is as shown in Table 33.1 (the exact sequence of events may, however, vary between fabricators).

33.6.1 Lifting equipment in fabrication workshops

Not all steel can be moved around a workshop on conveyor lines. It is not practical or cost-effective to do so and would restrict the flexibility for planning workshop activities. Much of the steel will be moved by cranes of one form or another.

Table 33.1   Sequence of activities in fabricating shops

Cleaning and cutting to length	Shot blasting Sawing	
Manufacture of attachments	Preparation of:	cleats brackets stiffening plates end-plates baseplates
Preparation of main components	Pre-assembly:	drilling punching coping cropping notching cutting of openings
Assembly of workpiece	Marshalling of components Setting-out of components Shop bolting Welding Cleaning up	
Quality control (assembly)	Check assembly	– dimensional – NDT – visual inspection
	Sign-off	
Surface treatment	Painting Metal spraying Galvanising	often by other specialists
Quality control (final product)	Conformity with specification Sign-off	
Transportation	Loading Despatch	

Craneage capacity can be a restriction on the work which a fabricator can undertake. Restricted headroom is another problem, particularly when considering the fabrication of deep roof trusses.

There are various types of crane in common use in most workshops:

- electric overhead travelling (EOT)
- goliath
- semi-goliath
- jib
- gantry.

EOT cranes are the main lifting vehicles in the larger fabrication workshops. Their capacity varies from fabricator to fabricator but rarely exceeds 30 tonnes SWL. EOT cranes run on rails supported on crane beams carried off either the main building stanchions or a separate gantry and are used for moving main components and finished workpieces between work centres. Heavy components need turning over for welding and this is achieved by the use of EOT cranes.

Goliath and semi-goliath cranes are heavy-duty lifting devices which tend to be used on a more local basis within the workshop than EOT cranes. The goliath is a

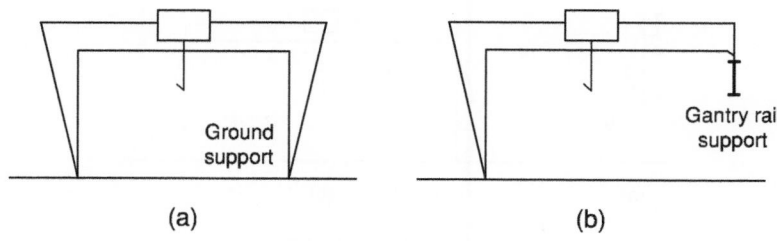

Figure 33.6   Goliath cranes: (a) goliath, (b) semi-goliath

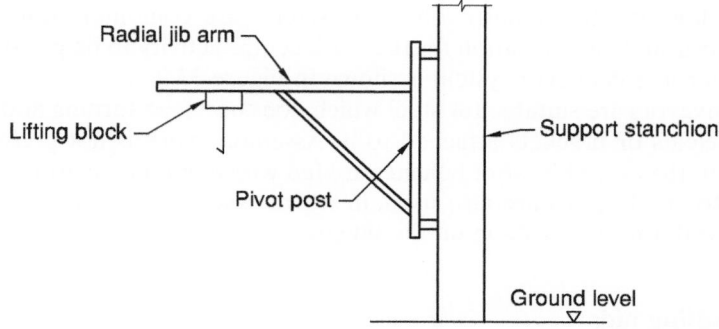

Figure 33.7   Jib crane

free-standing gantry supported on rails laid along the workshop floor. Semi-goliath cranes are ground rail supported on one side and gantry rail supported on the other. The lifting capacity of goliath-type cranes is normally less than that of an EOT crane though greater in some workshops where there are EOT crane support restrictions (Figure 33.6).

Jib cranes are light-duty (1–3 tonne capacity) cranes mounted from building side stanchions and operating on a radial arm with a reach of up to 5 m. They are useful for lifting or turning the lighter components (Figure 33.7).

Gantry cranes are purpose-built structures with limited lifting capacity (3–10 tonne). The lifting block travels along a gantry beam and is either underslung or top mounted. Gantry cranes are useful lifting devices where no building can be used to support an EOT crane, and some gantry cranes feature EOT cranes riding on rails alongside gantries. Stockyards normally are serviced by gantry cranes (Figure 33.8).

33.6.2 Conveyor systems

The use of factory automation is increasing the use of roller conveyor systems for moving main components. The roller conveyors can be designed so that they move

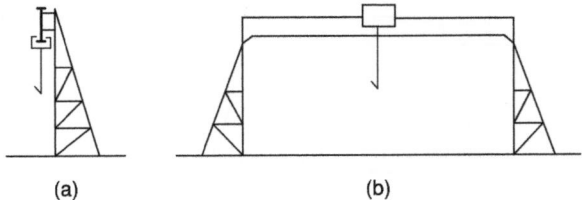

Figure 33.8   Gantry cranes: (a) single underslung, (b) EOT

steel between automated work centres either laterally across the workshop or lon-gitudinally along it. The modern conveyor systems are computer controlled and bring steel accurately into position for the work centre activity to be performed. An example of a roller conveyor system is shown in Figure 33.9.

Roller conveyors are suitable for steel which does not need turning and does not have many cleats or brackets attached to it. Assembly work is not performed on conveyors but the assembly work benches are fed with components from conveyors or cranes. Rolling buggies are also used, in which case fabrication operations are carried out with the steel section on the buggy.

33.6.3 Handling aids

Most steelwork assemblies can be picked up easily using chains or strops but the designer should be alert to the possibility of having to provide for temporary lifting brackets suitably stiffened. This may prove necessary where it is important that a component is lifted in a particular way either for stability reasons or for reasons of strength.

The designer should also be cautious in the provision of welded-on brackets. Flimsy outstands *will* get damaged, whether in the workshop, during transport or on site. Flimsy brackets should be supplied separately for bolting on at site.

Some components by nature of their design or geometry may require to be lifted in a specific way using strongbacks or lifting beams. The designer must make the fabricator aware of this requirement to avoid possible accidental damage or injury to workshop staff; steel is a very unforgiving material.

The fabricator will normally organise transport of steel from the workshop to site. There are loading restrictions on vehicles in terms of weight, width, height and length.

33.7 Quality management

The key activities within any quality management system are quality assurance and quality control. These activities are fully embodied in many industry-wide quality standards. While it is not a prerequisite that a manufacturer is registered under a

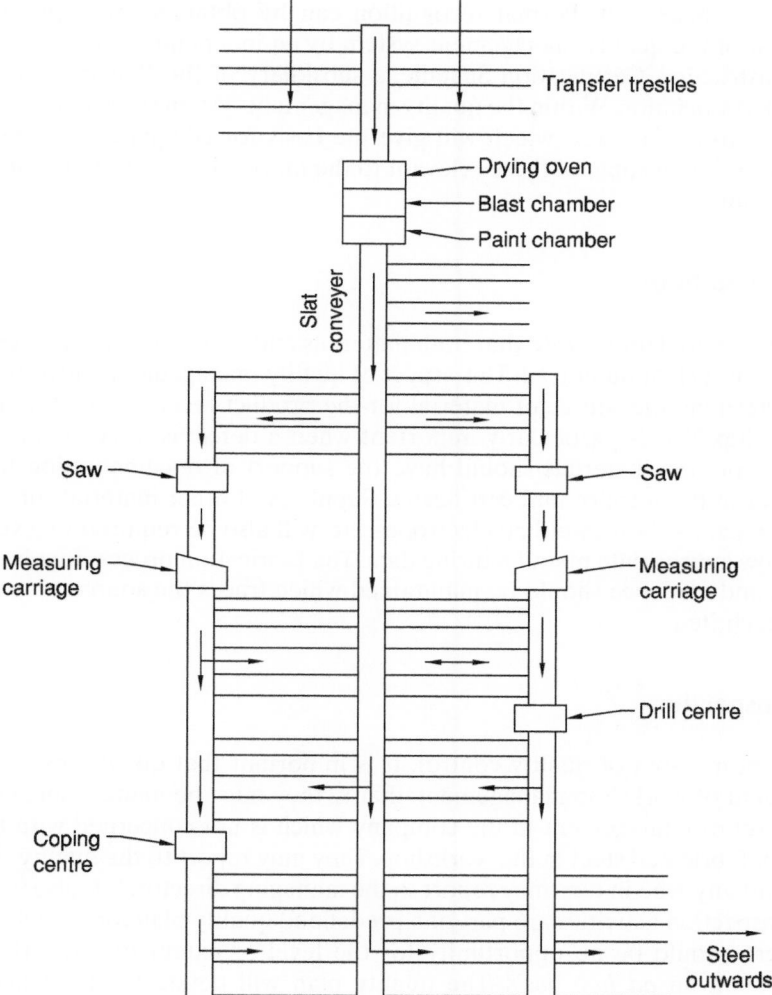

Figure 33.9   Roller conveyor system

standard-based scheme for appropriate levels of quality management to prevail, it does, however, allow specifiers to easily assess the relevance of the system under review. As far as design and manufacture of structural steelwork are concerned, there are a number of standards which apply, the most significant of these being BS EN 1090[1,2] and the National Structural Steelwork Specification[3]. These standards are the basis for the formulation and implementation of appropriate quality management systems.

While the use of these standards is commended to those setting up quality management systems, it does not imply any formal recognition that compliance with the standard has been met. Formal recognition can be obtained by application for assessment of the quality management system by an independent body, such as the Steel Construction Certification Scheme, a subsidiary of the British Construction Steelwork Association. Within the quality management system there must be written procedures to be followed which will give the basis for adequate QA and OC. A brief description of some of these relevant to the fabrication of structural steelwork is given below.

33.7.1 Traceability

It is necessary to demonstrate that both materials and manufacturing processes can be traced[4] by a clear audit trail. This aspect of quality management allows the specifier to determine the source of material for the product received and the origin of workmanship. This is particularly important when a defect is discovered. In terms of fabrication, the materials should have the support of documentation from the steelmaker in the form of mill certificates. Suppliers of other materials or consumable items such as bolts, welding electrodes, etc. will also be required to give details which show appropriate manufacturing data. The fabrication process involves workmanship, and evidence should be maintained which traces the source of workmanship on each item.

33.7.2 Inspection

This is a prime area of quality control. It is important that quality inspectors are independent of workshop management: that is, they must be made responsible to a higher level of management in the company which is not concerned with the production of fabricated steel in the workshop. They may report to the quality manager of the company, who in turn may report to the managing director. It is also important that all inspection activities are part of a predefined quality plan for the job in question. There should be no opportunity for the level or extent of inspection to be decided upon an *ad hoc* basis. The quality plan will define levels of inspection required and when the inspection is to take place, typically:

1. inspection of incoming materials or components
2. inspection of fabricated assemblies during and/or after fabrication
3. inspection of finished products prior to despatch
4. calibration of measuring equipment
5. certification of welders.

33.7.3 Defect feedback

At some stage or other, there will be occasions when defects are discovered in the fabricated item. This may be during inspection or on site. It is important that the defect is reported, and the quality management system must cater for this. The report should be formal and it must identify the defect and possible cause. This report is then acted upon to prevent recurrence of the defect. The audit trail of traceability and inspection should highlight the actual source of the defect, and this may be corrected by defined and planned actions.

33.7.4 Corrective action

All defects will cause corrective action to be taken. This action may take the form of revised procedures but more commonly will involve a process change. The most important corrective action is training as many defects stem from a lack of understanding on the part of someone in the production cycle. An important part of the prevention of recurrence of defects is trend analysis. Defects are recorded and trends studied which may highlight underlying problems which can be tackled. Occasionally, a finished product does not fully comply with the client's requirements and an approach may be made to determine if the product is suitable. While this process is going on the offending items must be put into quarantine areas which clearly distinguish them from others which do conform to requirements.

References to Chapter 33

1. British Standards Institution (2009) BS EN 1090-1. *Execution of steel structures and aluminium structures. Requirements for conformity assessment of structural components.* London, BSI.
2. British Standards Institution (2008) BS EN 1090-3 *Execution of steel structures and aluminium structures. Technical requirements for the execution of steel structures.* London, BSI.
3. British Constructional Steelwork Association (2010) *National Structural Steelwork Specification for Building Construction.* 5th edn. (CE Marking Version). London, BCSA/SCI.
4. British Constructional Steelwork Association (2009) *Inspection Documents.* Steel Industry Guidance Notes SN39. London, BCSA/SCI.

Further reading for Chapter 33

British Constructional Steelwork Association (2008) *Guide to the CE Marking of Structural steelwork.* BCSA publication No. 46/08. London, BCSA.
British Constructional Steelwork Association (2003) *Steel Buildings.* BCSA publication No. 35/03. London, BCSA.
British Constructional Steelwork Association Steel Industry Guidance Note SN17 (July 2007) *CE marking of steel products.* London, BCSA/SCI.
British Constructional Steelwork Association Steel Industry Guidance Note SN11 (January 2007) *Factors influencing steelwork prices.* London, BCSA/SCI.

Chapter 34
Erection

ALAN ROGAN

34.1 Introduction

Planning for the erection of any structure should commence at the design phase. The incorporation of buildability into the design phase may allow significant additional benefits. If a structure is to be erected quickly, to programme, and to the lowest possible cost, consideration must be given to the planning of the site works.

Work at a height and on site should be minimised, especially if this work could be carried out at the factory in ideal conditions. Where possible, frames and components should be designed to be built in the factory or assembled at low level for subsequent incorporation into the works. This will save time and cost as the work is not so weather-dependent and expensive temporary works are not required. Planners should:

- plan for repetition and standardisation
- plan for simplicity of assembly
- plan for ease of erection – keep it simple
- agree information handover dates and sign off design dates
- make allowance for trade interfaces
- allow realistic programmes for manufacture and erection
- recognise the complexity of the design process
- hold regular co-ordination meetings
- identify and fairly allocate risk
- consider long term stable relations with teamwork.

Attention to site construction methods are an essential part of any design. The Construction (Design and Management) (CDM) regulations[1] focuses on managing risks on site through effective planning and risk management, and good communication and teamwork to improve safety during construction work. The designer of the structure should take into account site access, materials handling, the construction

Steel Designers' Manual, Seventh Edition. Edited by Buick Davison and Graham W. Owens.
© 2012 Steel Construction Institute. Published 2012 by Blackwell Publishing Ltd.

sequence, and any limitations these may impose on the construction project. Utilising the expertise of specialist contractors at an early stage will maximise the speed of construction, reduce conflict, and give the opportunity of adding real value to the building.

34.2 Method statements, regulations and documentation

An erection method statement sets out a safe system of work for the delivery, erection, and completion of the intended structure thereby allowing the design team the opportunity to appraise this plan and make any appropriate observations or changes. The level of detail in a method statement will depend upon the size and complexity of the work. The BCSA have prepared a number of helpful publications[2,3] to guide designers in the production of suitable method statements.

There are many regulations that affect site erection of steelwork and these must be observed. The most practical way to ensure that the relevant regulations are observed is to follow approved codes of practice and guidance notes. In addition to the BCSA publications cited above, a list of regulations relevant to steel erection is included in Further Reading at the end of this chapter.

The European standard for the fabrication and erection of steel (and aluminium structures) is BS EN 1090-2:2008[4] 'Execution of steel structures and aluminium structures Part 2: Technical requirements for the execution of steel structures'. Execution is defined as 'all activities performed for the physical completion of the [structural steelwork], i.e. procurement, fabrication, welding, mechanical fastening, transportation, erection, surface treatment and the inspection and documentation thereof'. It has a very wide scope of application and requires specifiers to make a series of project or application-specific decisions before execution (fabrication and erection) can commence. The Standard introduces the concept of Execution Class. There are four execution classes which range from Execution Class 4 which is the most onerous through to Execution Class 1 which is the least onerous. Each Execution Class contains a set of requirements for fabrication and erection and these requirements may be applied to the structure as a whole, an individual component or a detail of a component. Those items, e.g. level of documentation and inspection, qualification of welding procedures and welding personnel, which are dependent on the choice of Execution Class, are itemised in Annex A.3 of BS EN 1090-2.

It is a design decision for the specifier to select the Execution Class required for the structure, an individual component or a particular detail of a component. The main reason for giving four execution classes is to provide a level of reliability against failure that is matched to the consequences of failure for the structure, the component or the detail. Execution Class is widely used throughout the Standard as a reliability differentiator for providing choice of quality, testing and qualification requirements.

Annex B of BS EN 1090-2 recommends that the choice of Execution Class is based on the 'service category' (SC) (SC1 – quasi-static, SC2 – fatigue) and the 'production category' (PC) (method of fabrication, PC1 or PC2, where structures/

components/details in PC2 are more difficult to produce than those in PC1). Most steel structures in the UK will include components in both production categories and most will be in SC1 (static) unless they are designed for fatigue (in which case they will be in SC2). Thus the default execution class for Building structures/components/details is likely to be Execution Class 2.

Section 9 of BS EN 1090-2 covers erection of structural steelwork in detail and should be consulted alongside the general description of best practice presented in this chapter.

34.3 Planning

34.3.1 Design information

Construction planning should start at the design phase since decisions at this stage will dictate the performance of the fabrication and erection teams. Site erection performance depends on many external and internal factors and it is essential these areas are properly managed. The site operation process is a complex inter-relationship of other disciplines which influence performance as shown in Figure 34.1.

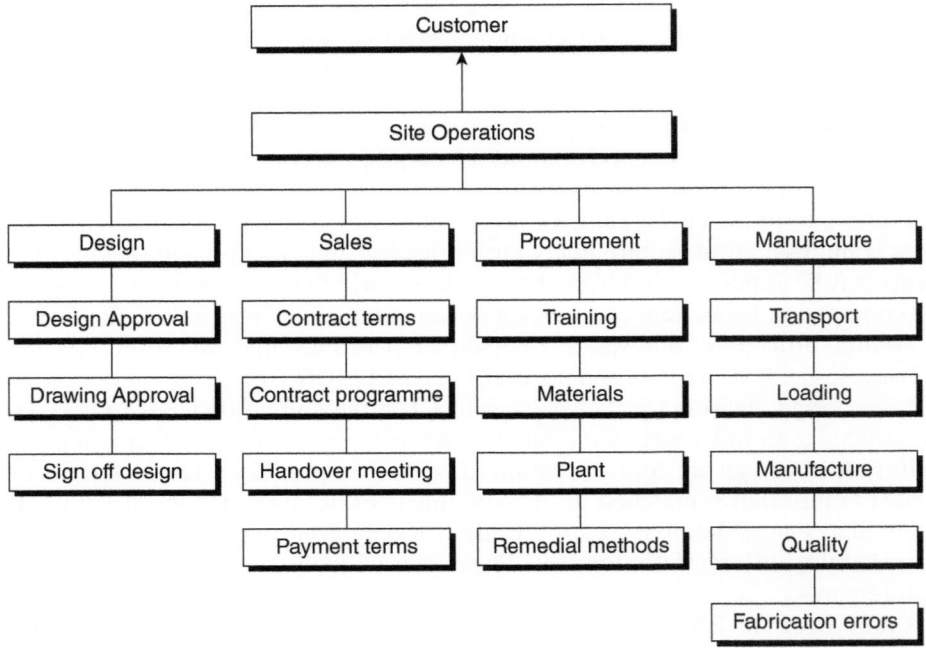

Figure 34.1 Chart showing the dependence of the erection process on other disciplines

34.3.2 Programming

Resources can be best utilised when the delivery and erection operations are well planned to follow a logical sequence. Typically, civil and other trades need to be sufficiently ahead of the steel erection (i.e. one week minimum) to allow for a reasonable flow of uninterrupted work. This will allow the steelwork contractor time to survey the foundation bolts prior to accepting them. It is not unusual for the civil contractor to have to carry out minor remedial work due to poor alignment of the holding-down bolts, and to this end it is advisable to allow the maximum time possible for handovers. Accurate positioning of foundations are essential, and tolerances defined in the specification and the National Structural Steelwork Specification (NSSS)[5] should be strictly adhered to.

Access restrictions or phasing of the works often govern the sequence of erection, which normally follows a grid pattern, split into zones. The siting of the erection crane will also dictate the construction pattern. Once the sequence and phasing has been agreed, the steelwork contractor can determine the resources required to meet the program.

34.3.3 Delivery and off-loading steelwork

A construction sequence programme is agreed prior to delivery of any materials to site, from which a delivery schedule may be produced. A more detailed piece-by-piece schedule for the steel erection method will need to be carried out to ensure smooth working on-site. Particular attention should be paid to the logistics of this exercise to prevent members being missed out, double handling of materials on site, lorries being under-utilised and the positioning of steel in the wrong location.

Delivery of the steelwork is normally in 20 ton lorry loads, often sub-divided into 4 bundles of 5 tons each to suit tower crane capacity, on a 'Just in time' delivery sequence in order to alleviate site congestion, minimise double handling, and on-site damage. Material needs to be stacked in the lorry for easy offloading, with items that are needed first readily available. Often the loading of the lorry can be dictated by crane off-loading capacity or stability of the load, as columns are usually required first yet they need to be at the bottom of the load because of their weight.

Shakedown areas (areas for separating and selecting steel bundles) are required on site to allow the sorting of the steelwork. The area should be firm and level with a plentiful supply of wooden sleepers or other suitable material. The steel members will have been allocated a unique code related to section, level, and piece number, this is usually stamped on the top surface to identify its orientation. In multi-storey high-rise buildings it is usual to create temporary shakeout platforms at various stages of the construction to minimise crane hook time.

Cold-formed roof members and steel decking packs also need to be placed in location as the work proceeds since crane access is often restricted later. Each bundle should be numbered and placed in convenient locations as subsequent relocation may prove difficult and costly.

34.3.4 Speed of construction and the use of sub-assemblies onsite

The speed of construction is dependent on the number and type of crane lifts and connection. The rate of erection will also vary dependent on the weight, size and location of the piece being erected. Erection speed is often dictated by the distance between the crane and hook, thereby emphasising the need to consider lifting bundles closer to the erection face and providing shake-out platforms. As a broad rule of thumb, an experienced erection crew working in ideal conditions at low level with average weight standard-sized components would be able to erect 40 to 60 pieces per day per hook.

As the designer may have been limited by the size and weight of components for transportation, it may prove advantageous to erect canopies, roof trusses, lattice girders and the like 'piece small' and then lift them as completed components. If the decision to sub-assemble has been made there will remain the further question of deciding whether an area in the stockyard is to be dedicated for this purpose or whether the work will be done behind the cranes on the erection front in order that the assembly can be lifted straight off the ground and into position as the crane is moved back.

The most common components to be assembled behind the crane are roof trusses or lattice girders which, because of their size, are almost always delivered to site 'piece small'. However, there is not always space for this to be done and other locations, such as the stockyard, have to be made available.

Where the potential sub-assembly area, and even the stockyard itself, is remote from the erection site, very careful investigations are needed before a decision to sub-assemble can be agreed. Local transportation size restrictions on the route to the erection cranes may rule this option out.

There are three factors which affect the practicability and economy of sub-assembling a unit on the ground:

1. the weight of the eventual assembly including any lifting beams required
2. the degree to which the unit is capable of being temporarily stiffened without unduly increasing its weight
3. the bulk of the unit, i.e. will it be possible to lift it to the height needed without fouling the crane jib?

Sub-assembly is only worthwhile if the unit can be lifted and bolted into place almost as easily as a single beam. In other words, the object of the exercise is to avoid carrying out operations at height which can easily be done at ground level.

34.3.5 Interface management

In order to minimise on-site disruption, the specialist/trade interface must be managed successfully. Failure to understand other trades requirements through unclear specifications and uncertain allocation of responsibility may result in conflict and disagreement. Thus, the development of open communication between all parties and the establishment of clear objectives during regular site meetings between over-lapping trades are necessary measures to ensure proper component, plant, and work area management. Failure to adequately address these issues can result in cladding fixings being left off or located in the wrong place, holding down bolts being incorrectly positioned, lift shaft tolerances proving incompatible with the main frame, and so on. Thus, co-operation between all parties is necessary to ensure enhanced efficiency; this may be achieved via an improved communication process, clear contractual definitions and the use of single-source suppliers.

34.3.6 Surveying and aligning the structure

Surveying should be carried out in accordance with BS 5964[6] (1990) Parts 1 to 3 Building Setting Out and Measurement, and adjusted for temperature outside the range of 5°C to 15°C. On a partly erected and unclad building frame the effects of temperature on the framework can exceed the effects of the wind. A tall framework will lean away from the sun as the sun moves round from east to west, thus checking the plumb of a building should only be done on a cloudy day or after the whole structure has been allowed to reach a uniform temperature (e.g. at night), and then only when the temperature is at or near the design mean figure. Tightening the bolts in the bracing when a building is at non-uniform temperature can lock in an error which may prove difficult to correct later. Accurate setting out is crucial for maintaining control of tolerances and achieving a structure which is acceptable to all the subsequent trades.

Further guidance on tolerances is provided in Chapter 32.

34.4 Site practices

34.4.1 Erection sequence

Columns are placed in grids to suit the grid/zone areas as previously agreed. The columns are lifted with a Dawson Ratchet (see Figure 34.2), nylon slings or chains.

Care must be taken to minimise damage to painted surfaces as repairs can often be difficult. Once placed into position the column is roughly positioned and plumbed. In extreme cases the columns may need guying off to ensure stability during the erection sequence. This procedure is repeated until a grid is formed. Beams are then placed in location and secured by two bolts at each end connection throughout the designated area. Once all of the beams are located, the remaining bolts are placed. Since many structures employ composite slab construction, an intricate part of the site erection is the placing and fixing of the decking bundles. A grid system should have been agreed for the placing of bundles which the erection crew situates in

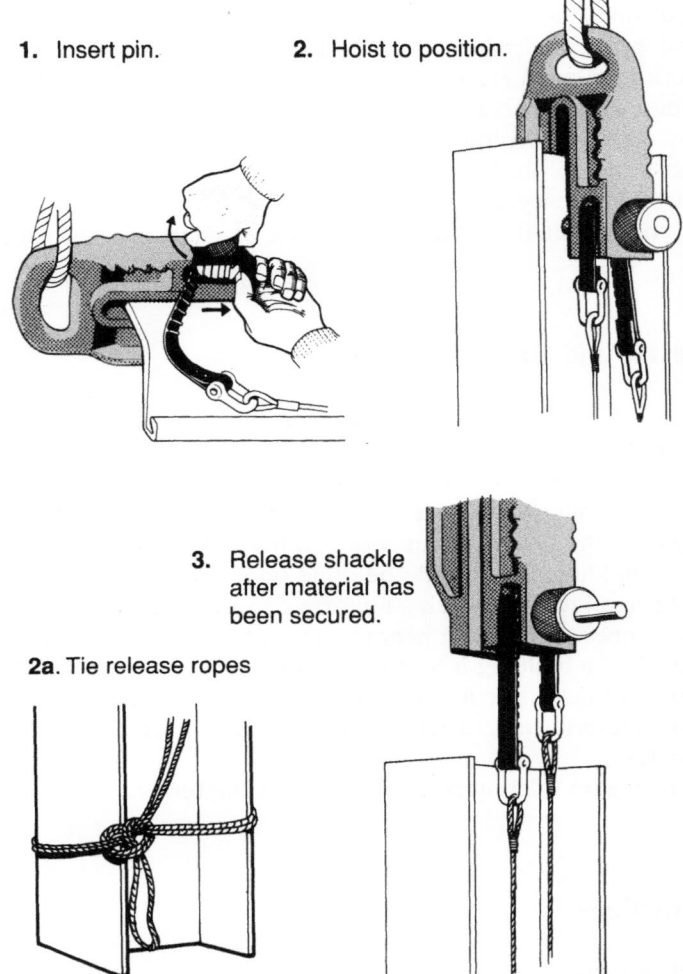

1. Insert pin.

2. Hoist to position.

3. Release shackle after material has been secured.

2a. Tie release ropes

Figure 34.2 Dawson Ratchet (Courtesy of Dawson Construction Plant Ltd.)

convenient locations to ensure that connections are not obstructed, nor the structure eccentrically loaded. Heavy wires are then placed and left in position. When this is achieved the erection crew can move onto the next area. The engineer with his line and levelling crew then follow on behind, pulling the structure to its correct alignment. Once this task is completed, the bolting-up crew can tighten all of the bolts and complete. The decking is then placed, followed by the stud welding. A typical flow chart of the process is shown in Figure 34.3.

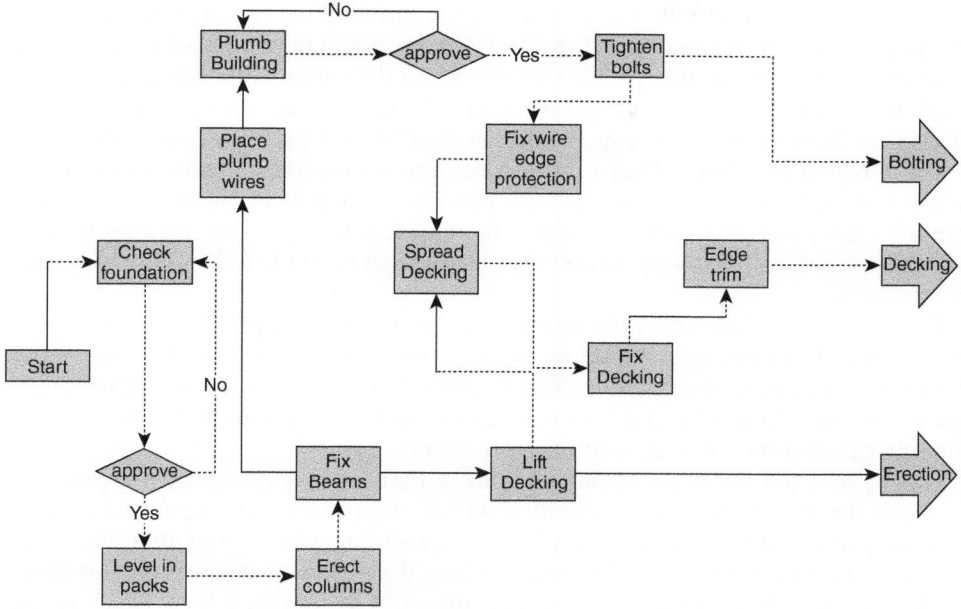

Figure 34.3 Flow chart of the erection process

34.4.2 Lining, levelling and plumbing

If, in spite of having taken care over the initial levelling of packers and of carefully positioning the column bases to the pre-set lines on the foundations, the structure still needs adjustment during lining, levelling and plumbing, then something may be wrong. The first thing to check is that no error has been made in the erection. If nothing is out of place then the drawings and the fabrication need to be checked. If there is an error it will be necessary to take careful note of all the circumstances and advise all concerned immediately.

It is necessary to do the lining and levelling check before final bolting up is done. In practice this means that it must be done immediately following erection since it is inefficient to slacken off bolts already tightened in order move the steelwork about. Erection cannot proceed until the braced bay is bolted up, and that cannot be done until it is level and plumb.

A supply of steel landing wedges and slip plates is needed to adjust the levels. They must be positioned in pairs opposite to each other on two sides of the base plate of the columns to be levelled. If the wedge is placed on one side only the column will be supported eccentrically and can in fact be brought down, especially if more than one column had been lifted on a series of wedges all driven in from the same direction. Alternatively toe jacks (jacks that can be used in tight areas) can be used in pairs to lift the columns. A temporary bench-mark should be established, and agreed with the client's representative, in the vicinity of the columns in a position where it cannot be disturbed. The level will then be used to check the final setting with the seating packers inserted, and the column landed back on them and bolted down with the holding-down bolts. The column can then be moved about in plan on these packers to bring it into line and bay length from its neighbours.

The position of a line, offset from the column centre-line in order to clear the columns, should be marked and agreed. This line is then used either to string and strain a piano wire or to set and sight a theodolite telescope. The readings from it will then give the amounts by which movements must be made to keep the structure in tolerance.

Running dimensions from the previous column will give required longitudinal movement. However, care should be taken to watch the plumb of the columns in this direction in view of the tendency of steelwork to 'grow' due to cumulative tolerances. Regard should be paid to column-to-column dimensions rather than to running dimensions from the end of the building.

Having levelled the bases of the columns in their correct positions, it is possible to check the plumb. As noted in Section 34.3.6, attention to temperature effects is necessary as it is futile to check the plumb of a building which is not all at the same temperature, and in the case of a long building, if the temperature is not standard.

The fact that the column bases are level does not mean that a level check is not needed on the various floors in a tall building. It is important that the levels of any one floor are checked for variations from a plane rather than on the basis of running vertical measurements from the base, since these are affected by temperature and the variable shortening effect on the lower columns as weight is added to them. Plumb can be most readily checked with a theodolite using its vertical axis and reading against a rule held to zero on the column centre-line. This eliminates the effect of rolling errors and is a check against the same centre-line used in the fabrication shops. If a theodolite cannot be used, a heavy plumb bob hung on a piano wire and provided with a simple damping arrangement, such as a bucket of water into which the bob is submerged, is a second-best alternative. Measurements are made from the wire to the centre lines in the same way as before. The disadvantage of using a plumb wire is that all the operatives have to climb on the steel to take and then to check the readings. Optical or laser plumbing units are available; these are particularly useful for checking multi-storey frames.

Adjustments, and in extreme cases provision for holding the framework in its correct position, are normally only necessary if the frame is not self-stable. For instance, if structural integrity depends on concrete diaphragm panels for stability, consideration should have been given at the design stage to one of two alternatives: either the concrete panels should be erected with the steel frame; or temporary bracing should have been provided as part of the original planning. To assure proper function, any bracing must be positioned so that it can be left in place until after the concrete panel has been placed and fixed.

If diagonal wires have to be used they should be tied off to the frame at node points rather than in mid-beam, in order to avoid bending members. Also their ends should be fixed using timber packers in the same way as is done when the pieces are slung for lifting. Turnbuckles or tirfor type pullers provide the effort necessary to tension the wires. The sequence of placing, tensioning and ultimately removing these temporary arrangements should form part of the method statement prepared for each particular situation. Once the components of the building have been pulled into position the final bolting up can be completed.

Portal frame construction in its temporary state often has the legs pre-set out of vertical which only becomes vertical after the final loading has taken place. It is the responsibility of the designer to calculate the extent of pre-set. This often proves a difficult process to assess. Further allowance must be made if the final position is needed for aesthetic purposes.

It is essential to ensure that the lining and levelling keep pace with the erection, since when additional floors are added it may become increasingly difficult to adjust, pull, or move the structure. As a rule of thumb, the lining and levelling of heavy sections should be within two to three bays of the erection face, whereas the lining and levelling of lighter sections should be within four to six bays of the erection face.

34.4.3 Tolerances

Prior to determining tolerances it is important to understand why they are needed. Rolling tolerances of the steel and manufacturing tolerances need to be accommodated. Tolerances must accommodate preceding work and subsequent building components. Tolerances must meet design requirements, architectural/aesthetic requirements and maximise buildability.

Tolerances can be categorised as structural (necessary for the integrity of the intended structure), architectural (necessary for the aesthetics of the structure), buildability (necessary to construct all of the components in a building). For a detailed coverage of tolerances in steel construction, refer to Chapter 32.

It is important to inform the designer of any deviation from specified tolerances, as failure to comply can often adversely affect both the structure and subsequent operations.

34.4.4 Holding-down bolts

To facilitate the erection of the structure, holding-down bolts are normally positioned by the civils contractor. Movement tolerances must be accommodated in the foundation to ensure correct alignment can take place (see Figure 34.4).

Conical sleeves maximise movement without reducing the area of anchorage. It is important to have the bolts set both vertical and loose in the sleeve. The key benefit of using sleeved cast-in bolts is to allow movement for the alignment of the structure to meet tolerances. After checking and confirming that the holding down

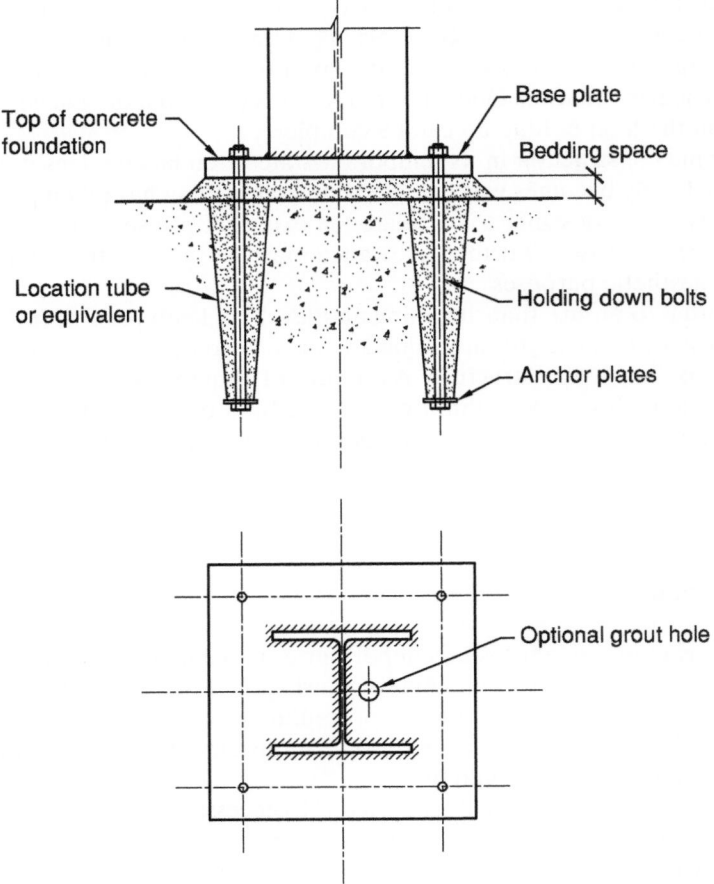

Figure 34.4   Holding-down bolts

bolts are acceptable, typical 100 mm × 100 mm steel packs (1 mm to 20 mm thick) or similar are placed on the base to the required level.

34.4.5 Site bolting

Wherever possible site splices should be designed to be bolted. This process is less affected by adverse weather than welding, uses simple equipment, and presents much less complex work access and subsequent inspection problems.

Bolts should be chosen to minimise size variation, and ease of repetition and selection. Design for Construction[7] makes the following recommendations when selecting bolts:

1. Pre-loaded bolts should only be used where relative movement of connected parts (slip) is unacceptable, or where there is a possibility of dynamic loading.
2. The use of different grade bolts of the same diameter on the same project should be avoided.
3. 'Just-in-time' delivery should be used, based on simple, easy to follow bolt lists. Where appropriate, bolts, nuts and washers should be supplied with a corrosion protection coating that does not require further on-site protection.
4. Bolt lengths should be rationalised. (Approximately 90% of simple connections could be made using M20 60mm long bolts.)
5. Bolts should be threaded full length where possible.
6. Connections should be standardised where possible.

Impact wrenches and powered nut runners speed the tightening of bolts. If small quantities of preloaded slip-resistant bolts are necessary, manual torque wrenches are suitable but where large quantities of bolts have to be tightened, an impact wrench is an essential tool. Where access is difficult, wrenches are made which run the nut up and are then used as manual ratchet wrenches to finally tighten the bolt.

Preloaded slip-resistant bolts must be tightened first to bring the plates together and thereafter be brought up to the required preload either by applying a further part turn of the nut or by using a wrench calibrated to indicate that the required torque has been reached. The manual wrench has a break action which gives the indication by means of a spring without endangering the erector by a sudden complete collapse. Alternatively, load-indicating washers may be used to demonstrate that the bolt has been tightened to the required tension.

34.5 Site fabrication and modifications

Special consideration must be given to the planning of the erection of a structural steel frame where site welding has been specified. Site welding has to be carried out in appropriate weather conditions, and in positions which make it more difficult and expensive than welding carried out in workshop conditions. In a workshop the work can be positioned to provide optimum conditions for the deposition of the weld metal, whereas most welds on-site are positional. Most site welding is carried out by manual metal arc and electrodes which is more flexible in use and is more able to lay good weld fillets in the vertical up-and-overhead positions.

Some means must be provided for temporarily aligning adjacent components which are to be welded together, and of holding them in position until they are welded. The methods adopted to cope with the need for alignment may have to carry the weight of the components, and in some cases a substantial load from the growing structure.

Safe means of access and safe working conditions must be provided for the welder and his equipment. The working platform may also have to incorporate weather

Figure 34.5   Run-off plates for a butt weld

protection since wind and rain and cold can all adversely affect the quality of the weld. The design of joint component weld preparation must take into account the position of these components in the structure. Provision must also be made for the necessary run-off coupon plates for butt welds and for their subsequent removal. Run-off plates are required in order to ensure that the full quality of the weld metal being deposited is maintained along the full design length of the weld (see Figure 34.5). The erection method and weld procedure statements for each joint must take all these factors into account (see Chapter 26).

Attention must be paid to the initial setting and positioning of the components in order that the inevitable weld shrinkage resulting from the cooling of joints does not result in loss of the required dimensional tolerances across the joints.

Health and Safety issues are clearly important when using welding and particular attention should be paid to fire precautions, the avoidance of eye damage, not just for the welders but for others working nearby, and the condition of the welding equipment.

34.6 Steel decking and shear connectors

34.6.1 Steel decking and shear connectors

Many steel framed commercial buildings in the UK use composite slab construction and so steel decking[8] and shear connectors now play an integral part in the steel erection programme. A key advantage of light-weight steel decking is that it can act as a working platform at each level. This eliminates the need for temporary platforms, and reduces the height an erector can fall – a critical safety issue in multi-storey construction.

Once the building or section has been lined, levelled and bolted up, then steel decking bundles previously positioned on the steel work can be spread. It is important to remember that the sheets are very light and should be protected from the wind. Since decking operatives are not normally steel erectors, they need safety wires to hook on to as they work. As the decking is spread, exposed edges must be protected since these are the vulnerable points where operatives can fall. The decking is then shot-fired to the steel using Hilti nails or similar. Once this is complete the edge trim is placed using shot-fired nails, the joints taped to minimise grout loss and then handed over to the welder for the shear connection installation. Shear connectors are usually placed and welded using automatic machines. Thus, provision should be made for a high current supply or adequate space for a portable generator. Most shear studs are 100 mm × 19 mm diameter and are delivered in barrels weighing about 100 kg. It is therefore important to allow for the lifting of studs and equipment to the various levels. The welder will then work methodically in lines, placing and testing the studs until each area is ready for inspection and handing over.

34.6.2 Cold-formed sections

Secondary items in many structures are cold-formed sections, used as roofing rails or cladding rails. Since roof and wall material is very light and easy to supply in large volumes, cold-formed sections are often delivered by 20 ton lorry. Care must be taken in the factory to ensure clear marking and banding of components to prevent wastage of man hours, days and even weeks of subsequent on-site sorting. Once the material has been off-loaded in bundles it will be placed in the rough location for installation. The material will need a large shake-out area to facilitate speed of erection. In situations of multi-storey high-rise construction, the material can be hoisted to the floor below the erection face, then shaken out and subsequently pulled up into position. This can often eliminate crucial tower crane hook time. Most manufacturers offer a total package of rails, sag rods and fastenings.

Selection of quick-fit materials will expedite site erection, thereby reducing construction time and man hours.

34.7 Cranes and craneage

34.7.1 Introduction

The choice of cranes and positioning may be dependent on many factors. Care needs to be taken to ensure good progress of the work. Often, cranes that work at maximum capacity in restricted areas actually slow production and are not efficient. A well-located crane with spare capacity can speed erection and more than compensate for the additional cost of hire.

Principal items to be considered when selecting a crane are:[7]

- site location – access and adjacent features
- duration of construction
- the weight of the pieces to be erected and their position relative to potential crane standing positions
- size of piece to be erected
- the need for tandem lifts
- maximum height of lifts
- number of pieces to be erected per week
- ground conditions
- the need to travel with loads
- the need for craneage to be spread over a number of locations
- organisation of off loading and stockyard areas dismantling.

On a large job, cranes of varying size and capacity perform various tasks, i.e. a heavy one for the main columns and a smaller one for the side framing posts and angles. The number of pieces to be lifted in any one working period determines the number of cranes required to achieve the desired work-rate. In practice, the choice of crane type and capacity is a compromise, with efforts being made to get as near to the optimum as possible.

In order to decide on a crane layout it is necessary to prepare a sketch or drawing of each critical lifting position showing, on a large scale, the position of the crane on the ground; the location of the hook on plan; the clearance between the jib and its load; and between the jib and the existing structure as the piece is being landed in its final position. These drawings will enable a check to be made of the match of each crane to its load; of the ability of the ground beneath the crane to sustain the crane and its load; and clearance to lift the component and place it in position. It is important to allow adequate clearance for the tail of the crane as it slews. A series of these drawings will rapidly confirm whether a particular crane is suitable. If there is an anomalous heavy lift in the series, the designer can be asked whether a splice position can be moved to reduce weight.

34.7.2 Types of crane

34.7.2.1 Mobile cranes

The mobile cranes include truck or wheel-mounted, and tracked cranes. Truck-mounted cranes are able to travel on the public roads under their own power and with their jibs shortened. Mobile cranes mounted on crawler tracks (see Figure 34.6) are not permitted to travel on the road under their own power.

A truck-mounted crane may prove a correct choice if a crane is only required on site for a short time, since it can come and go relatively easily. However, if a crane is to be on site for a longer period, the high transportation costs associated with crawler crane use can be justified. The real advantage of a crawler crane is that both its weight and the reaction from the load-lift are spread more widely by its tracks. However, it does not have the added stability of a truck-mounted crane with its outriggers set (see Figure 34.7).

Figure 34.6 Crawler crane (Courtesy of Baldwin Industrial Services/Chapman Brown Photography)

25 Tonne DEMAG AC 75 "CITY CLASS" Mobile Crane

Lifting Capacities Main Boom Extension

Telescopic Boom 20,7 - 25,0 m. On Fully Extended Outriggers 5,9 m. Working Range 360°

Radius m	Main Boom 20,7 m				Main Boom 25,0 m				Radius m
	7,1 m		13,0 m		7,1 m		13,0 m		
	0°	30°	0°	30°	0°	30°	0°	30°	
5	5,6	-	-	-	-	-	-	-	5
6	5,6	-	-	-	-	-	-	-	6
7	5,3	-	2,2	-	4,5	-	-	-	7
8	5,0	-	2,2	-	4,4	-	2,0	-	8
9	4,7	3,7	2,2	-	4,2	-	2,0	-	9
10	4,4	3,6	2,1	-	4,0	3,4	2,0	-	10
11	4,2	3,5	2,1	-	3,9	3,3	2,0	-	11
12	4,0	3,4	2,0	-	3,7	3,2	1,9	-	12
13	3,5	3,3	1,9	1,7	3,5	3,1	1,9	-	13
14	3,1	3,2	1,9	1,6	3,0	3,0	1,8	1,5	14
15	2,7	3,0	1,8	1,6	2,7	2,9	1,8	1,5	15
16	2,4	2,6	1,7	1,5	2,4	2,6	1,7	1,5	16
17	2,2	2,3	1,6	1,5	2,1	2,3	1,7	1,4	17
18	1,9	2,1	1,6	1,4	1,8	2,0	1,6	1,4	18
19	1,7	1,8	1,5	1,4	1,6	1,8	1,6	1,4	19
20	1,5	1,5	1,4	1,4	1,4	1,6	1,5	1,3	20
21	1,3	1,3	1,4	1,3	1,3	1,4	1,3	1,3	21
22	1,1	-	1,3	1,3	1,1	1,2	1,2	1,3	22
23	0,9	-	1,1	1,3	0,9	1,0	1,0	1,3	23
24	0,8	-	1,0	1,2	0,8	0,9	0,9	1,2	24
25	0,7	-	0,9	1,0	0,7	0,7	0,8	1,0	25
26	-	-	0,8	0,9	0,6	-	0,7	0,9	26
27	-	-	0,7	0,7	0,5	-	0,6	0,8	27
28	-	-	0,6	-	-	-	0,5	0,7	28
29	-	-	0,5	-	-	-	-	0,6	29

25 Tonne DEMAG AC 75 "CITY CLASS" Mobile Crane

Dimensions

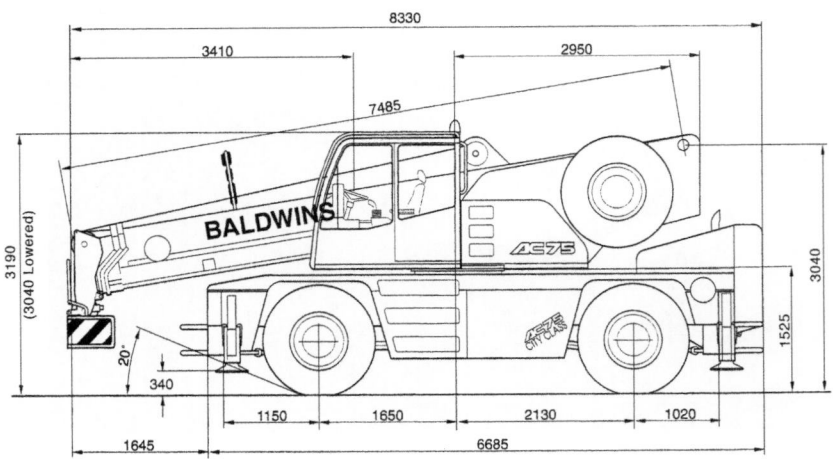

Figure 34.7   Typical mobile crane (Courtesy of Baldwin Industrial Services)

Most types of mobile crane form part of the erection fleet owned by steel erection and/or plant hire companies. The decision to hire from a plant hire company is often determined by the geography of the site, for example is the site close to a plant hire company's yard, or is it close to the contractor's own plant depot? Is the crane to be used for short or long duration? Is the lift for which the crane is required one of a series or simply a one-off? Mobile crane capacity normally ranges between 15 to 800 tonnes. Figures 34.7 and 34.8 provide performance data for a typical 25 and 800 tonne crane. It is important to note: **that the rated capacity of a mobile crane and the load which can be safely lifted are two very different things**. The maker's safe loading diagram for the crane must always be carefully consulted to see what load can be lifted with the jib length and the radius proposed. It is essential that a mobile crane is used on level ground so that no side loads are imposed on the jib structure.

34.7.2.2 Mobility for non-mobile cranes

Non-mobile cranes can be made mobile by mounting them on rails. This has two advantages: the positioning of the crane can be more easily dictated and controlled, and the loads transmitted by the rails to the ground act in a precisely known location. Many cranes have collapsed because of insufficient support underneath. However, most rail-mounted crane failures have occurred from overloading. Where the crane is to work over complex plant foundations the rail can be carried on a beam supported, if necessary, on piles especially driven for the purpose. If the rail is supported only by a beam on sleepers in direct contact with the ground, the load can be properly distributed by suitable spreaders. In either case conditions must be properly considered and designed for. Problems often occur when too much faith is invested in the capability of the ground to support a mobile crane and its outriggers.

34.7.2.3 Non-mobile cranes

Non-mobile cranes are generally larger than their non-mobile counterparts. They can reach a greater height, and are able to lift their rated loads at a greater radius.

There are two main types of non-mobile cranes, the tower crane and the (now rare) derrick. Due to their great size, the cranes must arrive on-site in pieces. Thus, the disadvantage of a non-mobile crane is that it has to be assembled on-site. Having been assembled, the crane must receive structural, winch and stability tests before being put into service.

A tower crane with sufficient height and lifting capacity (see Figure 34.9) has several advantages:

1. It requires only two rails for it to be 'mobile'. These two rails, although at a wide gauge, take up less ground space than a derrick.

800 Tonne LIEBHERR LTM 1800
Mobile Crane

Working Ranges Luffing Lattice Jib

Main Boom 83°. Luffing Lattice Jib 21,0 - 91,0 m. On Outriggers 13 m x 13 m. Working Range 360° Counterweight 153 t.

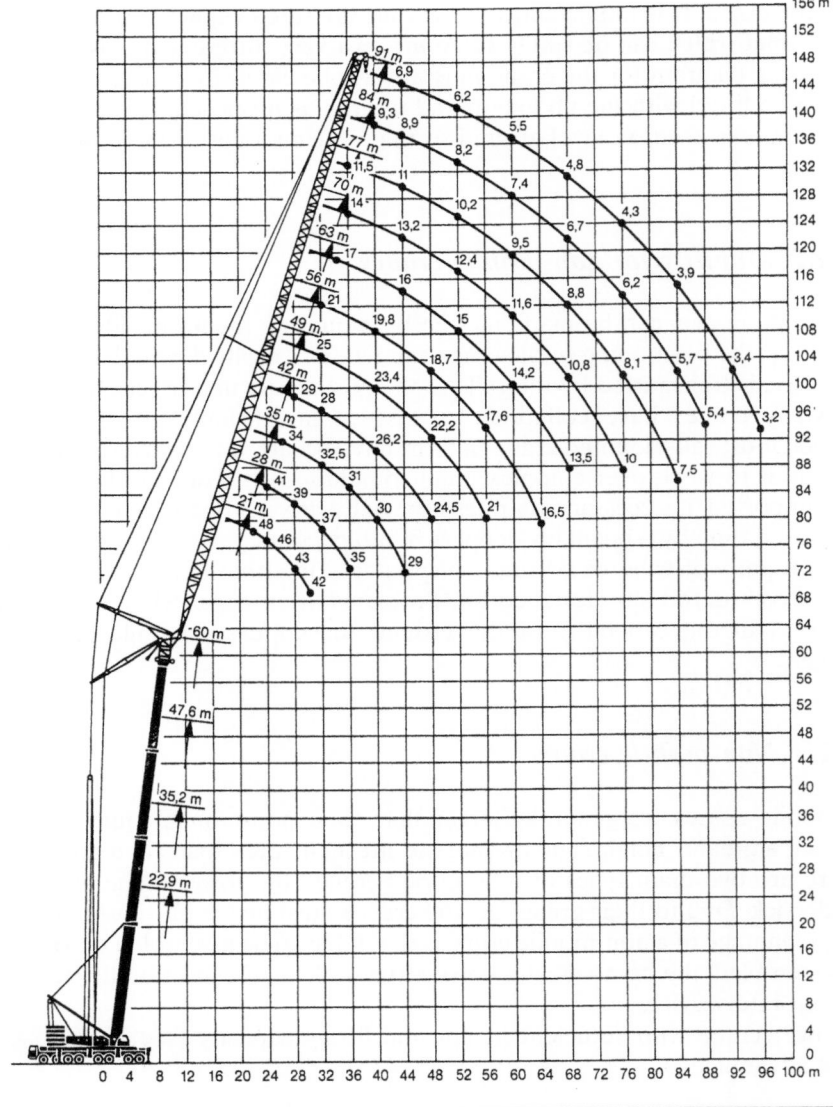

While every care has been taken in the preparation of the ratings given in this leaflet, and all reasonable steps have been taken to check the accuracy of the information, Baldwins cannot accept responsibility for any inaccuracy in respect of any matter arising out of, or in connection with the use of these tables.

Figure 34.8　Special 800 tonne Mobile crane (Courtesy of Baldwin Industrial Services)

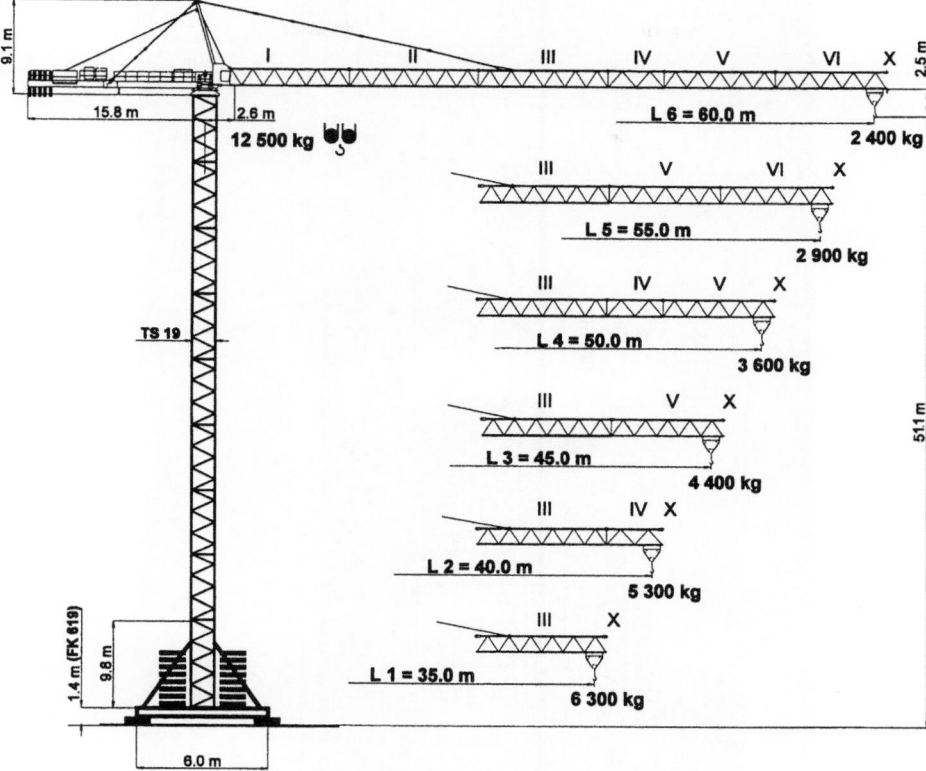

Figure 34.9 Tower crane (Courtesy of Delta Tower Cranes)

2. It carries most of its ballast at the top of the tower on the slewing jib/ counterbalance structure, and so very much less ballast is needed at the bottom. Indeed, in some cases, there is no need for any ballast at the tower base or portal.
3. Because the jib of a tower crane is often horizontal, with the luffing of a derrick jib replaced by a travelling crab, the crane can work much closer to the structure and can reach over to positions inaccessible to a luffing jib crane.
4. A tower crane is 'self-erecting' in the sense that, after initial assembly at or near ground level, the telescoping tower eliminates the need for secondary cranes.
5. As shown in Figure 34.10 a tower crane can be tied into the structure it is erecting, thus permitting its use at heights beyond its free standing capacity.

There exist several types of tower crane, e.g. articulated jib, luffing and saddle cranes as illustrated in Figure 34.11. It is essential that manufacturers or plant hirers are consulted in order to make the most appropriate choice of crane.

Figure 34.10   Citigroup Tower, London showing tower cranes tied into the building (Courtesy of Victor Buyck Hollandia)

34.7.2.4 Cranes for the stockyard

Stockyard cranes have to work hard. The tonnage per job has to be handled twice in the same period of time, often with many fewer cranes. It is therefore important that cranes be selected and sited carefully to ensure maximum efficiency.

34.7.3 Other solutions

If there is no suitable crane, or if there is no working place around or inside the building where a crane may be placed, then consideration must be given to a special

Figure 34.11 Tower cranes used in the construction of Citigroup Tower, London (Courtesy of Victor Buyck Hollandia)

mounting device for a standard crane, or even a special lifting device to do the work of a crane, designed to be supported on the growing structure under construction. In either event, close collaboration between the designer and erector members of the management team is of paramount importance. Conversely, inadequate communication may prove problematic. Once the decision to consider the use of a special lifting device has been made, a new range of options becomes available. The most important of these is the possibility of sub-assembling larger and heavier components thereby reducing the number of man hours worked at height. A major disadvantage of special lifting devices is that the apparatus being considered is often so specialised that it is unlikely to be of use on another job. Thus the whole cost is targeted at the one job for which it has been initially designed.

34.7.4 Crane layout

It is important to decide on the type, size and number of cranes that are required to carry out the work, since each has a designated range of positions relative to the work it is to perform. These positions are then co-ordinated into an overall plan which enables each crane to work without interfering with its neighbours, and at

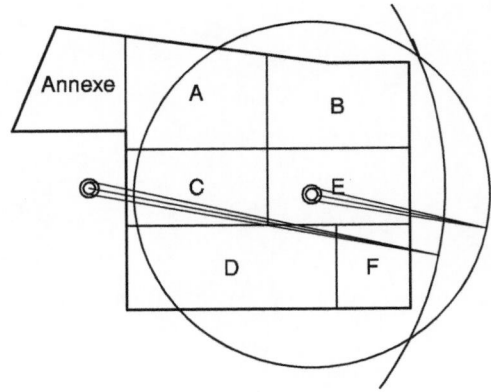

Figure 34.12   Typical crane layout

the same time enables each to work in a position where adequate support can safely be provided (see Figure 34.12). This plan will then form the basis of the erection method statement documentation.

A major factor in planning craneage is to ensure that access is both available and adequate to enable the necessary quantity and size of components to be moved. On large greenfield developments these movements may often have to take place along common access roads used by all contractors and along routes which may be subject to weight or size restrictions. On a tight urban site the access may be no more than a narrow one-way street subject to major traffic congestion.

34.7.5 The safe use of cranes

Mention has already been made of the UK Statutory Regulations. These lay down not only requirements for safe access and safe working but also a series of test requirements for cranes and other lifting appliances. It is the responsibility of management to ensure that plant put onto a site has a sufficient capacity to do the job for which it is intended, and that it remains in good condition during the course of the project. Shackles and slings must have test certificates showing when they were last tested. Cranes must be tested to an overload after they have been assembled. The crane test is to ensure that the winch capacity, as well as the resistance to overturning and the integrity of the structure, is adequate.

Although a crane may have been tested and used safely in many locations, its safety on a particular site is dependent upon adequate support under the tracks or outriggers. It is also important that the crane is sited on level ground, since an overload can easily be imposed, either directly or as a sideways twist to the jib, if the ground is not level.

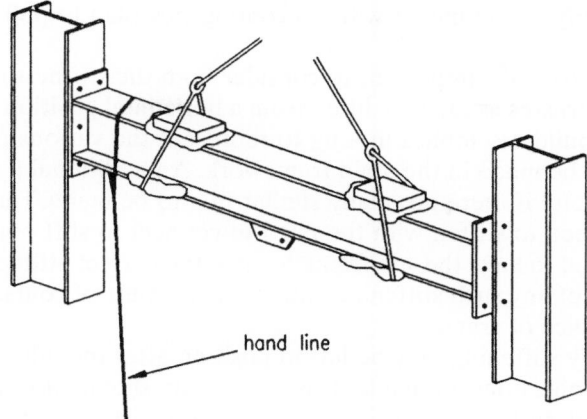

Figure 34.13 Typical slinging of a piece of steelwork

34.7.6 Slinging and lifting

Components, whether they are on transport or are lying in the stockyard, should always be landed on timber packers. The packers should be strong enough to support the weight of the steel placed above them, and thick enough to enable a sling to be slipped between each component.

When lifting a component for transport only, the aim is to have it hang horizontally. This means that it is necessary to estimate its centre of gravity. Although this calculation may be easy for a simple beam, it may prove more problematic for a complex component. The first lift should be made very slowly in order to check how it will behave, and also to check that the slings are properly bedded (see Figure 34.13).

Most steelwork arrives on-site with some or all of its paint treatment. Since the inevitable damage which slinging and handling can do to paintwork must be made good, it is therefore important to try to minimise that damage. The same measures that achieve this also ensure that the load will not slip as it is being lifted, and that the slings (chain or wire) are not themselves damaged as they bend sharply around the corners. Softwood packers should be used to ease these sharp corners.

Packers to prevent slipping are even more necessary if the piece being erected does not end up in a horizontal position. The aim should always be to sling the piece to hang at the same attitude that it will assume in its erected position. Pieces being lifted are usually controlled by a light hand line affixed to one end. This hand line is there to control the swing of the piece in the wind, and not to pull it into level. Wherever possible non-metallic slings should be used. They will reduce damage to paintwork and are less likely to slip than chain or wire slings.

In extreme cases two pieces may have to be erected simultaneously using two cranes. Staff, working back at the office, should account for this in the site erection method statement. It is too late to discover this omission when the erection is attempted with only one crane, or with no contingency plan to pull back the head of the column.

As discussed above, it is important to consider both the stiffness of large assemblies such as roof trusses as they are lifted from a horizontal position on the ground, and the need to build assemblies in a jig to represent the various points at which connection has to be made in the main framework. An additional jig for lifting can be particularly useful if there are many similar lifts to be made. This can be made to combine the need to stiffen with the need to connect to stiff points in the sub-frame, and the need to have the sub-frame hang in the correct attitude on the crane hook. The weight of any such stiffening and of any jig must of course be taken into account in the choice of crane.

Some temporary stiffening may be left in position after the initial erection until the permanent connections are made. This eventuality should also have been foreseen and sufficient stiffeners and lifting devices should be provided to avoid an unnecessary bottleneck caused by the shortage of a device for erection of the next sub-frame.

Where a particularly awkward or heavy lift has to be made, slinging and lifting can be made both quicker and safer if cleats for the slings have been incorporated in the fabrication. Each trial lift made after the first one wastes time until the piece hangs true. The drawing office should determine exactly where the centre of gravity is.

A chart giving details of standard hand signals is illustrated in Figure 34.14. Their use is essential when a banksman is employed to control the rear end of the transport thereby bringing the component to the hook as it is reversed. If the person directing the movements of the crane is out of sight of the crane driver, the banksman is needed to relay the signal. A clear system of signals should be agreed for the hand-over of crane control from the person on the ground to the person up on the steel who controls the actual landing of the component. A banksman may also be needed up on the steel if the crane driver cannot clearly see who is giving the control instructions: It is vital that there is no confusion over who is giving instructions to the crane driver.

34.8 Safety

The principal safely objectives when erecting steelwork are:

- stability of the part-erected structure
- safe lifting and placing of steel components safe access and working positions.

The most serious accidents that occur during the erection of structures are generally caused by falls from height, either from working positions or while gaining access to them. Other serious accidents can occur because of structural instability during erection and while handling, lifting and transporting materials. Failure to establish safe erection procedures and to implement them through effective site management can create unnecessary hazards, leading to risks being taken and hence

Figure 34.14 Standard hand signals for lifting

to accidents. Safe lifting and placing of steel components has been addressed in Section 33.8. Safe access and stability of the part-erected structure are covered in 33.9.1 and 33.9.2. These sections are edited extracts from the *BCSA Code of Practice for Erection of Multi-Storey Buildings*.[9] The reader is advised to consult this Code of Practice for a more detailed explanation.

34.8.1 The safety of the workforce

34.8.1.1 Temporary access during construction

The Steelwork Contractor must ensure that method statements and their associated risk assessments address the need to provide safe access and working positions. Installation of the permanent or temporary stair systems as soon as possible helps to eliminate some of the risks associated with temporary access.

Multi-storey and other high-rise structures are characterised by the fact that it is not always practicable to meet these requirements without the need for access over the steelwork using 'beam straddling'. The use of beam straddling should be avoided where the use of one of the following methods is practicable:

1. Using telescopic boom MEWPs ('cherry pickers') for both access and working positions: this method is preferred where it is practicable but it can only be used for the lower floor levels up to, say, 20 m – i.e. around the first splice level.
2. In some circumstances it is possible to use small boom MEWPs or 'spiders' for access and working positions: these can be temporarily positioned on the structure using support frames which can be re-located as the high-rise building sequence develops. The Steelwork Contractor must arrange for an engineer to evaluate the additional imposed loadings during the construction phase.
3. In other cases it may be possible to use such 'spiders' or alternatively scissor-lift MEWPs ('flying carpets') traversing the floor slab for both access and working positions; to do so, the permanent works designer must have included such construction loads in the initial design of the floor slab.
4. Using crane-mounted cradles or 'man baskets'. This method is often restricted by the availability of or access for suitable craneage. The use of man baskets on high-rise structures is also more likely to be restricted by the wind conditions.
5. When using MEWPs or baskets on site contractors should ensure their operators are secured to the anchorage point of the MEWP basket using safety harness and lanyard. It should be noted that the anchorage points on most types of MEWP are not designed for shock loading, therefore lanyards are provided as a fall restraint only.
6. Using mobile access towers (MATs): MATs must be constructed with due regard to stability and may be used only on a firm base. MATs can only be mounted directly on top of metal decking if suitable load spreaders (or 'elephant's feet') are used. If the MAT has wheels, they must be secured against movement when a person is working on the tower and movement of the tower must be only from the base.

7. Using scaffolding with suitable edge protection: this is rarely practicable or justi-fied for steel erection. However, the use of scaffolding may be practicable and justifiable in special situations, such as access to high-level elevations using scaf-folds cantilevered out from structurally finished floor levels or where a working platform is needed for, say, a later site welding task.

8. Using a Spandek system or similar span deck platforms for access: these are often used in the erection of multi-storey structures in conjunction with ladders whereby the span deck platform acts as a landing platform. The platforms are secured on the open steelwork, and as erection progresses they can be more readily relocated than scaffold platforms.

9. Using ladders for access to working positions on the steelwork: this is a very common method where MEWPs and man baskets are impracticable. Ladders must be properly positioned and tied, and they need to extend 1 m above the supporting steelwork. Landing platforms must be provided where access extends further than 9 m. Special care needs to be taken where ladders are positioned onto metal decking, as the surface of the decking has a low friction coefficient and is also profiled such that secure and even bearing may prove difficult to achieve. In such circumstances, the ladder should be footed by another worker or provided with a stability device whilst in use.

10. Using ladders for working positions: this is only justified for short duration tasks of less than 30 minutes in one location and for work that permits the user to have a minimum of three points of contact. Working off ladders should only be carried out if they are of suitable type, strength and length for the operation being carried out. Before work commences the ladders should be suitably tied at the top (or footed at all times during use where this is not practicable and the ladder does not exceed 4.6 m in length). Erectors must be instructed that, having used a ladder to reach a working position, they must immediately clip on to a suitable anchorage point on the steelwork before any work is commenced.

34.8.1.2 Beam straddling

This means of access may only be used by the Steelwork Contractor where there is no other practicable means of access, or a working position, other than by using the steelwork itself. A Task Specific Method Statement is required covering the means of access to the beam at height, the method of straddling the beam for access (at no time should erectors be allowed to 'walk beam top flanges' – where access is required beams should be crossed by 'straddling', i.e. the worker sitting on the top flange, using the bottom flange for a foothold and using both hands to grip the top flange in transit), and using the fall arrest system during access, and the method by which the erector is tied on during the work.

34.8.1.3 Exclusion zones and safe access routes

The Steelwork Contractor should try to ensure that others do not enter a hazardous area where steel erection work is taking place. Where reasonably practicable, the hazardous area should be established by the Principal Contractor as an exclusion zone within which only steel erection activity should be allowed.

For high-rise multi-storey structures, it is often impracticable to designate an exclusion zone solely in terms of a section of the building's plan area. Then, an additional segregation method is needed in terms of the building's height. This can be arranged if there are sufficient floors to act as a suitable safety barrier from falling materials between the erection activity and the following trades working below. With respect to falls of smaller objects, staggered construction sequences enable following trades to commence work activities under the protection of two structurally complete metal decked floors known as 'crash decks'. Fully concreted levels can be designed both to serve as 'crash decks' and to provide access for MEWPs operating over the slab.

34.8.1.4 Fall prevention and arrest

Falling from height is one of the most critical site hazards that the Steelwork Contractor has to address. If a person falls from a height, the resulting impact will generally result in a major or fatal injury. The only practical way to reduce the risk of falling from height is by risk control using fall protection. This protection can either be provided as:

- Fall prevention – which prevents a person from getting into a position from which they could fall, for example using guard rails, barriers and edge protection systems.
- Fall restraint – which restrains a worker from moving too far towards a position from which they could fall, for example using a system of fall-restraint lanyards attached to wires. The term 'fall restraint' used in the context of steel erection is a particular application of the term 'work restraint' used in HSE terminology.
- Fall arrest – which, should a person fall, 'arrests' the fall to limit its extent, for example using safety nets or a fall-arrest lanyard attached to a safe anchorage.

Further details are given in BCSA Code of Practice for Erection of Multi Storey Buildings.[9]

34.8.2 The safety of the structure

34.8.2.1 Bracing systems

Before erection is commenced, the Steelwork Contractor must ensure that the sequence included in the erection method statement has been reviewed by a suitably competent structural engineer who understands the means by which structural stability is achieved in the permanent works design. The engineer will also need to decide on the extent of any temporary bracing required and to design it for the interim loading conditions. The BCSA Guide to Steel Erection in Windy Conditions[10] provides recommendations for assessing the capacity of permanent or temporary bracing in terms of the wind loads to consider and the effect of column lack of verticality during erection.

The stability of multi-storey structures when completed is particularly affected by wind. The permanent works design will include a system by which wind loads are taken to the foundations. Generally this will be by using the floors as horizontal diaphragms to take the wind loads back to a vertical bracing system. This system will often be stair/lift cores of either concrete or steel. The ability of floors to act as diaphragms requires careful assessment as, for instance, precast planks may not be fully secured by grout immediately after installation and this could allow some adverse movement. In such cases temporary plan bracing or additional panels of temporary vertical bracing may need to be installed.

In the case of concrete cores, they need to be complete to the necessary number of floor levels before steel erection commences. The key issue is then to ensure that the connections between the steelwork and the concrete core are completed soon enough to be able to take the wind forces on the part-erected structure. In cases where the final connection is to be site welded, a 'temporary fix' will be needed prior to alignment of the steelwork.

In the case of steel vertical bracing systems, these can only be erected as the steel erection progresses generally. The key issue is then to ensure that the bracings are installed as soon as possible to be able to take the wind forces on the part-erected structure. Essentially this means that the steel in a braced core area should be erected first so that it can act as a stable 'box' to provide lateral support for subsequent erection. The stable box created needs to be able to provide lateral support in two orthogonal directions, which is more difficult to achieve if the vertical bracing panels are distributed around the building rather than being localised in core areas. In such cases, additional temporary bracing may be required to act as either vertical or horizontal bracing until the permanent works are complete.

In terms of permanent horizontal bracing, generally the floors will be either precast concrete planks or in situ concrete cast on metal decking. However, prior to placing of the precast planks or fixing of the metal decking, wind pressure acting directly on steel columns and beams can be taken straight into other steel members that frame into them. Hence, problems usually only occur where members are not framed in two directions, for example, where a column is supported by a single beam. If temporary bracing or temporary fixings are needed, they need to be retained until the permanent works are sufficiently complete, for example until after concrete floors are cast, to ensure that the frame remains both stable and held in position against movement under construction loads.

34.8.2.2 Temporary stability of columns

In the part-erected condition, the Steelwork Contractor must ensure that columns are stable against wind forces and the lack of verticality prior to alignment. Unless explicitly allowed in the erection method, columns will need to be tied in two directions within a working shift during which wind will be monitored. This means that the stability of a single 'flagpole' column may well be more influenced by its initial lack of verticality, and that:

- the bottom lift of column is designed with the base plate adequately sized to allow free-standing of the column until it is tied in
- column splices are adequately designed to allow the upper shaft to be erected safety until it is tied in.

The splice locations should be determined on three criteria as follows:

1. Subject to stability during erection, longer shafts with fewer splices are preferred. This often means that columns can be spliced every three floor levels.
2. Access methods for securing beams may limit the column shaft lengths to two levels. This depends on whether access is from the ground or from previously erected steelwork, by MEWP or ladder, as ladder access is limited to 9 m and MEWPs mounted aloft are often too small to be able to reach more than two levels.
3. To facilitate access to secure the splice during bolting up, the precise location should generally be 1100 mm above local finished floor level (i.e. about 1300 mm above any metal decking).

34.8.2.3 Typical erection sequence

By observing the following sequence, the Steelwork Contractor can generally ensure stability is maintained during erection:

- Erection should always commence from a braced bay (or a suitable structural core).
- A bay of steelwork should be erected including all floor members (up to first splice level -generally two or three floor levels) and bracings, ensuring that the structure is braced in both directions (using permanent and temporary bracing as necessary).
- Prior to the continued erection of the frame, the initial bay of steel should be fully lined, levelled, plumbed and bolted up to ensure a rigid structure is achieved as a 'stable box'.
- The remaining steelwork in the footprint area of the whole structure should then be erected to first splice. The steel should then be fully lined, levelled, plumbed and bolted up.
- On completion of the above, the bases of the columns should be grouted prior to commencement of steel erection for the next lift.
- Ensure all permanent and temporary bracing has been installed.
- Subsequent horizontal 'slices' of the structure are erected similarly, completing each up to the next column splice level before proceeding above.

In the case of multi-storey buildings with large plan areas and multiple cores, it may be possible to work on a series of areas divided vertically, especially if there are expansion joints between. Erection then proceeds as a 'carousel' whereby other trades such as decking and concreting follow round in sequence. This sequence will require additional temporary edge protection to phase edges.

34.8.2.4 *Temporary supports and temporary conditions*

Although much time and effort is invested in the design of a structure, the design of any temporary works on which that structure depends during construction must also be given very careful attention. The number of recorded collapses that take place after an initial failure in temporary supports bear testimony to the fact that this is not always the case. For example, a temporary support may only be designed to take a vertical load. In practice, the structure it is intended to support may move due to changes in temperature and wind loading, thereby imposing significant additional horizontal loads.

Sufficient consideration should be given to the foundations. Settlement in a trestle foundation can profoundly affect the stress distribution in the structure that it supports. Settlement under a crane outrigger from a load applied only momentarily can lead to the collapse of the crane and its load. The Code of Practice BS 5975[11] for falsework (which includes all temporary works, trestling, guy wires, etc., as well as temporary works associated with earthworks) deals with a wide range of falsework types and should be carefully read and observed. Particular attention should be paid to dealing with communication, co-ordination and supervision since failure in any of these areas can lead to a failure of the falsework itself.

34.9 Accidents

Clearly, the aim of all involved in the erection of steelwork is to complete the works without accident or injury. Following the guidance outlined in this chapter and in the cited documents will minimise the risk. However, the Steelwork Contractor should ensure that there are clear procedures for logging and reporting accidents and that First Aid, rescue and recovery plans are in place. Reference 9 gives detailed guidance.

There are many regulations that affect site erection of steelwork; the most important are:

Construction (Design & Management) Regulations (CDM)
Construction (Head Protection) Regulations
Construction (Health, Safety & Welfare) Regulations (CHSW)
Control of Substances Hazardous to Health Regulations (COSHH)
Control of Vibration at Work Regulations
Electricity at Work Regulations
Health and Safety (First Aid) Regulations
Lifting Operations and Lifting Equipment Regulations 1998
Highly Flammable Liquids and Liquefied Petroleum Gases Regulations
Lifting Operations and Lifting Equipment Regulations (LOLER)
Management of Health & Safety at Work Regulations (MHSW)
Manual Handling Operations Regulations
Noise at Work Regulations
Personal Protective Equipment at Work Regulations (PPE)
Provision and Use of Work Equipment Regulations (PUWER)
Reporting of Injuries, Diseases and Dangerous Occurrences Regulations (RIDDOR)

Workplace (Health, Safety & Welfare) Regulations
Work at Height Regulations
Most of the above can be accessed at http://www.hse.gov.uk/ or http://www.opsi.gov.uk/

References to Chapter 34

1. The Construction (Design & Management) Regulations 2007.
2. British Constructional Steelwork Association (2004) *BCSA Code of Practice for Erection of Low Rise Buildings*, BCSA Publication No. 36/04, London, BCSA.
3. British Constructional Steelwork Association (1999) *BCSA Guidance Notes on the Safer Erection of Steel-Framed Buildings*, Publication No. 11/99, London, BCSA.
4. British Standards Institution (2008) BS EN 1090 *Execution of steel structures and aluminium structures Part 2: Technical requirements for the execution of steel structures*. London, BSI.
5. British Constructional Steelwork Association (2010) *National Structural Steelwork Specification for Building Construction*, 5th edition (CE Marking Version) BCSA Publication No. 52/10. London, BCSA/SCI.
6. British Standards Institution (1990) BS 5964 *Building setting out and measurements. Part 1: Methods of measuring, planning and organisation and acceptance criteria. Part 2: Measuring stations and targets. Part 3: Check-lists for the procurement of surveys and measurement surveys*. London, BSI.
7. CIMSteel (1997) *Design for Construction*. Ascot, The Steel Construction Institute.
8. British Constructional Steelwork Association (2004) *BCSA Code of Practice for Metal Decking and Stud Welding*, Publication No. 37/04, London, BCSA.
9. British Constructional Steelwork Association (2006) *BCSA Code of Practice for Erection of Multi-Storey Buildings*, BCSA Publication No. 42/06, London, BCSA.
10. British Constructional Steelwork Association (2005) *BCSA Guide to Steel Erection in Windy Conditions*, BCSA Publication No. 39/05, London, BCSA.
11. British Standards Institution (2008) BS 5975 *Code of practice for temporary works procedures and the permissible stress design of falsework*. London, BSI.

Further reading for Chapter 34

British Standards Institution (2002) BS EN 1263 *Safety nets. Part 1: Safety requirements, test methods*. London, BSI.
British Standards Institution (2002) BS EN 1263 *Safety nets. Part 2: Safety requirements for the positioning limits*. London, BSI.
British Standards Institution (2003) BS EN 12811-1 *Temporary works equipment. Scaffolds. Performance requirements and general design*. London, BSI.
British Standards Institution (1990) BS 5974 *Code of practice for temporarily installed suspended scaffolds and access equipment*. London, BSI.
British Standards Institution (2006) BS 7121 *Code of practice for safe use of cranes*. Part 1: General. London, BSI.

Chapter 35
Fire protection and fire engineering

IAN SIMMS

35.1 Introduction

It is important that fire safety is considered in the early stages of the design process to avoid expensive problems later. There is no need for fire safety requirements to stifle aesthetic or functional aspects of building design, as a wide range of techniques are available to overcome most obstacles.

The fire resistance of the structural frame is only a small component of a fire safety strategy and other measures such as alarm systems, smoke control systems and sprinkler systems make a more effective contribution to the safety of building occupants and loss prevention.

The strength of all building materials will reduce as their temperature increases. Therefore structural members exposed to fire will weaken as the temperature rises until the member can no longer support the applied loading. The purpose of structural fire engineering is to determine when this will occur and make provision where necessary to elongate the life of a structural member in fire conditions, to ensure that the structure remains stable for a reasonable period.

In order to design structural members in fire conditions, the engineer must be able to predict the temperature of the member with time and the resistance of the member at a given temperature. Obviously the former also requires that the engineer can model the fire development in order to estimate the temperatures to which the structure will be exposed. These three components form the basis of structural fire engineering.

35.2 Building regulations

Buildings in the UK must be constructed in accordance with the requirements of Building Regulations. With respect to fire safety, the provisions of the UK Building Regulations are principally concerned with issues which affect life safety. The

Steel Designers' Manual, Seventh Edition. Edited by Buick Davison and Graham W. Owens.
© 2012 Steel Construction Institute. Published 2012 by Blackwell Publishing Ltd.

requirements as set out in the statutory instruments for England and Wales[1] are as follows.

- The building shall be designed and constructed so that there are appropriate provisions for the early warning of fire and appropriate means of escape, in case of fire, from the building to a place of safety outside the building, capable of being safely and effectively used at all material times.
- To inhibit the spread of fire within the building, the internal linings shall adequately resist the spread of flame over their surface and, if ignited, shall have a rate of heat release which is reasonable in the circumstances.
- The building shall be designed and constructed so that, in the event of fire, its stability will be maintained for a reasonable period.
- The external walls of the building shall adequately resist the spread of fire over the walls and from one building to another, having regard to the height, use and position of the building.[2]
- The building shall be designed and constructed so as to provide reasonable facilities to assist fire-fighters in the protection of life.

A second document, Approved Document B,[3] explains how the requirements of the Statutory Instrument (SI) can be fulfilled. For most buildings, following the recommendations of Approved Document B will provide a suitable and economic fire safety solution. However, for more unusual buildings with particular functional or aesthetic requirements, an advanced fire engineering solution may be required in order to achieve economic and practical solutions.

Similar requirements exist for Scotland[4] and Northern Ireland[5] with practical guidance being given in Technical Standards[6] and Technical Booklet E[7] respectively.

The fire resistance requirements for elements of structure are set out in Approved Document B.[3] The period of fire resistance required depends on building occupancy and building height. The fire resistance is expressed in units of time: either 30, 60, 90 or 120 minutes. These times are related to the performance of structural elements in standard fire resistance tests and are a convenient way of grading the performance of structural elements, but may not bear any relationship to the actual performance of the structural elements or to the time taken for occupants to escape from the building.

In accordance with Approved Document B, not all structural members are 'elements of structure', notable exceptions are members that only support a roof. Therefore single-storey buildings will not require fire resistance, except where space separation rules require fire resistance for the external walls of the building.

35.3 Fire engineering design codes

35.3.1 Introduction

The implementation of the Eurocodes is creating the opportunity for new approaches to fire engineering design.

BS EN 1990[8] states that fire design should be based on a consideration of fire development, heat transfer and mechanical behaviour. The required performance of the structure can be determined by global analysis, analysis of sub-assemblies or member analysis, as well as the use of tabular data and test results.

The behaviour of the structure exposed to fire should be assessed by considering an appropriate combination of actions and fire exposure. Appropriate combinations of actions are given in BS EN 1990,[8] and the fire exposure can be considered to be either nominal fire exposure, or modelled fire exposure as defined by BS EN 1991-1-2.[9]

Structural behaviour should be assessed using the thermal and structural models obtained from the relevant material code, BS EN 1992 to BS EN 1996. Where relevant to the specific material and method of analysis, thermal models may be based on the assumption of a uniform or non-uniform temperature within the cross-section and along the length of the member. Structural models may be confined to an analysis of individual members, or may account for interaction between the members. Models of mechanical behaviour should be non-linear.

The Eurocode for each structural material includes a part which relates to fire engineering design. Rules for calculating the resistance of steel and steel-concrete composite members are given in BS EN 1993-1-2[10] and BS EN 1994-1-2[11] respectively.

BS EN 1994 and 1993 allow the fire resistance of structural members to be assessed in terms of time, temperature or resistance. In most practical situations, it will be most convenient to use critical temperatures to describe the fire performance of structural members. Critical temperatures for protected and unprotected steel members are independent of time. Usually this eliminates the need to calculate the evolution of temperature in the member, which greatly simplifies the calculations. However, for structural members that are encased in concrete, the temperature distribution on the cross-section will need to be calculated using thermal analysis tools not provided in the code.

35.3.2 Thermal actions

The simplest form of thermal action is given by the nominal temperature time curves defined by BS EN 1991-1-2.[9] For building structures, the standard temperature-time curve is usually adopted. Nominal temperature-time curves do not necessarily reflect the real fire exposure conditions that a structure will experience. However they provide a convenient way in which to classify the behaviour of structural elements and to grade fire resistance requirements in Building Regulations. Further details of nominal temperature-time curves can be found on the Access Steel website.[12]

BS EN 1991-1-1 also provides simple fire models for localised and fully developed fires, which can be used to determine thermal actions by modelling the temperatures within the compartment, based on the: magnitude of the fire load, the thermal properties of the compartment boundaries, the compartment geometry, the rate of fire growth and the ventilation characteristics. Applying simple fire behaviour models is not straightforward, as at the design stage there may be uncertainty regarding some of the input data, and a range of likely fire scenarios will need to be considered before choosing the design fire. Examples of the use of the BS EN 1991-1-2[9] model for fully developed fires are available on Access Steel[13]. The UK National Annex to BS EN 1991-1-2[16] rejects the use of the localised fire model presented in BS EN 1991-1-2[9] Annex C of the code and recommends the use of an alternative plume model for localised fires.

35.3.3 Thermal analysis

BS EN 1993-1-2[10] and BS EN 1994-1-2[11] provide simplified methods for the thermal analysis of protected and unprotected steel members. However, the usefulness of the tools for analysing protected members is limited by the availability of thermal properties of fire protection products. As fire protection manufacturers in the UK have extensive data on their fire protection products, which allow the thickness of protection to be determined for a range of critical temperatures, the use of this method will generally be unnecessary.

The method for unprotected members is likely to be of more practical use, enabling the use of unprotected steelwork to be justified for a range of scenarios.

The heat transfer relationships given in the Eurocode are presented in a graphical format in two Access Steel resources for unprotected members[14] and protected members.[15]

Heat transfer analysis may also be carried out using the advanced models discussed in Section 35.6.

35.3.4 Load combinations

Fire is treated as an accidental design situation and as an ultimate limit state with members checked to the criterion of strength retention. BS EN 1990 gives the following equation to define the combination of actions in fire:

$$E_{d,fi} = G_{k,j} + \psi_{1,1}Q_{k,1} + \sum_{i>1} \psi_{2,1}Q_{k,i}$$

where
$\psi_{1,1}$ is the combination factor for the frequent value of the leading variable action
$\psi_{2,1}$ is the combination factor for the quasi-permanent value of the leading variable action.

The choice of $\psi_{1,1}$ or $\psi_{2,1}$ for use with the leading variable action in the above equation is a nationally determined parameter. The UK National Annex to BS EN 1991-1-2[16] states that the frequent value $\psi_{1,1}$ should be used in the UK.

As a simplification, BS EN 1991-1-2 4.3.2(2) permits the effect of actions in fire conditions to be derived from the design value for the effect of actions at room temperature by use of a reduction factor, as follows:

$$E_{\mathrm{d,fi,t}} = \eta_{\mathrm{fi}} E_{\mathrm{d}}$$

where:

η_{fi} is the reduction factor for the design load level for the fire situation

E_{d} is the design value for the effect of actions for room temperature design.

The calculation of the reduction factor will depend on the combination of actions used for room temperature design. Where the load combination given by BS EN 1990 Equation 6.10 is used for room temperature design, η_{fi} is given as follows:

$$\eta_{\mathrm{fi}} = \frac{G_{\mathrm{k}} + \psi_{1,1} Q_{\mathrm{k},1}}{\gamma_{\mathrm{G}} G_{\mathrm{k}} + \gamma_{\mathrm{Q},1} Q_{\mathrm{k},1}}$$

The value of η_{fi} will also be dependent on the relative magnitude of permanent and variable actions on the structural member being considered and the value of $\psi_{1,1}$ applicable to the building occupancy, as shown by Figure 35.1.

For use in tabular data BS EN 1991-1-2 4.3.3(1) also defines the load level of a structural member in fire conditions as:

$$E_{\mathrm{d,fi,t}} = \eta_{\mathrm{fi,t}} R_{\mathrm{d}}$$

where

R_{d} is the design value of resistance of the member at room temperature.

$\eta_{\mathrm{fi,t}}$ is the load level for fire design.

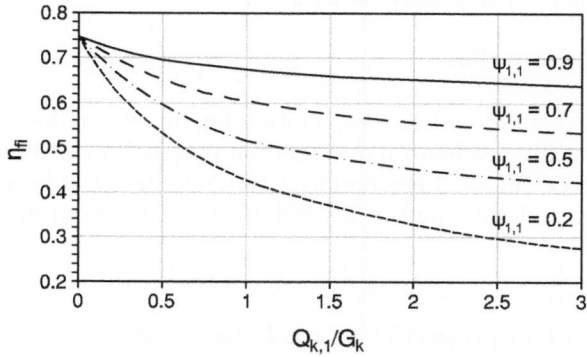

Figure 35.1   Variation of the reduction factor η_{fi} with proportion of permanent actions

35.3.5 Fire resistance testing

Fire resistance tests are conducted on single building elements with simple bound-ary conditions, in order to determine the fire rating of the element of construction. The general requirements for fire resistance tests are defined in BS EN 1363-1.[17]

The building elements are exposed to the standard temperature-time curve and their performance is assessed against the following three performance criteria.

- load bearing resistance, R
- integrity, E
- insulation, I.

For structural elements such as beams and columns, the fire performance is assessed against the load bearing criterion only. For separating elements such as composite floors, all three criteria will apply. Test procedures for specific elements of structure are given in the various parts of BS EN 1365.[18]

BS EN 1365 defines the scope of the application of fire test results for each type of structural element. This is referred to as 'Direct Application' and the deviations permitted between the test specimen and the structural element as constructed are very limited.

However, it is also possible to perform an 'Extended Application' which, with expert assessment, enables fire design data to be obtained for a wider range of applications. The process involves the use of heat transfer and structural models, whose performance is calibrated against the results of a fire resistance test. If ade-quate agreement between the models and the fire test can be achieved, the models may then be used to produce fire design data for a range of loadings and fire resist-ance periods which would be outside the scope of 'Direct Application'.

Extended application of fire resistance tests forms the basis of tabular design data for many products, such as composite slabs, 'Slimdek' and concrete filled hollow sections.

35.4 Structural performance in fire

35.4.1 Material properties

In order to use structural and thermal models for the evaluation of structural fire resistance, data on the mechanical and thermal properties of structural steel, con-crete and reinforcement steel are required. Suitable data can be found in BS EN 1993-1-2 for structural steel and in BS EN 1994-1-2 for concrete and reinforcing steel.

35.4.1.1 Mechanical properties of steel and concrete

BS EN 1993-1-2 provides mechanical and thermal property data for structural steel in the temperature range 20–1200°C that is appropriate for fire design purposes. The

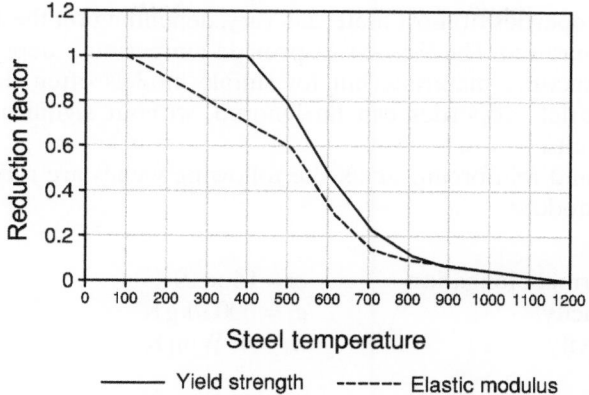

Figure 35.2   Variation of mechanical properties with temperature for structural steel (BS EN 1993-1-2)

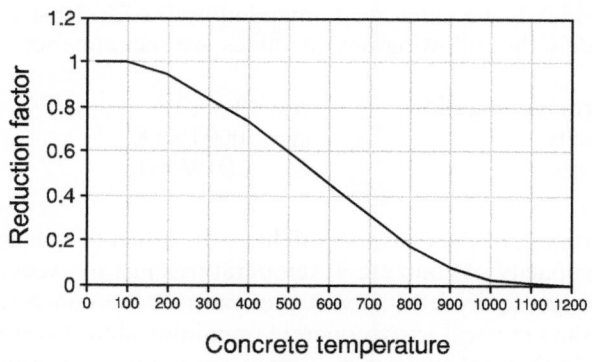

Figure 35.3   Strength reduction in NWC (BS EN 1994-1-2)

reduction in yield strength with temperature is given for a 2% proof strain. The variation of yield strength and the elastic modulus or slope of the linear elastic range with steel temperature is as shown in Figure 35.2. It should be noted that the mechanical properties of reinforcing steel differ from those of structural steel.

For concrete and reinforcement used in composite construction, mechanical and thermal material property data can be obtained from BS EN 1994-1-2 (Figure 35.3).[11]

35.4.1.2 Thermal properties of steel and concrete

For structural fire design, the main thermal properties of materials that need to be considered are thermal elongation, specific heat capacity and thermal conductivity.

The thermal properties of most materials vary, depending on the temperature at which they are measured. The Eurocodes provide temperature-dependent material properties for structural materials, but for simple models often the temperature variation in material properties can be ignored, without significant detrimental effects on the results.

For structural and reinforcing steels, the following values are recommended for use with simple models:

coefficient of thermal elongation	$\alpha = 14 \times 10^{-6}$
specific heat capacity	$c_a = 600 \, \text{J/kg K}$
thermal conductivity	$\lambda_a = 45 \, \text{W/m K}$

The thermal properties of concrete are more variable then those of steel and depend on the type of aggregates used. For normal weight concrete, Eurocodes give an upper and lower limit for thermal conductivity and permit national choice within these bounds. The UK National Annex to BS EN 1994-1-2[19] stipulates the use of the upper limit, which is the value recommended by BS EN 1994-1-2.

For simple models, the following design values are recommended:

coefficient of thermal elongation	$\alpha = 18 \times 10^{-6}$
specific heat capacity	$c_a = 1000 \, \text{J/kg K}$
thermal conductivity	$\lambda_a = 1.60 \, \text{W/m K}$

The moisture content of the concrete will have the effect of enhancing the apparent specific heat capacity of concrete at temperatures just in excess of 100°C. This is due to the energy required to convert the water in the concrete to steam and usually results in the temperature of concrete remaining almost constant for several minutes at 100°C. In thermal analysis, it is important not to overestimate the volume of water in the concrete or an unconservative result may be obtained from the analysis. BS EN 1994-1-2 recommends that the moisture content considered in analysis should not exceed 4% of the mass of the concrete.

35.4.2 Steel beams

The performance of steel beams in fire will depend on the magnitude of load that is applied to the steel section and how quickly the steel section heats up in the fire. Large sections, or lightly loaded sections, may be able to achieve 30 minutes fire resistance without applied fire protection but in most practical cases the rate of temperature rise in the section will have to be controlled, using fire protection material in order to provide adequate fire resistance. In order to determine the thickness of fire protection required, the critical temperature of the member must first be determined. The UK National Annex to BS EN 1993-1-2[20] provides critical temperature data for some common configurations of restrained steel sections in bending. These critical temperature data have been reproduced in Table 35.2.

Most rolled UKB sections subject to a load level of up to 0.6 will achieve 15 minutes fire resistance without the application of fire protection material. Therefore, one practical application of unprotected steelwork is in open-sided car parking structures. A detailed list of sections which can achieve 15 minutes fire resistance can be obtained from P186 *Design of Steel framed buildings without applied fire protection.*[21]

For Class 1, 2 and 3 steel sections restrained along their full length, the bending resistance can be calculated relatively simply using the following formula from BS EN 1993-1-2:

$$M_{\text{fi,t,Rd}} = k_{y,\theta} M_{\text{Rd}} \gamma_{\text{M,0}} / \gamma_{\text{M,fi}}$$

where:

M_{Rd} is the plastic moment resistance of the beam from room temperature design for a Class 1 or Class 2 cross-section or the elastic moment resistance from room temperature design for a Class 3 cross-section

$k_{y,\theta}$ is the reduction factor for the yield strength of steel at temperature, θ_{a}

$\gamma_{\text{M,fi}}$ is the partial factor for material strength in the fire situation.

The variation of this formula with steel temperature is shown in Figure 35.4.

The classification of the cross-section in fire conditions is conducted using the limits for Class 1 to 4 given in BS EN 1993-1-1[22] but with a reduced value of ε to account for the variation in the relationship between yield strength and elastic modulus at elevated temperature. The exact value of ε varies with steel temperature but BS EN 1993-1-2 permits the use of a single reduced value as follows:

$$\varepsilon = 0.85 \left[235/f_{\text{y}} \right]^{0.5}$$

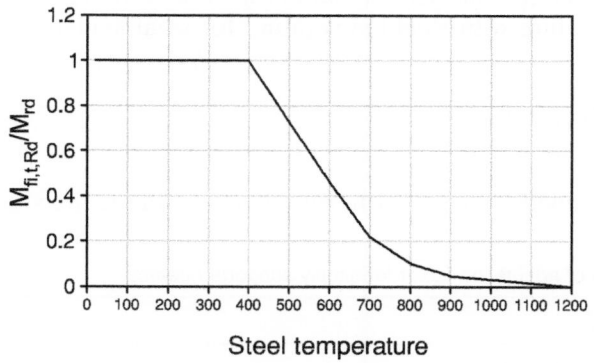

Figure 35.4 Bending resistance of a uniformly heated steel beam restrained along its full length

where
  f_y is the yield strength at room temperature.

Steel beams which support a concrete floor slab gain an advantage from the slab providing a heat sink and partially shielding the section from the fire. This results in a non-uniform temperature profile, which means that the top flange is cooler than other parts of the section. BS EN 1993-1-2 provides a simple calculation method which allows designers to take account of this non-uniform temperature distribution, without calculating the temperature distribution in the section.

$$M_{\text{fi,t,Rd}} = M_{\text{fi,}\theta\text{,Rd}}/\kappa_1\kappa_2$$

where
  $M_{\text{fi,}\theta\text{,Rd}}$ is the bending resistance of the beam for a uniform steel temperature, θ
  κ_1 is the adaptation factor for non-uniform temperature across the cross-section
  κ_2 is the adaptation factor for non-uniform temperature along the beam.

Values of κ_1 and κ_2 for simply supported beams are given in Table 35.1.

BS EN 1993-1-2 also provides a simple model to allow the calculation of the design lateral torsional buckling resistance for laterally unrestrained sections.

35.4.3 Columns

When calculating the fire performance of columns, buckling effects must be considered. As the reduction of the elastic modulus of steel occurs at a different rate from the reduction of strength, the resistance of the column at elevated temperature cannot be determined from the room temperature resistance by the application of a single factor, as is the case for restrained beams.

BS EN 1993-1-2 does, however, provide a method for calculating the resistance of columns in fire conditions. This equation can be used to determine the variation of critical temperature with steel temperature for axial loaded columns in braced frames.

$$N_{\text{b,fi,t,Rd}} = \chi_{\text{fi}} A k_{\text{y,}\theta} f_y / \gamma_{\text{M,fi}}$$

where
  χ_{fi} is the reduction factor for flexural buckling in the fire situation.

Table 35.1 Values of adaptation factor for simply supported beams

Exposure conditions	κ_1	κ_2
Beams exposed on four sides	1.0	1.0
Unprotected beam exposed on 3 sides supporting a concrete slab	0.7	1.0
Protected beam exposed on three sides supporting a concrete slab	0.85	1.0

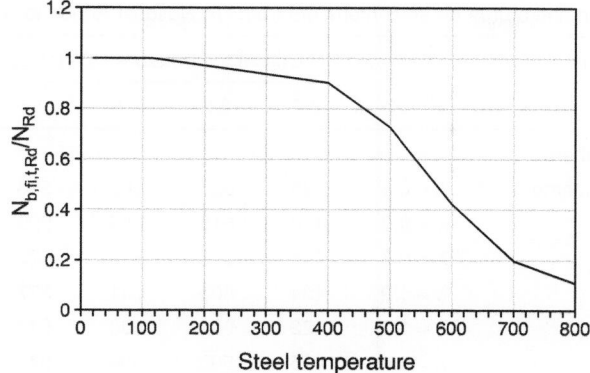

Figure 35.5   Variation of the design buckling resistance of a steel compression member with steel temperature

The reduction factor for flexural buckling in fire conditions is given as a function of the non-dimensional slenderness ratio, modified to account for the variation of yield strength and elastic modulus in fire conditions, as follows:

$$\lambda_\theta = \bar{\lambda}\left[k_{y,\theta}/k_{E,\theta}\right]^{0.5}$$

In general, the effective length of the column in fire conditions should be calculated as for normal design. However, for continuous columns in braced frames, BS EN 1993-1-2 permits the effective length of intermediate storeys to be taken as 0.5 times the system length, and for the top storey 0.7 times the system length.

Figure 35.5 shows the variation of resistance with temperature for an axially loaded column in a braced frame. The variation of critical temperature with non-dimensional slenderness and load level is shown in Table 35.2.

35.4.4   Composite beams

The majority of composite beams in the UK are downstand steel sections connected to composite concrete slabs via shear connectors, as shown by Figure 35.6. Unprotected, these beams achieve similar periods of fire resistance to steel sections. Most UKB sections will achieve 15 minutes fire resistance in this configuration, and heavy sections or lightly loaded sections will be able to achieve 30 minutes fire resistance.

BS EN 1994-1-2 covers the design of steel-concrete composite sections. For downstand composite steel beams, a simple critical temperature calculation is provided for steel sections up to 500 mm deep supporting concrete slabs not less than 120 mm deep. Using the load level for the composite beam in fire conditions, the critical

Table 35.2 Critical temperatures for steel members from UK National Annex to BS EN 1993-1-2[20]

Description of member		Critical temperature (°C) for utilisation factor μ_0					
		0.7	0.6	0.5	0.4	0.3	0.2
a) Compression members:							
Non-dimensional slenderness	$\lambda = 0.4$	485	526	562	598	646	694
	$\lambda = 0.6$	470	518	554	590	637	686
	$\lambda = 0.8$	451	510	546	583	627	678
	$\lambda = 1.0$	434	505	541	577	619	672
	$\lambda = 1.2$	422	502	538	573	614	668
	$\lambda = 1.4$	415	500	536	572	611	666
	$\lambda = 1.6$	411	500	535	571	610	665
b) Protected beam supporting concrete slabs or composite slabs		558	587	619	654	690	750
c) Unprotected beam supporting concrete slabs or composite slabs		594	621	650	670	717	775
d) Beam not supporting concrete slabs and members in tension		526	558	590	629	671	725

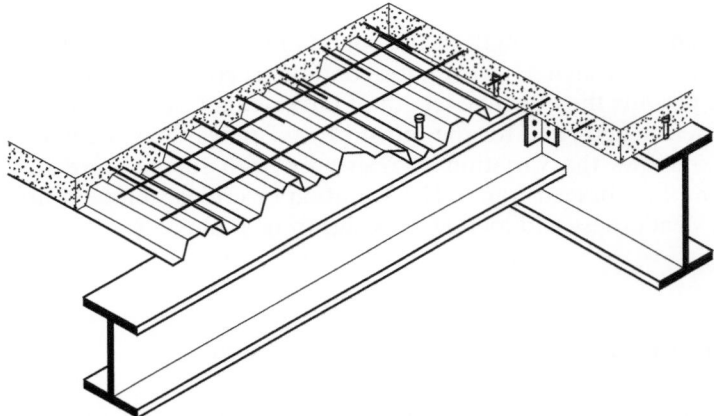

Figure 35.6 Downstand composite beam and composite slab

temperature is calculated from the reduction in steel strength by the following relationships:

R30 $0.9\eta_{fi,t} = f_{ay,\theta_{cr}} / f_{ay}$

>R30 $1.0\eta_{fi,t} = f_{ay,\theta_{cr}} / f_{ay}$

BS EN 1994-1-2 also provides a method of calculating the bending resistance of composite steel beams in fire conditions. The method is based on calculating the plastic moment resistance of the cross-section in a similar way to that used for room temperature design, taking account of the reduction in material properties with temperature. The shear connection between the steel section and the concrete slab is also calculated using the resistance equations given in BS EN 1994-1-1 taking account of the reduced strength of the shear studs and concrete. To simplify the analysis, BS EN 1994-1-2 permits the designer to assume that the design temperature of the shear studs is 80% of the top flange temperature and that the design temperature of the concrete is 40% of the top flange temperature of the steel section.

This model has been used to produce a table[23] of critical temperatures for composite beams using the calculation model from BS EN 1994-1-2. These critical temperatures are reproduced in Table 35.3 and are presented in terms of load level and degree of shear connection for room temperature design (see also Worked Example 1).

BS EN 1994-1-2 covers a number of other composite beam constructions with full or partial concrete encasement. Figure 35.7 shows a partially encased composite

Table 35.3   Critical temperatures for composite beams

Description of member		Critical temperatures (°C) for a load level of					
Construction	Degree of shear connection	0.7	0.6	0.5	0.4	0.3	0.2
Composite steel beams in bending	40%	558	588	628	666	698	763
	60%	556	586	619	655	691	752
	80%	545	577	608	647	684	744
	100%	536	567	598	639	679	738

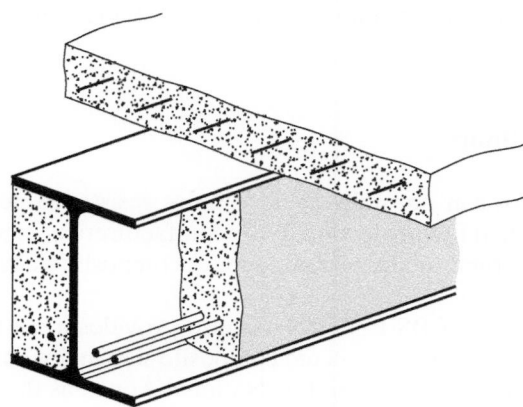

Figure 35.7   Partially encased composite beam

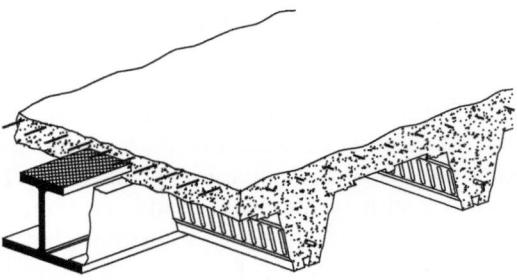

Figure 35.8   Slimdek system

beam, one of the more common systems. While their use is not widespread in the UK, they are used in other European countries and can usually achieve 60 to 120 minutes fire resistance without fire protection. The disadvantage of using these systems is that they tend to slow the construction process, with greater lead times being required. Encasing the sections in concrete also increases the self-weight of the member making them more cumbersome to lift and handle on site.[24]

35.4.5 Integrated beams

Integrating the steel beam within a floor slab can be an effective solution for periods of fire resistance up to 60 minutes. The Tata 'Slimdek' system, shown in Figure 35.8, is an example of this type of construction which can achieve 60 minutes fire resistance without the application of fire protection for load levels up to 0.6. For higher load levels, or for longer periods of fire resistance, fire protection is required. Shelf angle floors are another example of this type of construction. Further details for both construction types can be found in P186[21] and the Appendix (pp. 1324–7).

35.4.6 Composite floors

Composite floor slabs can normally achieve a fire resistance period of up to 120 minutes without applied fire protection. UK manufacturers of steel decking profiles normally provide information on the resistance of composite slabs constructed using their products.

Informative Annex D of BS EN 1994-1-2 also provides a method of calculating the moment resistance of an unprotected composite slab in fire conditions. However this method has been rejected by the UK National Annex as the geometry of most UK decking profiles is outside the field of application covered by the standard. For design temperature information reference may be made to BS5950-8[25] as an alternative to Annex D. The Eurocode also assumes that composite slabs will have reinforcement in the ribs of the profile which is not the case in most UK construction. Fire testing evidence in the UK supports the omission of bottom reinforcement for fire resistance periods of up to 120 minutes.

A simple design table is also provided in the fire datasheets in the Appendix (page 1323). This is likely to provide conservative results compared with manufacturers' design but is useful for initial design.

35.4.7 Developments in fire design

In September 1996, a programme of fire tests was completed in the UK at the Building Research Establishment's Cardington Laboratory. The tests were carried out on an eight-storey composite steel-framed building that had been designed and constructed as a typical multi-storey office building, see Figure 35.9. The purpose of the tests was to investigate the behaviour of a real structure under real fire

Figure 35.9   Cardington test frame

conditions and to collect data that would allow computer programs for the analysis of structures in fire to be verified. Details of the fire tests carried out on the frame and the data obtained from these tests are available in a British Steel publication.[26]

The experimental work at Cardington, and evidence from other real fires[27] in building structures, had served to illustrate that there are significant reserves of strength in composite steel concrete buildings, beyond those predicted by traditional design methods. In summary, performance of the assembled structure in fire exceeds the expectations created by standard fire tests on individual structural elements. Cardington demonstrated that it was possible to leave the composite steel beams that supported the concrete floor slab unprotected; work commenced to investigate suitable design models to allow structural engineers to justify the fire design of a floor slab supported by unprotected steel beams.

Researchers at the Building Research Establishment (BRE), with funding from the Steel Construction Institute, developed a simple design method for composite steel concrete floor slabs, following the experimental work at Cardington.[28,29] This BRE model has been validated against the Cardington large scale fire test results and previous experimental work conducted at room temperature.

Tata Steel distribute a design tool based on the design method developed by Bailey and Moore. The design tool is called TSLAB and is available free of charge from the Tata Steel website.

A SCI design guide P288[30] is also available which provides further guidance on use of the design tool and the application of the method to building structures. Using this design method it is possible to remove fire protection from 30 to 40% of the steel beams in a typical composite floor plate.

35.5 Fire protection materials

Fire protection materials are tested and assessed in accordance with BS EN 13381. For sprays and boards, referred to as non-reactive fire protection, the assessment is undertaken in accordance with the recommendations of BS EN 13381-4[31] and, for intumescent coatings, reactive fire protection, assessment is in accordance with BS EN 13381-8.[32]

The assessment provides a table of thickness versus section factor for each period of fire resistance and a particular value of steel temperature, known as the assessment temperature.[33] Multi-temperature assessments are usually provided, meaning that protection thicknesses are derived for a range of steel temperatures rather than a single value.

When specifying fire protection, the following information should be included in the specification:

- section factor of the structural member
- period of fire resistance specified by Approved Document B
- critical temperature of the structural member.

Further information on fire protection materials can be found in P197[34] and in the fire data sheets in the Appendix (pages 1310–1330).

35.5.1 Off-site application

The application of intumescent coating in workshop conditions prior to delivery to site is now a mature technology, with many coatings being developed specifically for this application. Further information on off-site application of intumescents can be found in ASFP TGD16.[35]

35.6 Advanced fire engineering

Advanced fire engineering[36] usually deviates from the prescriptive requirements of Approved Document B and seeks to demonstrate compliance with the performance-based requirements of the Statutory Instruments using advanced models. BS 7974[37] provides guidance on how the principles of fire safety engineering should be applied to the design of buildings. (Statutory Instruments are the legislative documents which form the legal basis of the Building Regulations.) A brief description of the design process is provided below.

35.6.1 Design process

A performance-based fire design procedure should be clearly documented so that the philosophy and assumptions can be clearly understood by a third party. The procedure may include the following main steps:

- review the architectural design of the building
- establish fire safety objectives
- identify fire hazards and possible consequences
- establish trial fire safety designs
- identify acceptance criteria and methods of analysis
- establish fire scenarios for analysis.

35.6.1.1 Review of architectural design

This review should aim to identify any architectural or client requirements that may be significant in the development of a fire safety solution. The review should consider the following aspects:

- the future use of the building
- the anticipated building contents
- anticipated permanent, variable and thermal actions
- the type of structure
- the building layout
- the presence of smoke ventilation systems or sprinkler systems
- the characteristics of the building's occupants
- the number of people likely to be in the building and their distribution
- the type of fire detection and alarm system
- the degree of building management throughout the life of the building.

35.6.1.2 Fire safety objectives

At an early stage of the design process, the fire safety objectives should be clearly identified. This process will ideally be undertaken in consultation with the client, regulatory authority and other stakeholders.

The main fire safety objectives that may be addressed are life safety, control of financial loss and environmental protection.

Life safety objectives are already set out in prescriptive regulations, but should include provisions to ensure that the building occupants can evacuate the building in reasonable safety, that fire-fighters can operate in reasonable safety and that collapse does not endanger people who are likely to be near the building.

The effects of fire on the continuing viability of a business can be substantial, and consideration should be given to minimising damage to the structure and fabric of the building, the building contents, the ongoing business viability and the corporate image. The level of precautions that are deemed necessary in a particular building will depend on the size and nature of the business undertaken there. In some cases it may be easy to relocate the business to alternative premises without serious disruption; in other cases, business may stop until the building is reinstated. Many businesses that experience fires in their premises go bankrupt before resuming business.

A large conflagration which releases hazardous materials into the environment may have a significant impact on the environment. The pollution may be airborne or waterborne as a result of the large volumes of water required for fire fighting operations.

35.6.1.3 Identification of fire hazards and possible consequences

A review of potential fire hazards may include consideration of ignition sources, the volume and distribution of combustible materials, activities undertaken in the building and any unusual factors. When evaluating the significance of these hazards, consideration of the likely consequences and their impact on achieving the fire safety objectives need to be considered.

35.6.1.4 Establish trial fire safety designs

In order to quantify the level of fire safety achieved, one or more trial safety strategies should be established for the building. These will normally be the most cost-effective strategies that satisfy the fire safety objectives.

The fire safety strategy is an integrated package of measures in the design of multi-storey buildings. The following should be considered when developing a fire safety strategy:

- automatic suppression measures (e.g. sprinkler systems) to limit the likelihood of the spread of fire and smoke
- automatic detection systems which provide an early warning of fire
- compartmentation of the building with fire-resisting construction and the provision of fire-resisting structural elements to ensure structural stability
- means of escape: provide adequate numbers of escape routes to ensure reasonable travel distances, and widths of escape routes for the number of people likely to occupy the building at any one time
- automatic systems such as self-closing fire doors or shutters to control the spread of smoke and flames
- automatic smoke control systems to ensure smoke-free escape routes
- alarm and warning systems to alert the building occupants
- evacuation strategy
- first aid and fire fighting equipment
- fire service facilities
- management of fire safety.

35.6.1.5 Establishing acceptance criteria and methods of analysis

Performance-based designs are based on the global analysis of a given fire strategy. Acceptance criteria must be established so that the performance of the building can be evaluated. The acceptance criteria should be agreed between the designers, regulators and clients. The evaluation of performance may be undertaken using comparative, deterministic or probabilistic approaches.

A comparative approach evaluates the level of fire safety obtained from performance-based design with that from a prescriptive approach, to ensure an equivalent level of fire safety is achieved. A deterministic approach aims to quantify the effects of a worst case fire scenario and to demonstrate that the effects will not exceed the acceptance criteria defined. A probabilistic approach aims to show that the fire safety strategy makes the likelihood of large losses occurring tolerably small.

35.6.1.6 Establishing fire scenarios

The number of possible fire scenarios in any building can become large, and resources to analyse all of them are usually not available. Therefore detailed analysis must be confined to the most significant fire scenarios, or the worst creditable cases as they are sometimes referred to. The failure of protection systems should also be included in the fire scenarios that are considered. In most buildings, more than one fire scenario will require detailed analysis.

35.6.2 Validation/verification of advanced models

Since a number of advanced calculation models are used to model fire action as well as the thermal and structural response of buildings, both validation and verification must be performed to ensure that a sensible solution is finally obtained.

Validation is the demonstration of the suitability of the design model or approach for the intended purpose, which includes those for predictions of fire severity, heat transfer and structural response.

Verification is the assessment of whether or not the design model has produced correct results. It includes a careful check on the input data, the consistency between the results obtained from the model and the results anticipated by qualitative analysis, and on the degrees of risk associated with possible errors. Advanced models should be verified against relevant test results and other calculation methods. They should be checked for compliance with normal engineering principles by the use of sensitivity studies.

In the context of validation and verification of models and results, ISO 16730:2008[38] provides a framework for assessment, verification and validation of all types of calculation methods used as tools for fire safety engineering. This international standard does not address specific fire models, but is intended to be applicable to both simple methods and advanced methods.

35.6.3 Regulatory approval

The complexity of obtaining regulatory approval will vary from country to country. However, regulatory bodies may require the designer to present a fire design in a form that may be easily checked by a third party with each design step clearly documented, including any assumptions and approximations that have been made. A checklist should be produced, which covers the overall design approach, the fire model, the heat transfer and the structural response.

35.6.4 The use of advanced models in structural fire engineering

The fire engineering calculation process involves three main steps as follows:

- fire behaviour modelling
- heat transfer modelling
- structural modelling.

Modelled fire behaviour needs to consider the worst possible fire scenarios. The fire behaviour model produces a set of thermal actions which are then used in the heat transfer analysis to establish the temperature-time history for the structural member. Using the characteristic permanent and variable actions from room temperature design, along with the combination of actions for accidental situations, a design load for the structure can be established. The structural behaviour can then be modelled using these thermal and mechanical actions. Usually, structural modelling for fire design is undertaken using finite element analysis. The scope and the complexity of the model will depend on the nature of the problem considered and will usually involve one storey and extend over one quarter or one half of the floor area, depending on the size of the building and the availability of lines of symmetry.

35.6.4.1 Fire behaviour models

Fire behaviour models can consider a fully developed fire or a localised fire. A localised fire model, or pre-flashover model, is useful in cases were the compartment geometry and/or the distribution of the combustible materials make it unlikely that flashover will occur. A fully developed fire, or post-flashover fire, describes the point at which all the available combustible material in the compartment is burning. The intensity of the fire will be controlled either by the quantity of the fuel or by the supply of oxygen to the fire.

Simple models for localised and fully developed fires are available. More advanced models such as zone models and computational fluid dynamics (CFD) models may also be used. Zone models are computer-based models that divide the fire compartment into separate zones within which conditions are assumed to be uniform. By considering the conservation of mass and energy within the compartment, the models calculate the evolution of temperature with time. The compartment can be one single zone or two zones, depending on whether a fully developed or localised fire is being modelled.

CFD modelling has been successfully used to model smoke movement and is now being applied to fire modelling. CFD models are based on the fundamental equations of fluid flow. They are demanding in terms of the input data required and the expertise required to assess the feasibility of the results.

35.6.4.2 Heat transfer models

Advanced heat transfer models are usually based on finite difference or finite element analysis techniques. Heat transfer to structural elements is dominated by radiation but the heat flux used for calculation normally includes both radiation and convection.

35.6.4.3 Structural models

Advanced structural models are used for either sub assemblies or whole frame behaviour. The advantage of using advanced models is that a more realistic analysis of building behaviour is obtained and redundancy and alternative load paths in the structural frame can be utilised in the fire condition to provide more economic and sustainable fire solutions.

35.7 Selection of an appropriate approach to fire protection and fire engineering for specific buildings

The approach to fire safety chosen for any project will depend mainly on economic considerations. For buildings that are commonplace and can be dealt with easily within the scope of building regulations, it is likely that the most economic approach in terms of design effort is to use pre-engineered data to obtain a solution. This will normally include the application of fire protection to steel members in accordance with fire protection manufacturers' recommendations. To justify using more advanced design methods and investing additional design effort, there will usually need to be a corresponding saving in the cost of construction.

For large or complex buildings the guidance and limitations imposed by building regulations may have an impact on the function of the building or may lead to excessive construction costs. In such circumstances in order to achieve the functionality required of the building, additional design effort may have to be investigated to demonstrate the fire safety performance of the completed structure and justify a departure from the requirements of legislation.

Further guidance on the selection of appropriate fire engineering strategies and their development is available from Access Steel documents.[39,40]

References to Chapter 35

1. The Building Regulations 2000 (SI 2000/2531). London, The Stationery Office.
2. The Steel Construction Institute (2002) *Single storey steel framed buildings in fire boundary conditions*, P313. Ascot, SCI.
3. Department of the Environment and The Welsh Office (2006) *The Building Regulations 2000, Approved Document B Fire Safety, Volume 2 Buildings other than dwelling houses*. 2006 edition, London, The Stationery Office.
4. Building Standards (Scotland) Regulations 1990, (Including amendments up to 2001). London, The Stationery Office.
5. The Building Regulations (Northern Ireland) 2000 (SR 2000/389). London, The Stationery Office.
6. Scottish Executive (2001) *Technical Standards: For compliance with the Building Standards (Scotland) Regulations 2001*. London, The Stationery Office.

7. Department of the Environment for Northern Ireland. *The Building Regulations (Northern Ireland) 2000, Technical Booklet E Fire Safety* (as amended 2000). London, The Stationery Office.

8. British Standards Institution (1990) BS EN 1990:2002, *Eurocode 0: Basis of structural design*. London, BSI.

9. British Standards Institution (2002) BS EN 1991-1-2:2002, *Eurocode 1: Actions on structures. Part 1-2: General actions – Actions on structures exposed to fire.* London, BSI.

10. British Standards Institution (2005) BS EN 1993-1-2:2005, *Eurocode 3: Design of steel structures. Part 1-2: General rules – structural fire design.* London, BSI.

11. British Standards Institution (2006) BS EN 1994-1-2:2006, *Eurocode 4: Design of composite steel and concrete structures. Part 1.2: General rules. Structural fire design.* London, BSI.

12. Access Steel (2008) *Nominal temperature-time curves*, SD007a-EN-EU, www.access-steel.com

13. Access Steel (2006) *Parametric fire curve for a fire compartment*, SX042a-EN-EU. www.access-steel.com

14. Access Steel (2008) *Nomogram for unprotected members*, SD004a-EN-EU,. www.access-steel.com

15. Access Steel (2008) *Nomogram for protected members*, SD005a-EN-EU,. www.access-steel.com

16. British Standards Institution (2002) NA to BS EN 1991-1-2, *UK National Annex to Eurocode 1: Actions on structures, Part 1.2 General actions – Actions on structures exposed to fire*. London, BSI.

17. British Standards Institution (1999) BS EN 1363-1 – *Fire resistance tests – Part 1: General requirements*. London, BSI.

18. British Standards Institution (2000) BS EN 1365, *Fire resistance tests for load-bearing elements*. London, BSI.

19. British Standards Institution (2005) NA to BS EN 1994-1-2, *UK National Annex to Eurocode 4: Design of composite steel and concrete structures, Part 1.2 General rules – Structural fire design*. London, BSI.

20. British Standards Institution (1993) NA to BS EN 1993-1-2, *UK National Annex to Eurocode 3: Design of steel structures, Part 1.2 General rules – Structural fire design*. London, BSI.

21. The Steel Construction Institute (1999) *Design of Steel framed buildings without applied fire protection*, P186. Ascot, SCI.

22. British Standards Institution (2005) BS EN 1993-1-1, *Eurocode 3: Design of steel structures, Part 1.1 General rules and rules for buildings*. London, BSI.

23. The Steel Construction Institute (2011) *Steel building design: Composite Design*, P359. Ascot, SCI.

24. Wang Y.C. (2002) *Steel and composite structures: Behaviour and design for fire safety*. Abingdon, Spon Press.

25. British Standards Institution (2003) BS5950-8:2003, *Structural use of steelwork in building: Code of practice for fire resistance design*. London, BSI.

26. British Steel (1999) *The behaviour of multi-storey steel framed buildings in fire*, A European joint research programme, Swinden Technology Centre.
27. The Steel Construction Institute (1991) *Investigation of Broadgate Phase 8 Fire*. Ascot, SCI.
28. Bailey C.G. and Moore D.B. (2000) The structural behaviour of steel frames with composite floor slabs subjected to fire: Part 1: Theory, *The Structural Engineer*, June 2000.
29. Bailey C.G. and Moore D.B. (2000) The structural behaviour of steel frames with composite floor slabs subjected to fire: Part 2: Design, *The Structural Engineer*, June 2000.
30. The Steel Construction Institute (2006) *Fire Safe Design: A new approach to multi-storey steel framed buildings* P288. Ascot, SCI.
31. British Standards Institution (2010) BS EN 13381-4, *Fire tests on elements of building construction. Method for determining the contribution to the fire resistance of structural members by applied fire protection to steel structural elements*. London, BSI.
32. British Standards Institution (2010) BS EN 13381-8 *Fire tests on elements of building construction. Method for determining the contribution to the fire resistance of structural members by applied fire protection to steel structural elements*. London, BSI.
33. Association for Specialist Fire Protection (2009) *Fire protection for structural steel in buildings*, 4th edn. Bordon, Hampshire, ASFP.
34. The Steel Construction Institute (1999) *Structural fire safety: A handbook for engineers and architects*, P197. Ascot, SCI.
35. Association of Specialist Fire Protection (2010) *Code of Practice for off-site applied thin film intumescent coatings, ASFP Technical Guidance Document 16*. Bordon, Hampshire, ASFP.
36. Institution of Structural Engineers (2007) *Guide to the advanced fire safety engineering of structures*. London, IStructE.
37. British Standards Institution (2001) BS7974:2001 *Application of fire safety engineering principles to the design of buildings: Code of practice*. London, BSI.
38. International Organization for Standardization (2008) *ISO 16730, Fire safety engineering – Assessment, verification and validation of calculation methods*. Geneva, ISO.
39. Access Steel (2008) *Selection of appropriate fire engineering strategy for multi-storey commercial and apartment buildings*, SS040a-EN-EU. www.access-steel.com
40. Access Steel (2008) *Fire safety strategy for multi-storey buildings for commercial and residential use*, SS008a-EN-EU. www.access-steel.com

 The Steel Construction Institute	Job No.			Sheet 1 of 7	Rev
Silwood Park, Ascot, Berks SL5 7QN Telephone: (01344) 623345 Fax: (01344) 622944	Job Title	Steel Designers Manual			
	Subject	Worked Example Fire Resistance – Composite Beams			
CALCULATION SHEET	Client		Made by	WIS	Date	March 2010
			Checked by		Date	

This worked example considers the fire resistance of a downstand steel beam designed to act compositely with a composite floor slab. The beam and floor slab have been designed in accordance with EN1994-1-1 for room temperature conditions and some of the design values used in the example come from the room temperature design. In the first instance the critical temperature for the beam is determined using Table 35.3. Using this critical temperature the resistance of the beam is then calculated using the design methods given in EN1994-1-2 Section 4.3.1.

Consider a simply supported composite beam shown in Figure 1. The beam supports a composite floor slab and is subject to a uniform load.

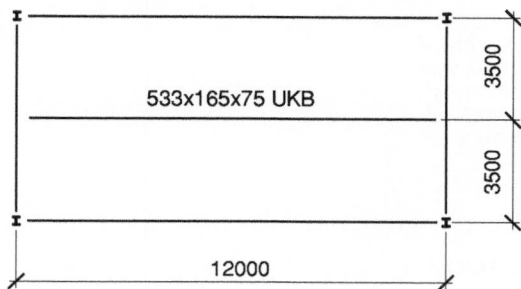

Figure 1   Floor plan

Details

Beam span	$L = 12\,m$
Beam spacing	$b = 2.5\,m$
Slab depth	$h_s = 130\,mm$
Steel decking	$0.9\,mm$ Tata CF60
Depth of concrete above profile	$h_c = 60\,mm$
Depth of decking profile	$h_p = 60\,mm$

Section properties

$533 \times 165 \times 75$ UKB Grade S275 steel

Section depth	$h_a = 829.1\,mm$
Flange width	$b = 165.9\,mm$
Web thickness	$w = 9.7\,mm$
Flange thickness	$t_f = 13.6\,mm$

Worked Example Fire Resistance – Composite Beams	Sheet 2 of 7	Rev

Shear connectors

Connector diameter	d	$= 19\,mm$
Overall height	h_{sc}	$= 100\,mm$
Ultimate tensile strength	f_u	$= 450\,N/mm^2$

Concrete

Normal Weight grade C25/30

Characteristic cylinder strength	f_{ck}	$= 25\,N/mm^2$
Characteristic cube strength	f_{cu}	$= 30\,N/mm^2$
Secant modulus of elasticity	E_{cm}	$= 31\,kNmm^2$

Reinforcement
A252 mesh

Bar diameter / spacing	$8\,mm$ @ $200\,mm$
Yield strength	f_{sj} $= 500\,N/mm^2$

Actions

Permanent actions

Slab self wt		$2.43\,kN/m^2$
Decking		$0.10\,kN/m^2$
Total	$g_{k,1}$	$= 2.53\,kN/m^2$
Beam self wt (allowance)	$g_{k,2}$	$= 0.80\,kN/m^2$
Ceiling and services	$g_{k,3}$	$= 0.70\,kN/m^2$

Variable actions

General use office floor area
BS EN 1991-1-1 (Category B1)

Imposed floor load (B1)	$q_{k,1}$	$= 2.5\,kN/m^2$

Moveable partitions BS EN 1991-1-1 6.3.1.2(8)

Allowance for moveable partitions	$q_{k,2}$	$= 0.9\,kN/m^2$

Room temperature design at the composite stage uses BS EN 1990 Eqn 6.10b

Load combinations

$$\sum_{j\geq1} G_{k,j} + \psi_{1,1}Q_{k,1}$$

Eqn 6.11(b)

$\psi_{1,1}$ 0.5 General office use

$G_{k,j}$ $= 2.53 + 0.8 + 0.7 = 4.03\,kN/m^2$

$Q_{k,1}$ $= 3.4\,kN/m^2$

F_d $= [4.03 + 0.5 \times 3.4] \times 3.5 = 20.06\,kN/m$

Applied bending moment for fire design is given by:

$$E_{fi,d,t} = \frac{20.06 \times 12^2}{8} = 361\,kNm$$

Worked Example Fire Resistance – Composite Beams	Sheet 3 of 7	Rev

Load level for fire design

$$\eta_{fi,t} = \frac{E_{fi,d,t}}{R_d} = \frac{361}{715.7}$$

EN 1994-1-2 4.1(7)

$\eta_{fi,t}$ = 0.50

R_d is the design resistance for room temperature design (M_{Rd} = 715.7 kNm)

$E_{fi,d,t}$ is the design effect of actions in the fire situation, at time t.

Degree of shear connection

Consider one shear connector per rib ($n_r = 1$). The degree of shear connection at room temperature can be determined from:

EN 1994-1-1 6.6.1.2(1)

$$\eta = \frac{N_c}{N_{c,f}}$$

N_c is the reduced force in the concrete flange

$N_{c,f}$ is the compressive force in the concrete flange at full shear connection

From room temperature design

N_c = 1242 kN

$N_{c,f}$ = 2618 kN

Effective width of concrete flange, b_{eff}

b_{eff} = 3.0 m

$$\eta = \frac{N_c}{N_{c,f}} = \frac{1242}{2618} = 0.47$$

Critical temperature

From Table 35.3 the critical temperature of a composite beam can be determined using the load level and the degree of shear connection.

Degree of shear connection $\eta = 0.47$

Load level $\eta_{fi,t} = 0.50$

Using linear interpolation the critical temperature can be determined.

Critical temperature $\theta_{cr} = 624°C$

Fire protection

The following data for needs to be supplied to a fire protection manufacturer to ensure that the correct thickness of fire protection is supplied.

$t_{ti,reqd}$ = 60 minutes

θ_{cr} = 624°C

Worked Example Fire Resistance – Composite Beams	Sheet 4 of 7	Rev

From Fire protection for structural steel in buildings[33]

Section Factor $A_m.V$ $= 160\,m^{-1}$

This information is sufficient for design purposes. However, in order to illustrate the calculation method given in EN1994-1-2 Section 4.3.1 and to demonstrate that the values given in Table 35.3 are conservative for design the bending resistance of the beam at this critical temperature has been calculated below.

Check bending resistance

Verify the bending resistance of a composite beam with a steel temperature, θ_a of 624°C

Check using the general rules given by EN 1994-1-2 4.3.1

<u>Effective width of concrete flange</u>

b_{eff} $= b_0 + \Sigma b_{e,i}$

For one stud $b_0 = 0$

$$b_{e,i} = \frac{L}{8} = \frac{12}{8} = 1.5\ m$$

b_{eff} $= 3.0\,m$

<u>Compressive resistance of concrete flange</u>

Design compressive strength of concrete

$$f_{cd} = \frac{f_{ck}}{\gamma_{fi,c}}$$

$\gamma_{fi,c}$ $= 1.0$ *EN 1992-1-2*

$$f_{cd} = \frac{25}{1.0} = 25.0\ N/mm^2$$

Temperature of concrete flange

θ_c $= 0.4\ \theta_a$ *EN 1994-1-2 4.3.4.2.5*

θ_c $= 250°C$

Strength retention factor for concrete

$k_{c,\theta}$ $= 0.9$ *EN 1994-1-2 3.2.2*

Compressive resistance of concrete flange

$N_{c,fi,Rd} = 0.85\,k_{c,\theta}\,f_{cd}\,b_{eff}h_c = 0.85 \times 0.9 \times 25.0 \times 3000 \times 70 \times 10^{-3}$ *EN 1994-1-2 4.3.1*

$N_{c,fi,Rd} = 4016.3\,kN$

Worked Example Fire Resistance – Composite Beams	Sheet 5 of 7	Rev

Resistance of steel section assuming uniform temperature

θ_a	$= 624°C$	
$k_{y,\theta}$	$= 0.41$	*EN 1994-1-2 3.2.1*

$$N_{fi,pl,a} = Ak_{y,\theta}\left(\frac{f_y}{\gamma_{M,fi,a}}\right) = 1073.4\ kN$$

Resistance of shear studs

The resistance of the shear connectors in fire conditions is obtained by applying strength reduction factors to the room temperature resistance values calculated using the equations given in EN1994-1-1.

EN 1994-1-2 4.3.4.2.5

The resistance of the headed shear stud is obtained from EN1994-1-1 as follows

$$P_{Rd} = k_t\frac{0.8 \times f_u \times \pi \times d^2 / 4}{\gamma_{M,fi,v}}$$

EN 1994-1-1 Eqn (6.18)

In fire conditions this value of P_{Rd} is modified as follows to account for the reduction in the strength of the stud due to increasing temperature.

$P_{fi,Rd} = 0.8\ k_{u,\theta}\ P_{Rd}$

EN 1994-1-2 4.3.4.2.5

The resistance of the concrete failure surface around the headed shear stud is given by the following equation.

$$P_{Rd} = k_t\frac{0.29 \times \alpha \times d^2\sqrt{f_{ck}E_{cm}}}{\gamma_{M,fi,v}}$$

EN1994-1-1 Eqn (6.19)

In fire conditions this is modified to account for the reduction in the strength of concrete with temperature as follows.

$P_{fi,Rd} = k_{u,\theta}\ P_{Rd}$

EN 1994-1-2 4.3.4.2.5

where

$$\alpha = 1.0 \quad as \quad \frac{h_{sc}}{d} < 4$$

For sheeting with ribs transverse to the beam, assume:

$k_t \qquad = 0.85$

EN1994-1-1 6.6.4.2

The room temperature resistance of the headed shear stud is the lesser of the following values.

$$P_{Rd} = 0.85\frac{0.8 \times 450 \times \pi \times \left(19^2 / 4\right)}{1.0} \times 10^{-3} = 86.8\ kN$$

Worked Example Fire Resistance – Composite Beams	Sheet 6 of 7	Rev

$$\frac{0.29 \times 1.0 \times 19^2 \times \sqrt{25 \times 31 \times 1000}}{1.0} \times 10^{-3} = 78.4\,kN$$

Considering the temperature of shear studs for fire conditions

$\theta_v \quad = 0.8\,\theta_a$ EN 1994-1-2 4.3.4.2.5

$\theta_v \quad = 500°C$

For this stud temperature the strength reduction is:

$k_{u,\theta} \quad = 0.78$ EN 1994-1-2 Table 3.2

Giving a resistance for the headed stud of:

$P_{fi,Rd} \quad = 0.8 \times 0.78 \times 86.8 = 54.2\,kN$

Considering the temperature of concrete for fire conditions

$\theta_c \quad = 0.4\,\theta_a$ EN 1994-1-2 4.3.4.2.5

$\theta_c \quad = 250°C$

For this design temperature the strength reduction in the concrete is:

$k_{c,\theta} \quad = 0.9$ En 1994-1-2 Table 3.3

Giving a resistance for the concrete failure surface of:

$P_{fi,Rd} \quad = 0.9 \times 78.4 = 70.6\,kN$

The design resistance is then the lesser of these two values

$P_{fi,Rd} \quad = 54.2\,kN$

The point of max moment occurs at mid span. Given that the ribs of the slab are at 300 mm centres, the number of studs between the support and point of max moment will be:

$$n_r = \frac{6}{0.3} = 20$$

$N_c \quad = 20 \times 54.2 = 1084\,kN$

Full shear resistance is achieved

Force in concrete

$N_{c,fi,Rd} = N_{fi,pl,a} = 1073.4\,kNm$

Depth of concrete

$$x_c = \frac{1073.4 \times 10^3}{0.85 \times 25.0 \times 3000} = 16.8\,mm$$

Worked Example Fire Resistance – Composite Beams	Sheet 7 of 7	Rev

Lever arm

$z \quad = h_s \, x_c/2 + h_d/2$

$z \quad = 130 - 16.8/2 + 529.1/2 = 386.2 \, mm$

Moment of resistance

$M_{fi,Rd} = N_{fi,pl,a} \times z$

$M_{fi,Rd} = 1073.4 \times 386.2 \times 10^{-3} = 414.5 \, kNm$

$E_{fi,d,t} \quad = 361 \, kNm$

$\therefore \; \theta_{cr} = 624°C$ is a conservative estimate of critical temperature.

Chapter 36
Corrosion and corrosion prevention

DAVID DEACON and ROGER HUDSON

36.1 Introduction

The specification of cost-effective protective treatments for structural steelwork should not present a major problem for most common applications if the factors that affect durability are appreciated. Primarily, it is important to recognise and define the corrosivity of the environment to which the structure is to be exposed to enable the specification of an appropriate protective system. Many structures are in relatively low risk category environments and therefore require minimal treatment. Conversely, a steel structure exposed to an aggressive environment needs to be protected with a durable system that may require maintenance for extended life.

The optimum protection treatment combines good surface preparation with suitable coating materials for a required durability at a minimum cost. Modern practices applied according to the relevant industry standards provide an opportunity to achieve the desired protection requirements for specific structures. There are many standards available to assist in the drafting of protection specifications. One of the most important is ISO 12944 *Paints and Varnishes – Corrosion Protection of Steel Structures by Protective Paint Systems*. This standard, which is published in eight parts, should be referred to when drafting protection specifications for steel structures. Part 5 of the series *Protective paint systems* contains a range of paint coatings and systems for different environmental categories that are defined in Part 2 *Classification of environments*. However, specifiers concerned with UK projects should be aware that not all of the paints listed are 'compliant' with current national environmental legislation, and further advice should be sought from the paint manufacturer.

---

[1]A comprehensive list of standards organised by topic is given at the end of the chapter.

Steel Designers' Manual, Seventh Edition. Edited by Buick Davison and Graham W. Owens.
© 2012 Steel Construction Institute. Published 2012 by Blackwell Publishing Ltd.

36.2 General corrosion

Most corrosion of steel can be considered as an electrochemical process which occurs in stages. Initial attack occurs at anodic areas on the surface, where ferrous ions go into solution. Electrons are released from the anode and move through the metallic structure to the adjacent cathodic sites on the surface, where they combine with oxygen and water to form hydroxyl ions. These react with the ferrous ions from the anode to produce ferrous hydroxide, which itself is further oxidised in air to produce hydrated ferric oxide: red rust (Figure 36.1).

The sum of these reactions is described by the following equation:

$$4Fe + 3O_2 + 2H_2O = 2Fe_2O_3H_2O$$

$$(iron/steel) + (oxygen) + (water) = rust$$

Two important points emerge:

1. For iron or steel to corrode it is necessary to have the *simultaneous* presence of water and oxygen; in the absence of either, corrosion does not occur.
2. All corrosion occurs at the anode; no corrosion occurs at the cathode.

However, after a period of time, polarisation effects such as the growth of corrosion products on the surface cause the corrosion process to be stifled. New, reactive

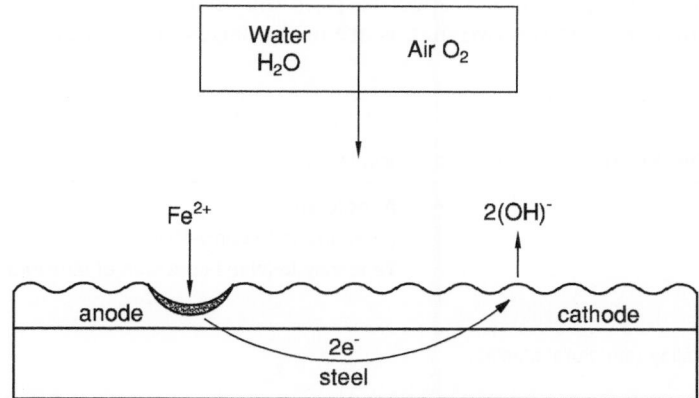

Reactions

At the anode	$Fe \longrightarrow Fe^{2+} + 2e^-$
At the cathode	$O_2 + 2H_2O + 4e^- \longrightarrow 4(OH)^-$
Combined	$Fe^{2+} + 2(OH)^- \longrightarrow Fe(OH)_2$ [ferrous hydroxide]
Further oxidation	$Fe(OH)_2 \longrightarrow Fe_2O_3H_2O$ [hydrated ferric oxide or rust]

Figure 36.1 Diagrammatic representation of the corrosion of steel

anodic sites may then be formed, thereby allowing further corrosion. Over long periods the loss of metal is reasonably uniform over the surface and so this case is usually described as *general corrosion*.

36.3 Other forms of corrosion

Various types of localised corrosion can also occur:

- **Pitting corrosion.** In some circumstances the attack on the original anodic area is not stifled and continues deep into the metal, forming a corrosion pit. Pitting more often occurs with mild steels immersed in water or buried in soil rather than those exposed in air.
- **Crevice corrosion.** Crevices can be formed by design-detailing, welding, surface debris, etc. Available oxygen in the crevice is quickly used by the corrosion process and, because of limited access, cannot be replaced. The entrance to the crevice becomes cathodic, since it can satisfy the oxygen-demanding cathode reaction. The tip of the crevice becomes a localised anode, and high corrosion rates occur at this point.
- **Bimetallic corrosion.** When two dissimilar metals are joined together in an electrolyte an electrical current passes between them and corrosion occurs on the anodic metal. Some metals (e.g. nickel and copper) cause steel to corrode preferentially whereas other metals corrode preferentially themselves, thereby protecting the steel. The tendency of dissimilar metals to bimetallic corrosion is partly dependent upon their respective positions in the galvanic series (Table 36.1): the further apart the two metals are in the series the greater the tendency.

Table 36.1   Bimetallic corrosion and structural steelwork

	Anodic end
Magnesium	(more prone to corrosion)
Zinc	Tendency to inhibit corrosion of structural steels
Aluminium	
Cadmium	
Carbon and low alloy (structural steels)	
Cast irons	
Lead	Tendency to accelerate corrosion of structural steels
Tin	
Copper	
Brasses	
Bronzes	
Nickel (passive)	
Titanium	
Stainless steels 430/304/316 (passive)	
	Cathodic end
	(less prone to corrosion)

Other aspects which influence bimetallic corrosion are the nature of the electrolyte and the respective surface areas of the anodic and cathodic metals. Bimetallic corrosion is most serious for immersed or buried structures but in less aggressive environments, e.g. stainless steel brick support angles attached to mild steel structural sections, the effect on the mild steel sections is practically minimal and no special precautions are required.

Further guidance for the avoidance of bimetallic corrosion can be found in BS PD 6484, *Commentary on corrosion at bimetallic contacts and its alleviation.*

36.4 Corrosion rates

The principal factors that determine the rate of corrosion of steel in air are:

- **Time of wetness.** This is the proportion of total time during which the surface is wet, due to rainfall, condensation, etc. It follows, therefore, that for unprotected steel in dry environments, e.g. inside heated buildings, corrosion will be negligible due to the low availability of water.
- **Atmospheric pollution.** The type and amount of atmospheric pollution and contaminants, e.g. sulphur dioxide, chlorides, dust, etc.
- **Sulphates.** These originate from sulphur dioxide gas, which is produced during the combustion of fossil fuels, e.g. sulphur-bearing oils and coal. The sulphur dioxide gas reacts with water or moisture in the atmosphere to form sulphurous and sulphuric acids. Industrial environments are a prime source of sulphur dioxide.
- **Chlorides.** These are mainly present in marine environments. The highest concentrations of chlorides are to be found in coastal regions, and there is a rapid reduction moving inland except where road de-icing salts are present.

Within a given local environment corrosion rates can vary markedly due to the effects of sheltering and prevailing winds. It is, therefore, the 'microclimate' immediately surrounding the structure which determines corrosion rates for practical purposes.

36.5 Effect of the environment

Corrosion rate data cannot be generalised; however, environments can be broadly classified and corresponding corrosion rates provide a useful indication. Atmospheric environment corrosivity categories, thickness loss data and typical examples are covered in ISO 12944-2. The 'C' category should be defined at the outset of drafting the specification for an appropriate coating system.

A range of UK environments and steel corrosion rates are considered as follows: (Note: Corrosion rates are usually expressed as μm/year; 1 μm = 0.001 mm.)

- **Rural atmospheric** – essentially inland, unpolluted environments; steel corrosion rates tend to be low, usually less than 50 μm/year.
- **Industrial atmospheric** – inland, polluted environments; corrosion rates are usually between 40 and 80 μm/year, depending upon level of SO_2.
- **Marine atmospheric** – in the UK a 2 km strip around the coast is broadly considered as being in a marine environment; corrosion rates are usually between 50 and 100 μm/year, largely dependent upon proximity to the sea.
- **Marine/industrial atmospheric** – polluted coastal environments which produce the highest corrosion rates e.g. between 50 and 150 μm/year.
- **Sea-water immersion** – in tidal waters four vertical zones are usually encountered:
  - the splash zone, immediately above the high-tide level, is usually the most corrosive zone with a mean corrosion rate of about 75 μm/year
  - the tidal zone, between high-tide and low-tide levels, is often covered with marine growths and exhibits low corrosion rates e.g. 35 μm/year
  - the low-water zone, a narrow band just below the low-water level, exhibits corrosion rates similar to the splash zone
  - the permanent immersion zone, from the low-water level down to bed level, exhibits low corrosion rates e.g. 35 μm/year.
- **Fresh-water immersion** – corrosion rates are lower in fresh water than in salt water e.g. 30–50 μm/year.
- **Soils** – the corrosion process is complex and very variable; various methods are used to assess the corrosivity of soils:
  - resistivity; generally high-resistance soils are least corrosive
  - Redox potential; to assess the soil's capability of anaerobic bacterial corrosion
  - pH; highly acidic soils (e.g. pH less than 4.0) can be corrosive
  - water content; corrosion depends upon the presence of moisture in the soil, the position of the water-table has an important bearing.

Long buried steel structures, e.g. pipelines, are most susceptible to corrosion. Steel piles driven into undisturbed soils are much less susceptible due to the low availability of oxygen.

36.6 Design and corrosion

In external or wet environments design can have an important bearing on the corrosion of steel structures. The prevention of corrosion should therefore be taken into account during the design stage of a project. The main points to be considered are:

- Entrapment of moisture and dirt:
  - avoid the creation of cavities, crevices, etc.
  - welded joints are preferable to bolted joints
  - avoid or seal lap joints
  - edge-seal HSFG faying surfaces
  - provide drainage holes for water, where necessary
  - seal box sections except where they are to be hot-dip galvanized
  - provide free circulation of air around the structure.
- Contact with other materials:
  - avoid bimetallic connections or insulate the contact surfaces (see BS PD 6484)
  - provide adequate depth of cover and quality of concrete
  - separate steel and timber by the use of coatings or sheet plastics.
- Coating application; design should ensure that the selected protective coatings can be applied efficiently:
  - provide vent-holes and drain-holes for items to be hot-dip galvanized (see ISO 14713-2)
  - provide adequate access for paint spraying, thermal (metal) spraying, etc. (see BS 4479: Part 7) both initially and during the lifetime maintenance.
- General factors.
  - large flat surfaces are easier to protect than more complicated shapes
  - provide access for subsequent maintenance
  - provide lifting lugs or brackets where possible to reduce damage during handling and erection.

ISO 12944-3 provides design guidance for the prevention of corrosion.

36.7 Surface preparation

Structural steel is a hot-rolled product. Sections leave the last rolling pass at about 1000°C and as they cool the steel surface reacts with oxygen in the atmosphere to produce mill-scale, a complex oxide which appears as a blue-grey tenacious scale completely covering the surface of the as-rolled steel section. Unfortunately, mill-scale is unstable. On weathering, water penetrates fissures in the scale, and rusting of the steel surface occurs. The mill-scale loses adhesion and begins to shed. Mill-scale is therefore an unsatisfactory base and needs to be removed before protective coatings are applied.

As mill-scale sheds, further rusting occurs. Rust is a hydrated oxide of iron which forms at ambient temperatures, producing a layer on the surface and which is itself an unsatisfactory base and also needs to be removed before protective coatings are applied.

Surface preparation of steel is therefore principally concerned with the removal of mill-scale and rust and is an essential process in corrosion protection treatments. Various methods of surface preparation are presented in ISO 8501 series of standards and are summarised as follows:

1. **Hand and power tool cleaning (St)**

   St 2: Thorough hand and power tool cleaning.

   St 3: Very thorough hand and power tool cleaning.

   Both manual and mechanical methods using scrapers, wire brushes etc. can remove about 30–50% of rust and scale. Disc and pencil grinders would be needed for severely pitted areas.

2. **Blast cleaning (Sa)**

   Sa 1: Light blast cleaning.

   Sa 2: Thorough blast cleaning.

   Sa 2½: Very thorough blast cleaning.

   Sa 3: Blast cleaning to visually clean steel.

   It is important to specify the steel condition (A, B, C, or D) before the contract is placed. Condition A or B should be specified for long-life coating systems since these are the two grades of new steel, without any pitting. Grades C and D are of pitted condition. The blast cleaning process involves the projection of abrasive particles (shot or grit) in a jet of compressed air or by centrifugal impellers at high velocities on to the steel surface. This process can be 100% efficient in the removal of rust and scale. The profile of the surface produced is dependent upon the size and shape of the abrasive used; angular grits produce angular surface profiles and round shot produce a rounded profile. The effectiveness of the surface preparation methods above are compared with the relevant photograph contained in the standard.

   Grit-blast abrasive can be either metallic (e.g. chilled iron grit) or non-metallic (e.g. slag grit). The latter are used only once and are referred to as expendable. They are used exclusively for site work. Metallic grits are expensive and are only used where they can be recycled. Grit blasting is always used for thermal (metal) sprayed coatings, where adhesion is at least partly dependent upon mechanical keying. It is also used for some paint coatings, particularly on site and for primers where adhesion may be a problem (e.g. zinc silicates and high-build solvent-free paint coatings).

   Shot-blast abrasives are always metallic, usually cast steel shot, and are used particularly in shot-blast plants, utilising impeller wheels and abrasive recycling. They are the preferred abrasive for most paints, particularly for thin film coatings (e.g. prefabrication primers).

   Blast-cleaned surfaces should be specified in terms of both visual surface cleanliness (ISO 8501) and surface roughness (ISO 8503). In addition acceptable levels of soluble salts, ISO 8502 and surface dust contaminants ISO 8502/3 should be stated.

3. **Wet (abrasive) blast cleaning**

   A further variation on the blast cleaning process is described as wet blasting. In this process, a small amount of water is entrained in the abrasive/compressed air stream. This is particularly useful in washing from the surface soluble iron salts, which are formed in the rust by atmospheric pollutants (e.g. chlorides and sulphates) during weathering. These are often located deep in corrosion pits on the steel surface and cannot be removed by conventional dry blast cleaning methods. Wet abrasive blasting has proved to be particularly useful on offshore structures and prior to maintenance painting for structures in heavily polluted environments. However, it is not entirely successful on pitted surfaces where it should be specified in conjunction with grinding.

4. **Acid pickling**

This process involves immersing the steel in a bath of suitable inhibited acids, which dissolve or remove mill-scale and rust but do not appreciably attack the exposed steel surface. It can be 100% effective. Acid pickling is normally used on structural steel intended for hot-dip galvanizing but is now rarely used as pre-treatment before painting.

36.8 Metallic coatings

There are four commonly used methods of applying metal coating to steel surfaces: hot-dip galvanizing, thermal (metal) spraying, electroplating and sherardizing. The latter two processes are not used in structural steelwork but are used for fittings, fasteners and other small items.

In general the corrosion protection afforded by metallic coatings is largely dependent upon the choice of coating metal and its thickness and is not greatly influenced by the method of application with the exception of thermal metal spray coatings as described in Section 36.8.2.

36.8.1 Hot-dip galvanizing

The most common method of applying a metal coating to structural steel is by hot-dip galvanizing.

The process involves the following stages:

1. Any surface oil or grease is removed by suitable degreasing agents.
2. The steel is cleaned of all rust and scale by acid pickling. This may be preceded by blast-cleaning to remove scale and roughen the surface but such surfaces are always subsequently pickled in inhibited hydrochloric acid.
3. The cleaned steel is then immersed in a fluxing agent to ensure good contact between the zinc and steel during immersion.
4. The cleaned and fluxed steel is dipped into a bath of molten zinc at a temperature of about 450°C at which the steel reacts with the molten zinc to form a series of zinc/iron alloys on its surface.
5. As the steel workpiece is removed from the bath a layer of relatively pure zinc is deposited on top of the alloy layers.

As the zinc solidifies it assumes a crystalline metallic lustre, usually referred to as spangling. The thickness of the galvanized coating is influenced by various factors including the size and thickness of the workpiece and the surface preparation of the steel. Thick steels and steels which have been abrasive blast cleaned tend to produce heavier coatings. Additionally, the steel composition has an effect on the coating produced. Steels containing silicon and phosphorus can have a marked effect on the thickness, structure and appearance of the coating. The thickness of the coating varies mainly with the silicon content of the steel and the bath immersion time. These thick coatings sometimes have a dull dark grey appearance and can be susceptible to mechanical damage.

Since hot-dip galvanizing is a dipping process, there is obviously some limitation on the size of components which can be galvanized. Double dipping, which involves dipping one end of the item before the other, can often be used when the length or width of the workpiece exceeds the size of the bath.

Some aspects of design need to take the galvanizing process into account, particularly filling, venting, draining and distortion. To enable a satisfactory coating, suitable holes must be provided in hollow articles (e.g. tubes and rectangular hollow sections) to allow access for the molten zinc, venting of hot gases to prevent explosions and the subsequent draining of zinc. Distortion of fabricated steel-work can be caused by differential thermal expansion and contraction and by the relief of unbalanced residual stresses during the galvanizing process. Further guidance on the design of articles to be hot-dip galvanized can be found ISO 14713-2.

ISO 1461 is the specification of hot-dip galvanized coating for structural steelwork. This requires, for sections not less than 6 mm thick, a minimum zinc coating weight of $610 \, g/m^2$, equivalent to a minimum average coating thickness of 85 μm.

For many applications, hot-dip galvanizing is used without further protection. However, to provide extra durability, particularly in certain atmospheric environments, or where there is a decorative requirement, paint coatings are applied. The combination of metal and paint coatings is usually referred to as a duplex coating. When applying paints to galvanized coatings, special surface preparation treatments should be used to ensure good adhesion. These include light blast cleaning to roughen the surface and provide a mechanical key, and the application of special etch primers or 'T' wash, which is an acidified solution designed to react with the surface and provide a visual indication of effectiveness.

36.8.2 Thermal (metal) spray coatings

An alternative method of applying a metallic coating to structural steelwork is by thermal (metal) spraying of either zinc or aluminium. The metal, in powder or wire form, is fed through a special spray-gun containing a heat source which can be either an oxy-gas flame or an electric arc. Molten globules of the metal are blown by a compressed air jet on to the previously blast-cleaned steel surface. No alloying occurs and the coating which is produced consists of overlapping platelets of metal and is porous. It is essential that the pores are subsequently sealed, either by applying a thin organic coating which soaks into the surface, or where further painting is not required by allowing the metal coating to weather, when corrosion products block the pores.

The adhesion of sprayed metal coatings to steel surfaces is considered to be essentially mechanical in nature. It is therefore necessary to apply the coating to a clean roughened surface for which blast-cleaning with a coarse grit abrasive is normally specified, usually chilled-iron grit, but for steels with a hardness exceeding 360 HV, alumina or silicon carbide grits may be necessary.

Typical specified coating thicknesses vary between 150–200 µm for aluminium and 100–150 µm for zinc.

Thermal (metal) spray coatings can be applied in the shops or at site, and there is no limitation on the size of the workpiece as there is with hot-dip galvanizing. Since the steel surface remains cool there are no distortion problems. Guidance on the design of articles to be thermally sprayed can be found in BS 4479-7. However, thermal spraying is considerably more expensive than hot-dip galvanizing.

For many applications thermal spray coatings are further protected by the subsequent application of paint coatings. A sealer is first applied, which fills the pores in the thermal spray coating and provides a smooth surface for application of the paint coating.

The protection of structural steelwork against atmospheric corrosion by thermal sprayed aluminium or zinc coatings is covered in ISO 2063.

Great care should be exercised in selecting metal spray for the main protective coating since the application process does not allow a uniform coating to be applied on certain design configurations.

36.9 Paint coatings

Painting is the principal method of protecting structural steelwork from corrosion.

36.9.1 Composition of paints and film formation

Paints are made by mixing and blending three main components:

1. Pigments: finely ground inorganic or organic powders which provide colour, opacity, film cohesion and sometimes corrosion inhibition or lamellar such as micaceous iron oxide (MIO) or aluminium flake to provide barrier protection.
2. Binders: usually resins or oils but can be organic chemicals, such as two-pack epoxy resins, or inorganic compounds such as soluble silicates. The binder is the film-forming component in the paint.
3. Solvents: used to dissolve the binder and to facilitate application of the paint. Solvents are usually organic liquids or water.

Paints are applied to steel surfaces by many methods but in all cases they produce a **wet film**. The thickness of the wet film can be measured, before the solvent evaporates, or the coating starts to cure, by using a comb-gauge.

As the solvent evaporates, film-formation occurs, leaving the binder and pigments on the surface as a **dry film**. The thickness of the dry film can be measured, usually with an electromagnetic gauge.

The relationship between the applied wet film thickness and the final **dry film thickness** (d.f.t.) is determined by the percentage volume solids of the paint, i.e.

d.f.t. = wet film thickness × % vol. solids.

In general, the corrosion protection afforded by a paint film is directly proportional to its dry film thickness.

36.9.2 Classification of paints

Since, in the broadest terms, a paint consists of a particular pigment, dispersed in a particular binder, dissolved in a particular solvent, the number of generic types of paint is limited. The most common methods of classifying paints are either by their pigmentation or by their binder-type.

Primers for steel are usually classified according to the main corrosion-inhibitive pigments used in their formulation, e.g. zinc phosphate, metallic zinc. Each of these inhibitive pigments can be incorporated into a range of binder resins, e.g. zinc phosphate alkyd primers or zinc phosphate epoxy primers.

Intermediate coats and finishing coats are usually classified according to their binders, e.g. epoxy build coats, vinyl finishes, urethane finishes, or by their pigment such as MIO.

36.9.3 Painting systems

Paints are usually applied one coat on top of another, each coat having a specific function or purpose.

The primer is applied directly on to the cleaned steel surface. Its purpose is to wet the surface and to provide good adhesion for subsequently applied coats. Primers for steel surfaces are also usually required to provide corrosion inhibition.

The intermediate coats (or undercoats) are applied to build the total film thickness of the system. This may involve the application of several coats.

The finishing coats provide the first line of defence against the environment and also determine the final appearance in terms of gloss, colour, etc.

The various superimposed coats within a painting system have, of course, to be compatible with one another. It is also important to apply extra coats on vulnerable parts of the structure to obtain the minimum thickness. These coats are known as stripe coats. As a first precaution, it is recommended that all paints within a system should normally be obtained from the same manufacturer.

36.9.4 Main generic types of paint and their properties

1. **Air-drying paints,** e.g. oil-based, alkyds. These materials dry and form a film by an oxidative process which involves absorption of oxygen from the atmosphere. They are therefore limited to relatively thin films. Once the film has formed it has limited solvent resistance and usually poor chemical resistance.

2. **One-pack chemical-resistant paints,** e.g. acrylated rubbers, vinyls. For these materials, film formation is by solvent evaporation and no oxidative process is involved. They can be applied as moderately thick films, of the order of 75 μm, although retention of solvent in the film can be a problem at the upper end of the range. The film formed remains relatively soft and has poor solvent resistance but good chemical resistance.

   Unmodified bituminous paints also dry by solvent evaporation. They are essentially solutions of either asphaltic bitumen or coal-tar pitch in organic solvents.
3. **Two-pack chemical-resistant paints,** e.g. epoxy, urethane. These materials are supplied as two separate components, usually referred to as the base and the curing agent. When the two components are mixed, immediately before use, a chemical reaction begins. These materials therefore have a limited 'pot-life' by which time the mixed coating must be applied. The polymerisation reaction continues after the paint has been applied and after the solvent has evaporated to produce a densely cross-linked film, which can be very hard and has good solvent and chemical resistance.

Liquid resins of low viscosity can be used in the formulation thereby avoiding the need for a solvent. Such coatings are referred to as 'solventless' or solvent-free and can be applied as very thick films.

A summary of the main generic types of paint and their properties is shown in Table 36.2.

Table 36.2 Main generic types of paint and their properties

	Cost	Tolerance of poor surface preparation	Chemical resistance	Solvent resistance	Over-coatability after ageing	Other Comments
Bituminous	Low	Good	Moderate	Poor	Good with coatings of same type	Limited to black and dark colours Thermoplastic
Alkyds	Low-medium	Moderate	Poor	Poor-moderate	Good	Good decorative properties
Acrylated-rubber	Medium	Poor	Good	Poor	Good	High-build films remain soft and are susceptible to 'sticking'
Vinyl	High	Poor	Good	Poor	Good	
Epoxy	Medium-high	V. poor	V. good	Good	Poor	Very susceptible to chalking in UV
Urethane	High	V. poor	V. good	Good	Poor	Can be more decorative than epoxies
Inorganic or organic silicate	High	V. poor	Moderate	Good	Moderate	May require special surface preparation

36.9.5 Prefabrication primers (also referred to as blast primers, shop-primers, weldable primers, temporary primers, holding primers, etc.)

These primers are used on structural steelwork, immediately after blast-cleaning, to hold the reactive blast-cleaned surface in a rust-free condition until final painting can be undertaken. They are mainly applied to steel plates and sections before fabrication. The main requirements of a blast primer are as follows:

1. The primer should be capable of airless-spray application to produce a very thin even coating. Dry-film thickness is usually limited to 15–25 μm. Below 15 μm the peaks of the blast profile are not protected and 'rust-rashing' occurs on weathering. Above 25 μm the primer affects the quality of the weld and produces excessive weld-fume.
2. The primer must dry very quickly, not only to protect the peaks of the blast profile, but also because priming is often done in-line with automatic blast-cleaning plant, which may be handling plates or sections at a pass-rate of 1–3 metres/minute. The interval between priming and handling is usually of the order of 1–10 minutes and hence the primer film must dry within this time.
3. Normal fabrication procedures, e.g. welding, gas-cutting, must not be significantly impeded by the coating and the primer should not cause excessive weld porosity.
4. Weld fumes emitted by the primer must not exceed the appropriate occupational exposure limits. Proprietary primers are tested and certificated by the Newcastle Occupational Health and Hygiene Service.
5. The primer coating should provide adequate protection. It should be noted that manufacturers may claim extended durability for their prefabrication primers, and suggested exposure periods of 6–12 months are not uncommon. In practice, such claims are rarely met except in the least arduous conditions, e.g. indoor storage. In aggressive conditions, durability can be measured in weeks rather than months.

Many proprietary blast-primers are available but they can be classified under the following main generic types:

1. **Etch primers** are based on polyvinyl butyral resin reinforced with a phenolic resin to uprate water resistance. These primers can be supplied in a single-pack or two-pack form and comprise a zinc chromate pigment.
2. **Epoxy primers** are two-pack materials utilising epoxy resins and usually either polyamide or polyamine curing agents. They are pigmented with a variety of inhibitive and non-inhibitive pigments. Zinc phosphate epoxy primers are the most frequently encountered and give the best durability within the group.
3. **Zinc epoxy primers** can be subdivided into zinc-rich and reduced-zinc types. Zinc-rich primers produce films which contain about 80% by weight of metallic zinc powder and the reduced-zinc as low as 55% by weight.

When exposed in either marine or highly industrial environments, zinc epoxy primers are prone to the formation of insoluble white zinc corrosion products, which must be removed from the surface before subsequent overcoating. All zinc epoxy primers produce zinc oxide fumes during welding and gas cutting and may be a health hazard and can cause weld porosity.

4. **Zinc silicate primers** produce a level of protection, which is comparable to the zinc-rich epoxy types and they suffer from the same drawbacks, e.g. formation of zinc salts and production of zinc oxide fumes during welding. They are however more expensive and usually are less convenient to use.

There are currently different categories of silicate primer based upon the binder (organic or inorganic) and the zinc content. Low-zinc primers in this group have been developed to improve weldability and minimise weld porosity. However, their durability is reduced. The organic silicate primers are the most suitable as prefabrication primers.

36.10 Application of paints

36.10.1 Methods of application

The method of application and the conditions under which paints are applied have a significant effect on the quality and durability of the coating.

The standard methods used for applying paints to structural steelwork are brush, roller, conventional air-spray, and airless spray, although other methods, e.g. dipping, can be used.

1. **Brush.** This is the simplest and also the slowest and therefore most expensive method. Nevertheless, it has certain advantages over the other methods, e.g. better wetting of the surface; can be used in restricted spaces; useful for small areas; less wastage and less contamination of surroundings.
2. **Roller.** This process is much quicker than brushing and is useful for large flat areas but demands suitable rheological properties of the paint. It should not be used for primer or build coat application and is banned in many protective coating specifications, in particular the UK Highways Agency bridge painting specifications with the exception of the final outer decorative coats.
3. **Air-spray.** The paint is atomised at the gun-nozzle by jets of compressed air; application rates are quicker than for brushing or rolling; paint wastage by over-spray is high and aeration of the wet film may cause porosity.
4. **Airless spray.** The paint is atomised at the gun-nozzle by very high hydraulic pressures; application rates are higher than for air-spray and overspray wastage is greatly reduced.

Airless spraying has become the most commonly used method of applying paint coatings to structural steelwork under controlled shop conditions. Brush application is more commonly used for site application, though spraying methods are also used.

36.10.2 Conditions for application

The principal conditions, which affect the application of paint coatings are temperature and humidity. These can be more easily controlled under shop conditions than on site.

1. **Temperature.** Air temperature and steel temperature affect solvent evaporation, brushing and spraying properties, drying and curing times, pot-life of two-pack materials, etc. Heating, if required, should only be by indirect methods.
2. **Humidity.** Paints should not be applied when there is condensation present on the steel surface or the relative humidity of the atmosphere is such that it will affect the application or drying of the coating. Normal practice is to measure the steel temperature with a contact thermometer and to ensure that it is maintained at least 3°C above dew-point.

36.11 Weather-resistant steels

Weather-resistant steels are high strength, low alloy weldable structural steels which possess good weathering resistance in many atmospheric conditions without the need for protective coatings. They contain up to 2.5% of alloying elements, e.g. chromium, copper, nickel and phosphorus. On exposure to air, under suitable conditions, they form an adherent protective rust layer. This acts as a protective film, which with time causes the corrosion rate to reduce until it reaches a low terminal level, usually between 2 and 3 years.

Conventional structural steels form rust layers that eventually become non-adherent and detach from the steel surface. The rate of corrosion progresses as a series of incremental curves approximating to a straight line, the slope of which is related to the aggressiveness of the environment. With weather-resistant steels, the rusting process is initiated in the same way but the alloying elements react with the environment to form an adherent, less porous rust layer. With time, this rust layer becomes protective and reduces the corrosion rate (see Figure 36.2).

Weather-resistant steels are specified in BS EN 10025 Part 5, and within this category Corten is one of the best known proprietary weather-resistant steels. These steels have mechanical properties comparable to those of grade S355 steels to BS EN 10025 Part 2.

36.11.1 Formation of the protective oxide layer

The time required for a weather-resistant steel to form a stable protective rust layer depends upon its orientation, the degree of atmospheric pollution and the frequency with which the surface is wetted and dried. The steel should be abrasive blast cleaned, to remove mill-scale, before exposure in order to provide a sound uniform surface for the formation of the oxide coatings.

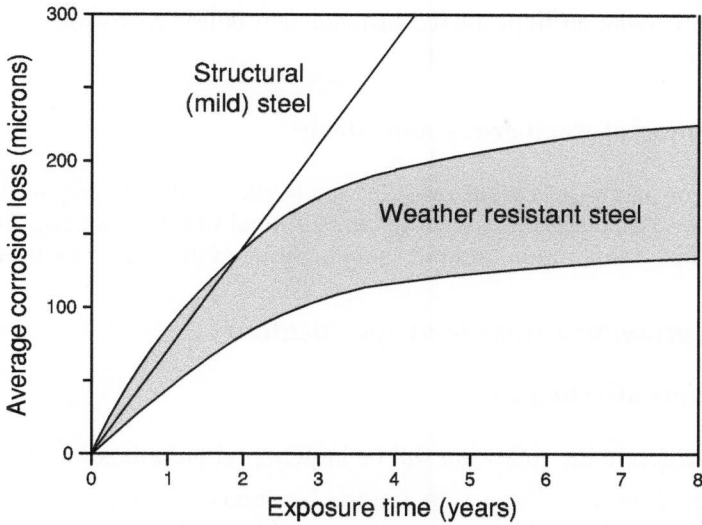

Figure 36.2 Typical corrosion losses of structural (mild) steel and weather-resistant steels in the UK

36.11.2 Precautions and limitations

The following points should be observed to maximise the benefits of using weather-resistant steels. Avoid:

- contact with absorbent surfaces, e.g. concrete and timber
- prolonged wet conditions
- burial in soils
- contact with dissimilar metals
- aggressive environments, e.g. marine atmospheres and road salts.

Drainage of corrosion products can be expected during the first years of exposure and can stain or streak adjacent materials, e.g. concrete piers. Provision should be made to divert corrosion products from vulnerable surfaces. Often the north faces of buildings experience long periods of wetness and do not favour the formation of a protective rust patina.

36.11.3 Welded and bolted connections

Weather-resistant steels can be welded by all the usual methods, e.g. manual metal arc, gas shielded, submerged arc and electrical resistance, including spot welding. Welding electrodes should be compatible with the welding process. For structural joints where high strength bolts are required, ASTM A325, Type 3 bolts (Corten X) must be used. Where lower strength bolts are satisfactory, these may be in Corten A or stainless steel. Galvanized, sherardized or electroplated nuts and bolts are not suitable for use in weather-resistant steel structures since, in time, the coatings will be consumed leaving an unprotected fastener that is less corrosion resistant than the surrounding weather-resistant steel.

36.11.4 Painting of weather-resistant steels

The practice of painting weather-resistant steels, should this be required, does not differ from those practices employed for conventional structural steels. They require the same surface preparation and the same painting systems may be used.

36.12 The protective treatment specification

36.12.1 Factors affecting choice

For a given structure the following will be largely predetermined:

1. the expected life of the structure and the feasibility of maintenance
2. the environment/s to which the steelwork will be subjected
3. the size and shape of the structural members
4. the shop-treatment facilities, which are available to the fabricator and/or the coatings sub-contractor
5. the site conditions, which will determine whether the steelwork can be treated after erection
6. the cost of providing protection.

These facts, and possibly others, have to be considered before making decisions on:

— the types of coating to be used
— the method of surface preparation
— the method/s of application
— the number of coats and the thickness of each coat.

In general, each case has to be decided on its own merits. However, the following points may be of assistance in making these decisions:

1. Protection requirements are minimal inside dry, heated buildings. Hidden steelwork in such situations requires no protection at all.
2. The durability of painting systems is increased several times over by using abrasive blast-cleaning rather than manual surface preparation.
3. Shot-blasting is preferred for most painting systems, except high build coatings, which require a higher grit profile.
4. Grit-blasting is essential for thermal (metal) spraying and some primers, e.g. zinc silicates.
5. If blast-cleaning is to be used, two alternative process routes are available, i.e.
   — blast/prime/fabricate/repair damage
   — fabricate/blast/prime.
   The former is usually cheaper but requires the use of a thinner, weldable prefabrication primer.

6. Prefabrication primers have to be applied to blast-cleaned surfaces as thin films, usually 25 µm maximum. Their durability is therefore limited and further shop-coating is often desirable.

7. Manual preparation methods are dependent upon weathering to loosen the mill-scale. These methods are therefore not usually appropriate for shop treatments. On site an adequate weathering period, usually several months, must be allowed and it is important that all the mill-scale has been removed.

8. Many modern primers based on synthetic resins are not compatible with manually prepared steel surfaces since they have a low tolerance to rust and scale.

9. Many oil-based and alkyd-based primers cannot be overcoated with finishing coats which contain strong solvents, e.g. acrylated rubbers, epoxies, bituminous coatings, etc.

10. Two-pack epoxies have poor resistance to UV radiation and are highly susceptible to 'chalking'. Overcoating problems can arise with two-pack epoxies unless they are overcoated before the prior coat is fully cured. This is particularly relevant when an epoxy system is to be applied partly in the shops and partly on site. Full curing generally takes 14–28 days depending on temperature of both the atmosphere and the substrate.

11. Steelwork, which is to be encased in concrete does not normally require any other protection, given an adequate depth of concrete cover and low moisture permeability.

12. Perimeter steelwork hidden in cavity walls falls into two categories:
    — Where an air gap (40 mm min.) exists between the steelwork and the outer brick or stone leaf, then adequate protection can be achieved by applying relatively simple painting systems.
    — Where the steelwork is in direct contact with the outer leaf, or is embedded in it, then the steel should be hot-dip galvanized and painted with a water-resistant coating, e.g. bitumen.

13. Where fire-protection systems are to be applied to the steelwork, consideration must be given to the question of compatibility between the corrosion-protection and the fire-protection systems.

14. New hot-dip galvanized surfaces can be difficult to paint and, unless special primers are used, adhesion problems can arise. Weathering the zinc surface before painting reduces this problem, but may not overcome it.

15. Thermal spraying produces a porous coating, which should be sealed by applying a thin low-viscosity sealer until all the porosity is satisfied. Further painting is then optional.

16. Particular attention should be paid to the treatment of weld areas. Flux residues and weld-spatter should be removed before application of coatings. In general, the objective should be to achieve the same standard of surface preparation and coating on the weld area as on the general surface.

17. Black-bolted joints require protection of the contact surfaces. This is normally restricted to the priming coat, which can be applied either in the shops or on site before the joint is assembled.

18. For high-strength friction-grip bolted joints, the faying surfaces must be free of any contaminant or coating which would reduce the slip-factor required on the joint. Some metal-spray coatings and some inorganic zinc silicate primers can be used but virtually all organic coatings adversely affect the slip-factor. Any coatings considered for this application should be carefully checked.

36.12.2 Writing the specification

The specification should 'say what it means and mean what it says'. It is intended to provide clear and precise instructions to the contractor on what is to be done and how it is to be done. It should be written in a logical sequence, starting with surface preparation, going through each paint or metal coating to be applied and finally dealing with specific areas, e.g. welds. It should also be as concise as possible, consistent with providing all the necessary information. The most important items of a specification are as follows:

1. The method of surface preparation and the standard required, which can often be specified by reference to an appropriate standard, e.g. ISO 8501–1, A or B Sa 2–3 qualities as well as the measured levels of salts or dirt.
2. The maximum interval between surface preparation and subsequent priming.
3. The types of paint to be used, supported by standards where these exist.
4. The method/s of application to be used.
5. The number of coats to be applied and the interval between coats, including stripe coats.
6. The wet and dry film thickness for each coat.
7. Where each coat is to be applied (e.g. shop or site) and the application conditions that are required, in terms of temperature, humidity, dew point, etc.
8. Details for treatment of welds, connections, etc.
9. Rectification procedures for damage, etc.

36.12.3 Quality of coating application

It is widely recognised that the successful performance of protective paint coatings is significantly influenced by the quality of the application. Modern coating materials require a detailed appreciation of their properties in terms of their preparation (mixing), correct environmental conditions at the time of application and the use of suitable equipment. For this reason, it is essential that all coating operatives are fully conversant with the materials that are to be applied to ensure that they achieve the maximum performance in service. To this end, the Institute of Corrosion, through its wholly owned subsidiary company Correx Ltd., has established a training and certification scheme Industrial Coating Applicator Training Scheme (ICATS) for coating applicators to be trained and certificated. Companies and individuals are registered with the scheme and specifiers should ensure that all coating contractors and their personnel are ICATS registered and qualified.

36.12.4 Inspection

Inspection is an essential activity to ensure that the requirements of the specification are being met. Ideally this inspection should be carried out throughout the course of the contract at each separate phase of the work, i.e. surface preparation, first coat, second coat, etc. A wide range of instruments are available to assess surface roughness and cleanliness on blast-cleaned steel, wet-film thickness on paint coatings, dry-film thickness on paint coatings and metal coatings and environmental conditions. It is strongly suggested that an independent coating inspector qualified through the Institute of Corrosion's training and certification scheme be employed.

36.12.5 Environmental protection

In addition to the requirements for corrosion protection, there is increasing pressure being introduced by legislation to use paints and coatings that are environmentally friendly and thereby minimise damage to the atmosphere. The use of coatings that do not contain high quantities of organic solvents and toxic or harmful substances has been encouraged and enforced by recent industry directives.

Following the introduction of the Environmental Protection Act (1990) the Secretary of State's Process Guidance Note PG6/23 Coating of Metal and Plastic was produced which had tables of maximum limits of volatile organic compounds (VOCs) for different coating types. The paint and coatings manufacturers responded by introducing 'compliant' coatings that contained low VOCs, high solids or water as the primary solvent or solvent-free.

Since October 2005, European harmonisation of national regulations has led to the following VOC directives within the EU:

1. The VOC Solvent Emission Directive (1999/13/EC), also known as the Installations VOC Directive, is applicable to installations and installation components using volatile organic compounds
2. The Paints Directive (2004/42/EC), or the Products VOC Directive, covers the volatile organic compounds contained in paints and varnishes.

Under The Solvent Emissions Directive (SED), the responsibility for compliance falls very much with the end user or applicator of the paint coating. The specifier then needs to be aware of the types of coating and their application to ensure compliance with the requirements of the legislation and consultation with the paint/coating industry and applicator should help.

Relevant standards

Surface preparation

British Standards Institution BS 7079: 2009 *General introduction to standards for preparation of steel substrates before application of paints and related products.* London, BSI.

International Organization for Standardization BS EN ISO 8501 (4 parts) *Preparation of steel substrates before application of paints and related products. Visual assessment of surface cleanliness.* London, BSI.

International Organization for Standardization BS EN ISO 8502 (9 parts) *Preparation of steel substrates before application of paints and related products. Tests for the assessment of surface cleanliness.* London, BSI.

International Organization for Standardization BS EN ISO 8503 (5 parts) *Preparation of steel substrates before application of paints and related products. Surface roughness characteristics of blast-cleaned steel substrates.* London, BSI.

International Organization for Standardization BS EN ISO 8504 (3 parts) *Preparation of steel substrates before application of paints and related products. Surface preparation methods.* London, BSI.

International Organization for Standardization BS EN ISO 11124 (4 parts) *Preparation of steel substrates before application of paints and related products. Specifications for metallic blast-cleaning abrasives.* London, BSI.

International Organization for Standardization BS EN ISO 11125 (7 parts) *Preparation of steel substrates before application of paints and related products. Test methods for metallic blast-cleaning abrasives.* London, BSI.

International Organization for Standardization BS EN ISO 11126 (8 parts) *Preparation of steel substrates before application of paints and related products. Specifications for non-metallic blast cleaning abrasives.* London, BSI.

International Organization for Standardization BS EN ISO 11127 (7 parts) *Preparation of steel substrates before application of paints and related products. Test methods for non-metallic blast-cleaning abrasives.* London, BSI.

Paints

International Organization for Standardization BS EN ISO 12944 (8 parts) *Paints and varnishes – Corrosion protection of steel structures by protective paint systems.* London, BSI.

International Organization for Standardization BS EN ISO 4618: 2006 *Paints and Varnishes. Terms and definitions.* London, BSI.

British Standards Institution BS 1070: 1993 *Black paint (tar based).* London, BSI.

British Standards Institution BS 2015: 1992 *Glossary of paint and related terms.* London, BSI.

British Standards Institution BS 3416: 1991 *Bitumen based coatings for cold application, suitable for use in contact with potable water.* London, BSI.

British Standards Institution BS EN 10300: 2005 *Steel tubes and fittings for onshore and offshore pipelines. Bitumen hot applied materials for external coating.* London, BSI.

British Standards Institution BS 4164: 2002 *Coal tar based hot applied coating materials for protecting iron and steel products, including suitable primers where required.* London, BSI.

British Standards Institution BS 4652: 1995 *Zinc rich priming paint (organic media)*. London, BSI.

British Standards Institution BS 6949: 1991 *Bitumen based coatings for cold application excluding use in contact with potable water*. London, BSI.

Metallic coatings

International Organization for Standardization BS EN ISO 1461: 2009 *Hot dip galvanized coatings on fabricated iron and steel articles – Specifications and test methods*. London, BSI.

International Organization for Standardization BS EN ISO 14713-1: 2009 *Zinc coatings. Guidelines and recommendations for the protection against corrosion of iron and steel in structures. General principles of design and corrosion resistance*. London, BSI.

International Organization for Standardization BS EN ISO 14713-2: 2009 *Zinc coatings. Guidelines and recommendations for the protection against corrosion of iron and steel in structures. Hot dip galvanizing*. London, BSI.

International Organization for Standardization BS EN ISO 14713-3: 2009 *Zinc coatings. Guidelines and recommendations for the protection against corrosion of iron and steel in structures. Sherardizing*. London, BSI.

British Standards Institution (1988) BS 4921: 1988 *Specification for sherardised coatings on iron or steel*. London, BSI.

International Organization for Standardization BS EN ISO 2081: 2008 *Metallic and other inorganic coatings. Electroplated coatings of zinc with supplementary treatments on iron or steel*. London, BSI.

International Organization for Standardization BS EN ISO 2082: 2008 *Metallic coatings. Electroplated coatings of cadmium with supplementary treatments on iron or steel*. London, BSI.

British Standards Institution BS 3083: 1988 *Hot dip zinc coated, hot dip aluminium/ zinc coated corrugated steel sheets for general purposes*. London, BSI.

British Standards Institution BS EN 10346: 2009 *Continuously hot-dip coated steel flat products. Technical delivery conditions*. London, BSI.

International Organization for Standardization BS EN ISO 2063: 2005 *Thermal spraying. Metallic and other inorganic coatings. Zinc, aluminium and their alloys*. London, BSI.

British Standards Institution (1990) BS 4479: Part 7: 1990 *Design of articles that are to be coated. Recommendations for thermally sprayed coatings*. London, BSI.

British Standards Institution BS 7371-12: 2008 *Coatings on metal fasteners. Requirements for imperial fasteners*. London, BSI.

British Standards Institution BS PD 6484: 1979 *Commentary on corrosion at bimetallic contacts and its alleviation*. London, BSI.

Appendix

Steel Designers' Manual, Seventh Edition. Edited by Buick Davison and Graham W. Owens.
© 2012 Steel Construction Institute. Published 2012 by Blackwell Publishing Ltd.

Elastic properties

Modulus of elasticity (Young's modulus)	$E = 205 \text{kN}/\text{mm}^2$
Poisson's ratio	$v = 0.30$
Coefficient of linear thermal expansion	$\alpha = 12 \times 10^{-6}{}^{\circ}\text{C}$

European standards for structural steels

Introduction

As part of the exercise towards the removal of technical barriers to trade, the European Committee for Iron and Steel Standardization (ECISS) has prepared a series of European Standards (ENs) for structural steels.

The changes since the 6th Edition of the Steel Designers' Manual are listed below.

BS EN 10025:1993 superseded by BS EN 10025 in the following parts:
   BS EN 10025-1:2004 Hot rolled products of structural steels. General technical delivery conditions
   BS EN 10025-2:2004 Hot rolled products of structural steels. Technical delivery conditions for non-alloy structural steels
   BS EN 10025-3:2004 Hot rolled products of structural steels. Technical delivery conditions for normalized/normalized rolled weldable fine grain structural steels
   BS EN 10025-4:2004 Hot rolled products of structural steels. Technical delivery conditions for thermomechanical rolled weldable fine grain structural steels
   BS EN 10025-5:2004 Hot rolled products of structual steels. Technical delivery conditons for structural steels with improved atmospheric corrosion resistance
   BS EN 10025-6:2004+A1:2009 Hot rolled products of structural steels. Technical delivery conditions for flat products of high yield strength structural steels in the quenched and tempered condition
BS EN 10113-1:1993 title withdrawn, replaced by parts of BS EN 10025
BS EN 10137-1:1996 title withdrawn, replaced by parts of BS EN 10025
BS EN 10137-2:1996 title withdrawn, replaced by parts of BS EN 10025
BS EN 10137-3:1996 title withdrawn 01 December 2004
BS EN 10155:1993 title withdrawn, replaced by parts of BS EN 10025
BS EN 10210:1994 title superseded by 2006 versions (2 parts)
BS 7668:1994 title replaced by BS 7668:2004 Weldable structural steels. Hot finished structural hollow sections in weather resistant steels. Specification

Table 1 summarizes the European and British standards which have superseded BS 4360.

Designation systems

The designation systems used in the EN are in accordance with EN 10027: Parts 1 and 2, together with ECISS Information Circular IC 10 (published by BSI as DD 214). These designations are totally different from the familiar BS 4360 designations. Tables 2a to 2d are intended to help users understand them.

Table 1 European and British Standards which have superseded BS 4360

Standard	Superseded BS 4360 grades
BS EN 10025: 1993	40 A, B, C, D; 43 A, B, C, D; 50 A, B, C, D, DD
BS EN 10113: Parts 1, 2 & 3: 1993	40 DD, E, EE; 43 DD, E, EE; 50 E, EE; 55 C, EE
BS EN 10137: Parts 1, 2 & 3: 1996	50 F and 55 F
BS EN 10155: 1993	WR 50 A, B, C
BS EN 10210: Part 1: 1994	Hot-finished structural hollow section grades – excluding weather resistant grades
BS 7668: 1994	Hot-finished weather resistant hollow section grades

Table 2a Symbols used in EN 10025

S. . . .	Structural steel
E. . . .	Engineering steel
.235 . . .	Minimum yield strength (R.) in N/mm2 @ 16mm
. . . JR..	Longitudinal Charpy V-notch impacts 27J @ +20°C
. . . J0..	Longitudinal Charpy V-notch impacts 27J @ 0°C
. . . J2..	Longitudinal Charpy V-notch impacts 27J @ –20°C
. . . K2..	Longitudinal Charpy V-notch impacts 40J @ –20°C
. . . . G1	Rimming steel (FU)
. . . . G2	Rimming steel not permitted (FN)
. . . . G3	FLAT products: Supply condition 'N', i.e. normalized or normalized rolled. LONG products: Supply condition at manufacturer's discretion
. . . . G4	ALL products: Supply condition at manufacturer's discretion

Examples: S235JRG1, S355K2G4

Table 2b Symbols used in EN 10155

S. . . .	Structural steel
.235 . . .	Minimum yield strength (R.) in N/mm2 @ 16mm
. . . J0..	Longitudinal Charpy V-notch impacts 27J @ 0°C
. . . J2..	Longitudinal Charpy V-notch impacts 27J @ –20°C
. . . K2..	Longitudinal Charpy V-notch impacts 40J @ –20°C
. . . . G1	FLAT products: Supply condition 'N', i.e. normalized or normalized rolled. LONG products: Supply condition at manufacturer's discretion
. . . . G2	All products: Supply condition at manufacturer's discretion
. . . . W	Weather resistant steel
. . . . P	High phosphorus grade

Examples: S235J0WP, S355K2G2W

Table 2c   Symbols used in EN 10113

S....	Structural steel
. 275..	Minimum yield strength (R.) in N/mm2 @ 16 mm
. . . . N.	Normalized or normalized rolled
. . . . M.	Thermomechanically rolled
. L	Charpy V-notch impacts down to −50°C

Examples: S275N, S355ML

Table 2d   Symbols used in EN 10137

S. . . .	Structural steel
.460..	Minimum yield strength (R.) in N/mm2 @ 16 mm
. . . . Q.	Quenched and tempered
. L	Charpy V-notch impacts down to −40°C
. . . . .L1	Charpy V-notch impacts down to −60°C

Examples: S460QL, S620QL1

Bending moment, shear and deflection

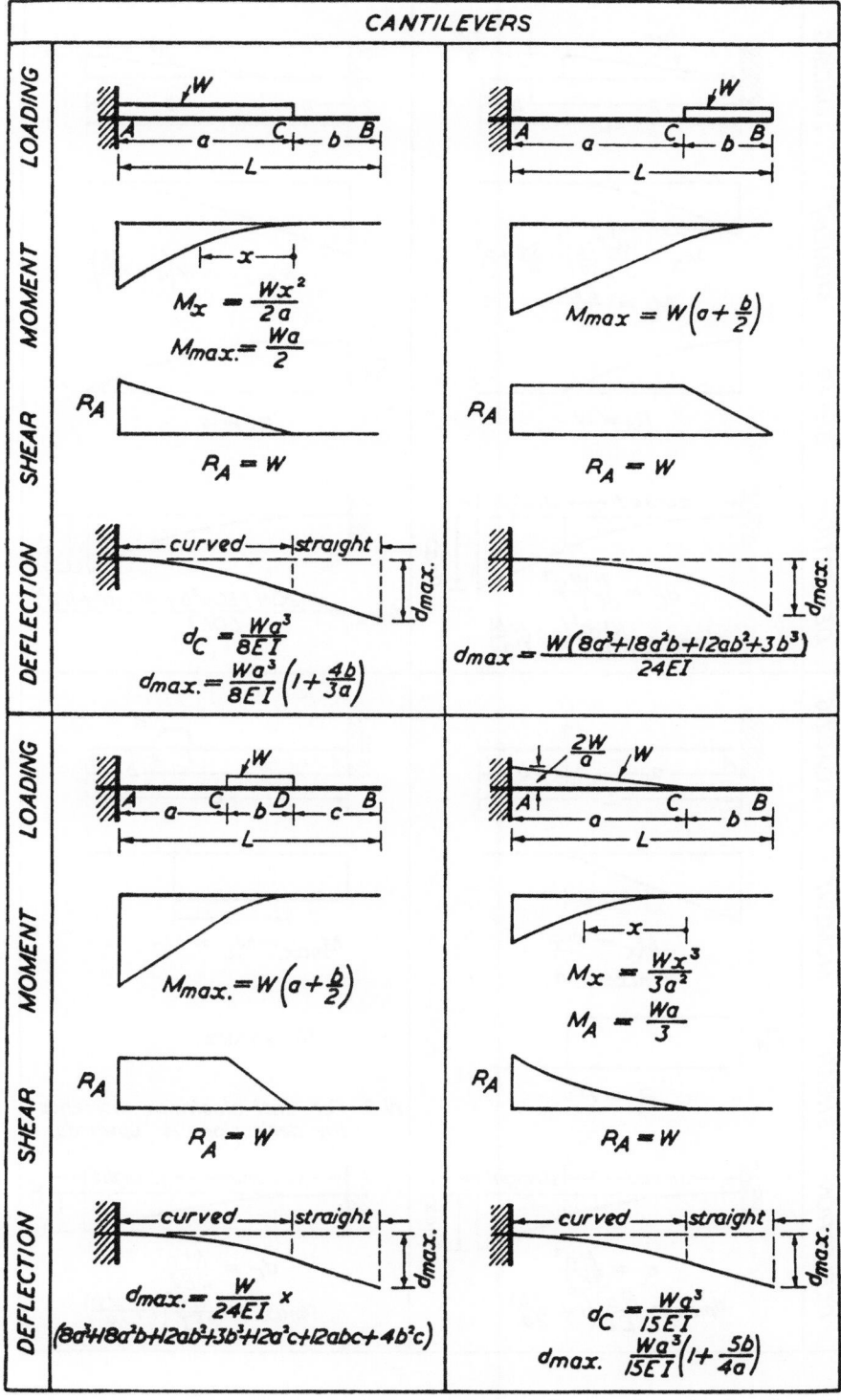

CANTILEVERS

LOADING

MOMENT

$$M_x = \frac{Wx^2}{2a}$$

$$M_{max.} = \frac{Wa}{2}$$

$$M_{max} = W\left(a + \frac{b}{2}\right)$$

SHEAR

$$R_A = W$$

$$R_A = W$$

DEFLECTION

curved | straight | $d_{max.}$

$$d_C = \frac{Wa^3}{8EI}$$

$$d_{max.} = \frac{Wa^3}{8EI}\left(1 + \frac{4b}{3a}\right)$$

$$d_{max} = \frac{W\left(8a^3 + 18a^2b + 12ab^2 + 3b^3\right)}{24EI}$$

LOADING

MOMENT

$$M_{max.} = W\left(a + \frac{b}{2}\right)$$

$$M_x = \frac{Wx^3}{3a^2}$$

$$M_A = \frac{Wa}{3}$$

SHEAR

$$R_A = W$$

$$R_A = W$$

DEFLECTION

curved | straight | $d_{max.}$

$$d_{max.} = \frac{W}{24EI} \times$$

$$(8a^3 + 18a^2b + 12ab^2 + 3b^3 + 12a^2c + 12abc + 4b^2c)$$

$$d_C = \frac{Wa^3}{15EI}$$

$$d_{max.} \frac{Wa^3}{15EI}\left(1 + \frac{5b}{4a}\right)$$

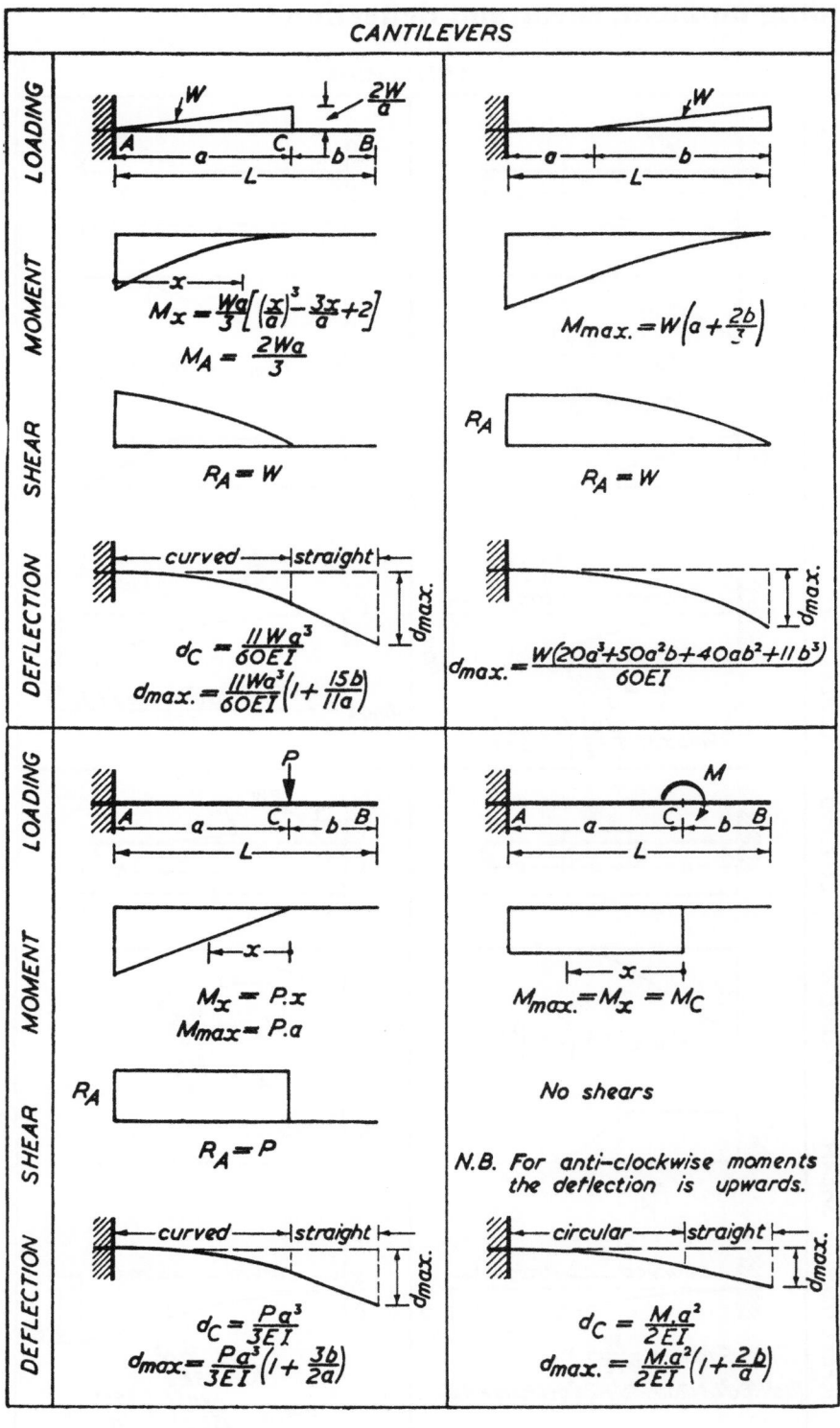

CANTILEVERS

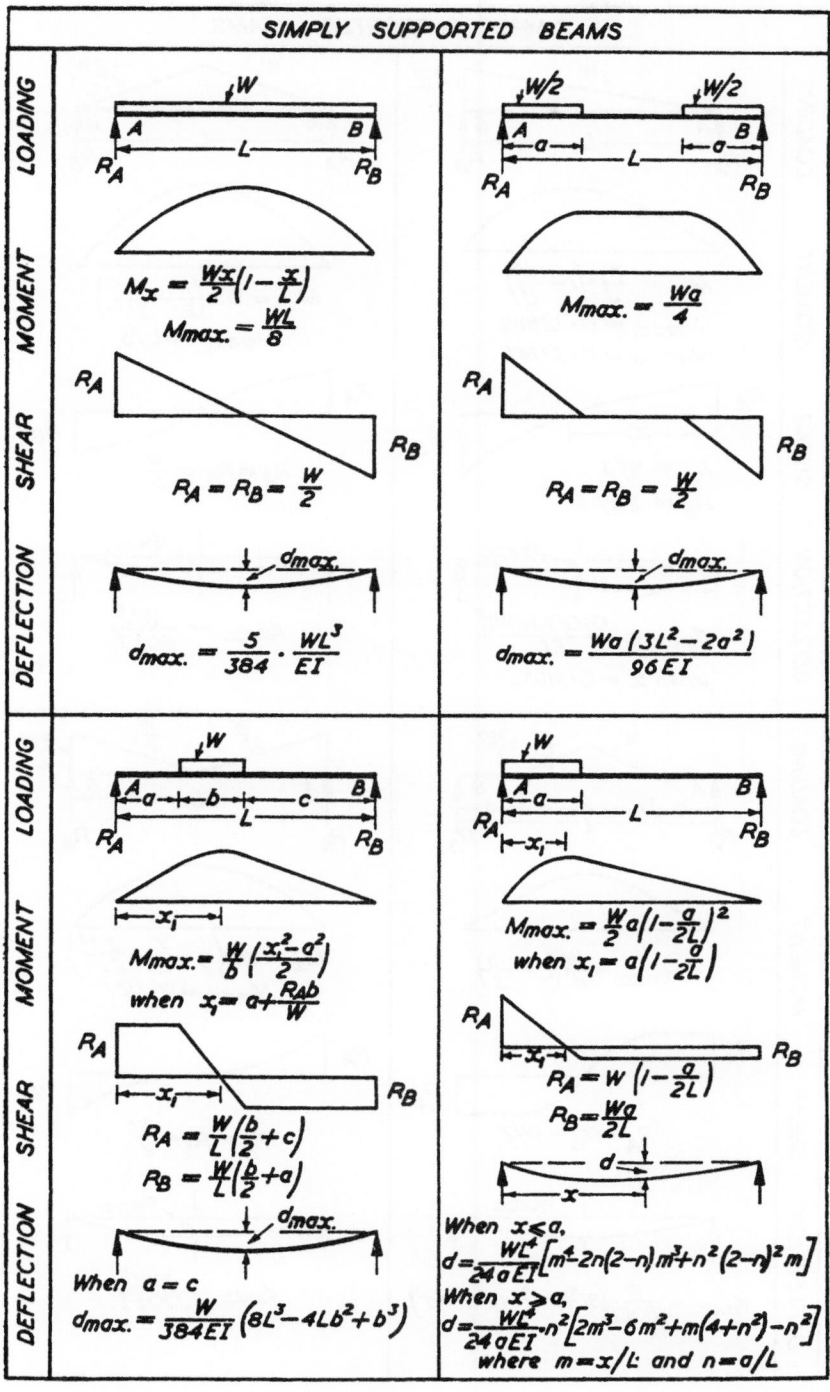

SIMPLY SUPPORTED BEAMS

LOADING / MOMENT / SHEAR / DEFLECTION (left-top panel)

$$M_x = \frac{Wx}{2}\left(1 - \frac{x}{L}\right)$$

$$M_{max.} = \frac{WL}{8}$$

$$R_A = R_B = \frac{W}{2}$$

$$d_{max.} = \frac{5}{384} \cdot \frac{WL^3}{EI}$$

(right-top panel)

$$M_{max.} = \frac{Wa}{4}$$

$$R_A = R_B = \frac{W}{2}$$

$$d_{max.} = \frac{Wa(3L^2 - 2a^2)}{96EI}$$

(left-bottom panel)

$$M_{max.} = \frac{W}{b}\left(\frac{x_1^2 - a^2}{2}\right)$$

$$\text{when } x_1 = a + \frac{R_A b}{W}$$

$$R_A = \frac{W}{L}\left(\frac{b}{2} + c\right)$$

$$R_B = \frac{W}{L}\left(\frac{b}{2} + a\right)$$

When $a = c$,
$$d_{max.} = \frac{W}{384EI}\left(8L^3 - 4Lb^2 + b^3\right)$$

(right-bottom panel)

$$M_{max.} = \frac{W}{2}a\left(1 - \frac{a}{2L}\right)^2$$

$$\text{when } x_1 = a\left(1 - \frac{a}{2L}\right)$$

$$R_A = W\left(1 - \frac{a}{2L}\right)$$

$$R_B = \frac{Wa}{2L}$$

When $x \leqslant a$,
$$d = \frac{WL^3}{24aEI}\left[m^4 - 2n(2-n)m^3 + n^2(2-n)^2 m\right]$$

When $x > a$,
$$d = \frac{WL^3}{24aEI} \cdot n^2\left[2m^3 - 6m^2 + m(4+n^2) - n^2\right]$$
$$\text{where } m = x/L \text{ and } n = a/L$$

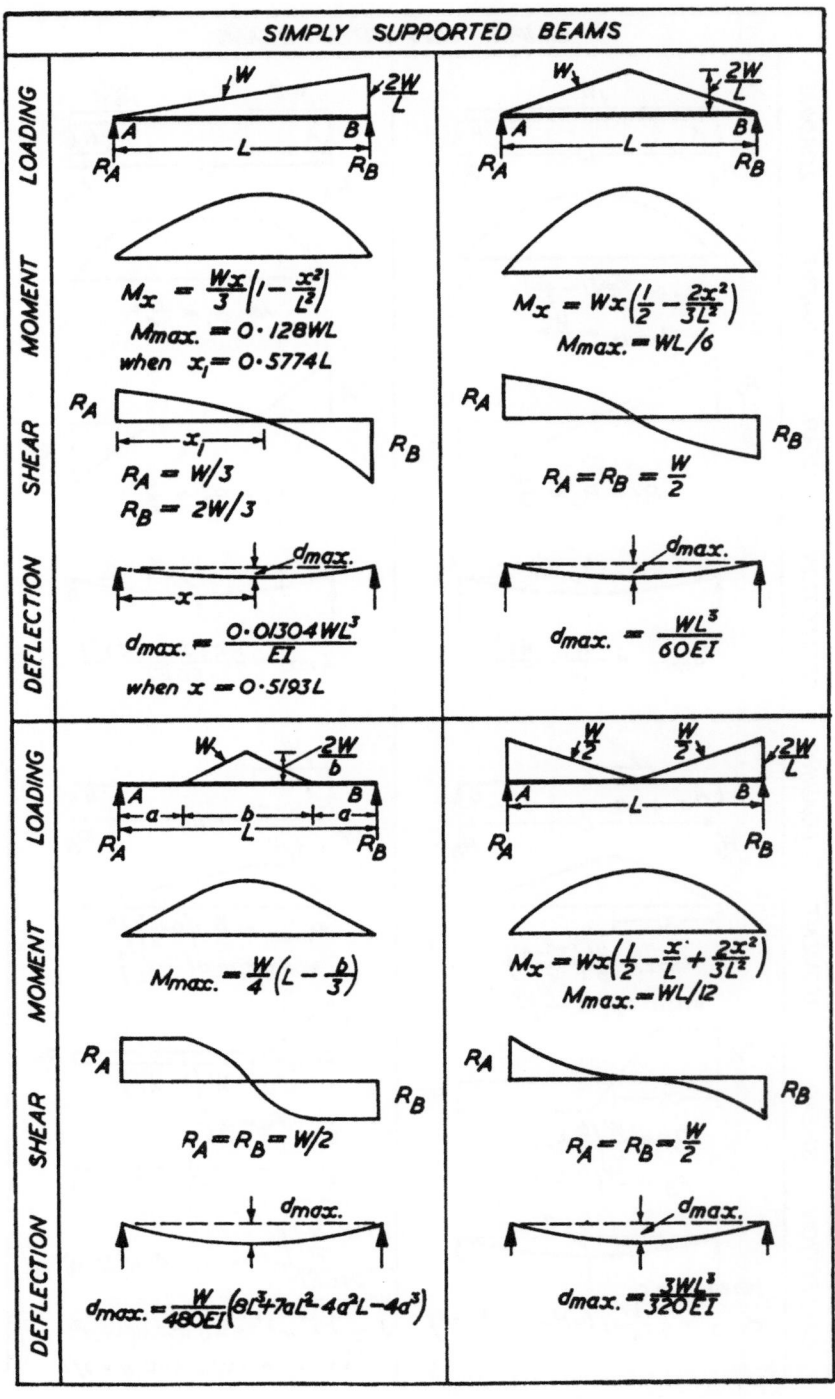

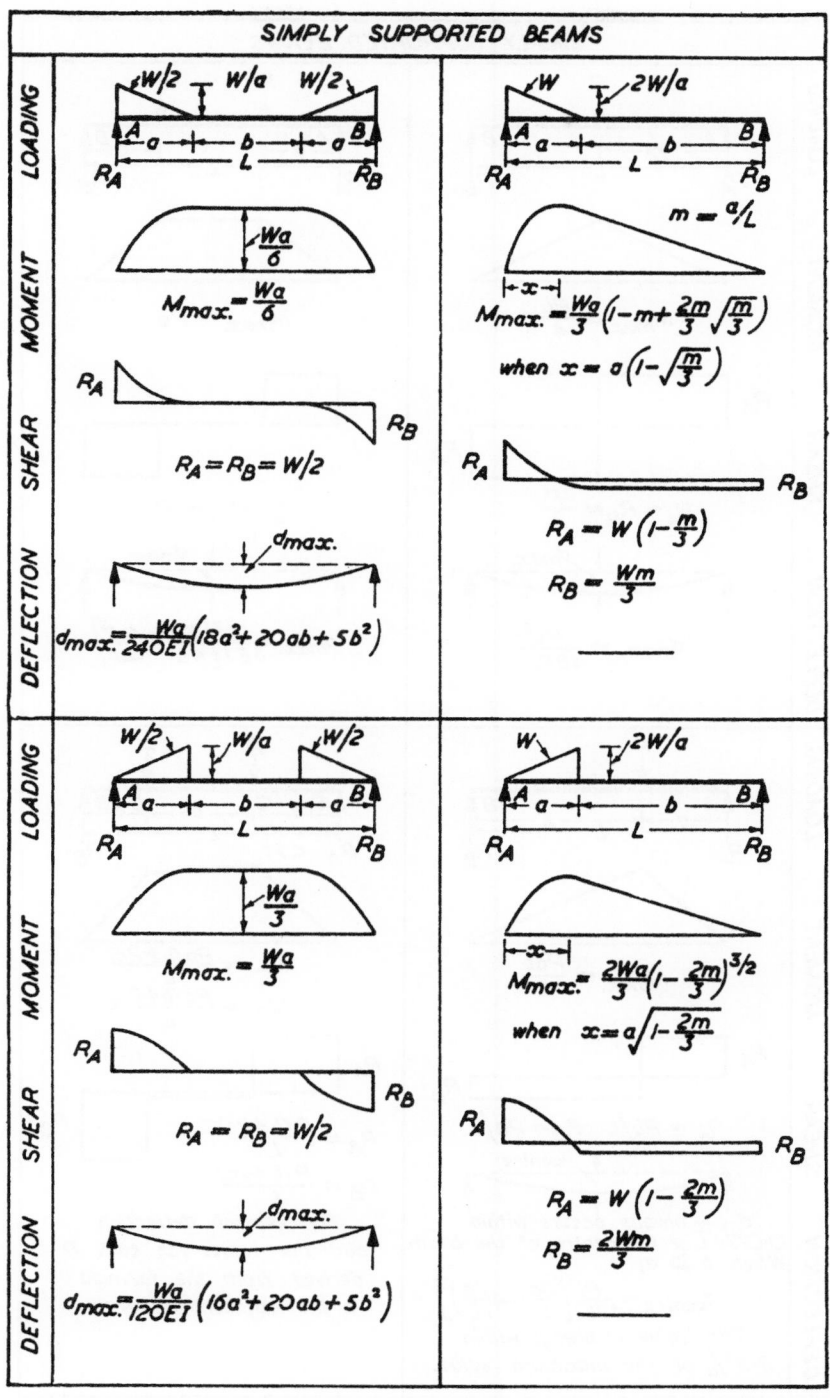

SIMPLY SUPPORTED BEAMS

First quadrant (top-left):

$M_{max.} = \dfrac{Wa}{6}$

$R_A = R_B = W/2$

$d_{max.} = \dfrac{Wa}{240EI}(18a^2 + 20ab + 5b^2)$

Second quadrant (top-right):

$m = a/L$

$M_{max.} = \dfrac{Wa}{3}\left(1 - m + \dfrac{2m}{3}\sqrt{\dfrac{m}{3}}\right)$

when $x = a\left(1 - \sqrt{\dfrac{m}{3}}\right)$

$R_A = W\left(1 - \dfrac{m}{3}\right)$

$R_B = \dfrac{Wm}{3}$

Third quadrant (bottom-left):

$M_{max.} = \dfrac{Wa}{3}$

$R_A = R_B = W/2$

$d_{max.} = \dfrac{Wa}{120EI}(16a^2 + 20ab + 5b^2)$

Fourth quadrant (bottom-right):

$M_{max.} = \dfrac{2Wa}{3}\left(1 - \dfrac{2m}{3}\right)^{3/2}$

when $x = a\sqrt{1 - \dfrac{2m}{3}}$

$R_A = W\left(1 - \dfrac{2m}{3}\right)$

$R_B = \dfrac{2Wm}{3}$

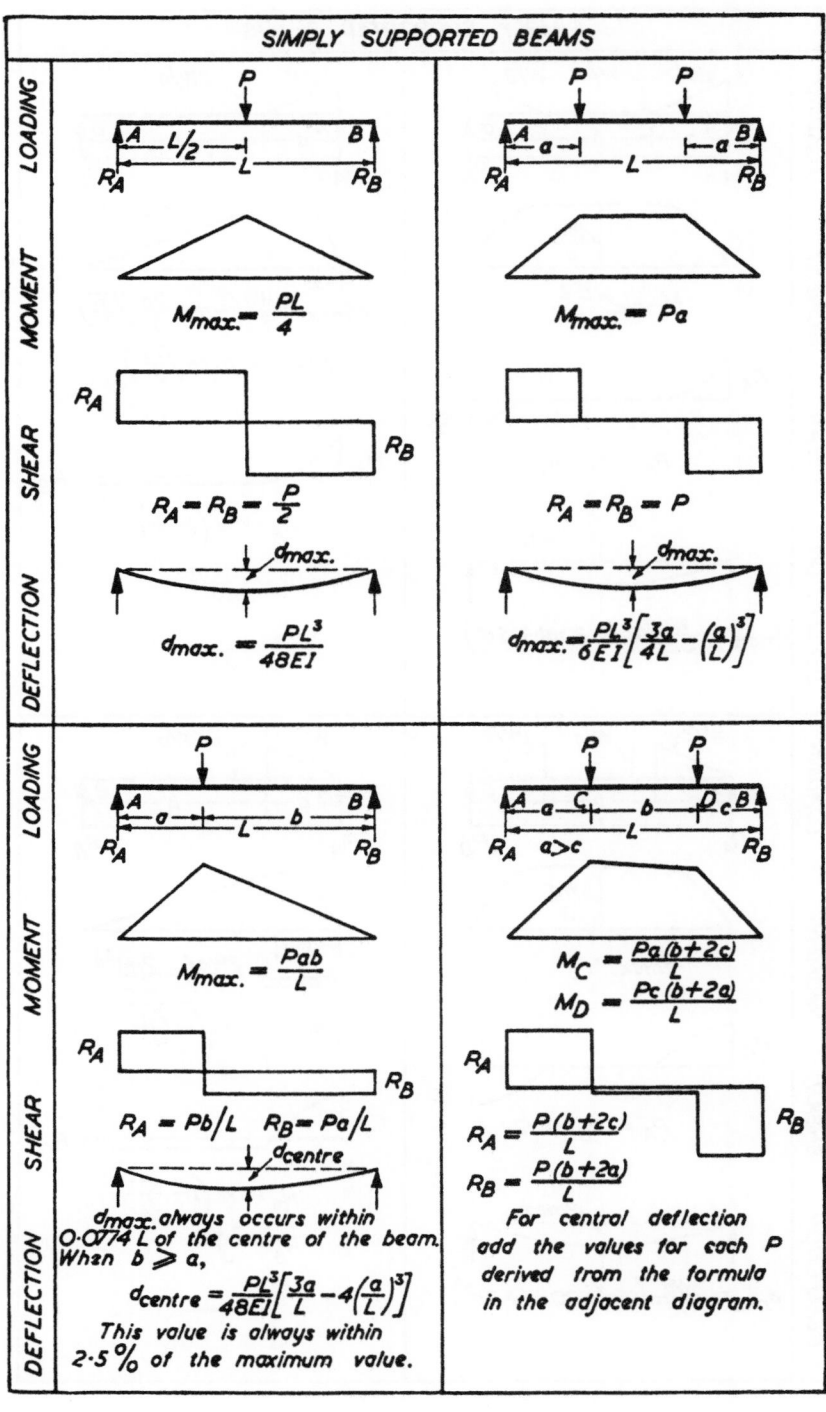

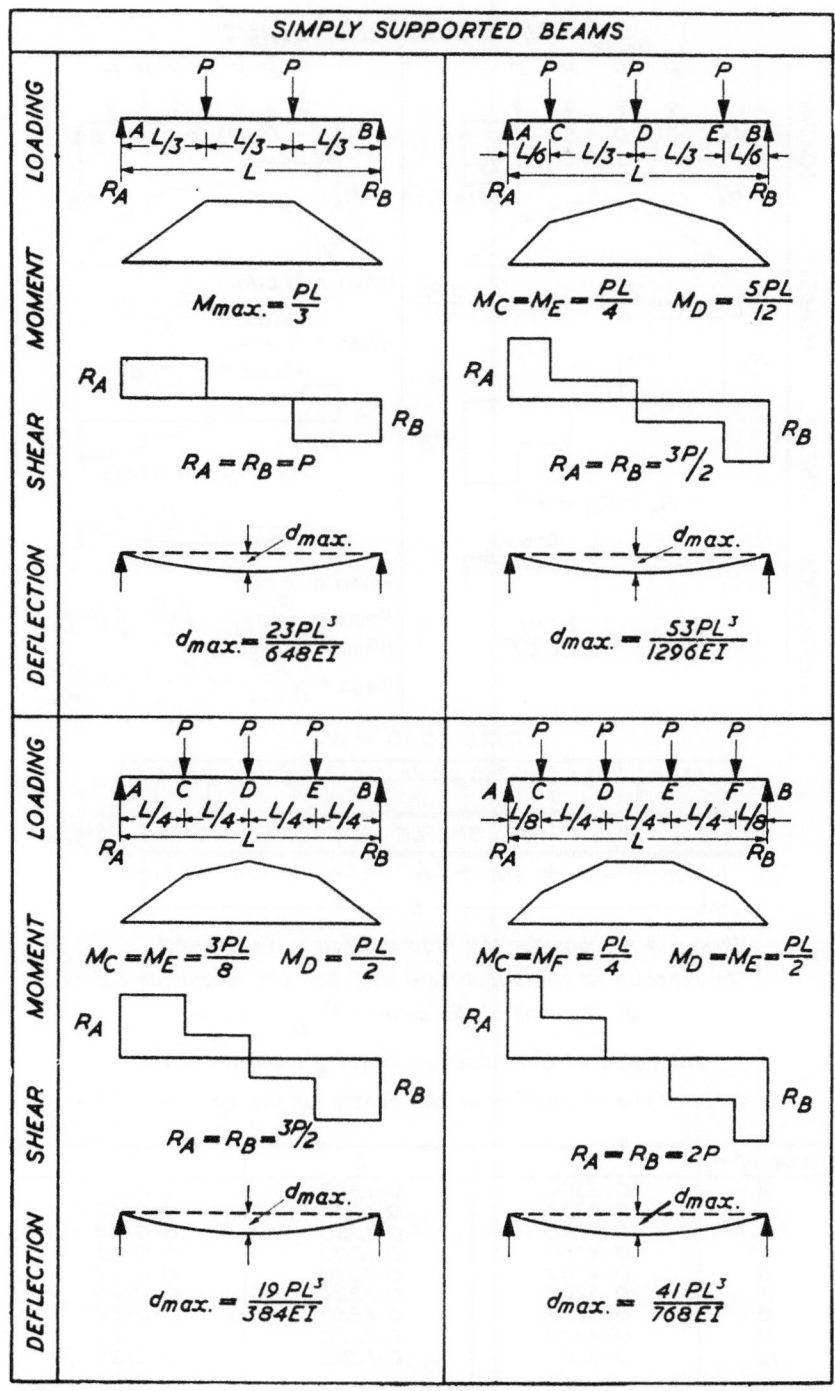

SIMPLY SUPPORTED BEAMS
LOADING
MOMENT
SHEAR
DEFLECTION
P P
A L/3 L/3 L/3 B
L
RA RB
Mmax. = PL/3
RA RB
RA = RB = P
dmax.
dmax. = 23PL³/648EI
P P P
A C D E B
L/6 L/3 L/3 L/6
L
RA RB
MC = ME = PL/4 MD = 5PL/12
RA RB
RA = RB = 3P/2
dmax.
dmax. = 53PL³/1296EI
P P P
A C D E B
L/4 L/4 L/4 L/4
L
RA RB
MC = ME = 3PL/8 MD = PL/2
RA RB
RA = RB = 3P/2
dmax.
dmax. = 19PL³/384EI
P P P P
A C D E F B
L/8 L/4 L/4 L/4 L/8
L
RA RB
MC = MF = PL/4 MD = ME = PL/2
RA RB
RA = RB = 2P
dmax.
dmax. = 41PL³/768EI

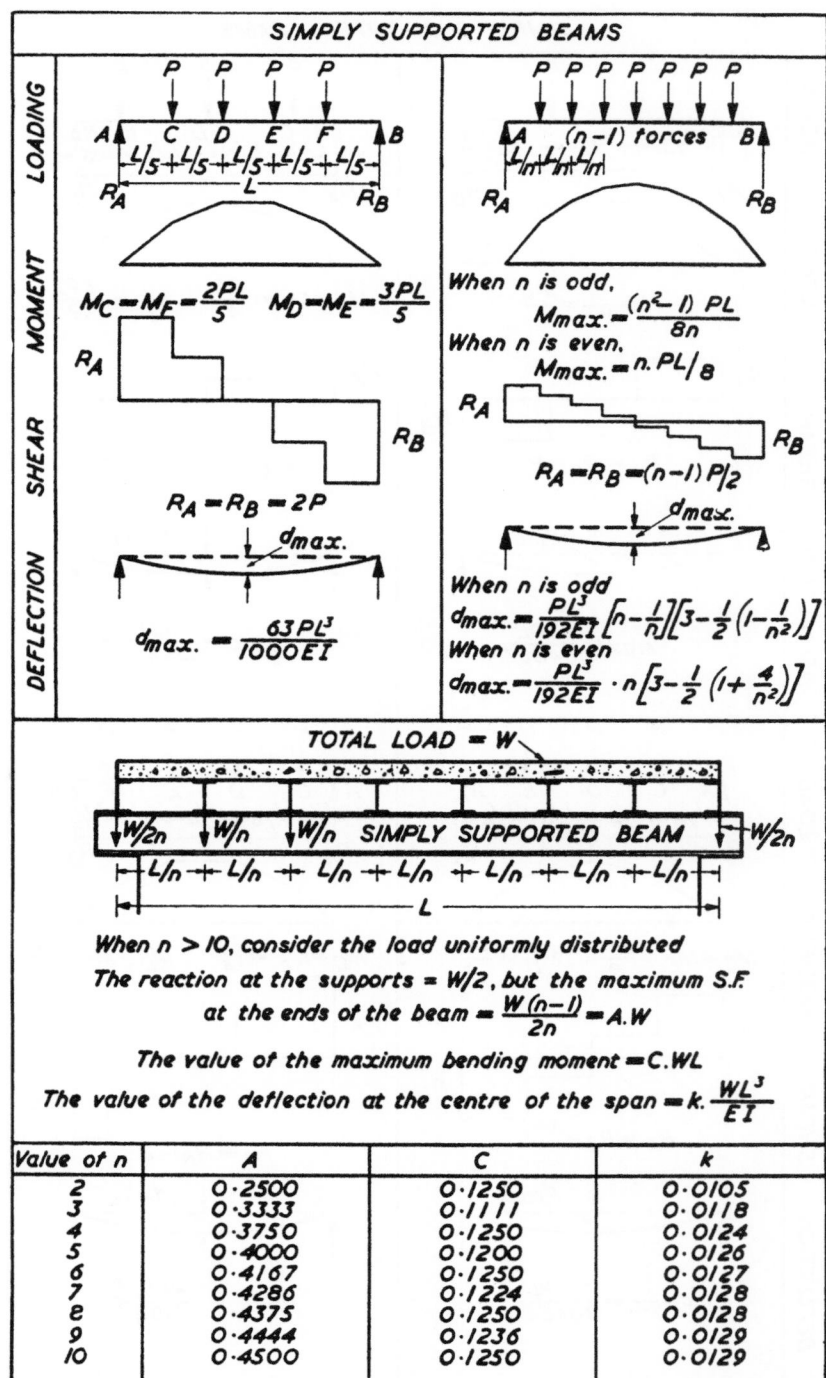

SIMPLY SUPPORTED BEAMS

LOADING

MOMENT

$M_C = M_F = \dfrac{2PL}{5}$ $M_D = M_E = \dfrac{3PL}{5}$

When n is odd,
$$M_{max.} = \frac{(n^2 - 1)\,PL}{8n}$$
When n is even,
$$M_{max.} = n.PL/8$$

SHEAR

$R_A = R_B = 2P$

$R_A = R_B = (n-1)\,P/2$

DEFLECTION

$d_{max.}$

$$d_{max.} = \frac{63\,PL^3}{1000\,EI}$$

When n is odd
$$d_{max.} = \frac{PL^3}{192EI}\left[n - \frac{1}{n}\right]\left[3 - \frac{1}{2}\left(1 - \frac{1}{n^2}\right)\right]$$
When n is even
$$d_{max.} = \frac{PL^3}{192EI}\cdot n\left[3 - \frac{1}{2}\left(1 + \frac{4}{n^2}\right)\right]$$

TOTAL LOAD = W

$\downarrow W/2n$ $\downarrow W/n$ $\downarrow W/n$ SIMPLY SUPPORTED BEAM $\downarrow W/2n$

When $n > 10$, consider the load uniformly distributed

The reaction at the supports $= W/2$, but the maximum S.F.
at the ends of the beam $= \dfrac{W(n-1)}{2n} = A.W$

The value of the maximum bending moment $= C.WL$

The value of the deflection at the centre of the span $= k.\dfrac{WL^3}{EI}$

Value of n	A	C	k
2	0·2500	0·1250	0·0105
3	0·3333	0·1111	0·0118
4	0·3750	0·1250	0·0124
5	0·4000	0·1200	0·0126
6	0·4167	0·1250	0·0127
7	0·4286	0·1224	0·0128
8	0·4375	0·1250	0·0128
9	0·4444	0·1236	0·0129
10	0·4500	0·1250	0·0129

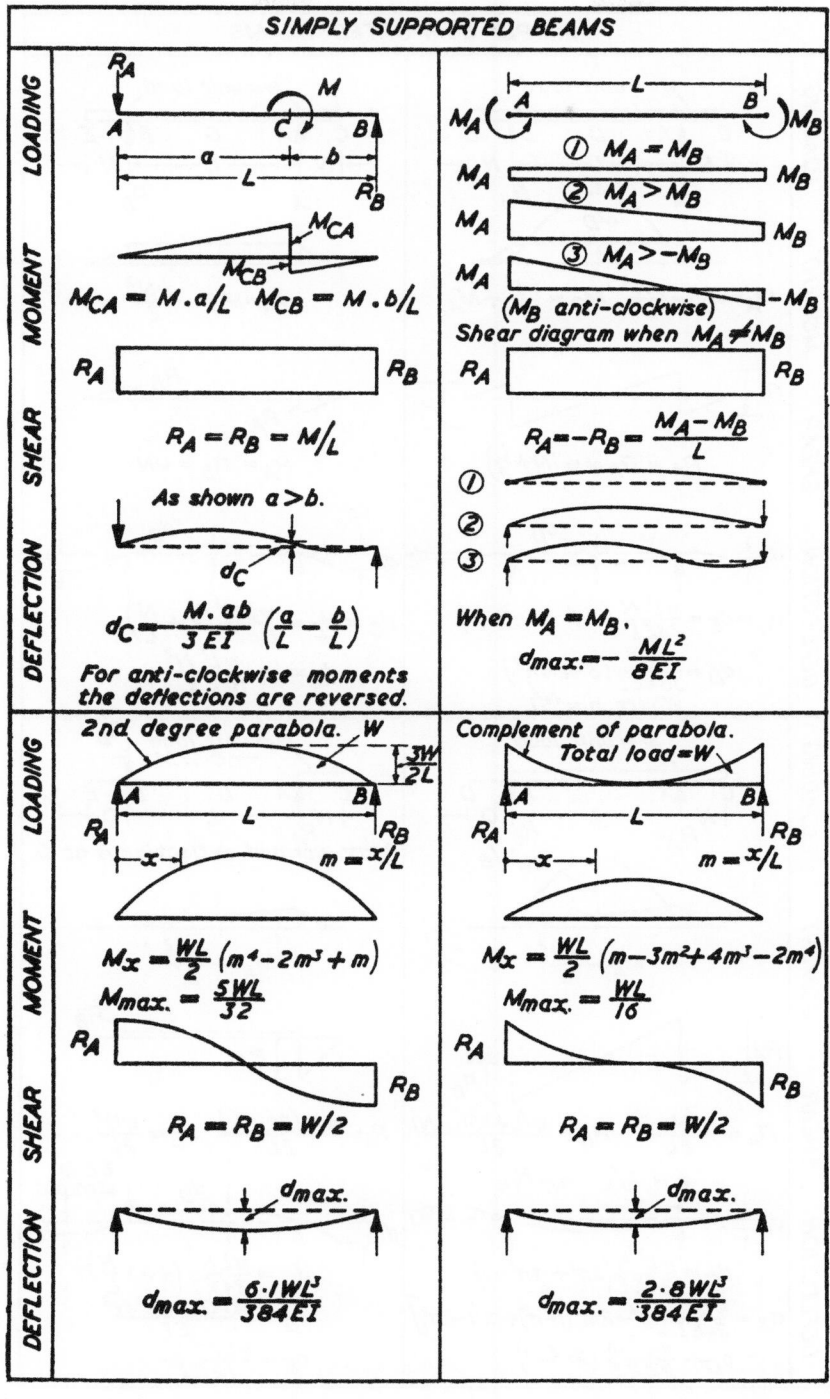

SIMPLY SUPPORTED BEAMS

LOADING

$M_{CA} = M \cdot a/L \qquad M_{CB} = M \cdot b/L$

① $M_A = M_B$

② $M_A > M_B$

③ $M_A > -M_B$

(M_B anti-clockwise)

Shear diagram when $M_A \neq M_B$

MOMENT

SHEAR

$R_A = R_B = M/L$

$R_A = -R_B = \dfrac{M_A - M_B}{L}$

DEFLECTION

As shown $a > b$.

$d_C = \dfrac{M \cdot ab}{3EI}\left(\dfrac{a}{L} - \dfrac{b}{L}\right)$

For anti-clockwise moments the deflections are reversed.

When $M_A = M_B$,

$d_{max.} = -\dfrac{ML^2}{8EI}$

LOADING

2nd degree parabola. W

$\dfrac{3W}{2L}$

$m = x/L$

Complement of parabola. Total load = W

$m = x/L$

MOMENT

$M_x = \dfrac{WL}{2}(m^4 - 2m^3 + m)$

$M_{max.} = \dfrac{5WL}{32}$

$M_x = \dfrac{WL}{2}(m - 3m^2 + 4m^3 - 2m^4)$

$M_{max.} = \dfrac{WL}{16}$

SHEAR

$R_A = R_B = W/2$

$R_A = R_B = W/2$

DEFLECTION

$d_{max.}$

$d_{max.} = \dfrac{6 \cdot 1 WL^3}{384EI}$

$d_{max.}$

$d_{max.} = \dfrac{2 \cdot 8 WL^3}{384EI}$

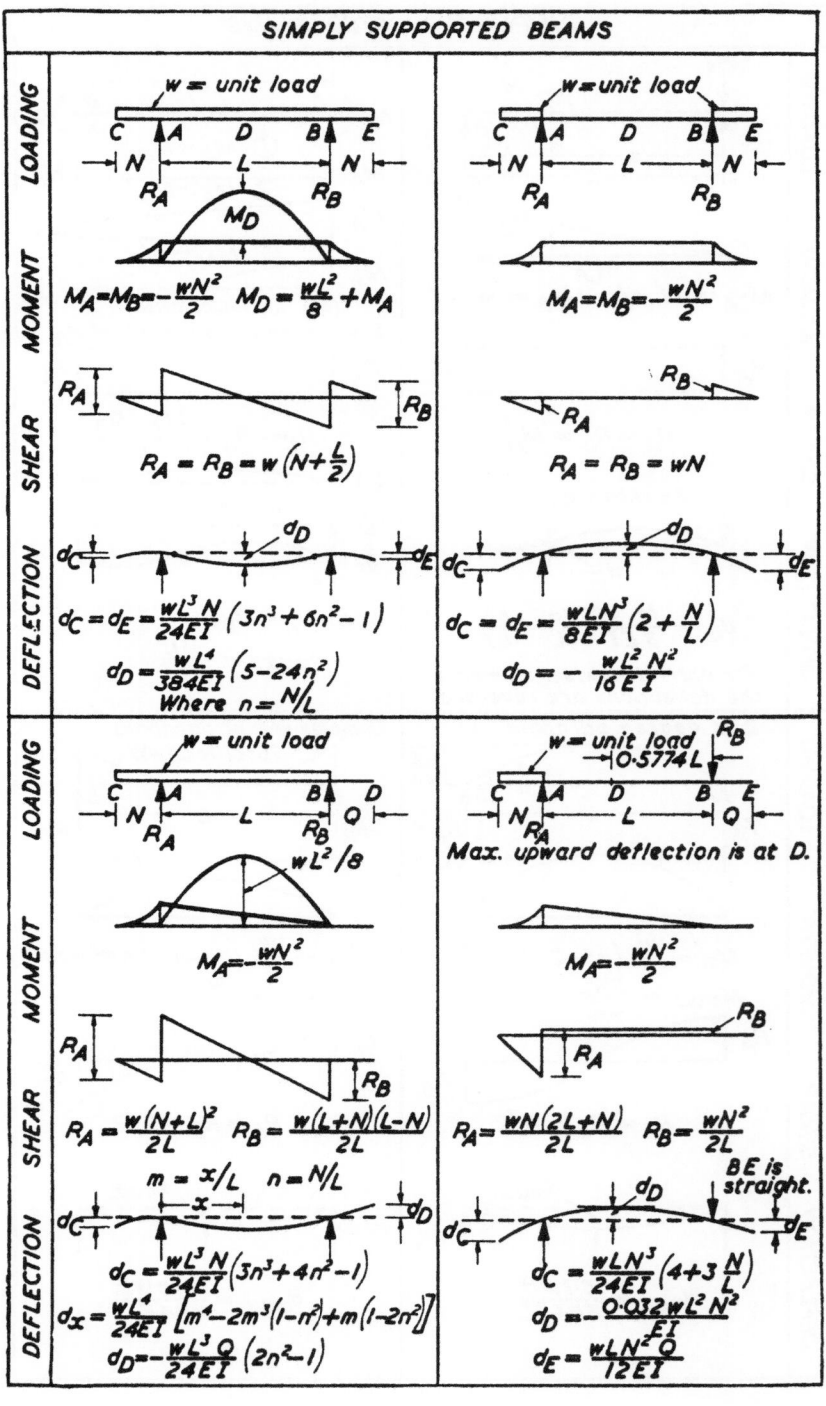

SIMPLY SUPPORTED BEAMS

LOADING (top left)

$w = $ unit load

R_A R_B

MOMENT

$M_A = M_B = -\dfrac{wN^2}{2}$ $M_D = \dfrac{wL^2}{8} + M_A$

SHEAR

$R_A = R_B = w\left(N + \dfrac{L}{2}\right)$

DEFLECTION

$d_C = d_E = \dfrac{wL^3 N}{24EI}\left(3n^3 + 6n^2 - 1\right)$

$d_D = \dfrac{wL^4}{384EI}\left(5 - 24n^2\right)$
Where $n = N/L$

LOADING (top right)

$w = $ unit load

R_A R_B

MOMENT

$M_A = M_B = -\dfrac{wN^2}{2}$

SHEAR

$R_A = R_B = wN$

DEFLECTION

$d_C = d_E = \dfrac{wLN^3}{8EI}\left(2 + \dfrac{N}{L}\right)$

$d_D = -\dfrac{wL^2 N^2}{16EI}$

LOADING (bottom left)

$w = $ unit load

R_A R_B

MOMENT

$wL^2/8$

$M_A = -\dfrac{wN^2}{2}$

SHEAR

$R_A = \dfrac{w(N+L)^2}{2L}$ $R_B = \dfrac{w(L+N)(L-N)}{2L}$

DEFLECTION

$m = x/L$ $n = N/L$

$d_C = \dfrac{wL^3 N}{24EI}\left(3n^3 + 4n^2 - 1\right)$

$d_x = \dfrac{wL^4}{24EI}\left[m^4 - 2m^3(1-n^2) + m(1-2n^2)\right]$

$d_D = -\dfrac{wL^3 Q}{24EI}\left(2n^2 - 1\right)$

LOADING (bottom right)

$w = $ unit load
$0.5774 L$ R_B

R_A

Max. upward deflection is at D.

MOMENT

$M_A = -\dfrac{wN^2}{2}$

SHEAR

$R_A = \dfrac{wN(2L+N)}{2L}$ $R_B = \dfrac{wN^2}{2L}$

DEFLECTION

BE is straight.

$d_C = \dfrac{wLN^3}{24EI}\left(4 + 3\dfrac{N}{L}\right)$

$d_D = -\dfrac{0.032\, wL^2 N^2}{EI}$

$d_E = \dfrac{wLN^2 Q}{12EI}$

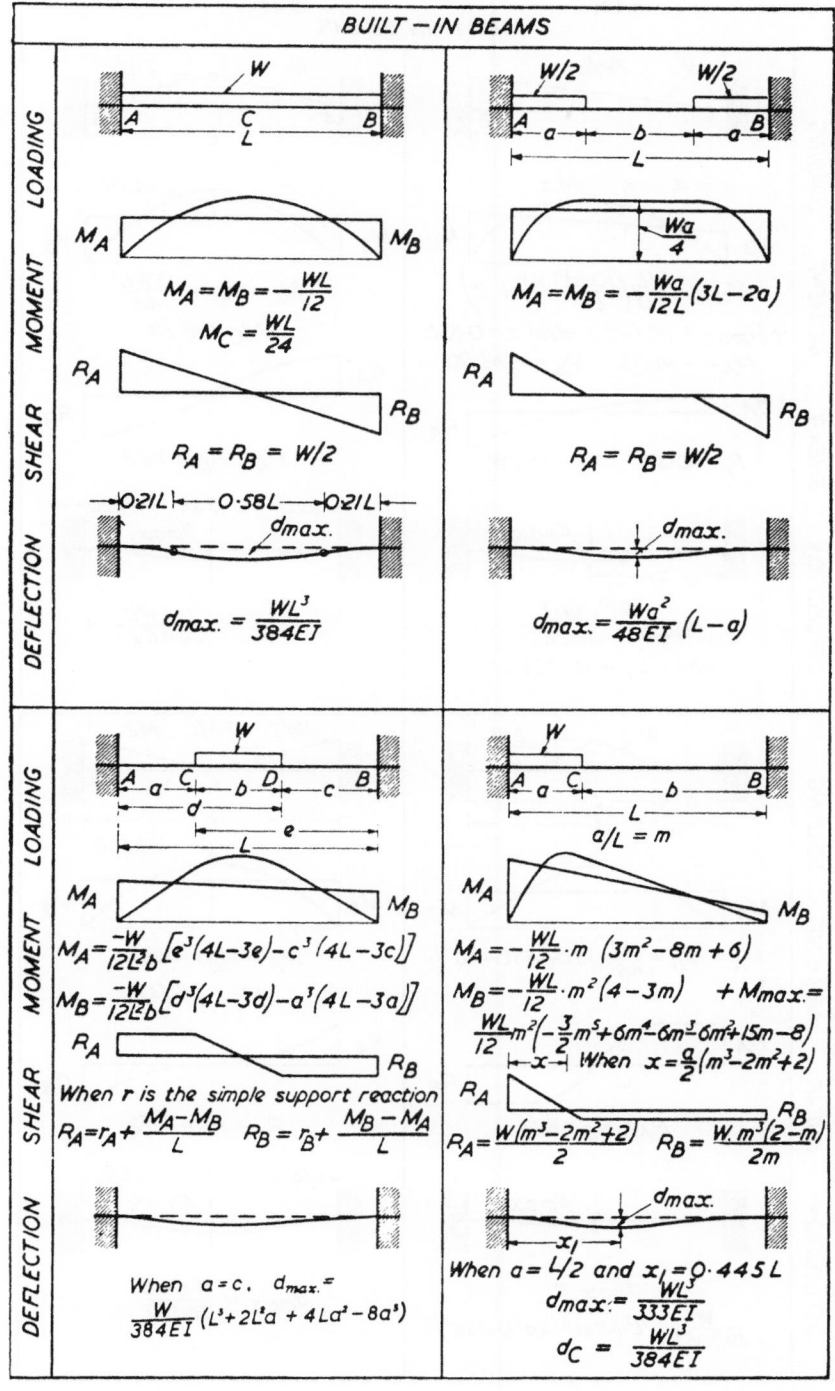

BUILT-IN BEAMS

LOADING

W
A — C — B, L

$W/2$ $W/2$
A — a — b — a — B, L

MOMENT

M_A M_B

$$M_A = M_B = -\frac{WL}{12}$$
$$M_C = \frac{WL}{24}$$

M_A $\frac{Wa}{4}$ M_B

$$M_A = M_B = -\frac{Wa}{12L}(3L - 2a)$$

SHEAR

R_A R_B

$$R_A = R_B = W/2$$

R_A R_B

$$R_A = R_B = W/2$$

DEFLECTION

$\vdash 0.21L \vdash$ — $0.58L$ — $\vdash 0.21L \vdash$
$d_{max.}$

$$d_{max.} = \frac{WL^3}{384EI}$$

$d_{max.}$

$$d_{max.} = \frac{Wa^2}{48EI}(L - a)$$

LOADING

W
A — a — C — b — D — c — B
d
e
L

W
A — a — C — b — B
L
$a/L = m$

MOMENT

M_A M_B

$$M_A = \frac{-W}{12L^2 b}\left[e^3(4L - 3e) - c^3(4L - 3c)\right]$$
$$M_B = \frac{-W}{12L^2 b}\left[d^3(4L - 3d) - a^3(4L - 3a)\right]$$

M_A M_B

$$M_A = -\frac{WL}{12} \cdot m\ (3m^2 - 8m + 6)$$
$$M_B = -\frac{WL}{12} \cdot m^2(4 - 3m)\ \ + M_{max.} =$$
$$\frac{WL}{12}m^2\left(-\frac{3}{2}m^5 + 6m^4 - 6m^3 + 6m^2 + 15m - 8\right)$$
$\vdash x \dashv$ When $x = \frac{a}{2}(m^3 - 2m^2 + 2)$

SHEAR

R_A R_B

When r is the simple support reaction
$$R_A = r_A + \frac{M_A - M_B}{L} \qquad R_B = r_B + \frac{M_B - M_A}{L}$$

R_A R_B

$$R_A = \frac{W(m^3 - 2m^2 + 2)}{2} \qquad R_B = \frac{W.m^3(2 - m)}{2m}$$

DEFLECTION

When $a = c$. $d_{max.} =$
$$\frac{W}{384EI}(L^3 + 2L^2 a + 4La^2 - 8a^3)$$

$d_{max.}$
x_1
When $a = L/2$ and $x_1 = 0.445L$
$$d_{max.} = \frac{WL^3}{333EI}$$
$$d_C = \frac{WL^3}{384EI}$$

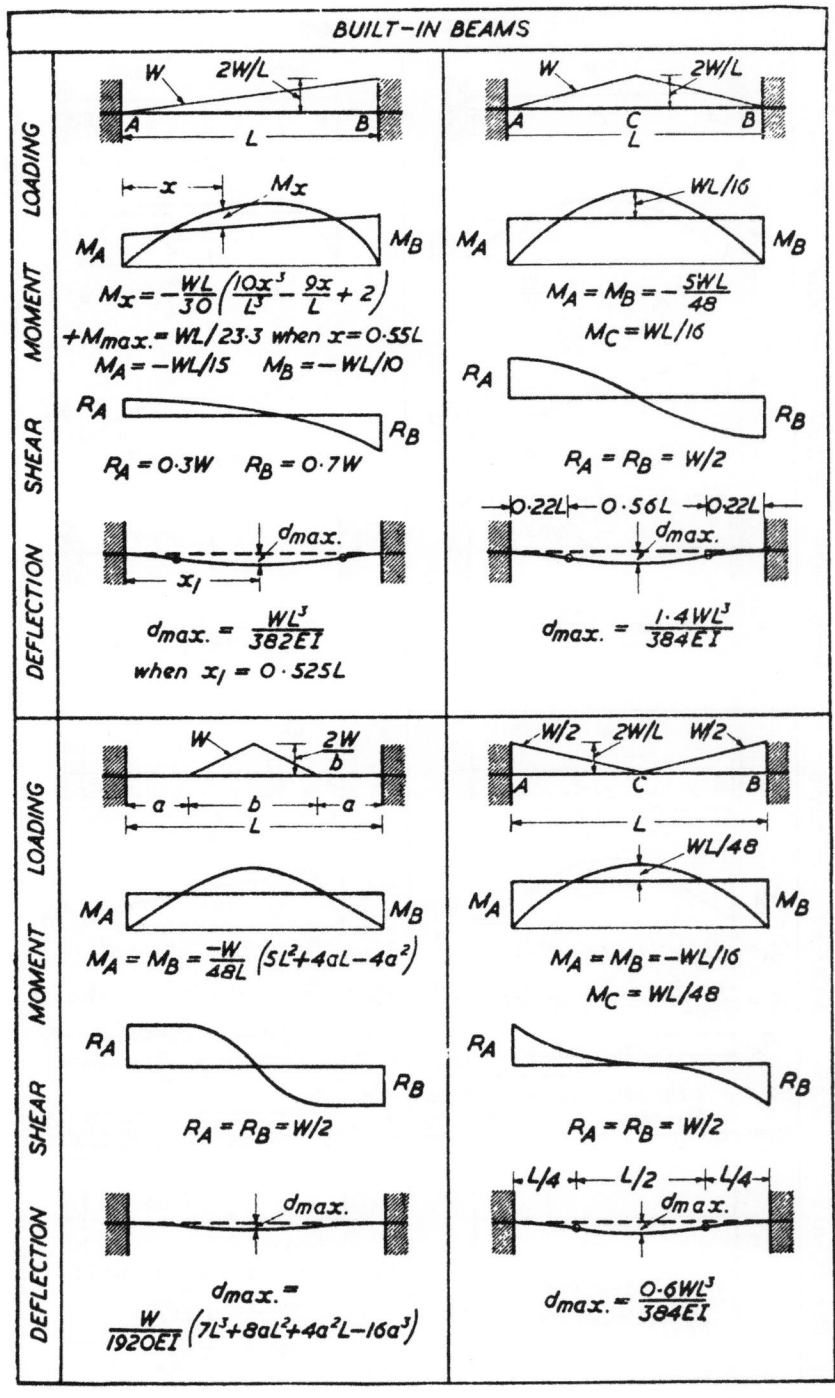

BUILT-IN BEAMS

LOADING / MOMENT / SHEAR / DEFLECTION

Top-left panel:

$$M_x = -\frac{WL}{30}\left(\frac{10x^3}{L^3} - \frac{9x}{L} + 2\right)$$

$+M_{max.} = WL/23.3$ when $x = 0.55L$

$M_A = -WL/15$ $M_B = -WL/10$

$R_A = 0.3W$ $R_B = 0.7W$

$$d_{max.} = \frac{WL^3}{382EI}$$

when $x_1 = 0.525L$

Top-right panel:

$$M_A = M_B = -\frac{5WL}{48}$$

$$M_C = WL/16$$

$R_A = R_B = W/2$

$$d_{max.} = \frac{1.4WL^3}{384EI}$$

Bottom-left panel:

$$M_A = M_B = \frac{-W}{48L}\left(5L^2 + 4aL - 4a^2\right)$$

$R_A = R_B = W/2$

$$d_{max.} = \frac{W}{1920EI}\left(7L^3 + 8aL^2 + 4a^2L - 16a^3\right)$$

Bottom-right panel:

$$M_A = M_B = -WL/16$$

$$M_C = WL/48$$

$R_A = R_B = W/2$

$$d_{max.} = \frac{0.6WL^3}{384EI}$$

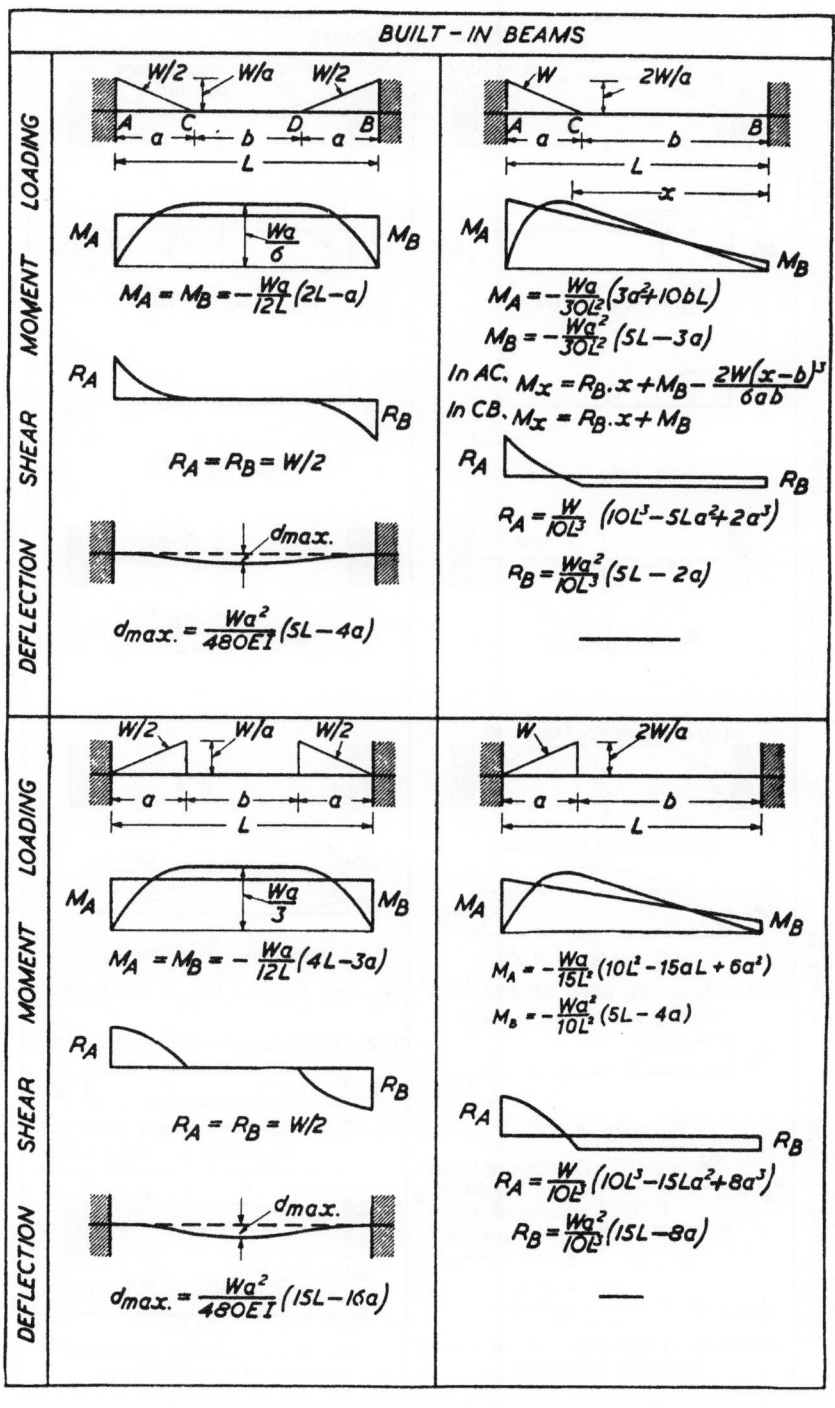

BUILT-IN BEAMS
LOADING
MOMENT
SHEAR
DEFLECTION
W/2
W/a
W/2
A
C
D
B
a
b
a
L
M_A
M_B
Wa/6
M_A = M_B = −Wa/12L (2L − a)
R_A
R_B
R_A = R_B = W/2
dmax.
dmax. = Wa²/480EI (5L − 4a)
W
2W/a
A
C
B
a
b
L
x
M_A
M_B
M_A = −Wa/30L² (3a² + 10bL)
M_B = −Wa²/30L² (5L − 3a)
In AC. M_x = R_B.x + M_B − 2W(x − b)³/6ab
In CB. M_x = R_B.x + M_B
R_A
R_B
R_A = W/10L³ (10L³ − 5La² + 2a³)
R_B = Wa²/10L³ (5L − 2a)
W/2
W/a
W/2
a
b
a
L
M_A
M_B
Wa/3
M_A = M_B = −Wa/12L (4L − 3a)
R_A
R_B
R_A = R_B = W/2
dmax.
dmax. = Wa²/480EI (15L − 16a)
W
2W/a
a
b
L
M_A
M_B
M_A = −Wa/15L² (10L² − 15aL + 6a²)
M_B = −Wa²/10L² (5L − 4a)
R_A
R_B
R_A = W/10L³ (10L³ − 15La² + 8a³)
R_B = Wa²/10L³ (15L − 8a)

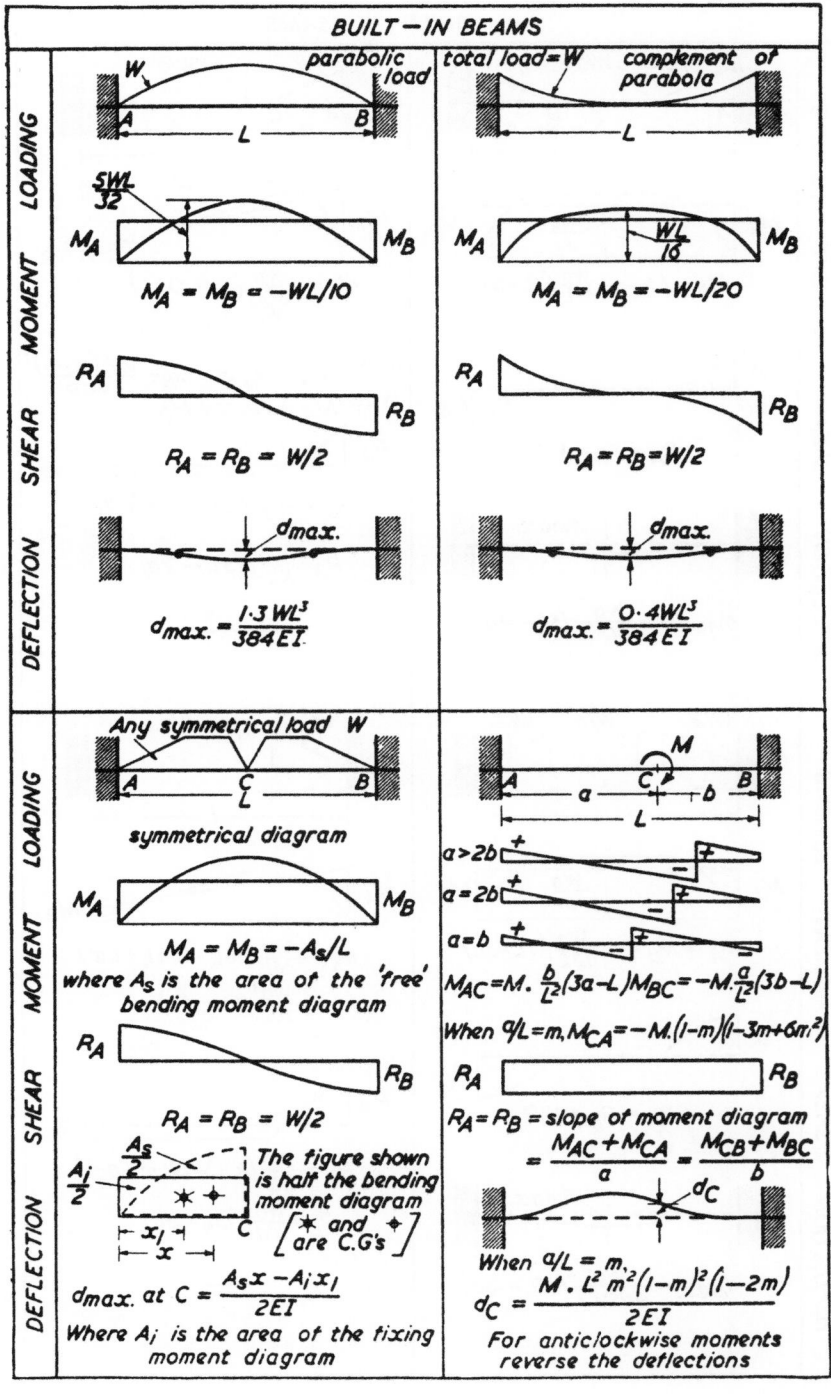

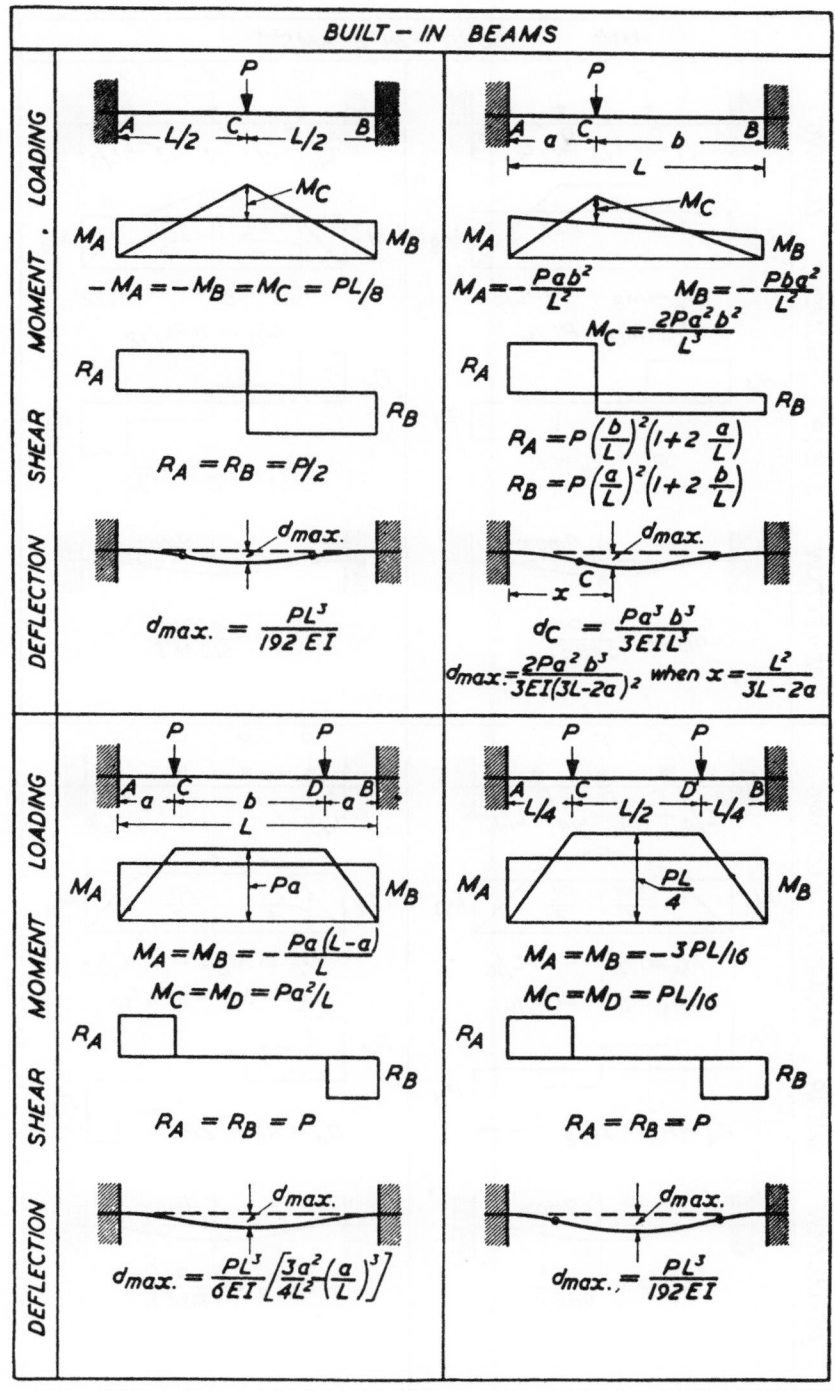

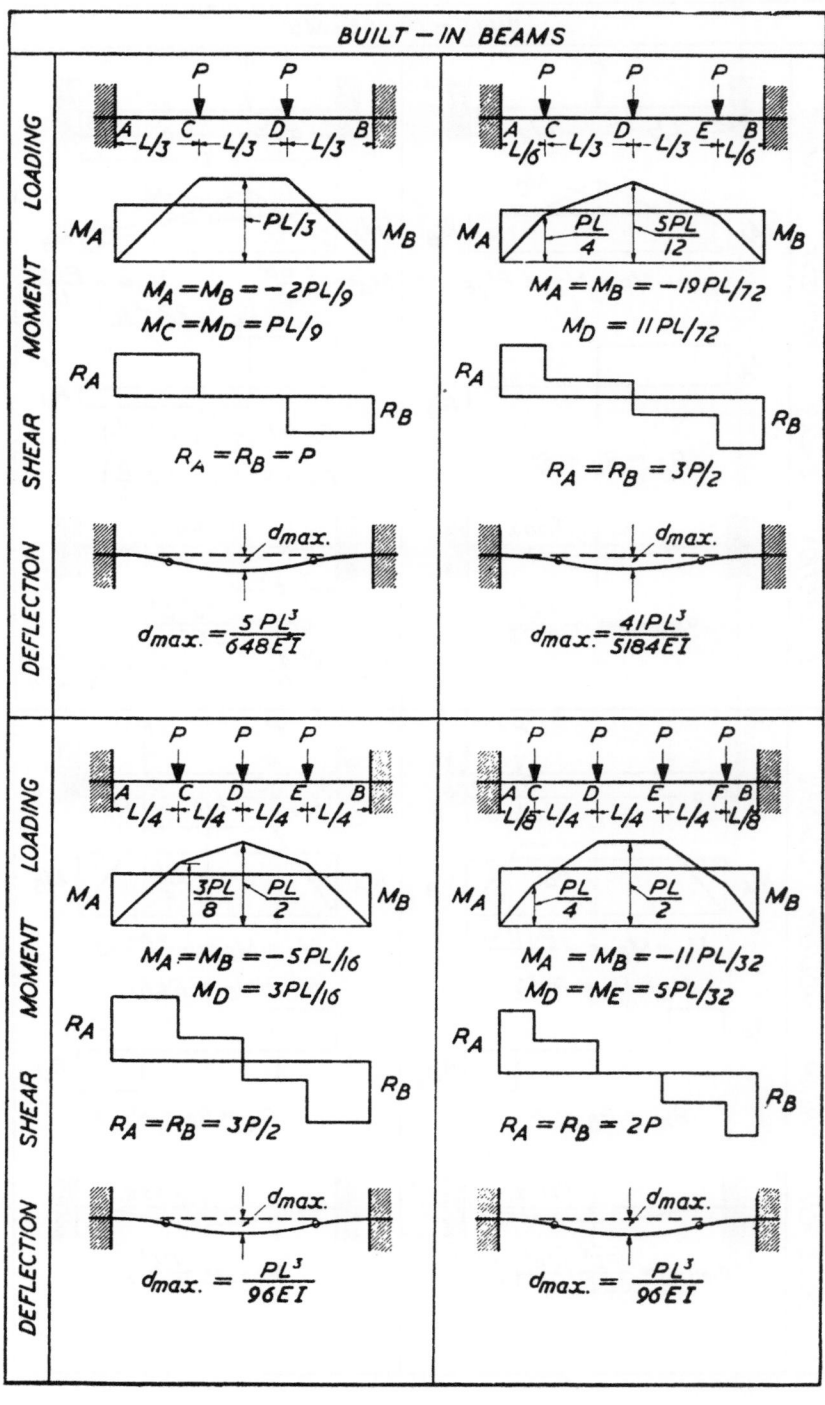

BUILT – IN BEAMS

LOADING / MOMENT / SHEAR / DEFLECTION

Top-left:

$M_A = M_B = -2PL/9$

$M_C = M_D = PL/9$

$R_A = R_B = P$

$d_{max.} = \dfrac{5PL^3}{648EI}$

Top-right:

$M_A = M_B = -19PL/72$

$M_D = 11PL/72$

$R_A = R_B = 3P/2$

$d_{max.} = \dfrac{41PL^3}{5184EI}$

Bottom-left:

$M_A = M_B = -5PL/16$

$M_D = 3PL/16$

$R_A = R_B = 3P/2$

$d_{max.} = \dfrac{PL^3}{96EI}$

Bottom-right:

$M_A = M_B = -11PL/32$

$M_D = M_E = 5PL/32$

$R_A = R_B = 2P$

$d_{max.} = \dfrac{PL^3}{96EI}$

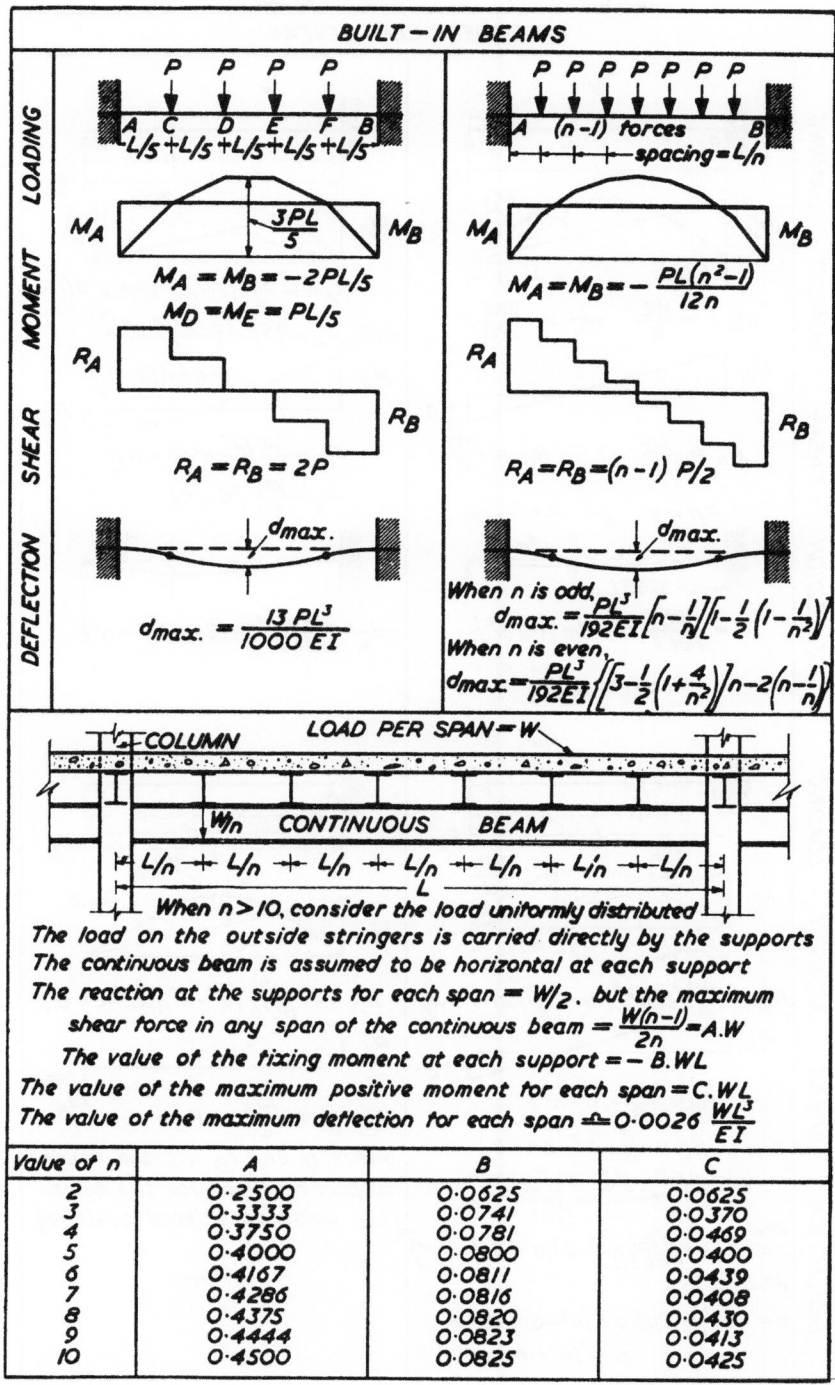

BUILT — IN BEAMS

$M_A = M_B = -2PL/5$

$M_D = M_E = PL/5$

$R_A = R_B = 2P$

$d_{max.} = \dfrac{13\,PL^3}{1000\,EI}$

$M_A = M_B = -\dfrac{PL(n^2-1)}{12n}$

$R_A = R_B = (n-1)\,P/2$

When n is odd,
$$d_{max.} = \dfrac{PL^3}{192EI}\left[n-\dfrac{1}{n}\right]\left[1-\dfrac{1}{2}\left(1-\dfrac{1}{n^2}\right)\right]$$
When n is even,
$$d_{max.} = \dfrac{PL^3}{192EI}\left\{\left[3-\dfrac{1}{2}\left(1+\dfrac{4}{n^2}\right)\right]n-2\left(n-\dfrac{1}{n}\right)\right\}$$

LOAD PER SPAN = W

CONTINUOUS BEAM

When n > 10, consider the load uniformly distributed

The load on the outside stringers is carried directly by the supports
The continuous beam is assumed to be horizontal at each support
The reaction at the supports for each span = W/2, but the maximum
shear force in any span of the continuous beam = $\dfrac{W(n-1)}{2n} = A.W$
The value of the fixing moment at each support = $- B.WL$
The value of the maximum positive moment for each span = $C.WL$
The value of the maximum deflection for each span $\simeq 0.0026\,\dfrac{WL^3}{EI}$

Value of n	A	B	C
2	0·2500	0·0625	0·0625
3	0·3333	0·0741	0·0370
4	0·3750	0·0781	0·0469
5	0·4000	0·0800	0·0400
6	0·4167	0·0811	0·0439
7	0·4286	0·0816	0·0408
8	0·4375	0·0820	0·0430
9	0·4444	0·0823	0·0413
10	0·4500	0·0825	0·0425

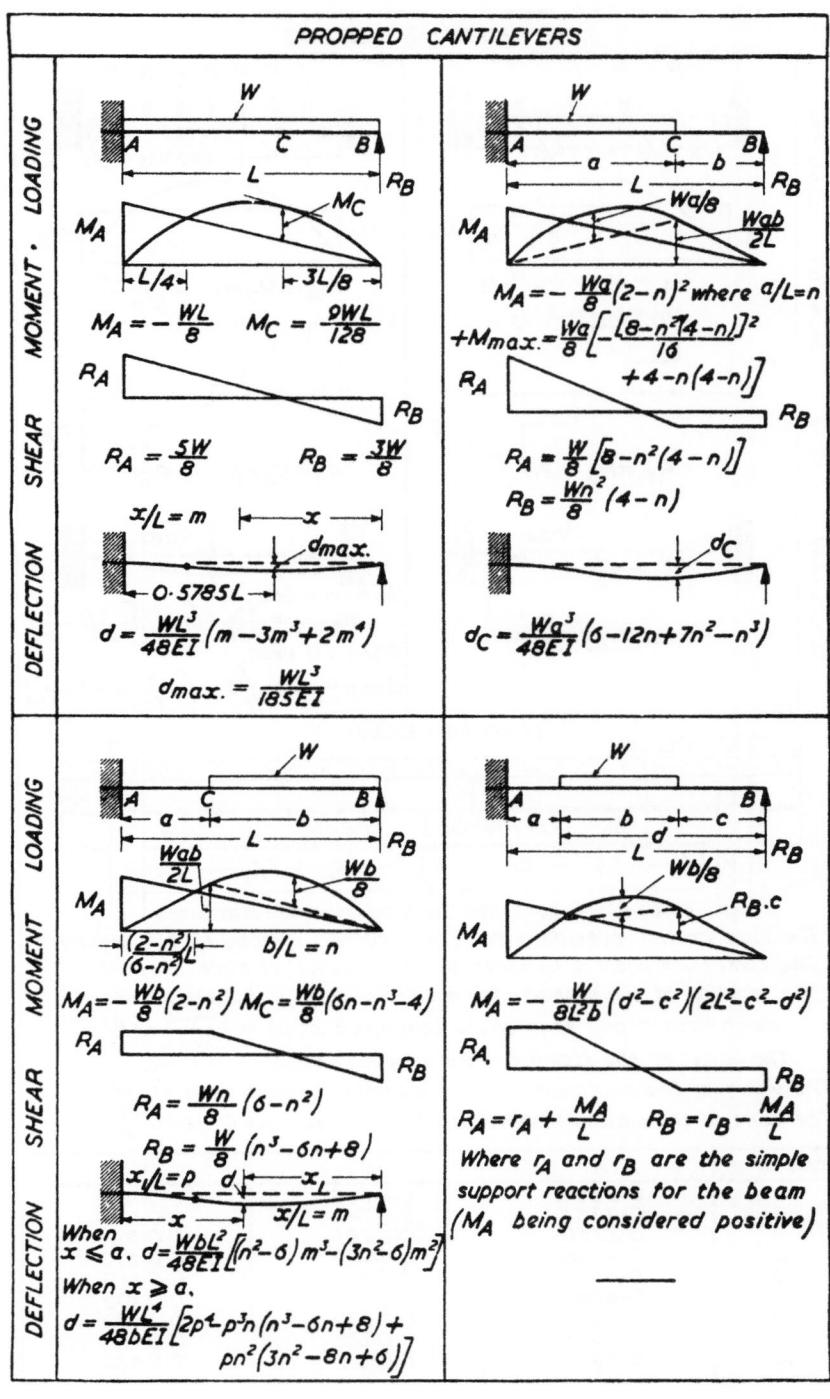

PROPPED CANTILEVERS

Top left — LOADING / MOMENT / SHEAR / DEFLECTION

$$M_A = -\frac{WL}{8} \qquad M_C = \frac{9WL}{128}$$

$$R_A = \frac{5W}{8} \qquad R_B = \frac{3W}{8}$$

$x/L = m$

$$d = \frac{WL^3}{48EI}(m - 3m^3 + 2m^4)$$

$$d_{max.} = \frac{WL^3}{185EI}$$

$0.5785L$

Top right

$$M_A = -\frac{Wa}{8}(2-n)^2 \text{ where } a/L = n$$

$$+M_{max.} = \frac{Wa}{8}\left[-\frac{[8-n^2(4-n)]^2}{16} + 4 - n(4-n) \right]$$

$$R_A = \frac{W}{8}[8 - n^2(4-n)]$$

$$R_B = \frac{Wn^2}{8}(4-n)$$

$$d_C = \frac{Wa^3}{48EI}(6 - 12n + 7n^2 - n^3)$$

Bottom left

$$M_A = -\frac{Wb}{8}(2-n^2) \qquad M_C = \frac{Wb}{8}(6n - n^3 - 4)$$

$b/L = n$, $\frac{(2-n^2)}{(6-n^2)}L$

$$R_A = \frac{Wn}{8}(6 - n^2)$$

$$R_B = \frac{W}{8}(n^3 - 6n + 8)$$

$x_1/L = p$, $x/L = m$

When $x \leqslant a$, $d = \frac{WbL^2}{48EI}[(n^2-6)m^3 - (3n^2-6)m^2]$

When $x \geqslant a$,

$$d = \frac{WL^4}{48bEI}\left[2p^4 - p^3n(n^3 - 6n + 8) + pn^2(3n^2 - 8n + 6) \right]$$

Bottom right

$$M_A = -\frac{W}{8L^2b}(d^2 - c^2)(2L^2 - c^2 - d^2)$$

$$R_A = r_A + \frac{M_A}{L} \qquad R_B = r_B - \frac{M_A}{L}$$

Where r_A and r_B are the simple support reactions for the beam (M_A being considered positive)

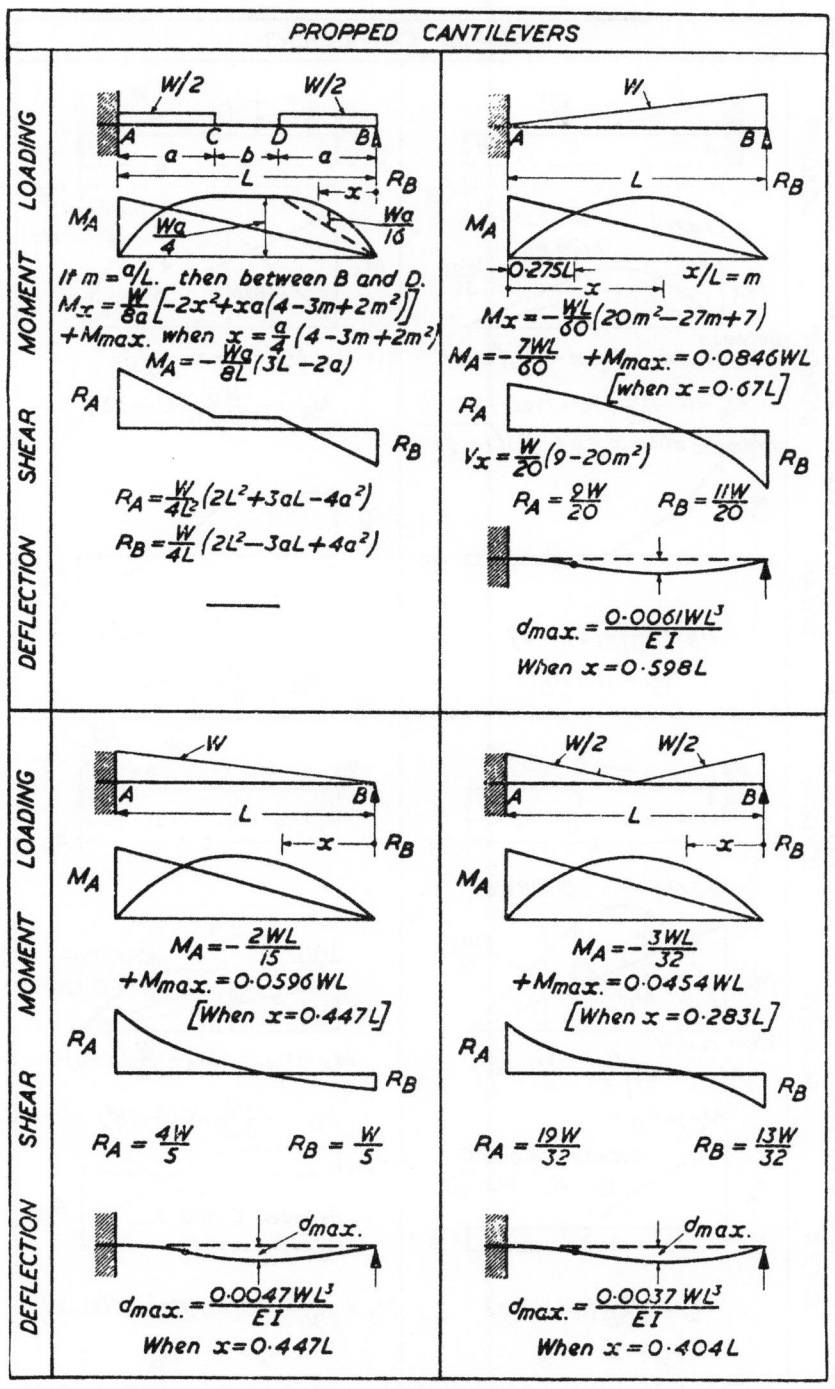

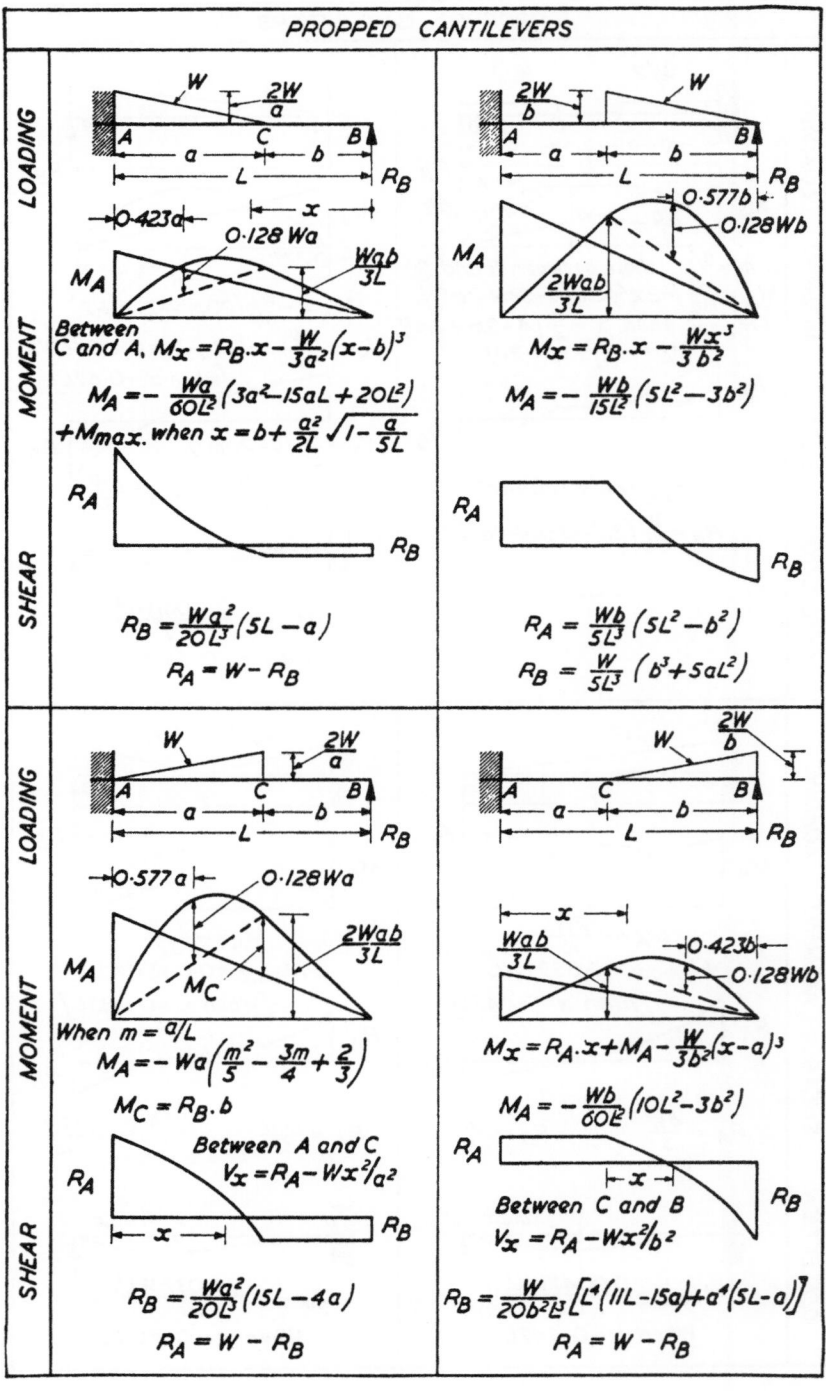

PROPPED CANTILEVERS

Top left panel:

LOADING: W, $\dfrac{2W}{a}$, A, C, B, a, b, L, R_B

MOMENT:

$0.423a$, x, $0.128\,Wa$, M_A, $\dfrac{Wab}{3L}$

Between C and A, $M_x = R_B \cdot x - \dfrac{W}{3a^2}(x-b)^3$

$M_A = -\dfrac{Wa}{60L^2}(3a^2 - 15aL + 20L^2)$

$+ M_{max.}$ when $x = b + \dfrac{a^2}{2L}\sqrt{1-\dfrac{a}{5L}}$

SHEAR: R_A, R_B

$R_B = \dfrac{Wa^2}{20L^3}(5L - a)$

$R_A = W - R_B$

Top right panel:

LOADING: $\dfrac{2W}{b}$, W, A, B, a, b, L, R_B

MOMENT:

M_A, $0.577b$, $0.128\,Wb$, $\dfrac{2Wab}{3L}$

$M_x = R_B \cdot x - \dfrac{Wx^3}{3b^2}$

$M_A = -\dfrac{Wb}{15L^2}(5L^2 - 3b^2)$

SHEAR: R_A, R_B

$R_A = \dfrac{Wb}{5L^3}(5L^2 - b^2)$

$R_B = \dfrac{W}{5L^3}(b^3 + 5aL^2)$

Bottom left panel:

LOADING: W, $\dfrac{2W}{a}$, A, C, B, a, b, L, R_B

MOMENT:

$0.577a$, $0.128\,Wa$, M_A, M_C, $\dfrac{2Wab}{3L}$

When $m = a/L$

$M_A = -Wa\left(\dfrac{m^2}{5} - \dfrac{3m}{4} + \dfrac{2}{3}\right)$

$M_C = R_B \cdot b$

SHEAR: R_A, R_B

Between A and C

$V_x = R_A - Wx^2/a^2$

x

$R_B = \dfrac{Wa^2}{20L^3}(15L - 4a)$

$R_A = W - R_B$

Bottom right panel:

LOADING: W, $\dfrac{2W}{b}$, A, C, B, a, b, L, R_B

MOMENT:

x, $\dfrac{Wab}{3L}$, $0.423b$, $0.128\,Wb$

$M_x = R_A \cdot x + M_A - \dfrac{W}{3b^2}(x-a)^3$

$M_A = -\dfrac{Wb}{60L^2}(10L^2 - 3b^2)$

R_A

x

Between C and B

$V_x = R_A - Wx^2/b^2$

$R_B = \dfrac{W}{20b^2L^3}\left[L^4(11L - 15a) + a^4(5L - a)\right]$

$R_A = W - R_B$

R_B

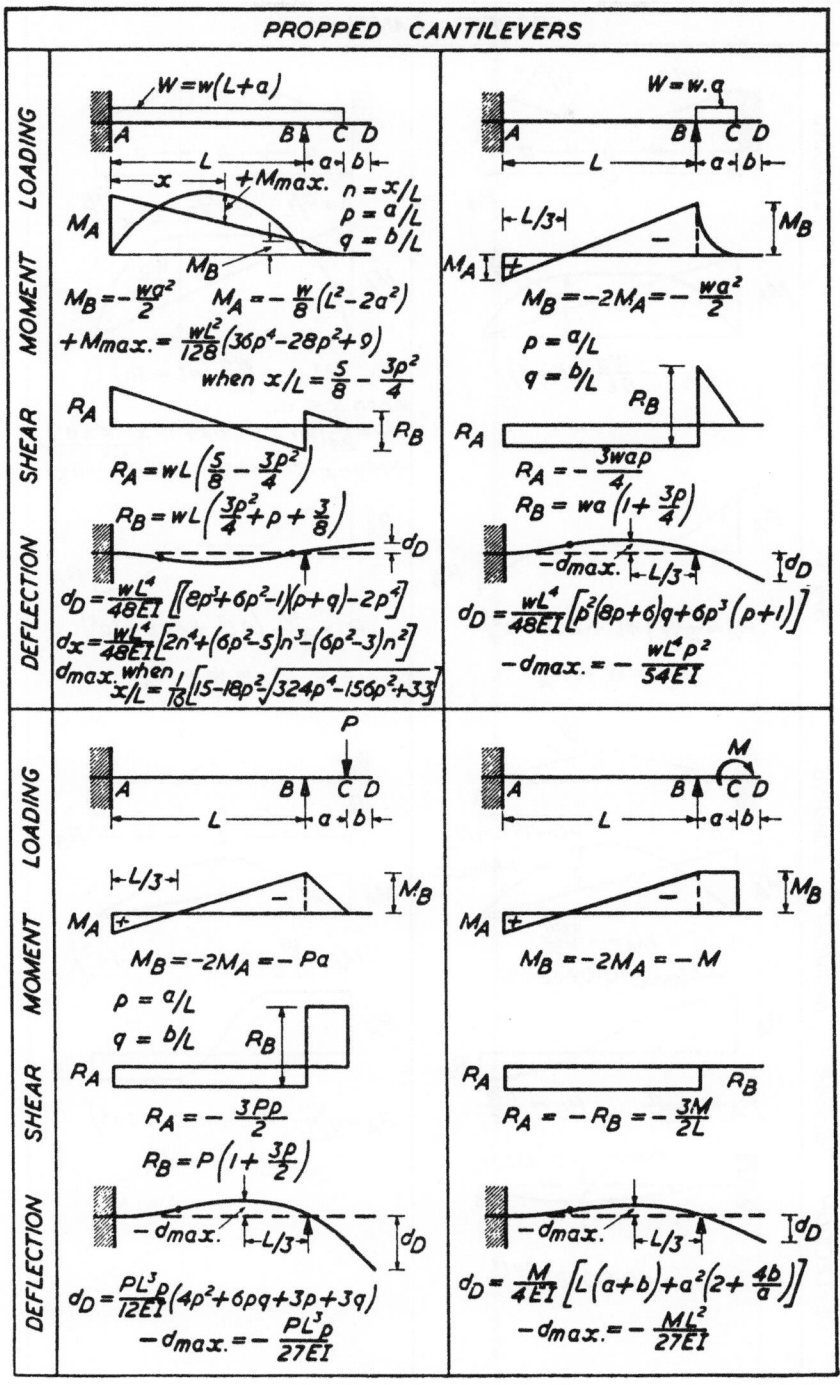

PROPPED CANTILEVERS

LOADING

$W = w(L+a)$

$W = w \cdot a$

MOMENT

$M_B = -\dfrac{wa^2}{2}$ $M_A = -\dfrac{w}{8}(L^2 - 2a^2)$

$+M_{max.} = \dfrac{wL^2}{128}(36p^4 - 28p^2 + 9)$ when $x/L = \dfrac{5}{8} - \dfrac{3p^2}{4}$

$M_B = -2M_A = -\dfrac{wa^2}{2}$

$p = a/L$

$q = b/L$

SHEAR

$R_A = wL\left(\dfrac{5}{8} - \dfrac{3p^2}{4}\right)$

$R_B = wL\left(\dfrac{3p^2}{4} + p + \dfrac{3}{8}\right)$

$R_A = -\dfrac{3wap}{4}$

$R_B = wa\left(1 + \dfrac{3p}{4}\right)$

DEFLECTION

$d_D = \dfrac{wL^4}{48EI}\left[(8p^4 + 6p^2 - 1)(p+q) - 2p^5\right]$

$d_x = \dfrac{wL^4}{48EIL}\left[2n^4 + (6p^2 - 5)n^3 - (6p^2 - 3)n^2\right]$

$d_{max.}$ when $x/L = \dfrac{1}{18}\left[15 - 18p^2 - \sqrt{324p^4 - 156p^2 + 33}\right]$

$d_D = \dfrac{wL^4}{48EI}\left[p^2(8p+6)q + 6p^3(p+1)\right]$

$-d_{max.} = -\dfrac{wL^4 p^2}{54EI}$

LOADING

P

M

MOMENT

$M_B = -2M_A = -Pa$

$p = a/L$

$q = b/L$

$M_B = -2M_A = -M$

SHEAR

$R_A = -\dfrac{3Pp}{2}$

$R_B = P\left(1 + \dfrac{3p}{2}\right)$

$R_A = -R_B = -\dfrac{3M}{2L}$

DEFLECTION

$d_D = \dfrac{PL^3 p}{12EI}(4p^2 + 6pq + 3p + 3q)$

$-d_{max.} = -\dfrac{PL^3 p}{27EI}$

$d_D = \dfrac{M}{4EI}\left[L(a+b) + a^2\left(2 + \dfrac{4b}{a}\right)\right]$

$-d_{max.} = -\dfrac{ML^2}{27EI}$

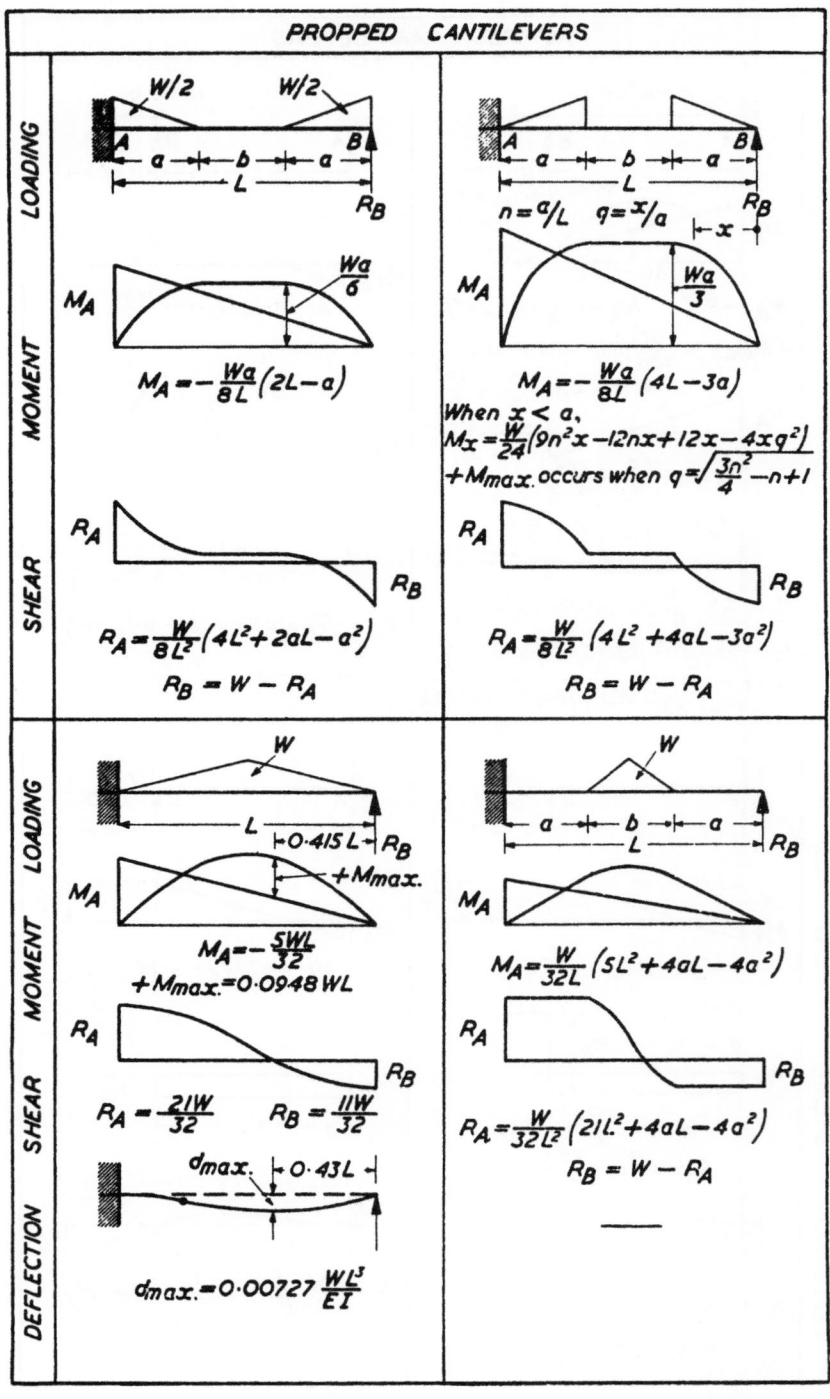

PROPPED CANTILEVERS

LOADING / MOMENT (top left)

$$M_A = -\frac{Wa}{8L}(2L-a)$$

Peak moment $\frac{Wa}{6}$

SHEAR (left)

$$R_A = \frac{W}{8L^2}(4L^2+2aL-a^2)$$
$$R_B = W - R_A$$

LOADING / MOMENT (top right)

$n = a/L \qquad q = x/a$

$$M_A = -\frac{Wa}{8L}(4L-3a)$$

Peak moment $\frac{Wa}{3}$

When $x < a$,
$$M_x = \frac{W}{24}(9n^2x - 12nx + 12x - 4xq^2)$$
$+M_{max.}$ occurs when $q = \sqrt{\frac{3n^2}{4} - n + 1}$

SHEAR (right)

$$R_A = \frac{W}{8L^2}(4L^2+4aL-3a^2)$$
$$R_B = W - R_A$$

LOADING / MOMENT (bottom left)

$\vdash 0.415L \dashv$ $+M_{max.}$

$$M_A = -\frac{5WL}{32}$$
$$+M_{max.} = 0.0948 WL$$

SHEAR (bottom left)

$$R_A = \frac{21W}{32} \qquad R_B = \frac{11W}{32}$$

DEFLECTION (bottom left)

$d_{max.}$ $\vdash 0.43L \dashv$

$$d_{max.} = 0.00727\frac{WL^3}{EI}$$

LOADING / MOMENT (bottom right)

$$M_A = \frac{W}{32L}(5L^2+4aL-4a^2)$$

SHEAR (bottom right)

$$R_A = \frac{W}{32L^2}(21L^2+4aL-4a^2)$$
$$R_B = W - R_A$$

———

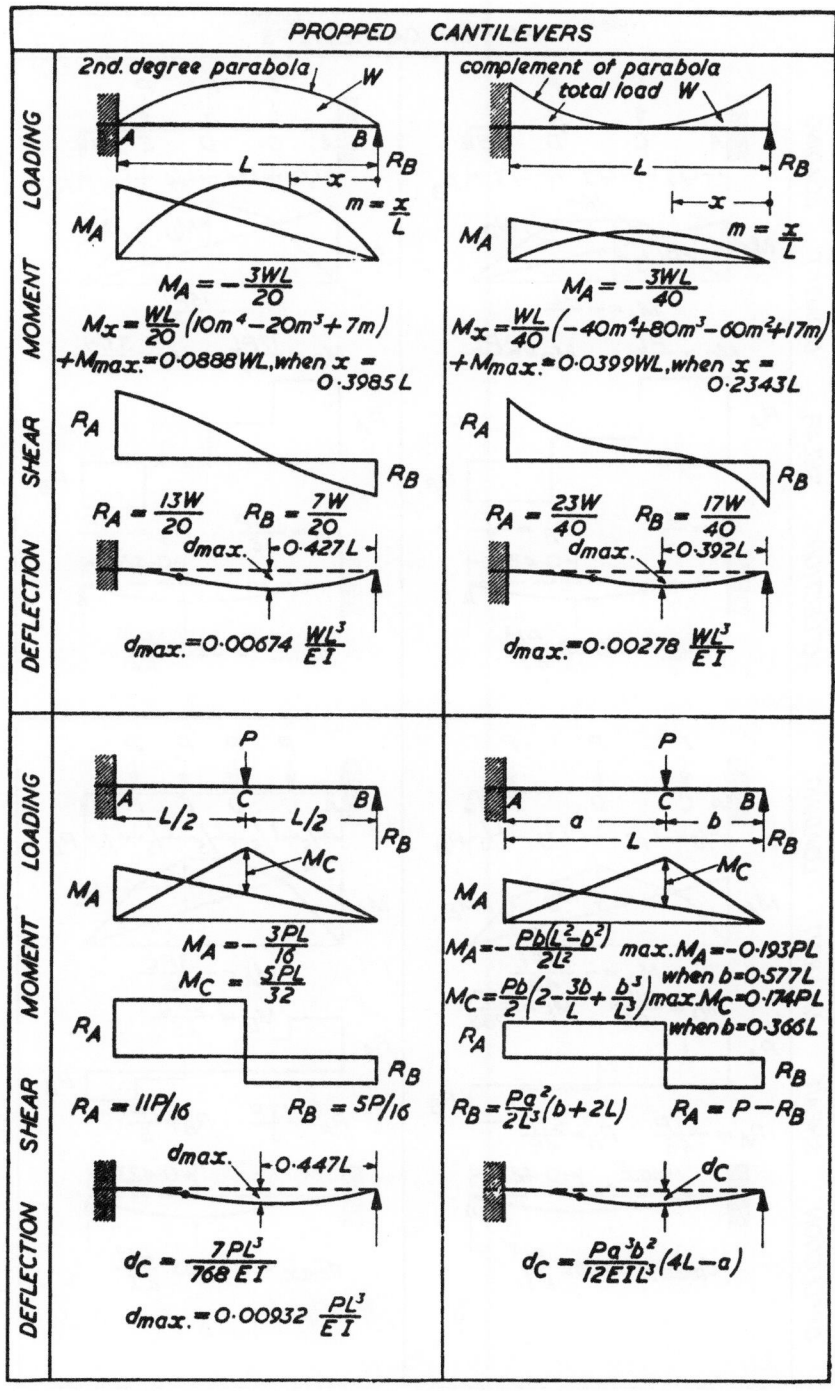

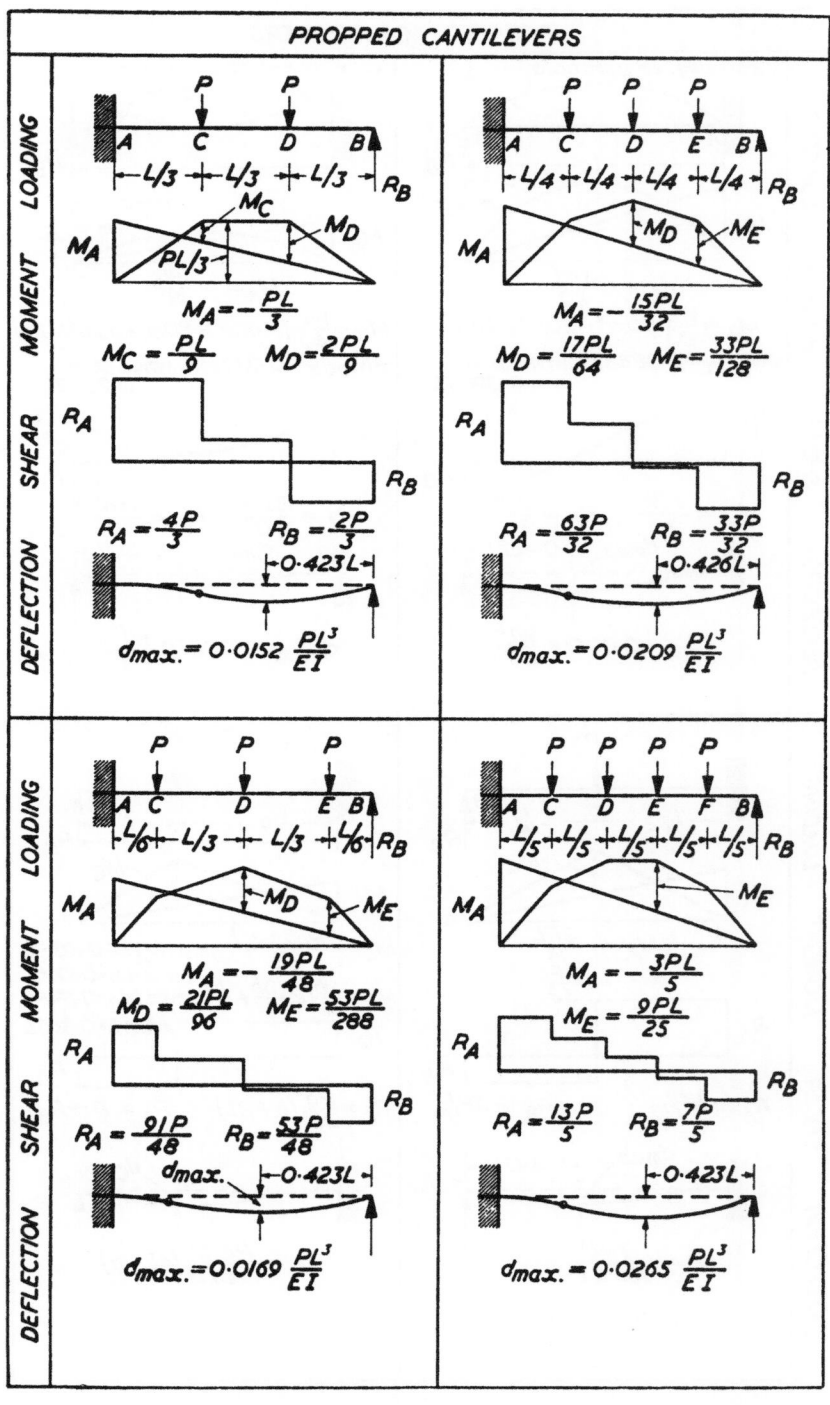

PROPPED CANTILEVERS

LOADING (top left)

$M_A = -\dfrac{PL}{3}$

$M_C = \dfrac{PL}{9}$ $M_D = \dfrac{2PL}{9}$

$R_A = \dfrac{4P}{3}$ $R_B = \dfrac{2P}{3}$

$d_{max.} = 0.0152\,\dfrac{PL^3}{EI}$

(top right)

$M_A = -\dfrac{15PL}{32}$

$M_D = \dfrac{17PL}{64}$ $M_E = \dfrac{33PL}{128}$

$R_A = \dfrac{63P}{32}$ $R_B = \dfrac{33P}{32}$

$d_{max.} = 0.0209\,\dfrac{PL^3}{EI}$

(bottom left)

$M_A = -\dfrac{19PL}{48}$

$M_D = \dfrac{21PL}{96}$ $M_E = \dfrac{53PL}{288}$

$R_A = \dfrac{91P}{48}$ $R_B = \dfrac{53P}{48}$

$d_{max.} = 0.0169\,\dfrac{PL^3}{EI}$

(bottom right)

$M_A = -\dfrac{3PL}{5}$

$M_E = \dfrac{9PL}{25}$

$R_A = \dfrac{13P}{5}$ $R_B = \dfrac{7P}{5}$

$d_{max.} = 0.0265\,\dfrac{PL^3}{EI}$

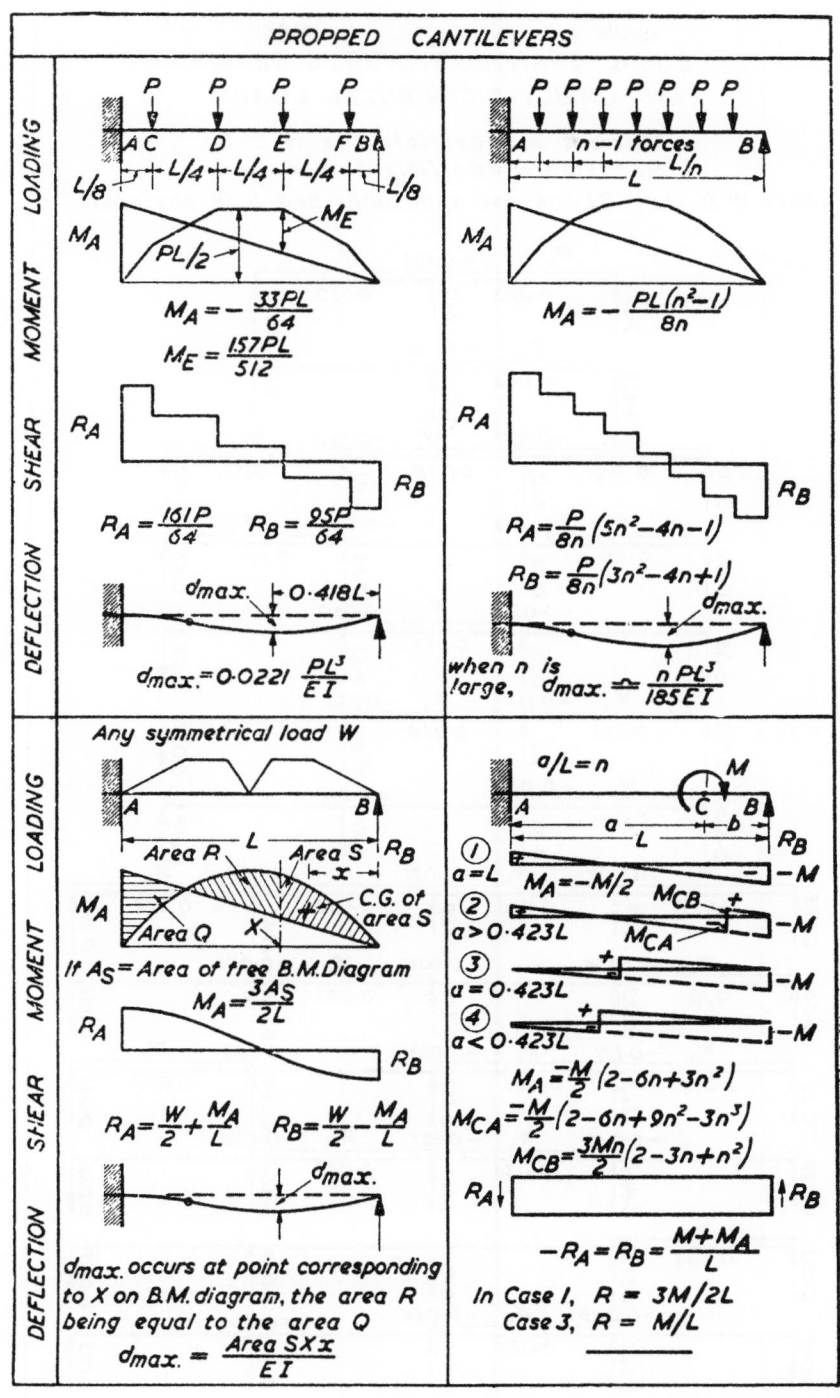

PROPPED CANTILEVERS

LOADING

MOMENT

$M_A = -\dfrac{33PL}{64}$

$M_E = \dfrac{157PL}{512}$

$M_A = -\dfrac{PL(n^2-1)}{8n}$

SHEAR

$R_A = \dfrac{161P}{64}$ $R_B = \dfrac{95P}{64}$

$R_A = \dfrac{P}{8n}(5n^2-4n-1)$

$R_B = \dfrac{P}{8n}(3n^2-4n+1)$

DEFLECTION

$d_{max} = 0.0221\,\dfrac{PL^3}{EI}$

when n is large, $d_{max} \simeq \dfrac{n\,PL^3}{185\,EI}$

Any symmetrical load W

If $A_S =$ Area of free B.M.Diagram

$M_A = \dfrac{3A_S}{2L}$

$R_A = \dfrac{W}{2} + \dfrac{M_A}{L}$ $R_B = \dfrac{W}{2} - \dfrac{M_A}{L}$

d_{max} occurs at point corresponding to X on B.M. diagram, the area R being equal to the area Q

$d_{max} = \dfrac{Area\ S \times x}{EI}$

$a/L = n$

① $a = L$

② $a > 0.423L$

③ $a = 0.423L$

④ $a < 0.423L$

$M_A = \dfrac{M}{2}(2-6n+3n^2)$

$M_{CA} = \dfrac{M}{2}(2-6n+9n^2-3n^3)$

$M_{CB} = \dfrac{3Mn}{2}(2-3n+n^2)$

$-R_A = R_B = \dfrac{M+M_A}{L}$

In Case I, $R = 3M/2L$

Case 3, $R = M/L$

EQUAL SPAN CONTINUOUS BEAMS
UNIFORMLY DISTRIBUTED LOADS

Moment = coefficient x W x L
Reaction = coefficient x W
where W is the U.D.L. on one span only and L is one span

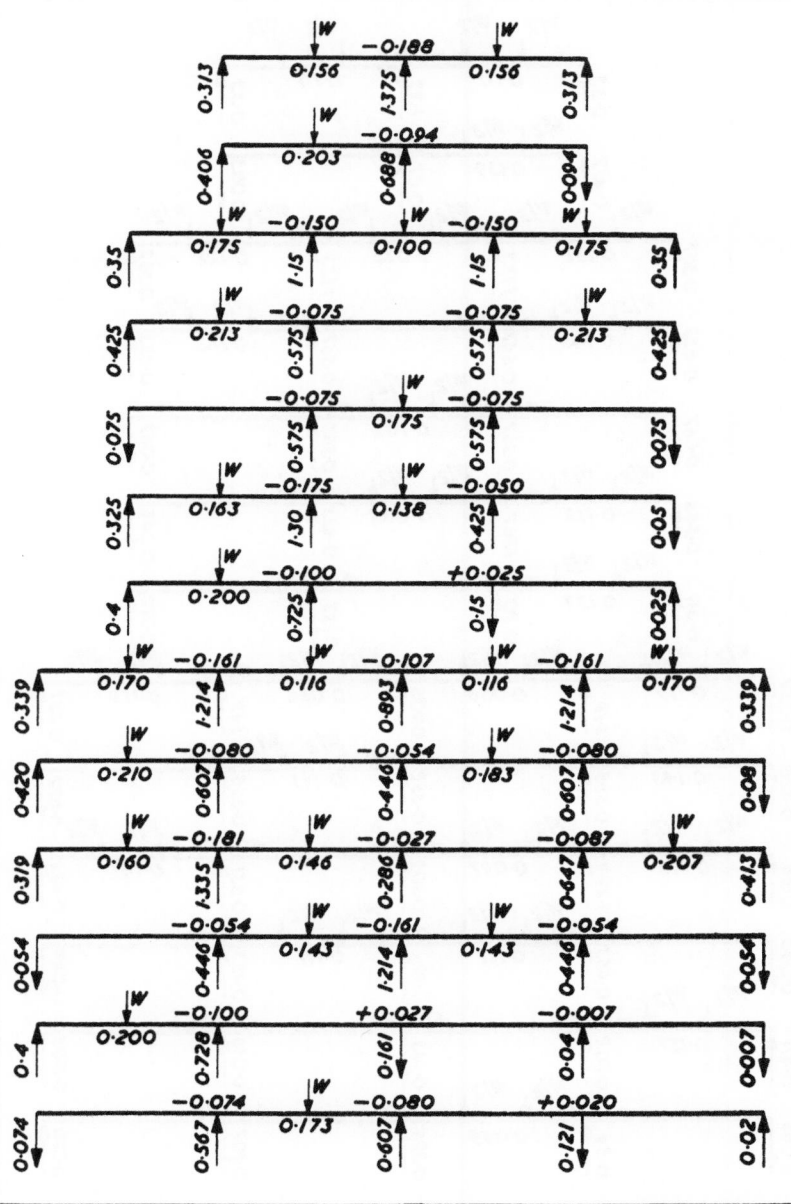

EQUAL SPAN CONTINUOUS BEAMS
CENTRAL POINT LOADS

Moment = coefficient x W x L
Reaction = coefficient x W
where W is the Load on one span only and L is one span

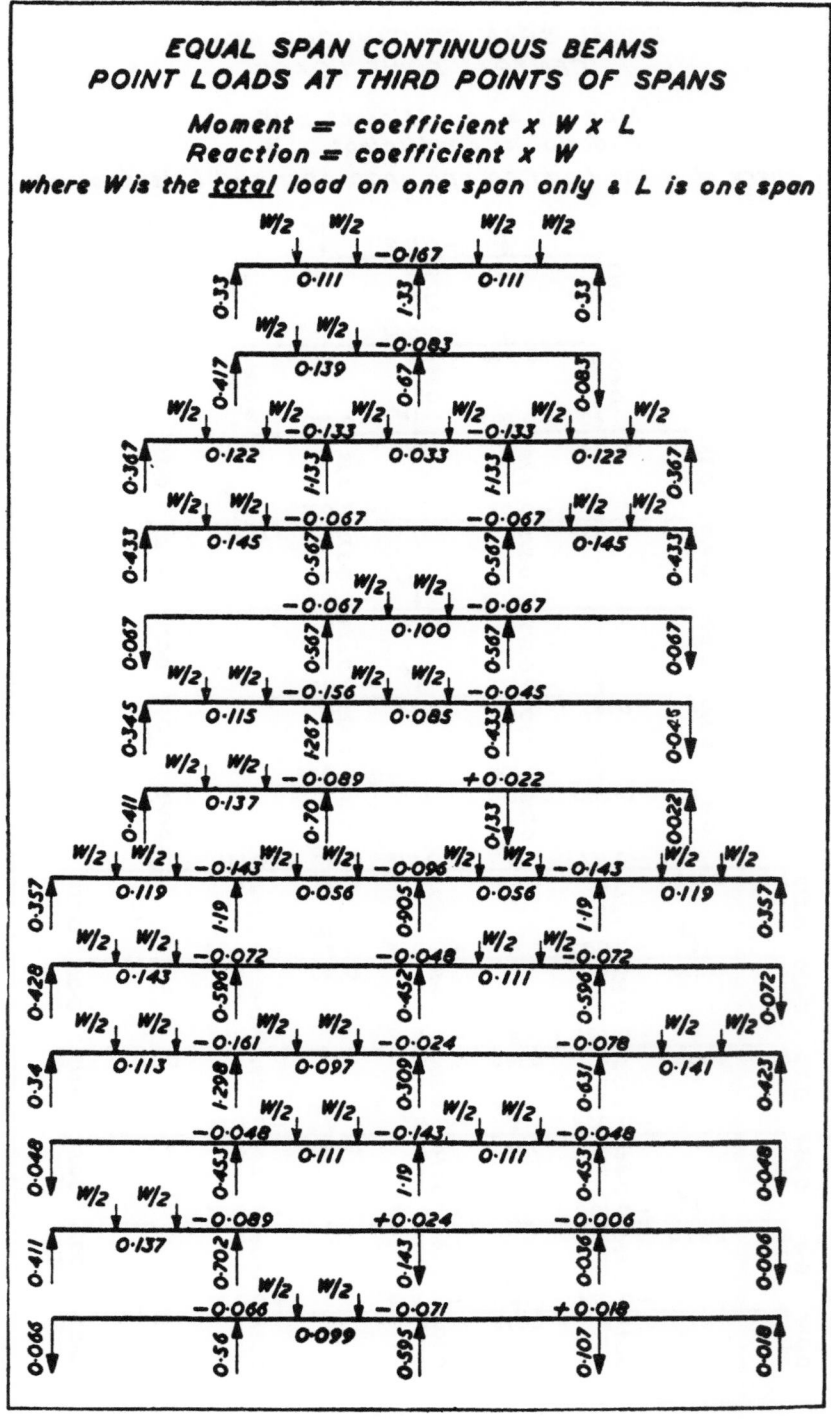

EQUAL SPAN CONTINUOUS BEAMS
POINT LOADS AT THIRD POINTS OF SPANS

Moment = coefficient x W x L
Reaction = coefficient x W
where W is the <u>total</u> load on one span only & L is one span

Second moments of area

SECOND MOMENTS OF AREA (cm^4)
OF TWO FLANGES
per millimetre of width

Distance d_w mm	THICKNESS OF EACH FLANGE IN MILLIMETRES									
	10	12	15	18	20	22	25	28	30	32
1000	510.1	614.5	772.7	932.8	1041	1149	1314	1480	1592	1705
1100	616.1	742.0	932.5	1125	1255	1385	1582	1782	1916	2051
1200	732.1	881.4	1107	1335	1489	1643	1876	2112	2270	2429
1300	858.1	1033	1297	1564	1743	1923	2195	2469	2654	2839
1400	994.1	1196	1502	1810	2017	2224	2539	2855	3068	3282
1500	1140	1372	1721	2074	2311	2548	2907	3269	3512	3756
1600	1296	1559	1956	2356	2625	2894	3301	3711	3986	4262
1700	1462	1759	2206	2656	2959	3262	3720	4181	4490	4800
1800	1638	1970	2471	2975	3313	3652	4164	4679	5024	5371
1900	1824	2193	2750	3311	3687	4064	4632	5204	5588	5973
2000	2020	2429	3045	3665	4081	4498	5126	5758	6182	6607
2100	2226	2676	3355	4037	4495	4953	5645	6340	6806	7273
2200	2442	2936	3680	4428	4929	5431	6189	6950	7460	7971
2300	2668	3207	4019	4836	5383	5931	6757	7588	8144	8702
2400	2904	3491	4374	5262	5857	6453	7351	8254	8858	9464
2500	3150	3786	4744	5706	6351	6997	7970	8947	9602	10258
2600	3406	4094	5129	6169	6865	7563	8614	9669	10376	11084
2700	3672	4413	5528	6649	7399	8150	9282	10419	11180	11943
2800	3948	4744	5943	7147	7953	8760	9976	11197	12014	12833
2900	4234	5088	6373	7663	8527	9392	10695	12003	12878	13755
3000	4530	5443	6818	8198	9121	10046	11439	12837	13772	14709
3100	4836	5811	7277	8750	9735	10722	12207	13699	14696	15696
3200	5152	6190	7752	9320	10369	11420	13001	14588	15650	16714
3300	5478	6582	8242	9908	11023	12139	13820	15506	16634	17764
3400	5814	6985	8747	10515	11697	12881	14664	16452	17648	18846
3500	6160	7401	9266	11139	12391	13645	15532	17426	18692	19961
3600	6516	7828	9801	11781	13105	14431	16426	18428	19766	21107
3700	6882	8267	10351	12441	13839	15239	17345	19458	20870	22285
3800	7258	8719	10916	13120	14593	16069	18289	20515	22004	23495
3900	7644	9182	11495	13816	15367	16920	19257	21601	23168	24738
4000	8040	9658	12090	14530	16161	17794	20251	22715	24362	26012
4100	8446	10145	12700	15262	16975	18690	21270	23857	25586	27318
4200	8862	10645	13325	16012	17809	19608	22314	25027	26840	28656
4300	9288	11156	13964	16781	18663	20548	23382	26225	28124	30027
4400	9724	11679	14619	17567	19537	21510	24476	27450	29438	31429
4500	10170	12215	15289	18371	20431	22494	25595	28704	30782	32863
4600	10626	12762	15974	19193	21345	23499	26739	29986	32156	34329
4700	11092	13322	16673	20034	22279	24527	27907	31296	33560	35827
4800	11568	13893	17388	20892	23233	25577	29101	32634	34994	37358
4900	12054	14477	18118	21768	24207	26649	30320	34000	36458	38920
5000	12550	15072	18863	22662	25201	27743	31564	35393	37952	40514

SECOND MOMENTS OF AREA (cm^4) OF TWO FLANGES

per millimetre of width

| THICKNESS OF EACH FLANGE IN MILLIMETRES | | | | | | | | | | Distance |
35	38	40	45	50	55	60	65	70	75	d, mm
1875	2048	2164	2459	2758	3064	3374	3691	4013	4341	**1000**
2255	2461	2600	2951	3308	3671	4040	4416	4797	5184	**1100**
2670	2913	3076	3489	3908	4334	4766	5205	5651	6103	**1200**
3120	3402	3592	4072	4558	5052	5552	6060	6575	7097	**1300**
3604	3930	4148	4700	5258	5825	6398	6980	7569	8166	**1400**
4124	4495	4744	5372	6008	6652	7304	7965	8633	9309	**1500**
4679	5099	5380	6090	6808	7535	8270	9014	9767	10528	**1600**
5269	5740	6056	6853	7658	8473	9296	10129	10971	11822	**1700**
5893	6420	6772	7661	8558	9466	10382	11309	12245	13191	**1800**
6553	7137	7528	8513	9508	10513	11528	12554	13589	14634	**1900**
7248	7892	8324	9411	10508	11616	12734	13863	15003	16153	**2000**
7978	8686	9160	10354	11558	12774	14000	15238	16487	17747	**2100**
8742	9517	10036	11342	12658	13987	15326	16678	18041	19416	**2200**
9542	10387	10952	12374	13808	15254	16712	18183	19665	21159	**2300**
10377	11294	11908	13452	15008	16577	18158	19752	21359	22978	**2400**
11247	12240	12904	14575	16258	17955	19664	21387	23123	24872	**2500**
12151	13223	13940	15743	17558	19388	21230	23087	24957	26841	**2600**
13091	14245	15016	16955	18908	20875	22856	24852	26861	28884	**2700**
14066	15304	16132	18213	20308	22418	24542	26681	28835	31003	**2800**
15076	16401	17288	19516	21758	24016	26288	28576	30879	33197	**2900**
16120	17537	18484	20864	23258	25669	28094	30536	32993	35466	**3000**
17200	18710	19720	22256	24808	27376	29960	32561	35177	37809	**3100**
18315	19922	20996	23694	26408	29139	31886	34650	37431	40228	**3200**
19465	21171	22312	25177	28058	30957	33872	36805	39755	42722	**3300**
20649	22459	23668	26705	29758	32830	35918	39025	42149	45291	**3400**
21869	23784	25064	28277	31508	34757	38024	41310	44613	47934	**3500**
23124	25147	26500	29895	33308	36740	40190	43659	47147	50653	**3600**
24414	26549	27976	31558	35158	38778	42416	46074	49751	53447	**3700**
25738	27988	29492	33266	37058	40871	44702	48554	52425	56316	**3800**
27098	29466	31048	35018	39008	43018	47048	51099	55169	59259	**3900**
28493	30981	32644	36816	41008	45221	49454	53708	57983	62278	**4000**
29923	32535	34280	38659	43058	47479	51920	56383	60867	65372	**4100**
31387	34126	35956	40547	45158	49792	54446	59123	63821	68541	**4200**
32887	35756	37672	42479	47308	52159	57032	61928	66845	71784	**4300**
34422	37423	39428	44457	49508	54582	59678	64797	69939	75103	**4400**
35992	39128	41224	46480	51758	57060	62384	67732	73103	78497	**4500**
37596	40872	43060	48548	54058	59593	65150	70732	76337	81966	**4600**
39236	42653	44936	50660	56408	62180	67976	73797	79641	85509	**4700**
40911	44473	46852	52818	58808	64823	70862	76926	83015	89128	**4800**
42621	46330	48808	55021	61258	67521	73808	80121	86459	92822	**4900**
44365	48226	50804	57269	63758	70274	76814	83381	89973	96591	**5000**

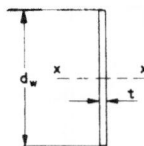

SECOND MOMENTS OF AREA (cm⁴)
OF RECTANGULAR PLATES
about axis x – x

Depth dₘ mm	THICKNESS t MILLIMETRES					
	3	4	5	6	8	10
25	.391	.521	.651	.781	1.04	1.30
50	3.13	4.17	5.21	6.25	8.33	10.4
75	10.5	14.1	17.6	21.1	28.1	35.2
100	25.0	33.3	41.7	50.0	66.7	83.3
125	48.8	65.1	81.4	97.7	130	163
150	84.4	113	141	169	225	281
175	134	179	223	268	357	447
200	200	267	333	400	533	667
225	285	380	475	570	759	949
250	391	521	651	781	1042	1302
275	520	693	867	1040	1386	1733
300	675	900	1125	1350	1800	2250
325	858	1144	1430	1716	2289	2861
350	1072	1429	1786	2144	2858	3573
375	1318	1758	2197	2637	3516	4395
400	1600	2133	2667	3200	4267	5333
425	1919	2559	3199	3838	5118	6397
450	2278	3038	3797	4556	6075	7594
475	2679	3572	4465	5359	7145	8931
500	3125	4167	5208	6250	8333	10417
525	3618	4823	6029	7235	9647	12059
550	4159	5546	6932	8319	11092	13865
575	4753	6337	7921	9505	12674	15842
600	5400	7200	9000	10800	14400	18000
625	6104	8138	10173	12207	16276	20345
650	6866	9154	11443	13731	18308	22885
675	7689	10252	12814	15377	20503	25629
700	8575	11433	14292	17150	22867	28583
725	9527	12703	15878	19054	25405	31757
750	10547	14063	17578	21094	28125	35156
775	11637	15516	19395	23274	31032	38790
800	12800	17067	21333	25600	34133	42667
825	14038	18717	23396	28076	37434	46793
850	15353	20471	25589	30706	40942	51177
875	16748	22331	27913	33496	44661	55827
900	18225	24300	30375	36450	48600	60750

SECOND MOMENTS OF AREA (cm⁴) OF RECTANGULAR PLATES

about axis x—x

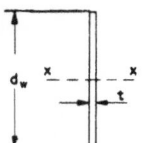

THICKNESS t MILLIMETRES						Depth
12	**15**	**18**	**20**	**22**	**25**	**d_w mm**
1.56	1.95	2.34	2.60	2.86	3.26	**25**
12.5	15.6	18.8	20.8	22.9	26.0	**50**
42.2	52.7	63 3	70.3	77.3	87.9	**75**
100	125	150	167	183	208	**100**
195	244	293	326	358	407	**125**
338	422	506	563	619	703	**150**
536	670	804	893	983	1117	**175**
800	1000	1200	1333	1467	1667	**200**
1139	1424	1709	1898	2088	2373	**225**
1563	1953	2344	2604	2865	3255	**250**
2080	2600	3120	3466	3813	4333	**275**
2700	3375	4050	4500	4950	5625	**300**
3433	4291	5149	5721	6293	7152	**325**
4288	5359	6431	7146	7860	8932	**350**
5273	6592	7910	8789	9668	10986	**375**
6400	8000	9600	10667	11733	13333	**400**
7677	9596	11515	12794	14074	15993	**425**
9113	11391	13669	15188	16706	18984	**450**
10717	13396	16076	17862	19648	22327	**475**
12500	15625	18750	20833	22917	26042	**500**
14470	18088	21705	24117	26529	30146	**525**
16638	20797	24956	27729	30502	34661	**550**
19011	23764	28516	31685	34853	39606	**575**
21600	27000	32400	36000	39600	45000	**600**
24414	30518	36621	40690	44759	50863	**625**
27463	34328	41194	45771	50348	57214	**650**
30755	38443	46132	51258	56384	64072	**675**
34300	42875	51450	57167	62883	71458	**700**
38108	47635	57162	63513	69864	79391	**725**
42188	52734	63281	70313	77344	87891	**750**
46548	58186	69823	77581	85339	96976	**775**
51200	64000	76800	85333	93867	106667	**800**
56152	70189	84227	93586	102945	116982	**825**
61413	76766	92119	102354	112590	127943	**850**
66992	83740	100488	111654	122819	139567	**875**
72900	91125	109350	121500	133650	151875	**900**

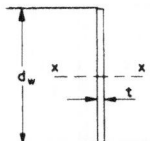

SECOND MOMENTS OF AREA (cm⁴)
OF RECTANGULAR PLATES
about axis x—x

Depth d, mm	THICKNESS t MILLIMETRES					
	3	4	5	6	8	10
1000	25000	33333	41667	50000	66667	83333
1100	33275	44367	55458	66550	88733	110917
1200	43200	57600	72000	86400	115200	144000
1300	54925	73233	91542	109850	146467	183083
1400	68600	91467	114333	137200	182933	228667
1500	84375	112500	140625	168750	225000	281250
1600	102400	136533	170667	204800	273067	341333
1700	122825	163767	204708	245650	327533	409417
1800	145800	194400	243000	291600	388800	486000
1900	171475	228633	285792	342950	457267	571583
2000	200000	266667	333333	400000	533333	666667
2100	231525	308700	385875	463050	617400	771750
2200	266200	354933	443667	532400	709867	887333
2300	304175	405567	506958	608350	811133	1013917
2400	345600	460800	576000	691200	921600	1152000
2500	390625	520833	651042	781250	1041667	1302083
2600	439400	585867	732333	878800	1171733	1464667
2700	492075	656100	820125	984150	1312200	1640250
2800	548800	731733	914667	1097600	1463467	1829333
2900	609725	812967	1016208	1219450	1625933	2032417
3000	675000	900000	1125000	1350000	1800000	2250000
3100	744775	993033	1241292	1489550	1986067	2482583
3200	819200	1092267	1365333	1638400	2184533	2730667
3300	898425	1197900	1497375	1796850	2395800	2994750
3400	982600	1310133	1637667	1965200	2620267	3275333
3500	1071875	1429167	1786458	2143750	2858333	3572917
3600	1166400	1555200	1944000	2332800	3110400	3888000
3700	1266325	1688433	2110542	2532650	3376867	4221083
3800	1371800	1829067	2286333	2743600	3658133	4572667
3900	1482975	1977300	2471625	2965950	3954600	4943250
4000	1600000	2133333	2666667	3200000	4266667	5333333
4100	1723025	2297333	2871708	3446050	4594733	5743417
4200	1852200	2469600	3087000	3704400	4939200	6174000
4300	1987675	2650233	3312792	3975350	5300467	6625583
4400	2129600	2839467	3549333	4259200	5678933	7098667
4500	2278125	3037500	3796875	4556250	6075000	7593750
4600	2433400	3244533	4055667	4866050	6489067	8111333
4700	2595575	3460767	4325958	5191150	6921533	8651917
4800	2764800	3686400	4608000	5529600	7372800	9216000
4900	2941225	3921633	4902042	5882450	7843267	9804083
5000	3125000	4166667	5208333	6250000	8333333	10416667

SECOND MOMENTS OF AREA (cm⁴)
OF RECTANGULAR PLATES
about axis x—x

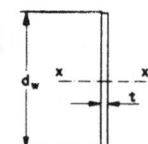

THICKNESS t MILLIMETRES						Depth
12	15	18	20	22	25	d_w mm
100000	125000	150000	166667	183333	208333	1000
133100	166375	199650	221833	244017	277292	1100
172800	216000	259200	288000	316800	360000	1200
219700	274625	329550	366167	402783	457708	1300
274400	343000	411600	457333	503067	571667	1400
337500	421875	506250	562500	618750	703125	1500
409600	512000	614400	682667	750933	853333	1600
491300	614125	736950	818833	900717	1023542	1700
583200	729000	874800	972000	1069200	1215000	1800
685900	857375	1028850	1143167	1257483	1428958	1900
800000	1000000	1200000	1333333	1466667	1666667	2000
926100	1157625	1389150	1543500	1697850	1929375	2100
1064800	1331000	1597200	1774667	1952133	2218333	2200
1216700	1520875	1825050	2027833	2230617	2534792	2300
1382400	1728000	2073600	2304000	2534400	2880000	2400
1562500	1953125	2343750	2604167	2864583	3255208	2500
1757600	2197000	2636400	2929333	3222267	3661667	2600
1968300	2460375	2952450	3280500	3608550	4100625	2700
2195200	2744000	3292800	3658667	4024533	4573333	2800
2438900	3048625	3658350	4064833	4471317	5081042	2900
2700000	3375000	4050000	4500000	4950000	5625000	3000
2979100	3723875	4468650	4965167	5461683	6206458	3100
3276800	4096000	4915200	5461333	6007467	6826667	3200
3593700	4492125	5390550	5989500	6588450	7486875	3300
3930400	4913000	5895600	6550667	7205733	8188333	3400
4287500	5359375	6431250	7145833	7860417	8932292	3500
4665600	5832000	6998400	7776000	8553600	9720000	3600
5065300	6331625	7597950	8442167	9286383	10552708	3700
5487200	6859000	8230800	9145333	10059867	11431667	3800
5931900	7414875	8897850	9886500	10875150	12358125	3900
6400000	8000000	9600000	10666667	11733333	13333333	4000
6892100	8615125	10338150	11486833	12635517	14358542	4100
7408800	9261000	11113200	12348000	13582800	15435000	4200
7950700	9938375	11926050	13251167	14576283	16563958	4300
8518400	10648000	12777600	14197333	15617067	17746667	4400
9112500	11390625	13668750	15187500	16706250	18984375	4500
9733600	12167000	14600400	16222667	17844933	20278333	4600
10382300	12977875	15573450	17303833	19034217	21629792	4700
11059200	13824000	16588800	18432000	20275200	23040000	4800
11764900	14706125	17647350	19608167	21568983	24510208	4900
12500000	15625000	18750000	20833333	22916667	26041667	5000

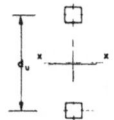

SECOND MOMENT
OF A PAIR OF UNIT AREAS

about axis x—x

Distance d_u mm	0	5	10	15	20	25	30	35	40	45
500	1250	1275	1301	1326	1352	1378	1405	1431	1458	1485
550	1513	1540	1568	1596	1625	1653	1682	1711	1741	1770
600	1800	1830	1861	1891	1922	1953	1985	2016	2048	2080
650	2113	2145	2178	2211	2245	2278	2312	2346	2381	2415
700	2450	2485	2521	2556	2592	2628	2665	2701	2738	2775
750	2813	2850	2888	2926	2965	3003	3042	3081	3121	3160
800	3200	3240	3281	3321	3362	3403	3445	3486	3528	3570
850	3613	3655	3698	3741	3785	3828	3872	3916	3961	4005
900	4050	4095	4141	4186	4232	4278	4325	4371	4418	4465
950	4513	4560	4608	4656	4705	4753	4802	4851	4901	4950
1000	5000	5050	5101	5151	5202	5253	5305	5356	5408	5460
1050	5513	5565	5618	5671	5725	5778	5832	5886	5941	5995
1100	6050	6105	6161	6216	6272	6328	6385	6441	6498	6555
1150	6613	6670	6728	6786	6845	6903	6962	7021	7081	7140
1200	7200	7260	7321	7381	7442	7503	7565	7626	7688	7750
1250	7813	7875	7938	8001	8065	8128	8192	8256	8321	8385
1300	8450	8515	8581	8646	8712	8778	8845	8911	8978	9045
1350	9113	9180	9248	9316	9385	9453	9522	9591	9661	9730
1400	9800	9870	9941	10011	10082	10153	10225	10296	10368	10440
1450	10513	10585	10658	10731	10805	10878	10952	11026	11101	11175
1500	11250	11325	11401	11476	11552	11628	11705	11781	11858	11935
1550	12013	12090	12168	12246	12325	12403	12482	12561	12641	12720
1600	12800	12880	12961	13041	13122	13203	13285	13366	13448	13530
1650	13613	13695	13778	13861	13945	14028	14112	14196	14281	14365
1700	14450	14535	14621	14706	14792	14878	14965	15051	15138	15225
1750	15313	15400	15488	15576	15665	15753	15842	15931	16021	16110
1800	16200	16290	16381	16471	16562	16653	16745	16836	16928	17020
1850	17113	17205	17298	17391	17485	17578	17672	17766	17861	17955
1900	18050	18145	18241	18336	18432	18528	18625	18721	18818	18915
1950	19013	19110	19208	19306	19405	19503	19602	19701	19801	19900
2000	20000	20100	20201	20301	20402	20503	20605	20706	20808	20910
2050	21013	21115	21218	21321	21425	21528	21632	21736	21841	21945
2100	22050	22155	22261	22366	22472	22578	22685	22791	22898	23005
2150	23113	23220	23328	23436	23545	23653	23762	23871	23981	24090
2200	24200	24310	24421	24531	24642	24753	24865	24976	25088	25200
2250	25313	25425	25538	25651	25765	25878	25992	26106	26221	26335
2300	26450	26565	26681	26796	26912	27028	27145	27261	27378	27495
2350	27613	27730	27848	27966	28085	28203	28322	28441	28561	28680
2400	28800	28920	29041	29161	29282	29403	29525	29646	29768	29890
2450	30013	30135	30258	30381	30505	30628	30752	30876	31001	31125
2500	31250	31375	31501	31626	31752	31878	32005	32131	32258	32385
2550	32513	32640	32768	32896	33025	33153	33282	33411	33541	33670
2600	33800	33930	34061	34191	34322	34453	34585	34716	34848	34980
2650	35113	35245	35378	35511	35645	35778	35912	36046	36181	36315
2700	36450	36585	36721	36856	36992	37128	37265	37401	37538	37675

Second moments are tabulated in cm⁴ and are for unit areas of 1 cm² each.

SECOND MOMENT
OF A PAIR OF UNIT AREAS

about axis x—x

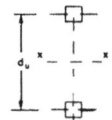

Distance d_u mm	0	5	10	15	20	25	30	35	40	45
2750	37813	37950	38088	38226	38365	38503	38642	38781	38921	39060
2800	39200	39340	39481	39621	39762	39903	40045	40186	40328	40470
2850	40613	40755	40898	41041	41185	41328	41472	41616	41761	41905
2900	42050	42195	42341	42486	42632	42778	42925	43071	43218	43365
2950	43513	43660	43808	43956	44105	44253	44402	44551	44701	44850
3000	45000	45150	45301	45451	45602	45753	45905	46056	46208	46360
3050	46513	46665	46818	46971	47125	47278	47432	47586	47741	47895
3100	48050	48205	48361	48516	48672	48828	48985	49141	49298	49455
3150	49613	49770	49928	50086	50245	50403	50562	50721	50881	51040
3200	51200	51360	51521	51681	51842	52003	52165	52326	52488	52650
3250	52813	52975	53138	53301	53465	53628	53792	53956	54121	54285
3300	54450	54615	54781	54946	55112	55278	55445	55611	55778	55945
3350	56113	56280	56448	56616	56785	56953	57122	57291	57461	57630
3400	57800	57970	58141	58311	58482	58653	58825	58996	59168	59340
3450	59513	59685	59858	60031	60205	60378	60552	60726	60901	61075
3500	61250	61425	61601	61776	61952	62128	62305	62481	62658	62835
3550	63013	63190	63368	63546	63725	63903	64082	64261	64441	64620
3600	64800	64980	65161	65341	65522	65703	65885	66066	66248	66430
3650	66613	66795	66978	67161	67345	67528	67712	67896	68081	68265
3700	68450	68635	68821	69006	69192	69378	69565	69751	69938	70125
3750	70313	70500	70688	70876	71065	71253	71442	71631	71821	72010
3800	72200	72390	72581	72771	72962	73153	73345	73536	73728	73920
3850	74113	74305	74498	74691	74885	75078	75272	75466	75661	75855
3900	76050	76245	76441	76636	76832	77028	77225	77421	77618	77815
3950	78013	78210	78408	78606	78805	79003	79202	79401	79601	79800
4000	80000	80200	80401	80601	80802	81003	81205	81406	81608	81810
4050	82013	82215	82418	82621	82825	83028	83232	83436	83641	83845
4100	84050	84255	84461	84666	84872	85078	85285	85491	85698	85905
4150	86113	86320	86528	86736	86945	87153	87362	87571	87781	87990
4200	88200	88410	88621	88831	89042	89253	89465	89676	89888	90100
4250	90313	90525	90738	90951	91165	91378	91592	91806	92021	92235
4300	92450	92665	92881	93096	93312	93528	93745	93961	94178	94395
4350	94613	94830	95048	95266	95485	95703	95922	96141	96361	96580
4400	96800	97020	97241	97461	97682	97903	98125	98346	98568	98790
4450	99013	99235	99458	99681	99905	100128	100352	100576	100801	101025
4500	101250	101475	101701	101926	102152	102378	102605	102831	103058	103285
4550	103513	103740	103968	104196	104425	104653	104882	105111	105341	105570
4600	105800	106030	106261	106491	106722	106953	107185	107416	107648	107880
4650	108113	108345	108578	108811	109045	109278	109512	109746	109981	110215
4700	110450	110685	110921	111156	111392	111628	111865	112101	112338	112575
4750	112813	113050	113288	113526	113765	114003	114242	114481	114721	114960
4800	115200	115440	115681	115921	116162	116403	116645	116886	117128	117370
4850	117613	117855	118098	118341	118585	118828	119072	119316	119561	119805
4900	120050	120295	120541	120786	121032	121278	121525	121771	122018	122265
4950	122513	122760	123008	123256	123505	123753	124002	124251	124501	124750

Second moments are tabulated in cm⁴ and are for unit areas of 1 cm² each.

GEOMETRICAL PROPERTIES OF PLANE SECTIONS

Section		Area	Position of Centroid	Moments of Inertia	Section Moduli
TRIANGLE		$A = \dfrac{bh}{2}$	$e_x = \dfrac{h}{3}$	$I_{XX} = bh^3/36$ $I_{YY} = hb^3/48$ $I_{aa} = bh^3/4$ $I_{bb} = bh^3/12$	Z_{XX} base $= bh^2/12$ apex $= bh^2/24$ $Z_{YY} = bh^2/24$
RECTANGLE		$A = bd$	$e_x = \dfrac{h}{2}$	$I_{XX} = bd^3/12$ $I_{YY} = db^3/12$ $I_{bb} = bd^3/3$	$Z_{XX} = bd^2/6$ $Z_{YY} = db^2/6$
RECTANGLE axis on diagonal		$A = bd$	$e_x = \dfrac{bd}{\sqrt{b^2+d^2}}$	$I_{XX} = \dfrac{b^3 d^3}{6(b^2+d^2)}$	$Z_{XX} = \dfrac{b^2 d^2}{6\sqrt{b^2+d^2}}$
RECTANGLE axis through C.G.		$A = bd$	$e_x = \dfrac{b.\sin\theta + d.\cos\theta}{2}$	$I_{XX} = \dfrac{bd(b^2\sin^2\theta + d^2\cos^2\theta)}{12}$	$Z_{XX} = \dfrac{bd(b^2\sin^2\theta + d^2\cos^2\theta)}{6(b.\sin\theta + d.\cos\theta)}$
SQUARE		$A = s^2$	$e_x = \dfrac{s}{2}$ $e_v = \dfrac{s}{\sqrt{2}}$	$I_{XX} = I_{YY} = s^4/12$ $I_{bb} = s^4/3$ $I_{VV} = s^4/12$	$Z_{XX} = Z_{YY} = \dfrac{s^3}{6}$ $Z_{VV} = \dfrac{s^3}{6\sqrt{2}}$
TRAPEZIUM		$A = \dfrac{d(a+b)}{2}$	$e_{x_1} = \dfrac{d(2a+b)}{3(a+b)}$	$I_{XX} = \dfrac{d^3(a^2+4ab+b^2)}{36(a+b)}$ $I_{YY} = \dfrac{d(a^3+a^2b+ab^2+b^3)}{48}$	$Z_{XX} = \dfrac{I_{XX}}{d-e_x}$ (two values) $Z_{YY} = \dfrac{2 I_{YY}}{b}$
DIAMOND		$A = \dfrac{bd}{2}$	$e_x = \dfrac{d}{2}$	$I_{XX} = \dfrac{bd^3}{48}$ $I_{YY} = \dfrac{db^3}{48}$	$Z_{XX} = \dfrac{bd^2}{24}$ $Z_{YY} = \dfrac{db^2}{24}$
HEXAGON		$A = 0.866d^2$	$e_x = 0.866s$ $= d/2$	$I_{XX} = I_{YY} = I_{VV}$ $= 0.0601d^4$	$Z_{XX} = 0.1203d^3$ $Z_{YY} = Z_{VV}$ $= 0.1042d^3$

GEOMETRICAL PROPERTIES OF PLANE SECTIONS

Section		Area	Position of Centroid	Moments of Inertia	Section Moduli
OCTAGON		$A = 0.8284 d^2$ $s = 0.4142 d$	$e_x = \dfrac{d}{2}$ $e_v = 0.541 d$	$I_{XX} = I_{YY} = I_{VV}$ $= 0.0547 d^4$	$Z_{XX} = Z_{YY}$ $= 0.1095 d^3$ $Z_{VV} = 0.1011 d^3$
POLYGON	 *n sides* *Regular figure*	$A = \dfrac{n s^2 \cot\theta}{4}$ $A = n r^2 \tan\theta$ $A = \dfrac{n R^2 \sin 2\theta}{2}$	$e = r$ or R depending on the axis and value of n	$I_1 = I_2$ $= \dfrac{A(6R^2 - s^2)}{24}$ $= \dfrac{A(12r^2 + s^2)}{48}$	$Z = \dfrac{I}{e}$
CIRCLE		$A = \pi r^2$ $A = 0.7854 d^2$	$e = r = \dfrac{d}{2}$	$I = \dfrac{\pi d^4}{64}$ $I = 0.7854 r^4$	$Z = \dfrac{\pi d^3}{32}$ $Z = 0.7854 r^3$
SEMI-CIRCLE		$A = 1.5708 r^2$	$e_x = 0.424 r$	$I_{XX} = 0.1098 r^4$ $I_{YY} = 0.3927 r^4$	Z_{XX} base $= 0.2587 r^3$ crown $= 0.1907 r^3$ $Z_{YY} = 0.3927 r^3$
SEGMENT		$A = $ $\dfrac{r^2}{2}\left(\dfrac{\pi\theta^\circ}{180^\circ} - \sin\theta\right)$	$e_0 = \dfrac{c^3}{12A}$ $e_x = e_0 - r.\cos\dfrac{\theta}{2}$	$I_{XX} = \dfrac{r^4}{16}\left(\dfrac{\pi\theta^\circ}{90^\circ} - \sin 2\theta\right)$ $- \dfrac{20r^4(1-\cos\theta)^3}{\pi\theta^\circ - 180^\circ \sin\theta}$ $I_{YY} = \dfrac{r^4}{48}\left(\dfrac{\pi\theta^\circ}{30^\circ} - 8\sin\theta + \sin 2\theta\right)$	Z_{XX} base $= I_{XX}/e_x$ crown $= \dfrac{I_{XX}}{b - e_x}$ $Z_{YY} = \dfrac{2I_{YY}}{c}$
SECTOR		$A = \dfrac{\theta^\circ}{360^\circ}\pi r^2$	$e_x = \dfrac{2}{3} r \dfrac{c}{a}$ $e_x = \dfrac{r^2 c}{3A}$	$I_{XX} = I_0 - \dfrac{360^\circ}{\theta^\circ\pi}\sin^2\dfrac{\theta}{2}\cdot\dfrac{4r^4}{9}$ $I_{YY} = \dfrac{r^4}{8}\left(\dfrac{\pi\theta^\circ}{180} - \sin\theta\right)$ $I_0 = \dfrac{r^4}{8}\left(\dfrac{\pi\theta^\circ}{180} + \sin\theta\right)$	Z_{XX} centre $= I_{XX}/e_x$ crown $= \dfrac{I_{XX}}{r - e_x}$ $Z_{YY} = \dfrac{2I_{YY}}{c}$
QUADRANT		$A = \dfrac{\pi r^2}{4}$	$e_x = 0.424 r$ $e_v = 0.6 r$ $e_u = 0.707 r$	$I_{XX} = I_{YY} = 0.0549 r^4$ $I_{bb} = 0.1963 r^4$ $I_{UU} = 0.0714 r^4$ $I_{VV} = 0.0384 r^4$	Minimum Values $Z_{XX} = Z_{YY}$ $= 0.0953 r^3$ $Z_{UU} = 0.1009 r^3$ $Z_{VV} = 0.064 r^3$
COMPLEMENT		$A = 0.2146 r^2$	$e_x = 0.777 r$ $e_v = 1.098 r$ $e_u = 0.707 r$ $e_a = 0.316 r$ $e_b = 0.391 r$	$I_{XX} = I_{YY} = 0.0076 r^4$ $I_{UU} = 0.012 r^4$ $I_{VV} = 0.0031 r^4$	Minimum Values $Z_{XX} = Z_{YY}$ $= 0.0097 r^3$ $Z_{UU} = 0.017 r^3$ $Z_{VV} = 0.0079 r^3$

GEOMETRICAL PROPERTIES OF PLANE SECTIONS

Section		Area	Position of Centroid	Moments of Inertia	Section Moduli
ELLIPSE		$A = \pi ab$	$e_x = a$ $e_y = b$	$I_{XX} = 0.7854ba^3$ $I_{YY} = 0.7854ab^3$	$Z_{XX} = 0.7854ba^2$ $Z_{YY} = 0.7854ab^2$
SEMI-ELLIPSE		$A = \dfrac{\pi ab}{2}$	$e_x = 0.424a$ $e_y = b$	$I_{XX} = 0.1098ba^3$ $I_{YY} = 0.3927ab^3$ $I_{base} = 0.3927ba^3$	$Z_{XX} - base$ $= 0.2587ba^2$ $Z_{XX} - crown$ $= 0.1907ba^2$ $Z_{YY} = 0.3927ab^2$
¼ ELLIPSE		$A = 0.7854ab$	$e_x = 0.424a$ $e_y = 0.424b$	$I_{XX} = 0.0549ba^3$ $I_{YY} = 0.0549ab^3$ $I_{b_1 a_1} = 0.1963ba^3$ $I_{b_1 c_1} = 0.1963ab^3$	$Z_{XX} - base$ $= 0.1293ba^2$ $Z_{XX} - crown$ $= 0.0953ba^2$ $Z_{YY} - base$ $= 0.1293ab^2$ $Z_{YY} - crown$ $= 0.0953ab^2$
COMPLEMENT		$A = 0.2146ab$	$e_x = 0.777a$ $e_y = 0.777b$	$I_{XX} = 0.0076ba^3$ $I_{YY} = 0.0076ab^3$	$Z_{XX} - base$ $= 0.0338ba^2$ $Z_{XX} - apex$ $= 0.0097ba^2$ $Z_{YY} - base$ $= 0.0338ab^2$ $Z_{YY} - apex$ $= 0.0097ab^2$
FULL PARABOLA		$A = \dfrac{4ab}{3}$	$e_x = \dfrac{2a}{5}$ $e_y = b$	$I_{XX} = 0.0914ba^3$ $I_{YY} = 0.2666ab^3$ $I_{base} = 0.3048ba^3$	$Z_{XX} - base$ $= 0.2286ba^2$ $Z_{XX} - crown$ $= 0.1524ba^2$ $Z_{YY} = 0.2666ab^2$
SEMI-PARABOLA		$A = \dfrac{2ab}{3}$	$e_x = \dfrac{2a}{5}$ $e_y = \dfrac{3b}{8}$	$I_{XX} = 0.0457ba^3$ $I_{YY} = 0.0396ab^3$ $I_{b_1 a_1} = 0.1524ba^3$ $I_{b_1 c_1} = 0.1333ab^3$	$Z_{XX} - base$ $= 0.1143ba^2$ $Z_{XX} - crown$ $= 0.076ba^2$ $Z_{YY} - base$ $= 0.1055ab^2$ $Z_{YY} - crown$ $= 0.0633ab^2$
COMPLEMENT		$A = \dfrac{ab}{3}$	$e_x = \dfrac{7a}{10}$ $e_y = \dfrac{3b}{4}$	$I_{XX} = 0.0176ba^3$ $I_{YY} = 0.0125ab^3$ $I_{a_1 b_1} = 0.181ba^3$ $I_{b_1 c_1} = 0.2ab^3$	$Z_{XX} - base$ $= 0.0587ba^2$ $Z_{XX} - apex$ $= 0.0252ba^2$ $Z_{YY} - base$ $= 0.05ab^2$ $Z_{YY} - apex$ $= 0.0167ab^2$
FILLET		$A = \dfrac{s^2}{6}$	$e_u = e_v = \dfrac{4s}{5}$	$I_{UU} = I_{VV} = 0.005245s^4$ $I_{ab} = 0.1119a^4$	$Z_{UU} = Z_{VV}$ $base$ $= 0.0262a^3$ $apex$ $= 0.0066a^3$

Plastic moduli

PLASTIC MODULUS OF TWO FLANGES

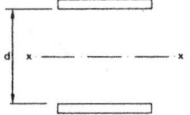

Dist d mm	Plastic Modulus Sxx(cm³) For Thickness t(mm)												
	15	20	25	30	35	40	45	50	55	60	65	70	75
1000	15.2	20.4	25.6	30.9	36.2	41.6	47.0	52.5	58.0	63.6	69.2	74.9	80.6
1100	16.7	22.4	28.1	33.9	39.7	45.6	51.5	57.5	63.5	69.6	75.7	81.9	88.1
1200	18.2	24.4	30.6	36.9	43.2	49.6	56.0	62.5	69.0	75.6	82.2	88.9	95.6
1300	19.7	26.4	33.1	39.9	46.7	53.6	60.5	67.5	74.5	81.6	88.7	95.9	103
1400	21.2	28.4	35.6	42.9	50.2	57.6	65.0	72.5	80.0	87.6	95.2	103	111
1500	22.7	30.4	38.1	45.9	53.7	61.6	69.5	77.5	85.5	93.6	102	110	118
1600	24.2	32.4	40.6	48.9	57.2	65.6	74.0	82.5	91.0	99.6	108	117	126
1700	25.7	34.4	43.1	51.9	60.7	69.6	78.5	87.5	96.5	106	115	124	133
1800	27.2	36.4	45.6	54.9	64.2	73.6	83.0	92.5	102	112	121	131	141
1900	28.7	38.4	48.1	57.9	67.7	77.6	87.5	97.5	108	118	128	138	148
2000	30.2	40.4	50.6	60.9	71.2	81.6	92.0	102	113	124	134	145	156
2100	31.7	42.4	53.1	63.9	74.7	85.6	96.5	107	119	130	141	152	163
2200	33.2	44.4	55.6	66.9	78.2	89.6	101	112	124	136	147	159	171
2300	34.7	46.4	58.1	69.9	81.7	93.6	106	117	130	142	154	166	178
2400	36.2	48.4	60.6	72.9	85.2	97.6	110	122	135	148	160	173	186
2500	37.7	50.4	63.1	75.9	88.7	102	115	127	141	154	167	180	193
2600	39.2	52.4	65.6	78.9	92.2	106	119	132	146	160	173	187	201
2700	40.7	54.4	68.1	81.9	95.7	110	124	137	152	166	180	194	208
2800	42.2	56.4	70.6	84.9	99.2	114	128	142	157	172	186	201	216
2900	43.7	58.4	73.1	87.9	103	118	133	147	163	178	193	208	223
3000	45.2	60.4	75.6	90.9	106	122	137	152	168	184	199	215	231
3100	46.7	62.4	78.1	93.9	110	126	142	157	174	190	206	222	238
3200	48.2	64.4	80.6	96.9	113	130	146	162	179	196	212	229	246
3300	49.7	66.4	83.1	99.9	117	134	151	167	185	202	219	236	253
3400	51.2	68.4	85.6	103	120	138	155	172	190	208	225	243	261
3500	52.7	70.4	88.1	106	124	142	160	177	196	214	232	250	268
3600	54.2	72.4	90.6	109	127	146	164	182	201	220	238	257	276
3700	55.7	74.4	93.1	112	131	150	169	187	207	226	245	264	283
3800	57.2	76.4	95.6	115	134	154	173	192	212	232	251	271	291
3900	58.7	78.4	98.1	118	138	158	178	197	218	238	258	278	298
4000	60.2	80.4	101	121	141	162	182	202	223	244	264	285	306
4100	61.7	82.4	103	124	145	166	187	207	229	250	271	292	313
4200	63.2	84.4	106	127	148	170	191	212	234	256	277	299	321
4300	64.7	86.4	108	130	152	174	196	217	240	262	284	306	328
4400	66.2	88.4	111	133	155	178	200	222	245	268	290	313	336
4500	67.7	90.4	113	136	159	182	205	227	251	274	297	320	343
4600	69.2	92.4	116	139	162	186	209	232	256	280	303	327	351
4700	70.7	94.4	118	142	166	190	214	237	262	286	310	334	358
4800	72.2	96.4	121	145	169	194	218	242	267	292	316	341	366
4900	73.7	98.4	123	148	173	198	223	247	273	298	323	348	373
5000	75.2	100.0	126	151	176	202	227	252	278	304	329	355	381

PLASTIC MODULUS OF RECTANGLES

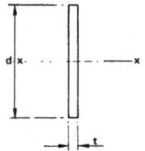

Depth d mm	Plastic Modulus S_{xx}(cm^3) For Thickness t(mm)									
	5	6	7	8	9	10	12.5	15	20	25
25	0.78	0.93	1.09	1.25	1.41	1.56	1.95	2.34	3.13	3.91
50	3.13	3.75	4.37	5.00	5.62	6.25	7.81	9.37	12.5	15.6
75	7.03	8.44	9.84	11.3	12.7	14.1	17.6	21.1	28.1	35.2
100	12.5	15.0	17.5	20.0	22.5	25.0	31.2	37.5	50.0	62.5
125	19.5	23.4	27.3	31.2	35.2	39.1	48.8	58.6	78.1	97.7
150	28.1	33.8	39.4	45.0	50.6	56.2	70.3	84.4	112	141
175	38.3	45.9	53.6	61.2	68.9	76.6	95.7	115	153	191
200	50.0	60.0	70.0	80.0	90.0	100.0	125	150	200	250
225	63.3	75.9	88.6	101	114	127	158	190	253	316
250	78.1	93.7	109	125	141	156	195	234	312	391
275	94.5	113	132	151	170	189	236	284	378	473
300	112	135	158	180	203	225	281	338	450	563
325	132	158	185	211	238	264	330	396	528	660
350	153	184	214	245	276	306	383	459	613	766
375	176	211	246	281	316	352	439	527	703	879
400	200	240	280	320	360	400	500	600	800	1000
425	226	271	316	361	406	452	564	677	903	1130
450	253	304	354	405	456	506	633	759	1010	1270
475	282	338	395	451	508	564	705	846	1130	1410
500	312	375	437	500	562	625	781	937	1250	1560
525	345	413	482	551	620	689	861	1030	1380	1720
550	378	454	529	605	681	756	945	1130	1510	1890
575	413	496	579	661	744	827	1030	1240	1650	2070
600	450	540	630	720	810	900	1130	1350	1800	2250
625	488	586	684	781	879	977	1220	1460	1950	2440
650	528	634	739	845	951	1060	1320	1580	2110	2640
675	570	683	797	911	1030	1140	1420	1710	2280	2850
700	613	735	858	980	1100	1230	1530	1840	2450	3060
725	657	788	920	1050	1180	1310	1640	1970	2630	3290
750	703	844	984	1120	1270	1410	1760	2110	2810	3520
775	751	901	1050	1200	1350	1500	1880	2250	3000	3750
800	800	960	1120	1280	1440	1600	2000	2400	3200	4000
825	851	1020	1190	1360	1530	1700	2130	2550	3400	4250
850	903	1080	1260	1440	1630	1810	2260	2710	3610	4520
875	957	1150	1340	1530	1720	1910	2390	2870	3830	4790
900	1010	1210	1420	1620	1820	2020	2530	3040	4050	5060

PLASTIC MODULUS OF RECTANGLES

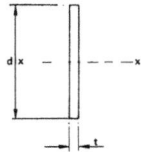

Depth d mm	Plastic Modulus S_{xx} (cm³) For Thickness t(mm)									
	5	6	7	8	9	10	12.5	15	20	25
1000	1250	1500	1750	2000	2250	2500	3120	3750	5000	6250
1100	1510	1810	2120	2420	2720	3020	3780	4540	6050	7560
1200	1800	2160	2520	2880	3240	3600	4500	5400	7200	9000
1300	2110	2530	2960	3380	3800	4220	5280	6340	8450	10600
1400	2450	2940	3430	3920	4410	4900	6130	7350	9800	12300
1500	2810	3370	3940	4500	5060	5620	7030	8440	11200	14100
1600	3200	3840	4480	5120	5760	6400	8000	9600	12800	16000
1700	3610	4330	5060	5780	6500	7220	9030	10800	14400	18100
1800	4050	4860	5670	6480	7290	8100	10100	12100	16200	20200
1900	4510	5410	6320	7220	8120	9020	11300	13500	18000	22600
2000	5000	6000	7000	8000	9000	10000	12500	15000	20000	25000
2100	5510	6620	7720	8820	9920	11000	13800	16500	22100	27600
2200	6050	7260	8470	9680	10900	12100	15100	18100	24200	30200
2300	6610	7930	9260	10600	11900	13200	16500	19800	26500	33100
2400	7200	8640	10100	11500	13000	14400	18000	21600	28800	36000
2500	7810	9370	10900	12500	14100	15600	19500	23400	31200	39100
2600	8450	10100	11800	13500	15200	16900	21100	25400	33800	42200
2700	9110	10900	12800	14600	16400	18200	22800	27300	36400	45600
2800	9800	11800	13700	15700	17600	19600	24500	29400	39200	49000
2900	10500	12600	14700	16800	18900	21000	26300	31500	42000	52600
3000	11200	13500	15700	18000	20200	22500	28100	33700	45000	56200
3100	12000	14400	16800	19200	21600	24000	30000	36000	48000	60100
3200	12800	15400	17900	20500	23000	25600	32000	38400	51200	64000
3300	13600	16300	19100	21800	24500	27200	34000	40800	54400	68100
3400	14400	17300	20200	23100	26000	28900	36100	43300	57800	72200
3500	15300	18400	21400	24500	27600	30600	38300	45900	61200	76600
3600	16200	19400	22700	25900	29200	32400	40500	48600	64800	81000
3700	17100	20500	24000	27400	30800	34200	42800	51300	68400	85600
3800	18000	21700	25300	28900	32500	36100	45100	54100	72200	90200
3900	19000	22800	26600	30400	34200	38000	47500	57000	76000	95100
4000	20000	24000	28000	32000	36000	40000	50000	60000	80000	100000
4100	21000	25200	29400	33600	37800	42000	52500	63000	84000	105000
4200	22100	26500	30900	35300	39700	44100	55100	66200	88200	110000
4300	23100	27700	32400	37000	41600	46200	57800	69300	92400	116000
4400	24200	29000	33900	38700	43600	48400	60500	72600	96800	121000
4500	25300	30400	35400	40500	45600	50600	63300	75900	101000	127000
4600	26500	31700	37000	42300	47600	52900	66100	79300	106000	132000
4700	27600	33100	38700	44200	49700	55200	69000	82800	110000	138000
4800	28800	34600	40300	46100	51800	57600	72000	86400	115000	144000
4900	30000	36000	42000	48000	54000	60000	75000	90000	120000	150000
5000	31200	37500	43700	50000	56200	62500	78100	93700	125000	156000

Formulae for rigid frames

Frame I

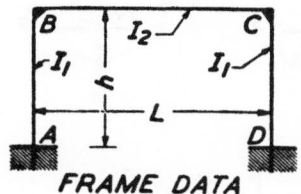

FRAME DATA

Coefficients:

$$k = \frac{I_2}{I_1} \cdot \frac{h}{L}$$

$$N_1 = k + 2 \qquad N_2 = 6k + 1$$

---

w per unit length

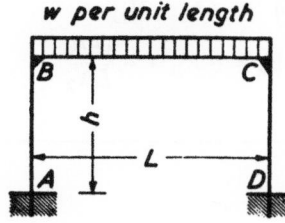

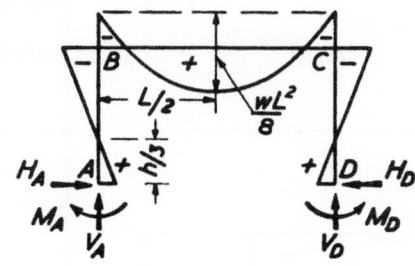

$$M_A = M_D = \frac{wL^2}{12N_1} \qquad M_B = M_C = -\frac{wL^2}{6N_1} = -2M_A$$

$$M_{\max} = \frac{wL^2}{8} + M_B \qquad V_A = V_D = \frac{wL}{2} \qquad H_A = H_D = \frac{3M_A}{h}$$

w per unit length

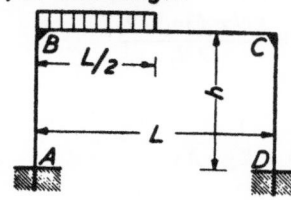

$$M_A = \frac{wL^2}{8}\left[\frac{1}{3N_1} - \frac{1}{8N_2}\right] \qquad M_B = -\frac{wL^2}{8}\left[\frac{2}{3N_1} + \frac{1}{8N_2}\right]$$

$$M_D = \frac{wL^2}{8}\left[\frac{1}{3N_1} + \frac{1}{8N_2}\right] \qquad M_C = -\frac{wL^2}{8}\left[\frac{2}{3N_1} - \frac{1}{8N_2}\right]$$

$$V_D = \frac{wL}{8}\left[1 - \frac{1}{4N_2}\right] \qquad V_A = \frac{wL}{2} - V_D \qquad H_A = H_D = \frac{wL^2}{8hN_1}$$

Extract: 'Kleinlogel, Rahmenformeln' 11. Auflage Berlin—Verlag von Wilhelm Ernst & Sohn.

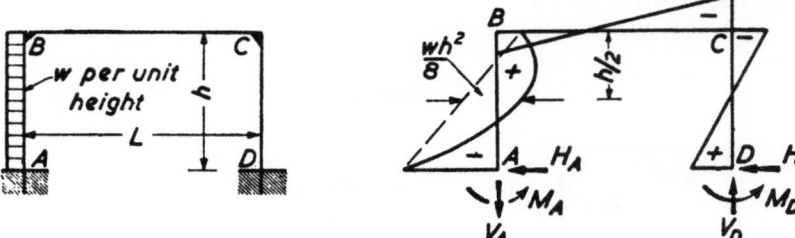

$$M_A = \frac{wh^2}{4}\left[-\frac{k+3}{6N_1} - \frac{4k+1}{N_2}\right] \qquad M_B = \frac{wh^2}{4}\left[-\frac{k}{6N_1} + \frac{2k}{N_2}\right]$$

$$M_D = \frac{wh^2}{4}\left[-\frac{k+3}{6N_1} + \frac{4k+1}{N_2}\right] \qquad M_C = \frac{wh^2}{4}\left[-\frac{k}{6N_1} - \frac{2k}{N_2}\right]$$

$$H_D = \frac{wh(2k+3)}{8N_1} \qquad H_A = -(wh - H_D) \qquad V_A = -V_D = -\frac{wh^2k}{LN_2}$$

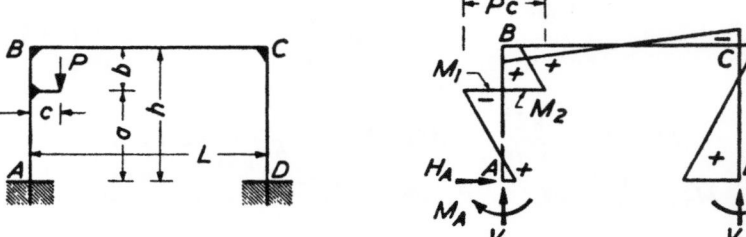

$$\text{Constants}: \quad a_1 = \frac{a}{h} \qquad b_1 = \frac{b}{h}$$

$$X_1 = \frac{Pc}{2N_1}[1 + 2b_1k - 3b_1^2(k+1)] \qquad X_2 = \frac{Pcka_1(3a_1 - 2)}{2N_1}$$

$$X_3 = \frac{3Pcka_1}{N_2}$$

$$M_A = +X_1 - \left(\frac{Pc}{2} - X_3\right) \qquad M_B = +X_2 + X_3$$

$$M_D = +X_1 + \left(\frac{Pc}{2} - X_3\right) \qquad M_C = +X_2 - X_3$$

$$H_A = H_D = \frac{Pc}{2h} + \frac{X_1 - X_2}{h} \qquad V_D = \frac{2X_3}{L} \qquad V_A = P - V_D$$

$$M_1 = M_A - H_A a \qquad M_2 = M_B + H_D b$$

Extract: 'Kleinlogel, Rahmenformeln' 11. Auflage Berlin—Verlag von Wilhelm Ernst & Sohn.

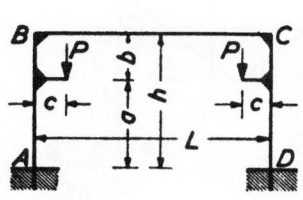
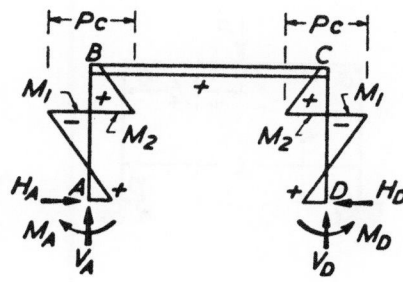

Constants: $a_1 = \dfrac{a}{h}$ $b_1 = \dfrac{b}{h}$

$$X_1 = \frac{Pc}{2N_1}[1 + 2b_1k - 3b_1^2(k + 1)] \qquad X_2 = \frac{Pcka_1(3a_1 - 2)}{2N_1}$$

$$M_A = M_D = \frac{Pc}{N_1}[1 + 2b_1k - 3b_1^2(k + 1)] = 2X_1$$

$$M_B = M_C = \frac{Pcka_1(3a_1 - 2)}{N_1} = 2X_2$$

$$V_A = V_D = P \qquad H_A = H_D = \frac{Pc + M_A - M_B}{h}$$

$$M_1 = M_A - H_A a \qquad M_2 = M_B + H_D b$$

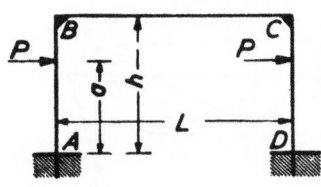
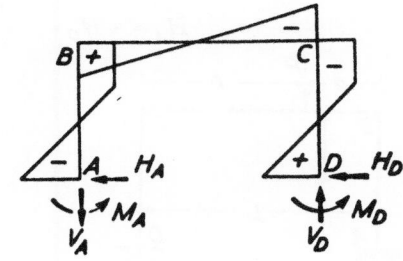

Constants: $a_1 = \dfrac{a}{h}$ $X_1 = \dfrac{3Paa_1k}{N_2}$

$$M_A = -Pa + X_1 \qquad\qquad M_B = X_1$$

$$M_D = +Pa - X_1 \qquad\qquad M_C = -X_1$$

$$V_A = -V_D = -\frac{2X_1}{L} \qquad H_A = -H_D = -P$$

Extract: 'Kleinlogel, Rahmenformeln' 11. Auflage Berlin—Verlag von Wilhelm Ernst & Sohn.

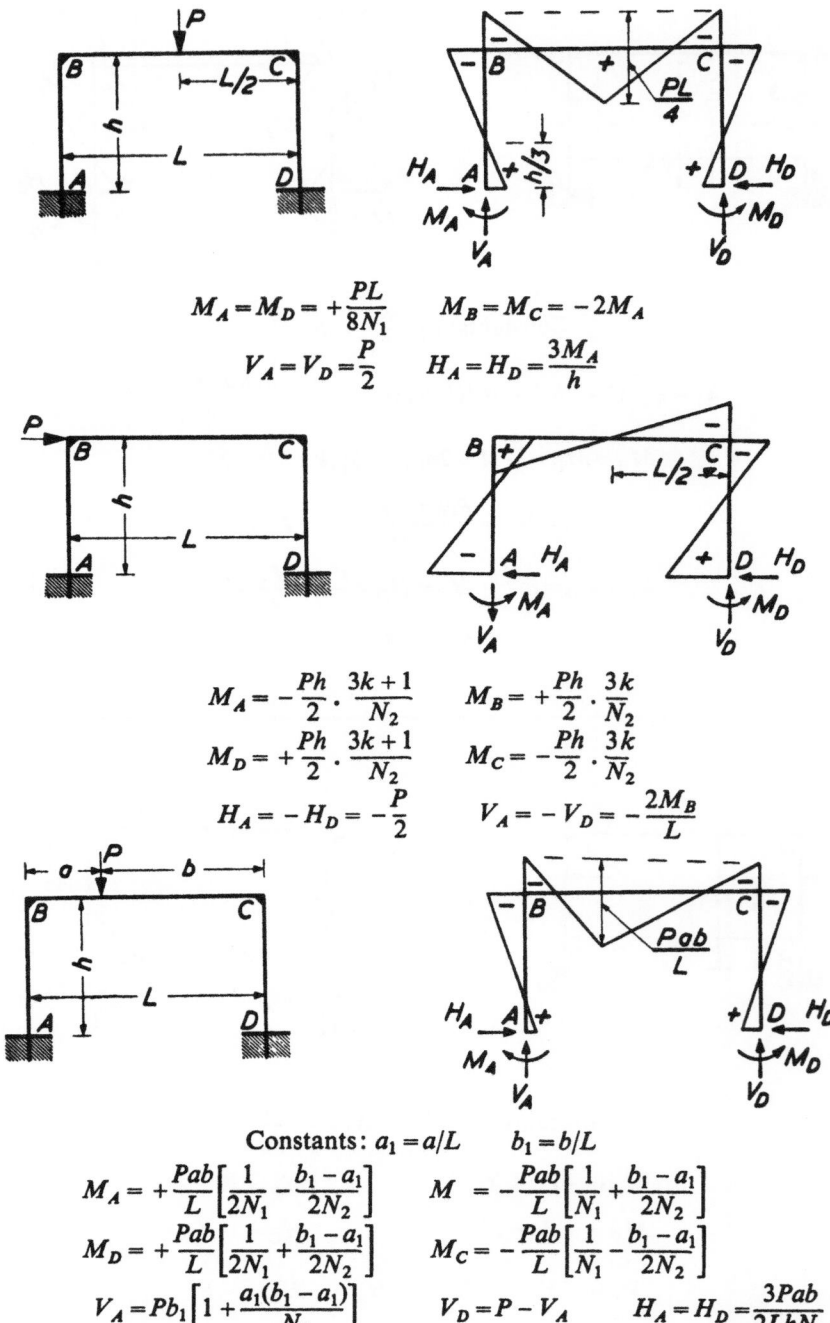

$$M_A = M_D = +\frac{PL}{8N_1} \qquad M_B = M_C = -2M_A$$

$$V_A = V_D = \frac{P}{2} \qquad H_A = H_D = \frac{3M_A}{h}$$

$$M_A = -\frac{Ph}{2} \cdot \frac{3k+1}{N_2} \qquad M_B = +\frac{Ph}{2} \cdot \frac{3k}{N_2}$$

$$M_D = +\frac{Ph}{2} \cdot \frac{3k+1}{N_2} \qquad M_C = -\frac{Ph}{2} \cdot \frac{3k}{N_2}$$

$$H_A = -H_D = -\frac{P}{2} \qquad V_A = -V_D = -\frac{2M_B}{L}$$

Constants: $a_1 = a/L \qquad b_1 = b/L$

$$M_A = +\frac{Pab}{L}\left[\frac{1}{2N_1} - \frac{b_1 - a_1}{2N_2}\right] \qquad M = -\frac{Pab}{L}\left[\frac{1}{N_1} + \frac{b_1 - a_1}{2N_2}\right]$$

$$M_D = +\frac{Pab}{L}\left[\frac{1}{2N_1} + \frac{b_1 - a_1}{2N_2}\right] \qquad M_C = -\frac{Pab}{L}\left[\frac{1}{N_1} - \frac{b_1 - a_1}{2N_2}\right]$$

$$V_A = Pb_1\left[1 + \frac{a_1(b_1 - a_1)}{N_2}\right] \qquad V_D = P - V_A \qquad H_A = H_D = \frac{3Pab}{2LhN_1}$$

Extract: 'Kleinlogel, Rahmenformeln' 11. Auflage Berlin—Verlag von Wilhelm Ernst & Sohn.

Frame II

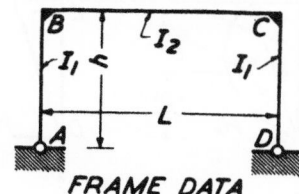

FRAME DATA

Coefficients:

$$k = \frac{I_2}{I_1} \cdot \frac{h}{L}$$

$$N = 2k + 3$$

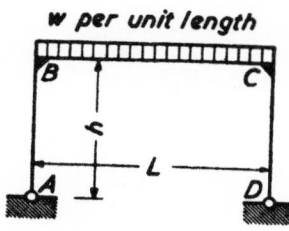

w per unit length

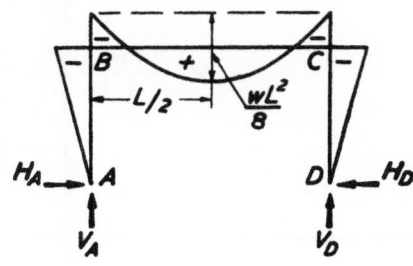

$$M_B = M_C = -\frac{wL^2}{4N} \qquad M_{max} = \frac{wL^2}{8} + M_B$$

$$V_A = V_D = \frac{wL}{2} \qquad H_A = H_D = -\frac{M_B}{h}$$

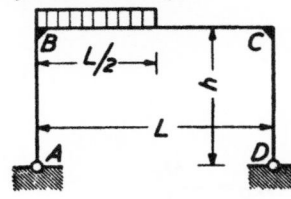

w per unit length

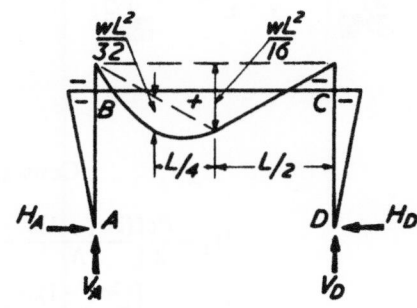

$$M_B = M_C = -\frac{wL^2}{8N}$$

$$V_A = \frac{3wL}{8} \qquad V_D = \frac{wL}{8} \qquad H_A = H_D = -\frac{M_B}{h}$$

Extract: 'Kleinlogel, Rahmenformeln' 11. Auflage Berlin—Verlag von Wilhelm Ernst & Sohn.

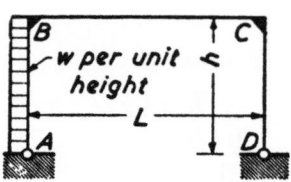
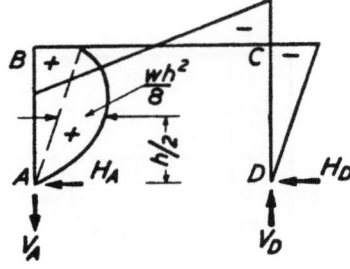

$$M_B = \frac{wh^2}{4}\left[-\frac{k}{2N}+1\right] \qquad H_D = -\frac{M_C}{h}$$

$$M_C = \frac{wh^2}{4}\left[-\frac{k}{2N}-1\right] \qquad H_A = -(wh - H_D)$$

$$V_A = -V_D = -\frac{wh^2}{2L}$$

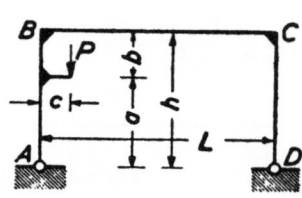
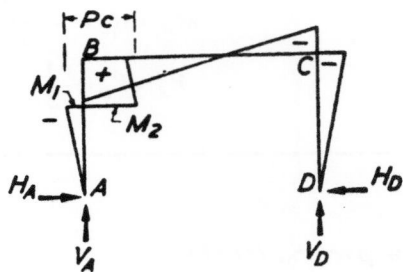

$$\text{Constant: } a_1 = \frac{a}{h}$$

$$M_B = \frac{Pc}{2}\left[\frac{(3a_1^2-1)k}{N}+1\right]$$
$$M_C = \frac{Pc}{2}\left[\frac{(3a_1^2-1)k}{N}-1\right] \qquad H_A = H_D = -\frac{M_C}{h}$$

$$V_D = \frac{Pc}{L} \qquad V_A = P - V_D$$

$$M_1 = -H_A a \qquad M_2 = Pc - H_A a$$

Extract: 'Kleinlogel, Rahmenformeln' 11. *Auflage Berlin—Verlag von Wilhelm Ernst & Sohn.*

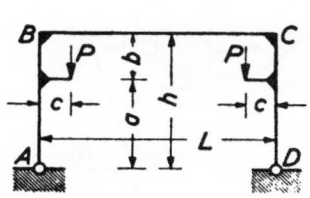
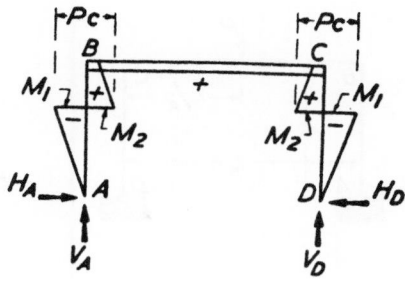

Constant: $a_1 = \dfrac{a}{h}$

$$M_B = M_C = \frac{Pc(3a_1^2 - 1)k}{N}$$

$$H_A = H_D = \frac{Pc - M_B}{h} \qquad V_A = V_D = P$$

$$M_1 = -H_A a \qquad M_2 = Pc - H_A a$$

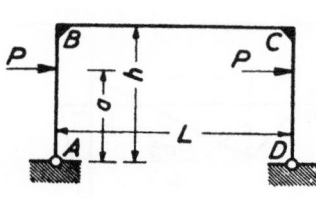
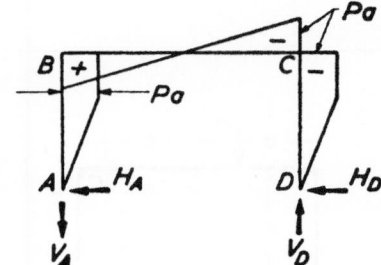

$$M_B = -M_C = Pa \qquad H_A = H_D = P$$

$$V_A = -V_D = -\frac{2Pa}{L}$$

Moment at loads $= \pm Pa$

Extract: '*Kleinlogel, Rahmenformeln*' 11. *Auflage Berlin—Verlag von Wilhelm Ernst & Sohn.*

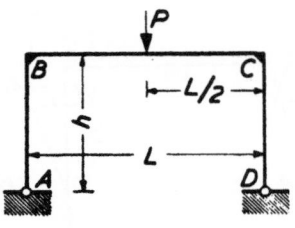
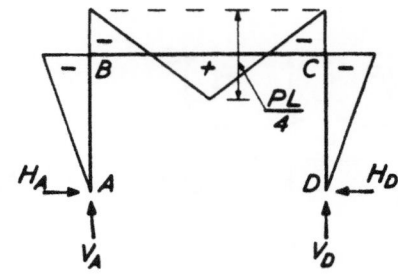

$$M_B = M_C = -\frac{3PL}{8N} \qquad V_A = V_D = \frac{P}{2} \qquad H_A = H_D = -\frac{{}^1M_B}{h}$$

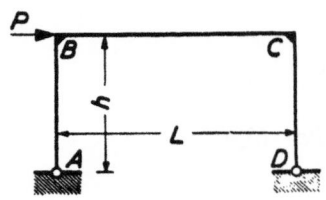
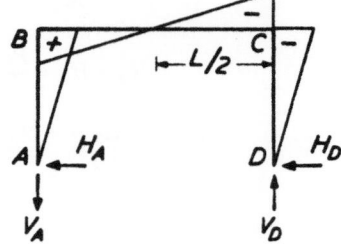

$$M_B = -M_C = +\frac{Ph}{2}$$

$$V_A = -V_D = -\frac{Ph}{L} \qquad H_A = -H_D = -\frac{P}{2}$$

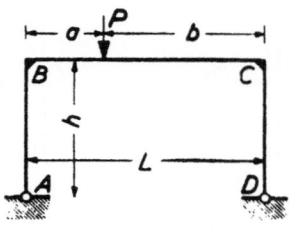
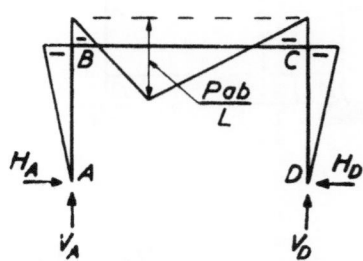

$$M_B = M_C = -\frac{Pab}{L} \cdot \frac{3}{2N}$$

$$V_A = \frac{Pb}{L} \qquad V_D = \frac{Pa}{L} \qquad H_A = H_D = -\frac{M_B}{h}$$

Extract: 'Kleinlogel, Rahmenformeln' 11. Auflage Berlin—Verlag von Wilhelm Ernst & Sohn.

Frame III

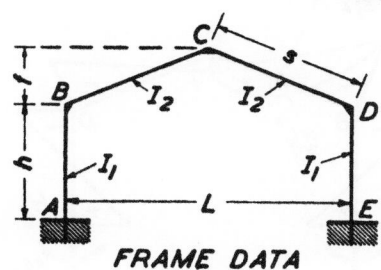

FRAME DATA

Coefficients:

$$k = \frac{I_2}{I_1} \cdot \frac{h}{s} \qquad \phi = \frac{f}{h}$$

$$m = 1 + \phi$$

$$B = 3k + 2 \qquad C = 1 + 2m$$

$$K_1 = 2(k + 1 + m + m^2) \qquad K_2 = 2(k + \phi^2)$$

$$R = \phi C - k \qquad N_1 = K_1 K_2 - R^2 \qquad N_2 = 3k + B$$

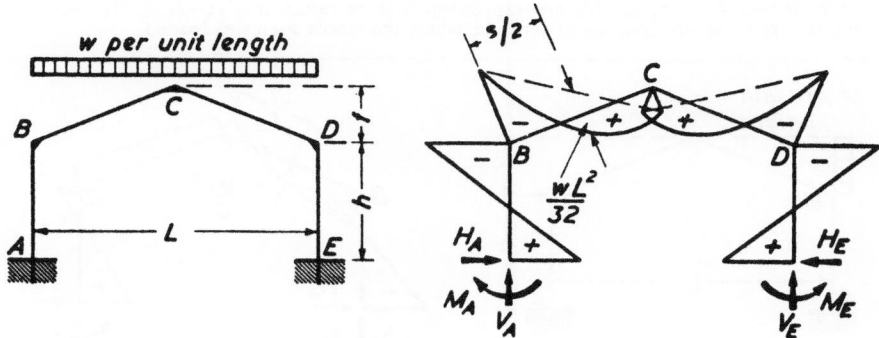

w per unit length

$\frac{wL^2}{32}$

$$M_A = M_E = \frac{wL^2}{16} \cdot \frac{k(8 + 15\phi) + \phi(6 - \phi)}{N_1}$$

$$M_B = M_D = -\frac{wL^2}{16} \cdot \frac{k(16 + 15\phi) + \phi^2}{N_1}$$

$$M_C = \frac{wL^2}{8} - \phi M_A + m M_B$$

$$V_A = V_E = \frac{wL}{2} \qquad H_A = H_E = \frac{M_A - M_B}{h}$$

Extract: 'Kleinlogel, Rahmenformeln' 11. Auflage Berlin—Verlag von Wilhelm Ernst & Sohn.

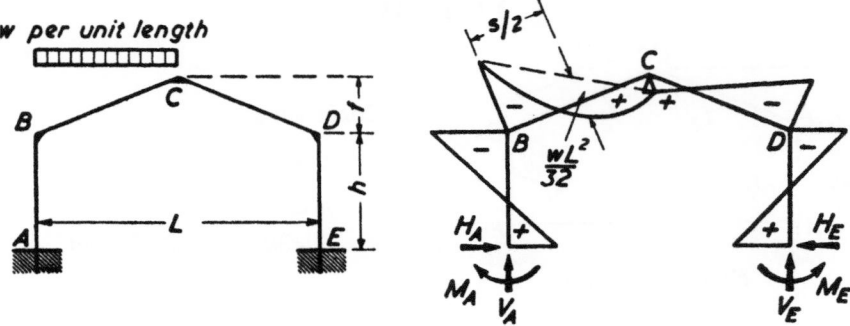

$$\text{Constants: } {}^*X_1 = \frac{wL^2}{32} \cdot \frac{k(8 + 15\phi) + \phi(6 - \phi)}{N_1}$$

$$^*X_2 = \frac{wL^2}{32} \cdot \frac{k(16 + 15\phi) + \phi^2}{N_1} \qquad X_3 = \frac{wL^2}{32 N_2}$$

$$M_A = +X_1 - X_3 \qquad M_B = -X_2 - X_3 \qquad M_E = +X_1 + X_3 \qquad M_D = -X_2 + X_3$$

$$^*M_C = \frac{wL^2}{16} - \phi X_1 - m X_2$$

$$V_E = \frac{wL}{8} - \frac{2X_3}{L} \qquad V_A = \frac{wL}{2} - V_E \qquad H_A = H_E = \frac{X_1 + X_2}{h}$$

* Note that X_1, $-X_2$ and M_C are respectively half the values of $M_A (=M_E)$, $M_B (=M_D)$ and M_C from the previous set of formulæ where the whole span was loaded.

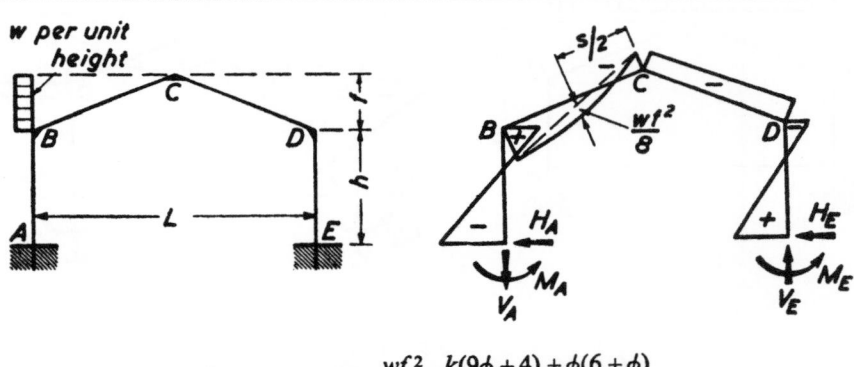

$$\text{Constants: } X_1 = \frac{wf^2}{8} \cdot \frac{k(9\phi + 4) + \phi(6 + \phi)}{N_1}$$

$$X_2 = \frac{wf^2}{8} \cdot \frac{k(8 + 9\phi) - \phi^2}{N_1} \qquad X_3 = \frac{wfh}{8} \cdot \frac{4B + \phi}{N_2}$$

$$M_A = -X_1 - X_3 \qquad M_B = +X_2 + \left(\frac{wfh}{2} - X_3\right)$$

$$M_E = -X_1 + X_3 \qquad M_D = +X_2 - \left(\frac{wfh}{2} - X_3\right)$$

$$M_C = -\frac{wf^2}{4} + \phi X_1 + m X_2$$

$$V_A = -V_E = -\frac{wfh(2 + \phi)}{2L} + \frac{2X_3}{L} \qquad H_E = \frac{wf}{2} - \frac{X_1 + X_2}{h} \qquad H_A = -(wf - H_E)$$

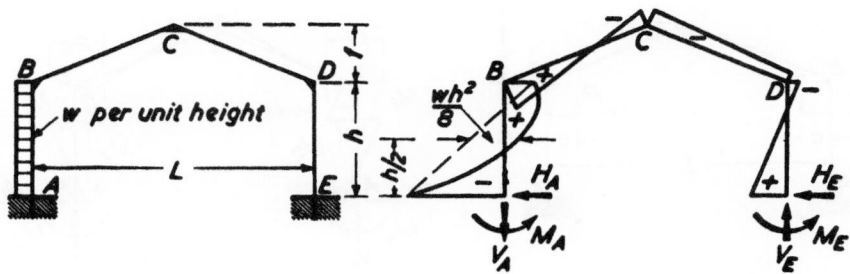

$$\text{Constants: } X_1 = \frac{wh^2}{8} \cdot \frac{k(k+6) + k\phi(15 + 16\phi) + 6\phi^2}{N_1}$$

$$X_2 = \frac{wh^2 k(9\phi + 8\phi^2 - k)}{8N_1} \qquad X_3 = \frac{wh^2(2k+1)}{2N_2}$$

$$M_A = -X_1 - X_3 \qquad M_B = +X_2 + \left(\frac{wh^2}{4} - X_3\right)$$

$$M_E = -X_1 + X_3 \qquad M_D = +X_2 - \left(\frac{wh^2}{4} - X_3\right)$$

$$M_C = -\frac{whf}{4} + \phi X_1 + m X_2$$

$$V_A = -V_E = -\frac{wh^2}{2L} + \frac{2X_3}{L} \qquad H_E = \frac{wh}{4} - \frac{X_1 + X_2}{h} \qquad H_A = -(wh - H_E)$$

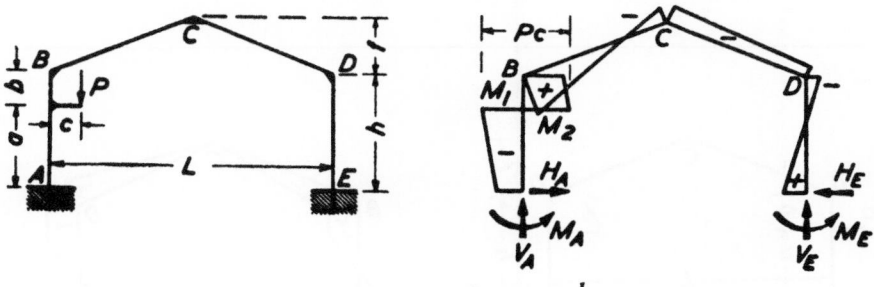

$$\text{Constants: } a_1 = \frac{a}{h} \qquad b_1 = \frac{b}{h}$$

$$Y_1 = Pc[2\phi^2 - (1 - 3b_1^2)k] \qquad Y_2 = Pc[\phi C - (3a_1^2 - 1)k]$$

$$X_1 = \frac{Y_1 K_1 - Y_2 R}{2N_1} \qquad X_2 = \frac{Y_2 K_2 - Y_1 R}{2N_1} \qquad X_3 = \frac{Pc}{2} \cdot \frac{B - 3(a_1 - b_1)k}{N_2}$$

$$M_A = -X_1 - X_3 \qquad M_B = +X_2 + \left(\frac{Pc}{2} - X_3\right)$$

$$M_E = -X_1 + X_3 \qquad M_D = +X_2 - \left(\frac{Pc}{2} - X_3\right) \qquad M_C = -\frac{\phi Pc}{2} + \phi X_1 + m X_2$$

$$M_1 = M_A - H_A a \qquad M_2 = M_B + H_E b$$

$$V_E = \frac{Pc - 2X_3}{L} \qquad V_A = P - V_E \qquad H_A = H_E = \frac{Pc}{2h} - \frac{X_1 + X_2}{h}$$

Extract: 'Kleinlogel, Rahmenformeln' 11. *Auflage Berlin—Verlag von Wilhelm Ernst & Sohn.*

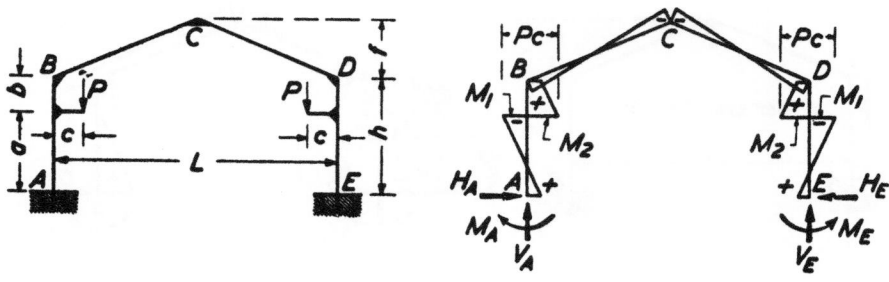

$$\text{Constants: } a_1 = \frac{a}{h} \qquad b_1 = \frac{b}{h}$$

$$Y_1 = Pc[2\phi^2 - (1 - 3b_1^2)k]$$

$$Y_2 = Pc[\phi C + (3a_1^2 - 1)k]$$

$$M_A = M_E = \frac{Y_2 R - Y_1 K_1}{N_1} \qquad M_B = M_D = \frac{Y_2 K_2 - Y_1 R}{N_1}$$

$$M_C = -\phi(Pc + M_A) + mM_B$$

$$V_A = V_D = P \qquad H_A = H_E = \frac{Pc + M_A - M_B}{h}$$

$$M_1 = M_A - H_A a \qquad M_2 = M_B + H_E b$$

$$\text{Constant: } X_1 = \frac{Pa(B + 3b_1 k)}{N_2}$$

$$M_A = -M_E = -X_1 \qquad M_B = -M_D = Pa - X_1 \qquad M_C = 0$$

$$V_A = -V_E = -2\left[\frac{Pa - X_1}{L}\right] \qquad H_A = -H_E = -P$$

Extract: 'Kleinlogel, Rahmenformeln' 11. Auflage Berlin—Verlag von Wilhelm Ernst & Sohn.

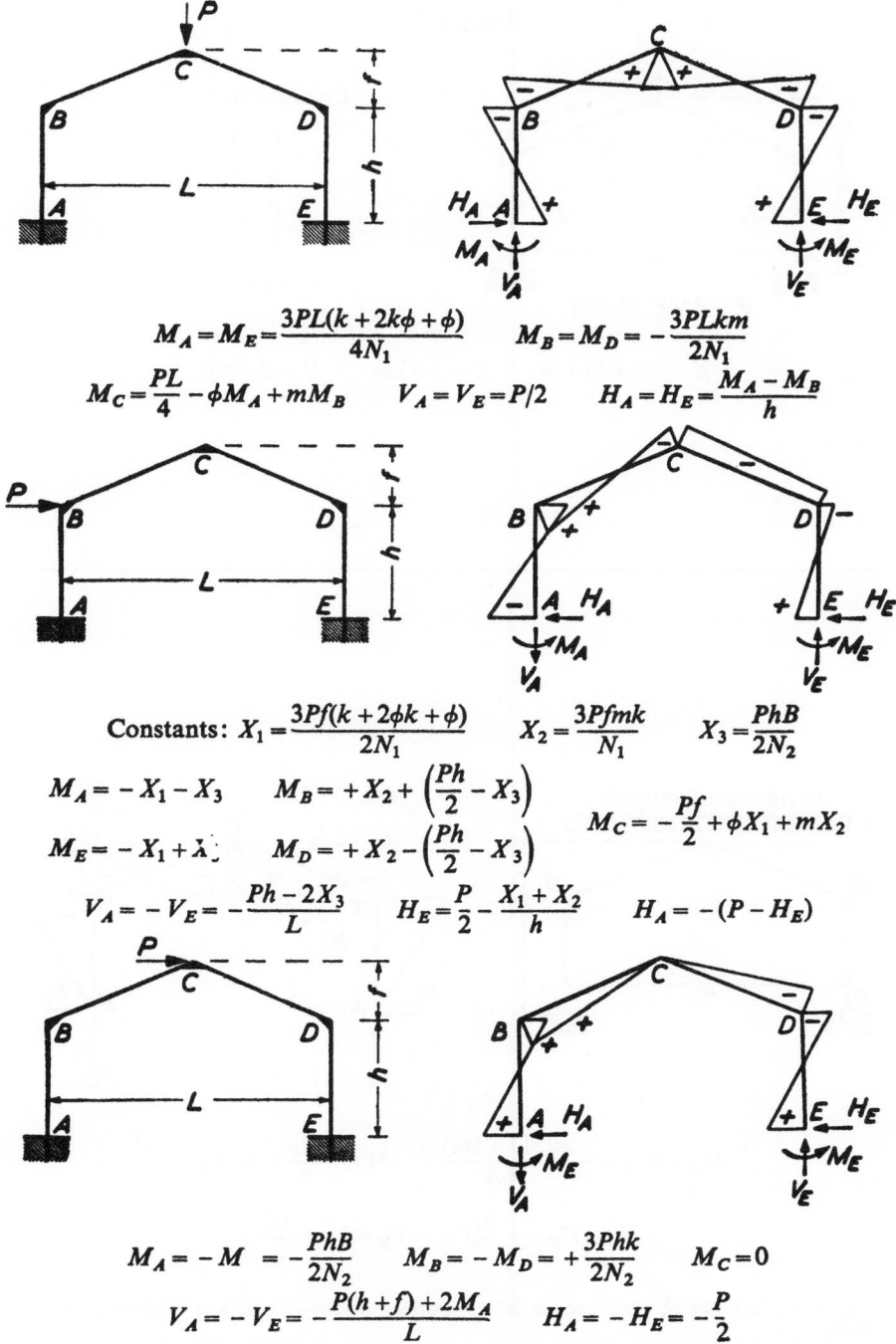

$$M_A = M_E = \frac{3PL(k + 2k\phi + \phi)}{4N_1} \qquad M_B = M_D = -\frac{3PLkm}{2N_1}$$

$$M_C = \frac{PL}{4} - \phi M_A + m M_B \qquad V_A = V_E = P/2 \qquad H_A = H_E = \frac{M_A - M_B}{h}$$

Constants: $X_1 = \dfrac{3Pf(k + 2\phi k + \phi)}{2N_1}$ $\qquad X_2 = \dfrac{3Pfmk}{N_1} \qquad X_3 = \dfrac{PhB}{2N_2}$

$$M_A = -X_1 - X_3 \qquad M_B = +X_2 + \left(\frac{Ph}{2} - X_3\right)$$

$$M_E = -X_1 + X_3 \qquad M_D = +X_2 - \left(\frac{Ph}{2} - X_3\right) \qquad M_C = -\frac{Pf}{2} + \phi X_1 + m X_2$$

$$V_A = -V_E = -\frac{Ph - 2X_3}{L} \qquad H_E = \frac{P}{2} - \frac{X_1 + X_2}{h} \qquad H_A = -(P - H_E)$$

$$M_A = -M = -\frac{PhB}{2N_2} \qquad M_B = -M_D = +\frac{3Phk}{2N_2} \qquad M_C = 0$$

$$V_A = -V_E = -\frac{P(h + f) + 2M_A}{L} \qquad H_A = -H_E = -\frac{P}{2}$$

Extract: 'Kleinlogel, Rahmenformeln' 11. Auflage Berlin—Verlag von Wilhelm Ernst & Sohn.

Frame IV

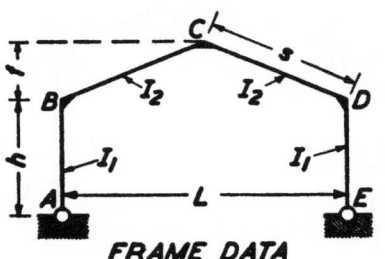

FRAME DATA

Coefficients:

$$k = \frac{I_2}{I_1} \cdot \frac{h}{s}$$

$$\phi = \frac{f}{h}$$

$$m = 1 + \phi$$

$$B = 2(k+1) + m \qquad C = 1 + 2m \qquad N = B + mC$$

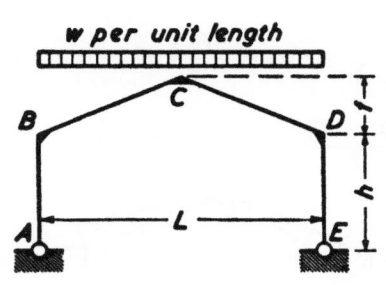

w per unit length

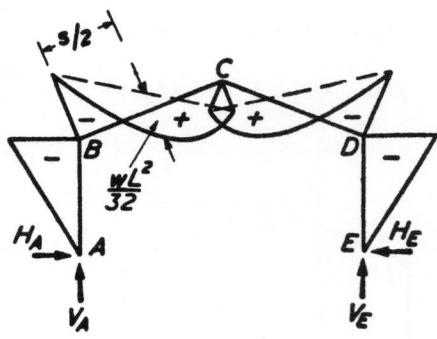

$$M_B = M_D = -\frac{wL^2(3+5m)}{16N} \qquad M_C = \frac{wL^2}{8} + mM_B$$

$$H_A = H_E = -\frac{M_B}{h} \qquad V_A = V_E = \frac{wL}{2}$$

Extract: 'Kleinlogel, Rahmenformeln' 11. *Auflage Berlin—Verlag von Wilhelm Ernst & Sohn.*

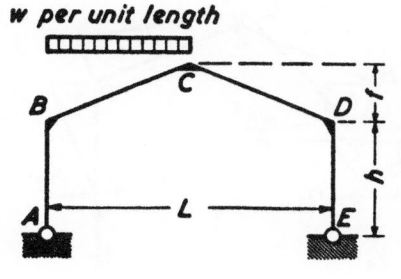

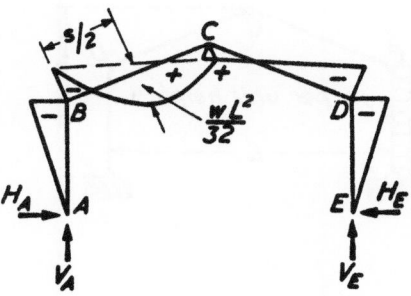

$$M_B = M_D = -\frac{wL^2(3+5m)}{32N} \qquad M_C = \frac{wL^2}{16} + mM_B$$

$$H_A = H_E = -\frac{M_B}{h} \qquad V_A = \frac{3wL}{8} \qquad V_E = \frac{wL}{8}$$

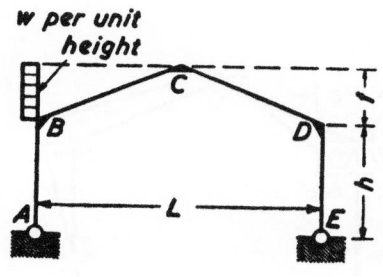

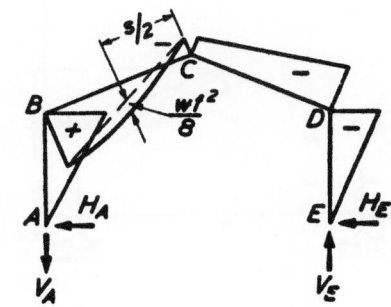

$$\text{Constant: } X = \frac{wf^2(C+m)}{8N}$$

$$M_B = +X + \frac{wfh}{2} \qquad M_C = -\frac{wf^2}{4} + mX$$

$$M_D = +X - \frac{wfh}{2} \qquad V_A = -V_E = -\frac{wfh(1+m)}{2L}$$

$$H_A = -\frac{X}{h} - \frac{wf}{2} \qquad H_E = -\frac{X}{h} + \frac{wf}{2}$$

Extract: 'Kleinlogel, Rahmenformeln' 11. Auflage Berlin—Verlag von Wilhelm Ernst & Sohn.

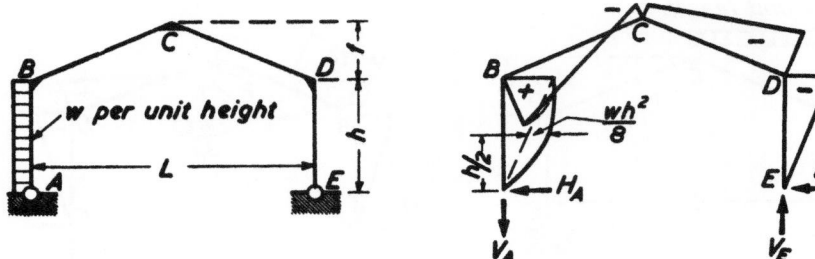

$$M_D = -\frac{wh^2}{8} \cdot \frac{2(B+C)+k}{N} \qquad M_B = \frac{wh^2}{2} + M_D$$

$$M_C = \frac{wh^2}{4} + mM_D$$

$$V_A = -V_E = -\frac{wh^2}{2L} \qquad H_E = -\frac{M_D}{h} \qquad H_A = -(wh - H_E)$$

$$\text{Constants: } a_1 = \frac{a}{h} \qquad X = \frac{Pc}{2} \cdot \frac{B+C-k(3a_1^2-1)}{N}$$

$$M_B = Pc - X \qquad M_D = -X \qquad M_C = \frac{Pc}{2} - mX$$

$$M_1 = -a_1 X \qquad M_2 = Pc - a_1 X$$

$$V_E = \frac{Pc}{L} \qquad V_A = P - V_E \qquad H_A = H_E = \frac{X}{h}$$

Extract: 'Kleinlogel, Rahmenformeln' 11. *Auflage Berlin—Verlag von Wilhelm Ernst & Sohn.*

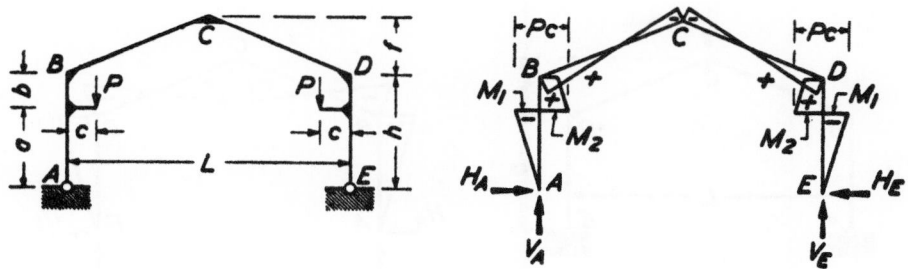

Constant: $a_1 = \dfrac{a}{h}$

$$M_B = M_D = Pc \cdot \frac{\phi C + k(3a_1^2 - 1)}{N} \qquad M_C = -\phi Pc + mM_B$$

$$H_A = H_E = \frac{Pc - M_B}{h} \qquad V_A = V_E = P$$

$$M_1 = -a_1(Pc - M_B) \qquad M_2 = (1 - a_1)Pc + a_1 M_B$$

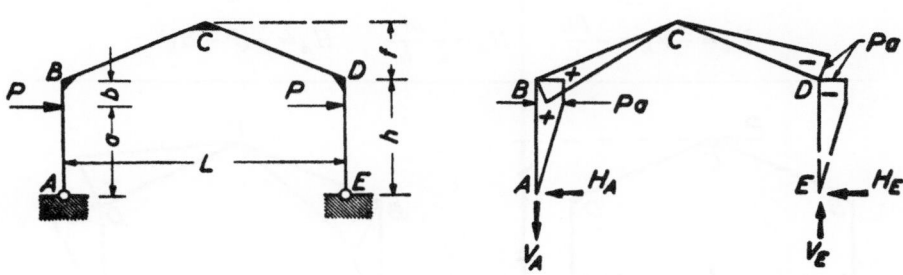

$$M_B = -M_D = Pa \qquad M_C = 0$$

$$H_A = -H_E = -P \qquad V_A = -V_E = -\frac{2Pa}{L}$$

Moment at loads $= \pm Pa$

Extract: 'Kleinlogel, Rahmenformeln' 11, Auflage Berlin—Verlag von Wilhelm Ernst & Sohn.

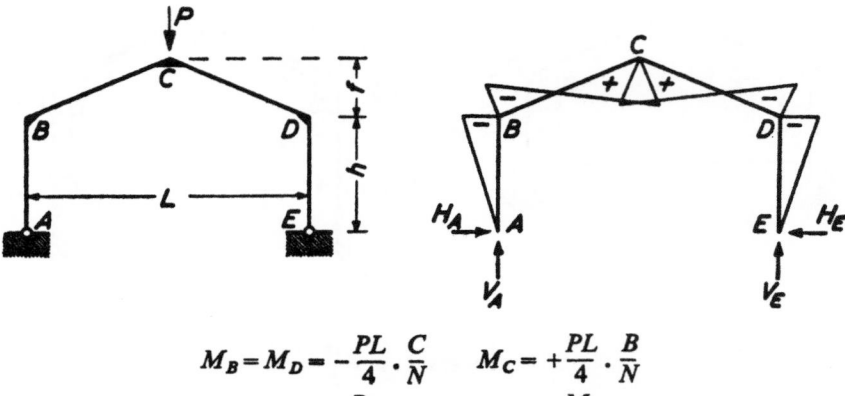

$$M_B = M_D = -\frac{PL}{4} \cdot \frac{C}{N} \qquad M_C = +\frac{PL}{4} \cdot \frac{B}{N}$$

$$V_A = V_E = \frac{P}{2} \qquad H_A = H_E = -\frac{M_B}{h}$$

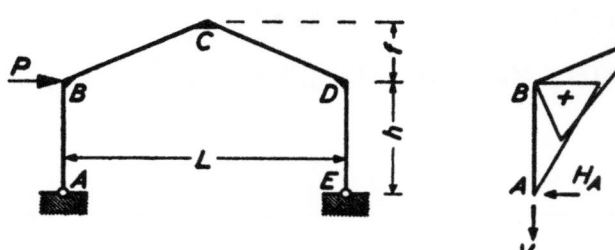

$$M_D = -\frac{Ph(B+C)}{2N} \qquad M_B = Ph + M_D \qquad M_C = \frac{Ph}{2} + mM_D$$

$$V_A = -V_E = -\frac{Ph}{L} \qquad H_E = -\frac{M_D}{h} \qquad H_A = -(P - H_E)$$

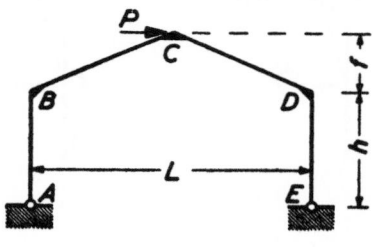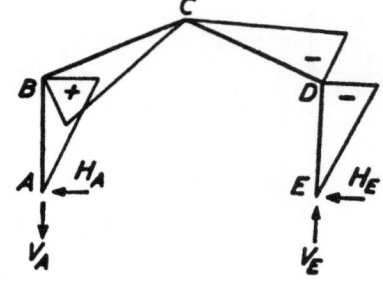

$$M_B = -M_D = +\frac{Ph}{2} \qquad M_C = 0 \qquad V_A = -V_E = -\frac{Phm}{L} \qquad H_A = -H_E = -\frac{P}{2}$$

Extract: 'Kleinlogel, Rahmenformeln' 11. *Auflage Berlin—Verlag von Wilhelm Ernst & Sohn.*

Explanatory notes on section dimensions and properties

1 General

This publication presents design data in tabular formats as assistance to engineers who are designing buildings in accordance with BS EN 1993-1-1: 2005, BS EN 1993-1-5: 2006 and BS EN 1993-1-8: 2005, and their respective National Annexes. Where these Parts do not give all the necessary expressions for the evaluation of data, reference is made to other published sources.

The symbols used are generally the same as those in these standards or the referred product standards. Where a symbol does not appear in the standards, a symbol has been chosen following the designation convention as closely as possible.

1.1 Material, section dimensions and tolerances

The structural sections referred to in this design guide are of weldable structural steels conforming to the relevant British Standards given in the table below:

Table 1.1 Structural steel products

Product	Technical delivery requirements		Dimensions	Tolerances
	Non alloy steels	Fine grain steels		
Universal beams, Universal columns, and universal bearing piles	BS EN 10025-2	BS EN 10025-3 BS EN 10025-4	BS 4-1	BS EN 10034
Joists			BS 4-1	BS 4-1 BS EN 10024
Parallel flange channels			BS 4-1	BS EN 10279
Angles			BS EN 10056-1	BS EN 10056-2
Structural tees cut from universal beams and universal columns			BS 4-1	–
ASB (asymmetric beams) Slimflor® beam	Generally BS EN 10025, but see note[b]		See note[a]	Generally BS EN 10034, but also see note[b]
Hot finished structural hollow sections	BS EN 10210-1		BS EN 10210-2	BS EN 10210-2
Cold formed hollow sections	BS EN 10219-1		BS EN 10219-2	BS EN 10219-2

Notes:
For full details of the British Standards, see the reference list at the end of the Appendix.
[a] See Tata publication, Advance™ Sections: CE marked structural sections.
[b] For further details, consult Tata.
Note that EN 1993 refers to the product standards by their CEN designation, e.g. EN 10025-2. The CEN standards are published in the UK by BSI with their prefix to the designation, e.g. BS EN 10025-2.

1.2 Dimensional units

The dimensions of sections are given in millimetres (mm).

1.3 Property units

Generally, the centimetre (cm) is used for the calculated properties but for surface areas and for the warping constant (I_w), the metre (m) and the decimetre (dm) respectively are used.

Note: $1\,\text{dm} = 0.1\,\text{m}$ $= 100\,\text{mm}$
$1\,\text{dm}^6 = 1 \times 10^{-6}\,\text{m}^6 = 1 \times 10^{12}\,\text{mm}^6$

1.4 Mass and force units

The units used are the kilogram (kg), the Newton (N) and the metre per second squared (m/s^2), so that $1\,\text{N} = 1\,\text{kg} \times 1\,\text{m}/\text{s}^2$. For convenience, a standard value of the acceleration due to gravity has been accepted as $9.80665\,\text{m}/\text{s}^2$. Thus, the force exerted by $1\,\text{kg}$ under the action of gravity is $9.80665\,\text{N}$ and the force exerted by 1 tonne (1000 kg) is 9.80665 kiloNewtons (kN).

2 Dimensions of section

2.1 Masses

The masses per metre have been calculated assuming that the density of steel is $7850\,\text{kg}/\text{m}^3$.

In all cases, including compound sections, the tabulated masses are for the steel section alone and no allowance has been made for connecting material or fittings.

2.2 Ratios for local buckling

The ratios of the flange outstand to thickness (c_f/t_f) and the web depth to thickness (c_w/t_w) are given for I-, H- and channel sections.

$$c_f = \frac{1}{2}[b - (t_w + 2r)]$$ for I- and H-sections

$$c_f = [b - (t_w + r)]$$ for channels

$$c_w = d = [h - 2(t_f + r)]$$ for I-, H- and channel sections

For circular hollow sections the ratios of the outside diameter to thickness (d/t) are given.

For square and rectangular hollow sections the ratios (c_f/t) and (c_w/t) are given where:

$$c_f = b - 3t \text{ and } c_w = h - 3t$$

For square hollow sections c_f and c_w are equal. Note that these relationships for c_f and c_w are applicable to both hot-finished and cold-formed sections.

The dimension c is not precisely defined in EN 1993-1-1 and the internal profile of the corners is not specified in either EN 10210-2 or EN 10219-2. The above expressions give conservative values of the ratio for both hot-finished and cold-formed sections.

2.3 Dimensions for detailing

The dimensions C, N and n have the meanings given in the figures at the heads of the tables and have been calculated according to the formulae below. The formulae for N and C make allowance for rolling tolerances, whereas the formulae for n make no such allowance.

2.3.1 UB sections, UC sections and bearing piles

$N = (b - t_w)/2 + 10\,\text{mm}$	(rounded to the nearest 2 mm above)
$N = (h - d)/2$	(rounded to the nearest 2 mm above)
$C = t_w/2 + 2\,\text{mm}$	(rounded to the nearest mm)

2.3.2 Joists

$N = (b - t_w)/2 + 6\,\text{mm}$	(rounded to the nearest 2 mm above)
$n = (h - d)/2$	(rounded to the nearest 2 mm above)
$C = t_w/2 + 2\,\text{mm}$	(rounded to the nearest mm)

Note: Flanges of BS 4-1 joists have an 8° taper.

2.3.3 Parallel flange channels

$N = (b - t_w) + 6\,\text{mm}$	(rounded up to the nearest 2 mm above)
$n = (h - d)/2$	(taken to the next higher multiple of 2 mm)
$C = t_w + 2\,\text{mm}$	(rounded up to the nearest mm)

3 Section properties

3.1 General

All section properties have been accurately calculated and rounded to three significant figures. They have been calculated from the metric dimensions given in the appropriate standards (see Section 1.1). For angles, BS EN 10056-1 assumes that the toe radius equals half the root radius.

3.2 Sections other than hollow sections

3.2.1 Second moment of area (I)

The second moment of area has been calculated taking into account all tapers, radii and fillets of the sections. Values are given about both the y-y and z-z axes.

3.2.2 Radius of gyration (i)

The radius of gyration is a parameter used in the calculation of buckling resistance and is derived as follows:

$$i = [I / A]^{1/2}$$

where:

   I is the second moment of area about the relevant axis
   A is the area of the cross section.

3.2.3 Elastic section modulus (W_{el})

The elastic section modulus is used to calculate the elastic design resistance for bending based on the yield strength of the section and the partial factor γ_M or to calculate the stress at the extreme fibre of the section due to a moment. It is derived as follows:

$$W_{el,y} = I_y / z$$
$$W_{el,z} = I_z / y$$

where:

   z, y are the distances to the extreme fibres of the section from the elastic y-y and z-z axes, respectively.

For parallel flange channels, the elastic section modulus about the minor (z-z) axis is given for the extreme fibre at the toe of the section only.

For angles, the elastic section moduli about both axes are given for the extreme fibres at the toes of the section only.

For asymmetric sections, the elastic section moduli are given at both top and bottom extreme fibres, as well as for the extreme lateral fibre.

3.2.4 Plastic section modulus (W_{pl})

The plastic section moduli about both y-y and z-z axes of the plastic cross sections are tabulated for all sections except angle sections.

3.2.5 Buckling parameter (U) and torsional index (X)

UB sections, UC sections, joists and parallel flange channels

The buckling parameter (U) and torsional index (X) used to determine the buckling resistance moment (see section 8.1) have been calculated using expressions in Access Steel document SN002 *Determination of non-dimensional slenderness of I and H sections*.

$$U = \left(\frac{W_{\mathrm{pl,y}}g}{A}\right)^{0.5} \times \left(\frac{I_z}{I_w}\right)^{0.25}$$

$$X = \sqrt{\frac{\pi^2 EAI_w}{20GI_T I_z}}$$

where:

$W_{\mathrm{pl,y}}$ is the plastic modulus about the major axis

$g \quad = \sqrt{1-\dfrac{I_z}{I_y}}$

I_y is the second moment of area about the major axis
I_z is the second moment of area about the minor axis
E =210 000 N/mm² is the modulus of elasticity
G is the shear modulus where $G = \dfrac{E}{2(1+v)}$
v is Poisson's ratio (=0.3)
A is the cross-sectional area
I_w is the warping constant
I_T is the torsional constant.

Tee sections and ASB sections

The buckling parameter (U) and the torsional index (X) used to determine the buckling resistance moment have been calculated using the following expressions:

$$U = [(4W_{\mathrm{pl,y}}^2 g / (A^2 h^2))]^{1/4}$$

$$X = 0.566 h [A / I_{\mathrm{T}}]^{1/2}$$

where:

$W_{\mathrm{pl,y}}$ is the plastic modulus about the major axis

$g \quad = \sqrt{1 - \dfrac{I_z}{I_y}}$

I_y is the second moment of area about the major axis
I_z is the second moment of area about the minor axis
A is the cross sectional area
h is the distance between shear centres of flanges (for T sections, h is the distance between the shear centre of the flange and the toe of the web)
I_{T} is the torsional constant.

3.2.6 Warping constant (Iw) and torsional constant (IT)

Rolled I-sections

The warping constant and St Venant torsional constant for rolled I-sections have been calculated using the formulae given in the SCI publication P057 *Design of members subject to combined bending and torsion.*

In Eurocode 3 terminology, these formulae are as follows:

$$I_{\mathrm{w}} = \frac{I_z h_{\mathrm{s}}^2}{4}$$

where:

I_z is the second moment of area about the minor axis
h_{s} is the distance between shear centres of flanges (i.e. $h_{\mathrm{s}} = h - t_{\mathrm{f}}$)

$$I_{\mathrm{T}} = \frac{2}{3} b t_{\mathrm{f}}^3 + \frac{1}{3}(h - 2t_{\mathrm{f}}) t_{\mathrm{w}}^3 + 2\alpha_1 D_1^4 - 0.420 t_{\mathrm{f}}^4$$

where:

$$\alpha_1 = -0.042 + 0.2204 \frac{t_{\mathrm{w}}}{t_{\mathrm{f}}} + 0.1355 \frac{r}{t_{\mathrm{f}}} - 0.0865 \frac{r t_{\mathrm{w}}}{t_{\mathrm{f}}^2} - 0.0725 \frac{t_{\mathrm{w}}^2}{t_{\mathrm{f}}^2}$$

$$D_1 = \frac{(t_{\mathrm{f}} + r)^2 + (r + 0.25 t_{\mathrm{w}}) t_{\mathrm{w}}}{2r + t_{\mathrm{f}}}$$

b is the width of the section
h is the depth of the section
t_{f} is the flange thickness
t_{w} is the web thickness
r is the root radius.

Tee sections

For Tee sections cut from UB and UC sections, the warping constant (I_w) and torsional constant (I_T) have been derived as given below.

$$I_w = \frac{1}{144}t_f^3 b^3 + \frac{1}{36}\left(h - \frac{t_f}{2}\right)^3 t_w^3$$

$$I_T = \frac{1}{3}bt_f^3 + \frac{1}{3}(h - t_f)t_w^3 + \alpha_1 D_1^4 - 0.21t_f^4 - 0.105t_w^4$$

where:

$$\alpha_1 = -0.042 + 0.2204\frac{t_w}{t_f} + 0.1355\frac{r}{t_f} - 0.0865\frac{t_w r}{t_f^2} - 0.0725\frac{t_w^2}{t_f^2}$$

D_1 is as defined above

Note: These formulae do not apply to tee sections cut from joists, which have tapered flanges. For such sections, expressions are given in SCI Publication 057.

Parallel flange channels

For parallel flange channels, the warping constant (I_w) and torsional constant (I_T) have been calculated as follows:

$$I_w = \frac{(h - t_f)^2}{4}\left[I_z - A\left(c_z - \frac{t_w}{2}\right)^2\left(\frac{(h - t_f)^2 A}{4I_y} - 1\right)\right]$$

$$I_T = \frac{2}{3}bt_f^3 + \frac{1}{3}(h - 2t_f)t_w^3 + 2\alpha_3 D_3^4 - 0.42t_f^4$$

where:

c_z is the distance from the back of the web to the centroidal axis

$$\alpha_3 = -0.0908 + 0.2621\frac{t_w}{t_f} + 0.1231\frac{r}{t_f} - 0.0752\frac{t_w r}{t_f^2} - 0.0945\left(\frac{t_w}{t_f}\right)^2$$

$$D_3 = 2\left[(3r + t_w + t_f) - \sqrt{2(2r + t_w)(2r + t_f)}\right]$$

Note: The formula for the torsional constant (I_T) is applicable to parallel flange channels only and does not apply to tapered flange channels.

Angles

For angles, the torsional constant (I_T) is calculated as follows:

$$I_T = \frac{1}{3}bt^3 + \frac{1}{3}(h - t)t^3 + \alpha_3 D_3^4 - 0.21t^4$$

where:
$$\alpha_3 = 0.0768 + 0.0479 \frac{r}{t}$$

$$D_3 = 2\left[(3r + 2t) - \sqrt{2(2r+t)^2} \right]$$

ASB sections

For ASB sections the warping constant (I_w) and torsional constant (I_T) are as given in Tata brochure, *Advance™ sections.*

3.2.7 Equivalent slenderness coefficient (θ) and monosymmetry index (ψ)

Angles

The buckling resistance moments for angles have not been included in the bending resistance tables of this publication as angles are predominantly used in compression and tension only. Where the designer wishes to use an angle section in bending, BS EN 1993-1-1, 6.3.2 enables the buckling resistance moment for angles to be determined. The procedure is quite involved.

As an alternative to the procedure in BS EN 1993-1-1, supplementary section properties have been included for angle sections in this publication which enable the designer to adopt a much simplified method for determining the buckling resistance moment. The method is based on that given in BS 5950-1:2000 Annex B.2.9. and makes use of the equivalent slenderness coefficient and the monosymmetry index.

The equivalent slenderness coefficient (ϕ) is tabulated for both equal and unequal angles. Two values of the equivalent slenderness coefficient are given for each unequal angle. The larger value is based on the major axis elastic section modulus ($W_{el,u}$) to the toe of the short leg and the lower value is based on the major axis elastic section modulus to the toe of the long leg.

The equivalent slenderness coefficient (ϕ) is calculated as follows:

$$\varphi = \frac{W_{el,u}g}{\sqrt{AI_T}}$$

where:
$W_{el,u}$ is the elastic section modulus about the major axis u-u

$$g = \sqrt{1 - \frac{I_v}{I_u}}$$

I_v is the second moment of area about the minor axis
I_u is the second moment of area about the major axis
A is the area of the cross-section
I_T is the torsional constant.

The monosymmetry index (ψ) is calculated as follows:

$$\psi = \left[2v_0 - \frac{\int v_i \left(u_i^2 + v_i^2 \right) dA}{I_u} \right] \frac{1}{t}$$

where:

u_i and v_i are the coordinates of an element of the cross-section

v_0 is the coordinate of the shear centre along the v-v axis, relative to the centroid

t is the thickness of the angle.

Tee sections

The monosymmetry index is tabulated for Tee sections cut from UBs and UCs. It has been calculated as:

$$\psi = \left(2z_0 - \frac{z_0 b^3 t_f / 12 + b t_f z_0^3 + \frac{t_w}{4} \left[(c - t_f)^4 - (h - c)^4 \right]}{I_y} \right) \frac{1}{(h - t_f / 2)}$$

where:

$z_0 = c - t_f / 2$

c is the distance from the outside of the flange to the centroid of the section

b is the flange width

t_f is the flange thickness

t_w is the web thickness

h is the depth of the section.

The above expression is based on BS 5950-1, Annex B.2.8.2.

ASB sections

The monosymmetry index is tabulated for ASB sections. It has been calculated using the equation in BS 5950-1, Annex B.2.4.1, re-expressed in BS EN 1993-1-1 nomenclature:

$$\psi = \frac{1}{h_s} \left(\frac{2(I_{zc} h_c - I_{zt} h_t)}{(I_{zc} + I_{zt})} - \frac{(I_{zc} h_c - I_{zt} h_t) + (b_c t_c h_c^3 - b_t t_t h_t^3) + \frac{t}{4}(d_c^4 - d_t^4)}{I_y} \right)$$

where:
$$h_s = \left(h - \frac{t_c + t_t}{2} \right)$$
$$d_c = h_c - t_c/2$$
$$d_t = h_t - t_t/2$$
$$I_{zc} = b_c^3 t_c /12$$
$$I_{zt} = b_t^3 t_t /12$$

h_c is the distance from the centre of the compression flange to the centroid of the section

h_t is the distance from the centre of the tension flange to the centroid of the section

b_c is the width of the compression flange

b_t is the width of the tension flange

t_c is the thickness of the compression flange

t_t is the thickness of the tension flange.

For ASB sections $t_c = t_t$ and this is shown as t_f in the tables.

3.3 Hollow sections

Section properties are given for both hot-finished and cold-formed hollow sections (but not for cold-formed elliptical hollow sections). For the same overall dimensions and wall thickness, the section properties for square and rectangular hot-finished and cold-formed sections are different because the corner radii are different.

3.3.1 Common properties

For general comment on second moment of area, radius of gyration, elastic and plastic modulus, see Sections 3.2.1, 3.2.2, 3.2.3 and 3.2.4.

For hot-finished square and rectangular hollow sections, the section properties have been calculated using corner radii of $1.5t$ externally and $1.0t$ internally, as specified by BS EN 10210-2.

For cold-formed square and rectangular hollow sections, the section properties have been calculated using the external corner radii of $2t$ if $t \leq 6$ mm, $2.5t$ if 6 mm $< t \leq 10$ mm and $3t$ if $t > 10$ mm, as specified by BS EN 10219-2. The internal corner radii used are $1.0t$ if $t \leq 6$ mm, $1.5t$ if 6 mm $< t \leq 10$ mm and $2t$ if $t > 10$ mm, as specified by BS EN 10219-2.

3.3.2 Plastic section modulus of hollow sections (W_{pl})

The plastic section moduli (W_{pl}) about both principal axes are given in the tables.

3.3.3 Torsional constant (I_T)

For circular hollow sections:

$$I_T = 2I$$

For square and rectangular hollow sections:

$$I_T = \frac{4A_h^2 t}{p} + \frac{t^3 p}{3}$$

For elliptical hollow sections:

$$I_T = \frac{4A_m^2 t}{U} + \frac{t^3 U}{3}$$

where:

I is the second moment of area of a CHS
t is the thickness of the section
p is the mean perimeter $= 2[(2b - t) + (2a - t)] - 2R_c(4 - \pi)$
A_h is the area enclosed by the mean perimeter $= (2b - t)(2a - t) - R_c^2(4 - \pi)$
a is half the outside dimension of the section on its major axis
b is half the outside dimension of the section on its minor axis
R_c is the average of the internal and external corner radii
A_m is the area enclosed by the mean perimeter of an elliptical hollow section

$$= \frac{\pi(2a - t)(2b - t)}{4}$$

$$U \quad = \frac{\pi}{2}(2a + 2b - 2t)\left(1 + 0.25\left(\frac{2a - 2b}{2a + 2b - 2t}\right)^2\right)$$

Note that the section height and width are described differently in the section property and resistance tables.

3.3.4 Torsional section modulus (W_t)

$$W_t = 2W_{el} \qquad \text{for circular hollow sections}$$

$$W_t = \frac{I_T}{\left(t + \dfrac{2A_h}{h}\right)} \qquad \text{for square and rectangular hollow sections}$$

$$W_t = \frac{I_T}{\left(t + \dfrac{2A_m}{2a}\right)} \qquad \text{for elliptical hollow sections}$$

where:

W_{el} is the elastic modulus and I_T, t, A_h, p, a and A_m are as defined in Section 3.3.3.

4 Effective section properties

4.1 General

In BS EN 1993-1-1:2005, effective section properties are required for the design of members with Class 4 cross-sections. In this publication, effective section properties are given for sections subject to compression only and bending only. Effective section properties depend on the grade of steel used and are given for rolled I-sections and angles in S275 and S355. Channels are not Class 4 and therefore no effective section properties are provided. For hot-finished and cold-formed hollow sections, effective section properties are only given for S355.

4.2 Effective section properties of members subject to compression

The effective cross-section properties of Class 4 cross-sections are based on the effective widths of the compression parts.

The effective cross–sectional area A_{eff} of Class 4 sections in compression is calculated in accordance with BS EN 1993-1-1, 6.2.2.5 and BS EN 1993–1–5:2006, 4.3 and 4.4.

The effective section properties tables list the sections that can be Class 4 and the identifier 'W', 'F' or 'WF' indicates whether the section is Class 4 due to the web, the flange or both. The effective area of the section is calculated from:

For UB, UC and joists: $A_{eff} = A - 4t_f(1 - \rho_f)c_f - t_w(1 - \rho_w)c_w$

For rectangular hollow sections and square hollow sections: $A_{eff} = A - 2t_f(1 - \rho_f)c_f - 2t_w(1 - \rho_w)c_w$

For parallel flange channels: $A_{eff} = A - 2t_f(1 - \rho_f)c_f - t_w(1 - \rho_w)c_w$

For equal angles: $A_{eff} = A - 2t(1 - \rho)h$

For unequal angles: $A_{eff} = A - t(1 - \rho)(h + b)$

For circular hollow sections: Effective areas are not tabulated for circular hollow sections in this publication. BS EN 1993-1-1 6.2.2.5(5) refers the reader to BS EN 1993-1-6.

For elliptical hollow sections: Effective areas are not tabulated in this publication, but may be calculated from Chan, T.M. and Gardner, L. Engineering Structures, **30**(2), 2008, 522–532.

$$A_{\text{eff}} = A \left(\frac{90t}{D_{\text{e}}} \frac{235}{f_{\text{y}}} \right)^{0.5}$$

where:

D_{e} is the equivalent diameter $= 2\dfrac{a^2}{b}$

a is half the outside dimension of the section on its major axis

b is half the outside dimension of the section on its minor axis.

Expressions for the reduction factors ρ_{f}, ρ_{w} and ρ are given in BS EN 1993-1-5, 4.4.

The ratio of effective area to gross area (A_{eff}/A) is also given in the tables to provide a guide as to how much of the section is effective. Note that although BS EN 1993-1-1 classifies some sections as Class 4, their effective area according to BS EN 1993-1-5 is equal to the gross area.

4.3 Effective section properties of members subject to pure bending

The effective cross-section properties of Class 4 cross-sections are based on the effective widths of the compression parts. The effective cross–sectional properties for Class 4 sections in bending have been calculated in accordance with BS EN 1993-1-1, 6.2.2.5 and BS EN 1993–1–5:2006, 4.3 and 4.4.

Cross-section properties are given for the effective second moment of area I_{eff} and the effective elastic section modulus $W_{\text{el,eff}}$. The identifier 'W', 'F' or 'WF' indicates whether the web, the flange or both controls the section Class 4 classification.

Equations for the effective section properties are not shown here because the process for determining these properties requires iteration. Also the equations are dependent on the classification status of each component part.

For the range of sections covered by this publication, only a selection of the hollow sections become Class 4 when subject to bending alone.

For cross-sections with a Class 3 web and Class 1 or 2 flanges, an effective plastic modulus $W_{\text{pl,eff}}$ can be calculated, following the recommendations given in BS EN 1993-1-1, 6.2.2.4 (1). This clause is applicable to open sections (UB, UC, joists and channels) and hollow sections.

For the range of sections covered by this publication, only a limited number of the hollow sections can be used with an effective plastic modulus $W_{\text{pl,eff}}$, when subject to bending alone.

5 Bolts and welds

5.1 Bolt resistances

The types of bolts covered are:

- Classes 4.6, 8.8 and 10.9, as specified in BS EN ISO 4014, BS EN ISO 4016, BS EN ISO 4017 and BS EN ISO 4018, assembled with a nut conforming to BS EN ISO 4032 or BS EN ISO 4034. Such bolts should be specified as also complying with BS EN 15048.
- Countersunk non-preloaded bolts as specified in BS 4933, assembled with a nut conforming to BS EN ISO 4032 or BS EN ISO 4034. Such bolts should be specified as also complying with BS EN 15048 and, for grade 8.8 and grade 10.9, with the mechanical property requirements of BS EN ISO 898-1.
- Preloaded bolts as specified in BS EN 14399. In the UK, either system HR bolts to BS EN 14399-3 or HRC bolts to EN 14399-10 should be used, with appropriate nuts and washers (including direct tension indicators to BS EN 14399-9, where required). Countersunk bolts to BS EN 14399-7 may alternatively be used. Bolts should be tightened in accordance with BS EN 1090-2.

(a) Non preloaded hexagon head bolts and countersunk bolts

For each grade:

- The first table gives the tensile stress area, the design tension resistance, the design shear resistance and the minimum thickness of ply passed through in order to avoid failure due to punching shear.
- The second table (and where applicable the third table) gives the design bearing resistance for the given bolt configurations.
- (i) The values of the tensile stress area A_s are those given in the relevant product standard.
- (ii) The design tension resistance of a bolt is given by:

$$F_{t,Rd} = \frac{k_2 f_{ub} A_s}{\gamma_{M2}}$$

where:

k_2 = 0.63 for countersunk bolts
= 0.9 for other bolts

f_{ub} is the ultimate tensile strength of the bolt from the relevant product standard

A_s is the tensile stress area of the bolt

γ_{M2} is the partial factor for bolts (γ_{M2} = 1.25 for Class 4.6 bolts and γ_{M2} = 1.25 for other classes of bolts, as given in the National Annex).

(iii) The shear resistance of the bolt is given by:

$$F_{v,Rd} = \frac{\alpha_v f_{ub} A_s}{\gamma_{M2}}$$

where:

α_v = 0.6 for Classes 4.6 and 8.8
  = 0.5 for Class 10.9
f_{ub} is the ultimate tensile strength of the bolt
A_s is the tensile stress area of the bolt.

(iv) The punching shear resistance is expressed in terms of the minimum thickness of the ply for which the design punching shear resistance would be equal to the design tension resistance. The value has been derived from the expression for the punching shear resistance given in BS EN 1993-1-8, table 3.4. The minimum thickness is given by:

$$t_{min} = \frac{B_{p,Rd}\gamma_{M2}}{0.6\pi d_m f_u}$$

where:

$B_{p,Rd} = F_{t,Rd}$
$F_{t,Rd}$ is the design tension resistance per bolt
$d_m \quad = \min\left(\left[\dfrac{e+s}{2}\right]_{head}; \left[\dfrac{e+s}{2}\right]_{nut}\right)$
e is the width across points of the bolt head or the nut
s is the width across flats of the bolt head or the nut
f_u is the ultimate tensile strength of the ply under the bolt head or nut
A_s is the tensile stress area of the bolt.

(v) The bearing resistance of a bolt adjacent to an edge, subject to force perpendicular to the edge is given by:

$$F_{b,Rd} = \frac{k_1 \alpha_b f_u dt}{\gamma_{M2}}$$

where:

$k_1 \quad = \min\left(\left(2.8\dfrac{e_2}{d_0} - 1.7\right); 2.5\right)$
e_2 is the edge distance from the centre of a fastener hole to the adjacent end of any part, measured at right angles to the direction of load transfer
d_0 is the hole diameter of the bolt
$\alpha_b \quad = \min\left(\alpha_d; \dfrac{f_{ub}}{f_u}; 1.0\right)$
f_{ub} is the ultimate tensile strength of the bolt
f_u is the ultimate tensile strength of ply passed through

$$\alpha_d = \frac{e_1}{3d_0}$$

e_1 is the end distance from the centre of a fastener hole to the adjacent end of any part, measured in the direction of load transfer.

Values of e_1, e_2, p_1 and p_2 given in the tables are the minimum distances needed in the connection configuration to achieve the bearing resistances given in each tables. Resistances have been calculated for values of $e_1 = 2d$ in the second table and $e_1 = 3d$ in the third table. Values of e_1 have then been rounded up to the nearest 5 mm.

For Class 4.6 bolts the values of e_1, e_2, p_1 and p_2 in the second table are based on typical connections used in the UK.

For Class 8.8 and 10.9 bolts the values of e_1, e_2, p_1 and p_2 in the second table are based on typical connections used in the UK. Values of e_1, e_2, p_1 and p_2 of the third table are based on the distances needed to maximise the bearing resistance of the bolts.

Clause 3.6.1(3) of BS EN 1993-1-8 states that where the threads do not comply with EN 1090 the relevant bolt resistances should be multiplied by a factor of 0.85.

Note 1 in table 3.4 of BS EN 1993-1-8 gives the reduction factors that should to be applied to the bearing resistance for oversize holes and slotted holes.

(b) Preloaded hexagon head bolts and countersunk bolts at serviceability and ultimate limit states

 (i) The tensile stress area A_s, tension resistance, shear resistance, punching shear resistance and bearing resistance are calculated as given above.
 (ii) The tension resistance has been calculated as for non-preloaded bolts.
 (iii) The bearing resistance has been calculated as for non-preloaded bolts.
 (iv) The slip resistance of the bolt is given by:

$$F_{s,Rd} = \frac{k_s n \mu}{\phantom{\gamma_{M3}}} F_{p,C} \qquad \text{at SLS}$$
$$F_{s,Rd} = \frac{k_s n \mu}{\gamma_{M3}} F_{p,C} \qquad \text{at ULS}$$

where:

k_s is taken as 1.0

n is the number of friction surfaces

μ is the slip factor

$F_{p,C} = 0.7 f_{ub} A_s$ is the preloading force

γ_{M3} is the partial factor for slip resistance (According to the National Annex, $\gamma_{M3} = 1.25$ at ultimate limit state, and $\gamma_{M3,ser} = 1.1$ at serviceability limit state).

Tables are provided for connections which are non-slip at the serviceability state and those that are non-slip at the ultimate limit state. In both cases, according to BS EN 1993-1-8, 3.4.1, bearing resistance must be checked, and resistance values are provided for this purpose.

Note that for connections which are non-slip at the serviceability limit state, the slip resistance must be equal to or greater than the design force due to the SLS values of actions (i.e. not the design force due to ULS actions). For connections which are non-slip at the serviceability limit state, the calculated resistances are based on a slip factor, μ of 0.5. Designers should ensure this value is appropriate for the surfaces to be fastened, or should calculate revised resistances.

5.2 Welds

Design resistances of fillet welds per unit length are tabulated. The design resistance of fillet welds will be sufficient if the following criteria are both satisfied:

$$\sqrt{\sigma_{\perp}^2 + 3\left(\tau_{\perp}^2 + \tau_{//}^2\right)} \leq \frac{f_u}{\beta_w \gamma_{M2}} \quad \text{and} \quad \sigma_{\perp} \leq 0.9 f_u / \gamma_{M2} \tag{4.1}$$

Each of these components of the stress in the weld can be expressed in terms of the longitudinal and transverse resistance of the weld per unit length as follows:

$$\sigma_{\perp} = \frac{F_{w,T,Ed} \sin \theta}{a}$$

$$\tau_{\perp} = \frac{F_{w,T,Ed} \cos \theta}{a}$$

$$\tau_{//} = \frac{F_{w,L,Ed}}{a}$$

Therefore:

$$\sqrt{\left(\frac{F_{w,T,Ed} \sin \theta}{a}\right)^2 + 3\left[\left(\frac{F_{w,T,Ed} \cos \theta}{a}\right)^2 + \left(\frac{F_{w,L,Ed}}{a}\right)^2\right]} \leq \frac{f_u}{\beta_w \gamma_{M2}}$$

$$\frac{1}{a}\sqrt{F_{w,T,Ed}^2\left(\sin^2 \theta + 3\cos^2 \theta\right) + 3F_{w,L,Ed}^2} \leq \frac{f_u}{\beta_w \gamma_{M2}}$$

$$\frac{1}{a}\sqrt{F_{w,T,Ed}^2\left(1 + 2\cos^2 \theta\right) + 3F_{w,L,Ed}^2} \leq \frac{f_u}{\beta_w \gamma_{M2}}$$

$$\frac{1}{a}\sqrt{\frac{F_{w,T,Ed}^2}{K} + F_{w,L,Ed}^2} \leq \frac{f_u}{\beta_w \gamma_{M2}} \quad \text{with} \quad K = \sqrt{\frac{3}{\left(1 + 2\cos^2 \theta\right)}}$$

From this equation, taking each of the components in turn as zero, the following expressions can be derived for the longitudinal and the transverse resistances of welds:

Design weld resistance, longitudinal: $F_{w,L,Rd} = f_{vw,d}a$
Design weld resistance, transverse: $F_{w,T,Rd} = KF_{w,L,Rd}$

where:

$\quad f_{vw,d}$ is the design shear strength of the weld $= \dfrac{f_u}{\sqrt{3}\beta_w \gamma_{M2}}$

$\quad f_u \quad = 410\,\text{N}/\text{mm}^2$ for S275
$\qquad\quad = 470\,\text{N}/\text{mm}^2$ for S355
$\qquad$ These values are valid for thicknesses up to 100 mm)

$\quad \beta_w \quad = 0.85$ for S275
$\qquad\quad = 0.9$ for S355

$\quad \gamma_{M2}$ is the partial factor for the resistance of welds ($\gamma_{M2} = 1.25$ according to the National Annex)

$\quad a \quad$ is the throat thickness of the fillet weld

$$K = \sqrt{\frac{3}{\left(1 + 2\cos^2\theta\right)}}$$

The above expression for transverse weld resistance is valid where the plates are at 90° and therefore $\theta = 45°$ and $K = 1.225$.

Tables of dimensions and gross section properties

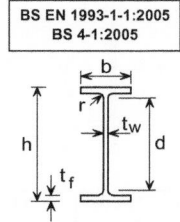

| BS EN 1993-1-1:2005 |
| BS 4-1:2005 |

UNIVERSAL BEAMS

Advance UKB

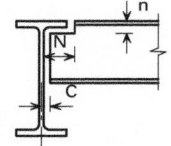

Dimensions

Section Designation	Mass per Metre	Depth of Section	Width of Section	Thickness		Root Radius	Depth between Fillets	Ratios for Local Buckling		Dimensions for Detailing			Surface Area	
				Web	Flange			Flange	Web	End Clearance	Notch		Per Metre	Per Tonne
		h	b	t_w	t_f	r	d	c_f / t_f	c_w / t_w	C	N	n		
	kg/m	mm	mm	mm	mm	mm	mm			mm	mm	mm	m²	m²
1016x305x487 +	486.7	1036.3	308.5	30.0	54.1	30.0	868.1	2.02	28.9	17	150	86	3.20	6.58
1016x305x437 +	437.0	1026.1	305.4	26.9	49.0	30.0	868.1	2.23	32.3	15	150	80	3.17	7.25
1016x305x393 +	392.7	1015.9	303.0	24.4	43.9	30.0	868.1	2.49	35.6	14	150	74	3.14	8.00
1016x305x349 +	349.4	1008.1	302.0	21.1	40.0	30.0	868.1	2.76	41.1	13	152	70	3.13	8.96
1016x305x314 +	314.3	999.9	300.0	19.1	35.9	30.0	868.1	3.08	45.5	12	152	66	3.11	9.89
1016x305x272 +	272.3	990.1	300.0	16.5	31.0	30.0	868.1	3.60	52.6	10	152	62	3.10	11.4
1016x305x249 +	248.7	980.1	300.0	16.5	26.0	30.0	868.1	4.30	52.6	10	152	56	3.08	12.4
1016x305x222 +	222.0	970.3	300.0	16.0	21.1	30.0	868.1	5.31	54.3	10	152	52	3.06	13.8
914x419x388	388.0	921.0	420.5	21.4	36.6	24.1	799.6	4.79	37.4	13	210	62	3.44	8.87
914x419x343	343.3	911.8	418.5	19.4	32.0	24.1	799.6	5.48	41.2	12	210	58	3.42	9.96
914x305x289	289.1	926.6	307.7	19.5	32.0	19.1	824.4	3.91	42.3	12	156	52	3.01	10.4
914x305x253	253.4	918.4	305.5	17.3	27.9	19.1	824.4	4.48	47.7	11	156	48	2.99	11.8
914x305x224	224.2	910.4	304.1	15.9	23.9	19.1	824.4	5.23	51.8	10	156	44	2.97	13.2
914x305x201	200.9	903.0	303.3	15.1	20.2	19.1	824.4	6.19	54.6	10	156	40	2.96	14.7
838x292x226	226.5	850.9	293.8	16.1	26.8	17.8	761.7	4.52	47.3	10	150	46	2.81	12.4
838x292x194	193.8	840.7	292.4	14.7	21.7	17.8	761.7	5.58	51.8	9	150	40	2.79	14.4
838x292x176	175.9	834.9	291.7	14.0	18.8	17.8	761.7	6.44	54.4	9	150	38	2.78	15.8
762x267x197	196.8	769.8	268.0	15.6	25.4	16.5	686.0	4.32	44.0	10	138	42	2.55	13.0
762x267x173	173.0	762.2	266.7	14.3	21.6	16.5	686.0	5.08	48.0	9	138	40	2.53	14.6
762x267x147	146.9	754.0	265.2	12.8	17.5	16.5	686.0	6.27	53.6	8	138	34	2.51	17.1
762x267x134	133.9	750.0	264.4	12.0	15.5	16.5	686.0	7.08	57.2	8	138	32	2.51	18.7
686x254x170	170.2	692.9	255.8	14.5	23.7	15.2	615.1	4.45	42.4	9	132	40	2.35	13.8
686x254x152	152.4	687.5	254.5	13.2	21.0	15.2	615.1	5.02	46.6	9	132	38	2.34	15.4
686x254x140	140.1	683.5	253.7	12.4	19.0	15.2	615.1	5.55	49.6	8	132	36	2.33	16.6
686x254x125	125.2	677.9	253.0	11.7	16.2	15.2	615.1	6.51	52.6	8	132	32	2.32	18.5
610x305x238	238.1	635.8	311.4	18.4	31.4	16.5	540.0	4.14	29.3	11	158	48	2.45	10.3
610x305x179	179.0	620.2	307.1	14.1	23.6	16.5	540.0	5.51	38.3	9	158	42	2.41	13.5
610x305x149	149.2	612.4	304.8	11.8	19.7	16.5	540.0	6.60	45.8	8	158	38	2.39	16.0
610x229x140	139.9	617.2	230.2	13.1	22.1	12.7	547.6	4.34	41.8	9	120	36	2.11	15.1
610x229x125	125.1	612.2	229.0	11.9	19.6	12.7	547.6	4.89	46.0	8	120	34	2.09	16.7
610x229x113	113.0	607.6	228.2	11.1	17.3	12.7	547.6	5.54	49.3	8	120	30	2.08	18.4
610x229x101	101.2	602.6	227.6	10.5	14.8	12.7	547.6	6.48	52.2	7	120	28	2.07	20.5
610x178x100 +	100.3	607.4	179.2	11.3	17.2	12.7	547.6	4.14	48.5	8	94	30	1.89	18.8
610x178x92 +	92.2	603.0	178.8	10.9	15.0	12.7	547.6	4.75	50.2	7	94	28	1.88	20.4
610x178x82 +	81.8	598.6	177.9	10.0	12.8	12.7	547.6	5.57	54.8	7	94	26	1.87	22.9
533x312x273 +	273.3	577.1	320.2	21.1	37.6	12.7	476.5	3.64	22.6	13	160	52	2.37	8.67
533x312x219 +	218.8	560.3	317.4	18.3	29.2	12.7	476.5	4.69	26.0	11	160	42	2.33	10.7
533x312x182 +	181.5	550.7	314.5	15.2	24.4	12.7	476.5	5.61	31.3	10	160	38	2.31	12.7
533x312x151 +	150.6	542.5	312.0	12.7	20.3	12.7	476.5	6.75	37.5	8	160	34	2.29	15.2

Advance and UKB are trademarks of Tata. A fuller description of the relationship between Universal Beams (UB) and the Advance range of sections manufactured by Tata is given in section 12.

+ These sections are in addition to the range of BS 4 sections.

FOR EXPLANATION OF TABLES SEE NOTE 2

BS EN 1993-1-1:2005
BS 4-1:2005

UNIVERSAL BEAMS

Advance UKB

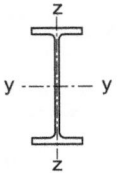

Properties

Section Designation	Second Moment of Area		Radius of Gyration		Elastic Modulus		Plastic Modulus		Buckling Parameter	Torsional Index	Warping Constant	Torsional Constant	Area of Section
	Axis y-y	Axis z-z	Axis y-y	Axis z-z	Axis y-y	Axis z-z	Axis y-y	Axis z-z	U	X	I_w	I_T	A
	cm^4	cm^4	cm	cm	cm^3	cm^3	cm^3	cm^3			dm^6	cm^4	cm^2
1016x305x487 +	1022000	26700	40.6	6.57	19700	1730	23200	2800	0.867	21.1	64.4	4300	620
1016x305x437 +	910000	23400	40.4	6.49	17700	1540	20800	2470	0.868	23.1	56.0	3190	557
1016x305x393 +	808000	20500	40.2	6.40	15900	1350	18500	2170	0.868	25.5	48.4	2330	500
1016x305x349 +	723000	18500	40.3	6.44	14300	1220	16600	1940	0.872	27.9	43.3	1720	445
1016x305x314 +	644000	16200	40.1	6.37	12900	1080	14800	1710	0.872	30.7	37.7	1260	400
1016x305x272 +	554000	14000	40.0	6.35	11200	934	12800	1470	0.872	35.0	32.2	835	347
1016x305x249 +	481000	11800	39.0	6.09	9820	784	11300	1240	0.861	39.9	26.8	582	317
1016x305x222 +	408000	9550	38.0	5.81	8410	636	9810	1020	0.850	45.7	21.5	390	283
914x419x388	720000	45400	38.2	9.59	15600	2160	17700	3340	0.885	26.7	88.9	1730	494
914x419x343	626000	39200	37.8	9.46	13700	1870	15500	2890	0.883	30.1	75.8	1190	437
914x305x289	504000	15600	37.0	6.51	10900	1010	12600	1600	0.867	31.9	31.2	926	368
914x305x253	436000	13300	36.8	6.42	9500	871	10900	1370	0.865	36.2	26.4	626	323
914x305x224	376000	11200	36.3	6.27	8270	739	9530	1160	0.860	41.3	22.1	422	286
914x305x201	325000	9420	35.7	6.07	7200	621	8350	982	0.853	46.9	18.4	291	256
838x292x226	340000	11400	34.3	6.27	7980	773	9160	1210	0.869	35.0	19.3	514	289
838x292x194	279000	9070	33.6	6.06	6640	620	7640	974	0.862	41.6	15.2	306	247
838x292x176	246000	7800	33.1	5.90	5890	535	6810	842	0.856	46.5	13.0	221	224
762x267x197	240000	8170	30.9	5.71	6230	610	7170	958	0.869	33.1	11.3	404	251
762x267x173	205000	6850	30.5	5.58	5390	514	6200	807	0.865	38.0	9.39	267	220
762x267x147	169000	5460	30.0	5.40	4470	411	5160	647	0.858	45.2	7.40	159	187
762x267x134	151000	4790	29.7	5.30	4020	362	4640	570	0.853	49.8	6.46	119	171
686x254x170	170000	6630	28.0	5.53	4920	518	5630	811	0.872	31.8	7.42	308	217
686x254x152	150000	5780	27.8	5.46	4370	455	5000	710	0.871	35.4	6.42	220	194
686x254x140	136000	5180	27.6	5.39	3990	409	4560	638	0.870	38.6	5.72	169	178
686x254x125	118000	4380	27.2	5.24	3480	346	3990	542	0.863	43.8	4.80	116	159
610x305x238	209000	15800	26.3	7.23	6590	1020	7490	1570	0.886	21.3	14.5	785	303
610x305x179	153000	11400	25.9	7.07	4930	743	5550	1140	0.885	27.7	10.2	340	228
610x305x149	126000	9310	25.7	7.00	4110	611	4590	937	0.886	32.7	8.17	200	190
610x229x140	112000	4510	25.0	5.03	3620	391	4140	611	0.875	30.6	3.99	216	178
610x229x125	98600	3930	24.9	4.97	3220	343	3680	535	0.875	34.0	3.45	154	159
610x229x113	87300	3430	24.6	4.88	2870	301	3280	469	0.870	38.0	2.99	111	144
610x229x101	75800	2910	24.2	4.75	2520	256	2880	400	0.863	43.0	2.52	77.0	129
610x178x100 +	72500	1660	23.8	3.60	2390	185	2790	296	0.854	38.7	1.44	95.0	128
610x178x92 +	64600	1440	23.4	3.50	2140	161	2510	258	0.850	42.7	1.24	71.0	117
610x178x82 +	55900	1210	23.2	3.40	1870	136	2190	218	0.843	48.5	1.04	48.8	104
533x312x273 +	199000	20600	23.9	7.69	6890	1290	7870	1990	0.891	15.9	15.0	1290	348
533x312x219 +	151000	15600	23.3	7.48	5400	982	6120	1510	0.884	19.8	11.0	642	279
533x312x182 +	123000	12700	23.1	7.40	4480	806	5040	1240	0.886	23.4	8.77	373	231
533x312x151 +	101000	10300	22.9	7.32	3710	659	4150	1010	0.885	27.8	7.01	216	192

Advance and UKB are trademarks of Tata. A fuller description of the relationship between Universal Beams (UB) and the Advance range of sections manufactured by Tata is given in section 12.

+ These sections are in addition to the range of BS 4 sections.

FOR EXPLANATION OF TABLES SEE NOTE 3

BS EN 1993-1-1:2005
BS 4-1:2005

UNIVERSAL BEAMS

Advance UKB

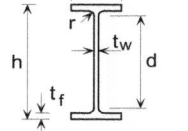

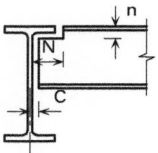

Dimensions

Section Designation	Mass per Metre	Depth of Section	Width of Section	Thickness Web	Thickness Flange	Root Radius	Depth between Fillets	Ratios for Local Buckling Flange	Ratios for Local Buckling Web	Dimensions for Detailing End Clearance	Dimensions for Detailing Notch N	Dimensions for Detailing Notch n	Surface Area Per Metre	Surface Area Per Tonne
	kg/m	h mm	b mm	t_w mm	t_f mm	r mm	d mm	c_f / t_f	c_w / t_w	C mm	N mm	n mm	m^2	m^2
533x210x138 +	138.3	549.1	213.9	14.7	23.6	12.7	476.5	3.68	32.4	9	110	38	1.90	13.7
533x210x122	122.0	544.5	211.9	12.7	21.3	12.7	476.5	4.08	37.5	8	110	34	1.89	15.5
533x210x109	109.0	539.5	210.8	11.6	18.8	12.7	476.5	4.62	41.1	8	110	32	1.88	17.2
533x210x101	101.0	536.7	210.0	10.8	17.4	12.7	476.5	4.99	44.1	7	110	32	1.87	18.5
533x210x92	92.1	533.1	209.3	10.1	15.6	12.7	476.5	5.57	47.2	7	110	30	1.86	20.2
533x210x82	82.2	528.3	208.8	9.6	13.2	12.7	476.5	6.58	49.6	7	110	26	1.85	22.5
533x165x85 +	84.8	534.9	166.5	10.3	16.5	12.7	476.5	3.96	46.3	7	90	30	1.69	19.9
533x165x75 +	74.7	529.1	165.9	9.7	13.6	12.7	476.5	4.81	49.1	7	90	28	1.68	22.5
533x165x66 +	65.7	524.7	165.1	8.9	11.4	12.7	476.5	5.74	53.5	6	90	26	1.67	25.4
457x191x161 +	161.4	492.0	199.4	18.0	32.0	10.2	407.6	2.52	22.6	11	102	44	1.73	10.7
457x191x133 +	133.3	480.6	196.7	15.3	26.3	10.2	407.6	3.06	26.6	10	102	38	1.70	12.8
457x191x106 +	105.8	469.2	194.0	12.6	20.6	10.2	407.6	3.91	32.3	8	102	32	1.67	15.8
457x191x98	98.3	467.2	192.8	11.4	19.6	10.2	407.6	4.11	35.8	8	102	30	1.67	17.0
457x191x89	89.3	463.4	191.9	10.5	17.7	10.2	407.6	4.55	38.8	7	102	28	1.66	18.6
457x191x82	82.0	460.0	191.3	9.9	16.0	10.2	407.6	5.03	41.2	7	102	28	1.65	20.1
457x191x74	74.3	457.0	190.4	9.0	14.5	10.2	407.6	5.55	45.3	7	102	26	1.64	22.1
457x191x67	67.1	453.4	189.9	8.5	12.7	10.2	407.6	6.34	48.0	6	102	24	1.63	24.3
457x152x82	82.1	465.8	155.3	10.5	18.9	10.2	407.6	3.29	38.8	7	84	30	1.51	18.4
457x152x74	74.2	462.0	154.4	9.6	17.0	10.2	407.6	3.66	42.5	7	84	28	1.50	20.2
457x152x67	67.2	458.0	153.8	9.0	15.0	10.2	407.6	4.15	45.3	7	84	26	1.50	22.3
457x152x60	59.8	454.6	152.9	8.1	13.3	10.2	407.6	4.68	50.3	6	84	24	1.49	24.9
457x152x52	52.3	449.8	152.4	7.6	10.9	10.2	407.6	5.71	53.6	6	84	22	1.48	28.3
406x178x85 +	85.3	417.2	181.9	10.9	18.2	10.2	360.4	4.14	33.1	7	96	30	1.52	17.8
406x178x74	74.2	412.8	179.5	9.5	16.0	10.2	360.4	4.68	37.9	7	96	28	1.51	20.4
406x178x67	67.1	409.4	178.8	8.8	14.3	10.2	360.4	5.23	41.0	6	96	26	1.50	22.3
406x178x60	60.1	406.4	177.9	7.9	12.8	10.2	360.4	5.84	45.6	6	96	24	1.49	24.8
406x178x54	54.1	402.6	177.7	7.7	10.9	10.2	360.4	6.86	46.8	6	96	22	1.48	27.3
406x140x53 +	53.3	406.6	143.3	7.9	12.9	10.2	360.4	4.46	45.6	6	78	24	1.35	25.3
406x140x46	46.0	403.2	142.2	6.8	11.2	10.2	360.4	5.13	53.0	5	78	22	1.34	29.1
406x140x39	39.0	398.0	141.8	6.4	8.6	10.2	360.4	6.69	56.3	5	78	20	1.33	34.1
356x171x67	67.1	363.4	173.2	9.1	15.7	10.2	311.6	4.58	34.2	7	94	26	1.38	20.6
356x171x57	57.0	358.0	172.2	8.1	13.0	10.2	311.6	5.53	38.5	6	94	24	1.37	24.1
356x171x51	51.0	355.0	171.5	7.4	11.5	10.2	311.6	6.25	42.1	6	94	22	1.36	26.7
356x171x45	45.0	351.4	171.1	7.0	9.7	10.2	311.6	7.41	44.5	6	94	20	1.36	30.2
356x127x39	39.1	353.4	126.0	6.6	10.7	10.2	311.6	4.63	47.2	5	70	22	1.18	30.2
356x127x33	33.1	349.0	125.4	6.0	8.5	10.2	311.6	5.82	51.9	5	70	20	1.17	35.4
305x165x54	54.0	310.4	166.9	7.9	13.7	8.9	265.2	5.15	33.6	6	90	24	1.26	23.3
305x165x46	46.1	306.6	165.7	6.7	11.8	8.9	265.2	5.98	39.6	5	90	22	1.25	27.1
305x165x40	40.3	303.4	165.0	6.0	10.2	8.9	265.2	6.92	44.2	5	90	20	1.24	30.8

Advance and UKB are trademarks of Tata. A fuller description of the relationship between Universal Beams (UB) and the Advance range of sections manufactured by Tata is given in section 12.

+ These sections are in addition to the range of BS 4 sections.

FOR EXPLANATION OF TABLES SEE NOTE 2

BS EN 1993-1-1:2005
BS 4-1:2005

UNIVERSAL BEAMS

Advance UKB

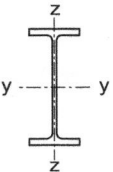

Properties

Section Designation	Second Moment of Area		Radius of Gyration		Elastic Modulus		Plastic Modulus		Buckling Parameter	Torsional Index	Warping Constant	Torsional Constant	Area of Section
	Axis y-y	Axis z-z	Axis y-y	Axis z-z	Axis y-y	Axis z-z	Axis y-y	Axis z-z	U	X	I_w	I_T	A
	cm^4	cm^4	cm	cm	cm^3	cm^3	cm^3	cm^3			dm^6	cm^4	cm^2
533x210x138 +	86100	3860	22.1	4.68	3140	361	3610	568	0.874	24.9	2.67	250	176
533x210x122	76000	3390	22.1	4.67	2790	320	3200	500	0.878	27.6	2.32	178	155
533x210x109	66800	2940	21.9	4.60	2480	279	2830	436	0.875	30.9	1.99	126	139
533x210x101	61500	2690	21.9	4.57	2290	256	2610	399	0.874	33.1	1.81	101	129
533x210x92	55200	2390	21.7	4.51	2070	228	2360	355	0.873	36.4	1.60	75.7	117
533x210x82	47500	2010	21.3	4.38	1800	192	2060	300	0.863	41.6	1.33	51.5	105
533x165x85 +	48500	1270	21.2	3.44	1820	153	2100	243	0.861	35.5	0.857	73.8	108
533x165x75 +	41100	1040	20.8	3.30	1550	125	1810	200	0.853	41.1	0.691	47.9	95.2
533x165x66 +	35000	859	20.5	3.20	1340	104	1560	166	0.847	47.0	0.566	32.0	83.7
457x191x161 +	79800	4250	19.7	4.55	3240	426	3780	672	0.881	16.5	2.25	515	206
457x191x133 +	63800	3350	19.4	4.44	2660	341	3070	535	0.879	19.6	1.73	292	170
457x191x106 +	48900	2510	19.0	4.32	2080	259	2390	405	0.876	24.4	1.27	146	135
457x191x98	45700	2350	19.1	4.33	1960	243	2230	379	0.881	25.8	1.18	121	125
457x191x89	41000	2090	19.0	4.29	1770	218	2010	338	0.878	28.3	1.04	90.7	114
457x191x82	37100	1870	18.8	4.23	1610	196	1830	304	0.879	30.8	0.922	69.2	104
457x191x74	33300	1670	18.8	4.20	1460	176	1650	272	0.877	33.8	0.818	51.8	94.6
457x191x67	29400	1450	18.5	4.12	1300	153	1470	237	0.873	37.8	0.705	37.1	85.5
457x152x82	36600	1180	18.7	3.37	1570	153	1810	240	0.872	27.4	0.591	89.2	105
457x152x74	32700	1050	18.6	3.33	1410	136	1630	213	0.872	30.1	0.518	65.9	94.5
457x152x67	28900	913	18.4	3.27	1260	119	1450	187	0.868	33.6	0.448	47.7	85.6
457x152x60	25500	795	18.3	3.23	1120	104	1290	163	0.868	37.5	0.387	33.8	76.2
457x152x52	21400	645	17.9	3.11	950	84.6	1100	133	0.859	43.8	0.311	21.4	66.6
406x178x85 +	31700	1830	17.1	4.11	1520	201	1730	313	0.880	24.4	0.728	93.0	109
406x178x74	27300	1550	17.0	4.04	1320	172	1500	267	0.882	27.5	0.608	62.8	94.5
406x178x67	24300	1360	16.9	3.99	1190	153	1350	237	0.880	30.4	0.533	46.1	85.5
406x178x60	21600	1200	16.8	3.97	1060	135	1200	209	0.880	33.7	0.466	33.3	76.5
406x178x54	18700	1020	16.5	3.85	930	115	1050	178	0.871	38.3	0.392	23.1	69.0
406x140x53 +	18300	635	16.4	3.06	899	88.6	1030	139	0.870	34.1	0.246	29.0	67.9
406x140x46	15700	538	16.4	3.03	778	75.7	888	118	0.871	39.0	0.207	19.0	58.6
406x140x39	12500	410	15.9	2.87	629	57.8	724	90.8	0.858	47.4	0.155	10.7	49.7
356x171x67	19500	1360	15.1	3.99	1070	157	1210	243	0.886	24.4	0.412	55.7	85.5
356x171x57	16000	1110	14.9	3.91	896	129	1010	199	0.882	28.8	0.330	33.4	72.6
356x171x51	14100	968	14.8	3.86	796	113	896	174	0.881	32.1	0.286	23.8	64.9
356x171x45	12100	811	14.5	3.76	687	94.8	775	147	0.874	36.8	0.237	15.8	57.3
356x127x39	10200	358	14.3	2.68	576	56.8	659	89.0	0.871	35.2	0.105	15.1	49.8
356x127x33	8250	280	14.0	2.58	473	44.7	543	70.2	0.863	42.1	0.081	8.79	42.1
305x165x54	11700	1060	13.0	3.93	754	127	846	196	0.889	23.6	0.234	34.8	68.8
305x165x46	9900	896	13.0	3.90	646	108	720	166	0.890	27.1	0.195	22.2	58.7
305x165x40	8500	764	12.9	3.86	560	92.6	623	142	0.889	31.0	0.164	14.7	51.3

Advance and UKB are trademarks of Tata. A fuller description of the relationship between Universal Beams (UB) and the Advance range of sections manufactured by Tata is given in section 12.

+ These sections are in addition to the range of BS 4 sections.

FOR EXPLANATION OF TABLES SEE NOTE 3

BS EN 1993-1-1:2005
BS 4-1:2005

UNIVERSAL BEAMS

Advance UKB

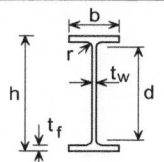

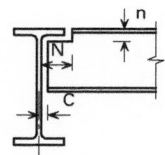

Dimensions

Section Designation	Mass per Metre	Depth of Section	Width of Section	Thickness		Root Radius	Depth between Fillets	Ratios for Local Buckling		Dimensions for Detailing			Surface Area	
				Web	Flange			Flange	Web	End Clearance	Notch		Per Metre	Per Tonne
		h	b	t_w	t_f	r	d	c_f/t_f	c_w/t_w	C	N	n		
	kg/m	mm	mm	mm	mm	mm	mm			mm	mm	mm	m²	m²
305x127x48	48.1	311.0	125.3	9.0	14.0	8.9	265.2	3.52	29.5	7	70	24	1.09	22.7
305x127x42	41.9	307.2	124.3	8.0	12.1	8.9	265.2	4.07	33.2	6	70	22	1.08	25.8
305x127x37	37.0	304.4	123.4	7.1	10.7	8.9	265.2	4.60	37.4	6	70	20	1.07	28.9
305x102x33	32.8	312.7	102.4	6.6	10.8	7.6	275.9	3.73	41.8	5	58	20	1.01	30.8
305x102x28	28.2	308.7	101.8	6.0	8.8	7.6	275.9	4.58	46.0	5	58	18	1.00	35.5
305x102x25	24.8	305.1	101.6	5.8	7.0	7.6	275.9	5.76	47.6	5	58	16	0.992	40.0
254x146x43	43.0	259.6	147.3	7.2	12.7	7.6	219.0	4.92	30.4	6	82	22	1.08	25.1
254x146x37	37.0	256.0	146.4	6.3	10.9	7.6	219.0	5.73	34.8	5	82	20	1.07	28.9
254x146x31	31.1	251.4	146.1	6.0	8.6	7.6	219.0	7.26	36.5	5	82	18	1.06	34.0
254x102x28	28.3	260.4	102.2	6.3	10.0	7.6	225.2	4.04	35.7	5	58	18	0.904	31.9
254x102x25	25.2	257.2	101.9	6.0	8.4	7.6	225.2	4.80	37.5	5	58	16	0.897	35.7
254x102x22	22.0	254.0	101.6	5.7	6.8	7.6	225.2	5.93	39.5	5	58	16	0.890	40.5
203x133x30	30.0	206.8	133.9	6.4	9.6	7.6	172.4	5.85	26.9	5	74	18	0.923	30.8
203x133x25	25.1	203.2	133.2	5.7	7.8	7.6	172.4	7.20	30.2	5	74	16	0.915	36.5
203x102x23	23.1	203.2	101.8	5.4	9.3	7.6	169.4	4.37	31.4	5	60	18	0.790	34.2
178x102x19	19.0	177.8	101.2	4.8	7.9	7.6	146.8	5.14	30.6	4	60	16	0.738	38.7
152x89x16	16.0	152.4	88.7	4.5	7.7	7.6	121.8	4.48	27.1	4	54	16	0.638	40.0
127x76x13	13.0	127.0	76.0	4.0	7.6	7.6	96.6	3.74	24.2	4	46	16	0.537	41.4

Advance and UKB are trademarks of Tata. A fuller description of the relationship between Universal Beams (UB) and the Advance range of sections manufactured by Tata is given in section 12.
FOR EXPLANATION OF TABLES SEE NOTE 2

BS EN 1993-1-1:2005
BS 4-1:2005

UNIVERSAL BEAMS

Advance UKB

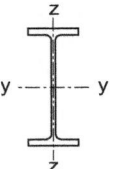

Properties

Section Designation	Second Moment of Area		Radius of Gyration		Elastic Modulus		Plastic Modulus		Buckling Parameter	Torsional Index	Warping Constant	Torsional Constant	Area of Section
	Axis y-y	Axis z-z	Axis y-y	Axis z-z	Axis y-y	Axis z-z	Axis y-y	Axis z-z	U	X	I_w	I_T	A
	cm^4	cm^4	cm	cm	cm^3	cm^3	cm^3	cm^3			dm^6	cm^4	cm^2
305x127x48	9570	461	12.5	2.74	616	73.6	711	116	0.873	23.3	0.102	31.8	61.2
305x127x42	8200	389	12.4	2.70	534	62.6	614	98.4	0.872	26.5	0.0846	21.1	53.4
305x127x37	7170	336	12.3	2.67	471	54.5	539	85.4	0.872	29.7	0.0725	14.8	47.2
305x102x33	6500	194	12.5	2.15	416	37.9	481	60.0	0.867	31.6	0.0442	12.2	41.8
305x102x28	5370	155	12.2	2.08	348	30.5	403	48.4	0.859	37.3	0.0349	7.40	35.9
305x102x25	4460	123	11.9	1.97	292	24.2	342	38.8	0.846	43.4	0.027	4.77	31.6
254x146x43	6540	677	10.9	3.52	504	92.0	566	141	0.891	21.1	0.103	23.9	54.8
254x146x37	5540	571	10.8	3.48	433	78.0	483	119	0.890	24.3	0.0857	15.3	47.2
254x146x31	4410	448	10.5	3.36	351	61.3	393	94.1	0.879	29.6	0.0660	8.55	39.7
254x102x28	4000	179	10.5	2.22	308	34.9	353	54.8	0.873	27.5	0.0280	9.57	36.1
254x102x25	3410	149	10.3	2.15	266	29.2	306	46.0	0.866	31.4	0.0230	6.42	32.0
254x102x22	2840	119	10.1	2.06	224	23.5	259	37.3	0.856	36.3	0.0182	4.15	28.0
203x133x30	2900	385	8.71	3.17	280	57.5	314	88.2	0.882	21.5	0.0374	10.3	38.2
203x133x25	2340	308	8.56	3.10	230	46.2	258	70.9	0.876	25.6	0.0294	5.96	32.0
203x102x23	2100	164	8.46	2.36	207	32.2	234	49.7	0.888	22.4	0.0154	7.02	29.4
178x102x19	1360	137	7.48	2.37	153	27.0	171	41.6	0.886	22.6	0.0099	4.41	24.3
152x89x16	834	89.8	6.41	2.10	109	20.2	123	31.2	0.890	19.5	0.00470	3.56	20.3
127x76x13	473	55.7	5.35	1.84	74.6	14.7	84.2	22.6	0.894	16.3	0.00200	2.85	16.5

Advance and UKB are trademarks of Tata. A fuller description of the relationship between Universal Beams (UB) and the Advance range of sections manufactured by Tata is given in section 12.

FOR EXPLANATION OF TABLES SEE NOTE 3

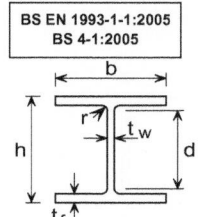

BS EN 1993-1-1:2005
BS 4-1:2005

UNIVERSAL COLUMNS

Advance UKC

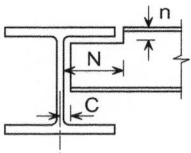

Dimensions

Section Designation	Mass per Metre	Depth of Section	Width of Section	Thickness Web	Thickness Flange	Root Radius	Depth between Fillets	Ratios for Local Buckling Flange	Ratios for Local Buckling Web	Dimensions for Detailing End Clearance	Dimensions for Detailing Notch N	Dimensions for Detailing Notch n	Surface Area Per Metre	Surface Area Per Tonne
		h	b	t_w	t_f	r	d	c_f/t_f	c_w/t_w	C	N	n		
	kg/m	mm	mm	mm	mm	mm	mm			mm	mm	mm	m²	m²
356x406x634	633.9	474.6	424.0	47.6	77.0	15.2	290.2	2.25	6.10	26	200	94	2.52	3.98
356x406x551	551.0	455.6	418.5	42.1	67.5	15.2	290.2	2.56	6.89	23	200	84	2.47	4.48
356x406x467	467.0	436.6	412.2	35.8	58.0	15.2	290.2	2.98	8.11	20	200	74	2.42	5.18
356x406x393	393.0	419.0	407.0	30.6	49.2	15.2	290.2	3.52	9.48	17	200	66	2.38	6.06
356x406x340	339.9	406.4	403.0	26.6	42.9	15.2	290.2	4.03	10.9	15	200	60	2.35	6.91
356x406x287	287.1	393.6	399.0	22.6	36.5	15.2	290.2	4.74	12.8	13	200	52	2.31	8.05
356x406x235	235.1	381.0	394.8	18.4	30.2	15.2	290.2	5.73	15.8	11	200	46	2.28	9.70
356x368x202	201.9	374.6	374.7	16.5	27.0	15.2	290.2	6.07	17.6	10	190	44	2.19	10.8
356x368x177	177.0	368.2	372.6	14.4	23.8	15.2	290.2	6.89	20.2	9	190	40	2.17	12.3
356x368x153	152.9	362.0	370.5	12.3	20.7	15.2	290.2	7.92	23.6	8	190	36	2.16	14.1
356x368x129	129.0	355.6	368.6	10.4	17.5	15.2	290.2	9.4	27.9	7	190	34	2.14	16.6
305x305x283	282.9	365.3	322.2	26.8	44.1	15.2	246.7	3.00	9.21	15	158	60	1.94	6.86
305x305x240	240.0	352.5	318.4	23.0	37.7	15.2	246.7	3.51	10.7	14	158	54	1.91	7.96
305x305x198	198.1	339.9	314.5	19.1	31.4	15.2	246.7	4.22	12.9	12	158	48	1.87	9.44
305x305x158	158.1	327.1	311.2	15.8	25.0	15.2	246.7	5.30	15.6	10	158	42	1.84	11.6
305x305x137	136.9	320.5	309.2	13.8	21.7	15.2	246.7	6.11	17.90	9	158	38	1.82	13.3
305x305x118	117.9	314.5	307.4	12.0	18.7	15.2	246.7	7.09	20.6	8	158	34	1.81	15.4
305x305x97	96.9	307.9	305.3	9.9	15.4	15.2	246.7	8.60	24.9	7	158	32	1.79	18.5
254x254x167	167.1	289.1	265.2	19.2	31.7	12.7	200.3	3.48	10.4	12	134	46	1.58	9.46
254x254x132	132.0	276.3	261.3	15.3	25.3	12.7	200.3	4.36	13.1	10	134	38	1.55	11.7
254x254x107	107.1	266.7	258.8	12.8	20.5	12.7	200.3	5.38	15.6	8	134	34	1.52	14.2
254x254x89	88.9	260.3	256.3	10.3	17.3	12.7	200.3	6.38	19.4	7	134	30	1.50	16.9
254x254x73	73.1	254.1	254.6	8.6	14.2	12.7	200.3	7.77	23.3	6	134	28	1.49	20.4
203x203x127 +	127.5	241.4	213.9	18.1	30.1	10.2	160.8	2.91	8.88	11	108	42	1.28	10.0
203x203x113 +	113.5	235.0	212.1	16.3	26.9	10.2	160.8	3.26	9.87	10	108	38	1.27	11.2
203x203x100 +	99.6	228.6	210.3	14.5	23.7	10.2	160.8	3.70	11.1	9	108	34	1.25	12.6
203x203x86	86.1	222.2	209.1	12.7	20.5	10.2	160.8	4.29	12.7	8	110	32	1.24	14.4
203x203x71	71.0	215.8	206.4	10.0	17.3	10.2	160.8	5.09	16.1	7	110	28	1.22	17.2
203x203x60	60.0	209.6	205.8	9.4	14.2	10.2	160.8	6.20	17.1	7	110	26	1.21	20.2
203x203x52	52.0	206.2	204.3	7.9	12.5	10.2	160.8	7.04	20.4	6	110	24	1.20	23.1
203x203x46	46.1	203.2	203.6	7.2	11.0	10.2	160.8	8.00	22.3	6	110	22	1.19	25.8
152x152x51 +	51.2	170.2	157.4	11.0	15.7	7.6	123.6	4.18	11.2	8	84	24	0.935	18.3
152x152x44 +	44.0	166.0	155.9	9.5	13.6	7.6	123.6	4.82	13.0	7	84	22	0.924	21.0
152x152x37	37.0	161.8	154.4	8.0	11.5	7.6	123.6	5.70	15.5	6	84	20	0.912	24.7
152x152x30	30.0	157.6	152.9	6.5	9.4	7.6	123.6	6.98	19.0	5	84	18	0.901	30.0
152x152x23	23.0	152.4	152.2	5.8	6.8	7.6	123.6	9.65	21.3	5	84	16	0.889	38.7

Advance and UKC are trademarks of Tata. A fuller description of the relationship between Universal Columns (UC) and the Advance range of sections manufactured by Tata is given in section 12.

+ These sections are in addition to the range of BS 4 sections.

FOR EXPLANATION OF TABLES SEE NOTE 2

BS EN 1993-1-1:2005
BS 4-1:2005

UNIVERSAL COLUMNS

Advance UKC

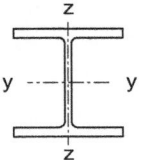

Properties

Section Designation	Second Moment of Area		Radius of Gyration		Elastic Modulus		Plastic Modulus		Buckling Parameter	Torsional Index	Warping Constant	Torsional Constant	Area of Section
	Axis y-y	Axis z-z	Axis y-y	Axis z-z	Axis y-y	Axis z-z	Axis y-y	Axis z-z					
	cm^4	cm^4	cm	cm	cm^3	cm^3	cm^3	cm^3	U	X	I_w dm^6	I_T cm^4	A cm^2
356x406x634	275000	98100	18.4	11.0	11600	4630	14200	7110	0.843	5.46	38.8	13700	808
356x406x551	227000	82700	18.0	10.9	9960	3950	12100	6060	0.841	6.05	31.1	9240	702
356x406x467	183000	67800	17.5	10.7	8380	3290	10000	5030	0.839	6.85	24.3	5810	595
356x406x393	147000	55400	17.1	10.5	7000	2720	8220	4150	0.837	7.86	18.9	3550	501
356x406x340	123000	46900	16.8	10.4	6030	2330	7000	3540	0.836	8.84	15.5	2340	433
356x406x287	99900	38700	16.5	10.3	5070	1940	5810	2950	0.835	10.17	12.3	1440	366
356x406x235	79100	31000	16.3	10.2	4150	1570	4690	2380	0.834	12.04	9.54	812	299
356x368x202	66300	23700	16.1	9.60	3540	1260	3970	1920	0.844	13.35	7.16	558	257
356x368x177	57100	20500	15.9	9.54	3100	1100	3460	1670	0.844	15.00	6.09	381	226
356x368x153	48600	17600	15.8	9.49	2680	948	2960	1430	0.844	17.01	5.11	251	195
356x368x129	40200	14600	15.6	9.43	2260	793	2480	1200	0.844	19.81	4.18	153	164
305x305x283	78900	24600	14.8	8.27	4320	1530	5110	2340	0.855	7.64	6.35	2030	360
305x305x240	64200	20300	14.5	8.15	3640	1280	4250	1950	0.854	8.73	5.03	1270	306
305x305x198	50900	16300	14.2	8.04	3000	1040	3440	1580	0.854	10.23	3.88	734	252
305x305x158	38700	12600	13.9	7.90	2370	808	2680	1230	0.851	12.46	2.87	378	201
305x305x137	32800	10700	13.7	7.83	2050	692	2300	1050	0.851	14.13	2.39	249	174
305x305x118	27700	9060	13.6	7.77	1760	589	1960	895	0.850	16.14	1.98	161	150
305x305x97	22200	7310	13.4	7.69	1450	479	1590	726	0.850	19.19	1.56	91.2	123
254x254x167	30000	9870	11.9	6.81	2080	744	2420	1140	0.851	8.48	1.63	626	213
254x254x132	22500	7530	11.6	6.69	1630	576	1870	878	0.850	10.32	1.19	319	168
254x254x107	17500	5930	11.3	6.59	1310	458	1480	697	0.848	12.38	0.898	172	136
254x254x89	14300	4860	11.2	6.55	1100	379	1220	575	0.850	14.46	0.717	102	113
254x254x73	11400	3910	11.1	6.48	898	307	992	465	0.849	17.24	0.562	57.6	93.1
203x203x127 +	15400	4920	9.75	5.50	1280	460	1520	704	0.854	7.38	0.549	427	162
203x203x113 +	13300	4290	9.59	5.45	1130	404	1330	618	0.853	8.11	0.464	305	145
203x203x100 +	11300	3680	9.44	5.39	988	350	1150	534	0.852	9.02	0.386	210	127
203x203x86	9450	3130	9.28	5.34	850	299	977	456	0.850	10.20	0.318	137	110
203x203x71	7620	2540	9.18	5.30	706	246	799	374	0.853	11.90	0.250	80.2	90.4
203x203x60	6120	2060	8.96	5.20	584	201	656	305	0.846	14.10	0.197	47.2	76.4
203x203x52	5260	1780	8.91	5.18	510	174	567	264	0.848	15.80	0.167	31.8	66.3
203x203x46	4570	1550	8.82	5.13	450	152	497	231	0.847	17.70	0.143	22.2	58.7
152x152x51 +	3230	1020	7.04	3.96	379	130	438	199	0.848	10.10	0.061	48.8	65.2
152x152x44 +	2700	860	6.94	3.92	326	110	372	169	0.848	11.50	0.050	31.7	56.1
152x152x37	2210	706	6.85	3.87	273	91.5	309	140	0.848	13.30	0.040	19.2	47.1
152x152x30	1750	560	6.76	3.83	222	73.3	248	112	0.849	16.00	0.031	10.5	38.3
152x152x23	1250	400	6.54	3.70	164	52.6	182	80.1	0.840	20.70	0.021	4.63	29.2

Advance and UKC are trademarks of Tata. A fuller description of the relationship between Universal Columns (UC) and the Advance range of sections manufactured by Tata is given in section 12.

+ These sections are in addition to the range of BS 4 sections.

FOR EXPLANATION OF TABLES SEE NOTE 3

JOISTS

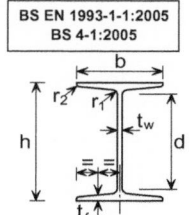

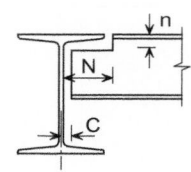

Dimensions

Section Designation	Mass per Metre	Depth of Section	Width of Section	Thickness		Radii		Depth between Fillets	Ratios for Local Buckling		Dimensions for Detailing			Surface Area	
				Web	Flange	Root	Toe		Flange	Web	End Clearance	Notch		Per Metre	Per Tonne
		h	b	t_w	t_f	r_1	r_2	d	c_f / t_f	c_w / t_w	C	N	n		
	kg/m	mm	mm	mm	mm	mm	mm	mm			mm	mm	mm	m²	m²
254x203x82	82.0	254.0	203.2	10.2	19.9	19.6	9.7	166.6	3.86	16.3	7	104	44	1.21	14.8
254x114x37	37.2	254.0	114.3	7.6	12.8	12.4	6.1	199.3	3.20	26.2	6	60	28	0.899	24.2
203x152x52	52.3	203.2	152.4	8.9	16.5	15.5	7.6	133.2	3.41	15.0	6	78	36	0.932	17.8
152x127x37	37.3	152.4	127.0	10.4	13.2	13.5	6.6	94.3	3.39	9.07	7	66	30	0.737	19.8
127x114x29	29.3	127.0	114.3	10.2	11.5	9.9	4.8	79.5	3.67	7.79	7	60	24	0.646	22.0
127x114x27	26.9	127.0	114.3	7.4	11.4	9.9	5.0	79.5	3.82	10.7	6	60	24	0.650	24.2
127x76x16	16.5	127.0	76.2	5.6	9.6	9.4	4.6	86.5	2.70	15.4	5	42	22	0.512	31.0
114x114x27	27.1	114.3	114.3	9.5	10.7	14.2	3.2	60.8	3.57	6.40	7	60	28	0.618	22.8
102x102x23	23.0	101.6	101.6	9.5	10.3	11.1	3.2	55.2	3.39	5.81	7	54	24	0.549	23.9
102x44x7	7.5	101.6	44.5	4.3	6.1	6.9	3.3	74.6	2.16	17.3	4	28	14	0.350	46.6
89x89x19	19.5	88.9	88.9	9.5	9.9	11.1	3.2	44.2	2.89	4.65	7	46	24	0.476	24.4
76x76x15	15.0	76.2	80.0	8.9	8.4	9.4	4.6	38.1	3.11	4.28	6	42	20	0.419	27.9
76x76x13	12.8	76.2	76.2	5.1	8.4	9.4	4.6	38.1	3.11	7.47	5	42	20	0.411	32.1

FOR EXPLANATION OF TABLES SEE NOTE 2

BS EN 1993-1-1:2005
BS 4-1:2005

JOISTS

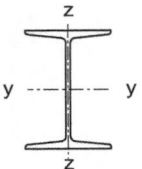

Properties

Section Designation	Second Moment of Area		Radius of Gyration		Elastic Modulus		Plastic Modulus		Buckling Parameter	Torsional Index	Warping Constant	Torsional Constant	Area of Section
	Axis y-y	Axis z-z	Axis y-y	Axis z-z	Axis y-y	Axis z-z	Axis y-y	Axis z-z	U	X	I_w	I_T	A
	cm^4	cm^4	cm	cm	cm^3	cm^3	cm^3	cm^3			dm^6	cm^4	cm^2
254x203x82	12000	2280	10.7	4.67	947	224	1080	371	0.888	11.0	0.312	152	105
254x114x37	5080	269	10.4	2.39	400	47.1	459	79.1	0.884	18.7	0.0392	25.2	47.3
203x152x52	4800	816	8.49	3.50	472	107	541	176	0.890	10.7	0.0711	64.8	66.6
152x127x37	1820	378	6.19	2.82	239	59.6	279	99.8	0.867	9.3	0.0183	33.9	47.5
127x114x29	979	242	5.12	2.54	154	42.3	181	70.8	0.853	8.8	0.00807	20.8	37.4
127x114x27	946	236	5.26	2.63	149	41.3	172	68.2	0.868	9.3	0.00788	16.9	34.2
127x76x16	571	60.8	5.21	1.70	90.0	16.0	104	26.4	0.890	11.8	0.00210	6.72	21.1
114x114x27	736	224	4.62	2.55	129	39.2	151	65.8	0.839	7.9	0.00601	18.9	34.5
102x102x23	486	154	4.07	2.29	95.6	30.3	113	50.6	0.836	7.4	0.00321	14.2	29.3
102x44x7	153	7.82	4.01	0.907	30.1	3.51	35.4	6.03	0.872	14.9	0.000178	1.25	9.50
89x89x19	307	101	3.51	2.02	69.0	22.8	82.7	38.0	0.829	6.6	0.00158	11.5	24.9
76x76x15	172	60.9	3.00	1.78	45.2	15.2	54.2	25.8	0.820	6.4	0.000700	6.83	19.1
76x76x13	158	51.8	3.12	1.79	41.5	13.6	48.7	22.4	0.853	7.2	0.000595	4.59	16.2

FOR EXPLANATION OF TABLES SEE NOTE 3

BS EN 1993-1-1:2005
BS 4-1:2005

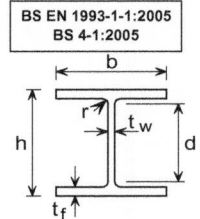

UNIVERSAL BEARING PILES

Advance UKBP

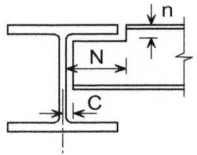

Dimensions

Section Designation	Mass per Metre	Depth of Section	Width of Section	Thickness		Root Radius	Depth between Fillets	Ratios for Local Buckling		Dimensions for Detailing			Surface Area	
				Web	Flange			Flange	Web	End Clearance	Notch		Per Metre	Per Tonne
		H	b	t_w	t_f	r	d	c_f/t_f	c_w/t_w	C	N	n		
	kg/m	mm	mm	mm	mm	mm	mm			mm	mm	mm	m^2	m^2
356x368x174	173.9	361.4	378.5	20.3	20.4	15.2	290.2	8.03	14.3	12	190	36	2.17	12.5
356x368x152	152.0	356.4	376.0	17.8	17.9	15.2	290.2	9.16	16.3	11	190	34	2.16	14.2
356x368x133	133.0	352.0	373.8	15.6	15.7	15.2	290.2	10.44	18.6	10	190	32	2.14	16.1
356x368x109	108.9	346.4	371.0	12.8	12.9	15.2	290.2	12.71	22.7	8	190	30	2.13	19.5
305x305x223	222.9	337.9	325.7	30.3	30.4	15.2	246.7	4.36	8.14	17	158	46	1.89	8.49
305x305x186	186.0	328.3	320.9	25.5	25.6	15.2	246.7	5.18	9.67	15	158	42	1.86	10.0
305x305x149	149.1	318.5	316.0	20.6	20.7	15.2	246.7	6.40	12.0	12	158	36	1.83	12.3
305x305x126	126.1	312.3	312.9	17.5	17.6	15.2	246.7	7.53	14.1	11	158	34	1.82	14.4
305x305x110	110.0	307.9	310.7	15.3	15.4	15.2	246.7	8.60	16.1	10	158	32	1.80	16.4
305x305x95	94.9	303.7	308.7	13.3	13.3	15.2	246.7	9.96	18.5	9	158	30	1.79	18.9
305x305x88	88.0	301.7	307.8	12.4	12.3	15.2	246.7	10.77	19.9	8	158	28	1.78	20.3
305x305x79	78.9	299.3	306.4	11.0	11.1	15.2	246.7	11.94	22.4	8	158	28	1.78	22.5
254x254x85	85.1	254.3	260.4	14.4	14.3	12.7	200.3	7.71	13.9	9	134	28	1.50	17.6
254x254x71	71.0	249.7	258.0	12.0	12.0	12.7	200.3	9.19	16.7	8	134	26	1.49	20.9
254x254x63	63.0	247.1	256.6	10.6	10.7	12.7	200.3	10.31	18.9	7	134	24	1.48	23.5
203x203x54	53.9	204.0	207.7	11.3	11.4	10.2	160.8	7.72	14.2	8	110	22	1.20	22.2
203x203x45	44.9	200.2	205.9	9.5	9.5	10.2	160.8	9.26	16.9	7	110	20	1.19	26.4

Advance and UKBP are trademarks of Tata. A fuller description of the relationship between Universal Bearing Piles (UBP) and the Advance range of sections manufactured by Tata is given in section 12.

FOR EXPLANATION OF TABLES SEE NOTE 2

BS EN 1993-1-1:2005
BS 4-1:2005

UNIVERSAL BEARING PILES

Advance UKBP

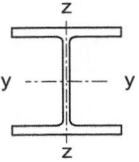

Properties

Section Designation	Second Moment of Area		Radius of Gyration		Elastic Modulus		Plastic Modulus		Buckling Parameter	Torsional Index	Warping Constant	Torsional Constant	Area of Section
	Axis y-y	Axis z-z	Axis y-y	Axis z-z	Axis y-y	Axis z-z	Axis y-y	Axis z-z	U	X	I_w	I_T	A
	cm^4	cm^4	cm	cm	cm^3	cm^3	cm^3	cm^3			dm^6	cm^4	cm^2
356x368x174	51000	18500	15.2	9.13	2820	976	3190	1500	0.822	15.8	5.37	330	221
356x368x152	44000	15900	15.1	9.05	2470	845	2770	1290	0.821	17.9	4.55	223	194
356x368x133	38000	13700	15.0	8.99	2160	732	2410	1120	0.823	20.1	3.87	151	169
356x368x109	30600	11000	14.9	8.90	1770	592	1960	903	0.822	24.2	3.05	84.6	139
305x305x223	52700	17600	13.6	7.87	3120	1080	3650	1680	0.827	9.5	4.15	943	284
305x305x186	42600	14100	13.4	7.73	2600	881	3000	1370	0.827	11.1	3.24	560	237
305x305x149	33100	10900	13.2	7.58	2080	691	2370	1070	0.828	13.5	2.42	295	190
305x305x126	27400	9000	13.1	7.49	1760	575	1990	885	0.829	15.7	1.95	182	161
305x305x110	23600	7710	13.0	7.42	1530	496	1720	762	0.830	17.7	1.65	122	140
305x305x95	20000	6530	12.9	7.35	1320	423	1470	648	0.829	20.2	1.38	80.0	121
305x305x88	18400	5980	12.8	7.31	1220	389	1360	595	0.831	21.6	1.25	64.2	112
305x305x79	16400	5330	12.8	7.28	1100	348	1220	531	0.833	23.8	1.11	46.9	100
254x254x85	12300	4220	10.6	6.24	966	324	1090	498	0.826	15.6	0.607	81.8	108
254x254x71	10100	3440	10.6	6.17	807	267	904	409	0.826	18.4	0.486	48.4	90.4
254x254x63	8860	3020	10.5	6.13	717	235	799	360	0.828	20.4	0.421	34.3	80.2
203x203x54	5030	1710	8.55	4.98	493	164	557	252	0.827	15.8	0.158	32.7	68.7
203x203x45	4100	1380	8.46	4.92	410	134	459	206	0.827	18.6	0.126	19.2	57.2

Advance and UKBP are trademarks of Tata. A fuller description of the relationship between Universal Bearing Piles (UBP) and the Advance range of sections manufactured by Tata is given in section 12.

FOR EXPLANATION OF TABLES SEE NOTE 3

BS EN 1993-1-1:2005
BS EN 10210-2:2006

HOT-FINISHED
CIRCULAR HOLLOW SECTIONS

Celsius® CHS

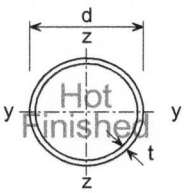

Dimensions and properties

Section Designation		Mass per Metre	Area of Section	Ratio for Local Buckling	Second Moment of Area	Radius of Gyration	Elastic Modulus	Plastic Modulus	Torsional Constants		Surface Area	
Outside Diameter	Thickness										Per Metre	Per Tonne
d	t		A	d/t	I	i	W_{el}	W_{pl}	I_T	W_t		
mm	mm	kg/m	cm²		cm⁴	cm	cm³	cm³	cm⁴	cm³	m²	m²
26.9	3.2	1.87	2.38	8.41	1.70	0.846	1.27	1.81	3.41	2.53	0.085	45.2
33.7	2.6	1.99	2.54	13.0	3.09	1.10	1.84	2.52	6.19	3.67	0.106	53.1
	3.2	2.41	3.07	10.5	3.60	1.08	2.14	2.99	7.21	4.28	0.106	44.0
	4.0	2.93	3.73	8.43	4.19	1.06	2.49	3.55	8.38	4.97	0.106	36.1
42.4	2.6	2.55	3.25	16.3	6.46	1.41	3.05	4.12	12.9	6.10	0.133	52.2
	3.2	3.09	3.94	13.3	7.62	1.39	3.59	4.93	15.2	7.19	0.133	43.1
	4.0	3.79	4.83	10.6	8.99	1.36	4.24	5.92	18.0	8.48	0.133	35.2
	5.0	4.61	5.87	8.48	10.5	1.33	4.93	7.04	20.9	9.86	0.133	28.9
48.3	3.2	3.56	4.53	15.1	11.6	1.60	4.80	6.52	23.2	9.59	0.152	42.6
	4.0	4.37	5.57	12.1	13.8	1.57	5.70	7.87	27.5	11.4	0.152	34.7
	5.0	5.34	6.80	9.66	16.2	1.54	6.69	9.42	32.3	13.4	0.152	28.4
60.3	3.2	4.51	5.74	18.8	23.5	2.02	7.78	10.4	46.9	15.6	0.189	42.0
	4.0	5.55	7.07	15.1	28.2	2.00	9.34	12.7	56.3	18.7	0.189	34.1
	5.0	6.82	8.69	12.1	33.5	1.96	11.1	15.3	67.0	22.2	0.189	27.8
76.1	2.9	5.24	6.67	26.2	44.7	2.59	11.8	15.5	89.5	23.5	0.239	45.7
	3.2	5.75	7.33	23.8	48.8	2.58	12.8	17.0	97.6	25.6	0.239	41.6
	4.0	7.11	9.06	19.0	59.1	2.55	15.5	20.8	118	31.0	0.239	33.6
	5.0	8.77	11.2	15.2	70.9	2.52	18.6	25.3	142	37.3	0.239	27.3
88.9	3.2	6.76	8.62	27.8	79.2	3.03	17.8	23.5	158	35.6	0.279	41.3
	4.0	8.38	10.7	22.2	96.3	3.00	21.7	28.9	193	43.3	0.279	33.3
	5.0	10.3	13.2	17.8	116	2.97	26.2	35.2	233	52.4	0.279	27.0
	6.3	12.8	16.3	14.1	140	2.93	31.5	43.1	280	63.1	0.279	21.8
114.3	3.2	8.77	11.2	35.7	172	3.93	30.2	39.5	345	60.4	0.359	41.0
	3.6	9.83	12.5	31.8	192	3.92	33.6	44.1	384	67.2	0.359	36.5
	4.0	10.9	13.9	28.6	211	3.90	36.9	48.7	422	73.9	0.359	33.0
	5.0	13.5	17.2	22.9	257	3.87	45.0	59.8	514	89.9	0.359	26.6
	6.3	16.8	21.4	18.1	313	3.82	54.7	73.6	625	109	0.359	21.4
139.7	5.0	16.6	21.2	27.9	481	4.77	68.8	90.8	961	138	0.439	26.4
	6.3	20.7	26.4	22.2	589	4.72	84.3	112	1180	169	0.439	21.2
	8.0	26.0	33.1	17.5	720	4.66	103	139	1440	206	0.439	16.9
	10.0	32.0	40.7	14.0	862	4.60	123	169	1720	247	0.439	13.7
168.3	5.0	20.1	25.7	33.7	856	5.78	102	133	1710	203	0.529	26.3
	6.3	25.2	32.1	26.7	1050	5.73	125	165	2110	250	0.529	21.0
	8.0	31.6	40.3	21.0	1300	5.67	154	206	2600	308	0.529	16.7
	10.0	39.0	49.7	16.8	1560	5.61	186	251	3130	372	0.529	13.5
	12.5	48.0	61.2	13.5	1870	5.53	222	304	3740	444	0.529	11.0
193.7	5.0	23.3	29.6	38.7	1320	6.67	136	178	2640	273	0.609	26.2
	6.3	29.1	37.1	30.7	1630	6.63	168	221	3260	337	0.609	20.9
	8.0	36.6	46.7	24.2	2020	6.57	208	276	4030	416	0.609	16.6
	10.0	45.3	57.7	19.4	2440	6.50	252	338	4880	504	0.609	13.4
	12.5	55.9	71.2	15.5	2930	6.42	303	411	5870	606	0.609	10.9

Celsius® is a trademark of Tata. A fuller description of the relationship between Hot Finished Circular Hollow Sections (HFCHS) and the Celsius® range of sections manufactured by Tata is given in section 12.

░░░░ Check availability

FOR EXPLANATION OF TABLES SEE NOTES 2 AND 3

BS EN 1993-1-1:2005
BS EN 10210-2:2006

HOT-FINISHED
CIRCULAR HOLLOW SECTIONS

Celsius® CHS

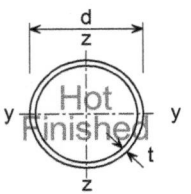

Dimensions and properties

Section Designation		Mass per Metre	Area of Section	Ratio for Local Buckling	Second Moment of Area	Radius of Gyration	Elastic Modulus	Plastic Modulus	Torsional Constants		Surface Area	
Outside Diameter	Thickness										Per Metre	Per Tonne
d	t		A	d/t	I	i	W_{el}	W_{pl}	I_T	W_t		
mm	mm	kg/m	cm²		cm⁴	cm	cm³	cm³	cm⁴	cm³	m²	m²
219.1	5.0	26.4	33.6	43.8	1930	7.57	176	229	3860	352	0.688	26.1
	6.3	33.1	42.1	34.8	2390	7.53	218	285	4770	436	0.688	20.8
	8.0	41.6	53.1	27.4	2960	7.47	270	357	5920	540	0.688	16.5
	10.0	51.6	65.7	21.9	3600	7.40	328	438	7200	657	0.688	13.3
	12.5	63.7	81.1	17.5	4350	7.32	397	534	8690	793	0.688	10.8
	14.2	71.8	91.4	15.4	4820	7.26	440	597	9640	880	0.688	9.59
	16.0	80.1	102	13.7	5300	7.20	483	661	10600	967	0.688	8.59
244.5	8.0	46.7	59.4	30.6	4160	8.37	340	448	8320	681	0.768	16.5
	10.0	57.8	73.7	24.5	5070	8.30	415	550	10100	830	0.768	13.3
	12.5	71.5	91.1	19.6	6150	8.21	503	673	12300	1010	0.768	10.7
	14.2	80.6	103	17.2	6840	8.16	559	754	13700	1120	0.768	9.52
	16.0	90.2	115	15.3	7530	8.10	616	837	15100	1230	0.768	8.52
273.0	6.3	41.4	52.8	43.3	4700	9.43	344	448	9390	688	0.858	20.7
	8.0	52.3	66.6	34.1	5850	9.37	429	562	11700	857	0.858	16.4
	10.0	64.9	82.6	27.3	7150	9.31	524	692	14300	1050	0.858	13.2
	12.5	80.3	102	21.8	8700	9.22	637	849	17400	1270	0.858	10.7
	14.2	90.6	115	19.2	9700	9.16	710	952	19400	1420	0.858	9.46
	16.0	101	129	17.1	10700	9.10	784	1060	21400	1570	0.858	8.46
323.9	6.3	49.3	62.9	51.4	7930	11.2	490	636	15900	979	1.02	20.6
	8.0	62.3	79.4	40.5	9910	11.2	612	799	19800	1220	1.02	16.3
	10.0	77.4	98.6	32.4	12200	11.1	751	986	24300	1500	1.02	13.1
	12.5	96.0	122	25.9	14800	11.0	917	1210	29700	1830	1.02	10.6
	14.2	108	138	22.8	16600	11.0	1030	1360	33200	2050	1.02	9.38
	16.0	121	155	20.2	18400	10.9	1140	1520	36800	2270	1.02	8.38
355.6	14.2	120	152	25.0	22200	12.1	1250	1660	44500	2500	1.12	9.34
	16.0	134	171	22.2	24700	12.0	1390	1850	49300	2770	1.12	8.34
406.4	6.3	62.2	79.2	64.5	15800	14.1	780	1010	31700	1560	1.28	20.5
	8.0	78.6	100	50.8	19900	14.1	978	1270	39700	1960	1.28	16.2
	10.0	97.8	125	40.6	24500	14.0	1210	1570	49000	2410	1.28	13.1
	12.5	121	155	32.5	30000	13.9	1480	1940	60100	2960	1.28	10.5
	14.2	137	175	28.6	33700	13.9	1660	2190	67400	3320	1.28	9.30
	16.0	154	196	25.4	37400	13.8	1840	2440	74900	3690	1.28	8.29
457.0	8.0	88.6	113	57.1	28400	15.9	1250	1610	56900	2490	1.44	16.2
	10.0	110	140	45.7	35100	15.8	1540	2000	70200	3070	1.44	13.0
	12.5	137	175	36.6	43100	15.7	1890	2470	86300	3780	1.44	10.5
	14.2	155	198	32.2	48500	15.7	2120	2790	96900	4240	1.44	9.26
	16.0	174	222	28.6	54000	15.6	2360	3110	108000	4720	1.44	8.25
508.0	10.0	123	156	50.8	48500	17.6	1910	2480	97000	3820	1.60	13.0
	12.5	153	195	40.6	59800	17.5	2350	3070	120000	4710	1.60	10.4
	14.2	173	220	35.8	67200	17.5	2650	3460	134000	5290	1.60	9.23
	16.0	194	247	31.8	74900	17.4	2950	3870	150000	5900	1.60	8.22

Celsius® is a trademark of Tata. A fuller description of the relationship between Hot Finished Circular Hollow Sections (HFCHS) and the Celsius® range of sections manufactured by Tata is given in section 12.

▭ Check availability

FOR EXPLANATION OF TABLES SEE NOTES 2 AND 3

BS EN 1993-1-1:2005
BS EN 10210-2:2006

HOT-FINISHED
SQUARE HOLLOW SECTIONS

Celsius® SHS

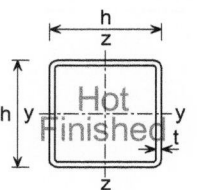

Dimensions and properties

Section Designation		Mass per Metre	Area of Section	Ratio for Local Buckling	Second Moment of Area	Radius of Gyration	Elastic Modulus	Plastic Modulus	Torsional Constants		Surface Area	
Size	Thickness										Per Metre	Per Tonne
$h \times h$	t		A	c/t [(1)]	I	i	W_{el}	W_{pl}	I_T	W_t		
mm	mm	kg/m	cm²		cm⁴	cm	cm³	cm³	cm⁴	cm³	m²	m²
40 x 40	3.0	3.41	4.34	10.3	9.78	1.50	4.89	5.97	15.7	7.10	0.152	44.7
	3.2	3.61	4.60	9.50	10.2	1.49	5.11	6.28	16.5	7.42	0.152	42.0
	4.0	4.39	5.59	7.00	11.8	1.45	5.91	7.44	19.5	8.54	0.150	34.1
	5.0	5.28	6.73	5.00	13.4	1.41	6.68	8.66	22.5	9.60	0.147	27.8
50 x 50	3.0	4.35	5.54	13.7	20.2	1.91	8.08	9.70	32.1	11.8	0.192	44.2
	3.2	4.62	5.88	12.6	21.2	1.90	8.49	10.2	33.8	12.4	0.192	41.5
	4.0	5.64	7.19	9.50	25.0	1.86	9.99	12.3	40.4	14.5	0.190	33.6
	5.0	6.85	8.73	7.00	28.9	1.82	11.6	14.5	47.6	16.7	0.187	27.3
	6.3	8.31	10.6	4.94	32.8	1.76	13.1	17.0	55.2	18.8	0.184	22.1
60 x 60	3.0	5.29	6.74	17.0	36.2	2.32	12.1	14.3	56.9	17.7	0.232	43.9
	3.2	5.62	7.16	15.8	38.2	2.31	12.7	15.2	60.2	18.6	0.232	41.2
	4.0	6.90	8.79	12.0	45.4	2.27	15.1	18.3	72.5	22.0	0.230	33.3
	5.0	8.42	10.7	9.00	53.3	2.23	17.8	21.9	86.4	25.7	0.227	27.0
	6.3	10.3	13.1	6.52	61.6	2.17	20.5	26.0	102	29.6	0.224	21.7
	8.0	12.5	16.0	4.50	69.7	2.09	23.2	30.4	118	33.4	0.219	17.5
70 x 70	3.6	7.40	9.42	16.4	68.6	2.70	19.6	23.3	108	28.7	0.271	36.6
	5.0	9.99	12.7	11.0	88.5	2.64	25.3	30.8	142	36.8	0.267	26.7
	6.3	12.3	15.6	8.11	104	2.58	29.7	36.9	169	42.9	0.264	21.5
	8.0	15.0	19.2	5.75	120	2.50	34.2	43.8	200	49.2	0.259	17.3
80 x 80	3.6	8.53	10.9	19.2	105	3.11	26.2	31.0	164	38.5	0.311	36.4
	4.0	9.41	12.0	17.0	114	3.09	28.6	34.0	180	41.9	0.310	32.9
	5.0	11.6	14.7	13.0	137	3.05	34.2	41.1	217	49.8	0.307	26.6
	6.3	14.2	18.1	9.70	162	2.99	40.5	49.7	262	58.7	0.304	21.3
	8.0	17.5	22.4	7.00	189	2.91	47.3	59.5	312	68.3	0.299	17.1
90 x 90	3.6	9.66	12.3	22.0	152	3.52	33.8	39.7	237	49.7	0.351	36.3
	4.0	10.7	13.6	19.5	166	3.50	37.0	43.6	260	54.2	0.350	32.8
	5.0	13.1	16.7	15.0	200	3.45	44.4	53.0	316	64.8	0.347	26.4
	6.3	16.2	20.7	11.3	238	3.40	53.0	64.3	382	77.0	0.344	21.2
	8.0	20.1	25.6	8.25	281	3.32	62.6	77.6	459	90.5	0.339	16.9

Celsius® is a trademark of Tata. A fuller description of the relationship between Hot Finished Square Hollow Sections (HFSHS) and the Celsius® range of sections manufactured by Tata is given in section 12.

(1) For local buckling calculation c = h - 3t.

FOR EXPLANATION OF TABLES SEE NOTES 2 AND 3

BS EN 1993-1-1:2005
BS EN 10210-2:2006

HOT-FINISHED
SQUARE HOLLOW SECTIONS

Celsius® SHS

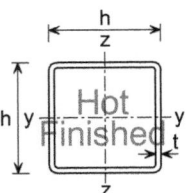

Dimensions and properties

Section Designation		Mass per Metre	Area of Section	Ratio for Local Buckling	Second Moment of Area	Radius of Gyration	Elastic Modulus	Plastic Modulus	Torsional Constants		Surface Area	
Size	Thickness										Per Metre	Per Tonne
$h \times h$	t		A	c/t [(1)]	I	i	W_{el}	W_{pl}	I_T	W_t		
mm	mm	kg/m	cm²		cm⁴	cm	cm³	cm³	cm⁴	cm³	m²	m²
100 x 100	4.0	11.9	15.2	22.0	232	3.91	46.4	54.4	361	68.2	0.390	32.7
	5.0	14.7	18.7	17.0	279	3.86	55.9	66.4	439	81.8	0.387	26.3
	6.3	18.2	23.2	12.9	336	3.80	67.1	80.9	534	97.8	0.384	21.1
	8.0	22.6	28.8	9.50	400	3.73	79.9	98.2	646	116	0.379	16.8
	10.0	27.4	34.9	7.00	462	3.64	92.4	116	761	133	0.374	13.6
120 x 120	5.0	17.8	22.7	21.0	498	4.68	83.0	97.6	777	122	0.467	26.2
	6.3	22.2	28.2	16.0	603	4.62	100	120	950	147	0.464	20.9
	8.0	27.6	35.2	12.0	726	4.55	121	146	1160	176	0.459	16.6
	10.0	33.7	42.9	9.00	852	4.46	142	175	1380	206	0.454	13.5
	12.5	40.9	52.1	6.60	982	4.34	164	207	1620	236	0.448	11.0
140 x 140	5.0	21.0	26.7	25.0	807	5.50	115	135	1250	170	0.547	26.1
	6.3	26.1	33.3	19.2	984	5.44	141	166	1540	206	0.544	20.8
	8.0	32.6	41.6	14.5	1200	5.36	171	204	1890	249	0.539	16.5
	10.0	40.0	50.9	11.0	1420	5.27	202	246	2270	294	0.534	13.4
	12.5	48.7	62.1	8.20	1650	5.16	236	293	2700	342	0.528	10.8
150 x 150	5.0	22.6	28.7	27.0	1000	5.90	134	156	1550	197	0.587	26.0
	6.3	28.1	35.8	20.8	1220	5.85	163	192	1910	240	0.584	20.8
	8.0	35.1	44.8	15.8	1490	5.77	199	237	2350	291	0.579	16.5
	10.0	43.1	54.9	12.0	1770	5.68	236	286	2830	344	0.574	13.3
	12.5	52.7	67.1	9.00	2080	5.57	277	342	3380	402	0.568	10.8
160 x 160	5.0	24.1	30.7	29.0	1230	6.31	153	178	1890	226	0.627	26.0
	6.3	30.1	38.3	22.4	1500	6.26	187	220	2330	275	0.624	20.7
	8.0	37.6	48.0	17.0	1830	6.18	229	272	2880	335	0.619	16.5
	10.0	46.3	58.9	13.0	2190	6.09	273	329	3480	398	0.614	13.3
	12.5	56.6	72.1	9.80	2580	5.98	322	395	4160	467	0.608	10.7
	14.2	63.3	80.7	8.27	2810	5.90	351	436	4580	508	0.603	9.53
180 x 180	6.3	34.0	43.3	25.6	2170	7.07	241	281	3360	355	0.704	20.7
	8.0	42.7	54.4	19.5	2660	7.00	296	349	4160	434	0.699	16.4
	10.0	52.5	66.9	15.0	3190	6.91	355	424	5050	518	0.694	13.2
	12.5	64.4	82.1	11.4	3790	6.80	421	511	6070	613	0.688	10.7
	14.2	72.2	92.0	9.68	4150	6.72	462	566	6710	670	0.683	9.46
	16.0	80.2	102	8.25	4500	6.64	500	621	7340	724	0.679	8.46
200 x 200	5.0	30.4	38.7	37.0	2450	7.95	245	283	3760	362	0.787	25.9
	6.3	38.0	48.4	28.7	3010	7.89	301	350	4650	444	0.784	20.6
	8.0	47.7	60.8	22.0	3710	7.81	371	436	5780	545	0.779	16.3
	10.0	58.8	74.9	17.0	4470	7.72	447	531	7030	655	0.774	13.2
	12.5	72.3	92.1	13.0	5340	7.61	534	643	8490	778	0.768	10.6
	14.2	81.1	103	11.1	5870	7.54	587	714	9420	854	0.763	9.41
	16.0	90.3	115	9.50	6390	7.46	639	785	10300	927	0.759	8.40

Celsius® is a trademark of Tata. A fuller description of the relationship between Hot Finished Square Hollow Sections (HFSHS) and the Celsius® range of sections manufactured by Tata is given in section 12.
(1) For local buckling calculation c = h - 3t.
        Check availability
FOR EXPLANATION OF TABLES SEE NOTES 2 AND 3

BS EN 1993-1-1:2005
BS EN 10210-2:2006

HOT-FINISHED
SQUARE HOLLOW SECTIONS

Celsius® SHS

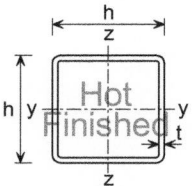

Dimensions and properties

Section Designation		Mass per	Area of	Ratio for	Second Moment	Radius of	Elastic Modulus	Plastic Modulus	Torsional Constants		Surface Area	
Size	Thickness	Metre	Section	Local Buckling	of Area	Gyration					Per Metre	Per Tonne
h x h	t		A	c/t (1)	I	i	W_{el}	W_{pl}	I_T	W_t		
mm	mm	kg/m	cm²		cm⁴	cm	cm³	cm³	cm⁴	cm³	m²	m²
250 x 250	6.3	47.9	61.0	36.7	6010	9.93	481	556	9240	712	0.984	20.5
	8.0	60.3	76.8	28.3	7460	9.86	596	694	11500	880	0.979	16.3
	10.0	74.5	94.9	22.0	9060	9.77	724	851	14100	1070	0.974	13.1
	12.5	91.9	117	17.0	10900	9.66	873	1040	17200	1280	0.968	10.5
	14.2	103	132	14.6	12100	9.58	967	1160	19100	1410	0.963	9.31
	16.0	115	147	12.6	13300	9.50	1060	1280	21100	1550	0.959	8.31
260 x 260	6.3	49.9	63.5	38.3	6790	10.3	522	603	10400	773	1.02	20.5
	8.0	62.8	80.0	29.5	8420	10.3	648	753	13000	956	1.02	16.2
	10.0	77.7	98.9	23.0	10200	10.2	788	924	15900	1160	1.01	13.1
	12.5	95.8	122	17.8	12400	10.1	951	1130	19400	1390	1.01	10.5
	14.2	108	137	15.3	13700	9.99	1060	1260	21700	1540	1.00	9.30
	16.0	120	153	13.3	15100	9.91	1160	1390	23900	1690	0.999	8.29
300 x 300	6.3	57.8	73.6	44.6	10500	12.0	703	809	16100	1040	1.18	20.5
	8.0	72.8	92.8	34.5	13100	11.9	875	1010	20200	1290	1.18	16.2
	10.0	90.2	115	27.0	16000	11.8	1070	1250	24800	1580	1.17	13.0
	12.5	112	142	21.0	19400	11.7	1300	1530	30300	1900	1.17	10.5
	14.2	126	160	18.1	21600	11.6	1440	1710	33900	2110	1.16	9.25
	16.0	141	179	15.8	23900	11.5	1590	1900	37600	2330	1.16	8.25
350 x 350	8.0	85.4	109	40.8	21100	13.9	1210	1390	32400	1790	1.38	16.2
	10.0	106	135	32.0	25900	13.9	1480	1720	39900	2190	1.37	13.0
	12.5	131	167	25.0	31500	13.7	1800	2110	48900	2650	1.37	10.4
	14.2	148	189	21.6	35200	13.7	2010	2360	54900	2960	1.36	9.21
	16.0	166	211	18.9	38900	13.6	2230	2630	61000	3260	1.36	8.20
400 x 400	10.0	122	155	37.0	39100	15.9	1960	2260	60100	2900	1.57	12.9
	12.5	151	192	29.0	47800	15.8	2390	2780	73900	3530	1.57	10.4
	14.2	170	217	25.2	53500	15.7	2680	3130	83000	3940	1.56	9.18
	16.0	191	243	22.0	59300	15.6	2970	3480	92400	4360	1.56	8.17
	20.0 ^	235	300	17.0	71500	15.4	3580	4250	112000	5240	1.55	6.58

Celsius® is a trademark of Tata. A fuller description of the relationship between Hot Finished Square Hollow Sections (HFSHS) and the Celsius® range of sections manufactured by Tata is given in section 12.

(1) For local buckling calculation c = h - 3t.

^ SAW process (single longitudinal seam weld, slightly proud)

▨ Check availability

FOR EXPLANATION OF TABLES SEE NOTES 2 AND 3

BS EN 1993-1-1:2005
BS EN 10210-2:2006

HOT-FINISHED
RECTANGULAR HOLLOW SECTIONS

Celsius® RHS

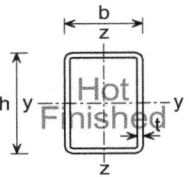

Dimensions and properties

Section Designation		Mass per Metre	Area of Section	Ratios for Local Buckling		Second Moment of Area		Radius of Gyration		Elastic Modulus		Plastic Modulus		Torsional Constants		Surface Area	
Size	Thickness					Axis y-y	Axis z-z	Axis y-y	Axis z-z	Axis y-y	Axis z-z	Axis y-y	Axis z-z			Per Metre	Per Tonne
h x b	t		A	c_w/t [1]	c_f/t [1]									I_T	W_t		
mm	mm	kg/m	cm²			cm⁴	cm⁴	cm	cm	cm³	cm³	cm³	cm³	cm⁴	cm³	m²	m²
50x30	3.2	3.61	4.60	12.6	6.38	14.2	6.20	1.76	1.16	5.68	4.13	7.25	5.00	14.2	6.80	0.152	42.1
60x40	3.0	4.35	5.54	17.0	10.3	26.5	13.9	2.18	1.58	8.82	6.95	10.9	8.19	29.2	11.2	0.192	44.1
	4.0	5.64	7.19	12.0	7.00	32.8	17.0	2.14	1.54	10.9	8.52	13.8	10.3	36.7	13.7	0.190	33.7
	5.0	6.85	8.73	9.00	5.00	38.1	19.5	2.09	1.50	12.7	9.77	16.4	12.2	43.0	15.7	0.187	27.3
80x40	3.2	5.62	7.16	22.0	9.50	57.2	18.9	2.83	1.63	14.3	9.46	18.0	11.0	46.2	16.1	0.232	41.3
	4.0	6.90	8.79	17.0	7.00	68.2	22.2	2.79	1.59	17.1	11.1	21.8	13.2	55.2	18.9	0.230	33.3
	5.0	8.42	10.7	13.0	5.00	80.3	25.7	2.74	1.55	20.1	12.9	26.1	15.7	65.1	21.9	0.227	27.0
	6.3	10.3	13.1	9.70	3.35	93.3	29.2	2.67	1.49	23.3	14.6	31.1	18.4	75.6	24.8	0.224	21.7
	8.0	12.5	16.0	7.00	2.00	106	32.1	2.58	1.42	26.5	16.1	36.5	21.2	85.8	27.4	0.219	17.5
90x50	3.6	7.40	9.42	22.0	10.9	98.3	38.7	3.23	2.03	21.8	15.5	27.2	18.0	89.4	25.9	0.271	36.6
	5.0	9.99	12.7	15.0	7.00	127	49.2	3.16	1.97	28.3	19.7	36.0	23.5	116	32.9	0.267	26.7
	6.3	12.3	15.6	11.3	4.94	150	57.0	3.10	1.91	33.3	22.8	43.2	28.0	138	38.1	0.264	21.5
100x50	3.0	6.71	8.54	30.3	13.7	110	36.8	3.58	2.08	21.9	14.7	27.3	16.8	88.4	25.0	0.292	43.5
	3.2	7.13	9.08	28.3	12.6	116	38.8	3.57	2.07	23.2	15.5	28.9	17.7	93.4	26.4	0.292	41.0
	4.0	8.78	11.2	22.0	9.50	140	46.2	3.53	2.03	27.9	18.5	35.2	21.5	113	31.4	0.290	33.0
	5.0	10.8	13.7	17.0	7.00	167	54.3	3.48	1.99	33.3	21.7	42.6	25.8	135	36.9	0.287	26.6
	6.3	13.3	16.9	12.9	4.94	197	63.0	3.42	1.93	39.4	25.2	51.3	30.8	160	42.9	0.284	21.4
	8.0	16.3	20.8	9.50	3.25	230	71.7	3.33	1.86	46.0	28.7	61.4	36.3	186	48.9	0.279	17.1
100x60	3.6	8.53	10.9	24.8	13.7	145	64.8	3.65	2.44	28.9	21.6	35.6	24.9	142	35.6	0.311	36.5
	5.0	11.6	14.7	17.0	9.00	189	83.6	3.58	2.38	37.8	27.9	47.4	32.9	188	45.9	0.307	26.5
	6.3	14.2	18.1	12.9	6.52	225	98.1	3.52	2.33	45.0	32.7	57.3	39.5	224	53.8	0.304	21.4
	8.0	17.5	22.4	9.50	4.50	264	113	3.44	2.25	52.8	37.8	68.7	47.1	265	62.2	0.299	17.1
120x60	3.6	9.66	12.3	30.3	13.7	227	76.3	4.30	2.49	37.9	25.4	47.2	28.9	183	43.3	0.351	36.3
	5.0	13.1	16.7	21.0	9.00	299	98.8	4.23	2.43	49.9	32.9	63.1	38.4	242	56.0	0.347	26.5
	6.3	16.2	20.7	16.0	6.52	358	116	4.16	2.37	59.7	38.8	76.7	46.3	290	65.9	0.344	21.2
	8.0	20.1	25.6	12.0	4.50	425	135	4.08	2.30	70.8	45.0	92.7	55.4	344	76.6	0.339	16.9
120x80	5.0	14.7	18.7	21.0	13.0	365	193	4.42	3.21	60.9	48.2	74.6	56.1	401	77.9	0.387	26.3
	6.3	18.2	23.2	16.0	9.70	440	230	4.36	3.15	73.3	57.6	91.0	68.2	487	92.9	0.384	21.1
	8.0	22.6	28.8	12.0	7.00	525	273	4.27	3.08	87.5	68.1	111	82.6	587	110	0.379	16.8
	10.0	27.4	34.9	9.00	5.00	609	313	4.18	2.99	102	78.1	131	97.3	688	126	0.374	13.6
150x100	5.0	18.6	23.7	27.0	17.0	739	392	5.58	4.07	98.5	78.5	119	90.1	807	127	0.487	26.2
	6.3	23.1	29.5	20.8	12.9	898	474	5.52	4.01	120	94.8	147	110	986	153	0.484	21.0
	8.0	28.9	36.8	15.8	9.50	1090	569	5.44	3.94	145	114	180	135	1200	183	0.479	16.6
	10.0	35.3	44.9	12.0	7.00	1280	665	5.34	3.85	171	133	216	161	1430	214	0.474	13.4
	12.5	42.8	54.6	9.00	5.00	1490	763	5.22	3.74	198	153	256	190	1680	246	0.468	10.9
150x125	4.0	16.6	21.2	34.5	28.3	714	539	5.80	5.04	95.2	86.3	112	98.9	949	133	0.540	32.5
	5.0	20.6	26.2	27.0	22.0	870	656	5.76	5.00	116	105	138	121	1160	162	0.537	26.1
	6.3	25.6	32.6	20.8	16.8	1060	798	5.70	4.94	141	128	169	149	1430	196	0.534	20.9
	8.0	32.0	40.8	15.8	12.6	1290	966	5.62	4.87	172	155	208	183	1750	237	0.529	16.5
	10.0	39.2	49.9	12.0	9.50	1530	1140	5.53	4.78	204	183	251	221	2100	279	0.524	13.4
	12.5	47.7	60.8	9.00	7.00	1780	1330	5.42	4.67	238	212	299	262	2490	324	0.518	10.9

Celsius® is a trademark of Tata. A fuller description of the relationship between Hot Finished Rectangular Hollow Sections (HFRHS) and the Celsius® range of sections manufactured by Tata is given in section 12.

(1) For local buckling calculation $c_w = h - 3t$ and $c_f = b - 3t$.

░░░░░ Check availability

FOR EXPLANATION OF TABLES SEE NOTES 2 AND 3

BS EN 1993-1-1:2005
BS EN 10210-2:2006

HOT-FINISHED
RECTANGULAR HOLLOW SECTIONS

Celsius® RHS

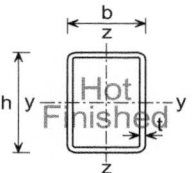

Dimensions and properties

Section Designation		Mass per Metre	Area of Section	Ratios for Local Buckling		Second Moment of Area		Radius of Gyration		Elastic Modulus		Plastic Modulus		Torsional Constants		Surface Area	
Size	Thickness					Axis y-y	Axis z-z	Axis y-y	Axis z-z	Axis y-y	Axis z-z	Axis y-y	Axis z-z			Per Metre	Per Tonne
h x b	t	Metre	A	c_w/t [(1)]	c_f/t [(1)]									I_T	W_t		
mm	mm	kg/m	cm²			cm⁴	cm⁴	cm	cm	cm³	cm³	cm³	cm³	cm⁴	cm³	m²	m²
160x80	4.0	14.4	18.4	37.0	17.0	612	207	5.77	3.35	76.5	51.7	94.7	58.3	493	88.1	0.470	32.6
	5.0	17.8	22.7	29.0	13.0	744	249	5.72	3.31	93.0	62.3	116	71.1	600	106	0.467	26.2
	6.3	22.2	28.2	22.4	9.70	903	299	5.66	3.26	113	74.8	142	86.8	730	127	0.464	20.9
	8.0	27.6	35.2	17.0	7.00	1090	356	5.57	3.18	136	89.0	175	106	883	151	0.459	16.6
	10.0	33.7	42.9	13.0	5.00	1280	411	5.47	3.10	161	103	209	125	1040	175	0.454	13.5
200x100	5.0	22.6	28.7	37.0	17.0	1500	505	7.21	4.19	149	101	185	114	1200	172	0.587	26.0
	6.3	28.1	35.8	28.7	12.9	1830	613	7.15	4.14	183	123	228	140	1480	208	0.584	20.8
	8.0	35.1	44.8	22.0	9.50	2230	739	7.06	4.06	223	148	282	172	1800	251	0.579	16.5
	10.0	43.1	54.9	17.0	7.00	2660	869	6.96	3.98	266	174	341	206	2160	295	0.574	13.3
	12.5	52.7	67.1	13.0	5.00	3140	1000	6.84	3.87	314	201	408	245	2540	341	0.568	10.8
200x120	5.0	24.1	30.7	37.0	21.0	1690	762	7.40	4.98	168	127	205	144	1650	210	0.627	26.0
	6.3	30.1	38.3	28.7	16.0	2070	929	7.34	4.92	207	155	253	177	2030	255	0.624	20.7
	8.0	37.6	48.0	22.0	12.0	2530	1130	7.26	4.85	253	188	313	218	2500	310	0.619	16.5
	10.0	46.3	58.9	17.0	9.00	3030	1340	7.17	4.76	303	223	379	263	3000	367	0.614	13.3
	14.2	63.3	80.7	11.1	5.45	3910	1690	6.96	4.58	391	282	503	346	3920	464	0.603	9.53
200x150	8.0	41.4	52.8	22.0	15.8	2970	1890	7.50	5.99	297	253	359	294	3640	398	0.679	16.4
	10.0	51.0	64.9	17.0	12.0	3570	2260	7.41	5.91	357	302	436	356	4410	475	0.674	13.2
250x120	10.0	54.1	68.9	22.0	9.00	5310	1640	8.78	4.88	425	273	539	318	4090	468	0.714	13.2
	12.5	66.4	84.6	17.0	6.60	6330	1930	8.65	4.77	506	321	651	381	4880	549	0.708	10.7
	14.5	74.5	94.9	14.6	5.45	6960	2090	8.56	4.70	556	349	722	421	5360	597	0.703	9.44
250x150	5.0	30.4	38.7	47.0	27.0	3360	1530	9.31	6.28	269	204	324	228	3280	337	0.787	25.9
	6.3	38.0	48.4	36.7	20.8	4140	1870	9.25	6.22	331	250	402	283	4050	413	0.784	20.6
	8.0	47.7	60.8	28.3	15.8	5110	2300	9.17	6.15	409	306	501	350	5020	506	0.779	16.3
	10.0	58.8	74.9	22.0	12.0	6170	2760	9.08	6.06	494	367	611	426	6090	605	0.774	13.2
	12.5	72.3	92.1	17.0	9.00	7390	3270	8.96	5.96	591	435	740	514	7330	717	0.768	10.6
	14.2	81.1	103	14.6	7.56	8140	3580	8.87	5.88	651	477	823	570	8100	784	0.763	9.41
	16.0	90.3	115	12.6	6.38	8880	3870	8.79	5.80	710	516	906	625	8870	849	0.759	8.41
250x200	10.0	66.7	84.9	22.0	17.0	7610	5370	9.47	7.95	609	537	731	626	9890	835	0.874	13.1
	12.5	82.1	105	17.0	13.0	9150	6440	9.35	7.85	732	644	888	760	12000	997	0.868	10.6
	14.2	92.3	118	14.6	11.1	10100	7100	9.28	7.77	809	710	990	846	13300	1100	0.863	9.35
260x140	5.0	30.4	38.7	49.0	25.0	3530	1350	9.55	5.91	272	193	331	216	3080	326	0.787	25.9
	6.3	38.0	48.4	38.3	19.2	4360	1660	9.49	5.86	335	237	411	267	3800	399	0.784	20.6
	8.0	47.7	60.8	29.5	14.5	5370	2030	9.40	5.78	413	290	511	331	4700	488	0.779	16.3
	10.0	58.8	74.9	23.0	11.0	6490	2430	9.31	5.70	499	347	624	402	5700	584	0.774	13.2
	12.5	72.3	92.1	17.8	8.20	7770	2880	9.18	5.59	597	411	756	485	6840	690	0.768	10.6
	14.2	81.1	103	15.3	6.86	8560	3140	9.10	5.52	658	449	840	537	7560	754	0.763	9.41
	16.0	90.3	115	13.3	5.75	9340	3400	9.01	5.44	718	486	925	588	8260	815	0.759	8.41
300x100	8.0	47.7	60.8	34.5	9.50	6310	1080	10.2	4.21	420	216	546	245	3070	387	0.779	16.3
	10.0	58.8	74.9	27.0	7.00	7610	1280	10.1	4.13	508	255	666	296	3680	458	0.774	13.2
	14.2	81.1	103	18.1	4.04	10000	1610	9.85	3.94	669	321	896	390	4760	578	0.763	9.41

Celsius® is a trademark of Tata. A fuller description of the relationship between Hot Finished Rectangular Hollow Sections (HFRHS) and the Celsius® range of sections manufactured by Tata is given in section 12.
(1) For local buckling calculation $c_w = h - 3t$ and $c_f = b - 3t$.
▓▓▓ Check availability
FOR EXPLANATION OF TABLES SEE NOTES 2 AND 3

BS EN 1993-1-1:2005
BS EN 10210-2:2006

HOT-FINISHED
RECTANGULAR HOLLOW SECTIONS

Celsius® RHS

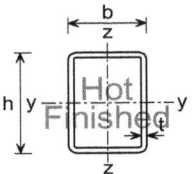

Dimensions and properties

Section Designation		Mass per Metre	Area of Section	Ratios for Local Buckling		Second Moment of Area		Radius of Gyration		Elastic Modulus		Plastic Modulus		Torsional Constants		Surface Area	
Size	Thickness					Axis y-y	Axis z-z	Axis y-y	Axis z-z	Axis y-y	Axis z-z	Axis y-y	Axis z-z			Per Metre	Per Tonne
$h \times b$	t		A	c_w/t [1]	c_f/t [1]									I_T	W_t		
mm	mm	kg/m	cm²			cm⁴	cm⁴	cm	cm	cm³	cm³	cm³	cm³	cm⁴	cm³	m²	m²
300x150	8.0	54.0	68.8	34.5	15.8	8010	2700	10.8	6.27	534	360	663	407	6450	613	0.879	16.3
	10.0	66.7	84.9	27.0	12.0	9720	3250	10.7	6.18	648	433	811	496	7840	736	0.874	13.1
	12.5	82.1	105	21.0	9.00	11700	3860	10.6	6.07	779	514	986	600	9450	874	0.868	10.6
	14.2	92.3	118	18.1	7.56	12900	4230	10.5	6.00	862	564	1100	666	10500	959	0.863	9.35
	16.0	103	131	15.8	6.38	14200	4600	10.4	5.92	944	613	1210	732	11500	1040	0.859	8.34
300x200	6.3	47.9	61.0	44.6	28.7	7830	4190	11.3	8.29	522	419	624	472	8480	681	0.984	20.5
	8.0	60.3	76.8	34.5	22.0	9720	5180	11.3	8.22	648	518	779	589	10600	840	0.979	16.2
	10.0	74.5	94.9	27.0	17.0	11800	6280	11.2	8.13	788	628	956	721	12900	1020	0.974	13.1
	12.5	91.9	117	21.0	13.0	14300	7540	11.0	8.02	952	754	1170	877	15700	1220	0.968	10.5
	14.2	103	132	18.1	11.1	15800	8330	11.0	7.95	1060	833	1300	978	17500	1340	0.963	9.35
	16.0	115	147	15.8	9.50	17400	9110	10.9	7.87	1160	911	1440	1080	19300	1470	0.959	8.34
300x250	5.0	42.2	53.7	57.0	47.0	7410	5610	11.7	10.2	494	449	575	508	9770	697	1.09	25.8
	6.3	52.8	67.3	44.6	36.7	9190	6950	11.7	10.2	613	556	716	633	12200	862	1.08	20.5
	8.0	66.5	84.8	34.5	28.3	11400	8630	11.6	10.1	761	690	896	791	15200	1070	1.08	16.2
	10.0	82.4	105	27.0	22.0	13900	10500	11.5	10.0	928	840	1100	971	18600	1300	1.07	13.0
	12.5	102	130	21.0	17.0	16900	12700	11.4	9.89	1120	1010	1350	1190	22700	1560	1.07	10.5
	14.2	115	146	18.1	14.6	18700	14100	11.3	9.82	1250	1130	1510	1330	25400	1730	1.06	9.22
	16.0	128	163	15.8	12.6	20600	15500	11.2	9.74	1380	1240	1670	1470	28100	1900	1.06	8.28
350x150	5.0	38.3	48.7	67.0	27.0	7660	2050	12.5	6.49	437	274	543	301	5160	477	0.987	25.8
	6.3	47.9	61.0	52.6	20.8	9480	2530	12.5	6.43	542	337	676	373	6390	586	0.984	20.5
	8.0	60.3	76.8	40.8	15.8	11800	3110	12.4	6.36	673	414	844	464	7930	721	0.979	16.2
	10.0	74.5	94.9	32.0	12.0	14300	3740	12.3	6.27	818	498	1040	566	9630	867	0.974	13.1
	12.5	91.9	117	25.0	9.00	17300	4450	12.2	6.17	988	593	1260	686	11600	1030	0.968	10.5
	14.2	103	132	21.6	7.56	19200	4890	12.1	6.09	1100	652	1410	763	12900	1130	0.963	9.35
	16.0	115	147	18.9	6.38	21100	5320	12.0	6.01	1210	709	1560	840	14100	1230	0.959	8.34
350x250	5.0	46.1	58.7	67.0	47.0	10600	6360	13.5	10.4	607	509	716	569	12200	817	1.19	25.8
	6.3	57.8	73.6	52.6	36.7	13200	7890	13.4	10.4	754	631	892	709	15200	1010	1.18	20.4
	8.0	72.8	92.8	40.8	28.3	16400	9800	13.3	10.3	940	784	1120	888	19000	1250	1.18	16.2
	10.0	90.2	115	32.0	22.0	20100	11900	13.2	10.2	1150	955	1380	1090	23400	1530	1.17	13.0
	12.5	112	142	25.0	17.0	24400	14400	13.1	10.1	1400	1160	1690	1330	28500	1840	1.17	10.4
	14.2	126	160	21.6	14.6	27200	16000	13.0	10.0	1550	1280	1890	1490	31900	2040	1.16	9.21
	16.0	141	179	18.9	12.6	30000	17700	12.9	9.93	1720	1410	2100	1660	35300	2250	1.16	8.23
400x120	5.0	39.8	50.7	77.0	21.0	9520	1420	13.7	5.30	476	237	612	259	4090	430	1.03	25.9
	6.3	49.9	63.5	60.5	16.0	11800	1740	13.6	5.24	590	291	762	320	5040	527	1.02	20.4
	8.0	62.8	80.0	47.0	12.0	14600	2130	13.5	5.17	732	356	952	397	6220	645	1.02	16.2
	10.0	77.7	98.9	37.0	9.00	17800	2550	13.4	5.08	891	425	1170	483	7510	771	1.01	13.0
	12.5	95.8	122	29.0	6.60	21600	3010	13.3	4.97	1080	502	1430	583	8980	911	1.01	10.5
	14.2	108	137	25.2	5.45	23900	3290	13.2	4.89	1200	549	1590	646	9890	996	1.00	9.26
	16.0	120	153	22.0	4.50	26300	3560	13.1	4.82	1320	593	1760	709	10800	1080	0.999	8.33

Celsius® is a trademark of Tata. A fuller description of the relationship between Hot Finished Rectangular Hollow Sections (HFRHS) and the Celsius® range of sections manufactured by Tata is given in section 12.

(1) For local buckling calculation $c_w = h - 3t$ and $c_f = b - 3t$.

▨ Check availability

FOR EXPLANATION OF TABLES SEE NOTES 2 AND 3

BS EN 1993-1-1:2005
BS EN 10210-2:2006

HOT-FINISHED
RECTANGULAR HOLLOW SECTIONS

Celsius® RHS

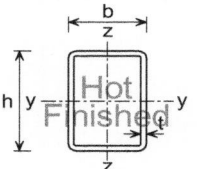

Dimensions and properties

Section Designation		Mass per Metre	Area of Section	Ratios for Local Buckling		Second Moment of Area		Radius of Gyration		Elastic Modulus		Plastic Modulus		Torsional Constants		Surface Area	
Size	Thickness					Axis y-y	Axis z-z	Axis y-y	Axis z-z	Axis y-y	Axis z-z	Axis y-y	Axis z-z			Per Metre	Per Tonne
h x b mm	t mm	kg/m	A cm²	c_w/t (1)	c_f/t (1)	cm⁴	cm⁴	cm	cm	cm³	cm³	cm³	cm³	I_T cm⁴	W_t cm³	m²	m²
400x150	5.0	42.2	53.7	77.0	27.0	10700	2320	14.1	6.57	534	309	671	337	6130	547	1.09	25.8
	6.3	52.8	67.3	60.5	20.8	13300	2850	14.0	6.51	663	380	836	418	7600	673	1.08	20.5
	8.0	66.5	84.8	47.0	15.8	16500	3510	13.9	6.43	824	468	1050	521	9420	828	1.08	16.2
	10.0	82.4	105	37.0	12.0	20100	4230	13.8	6.35	1010	564	1290	636	11500	998	1.07	13.0
	12.5	102	130	29.0	9.00	24400	5040	13.7	6.24	1220	672	1570	772	13800	1190	1.07	10.5
	14.2	115	146	25.2	7.56	27100	5550	13.6	6.16	1360	740	1760	859	15300	1310	1.06	9.22
	16.0	128	163	22.0	6.38	29800	6040	13.5	6.09	1490	805	1950	947	16800	1430	1.06	8.28
400x200	8.0	72.8	92.8	47.0	22.0	19600	6660	14.5	8.47	978	666	1200	743	15700	1140	1.18	16.2
	10.0	90.2	115	37.0	17.0	23900	8080	14.4	8.39	1200	808	1480	911	19300	1380	1.17	13.0
	12.5	112	142	29.0	13.0	29100	9740	14.3	8.28	1450	974	1810	1110	23400	1660	1.17	10.4
	14.2	126	160	25.2	11.1	32400	10800	14.2	8.21	1620	1080	2030	1240	26100	1830	1.16	9.21
	16.0	141	179	22.0	9.50	35700	11800	14.1	8.13	1790	1180	2260	1370	28900	2010	1.16	8.23
400x300	8.0	85.4	109	47.0	34.5	25700	16500	15.4	12.3	1290	1100	1520	1250	31000	1750	1.38	16.2
	10.0	106	135	37.0	27.0	31500	20200	15.3	12.2	1580	1350	1870	1540	38200	2140	1.37	12.9
	12.5	131	167	29.0	21.0	38500	24600	15.2	12.1	1920	1640	2300	1880	46800	2590	1.37	10.5
	14.2	148	189	25.2	18.1	43000	27400	15.1	12.1	2150	1830	2580	2110	52500	2890	1.36	9.19
	16.0	166	211	22.0	15.8	47500	30300	15.0	12.0	2380	2020	2870	2350	58300	3180	1.36	8.19
450x250	8.0	85.4	109	53.3	28.3	30100	12100	16.6	10.6	1340	971	1620	1080	27100	1630	1.38	16.2
	10.0	106	135	42.0	22.0	36900	14800	16.5	10.5	1640	1190	2000	1330	33300	1990	1.37	12.9
	12.5	131	167	33.0	17.0	45000	18000	16.4	10.4	2000	1440	2460	1630	40700	2410	1.37	10.5
	14.2	148	189	28.7	14.6	50300	20000	16.3	10.3	2240	1600	2760	1830	45600	2680	1.36	9.19
	16.0	166	211	25.1	12.6	55700	22000	16.2	10.2	2480	1760	3070	2030	50500	2950	1.36	8.19
500x200	8.0	85.4	109	59.5	22.0	34000	8140	17.7	8.65	1360	814	1710	896	21100	1430	1.38	16.2
	10.0	106	135	47.0	17.0	41800	9890	17.6	8.56	1670	989	2110	1100	25900	1740	1.37	12.9
	12.5	131	167	37.0	13.0	51000	11900	17.5	8.45	2040	1190	2590	1350	31500	2100	1.37	10.5
	14.2	148	189	32.2	11.1	56900	13200	17.4	8.38	2280	1320	2900	1510	35200	2320	1.36	9.19
	16.0	166	211	28.3	9.50	63000	14500	17.3	8.30	2520	1450	3230	1670	38900	2550	1.36	8.19
500x300	8.0	97.9	125	59.5	34.5	43700	20000	18.7	12.6	1750	1330	2100	1480	42600	2200	1.58	16.1
	10.0	122	155	47.0	27.0	53800	24400	18.6	12.6	2150	1630	2600	1830	52500	2700	1.57	12.9
	12.5	151	192	37.0	21.0	65800	29800	18.5	12.5	2630	1990	3200	2240	64400	3280	1.57	10.4
	14.2	170	217	32.2	18.1	73700	33200	18.4	12.4	2950	2220	3590	2520	72200	3660	1.56	9.18
	16.0	191	243	28.3	15.8	81800	36800	18.3	12.3	3270	2450	4010	2800	80300	4040	1.56	8.17
	20.0 ^	235	300	22.0	12.0	98800	44100	18.2	12.1	3950	2940	4890	3410	97400	4840	1.55	6.58

Celsius® is a trademark of Tata. A fuller description of the relationship between Hot Finished Rectangular Hollow Sections (HFRHS) and the Celsius® range of sections manufactured by Tata is given in section 12.

(1) For local buckling calculation $c_w = h - 3t$ and $c_f = b - 3t$.

^ SAW process (single longitudinal seam weld, slightly proud)

▨ Check availability

FOR EXPLANATION OF TABLES SEE NOTES 2 AND 3

BS EN 1993-1-1:2005
BS EN 10210-2:2006

HOT-FINISHED
ELLIPTICAL HOLLOW SECTIONS

Celsius® OHS

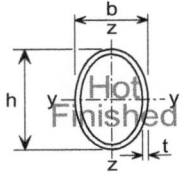

Dimensions and properties

Section Designation		Mass per Metre	Area of Section	Second Moment of Area		Radius of Gyration		Elastic Modulus		Plastic Modulus		Torsional Constants		Surface Area	
Size	Thickness			Axis y-y	Axis z-z	Axis y-y	Axis z-z	Axis y-y	Axis z-z	Axis y-y	Axis z-z			Per Metre	Per Tonne
h x b	t		A									I_T	W_t		
mm	mm	kg/m	cm^2	cm^4	cm^4	cm	cm	cm^3	cm^3	cm^3	cm^3	cm^4	cm^3	m^2	m^2
150 x 75	4.0	10.7	13.6	301	101	4.70	2.72	40.1	26.9	56.1	34.4	303	60.1	0.363	33.9
150 x 75	5.0	13.3	16.9	367	122	4.66	2.69	48.9	32.5	68.9	42.0	367	72.2	0.363	27.4
150 x 75	6.3	16.5	21.0	448	147	4.62	2.64	59.7	39.1	84.9	51.5	443	86.3	0.363	22.0
200 x 100	5.0	17.9	22.8	897	302	6.27	3.64	89.7	60.4	125	76.8	905	135	0.484	27.1
200 x 100	6.3	22.3	28.4	1100	368	6.23	3.60	110	73.5	155	94.7	1110	163	0.484	21.7
200 x 100	8.0	28.0	35.7	1360	446	6.17	3.54	136	89.3	193	117	1350	197	0.484	17.3
200 x 100	10.0	34.5	44.0	1640	529	6.10	3.47	164	106	235	141	1610	232	0.484	14.0
250 x 125	6.3	28.2	35.9	2210	742	7.84	4.55	176	119	246	151	2220	265	0.605	21.5
250 x 125	8.0	35.4	45.1	2730	909	7.78	4.49	219	145	307	188	2730	323	0.605	17.1
250 x 125	10.0	43.8	55.8	3320	1090	7.71	4.42	265	174	376	228	3290	385	0.605	13.8
250 x 125	12.5	53.9	68.7	4000	1290	7.63	4.34	320	207	458	276	3920	453	0.605	11.2
300 x 150	8.0	42.8	54.5	4810	1620	9.39	5.44	321	215	449	275	4850	481	0.726	17.0
300 x 150	10.0	53.0	67.5	5870	1950	9.32	5.37	391	260	551	336	5870	577	0.726	13.7
300 x 150	12.5	65.5	83.4	7120	2330	9.24	5.29	475	311	674	409	7050	686	0.726	11.1
300 x 150	16.0	82.5	105	8730	2810	9.12	5.17	582	374	837	503	8530	818	0.726	8.81
400 x 200	8.0	57.6	73.4	11700	3970	12.6	7.35	584	397	811	500	11900	890	0.969	16.8
400 x 200	10.0	71.5	91.1	14300	4830	12.5	7.28	717	483	1000	615	14500	1080	0.969	13.5
400 x 200	12.5	88.6	113	17500	5840	12.5	7.19	877	584	1230	753	17600	1300	0.969	10.9
400 x 200	16.0	112	143	21700	7140	12.3	7.07	1090	714	1540	936	21600	1580	0.969	8.64
500 x 250	10.0	90	115	28539	9682	15.8	9.2	1142	775	1585	976	28950	1739	1.21	13.5
500 x 250	12.5	112	142	35000	11800	15.7	9.10	1400	943	1960	1200	35300	2110	1.21	10.8
500 x 250	16.0	142	180	43700	14500	15.6	8.98	1750	1160	2460	1500	43700	2590	1.21	8.55

Celsius® is a trademark of Tata. A fuller description of the relationship between Hot Finished Elliptical Hollow Sections (HFEHS) and the Celsius® range of sections manufactured by Tata is given in section 12.
FOR EXPLANATION OF TABLES SEE NOTES 2 AND 3

BS EN 1993-1-1:2005
BS EN 10219-2:2006

COLD-FORMED
CIRCULAR HOLLOW SECTIONS

Hybox® CHS

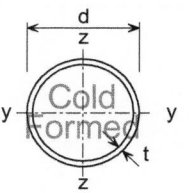

Dimensions and properties

Section Designation		Mass per Metre	Area of Section	Ratio for Local Buckling	Second Moment of Area	Radius of Gyration	Elastic Modulus	Plastic Modulus	Torsional Constants		Surface Area	
Outside Diameter	Thickness										Per Metre	Per Tonne
d	t		A	d/t	I	i	W_{el}	W_{pl}	I_T	W_t		
mm	mm	kg/m	cm²		cm⁴	cm	cm³	cm³	cm⁴	cm³	m²	m²
33.7	3.0	2.27	2.89	11.2	3.44	1.09	2.04	2.84	6.88	4.08	0.106	46.6
42.4	4.0	3.79	4.83	10.6	8.99	1.36	4.24	5.92	18.0	8.48	0.133	35.2
48.3	3.0	3.35	4.27	16.1	11.0	1.61	4.55	6.17	22.0	9.11	0.152	45.3
48.3	3.5	3.87	4.93	13.8	12.4	1.59	5.15	7.04	24.9	10.3	0.152	39.2
48.3	4.0	4.37	5.57	12.1	13.8	1.57	5.70	7.87	27.5	11.4	0.152	34.7
60.3	3.0	4.24	5.40	20.1	22.2	2.03	7.37	9.86	44.4	14.7	0.189	44.7
60.3	4.0	5.55	7.07	15.1	28.2	2.00	9.34	12.7	56.3	18.7	0.189	34.1
76.1	3.0	5.41	6.89	25.4	46.1	2.59	12.1	16.0	92.2	24.2	0.239	44.2
76.1	4.0	7.11	9.06	19.0	59.1	2.55	15.5	20.8	118	31.0	0.239	33.6
88.9	3.0	6.36	8.10	29.6	74.8	3.04	16.8	22.1	150	33.6	0.279	43.9
88.9	3.5	7.37	9.39	25.4	85.7	3.02	19.3	25.5	171	38.6	0.279	37.9
88.9	4.0	8.38	10.7	22.2	96.3	3.00	21.7	28.9	193	43.3	0.279	33.3
88.9	5.0	10.3	13.2	17.8	116	2.97	26.2	35.2	233	52.4	0.279	27.0
88.9	6.3	12.8	16.3	14.1	140	2.93	31.5	43.1	280	63.1	0.279	21.8
114.3	3.0	8.23	10.5	38.1	163	3.94	28.4	37.2	325	56.9	0.359	43.6
114.3	3.5	9.56	12.2	32.7	187	3.92	32.7	43.0	374	65.5	0.359	37.5
114.3	4.0	10.9	13.9	28.6	211	3.90	36.9	48.7	422	73.9	0.359	33.0
114.3	5.0	13.5	17.2	22.9	257	3.87	45.0	59.8	514	89.9	0.359	26.6
114.3	6.0	16.0	20.4	19.1	300	3.83	52.5	70.4	600	105	0.359	22.4
139.7	4.0	13.4	17.1	34.9	393	4.80	56.2	73.7	786	112	0.439	32.8
139.7	5.0	16.6	21.2	27.9	481	4.77	68.8	90.8	961	138	0.439	26.4
139.7	6.0	19.8	25.2	23.3	564	4.73	80.8	107	1130	162	0.439	22.2
139.7	8.0	26.0	33.1	17.5	720	4.66	103	139	1440	206	0.439	16.9
139.7	10.0	32.0	40.7	14.0	862	4.60	123	169	1720	247	0.439	13.7
168.3	4.0	16.2	20.6	42.1	697	5.81	82.8	108	1390	166	0.529	32.6
168.3	5.0	20.1	25.7	33.7	856	5.78	102	133	1710	203	0.529	26.3
168.3	6.0	24.0	30.6	28.1	1010	5.74	120	158	2020	240	0.529	22.0
168.3	8.0	31.6	40.3	21.0	1300	5.67	154	206	2600	308	0.529	16.7
168.3	10.0	39.0	49.7	16.8	1560	5.61	186	251	3130	372	0.529	13.5
168.3	12.5	48.0	61.2	13.5	1870	5.53	222	304	3740	444	0.529	11.0
193.7	4.0	18.7	23.8	48.4	1070	6.71	111	144	2150	222	0.609	32.5
193.7	4.5	21.0	26.7	43.0	1200	6.69	124	161	2400	247	0.609	29.0
193.7	5.0	23.3	29.6	38.7	1320	6.67	136	178	2640	273	0.609	26.2
193.7	6.0	27.8	35.4	32.3	1560	6.64	161	211	3120	322	0.609	21.9
193.7	8.0	36.6	46.7	24.2	2020	6.57	208	276	4030	416	0.609	16.6
193.7	10.0	45.3	57.7	19.4	2440	6.50	252	338	4880	504	0.609	13.4
193.7	12.5	55.9	71.2	15.5	2930	6.42	303	411	5870	606	0.609	10.9
219.1	4.5	23.8	30.3	48.7	1750	7.59	159	207	3490	319	0.688	28.9
219.1	5.0	26.4	33.6	43.8	1930	7.57	176	229	3860	352	0.688	26.1
219.1	6.0	31.5	40.2	36.5	2280	7.54	208	273	4560	417	0.688	21.8
219.1	8.0	41.6	53.1	27.4	2960	7.47	270	357	5920	540	0.688	16.5
219.1	10.0	51.6	65.7	21.9	3600	7.40	328	438	7200	657	0.688	13.3
219.1	12.0	61.3	78.1	18.3	4200	7.33	383	515	8400	767	0.688	11.2
219.1	12.5	63.7	81.1	17.5	4350	7.32	397	534	8690	793	0.688	10.8
219.1	16.0	80.1	102	13.7	5300	7.20	483	661	10600	967	0.688	8.59

Hybox® is a trademark of Tata. A fuller description of the relationship between Cold Formed Circular Hollow Sections (CFCHS) and the Hybox® range of sections manufactured by Tata is given in section 12.

FOR EXPLANATION OF TABLES SEE NOTES 2 AND 3

BS EN 1993-1-1:2005
BS EN 10219-2:2006

COLD-FORMED
CIRCULAR HOLLOW SECTIONS

Hybox® CHS

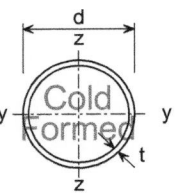

Dimensions and properties

Section Designation		Mass per Metre	Area of Section	Ratio for Local Buckling	Second Moment of Area	Radius of Gyration	Elastic Modulus	Plastic Modulus	Torsional Constants		Surface Area	
Outside Diameter	Thickness										Per Metre	Per Tonne
d	t		A	d/t	I	i	W_{el}	W_{pl}	I_T	W_t		
mm	mm	kg/m	cm²		cm⁴	cm	cm³	cm³	cm⁴	cm³	m²	m²
244.5	5.0	29.5	37.6	48.9	2700	8.47	221	287	5400	441	0.768	26.0
244.5	6.0	35.3	45.0	40.8	3200	8.43	262	341	6400	523	0.768	21.8
244.5	8.0	46.7	59.4	30.6	4160	8.37	340	448	8320	681	0.768	16.5
244.5	10.0	57.8	73.7	24.5	5070	8.30	415	550	10100	830	0.768	13.3
244.5	12.0	68.8	87.7	20.4	5940	8.23	486	649	11900	972	0.768	11.2
244.5	12.5	71.5	91.1	19.6	6150	8.21	503	673	12300	1010	0.768	10.7
244.5	16.0	90.2	115	15.3	7530	8.10	616	837	15100	1230	0.768	8.52
273.0	5.0	33.0	42.1	54.6	3780	9.48	277	359	7560	554	0.858	26.0
273.0	6.0	39.5	50.3	45.5	4490	9.44	329	428	8970	657	0.858	21.7
273.0	8.0	52.3	66.6	34.1	5850	9.37	429	562	11700	857	0.858	16.4
273.0	10.0	64.9	82.6	27.3	7150	9.31	524	692	14300	1050	0.858	13.2
273.0	12.0	77.2	98.4	22.8	8400	9.24	615	818	16800	1230	0.858	11.1
273.0	12.5	80.3	102	21.8	8700	9.22	637	849	17400	1270	0.858	10.7
273.0	16.0	101	129	17.1	10700	9.10	784	1060	21400	1570	0.858	8.46
323.9	5.0	39.3	50.1	64.8	6370	11.3	393	509	12700	787	1.02	25.9
323.9	6.0	47.0	59.9	54.0	7570	11.2	468	606	15100	935	1.02	21.6
323.9	8.0	62.3	79.4	40.5	9910	11.2	612	799	19800	1220	1.02	16.3
323.9	10.0	77.4	98.6	32.4	12200	11.1	751	986	24300	1500	1.02	13.1
323.9	12.0	92.3	118	27.0	14300	11.0	884	1170	28600	1770	1.02	11.0
323.9	12.5	96.0	122	25.9	14800	11.0	917	1210	29700	1830	1.02	10.6
323.9	16.0	121	155	20.2	18400	10.9	1140	1520	36800	2270	1.02	8.38
355.6	5.0	43.2	55.1	71.1	8460	12.4	476	615	16900	952	1.12	25.8
355.6	6.0	51.7	65.9	59.3	10100	12.4	566	733	20100	1130	1.12	21.6
355.6	8.0	68.6	87.4	44.5	13200	12.3	742	967	26400	1490	1.12	16.3
355.6	10.0	85.2	109	35.6	16200	12.2	912	1200	32400	1830	1.12	13.1
355.6	12.0	102	130	29.6	19100	12.2	1080	1420	38300	2150	1.12	11.0
355.6	12.5	106	135	28.4	19900	12.1	1120	1470	39700	2230	1.12	10.6
355.6	16.0	134	171	22.2	24700	12.0	1390	1850	49300	2770	1.12	8.34
406.4	6.0	59.2	75.5	67.7	15100	14.2	745	962	30300	1490	1.28	21.5
406.4	8.0	78.6	100	50.8	19900	14.1	978	1270	39700	1960	1.28	16.2
406.4	10.0	97.8	125	40.6	24500	14.0	1210	1570	49000	2410	1.28	13.1
406.4	12.0	117	149	33.9	28900	14.0	1420	1870	57900	2850	1.28	10.9
406.4	12.5	121	155	32.5	30000	13.9	1480	1940	60100	2960	1.28	10.5
406.4	16.0	154	196	25.4	37400	13.8	1840	2440	74900	3690	1.28	8.29
457.0	6.0	66.7	85.0	76.2	21600	15.9	946	1220	43200	1890	1.44	21.5
457.0	8.0	88.6	113	57.1	28400	15.9	1250	1610	56900	2490	1.44	16.2
457.0	10.0	110	140	45.7	35100	15.8	1540	2000	70200	3070	1.44	13.0
457.0	12.0	132	168	38.1	41600	15.7	1820	2380	83100	3640	1.44	10.9
457.0	12.5	137	175	36.6	43100	15.7	1890	2470	86300	3780	1.44	10.5
457.0	16.0	174	222	28.6	54000	15.6	2360	3110	108000	4720	1.44	8.25
508.0	6.0	74.3	94.6	84.7	29800	17.7	1170	1510	59600	2350	1.60	21.5
508.0	8.0	98.6	126	63.5	39300	17.7	1550	2000	78600	3090	1.60	16.2
508.0	10.0	123	156	50.8	48500	17.6	1910	2480	97000	3820	1.60	13.0
508.0	12.0	147	187	42.3	57500	17.5	2270	2950	115000	4530	1.60	10.9
508.0	12.5	153	195	40.6	59800	17.5	2350	3070	120000	4710	1.60	10.4
508.0	16.0	194	247	31.8	74900	17.4	2950	3870	150000	5900	1.60	8.22

Hybox® is a trademark of Tata. A fuller description of the relationship between Cold Formed Circular Hollow Sections (CFCHS) and the Hybox® range of sections manufactured by Tata is given in section 12.
FOR EXPLANATION OF TABLES SEE NOTES 2 AND 3

BS EN 1993-1-1:2005
BS EN 10219-2:2006

COLD-FORMED
SQUARE HOLLOW SECTIONS

Hybox® SHS

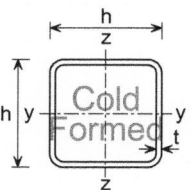

Dimensions and properties

Section Designation		Mass per Metre	Area of Section	Ratio for Local Buckling	Second Moment of Area	Radius of Gyration	Elastic Modulus	Plastic Modulus	Torsional Constants		Surface Area	
Size	Thickness										Per Metre	Per Tonne
h x h	t	Metre	A	c/t [(1)]	I	i	W_{el}	W_{pl}	I_T	W_t		
mm	mm	kg/m	cm²		cm⁴	cm	cm³	cm³	cm⁴	cm³	m²	m²
25x25	2.0	1.36	1.74	7.50	1.48	0.924	1.19	1.47	2.53	1.80	0.093	68.3
	2.5	1.64	2.09	5.00	1.69	0.899	1.35	1.71	2.97	2.07	0.091	55.7
	3.0	1.89	2.41	3.33	1.84	0.874	1.47	1.91	3.33	2.27	0.090	47.4
30x30	2.5	2.03	2.59	7.00	3.16	1.10	2.10	2.61	5.40	3.20	0.111	54.8
	3.0	2.36	3.01	5.00	3.50	1.08	2.34	2.96	6.15	3.58	0.110	46.5
40x40	2.0	2.31	2.94	15.0	6.94	1.54	3.47	4.13	11.3	5.23	0.153	66.4
	2.5	2.82	3.59	11.0	8.22	1.51	4.11	4.97	13.6	6.21	0.151	53.7
	3.0	3.30	4.21	8.33	9.32	1.49	4.66	5.72	15.8	7.07	0.150	45.3
	4.0	4.20	5.35	5.00	11.1	1.44	5.54	7.01	19.4	8.48	0.146	34.8
50x50	2.5	3.60	4.59	15.0	16.9	1.92	6.78	8.07	27.5	10.2	0.191	53.1
	3.0	4.25	5.41	11.7	19.5	1.90	7.79	9.39	32.1	11.8	0.190	44.7
	4.0	5.45	6.95	7.50	23.7	1.85	9.49	11.7	40.4	14.4	0.186	34.2
	5.0	6.56	8.36	5.00	27.0	1.80	10.8	13.7	47.5	16.6	0.183	27.9
60x60	3.0	5.19	6.61	15.0	35.1	2.31	11.7	14.0	57.1	17.7	0.230	44.3
	4.0	6.71	8.55	10.0	43.6	2.26	14.5	17.6	72.6	22.0	0.226	33.7
	5.0	8.13	10.4	7.00	50.5	2.21	16.8	20.9	86.4	25.6	0.223	27.4
70x70	2.5	5.17	6.59	23.0	49.4	2.74	14.1	16.5	78.5	21.2	0.271	52.5
	3.0	6.13	7.81	18.3	57.5	2.71	16.4	19.4	92.4	24.7	0.270	44.0
	3.5	7.06	8.99	15.0	65.1	2.69	18.6	22.2	106	28.0	0.268	38.0
	4.0	7.97	10.1	12.5	72.1	2.67	20.6	24.8	119	31.1	0.266	33.4
	5.0	9.70	12.4	9.00	84.6	2.62	24.2	29.6	142	36.7	0.263	27.1
80x80	3.0	7.07	9.01	21.7	87.8	3.12	22.0	25.8	140	33.0	0.310	43.8
	3.5	8.16	10.4	17.9	99.8	3.10	25.0	29.5	161	37.6	0.308	37.7
	4.0	9.22	11.7	15.0	111	3.07	27.8	33.1	180	41.8	0.306	33.2
	5.0	11.3	14.4	11.0	131	3.03	32.9	39.7	218	49.7	0.303	26.9
	6.0	13.2	16.8	8.33	149	2.98	37.3	45.8	252	56.6	0.299	22.7
90x90	3.0	8.01	10.2	25.0	127	3.53	28.3	33.0	201	42.5	0.350	43.6
	3.5	9.26	11.8	20.7	145	3.51	32.2	37.9	232	48.5	0.348	37.6
	4.0	10.5	13.3	17.5	162	3.48	36.0	42.6	261	54.2	0.346	33.0
	5.0	12.8	16.4	13.0	193	3.43	42.9	51.4	316	64.7	0.343	26.7
	6.0	15.1	19.2	10.0	220	3.39	49.0	59.5	368	74.2	0.339	22.5
100x100	3.0	8.96	11.4	28.3	177	3.94	35.4	41.2	279	53.2	0.390	43.5
	4.0	11.7	14.9	20.0	226	3.89	45.3	53.3	362	68.1	0.386	32.9
	5.0	14.4	18.4	15.0	271	3.84	54.2	64.6	441	81.7	0.383	26.6
	6.0	17.0	21.6	11.7	311	3.79	62.3	75.1	514	94.1	0.379	22.3
	8.0	21.4	27.2	7.50	366	3.67	73.2	91.1	645	114	0.366	17.1
120x120	4.0	14.2	18.1	25.0	402	4.71	67.0	78.3	637	101	0.466	32.7
	5.0	17.5	22.4	19.0	485	4.66	80.9	95.4	778	122	0.463	26.4
	6.0	20.7	26.4	15.0	562	4.61	93.7	112	913	141	0.459	22.1
	8.0	26.4	33.6	10.0	677	4.49	113	138	1160	175	0.446	16.9
	10.0	31.8	40.6	7.00	777	4.38	129	162	1380	203	0.437	13.7

Hybox® is a trademark of Tata. A fuller description of the relationship between Cold Formed Square Hollow Sections (CFSHS) and the Hybox® range of sections manufactured by Tata is given in section 12.

(1) For local buckling calculation c = h - 3t.

FOR EXPLANATION OF TABLES SEE NOTES 2 AND 3

BS EN 1993-1-1:2005
BS EN 10219-2:2006

COLD-FORMED
SQUARE HOLLOW SECTIONS

Hybox® SHS

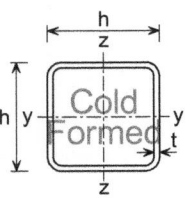

Dimensions and properties

Section Designation		Mass per Metre	Area of Section	Ratio for Local Buckling	Second Moment of Area	Radius of Gyration	Elastic Modulus	Plastic Modulus	Torsional Constants		Surface Area	
Size	Thickness										Per Metre	Per Tonne
h x h	t		A	c/t (1)	I	i	W_{el}	W_{pl}	I_T	W_t		
mm	mm	kg/m	cm²		cm⁴	cm	cm³	cm³	cm⁴	cm³	m²	m²
140x140	4.0	16.8	21.3	30.0	652	5.52	93.1	108	1020	140	0.546	32.6
	5.0	20.7	26.4	23.0	791	5.48	113	132	1260	170	0.543	26.2
	6.0	24.5	31.2	18.3	920	5.43	131	155	1480	198	0.539	22.0
	8.0	31.4	40.0	12.5	1130	5.30	161	194	1900	248	0.526	16.7
	10.0	38.1	48.6	9.00	1310	5.20	187	230	2270	291	0.517	13.6
150x150	4.0	18.0	22.9	32.5	808	5.93	108	125	1270	162	0.586	32.5
	5.0	22.3	28.4	25.0	982	5.89	131	153	1550	197	0.583	26.2
	6.0	26.4	33.6	20.0	1150	5.84	153	180	1830	230	0.579	21.9
	8.0	33.9	43.2	13.8	1410	5.71	188	226	2360	289	0.566	16.7
	10.0	41.3	52.6	10.0	1650	5.61	220	269	2840	341	0.557	13.5
160x160	4.0	19.3	24.5	35.0	987	6.34	123	143	1540	185	0.626	32.5
	5.0	23.8	30.4	27.0	1200	6.29	150	175	1900	226	0.623	26.1
	6.0	28.3	36.0	21.7	1410	6.25	176	206	2240	264	0.619	21.9
	8.0	36.5	46.4	15.0	1740	6.12	218	260	2900	334	0.606	16.6
	10.0	44.4	56.6	11.0	2050	6.02	256	311	3490	395	0.597	13.4
180x180	5.0	27.0	34.4	31.0	1740	7.11	193	224	2720	290	0.703	26.1
	6.0	32.1	40.8	25.0	2040	7.06	226	264	3220	340	0.699	21.8
	8.0	41.5	52.8	17.5	2550	6.94	283	336	4190	432	0.686	16.5
	10.0	50.7	64.6	13.0	3020	6.84	335	404	5070	515	0.677	13.4
	12.0	58.5	74.5	10.0	3320	6.68	369	454	5870	584	0.658	11.3
	12.5	60.5	77.0	9.40	3410	6.65	378	467	6050	600	0.656	10.8
200x200	5.0	30.1	38.4	35.0	2410	7.93	241	279	3760	362	0.783	26.0
	6.0	35.8	45.6	28.3	2830	7.88	283	330	4460	426	0.779	21.8
	8.0	46.5	59.2	20.0	3570	7.76	357	421	5820	544	0.766	16.5
	10.0	57.0	72.6	15.0	4250	7.65	425	508	7070	651	0.757	13.3
	12.0	66.0	84.1	11.7	4730	7.50	473	576	8230	743	0.738	11.2
	12.5	68.3	87.0	11.0	4860	7.47	486	594	8500	765	0.736	10.8
250x250	6.0	45.2	57.6	36.7	5670	9.92	454	524	8840	681	0.979	21.6
	8.0	59.1	75.2	26.3	7230	9.80	578	676	11600	878	0.966	16.3
	10.0	72.7	92.6	20.0	8710	9.70	697	822	14200	1060	0.957	13.2
	12.0	84.8	108	15.8	9860	9.55	789	944	16700	1230	0.938	11.1
	12.5	88.0	112	15.0	10200	9.52	813	975	17300	1270	0.936	10.6
300x300	6.0	54.7	69.6	45.0	9960	12.0	664	764	15400	997	1.18	21.6
	8.0	71.6	91.2	32.5	12800	11.8	853	991	20300	1290	1.17	16.3
	10.0	88.4	113	25.0	15500	11.7	1040	1210	25000	1570	1.16	13.1
	12.0	104	132	20.0	17800	11.6	1180	1400	29500	1830	1.14	11.0
	12.5	108	137	19.0	18300	11.6	1220	1450	30600	1890	1.14	10.6
350x350	8.0	84.2	107	38.8	20700	13.9	1180	1370	32600	1790	1.37	16.2
	10.0	104	133	30.0	25200	13.8	1440	1680	40100	2180	1.36	13.0
	12.0	123	156	24.2	29100	13.6	1660	1950	47600	2550	1.34	10.9
	12.5	127	162	23.0	30000	13.6	1720	2020	49400	2640	1.34	10.5
400x400	8.0	96.7	123	45.0	31300	15.9	1560	1800	48900	2360	1.57	16.2
	10.0	120	153	35.0	38200	15.8	1910	2210	60400	2890	1.56	13.0
	12.0	141	180	28.3	44300	15.7	2220	2590	71800	3400	1.54	10.9
	12.5	147	187	27.0	45900	15.7	2290	2680	74600	3520	1.54	10.5

Hybox® is a trademark of Tata. A fuller description of the relationship between Cold Formed Square Hollow Sections (CFSHS) and the Hybox® range of sections manufactured by Tata is given in section 12.
(1) For local buckling calculation c = h - 3t.
FOR EXPLANATION OF TABLES SEE NOTES 2 AND 3

BS EN 1993-1-1:2005
BS EN 10219-2:2006

COLD-FORMED
RECTANGULAR HOLLOW SECTIONS

Hybox® RHS

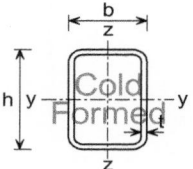

Dimensions and properties

Section Designation		Mass per	Area of	Ratios for Local Buckling		Second Moment of Area		Radius of Gyration		Elastic Modulus		Plastic Modulus		Torsional Constants		Surface Area	
Size	Thickness	Metre	Section			Axis y-y	Axis z-z	Axis y-y	Axis z-z	Axis y-y	Axis z-z	Axis y-y	Axis z-z			Per Metre	Per Tonne
h x b mm	t mm	kg/m	A cm²	c_w/t [(1)]	c_f/t [(1)]	cm⁴	cm⁴	cm	cm	cm³	cm³	cm³	cm³	I_T cm⁴	W_t cm³	m²	m²
50 x 25	2.0	2.15	2.74	20.0	7.50	8.38	2.81	1.75	1.01	3.35	2.25	4.26	2.62	7.06	3.92	0.143	66.6
50 x 25	3.0	3.07	3.91	11.7	3.33	11.2	3.67	1.69	0.969	4.47	2.93	5.86	3.56	9.64	5.18	0.140	45.5
50 x 30	2.5	2.82	3.59	15.0	7.00	11.3	5.05	1.77	1.19	4.52	3.37	5.70	3.98	11.7	5.72	0.151	53.7
	3.0	3.30	4.21	11.7	5.00	12.8	5.70	1.75	1.16	5.13	3.80	6.57	4.58	13.5	6.49	0.150	45.3
	4.0	4.20	5.35	7.50	2.50	15.3	6.69	1.69	1.12	6.10	4.46	8.05	5.58	16.5	7.71	0.146	34.8
60 x 30	3.0	3.77	4.81	15.0	5.00	20.5	6.80	2.06	1.19	6.83	4.53	8.82	5.39	17.5	7.95	0.170	45.0
	4.0	4.83	6.15	10.0	2.50	24.7	8.06	2.00	1.14	8.23	5.37	10.9	6.62	21.5	9.52	0.166	34.5
60 x 40	3.0	4.25	5.41	15.0	8.33	25.4	13.4	2.17	1.58	8.46	6.72	10.5	7.94	29.3	11.2	0.190	44.7
	4.0	5.45	6.95	10.0	5.00	31.0	16.3	2.11	1.53	10.3	8.14	13.2	9.89	36.7	13.7	0.186	34.2
	5.0	6.56	8.36	7.00	3.00	35.3	18.4	2.06	1.48	11.8	9.21	15.4	11.5	42.8	15.6	0.183	27.9
70 x 40	3.0	4.72	6.01	18.3	8.33	37.3	15.5	2.49	1.61	10.7	7.75	13.4	9.05	36.5	13.2	0.210	44.5
	4.0	6.08	7.75	12.5	5.00	46.0	18.9	2.44	1.56	13.1	9.44	16.8	11.3	45.8	16.2	0.206	33.9
70 x 50	3.0	5.19	6.61	18.3	11.7	44.1	26.1	2.58	1.99	12.6	10.4	15.4	12.2	53.6	17.1	0.230	44.3
	4.0	6.71	8.55	12.5	7.50	54.7	32.2	2.53	1.94	15.6	12.9	19.5	15.4	68.1	21.2	0.226	33.7
80 x 40	3.0	5.19	6.61	21.7	8.33	52.3	17.6	2.81	1.63	13.1	8.78	16.5	10.2	43.9	15.3	0.230	44.3
	4.0	6.71	8.55	15.0	5.00	64.8	21.5	2.75	1.59	16.2	10.7	20.9	12.8	55.2	18.8	0.226	33.7
	5.0	8.13	10.4	11.0	3.00	75.1	24.6	2.69	1.54	18.8	12.3	24.7	15.0	65.0	21.7	0.223	27.4
80 x 50	3.0	5.66	7.21	21.7	11.7	61.1	29.4	2.91	2.02	15.3	11.8	18.8	13.6	65.0	19.7	0.250	44.1
	4.0	7.34	9.35	15.0	7.50	76.4	36.5	2.86	1.98	19.1	14.6	24.0	17.2	82.7	24.6	0.246	33.6
	5.0	8.91	11.4	11.0	5.00	89.2	42.3	2.80	1.93	22.3	16.9	28.5	20.5	98.4	28.7	0.243	27.2
80 x 60	3.0	6.13	7.81	21.7	15.0	70.0	44.9	3.00	2.40	17.5	15.0	21.2	17.4	88.3	24.1	0.270	44.0
	4.0	7.97	10.1	15.0	10.0	87.9	56.1	2.94	2.35	22.0	18.7	27.0	22.1	113	30.3	0.266	33.4
	5.0	9.70	12.4	11.0	7.00	103	65.7	2.89	2.31	25.8	21.9	32.2	26.4	136	35.7	0.263	27.1
90 x 50	3.0	6.13	7.81	25.0	11.7	81.9	32.7	3.24	2.05	18.2	13.1	22.6	15.0	76.7	22.4	0.270	44.0
	4.0	7.97	10.1	17.5	7.50	103	40.7	3.18	2.00	22.8	16.3	28.8	19.1	97.7	28.0	0.266	33.4
	5.0	9.70	12.4	13.0	5.00	121	47.4	3.12	1.96	26.8	18.9	34.4	22.7	116	32.7	0.263	27.1
100 x 40	3.0	6.13	7.81	28.3	8.33	92.3	21.7	3.44	1.67	18.5	10.8	23.7	12.4	59.0	19.4	0.270	44.0
	4.0	7.97	10.1	20.0	5.00	116	26.7	3.38	1.62	23.1	13.3	30.3	15.7	74.5	24.0	0.266	33.4
	5.0	9.70	12.4	15.0	3.00	136	30.8	3.31	1.58	27.1	15.4	36.1	18.5	87.9	27.9	0.263	27.1
100 x 50	3.0	6.60	8.41	28.3	11.7	106	36.1	3.56	2.07	21.3	14.4	26.7	16.4	88.6	25.0	0.290	43.9
	4.0	8.59	10.9	20.0	7.50	134	44.9	3.50	2.03	26.8	18.0	34.1	20.9	113	31.3	0.286	33.3
	5.0	10.5	13.4	15.0	5.00	158	52.5	3.44	1.98	31.6	21.0	40.8	25.0	135	36.8	0.283	27.0
	6.0	12.3	15.6	11.7	3.33	179	58.7	3.38	1.94	35.8	23.5	46.9	28.5	154	41.4	0.279	22.8
100 x 60	3.0	7.07	9.01	28.3	15.0	121	54.6	3.66	2.46	24.1	18.2	29.6	20.8	122	30.6	0.310	43.8
	3.5	8.16	10.4	23.6	12.1	137	61.9	3.63	2.44	27.4	20.6	33.8	23.8	139	34.8	0.308	37.7
	4.0	9.22	11.7	20.0	10.0	153	68.7	3.60	2.42	30.5	22.9	37.9	26.6	156	38.7	0.306	33.2
	5.0	11.3	14.4	15.0	7.00	181	80.8	3.55	2.37	36.2	26.9	45.6	31.9	188	45.8	0.303	26.9
	6.0	13.2	16.8	11.7	5.00	205	91.2	3.49	2.33	41.1	30.4	52.5	36.6	216	51.9	0.299	22.7
100 x 80	3.0	8.01	10.2	28.3	21.7	149	106	3.82	3.22	29.8	26.4	35.4	30.4	196	41.9	0.350	43.6
	4.0	10.5	13.3	20.0	15.0	189	134	3.77	3.17	37.9	33.5	45.6	39.2	254	53.4	0.346	33.0
	5.0	12.8	16.4	15.0	11.0	226	160	3.72	3.12	45.2	39.9	55.1	47.2	308	63.7	0.343	26.7

Hybox® is a trademark of Tata. A fuller description of the relationship between Cold Formed Rectangular Hollow Sections (CFRHS) and the Hybox® range of sections manufactured by Tata is given in section 12.
(1) For local buckling calculation $c_w = d - 3t$ and $c_f = h - 3t$.
FOR EXPLANATION OF TABLES SEE NOTES 2 AND 3

BS EN 1993-1-1:2005
BS EN 10219-2:2006

COLD-FORMED
RECTANGULAR HOLLOW SECTIONS

Hybox® RHS

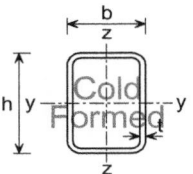

Dimensions and properties

Section Designation		Mass per Metre	Area of Section	Ratios for Local Buckling		Second Moment of Area		Radius of Gyration		Elastic Modulus		Plastic Modulus		Torsional Constants		Surface Area	
Size	Thickness					Axis y-y	Axis z-z	Axis y-y	Axis z-z	Axis y-y	Axis z-z	Axis y-y	Axis z-z			Per Metre	Per Tonne
h x b mm	t mm	kg/m	A cm²	c_w/t [(1)]	c_f/t [(1)]	cm⁴	cm⁴	cm	cm	cm³	cm³	cm³	cm³	I_T cm⁴	W_t cm³	m²	m²
120 x 40	3.0	7.07	9.01	35.0	8.33	148	25.8	4.05	1.69	24.7	12.9	32.2	14.6	74.6	23.5	0.310	43.8
	4.0	9.22	11.7	25.0	5.00	187	31.9	3.99	1.65	31.1	15.9	41.2	18.5	94.2	29.2	0.306	33.2
	5.0	11.3	14.4	19.0	3.00	221	36.9	3.92	1.60	36.8	18.5	49.4	22.0	111	34.1	0.303	26.9
120 x 60	3.0	8.01	10.2	35.0	15.0	189	64.4	4.30	2.51	31.5	21.5	39.2	24.2	156	37.1	0.350	43.6
	3.5	9.26	11.8	29.3	12.1	216	73.1	4.28	2.49	35.9	24.4	44.9	27.7	179	42.2	0.348	37.6
	4.0	10.5	13.3	25.0	10.0	241	81.2	4.25	2.47	40.1	27.1	50.5	31.1	201	47.0	0.346	33.0
	5.0	12.8	16.4	19.0	7.00	287	96.0	4.19	2.42	47.8	32.0	60.9	37.4	242	55.8	0.343	26.7
	6.0	15.1	19.2	15.0	5.00	328	109	4.13	2.38	54.7	36.3	70.6	43.1	280	63.6	0.339	22.5
120 x 80	4.0	11.7	14.9	25.0	15.0	295	157	4.44	3.24	49.1	39.3	59.8	45.2	331	64.9	0.386	32.9
	5.0	14.4	18.4	19.0	11.0	353	188	4.39	3.20	58.9	46.9	72.4	54.7	402	77.8	0.383	26.6
	6.0	17.0	21.6	15.0	8.33	406	215	4.33	3.15	67.7	53.8	84.3	63.5	469	89.4	0.379	22.3
	8.0	21.4	27.2	10.0	5.00	476	252	4.18	3.04	79.3	62.9	102	76.9	584	108	0.366	17.1
140 x 80	3.0	9.90	12.6	41.7	21.7	334	141	5.15	3.35	47.8	35.3	58.2	39.6	317	59.7	0.430	43.4
	4.0	13.0	16.5	30.0	15.0	430	180	5.10	3.30	61.4	45.1	75.5	51.3	412	76.5	0.426	32.8
	5.0	16.0	20.4	23.0	11.0	517	216	5.04	3.26	73.9	54.0	91.8	62.2	501	91.8	0.423	26.5
	6.0	18.9	24.0	18.3	8.33	597	248	4.98	3.21	85.3	62.0	107	72.4	584	106	0.419	22.2
	8.0	23.9	30.4	12.5	5.00	708	293	4.82	3.10	101	73.3	131	88.4	731	129	0.406	17.0
	10.0	28.7	36.6	9.00	3.00	804	330	4.69	3.01	115	82.6	152	103	851	147	0.397	13.8
150 x 100	3.0	11.3	14.4	45.0	28.3	461	248	5.65	4.15	61.4	49.5	73.5	55.8	507	81.4	0.490	43.3
	4.0	14.9	18.9	32.5	20.0	595	319	5.60	4.10	79.3	63.7	95.7	72.5	662	105	0.486	32.7
	5.0	18.3	23.4	25.0	15.0	719	384	5.55	4.05	95.9	76.8	117	88.3	809	127	0.483	26.3
	6.0	21.7	27.6	20.0	11.7	835	444	5.50	4.01	111	88.8	137	103	948	147	0.479	22.1
	8.0	27.7	35.2	13.8	7.50	1010	536	5.35	3.90	134	107	169	128	1210	182	0.466	16.8
	10.0	33.4	42.6	10.0	5.00	1160	614	5.22	3.80	155	123	199	150	1430	211	0.457	13.7
160 x 80	4.0	14.2	18.1	35.0	15.0	598	204	5.74	3.35	74.7	50.9	92.9	57.4	494	88.0	0.466	32.7
	5.0	17.5	22.4	27.0	11.0	722	244	5.68	3.30	90.2	61.0	113	69.7	601	106	0.463	26.4
	6.0	20.7	26.4	21.7	8.33	836	281	5.62	3.26	105	70.2	132	81.3	702	122	0.459	22.1
	8.0	26.4	33.6	15.0	5.00	1000	335	5.46	3.16	125	83.7	163	100	882	150	0.446	16.9
180 x 80	4.0	15.5	19.7	40.0	15.0	802	227	6.37	3.39	89.1	56.7	112	63.5	578	99.6	0.506	32.7
	5.0	19.1	24.4	31.0	11.0	971	272	6.31	3.34	108	68.1	137	77.2	704	120	0.503	26.3
	6.0	22.6	28.8	25.0	8.33	1130	314	6.25	3.30	125	78.5	160	90.2	823	139	0.499	22.1
	8.0	28.9	36.8	17.5	5.00	1360	377	6.08	3.20	151	94.1	198	111	1040	170	0.486	16.8
	10.0	35.0	44.6	13.0	3.00	1570	429	5.94	3.10	174	107	234	131	1210	196	0.477	13.6
180 x 100	4.0	16.8	21.3	40.0	20.0	926	374	6.59	4.18	103	74.8	126	84.0	854	127	0.546	32.6
	5.0	20.7	26.4	31.0	15.0	1120	452	6.53	4.14	125	90.4	154	103	1050	154	0.543	26.2
	6.0	24.5	31.2	25.0	11.7	1310	524	6.48	4.10	146	105	181	120	1230	179	0.539	22.0
	8.0	31.4	40.0	17.5	7.50	1600	637	6.32	3.99	178	127	226	150	1570	222	0.526	16.7
	10.0	38.1	48.6	13.0	5.00	1860	736	6.19	3.89	207	147	268	177	1860	260	0.517	13.6

Hybox® is a trademark of Tata. A fuller description of the relationship between Cold Formed Rectangular Hollow Sections (CFRHS) and the Hybox® range of sections manufactured by Tata is given in section 12.
(1) For local buckling calculation $c_w = d - 3t$ and $c_f = h - 3t$.
FOR EXPLANATION OF TABLES SEE NOTES 2 AND 3

BS EN 1993-1-1:2005
BS EN 10219-2:2006

COLD-FORMED
RECTANGULAR HOLLOW SECTIONS

Hybox® RHS

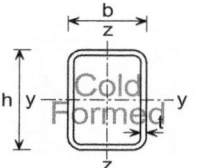

Dimensions and properties

Section Designation		Mass per Metre	Area of Section	Ratios for Local Buckling		Second Moment of Area		Radius of Gyration		Elastic Modulus		Plastic Modulus		Torsional Constants		Surface Area	
Size	Thickness					Axis y-y	Axis z-z	Axis y-y	Axis z-z	Axis y-y	Axis z-z	Axis y-y	Axis z-z			Per Metre	Per Tonne
h x b	t		A	c_w/t [(1)]	c_f/t [(1)]									I_T	W_t		
mm	mm	kg/m	cm²			cm⁴	cm⁴	cm	cm	cm³	cm³	cm³	cm³	cm⁴	cm³	m²	m²
200 x 100	4.0	18.0	22.9	45.0	20.0	1200	411	7.23	4.23	120	82.2	148	91.7	985	142	0.586	32.5
	5.0	22.3	28.4	35.0	15.0	1460	497	7.17	4.19	146	99.4	181	112	1210	172	0.583	26.2
	6.0	26.4	33.6	28.3	11.7	1700	577	7.12	4.14	170	115	213	132	1420	200	0.579	21.9
	8.0	33.9	43.2	20.0	7.50	2090	705	6.95	4.04	209	141	267	165	1810	250	0.566	16.7
	10.0	41.3	52.6	15.0	5.00	2440	818	6.82	3.94	244	164	318	195	2150	292	0.557	13.5
200 x 120	4.0	19.3	24.5	45.0	25.0	1350	618	7.43	5.02	135	103	164	115	1350	172	0.626	32.5
	5.0	23.8	30.4	35.0	19.0	1650	750	7.37	4.97	165	125	201	141	1650	210	0.623	26.1
	6.0	28.3	36.0	28.3	15.0	1930	874	7.32	4.93	193	146	237	166	1950	245	0.619	21.9
	8.0	36.5	46.4	20.0	10.0	2390	1080	7.17	4.82	239	180	298	209	2510	308	0.606	16.6
	10.0	44.4	56.6	15.0	7.00	2810	1260	7.04	4.72	281	210	356	250	3010	364	0.597	13.4
200 x 150	4.0	21.2	26.9	45.0	32.5	1580	1020	7.67	6.16	158	136	187	154	1940	219	0.686	32.4
	5.0	26.2	33.4	35.0	25.0	1940	1250	7.62	6.11	193	166	230	189	2390	267	0.683	26.1
	6.0	31.1	39.6	28.3	20.0	2270	1460	7.56	6.06	227	194	271	223	2830	313	0.679	21.8
	8.0	40.2	51.2	20.0	13.8	2830	1820	7.43	5.95	283	242	344	283	3670	396	0.666	16.5
	10.0	49.1	62.6	15.0	10.0	3350	2140	7.31	5.85	335	286	413	339	4430	471	0.657	13.4
250 x 150	5.0	30.1	38.4	45.0	25.0	3300	1510	9.28	6.27	264	201	320	225	3290	337	0.783	26.0
	6.0	35.8	45.6	36.7	20.0	3890	1770	9.23	6.23	311	236	378	266	3890	396	0.779	21.8
	8.0	46.5	59.2	26.3	13.8	4890	2220	9.08	6.12	391	296	482	340	5050	504	0.766	16.5
	10.0	57.0	72.6	20.0	10.0	5830	2630	8.96	6.02	466	351	582	409	6120	602	0.757	13.3
	12.0	66.0	84.1	15.8	7.50	6460	2930	8.77	5.90	517	390	658	463	7090	684	0.738	11.2
	12.5	68.3	87.0	15.0	7.00	6630	3000	8.73	5.87	531	400	678	477	7320	704	0.736	10.8
300 x 100	6.0	35.8	45.6	45.0	11.7	4780	842	10.2	4.30	318	168	411	188	2400	306	0.779	21.8
	8.0	46.5	59.2	32.5	7.50	5980	1050	10.0	4.20	399	209	523	238	3080	385	0.766	16.5
	10.0	57.0	72.6	25.0	5.00	7110	1220	9.90	4.11	474	245	631	285	3680	455	0.757	13.3
	12.5	68.3	87.0	19.0	3.00	8010	1370	9.59	3.97	534	275	732	330	4290	521	0.736	10.8
300 x 200	6.0	45.2	57.6	45.0	28.3	7370	3960	11.3	8.29	491	396	588	446	8120	651	0.979	21.6
	8.0	59.1	75.2	32.5	20.0	9390	5040	11.2	8.19	626	504	757	574	10600	838	0.966	16.3
	10.0	72.7	92.6	25.0	15.0	11300	6060	11.1	8.09	754	606	921	698	13000	1010	0.957	13.2
	12.0	84.8	108	20.0	11.7	12800	6850	10.9	7.96	853	685	1060	801	15200	1170	0.938	11.1
	12.5	88.0	112	19.0	11.0	13200	7060	10.8	7.94	879	706	1090	828	15800	1200	0.936	10.6
400 x 200	8.0	71.6	91.2	45.0	20.0	19000	6520	14.4	8.45	949	652	1170	728	15800	1130	1.17	16.3
	10.0	88.4	113	35.0	15.0	23000	7860	14.3	8.36	1150	786	1430	888	19400	1370	1.16	13.1
	12.0	104	132	28.3	11.7	26200	8980	14.1	8.24	1310	898	1660	1030	22800	1590	1.14	11.0
	12.5	108	137	27.0	11.0	27100	9260	14.1	8.22	1360	926	1710	1060	23600	1640	1.14	10.6
450 x 250	8.0	84.2	107	51.3	26.3	29300	11900	16.5	10.5	1300	953	1590	1060	27200	1630	1.37	16.2
	10.0	104	133	40.0	20.0	35700	14500	16.4	10.4	1590	1160	1950	1300	33500	1980	1.36	13.0
	12.0	123	156	32.5	15.8	41100	16700	16.2	10.3	1830	1330	2260	1520	39600	2310	1.34	10.9
	12.5	127	162	31.0	15.0	42500	17200	16.2	10.3	1890	1380	2350	1570	41100	2390	1.34	10.5
500 x 300	8.0	96.7	123	57.5	32.5	42800	19600	18.6	12.6	1710	1310	2060	1460	42800	2200	1.57	16.2
	10.0	120	153	45.0	25.0	52300	23900	18.5	12.5	2090	1600	2540	1790	52700	2690	1.56	13.0
	12.0	141	180	36.7	20.0	60600	27700	18.3	12.4	2420	1850	2960	2090	62600	3160	1.54	10.9
	12.5	147	187	35.0	19.0	62700	28700	18.3	12.4	2510	1910	3070	2170	65000	3270	1.54	10.5

Hybox® is a trademark of Tata. A fuller description of the relationship between Cold Formed Rectangular Hollow Sections (CFRHS) and the Hybox® range of sections manufactured by Tata is given in section 12.

(1) For local buckling calculation c_w = d - 3t and c_f = h - 3t.

FOR EXPLANATION OF TABLES SEE NOTES 2 AND 3

**BS EN 1993-1-1:2005
Tata ASB**

ASB (ASYMMETRIC BEAMS)

Dimensions and properties

Section Designation	Mass per Metre	Depth of Section	Width of Flange		Thickness		Root Radius	Depth between Fillets	Ratios for Local Buckling			Second Moment of Area		Surface Area	
			Top	Bottom	Web	Flange			Flanges		Web	Axis y-y	Axis z-z	Per Metre	Per Tonne
		h	b_t	b_b	t_w	t_f	r	d	c_{fl}/t_f	c_{fb}/t_f	c_w/t_w				
	kg/m	mm	mm	mm	mm	mm	mm	mm				cm^4	cm^4	m^2	m^2
300 ASB 249 ^	249	342	203	313	40.0	40.0	27.0	208	1.36	2.74	5.20	52900	13200	1.59	6.38
300 ASB 196	196	342	183	293	20.0	40.0	27.0	208	1.36	2.74	10.4	45900	10500	1.55	7.93
300 ASB 185 ^	185	320	195	305	32.0	29.0	27.0	208	1.88	3.78	6.50	35700	8750	1.53	8.29
300 ASB 155	155	326	179	289	16.0	32.0	27.0	208	1.70	3.42	13.0	34500	7990	1.51	9.71
300 ASB 153 ^	153	310	190	300	27.0	24.0	27.0	208	2.27	4.56	7.70	28400	6840	1.50	9.81
280 ASB 136 ^	136	288	190	300	25.0	22.0	24.0	196	2.66	5.16	7.84	22200	6260	1.46	10.7
280 ASB 124	124	296	178	288	13.0	26.0	24.0	196	2.25	4.37	15.1	23500	6410	1.46	11.8
280 ASB 105	105	288	176	286	11.0	22.0	24.0	196	2.66	5.16	17.8	19200	5300	1.44	13.7
280 ASB 100 ^	100	276	184	294	19.0	16.0	24.0	196	3.66	7.09	10.3	15500	4250	1.43	14.2
280 ASB 74	73.6	272	175	285	10.0	14.0	24.0	196	4.18	8.11	19.6	12200	3330	1.40	19.1

^ Sections are fire engineered with thick webs.

FOR EXPLANATION OF TABLES SEE NOTES 2 AND 3

ASB (ASYMMETRIC BEAMS)

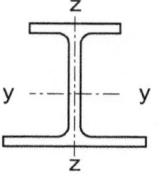

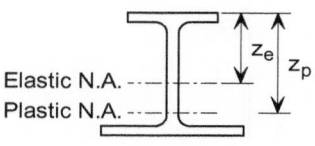

Elastic N.A.

Plastic N.A.

Properties (Continued)

Section Designation	Radius of Gyration		Elastic Modulus			Neutral Axis Position		Plastic Modulus		Buckling Parameter	Torsional Index	Mono-symmetry index*	Warping Constant	Torsional Constant	Area of Section
	Axis y-y	Axis z-z	Axis y-y Top	Axis y-y Bottom	Axis z-z	Elastic	Plastic	Axis y-y	Axis z-z						
	cm	cm	cm³	cm³	cm³	z_e cm	z_p cm	cm³	cm³	U	X	⊠	I_w dm⁶	I_T cm⁴	A cm²
300 ASB 249 ^	12.9	6.40	2760	3530	843	19.2	22.6	3760	1510	0.820	6.80	0.663	2.00	2000	318
300 ASB 196	13.6	6.48	2320	3180	714	19.8	28.1	3060	1230	0.840	7.86	0.895	1.50	1180	249
300 ASB 185 ^	12.3	6.10	1980	2540	574	18.0	21.0	2660	1030	0.820	8.56	0.662	1.20	871	235
300 ASB 155	13.2	6.35	1830	2520	553	18.9	27.3	2360	950	0.840	9.40	0.868	1.07	620	198
300 ASB 153 ^	12.1	5.93	1630	2090	456	17.4	20.4	2160	817	0.820	9.97	0.643	0.895	513	195
280 ASB 136 ^	11.3	6.00	1370	1770	417	16.3	19.2	1810	741	0.810	10.2	0.628	0.710	379	174
280 ASB 124	12.2	6.37	1360	1900	445	17.3	25.7	1730	761	0.830	10.5	0.807	0.721	332	158
280 ASB 105	12.0	6.30	1150	1610	370	16.8	25.3	1440	633	0.830	12.1	0.777	0.574	207	133
280 ASB 100 ^	11.0	5.76	995	1290	289	15.6	18.4	1290	511	0.810	13.2	0.616	0.451	160	128
280 ASB 74	11.4	5.96	776	1060	234	15.7	21.3	978	403	0.830	16.7	0.699	0.338	72.0	93.7

^ Sections are fire engineered with thick webs.

* Monosymmetry index is positive when the wide flange is in compression and negative when the narrow flange is in compression

FOR EXPLANATION OF TABLES SEE NOTES 2 AND 3

BS EN 1993-1-1:2005
BS 4-1:2005

PARALLEL FLANGE CHANNELS

Advance UKPFC

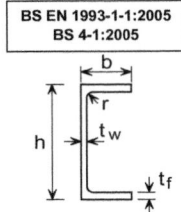

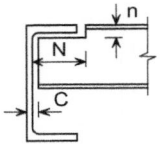

Dimensions

Section Designation	Mass per Metre	Depth of Section	Width of Section	Thickness		Root Radius	Depth between Fillets	Ratios for Local Buckling		Distance	Dimensions for Detailing			Surface Area	
				Web	Flange			Flange	Web		End Clearance	Notch		Per Metre	Per Tonne
		h	b	t_w	t_f	r	d	c_f/t_f	c_w/t_w	e_o	C	N	n		
	kg/m	mm	mm	mm	mm	mm	mm			cm	mm	mm	mm	m^2	m^2
430x100x64	64.4	430	100	11.0	19.0	15	362	3.89	32.9	3.27	13	96	36	1.23	19.0
380x100x54	54.0	380	100	9.5	17.5	15	315	4.31	33.2	3.48	12	98	34	1.13	20.9
300x100x46	45.5	300	100	9.0	16.5	15	237	4.61	26.3	3.68	11	98	32	0.969	21.3
300x90x41	41.4	300	90	9.0	15.5	12	245	4.45	27.2	3.18	11	88	28	0.932	22.5
260x90x35	34.8	260	90	8.0	14.0	12	208	5.00	26.0	3.32	10	88	28	0.854	24.5
260x75x28	27.6	260	75	7.0	12.0	12	212	4.67	30.3	2.62	9	74	26	0.796	28.8
230x90x32	32.2	230	90	7.5	14.0	12	178	5.04	23.7	3.46	10	90	28	0.795	24.7
230x75x26	25.7	230	75	6.5	12.5	12	181	4.52	27.8	2.78	9	76	26	0.737	28.7
200x90x30	29.7	200	90	7.0	14.0	12	148	5.07	21.1	3.60	9	90	28	0.736	24.8
200x75x23	23.4	200	75	6.0	12.5	12	151	4.56	25.2	2.91	8	76	26	0.678	28.9
180x90x26	26.1	180	90	6.5	12.5	12	131	5.72	20.2	3.64	9	90	26	0.697	26.7
180x75x20	20.3	180	75	6.0	10.5	12	135	5.43	22.5	2.87	8	76	24	0.638	31.4
150x90x24	23.9	150	90	6.5	12.0	12	102	5.96	15.7	3.71	9	90	26	0.637	26.7
150x75x18	17.9	150	75	5.5	10.0	12	106	5.75	19.3	2.99	8	76	24	0.579	32.4
125x65x15	14.8	125	65	5.5	9.5	12	82.0	5.00	14.9	2.56	8	66	22	0.489	33.1
100x50x10	10.2	100	50	5.0	8.5	9	65.0	4.24	13.0	1.94	7	52	18	0.382	37.5

Advance and UKPFC are trademarks of Tata. A fuller description of the relationship between Parallel Flange Channels (PFC) and the Advance range of sections manufactured by Tata is given in section 12.

e_0 is the distance from the centre of the web to the shear centre

FOR EXPLANATION OF TABLES SEE NOTE 2

BS EN 1993-1-1:2005
BS 4-1:2005

PARALLEL FLANGE CHANNELS

Advance UKPFC

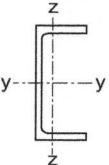

Properties

Section Designation	Second Moment of Area		Radius of Gyration		Elastic Modulus		Plastic Modulus		Buckling Parameter	Torsional Index	Warping Constant	Torsional Constant	Area of Section
	Axis y-y	Axis z-z	Axis y-y	Axis z-z	Axis y-y	Axis z-z	Axis y-y	Axis z-z	U	X	I_w	I_T	A
	cm^4	cm^4	cm	cm	cm^3	cm^3	cm^3	cm^3			dm^6	cm^4	cm^2
430x100x64	21900	722	16.3	2.97	1020	97.9	1220	176	0.917	22.5	0.219	63.0	82.1
380x100x54	15000	643	14.8	3.06	791	89.2	933	161	0.933	21.2	0.150	45.7	68.7
300x100x46	8230	568	11.9	3.13	549	81.7	641	148	0.944	17.0	0.0813	36.8	58.0
300x90x41	7220	404	11.7	2.77	481	63.1	568	114	0.934	18.3	0.0581	28.8	52.7
260x90x35	4730	353	10.3	2.82	364	56.3	425	102	0.943	17.2	0.0379	20.6	44.4
260x75x28	3620	185	10.1	2.30	278	34.4	328	62.0	0.932	20.5	0.0203	11.7	35.1
230x90x32	3520	334	9.27	2.86	306	55.0	355	98.9	0.949	15.1	0.0279	19.3	41.0
230x75x26	2750	181	9.17	2.35	239	34.8	278	63.2	0.945	17.3	0.0153	11.8	32.7
200x90x30	2520	314	8.16	2.88	252	53.4	291	94.5	0.952	12.9	0.0197	18.3	37.9
200x75x23	1960	170	8.11	2.39	196	33.8	227	60.6	0.956	14.7	0.0107	11.1	29.9
180x90x26	1820	277	7.40	2.89	202	47.4	232	83.5	0.950	12.8	0.0141	13.3	33.2
180x75x20	1370	146	7.27	2.38	152	28.8	176	51.8	0.945	15.3	0.00754	7.34	25.9
150x90x24	1160	253	6.18	2.89	155	44.4	179	76.9	0.937	10.8	0.00890	11.8	30.4
150x75x18	861	131	6.15	2.40	115	26.6	132	47.2	0.945	13.1	0.00467	6.10	22.8
125x65x15	483	80.0	5.07	2.06	77.3	18.8	89.9	33.2	0.942	11.1	0.00194	4.72	18.8
100x50x10	208	32.3	4.00	1.58	41.5	9.89	48.9	17.5	0.942	10.0	0.000491	2.53	13.0

Advance and UKPFC are trademarks of Tata. A fuller description of the relationship between Parallel Flange Channels (PFC) and the Advance range of sections manufactured by Tata is given in section 12.

FOR EXPLANATION OF TABLES SEE NOTE 3

| BS EN 1993-1-1:2005 |
| BS 4-1:2005 |

TWO PARALLEL FLANGE CHANNELS LACED

TWO Advance UKPFC LACED

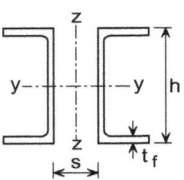

Dimensions and properties

Composed of Two Channels	Total Mass per Metre	Total Area	Space between Webs	Second Moment of Area		Radius of Gyration		Elastic Modulus		Plastic Modulus	
				Axis y-y	Axis z-z	Axis y-y	Axis z-z	Axis y-y	Axis z-z	Axis y-y	Axis z-z
			s								
	kg/m	cm^2	mm	cm^4	cm^4	cm	cm	cm^3	cm^3	cm^3	cm^3
430x100x64	129	164	270	43900	44100	16.3	16.4	2040	1880	2440	2650
380x100x54	108	137	235	30100	30400	14.8	14.9	1580	1400	1870	2000
300x100x46	91.1	116	170	16500	16600	11.9	12.0	1100	898	1280	1340
300x90x41	82.8	105	175	14400	14400	11.7	11.7	962	811	1140	1200
260x90x35	69.7	88.8	145	9460	9560	10.3	10.4	727	588	849	886
260x75x28	55.2	70.3	155	7240	7190	10.1	10.1	557	472	656	692
230x90x32	64.3	81.9	120	7040	7190	9.27	9.37	612	479	709	731
230x75x26	51.3	65.4	135	5500	5720	9.17	9.35	478	401	557	592
200x90x30	59.4	75.7	90.0	5050	5030	8.16	8.15	505	372	583	577
200x75x23	46.9	59.7	105	3930	3910	8.11	8.09	393	306	454	462
180x90x26	52.1	66.4	75.0	3640	3730	7.40	7.49	404	292	464	459
180x75x20	40.7	51.8	90.0	2740	2770	7.27	7.31	304	231	352	358
150x90x24	47.7	60.8	45.0	2320	2380	6.18	6.26	310	212	357	338
150x75x18	35.7	45.5	65.0	1720	1810	6.15	6.30	230	168	264	265
125x65x15	29.5	37.6	50.0	966	1010	5.07	5.18	155	112	180	178
100x50x10	20.4	26.0	40.0	415	427	4.00	4.05	83.1	61.0	97.7	97.1

Advance and UKPFC are trademarks of Tata. A fuller description of the relationship between Parallel Flange Channels (PFC) and the Advance range of sections manufactured by Tata is given in section 12.

FOR EXPLANATION OF TABLES SEE NOTES 2 AND 3

BS EN 1993-1-1:2005
BS 4-1:2005

TWO PARALLEL FLANGE CHANNELS BACK TO BACK

TWO Advance UKPFC BACK TO BACK

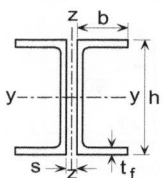

Dimensions and properties

Composed of Two Channels	Total Mass per Metre	Total Area	Properties about Axis y-y				Radius of Gyration i_z about Axis z-z (cm)				
			I_y	i_y	$W_{el.y}$	$W_{pl.y}$	Space between webs, s (mm)				
	kg/m	cm^2	cm^4	cm	cm^3	cm^3	0	8	10	12	15
430x100x64	129	164	43900	16.3	2040	2440	3.96	4.23	4.31	4.38	4.49
380x100x54	108	137	30100	14.8	1580	1870	4.14	4.42	4.49	4.57	4.68
300x100x46	91.1	116	16500	11.9	1100	1280	4.37	4.66	4.73	4.81	4.92
300x90x41	82.8	105	14400	11.7	962	1140	3.80	4.08	4.16	4.23	4.35
260x90x35	69.7	88.8	9460	10.3	727	849	3.93	4.22	4.29	4.37	4.48
260x75x28	55.2	70.3	7240	10.1	557	656	3.11	3.40	3.47	3.55	3.66
230x90x32	64.3	81.9	7040	9.27	612	709	4.09	4.38	4.46	4.53	4.65
230x75x26	51.3	65.4	5500	9.17	478	557	3.29	3.58	3.66	3.73	3.85
200x90x30	59.4	75.7	5050	8.16	505	583	4.25	4.55	4.63	4.71	4.83
200x75x23	46.9	59.7	3930	8.11	393	454	3.44	3.74	3.82	3.89	4.01
180x90x26	52.1	66.4	3640	7.40	404	464	4.29	4.59	4.67	4.75	4.87
180x75x20	40.7	51.8	2740	7.27	304	352	3.39	3.68	3.76	3.84	3.95
150x90x24	47.7	60.8	2320	6.18	310	357	4.39	4.69	4.77	4.85	4.98
150x75x18	35.7	45.5	1720	6.15	230	264	3.52	3.82	3.90	3.98	4.10
125x65x15	29.5	37.6	966	5.07	155	180	3.05	3.36	3.44	3.52	3.64
100x50x10	20.4	26.0	415	4.00	83.1	97.7	2.34	2.65	2.73	2.82	2.94

Advance and UKPFC are trademarks of Tata. A fuller description of the relationship between Parallel Flange Channels (PFC) and the Advance range of sections manufactured by Tata is given in section 12.

Properties about y axis:

$I_z = (\text{Total Area}).(i_z)^2$

$W_{el.z} = I_z/(b+0.5s)$

where s is the space between webs.

FOR EXPLANATION OF TABLES SEE NOTES 2 AND 3

BS EN 1993-1-1:2005
BS EN 10056-1:1999

EQUAL ANGLES

Advance UKA - Equal Angles

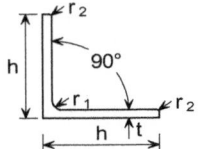

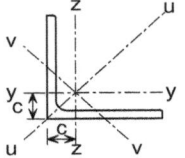

Dimensions and properties

Section Designation		Mass per Metre	Radius		Area of Section	Distance to centroid	Second Moment of Area			Radius of Gyration			Elastic Modulus	Torsional Constant	Equivalent Slenderness Coefficient
Size	Thickness		Root	Toe			Axis y-y, z-z	Axis u-u	Axis v-v	Axis y-y, z-z	Axis u-u	Axis v-v	Axis y-y, z-z		
$h \times h$	t		r_1	r_2		c								I_T	ϕ_a
mm	mm	kg/m	mm	mm	cm²	cm	cm⁴	cm⁴	cm⁴	cm	cm	cm	cm³	cm⁴	
200x200	24	71.1	18.0	9.00	90.6	5.84	3330	5280	1380	6.06	7.64	3.90	235	182	2.50
	20	59.9	18.0	9.00	76.3	5.68	2850	4530	1170	6.11	7.70	3.92	199	107	3.05
	18	54.3	18.0	9.00	69.1	5.60	2600	4150	1050	6.13	7.75	3.90	181	78.9	3.43
	16	48.5	18.0	9.00	61.8	5.52	2340	3720	960	6.16	7.76	3.94	162	56.1	3.85
150x150	18 +	40.1	16.0	8.00	51.2	4.38	1060	1680	440	4.55	5.73	2.93	99.8	58.6	2.48
	15	33.8	16.0	8.00	43.0	4.25	898	1430	370	4.57	5.76	2.93	83.5	34.6	3.01
	12	27.3	16.0	8.00	34.8	4.12	737	1170	303	4.60	5.80	2.95	67.7	18.2	3.77
	10	23.0	16.0	8.00	29.3	4.03	624	990	258	4.62	5.82	2.97	56.9	10.8	4.51
120x120	15 +	26.6	13.0	6.50	34.0	3.52	448	710	186	3.63	4.57	2.34	52.8	27.0	2.37
	12	21.6	13.0	6.50	27.5	3.40	368	584	152	3.65	4.60	2.35	42.7	14.2	2.99
	10	18.2	13.0	6.50	23.2	3.31	313	497	129	3.67	4.63	2.36	36.0	8.41	3.61
	8 +	14.7	13.0	6.50	18.8	3.24	259	411	107	3.71	4.67	2.38	29.5	4.44	4.56
100x100	15 +	21.9	12.0	6.00	28.0	3.02	250	395	105	2.99	3.76	1.94	35.8	22.3	1.92
	12	17.8	12.0	6.00	22.7	2.90	207	328	85.7	3.02	3.80	1.94	29.1	11.8	2.44
	10	15.0	12.0	6.00	19.2	2.82	177	280	73.0	3.04	3.83	1.95	24.6	6.97	2.94
	8	12.2	12.0	6.00	15.5	2.74	145	230	59.9	3.06	3.85	1.96	19.9	3.68	3.70
90x90	12 +	15.9	11.0	5.50	20.3	2.66	149	235	62.0	2.71	3.40	1.75	23.5	10.5	2.17
	10	13.4	11.0	5.50	17.1	2.58	127	201	52.6	2.72	3.42	1.75	19.8	6.20	2.64
	8	10.9	11.0	5.50	13.9	2.50	104	166	43.1	2.74	3.45	1.76	16.1	3.28	3.33
	7	9.61	11.0	5.50	12.2	2.45	92.6	147	38.3	2.75	3.46	1.77	14.1	2.24	3.80
80x80	10	11.9	10.0	5.00	15.1	2.34	87.5	139	36.4	2.41	3.03	1.55	15.4	5.45	2.33
	8	9.63	10.0	5.00	12.3	2.26	72.2	115	29.9	2.43	3.06	1.56	12.6	2.88	2.94
75x75	8	8.99	9.00	4.50	11.4	2.14	59.1	93.8	24.5	2.27	2.86	1.46	11.0	2.65	2.76
	6	6.85	9.00	4.50	8.73	2.05	45.8	72.7	18.9	2.29	2.89	1.47	8.41	1.17	3.70
70x70	7	7.38	9.00	4.50	9.40	1.97	42.3	67.1	17.5	2.12	2.67	1.36	8.41	1.69	2.92
	6	6.38	9.00	4.50	8.13	1.93	36.9	58.5	15.3	2.13	2.68	1.37	7.27	1.09	3.41
65x65	7	6.83	9.00	4.50	8.73	2.05	33.4	53.0	13.8	1.96	2.47	1.26	7.18	1.58	2.67
60x60	8	7.09	8.00	4.00	9.03	1.77	29.2	46.1	12.2	1.80	2.26	1.16	6.89	2.09	2.14
	6	5.42	8.00	4.00	6.91	1.69	22.8	36.1	9.44	1.82	2.29	1.17	5.29	0.922	2.90
	5	4.57	8.00	4.00	5.82	1.64	19.4	30.7	8.03	1.82	2.30	1.17	4.45	0.550	3.48
50x50	6	4.47	7.00	3.50	5.69	1.45	12.8	20.3	5.34	1.50	1.89	0.968	3.61	0.755	2.38
	5	3.77	7.00	3.50	4.80	1.40	11.0	17.4	4.55	1.51	1.90	0.973	3.05	0.450	2.88
	4	3.06	7.00	3.50	3.89	1.36	8.97	14.2	3.73	1.52	1.91	0.979	2.46	0.240	3.57
45x45	5	3.06	7.00	3.50	3.90	1.25	7.14	11.4	2.94	1.35	1.71	0.870	2.20	0.304	2.84
40x40	5	2.97	6.00	3.00	3.79	1.16	5.43	8.60	2.26	1.20	1.51	0.773	1.91	0.352	2.26
	4	2.42	6.00	3.00	3.08	1.12	4.47	7.09	1.86	1.21	1.52	0.777	1.55	0.188	2.83
35x35	4	2.09	5.00	2.50	2.67	1.00	2.95	4.68	1.23	1.05	1.32	0.678	1.18	0.158	2.50
30x30	4	1.78	5.00	2.50	2.27	0.878	1.80	2.85	0.754	0.892	1.12	0.577	0.850	0.137	2.07
	3	1.36	5.00	2.50	1.74	0.835	1.40	2.22	0.585	0.899	1.13	0.581	0.649	0.0613	2.75
25x25	4	1.45	3.50	1.75	1.85	0.762	1.02	1.61	0.430	0.741	0.931	0.482	0.586	0.1070	1.75
	3	1.12	3.50	1.75	1.42	0.723	0.803	1.27	0.334	0.751	0.945	0.484	0.452	0.0472	2.38
20x20	3	0.882	3.50	1.75	1.12	0.598	0.392	0.618	0.165	0.590	0.742	0.383	0.279	0.0382	1.81

Advance and UKA are trademarks of Tata. A fuller description of the relationship between Angles and the Advance range of sections manufactured by Tata is given in section 12.
+ These sections are in addition to the range of BS EN 10056-1 sections.
c is the distance from the back of the leg to the centre of gravity.
FOR EXPLANATION OF TABLES SEE NOTES 2 AND 3

BS EN 1993-1-1:2005
BS EN 10056-1:1999

UNEQUAL ANGLES

Advance UKA - Unequal Angles

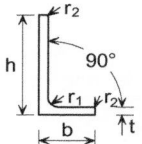

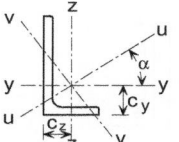

Dimensions and properties

Section Designation		Mass per	Radius		Dimension		Second Moment of Area				Radius of Gyration			
Size	Thickness	Metre	Root	Toe			Axis y-y	Axis z-z	Axis u-u	Axis v-v	Axis y-y	Axis z-z	Axis u-u	Axis v-v
h x b	t		r_1	r_2	c_y	c_z								
mm	mm	kg/m	mm	mm	cm	cm	cm^4	cm^4	cm^4	cm^4	cm	cm	cm	cm
200x150	18 +	47.1	15.0	7.50	6.33	3.85	2380	1150	2920	623	6.29	4.37	6.97	3.22
	15	39.6	15.0	7.50	6.21	3.73	2020	979	2480	526	6.33	4.40	7.00	3.23
	12	32.0	15.0	7.50	6.08	3.61	1650	803	2030	430	6.36	4.44	7.04	3.25
200x100	15	33.8	15.0	7.50	7.16	2.22	1760	299	1860	193	6.40	2.64	6.59	2.12
	12	27.3	15.0	7.50	7.03	2.10	1440	247	1530	159	6.43	2.67	6.63	2.14
	10	23.0	15.0	7.50	6.93	2.01	1220	210	1290	135	6.46	2.68	6.65	2.15
150x90	15	33.9	12.0	6.00	5.21	2.23	761	205	841	126	4.74	2.46	4.98	1.93
	12	21.6	12.0	6.00	5.08	2.12	627	171	694	104	4.77	2.49	5.02	1.94
	10	18.2	12.0	6.00	5.00	2.04	533	146	591	88.3	4.80	2.51	5.05	1.95
150x75	15	24.8	12.0	6.00	5.52	1.81	713	119	753	78.6	4.75	1.94	4.88	1.58
	12	20.2	12.0	6.00	5.40	1.69	588	99.6	623	64.7	4.78	1.97	4.92	1.59
	10	17.0	12.0	6.00	5.31	1.61	501	85.6	531	55.1	4.81	1.99	4.95	1.60
125x75	12	17.8	11.0	5.50	4.31	1.84	354	95.5	391	58.5	3.95	2.05	4.15	1.61
	10	15.0	11.0	5.50	4.23	1.76	302	82.1	334	49.9	3.97	2.07	4.18	1.61
	8	12.2	11.0	5.50	4.14	1.68	247	67.6	274	40.9	4.00	2.09	4.21	1.63
100x75	12	15.4	10.0	5.00	3.27	2.03	189	90.2	230	49.5	3.10	2.14	3.42	1.59
	10	13.0	10.0	5.00	3.19	1.95	162	77.6	197	42.2	3.12	2.16	3.45	1.59
	8	10.6	10.0	5.00	3.10	1.87	133	64.1	162	34.6	3.14	2.18	3.47	1.60
100x65	10 +	12.3	10.0	5.00	3.36	1.63	154	51.0	175	30.1	3.14	1.81	3.35	1.39
	8 +	9.94	10.0	5.00	3.27	1.55	127	42.2	144	24.8	3.16	1.83	3.37	1.40
	7 +	8.77	10.0	5.00	3.23	1.51	113	37.6	128	22.0	3.17	1.83	3.39	1.40
100x50	8 8.97		8.00	4.00	3.60	1.13	116	19.7	123	12.8	3.19	1.31	3.28	1.06
	6 6.84		8.00	4.00	3.51	1.05	89.9	15.4	95.4	9.92	3.21	1.33	3.31	1.07
80x60	7 7.36		8.00	4.00	2.51	1.52	59.0	28.4	72.0	15.4	2.51	1.74	2.77	1.28
80x40	8 7.07		7.00	3.50	2.94	0.963	57.6	9.61	60.9	6.34	2.53	1.03	2.60	0.838
	6 5.41		7.00	3.50	2.85	0.884	44.9	7.59	47.6	4.93	2.55	1.05	2.63	0.845
75x50	8 7.39		7.00	3.50	2.52	1.29	52.0	18.4	59.6	10.8	2.35	1.40	2.52	1.07
	6 5.65		7.00	3.50	2.44	1.21	40.5	14.4	46.6	8.36	2.37	1.42	2.55	1.08
70x50	6 5.41		7.00	3.50	2.23	1.25	33.4	14.2	39.7	7.92	2.20	1.43	2.40	1.07
65x50	5 4.35		6.00	3.00	1.99	1.25	23.2	11.9	28.8	6.32	2.05	1.47	2.28	1.07
60x40	6 4.46		6.00	3.00	2.00	1.01	20.1	7.12	23.1	4.16	1.88	1.12	2.02	0.855
	5 3.76		6.00	3.00	1.96	0.972	17.2	6.11	19.7	3.54	1.89	1.13	2.03	0.860
60x30	5 3.36		5.00	2.50	2.17	0.684	15.6	2.63	16.5	1.71	1.91	0.784	1.97	0.633
50x30	5 2.96		5.00	2.50	1.73	0.741	9.36	2.51	10.3	1.54	1.57	0.816	1.65	0.639
45x30	4 2.25		4.50	2.25	1.48	0.740	5.78	2.05	6.65	1.18	1.42	0.850	1.52	0.640
40x25	4 1.93		4.00	2.00	1.36	0.623	3.89	1.16	4.35	0.700	1.26	0.687	1.33	0.534
40x20	4 1.77		4.00	2.00	1.47	0.480	3.59	0.600	3.80	0.393	1.26	0.514	1.30	0.417
30x20	4 1.46		4.00	2.00	1.03	0.541	1.59	0.553	1.81	0.330	0.925	0.546	0.988	0.421
	3 1.12		4.00	2.00	0.990	0.502	1.25	0.437	1.43	0.256	0.935	0.553	1.00	0.424

Advance and UKA are trademarks of Tata. A fuller description of the relationship between Angles and the Advance range of sections manufactured by Tata is given in section 12.

+ These sections are in addition to the range of BS EN 10056-1 sections.

c_x is the distance from the back of the short leg to the centre of gravity.

c_y is the distance from the back of the long leg to the centre of gravity.

FOR EXPLANATION OF TABLES SEE NOTES 2 AND 3

BS EN 1993-1-1:2005
BS EN 10056-1:1999

UNEQUAL ANGLES

Advance UKA - Unequal Angles

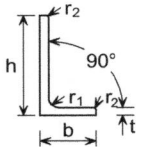

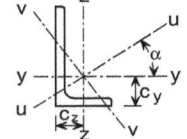

Dimensions and properties (continued)

Section Designation		Elastic Modulus		Angle Axis y-y to	Torsional Constant	Equivalent Slenderness Coefficient		Mono-symmetry Index	Area of Section
Size	Thickness	Axis y-y	Axis z-z	Axis u-u		Min	Max		
h x b	t			Tan α	I_T	ϕ_a	ϕ_a	ψ_a	
mm	mm	cm³	cm³		cm⁴				cm²
200x150	18 +	174	103	0.549	67.9	2.93	3.72	4.60	60.0
	15	147	86.9	0.551	39.9	3.53	4.50	5.55	50.5
	12	119	70.5	0.552	20.9	4.43	5.70	6.97	40.8
200x100	15	137	38.5	0.260	34.3	3.54	5.17	9.19	43.0
	12	111	31.3	0.262	18.0	4.42	6.57	11.5	34.8
	10	93.2	26.3	0.263	10.66	5.26	7.92	13.9	29.2
150x90	15	77.7	30.4	0.354	26.8	2.58	3.59	5.96	33.9
	12	63.3	24.8	0.358	14.1	3.24	4.58	7.50	27.5
	10	53.3	21.0	0.360	8.30	3.89	5.56	9.03	23.2
150x75	15	75.2	21.0	0.253	25.1	2.62	3.74	6.84	31.7
	12	61.3	17.1	0.258	13.2	3.30	4.79	8.60	25.7
	10	51.6	14.5	0.261	7.80	3.95	5.83	10.4	21.7
125x75	12	43.2	16.9	0.354	11.6	2.66	3.73	6.23	22.7
	10	36.5	14.3	0.357	6.87	3.21	4.55	7.50	19.1
	8	29.6	11.6	0.360	3.62	4.00	5.75	9.43	15.5
100x75	12	28.0	16.5	0.540	10.05	2.10	2.64	3.46	19.7
	10	23.8	14.0	0.544	5.95	2.54	3.22	4.17	16.6
	8	19.3	11.4	0.547	3.13	3.18	4.08	5.24	13.5
100x65	10 +	23.2	10.5	0.410	5.61	2.52	3.43	5.45	15.6
	8 +	18.9	8.54	0.413	2.96	3.14	4.35	6.86	12.7
	7 +	16.6	7.53	0.415	2.02	3.58	5.00	7.85	11.2
100x50	8 18.2		5.08	0.258	2.61	3.30	4.80	8.61	11.4
	6 13.8		3.89	0.262	1.14	4.38	6.52	11.6	8.71
80x60	7 10.7		6.34	0.546	1.66	2.92	3.72	4.78	9.38
80x40	8 11.4		3.16	0.253	2.05	2.61	3.73	6.85	9.01
	6 8.73		2.44	0.258	0.899	3.48	5.12	9.22	6.89
75x50	8 10.4		4.95	0.430	2.14	2.36	3.18	4.92	9.41
	6 8.01		3.81	0.435	0.935	3.18	4.34	6.60	7.19
70x50	6 7.01		3.78	0.500	0.899	2.96	3.89	5.44	6.89
65x50	5 5.14		3.19	0.577	0.498	3.38	4.26	5.08	5.54
60x40	6 5.03		2.38	0.431	0.735	2.51	3.39	5.26	5.68
	5 4.25		2.02	0.434	0.435	3.02	4.11	6.34	4.79
60x30	5 4.07		1.14	0.257	0.382	3.15	4.56	8.26	4.28
50x30	5 2.86		1.11	0.352	0.340	2.51	3.52	5.99	3.78
45x30	4 1.91		0.910	0.436	0.166	2.85	3.87	5.92	2.87
40x25	4 1.47		0.619	0.380	0.142	2.51	3.48	5.75	2.46
40x20	4 1.42		0.393	0.252	0.131	2.57	3.68	6.86	2.26
30x20	4 0.807		0.379	0.421	0.1096	1.79	2.39	3.95	1.86
	3 0.621		0.292	0.427	0.0486	2.40	3.28	5.31	1.43

Advance and UKA are trademarks of Tata. A fuller description of the relationship between Angles and the Advance range of sections manufactured by Tata is given in section 12.

+ These sections are in addition to the range of BS EN 10056-1 sections.

FOR EXPLANATION OF TABLES SEE NOTES 2 AND 3

BS EN 1993-1-1:2005
BS EN 10056-1:1999

EQUAL ANGLES BACK TO BACK

Advance UKA - Equal Angles BACK TO BACK

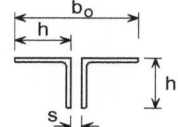

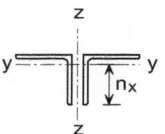

Dimensions and properties

Composed of Two Angles		Total Mass per Metre	Distance	Total Area	Properties about Axis y-y			Radius of Gyration i_z about Axis z-z (cm)				
h x h	t		n_y		I_y	i_y	$W_{el,y}$	Space between angles, s, (mm)				
mm	mm	kg/m	cm	cm^2	cm^4	cm	cm^3	0	8	10	12	15
200x200	24	142	14.2	181	6660	6.06	470	8.42	8.70	8.77	8.84	8.95
	20	120	14.3	153	5700	6.11	398	8.34	8.62	8.69	8.76	8.87
	18	109	14.4	138	5200	6.13	362	8.31	8.58	8.65	8.72	8.83
	16	97.0	14.5	124	4680	6.16	324	8.27	8.54	8.61	8.68	8.79
150x150	18 +	80.2	10.6	102	2120	4.55	200	6.32	6.60	6.67	6.75	6.86
	15	67.6	10.8	86.0	1800	4.57	167	6.24	6.52	6.59	6.66	6.77
	12	54.6	10.9	69.6	1470	4.60	135	6.18	6.45	6.52	6.59	6.70
	10	46.0	11.0	58.6	1250	4.62	114	6.13	6.40	6.47	6.54	6.64
120x120	15 +	53.2	8.48	68.0	896	3.63	106	5.06	5.34	5.42	5.49	5.60
	12	43.2	8.60	55.0	736	3.65	85.4	4.99	5.27	5.35	5.42	5.53
	10	36.4	8.69	46.4	626	3.67	72.0	4.94	5.22	5.29	5.36	5.47
	8 +	29.4	8.76	37.6	518	3.71	59.0	4.93	5.20	5.27	5.34	5.45
100x100	15 +	43.8	6.98	56.0	500	2.99	71.6	4.25	4.54	4.62	4.69	4.81
	12	35.6	7.10	45.4	414	3.02	58.2	4.19	4.47	4.55	4.62	4.74
	10	30.0	7.18	38.4	354	3.04	49.2	4.14	4.43	4.50	4.57	4.69
	8	24.4	7.26	31.0	290	3.06	39.8	4.11	4.38	4.46	4.53	4.64
90x90	12 +	31.8	6.34	40.6	298	2.71	47.0	3.80	4.09	4.16	4.24	4.36
	10	26.8	6.42	34.2	254	2.72	39.6	3.75	4.04	4.11	4.19	4.30
	8	21.8	6.50	27.8	208	2.74	32.2	3.71	3.99	4.06	4.13	4.25
	7	19.2	6.55	24.4	185	2.75	28.2	3.69	3.96	4.04	4.11	4.22
80x80	10	23.8	5.66	30.2	175	2.41	30.8	3.36	3.65	3.72	3.80	3.92
	8	19.3	5.74	24.6	144	2.43	25.2	3.31	3.60	3.67	3.75	3.86
75x75	8	18.0	5.36	22.8	118	2.27	22.0	3.12	3.41	3.49	3.56	3.68
	6	13.7	5.45	17.5	91.6	2.29	16.8	3.07	3.35	3.43	3.50	3.62
70x70	7	14.8	5.03	18.8	84.6	2.12	16.8	2.89	3.18	3.26	3.33	3.45
	6	12.8	5.07	16.3	73.8	2.13	14.5	2.87	3.16	3.23	3.31	3.42
65x65	7	13.7	4.45	17.5	66.8	1.96	14.4	2.83	3.14	3.21	3.29	3.42
60x60	8	14.2	4.23	18.1	58.4	1.80	13.8	2.52	2.82	2.90	2.97	3.10
	6	10.8	4.31	13.8	45.6	1.82	10.6	2.48	2.77	2.85	2.92	3.04
	5	9.14	4.36	11.6	38.8	1.82	8.90	2.45	2.74	2.81	2.89	3.01
50x50	6	8.94	3.55	11.4	25.6	1.50	7.22	2.09	2.38	2.46	2.54	2.66
	5	7.54	3.60	9.60	22.0	1.51	6.10	2.06	2.35	2.43	2.51	2.63
	4	6.12	3.64	7.78	17.9	1.52	4.92	2.04	2.32	2.40	2.48	2.60

Advance and UKA are trademarks of Tata. A fuller description of the relationship between Angles and the Advance range of sections manufactured by Tata is given in section 12.

+ These sections are in addition to the range of BS EN 10056-1 sections.

Properties about y-y axis:

$I_z = \text{(Total Area)}.(i_z)^2$

$W_{el,z} = I_z / (0.5 b_o)$

FOR EXPLANATION OF TABLES SEE NOTES 2 AND 3

BS EN 1993-1-1:2005
BS EN 10056-1:1999

UNEQUAL ANGLES BACK TO BACK

Advance UKA - Unequal Angles BACK TO BACK

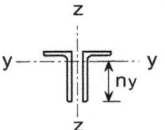

Dimensions and properties

Composed of Two Angles		Total Mass per Metre	Distance	Total Area	Properties about Axis y-y			Radius of Gyration i_z about Axis z-z (cm)				
h x b	t		n_y		I_y	i_y	$W_{el,y}$	Space between angles, s, (mm)				
mm	mm	kg/m	cm	cm²	cm⁴	cm	cm³	0	8	10	12	15
200x150	18 +	94.2	13.7	120	4750	6.29	348	5.84	6.11	6.18	6.25	6.36
	15	79.2	13.8	101	4040	6.33	294	5.77	6.04	6.11	6.18	6.28
	12	64.0	13.9	81.6	3300	6.36	238	5.72	5.98	6.05	6.12	6.22
200x100	15	67.5	12.8	86.0	3520	6.40	274	3.45	3.72	3.79	3.86	3.97
	12	54.6	13.0	69.6	2880	6.43	222	3.39	3.65	3.72	3.79	3.90
	10	46.0	13.1	58.4	2440	6.46	186	3.35	3.61	3.67	3.74	3.85
150x90	15	53.2	9.79	67.8	1522	4.74	155	3.32	3.60	3.67	3.75	3.86
	12	43.2	9.92	55.0	1250	4.77	127	3.27	3.55	3.62	3.69	3.80
	10	36.4	10.0	46.4	1070	4.80	107	3.23	3.50	3.57	3.64	3.75
150x75	15	49.6	9.48	63.4	1430	4.75	150	2.65	2.94	3.01	3.09	3.21
	12	40.4	9.60	51.4	1180	4.78	123	2.59	2.87	2.94	3.02	3.14
	10	34.0	9.69	43.4	1000	4.81	103	2.56	2.83	2.90	2.97	3.08
125x75	12	35.6	8.19	45.4	708	3.95	86.4	2.76	3.04	3.11	3.19	3.30
	10	30.0	8.27	38.2	604	3.97	73.0	2.72	2.99	3.07	3.14	3.26
	8	24.4	8.36	31.0	494	4.00	59.2	2.68	2.95	3.02	3.09	3.20
100x75	12	30.8	6.73	39.4	378	3.10	56.0	2.95	3.24	3.31	3.39	3.51
	10	26.0	6.81	33.2	324	3.12	47.6	2.91	3.19	3.27	3.34	3.46
	8	21.2	6.90	27.0	266	3.14	38.6	2.87	3.15	3.22	3.29	3.41
100x65	10 +	24.6	6.64	31.2	308	3.14	46.4	2.43	2.72	2.79	2.87	2.99
	8 +	19.9	6.73	25.4	254	3.16	37.8	2.39	2.67	2.74	2.82	2.93
	7 +	17.5	6.77	22.4	226	3.17	33.2	2.37	2.65	2.72	2.79	2.91
100x50	8	17.9	6.40	22.8	232	3.19	36.4	1.73	2.02	2.09	2.17	2.29
	6	13.7	6.49	17.4	180	3.21	27.6	1.69	1.97	2.04	2.12	2.24
80x60	7	14.7	5.49	18.8	118	2.51	21.4	2.31	2.59	2.67	2.74	2.86
80x40	8	14.1	5.06	18.0	115	2.53	22.8	1.41	1.71	1.79	1.87	2.00
	6	10.8	5.15	13.8	89.8	2.55	17.5	1.37	1.66	1.74	1.82	1.94
75x50	8	14.8	4.98	18.8	104	2.35	20.8	1.90	2.19	2.27	2.35	2.47
	6	11.3	5.06	14.4	81.0	2.37	16.0	1.86	2.14	2.22	2.30	2.42
70x50	6	10.8	4.77	13.8	66.8	2.20	14.0	1.90	2.19	2.26	2.34	2.46
65x50	5	8.70	4.51	11.1	46.4	2.05	10.3	1.93	2.21	2.28	2.36	2.48
60x40	6	8.92	4.00	11.4	40.2	1.88	10.1	1.51	1.80	1.88	1.96	2.09
	5	7.52	4.04	9.58	34.4	1.89	8.50	1.49	1.78	1.86	1.94	2.06

Advance and UKA are trademarks of Tata. A fuller description of the relationship between Angles and the Advance range of sections manufactured by Tata is given in section 12.

+ These sections are in addition to the range of BS EN 10056-1 sections.

Properties about y-y axis:

$I_z = (Total\ Area).(i_z)^2$

$W_{el,z} = I_z / (0.5b_o)$

FOR EXPLANATION OF TABLES SEE NOTES 2 AND 3

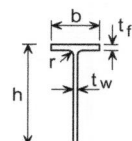

STRUCTURAL TEES CUT FROM UNIVERSAL BEAMS

Advance UKT split from Advance UKB

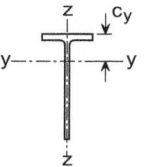

BS EN 1993-1-1:2005
BS 4-1:2005

Dimensions and properties

Section Designation	Cut from Universal Beam Section Designation	Mass per Metre	Width of Section	Depth of Section	Thickness Web	Thickness Flange	Root Radius	Ratios for Local Buckling Flange	Ratios for Local Buckling Web	Dimension	Second Moment of Area Axis y-y	Second Moment of Area Axis z-z
		kg/m	b mm	h mm	t_w mm	t_f mm	r mm	c_f/t_f	c_w/t_w	c_y cm	cm^4	cm^4
305x457x127	914x305x253	126.7	305.5	459.1	17.3	27.9	19.1	5.47	26.5	12.0	32700	6650
305x457x112	914x305x224	112.1	304.1	455.1	15.9	23.9	19.1	6.36	28.6	12.1	29100	5620
305x457x101	914x305x201	100.4	303.3	451.4	15.1	20.2	19.1	7.51	29.9	12.5	26400	4710
292x419x113	838x292x226	113.3	293.8	425.4	16.1	26.8	17.8	5.48	26.4	10.8	24600	5680
292x419x97	838x292x194	96.9	292.4	420.3	14.7	21.7	17.8	6.74	28.6	11.1	21300	4530
292x419x88	838x292x176	87.9	291.7	417.4	14.0	18.8	17.8	7.76	29.8	11.4	19600	3900
267x381x99	762x267x197	98.4	268.0	384.8	15.6	25.4	16.5	5.28	24.7	9.89	17500	4090
267x381x87	762x267x173	86.5	266.7	381.0	14.3	21.6	16.5	6.17	26.6	9.98	15500	3430
267x381x74	762x267x147	73.5	265.2	376.9	12.8	17.5	16.5	7.58	29.4	10.2	13200	2730
267x381x67	762x267x134	66.9	264.4	374.9	12.0	15.5	16.5	8.53	31.2	10.3	12100	2390
254x343x85	686x254x170	85.1	255.8	346.4	14.5	23.7	15.2	5.40	23.9	8.67	12100	3320
254x343x76	686x254x152	76.2	254.5	343.7	13.2	21.0	15.2	6.06	26.0	8.61	10800	2890
254x343x70	686x254x140	70.0	253.7	341.7	12.4	19.0	15.2	6.68	27.6	8.63	9910	2590
254x343x63	686x254x125	62.6	253.0	338.9	11.7	16.2	15.2	7.81	29.0	8.85	8980	2190
305x305x119	610x305x238	119.0	311.4	317.9	18.4	31.4	16.5	4.96	17.3	7.11	12400	7920
305x305x90	610x305x179	89.5	307.1	310.0	14.1	23.6	16.5	6.51	22.0	6.69	9040	5700
305x305x75	610x305x149	74.6	304.8	306.1	11.8	19.7	16.5	7.74	25.9	6.45	7410	4650
229x305x70	610x229x140	69.9	230.2	308.5	13.1	22.1	12.7	5.21	23.5	7.61	7740	2250
229x305x63	610x229x125	62.5	229.0	306.0	11.9	19.6	12.7	5.84	25.7	7.54	6900	1970
229x305x57	610x229x113	56.5	228.2	303.7	11.1	17.3	12.7	6.60	27.4	7.58	6270	1720
229x305x51	610x229x101	50.6	227.6	301.2	10.5	14.8	12.7	7.69	28.7	7.78	5690	1460
178x305x50 +	610x178x100	50.1	179.2	303.7	11.3	17.2	12.7	5.21	26.9	8.57	5890	829
178x305x46 +	610x178x92	46.1	178.8	301.5	10.9	15.0	12.7	5.96	27.7	8.78	5450	718
178x305x41 +	610x178x82	40.9	177.9	299.3	10.0	12.8	12.7	6.95	29.9	8.88	4840	603
312x267x136 +	533x312x272	136.6	320.2	288.8	21.1	37.6	12.7	4.26	13.7	6.28	10600	10300
312x267x110 +	533x312x219	109.4	317.4	280.4	18.3	29.2	12.7	5.43	15.3	6.09	8530	7790
312x267x91 +	533x312x182	90.7	314.5	275.6	15.2	24.4	12.7	6.44	18.1	5.78	6890	6330
312x267x75 +	533x312x151	75.3	312.0	271.5	12.7	20.3	12.7	7.68	21.4	5.54	5620	5140

Advance, UKT and UKB are trademarks of Tata. A fuller description of the relationship between Structural Tees and the Advance range of sections manufactured by Tata is given in section 12.

+ These sections are in addition to the range of BS 4 sections

FOR EXPLANATION OF TABLES SEE NOTES 2 AND 3

BS EN 1993-1-1:2005
BS 4-1:2005

STRUCTURAL TEES CUT FROM UNIVERSAL BEAMS

Advance UKT split from Advance UKB

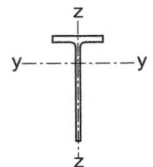

Properties (continued)

Section Designation	Radius of Gyration Axis y-y	Radius of Gyration Axis z-z	Elastic Modulus Axis y-y Flange	Elastic Modulus Axis y-y Toe	Elastic Modulus Axis z-z	Plastic Modulus Axis y-y	Plastic Modulus Axis z-z	Buckling Parameter U	Torsional Index X	Mono-symmetry Index ψ	Warping Constant (*) I_w	Torsional Constant I_t	Area of Section A
	cm	cm	cm³	cm³	cm³	cm³	cm³				cm⁶	cm⁴	cm²
305x457x127	14.2	6.42	2720	965	435	1730	685	0.656	18.1	0.749	17000	313	161
305x457x112	14.3	6.27	2400	871	369	1570	582	0.666	20.6	0.753	12400	211	143
305x457x101	14.4	6.07	2110	808	311	1460	491	0.685	23.4	0.759	9820	146	128
292x419x113	13.1	6.27	2280	776	387	1380	606	0.640	17.5	0.742	11500	257	144
292x419x97	13.1	6.06	1930	689	310	1240	487	0.660	20.8	0.747	7830	153	123
292x419x88	13.2	5.90	1720	644	267	1160	421	0.675	23.2	0.751	6320	111	112
267x381x99	11.8	5.71	1770	613	305	1090	479	0.641	16.6	0.741	7620	202	125
267x381x87	11.9	5.58	1550	550	257	986	404	0.654	19.0	0.745	5450	134	110
267x381x74	11.9	5.40	1300	481	206	867	324	0.670	22.6	0.749	3600	79.5	93.6
267x381x67	11.9	5.30	1180	445	181	806	285	0.679	24.9	0.753	2850	59.2	85.3
254x343x85	10.5	5.53	1390	464	259	826	406	0.624	15.9	0.731	4720	154	108
254x343x76	10.5	5.46	1250	417	227	743	355	0.627	17.7	0.732	3420	110	97.0
254x343x70	10.5	5.39	1150	388	204	691	319	0.633	19.3	0.734	2720	84.3	89.2
254x343x63	10.6	5.24	1010	358	173	643	271	0.651	21.9	0.740	2090	57.9	79.7
305x305x119	9.03	7.23	1740	501	509	894	787	0.483	10.6	0.662	11300	391	152
305x305x90	8.91	7.07	1350	372	371	656	572	0.484	13.8	0.664	4710	170	114
305x305x75	8.83	7.00	1150	307	305	538	469	0.483	16.4	0.666	2690	99.8	95.0
229x305x70	9.32	5.03	1020	333	196	592	306	0.613	15.3	0.727	2560	108	89.1
229x305x63	9.31	4.97	915	299	172	531	268	0.617	17.1	0.728	1840	76.9	79.7
229x305x57	9.33	4.88	826	275	150	489	235	0.626	19.0	0.731	1400	55.5	72.0
229x305x51	9.40	4.76	732	255	128	456	200	0.644	21.6	0.736	1080	38.3	64.4
178x305x50 +	9.60	3.60	688	270	92.5	490	148	0.694	19.4	0.768	1230	47.3	63.9
178x305x46 +	9.64	3.50	621	255	80.3	468	129	0.710	21.5	0.774	1050	35.3	58.7
178x305x41 +	9.64	3.40	545	230	67.8	425	109	0.722	24.3	0.778	780	24.3	52.1
312x267x136 +	7.81	7.69	1690	469	644	857	993	0.247	7.96	0.613	17300	642	174
312x267x110 +	7.82	7.48	1400	389	491	696	757	0.332	9.93	0.617	8730	320	139
312x267x91 +	7.72	7.40	1190	317	403	562	619	0.324	11.7	0.618	4920	186	116
312x267x75 +	7.65	7.32	1010	260	330	458	505	0.326	14.0	0.619	2780	108	95.9

Advance, UKT and UKB are trademarks of Tata. A fuller description of the relationship between Structural Tees and the Advance range of sections manufactured by Tata is given in section 12.

+ These sections are in addition to the range of BS 4 sections

(*) Note units are cm⁶ and not dm⁶.

FOR EXPLANATION OF TABLES SEE NOTES 2 AND 3

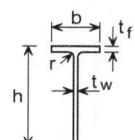

BS EN 1993-1-1:2005
BS 4-1:2005

STRUCTURAL TEES CUT FROM UNIVERSAL BEAMS

Advance UKT split from Advance UKB

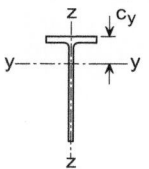

Dimensions and properties

Section Designation	Cut from Universal Beam Section Designation	Mass per Metre	Width of Section	Depth of Section	Thickness Web	Thickness Flange	Root Radius	Ratios for Local Buckling Flange	Ratios for Local Buckling Web	Dimension	Second Moment of Area Axis y-y	Second Moment of Area Axis z-z
		kg/m	b mm	h mm	t_w mm	t_f mm	r mm	c_f/t_f	c_w/t_w	c_y cm	cm^4	cm^4
210x267x69 +	533x210x138	69.1	213.9	274.5	14.7	23.6	12.7	4.53	18.7	6.94	5990	1930
210x267x61	533x210x122	61.0	211.9	272.2	12.7	21.3	12.7	4.97	21.4	6.66	5160	1690
210x267x55	533x210x109	54.5	210.8	269.7	11.6	18.8	12.7	5.61	23.3	6.61	4600	1470
210x267x51	533x210x101	50.5	210.0	268.3	10.8	17.4	12.7	6.03	24.8	6.53	4250	1350
210x267x46	533x210x92	46.0	209.3	266.5	10.1	15.6	12.7	6.71	26.4	6.55	3880	1190
210x267x41	533x210x82	41.1	208.8	264.1	9.6	13.2	12.7	7.91	27.5	6.75	3530	1000
165x267x43 +	533x165x85	42.3	166.5	267.1	10.3	16.5	12.7	5.05	25.9	7.23	3750	637
165x267x37 +	533x165x75	37.3	165.9	264.5	9.7	13.6	12.7	6.10	27.3	7.46	3350	520
165x267x33 +	533x165x66	32.8	165.1	262.4	8.9	11.4	12.7	7.24	29.5	7.59	2960	429
191x229x81 +	457x191x161	80.7	199.4	246.0	18.0	32.0	10.2	3.12	13.7	6.22	5160	2130
191x229x67 +	457x191x133	66.6	196.7	240.3	15.3	26.3	10.2	3.74	15.7	5.96	4180	1670
191x229x53 +	457x191x106	52.9	194.0	234.6	12.6	20.6	10.2	4.71	18.6	5.73	3260	1260
191x229x49	457x191x98	49.1	192.8	233.5	11.4	19.6	10.2	4.92	20.5	5.53	2970	1170
191x229x45	457x191x89	44.6	191.9	231.6	10.5	17.7	10.2	5.42	22.1	5.47	2680	1040
191x229x41	457x191x82	41.0	191.3	229.9	9.9	16.0	10.2	5.98	23.2	5.47	2470	935
191x229x37	457x191x74	37.1	190.4	228.4	9.0	14.5	10.2	6.57	25.4	5.38	2220	836
191x229x34	457x191x67	33.5	189.9	226.6	8.5	12.7	10.2	7.48	26.7	5.46	2030	726
152x229x41	457x152x82	41.0	155.3	232.8	10.5	18.9	10.2	4.11	22.2	5.96	2600	592
152x229x37	457x152x74	37.1	154.4	230.9	9.6	17.0	10.2	4.54	24.1	5.88	2330	523
152x229x34	457x152x67	33.6	153.8	228.9	9.0	15.0	10.2	5.13	25.4	5.91	2120	456
152x229x30	457x152x60	29.9	152.9	227.2	8.1	13.3	10.2	5.75	28.0	5.84	1880	397
152x229x26	457x152x52	26.1	152.4	224.8	7.6	10.9	10.2	6.99	29.6	6.04	1670	322
178x203x43 +	406x178x85	42.6	181.9	208.6	10.9	18.2	10.2	5.00	19.1	4.91	2030	915
178x203x37	406x178x74	37.1	179.5	206.3	9.5	16.0	10.2	5.61	21.7	4.76	1740	773
178x203x34	406x178x67	33.5	178.8	204.6	8.8	14.3	10.2	6.25	23.3	4.73	1570	682
178x203x30	406x178x60	30.0	177.9	203.1	7.9	12.8	10.2	6.95	25.7	4.64	1400	602
178x203x27	406x178x54	27.0	177.7	201.2	7.7	10.9	10.2	8.15	26.1	4.83	1290	511
140x203x27 +	406x140x53	26.6	143.3	203.3	7.9	12.9	10.2	5.55	25.7	5.16	1320	317
140x203x23	406x140x46	23.0	142.2	201.5	6.8	11.2	10.2	6.35	29.6	5.02	1120	269
140x203x20	406x140x39	19.5	141.8	198.9	6.4	8.6	10.2	8.24	31.1	5.32	979	205

Advance, UKT and UKB are trademarks of Tata. A fuller description of the relationship between Structural Tees and the Advance range of sections manufactured by Tata is given in section 12.

+ These sections are in addition to the range of BS 4 sections

FOR EXPLANATION OF TABLES SEE NOTES 2 AND 3

BS EN 1993-1-1:2005
BS 4-1:2005

STRUCTURAL TEES CUT FROM UNIVERSAL BEAMS

Advance UKT split from Advance UKB

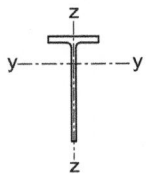

Properties (continued)

Section Designation	Radius of Gyration		Elastic Modulus			Plastic Modulus		Buckling Parameter	Torsional Index	Mono-symmetry Index	Warping Constant (*)	Torsional Constant	Area of Section
	Axis y-y	Axis z-z	Axis y-y Flange	Axis y-y Toe	Axis z-z	Axis y-y	Axis z-z	U	X	ψ	I_w	I_t	A
	cm	cm	cm^3	cm^3	cm^3	cm^3	cm^3				cm^6	cm^4	cm^2
210x267x69 +	8.24	4.68	862	292	181	520	284	0.609	12.5	0.719	2490	125	88.1
210x267x61	8.15	4.67	775	251	160	446	250	0.600	13.8	0.719	1660	88.9	77.7
210x267x55	8.14	4.60	697	226	140	401	218	0.605	15.5	0.721	1200	63.0	69.4
210x267x51	8.12	4.57	650	209	128	371	200	0.606	16.6	0.722	951	50.3	64.3
210x267x46	8.14	4.51	593	193	114	343	178	0.613	18.3	0.724	737	37.7	58.7
210x267x41	8.21	4.38	523	179	96.1	320	150	0.634	20.8	0.730	565	25.7	52.3
165x267x43 +	8.34	3.44	519	192	76.6	346	122	0.672	17.7	0.758	670	36.8	54.0
165x267x37 +	8.39	3.30	449	176	62.7	321	100	0.693	20.6	0.765	514	23.9	47.6
165x267x33 +	8.41	3.20	390	159	52.0	291	83.1	0.708	23.6	0.771	378	15.9	41.9
191x229x81 +	7.09	4.55	830	281	213	507	336	0.573	8.24	0.699	3780	256	103
191x229x67 +	7.01	4.44	702	231	170	414	267	0.576	9.82	0.702	2130	146	84.9
191x229x53 +	6.96	4.32	569	184	130	328	203	0.583	12.2	0.706	1070	72.6	67.4
191x229x49	6.88	4.33	536	167	122	296	189	0.573	12.9	0.705	835	60.5	62.6
191x229x45	6.87	4.29	491	152	109	269	169	0.576	14.1	0.706	628	45.2	56.9
191x229x41	6.88	4.23	452	141	97.8	250	152	0.583	15.5	0.709	494	34.5	52.2
191x229x37	6.86	4.20	413	127	87.8	225	136	0.583	16.9	0.709	365	25.8	47.3
191x229x34	6.90	4.12	372	118	76.5	209	119	0.597	18.9	0.713	280	18.5	42.7
152x229x41	7.05	3.37	436	150	76.3	267	120	0.634	13.7	0.740	534	44.5	52.3
152x229x37	7.03	3.33	397	135	67.8	242	107	0.636	15.1	0.742	396	32.9	47.2
152x229x34	7.04	3.27	359	125	59.3	223	93.3	0.646	16.8	0.745	305	23.8	42.8
152x229x30	7.02	3.23	322	111	52.0	199	81.5	0.648	18.8	0.746	217	16.9	38.1
152x229x26	7.08	3.11	276	102	42.3	183	66.6	0.671	22.0	0.753	161	10.7	33.3
178x203x43 +	6.11	4.11	413	127	101	226	157	0.556	12.2	0.694	538	46.3	54.3
178x203x37	6.06	4.04	365	109	86.1	194	133	0.555	13.8	0.696	350	31.3	47.2
178x203x34	6.07	3.99	332	100	76.3	177	118	0.561	15.2	0.698	262	23.0	42.8
178x203x30	6.04	3.97	301	89.0	67.6	157	104	0.561	16.9	0.699	186	16.6	38.3
178x203x27	6.13	3.85	268	84.6	57.5	150	89.1	0.588	19.2	0.705	146	11.5	34.5
140x203x27 +	6.23	3.06	256	87.0	44.3	155	69.5	0.636	17.1	0.739	148	14.4	34.0
140x203x23	6.19	3.03	224	74.2	37.8	132	59.0	0.633	19.5	0.740	93.7	9.49	29.3
140x203x20	6.28	2.87	184	67.2	28.9	121	45.4	0.668	23.8	0.750	66.3	5.33	24.8

Advance, UKT and UKB are trademarks of Tata. A fuller description of the relationship between Structural Tees and the Advance range of sections manufactured by Tata is given in section 12.

+ These sections are in addition to the range of BS 4 sections

(*) Note units are cm^6 and not dm^6.

FOR EXPLANATION OF TABLES SEE NOTES 2 AND 3

BS EN 1993-1-1:2005
BS 4-1:2005

STRUCTURAL TEES CUT FROM UNIVERSAL BEAMS

Advance UKT split from Advance UKB

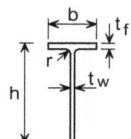

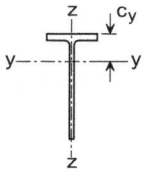

Dimensions and properties

Section Designation	Cut from Universal Beam Section Designation	Mass per Metre	Width of Section	Depth of Section	Thickness		Root Radius	Ratios for Local Buckling		Dimension	Second Moment of Area	
					Web	Flange		Flange	Web		Axis y-y	Axis z-z
		kg/m	b mm	h mm	t_w mm	t_f mm	r mm	c_f/t_f	c_w/t_w	c_y cm	cm^4	cm^4
171x178x34	356x171x67	33.5	173.2	181.6	9.1	15.7	10.2	5.52	20.0	4.00	1150	681
171x178x29	356x171x57	28.5	172.2	178.9	8.1	13.0	10.2	6.62	22.1	3.97	986	554
171x178x26	356x171x51	25.5	171.5	177.4	7.4	11.5	10.2	7.46	24.0	3.94	882	484
171x178x23	356x171x45	22.5	171.1	175.6	7.0	9.7	10.2	8.82	25.1	4.05	798	406
127x178x20	356x127x39	19.5	126.0	176.6	6.6	10.7	10.2	5.89	26.8	4.43	728	179
127x178x17	356x127x33	16.5	125.4	174.4	6.0	8.5	10.2	7.38	29.1	4.56	626	140
165x152x27	305x165x54	27.0	166.9	155.1	7.9	13.7	8.9	6.09	19.6	3.21	642	531
165x152x23	305x165x46	23.0	165.7	153.2	6.7	11.8	8.9	7.02	22.9	3.07	536	448
165x152x20	305x165x40	20.1	165.0	151.6	6.0	10.2	8.9	8.09	25.3	3.03	468	382
127x152x24	305x127x48	24.0	125.3	155.4	9.0	14.0	8.9	4.48	17.3	3.94	662	231
127x152x21	305x127x42	20.9	124.3	153.5	8.0	12.1	8.9	5.14	19.2	3.87	573	194
127x152x19	305x127x37	18.5	123.4	152.1	7.1	10.7	8.9	5.77	21.4	3.78	501	168
102x152x17	305x102x33	16.4	102.4	156.3	6.6	10.8	7.6	4.74	23.7	4.14	487	97.1
102x152x14	305x102x28	14.1	101.8	154.3	6.0	8.8	7.6	5.78	25.7	4.20	420	77.7
102x152x13	305x102x25	12.4	101.6	152.5	5.8	7.0	7.6	7.26	26.3	4.43	377	61.5
146x127x22	254x146x43	21.5	147.3	129.7	7.2	12.7	7.6	5.80	18.0	2.64	343	339
146x127x19	254x146x37	18.5	146.4	127.9	6.3	10.9	7.6	6.72	20.3	2.55	292	285
146x127x16	254x146x31	15.5	146.1	125.6	6.0	8.6	7.6	8.49	20.9	2.66	259	224
102x127x14	254x102x28	14.1	102.2	130.1	6.3	10.0	7.6	5.11	20.7	3.24	277	89.3
102x127x13	254x102x25	12.6	101.9	128.5	6.0	8.4	7.6	6.07	21.4	3.32	250	74.3
102x127x11	254x102x22	11.0	101.6	126.9	5.7	6.8	7.6	7.47	22.3	3.45	223	59.7
133x102x15	203x133x30	15.0	133.9	103.3	6.4	9.6	7.6	6.97	16.1	2.11	154	192
133x102x13	203x133x25	12.5	133.2	101.5	5.7	7.8	7.6	8.54	17.8	2.10	131	154

Advance, UKT and UKB are trademarks of Tata. A fuller description of the relationship between Structural Tees and the Advance range of sections manufactured by Tata is given in section 12.

FOR EXPLANATION OF TABLES SEE NOTES 2 AND 3

BS EN 1993-1-1:2005
BS 4-1:2005

STRUCTURAL TEES CUT FROM UNIVERSAL BEAMS

Advance UKT split from Advance UKB

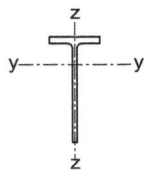

Properties (continued)

Section Designation	Radius of Gyration		Elastic Modulus			Plastic Modulus		Buckling Parameter	Torsional Index	Mono-symmetry Index	Warping Constant (*)	Torsional Constant	Area of Section
	Axis y-y	Axis z-z	Axis		Axis z-z	Axis y-y	Axis z-z						
			y-y Flange	y-y Toe	z-z	y-y	z-z	U	X	ψ	I_w	I_t	A
	cm	cm	cm^3	cm^3	cm^3	cm^3	cm^3				cm^6	cm^4	cm^2
171x178x34	5.20	3.99	288	81.5	78.6	145	121	0.500	12.2	0.672	249	27.8	42.7
171x178x29	5.21	3.91	248	70.9	64.4	125	99.4	0.514	14.4	0.676	154	16.6	36.3
171x178x26	5.21	3.86	224	63.9	56.5	113	87.1	0.521	16.1	0.677	110	11.9	32.4
171x178x23	5.28	3.76	197	59.1	47.4	104	73.3	0.546	18.4	0.683	79.2	7.90	28.7
127x178x20	5.41	2.68	164	55.0	28.4	98.0	44.5	0.632	17.6	0.739	57.1	7.53	24.9
127x178x17	5.45	2.58	137	48.6	22.3	87.2	35.1	0.655	21.1	0.746	38.0	4.38	21.1
165x152x27	4.32	3.93	200	52.2	63.7	92.8	97.8	0.389	11.8	0.636	128	17.3	34.4
165x152x23	4.27	3.91	174	43.7	54.1	77.1	82.8	0.380	13.6	0.636	78.6	11.1	29.4
165x152x20	4.27	3.86	155	38.6	46.3	67.6	70.9	0.393	15.5	0.638	52.0	7.35	25.7
127x152x24	4.65	2.74	168	57.1	36.8	102	58.0	0.602	11.7	0.714	104	15.8	30.6
127x152x21	4.63	2.70	148	49.9	31.3	88.9	49.2	0.606	13.3	0.716	69.2	10.5	26.7
127x152x19	4.61	2.67	132	43.8	27.2	77.9	42.7	0.606	14.9	0.718	47.4	7.36	23.6
102x152x17	4.82	2.15	118	42.3	19.0	75.8	30.0	0.656	15.8	0.749	36.8	6.08	20.9
102x152x14	4.84	2.08	100.0	37.4	15.3	67.5	24.2	0.673	18.7	0.756	25.2	3.69	17.9
102x152x13	4.88	1.97	85.0	34.8	12.1	63.9	19.4	0.705	21.8	0.766	20.4	2.37	15.8
146x127x22	3.54	3.52	130	33.2	46.0	59.5	70.5	0.202	10.6	0.613	64.9	11.9	27.4
146x127x19	3.52	3.48	115	28.5	39.0	50.7	59.7	0.233	12.2	0.616	41.0	7.65	23.6
146x127x16	3.61	3.36	97.4	26.2	30.6	46.0	47.1	0.376	14.8	0.623	24.5	4.26	19.8
102x127x14	3.92	2.22	85.5	28.3	17.5	50.4	27.4	0.607	13.8	0.720	21.0	4.77	18.0
102x127x13	3.95	2.15	75.3	26.2	14.6	46.9	23.0	0.628	15.8	0.727	15.9	3.20	16.0
102x127x11	3.99	2.06	64.5	24.1	11.7	43.5	18.6	0.656	18.2	0.736	12.0	2.06	14.0
133x102x15	2.84	3.17	73.1	18.8	28.7	33.5	44.1	-	-	0.569	21.7	5.13	19.1
133x102x13	2.86	3.10	62.4	16.2	23.1	28.7	35.5	-	-	0.572	12.6	2.97	16.0

Advance, UKT and UKB are trademarks of Tata. A fuller description of the relationship between Structural Tees and the Advance range of sections manufactured by Tata is given in section 12.

(*) Note units are cm^6 and not dm^6.

– Indicates that no values of U and X are given, as lateral torsional buckling due to bending about the x-x axis is not possible, because the second moment of area about the z-z axis exceeds the second moment of area about the y-y axis.

FOR EXPLANATION OF TABLES SEE NOTES 2 AND 3

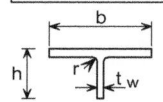

STRUCTURAL TEES CUT FROM UNIVERSAL COLUMNS

Advance UKT split from Advance UKC

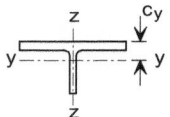

Dimensions

Section Designation	Cut from Universal Beam Section Designation	Mass per Metre	Width of Section	Depth of Section	Thickness		Root Radius	Ratios for Local Buckling		Dimension
			b	h	Web t_w	Flange t_f	r	Flange f/t_f	Web c_w/t_w	c_y
		kg/m	mm	mm	mm	mm	mm			cm
406x178x118	356x406x235	117.5	394.8	190.4	18.4	30.2	15.2	6.54	10.3	3.40
368x178x101	356x368x202	100.9	374.7	187.2	16.5	27.0	15.2	6.94	11.3	3.29
368x178x89	356x368x177	88.5	372.6	184.0	14.4	23.8	15.2	7.83	12.8	3.09
368x178x77	356x368x153	76.5	370.5	180.9	12.3	20.7	15.2	8.95	14.7	2.88
368x178x65	356x368x129	64.5	368.6	177.7	10.4	17.5	15.2	10.5	17.1	2.69
305x152x79	305x305x158	79.0	311.2	163.5	15.8	25.0	15.2	6.22	10.3	3.04
305x152x69	305x305x137	68.4	309.2	160.2	13.8	21.7	15.2	7.12	11.6	2.86
305x152x59	305x305x118	58.9	307.4	157.2	12.0	18.7	15.2	8.22	13.1	2.69
305x152x49	305x305x97	48.4	305.3	153.9	9.9	15.4	15.2	9.91	15.5	2.50
254x127x84	254x254x167	83.5	265.2	144.5	19.2	31.7	12.7	4.18	7.53	3.07
254x127x66	254x254x132	66.0	261.3	138.1	15.3	25.3	12.7	5.16	9.03	2.70
254x127x54	254x254x107	53.5	258.8	133.3	12.8	20.5	12.7	6.31	10.4	2.45
254x127x45	254x254x89	44.4	256.3	130.1	10.3	17.3	12.7	7.41	12.6	2.21
254x127x37	254x254x73	36.5	254.6	127.0	8.6	14.2	12.7	8.96	14.8	2.05
203x102x64 +	203x203x127	63.7	213.9	120.7	18.1	30.1	10.2	3.55	6.67	2.73
203x102x57 +	203x203x113	56.7	212.1	117.5	16.3	26.9	10.2	3.94	7.21	2.56
203x102x50 +	203x203x100	49.8	210.3	114.3	14.5	23.7	10.2	4.44	7.88	2.38
203x102x43	203x203x86	43.0	209.1	111.0	12.7	20.5	10.2	5.10	8.74	2.20
203x102x36	203x203x71	35.5	206.4	107.8	10.0	17.3	10.2	5.97	10.8	1.95
203x102x30	203x203x60	30.0	205.8	104.7	9.4	14.2	10.2	7.25	11.1	1.89
203x102x26	203x203x52	26.0	204.3	103.0	7.9	12.5	10.2	8.17	13.0	1.75
203x102x23	203x203x46	23.0	203.6	101.5	7.2	11.0	10.2	9.25	14.1	1.69
152x76x26 +	152x152x51	25.6	157.4	85.1	11.0	15.7	7.6	5.01	7.74	1.79
152x76x22 +	152x152x44	22.0	155.9	83.0	9.5	13.6	7.6	5.73	8.74	1.66
152x76x19	152x152x37	18.5	154.4	80.8	8.0	11.5	7.6	6.71	10.1	1.53
152x76x15	152x152x30	15.0	152.9	78.7	6.5	9.4	7.6	8.13	12.1	1.41
152x76x12	152x152x23	11.5	152.2	76.1	5.8	6.8	7.6	11.2	13.1	1.39

Advance, UKT and UKC are trademarks of Tata. A fuller description of the relationship between Structural Tees and the Advance range of sections manufactured by Tata is given in section 12.

+ These sections are in addition to the range of BS 4 sections

FOR EXPLANATION OF TABLES SEE NOTES 2 AND 3

BS EN 1993-1-1:2005
BS 4-1:2005

STRUCTURAL TEES CUT FROM UNIVERSAL COLUMNS

Advance UKT split from Advance UKC

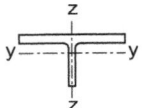

Properties

Section Designation	Second Moment of Area		Radius of Gyration		Elastic Modulus			Plastic Modulus		Mono-symmetry Index	Warping Constant (*)	Torsional Constant	Area of Section
	Axis y-y	Axis z-z	Axis y-y	Axis z-z	Axis y-y Flange	Axis y-y Toe	Axis z-z	Axis y-y	Axis z-z	ψ	I_w	I_t	A
	cm^4	cm^4	cm	cm	cm^3	cm^3	cm^3	cm^3	cm^3		cm^6	cm^4	cm^2
406x178x118	2860	15500	4.37	10.2	843	183	785	367	1190	0.165	12700	405	150
368x178x101	2460	11800	4.38	9.60	749	160	632	312	960	0.216	7840	278	129
368x178x89	2090	10300	4.30	9.54	676	136	551	263	835	0.212	5270	190	113
368x178x77	1730	8780	4.22	9.49	601	114	474	216	717	0.209	3390	125	97.4
368x178x65	1420	7310	4.16	9.43	527	94.1	396	175	600	0.207	2010	76.2	82.2
305x152x79	1530	6280	3.90	7.90	503	115	404	225	615	0.268	3650	188	101
305x152x69	1290	5350	3.84	7.83	450	97.7	346	188	526	0.263	2340	124	87.2
305x152x59	1080	4530	3.79	7.77	401	82.8	295	156	448	0.262	1470	80.3	75.1
305x152x49	858	3650	3.73	7.69	343	66.5	239	123	363	0.258	806	45.5	61.7
254x127x84	1200	4930	3.36	6.81	391	105	372	220	569	0.261	4540	312	106
254x127x66	871	3770	3.22	6.69	323	78.3	288	159	439	0.250	2200	159	84.1
254x127x54	676	2960	3.15	6.59	276	62.1	229	122	348	0.245	1150	85.9	68.2
254x127x45	524	2430	3.04	6.55	237	48.5	190	94.0	288	0.242	660	51.1	56.7
254x127x37	417	1950	2.99	6.48	204	39.2	153	74.0	233	0.236	359	28.8	46.5
203x102x64 +	637	2460	2.80	5.50	233	68.2	230	145	352	0.279	2050	212	81.2
203x102x57 +	540	2140	2.73	5.45	211	58.8	202	123	309	0.270	1430	152	72.3
203x102x50 +	453	1840	2.67	5.39	190	50.0	175	103	267	0.266	951	104	63.4
203x102x43	373	1560	2.61	5.34	169	41.9	150	84.6	228	0.257	605	68.1	54.8
203x102x36	280	1270	2.49	5.30	143	31.8	123	63.6	187	0.254	343	40.0	45.2
203x102x30	244	1030	2.53	5.20	129	28.4	100	54.3	153	0.245	195	23.5	38.2
203x102x26	200	889	2.46	5.18	115	23.4	87.0	44.5	132	0.243	128	15.8	33.1
203x102x23	177	774	2.45	5.13	105	20.9	76.0	39.0	115	0.242	87.2	11.0	29.4
152x76x26 +	141	511	2.08	3.96	79.0	21.0	64.9	41.4	99.5	0.281	122	24.3	32.6
152x76x22 +	116	430	2.04	3.92	70.0	17.5	55.2	34.0	84.4	0.281	76.7	15.8	28.0
152x76x19	93.1	353	1.99	3.87	60.7	14.2	45.7	27.1	69.8	0.277	44.9	9.54	23.5
152x76x15	72.2	280	1.94	3.83	51.4	11.2	36.7	20.9	55.8	0.269	23.7	5.24	19.1
152x76x12	58.5	200	2.00	3.70	41.9	9.41	26.3	16.9	40.1	0.278	9.78	2.30	14.6

Advance, UKT and UKC are trademarks of Tata. A fuller description of the relationship between Structural Tees and the Advance range of sections manufactured by Tata is given in section 12.

+ These sections are in addition to the range of BS 4 sections

(*) Note units are cm^6 and not dm^6.

Values of U and X are not given, as lateral torsional buckling due to bending about the y-y axis is not possible, because the second moment of area about the z-z axis exceeds the second moment of area about the y-y axis.

FOR EXPLANATION OF TABLES SEE NOTES 2 AND 3

BS EN 1993-1-1:2005
BS 4-1:2005

EFFECTIVE SECTION PROPERTIES

UNIVERSAL BEAMS
Advance UKB

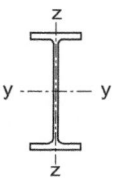

Classification and effective area for sections subject to axial compression

Section Designation	S275 / Advance275					S355 / Advance355				
	Classification		Gross Area A cm^2	Effective Area A$_{eff}$ cm^2	A$_{eff}$/A	Classification		Gross Area A cm^2	Effective Area A$_{eff}$ cm^2	A$_{eff}$/A
1016x305x393 +	Not class 4		500	500	1.00	Class 4	W	500	**488**	0.976
1016x305x349 +	Class 4	W	445	**432**	0.970	Class 4	W	445	**418**	0.940
1016x305x314 +	Class 4	W	400	**379**	0.947	Class 4	W	400	**366**	0.916
1016x305x272 +	Class 4	W	347	**317**	0.913	Class 4	W	347	**306**	0.883
1016x305x249 +	Class 4	W	317	**287**	0.905	Class 4	W	317	**276**	0.872
1016x305x222 +	Class 4	W	283	**251**	0.888	Class 4	W	283	**241**	0.853
914x419x388	Not class 4		494	494	1.00	Class 4	W	494	**478**	0.968
914x419x343	Class 4	W	437	**426**	0.974	Class 4	W	437	**414**	0.948
914x305x289	Class 4	W	368	**354**	0.962	Class 4	W	368	**342**	0.929
914x305x253	Class 4	W	323	**301**	0.932	Class 4	W	323	**290**	0.899
914x305x224	Class 4	W	286	**259**	0.907	Class 4	W	286	**250**	0.874
914x305x201	Class 4	W	256	**227**	0.887	Class 4	W	256	**218**	0.852
838x292x226	Class 4	W	289	**271**	0.936	Class 4	W	289	**261**	0.904
838x292x194	Class 4	W	247	**224**	0.908	Class 4	W	247	**216**	0.875
838x292x176	Class 4	W	224	**200**	0.891	Class 4	W	224	**192**	0.856
762x267x197	Class 4	W	251	**239**	0.953	Class 4	W	251	**231**	0.922
762x267x173	Class 4	W	220	**204**	0.929	Class 4	W	220	**197**	0.896
762x267x147	Class 4	W	187	**168**	0.896	Class 4	W	187	**161**	0.862
762x267x134	Class 4	W	171	**149**	0.871	Class 4	W	171	**143**	0.839
686x254x170	Class 4	W	217	**209**	0.963	Class 4	W	217	**202**	0.933
686x254x152	Class 4	W	194	**182**	0.941	Class 4	W	194	**176**	0.909
686x254x140	Class 4	W	178	**164**	0.924	Class 4	W	178	**159**	0.892
686x254x125	Class 4	W	159	**144**	0.905	Class 4	W	159	**139**	0.872
610x305x179	Not class 4		228	228	1.00	Class 4	W	228	**220**	0.965
610x305x149			190	**182**	0.956	Class 4	W	190	**177**	0.931
610x229x140	Class 4	W	178	**172**	0.968	Class 4	W	178	**167**	0.937
610x229x125	Class 4	W	159	**150**	0.945	Class 4	W	159	**145**	0.914
610x229x113	Class 4	W	144	**133**	0.926	Class 4	W	144	**129**	0.895
610x229x101	Class 4	W	129	**117**	0.904	Class 4	W	129	**113**	0.872
610x178x100 +	Class 4	W	128	**118**	0.921	Class 4	W	128	**113**	0.885
610x178x92 +	Class 4	W	117	**105**	0.900	Class 4	W	117	**101**	0.864
610x178x82 +	Class 4	W	104	**90.7**	0.872	Class 4	W	104	**86.9**	0.835
533x312x150 +	Not class 4		192	192	1.00	Class 4	W	192	**186**	0.970
533x210x122	Not class 4		155	155	1.00	Class 4	W	155	**149**	0.963
533x210x109	Class 4	W	139	**135**	0.972	Class 4	W	139	**131**	0.942
533x210x101	Class 4	W	129	**123**	0.956	Class 4	W	129	**119**	0.926
533x210x92	Class 4	W	117	**109**	0.934	Class 4	W	117	**106**	0.905
533x210x82	Class 4	W	105	**96.4**	0.918	Class 4	W	105	**93.1**	0.887

Advance and UKB are trademarks of Tata. A fuller description of the relationship between Universal Beams (UB) and the Advance range of sections manufactured by Tata is given in note 12.

+ These sections are in addition to the range of BS 4-1 sections

W indicates that the section classification is controlled by the web.

Values of A$_{eff}$ not in **bold type** are the same as the gross area.

Only the sections which can be class 4 under axial compression are given in the table.

FOR EXPLANATION OF TABLES SEE NOTE 4

BS EN 1993-1-1:2005
BS 4-1:2005

EFFECTIVE SECTION PROPERTIES

UNIVERSAL BEAMS
Advance UKB

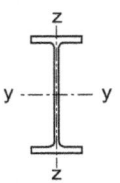

Classification and effective area for sections subject to axial compression

Section Designation	S275 / Advance275					S355 / Advance355				
	Classification		Gross Area A cm²	Effective Area A_{eff} cm²	A_{eff}/A	Classification		Gross Area A cm²	Effective Area A_{eff} cm²	A_{eff}/A
533x165x85 +	Class 4	W	108	**101**	0.937	Class 4	W	108	**97.6**	0.903
533x165x74 +	Class 4	W	95.2	**86.8**	0.911	Class 4	W	95.2	**83.5**	0.877
533x165x66 +	Class 4	W	83.7	**73.9**	0.883	Class 4	W	83.7	**70.9**	0.848
457x191x98	Not class 4		125	125	1.00	Class 4	W	125	**122**	0.975
457x191x89	Not class 4		114	114	1.00	Class 4	W	114	**109**	0.957
457x191x82	Class 4	W	104	**101**	0.968	Class 4	W	104	**97.8**	0.940
457x191x74	Class 4	W	94.6	**89.6**	0.947	Class 4	W	94.6	**86.9**	0.919
457x191x67	Class 4	W	85.5	**79.7**	0.932	Class 4	W	85.5	**77.2**	0.903
457x152x82	Not class 4		105	105	1.00	Class 4	W	105	**100**	0.954
457x152x74	Class 4	W	94.5	**91.0**	0.963	Class 4	W	94.5	**88.1**	0.932
457x152x67	Class 4	W	85.6	**80.6**	0.942	Class 4	W	85.6	**77.9**	0.911
457x152x60	Class 4	W	76.2	**69.7**	0.915	Class 4	W	76.2	**67.4**	0.884
457x152x52	Class 4	W	66.6	**59.4**	0.892	Class 4	W	66.6	**57.3**	0.860
406x178x74	Not class 4		94.5	94.5	1.00	Class 4	W	94.5	**90.8**	0.961
406x178x67	Class 4	W	85.5	**83.0**	0.970	Class 4	W	85.5	**80.7**	0.944
406x178x60	Class 4	W	76.5	**72.5**	0.948	Class 4	W	76.5	**70.4**	0.921
406x178x54	Class 4	W	69.0	**64.7**	0.938	Class 4	W	69.0	**62.7**	0.909
406x140x53 +	Class 4	W	67.9	**63.9**	0.941	Class 4	W	67.9	**61.8**	0.911
406x140x46	Class 4	W	58.6	**53.1**	0.906	Class 4	W	58.6	**51.4**	0.876
406x140x39	Class 4	W	49.7	**43.7**	0.880	Class 4	W	49.7	**42.1**	0.848
356x171x67	Not class 4		85.5	85.5	1.00	Class 4	W	85.5	**84.1**	0.983
356x171x57	Not class 4		72.6	72.6	1.00	Class 4	W	72.6	**69.7**	0.960
356x171x51	Class 4	W	64.9	**62.7**	0.966	Class 4	W	64.9	**61.0**	0.940
356x171x45	Class 4	W	57.3	**54.5**	0.952	Class 4	W	57.3	**53.0**	0.924
356x127x39	Class 4	W	49.8	**46.5**	0.934	Class 4	W	49.8	**45.0**	0.904
356x127x33	Class 4	W	42.1	**38.1**	0.905	Class 4	W	42.1	**36.8**	0.874
305x165x46	Class 4	W	58.7	**57.6**	0.982	Class 4	W	58.7	**56.3**	0.960
305x165x40	Class 4	W	51.3	**49.4**	0.962	Class 4	W	51.3	**48.2**	0.940
305x127x37	Not class 4		47.2	47.2	1.00	Class 4	W	47.2	**45.3**	0.960
305x102x33	Class 4	W	41.8	**40.1**	0.960	Class 4	W	41.8	**38.8**	0.929
305x102x28	Class 4	W	35.9	**33.5**	0.933	Class 4	W	35.9	**32.3**	0.900
305x102x25	Class 4	W	31.6	**29.0**	0.917	Class 4	W	31.6	**27.8**	0.880
254x146x37	Not class 4		47.2	47.2	1.00	Class 4	W	47.2	**46.4**	0.983
254x146x31	Not class 4		39.7	39.7	1.00	Class 4	W	39.7	**38.6**	0.971
254x102x28	Not class 4		36.1	36.1	1.00	Class 4	W	36.1	**35.0**	0.971
254x102x25	Not class 4		32.0	32.0	1.00	Class 4	W	32.0	**30.6**	0.957
254x102x22	Class 4	W	28.0	**27.2**	0.973	Class 4	W	28.0	**26.3**	0.940

Advance and UKB are trademarks of Tata. A fuller description of the relationship between Universal Beams (UB) and the Advance range of sections manufactured by Tata is given in note 12.

+ These sections are in addition to the range of BS 4-1 sections

W indicates that the section classification is controlled by the web.

Values of A_{eff} not in **bold type** are the same as the gross area.

Only the sections which can be class 4 under axial compression are given in the table.

FOR EXPLANATION OF TABLES SEE NOTE 4

EFFECTIVE SECTION PROPERTIES

Classification and effective area for sections subject to axial compression

HOLLOW SECTIONS - S275

There are no effective property tables given for hot finished hollow sections in S275, because hot finished hollow sections are normally available in S355 only.

There are no effective property tables given for cold-formed hollow sections in S275, because cold-formed hollow sections are normally available in S235 and S355 only. No effective property tables are given for S235 either, because sections available may not be manufactured to BS EN 10219-2: 2006.

BS EN 1993-1-1:2005
BS EN 10210-2:2005

EFFECTIVE SECTION PROPERTIES   S355 / Celsius® 355

HOT-FINISHED
SQUARE HOLLOW SECTIONS

Celsius® SHS

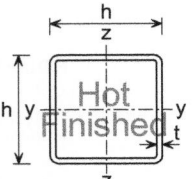

Classification and effective area for sections subject to axial compression

Designation			S355 / Celsius®355			
Size	Wall	Classification	Gross	Effective		
	Thickness		Area	Area		
h x h	t		A	A_{eff}		A_{eff}/A
mm	mm		cm^2	cm^2		
200x200	5.0	Class 4	38.7	**35.2**		0.910
250x250	6.3	Class 4	61.0	**55.8**		0.915
260x260	6.3	Class 4	63.5	**56.6**		0.892
300x300	6.3	Class 4	73.6	**59.4**		0.807
	8.0	Class 4	92.8	**87.9**		0.947
350x350	8.0	Class 4	109	**93.5**		0.858
400x400	10.0	Class 4	155	**141**		0.910

Celsius® is a trademark of Tata. A fuller description of the relationship between Hot Finished Square Hollow Sections (HFSHS) and the Celsius® range of sections manufactured by Tata is given in note 12.

Values of A_{eff} not in **bold type** are the same as the gross area.

Only the sections which can be class 4 under axial compression are given in the table.

FOR EXPLANATION OF TABLES SEE NOTE 4.

BS EN 1993-1-1:2005
BS EN 10219-2:2005

EFFECTIVE SECTION PROPERTIES   S355 / Hybox® 355

COLD-FORMED
SQUARE HOLLOW SECTIONS

Hybox® SHS

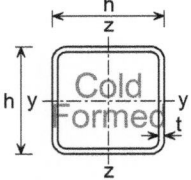

Classification and effective area for sections subject to axial compression

Designation			S355 / Hybox®355			
Size	Wall	Classification	Gross	Effective		
	Thickness		Area	Area		
h x h	t		A	A_{eff}		A_{eff}/A
mm	mm		cm^2	cm^2		
150x150	4.0	Class 4	22.9	**21.7**		0.947
160x160	4.0	Class 4	24.5	**22.3**		0.909
200x200	5.0	Class 4	38.4	**34.9**		0.909
250x250	6.0	Class 4	57.6	**51.0**		0.885
300x300	6.0	Class 4	69.6	**54.1**		0.777
	8.0	Class 4	91.2	**86.3**		0.947
350x350	8.0	Class 4	107	**91.5**		0.855
400x400	8.0	Class 4	123	**95.4**		0.776
	10.0	Class 4	153	**139**		0.909

Hybox® is a trademark of Tata. A fuller description of the relationship between Cold Formed Square Hollow Sections (CFSHS) and the Hybox® range of sections manufactured by Tata is given in note 12.

Values of A_{eff} not in **bold type** are the same as the gross area.

Only the sections which can be class 4 under axial compression are given in the table.

FOR EXPLANATION OF TABLES SEE NOTE 4.

BS EN 1993-1-1:2005
BS EN 10210-2:2005

EFFECTIVE SECTION PROPERTIES   S355 / Celsius®355

HOT-FINISHED
RECTANGULAR HOLLOW SECTIONS

Celsius® RHS

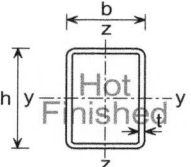

Classification and effective area for sections subject to axial compression

Designation		Classification		S355 / Celsius®355		
Size	Wall Thickness			Gross Area	Effective Area	
h x b	t			A	A_{eff}	A_{eff}/A
mm	mm			cm^2	cm^2	
150x125	4.0	Class 4	W	21.2	**20.6**	0.971
160x80	4.0	Class 4	W	18.4	**17.3**	0.939
200x100	5.0	Class 4	W	28.7	**27.0**	0.939
200x120	5.0	Class 4	W	30.7	**29.0**	0.943
250x150	5.0	Class 4	W	38.7	**33.3**	0.861
	6.3	Class 4	W	48.4	**45.8**	0.946
260x140	5.0	Class 4	W	38.7	**32.5**	0.840
	6.3	Class 4	W	48.4	**45.0**	0.929
300x100	8.0	Class 4	W	60.8	**58.4**	0.960
300x150	8.0	Class 4	W	68.8	**66.4**	0.965
300x200	6.3	Class 4	W	61.0	**53.9**	0.884
	8.0	Class 4	W	76.8	**74.4**	0.968
300x250	5.0	Class 4	F,W	53.7	**38.8**	0.722
	6.3	Class 4	F,W	67.3	**57.6**	0.856
	8.0	Class 4	W	84.8	**82.4**	0.971
350x150	5.0	Class 4	W	48.7	**34.8**	0.715
	6.3	Class 4	W	61.0	**48.9**	0.801
	8.0	Class 4	W	76.8	**69.0**	0.899
350x250	5.0	Class 4	F,W	58.7	**39.4**	0.671
	6.3	Class 4	F,W	73.6	**58.9**	0.800
	8.0	Class 4	W	92.8	**85.0**	0.916
400x120	5.0	Class 4	W	50.7	**32.3**	0.636
	6.3	Class 4	W	63.5	**46.0**	0.724
	8.0	Class 4	W	80.0	**66.2**	0.827
	10.0	Class 4	W	98.9	**91.9**	0.930
400x150	5.0	Class 4	W	53.7	**35.3**	0.657
	6.3	Class 4	W	67.3	**49.8**	0.740
	8.0	Class 4	W	84.8	**71.0**	0.837
	10.0	Class 4	W	105	**98.0**	0.934

Celsius® is a trademark of Tata. A fuller description of the relationship between Hot Finished Rectangular Hollow Sections (HFRHS) and the Celsius® range of sections manufactured by Tata is given in note 12.

W indicates that the section classification is controlled by the web.

F indicates that the section classification is controlled by the flange.

F,W indicates that the section classification is controlled by both flange and web.

Values of A_{eff} not in **bold type** are the same as the gross area.

Only the sections which can be class 4 under axial compression are given in the table.

FOR EXPLANATION OF TABLES SEE NOTE 4.

BS EN 1993-1-1:2005
BS EN 10210-2:2005

EFFECTIVE SECTION PROPERTIES   S355 / Celsius® 355

**HOT-FINISHED
RECTANGULAR HOLLOW SECTIONS**

Celsius® RHS

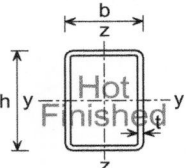

Classification and effective area for sections subject to axial compression

Designation		S355 / Celsius®355				
Size	Wall Thickness	Classification		Gross Area	Effective Area	
h x b	t			A	A_{eff}	A_{eff}/A
mm	mm			cm²	cm²	
400x200	8.0	Class 4	W	92.8	**79.0**	0.851
	10.0	Class 4	W	115	**108**	0.939
400x300	8.0	Class 4	F,W	109	**92.8**	0.851
	10.0	Class 4	W	135	**128**	0.948
450x250	8.0	Class 4	W	109	**88.7**	0.814
	10.0	Class 4	W	135	**121**	0.897
500x200	8.0	Class 4	W	109	**81.9**	0.751
	10.0	Class 4	W	135	**113**	0.840
	12.5	Class 4	W	167	**156**	0.935
500x300	8.0	Class 4	F,W	125	**95.4**	0.764
	10.0	Class 4	W	155	**133**	0.861
	12.5	Class 4	W	192	**181**	0.943

Celsius® is a trademark of Tata. A fuller description of the relationship between Hot Finished Rectangular Hollow Sections (HFRHS) and the Celsius® range of sections manufactured by Tata is given in note 12.

W indicates that the section classification is controlled by the web.

F indicates that the section classification is controlled by the flange.

F,W indicates that the section classification is controlled by both flange and web.

Values of A_{eff} not in **bold type** are the same as the gross area.

Only the sections which can be class 4 under axial compression are given in the table.

FOR EXPLANATION OF TABLES SEE NOTE 4.

BS EN 1993-1-1:2005
BS EN 10219-2:2005

EFFECTIVE SECTION PROPERTIES

S355 / Hybox®355

COLD-FORMED
RECTANGULAR HOLLOW SECTIONS

Hybox® RHS

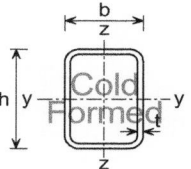

Classification and effective area for sections subject to axial compression

Designation		S355 / Hybox®355				
Size	Wall Thickness	Classification		Gross Area	Effective Area	
h x b mm	t mm			A cm²	A_{eff} cm²	A_{eff}/A
120x40	3.0	Class 4	W	9.01	**8.38**	0.930
120x60	3.0	Class 4	W	10.2	**9.57**	0.938
140x80	3.0	Class 4	W	12.6	**11.1**	0.883
150x100	3.0	Class 4	W	14.4	**12.5**	0.865
	4.0	Class 4	W	18.9	**18.3**	0.968
160x80	4.0	Class 4	W	18.1	**17.0**	0.938
180x80	4.0	Class 4	W	19.7	**17.5**	0.887
180x100	4.0	Class 4	W	21.3	**19.1**	0.895
200x100	4.0	Class 4	W	22.9	**19.4**	0.849
	5.0	Class 4	W	28.4	**26.7**	0.939
200x120	4.0	Class 4	W	24.5	**21.0**	0.859
	5.0	Class 4	W	30.4	**28.7**	0.943
200x150	4.0	Class 4	F,W	26.9	**22.8**	0.849
	5.0	Class 4	W	33.4	**31.7**	0.948
250x150	5.0	Class 4	W	38.4	**33.0**	0.860
	6.0	Class 4	W	45.6	**42.3**	0.927
300x100	6.0	Class 4	W	45.6	**37.8**	0.830
	8.0	Class 4	W	59.2	**56.8**	0.959
300x200	6.0	Class 4	W	57.6	**49.8**	0.865
	8.0	Class 4	W	75.2	**72.8**	0.968
400x200	8.0	Class 4	W	91.2	**77.4**	0.849
	10.0	Class 4	W	113	**106**	0.938
450x250	8.0	Class 4	W	107	**86.7**	0.810
	10.0	Class 4	W	133	**119**	0.895
	12.0	Class 4	W	156	**151**	0.965
500x300	8.0	Class 4	F,W	123	**93.4**	0.760
	10.0	Class 4	W	153	**131**	0.859
	12.0	Class 4	W	180	**167**	0.926
	12.5	Class 4	W	187	**176**	0.942

Hybox® is a trademark of Tata. A fuller description of the relationship between Cold Formed Rectangular Hollow Sections (CFRHS) and the Hybox® range of sections manufactured by Tata is given in note 12.

W indicates that the section classification is controlled by the web.

F indicates that the section classification is controlled by the flange.

F,W indicates that the section classification is controlled by both flange and web

Values of A_{eff} not in **bold type** are the same as the gross area.

Only the sections which can be class 4 under axial compression are given in the table.

FOR EXPLANATION OF TABLES SEE NOTE 4.

BS EN 1993-1-1:2005
BS EN 10056-1:1999

EFFECTIVE SECTION PROPERTIES

EQUAL ANGLES
Advance UKA - Equal Angles

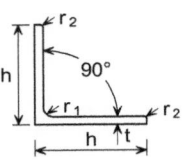

Classification and effective area for sections subject to axial compression

Designation		S275 / Advance275				S355 / Advance355			
Size	Thickness	Classification	Gross Area	Effective Area		Classification	Gross Area	Effective Area	
h x h	t		A	A_{eff}	A_{eff}/A		A	A_{eff}	A_{eff}/A
mm	mm		cm^2	cm^2			cm^2	cm^2	
200x200	20.0	Not class 4	76.3	76.3	1.00	Class 4	76.3	76.3	1.00
	18.0	Class 4	69.1	69.1	1.00	Class 4	69.1	69.1	1.00
	16.0	Class 4	61.8	61.8	1.00	Class 4	61.8	**57.7**	0.934
150x150	15.0	Not class 4	43.0	43.0	1.00	Class 4	43.0	43.0	1.00
	12.0	Class 4	34.8	34.8	1.00	Class 4	34.8	**32.5**	0.934
	10.0	Class 4	29.3	**26.3**	0.898	Class 4	29.3	**23.8**	0.814
120x120	12.0	Not class 4	27.5	27.5	1.00	Class 4	27.5	27.5	1.00
	10.0	Class 4	23.2	23.2	1.00	Class 4	23.2	**22.3**	0.962
	8.0 +	Class 4	18.8	**16.9**	0.898	Class 4	18.8	**15.3**	0.814
100x100	10.0	Not class 4	19.2	19.2	1.00	Class 4	19.2	19.2	1.00
	8.0	Class 4	15.5	15.5	1.00	Class 4	15.5	**14.5**	0.934
90x90	8.0	Class 4	13.9	13.9	1.00	Class 4	13.9	13.9	1.00
	7.0	Class 4	12.2	12.2	1.00	Class 4	12.2	**11.2**	0.915
80x80	8.0	Not class 4	12.3	12.3	1.00	Class 4	12.3	12.3	1.00
75x75	8.0	Not class 4	11.4	11.4	1.00	Class 4	11.4	11.4	1.00
	6.0	Class 4	8.73	8.73	1.00	Class 4	8.73	**8.15**	0.934
70x70	7.0	Not class 4	9.40	9.40	1.00	Class 4	9.40	9.40	1.00
	6.0	Class 4	8.13	8.13	1.00	Class 4	8.13	**7.98**	0.981
60x60	6.0	Not class 4	6.91	6.91	1.00	Class 4	6.91	6.91	1.00
	5.0	Class 4	5.82	5.82	1.00	Class 4	5.82	**5.60**	0.962
50x50	5.0	Not class 4	4.80	4.80	1.00	Class 4	4.80	4.80	1.00
	4.0	Class 4	3.89	3.89	1.00	Class 4	3.89	**3.63**	0.934
45x45	4.5	Not class 4	3.90	3.90	1.00	Class 4	3.90	3.90	1.00
40x40	4.0	Not class 4	3.08	3.08	1.00	Class 4	3.08	3.08	1.00
30x30	3.0	Not class 4	1.74	1.74	1.00	Class 4	1.74	1.74	1.00

Advance and UKA are trademarks of Tata. A fuller description of the relationship between Equal Angles and the Advance range of sections manufactured by Tata is given in note 12.

+ These sections are in addition to the range of BS EN 10056-1 sections.

Values of A_{eff} not in **bold type** are the same as the gross area.

Only the sections which can be class 4 under axial compression are given in the table.

FOR EXPLANATION OF TABLES SEE NOTE 4.

BS EN 1993-1-1:2005
BS EN 10056-1:1999

EFFECTIVE SECTION PROPERTIES

UNEQUAL ANGLES
Advance UKA - Unequal Angles

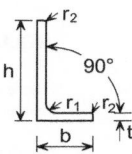

Classification and effective area for sections subject to axial compression

Designation		S275 / Advance275				S355 / Advance355			
Size	Thickness	Classification	Gross Area	Effective Area		Classification	Gross Area	Effective Area	
h x b	t		A	A_{eff}	A_{eff}/A		A	A_{eff}	A_{eff}/A
mm	mm		cm^2	cm^2			cm^2	cm^2	
200x150	18.0 +	Not class 4	60.1	60.1	1.00	Class 4	60.1	60.1	1.00
	15.0	Class 4	50.5	**49.3**	0.977	Class 4	50.5	**44.9**	0.889
	12.0	Class 4	40.8	**33.8**	0.827	Class 4	40.8	**30.5**	0.746
200x100	15.0	Not class 4	43.0	43.0	1.00	Class 4	43.0	**38.2**	0.889
	12.0	Class 4	34.8	**28.8**	0.827	Class 4	34.8	**25.9**	0.745
	10.0	Class 4	29.2	**20.8**	0.714	Class 4	29.2	**18.7**	0.640
150x90	12.0	Not class 4	27.5	27.5	1.00	Class 4	27.5	**25.7**	0.933
	10.0	Class 4	23.2	**20.8**	0.897	Class 4	23.2	**18.8**	0.812
150x75	12.0	Not class 4	25.7	25.7	1.00	Class 4	25.7	**24.0**	0.933
	10.0	Class 4	21.7	**19.5**	0.896	Class 4	21.7	**17.6**	0.812
125x75	10.0	Not class 4	19.1	19.1	1.00	Class 4	19.1	**17.8**	0.933
	8.0	Class 4	15.5	**13.5**	0.869	Class 4	15.5	**12.2**	0.786
100x75	8.0	Class 4	13.5	13.5	1.00	Class 4	13.5	**12.6**	0.934
100x65	8.0 +	Not class 4	12.7	12.7	1.00	Class 4	12.7	**11.9**	0.933
	7.0 +	Class 4	11.2	**10.4**	0.930	Class 4	11.2	**9.46**	0.844
100x50	8.0	Not class 4	11.4	11.4	1.00	Class 4	11.4	**10.6**	0.933
	6.0	Class 4	8.71	**7.20**	0.827	Class 4	8.71	**6.49**	0.746
80x60	7.0	Not class 4	9.38	9.38	1.00	Class 4	9.38	**9.33**	0.995
80x40	6.0	Not class 4	6.89	6.89	1.00	Class 4	6.89	**6.12**	0.889
75x50	6.0	Not class 4	7.19	7.19	1.00	Class 4	7.19	**6.71**	0.933
70x50	6.0	Not class 4	6.89	6.89	1.00	Class 4	6.89	**6.76**	0.981
65x50	5.0	Class 4	5.54	**5.51**	0.994	Class 4	5.54	**5.02**	0.907
60x40	5.0	Not class 4	4.79	4.79	1.00	Class 4	4.79	**4.60**	0.961
45x30	4.0	Not class 4	2.87	2.87	1.00	Class 4	2.87	2.87	1.00

Advance and UKA are trademarks of Tata. A fuller description of the relationship between Unequal Angles and the Advance range of sections manufactured by Tata is given in note 12.

Classification done when section loaded to capacity.

+ These sections are in addition to the range of BS EN 10056-1 sections.

Values of A_{eff} not in **bold type** are the same as the gross area.

Only the sections which can be class 4 under axial compression are given in the table.

FOR EXPLANATION OF TABLES SEE NOTE 4

EFFECTIVE SECTION PROPERTIES

Classification and effective section properties for sections subject to bending

HOLLOW SECTIONS - S275

There are no effective property tables given for hot finished hollow sections in S275, because hot finished hollow sections are normally available in S355 only.

There are no effective property tables given for cold-formed hollow sections in S275, because normally cold-formed hollow sections are normally available in S235 and S355 only. No effective property tables are given for S235 either, because sections available may not be manufactured to BS EN 10219-2: 2006.

BS EN 1993-1-1:2005
BS EN 10210-2:2005

EFFECTIVE SECTION PROPERTIES   S355 / Celsius® 355

HOT-FINISHED
SQUARE HOLLOW SECTIONS

Celsius® SHS

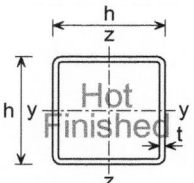

Classification and effective section properties for sections subject to bending about y-y axis

Designation		S355 / Celsius®355					
Size	Wall Thickness	Classification			Properties		
$h \times h$	t				$I_{eff,y}$	$W_{el,eff,y}$	$W_{pl,eff,y}$
mm	mm				cm^4	cm^3	cm^3
200 x 200	5.0	Class 4	F		2360	231	*
250 x 250	6.3	Class 4	F		5817	456	*
260 x 260	6.3	Class 4	F		6503	487	*
300 x 300	6.3	Class 4	F		9743	619	*
300 x 300	8.0	Class 4	F		12865	847	*
350 x 350	8.0	Class 4	F		19952	1100	*
400 x 400	10.0	Class 4	F		37772	1847	*

Celsius® is a trademark of Tata. A fuller description of the relationship between Hot Finished Square Hollow Sections (HFSHS) and the Celsius® range of sections manufactured by Tata is given in note 12.

F indicates that the section classification is controlled by the flange.

$W_{el,eff,y}$ is the minimum value of the effective elastic modulus about the y-y axis for a class 4 section.

$W_{pl,eff,y}$ is the effective plastic modulus about the y-y axis for a class 3 section.

* Indicates that this parameter is not applicable to the section

Only the sections which can be class 3 or class 4 when subject to pure bending are given in the table.

FOR EXPLANATION OF TABLES SEE NOTE 4.

BS EN 1993-1-1:2005
BS EN 10219-2:2005

EFFECTIVE SECTION PROPERTIES

S355 / Hybox®355

COLD-FORMED
SQUARE HOLLOW SECTIONS

Hybox® SHS

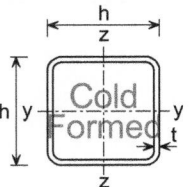

Classification and effective section properties for sections subject to bending about y-y axis

Designation				S355 / Hybox®355			
Size	Wall Thickness	Classification			Properties		
h x h	t			$I_{eff,y}$	$W_{el,eff,y}$	$W_{pl,eff,y}$	
mm	mm			cm⁴	cm³	cm³	
150 x 150	4.0	Class 4	F	792	104	*	
160 x 160	4.0	Class 4	F	952	116	*	
200 x 200	5.0	Class 4	F	2325	227	*	
250 x 250	6.0	Class 4	F	5418	421	*	
300 x 300	6.0	Class 4	F	9075	572	*	
300 x 300	8.0	Class 4	F	12537	825	*	
350 x 350	8.0	Class 4	F	19503	1075	*	
400 x 400	8.0	Class 4	F	28460	1345	*	
400 x 400	10.0	Class 4	F	36860	1802	*	

Hybox® is a trademark of Tata. A fuller description of the relationship between Cold Formed Square Hollow Sections (CFSHS) and the Hybox® range of sections manufactured by Tata is given in note 12.

F indicates that the section classification is controlled by the flange.

$W_{el,eff,y}$ is the minimum value of the effective elastic modulus about the y-y axis for a class 4 section.

$W_{pl,eff,y}$ is the effective plastic modulus about the y-y axis for a class 3 section.

* Indicates that this parameter is not applicable to the section

FOR EXPLANATION OF TABLES SEE NOTE 4.

BS EN 1993-1-1:2005
BS EN 10210-2:2005

EFFECTIVE SECTION PROPERTIES   **S355 / Celsius® 355**

HOT-FINISHED
RECTANGULAR HOLLOW SECTIONS

Celsius® RHS

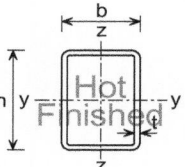

Classification and effective section properties for sections subject to bending about y-y axis

Designation		S355 / Celsius®355				
Size	Wall	Classification		Properties		
	Thickness					
h x b	t			$I_{eff,y}$	$W_{el,eff,y}$	$W_{pl,eff,y}$
mm	mm			cm^4	cm^3	cm^3
300 x 250	5.0	Class 4	F	6792	430	*
300 x 250	6.3	Class 4	F	8902	582	*
350 x 250	5.0	Class 4	F	9790	534	*
350 x 250	6.3	Class 4	F	12812	719	*
400 x 120	5.0	Class 3	W	-	-	566
400 x 150	5.0	Class 3	W	-	-	626
400 x 300	8.0	Class 4	F	25235	1248	*
500 x 300	8.0	Class 4	F	42983	1703	*

Celsius® is a trademark of Tata. A fuller description of the relationship between Hot Finished Rectangular Hollow Sections (HFRHS) and the Celsius® range of sections manufactured by Tata is given in note 12.

- Indicates that the effective section property is not applicable due to the section classification and the gross property should be used.

W indicates that the section classification is controlled by the web.

F indicates that the section classification is controlled by the flange.

$W_{el,eff,y}$ is the minimum value of the effective elastic modulus about the y-y axis for a class 4 section.

$W_{pl,eff,y}$ is the effective plastic modulus about the y-y axis for a class 3 section.

* Indicates that this parameter is not applicable to the section

⬛ Check availability

Only the sections which can be class 3 or class 4 when subject to pure bending are given in the table.

FOR EXPLANATION OF TABLES SEE NOTE 4.

BS EN 1993-1-1:2005
BS EN 10210-2:2005

EFFECTIVE SECTION PROPERTIES   S355 / Celsius® 355

HOT-FINISHED
RECTANGULAR HOLLOW SECTIONS

Celsius® RHS

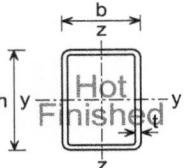

Classification and effective section properties for sections subject to bending about z-z axis

Designation		S355 / Celsius® 355				
Size	Wall Thickness	Classification		Properties		
h x b mm	t mm			$I_{eff,z}$ cm^4	$W_{el,eff,z}$ cm^3	$W_{pl,eff,z}$ cm^3
300 x 250	5.0	Class 4	F	4828	353	*
300 x 250	6.3	Class 4	F	6394	485	*
350 x 250	5.0	Class 4	F	5179	366	*
350 x 250	6.3	Class 4	F	6903	508	*
400 x 120	5.0	Class 4	F	1051	144	-
400 x 150	5.0	Class 4	F	1731	192	-
400 x 300	8.0	Class 4	F	14969	936	*
500 x 300	8.0	Class 4	F	16709	996	*

Celsius® is a trademark of Tata. A fuller description of the relationship between Hot Finished Rectangular Hollow Sections (HFRHS) and the Celsius® range of sections manufactured by Tata is given in note 12.

- Indicates that the effective section property is not applicable due to the section classification and the gross property should be used.

F indicates that the section classification is controlled by the flange.

$W_{el,eff,z}$ is the minimum value of the effective elastic modulus about the z-z axis for a class 4 section.

$W_{pl,eff,z}$ is not applicable for RHS.

Only the sections which can be class 3 or class 4 when subject to pure bending are given in the table.

* Indicates that this parameter is not applicable to the section

▨▨▨▨▨▨   Check availability

FOR EXPLANATION OF TABLES SEE NOTE 4.

BS EN 1993-1-1:2005
BS EN 10219-2:2005

EFFECTIVE SECTION PROPERTIES   **S355 / Hybox®355**

**COLD-FORMED
RECTANGULAR HOLLOW SECTIONS**

Hybox® RHS

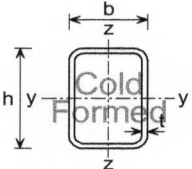

Classification and effective section properties for sections subject to bending about y-y axis

Designation		S355 / Hybox®355				
Size	Wall Thickness	Classification		Properties		
h x b mm	t mm			$I_{eff,y}$ cm^4	$W_{el,eff,y}$ cm^3	$W_{pl,eff,y}$ cm^3
200 x 150	4.0	Class 4	F	1554	154	*
500 x 300	8.0	Class 4	F	42060	1666	*

Hybox® is a trademark of Tata. A fuller description of the relationship between Cold Formed Rectangular Hollow Sections (CFRHS) and the Hybox® range of sections manufactured by Tata is given in note 12.

F indicates that the section classification is controlled by the flange.

$W_{el,eff,y}$ is the minimum value of the effective elastic modulus about the y-y axis for a class 4 section.

$W_{pl,eff,y}$ is the effective plastic modulus about the y-y axis for a class 3 section.

* Indicates that this parameter is not applicable to the section

Only the sections which can be class 3 or class 4 when subject to pure bending are given in the table.

FOR EXPLANATION OF TABLES SEE NOTE 4.

BS EN 1993-1-1:2005
BS EN 10219-2:2005

EFFECTIVE SECTION PROPERTIES   S355 / Hybox®355

COLD-FORMED
RECTANGULAR HOLLOW SECTIONS

Hybox® RHS

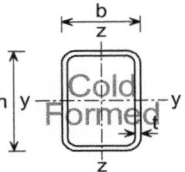

Classification and effective section properties for sections subject to bending about z-z axis

Designation		Classification		S355 / Hybox®355 Properties		
Size h x b mm	Wall Thickness t mm			$I_{eff,z}$ cm^4	$W_{el,eff,z}$ cm^3	$W_{pl,eff,z}$ cm^3
200 x 150	4.0	Class 4	F	923	115	*
500 x 300	8.0	Class 4	F	16375	974	*

Hybox® is a trademark of Tata. A fuller description of the relationship between Cold Formed Rectangular Hollow Sections (CFRHS) and the Hybox® range of sections manufactured by Tata is given in note 12.

F indicates that the section classification is controlled by the flange.

$W_{el,eff,z}$ is the minimum value of the effective elastic modulus about the z-z axis for a class 4 section.

$W_{pl,eff,z}$ is not applicable for RHS.

* Indicates that this parameter is not applicable to the section

Only the sections which can be class 3 or class 4 when subject to pure bending are given in the table.

FOR EXPLANATION OF TABLES SEE NOTE 4.

Back marks in channel flanges

RSC	Nominal flange width (mm)	Back mark (mm)	Edge dist. (mm)	Recommended diameter (mm)
	102	55	47	24
	89	55	34	20
	76	45	31	20
	64	35	29	16
	51	30	21	10
	38	22	–	–

Back marks in angles

Nominal leg (mm)	S_1 (mm)	S_2 (mm)	S_3 (mm)	S_4 (mm)	S_5 (mm)	S_6 (mm)	Nominal leg (mm)	S_1 (mm)
250							75	45 (20)
200		75 (30)	75 (30)	55 (20)	55 (20)	55 (20)	70	40 (20)
150		55 (20)	55 (20)				65	35 (20)
125		45 (20)	50 (20)				60	35 (16)
120		45 (16)	50 (16)				50	28 (12)
100	55 (24)						45	25
90	50 (24)						40	23
80	45 (20)						30	20
							25	15

Maximum recommended bolt sizes are given in brackets

This table is reproduced from BCSA Publication No. 5/79, *Metric Practice for Structural Steelworks*, 3rd edn, 1979.

Cross centres through flanges

Flange width (mm)	Minimum for accessibility (mm)	Maximum for edge dist. (mm)	S_1 (mm)	S_2 (mm)	S_3 (mm)	S_4 (mm)
Joists						
44	27 (5)	30	30			
64	38 (10)	39	40			
76	48 (10)	51	48			
89	54 (12)	59	56			
102	60 (16)	62	60			
114	66 (16)	74	70			
127	72 (20)	77	75			
152	75 (20)	102	90			
203	91 (24)	143	140			
UCs						
152	65 (24)	92	90			
203	75 (24)	143	140			
254	87 (24)	194	140			
305	100 (24)	245	140	120 (24)	60 (24)	240 (24)
368	88 (24)	308	140	140 (24)	75 (24)	290 (24)
406	120 (24)	346	140	140 (24)	75 (24)	290 (24)
UBs						
102	50 (16)	62	54			
127	62 (20)	77	70			
133	57 (20)	83	70			
140	69 (24)	80	70			
146	64 (24)	86	70			
152	73 (24)	92	90			
165	67 (24)	105	90			
171	72 (24)	111	90			
178	72 (24)	118	90			
191	74 (24)	131	90			
210	80 (24)	150	140			
229	80 (24)	169	140			
254	87 (24)	194	140			
267	91 (24)	207	140	90 (20)	50 (20)	190 (20)
292	94 (24)	232	140	100 (24)	60 (24)	220 (24)
305	100 (24)	245	140	120 (24)	60 (24)	240 (24)
419	112 (24)	359	140	140 (24)	75 (24)	290 (24)

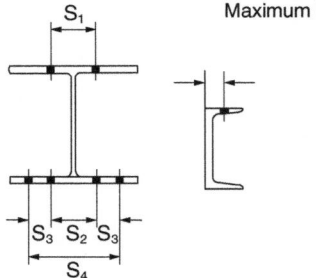

Maximum bolt diameters for dimensions shown are given in brackets

Bolt and Weld Data for S275

BS EN 1993-1-8:2005 BS EN ISO 4016 BS EN ISO 4018

BOLT RESISTANCES   S275

Non Preloaded bolts

Class 4.6 hexagon head bolts

Diameter of Bolt	Tensile Stress Area	Tension Resistance	Shear Resistance		Bolts in tension
			Single Shear	Double Shear	Min thickness for punching shear
d mm	A_s mm^2	$F_{t,Rd}$ kN	$F_{v,Rd}$ kN	$2 \times F_{v,Rd}$ kN	t_{min} mm
12	84.3	24.3	13.8	27.5	2.1
16	157	45.2	30.1	60.3	3.2
20	245	70.6	47.0	94.1	3.9
24	353	102	67.8	136	4.7
30	561	162	108	215	5.8

Diameter of Bolt	Minimum				Bearing Resistance (kN)										
	Edge distance	End distance	Pitch	Gauge						Thickness in mm of ply, t.					
d mm	e_2 mm	e_1 mm	p_1 mm	p_2 mm	5	6	7	8	9	10	12	15	20	25	30
12	20	25	35	40	25.9	31.0	36.2	41.4	46.6	51.7	62.1	77.6	103	129	155
16	25	35	50	50	34.0	40.8	47.7	54.5	61.3	68.1	81.7	102	136	170	204
20	30	40	60	60	**42.1**	50.5	58.9	67.4	75.8	84.2	*101*	*126*	*168*	*211*	*253*
24	35	50	70	70	**50.1**	**60.1**	70.2	80.2	90.2	100	120	*150*	*200*	*251*	*301*
30	45	60	85	90	**63.2**	**75.8**	**88.4**	**101**	114	126	152	189	*253*	*316*	*379*

For M12 bolts the design shear resistance $F_{v,Rd}$ has been calculated as 0.85 times the value given in BS EN 1993-1-8, Table 3.4 (Cl3.6.1(5))

See clause 3.7(1) of BS EN 1993-1-8: 2005 for calculation of the design resistance of a group of fasteners.

For bolts with cut threads that do not comply with EN 1090, the given values for tension and shear should be multiplied by 0.85

Values of bearing resistance in **bold** are less than the single shear resistance of the bolt.

Values of bearing resistance in *italic* are greater than the double shear resistance of the bolt.

Bearing values assume standard clearance holes.

If oversize or short slotted holes are used, bearing values should be multiplied by 0.8.

If long slotted or kidney shaped holes are used, bearing values should be multiplied by 0.6.

In single lap joints with only one bolt row, the design bearing resistance for each bolt should be limited to $1.5f_u d\, t/\gamma_{M2}$

FOR EXPLANATION OF TABLES SEE NOTE 5.

BS EN 1993-1-8:2005
BS EN ISO 4014
BS EN ISO 4017

BOLT RESISTANCES

S275

Non Preloaded bolts

Class 8.8 hexagon head bolts

Diameter of Bolt	Tensile Stress Area	Tension Resistance	Shear Resistance		Bolts in tension
			Single Shear	Double Shear	Min thickness for punching shear
d	A_s	$F_{t,Rd}$	$F_{v,Rd}$	$2 \times F_{v,Rd}$	t_{min}
mm	mm^2	kN	kN	kN	mm
12	84.3	48.6	27.5	55.0	4.3
16	157	90.4	60.3	121	6.3
20	245	141	94.1	188	7.8
24	353	203	136	271	9.4
30	561	323	215	431	11.6

Diameter of Bolt	Minimum				Bearing Resistance (kN)											
	Edge distance	End distance	Pitch	Gauge												
d	e_2	e_1	p_1	p_2	Thickness in mm of ply, t.											
mm	mm	mm	mm	mm	5	6	7	8	9	10	12	15	20	25	30
12	20	25	35	40	25.9	31.0	36.2	41.4	46.6	51.7	62.1	77.6	103	129	155
16	25	35	50	50	34.0	40.8	47.7	54.5	61.3	68.1	81.7	102	136	170	204
20	30	40	60	60	42.1	50.5	58.9	67.4	75.8	84.2	101	126	168	211	253
24	35	50	70	70	50.1	60.1	70.2	80.2	90.2	100	20	150	200	251	301
30	45	60	85	90	63.2	75.8	88.4	101	114	126	152	89	253	316	379

Diameter of Bolt	Minimum				Bearing Resistance (kN)											
	Edge distance	End distance	Pitch	Gauge												
d	e_2	e_1	p_1	p_2	Thickness in mm of ply, t.											
mm	mm	mm	mm	mm	5	6	7	8	9	10	12	15	20	25	30
12	25	40	50	45	42.2	50.6	59.0	67.5	75.9	84.3	101	127	169	211	253
16	30	50	65	55	58.3	70.0	81.6	93.3	105	117	140	175	233	292	350
20	35	60	80	70	74.5	89.5	104	119	134	149	179	224	298	373	447
24	40	75	95	80	90.8	109	127	145	163	182	218	272	363	454	545
30	50	90	115	100	112	134	157	179	201	224	268	335	447	559	671

For M12 bolts the design shear resistance $F_{v,Rd}$ has been calculated as 0.85 times the value given in BS EN 1993-1-8, Table 3.4 (Cl3.6.1(5))
See clause 3.7(1) of BS EN 1993-1-8: 2005 for calculation of the design resistance of a group of fasteners.
For bolts with cut threads that do not comply with EN 1090, the given values for tension and shear should be multiplied by 0.85
Values of bearing resistance in bold are less than the single shear resistance of the bolt.
Values of bearing resistance in italic are greater than the double shear resistance of the bolt.
Bearing values assume standard clearance holes.
If oversize or short slotted holes are used, bearing values should be multiplied by 0.8.
If long slotted or kidney shaped holes are used, bearing values should be multiplied by 0.6.
In single lap joints with only one bolt row, the design bearing resistance for each bolt should be limited to $1.5 f_u$ d t/γ_{M2}
FOR EXPLANATION OF TABLES SEE NOTE 5.

| BS EN 1993-1-8:2005 |
| BS EN ISO 4014 |
| BS EN ISO 4017 |

BOLT RESISTANCES

S275

Non Preloaded bolts

Class 10.9 hexagon head bolts

Diameter of Bolt	Tensile Stress Area	Tension Resistance	Shear Resistance		Bolts in tension
			Single Shear	Double Shear	Min thickness for punching shear
d	A_s	$F_{t,Rd}$	$F_{v,Rd}$	$2 \times F_{v,Rd}$	t_{min}
mm	mm^2	kN	kN	kN	mm
12	84.3	60.7	28.7	57.3	5.3
16	157	113	62.8	126	7.9
20	245	176	98.0	196	9.8
24	353	254	141	282	11.7
30	561	404	224	449	14.5

Diameter of Bolt	Minimum				Bearing Resistance (kN)										
	Edge distance	End distance	Pitch	Gauge	Thickness in mm of ply, t.										
d	e_2	e_1	p_1	p_2	5	6	7	8	9	10	12	15	20	25	30
mm	mm	mm	mm	mm											
12	20	25	35	40	25.9	31.0	36.2	41.4	46.6	51.7	62.1	77.6	103	129	155
16	25	35	50	50	34.0	40.8	47.7	54.5	61.3	68.1	81.7	102	136	170	204
20	30	40	60	60	42.1	50.5	58.9	67.4	75.8	84.2	101	126	168	211	253
24	35	50	70	70	50.1	60.1	70.2	80.2	90.2	100	120	150	200	251	301
30	45	60	85	90	63.2	75.8	88.4	101	114	126	152	189	253	316	379

Diameter of Bolt	Minimum				Bearing Resistance (kN)										
	Edge distance	End distance	Pitch	Gauge	Thickness in mm of ply, t.										
d	e_2	e_1	p_1	p_2	5	6	7	8	9	10	12	15	20	25	30
mm	mm	mm	mm	mm											
12	25	40	50	45	42.2	50.6	59.0	67.5	75.9	84.3	101	127	169	211	253
16	30	50	65	55	58.3	70.0	81.6	93.3	105	117	140	175	233	292	350
20	35	60	80	70	74.5	89.5	104	119	134	149	179	224	298	373	447
24	40	75	95	80	90.8	109	127	145	163	182	218	272	363	454	545
30	50	90	115	100	112	134	157	179	201	224	268	335	447	559	671

For M12 bolts the design shear resistance $F_{v,Rd}$ has been calculated as 0.85 times the value given in BS EN 1993-1-8, Table 3.4 (Cl3.6.1(5))

See clause 3.7(1) of BS EN 1993-1-8: 2005 for calculation of the design resistance of a group of fasteners.

For bolts with cut threads that do not comply with EN 1090, the given values for tension and shear should be multiplied by 0.85

Values of bearing resistance in bold are less than the single shear resistance of the bolt.

Values of bearing resistance in italic are greater than the double shear resistance of the bolt.

Bearing values assume standard clearance holes.

If oversize or short slotted holes are used, bearing values should be multiplied by 0.8.

If long slotted or kidney shaped holes are used, bearing values should be multiplied by 0.6.

In single lap joints with only one bolt row, the design bearing resistance for each bolt should be limited to $1.5f_u d\, t/\gamma_{M2}$

FOR EXPLANATION OF TABLES SEE NOTE 5.

BS EN 1993-1-8:2005
BS EN ISO 4016
BS EN ISO 4018

BOLT RESISTANCES

S275

Non Preloaded bolts

Class 4.6 countersunk bolts

Diameter of Bolt	Tensile Stress Area	Tension Resistance	Shear Resistance		Bolts in tension
			Single Shear	Double Shear	Min thickness for punching shear
d	A_s	$F_{t,Rd}$	$F_{v,Rd}$	$2 \times F_{v,Rd}$	t_{min}
mm	mm^2	kN	kN	kN	mm
12	84.3	17.0	13.8	27.5	1.5
16	157	31.7	30.1	60.3	2.2
20	245	49.4	47.0	94.1	2.7
24	353	71.2	67.8	136	3.3
30	561	113	108	215	4.1

Diameter of Bolt	Minimum				Bearing Resistance (kN)											
	Edge distance	End distance	Pitch	Gauge												
d	e_2	e_1	p_1	p_2	Thickness in mm of ply, t.											
mm	mm	mm	mm	mm	5	6	7	8	9	10	12	15	20	25	30
12	20	25	35	40	**10.3**	15.5	20.7	25.9	*31.0*	*36.2*	*46.6*	*62.1*	*87.9*	*114*	*140*
16	25	35	50	50	**6.81**	**13.6**	**20.4**	**27.2**	34.0	40.8	54.5	*74.9*	*109*	*143*	*177*
20	30	40	60	60	0	**8.42**	**16.8**	**25.3**	**33.7**	42.1	58.9	84.2	*126*	*168*	*211*
24	35	50	70	70	0	0	**10.0**	**20.0**	**30.1**	**40.1**	**60.1**	90.2	*140*	*190*	*241*
30	45	60	85	90	0	0	0	**6.32**	**18.9**	**31.6**	**56.8**	**94.7**	158	*221*	*284*

For M12 bolts the design shear resistance $F_{v,Rd}$ has been calculated as 0.85 times the value given in BS EN 1993-1-8, Table 3.4 (Cl3.6.1(5))

See clause 3.7(1) of BS EN 1993-1-8: 2005 for calculation of the design resistance of a group of fasteners.

For bolts with cut threads that do not comply with EN 1090, the given values for tension and shear should be multiplied by 0.85

Values of bearing resistance in **bold** are less than the single shear resistance of the bolt.

Values of bearing resistance in *italic* are greater than the double shear resistance of the bolt.

Bearing values assume standard clearance holes.

If oversize or short slotted holes are used, bearing values should be multiplied by 0.8.

If long slotted or kidney shaped holes are used, bearing values should be multiplied by 0.6.

In single lap joints with only one bolt row, the design bearing resistance for each bolt should be limited to 1.5f_u d t/γ_{M2}

FOR EXPLANATION OF TABLES SEE NOTE 5.

| BS EN 1993-1-8:2005 |
| BS EN ISO 4014 |
| BS EN ISO 4017 |

BOLT RESISTANCES

S275

Non Preloaded bolts

Class 8.8 countersunk bolts

Diameter of Bolt	Tensile Stress Area	Tension Resistance	Shear Resistance		Bolts in tension
			Single Shear	Double Shear	Min thickness for punching shear
d	A_s	$F_{t,Rd}$	$F_{v,Rd}$	$2 \times F_{v,Rd}$	t_{min}
mm	mm^2	kN	kN	kN	mm
12	84.3	34.0	27.5	55.0	3.0
16	157	63.3	60.3	121	4.4
20	245	98.8	94.1	188	5.5
24	353	142	136	271	6.6
30	561	226	215	431	8.1

Diameter of Bolt	Minimum				Bearing Resistance (kN)										
	Edge distance	End distance	Pitch	Gauge											
d	e_2	e_1	p_1	p_2	Thickness in mm of ply, t.										
mm	mm	mm	mm	mm	5	6	7	8	9	10	12	15	20	25	30
12	20	25	35	40	10.3	15.5	20.7	25.9	31.0	36.2	46.6	62.1	87.9	114	140
16	25	35	50	50	6.81	13.6	20.4	27.2	34.0	40.8	54.5	74.9	109	143	177
20	30	40	60	60	0	8.42	16.8	25.3	33.7	42.1	58.9	84.2	126	168	211
24	35	50	70	70	0	0	10.0	20.0	30.1	40.1	60.1	90.2	140	190	241
30	45	60	85	90	0	0	0	6.32	18.9	31.6	56.8	94.7	158	221	284

Diameter of Bolt	Minimum				Bearing Resistance (kN)										
	Edge distance	End distance	Pitch	Gauge											
d	e_2	e_1	p_1	p_2	Thickness in mm of ply, t.										
mm	mm	mm	mm	mm	5	6	7	8	9	10	12	15	20	25	30
12	25	45	55	45	19.7	29.5	39.4	49.2	59.0	68.9	88.6	118	167	216	266
16	30	55	70	55	13.1	26.2	39.4	52.5	65.6	78.7	105	144	210	276	341
20	35	70	85	70	0	16.4	32.8	49.2	65.6	82.0	115	164	246	328	410
24	40	80	100	80	0	0	19.7	39.4	59.0	78.7	118	177	276	374	472
30	50	100	125	100	0	0	0	12.3	36.9	61.5	111	185	308	431	554

For M12 bolts the design shear resistance $F_{v,Rd}$ has been calculated as 0.85 times the value given in BS EN 1993-1-8, Table 3.4 (Cl3.6.1(5))

See clause 3.7(1) of BS EN 1993-1-8: 2005 for calculation of the design resistance of a group of fasteners.

For bolts with cut threads that do not comply with EN 1090, the given values for tension and shear should be multiplied by 0.85

Values of bearing resistance in **bold** are less than the single shear resistance of the bolt.

Values of bearing resistance in *italic* are greater than the double shear resistance of the bolt.

Bearing values assume standard clearance holes.

If oversize or short slotted holes are used, bearing values should be multiplied by 0.8.

If long slotted or kidney shaped holes are used, bearing values should be multiplied by 0.6.

In single lap joints with only one bolt row, the design bearing resistance for each bolt should be limited to $1.5f_u d\ t/\gamma_{M2}$

FOR EXPLANATION OF TABLES SEE NOTE 5.

BS EN 1993-1-8:2005
BS EN ISO 4014
BS EN ISO 4017

BOLT RESISTANCES

S275

Non Preloaded bolts

Class 10.9 countersunk bolts

Diameter of Bolt	Tensile Stress Area	Tension Resistance	Shear Resistance		Bolts in tension
			Single Shear	Double Shear	Min thickness for punching shear
d mm	A_s mm^2	$F_{t,Rd}$ kN	$F_{v,Rd}$ kN	$2 \times F_{v,Rd}$ kN	t_{min} mm
12	84.3	42.5	28.7	57.3	3.7
16	157	79.1	62.8	126	5.5
20	245	123	98.0	196	6.9
24	353	178	141	282	8.2
30	561	283	224	449	10.2

Diameter of Bolt	Minimum				Bearing Resistance (kN)										
	Edge distance	End distance	Pitch	Gauge	Thickness in mm of ply, t.										
d mm	e_2 mm	e_1 mm	p_1 mm	p_2 mm	5	6	7	8	9	10	12	15	20	25	30
12	20	25	35	40	10.3	15.5	20.7	25.9	31.0	36.2	46.6	62.1	87.9	114	140
16	25	35	50	50	6.81	13.6	20.4	27.2	34.0	40.8	54.5	74.9	109	143	177
20	30	40	60	60	0	8.42	16.8	25.3	33.7	42.1	58.9	84.2	126	168	211
24	35	50	70	70	0	0	10.0	20.0	30.1	40.1	60.1	90.2	140	190	241
30	45	60	85	90	0	0	0	6.32	18.9	31.6	56.8	94.7	158	221	284

Diameter of Bolt	Minimum				Bearing Resistance (kN)										
	Edge distance	End distance	Pitch	Gauge	Thickness in mm of ply, t.										
d mm	e_2 mm	e_1 mm	p_1 mm	p_2 mm	5	6	7	8	9	10	12	15	20	25	30
12	25	45	55	45	19.7	29.5	39.4	49.2	59.0	68.9	88.6	118	167	216	266
16	30	55	70	55	13.1	26.2	39.4	52.5	65.6	78.7	105	144	210	276	341
20	35	70	85	70	0	16.4	32.8	49.2	65.6	82.0	115	164	246	328	410
24	40	80	100	80	0	0	19.7	39.4	59.0	78.7	118	177	276	374	472
30	50	100	125	100	0	0	0	12.3	36.9	61.5	111	185	308	431	554

For M12 bolts the design shear resistance $F_{v,Rd}$ has been calculated as 0.85 times the value given in BS EN 1993-1-8, Table 3.4 (Cl3.6.1(5))

See clause 3.7(1) of BS EN 1993-1-8: 2005 for calculation of the design resistance of a group of fasteners.

For bolts with cut threads that do not comply with EN 1090, the given values for tension and shear should be multiplied by 0.85

Values of bearing resistance in **bold** are less than the single shear resistance of the bolt.

Values of bearing resistance in *italic* are greater than the double shear resistance of the bolt.

Bearing values assume standard clearance holes.

If oversize or short slotted holes are used, bearing values should be multiplied by 0.8.

If long slotted or kidney shaped holes are used, bearing values should be multiplied by 0.6.

In single lap joints with only one bolt row, the design bearing resistance for each bolt should be limited to $1.5f_u \, d \, t/\gamma_{M2}$

FOR EXPLANATION OF TABLES SEE NOTE 5.

| BS EN 1993-1-8:2005 |
| BS EN 14399:2005 |
| EN 1090:2008 |

BOLT RESISTANCES

S275

Preloaded bolts at serviceability limit state

Class 8.8 hexagon head bolts

Diameter of Bolt	Tensile Stress Area	Tension Resistance	Shear Resistance		Bolts in tension	Slip Resistance $\mu = 0.5$	
			Single Shear	Double Shear	Min thickness for punching shear	Single Shear	Double Shear
d	A_s	$F_{t,Rd}$	$F_{v,Rd}$	$2 \times F_{v,Rd}$	t_{min}		
mm	mm^2	kN	kN	kN	mm	kN	kN
12	84.3	48.6	27.5	55.0	3.71	21.5	42.9
16	157	90.4	60.3	121	5.59	40.0	79.9
20	245	141	94.1	188	7.36	62.4	125
24	353	203	136	271	8.22	89.9	180
30	561	323	215	431	10.7	143	286

Diameter of Bolt	Minimum				Bearing Resistance (kN)										
	Edge distance	End distance	Pitch	Gauge	Thickness in mm of ply, t.										
d	e_2	e_1	p_1	p_2	5	6	7	8	9	10	12	15	20	25	30
mm	mm	mm	mm	mm											
12	20	40	50	40	38.8	46.6	54.3	62.1	69.8	77.6	93.1	116	155	194	233
16	25	50	65	50	51.1	61.3	71.5	81.7	91.9	102	123	153	204	255	306
20	30	60	80	60	63.2	75.8	88.4	101	114	126	152	189	253	316	379
24	35	75	95	70	75.2	90.2	105	120	135	150	180	226	301	376	451
30	45	90	115	90	94.7	114	133	152	171	189	227	284	379	474	568

For M12 bolts the design shear resistance $F_{v,Rd}$ has been calculated as 0.85 times the value given in BS EN 1993-1-8, Table 3.4 (Cl3.6.1(5))

See clause 3.7(1) of BS EN 1993-1-8: 2005 for calculation of the design resistance of a group of fasteners.

For bolts with cut threads that do not comply with EN 1090, the given values for tension and shear should be multiplied by 0.85

Values of bearing resistance in **bold** are less than the single shear resistance of the bolt.

Values of bearing resistance in *italic* are greater than the double shear resistance of the bolt.

The tension resistance, the shear resistance and the bearing resistance should be greater than the design ultimate force

The slip resistance should be greater than the design force at serviceability

In single lap joints with only one bolt row, the design bearing resistance for each bolt should be limited to $1.5f_u d\ t/\gamma_{M2}$

Values have been calculated assuming $k_s=1$ and $\mu=0.5$. See BS EN 1993-1-8, section 3.9 for other values of k_s and μ

FOR EXPLANATION OF TABLES SEE NOTE 5.

BS EN 1993-1-8:2005
BS EN 14399:2005
EN 1090:2008

BOLT RESISTANCES

S275

Preloaded bolts at serviceability limit state

Class 10.9 hexagon head bolts

Diameter of Bolt	Tensile Stress Area	Tension Resistance	Shear Resistance		Bolts in tension	Slip Resistance $\mu = 0.5$	
			Single Shear	Double Shear	Min thickness for punching shear	Single Shear	Double Shear
d	A_s	$F_{t,Rd}$	$F_{v,Rd}$	$2 \times F_{v,Rd}$	t_{min}		
mm	mm^2	kN	kN	kN	mm	kN	kN
16	157	113	62.8	126	6.99	50.0	99.9
20	245	176	98.0	196	9.20	78.0	156
24	353	254	141	282	10.3	112	225
30	561	404	224	449	13.3	179	357

Diameter of Bolt	Minimum				Bearing Resistance (kN)										
	Edge distance	End distance	Pitch	Gauge	Thickness in mm of ply, t.										
d	e_2	e_1	p_1	p_2	5	6	7	8	9	10	12	15	20	25	30
mm	mm	mm	mm	mm											
16	25	50	65	50	51.1	61.3	71.5	81.7	91.9	102	123	153	204	255	306
20	30	60	80	60	63.2	75.8	88.4	101	114	126	152	189	253	316	379
24	35	75	95	70	75.2	90.2	105	120	135	150	180	226	301	376	451
30	45	90	115	90	94.7	114	133	152	171	189	227	284	379	474	568

For M12 bolts the design shear resistance $F_{v,Rd}$ has been calculated as 0.85 times the value given in BS EN 1993-1-8, Table 3.4 (Cl3.6.1(5))

See clause 3.7(1) of BS EN 1993-1-8: 2005 for calculation of the design resistance of a group of fasteners.

For bolts with cut threads that do not comply with EN 1090, the given values for tension and shear should be multiplied by 0.85

Values of bearing resistance in **bold** are less than the single shear resistance of the bolt.

Values of bearing resistance in *italic* are greater than the double shear resistance of the bolt.

The tension resistance, the shear resistance and the bearing resistance should be greater than the design ultimate force

The slip resistance should be greater than the design force at serviceability

In single lap joints with only one bolt row, the design bearing resistance for each bolt should be limited to 1.5f$_u$ d t/γ_{M2}

Values have been calculated assuming k$_s$=1 and μ=0.5. See BS EN 1993-1-8, section 3.9 for other values of k$_s$ and μ

FOR EXPLANATION OF TABLES SEE NOTE 5.

| BS EN 1993-1-8:2005 |
| BS EN 14399:2005 |
| EN 1090:2008 |

BOLT RESISTANCES

S275

Preloaded bolts at ultimate limit state

Class 8.8 hexagon head bolts

Diameter of Bolt	Tensile Stress Area	Bolts in tension		Slip Resistance							
		Tension Resistance	Min thcks for punching shear	$\mu = 0.2$		$\mu = 0.3$		$\mu = 0.4$		$\mu = 0.5$	
				Single Shear	Double Shear	Single Shear	Double Shear	Single Shear	Double Shear	Single Shear	Double Shear
d	A_s	$F_{t,Rd}$	t_{min}								
mm	mm^2	kN	mm	kN	kN	kN	kN	kN	kN	kN	kN
12	84.3	48.6	3.71	7.55	15.1	11.3	22.7	15.1	30.2	18.9	37.8
16	157	90.4	5.59	14.1	28.1	21.1	42.2	28.1	56.3	35.2	70.3
20	245	141	7.36	22.0	43.9	32.9	65.9	43.9	87.8	54.9	110
24	353	203	8.22	31.6	63.3	47.4	94.9	63.3	127	79.1	158
30	561	323	10.7	50.3	101	75.4	151	101	201	126	251

Diameter of Bolt	Minimum				Bearing Resistance (kN)										
	Edge distance	End distance	Pitch	Gauge											
					Thickness in mm of ply, t.										
d	e_2	e_1	p_1	p_2	5	6	7	8	9	10	12	15	20	25	30
mm	mm	mm	mm	mm											
12	20	40	50	40	38.8	46.6	54.3	62.1	69.8	77.6	93.1	116	155	194	233
16	25	50	65	50	51.1	61.3	71.5	81.7	91.9	102	123	153	204	255	306
20	30	60	80	60	63.2	75.8	88.4	101	114	126	152	189	253	316	379
24	35	75	95	70	75.2	90.2	105	120	135	150	180	226	301	376	451
30	45	90	115	90	94.7	114	133	152	171	189	227	284	379	474	568

For bolts with cut threads that do not comply with EN 1090, the given values for tension resistance and minimum thickness to avoid punching should be multiplied by 0.85

Values of bearing resistance in **bold** are less than the single shear resistance of the bolt.

Values of bearing resistance in *italic* are greater than the double shear resistance of the bolt.

In single lap joints with only one bolt row, the design bearing resistance for each bolt should be limited to $1.5 f_u d\, t/\gamma_{M2}$

FOR EXPLANATION OF TABLES SEE NOTE 5.

BS EN 1993-1-8:2005
BS EN 14399:2005
EN 1090:2008

BOLT RESISTANCES

S275

Preloaded bolts at ultimate limit state

Class 10.9 hexagon head bolts

Diameter of Bolt	Tensile Stress Area	Bolts in tension		Slip Resistance							
		Tension Resistance	Min thcks for punching shear	$\mu = 0.2$		$\mu = 0.3$		$\mu = 0.4$		$\mu = 0.5$	
				Single Shear	Double Shear	Single Shear	Double Shear	Single Shear	Double Shear	Single Shear	Double Shear
d	A_s	$F_{t,Rd}$	t_{min}								
mm	mm^2	kN	mm	kN	kN	kN	kN	kN	kN	kN	kN
16	157	113	6.99	17.6	35.2	26.4	52.8	35.2	70.3	44.0	87.9
20	245	176	9.20	27.4	54.9	41.2	82.3	54.9	110	68.6	137
24	353	254	10.3	39.5	79.1	59.3	119	79.1	158	98.8	198
30	561	404	13.3	62.8	126	94.2	188	126	251	157	314

Diameter of Bolt	Minimum				Bearing Resistance (kN)											
	Edge distance	End distance	Pitch	Gauge												
					Thickness in mm of ply, t.											
d	e_2	e_1	p_1	p_2	5	6	7	8	9	10	12	15	20	25	30	
mm	mm	mm	mm	mm												
16	25	50	65	50	51.1	61.3	71.5	81.7	91.9	102	123	*153*	*204*	*255*	*306*	
20	30	60	80	60	63.2	75.8	**88.4**	101	114	126	152	189	*253*	*316*	*379*	
24	35	75	95	70	75.2	90.2	**105**	**120**	**135**	150	180	226	*301*	*376*	*451*	
30	45	90	115	90	**94.7**	**114**	**133**	**152**	**171**	**189**	227	284	379	*474*	*568*	

For bolts with cut threads that do not comply with EN 1090, the given values for tension resistance and minimum thickness to avoid punching should be multiplied by 0.85

Values of bearing resistance in **bold** are less than the single shear resistance of the bolt.

Values of bearing resistance in *italic* are greater than the double shear resistance of the bolt.

In single lap joints with only one bolt row, the design bearing resistance for each bolt should be limited to $1.5 f_u d\, t / \gamma_{M2}$

FOR EXPLANATION OF TABLES SEE NOTE 5.

| BS EN 1993-1-8:2005 |
| BS EN 14399:2005 |
| EN 1090:2008 |

BOLT RESISTANCES

S275

Preloaded bolts at serviceability limit state

Class 8.8 countersunk bolts

Diameter of Bolt	Tensile Stress Area	Tension Resistance	Shear Resistance		Bolts in tension	Slip Resistance $\mu = 0.5$	
			Single Shear	Double Shear	Min thickness for punching shear	Single Shear	Double Shear
d	A_s		$F_{v,Rd}$	$2 \times F_{v,Rd}$	t_{min}		
mm	mm^2	kN	kN	kN	mm	kN	kN
12	84.3	34.0	27.5	55.0	2.60	21.5	42.9
16	157	63.3	60.3	121	3.91	40.0	79.9
20	245	98.8	94.1	188	5.15	62.4	125
24	353	142	136	271	5.76	89.9	180
30	561	226	215	431	7.47	143	286

Diameter of Bolt	Minimum				Bearing Resistance (kN)											
	Edge distance	End distance	Pitch	Gauge						Thickness in mm of ply, t.						
d	e_2	e_1	p_1	p_2	5	6	7	8	9	10	12	15	20	25	30
mm	mm	mm	mm	mm											
12	20	40	50	40	**15.5**	**23.3**	31.0	38.8	46.6	54.3	69.8	*93.1*	*132*	*171*	*210*
16	25	50	65	50	**10.2**	**20.4**	**30.6**	**40.8**	51.1	61.3	81.7	112	*163*	*214*	*265*
20	30	60	80	60	**0**	**12.6**	**25.3**	**37.9**	**50.5**	63.2	88.4	126	*189*	*253*	*316*
24	35	75	95	70	**0**	**0**	**15.0**	**30.1**	**45.1**	**60.1**	90.2	135	211	*286*	*361*
30	45	90	115	90	**0**	**0**	**0**	**9.47**	**28.4**	**47.4**	**85.3**	**142**	237	332	426

For M12 bolts the design shear resistance $F_{v,Rd}$ has been calculated as 0,85 times the value given in BS EN 1993-1-8, Table 3.4 (Cl3.6.1(5))

See clause 3.7(1) of BS EN 1993-1-8: 2005 for calculation of the design resistance of a group of fasteners.

For bolts with cut threads that do not comply with EN 1090, the given values for tension and shear should be multiplied by 0.85

Values of bearing resistance in **bold** are less than the single shear resistance of the bolt.

Values of bearing resistance in *italic* are greater than the double shear resistance of the bolt.

The tension resistance, the shear resistance and the bearing resistance should be greater than the design ultimate force

The slip resistance should be greater than the design force at serviceability

In single lap joints with only one bolt row, the design bearing resistance for each bolt should be limited to 1.5f$_u$ d t/γ_{M2}

Values have been calculated assuming k_s=1 and μ=0.5. See BS EN 1993-1-8, section 3.9 for other values of k_s and μ

FOR EXPLANATION OF TABLES SEE NOTE 5.

| BS EN 1993-1-8:2005 |
| BS EN 14399:2005 |
| EN 1090:2008 |

BOLT RESISTANCES

S275

Preloaded bolts at serviceability limit state

Class 10.9 countersunk bolts

Diameter of Bolt	Tensile Stress Area	Tension Resistance	Shear Resistance		Bolts in tension	Slip Resistance $\mu = 0.5$	
			Single Shear	Double Shear	Min thickness for punching shear	Single Shear	Double Shear
d	A_s		$F_{v,Rd}$	$2 \times F_{v,Rd}$	t_{min}		
mm	mm^2	kN	kN	kN	mm	kN	kN
16	157	79.1	62.8	126	4.89	50.0	99.9
20	245	123	98.0	196	6.44	78.0	156
24	353	178	141	282	7.19	112	225
30	561	283	224	449	9.33	179	357

Diameter of Bolt	Minimum				Bearing Resistance (kN)											
	Edge distance	End distance	Pitch	Gauge	Thickness in mm of ply, t.											
d	e_2	e_1	p_1	p_2	5	6	7	8	9	10	12	15	20	25	30
mm	mm	mm	mm	mm											
16	25	50	65	50	10.2	20.4	30.6	40.8	51.1	61.3	81.7	112	*163*	*214*	*265*
20	30	60	80	60	0	12.6	25.3	37.9	50.5	63.2	88.4	126	189	*253*	*316*
24	35	75	95	70	0	0	15.0	30.1	45.1	60.1	90.2	135	211	*286*	*361*
30	45	90	115	90	0	0	0	9.47	28.4	47.4	85.3	142	237	332	426

For M12 bolts the design shear resistance $F_{v,Rd}$ has been calculated as 0,85 times the value given in BS EN 1993-1-8, Table 3.4 (Cl3.6.1(5))

See clause 3.7(1) of BS EN 1993-1-8: 2005 for calculation of the design resistance of a group of fasteners.

For bolts with cut threads that do not comply with EN 1090, the given values for tension and shear should be multiplied by 0.85

Values of bearing resistance in **bold** are less than the single shear resistance of the bolt.

Values of bearing resistance in *italic* are greater than the double shear resistance of the bolt.

The tension resistance, the shear resistance and the bearing resistance should be greater than the design ultimate force

The slip resistance should be greater than the design force at serviceability

In single lap joints with only one bolt row, the design bearing resistance for each bolt should be limited to 1.5f_u d t/γ_{M2}

Values have been calculated assuming k_s=1 and μ=0.5. See BS EN 1993-1-8, section 3.9 for other values of k_s and μ

FOR EXPLANATION OF TABLES SEE NOTE 5.

BS EN 1993-1-8:2005
BS EN 14399:2005
EN 1090:2008

BOLT RESISTANCES

S275

Preloaded bolts at ultimate limit state

Class 8.8 countersunk bolts

Diameter of Bolt	Tensile Stress Area	Bolts in tension		Slip Resistance							
		Tension Resistance	Min thcks for punching shear	$\mu = 0.2$		$\mu = 0.3$		$\mu = 0.4$		$\mu = 0.5$	
				Single Shear	Double Shear	Single Shear	Double Shear	Single Shear	Double Shear	Single Shear	Double Shear
d	A_s	$F_{t,Rd}$	t_{min}								
mm	mm^2	kN	mm	kN	kN	kN	kN	kN	kN	kN	kN
12	84.3	34.0	2.60	7.55	15.1	11.3	22.7	15.1	30.2	18.9	37.8
16	157	63.3	3.91	14.1	28.1	21.1	42.2	28.1	56.3	35.2	70.3
20	245	98.8	5.15	22.0	43.9	32.9	65.9	43.9	87.8	54.9	110
24	353	142	5.76	31.6	63.3	47.4	94.9	63.3	127	79.1	158
30	561	226	7.47	50.3	101	75.4	151	101	201	126	251

Diameter of Bolt	Edge distance	End distance	Pitch	Gauge	Bearing Resistance (kN)										
					Thickness in mm of ply, t.										
d	e_2	e_1	p_1	p_2	5	6	7	8	9	10	12	15	20	25	30
mm	mm	mm	mm	mm											
12	20	40	50	40	15.5	23.3	31.0	38.8	46.6	54.3	69.8	93.1	*132*	*171*	*210*
16	25	50	65	50	10.2	20.4	30.6	40.8	51.1	61.3	81.7	112	*163*	*214*	*265*
20	30	60	80	60	0	12.6	25.3	37.9	50.5	63.2	88.4	126	*189*	*253*	*316*
24	35	75	95	70	0	0	15.0	30.1	45.1	60.1	90.2	135	211	*286*	*361*
30	45	90	115	90	0	0	0	9.47	28.4	47.4	85.3	142	237	332	*426*

For bolts with cut threads that do not comply with EN 1090, the given values for tension resistance and minimum thickness to avoid punching should be multiplied by 0.85

Values of bearing resistance in **bold** are less than the single shear resistance of the bolt.

Values of bearing resistance in *italic* are greater than the double shear resistance of the bolt.

In single lap joints with only one bolt row, the design bearing resistance for each bolt should be limited to $1.5f_u\, d\, t/\gamma_{M2}$

FOR EXPLANATION OF TABLES SEE NOTE 5.

BS EN 1993-1-8:2005
BS EN 14399:2005
EN 1090:2008

BOLT RESISTANCES

S275

Preloaded bolts at ultimate limit state

Class 10.9 countersunk bolts

Diameter of Bolt	Tensile Stress Area	Bolts in tension		Slip Resistance							
		Tension Resistance	Min thcks for punching shear	$\mu = 0.2$		$\mu = 0.3$		$\mu = 0.4$		$\mu = 0.5$	
				Single Shear	Double Shear	Single Shear	Double Shear	Single Shear	Double Shear	Single Shear	Double Shear
d	A_s	$F_{t,Rd}$	t_{min}								
mm	mm^2	kN	mm	kN	kN	kN	kN	kN	kN	kN	kN
16	157	79.1	4.89	17.6	35.2	26.4	52.8	35.2	70.3	44.0	87.9
20	245	123	6.44	27.4	54.9	41.2	82.3	54.9	110	68.6	137
24	353	178	7.19	39.5	79.1	59.3	119	79.1	158	98.8	198
30	561	283	9.33	62.8	126	94.2	188	126	251	157	314

Diameter of Bolt	Edge distance	End distance	Pitch	Gauge	Bearing Resistance (kN)										
					Thickness in mm of ply, t.										
d	e_2	e_1	p_1	p_2	5	6	7	8	9	10	12	15	20	25	30
mm	mm	mm	mm	mm											
16	25	50	65	50	10.2	20.4	30.6	40.8	51.1	61.3	81.7	112	*163*	*214*	*265*
20	30	60	80	60	0	12.6	25.3	37.9	50.5	63.2	88.4	126	189	*253*	*316*
24	35	75	95	70	0	0	15.0	30.1	45.1	60.1	90.2	135	211	*286*	*361*
30	45	90	115	90	0	0	0	9.47	28.4	47.4	85.3	142	237	332	*426*

For bolts with cut threads that do not comply with EN 1090, the given values for tension resistance and minimum thickness to avoid punching should be multiplied by 0.85

Values of bearing resistance in **bold** are less than the single shear resistance of the bolt.

Values of bearing resistance in *italic* are greater than the double shear resistance of the bolt.

In single lap joints with only one bolt row, the design bearing resistance for each bolt should be limited to $1.5f_u d \, t/\gamma_{M2}$

FOR EXPLANATION OF TABLES SEE NOTE 5.

FILLET WELDS

Design weld resistances

Leg Length	Throat Thickness	Longitudinal resistance	Transverse resistance
s	a	$F_{w,L,Rd}$	$F_{w,T,Rd}$
mm	mm	kN/mm	kN/mm
3.0	2.1	0.47	0.57
4.0	2.8	0.62	0.76
5.0	3.5	0.78	0.96
6.0	4.2	0.94	1.15
8.0	5.6	1.25	1.53
10.0	7.0	1.56	1.91
12.0	8.4	1.87	2.29
15.0	10.5	2.34	2.87
18.0	12.6	2.81	3.44
20.0	14.0	3.12	3.82
22.0	15.4	3.43	4.20
25.0	17.5	3.90	4.78

FOR EXPLANATION OF TABLES SEE NOTE 5.2.

Bolt and Weld Data for S355

| BS EN 1993-1-8:2005
BS EN ISO 4016
BS EN ISO 4018 | **BOLT RESISTANCES** | | | S355 |

Non Preloaded bolts

Class 4.6 hexagon head bolts

Diameter of Bolt	Tensile Stress Area	Tension Resistance	Shear Resistance		Bolts in tension
			Single Shear	Double Shear	Min thickness for punching shear
d	A_s	$F_{t,Rd}$	$F_{v,Rd}$	$2 \times F_{v,Rd}$	t_{min}
mm	mm^2	kN	kN	kN	mm
12	84.3	24.3	13.8	27.5	1.9
16	157	45.2	30.1	60.3	2.8
20	245	70.6	47.0	94.1	3.4
24	353	102	67.8	136	4.1
30	561	162	108	215	5.1

Diameter of Bolt	Minimum				Bearing Resistance (kN)											
	Edge distance	End distance	Pitch	Gauge												
d	e_2	e_1	p_1	p_2	Thickness in mm of ply, t											
mm	mm	mm	mm	mm	5	6	7	8	9	10	12	15	20	25	30
12	20	25	35	40	29.7	35.6	41.5	47.4	53.4	59.3	71.2	89.0	119	148	178
16	25	35	50	50	39.0	46.8	54.6	62.4	70.2	78.0	93.6	117	156	195	234
20	30	40	60	60	48.3	57.9	67.6	77.2	86.9	96.5	116	145	193	241	290
24	35	50	70	70	**57.5**	68.9	80.4	91.9	103	115	*138*	*172*	*230*	*287*	*345*
30	45	60	85	90	**72.4**	**86.9**	**101**	116	130	145	*174*	*217*	*290*	*362*	*434*

For M12 bolts the design shear resistance $F_{v,Rd}$ has been calculated as 0.85 times the value given in BS EN 1993-1-8, Table 3.4 (Cl3.6.1(5))

See clause 3.7(1) of BS EN 1993-1-8: 2005 for calculation of the design resistance of a group of fasteners.

For bolts with cut threads that do not comply with EN 1090, the given values for tension and shear should be multiplied by 0.85

Values of bearing resistance in **bold** are less than the single shear resistance of the bolt.

Values of bearing resistance in *italic* are greater than the double shear resistance of the bolt.

Bearing values assume standard clearance holes.

If oversize or short slotted holes are used, bearing values should be multiplied by 0.8.

If long slotted or kidney shaped holes are used, bearing values should be multiplied by 0.6.

In single lap joints with only one bolt row, the design bearing resistance for each bolt should be limited to $1.5f_u \, d \, t/\gamma_{M2}$

FOR EXPLANATION OF TABLES SEE NOTE 5.

BS EN 1993-1-8:2005
BS EN ISO 4014
BS EN ISO 4017

BOLT RESISTANCES

S355

Non Preloaded bolts

Class 8.8 hexagon head bolts

Diameter of Bolt	Tensile Stress Area	Tension Resistance	Shear Resistance		Bolts in tension
			Single Shear	Double Shear	Min thickness for punching shear
d	A_s	$F_{t,Rd}$	$F_{v,Rd}$	$2 \times F_{v,Rd}$	t_{min}
mm	mm^2	kN	kN	kN	mm
12	84.3	48.6	27.5	55.0	3.7
16	157	90.4	60.3	121	5.5
20	245	141	94.1	188	6.8
24	353	203	136	271	8.2
30	561	323	215	431	10.1

Diameter of Bolt	Minimum				Bearing Resistance (kN)										
	Edge distance	End distance	Pitch	Gauge											
d	e_2	e_1	p_1	p_2					Thickness in mm of ply, t						
mm	mm	mm	mm	mm	5	6	7	8	9	10	12	15	20	25	30
12	20	25	35	40	29.7	35.6	41.5	47.4	53.4	59.3	71.2	89.0	119	148	178
16	25	35	50	50	39.0	46.8	54.6	62.4	70.2	78.0	93.6	117	156	195	234
20	30	40	60	60	48.3	57.9	67.6	77.2	86.9	96.5	116	145	193	241	290
24	35	50	70	70	57.5	68.9	80.4	91.9	103	115	138	172	230	287	345
30	45	60	85	90	72.4	86.9	101	116	130	145	174	217	290	362	434

Diameter of Bolt	Minimum				Bearing Resistance (kN)										
	Edge distance	End distance	Pitch	Gauge											
d	e_2	e_1	p_1	p_2					Thickness in mm of ply, t						
mm	mm	mm	mm	mm	5	6	7	8	9	10	12	15	20	25	30
12	25	40	50	45	48.3	58.0	67.7	77.3	87.0	96.7	116	145	193	242	290
16	30	50	65	55	66.8	80.2	93.6	107	120	134	160	201	267	334	401
20	35	60	80	70	85.5	103	120	137	154	171	205	256	342	427	513
24	40	75	95	80	104	125	146	167	187	208	250	312	416	521	625
30	50	90	115	100	128	154	179	205	231	256	308	385	513	641	769

For M12 bolts the design shear resistance $F_{v,Rd}$ has been calculated as 0.85 times the value given in BS EN 1993-1-8, Table 3.4 (Cl3.6.1(5))

See clause 3.7(1) of BS EN 1993-1-8: 2005 for calculation of the design resistance of a group of fasteners.

For bolts with cut threads that do not comply with EN 1090, the given values for tension and shear should be multiplied by 0.85

Values of bearing resistance in **bold** are less than the single shear resistance of the bolt.

Values of bearing resistance in *italic* are greater than the double shear resistance of the bolt.

Bearing values assume standard clearance holes.

If oversize or short slotted holes are used, bearing values should be multiplied by 0.8.

If long slotted or kidney shaped holes are used, bearing values should be multiplied by 0.6.

In single lap joints with only one bolt row, the design bearing resistance for each bolt should be limited to $1.5f_u \, d \, t/\gamma_{M2}$

FOR EXPLANATION OF TABLES SEE NOTE 5.

BS EN 1993-1-8:2005
BS EN ISO 4014
BS EN ISO 4017

BOLT RESISTANCES   S355

Non Preloaded bolts

Class 10.9 hexagon head bolts

Diameter of Bolt	Tensile Stress Area	Tension Resistance	Shear Resistance		Bolts in tension
			Single Shear	Double Shear	Min thickness for punching shear
d	A_s	$F_{t,Rd}$	$F_{v,Rd}$	$2 \times F_{v,Rd}$	t_{min}
mm	mm²	kN	kN	kN	mm
12	84.3	60.7	28.7	57.3	4.6
16	157	113	62.8	126	6.9
20	245	176	98.0	196	8.5
24	353	254	141	282	10.2
30	561	404	224	449	12.7

Diameter of Bolt	Minimum				Bearing Resistance (kN)										
	Edge distance	End distance	Pitch	Gauge	Thickness in mm of ply, t										
d	e_2	e_1	p_1	p_2	5	6	7	8	9	10	12	15	20	25	30
mm	mm	mm	mm	mm											
12	20	25	35	40	29.7	35.6	41.5	47.4	53.4	59.3	71.2	89.0	119	148	178
16	25	35	50	50	39.0	46.8	54.6	62.4	70.2	78.0	93.6	117	156	195	234
20	30	40	60	60	48.3	57.9	67.6	77.2	86.9	96.5	116	145	193	241	290
24	35	50	70	70	57.5	68.9	80.4	91.9	103	115	138	172	230	287	345
30	45	60	85	90	72.4	86.9	101	116	130	145	174	217	290	362	434

Diameter of Bolt	Minimum				Bearing Resistance (kN)										
	Edge distance	End distance	Pitch	Gauge	Thickness in mm of ply, t										
d	e_2	e_1	p_1	p_2	5	6	7	8	9	10	12	15	20	25	30
mm	mm	mm	mm	mm											
12	25	40	50	45	48.3	58.0	67.7	77.3	87.0	96.7	116	145	193	242	290
16	30	50	65	55	66.8	80.2	93.6	107	120	134	160	201	267	334	401
20	35	60	80	70	85.5	103	120	137	154	171	205	256	342	427	513
24	40	75	95	80	104	125	146	167	187	208	250	312	416	521	625
30	50	90	115	100	128	154	179	205	231	256	308	385	513	641	769

For M12 bolts the design shear resistance $F_{v,Rd}$ has been calculated as 0.85 times the value given in BS EN 1993-1-8, Table 3.4 (Cl3.6.1(5))
See clause 3.7(1) of BS EN 1993-1-8: 2005 for calculation of the design resistance of a group of fasteners.
For bolts with cut threads that do not comply with EN 1090, the given values for tension and shear should be multiplied by 0.85
Values of bearing resistance in **bold** are less than the single shear resistance of the bolt.
Values of bearing resistance in *italic* are greater than the double shear resistance of the bolt.
Bearing values assume standard clearance holes.
If oversize or short slotted holes are used, bearing values should be multiplied by 0.8.
If long slotted or kidney shaped holes are used, bearing values should be multiplied by 0.6.
In single lap joints with only one bolt row, the design bearing resistance for each bolt should be limited to $1.5 f_u d\, t / \gamma_{M2}$
FOR EXPLANATION OF TABLES SEE NOTE 5.

BS EN 1993-1-8:2005
BS EN ISO 4016
BS EN ISO 4018

BOLT RESISTANCES

S355

Non Preloaded bolts

Class 4.6 countersunk bolts

Diameter of Bolt	Tensile Stress Area	Tension Resistance	Shear Resistance		Bolts in tension
			Single Shear	Double Shear	Min thickness for punching shear
d mm	A_s mm^2	$F_{t,Rd}$ kN	$F_{v,Rd}$ kN	$2 \times F_{v,Rd}$ kN	t_{min} mm
12	84.3	17.0	13.8	27.5	1.3
16	157	31.7	30.1	60.3	1.9
20	245	49.4	47.0	94.1	2.4
24	353	71.2	67.8	136	2.9
30	561	113	108	215	3.6

Diameter of Bolt	Minimum				Bearing Resistance (kN)											
	Edge distance	End distance	Pitch	Gauge	Thickness in mm of ply, t											
d mm	e_2 mm	e_1 mm	p_1 mm	p_2 mm	5	6	7	8	9	10	12	15	20	25	30
12	20	25	35	40	11.9	17.8	23.7	29.7	35.6	41.5	53.4	71.2	101	130	160
16	25	35	50	50	7.80	15.6	23.4	31.2	39.0	46.8	62.4	85.8	125	164	203
20	30	40	60	60	0	9.65	19.3	29.0	38.6	48.3	67.6	96.5	145	193	241
24	35	50	70	70	0	0	11.5	23.0	34.5	46.0	68.9	103	161	218	276
30	45	60	85	90	0	0	0	7.24	21.7	36.2	65.2	109	181	253	326

For M12 bolts the design shear resistance $F_{v,Rd}$ has been calculated as 0.85 times the value given in BS EN 1993-1-8, Table 3.4 (Cl3.6.1(5))

See clause 3.7(1) of BS EN 1993-1-8: 2005 for calculation of the design resistance of a group of fasteners.

For bolts with cut threads that do not comply with EN 1090, the given values for tension and shear should be multiplied by 0.85

Values of bearing resistance in **bold** are less than the single shear resistance of the bolt.

Values of bearing resistance in *italic* are greater than the double shear resistance of the bolt.

Bearing values assume standard clearance holes.

If oversize or short slotted holes are used, bearing values should be multiplied by 0.8.

If long slotted or kidney shaped holes are used, bearing values should be multiplied by 0.6.

In single lap joints with only one bolt row, the design bearing resistance for each bolt should be limited to $1.5f_u$ d t/γ_{M2}

FOR EXPLANATION OF TABLES SEE NOTE 5.

BS EN 1993-1-8:2005
BS EN ISO 4014
BS EN ISO 4017

BOLT RESISTANCES

S355

Non Preloaded bolts

Class 8.8 countersunk bolts

Diameter of Bolt	Tensile Stress Area	Tension Resistance	Shear Resistance		Bolts in tension
			Single Shear	Double Shear	Min thickness for punching shear
d	A_s	$F_{t,Rd}$	$F_{v,Rd}$	$2 \times F_{v,Rd}$	t_{min}
mm	mm^2	kN	kN	kN	mm
12	84.3	34.0	27.5	55.0	2.6
16	157	63.3	60.3	121	3.9
20	245	98.8	94.1	188	4.8
24	353	142	136	271	5.7
30	561	226	215	431	7.1

Diameter of Bolt	Minimum				Bearing Resistance (kN)											
	Edge distance	End distance	Pitch	Gauge												
d	e_2	e_1	p_1	p_2	Thickness in mm of ply, t											
mm	mm	mm	mm	mm	5	6	7	8	9	10	12	15	20	25	30
12	20	25	35	40	**11.9**	**17.8**	**23.7**	**29.7**	35.6	41.5	53.4	71.2	*101*	*130*	*160*
16	25	35	50	50	**7.80**	**15.6**	**23.4**	**31.2**	**39.0**	**46.8**	62.4	85.8	*125*	*164*	*203*
20	30	40	60	60	0	**9.65**	**19.3**	**29.0**	**38.6**	**48.3**	**67.6**	96.5	145	*193*	*241*
24	35	50	70	70	0	0	**11.5**	**23.0**	**34.5**	**46.0**	**68.9**	103	161	218	*276*
30	45	60	85	90	0	0	0	**7.24**	**21.7**	**36.2**	**65.2**	109	181	253	326

Diameter of Bolt	Minimum				Bearing Resistance (kN)											
	Edge distance	End distance	Pitch	Gauge												
d	e_2	e_1	p_1	p_2	Thickness in mm of ply, t											
mm	mm	mm	mm	mm	5	6	7	8	9	10	12	15	20	25	30
12	25	45	55	45	*22.6*	*33.8*	*45.1*	**56.4**	**67.7**	**79.0**	102	135	*192*	*248*	*305*
16	30	55	70	55	*15.0*	*30.1*	*45.1*	60.2	75.2	90.2	120	165	*241*	*316*	*391*
20	35	70	85	70	0	*18.8*	*37.6*	56.4	75.2	94.0	132	188	*282*	*376*	*470*
24	40	80	100	80	0	0	*22.6*	45.1	67.7	90.2	135	203	316	*429*	*541*
30	50	100	125	100	0	0	0	*14.1*	42.3	70.5	127	212	353	**494**	635

For M12 bolts the design shear resistance $F_{v,Rd}$ has been calculated as 0.85 times the value given in BS EN 1993-1-8, Table 3.4 (Cl3.6.1(5))

See clause 3.7(1) of BS EN 1993-1-8: 2005 for calculation of the design resistance of a group of fasteners.

For bolts with cut threads that do not comply with EN 1090, the given values for tension and shear should be multiplied by 0.85

Values of bearing resistance in **bold** are less than the single shear resistance of the bolt.

Values of bearing resistance in *italic* are greater than the double shear resistance of the bolt.

Bearing values assume standard clearance holes.

If oversize or short slotted holes are used, bearing values should be multiplied by 0.8.

If long slotted or kidney shaped holes are used, bearing values should be multiplied by 0.6.

In single lap joints with only one bolt row, the design bearing resistance for each bolt should be limited to $1.5f_u \, d \, t/\gamma_{M2}$

FOR EXPLANATION OF TABLES SEE NOTE 5.

BS EN 1993-1-8:2005
BS EN ISO 4014
BS EN ISO 4017

BOLT RESISTANCES

S355

Non Preloaded bolts

Class 10.9 countersunk bolts

Diameter of Bolt	Tensile Stress Area	Tension Resistance	Shear Resistance		Bolts in tension
			Single Shear	Double Shear	Min thickness for punching shear
d	A_s	$F_{t,Rd}$	$F_{v,Rd}$	$2 \times F_{v,Rd}$	t_{min}
mm	mm^2	kN	kN	kN	mm
12	84.3	42.5	28.7	57.3	3.2
16	157	79.1	62.8	126	4.8
20	245	123	98.0	196	6.0
24	353	178	141	282	7.2
30	561	283	224	449	8.9

Diameter of Bolt	Minimum				Bearing Resistance (kN)											
	Edge distance	End distance	Pitch	Gauge												
d	e_2	e_1	p_1	p_2					Thickness in mm of ply, t							
mm	mm	mm	mm	mm	5	6	7	8	9	10	12	15	20	25	30
12	20	25	35	40	11.9	17.8	23.7	29.7	35.6	41.5	53.4	71.2	101	130	160
16	25	35	50	50	7.80	15.6	23.4	31.2	39.0	46.8	62.4	85.8	125	164	203
20	30	40	60	60	0	9.65	19.3	29.0	38.6	48.3	67.6	96.5	145	193	241
24	35	50	70	70	0	0	11.5	23.0	34.5	46.0	68.9	103	161	218	276
30	45	60	85	90	0	0	0	7.24	21.7	36.2	65.2	109	181	253	326

Diameter of Bolt	Minimum				Bearing Resistance (kN)											
	Edge distance	End distance	Pitch	Gauge												
	e_2	e_1	p_1	p_2					Thickness in mm of ply, t							
mm	mm	mm	mm	mm	5	6	7	8	9	10	12	15	20	25	30
12	25	45	55	45	22.6	33.8	45.1	56.4	67.7	79.0	102	135	192	248	305
16	30	55	70	55	15.0	30.1	45.1	60.2	75.2	90.2	120	165	241	316	391
20	35	70	85	70	0	18.8	37.6	56.4	75.2	94.0	132	188	282	376	470
24	40	80	100	80	0	0	22.6	45.1	67.7	90.2	135	203	316	429	541
30	50	100	125	100	0	0	0	14.1	42.3	70.5	127	212	353	494	635

For M12 bolts the design shear resistance $F_{v,Rd}$ has been calculated as 0.85 times the value given in BS EN 1993-1-8, Table 3.4 (Cl3.6.1(5))

See clause 3.7(1) of BS EN 1993-1-8: 2005 for calculation of the design resistance of a group of fasteners.

For bolts with cut threads that do not comply with EN 1090, the given values for tension and shear should be multiplied by 0.85

Values of bearing resistance in **bold** are less than the single shear resistance of the bolt.

Values of bearing resistance in *italic* are greater than the double shear resistance of the bolt.

Bearing values assume standard clearance holes.

If oversize or short slotted holes are used, bearing values should be multiplied by 0.8.

If long slotted or kidney shaped holes are used, bearing values should be multiplied by 0.6.

In single lap joints with only one bolt row, the design bearing resistance for each bolt should be limited to $1.5 f_u d \, t / \gamma_{M2}$

FOR EXPLANATION OF TABLES SEE NOTE 5.

BS EN 1993-1-8:2005
BS EN 14399:2005
EN 1090:2008

BOLT RESISTANCES

S355

Preloaded bolts at serviceability limit state

Class 8.8 hexagon head bolts

Diameter of Bolt	Tensile Stress Area	Tension Resistance	Shear Resistance		Bolts in tension	Slip Resistance $\mu = 0.5$	
			Single Shear	Double Shear	Min thickness for punching shear	Single Shear	Double Shear
d	A_s	$F_{t,Rd}$	$F_{v,Rd}$	$2 \times F_{v,Rd}$	t_{min}		
mm	mm^2	kN	kN	kN	mm	kN	kN
12	84.3	48.6	27.5	55.0	3.24	21.5	42.9
16	157	90.4	60.3	121	4.88	40.0	79.9
20	245	141	94.1	188	6.42	62.4	125
24	353	203	136	271	7.17	89.9	180
30	561	323	215	431	9.30	143	286

Diameter of Bolt	Minimum				Bearing Resistance (kN)											
	Edge distance	End distance	Pitch	Gauge	Thickness in mm of ply, t											
d	e_2	e_1	p_1	p_2	5	6	7	8	9	10	12	15	20	25	30
mm	mm	mm	mm	mm											
12	20	40	50	40	44.5	53.4	62.3	71.2	80.1	89.0	107	133	178	222	267
16	25	50	65	50	58.5	70.2	81.9	93.6	105	117	140	176	234	293	351
20	30	60	80	60	72.4	86.9	101	116	130	145	174	217	290	362	434
24	35	75	95	70	86.2	103	121	138	155	172	207	259	345	431	517
30	45	90	115	90	109	130	152	174	195	217	261	326	434	543	652

For M12 bolts the design shear resistance $F_{v,Rd}$ has been calculated as 0.85 times the value given in BS EN 1993-1-8, Table 3.4 (Cl3.6.1(5))

See clause 3.7(1) of BS EN 1993-1-8: 2005 for calculation of the design resistance of a group of fasteners.

For bolts with cut threads that do not comply with EN 1090, the given values for tension and shear should be multiplied by 0.85

Values of bearing resistance in **bold** are less than the single shear resistance of the bolt.

Values of bearing resistance in *italic* are greater than the double shear resistance of the bolt.

The tension resistance, the shear resistance and the bearing resistance should be greater than the design ultimate force

The slip resistance should be greater than the design force at serviceability

In single lap joints with only one bolt row, the design bearing resistance for each bolt should be limited to 1.5f_u d t/γ_{M2}

Values have been calculated assuming k_s=1 and μ=0.5. See BS EN 1993-1-8, section 3.9 for other values of k_s and μ

FOR EXPLANATION OF TABLES SEE NOTE 5.

BS EN 1993-1-8:2005
BS EN 14399:2005
EN 1090:2008

BOLT RESISTANCES

S355

Preloaded bolts at serviceability limit state

Class 10.9 hexagon head bolts

Diameter of Bolt	Tensile Stress Area	Tension Resistance	Shear Resistance		Bolts in tension	Slip Resistance $\mu = 0.5$	
			Single Shear	Double Shear	Min thickness for punching shear	Single Shear	Double Shear
d	A_s	$F_{t,Rd}$	$F_{v,Rd}$	$2 \times F_{v,Rd}$	t_{min}		
mm	mm^2	kN	kN	kN	mm	kN	kN
16	157	113	62.8	126	6.10	50.0	99.9
20	245	176	98.0	196	8.03	78.0	156
24	353	254	141	282	8.97	112	225
30	561	404	224	449	11.6	179	357

Diameter of Bolt	Minimum				Bearing Resistance (kN)											
	Edge distance	End distance	Pitch	Gauge	Thickness in mm of ply, t											
d	e_2	e_1	p_1	p_2	5	6	7	8	9	10	12	15	20	25	30
mm	mm	mm	mm	mm											
16	25	50	65	50	58.5	70.2	81.9	93.6	105	117	140	176	234	293	351
20	30	60	80	60	72.4	86.9	101	116	130	145	174	217	290	362	434
24	35	75	95	70	86.2	103	121	138	155	172	207	259	345	431	517
30	45	90	115	90	109	130	152	174	195	217	261	326	434	543	652

For M12 bolts the design shear resistance $F_{v,Rd}$ has been calculated as 0.85 times the value given in BS EN 1993-1-8, Table 3.4 (Cl3.6.1(5))

See clause 3.7(1) of BS EN 1993-1-8: 2005 for calculation of the design resistance of a group of fasteners.

For bolts with cut threads that do not comply with EN 1090, the given values for tension and shear should be multiplied by 0.85

Values of bearing resistance in **bold** are less than the single shear resistance of the bolt.

Values of bearing resistance in *italic* are greater than the double shear resistance of the bolt.

The tension resistance, the shear resistance and the bearing resistance should be greater than the design ultimate force

The slip resistance should be greater than the design force at serviceability

In single lap joints with only one bolt row, the design bearing resistance for each bolt should be limited to 1.5f_u d t/γ_{M2}

Values have been calculated assuming k_s=1 and μ=0.5. See BS EN 1993-1-8, section 3.9 for other values of k_s and μ

FOR EXPLANATION OF TABLES SEE NOTE 5.

BS EN 1993-1-8:2005
BS EN 14399:2005
EN 1090:2008

BOLT RESISTANCES   S355

Preloaded bolts at ultimate limit state

Class 8.8 hexagon head bolts

Diameter	Tensile	Bolts in tension		Slip Resistance							
of	Stress	Tension	Min thcks	$\mu = 0.2$		$\mu = 0.3$		$\mu = 0.4$		$\mu = 0.5$	
Bolt	Area	Resistance	for punching shear	Single Shear	Double Shear	Single Shear	Double Shear	Single Shear	Double Shear	Single Shear	Double Shear
d	A_s	$F_{t,Rd}$	t_{min}								
mm	mm^2	kN	mm	kN	kN	kN	kN	kN	kN	kN	kN
12	84.3	48.6	3.24	7.55	15.1	11.3	22.7	15.1	30.2	18.9	37.8
16	157	90.4	4.88	14.1	28.1	21.1	42.2	28.1	56.3	35.2	70.3
20	245	141	6.42	22.0	43.9	32.9	65.9	43.9	87.8	54.9	110
24	353	203	7.17	31.6	63.3	47.4	94.9	63.3	127	79.1	158
30	561	323	9.30	50.3	101	75.4	151	101	201	126	251

Diameter	Minimum				Bearing Resistance (kN)										
of	Edge distance	End distance	Pitch	Gauge											
Bolt					Thickness in mm of ply, t										
d	e_2	e_1	p_1	p_2	5	6	7	8	9	10	12	15	20	25	30
mm	mm	mm	mm	mm											
12	20	40	50	40	44.5	53.4	62.3	71.2	80.1	89.0	107	133	178	222	267
16	25	50	65	50	**58.5**	70.2	81.9	93.6	105	117	140	176	234	293	351
20	30	60	80	60	**72.4**	**86.9**	101	116	130	145	174	217	*290*	*362*	*434*
24	35	75	95	70	**86.2**	**103**	**121**	138	155	172	207	259	*345*	*431*	*517*
30	45	90	115	90	**109**	**130**	**152**	**174**	**195**	217	261	326	*434*	*543*	*652*

For bolts with cut threads that do not comply with EN 1090, the given values for tension resistance and minimum thickness to avoid punching should be multiplied by 0.85

Values of bearing resistance in **bold** are less than the single shear resistance of the bolt.

Values of bearing resistance in *italic* are greater than the double shear resistance of the bolt.

In single lap joints with only one bolt row, the design bearing resistance for each bolt should be limited to $1.5f_u\, d\, t/\gamma_{M2}$

FOR EXPLANATION OF TABLES SEE NOTE 5.

BS EN 1993-1-8:2005
BS EN 14399:2005
EN 1090:2008

BOLT RESISTANCES

S355

Preloaded bolts at ultimate limit state

Class 10.9 hexagon head bolts

Diameter	Tensile	Bolts in tension		Slip Resistance							
of	Stress	Tension	Min thcks	$\mu = 0.2$		$\mu = 0.3$		$\mu = 0.4$		$\mu = 0.5$	
Bolt	Area	Resistance	for punching	Single	Double	Single	Double	Single	Double	Single	Double
			shear	Shear	Shear	Shear	Shear	Shear	Shear	Shear	Shear
d	A_s	$F_{t,Rd}$	t_{min}								
mm	mm^2	kN	mm	kN	kN	kN	kN	kN	kN	kN	kN
16	157	113	6.10	17.6	35.2	26.4	52.8	35.2	70.3	44.0	87.9
20	245	176	8.03	27.4	54.9	41.2	82.3	54.9	110	68.6	137
24	353	254	8.97	39.5	79.1	59.3	119	79.1	158	98.8	198
30	561	404	11.6	62.8	126	94.2	188	126	251	157	314

Diameter	Minimum				Bearing Resistance (kN)										
of	Edge	End	Pitch	Gauge											
Bolt	distance	distance													
d	e_2	e_1	p_1	p_2						Thickness in mm of ply, t					
mm	mm	mm	mm	mm	5	6	7	8	9	10	12	15	20	25	30
16	25	50	65	50	**58.5**	70.2	81.9	93.6	105	117	*140*	*176*	*234*	*293*	*351*
20	30	60	80	60	**72.4**	**86.9**	101	116	130	145	174	*217*	*290*	*362*	*434*
24	35	75	95	70	**86.2**	**103**	**121**	138	155	172	207	259	*345*	*431*	*517*
30	45	90	115	90	**109**	**130**	**152**	**174**	**195**	**217**	261	326	434	*543*	*652*

For bolts with cut threads that do not comply with EN 1090, the given values for tension resistance and minimum thickness to avoid punching should be multiplied by 0.85

Values of bearing resistance in **bold** are less than the single shear resistance of the bolt.

Values of bearing resistance in *italic* are greater than the double shear resistance of the bolt.

In single lap joints with only one bolt row, the design bearing resistance for each bolt should be limited to $1.5f_u\, d\, t/\gamma_{M2}$

FOR EXPLANATION OF TABLES SEE NOTE 5.

| BS EN 1993-1-8:2005 |
| BS EN 14399:2005 |
| EN 1090:2008 |

BOLT RESISTANCES

S355

Preloaded bolts at serviceability limit state

Class 8.8 countersunk bolts

Diameter of Bolt	Tensile Stress Area	Tension Resistance	Shear Resistance		Bolts in tension	Slip Resistance $\mu = 0.5$	
			Single Shear	Double Shear	Min thickness for punching shear	Single Shear	Double Shear
d	A_s		$F_{v,Rd}$	$2 \times F_{v,Rd}$	t_{min}		
mm	mm^2	kN	kN	kN	mm	kN	kN
12	84.3	34.0	27.5	55.0	2.27	21.5	42.9
16	157	63.3	60.3	121	3.41	40.0	79.9
20	245	98.8	94.1	188	4.50	62.4	125
24	353	142	136	271	5.02	89.9	180
30	561	226	215	431	6.51	143	286

Diameter of Bolt	Minimum				Bearing Resistance (kN)										
	Edge distance	End distance	Pitch	Gauge	Thickness in mm of ply, t										
d	e_2	e_1	p_1	p_2	5	6	7	8	9	10	12	15	20	25	30
mm	mm	mm	mm	mm											
12	20	40	50	40	17.8	26.7	35.6	44.5	53.4	62.3	80.1	107	151	196	240
16	25	50	65	50	11.7	23.4	35.1	46.8	58.5	70.2	93.6	129	187	246	304
20	30	60	80	60	0	14.5	29.0	43.4	57.9	72.4	101	145	217	290	362
24	35	75	95	70	0	0	17.2	34.5	51.7	68.9	103	155	241	327	414
30	45	90	115	90	0	0	0	10.9	32.6	54.3	97.7	163	272	380	489

For M12 bolts the design shear resistance $F_{v,Rd}$ has been calculated as 0,85 times the value given in BS EN 1993-1-8, Table 3.4 (Cl3.6.1(5))

See clause 3.7(1) of BS EN 1993-1-8: 2005 for calculation of the design resistance of a group of fasteners.

For bolts with cut threads that do not comply with EN 1090, the given values for tension and shear should be multiplied by 0.85

Values of bearing resistance in **bold** are less than the single shear resistance of the bolt.

Values of bearing resistance in *italic* are greater than the double shear resistance of the bolt.

The tension resistance, the shear resistance and the bearing resistance should be greater than the design ultimate force

The slip resistance should be greater than the design force at serviceability

In single lap joints with only one bolt row, the design bearing resistance for each bolt should be limited to 1.5f_u d t/γ_{M2}

Values have been calculated assuming k_s=1 and μ=0.5. See BS EN 1993-1-8, section 3.9 for other values of k_s and μ

FOR EXPLANATION OF TABLES SEE NOTE 5.

BS EN 1993-1-8:2005
BS EN 14399:2005
EN 1090:2008

BOLT RESISTANCES

S355

Preloaded bolts at serviceability limit state

Class 10.9 countersunk bolts

Diameter of Bolt	Tensile Stress Area	Tension Resistance	Shear Resistance		Bolts in tension	Slip Resistance $\mu = 0.5$	
			Single Shear	Double Shear	Min thickness for punching shear	Single Shear	Double Shear
d	A_s		$F_{v,Rd}$	$2 \times F_{v,Rd}$	t_{min}		
mm	mm^2	kN	kN	kN	mm	kN	kN
16	157	79.1	62.8	126	4.27	50.0	99.9
20	245	123	98.0	196	5.62	78.0	156
24	353	178	141	282	6.28	112	225
30	561	283	224	449	8.14	179	357

Diameter of Bolt	Minimum				Bearing Resistance (kN)											
	Edge distance	End distance	Pitch	Gauge	Thickness in mm of ply, t											
d	e_2	e_1	p_1	p_2	5	6	7	8	9	10	12	15	20	25	30
mm	mm	mm	mm	mm											
16	25	50	65	50	11.7	23.4	35.1	46.8	58.5	70.2	93.6	*129*	*187*	*246*	*304*
20	30	60	80	60	0	14.5	29.0	43.4	57.9	72.4	101	145	*217*	*290*	*362*
24	35	75	95	70	0	0	17.2	34.5	51.7	68.9	103	155	241	*327*	*414*
30	45	90	115	90	0	0	0	10.9	32.6	54.3	97.7	**163**	272	380	*489*

For M12 bolts the design shear resistance $F_{v,Rd}$ has been calculated as 0,85 times the value given in BS EN 1993-1-8, Table 3.4 (Cl3.6.1(5))

See clause 3.7(1) of BS EN 1993-1-8: 2005 for calculation of the design resistance of a group of fasteners.

For bolts with cut threads that do not comply with EN 1090, the given values for tension and shear should be multiplied by 0.85

Values of bearing resistance in **bold** are less than the single shear resistance of the bolt.

Values of bearing resistance in *italic* are greater than the double shear resistance of the bolt.

The tension resistance, the shear resistance and the bearing resistance should be greater than the design ultimate force

The slip resistance should be greater than the design force at serviceability

In single lap joints with only one bolt row, the design bearing resistance for each bolt should be limited to $1.5f_u d \, t/\gamma_{M2}$

Values have been calculated assuming k_s=1 and μ=0.5. See BS EN 1993-1-8, section 3.9 for other values of k_s and μ

FOR EXPLANATION OF TABLES SEE NOTE 5.

BS EN 1993-1-8:2005
BS EN 14399:2005
EN 1090:2008

BOLT RESISTANCES

S355

Preloaded bolts at ultimate limit state

Class 8.8 countersunk bolts

Diameter of Bolt	Tensile Stress Area	Bolts in tension		Slip Resistance							
		Tension Resistance	Min thcks for punching shear	μ = 0.2		μ = 0.3		μ = 0.4		μ = 0.5	
				Single Shear	Double Shear	Single Shear	Double Shear	Single Shear	Double Shear	Single Shear	Double Shear
d	A_s	$F_{t,Rd}$	t_{min}								
mm	mm²	kN	mm	kN	kN	kN	kN	kN	kN	kN	kN
12	84.3	34.0	2.27	7.55	15.1	11.3	22.7	15.1	30.2	18.9	37.8
16	157	63.3	3.41	14.1	28.1	21.1	42.2	28.1	56.3	35.2	70.3
20	245	98.8	4.50	22.0	43.9	32.9	65.9	43.9	87.8	54.9	110
24	353	142	5.02	31.6	63.3	47.4	94.9	63.3	127	79.1	158
30	561	226	6.51	50.3	101	75.4	151	101	201	126	251

Diameter of Bolt	Edge distance	End distance	Pitch	Gauge	Bearing Resistance (kN)											
					Thickness in mm of ply, t											
d	e_2	e_1	p_1	p_2	5	6	7	8	9	10	12	15	20	25	30
mm	mm	mm	mm	mm											
12	20	40	50	40	17.8	26.7	35.6	44.5	53.4	62.3	80.1	107	151	196	240
16	25	50	65	50	11.7	23.4	35.1	46.8	58.5	70.2	93.6	129	187	246	304
20	30	60	80	60	0	14.5	29.0	43.4	57.9	72.4	101	145	217	290	362
24	35	75	95	70	0	0	17.2	34.5	51.7	68.9	103	155	241	327	414
30	45	90	115	90	0	0	0	10.9	32.6	54.3	97.7	163	272	380	489

For bolts with cut threads that do not comply with EN 1090, the given values for tension resistance and minimum thickness to avoid punching should be multiplied by 0.85

Values of bearing resistance in **bold** are less than the single shear resistance of the bolt.

Values of bearing resistance in *italic* are greater than the double shear resistance of the bolt.

In single lap joints with only one bolt row, the design bearing resistance for each bolt should be limited to $1.5 f_u d\, t/\gamma_{M2}$

FOR EXPLANATION OF TABLES SEE NOTE 5.

BS EN 1993-1-8:2005
BS EN 14399:2005
EN 1090:2008

BOLT RESISTANCES

S355

Preloaded bolts at ultimate limit state

Class 10.9 countersunk bolts

Diameter of Bolt	Tensile Stress Area	Bolts in tension		Slip Resistance							
		Tension Resistance	Min thcks for punching shear	$\mu = 0.2$		$\mu = 0.3$		$\mu = 0.4$		$\mu = 0.5$	
				Single Shear	Double Shear	Single Shear	Double Shear	Single Shear	Double Shear	Single Shear	Double Shear
d	A_s	$F_{t,Rd}$	t_{min}								
mm	mm^2	kN	mm	kN	kN	kN	kN	kN	kN	kN	kN
16	157	79.1	4.27	17.6	35.2	26.4	52.8	35.2	70.3	44.0	87.9
20	245	123	5.62	27.4	54.9	41.2	82.3	54.9	110	68.6	137
24	353	178	6.28	39.5	79.1	59.3	119	79.1	158	98.8	198
30	561	283	8.14	62.8	126	94.2	188	126	251	157	314

Diameter of Bolt	Edge distance	End distance	Pitch	Gauge	Bearing Resistance (kN)										
					Thickness in mm of ply, t										
d	e_2	e_1	p_1	p_2	5	6	7	8	9	10	12	15	20	25	30
mm	mm	mm	mm	mm											
16	25	50	65	50	**11.7**	**23.4**	**35.1**	**46.8**	**58.5**	70.2	93.6	*129*	*187*	*246*	*304*
20	30	60	80	60	0	**14.5**	**29.0**	**43.4**	**57.9**	**72.4**	101	145	*217*	*290*	*362*
24	35	75	95	70	0	0	**17.2**	**34.5**	**51.7**	**68.9**	**103**	155	*241*	*327*	*414*
30	45	90	115	90	0	0	0	**10.9**	**32.6**	**54.3**	**97.7**	**163**	272	*380*	*489*

For bolts with cut threads that do not comply with EN 1090, the given values for tension resistance and minimum thickness to avoid punching should be multiplied by 0.85

Values of bearing resistance in **bold** are less than the single shear resistance of the bolt.

Values of bearing resistance in *italic* are greater than the double shear resistance of the bolt.

In single lap joints with only one bolt row, the design bearing resistance for each bolt should be limited to $1.5 f_u \, d \, t / \gamma_{M2}$

FOR EXPLANATION OF TABLES SEE NOTE 5.

FILLET WELDS   **S355**

Design weld resistances

Leg Length s mm	Throat Thickness a mm	Longitudinal resistance $F_{w,L,Rd}$ kN/mm	Transverse resistance $F_{w,T,Rd}$ kN/mm
3.0	2.1	0.51	0.62
4.0	2.8	0.68	0.83
5.0	3.5	0.84	1.03
6.0	4.2	1.01	1.24
8.0	5.6	1.35	1.65
10.0	7.0	1.69	2.07
12.0	8.4	2.03	2.48
15.0	10.5	2.53	3.10
18.0	12.6	3.04	3.72
20.0	14.0	3.38	4.14
22.0	15.4	3.71	4.55
25.0	17.5	4.22	5.17

FOR EXPLANATION OF TABLES SEE NOTE 5.2.

Extracts from Concise Eurocodes

IMPORTANT NOTE: The following extracts from the Concise Eurocode, [1] ***itself drawn from BS EN 1993-1-1:2000, are designed to be an aide memoire for some of the latter's key numerical content. They are not complete and should not be used without reference to the full standard to ensure overall compliance with that design standard.***

They are reproduced with the kind permission of the British Standards Institution.

British Standards can be obtained from BSI Customer Services, 389 Chiswick High Road, London W4 4AL, United Kingdom (Tel +44(0)207 8996 9001).

Section, figure and table numbers relate to Steel Building Design: Concise Eurocodes. [1]

1.5 Terminology

The Eurocodes contain different terms from those familiar to UK designers. Some important changes are given below.

Table 2.1   Eurocode terms

Eurocode term	UK term
Actions	Loads
Permanent action	Dead load
Variable action	Imposed, or live load
Design value of actions	Ultimate loads
Verification	Check
Effects	Internal bending moments and forces which result from the application of the actions
Resistance	Capacity, or Resistance
Effects of deformed geometry	Second-order effects

2.3.3 Combination of actions at ULS for persistent or transient design situations

The combinations of actions may either be expressed as:

$$\sum_{j\geq 1} \gamma_{G,j} G_{k,j} + \gamma_{Q,1} Q_{k,1} + \sum_{i\geq 1} \gamma_{Q,i} \psi_{0,i} Q_{k,i}$$

---

[1]Brettle, M.E. and Brown, D.G. *Steel Building Design: Concise Eurocodes. P362* The Steel Construction Institute, 2009.

Table 2.2   Partial factor for actions (γ_F)

Ultimate Limit State	Permanent Actions $\gamma_{G,j}$		Leading or Main Variable Action $\gamma_{Q,1}$	Accompanying Variable Action $\gamma_{Q,i}$
	Unfavourable	Favourable		
EQU	1.1	0.9	1.5	1.5
STR	1.35	1.0	1.5	1.5

Note: When variable actions are favourable Q_k should be taken as zero:

Table 2.3   Values of ψ factors for buildings

Action	ψ_0	ψ_1	ψ_2
Imposed loads in buildings, category (see EN 1991-1-1)			
Category A: domestic, residential areas	0.7	0.5	0.3
Category B: office areas	0.7	0.5	0.3
Category C: congregation areas	0.7	0.7	0.6
Category D: shopping areas	0.7	0.7	0.6
Category E: storage areas	1.0	0.9	0.8
Category H: roofs[a]	0.7	0	0
Snow loads on buildings (see EN 1991-3)			
– for sites located at altitude H > 1000 m a.s.l.	0.70	0.50	0.20
– for sites located at altitude H ≤ 1000 m a.s.l.	0.50	0.20	0
Wind loads on buildings (see (EN 1991-1-4)	0.5	0.2	0
Temperature (non-fire) in buildings (see EN 1991-1-5)	0.6	0.5	0

[a]On roofs, imposed loads should not be combined with either wind loads or snow loads – see BS EN 1991-1-1 Clause 3.1(4)

2.5 General requirements for the design of joints

The values for the partial factors (γ_{Mi}) are as follows:

- Resistance of bolts γ_{M2} 1.25
- Resistance of welds γ_{M2} 1.25
- Resistance of plates in bearing γ_{M2} 1.25
- Slip resistance   at ULS γ_{M3} 1.25
                       at SLS $\gamma_{M3,ser}$ 1.10
- Preload of high strength bolts γ_{M7} 1.10

When determining the tying resistance for structural integrity verifications, the following value for the partial factor (γ_{Mi}) should be used:

- Tying resistance γ_{Mu} 1.10

3.2 Densities of construction materials

Table 3.1 Densities of construction materials

Material	Density (kN/m³)
Concrete*	
normal weight	24
lightweight, density class LC 2.0	20
lightweight, density class LC 1.8	18
Steel	77

*Add 1 kN/m³ for reinforcement and 1 kN/m³ for unhardened concrete.

3.3.2 Characteristic values of imposed loads

1. Imposed load categories and minimum imposed floor loads are given in Table 3.2.
2. Minimum imposed roof loads are given in Table 3.3. See Section 3.4 for snow loads.

Table 3.2 Categories of loaded areas and minimum imposed floor loads

Category	Example	q_k (kN/m²)
A1	All areas within self-contained single family dwellings or modular student accommodation[1] Communal areas (including kitchens) in blocks of flats that are no more than 2 storeys and only 4 dwellings per floor are accessible from a single staircase.	1.5
A2	Bedrooms and dormitories except those in A1 and A3	1.5
A3	Bedrooms in hotels and motels; hospital wards; toilet areas	2.0
B1	General office use other than in B2	2.5
B2	Office areas at or below ground floor level	3.0
C31	Corridors, hallways, aisles which are not subjected to crowds or wheeled vehicles and communal areas in blocks of flats not covered by A1	3.0
C51	Areas susceptible to large crowds	5.0
C52	Stages in public assembly areas (see Note 5)	7.5
D	Areas in general retail shops and department stores	4.0

[1] Each module has a secure door and there are not more than six single bedrooms and an internal corridor.

Minimum imposed floor loads for other areas in categories A to D are given in Table NA.3 of BS EN 1991-1-1 and in Table NA.5 for category E.

4.1.2 Material properties for hot-rolled steel

Table 4.1 Nominal values of yield strength (f_y) and ultimate tensile strength (f_u)

Steel grade and sub-grade	f_y (N/mm^2) Nominal thickness of element t (mm)				f_u (N/mm^2) Nominal thickness t (mm)
	$t \leq 16$	$16 < t \leq 40$	$40 < t \leq 63$	$63 < t \leq 80$	$3 \leq t \leq 100$
S275JR S275J0 S275J2	275	265	255	245	410
S355JR S355J0 S355J2 S355K2	355	345	335	325	470
S355J0H S355J2H S355K2H	355	345	335	325	470

Note 1: As stated in the National Annex to BS EN 1993-1-1, NA.2.4, the ultimate strength f_u should be taken as the lowest value of the range given (in the Product Standard). This minimum value is quoted above.
Note 2: Although not stated in the Eurocodes, for rolled sections, t may be taken as the flange thickness

Table 4.4 Maximum thickness for internal and external steelwork in buildings[1]

Detail type		Tensile stress level, $\sigma_{Ed}/f_y(t)$ [2]									
Description	ΔT_{RD}	Comb. 1	Comb. 2	Comb. 3	Comb. 4	Comb. 5	Comb. 6	Comb. 7	Comb. 8	Comb. 9	Comb. 10
Plain material	+30°	≤0	0.15	0.3	≥0.5						
Bolted	+20°		≤0	0.15	0.3	≥0.5					
Welded – moderate	0°				≤0	0.15	0.3	≥0.5			
Welded – severe	−20°						≤0	0.15	0.3	≥0.5	
Welded – very severe	−30°							≤0	0.15	0.3	≥0.5

Steel grade   **Subgrade**   **Maximum thickness (mm) according to combination of stress level and detail type**

Steel grade	Subgrade	Comb. 1	Comb. 2	Comb. 3	Comb. 4	Comb. 5	Comb. 6	Comb. 7	Comb. 8	Comb. 9	Comb. 10
Internal steelwork $T_{md}=-5°C$											
S275	JR	122.5	102.5	85	70	60	50	40	32.5	27.5	**22.5**
	JO	192.5	172.5	147.5	122.5	102.5	85	70	60	50	**40**
	J2	200	200	192.5	172.5	147.5	122.5	102.5	85	70	**60**
	M, N	200	200	200	192.5	172.5	147.5	122.5	102.5	85	**70**
	ML, NL	200	200	200	200	200	192.5	172.5	147.5	122.5	**102.5**
S355	JR	82.5	67.5	55	45	37.5	30	22.5	17.5	15	**12.5**
	JO	142.5	120	100	82.5	67.5	55	45	37.5	30	**22.5**
	J2	190	167.5	142.5	120	100	82.5	67.5	55	45	**37.5**
	K2, M, N	200	190	167.5	142.5	120	100	82.5	67.5	55	**45**
	ML, NL	200	200	200	190	167.5	142.5	120	100	82.5	**67.5**
External Steelwork $T_{md}=-15°C$											
S275	JR	70	60	50	40	32.5	27.5	22.5	17.5	12.5	**10**
	JO	172.5	147.5	122.5	102.5	85	70	60	50	40	**32.5**
	J2	200	192.5	172.5	147.5	122.5	102.5	85	70	60	**50**
	M, N	200	200	192.5	172.5	147.5	122.5	102.5	85	70	**60**
	ML, NL	200	200	200	200	192.5	172.5	147.5	122.5	102.5	**85**
S355	JR	45	37.5	30	22.5	17.5	15	12.5	10	7.5	**5**
	JO	120	100	82.5	67.5	55	45	37.5	30	22.5	**17.5**
	J2	167.5	142.5	120	100	82.5	67.5	55	45	37.5	**30**
	K2, M, N	190	167.5	142.5	120	100	82.5	67.5	55	45	**37.5**
	ML, NL	200	200	190	167.5	142.5	120	100	82.5	67.5	**55**

Notes:
[1] This Table is based on the following conditions:
i) $\Delta T_{Rg}=0$
ii) $\Delta T_{\varepsilon}=0$
If either of conditions i) or ii) are not complied with, an appropriate adjustment towards the right side of the table should be made. In accordance with BS EN 1993-1-1
[2] $f_y(t)$ should be obtained from the National Annex to BS EN 1993-1-1, Section 2.4.

4.4 Concrete

Table 4.6 Normal concrete material properties

		C25/30	C30/37	C35/45	C40/50
Cylinder strength (MPa)	f_{ck}	25	30	35	40
Cube strength (MPa)	$f_{ck,cube}$	30	37	45	50
Secant modulus of elasticity (GPa)	E_{cm}	31	33	34	35

Table 4.4 Lightweight concrete material properties

		LC25/28	LC30/33	LC35/38
Cylinder strength (MPa)	f_{ck}	25	30	35
Cube strength (MPa)	$f_{ck,cube}$	28	33	38
Secant modulus of elasticity (GPa) E_{lcm}	E_{lcm}	$E_{lcm} = E_{cm}\eta_E$		

Notes:
E_{cm} is given by Table 4.6 for the corresponding value of f_{ck}
$\eta_E = (p/2200)^2 <$ where p denotes the oven-dry density in accordance with BS EN 206-1 Section 4

5.3.2 Imperfections for global analysis of frames

The initial sway imperfections (See Figure 5.1) may be determined from:

$$\phi = \phi_0\alpha_h\alpha_m$$

where:

ϕ_0 $= 1/200$

α_h is the reduction factor for height h applicable to columns: $= \dfrac{2}{\sqrt{h}}$ but

$\dfrac{2}{3} \leq \alpha_h \leq 1.0$

h is the height of the structure in metres

α_m is the reduction factor for the number of columns in a row $= \sqrt{0.5(1+1/m)}$

m is the number of vertical members contributing to the horizontal force on the bracing system.

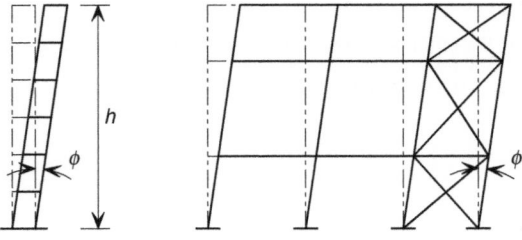

Figure 5.1 Equivalent sway imperfections

Sway imperfections may be disregarded where:

$$H_{Ed} \geq 0.15 V_{Ed}$$

For the determination of horizontal forces to floor diaphragms, the configuration of imperfections as given in Figure 5.2 should be applied, where ϕ is a sway imperfection defined above assuming a single storey with height of the structure in metres, h.

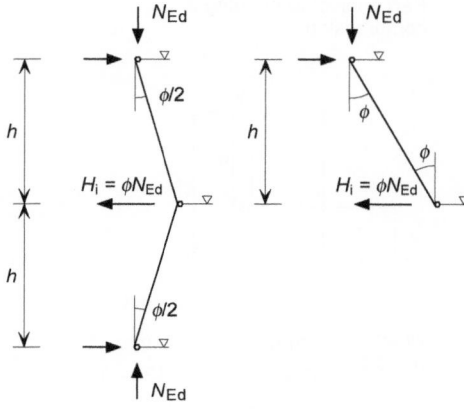

Figure 5.2   Configuration of sway imperfection ϕ for horizontal forces on floor diaphragm

5.5 Classification of cross-sections

Table 5.1　(Sheet 1 of 2): Maximum c/t ratios for compression parts

Internal compression parts

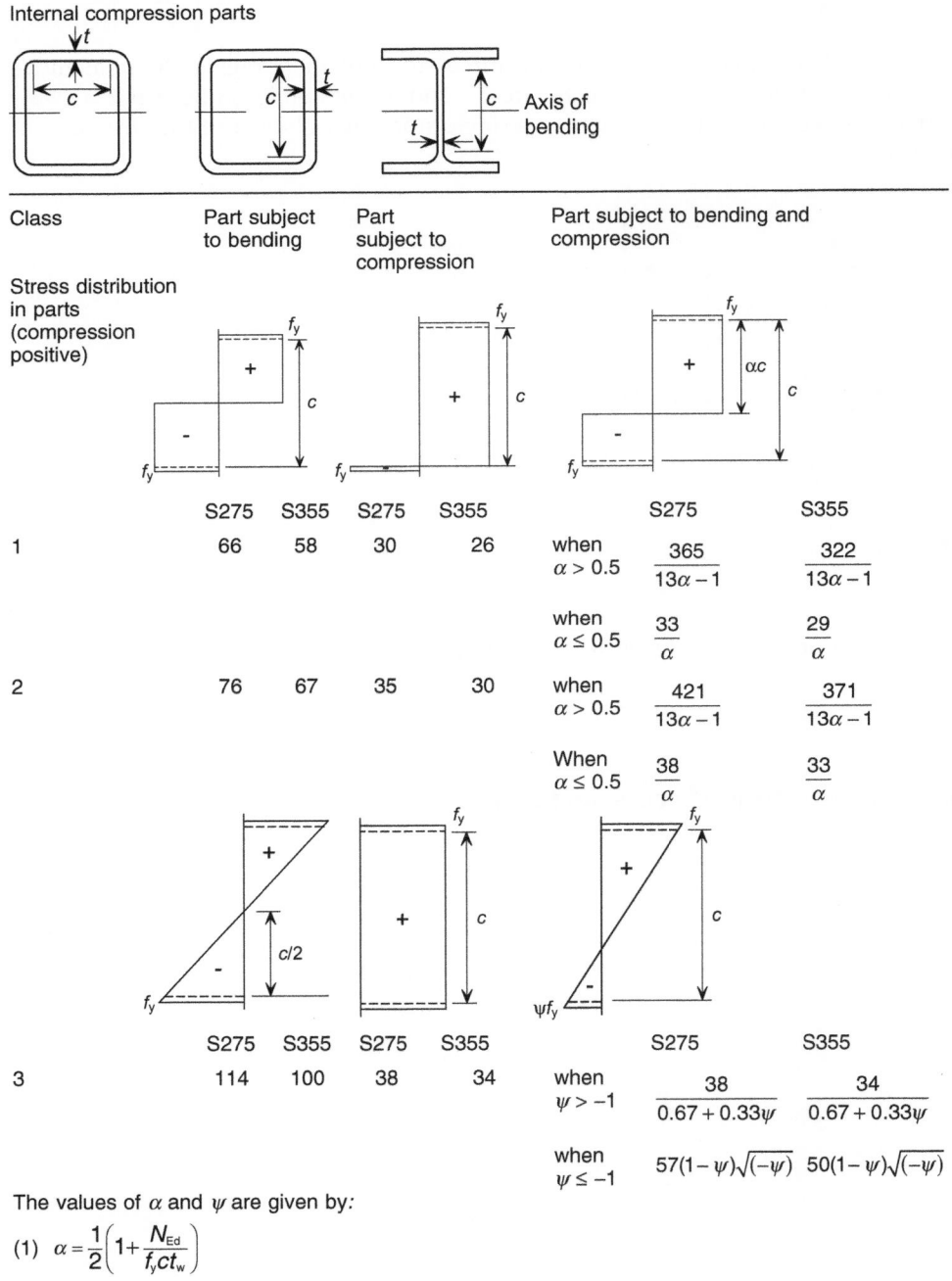

Class	Part subject to bending		Part subject to compression		Part subject to bending and compression	
Stress distribution in parts (compression positive)						
	S275	S355	S275	S355	S275	S355
1	66	58	30	26	when $\alpha > 0.5$ $\dfrac{365}{13\alpha - 1}$	$\dfrac{322}{13\alpha - 1}$
					when $\alpha \leq 0.5$ $\dfrac{33}{\alpha}$	$\dfrac{29}{\alpha}$
2	76	67	35	30	when $\alpha > 0.5$ $\dfrac{421}{13\alpha - 1}$	$\dfrac{371}{13\alpha - 1}$
					When $\alpha \leq 0.5$ $\dfrac{38}{\alpha}$	$\dfrac{33}{\alpha}$
	S275	S355	S275	S355	S275	S355
3	114	100	38	34	when $\psi > -1$ $\dfrac{38}{0.67 + 0.33\psi}$	$\dfrac{34}{0.67 + 0.33\psi}$
					when $\psi \leq -1$ $57(1-\psi)\sqrt{(-\psi)}$	$50(1-\psi)\sqrt{(-\psi)}$

The values of α and ψ are given by:

(1)　$\alpha = \dfrac{1}{2}\left(1 + \dfrac{N_{Ed}}{f_y c t_w}\right)$

(2)　$\psi = \dfrac{2N_{Ed}}{A f_y} - 1$

where N_{Ed} is positive in compression

Table 5.1 (Sheet 2 of 2): Maximum c/t ratios for compression parts

Outstand flanges

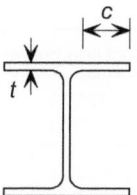 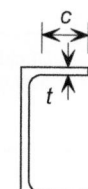

Rolled sections

Class Stress distribution in parts (compression positive	Part subject to compression 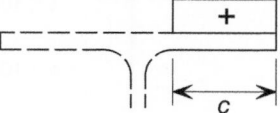	
	S275	S355
1	8	7
2	9	8
3	12	11

Angles

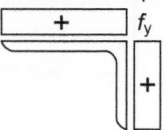

Class Stress distribution across section (compression positive)	Section in compression	
3	S275	S355
	$h/t \le 13 : \dfrac{b+h}{2t} \le 10$	$h/t \le 12 : \dfrac{b+h}{2t} \le 9$

Tubular sections

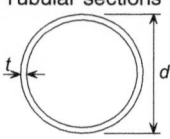

Class	Section in bending and/or compression S355
1	$d/t \le 33$
2	$d/t \le 46$
3	$d/t \le 59$

6.3 Buckling resistance of members

Table 6.1 Selection of flexural buckling curve for a cross-section

Cross-section	Limits		Buckling about axis	Buckling curve S 275 S 355
Rolled sections	$h/b > 1.2$	$t_f \leq 40\,\text{mm}$	$y - y$ $z - z$	a b
		$40\,\text{mm} < t_f \leq 100$	$y - y$ $z - z$	b c
	$h/b \leq 1.2$	$t_f \leq 100\,\text{mm}$	$y - y$ $z - z$	b c
		$t_f > 100\,\text{mm}$	$y - y$ $z - z$	d d
U-, T- and solid sections			any	c
L-sections			any	b
Hollow sections	Hot-finished		any	a
	Cold-formed		any	c

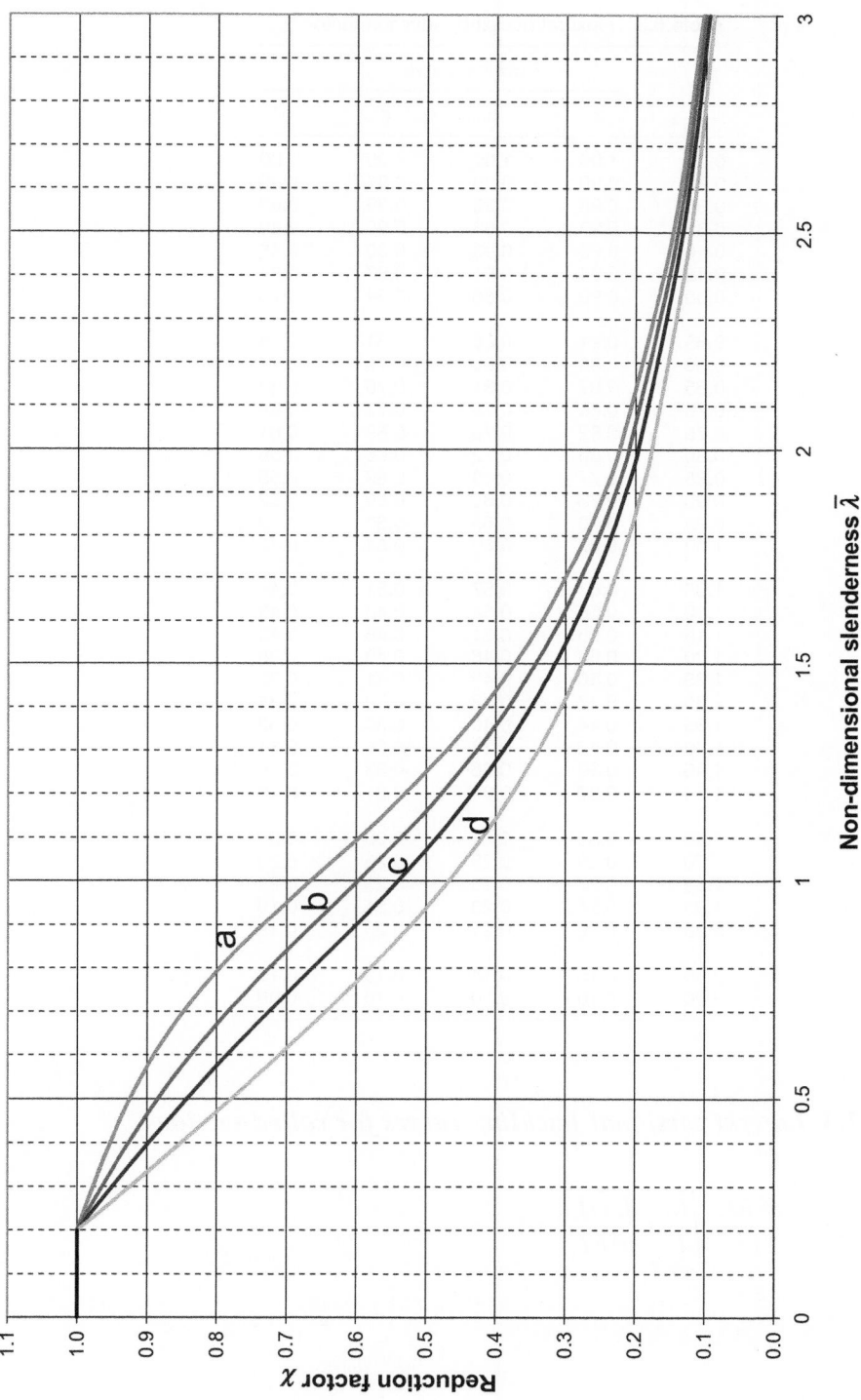

Figure 6.1   Buckling curves

Table 6.3   Flexural buckling reduction factor, χ

$\bar{\lambda}$	Buckling curve			
	a	b	c	d
0.20	1.00	1.00	1.00	1.00
0.25	0.99	0.98	0.97	0.96
0.30	0.98	0.96	0.95	0.92
0.35	0.97	0.95	0.92	0.89
0.40	0.95	0.93	0.90	0.85
0.45	0.94	0.91	0.87	0.81
0.50	0.92	0.88	0.84	0.78
0.55	0.91	0.86	0.81	0.74
0.60	0.89	0.84	0.79	0.71
0.65	0.87	0.81	0.76	0.68
0.70	0.85	0.78	0.72	0.64
0.75	0.82	0.75	0.69	0.61
0.80	0.80	0.72	0.66	0.58
0.85	0.77	0.69	0.63	0.55
0.90	0.73	0.66	0.60	0.52
0.95	0.70	0.63	0.57	0.49
1.00	0.67	0.60	0.54	0.47
1.05	0.63	0.57	0.51	0.44
1.10	0.60	0.54	0.48	0.42
1.15	0.56	0.51	0.46	0.40
1.20	0.53	0.48	0.43	0.38
1.25	0.50	0.45	0.41	0.36
1.30	0.47	0.43	0.39	0.34
1.35	0.44	0.40	0.37	0.32
1.40	0.42	0.38	0.35	0.31
1.45	0.39	0.36	0.33	0.29
1.50	0.37	0.34	0.31	0.28
1.60	0.33	0.31	0.28	0.25
1.70	0.30	0.28	0.26	0.23
1.80	0.27	0.25	0.23	0.21
1.90	0.24	0.23	0.21	0.19
2.00	0.22	0.21	0.20	0.18
2.50	0.15	0.14	0.13	0.12
3.00	0.10	0.10	0.10	0.09

6.3.2.3 *Lateral torsional buckling curves for rolled sections*

$$M_{cr} = C_1 \frac{\pi^2 EI_z}{L^2} \sqrt{\frac{I_w}{I_z} + \frac{L^2 GI_t}{\pi^2 EI_z}}$$

Table 6.4   Values of $\dfrac{1}{\sqrt{C_1}}$ and C_1 for various moment conditions (load is not destabilising)

End Moment Loading	ψ	$\dfrac{1}{\sqrt{C_1}}$	C_1
	+1.00	1.00	1.00
	+0.75	0.92	1.17
	+0.50	0.86	1.36
	+0.25	0.80	1.56
$-1 \le \psi \le +1$	0.00	0.75	1.77
	−0.25	0.71	2.00
	−0.50	0.67	2.24
	−0.75	0.63	2.49
	−1.00	0.60	2.76

Intermediate Transverse Loading	$\dfrac{1}{\sqrt{C_1}}$	C_1
	0.94	1.13
	0.62	2.60
	0.86	1.35
	0.77	1.69

Table 6.6   Recommendations for the selection of lateral torsional buckling curve

Cross-section	Limits	Buckling curve
Rolled doubly symmetric I- and H- sections, and hot-finished hollow sections	$h/b \le 2$	b
	$2 < h/b \le 3.1$	c
	$h/b > 3.1$	d
Angles (for moments in the major principal plane)		D
All other hot-rolled sections		D
Cold-formed hollow sections	$h/b \le 2$	c
	$h/b > 2$	d

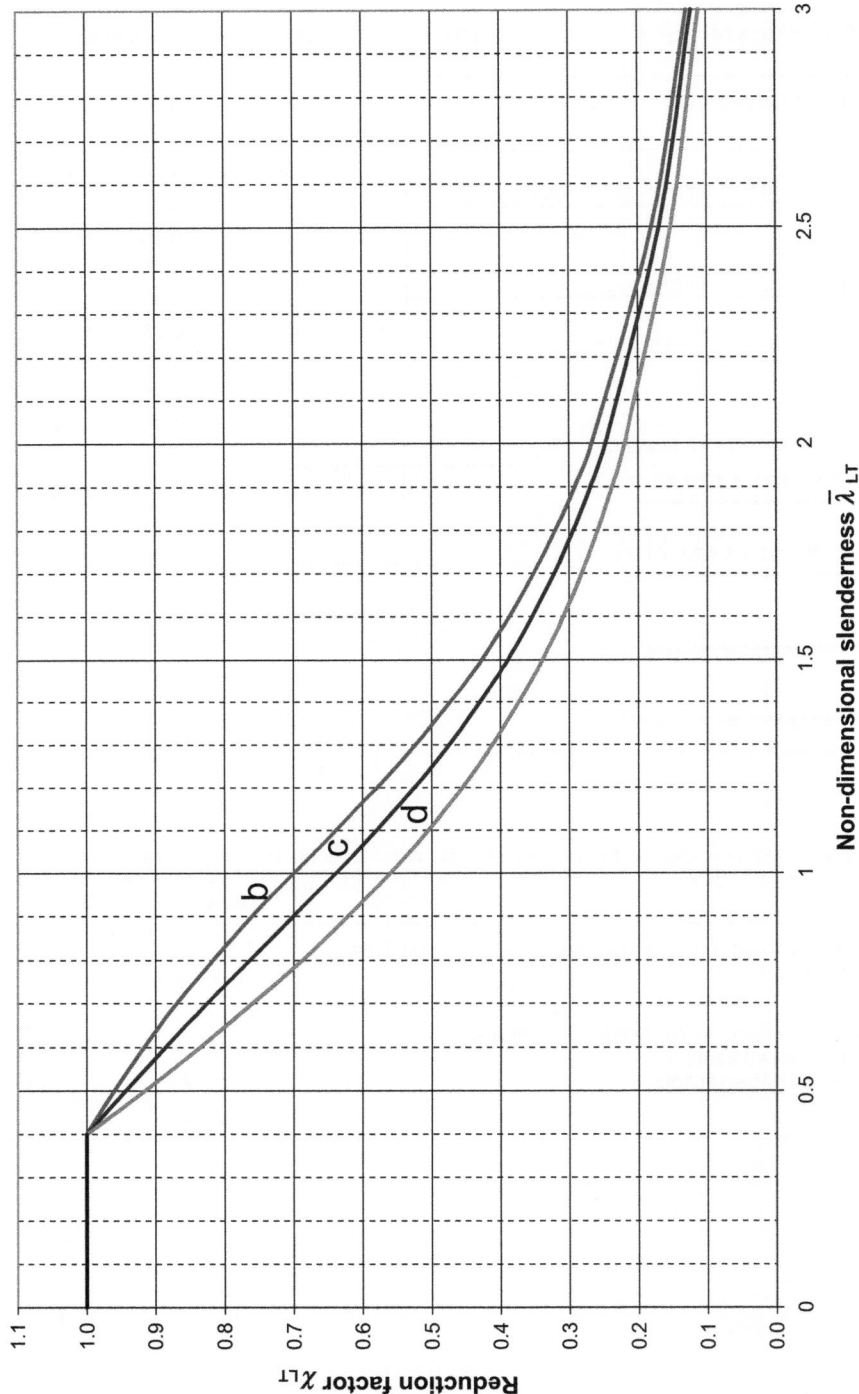

Figure 6.2   Lateral torsional buckling curves for rolled sections

Table 6.7 Lateral torsional buckling reduction factors, χ_{LT}

$\bar{\lambda}_{LT}$	Rolled I, H Sections		
	$h/b \leq 2$	$2 < h/b \leq 3.1$	$h/b > 3.1$
0.20	1.00	1.00	1.00
0.25	1.00	1.00	1.00
0.30	1.00	1.00	1.00
0.35	1.00	1.00	1.00
0.40	1.00	1.00	1.00
0.45	0.98	0.97	0.96
0.50	0.96	0.94	0.92
0.55	0.94	0.92	0.88
0.60	0.92	0.89	0.84
0.65	0.89	0.86	0.80
0.70	0.87	0.83	0.76
0.75	0.84	0.79	0.72
0.80	0.82	0.76	0.69
0.85	0.79	0.73	0.65
0.90	0.76	0.70	0.62
0.95	0.73	0.67	0.59
1.00	0.70	0.64	0.56
1.05	0.67	0.61	0.53
1.10	0.64	0.58	0.50
1.15	0.61	0.55	0.48
1.20	0.58	0.52	0.46
1.25	0.55	0.50	0.43
1.30	0.52	0.47	0.41
1.35	0.50	0.45	0.39
1.40	0.47	0.43	0.37
1.45	0.45	0.41	0.36
1.50	0.43	0.39	0.34
1.60	0.39	0.35	0.31
1.70	0.35	0.32	0.28
1.80	0.32	0.29	0.26
1.90	0.29	0.27	0.24
2.00	0.27	0.25	0.22
2.50	0.18	0.17	0.15
3.00	0.13	0.12	0.11

7.2 Vertical deflections

7.2.1 Steel members

Table 7.1 Suggested limits for vertical deflection due to characteristic combination (variable actions only)

Vertical deflection	
Cantilevers	Length/180
Beams carrying plaster or other brittle finish	Span/360
Other beams (except purlins and sheeting rails)	Span/200
Purlins and sheeting rails	To suit the characteristics of the particular cladding

7.2.3 Steel and concrete composite floors

The deflection limits that should be used for composite floors during execution are:

When ponding is not explicitly considered: $\dfrac{\text{effective span}}{180}$ but $\leq 20\,\text{mm}$

When ponding is explicitly considered: $\dfrac{\text{effective span}}{130}$ but $\leq 30\,\text{mm}$

At the normal stage, the deflection can be calculated using the average of the cracked and uncracked second moment of areas (BS EN 1994-1-1, 9.8.2(5)), and should be limited to effective span/500 for the quasi-permanent actions (BS EN 1992-1-1, 7.4.1(5)).

7.3 Horizontal deflections

Table 7.2 Suggested limits for horizontal deflection

Horizontal deflection	
Tops of columns in single-storey buildings except portal frames	Height/300
Columns in portal frame buildings, not supporting crane runways	To suit the characteristics of the particular cladding
In each storey of a building with more than one storey	Height of that storey/300

8.2 Bolted connections

8.2.3 Categories of bolted connections

Table 8.2 Categories of bolted connections

Category	Criteria	Remarks
Shear Connections		
A Bearing type	$F_{v,Ed} \leq F_{v,Rd}$ $F_{v,Ed} \leq F_{b,Rd}$	No preloading required. Bolt classes 4.6 and 8.8.
B Slip-resistant at serviceability limit state	$F_{v,Ed,ser} \leq F_{s,Rd,ser}$ $F_{v,Ed} \leq F_{v,Rd}$ $F_{v,Ed} \leq F_{b,Rd}$	Preloaded 8.8 bolts should be used. For slip resistance at serviceability see BS EN 1993-1-8 Section 8.2.8
C Slip-resistant at ultimate limit state	$F_{v,Ed} \leq F_{s,Rd}$ $F_{v,Ed} \leq F_{b,Rd}$ $F_{v,Ed} \leq N_{net,Rd}$	Preloaded 8.8 bolts should be used. For slip resistance at serviceability see BS EN 1993-1-8 Section 8.2.8 See note to BS EN 1993-1-8 Section 5.2.1
Tension Connections		
D Non-preloaded	$F_{t,Ed} \leq F_{t,Rd}$ $F_{t,Ed} \leq B_{p,Rd}$	No preloading required. Bolt classes 4.6 and 8.8.
E Preloaded	$F_{t,Ed} \leq F_{t,Rd}$ $F_{t,Ed} \leq B_{p,Rd}$	Preloaded 8.8 bolts should be used.

The design tensile force $F_{t,Ed}$ should include any force due to prying action, see BS EN 1993-1-8 Section 3.11. Bolts subject to both shear force and tensile force should also satisfy the criteria given in Table 8.4

For wind and stability bracings, bolts in Category A connections may be used.

Category D connections may be used in connections designed to resist normal wind loads.

8.2.4 Positioning holes for bolts

Table 8.3 Minimum and maximum spacing, end and edge distances

Distances and spacings	Minimum	Maximum[1] [2] [3]	
		Steel exposed to the weather or other corrosive influences	Steel not exposed to the weather or other corrosive influences
Edge distance e_1	$1.2d_0$	$4t + 40$ mm	
Edge distance e_2	$1.2d_0$	$4t + 40$ mm	
Spacing p_1	$2.2d_0$	The smaller of $14t$ or 200 mm	The smaller of $14t$ or 200 mm
Spacing p_2	$2.4d_0$	The smaller of $14t$ or 200 mm	The smaller of $14t$ or 200 mm

[1] Maximum values for spacings, edge and end distances are unlimited, except in the following cases:
– for compression members in order to avoid local buckling and to prevent corrosion in exposed members and;
– for exposed tension members to prevent corrosion.
[2] The local buckling resistance of the plate in compression between fasteners should be calculated according to BS EN 1993-1-1 Clause 6.3.1.1 using $0.6p_1$ as buckling length. Local buckling between the fasteners need not be checked if p_1/t is smaller than 9ε. The edge distance should not exceed the local buckling requirements for an outstand element in the compression members, see Table 5.1. The end distance is not affected by this requirement.
[3] t is the thickness of the thinner outer connected part.

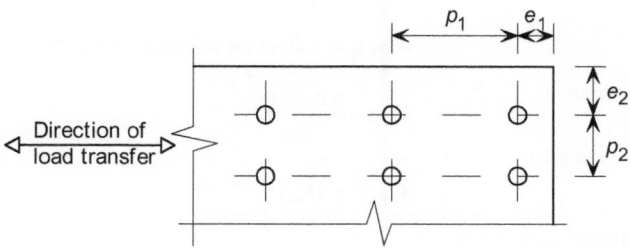

Figure 8.1 Symbols for spacing of fasteners

8.2.5 Design resistance of individual bolts

Table 8.3 Design resistance for individual fasteners subject to shear and/or tension

Failure mode	Design resistance
Shear resistance per shear plane	Where the shear plane passes through the threaded portion of the bolt: $$F_{v,Rd} = \frac{0.6f_{ub}A_s}{\gamma_{M2}}$$ A_s is the tensile stress area of the bolt Where the shear plane passes through the unthreaded portion of the bolt: $$F_{v,Rd} = \frac{0.6f_{ub}A}{\gamma_{M2}}$$ A is the gross cross-sectional area of the bolt.
Bearing resistance [1),2),3)]	$$F_{b,Rd} = \frac{k_1 \alpha_b f_u dt}{\gamma_{M2}}$$ Where d is the nominal diameter of the bolt, α_b is the smallest of α_d, $\frac{f_{ub}}{f_u}$ or 1.0 For end bolts: $\alpha_d = \dfrac{e_1}{3d_0}$ For inner bolts $\alpha_d = \dfrac{p_1}{3d_0} - \dfrac{1}{4}$ For edge bolts k_1 is the smallest of $2.8\dfrac{e_2}{d_0} - 1.7$ or 2.5 For inner bolts k_1 is the smallest of $1.4\dfrac{p_2}{d_0} - 1.7$ or 2.5
Tension resistance [2)]	$$F_{t,Rd} = \frac{k_2 f_{ub} A_s}{\gamma_{M2}}$$ Where $k_2 = 0.63$ for countersunk bolts otherwise $k_2 = 0.9$
Punching shear resistance	$$B_{p,Rd} = \frac{0.6\pi d_m t_p f_u}{\gamma_{M2}}$$
Combined shear and tension	$$\frac{F_{v,Ed}}{F_{v,Rd}} + \frac{F_{t,Ed}}{1.4F_{t,Rd}} \leq 1.0$$

[1)] The bearing resistance $F_{b,Rd}$ for bolts
– In oversized holes is 0.8 times the bearing resistance for bolts in normal holes.
[2)] For a countersunk bolt:
– The bearing resistance $F_{b,Rd}$ should be based on a plate thickness t equal to the thickness of the connected plate minus half the depth of the countersinking.
– For the determination of the tension resistance, $F_{b,Rd}$ the angle and depth of countersinking should conform with the reference standards listed in BS EN 1993-1-8 Clause 1.2.4. Otherwise the tension resistance $F_{t,Rd}$ should be adjusted accordingly.
[3)] When the load on a bolt is not parallel to the edge, the bearing resistance may be verified separately for the bolt load components parallel and normal to the end.

8.3 Welded connections

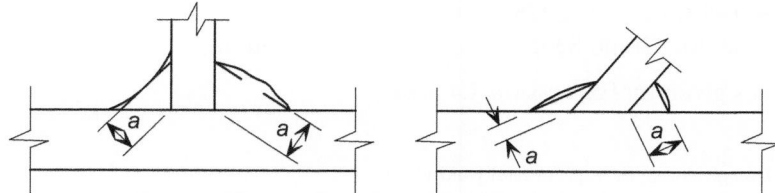

Figure 8.6 Throat thickness of a fillet weld

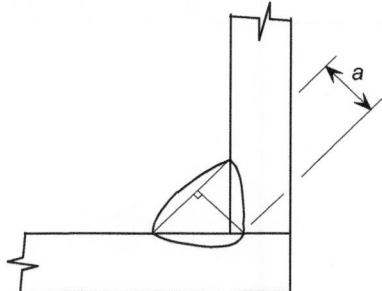

Figure 8.7 Throat thickness of a deep penetration fillet weld

8.3.3.3 *Design resistance of fillet welds*

The design resistance of a fillet weld may be assumed to be adequate if, at every point along its length, the resultant of all the forces per unit length transmitted by the weld satisfy the following criterion:

$$F_{w,Ed} \leq F_{w,Rd}$$

where:

$F_{w,Ed}$ is the design value of the weld force per unit length
$F_{w,Rd}$ is the design weld resistance per unit length.

Independent of the orientation of the weld throat plane to the applied force, the design resistance per unit length $F_{w,Rd}$ should be determined from:

$$F_{w,Rd} = f_{vw,d}a$$

where:

$F_{vw,d}$ is the design shear strength of the weld.

The design shear strength $f_{vw,d}$ of the weld should be determined from:

$$f_{vw,d} = \frac{f_u/\sqrt{3}}{\beta_w \gamma_{M2}}$$

where:

f_u is the nominal ultimate tensile strength of the weaker part joined;

β_w = 0.85 for Grade S275

β_w = 0.9 for Grade S355

The values given by Expression 4.4 are:

Table 8.4   Design shear strength of fillet weld ($f_{vw,d}$)

Steel grade	f_u (N/mm^2)	Thickness of weaker jointed part	$f_{vw,d}$ (N/mm^2)
S275	410	3 mm ≤ t_p ≤ 100 mm	223
S355	470	3 mm ≤ t_p ≤ 100 mm	241

Floor plates

The following traditional approaches to the design of floor plates have been retained from earlier editions of The Steel Designers' Manual. They are based on Pounder's Formula and are still appropriate for initial design. More details of Pounder's Formula may be found at

www.civl.port.ac.uk/britishsteel/media/BSCM%20HTML%20Docs/Notes%20 to%20durbar%20floor%20plate%20tables.html

Ultimate load capacity (kN/m²) for floor plates simply supported on two edges stressed to 275 n/mm²

Thickness on plain mm	Span (mm)							
	600	800	1000	1200	1400	1600	1800	2000
4.5	20.48	11.62	7.45	5.17	3.80	2.95	2.28	1.87
6.0	36.77	20.68	13.28	9.20	6.73	5.20	4.07	3.30
8.0	65.40	36.87	23.48	16.38	11.97	9.23	7.23	5.93
10.0	102.03	57.42	36.67	25.55	18.70	14.45	11.30	9.25
12.5	159.70	89.85	57.40	39.98	29.27	22.62	17.68	14.50

Stiffeners should be used for spans in excess of 1100 mm to avoid excessive deflections.

Ultimate load capacity (kN/m²) for floor plates simply supported on all four edges stressed to 275 N/mm² (Values obtained using Pounder's Formula allowing corners to lift)

Thickness on plain mm	Breadth B mm	Length (mm)							
		600	800	1000	1200	1400	1600	1800	2000
4.5	600	34.9	25.5	22.7	21.7	21.2	21.0	20.8	20.8
	800		19.6	15.1	13.4	12.6	12.2	12.0	11.8
	1000			12.6	10.0	8.8	8.3	7.9	7.7
	1200				8.7	7.1	6.3	5.9	5.6
	1400					6.4	5.3	4.8	4.4
	1600						4.9	4.1	3.7
	1800							3.8	3.3
6.0	600	62.1	45.3	40.4	38.5	37.7	37.3	37.0	36.9
	800		34.9	26.8	23.7	22.3	21.7	21.3	21.1
	1000			22.4	17.8	15.8	14.8	14.2	13.9
	1200				15.5	12.7	11.3	10.6	10.1
	1400					11.4	9.5	8.5	7.9
	1600						8.7	7.4	6.7
	1800							6.9	5.9
8.0	600	110	80.6	71.1	68.4	67.0	66.2	65.8	65.6
	800		62.1	47.7	42.2	39.7	38.5	37.8	37.4
	1000			39.7	31.7	28.1	26.2	25.2	24.6
	1200				27.6	22.6	20.1	18.8	17.9
	1400					20.3	17.0	15.2	14.1
	1600						15.5	13.3	11.9
	1800							12.3	10.6
10.0	600	172*	126*	112*	107*	105*	103*	103*	103*
	800		97.0	74.5	65.9	62.1	60.1	59.1	58.5
	1000			62.1	49.5	43.9	41.0	39.4	38.5
	1200				43.1	35.4	31.5	29.3	28.0
	1400					31.7	26.6	23.8	22.1
	1600						24.3	20.7	18.6
	1800							19.2	16.6
12.5	600	269*	197*	175*	167*	163*	162*	161*	160*
	800		152	116*	103*	97.0*	94.0*	92.3*	91.4*
	1000			97.0	77.4	68.5	64.1	61.6	60.1
	1200				67.4	55.3	49.2	45.8	43.8
	1400					49.5	41.5	37.1	34.5
	1600						37.9	32.4	29.1
	1800							29.9	25.9

Ultimate load capacity (kN/m²) for floor plates fixed on all four edges stressed to 275 N/mm²

Thickness on plain mm	Breadth B mm	Length (mm)							
		600	800	1000	1200	1400	1600	1800	2000
4.5	600	47.7*	36.8*	33.5*	32.2*	31.6*	31.4*	31.2*	31.1*
	800		26.8	21.5*	19.5*	18.6*	18.1*	17.9*	17.7*
	1000			17.2*	14.2*	12.9*	12.2*	11.8*	11.6*
	1200				11.9*	10.1	9.1	8.6	8.3
	1400					8.7	7.5	6.9	6.5
	1600						6.7	5.8	5.3
	1800							5.3	4.7
6.0	600	84.8*	65.4*	59.5*	57.3*	56.2*	55.7*	55.5*	55.3*
	800		47.7*	38.3*	34.7*	33.1*	32.2*	31.7*	31.5*
	1000			30.5*	25.3*	22.9*	21.7*	21.0*	20.6*
	1200				21.2*	18.0*	16.3*	15.4*	14.9*
	1400					15.6*	13.4*	12.3*	11.6
	1600						11.9	10.4	9.5
	1800							9.4	8.3
8.0	600	151*	116*	106*	102*	100*	99.1*	98.6*	98.3*
	800		68.1*	61.7*	58.8*	57.3*	56.4*	55.9*	55.3
	1000			54.3*	44.9*	40.7*	38.6*	37.4*	36.7*
	1200				37.7*	31.9*	29.0*	27.4*	26.5*
	1400					27.7*	23.9*	21.8*	20.6*
	1600						21.2*	18.6*	17.0*
	1800							16.9*	14.8*
10.0	600	236*	182*	165*	159*	156*	155*	154*	154*
	800		132*	106*	96.4*	91.8*	89.5*	88.2*	87.4*
	1000			84.8*	70.2*	63.7*	60.3*	58.4*	57.3*
	1200				58.9*	49.9*	45.4*	42.9*	41.3*
	1400					43.3*	37.3*	34.1*	32.2*
	1600						33.1*	29.0*	26.6*
	1800							26.2*	23.2*
12.5	600	368*	284*	258*	249*	244*	242*	241*	240*
	800		207*	166*	151*	144*	140*	138*	137*
	1000			132*	110*	99.5*	94.2*	91.2*	89.5*
	1200				92.0*	77.9*	70.9*	67.0*	64.6*
	1400					67.6*	58.3*	53.3*	50.3*
	1600						51.8*	45.3*	41.6*
	1800							40.9*	36.2*

Note on tables:

Values without an asterisk cause deflection greater than B / 100 at serviceability, assuming that the only dead load present is due to self-weight.

Fire resistance

STEELWORK IN FIRE
INFORMATION SHEET

This series of information sheets is intended to illustrate methods of achieving fire resistance in steel structures. It should not be used for design without consulting detailed design guidance referenced below.

SPRAYED PROTECTION

UP TO 4 HRS

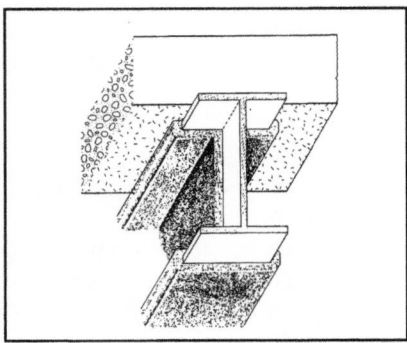

METHOD

Fire protective insulation can be applied by spraying to almost any type of steel member. Most products can achieve up to 4 hours rating.

PRINCIPLE

Insulation reduces the heating rate of a steel member so that its limiting temperature is not exceeded for the required fire resistance period. The protection material thickness necessary depends on the section factor (Hp/A) of the member and the fire rating required.

ADVANTAGES

a) Low cost
b) Rapid application
c) Easy to cover complex details
d) Often applied to non-primed steelwork
e) Some products may be suitable for external use

LIMITATIONS (check with manufacturer)

a) Appearance may be inadequate for visible members
b) Overspray may need masking or shielding
c) Primer, if used, must be compatible

FOR MORE DETAILED INFORMATION SEE:-
"Fire protection of Structural Steel in Building"
Published jointly by:
ASFP and The Steel Construction Institute

Sheet Code
ISF/No.01
January 1997

PROTECTION THICKNESS

Thickness recommendations given in "Fire Protection of Structural Steel in Building" have normally been derived from fire tests on orthodox H or I rolled sections. For other sections the recommended thickness for a given section factor and fire rating should be modified as follows:

CASTELLATED SECTIONS

The thickness of fire protection material on a castellated section should be 20% greater than that required for the section from which it was cut.

HOLLOW SECTIONS

For spray applied fire protection materials the recommended thickness (t) should be increased as follows

For section factor (Hp/A) less than 250
modified thickness = t [1+(Hp/A) / 1000]

For section factor (Hp/A) 250 or over
modified thickness = 1.25 x t

STEELWORK IN FIRE
INFORMATION SHEET

This series of information sheets is intended to illustrate methods of achieving fire resistance in steel structures. It should not be used for design without consulting detailed design guidance referenced below.

BOARD PROTECTION

UP TO 4 HRS

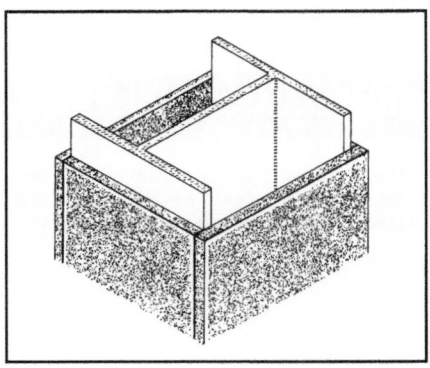

METHOD

Fire protective insulation can be applied by fixing boards to almost any type of steel member. Most products can achieve up to 4 hour rating. Fixing methods vary.

PRINCIPLE

Insulation reduces the heating rate of a steel member so that its limiting temperature is not exceeded during the required fire resistance period. The protection board thickness necessary depends on the section factor (Hp/A) of the member and the fire rating required.

ADVANTAGES

a) Boxed appearance suitable for visible members
b) Clean dry fixing
c) Factory manufactured, guaranteed thickness
d) Often applied to non-primed steelwork
e) Some products may be suitable for external use

LIMITATIONS (check with manufacturer)

a) Require fitting around complex details
b) May be more expensive and slower to fix than sprays

FOR MORE DETAILED INFORMATION SEE:-
"Fire protection of Structural Steel in Building"
Publication jointly by:
ASFP and The Steel Construction Institute

Sheet Code
ISF/No.02
January 1997

PROTECTION THICKNESS

Thickness recommendations given in "Fire Protection of Structural Steel in Building" have normally been derived from fire tests on orthodox H or I rolled sections. For other sections the recommended thickness for a given section factor and fire rating should be modified as follows.

CASTELLATED SECTIONS

The thickness of fire protection material on a castellated section should be 20% greater than that required for the section from which it was cut.

STEELWORK IN FIRE
INFORMATION SHEET

This series of information sheets is intended to illustrate methods of achieving fire resistance in steel structures. It should not be used for design without consulting detailed design guidance referenced below.

THIN FILM INTUMESCENT COATINGS

UP TO 2 HRS

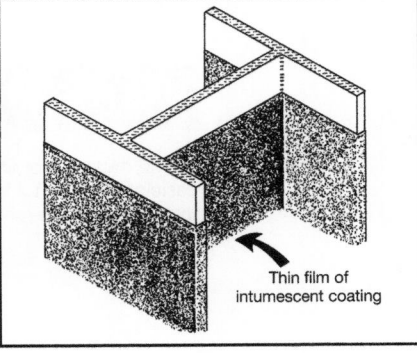

Thin film of intumescent coating

METHOD

Most thin film intumescent coatings can be applied by spray, brush or roller and can achieve up to 1 hour fire resistance on fully exposed steel members. Some products can achieve up to 2 hours fire resistance on some section sizes.

PRINCIPLE

Insulation is created by swelling of the coating at elevated temperatures to generate a foam like char. This reduces the heating rate so that the limiting temperature of the steel member is not exceeded during the required fire resistance period. The coating thickness necessary depends on the section factor (Hp/A) and the fire rating required.

ADVANTAGES

a) Decorative finish
b) Rapid application
c) Easy to cover complex details
d) Easy post protection fixings to steelwork eg service hangers

LIMITATIONS (check with manufacturer)

a) May be suitable for dry internal environments only
b) May be more expensive than sprayed insulation
c) May require blast cleaned surface and compatible primer

FOR MORE DETAILED INFORMATION SEE:-
"Fire protection of Structural Steel in Building"
Publication jointly by:
ASFP and The Steel Construction Institute

Sheet Code
ISF/No.03
January 1997

PROTECTION THICKNESS

Thickness recommendations given in "Fire Protection of Structural Steel in Building" have normally been derived from fire tests on orthodox H or I rolled sections. For other sections the recommended thickness for a given section factor and fire rating should be modified as follows:

CASTELLATED SECTIONS

The thickness of fire protection material on a castellated section should be 20% greater than that required for the section from which it was cut.

HOLLOW SECTIONS

For intumescent materials applied to hollow sections the manufacturers should have carried out separate tests and appraisal

STEELWORK IN FIRE
INFORMATION SHEET

This series of information sheets is intended to illustrate methods of achieving fire resistance in steel structures. It should not be used for design without consulting detailed design guidance referenced below.

BLOCK - FILLED COLUMNS

30 MINUTES

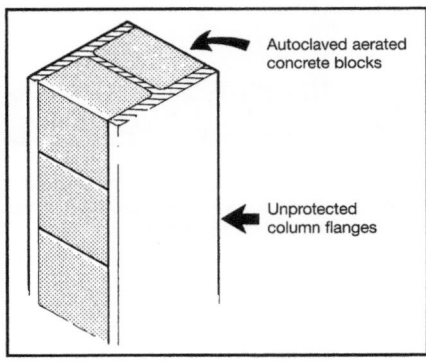

Autoclaved aerated concrete blocks

Unprotected column flanges

METHOD

Unprotected universal sections with section factors up to 69m^{-1} (see overleaf) can attain 30 minutes fire resistance by fitting autoclaved aerated concrete blocks between the flanges tied to the web at approximately 1m intervals

PRINCIPLE

Partial exposure of steel members affects fire resistance in two ways-

Firstly the reduction of exposed surface area reduces the rate of heating by radiation and thus increases the time to reach failure temperature.

Secondary, if the exposure creates both hot and cold regions in the cross section, plastic yielding occurs in the hot region and load is transferred to the stronger cooler region. Thus a non-uniformly heated section has a higher fire resistance than one heated evenly.

ADVANTAGES

a) Reduced cost - compared with total encasement with insulation
b) More slender finished columns occupy less floor space
c) Good durability - high resistance to impact and abrasion damage

LIMITATIONS

With unprotected steel the method is limited to 30 minutes fire rating.

When higher ratings are required exposed steel must be treated with the full insulation or intumescent coating thickness recommended for the higher rating.

This method should not be used when the blockwork also forms a separating wall. In this case the column will be heated on one side only and thermal bowing may cause the wall to crack or collapse. In such cases the flange(s) should be protected. Alternatively, if the limit of wall deformation is known, the bowing can be calculated to ensure no integrity failure.

FOR MORE DETAILED INFORMATION SEE:-
BRE digest 317, Building Research Establishment, Garston, Watford WD2 7SR

Sheet Code
ISF/No.04
January 1997

METHODS OF ACHIEVING 30 MINUTES FIRE RESISTANCE

COLUMN SECTION - AXIALLY LOADED [1] FREE STANDING

SERIAL SIZE mm	MASS/METRE kg	PROTECTION METHOD RECOMMENDED
305 x 406	393 and over	No fire protection required
356 x 406 305 x 305 254 x 254 203 x 203 203 x 203	340 and under All weights All weights 52 and over 46[2]	Block filling with autoclaved aerated concrete blocks
152 x 203	All weights	Apply fire protection material as per manufacturer's recommendations

BEAM SECTIONS ACTING AS PORTAL FRAME STANCHIONS [1]

914 x 419 914 x 305 *610 x 305	All weights 289 238	No fire protection required
*914 x 305 838 x 292 762 x 267 686 x 254 *610 x 305 610 x 229 533 x 210 457 x 191 457 x 152 406 x 178 356 x 171 305 x 165 305 x 127 254 x146	252 and under All weights All weights All weights 179 and under All weights All weights All weights 60 and over 60 and over 57 and over 54 48 43	Block filling with autoclaved aerated concrete blocks
Other beam sizes		Apply fire protection material as per manufacturer's recommendations

Notes:
1) This table applies to sections designed to BS 5950: Part 1:1990 provided the load factor (γf) does not exceed 1.5
2) To achieve 30 min fire resistance, a 203 x 203 x 46 kg/m column with blocked in webs should be loaded only up to 80% of the maximum allowable per BS 449:Part 2:1969 or BS 5950:Part 1:1990
*3) The table revises BRE Digest 317 (1986) in accordance with BS 5950:Part 8:1990

STEELWORK IN FIRE

INFORMATION SHEET

This series of information sheets is intended to illustrate methods of achieving fire resistance in steel structures. It should not be used for design without consulting detailed design guidance referenced below.

CONCRETE FILLED HOLLOW COLUMNS

UP TO 2 HRS

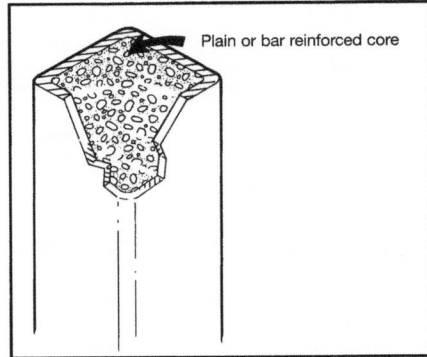

Plain or bar reinforced core

METHOD
Unprotected square or rectangular hollow sections can attain up to 120 minutes fire resistance by filling with plain or bar reinforced concrete.

PRINCIPLE
Heat flows through the steel wall into the concrete core which being a poor conductor heats up slowly.

As the temperature increases the steel yield strength reduces and the load is progressively transferred into the concrete core
The steel acts as a restraint to the concrete preventing spalling and hence the rate of degrading of the concrete.

ADVANTAGES

a) Steel acts as a permanent shuttering.
b) More slender finished columns occupy less floor space.
c) Good durability – high resistance to impact and abrasion damage.

LIMITATIONS

a) A minimum column size of 200 mm × 200 mm is required for bar reinforced sections.
b) Unreinforced sections can only achieve 30 minutes fire resistance.

FOR MORE DETAILED INFORMATION SEE:-
BS EN 1994-1-2

The fire resistance of concrete filled tubes to Eurocode 4, Technical Report, P259, 2000

Design guide for structural hollow section columns exposed to fire, Design Guide No 4, CIDECT, 1996

Sheet Code
ISF/No.05
January 1997

CONCRETE FILLED RECTANGULAR HOLLOW SECTIONS

The fire resistance of externally unprotected concrete filled hollow sections is dependent on three main variables.

- The concrete strength selected
- The ratio of axial load and moment
- The addition of fibre or bar reinforcement

CONCRETE STRENGTH

The core capacity and hence its fire resistance is directly related to the concrete strength.

AXIAL LOAD AND MOMENT

Plain concrete does not perform well in tension and when subject to axial load and moment it is necessary to produce a resultant compressive stress in the core.

COMBINED PROTECTION

As an alternative the concrete filled section can be designed for full factored loads and provided with external fire protection.

The thickness of the external fire protection is assessed as far for an unfilled section, and, due to the effect of the core, the thickness can be reduced.

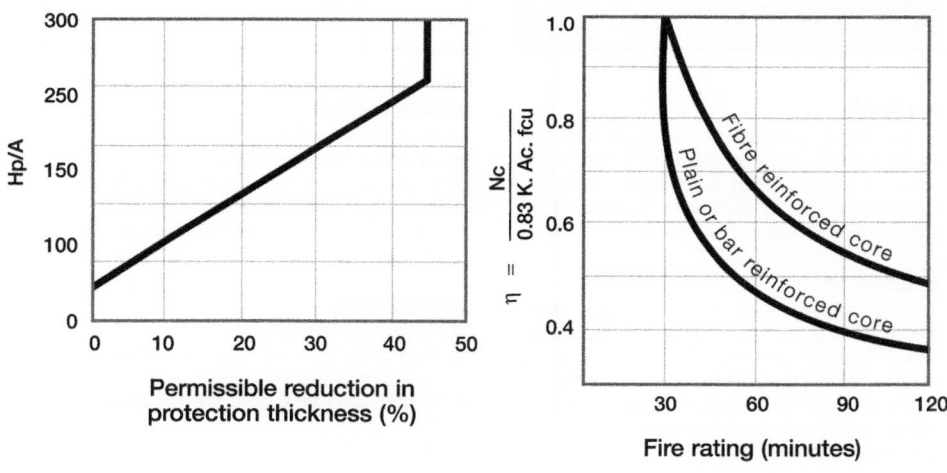

STEELWORK IN FIRE
INFORMATION SHEET

This series of information sheets is intended to illustrate methods of achieving fire resistance in steel structures. It should not be used for design without consulting detailed design guidance referenced below.

COMPOSITE SLABS WITH PROFILED METAL DECK WITH UNFILLED VOIDS

UP TO 2 HRS

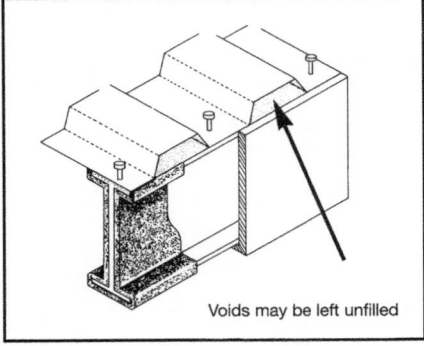

Voids may be left unfilled

METHOD

In composite construction using profiled metal deck floors it is unnecessary to fill the deck voids above the top flange for any fire resistance period using dovetail deck, or up to 90 minutes using trapezoidal deck (see overleaf)

PRINCIPLE

In a composite beam/slab member the neutral axis in bending lies in, or close to, the beam top flange. Thus the top flange makes little significant contribution to the structural behaviour of the total composite system and it's temperature can be allowed to increase with little detriment to performance in fire.

ADVANTAGES

a) Saving in time on site
b) Saving in cost for filling voids
c) It is unnecessary to build up the full thickness of protection on toes of upper flange
d) Void filling is unnecessary when using dovetail deck

LIMITATIONS

Voids must be filled where:-
a) Trapezoidal deck is used for fire ratings over 90 minutes
b) Trapezoidal deck is used in non-composite construction
c) Any type of deck crosses a fire separating wall

FOR MORE DETAILED INFORMATION SEE:-
Technical Report 109
"Fire resistance of composite beams"
The British Steel Construction Institute

Sheet Code
ISF/No.06
January 1997

COMPOSITE BEAMS - UNFILLED VOIDS

TRAPEZOIDAL DECK

Construction	Fire Protection On Beam	Fire Resistance (minutes)		
		Up to 60	90	Over 90
Composite Beams	BOARD or SPRAY	No Increase in thickness*	Increase thickness* by 10% (or use thickness* appropriate to beam Hp/A + 15% whichever is less)	Fill voids
	INTUMESCENT	Increase thickness* by 20% (or use thickness* appropriate to beam Hp/A + 30% whichever is less)	Increase thickness* by 30% (or use thickness* appropriate to beam Hp/A + 50% whichever is less)	Fill voids
Non-Composite Beams	All types	Fill voids		

DOVETAIL DECK

Construction	Fire Protection On Beam	Fire Resistance (minutes)
Composite or Non-composite Beams	All Types	Voids may be left unfilled for all fire resistance periods.

* Thickness is the board, spray or intumescent thickness given for 30, 60 or 90 minutes rating in "Fire Protection for Structural Steel in Buildings" published by ASFP and The Steel Construction Institute

STEELWORK IN FIRE
INFORMATION SHEET

This series of information sheets is intended to illustrate methods of achieving fire resistance in steel structures.
It should not be used for design without consulting detailed design guidance referenced below.

COMPOSITE SLABS WITH PROFILED METAL DECK　　UP TO 2 HRS

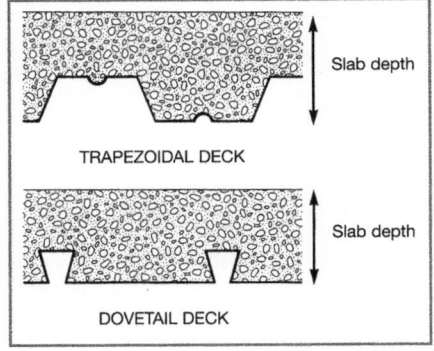

TRAPEZOIDAL DECK

Slab depth

DOVETAIL DECK

Slab depth

METHOD
Fire resistance of composite slabs up to 90 mins can be achieved using normal A142 mesh reinforcement. This can be increased to 120 mins if heavier mesh is used and the slab depth increased (see overleaf).

Other cases outside the limit overleaf can be evaluated by the "Fire Engineering Method" (See below)

PRINCIPLE
Mesh reinforcement, which is not designed to act structurally under normal conditions, makes a significant contribution to structural continuity in fire.

ADVANTAGES
a) Standard mesh, without additional reinforcing bars, may be used
b) No fire protection is required on the deck soffit

LIMITATIONS
a) Applies only to slabs designed to BS5950 Part 4
b) Mesh overlaps should exceed 50 times bar diameters
c) Mesh bar ductility should exceed 12% elongation in tension (to BS 4449)
d) Mesh should lie between 20 & 45mm from slab upper surface
e) Imposed load should not exceed 6.7kN/m^2 (including finishes)

FOR MORE DETAILED INFORMATION SEE:-
SCI Technical Report 056
"Fire resistance of composite floors with steel decking".
The Steel Construction Institute and
CIRIA Special publication 42 CIRIA

Sheet Code
ISF/No.07
January 1997

FIRE RESISTANT COMPOSITE SLABS

TRAPEZOIDAL DECK

60mm max

| Maximum Span (m) | Fire Rating (h) | Minimum Dimensions | | | Mesh Size |
| | | Sheet thickness | Slab depth (mm) | | |
			NWC [2]	LWC [3]	
2.7	1	0.8	130	120	A142
3.0	1	0.9	130	120	A142
	1.5	0.9	140	130	A142
	2	0.9	155	140	A193
3.6	1	1.0	130	120	A193
	1.5	1.2	140	130	A193
	2	1.2	155	140	A252

DOVETAIL DECK

51mm max

| Maximum Span (m) | Fire Rating (h) | Minimum Dimensions | | | Mesh Size |
| | | Sheet thickness | Slab depth (mm) | | |
			NWC [2]	LWC [3]	
2.5	1	0.8	100	100	A142
	1.5	0.8	110	105	A142
3.0	1	0.9	120	110	A142
	1.5	0.9	130	120	A142
	2	0.9	140	130	A193
3.6	1	1.0	125	120	A193
	1.5	1.2	135	125	A193
	2	1.2	145	130	A252

1) Imposed load not exceeding $5kN/M^2$ ($+ 1.7kN/m^2$ ceiling and services)
2) NWC = Normal weight concrete
3) LWC = Light weight concrete

NOTE: Minimum slab depths given in BS 5950 part 8 are to safety the insulation criterion only. Figures given in the table above incorporate a strength criterion also and thus may exceed the minimum depth given in the code.

STEELWORK IN FIRE
INFORMATION SHEET

This series of information sheets is intended to illustrate methods of achieving fire resistance in steel structures. It should not be used for design without consulting detailed design guidance referenced below.

SLIMDEK BEAMS
WITH DEEP DECK

UP TO 1 HOUR UNPROTECTED

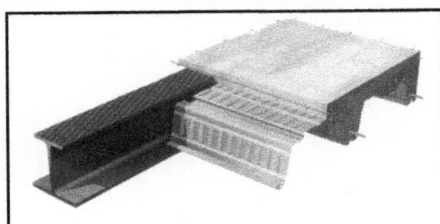

METHOD

The SLIMDEK system consists of an asymmetric beam, with a narrow upper flange and a 225mm deep deck positioned on the outstand of the lower flange. The floor is formed from in-situ applied concrete. This arrangement, shown above, can be designed to provide 60 minutes fire resistance in most cases without the need for applied fire protection.

PRINCIPLE

The section is protected form the effects of fire by the insulating concrete floor. Thus only the bottom flange is directly exposed in fire. Composite action, which develops as a consequence of the raised pattern on the upper flange compensates for much of the loss of strength in the steel at high temperatures

ADVANTAGES

- Fire resistance periods of 60 minutes can be achieved in most instances without any restrictions in loadings.

- Flat slab construction.

*SLIMDEK is a Registered Trade Mark of British Steel plc.

- Clear service runs.

- Reduced construction runs and building heights.

- Services can be passed through prepared openings in the rib of the decking to further reduce floor depth.

LIMITATIONS

a) For fire resistance periods greater than 60 minutes, the exposed lower flange requires fire protection.

b) Where holes are cut in the beam to allow services to pass through, the exposed bottom flange will generally require fire protection.

FOR MORE DETAILED INFORMATION SEE:-
SCI Publication 175 "Design of Asymmetric Slimflor Beams using Deep Composite Decking"
The Steel Construction Institute

Sheet Code
ISF/No.9
April 1997

Table 1

Summary of recommendations

Fire Resistance (Minutes)	Design Type	
	Without holes or action	**With service holes**
30 Minutes	No protection required	No protection required
60 Minutes	No protection required in most circumstances (see table 2)	Protect bottom flange
Greater than 60	Protect bottom flange	

Table 2

Load table for ASB beams for 60 minutes fire resistance (Concrete grade 30, steel grade S355)

Section	Span of Beam (mm)	Effective width of slab (mm)	Moment resistance at ultimate limit state (kNm)	Moment resistance at fire resistance of 60 mins. (kNm)	Maximum Load Ratio
280 ASB 100	5500	688	554	257	0.48
	6000	750	562	258	0.47
	6500	813	570	260	0.47
280 ASB 136	5500	688	726	376	0.52
	6000	750	736	378	0.51
	6500	813	745	381	0.51
	7000	875	754	383	0.51
300 ASB 153	6000	750	870	472	0.51
	6500	813	880	475	0.54
	7000	875	890	478	0.54
	7500	938	900	481	0.54
300 ASB 153 (slab flush with top flange, and lightweight concrete used)	6000	750	835	440	0.53
	6500	813	842	442	0.53
	7000	875	849	443	0.52
	7500	938	856	445	0.52

STEELWORK IN FIRE
INFORMATION SHEET

This series of information sheets is intended to illustrate methods of achieving fire resistance in steel structures. It should not be used for design without consulting detailed design guidance referenced below.

SHELF ANGLE FLOOR BEAMS

UP TO 1 HOUR UNPROTECTED

Concrete floor slab

METHOD

Shelf Angle Floor Beams consist of Universal Beams with angles bolted or welded to the web. The floor is formed from concrete floor slabs which sit on the outstand of the angle. The gap between the web and the floor slab is filled with grout or concrete to ensure that an effective heat sink is created around the section. This arrangement can be designed to provide 30 or 60 minutes fire resistance in many cases without the need for applied fire protection.

PRINCIPLE

The section is partly protected from the effects of fire by the insulating concrete floor and infill. Thus only part of the web and the bottom flange is exposed to the fire. The angles, which are ignored in cold design, are considered in fire and provide additional capacity. As the angles are placed further down the web, the insulated area, and thus the fire resistance is increased. Where fire protection is required, the reduced exposed perimeter leads in turn to reduced fire protection thicknesses.

ADVANTAGES

- Fire resistance periods of 30 minutes can be achieved in most cases without fire protection to the exposed web and bottom flange.

- Fire resistance periods of 60 minutes can be achieved in some instances depending on the load and the exposed area of the beam.

- Reduced construction runs and building height

- Where the exposed steelwork requires fire protection, reduced thicknesses are possible

LIMITATIONS

a) The angle must be 125 x 75 x 12 Grade S355, short leg attached vertically to the beam as shown overleaf.

b) For 60 minutes fire resistance at high loads, the required depth of floor slab may be unavailable or uneconomic.

c) For fire resistance periods over 60 minutes, fire protection to the exposed perimeter will always be necessary.

FOR MORE DETAILED INFORMATION SEE:-
SCI Publication 126 "The Fire Resistance of Shelf Angle
Floor Beams to BS5950 Part 8" The Steel Construction Institute

Sheet Code
ISF/No.10
April 1997

Typical recommendations for fire resistance of Shelf Angle Floor Beams, taken from the Steel Construction Institute publication, are given in Table 1. These can be expanded to take into account other grades of steel and fire resistance periods. The allowable angle connection force must also be considered.

Table 1

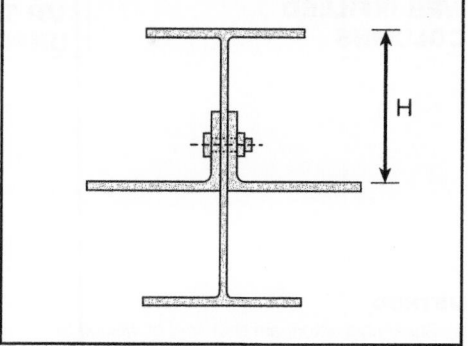

Fire Resistance	60 minutes

Beam Grade	S355
Angle Grade	S355

Section Size	M_p (KNm)	H(mm), Position of angle below top of beam for load ratio[1] of						
		0.4	0.45	0.5	0.55	0.6	0.65	0.7
305 x 102 x 25 UB	120	129	144	158	172	184	196	208
305 x 102 x 28 UB	145	137	152	167	180	193	205	217
305 x 102 x 33 UB	170	144	159	174	188	201	214	227
305 x 127 x 37 UB	192	145	160	174	188	202	209	222
305 x 127 x 42 UB	217	150	166	181	196	207	216	230
305 x 127 x 48 UB	251	158	174	190	206	212	227	235

[1]Load Ratio is defined in BS5950 Part 8 is the ratio of the load at the fire limit state to the cold capacity of the section.

STEELWORK IN FIRE
INFORMATION SHEET

This series of information sheets is intended to illustrate methods of achieving fire resistance in steel structures. It should not be used for design without consulting detailed design guidance referenced below.

WEB INFILLED COLUMNS

UP TO 1 HOUR UNPROTECTED

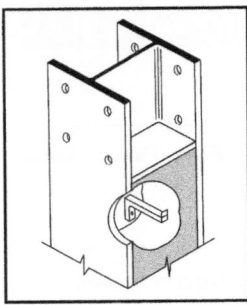

METHOD

Shear connectors are shot fired or welded to the web of the column. A web stiffening plate is welded to the column below the connection zone to contain the concrete. The area below the stiffener and between the flanges is then filled with concrete.

PRINCIPLE

In cold design any beneficial effects of the concrete are ignored. In fire however, as the steel becomes hot, the load is transferred to the concrete. Load transfer is accommodated both by the shear studs and the welded stiffener plate. The unconcreted part of the column and the connections are protected by the same system used to protect the steel beam.

ADVANTAGES

- Fire resistance periods of 60 minutes can be achieved in most cases without fire protection to the exposed flanges.

- For higher periods of fire resistance reduced fire protection thicknesses are required.

- The complete column can be constructed off-site.

- The complete section takes up no space not already occupied by the steel column.

LIMITATIONS

a) The method should not be used where a high specification is required and the column is exposed.

b) Although the system can be used in simple construction where moments are relatively small, it is not suitable for columns in moment resisting frames.

c) For fire resistance periods over 60 minutes, fire protection to the exposed perimeter will always be necessary.

d) The method can only be used for 203x203x46 UCs and above in size.

FOR MORE DETAILED INFORMATION SEE:-
SCI Publication 124 "The Fire Resistance of Web Infilled Steel Columns"
The Steel Construction Institute

Sheet Code
ISF/No.11
April 1997

Typical recommendations for fire resistance of Web Infilled Columns, taken from the Steel Construction Institute publication, are given in Table 1

Table 1

Fire Resistance   60 minutes

Column Grade   S275

Section Size	Moment Capacity (kNm)		Load Ratio[1]	Compressive Capacity (kN) at effective lengths used for normal design				
	M_{fx}	M_{fy}		2500mm	3000mm	3500mm	4000mm	4500mm
203 x 203 x 46	40.3	22.9	0.57	744	682	616	551	488
203 x 203 x 52	47.0	25.4	0.54	803	738	669	600	533
203 x 203 x 60	56.7	29.0	0.52	882	881	735	659	586
203 x 203 x 71	69.9	34.6	0.49	974	899	820	740	662

[1]Load Ratio is defined in BS5950 Part 8 as the ratio of the load at the fire limit state to the cold capacity of the section.

M_{fx} = the moment capacity in fire conditions for bending about the x - x axis.

M_{fy} = the moment capacity in fire conditions for bending about the y - y axis.

Section factors for fire design

Universal beams							Section factor $\frac{A_m}{V}$			
							Profile		Box	
							3 sides	4 sides	3 sides	4 sides
Designation		Depth of section D	Width of section B	Thickness		Area of section				
Serial size	Mass per metre			Web t	Flange T					
mm	kg	mm	mm	mm	mm	cm²	m⁻¹	m⁻¹	m⁻¹	m⁻¹
914×419	388	920.5	420.5	21.5	36.6	494.4	60	70	45	55
	343	911.4	418.5	19.4	32.0	437.4	70	80	50	60
914×305	289	926.6	307.8	19.6	32.0	368.8	75	80	60	65
	253	918.5	305.5	17.3	27.9	322.8	85	95	65	75
	224	910.3	304.1	15.9	23.9	285.2	95	105	75	85
	201	903	303.4	15.2	20.2	256.4	105	115	80	95
838×292	226	850.9	293.8	16.1	26.8	288.7	85	95	70	80
	194	840.7	292.4	14.7	21.7	247.1	100	115	80	90
	176	834.9	291.6	14	18.8	224.1	110	125	90	100
762×267	197	769.6	268	15.6	25.4	250.7	90	100	70	85
	173	762	266.7	14.3	21.6	220.4	105	115	80	95
	147	753.9	265.3	12.9	17.5	188.0	120	135	95	110
686×254	170	692.9	255.8	14.5	23.7	216.5	95	110	75	90
	152	687.6	254.5	13.2	21.0	193.8	110	120	85	95
	140	683.5	253.7	12.4	19.0	178.6	115	130	90	105
	125	677.9	253	11.7	16.2	159.6	130	145	100	115
610×305	238	633	311.5	18.6	31.4	303.7	70	80	50	60
	179	617.5	307	14.1	23.6	227.9	90	105	70	80
	149	609.6	304.8	11.9	19.7	190.1	110	125	80	95
610×229	140	617	230.1	13.1	22.1	178.3	105	120	80	95
	125	611.9	229	11.9	19.6	159.5	115	130	90	105
	113	607.3	228.2	11.2	17.3	144.4	130	145	100	115
	101	602.2	227.6	10.6	14.8	129.1	145	160	110	130
533×210	122	544.6	211.9	12.8	21.3	155.7	110	120	85	95
	109	539.5	210.7	11.6	18.8	138.5	120	135	95	110
	101	536.7	210.1	10.9	17.4	129.7	130	145	100	115
	92	533.1	209.3	10.2	15.6	117.7	140	160	110	125
	82	528.3	208.7	9.6	13.2	104.4	155	175	120	140
457×191	98	467.4	192.8	11.4	19.6	125.2	120	135	90	105
	89	463.6	192	10.6	17.7	113.9	130	145	100	115
	82	460.2	191.3	9.9	16.0	104.5	140	160	105	125
	74	457.2	190.5	9.1	14.5	94.98	155	175	115	135
	67	453.6	189.9	8.5	12.7	85.44	170	190	130	150
457×152	82	465.1	153.5	10.7	18.9	104.4	130	145	105	120
	74	461.3	152.7	9.9	17.0	94.99	140	155	115	130
	67	457.2	151.9	9.1	15.0	85.41	155	175	125	145
	60	454.7	152.9	8.0	13.3	75.93	175	195	140	160
	52	449.8	152.4	7.6	10.9	66.49	200	220	160	180
406×178	74	412.8	179.7	9.7	16.0	94.95	140	160	105	125
	67	409.4	178.8	8.8	14.3	85.49	155	175	115	140
	60	406.4	177.8	7.8	12.8	76.01	175	195	130	155
	54	402.6	177.6	7.6	10.9	68.42	190	215	145	170
406×140	46	402.3	142.4	6.9	11.2	58.96	205	230	160	185
	39	397.3	141.8	6.3	8.6	49.40	240	270	190	220
356×171	67	364	173.2	9.1	15.7	85.42	140	160	105	125
	57	358.6	172.1	8	13.0	72.18	165	190	125	145
	51	355.6	171.5	7.3	11.5	64.58	185	210	135	165
	45	352	171	6.9	9.7	56.96	210	240	155	185
356×127	39	352.8	126	6.5	10.7	49.40	215	240	170	195
	33	248.5	125.4	5.9	8.5	41.83	250	280	195	225
305×165	54	310.9	166.8	7.7	13.7	68.38	160	185	115	140
	46	307.1	165.7	6.7	11.8	58.90	185	210	130	160
	40	303.8	165.1	6.1	10.2	51.50	210	240	150	180
305×127	48	310.4	125.2	9.9	14.0	60.83	160	180	125	145
	42	306.6	124.3	8	12.1	53.18	180	205	140	160
	37	303.8	123.5	7.2	10.7	47.47	200	225	155	180
305×102	33	312.7	102.4	6.6	10.8	41.77	215	240	175	200
	28	308.9	101.9	6.1	8.9	36.30	245	275	200	225
	25	304.8	101.6	5.8	6.8	31.39	285	315	225	260
254×146	43	259.6	147.3	7.3	12.7	55.10	170	195	120	150
	37	256	146.4	6.4	10.9	47.45	195	225	140	170
	31	251.5	146.1	6.1	8.6	40.00	230	265	160	200
254×102	28	260.4	102.1	6.4	10.0	36.19	220	250	170	200
	25	257	101.9	6.1	8.4	32.17	245	280	190	225
	22	254	101.6	5.8	6.8	28.42	275	315	215	250
203×133	30	206.8	133.8	6.3	9.6	38.00	210	245	145	180
	25	203.2	133.4	5.8	7.8	32.31	240	285	165	210
203×102	23	203.2	101.6	5.2	9.3	29	235	270	175	210
178×102	19	177.8	101.6	4.7	7.9	24.2	265	305	190	230
152×89	16	152.4	88.9	4.6	7.7	20.5	270	310	190	235
127×76	13	127	76.2	4.2	7.6	16.8	275	320	195	240

Universal columns		Depth of section D	Width of section B	Thickness		Area of section	Section factor $\frac{A_m}{V}$			
							Profile		Box	
Designation							3 sides	4 sides	3 sides	4 sides
Serial size	Mass per metre			Web t	Flange T					
mm	kg	mm	mm	mm	mm	cm²	m⁻¹	m⁻¹	m⁻¹	m⁻¹
356 × 406	634	474.7	424.1	47.6	77.0	808.1	25	30	15	20
	551	455.7	418.5	42.0	67.5	701.8	30	35	20	25
	467	436.6	412.4	35.9	58.0	595.5	35	40	20	30
	393	419.1	407.0	30.6	49.2	500.9	40	45	25	35
	340	406.4	403.0	26.5	42.9	432.7	45	55	30	35
	287	393.7	399.0	22.6	36.5	366.0	50	65	30	45
	235	381.0	395.0	18.5	30.2	299.8	65	75	40	50
356 × 368	202	374.7	374.4	16.8	27.0	257.9	70	85	45	60
	177	368.3	372.1	14.5	23.8	225.7	80	95	50	65
	153	362.0	370.2	12.6	20.7	195.2	90	110	55	75
	129	355.6	368.3	10.7	17.5	164.9	105	130	65	90
305 × 305	283	365.3	321.8	26.9	44.1	360.4	45	55	30	40
	240	352.6	317.9	23.0	37.7	305.6	50	60	35	45
	198	339.9	314.1	19.2	31.4	252.3	60	75	40	50
	158	327.2	310.6	15.7	25.0	201.2	75	90	50	65
	137	320.5	308.7	13.8	21.7	174.6	85	105	55	70
	118	314.5	306.8	11.9	18.7	149.8	100	120	60	85
	97	307.8	304.8	9.9	15.4	123.3	120	145	75	100
254 × 254	167	289.1	264.5	19.2	31.7	212.4	60	75	40	50
	132	276.4	261.0	15.6	25.3	167.7	75	90	50	65
	107	266.7	258.3	13.0	20.5	136.6	90	110	60	75
	89	260.4	255.9	10.5	17.3	114.0	110	130	70	90
	73	254.0	254.0	8.6	14.2	92.9	130	160	80	110
203 × 203	86	222.3	208.8	13.0	20.5	110.1	95	110	60	80
	71	215.9	206.2	10.3	17.3	91.1	110	135	70	95
	60	209.6	205.2	9.3	14.2	75.8	130	160	80	110
	52	206.2	203.9	8.0	12.5	66.4	150	180	95	125
	46	203.2	203.2	7.3	11.0	58.8	165	200	105	140
152 × 152	37	161.8	154.4	8.1	11.5	47.4	160	190	100	135
	30	157.5	152.9	6.6	9.4	38.2	195	235	120	160
	23	152.4	152.4	6.1	6.8	29.8	245	300	155	205

Circular hollow sections				Section factor $\frac{A_m}{V}$ Profile or Box
Designation		**Mass per metre**	**Area of section**	
Outside diameter D	Thickness t			
mm	mm	kg	cm²	m⁻¹
21.3	3.2	1.43	1.82	370
26.9	3.2	1.87	2.38	355
33.7	2.6	1.99	2.54	415
	3.2	2.41	3.07	345
	4.0	2.93	3.73	285
42.4	2.6	2.55	3.25	410
	3.2	3.09	3.94	340
	4.0	3.79	4.83	275
48.3	3.2	3.56	4.53	335
	4.0	4.37	5.57	270
	5.0	5.34	6.80	225
60.3	3.2	4.51	5.74	330
	4.0	5.55	7.07	270
	5.0	6.82	8.69	220
76.1	3.2	5.75	7.33	325
	4.0	7.11	9.06	265
	5.0	8.77	11.2	215
88.9	3.2	6.76	8.62	325
	4.0	8.38	10.70	260
	5.0	10.3	13.2	210
114.3	3.6	9.83	12.5	285
	5.0	13.5	17.2	210
	6.3	16.8	21.4	170
139.7	5.0	16.6	21.2	205
	6.3	20.7	26.4	165
	8.0	26.0	33.1	135
	10.0	32.0	40.7	110
168.3	5.0	20.1	25.7	205
	6.3	25.2	37.1	165
	8.0	31.6	40.3	130
	10.0	39.0	49.7	105
193.7	5.0	23.3	29.6	205
	6.3	29.1	37.1	165
	8.0	36.6	46.7	130
	10.0	45.3	57.7	105
	12.5	55.9	71.2	85
	16.0	70.1	89.3	70
219.1	5.0	26.4	33.6	205
	6.3	33.1	42.1	165
	8.0	41.6	53.1	130
	10.0	51.6	65.7	105
	12.5	63.7	81.1	85
	16.0	80.1	102	65
	20.0	98.2	125	55

				Section factor $\frac{A_m}{V}$ Profile or Box
Designation		**Mass per metre**	**Area of section**	
Outside diameter D	Thickness t			
mm	mm	kg	cm²	m⁻¹
244.5	6.3	37.0	47.1	165
	8.0	46.7	59.4	130
	10.0	57.8	73.7	105
	12.5	71.5	91.1	85
	16.0	90.2	115	65
	20.0	111	141	55
273.0	6.3	41.4	52.8	160
	8.0	52.3	66.6	130
	10.0	64.9	82.6	105
	12.5	80.3	102	85
	16.0	101	129	65
	20.0	125	159	55
	25.0	153	195	45
323.9	6.3	49.3	62.9	160
	8.0	62.3	79.4	130
	10.0	77.4	98.6	105
	12.5	96.0	122	85
	16.0	121	155	65
	20.0	150	191	55
	25.0	184	235	45
355.6	8.0	68.6	87.4	130
	10.0	85.2	109	100
	12.5	106	135	85
	16.0	134	171	65
	20.0	166	211	55
	25.0	204	260	45
406.4	10.0	97.8	125	100
	12.5	121	155	80
	16.0	154	196	65
	20.0	191	243	55
	25.0	235	300	45
	32.0	295	376	35
457.0	10.0	110	140	105
	12.5	137	175	80
	16.0	174	222	65
	20.0	216	275	50
	25.0	266	339	40
	32.0	335	427	35
	40.0	411	524	25
508.0	10.0	123	156	100
	12.5	153	195	80
	16.0	194	247	65

Rectangular hollow sections				Section factor $\frac{A_m}{V}$		
				3 sides		4 sides
Designation		Mass per metre	Area of section			
Size D × B	Thickness t					
mm	mm	kg	cm²	m⁻¹	m⁻¹	m⁻¹
50×25	2.5	2.72	3.47	360	290	430
	3.0	3.22	4.10	305	245	365
	3.2	3.41	4.34	290	230	345
50×30	2.5	2.92	3.72	350	295	430
	3.0	3.45	4.40	295	250	365
	3.2	3.66	4.66	280	235	345
	4.0	4.46	5.68	230	195	280
	5.0	5.40	6.88	190	160	235
60×40	2.5	3.71	4.72	340	295	425
	3.0	4.39	5.60	285	250	355
	3.2	4.66	5.94	270	235	335
	4.0	5.72	7.28	220	190	275
	5.0	6.97	8.88	180	160	225
	6.3	8.49	10.8	150	130	185
80×40	3.0	5.34	6.80	295	235	355
	3.2	5.67	7.22	275	220	330
	4.0	6.97	8.88	225	180	270
	5.0	8.54	10.9	185	145	220
	6.3	10.5	13.3	150	120	180
	8.0	12.8	16.3	125	100	145
90×50	3.0	6.28	8.00	290	240	350
	3.6	7.46	9.50	240	200	295
	5.0	10.1	12.9	180	145	215
	6.3	12.5	15.9	145	120	175
	8.0	15.3	19.5	120	95	145
100×50	3.0	6.75	8.60	290	235	350
	3.2	7.18	9.14	275	220	330
	4.0	8.86	11.3	220	175	265
	5.0	10.9	13.9	180	145	215
	6.3	13.4	17.1	145	115	175
	8.0	16.6	21.1	120	95	140
100×60	3.0	7.22	9.20	285	240	350
	3.6	8.59	10.9	240	200	295
	5.0	11.7	14.9	175	150	215
	6.3	14.4	18.4	140	120	175
	8.0	17.8	22.7	115	95	140
120×60	3.6	9.72	12.4	240	195	290
	5.0	13.3	16.9	180	140	215
	6.3	16.4	20.9	145	115	170
	8.0	20.4	25.9	115	95	140
120×80	5.0	14.8	18.9	170	150	210
	6.3	18.4	23.4	135	120	170
	8.0	22.9	29.1	110	95	135
	10.0	27.9	35.5	90	80	115
150×100	5.0	18.7	23.9	165	145	210
	6.3	23.8	29.7	135	120	170
	8.0	29.1	37.1	110	95	135
	10.0	35.7	45.5	90	75	110
	12.5	43.6	55.5	70	65	90
160×80	5.0	18.0	22.9	175	140	210
	6.3	22.3	28.5	140	110	170
	8.0	27.9	35.5	115	90	135
	10.0	34.2	43.5	90	75	110
	12.5	41.6	53.0	75	60	90
200×100	5.0	22.7	28.9	175	140	210
	6.3	28.3	36.0	140	110	165
	8.0	35.4	45.1	110	90	135
	10.0	43.6	55.5	90	70	110
	12.5	53.4	68.0	75	60	90
	16.0	66.4	84.5	60	45	70
250×150	6.3	38.2	48.6	135	115	165
	8.0	48.0	61.1	105	90	130
	10.0	59.3	75.5	85	75	105
	12.5	73.0	93.0	70	60	85
	16.0	91.5	117	55	45	70
300×200	6.3	48.1	61.2	130	115	165
	8.0	60.5	77.1	105	90	130
	10.0	75.0	95.5	85	75	105
	12.5	92.6	118	70	60	85
	16.0	117	149	55	45	65
400×200	10.0	90.7	116	85	70	105
	12.5	112	143	70	55	85
	16.0	142	181	55	45	65
450×250	10.0	106	136	85	70	105
	12.5	132	168	70	55	85
	16.0	167	213	55	45	65

Rectangular hollow sections (square)				Section factor $\frac{A_m}{V}$	
Designation		Mass per metre	Area of section	3 sides	4 sides
Size D × D	Thickness t				
mm	mm	kg	cm²	m⁻¹	m⁻¹
20×20	2.0	1.12	1.42	425	565
	2.5	1.35	1.72	350	465
25×25	2.0	1.43	1.82	410	550
	2.5	1.74	2.22	340	450
	3.0	2.04	2.60	290	385
	3.2	2.15	2.74	275	365
30×30	2.5	2.14	2.72	330	440
	3.0	2.51	3.20	280	375
	3.2	2.65	3.38	265	355
40×40	2.5	2.92	3.72	325	430
	3.0	3.45	4.40	275	365
	3.2	3.66	4.66	260	345
	4.0	4.46	5.68	210	280
	5.0	5.40	6.88	175	235
50×50	2.5	3.71	4.72	320	425
	3.0	4.39	5.60	270	355
	3.2	4.66	5.94	255	335
	4.0	5.72	7.28	205	275
	5.0	6.97	8.88	170	225
	6.3	8.49	10.8	140	185
60×60	3.0	5.34	6.80	265	355
	3.2	5.67	7.22	250	330
	4.0	6.97	8.88	205	270
	5.0	8.54	10.9	165	220
	6.3	10.5	13.3	135	180
	8.0	12.8	16.3	110	145
70×70	3.0	6.28	8.00	260	350
	3.6	7.46	9.50	220	295
	5.0	10.1	12.9	165	215
	6.3	12.5	15.9	130	175
	8.0	15.3	19.5	110	145
80×80	3.0	7.22	9.20	260	350
	3.6	8.59	10.9	220	295
	5.0	11.7	14.9	160	215
	6.3	14.4	18.4	130	175
	8.0	17.8	22.7	105	140
90×90	3.6	9.72	12.4	220	290
	5.0	13.3	16.9	160	215
	6.3	16.4	20.9	130	170
	8.0	20.4	25.9	105	140
100×100	4.0	12.0	15.3	195	260
	5.0	14.8	18.9	160	210
	6.3	18.4	23.4	130	170
	8.0	22.9	29.1	105	135
	10.0	27.9	35.5	85	115

Designation		Mass per metre	Area of section	Section factor $\frac{A_m}{V}$ 3 sides	4 sides
Size D × D	Thickness t				
mm	mm	kg	cm²	m⁻¹	m⁻¹
120×120	5.0	18.0	22.9	155	210
	6.3	22.3	28.5	125	170
	8.0	27.9	35.5	100	135
	10.0	34.2	43.5	85	110
	12.5	41.6	53.0	70	90
140×140	5.0	21.1	26.9	155	210
	6.3	26.3	33.5	125	165
	8.0	32.9	41.9	100	135
	10.0	40.4	51.5	80	110
	12.5	49.5	63.0	65	90
150×150	5.0	22.7	28.9	155	210
	6.3	28.3	36.0	125	165
	8.0	35.4	45.1	100	135
	10.0	43.6	55.5	80	110
	12.5	53.4	68.0	65	90
	16.0	66.4	84.5	55	70
180×180	6.3	34.2	43.6	125	165
	8.0	43.0	54.7	100	130
	10.0	53.0	67.5	80	105
	12.5	65.2	83.0	65	85
	16.0	81.4	104	50	70
200×200	6.3	38.2	48.6	125	165
	8.0	48.0	61.1	100	130
	10.0	59.3	75.5	80	105
	12.5	73.0	93.0	65	85
	16.0	91.5	117	50	70
250×250	6.3	48.1	61.2	125	165
	8.0	60.5	77.1	95	130
	10.0	75.0	95.5	80	105
	12.5	92.6	118	65	85
	16.0	117	149	50	65
300×300	10.0	90.7	116	80	105
	12.5	112	143	65	85
	16.0	142	181	50	65
350×350	10.0	106	136	75	105
	12.5	132	168	65	85
	16.0	167	213	50	65
400×400	10.0	122	156	75	105
	12.5	152	193	60	85
	16.0	192	245	50	65

Corrosion resistance

Basic data on corrosion

Atmospheric corrosivity categories and examples of typical environments (ISO 12944 Part 2).

Corrosivity category and risk	Mass loss per unit surface/thickness loss (see Note 1)			Examples of typical environments in a temperate climate (informative only)	
	Low-carbon steel Thickness loss μm	Exterior		Interior	
C1 very low	≤1.3	–		Heated buildings with clean atmospheres, e.g. offices, shops, schools, hotels	
C2 low	>1.3–25	Atmospheres with low level of pollution. Mostly rural areas		Unheated buildings where condensation may occur, e.g. depots, sports halls	
C3 medium	>25–60	Urban and industrial atmospheres, moderate sulphur dioxide pollution. Coastal area with low salinity		Production rooms with high humidity and some air pollution, e.g. food-processing plants, laundries, breweries, dairies	
C4 high	>50–80	Industrial areas and coastal areas with moderate salinity		Chemical plants, swimming pools, coastal, ship and boatyards	
C5-I very high (industrial)	>80–200	Industrial areas with high humidity and aggressive atmosphere		Buildings or areas with almost permanent condensation and high pollution	
C5-M very high (marine)	>80–200	Coastal and offshore areas with high salinity		Buildings or areas with almost permanent condensation and with high pollution	

1. The thickness loss values are after the first year of exposure. Losses may reduce over subsequent years.
2. The loss values used for the corrosivity categories are identical to those given in ISO 9223.
3. In coastal areas in hot, humid zones, the thickness losses can exceed the limits of category C5-M. Special precautions must therefore be taken when selecting protective paint systems for structures in such areas.

1 μm = 0.001 mm

Main generic types of paint and their properties

Binder	System cost	Tolerance of poor surface	Chemical resistance	Solvent resistance	Water resistance	Overcoating after ageing	Comments
Black coatings (based on tar products)	Low	Good	Moderate	Poor	Good	Very Good with coatings of same type	Limited to black or dark colours. May soften in hot conditions
Alkyds	Low – medium	Moderate	Poor	Poor – moderate	Moderate	Good	Good decorative properties. High solvent levels
Acrylated rubbers	Medium – high	Poor	Good	Poor	Good	Good	High build films that remain soft and are susceptible to sticking
Epoxy Surface tolerant	Medium – high	Good	Good	Good	Good	Good	Can be applied to a range of surfaces and coatings*
High performance	Medium – high	Very poor	Very good	Good	Very good	Poor	Susceptible to 'chalking' in U.V. light
Urethane and polyurethane	High	Very poor	Very good	Good	Very good	Poor	Can be more decorative than epoxies
Organic silicate and inorganic silicate	High	Very poor	Moderate	Good	Good	Moderate	May require special surface preparation

*Widely used for maintenance painting

Details should be designed to enhance durability by avoiding water entrapment.

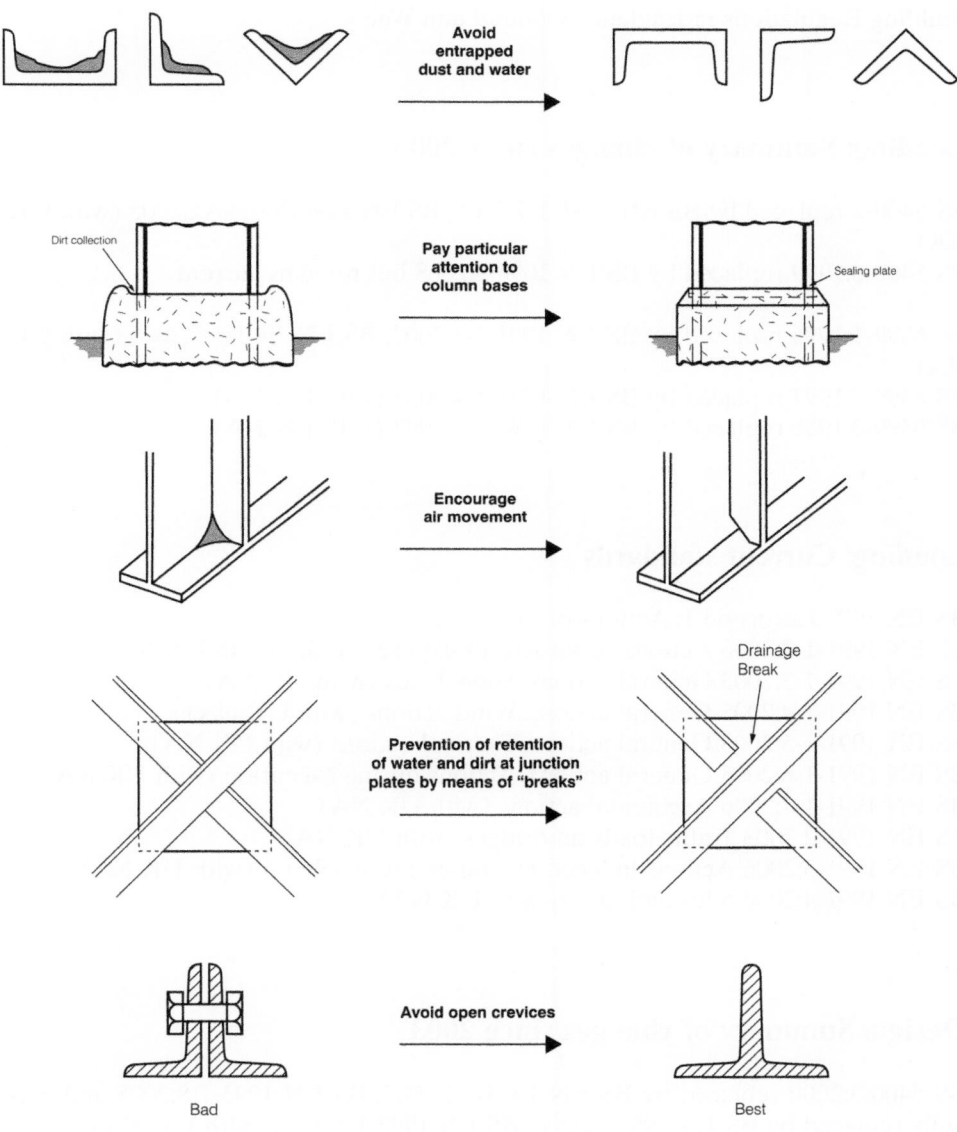

British and European Standards for steelwork

In March 2010 BSI withdrew all British Standards that conflict with the Eurocodes. However, British Standards can continue to be used as long as they satisfy the Building Regulations in England, Scotland and Wales.

Loading: Summary of changes since 2003

BS 5400-2 replaced by BS EN 1991-1-7:2006, BS EN 1990:2002+A1:2005 (with UK NA)
BS 5400-6:1999 replaced by BS EN 1090-2:2008 but remains current

BS 6399-1:1996 replaced by BS EN 1991-1-1:2002, BS EN 1991-1-7:2006 (with UK NA)
BS 6399-2:1997 replaced by BS EN 1991-1-4:2005 (with UK NA)
BS 6399-3:1988 replaced by BS EN 1991-1-3:2003 (with UK NA)

Loading: Current standards

BS EN 1991 Eurocode 1: Actions on structures
BS EN 1991-1-2:2006 Actions on structures exposed to fire (with UK NA)
BS EN 1991-1-3:2003 General actions. Snow loads (with UK NA)
BS EN 1991-1-4:2005 General actions. Wind actions (with UK NA)
BS EN 1991-1-5:2003 General actions. Thermal actions (with UK NA)
BS EN 1991-1-6:2005 General actions. Actions during execution (with UK NA)
BS EN 1991-1-7:2006 Accidental actions (with UK NA)
BS EN 1991-2:2003 Traffic loads on bridges (with UK NA)
BS EN 1991-3:2006 Actions induced by cranes and machines (with UK NA)
BS EN 1991-4:2006 Silos and tanks (with UK NA)

Design: Summary of changes since 2003

BS 5400-3:2000 replaced by BS EN 1993-1-1:2005, BS EN 1993-1-8:2005 and partially replaced by BS EN 1993-2:2006, BS EN 1993-1-5:2006 (with UK NA's)
BS 5400-5:2005 replaced by BS EN 1994-2:2005 (with UK NA)
BS 5400-9:1983 replaced by BS EN 1337-2:2004, BS EN 1337-7:2004, BS EN 1337-3:2005, BS EN 1337-5:2005, BS EN 1337-8:2007
BS 5400-10:1980 replaced by BS EN 1993-1-9:2005 (with UK NA)

BS 5400-10C:1999 Steel, concrete and composite bridges. Charts for classification of details for fatigue replaced by BS EN 1993-1-9 Tables 8.1 to 8.10

BS 5427-1:1996 Code of practice for the use of profiled sheet for roof and wall cladding on buildings. Design

BS 5950-1:2000 replaced by BS EN 1993-6:2007, BS EN 1993-1-1:2005, BS EN 1993-1-8:2005, BS EN 1993-5:2007 and partially replaced by BS EN 1993-1-5:2006 (with UK NA's)
BS 5950-3.1:1990 replaced by BS EN 1994-1-1:2004 (with UK NA)
BS 5950-4:1994 replaced by BS EN 1994-1-1:2004 (with UK NA)
BS 5950-5:1998 replaced by BS EN 1994-2:2005 (with UK NA)
BS 5950-6:1995 replaced by BS EN 1090-2:2008
BS 5950-8:2003 replaced by BS EN 1993-1-2:2005 (with UK NA)
BS 5950-9:1994 partially replaced by BS EN 1993-1-3:2006 (with UK NA)

Design: Current standards

BS EN 1991 Eurocode 1: Actions on structures
BS EN 1991-3:2006 Actions induced by cranes and machines (with UK NA)
BS EN 1991-1-7:2006 General actions. Accidental actions (with UK NA)
BS EN 1993 Eurocode 3: Design of steel structures
BS EN 1993-1-1:2005 General rules and rules for buildings (with UK NA)
BS EN 1993-1-2:2005 General rules. Structural fire design (with UK NA)
BS EN 1993-1-3:2006 General rules. Supplementary rules for cold-formed members and sheeting (with UK NA)
BS EN 1993-1-6:2007 Strength and Stability of Shell Structures (with UK NA)
BS EN 1993-1-7:2007 Plated structures subject to out of plane loading
BS EN 1993-4-1:2007 Silos (with UK NA)
BS EN 1993-4-2:2007 Tanks (with UK NA)
BS EN 1993-4-3:2007 Pipelines (with UK NA)

BS EN 1994 Eurocode 4: Design of composite steel and concrete structures
BS EN 1994-1-1:2004 General rules and rules for buildings (with UK NA)

BS EN 1998 Eurocode 8: Design of structures for earthquake resistance
BS EN 1998-1:2004 General rules, seismic actions and rules for buildings (with UK NA)
BS EN 1998-2:2005+A1:2009 Bridges (with UK NA)
BS EN 1998-3:2005 Assessment and retrofitting of buildings
BS EN 1998-4:2006 Silos, tanks and pipelines (with UK NA)
BS EN 1998-5:2004 Foundations, retaining structures and geotechnical aspects (with UK NA)
BS EN 1998-6:2005 Towers, masts and chimneys (with UK NA)

Steel fabrication and erection: Summary of changes since 2003

BS 4604-1:1970 replaced by BS EN 1993-1-8:2005 (with UK NA)
BS 4604-2:1970 replaced by BS EN 1993-1-8:2005 (with UK NA)

BS 5400-6:1999 replaced by BS EN 1090-2:2008 but remains current

BS 5950-2:2001 replaced by BS EN 1090-2:2008 but remains current

Steel fabrication and erection: Current standards

BS EN 1090 Execution of steel structures and aluminium structures
BS EN 1090-1:2009 Requirements for conformity assessment of structural components
BS EN 1090-2:2008 Technical requirements for the execution of steel structures
BS EN 1090-3:2008 Technical requirements for aluminium structures

Foundations and piling: Summary of changes since 2003

BS 449-2:1969 replaced by BS EN 1993-6:2007, BS EN 1993-1-1:2005, BS EN 1993-1-8:2005, BS EN 1993-5:2007 and partially replaced by BS EN 1993-1-5:2006 (with UK NA's)

BS 5400-1:1988 replaced by BS EN 1991-1-7:2006, BS EN 1990:2002+A1:2005 (with UK NA's)

BS 5493:1977 partially replaced by Parts 1 to 8 of BS EN ISO 12944 and BS EN ISO 14713:1999

BS 5950-1:2000 replaced by BS EN 1993-6:2007, BS EN 1993-1-1:2005, BS EN 1993-1-8:2005, BS EN 1993-5:2007and partially replaced by BS EN 1993-1-5:2006 (with UK NA's)

BS 8002:1994 replaced by BS EN 1997-1:2004 (with UK NA)

BS 8004:1986 replaced by BS EN 1997-1:2004 (with UK NA)

BS 8081:1989 partially replaced by BS EN 1537:2000

Foundations and piling: Current standards

BS 4-1:2005 Structural steel sections. Specification for hot-rolled sections
BS EN 10248-1:1996 Hot rolled sheet piling of non alloy steels. Technical delivery conditions

BS EN 10248-2:1996 Hot rolled sheet piling of non alloy steels. Tolerances on shape and dimensions

BS EN 12063:1999 Execution of special geotechnical work. Sheet pile walls

Structural steel: Current standards

BS 7668:2004 Weldable structural steels. Hot finished structural hollow sections in weather resistant steels. Specification

BS EN 10025 Hot rolled products of structural steels.
BS EN 10025-1:2004 General technical delivery conditions
BS EN 10025-3:2004 Technical delivery conditions for normalized/normalized rolled weldable fine grain structural steels
BS EN 10025-4:2004 Technical delivery conditions for thermomechanical rolled weldable fine grain structural steels
BS EN 10025-5:2004 Technical delivery conditions for structural steels with improved atmospheric corrosion resistance

BS EN 10025-6:2004+A1:2009 Technical delivery conditions for flat products of high yield strength structural steels in the quenched and tempered condition

BS EN 10029:1991 Specification for tolerances on dimensions, shape and mass for hot rolled steel plates 3 mm thick or above

BS EN 10111:2008 Continuously hot rolled low carbon steel sheet and strip for cold forming. Technical delivery conditions

BS EN 10130:2006 Cold rolled low carbon steel flat products for cold forming. Technical delivery conditions

BS EN 10139:1998 Cold rolled uncoated mild steel narrow strip for cold forming. Technical delivery conditions

BS EN 10164:2004 Steel products with improved deformation properties perpendicular to the surface of the product. Technical delivery conditions

BS EN 10210 Hot finished structural hollow sections of non-alloy and fine grain steels
BS EN 10210-1:2006 Technical delivery requirements
BS EN 10210-2:2006 Tolerances, dimensions and sectional properties

BS EN 10219 Cold formed welded structural hollow sections of non-alloy and fine grain steels

BS EN 10219-1:2006 Technical delivery requirements
BS EN 10219-2:2006 Tolerances, dimensions and sectional properties

BS EN 10268:2006 Cold rolled steel flat products with high yield strength for cold forming. Technical delivery conditions

BS EN 10273:2007 Hot rolled weldable steel bars for pressure purposes with specified elevated temperature properties

Steel products: Current standards

BS 4-1:2005 Structural steel sections. Specification for hot-rolled sections
BS EN 10029:1991 Specification for tolerances on dimensions, shape and mass for hot rolled steel plates 3 mm thick or above
BS EN 10051:1991+A1:1997 Continuously hot-rolled uncoated plate, sheet and strip of non-alloy and alloy steels. Tolerances on dimensions and shape

BS EN 10055:1996 Hot rolled steel equal flange tees with radiused root and toes. Dimensions and tolerances on shape and dimensions
BS EN 10056 Specification for structural steel equal and unequal angles
BS EN 10056-1:1999 Dimensions
BS EN 10056-2:1993 Tolerances on shape and dimensions

BS EN 10067:1997 Hot rolled bulb flats. Dimensions and tolerances on shape, dimensions and mass

BS EN 10163 Delivery requirements for surface condition of hot-rolled steel plates, wide flats and sections
BS EN 10163-1:2004 General requirements
BS EN 10163-2:2004 Plate and wide flats
BS EN 10163-3:2004 Sections

BS EN 10210-2:2006 Hot finished structural hollow sections of non-alloy and fine grain steels. Tolerances, dimensions and sectional properties

BS EN 10219-2:2006 Cold formed welded structural hollow sections of non-alloy and fine grain steels. Tolerances, dimensions and sectional properties

BS EN 10084:2008 Case hardening steels. Technical delivery conditions

Cold-rolled thin gauge sections and sheets: Summary of changes since 2003

BS 5950 Structural use of steelwork in building
BS 5950-5:1998 replaced by BS EN 1994-2:2005 (with UK NA)

BS 5950-6:1995 replaced by BS EN 1090-2:2008
BS 5950-9:1994 partially replaced by BS EN 1993-1-3:2006 (with UK NA)

Cold-rolled thin gauge sections and sheets: Current standards

BS EN 10031:2003 Semi finished products for forging. Tolerances on dimensions shape and mass

BS EN 10048:1997 Hot rolled narrow steel strip. Tolerances on dimensions and shape

BS EN 10139:1998 Cold rolled uncoated mild steel narrow strip for cold forming. Technical delivery conditions

BS EN 10140:2006 Cold rolled narrow steel strip. Tolerances on dimensions and shape
BS EN 10143:2006 Continuously hot-dip coated steel sheet and strip. Tolerances on dimensions and shape

BS EN 10149 Specification for hot-rolled flat products made of high yield strength steels for cold forming
BS EN 10149-1:1996 General delivery conditions
BS EN 10149-2:1996 Delivery conditions for thermomechanically rolled steels
BS EN 10149-3:1996 Delivery conditions for normalized or normalized rolled steels

BS EN 10162:2003 Cold rolled steel sections. Technical delivery conditions. Dimensional and cross-sectional tolerances

BS EN 10169-2:2006 Continuously organic coated (coil coated) steel flat products. Products for building exterior applications

BS EN 10328:2005 Iron and steel. Determination of the conventional depth and hardening after surface heating

BS EN 10346:2009 Continuously hot-dip coated steel flat products. Technical delivery conditions

BS ISO 4997:2007 Cold-reduced carbon steel sheet of structural quality

BS ISO 4999:2005 Continuous hot-dip terne (lead alloy) coated cold-reduced carbon steel sheet of commercial, drawing and structural qualities

ISO 4495:2008 Hot-rolled steel sheet of structural quality

ISO 5951:2008 Hot-rolled steel sheet of higher yield strength with improved formability

ISO 6316:2008 Hot-rolled steel strip of structural quality

ISO 16162:2005 Continuously cold-rolled steel sheet products – Dimensional and shape tolerances

ISO 16163:2005 Continuously hot-dipped coated steel sheet products – Dimensional and shape tolerances

Stainless steels: Current standards

BS EN 1011-3:2000 Welding. Recommendations for welding of metallic materials. Arc welding of stainless steels

BS EN 10088 Stainless steels
BS EN 10088-1:2005 List of stainless steels
BS EN 10088-2:2005 Technical delivery conditions for sheet/plate and strip of corrosion resisting steels for general purposes
BS EN 10088-3:2005 Technical delivery conditions for semi-finished products, bars, rods, wire, sections and bright products of corrosion resisting steels for general purposes
BS EN 10088-4:2009 Technical delivery conditions for sheet/plate and strip of corrosion resisting steels for construction purposes
BS EN 10088-5:2009 Technical delivery conditions for bars, rods, wire, sections and bright products of corrosion resisting steels for construction purposes

BS EN ISO 3506 Mechanical properties of corrosion-resistant stainless steel fasteners
BS EN ISO 3506-1:2009 Bolts, screws and studs
BS EN ISO 3506-2:2009 Nuts
BS EN ISO 3506-3:2009 Set screws and similar fasteners not under tensile stress
BS EN ISO 3506-4:2009 Tapping screws

Castings and forgings: Current standards

BS EN 10293:2005 Steel castings for general engineering uses

BS EN 10088 Stainless steels
BS EN 10088-1:2005 List of stainless steels
BS EN 10088-2:2005 Technical delivery conditions for sheet/plate and strip of corrosion resisting steels for general purposes

BS EN 10088-3:2005 Technical delivery conditions for semi-finished products, bars, rods, wire, sections and bright products of corrosion resisting steels for general purposes
BS EN 10088-4:2009 Technical delivery conditions for sheet/plate and strip of corrosion resisting steels for construction purposes
BS EN 10088-5:2009 Technical delivery conditions for bars, rods, wire, sections and bright products of corrosion resisting steels for construction purposes

BS EN 1560:1997 Founding. Designation system for cast iron. Material symbols and material numbers

BS EN 1561:1997 Founding. Grey cast irons

BS EN 1563:1997 Founding. Spheroidal graphite cast iron

Steel construction components: Current standards

BS EN 10162:2003 Cold rolled steel sections. Technical delivery conditions. Dimensional and cross-sectional tolerances

BS 5427-1:1996 Code of practice for the use of profiled sheet for roof and wall cladding on buildings. Design

BS EN 1337 Structural bearings (several parts)

BS EN 1462:2004 Brackets for eaves gutters. Requirements and testing

Welding materials and processes: Current standards

BS 499 Welding terms and symbols
BS 499-1:2009 Glossary for welding, brazing and thermal cutting
BS 499-2C:1999 European arc welding symbols in chart form

BS EN ISO 4063:2009 Welding and allied processes. Nomenclature of processes and reference numbers

BS EN ISO 9692 Welding and allied processes. Joint preparation.
BS EN ISO 9692-1:2003 Recommendations for joint preparation. Manual metal-arc welding, gas-shielded metal-arc welding, gas welding, TIG welding and beam welding of steels
BS EN ISO 9692-2:1998 Submerged arc welding of steels

Processes and consumables: Current standards

BS EN 756:2004 Welding consumables. Solid wires, solid wire-flux and tubular cored electrode-flux combinations for submerged arc welding of non alloy and fine grain steels.
Classification

BS EN 757:1997 Welding consumables. Covered electrodes for manual metal arc welding of high strength steels. Classification

BS EN ISO 2560:2009 Welding consumables. Covered electrodes for manual metal arc welding of non-alloy and fine grain steels. Classification

BS EN ISO 14341:2008 Welding consumables. Wire electrodes and deposits for gas shielded metal arc welding of non alloy and fine grain steels. Classification
BS EN ISO 17632:2008 Welding consumables. Tubular cored electrodes for gas shielded and non-gas shielded metal arc welding of non-alloy and fine grain steels. Classification

Testing and examination: Current standards

BS EN 875:1995 Destructive tests on welds in metallic materials. Impact tests. Test specimen location, notch orientation and examination

BS EN 895:1995 Destructive tests on welds in metallic materials. Transverse tensile test
BS EN 876:1995 Destructive tests on welds in metallic materials. Longitudinal tensile test on weld metal in fusion welded joints

BS EN 910:1996 Destructive tests on welds in metallic materials. Bend tests
BS EN 1320:1997 Destructive tests on welds in metallic materials. Fracture tests
BS EN 1321:1997 Destructive test on welds in metallic materials. Macroscopic and microscopic examination of welds

BS EN 1043-1:1996 Destructive tests on welds in metallic materials. Hardness testing. Hardness test on arc welded joints

BS EN 1043-2:1997 Destructive tests on welds in metallic materials. Hardness testing. Micro hardness testing on welded joints

BS EN 1435:1997 Non-destructive examination of welds. Radiographic examination of welded joints

BS EN 1713:1998 Non-destructive testing of welds. Ultrasonic testing. Characterization of indications in welds

BS EN 1714:1998 Non destructive testing of welded joints. Ultrasonic testing of welded joints

BS EN 12062:1998 Non-destructive examination of welds. General rules for metallic materials

BS EN ISO 5817:2007 Welding. Fusion-welded joints in steel, nickel, titanium and their alloys (beam welding excluded). Quality levels for imperfections

BS EN ISO 9018:2003 Destructive tests on welds in metallic materials. Tensile test on cruciform and lapped joints

Bolts and fasteners: Summary of changes since 2003

BS 4395-1:1969 & BS 4395-2:1969 replaced by Parts 1-8 and 10 of BS EN 14399 but remains current

BS 4604-1:1970 & BS 4604-2:1970 replaced by BS EN 1993-1-8:2005 (with UK NA)

BS 7644-1:1993 & BS 7644-2:1993 replaced by BS EN 14399-9:2009 but remains current

Fire resistance: Current standards

BS 476 Fire tests on building materials and structures. Guide to the principles, selection, role and application of fire testing and their outputs
BS 476-20:1987 Method for determination of the fire resistance of elements of construction (general principles)
BS 476-21:1987 Methods for determination of the fire resistance of loadbearing elements of construction
BS 476-22:1987 Methods for determination of the fire resistance of non-loadbearing elements of construction
BS 476-23:1987 Methods for determination of the contribution of components to the fire resistance of a structure

BS 9999:2008 Code of practice for fire safety in the design, management and use of buildings
BS 5950-8:2003 Structural use of steelwork in building. Code of practice for fire resistant design

BS 8202 Coatings for fire protection of building elements
BS 8202-1:1995 Code of practice for the selection and installation of sprayed mineral coatings

BS 8202-2:1992 Code of practice for the use of intumescent coating systems to metallic substrates for providing fire resistance

Corrosion prevention and coatings: Current standards

BS 2569-2:1965 Specification for sprayed metal coatings. Protection of iron and steel against corrosion and oxidation at elevated temperatures

BS 4652:1995 Specification for zinc-rich priming paint (organic media)

BS 4921:1988 Specification for sherardized coatings on iron or steel

BS 5493:1977 partially replaced by Parts 1 to 8 of BS EN ISO 12944 and BS EN ISO 14713:1999

BS 7079:2009 General introduction to standards for preparation of steel substrates before application of paints and related products

Quality assurance: Current standards

BS EN ISO 9000:2005 Quality management systems. Fundamentals and vocabulary

BS EN ISO 9001:2008 Quality management systems. Requirements

Environmental: Current standards

BS 6187:2000 Code of practice for demolition

BS EN ISO 14001:2004 Environmental management systems. Requirements with guidance for use

BS EN ISO 19011:2002 Guidelines for quality and/or environmental management systems auditing

BS ISO 14004:2004 Environmental management systems. General guidelines on principles, systems and supporting techniques

Index